A COLLECTION OF TECHNICAL PAPERS

AIAA/ASME
Adaptive Structures Forum

April 21-22, 1994/ Hilton Head, SC

Copyright © by
American Institute of Aeronautics and Astronautics

All rights reserved. No part of this volume may be reproduced or transmitted in any form or by any means, electronic or mechanical, including photocopying and recording, for any purpose without permission from the publisher.

ISBN   1-56347-102-7

Printed in the U.S.A.

AIAA / ASME Adaptive Structures Forum

Hilton Head, SC
April 21–22, 1994

General Chair
Ben Wada
Jet Propulsion Laboratory
Pasadena, CA

Cochair
Brantley Hanks
NASA Langley
Hampton, VA

Technical Chair
Andrew Bicos
McDonnell Douglas Aerospace
Huntington Beach, CA

―――――――――――― Conference Committee ――――――――――――

AIAA Materials TC
Joseph Stoyack
Loral Vought Systems Corp.
Dallas, TX

AIAA Structures TC
David Jensen
Brigham Young University
Provo, UT

AIAA Structures TC
M. Natori
ISAS
Kanagawa, Japan

AIAA Structural
Dynamics TC
Edward White
McDonnell Douglas
Aerospace
St. Louis, MO

AIAA Structural
Dynamics TC
Yuji Matsuzaki
Nagoya University
Nagoya, Japan

AIAA Multidisciplinary
Design Optimization TC
Mokhtar Salama
Jet Propulsion
Laboratory
Pasadena, CA

Alok Das (ASME)
Phillips Laboratory
Kirtland AFB, NM

Ephrahim Garcia
(ASME)
Vanderbilt University
Nashville, TN

David Martinez (ASME)
Sandia National
Laboratory
Albuquerque, NM

Mike Obal (ASME)
Ballistic Missile
Defense Organization
Washington, DC

Craig Rogers (ASME)
VPI & SU
Blacksburg, VA

AIAA/ASME
Adaptive Structures Forum

Hilton Head, SC/April 21-22, 1994

Table of Contents

Paper No.	Title and Author	Page No.

1 System Identification

94-1730	**Optimal Sensor Placement for Modal Identification Using System-Realization Methods** D. Kammer and R. Brillhart	1
94-1731	**A Time and Frequency Domain Procedure fior Identification of Structural Dynamic Models** L. Peterson and K. Alvin	14
94-1733	**A Comparison and Enhancement of Flexible Structure System Identification Methods for Control** L. Hufford and W. Robinson	25
94-1734	**Experiment Dynamic Characterization of a Reconfigurable Adaptive Precision Truss** J. Hinkle and L. Peterson	35
94-1735	**Interval Model Identification for Flexible Structures with Certain Parameters** J-S. Lew, T. Link, E. Garcia and L. Keel	42

2 Piezoelectric Actuators and Sensors

94-1736	**Piezoelectric Actuator Capable of Inducing Torsional and Bending Control Loads with no Structural Coupling** G. Kawiecki and W. Smith	49
94-1737	**Collocated Independent Modal Control with Self-Sensing Orthogonal Piezoelectric Actuator (Theory and Equipment)** H. Tzou and J. Holkamp	57
94-1738	**A Comparison of Actuation Techniques for Aircraft Cabin Noise Control** D. Rosetti and M. Norris	65
94-1739	**Determination of the Power Consumption of Stacked PZT Actuator-Driven Underwater Plates for Active Structural Acoustic Control** C. Liang, F. Sun, C. Rogers and Y. Gu	73
94-1740	**Modal Analysis Using Piezoelectric Sensors** J. Callahan and H. Baruh	83
94-1741	**Modal Filtering Using Lineal Sensors** E. Lang and T. Chou	95

NA = Not Available   WD = Withdrawn

94-1742	**Development of a Linear Piezoelectric Motor** D. Newton, E. Garcia and G. Horner	101

3 Adaptive Control of Structures I

94-1743	**Nonlinear Flutter Control of Composite Panels with Embedded Piezoelectric Materials Using Finite Element Method** R. Zhou, M. Hsiao, D. Xue, Z. Lai, C. Mei, and J-K. Huang	107
94-1744	**Flutter Control of an Adaptive Composite Panel** A. Suleman and V. Venkayya	118
94-1745	**Active Flutter Suppression of Composite Plate with Piezoelectric Actuators** C. Nam, W. Kim and S. Oh	127
94-1746	**Flutter Suppression of an Airfoil with Unsteady Flow Forces Using a Piezoelectric Active Strut** G. Flowers and G. Prasad	NA
94-1747	**Structure-Control Interaction and the Design of Piezoceramic Actuated Adaptive Airfoils** D. Steadman, S. Griffin, and S. Hanagud	135
94-1748	**Seismic Vibration Suppression by the Axial-Force-Rated Actuators** V. Chawla, M. Sener, S. Utku and B. Wada	143
94-1749	**Active Base Isolation in Buildings Subjected to Earthquake Excitation** P. Jalihal, S. Utku and B. Wada	WD

4 Damage Detection

94-1750	**A Nondestructive Laser Test Method for Crack Detection in Structures** O. Sensburg, H. Honlinger, G. Tomlinson and K. Worden	NA
94-1751	**An Artificial Neural Network Approach to Structural Damage Detection Using Frequency Response Functions** C. Povich and T. Lim	151
94-1752	**Damage Detection in Adaptive Structures Using Neural Networks** R. Manning	160
94-1753	**A Neural Network Approach for Damage Detection and Identification of Structures** J. Rhim and S. Lee	173
94-1754	**Detection Delaminations in Composite Beams Using Piezoelectric Sensors** D. Saravanos, V. Birman and D. Hopkins	181

NA = Not Available   WD = Withdrawn

94-1755	**Location and Estimation of Damage in a Beam Using Identification Algorithms**
	D. Lindner and G. Kirby 192
94-1756	**Mechanics for the Coupled Dynamic Response of Active/Sensory Composite Structures**
	D. Saravanos, P. Heyliger and D. Hopkins 199

5 Actuators for Adaptive Structures

94-1757	**The Design of Extended Bandwidth Shape Memory Alloy Actuators**
	J. Ditman, L. Bergman and T. Tsao 210
94-1758	**Shape Memory Ceramic Actuation of Adaptive Structures**
	K. Ghandi and N. Hagood 221
94-1759	**Structural Damping and Self-Sensing Actuation in Tertenol-D Magnetostrictive Materials**
	R. Fenn and M. Gerver 232
94-1760	**Torsional Actuation with Extension—Torsion Composite Coupling and Magnetorestrictive Actuators**
	C. Bothwell, R. Chandra and I. Chopra 241
94-1761	**Active Damping of a Flexible Beam Using ER Fluid Actuators**
	N. Wereley 253
94-1762	**Coupled Electro-Mechanical Impedance Modeling to Predict Power Requirment and Energy Efficiency of Piezoelectric Actuators Integrated with Plate-Like Structures**
	S. Zhou, C. liang and C. Rogers 259
94-1763	**The Effects of Shaped Piezoelectric Actuators on the Excitation of Beams**
	G. Diehl and H. Cudney 270

6 Adaptive Control of Structures II

94-1764	**Adaptive Airfoils for Helicopters**
	R. Roglin, S. Kondor and S. Hanagud 279
94-1765	**Active Control of Helicopter Rotor Blades with Induced Strain Actuators**
	V. Giurgiutiu, Z. Chaundry, and C. Rogers 288
94-1766	**Designing Efficient Helicopter Individual Blade Controllers Using Smart Structures**
	F. Nitzsche 298
94-1767	**Design and Testing of a Helicopter Rotor Model with Smart Trailing Edge Flaps**
	C. Waltz and I. Chopra 309
94-1768	**Active Closed Cell Beam Shape Control**
	S. Ehlers 320

NA = Not Available   WD = Withdrawn

94-1769	**Advanced Control Law for Vibration Suppression with a TYPE-II Variable Stiffness Member** K. Minesugi and J. Onoda	333
94-1770	**Semi-Active Vibration Suppression by Variable-Damping Members** J. Onoda and K. Minesugi	340

7 Adaptive Shape Control

94-1771	**On-Orbit Shape Correction of Inflatable Structures** M. Salama, C. Kuo, J. Garba, B. Wada and M. Thomas	348
94-1772	**Static Shape Control of Space Trusses with Partial Measurements** H. Furuya and R. Haftka	356
94-1773	**Thermal Gradients for Delamination Control** S. Hanagud, S. Atluri, L. Gummadi and C. McColl	368
94-1774	**Optimizing Induced Strain Actuators for Maximum Panel Deflection** T. Leeks and T. Weisshaar	378
94-1775	**Maximizing Deflections Produced by Induced Beam-Like Actuators for Transonic Drag Reduction** M. Muller and T. Weisshaar	388
94-1776	**Identification of Nonlinear Joints in a Truss Structure** R. Bruno	402
94-1777	**Development of Intelligent Structures Using Multiobjective Optimization and Simulated Annealing** C. Seeley and A. Chattopadhyay	411

8 Modeling of Adaptive Structures

94-1778	**A Mechanical Approach on Interfacial Stress Alleviation in an Integrated Induced Strain Actuator/Substructure System** M. Lin and C. Rogers	422
94-1779	**Modeling Considerations for In-Phase Actuation of Actuators Bonded to Shell Structures** F. Lalande, Z. Chaudhry, and C. Rogers	429
94-1780	**A New Modeling Technique for Piezoelectrically Actuated Composite Beams** M. Shen	NA
94-1781	**Modeling Piezoceramic Actuation of Beams in Torsion** C. Park and I. Chopra	438
94-1782	**Two-Dimensional Finite Element Analysis of Laminated Composite Plates Containing Distributed Piezoelectric Actuators and Sensors** D. Detwiler, M. Shen and V. Venkayya	451

NA = Not Available   WD = Withdrawn

94-1783	**Adaptive Electrostatic Structures: A Fundamental Study of the Electrostatic Oscillator** L. Silverberg and R. McDaniel	461
94-1784	**An Object Oriented Approach to the Motion Control of a Free-Floating Variable Geometry Truss** S. Huang, M. Natori, K. Miura, S. Nakai and H. Katsukura	465

9 Adaptive Control of Structures III

94-1785	**A Biologically Inspired Controller for Sound and Vibration Applications** J. Carneal and C. Fuller	474
94-1786	**Neural Network Based Time-Optimal Control of a Magnetically Levitated Precision Positioning System** J. Redmond and S. Tucker	485
94-1787	**Discrete-Time Implementation of Positive Position Feedback: Analysis and Design Approaches with an Experiment** G. Fagan and H. Robertshaw	494
94-1788	**Spillover and Energy Considerations in the Output Feedback Control of Adaptive Structures** C. Baycan, S. Utku and B. Wada	506
94-1789	**Direct Rate Feedback for Piezostructures Using Sensoriactuators with Feedthrough Dynamics** D. Cole and H. Robertshaw	514
94-1790	**A Self-Tuning Piezoelectric Vibration Absorber** J. Hollkamp and T. Starchville Jr.	521
94-1791	**Adaptive Structures and Active Control of Delaminations** S. Hanagud	NA

NA = Not Available   WD = Withdrawn

OPTIMAL SENSOR PLACEMENT FOR MODAL IDENTIFICATION USING SYSTEM-REALIZATION METHODS

Daniel C. Kammer[*]
Department of Engineering Mechanics
University of Wisconsin
Madison, WI 53706
(608) 262-5724

Ralph D. Brillhart[**]
Structural Dynamics Research Corporation
Engineering Services Division
11995 El Camino Real, Suite 200
San Diego, CA 92130

Abstract

The relationship between a previously published sensor placement technique, called Effective Independence, and system-realization methods of modal identification is presented. The sensor placement method maximizes spatial independence and signal strength of targeted mode shapes by maximizing the determinant of an associated Fisher information matrix. It is shown that the sensor placement method also enhances modal identification using system realization techniques by minimizing the size of the required test data matrix, maximizing the modal observability, enhancing the separation of target modes from computational or noise modes, and optimizing the estimation of the discrete system plant matrix. Three currently popular system-realization methods are considered in the analysis, including the Eigensystem Realization Algorithm, the Q-Markov COVER method, and the Polyreference method. A numerical example is also presented using the polyreference modal identification technique in conjunction with several sensor configurations selected using differing placement methods. The corresponding test data is from a modal survey performed on the Controls-Structures-Interaction Evolutionary Model testbed at the NASA LaRC Space Structures Research Laboratory. It is shown that the Effective Independence sensor configuration provides superior modal identification results as predicted by the analytical work.

Introduction

Sensor placement is an important issue which must be addressed by engineers working problems in identification, analysis, control, and health monitoring of large flexible structures. It has long been a critical issue in control dynamics because there is usually a small number of sensors available to the controls engineer and thus there has been a great deal of literature devoted to this subject [1-7]. Sensor placement for structural system identification, mainly for control purposes, has also received much attention [8-14]. Until recently, optimal sensor placement for modal identification has not been studied extensively because, for a typical modern ground vibration test, a large number of sensors are available to the test engineer. If the appropriate dynamic data is not being collected during the course of the test, sensors can either be repositioned or more sensors can be easily added. However, there are currently important modal identification problems in which the number of available sensors is very limited and their positions are essentially fixed once in service. In this type of situation, sensor placement becomes a critical issue which governs the success or failure of the modal identification experiment. Examples include on-orbit modal identification of large space structures, health monitoring of flexible structures in hostile environments, and permanent placement of sensors in smart structures. There has thus been a recent interest in optimal placement of a small number of sensors for modal identification [15-19].

This paper considers the Effective Independence (EfI) sensor placement method proposed by Kammer [16, 17]. The method was developed based upon the needs of a structural dynamicist who must use the modal test parameters to validate a finite element model (FEM) using test-analysis correlation and model updating techniques. In order for the model validation analysis to be successful, the test and FEM mode shapes partitioned to the sensor degrees of freedom must be spatially independent. The EfI sensor placement method maximizes both spatial independence and signal strength of the targeted mode shapes by maximizing the determinant of an associated Fisher information matrix. While the EfI method has been shown to place sensors to the benefit of post-test correlation and model updating, it is not clear that the method enhances the modal identification process itself which is of prime importance to the test engineer. In addition, the method was derived using an input-output equation corresponding to displacement sensors. Therefore, there is a question concerning the use of the method in conjunction with velocity sensors or accelerometers.

The contribution of this paper is the presentation of the relationship between the EfI sensor placement technique and system-realization methods of modal identification. It will be shown that sensor placement using the EfI method enhances modal identification. Three currently popular system-realization methods are considered, includ-

---

[*] Assistant Professor, Senior Member AIAA
[**] Director, Test Projects

Copyright © 1993 by Daniel C. Kammer.
Published by the American Institute of Aeronautics and Astronautics, Inc. with permission.

the Q-Markov COVER (QMC) method [21], and the Polyreference method [22]. Initially, the EfI sensor placement technique will be summarized, the theory relating the EfI method and the system-realization approach to modal identification will then be presented with emphasis on ERA and QMC. Finally, a numerical example will be presented using the polyreference modal identification technique in conjunction with several sensor configurations selected using differing placement techniques.

Sensor Placement Theory

The objective of the EfI sensor placement method is to place sensors such that the spatial independence and signal strength of a desired set of target modes are maximized to enhance post-test correlation and model updating analyses. The real target mode shapes are not known prior to the test, therefore sensor placement must be based upon a pretest FEM. This problem can be cast in the form of a state estimation problem with corresponding output equation given by

$$y_s = \Phi_{fs} q + w \qquad (1)$$

where y_s is a vector of sensor outputs at a large candidate set of sensor locations, Φ_{fs} is a matrix of FEM target modes Φ_f partitioned to the corresponding sensor location degrees of freedom, q is the target mode response vector, and w is a sensor noise vector. It is assumed that the target mode set corresponds to the modes which significantly participate in the sensor output. Spatial independence of the target mode partitions implies that at any time t, the sensor output can be sampled and the target modal response can be estimated. The noise w is assumed to be a stationary random observation disturbance possessing zero mean and covariance intensity matrix R satisfying the relation

$$E\left[w(t)w(\tau)^T\right] = R\delta(t - \tau) \qquad (2)$$

in which E is the expectation operator and δ represents the delta function. The only restriction on the intensity matrix R is that it is positive definite, implying that none of the sensors is perfect.

Using an efficient unbiased estimator to estimate the target modal response results in an estimate error covariance matrix given by

$$J = E\left[(q - \hat{q})(q - \hat{q})^T\right] = \left[\Phi_{fs}^T R^{-1} \Phi_{fs}\right]^{-1} = Q^{-1} \qquad (3)$$

where Q represents the corresponding Fisher information matrix [23]. Using the definition $[R^{-1/2} R^{-1/2}] = R^{-1}$, the information matrix can be alternatively expressed in the form

$$Q = \Phi_{fs}^T R^{-1} \Phi_{fs} = \Phi_{fs}^T \left[R^{-1/2} R^{-1/2}\right] \Phi_{fs} = \overline{\Phi}_{fs}^T \overline{\Phi}_{fs} \qquad (4)$$

in which

$$\overline{\Phi}_{fs} = R^{-1/2} \Phi_{fs} \qquad (5)$$

represents noise-modified target mode shapes. If the noise is uncorrelated between sensors, the covariance matrix R is diagonal. If, in addition, each sensor possesses the same noise statistics, then sensor noise has no impact upon the sensor placement strategy. The corresponding information matrix can be expressed in terms of the target mode partitions in the form $Q = \Phi_{fs}^T \Phi_{fs}$. This is also the appropriate form of the information matrix when there is no a priori information concerning the sensor noise statistics. This form of the information matrix will be used in subsequent sections to study the effect of the EfI sensor placement method upon modal identification.

It is well known [24] that minimization of the estimate error covariance matrix or maximization of the information matrix Q results in the best estimate of the modal response $\hat{q}$. Sensors should therefore be placed such that Q is maximized in the appropriate norm. Maximization of the information matrix determinant has been a commonly used criterion for optimal parameter estimation [24]. Examination of the form of the information matrix presented in equation (4) indicates that if the target modal partitions are not linearly independent, the determinant will be zero. Therefore, maximizing the information matrix determinant will maximize the spatial independence of the target modal partitions as desired. It will also maximize the signal strength of the target modal responses in the sensor output which is very desirable in the presence of noise [17]. Thus, the determinant is also the appropriate measure of the size of the Fisher information matrix for the purpose of sensor placement using a state estimation formulation. It is important to note that if the noise modified mode shapes are linearly independent, the original target modes will also be independent because multiplication by positive definite $R^{-1/2}$ in equation (5) does not affect rank.

The EfI method of sensor placement determines the contribution of each candidate sensor location to the eigenvalues of the information matrix by first computing

$$G = \left[\overline{\Phi}_{fs} \Psi\right]^{\wedge 2} \qquad (6)$$

in which Ψ is a matrix containing the eigenvectors of the information matrix and $^{\wedge 2}$ denotes a term-by-term square of the resulting matrix in the brackets. For k target modes, there will be k eigenvalues and orthonormal eigenvectors. The eigenvectors represent k orthogonal directions in target mode identification space. Each row of G corresponds to a candidate sensor location, while each column corresponds to an eigenvector. Each of the columns of matrix G adds to the corresponding eigenvalue of the information matrix. If G is post-multiplied by the inverse of the eigenvalue matrix, producing

$$F_E = [\overline{\Phi}_{fs}\Psi]^{\wedge 2}\,\lambda^{-1} \quad (7)$$

each column then adds to unity. The ith term in the jth column of F_E represents the fractional contribution of the ith candidate sensor location to the jth eigenvalue of Q. Finally, adding the columns of F_E produces an n_S dimensional vector called the Effective Independence Distribution

$$E_D = [\overline{\Phi}_{fs}\Psi]^{\wedge 2}\,\lambda^{-1}\{1\}_k \quad (8)$$

where $\{1\}_k$ is a k dimensional column vector of 1's and n_S is the number of candidate sensors. It is important to note that the sum of the terms within E_D corresponds to the number of target modes, k.

It has been shown in Ref. [16] that vector E_D is the diagonal of the matrix

$$\overline{E} = \overline{\Phi}_{fs}\Psi\lambda^{-1}\Psi^T\overline{\Phi}_{fs}^T = \overline{\Phi}_{fs}\left[\overline{\Phi}_{fs}^T\overline{\Phi}_{fs}\right]^{-1}\overline{\Phi}_{fs}^T \quad (9)$$

The right portion of this expression represents an orthogonal projector [25] onto the column space of the noise-modified target modes. It is well known that an orthogonal projector is idempotent, meaning its trace is equal to its rank, which in this case is the number of target modes, k. Therefore, each term within vector E_D represents the contribution of the corresponding candidate sensor location to the rank of the information matrix and thus the linear independence of the target modes. More importantly, it has been shown in Refs. [26] and [27] that the Effective Independence of the ith sensor, E_{Di}, is related to the determinant of the information matrix by the expression

$$E_{Di} = \frac{|Q| - |Q_{Ti}|}{|Q|} \quad (10)$$

in which Q_{Ti} represents the information matrix with the ith candidate sensor location deleted from the target modes. Therefore, E_{Di} corresponds to the fractional change in the determinant of the information matrix if the ith candidate sensor location is deleted. The values of E_{Di} thus satisfy the relation

$$0 \le E_{Di} \le 1.0 \quad (11)$$

where a value of 0.0 indicates that the corresponding candidate sensor location will contribute nothing to the identification of the target modes, while a value of 1.0 indicates that the location is absolutely vital to target mode identification.

The Effective Independence method ranks the candidate sensor locations based upon the corresponding Effective Independence Distribution vector E_D given in Eq. (8). The entries in E_D are sorted by magnitude and the lowest ranked sensor is deleted. The remaining sensor locations are then reranked. In an iterative manner the large candidate set of sensor locations can be quickly reduced to the desired number. As each sensor is deleted, the determinant of the information matrix is reduced. At each iteration, according to Eq. (10), the deletion of the lowest ranked sensor location will produce the smallest change in the information matrix determinant. Therefore, even though the Effective Independence method is suboptimal due to its iterative nature, for a given number of allowed sensors, it tends to produce a sensor configuration which maximizes the determinant of the information matrix. Details of the method can be found in Refs. [16, 17, 27].

Effects of EfI on the Observability Matrix

The realization methods for modal identification which are considered in this paper rely on a generalized observability matrix V_p. For the system

$$\dot{z} = Az + Bu$$
$$y_s = Cz + Nu \quad (12)$$

the generalized observability matrix is given by

$$V_p = \begin{bmatrix} C \\ CA \\ \vdots \\ CA^{p-1} \end{bmatrix} \quad (13)$$

It is assumed in this paper that the structure being considered is lightly damped such that proportional damping is a good model for energy dissipation effects. It is also assumed that there are no rigid body modes in the target mode set. In target normal mode coordinates, the matrices in Eqs. (12) are given by

$$z = \{q^T\ \dot{q}^T\}^T \quad A = \begin{bmatrix} 0 & I \\ -\omega^2 & -2\varsigma\omega \end{bmatrix} \quad B = \begin{Bmatrix} 0 \\ \Phi_{fa}^T \end{Bmatrix} \quad (14)$$

in which ω is a diagonal matrix of modal frequencies, ς is a diagonal matrix of modal damping coefficients, and Φ_{fa} is the matrix of FEM target modes partitioned to the actuator locations. In order to identify the k responding target modes, V_p must be full column rank, i.e. $rk(V_p)=2k$. In modal coordinates, V_p can be decomposed into the product of two matrices

$$V_p = S_p(\Phi_{fs})Z_p(\omega,\varsigma) \quad (15)$$

where S is a $(pn_S \times pk)$ matrix function of Φ_{fs} and Z is a $(pk \times 2k)$ matrix function of ω and ς. An optimum sensor configuration will minimize the required size of the generalized observability matrix while still maintaining its rank. The smallest possible observability matrix which still has rank $2k$ is given by

$$V_2 = \begin{bmatrix} C \\ CA \end{bmatrix} \quad (16)$$

While in practice, a much larger observability matrix would be used due to the presence of noise, etc., this minimum sized observability matrix will be used in this paper to demonstrate the effects of the EfI sensor placement strategy.

<u>Acceleration Sensors</u>

In the case of accelerometers, the output influence matrix is given by

$$C = \begin{bmatrix} -\Phi_{fs}\omega^2 & -2\Phi_{fs}\varsigma\omega \end{bmatrix} \quad (17)$$

which results in the minimum observability matrix

$$V_2 = \begin{bmatrix} -\Phi_{fs}\omega^2 & -2\Phi_{fs}\varsigma\omega \\ 2\Phi_{fs}\varsigma\omega^3 & -\Phi_{fs}\omega^2 + 4\Phi_{fs}\varsigma^2\omega^2 \end{bmatrix} = S_2 Z_2 \quad (18)$$

where S_2 and Z_2 are ($2n_s$ x $2k$) and ($2k$ x $2k$) matrices, respectively, given by

$$S_2 = \begin{bmatrix} \Phi_{fs} & 0 \\ 0 & \Phi_{fs} \end{bmatrix} \text{ and } Z_2 = \begin{bmatrix} -\omega^2 & -2\varsigma\omega \\ 2\varsigma\omega^3 & -\omega^2 + 4\varsigma^2\omega^2 \end{bmatrix} \quad (19)$$

According to Sylvester's inequality [28], V_2 will have rank $2k$ if S_2 is full column rank and Z_2 is nonsingular.

The determinant of Z_2 can be computed using the well known expression

$$|Z_2| = \begin{vmatrix} Z_{11} & Z_{12} \\ Z_{21} & Z_{22} \end{vmatrix} = |Z_{11}||Z_{22} - Z_{21} Z_{11}^{-1} Z_{12}| \quad (20)$$

Substituting the partitions from Eq. (19) produces

$$|Z_2| = |-\omega^2||-\omega^2 + 4\varsigma^2\omega^2 - 2\varsigma\omega^3(-\omega^2)^{-1}(-2\varsigma\omega)| = |\omega^4| \quad (21)$$

which indicates that Z_2 is nonsingular regardless of the damping values and sensor placement. It is obvious from Eq. (19) that spatial matrix S_2 will be full column rank, and thus the system will be observable, if and only if the target mode shape partitions Φ_{fs} are full column rank. Therefore at least $2k$ sensors must be placed such that the target mode partitions are linearly independent. This is precisely the objective of the EfI sensor placement methodology described earlier. The EfI method thus results in a minimization of the size of the generalized observability matrix required to produce a rank of $2k$. It is also important to note that the observability matrix is affected by sensor placement only through the spatially dependent matrix S_p.

The optimum sensor configuration should not only render the system observable, but it should also enhance the degree of observability. An observability grammian can be generated in the form

$$W_o = V_2^T V_2 = Z_2^T S_2^T S_2 Z_2 \quad (22)$$

Associated with the observability grammian is a hyperellipsoid in observability space given by

$$x^T W_o^{-1} x = 1 \quad (23)$$

The volume of this hyperellipsoid is proportional to $|W_o|^{1/2}$, therefore the determinant of the observability grammian is often used as a measure of the degree of observability of a system [29]. Sensors should be placed such that the determinant of the observability grammian is maximized. Using Eq. (22), the determinant can be written as

$$|W_o| = |Z_2^T||S_2^T S_2||Z_2| = |Z_2|^2 |S_2^T S_2| \quad (24)$$

It has already been shown that sensor placement has no effect upon the matrix Z_2 and thus also its determinant. On the other hand, the determinant of the matrix product $S_2^T S_2$ is given by

$$|S_2^T S_2| = |\Phi_{fs}^T \Phi_{fs}|^2 = |Q|^2 \quad (25)$$

where Q is the corresponding information matrix discussed earlier. Maximization of the determinant of the information matrix thus results in the maximization of the degree of system observability. As described in the previous section, the EfI sensor placement methodology maximizes the information matrix determinant. Thus, the EfI sensor placement technique not only results in a minimum sized generalized observability matrix, it also maximizes the observability of the system.

<u>Displacement Sensors</u>

For the case of displacement sensors, the output influence matrix is given by

$$C = \begin{bmatrix} \Phi_{fs} & 0 \end{bmatrix} \quad (26)$$

in which 0 is an (n_s x k) null matrix. Using Eqs. (14) and (16), the corresponding minimum generalized observability matrix V_2 is of the form

$$V_2 = \begin{bmatrix} \Phi_{fs} & 0 \\ 0 & \Phi_{fs} \end{bmatrix} = S_2 \quad (27)$$

Thus placing displacement sensors such that the target mode shape partitions are full column rank produces the required rank of $2k$ for the generalized observability matrix and maximization of the determinant of the information matrix maximizes the degree of system observability as in the case of accelerometers.

Velocity Sensors

In the case of velocity sensors, the output influence matrix is of the form

$$C = \begin{bmatrix} 0 & \Phi_{fs} \end{bmatrix} \quad (28)$$

The minimum generalized observability matrix is now given by

$$V_2 = \begin{bmatrix} \Phi_{fs} & 0 \\ 0 & \Phi_{fs} \end{bmatrix} A = S_2 A \quad (29)$$

Assuming that there are no rigid body modes, the system matrix A is nonsingular. The observability matrix will thus have rank $2k$ if and only if the matrix S_2 is full column rank as in the previous two cases and the degree of observability will be maximized by maximizing the determinant of the information matrix.

These results show that maximizing the determinant of the Fisher information matrix $Q = \Phi_{fs}^T \Phi_{fs}$ is the proper objective for sensor placement regardless of the type of sensor being used. This coincides with the design objective of the EfI sensor placement methodology.

Effect of Discretization

So far, only continuous time systems have been considered in this analysis. In reality, the sensors will be sampled at discrete time intervals of length T. Therefore, it is important to consider the effects of discretization upon the results that have already been derived. The discrete version of Eq. (12) is of the form

$$z(j+1) = A_D z(j) + B_D u(j)$$

$$y_s(j) = C z(j) + N u(j) \quad (30)$$

where A_D and B_D are the discrete versions of the system and output influence matrices, respectively, given by

$$A_D = e^{AT} \quad B_D = \left(\int_0^T e^{At} dt\right) B \quad (31)$$

For the type of structural system being considered, the form of the discrete system matrix can be found in Skelton [30] as

$$A_D = \begin{bmatrix} A_{D11} & A_{D12} \\ A_{D21} & A_{D22} \end{bmatrix} \quad (32)$$

in which

$$A_{D11} = e^{-\varsigma\omega T}\left[\cos\omega\left(I-\varsigma^2\right)^{1/2}T + \varsigma\left(I-\varsigma^2\right)^{-1/2}\sin\omega\left(I-\varsigma^2\right)^{1/2}T\right]$$

$$A_{D12} = e^{-\varsigma\omega T}\omega^{-1}\left(I-\varsigma^2\right)^{-1/2}\sin\omega\left(I-\varsigma^2\right)^{1/2}T$$

(33)

$$A_{D21} = -e^{-\varsigma\omega T}\omega\left(I-\varsigma^2\right)^{-1/2}\sin\omega\left(I-\varsigma^2\right)^{1/2}T$$

$$A_{D22} = e^{-\varsigma\omega T}\left[\cos\omega\left(I-\varsigma^2\right)^{1/2}T - \varsigma\left(I-\varsigma^2\right)^{-1/2}\sin\omega\left(I-\varsigma^2\right)^{1/2}T\right]$$

Accelerometers are assumed for sensors in this derivation, but analogous results are obtained for displacement or velocity sensors. The corresponding discrete minimum generalized observability matrix is given by

$$V_{D2} = \begin{bmatrix} C \\ CA_D \end{bmatrix} =$$

$$\begin{bmatrix} -\Phi_{fs}\omega^2 & -2\Phi_{fs}\varsigma\omega \\ -\Phi_{fs}\omega^2 A_{D11} - 2\Phi_{fs}\varsigma\omega A_{D21} & -\Phi_{fs}\omega^2 A_{D12} - 2\Phi_{fs}\varsigma\omega A_{D22} \end{bmatrix}$$

(34)

As in the continuous time case, this can be factored into the product

$$V_{D2} = S_2 Z_{D2} = \begin{bmatrix} \Phi_{fs} & 0 \\ 0 & \Phi_{fs} \end{bmatrix}$$

$$\begin{bmatrix} -\omega^2 & -2\varsigma\omega \\ -\omega^2 A_{D11} - 2\varsigma\omega A_{D21} & -\omega^2 A_{D12} - 2\varsigma\omega A_{D22} \end{bmatrix}$$

(35)

where it can be seen that discretization does not affect the spatially dependent matrix S_2. Using Eqs. (20) and (33), the determinant of Z_{D2} can be shown to be

$$|Z_{D2}| = \left|e^{-\varsigma\omega T}\omega^3\left(I-\varsigma^2\right)^{-1/2}\sin\omega\left(I-\varsigma^2\right)^{1/2}T\right| \quad (36)$$

Once again assuming no rigid body modes and an underdamped system, the matrices $e^{-\varsigma\omega T}$, ω^3, and $(I-\varsigma^2)^{-1/2}$, are positive definite. Therefore, Z_{D2} will be nonsingular if and only if the sampling period is chosen such that

$$\omega_i\left(I-\varsigma_i^2\right)^{1/2}T \neq n\pi \quad n = \pm 1, 2, \cdots$$

where ω_i and ζ_i are the frequency and damping coefficient of the ith mode, respectively. Because the eigenvalues occur only in complex conjugate pairs, this condition corresponds to the well known requirement for the preservation of system observability under discretization [31]. Assuming that the sampling period is chosen correctly, the generalized observability matrix will be of rank $2k$ if and only if matrix S_2 is full column rank as in the continuous time case. Discretization, therefore, has no effect on the optimal sensor placement problem.

Nonproportional Damping

To this point it has been assumed that the structure is proportionally damped such that the normal modes decouple the system. In general, most structural systems are not proportionally damped. Therefore, it is important to examine the effects of nonproportional damping on the sensor placement problem for modal identification. For simplicity, the continuous time system will again be considered. In normal mode coordinates, the system matrix from Eq. (12) is given by

$$A = \begin{bmatrix} 0 & I \\ -\omega^2 & -D_q \end{bmatrix} \quad (37)$$

The normal modes decouple the mass and stiffness matrices, but the modal damping matrix D_q, given by

$$D_q = \Phi_f^T D \Phi_f \quad (38)$$

is in general fully populated due to the difference between the normal modes and the nonproportionally damped mode shapes.

Once again, accelerometers will be used in this analysis, but analogous results are obtained for displacement or velocity sensors. The corresponding output influence matrix in normal mode coordinates is given by

$$C_N = \begin{bmatrix} -\Phi_{fs}\omega^2 & -\Phi_{fs}D_q \end{bmatrix} \quad (39)$$

which results in the minimum generalized observability matrix

$$V_{N2} = \begin{bmatrix} -\Phi_{fs}\omega^2 & -\Phi_{fs}D_q \\ \Phi_{fs}D_q\omega^2 & -\Phi_{fs}\omega^2 + \Phi_{fs}D_q^2 \end{bmatrix} \quad (40)$$

As in the case of proportional damping, this matrix can be decomposed as follows

$$V_{N2} = S_2 Z_{N2} = \begin{bmatrix} \Phi_{fs} & 0 \\ 0 & \Phi_{fs} \end{bmatrix} \begin{bmatrix} -\omega^2 & -D_q \\ D_q\omega^2 & -\omega^2 + D_q^2 \end{bmatrix} \quad (41)$$

The determinant of Z_{N2} can be computed using Eq. (20) resulting in

$$|Z_{N2}| = \left|-\omega^2\right|\left|-\omega^2 + D_q^2 - D_q\omega^2\left(-\omega^2\right)^{-1}\left(-D_q\right)\right| = \left|\omega^4\right| \quad (42)$$

as in Eq. (21) for the proportionally damped case. The result is independent of the damping. As in previous cases, the observability of the system hinges on the matrix S_2. Maximization of the system observability reduces to placement of the sensors such that the Fisher information matrix determinant is maximized. The important thing to note is that the information matrix can be based solely on the normal modes of the system, regardless of the type of damping.

Effects of EfI on ERA and Q-Markov COVER

The relationship between the EfI sensor placement technique and the Eigen-System Realization method [20] of modal identification will be investigated first. Considering a discrete time system, the ERA method is based upon a block Hankel data matrix which can be written in the form

$$H_{pd}(j) = V_p A_D^j W_d \quad (43)$$

in which j represents the jth time step and W_d is a generalized controllability matrix given by

$$W_d = \begin{bmatrix} B_D & A_D B_D & \cdots & A_D^{d-1} B_D \end{bmatrix} \quad (44)$$

It is assumed that the actuator locations are fixed such that all the responding target modes are controllable implying that the rank of W_d is $2k$. It is important to note that sensor placement has no effect upon the controllability of the system.

The ERA method provides a minimum realization of the system which relies upon the singular value decomposition of the block Hankel matrix evaluated at $j=0$, which for the minimum generalized observability matrix is given by

$$H_{2d}(0) = V_2 W_d = UOY^T \quad (45)$$

in which U and V are orthonormal matrices and O is the corresponding matrix of singular values. In order to identify the $2k$ target modes, the block Hankel matrix must have rank $2k$. In general, the test data will contain noise, therefore, the rank of $H_{2d}(0)$ will be larger than $2k$ and there will be more than $2k$ positive singular values. In order to distinguish between target mode singular values and noise or computational singular values, the objective of the sensor placement method should be to maximize the magnitudes of the singular values of $H_{2d}(0)$ which correspond to the structural target modes. The minimum realization is generated by partitioning the matrix of singular values to the submatrix O_r which corresponds to the responding target modes. The corresponding minimum rank block Hankel matrix is then given by

$$H_{2d}(0) = V_2 W_d = U_r O_r Y_r^T \quad (46)$$

in which U_r and Y_r represent the first $2k$ columns of U and Y, respectively.

The effect of sensor placement upon the singular values of H_{2d} can be investigated by postmultiplying Eq. (46) by Y_r and taking advantage of the orthonormality of its columns, yielding

$$U_r O_r = V_2 W_d Y_r \quad (47)$$

Premultiplying each side of Eq. (47) by its transpose produces

$$O_r^T U_r^T U_r O_r = Y_r^T W_d^T V_2^T V_2 W_d Y_r \quad (48)$$

which, when utilizing the orthonormality of the columns of U_r, the diagonal form of O_r, and the definition of the observability grammian, becomes

$$O_r^2 = Y_r^T W_d^T W_o W_d Y_r \quad (49)$$

Noting that $W_d Y_r$ is a $(2k \times 2k)$ full rank matrix, the determinant of each side of Eq. (49) can be taken to produce

$$|O_r|^2 = |Y_r^T W_d^T||W_o||W_d Y_r| = |W_o||W_d Y_r Y_r^T W_d^T| \quad (50)$$

The columns of Y_r span the row space of the generalized controllability matrix W_d, therefore, $Y_r Y_r^T$ is an orthogonal projector [25] onto the column space of W_d^T. Using the fact that $Y_r Y_r^T W_d^T = W_d^T$, Eq. (50) can be expressed as

$$|O_r|^2 = |W_o||W_d W_d^T| = |W_o||W_c| \quad (51)$$

The geometric mean of the Hankel matrix singular values can then be expressed as

$$|O_r|^{1/2k} = |W_o|^{1/4k} |W_c|^{1/4k} \quad (52)$$

Therefore, for fixed actuator locations, maximizing the determinant of the observability grammian maximizes the geometric mean of the Hankel matrix singular values. As discussed in the previous section, the EfI sensor placement methodology maximizes the determinant of the Fisher information matrix which maximizes the determinant of the observability grammian. Thus the EfI sensor placement approach enhances the separation of target mode singular values from computational singular values.

As shown by Peterson, the ERA method [32, 33] uses a least squares approach to estimate the discrete time system matrix A_D from the time shifted block Hankel matrix

$$H_{2d}(1) = V_2 A_D W_d \quad (53)$$

The system matrix can then be estimated as

$$\hat{A}_D = V_2^+ H(1) W_d^+ \quad (54)$$

where the generalized inverses V_2^+ and W_d^+ are computed using the singular value decomposition of the Hankel matrix $H(0)$ in which

$$V_2 = U_r O_r^{1/2} \quad W_d = O_r^{1/2} Y_r^T \quad (55)$$

In order to investigate the effect of sensor placement, the least squares solution in Eq. (54) will be studied as a two step process. In the first step, assume that the generalized inverse of W_d has been generated and used to postmultiply Eq. (53) to form the expression

$$H(1) W_d^+ = V_2 A_D \quad (56)$$

For fixed actuator locations, sensor placement has no impact upon the generalized inverse of the controllability matrix. Equation (56) represents a straightforward least squares estimation problem in which the discrete system matrix can be estimated as

$$\hat{A}_D = \left(V_2^T V_2\right)^{-1} V_2^T H(1) W_d^+ = \left(W_o\right)^{-1} V_2^T H(1) W_d^+ \quad (57)$$

where the Moore-Penrose form of the generalized inverse [25] has been used to invert V_2 to illustrate the effect of sensor placement. For this estimation problem, the observability grammian is the corresponding Fisher information matrix. Therefore, the best estimate for A_D is obtained by maximizing the determinant of the observability grammian. As previous analysis shows, sensor placement using the EfI method produces this result. Therefore, the EfI sensor placement methodology not only produces larger Hankel matrix singular values which enhances system order determination, it also provides a better estimate of the discrete time system matrix.

The effect of the EfI sensor placement methodology upon the Q-Markov COVER (QMC) can also be easily investigated. In contrast with the ERA approach, the Q-Markov COVER (QMC) uses a data matrix of the form

$$D_p = V_p X V_p^T \quad (58)$$

where X is the state covariance matrix. Minimization of the generalized observability matrix also minimizes the size of the data matrix used in QMC. Peterson [32] showed that in the limit as the integer d becomes very large, the data matrix D_p can be expressed in the form

$$D_p = H_{pd}(0) H_{pd}(0)^T \quad (59)$$

The QMC method determines the system order by performing a singular value decomposition of the data matrix D_p. It is obvious from the form of Eq. (59) that the singular values of D_p are just the squares of the singular values of the block Hankel matrix $H_{pd}(0)$.

Maximizing the singular values of the Hankel matrix for the ERA approach also maximizes the singular values of the data matrix D_p used in QMC. The EfI sensor placement method therefore minimizes the size of the required data matrix and enhances system order determination using either the ERA or QMC techniques.

In order to determine the discrete system matrix, the QMC method uses a time shifted generalized observability matrix. The minimum sized observability matrix required is

$$V_3 = \begin{bmatrix} C \\ CA_D \\ CA_D^2 \end{bmatrix} \quad (60)$$

in which $p=3$. Using Peterson's notation [32], the first $p-1$ row blocks and last $p-1$ row blocks are defined as

$$V_3^1 = \begin{bmatrix} C \\ CA_D \end{bmatrix} \quad V_3^2 = \begin{bmatrix} CA_D \\ CA_D^2 \end{bmatrix} \quad (61)$$

Note that V_3^1 is just V_2 from the ERA analysis. These time shifted observability matrices are related by the expression

$$V_3^1 A_D = V_3^2 \quad (62)$$

The least squares estimate of the discrete system matrix is then given by

$$\hat{A}_D = \left[V_2^T V_2\right]^{-1} V_2^T V_3^2 = W_o^{-1} V_2^T V_3^2 \quad (63)$$

The form of Eq. (63) indicates that the observability grammian is the corresponding Fisher information matrix. As in previous analysis, maximizing the information matrix determinant provides the best estimate of the system matrix. Therefore, the EfI sensor placement methodology enhances the estimation of the discrete system matrix for both the ERA and QMC modal identification techniques. Further details on the generation of the generalized observability matrix from the data matrix D_p can be found in Refs. [32, 33].

Numerical Example using Polyreference

In order to further illustrate the advantages of using the EfI sensor placement method in conjunction with system-realization based modal identification techniques, the sensor placement approach will be applied in combination with the polyreference modal identification method to actual test data. The polyreference method is also a realization technique which is based upon a block Hankel matrix of Markov parameters and its singular value decomposition. However, the technique provides an observable canonical-form realization where the system is observable but not necessarily controllable. Therefore, the resulting identified system is not a minimum realization. Further technical details concerning the formulation of the polyreference technique and its relation to realization theory can be found in Refs. [22, 34].

The example test data selected for this paper is based on a modal survey performed on the Controls-Structures Interaction Evolutionary Model (CEM) testbed at NASA LaRC Space Structures Research Laboratory. Test data was obtained and stored in the form of frequency response functions for 103 sensor locations which contained a subset of 41 low frequency accelerometers selected based upon kinetic energy and engineering judgment. Sixty seven modes were identified using direct parameter estimation and polyreference with the data from all 103 sensor locations. Twenty seven analytical modes were predicted between 0.0 and 34.0 Hz. Twenty four of these modes were accurately correlated with the test modes. Of these twenty four modes, sixteen were used as target modes in this investigation ranging in frequency from 3.1 to 31.0 Hz. The first seven analytical modes were excluded because their low frequencies would give the 41 low frequency accelerometers an unfair advantage. Modes 9 and 11 were excluded because they are cable modes which are not observable from the 41 low frequency accelerometer locations. Mode 17 was excluded because it had poor test-analysis correlation. The target mode set therefore included analytical modes 8, 10, 12-16, and 18-26. The corresponding frequencies are listed in Table 1.

The 103 sensor locations were used as the initial candidate set for the EfI sensor placement process. In order to exaggerate the difficulties associated with modal identification using a small number of sensors and the need for systematic sensor placement strategies, the minimum allowable number of sensors were used to identify the target modes, which in this case was sixteen. The EfI sensor placement methodology discussed previously was applied to the initial candidate sensor set. Eighty seven iterations were performed to reduce the candidate sensor set to the best sixteen locations. Figure 1 illustrates the Fisher information matrix determinant versus iteration. It is easy to see that the EfI method maintains the magnitude of the determinant as candidate sensor locations are deleted. The best sixteen EfI sensor locations, designated as Set 1, are illustrated in Fig. 2. The information matrix determinant for the EfI sensor configuration has a value of 2.73×10^{17}.

A second and third set of sixteen sensor locations were selected for comparison purposes using modal kinetic energy which is commonly used to place sensors for modal identification. Set 2 was selected from the 41 low frequency accelerometer locations. The sixteen locations possessing the largest modal kinetic energy averaged over all sixteen target modes were selected for this sensor configuration. Set 2, illustrated in Fig. 3, has an information matrix determinant value of 5.06×10^8. Set 3 was selected from the initial candidate set of 103 sensor locations. The sixteen locations possessing the largest modal kinetic energy averaged over all sixteen target modes were selected for this sensor configuration. Set 3, illustrated in Fig. 4, has an information matrix determinant value of 4.23×10^{-18}. The EfI sensor configuration clearly has the largest information matrix determinant and is thus the superior sensor configuration for modal identification.

Subsets of the test data were selected to correspond to the three sensor configurations. Each of the subsets was used to compute the multivariate mode indicator function (MMIF) [35] to see whether all sixteen of the target modes were indicated. All of the excitation references (four) used in conducting the multi-input modal survey were employed in the MMIF generation. For comparison, the full 103 and the 41 low frequency transducer sets were also evaluated. The 103 sensor set was considered to be the "true" structural representation since it contained the most information. There were noticeable differences in the quality of the MMIF which resulted from the various sixteen sensor subsets of data. "Noisier" MMIF resulted at low frequency for Set 1 and Set 3, which included a mix of low frequency and other transducers. This was believed to be the result of poorer physical characteristics at low frequencies for the other transducers contained in these sets. Set 2, which comprised only low frequency transducers, yielded a MMIF which was smoother and similar to that obtained from the 41 transducer set.

After evaluating the MMIF, a modal parameter extraction was performed using the polyreference method. As part of this process, a correlation matrix was generated using each of the subsets of data. The same size correlation matrix was used in all cases so that there would be no bias due to matrix size limitations. Selection of the actual number of poles used for the final modal data extraction was made based on the proper identification of the target mode frequencies. Once an appropriate number of poles was selected, the modal parameters were extracted. Mode shapes were generated for each of the subsets of data in order to determine how effective each sensor set was at identifying linearly independent target modes. The modal assurance criterion (MAC) [36] was used as the basis to assess the quality of the extracted mode shape data for each sensor set. The ijth term in the *MAC* matrix is given by the relation

$$MAC_{ij} = \frac{\left(\Phi_i^T \Phi_j\right)^2}{\|\Phi_i\|^2 \|\Phi_j\|^2}$$

If the two mode shapes Φ_i and Φ_j are from the same set, the computation is referred to as a self-MAC, otherwise it is referred to as a cross-MAC. Ideally, off-diagonal terms should be less than 0.10, and, in the cross-MAC, the diagonal terms should also be greater than 0.90 for good agreement.

Initially, the self-MAC computation was made for the complete 103 sensor set modes. There were no off-diagonal terms greater than 0.10 and their RMS value was 0.02 indicating clean linearly independent test mode shapes. The 103 sensor set modes were then partitioned to each of the three sensor subset degrees of freedom and used as a baseline for cross-MAC comparisons with the modes obtained using the data from the three sensor subsets. The cross-MAC computations for all three sensor sets predict diagonal values of approximately 1.0 indicating that, for the amount of test data used, each of the sensor configurations was able to accurately identify the sixteen target test modes. However, the cross-MAC computations for the three sensor sets differed significantly in the off-diagonal terms. Figure 5 shows the cross-MAC results for sensor Set 1, which was derived using the EfI technique. The off-diagonal terms have an RMS value of 0.04, with 12 being over 0.10. The cross-MAC comparison for sensor Set 2, which was selected from the 41 low frequency sensor locations using target modal kinetic energy, was poorer in the off-diagonal terms and the order of target modes 5 and 6 was switched as illustrated in Fig. 6. The off-diagonal terms have an RMS value of 0.08, with 25 being greater than 0.10. The worst off-diagonal comparison was obtained using Set 3 as shown in Fig. 7. This set was selected from the full 103 available sensor locations based upon target modal kinetic energy. The RMS value of the off-diagonal terms is 0.17, with 42 terms larger than 0.10. Both Sets 2 and 3 produce mode shapes which are highly coupled while Set 1 produces modes which more accurately approximate the coupling level produced by the test modes at all 103 sensor locations.

It is clear from these computations that the sensor set selected based upon Effective Independence provides more linearly independent test mode shapes which is essential for accurate test/analysis correlation and finite element model updating. The theory presented in this paper indicates that this is also a vital characteristic when using small amounts of test data to extract the target modes. This will be investigated further in the future using the ERA modal identification technique.

Conclusion

In cases where limited numbers of sensors are available for modal identification, it is vital that a systematic method is used for selecting their locations. A method called Effective Independence has been devised to place sensors such that the locations maintain the linear independence of the modal partitions. This is required for post-test correlation and finite element model updating. This paper has shown analytically that the EfI method of sensor placement also enhances the modal identification process itself using system realization methods. It was shown that maximizing the independence of the target modes minimizes the size of the observability matrix required to render the system observable using displacement, velocity, or acceleration sensors. This also minimizes the size of the required test data matrices used in the modal identification process. By maximizing the determinant of the Fisher information matrix, the EfI method was shown to maximize the observability of the system. The effects of discretization and nonproportional damping were examined and the EfI method was found to produce the same results. It was also shown that maximizing the determinant of the Fisher information matrix maximizes the singular values of the data matrices in both the ERA and QMC methods. The singular values are used to differentiate between target modes and computational modes. Maximizing the determinant also enhances the estimation of the discrete system matrix in both methods. A numerical example using test data from the Controls-Structures Interaction Evolutionary Model (CEM) testbed at NASA LaRC Space Structures Research Laboratory showed that a minimum sized set of sixteen sensors derived using EfI is capable of accurately identifying sixteen target modes generated using a complete set of 103 sensors. In addition, the linear independence of the target modes derived using the EfI sensor set was comparable to the linear independence of the complete 103 sensor target modes. It is believed that the Effective Independence method provides an efficient technique for selecting a sensor configuration which not only provides the modal data required by structural dynamicists, but also produces an enhanced modal identification.

Acknowledgment

The authors wish to thank NASA LaRC for the use of the data in this paper. Specific thanks go to Ken Elliott, who conducted the original testing and provided the modal data.

References

[1] Yu, T. K. and J. H. Seinfeld, "Observability and Optimal Measurement Location in Linear Distributed Parameter Systems," *International Journal of Control*, Vol. 18, No. 4, 1973, pp. 785-799.

[2] Juang, J. N. and G. Rodriguez, "Formulation and Application of Large Structure Sensor and Actuator Placement," *2nd VPI&SU/AIAA Symposium on Dynamics and Control of Large Flexible Spacecraft*, 1979, pp. 247-262.

[3] Norris, G. A. and R. E. Skelton, "Selection of Dynamic Sensors and Actuators in the Control of Linear Systems," *Journal of Dynamic Systems, Measurement, and Control*, Vol. 111, 1989, pp. 389-397.

[4] Baruh, H. and K. Choe, "Sensor Placement in Structural Control," *Journal of Guidance, Control, and Dynamics*, Vol. 13, No. 3, 1990, pp. 524-533.

[5] DeLorenzo, M. L., "Sensor and Actuator Selection for Large Space Structure Control," *Journal of Guidance, Control, and Dynamics*, Vol. 13, No. 2, 1990, pp. 249-257.

[6] Maghami, P. G. and S. M. Joshi, "Sensor-Actuator Placement for Flexible Structures with Actuator Dynamics," *submitted to the Journal of Guidance, Control, and Dynamics*, 1991.

[7] Lim, K. B., "Method for Optimal Actuator and Sensor Placement for Large Flexible Structures," *Journal of Guidance, Control, and Dynamics*, Vol. 15, No. 1, 1992, pp. 49-57.

[8] Goodson, R. E. and M. P. Polis, "Identification of Parameters in Distributed Systems," *Distributed Parameter Systems*, edited by W. H. Ray and D. G. Lainiotis, Marcel Dekker, New York, 1978.

[9] Gupta, N. K. and W. E. Hall, "Design and Evaluation of Sensor Systems for State and Parameter Estimation," *Journal of Guidance, Control, and Dynamics*, Vol. 1, No. 6, 1978, pp. 397-403.

[10] Le Pourhiet, A. and L. Le Letty, "Optimization of Sensor Locations in Distributed Parameter System Identification," *Identification and System Parameter Estimation*, edited by Rajbman, North-Holland Co., Amsterdam, 1978, pp. 1581-1592.

[11] Shah, P. C., and F. E. Udwadia, "A Methodology for Optimal Sensor Locations for Identification of Dynamic Systems," *Journal of Applied Mechanics*, Vol. 45, 1978, pp. 188-196.

[12] Qureshi, Z. H., T. S. Ng and G. C. Goodwin, "Optimum Experimental Design for Identification of Distributed Parameter Systems," *International Journal of Control*, Vol. 31, 1980, pp. 21.

[13] Sirlin, S. W., R. W. Longman, and J. N. Juang, "Identifiability of Conservative Linear Mechanical Systems," Vol. 33, No. 1, 1985, pp. 95-118.

[14] Udwadia, F. E. and J. A. Garba, "Optimal Sensor Locations for Structural Identification," *JPL Workshop on Identification and Control of Flexible Space Structures*, 1985, pp. 247-261.

[15] Salama, M., T. Rose and J. Garba, "Optimal Placement of Excitation and Sensors for Verification of Large Dynamical Systems," *AIAA/ASME/ASCE/AHS 28th Structures, Structural Dynamics, and Materials Conference*, Monterey, CA, 1987, pp. 1024-1031.

[16] Kammer, D. C., "Sensor Placement for On-Orbit Modal Identification and Correlation of Large Space Structures," *Journal of Guidance, Control, and Dynamics*, Vol. 14, No. 2, 1991, pp. 251-259.

[17] Kammer, D. C., "Effects of Noise on Sensor placement for On-Orbit Modal Identification of Large Space Structures," Vol. 114, No. 3, 1992, pp. 436-443.

[18] Lim, T. W., "Sensor Placement for On-Orbit Modal Testing," Vol. 29, No. 2, 1992, pp. 239-246.

[19] Lim, T. W., "Actuator/Sensor Placement for Modal Parameter Identification of Flexible Structures," Vol. 8, No. 1, 1993, pp. 1-13.

[20] Juang, J. and R. S. Pappa, "An Eigensystem Realization Algorithm for Modal Identification and Model Reduction," *Journal of Guidance, Control, and Dynamics*, Vol. 8, No. 5, 1985, pp. 620-627.

[21] King, A. M., U. B. Desai and R. E. Skelton, "A Generalized Approach to q-Markov Covariance Equivalent Realizations for Discrete Systems," Vol. 24, No. 4, 1988, pp. 507-515.

[22] Vold, H., J. Kundrat, G. T. Rocklin and R. Russel, "A Multi-Input Modal Estimation Algorithm for Minicomputers," Society of Automotive Engineers, No. 820194, 1982.

[23] Middleton, D., *An Introduction to Statistical Communication Theory*, McGraw-Hill, New York, NY, 1960.

[24] Fedorov, V. V., *Theory of Optimal Experiments*, Academic, New York, NY, 1972.

[25] Ben-Israel, A. and T. N. E. Greville, *Generalized Inverses: Theory and Application*, John Wiley and Sons, New York, NY, 1974.

[26] Poston, W. L. and R. H. Tolson, "Maximizing the Determinant of the Information Matrix with the Effective Independence Method," Vol. 15, No. 6, 1992, pp. 1513-1514.

[27] Kammer, D. C. and L. Yao, "Enhancement of On-Orbit Modal Identification of Large Space Structures through Sensor Placement," accepted for publication in the *Journal of Sound and Vibration*, 1993.

[28] Horn, R. A. and C. R. Johnson, *Matrix Analysis*, Cambridge University Press, New York, 1992.

[29] Guangqian, X. and P. M. Bainum, "Actuator Placement Using Degree of Controllability for Discrete-Time Systems," *Journal of Dynamic Systems, Measurement, and Control*, Vol. 114, No. 3, 1992, pp. 508-516.

[30] Skelton, R. E., *Dynamic Systems Control: Linear Systems Analysis and Synthesis*, John Wiley & Sons, New York, 1988.

[31] Ogata, K., *Discrete-Time Control Systems*, Prentice-Hall, Inc., Englewood Cliffs, 1987.

[32] Peterson, L. D. and S. J. Bullock, "A Comparison of the Eigensystem Realization Algorithm and the Q-Markov COVER," *32nd AIAA Structures, Structural Dynamics, and Materials Conference*, Baltimore, MD, 1991, pp. 2169-2178.

[33] Peterson, L. D., "Efficient Computation of the Eigensystem Realization Algorithm," *33rd AIAA Structures, Structural Dynamics, and Materials Conference*, Dallas, TX, 1992, pp. 1122-1131.

[34] Juang, J. N., "Mathematical Correlation of Modal-Parameter-Identification Methods via System-Realization Theory," *International Journal of Analytical and Experimental Modal Analysis*, 1987, pp. 1-18.

[35] Williams, R., J. Crowley and H. Vold, "The Multivariate Mode Indicator Function in Modal Analysis," *Third International Modal Analysis Conference*, 1985.

[36] Allemang, R. J. and D. L. Brown, "A Correlation Coefficient for Modal Vector Analysis," *1st International Modal Analysis Conference*, Orlando, FL, 1982.

Table 1. Target mode natural frequencies.

Analytical Mode	Freq. (Hz.)	Test Mode	Freq. (Hz.)
8	3.125	12	3.249
10	3.683	13	3.796
12	6.122	16	6.401
13	7.827	19	8.001
14	9.451	25	9.919
15	9.769	26	9.975
16	11.853	34	12.310
18	13.418	36	13.656
19	13.572	37	13.909
20	14.113	38	14.508
21	14.715	39	15.066
22	16.556	43	17.151
23	20.114	50	21.000
24	23.258	56	24.416
25	30.724	67	31.784
26	31.011	64	31.302

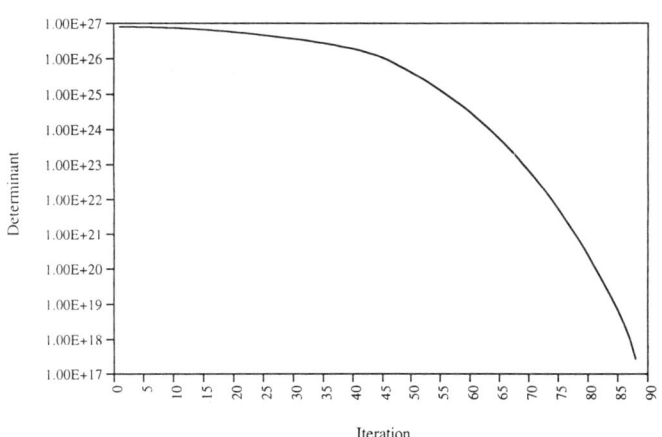

Fig. 1. Fisher information matrix determinant by iteration for EfI sensor placement method.

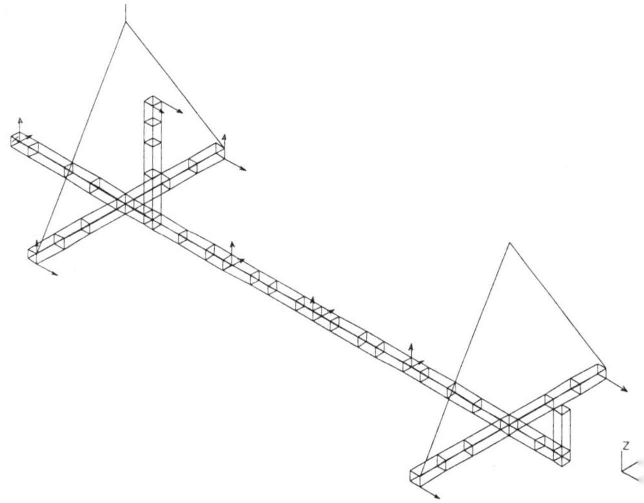

Fig. 3. Sensor set 2 selected based on modal kinetic energy from 41 low frequency accelerometers.

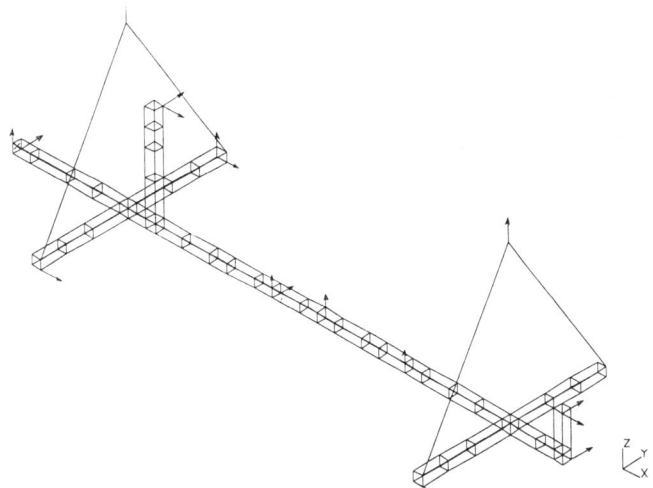

Fig. 2. Sensor set 1 selected using EfI from 103 sensor candidate set.

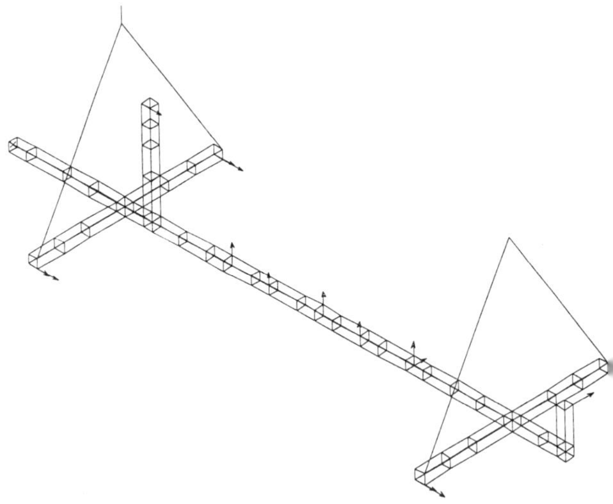

Fig. 4. Sensor set 3 selected based on modal kinetic energy from 103 sensor candidate set.

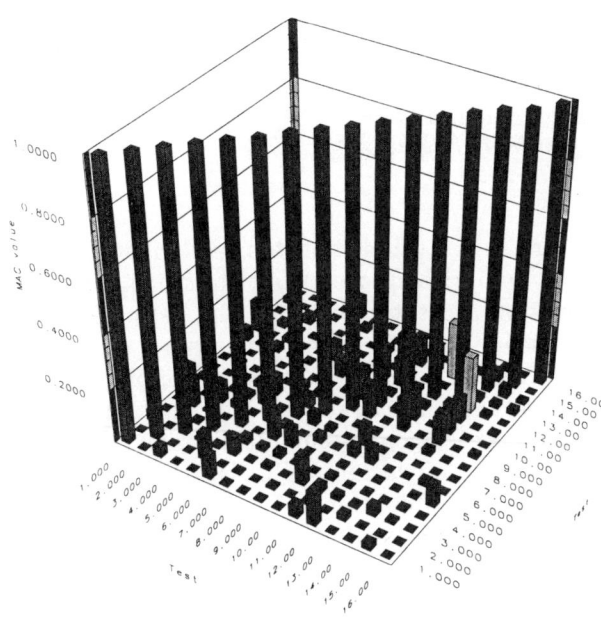

Fig. 5. Cross-MAC for sensor set 1 selected using EfI.

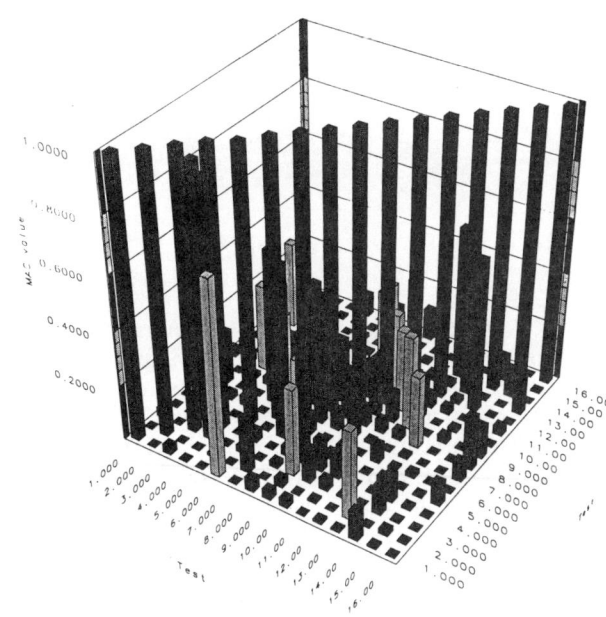

Fig. 7. Cross-MAC for sensor set 3 selected using kinetic energy from 103 candidate sensor set.

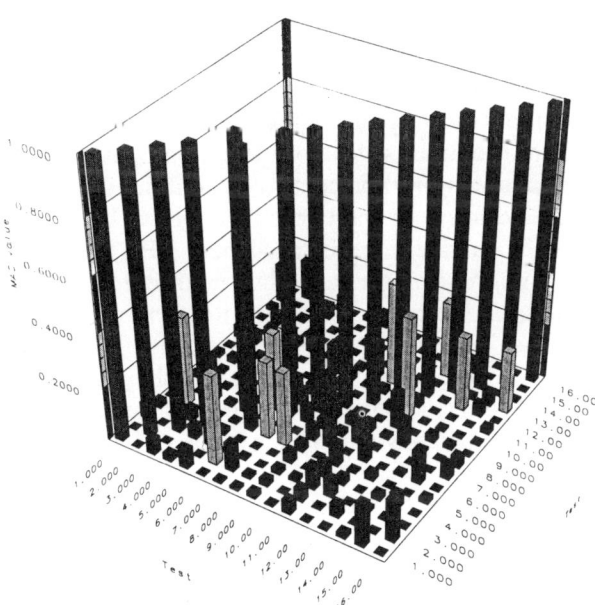

Fig. 6. Cross-MAC for sensor set 2 selected using kinetic energy from 41 low frequency accelerometers.

A TIME AND FREQUENCY DOMAIN PROCEDURE FOR IDENTIFICATION OF STRUCTURAL DYNAMIC MODELS

L.D. Peterson[1] and K.F. Alvin[2]
University of Colorado
Center for Aerospace Structures and
Department of Aerospace Engineering Sciences
Boulder, Colorado 80309-0429

ABSTRACT

A system identification procedure is presented which uses a time domain least squares curve fit to extract system poles and a frequency domain curve fit to extract system modal vectors. The time domain pole extraction relies on an asymptotically large Hankel matrix to converge the modal frequencies and damping ratios. The modal vectors are then extracted using the converged system poles in a linear least squares curve fit in the frequency domain. An important component of the frequency domain curve fit is that it includes an improper term as an unknown of the curve fit. This compensates for the fact that most modal vibration data is significantly influenced by modes outside the bandwidth of the data collection procedure. The resulting minimal order model achieves excellent time and frequency domain accuracy without the need for a nonlinear regression. An application to experimental test data is presented to demonstrate the accuracy of the method.

INTRODUCTION

Obtaining accurate estimates of modal parameters from test data is an essential part of designing and analyzing aerospace structures. Finite element models are often uncertain in the characteristics of critical structural parameters, such as interface compliances, boundary conditions, and material damping. Experimental modal parameters can be directly compared to the modes of the finite element model to validate the uncertain components of the model. For this reason, the validity of an analytical structural model often relies on the accuracy of experimentally derived modes.

As both the precision required of structural models and the modal complexity of the structures themselves have increased, it has become more challenging to obtain accurate estimates of modal parameters. A central part of this challenge is improving the numerical procedure with which the modal parameters are extracted from the data. Most current research has focused primarily on state-space system realization procedures to accomplish the modal parameter extraction. Most notable among these is the Eigensystem Realization Algorithm (ERA) [1], and its variants ERA/DC[2] and OKID[3]. The Q-Markov Covariance Equivalent Realization (QMC)[4] is a similar, but related algorithm[5], and the Polyreference algorithm,[6] which is more widely used within the modal test community, has similar algorithmic properties.[7]

These algorithms all attempt to find a discrete state-space model of the multi-input multi-output (MIMO) structural test data by minimizing particular error norms in the time domain. Although each performs well on simulated data sets, none has been shown to have a distinct advantage when applied to actual modal test data. In fact, the accuracy of the modal identification often depends more on the experience of the analyst than on the details of the algorithm. The major shortcoming is that the high time domain accuracy of the curve fit is not well reflected in the frequency domain accuracy. ERA, ERA/DC, QMC, and Polyreference models usually retain large errors in the frequency domain near the Frequency Response Function (FRF) zeros. This is an error which translates to inaccuracies in the modal vectors, and reflects poor knowledge of the stiffness and mass distribution within the structure.

In Reference [8], we examined the possibility that increasing the data set size would alleviate this problem. We found, however, that even an order of magnitude increase in the Hankel matrix size does not significantly improve the error near the zeros. It does, however, converge the poles of the structure very tightly. This feature is exploited in this paper in a frequency domain curve fit adjustment of the ERA modal vectors using the converge values of the structural poles.

Several researchers have also investigate similar methods. Jacques and Miller in Reference [9] begin with a set of ERA poles determined by overspecifying

---
1. Assistant Professor, Senior Member AIAA, Associate Member ASME
   ldpeter@Colorado.EDU
2. Post Doctoral Research Associate Member AIAA

Copyright © by L.D. Peterson and K.F. Alvin, 1994
Printed by the American Institute of Aeronautics and Astronautics with permission.

the model order sufficiently to obtain close estimates of the frequency and damping ratios for the identified modes. They then begin to remove states from the ERA model while adjusting the system residues (modal vectors) using a nonlinear least squares curve fit to the logarithmic error in the frequency domain. Once this has converged, their procedure concludes with a nonlinear least squares curve fit to the absolute error in the frequency domain. This last nonlinear regression includes a readjustment of the system poles. A similar type of frequency domain adjustment was also developed for Polyreference by Mayes in Reference [10]. In this approach, Polyreference is used to determine modal frequencies and damping ratios. After this initial curve fit, the modal residues (modal vectors) are determined using a mode-by-mode curve fit near each individual modal resonance.

In contrast to these frequency domain adjustment methods, Horta and Juang developed a matrix fraction decomposition (MFD) approach in Reference [11]. This method parameterizes the system model using the matrix fraction decomposition in the z-domain. It then fits either the FRF or the auto/cross spectra data using a linear least squares curve fit. This results in a large order state-space model which must be reduced. Rather than use more direct model reduction techniques applied to the system matrices themselves, the authors in [11] found it more computationally efficient to generate the Markov sequence of the frequency domain curve fit and pass it to ERA/DC to obtain a minimal order realization. In this way, the MFD method preconditions the vibration data using a frequency domain curve fit before ERA/DC is applied. In the sense that the MFD method is a preconditioning of the data, it is similar in its point of view to that of OKID,[3] which begins with a time domain preconditioning.

All these methods report similar and, indeed, improved levels of system model accuracy in the frequency domain. The Jacques and Miller method [9] has the advantage of using good initial pole estimates, but requires a nonlinear regression. Nonlinear regression can lead to multiply-valued numerical searches unless the starting point is very close to the actual poles. The Mayes procedure [10] uses only a well-determined linear regression, but it requires a mode by mode adjustment of the data. The Horta and Juang method [11] has the advantage of using only an overdetermined linear regression, but it requires a model reduction step using synthesized Markov parameters as analyzed by ERA.

In contrast, the procedure we present in this paper achieves high frequency domain accuracy using only linear regressions without complex model reduction procedures. As shown in Figure (1), the proposed

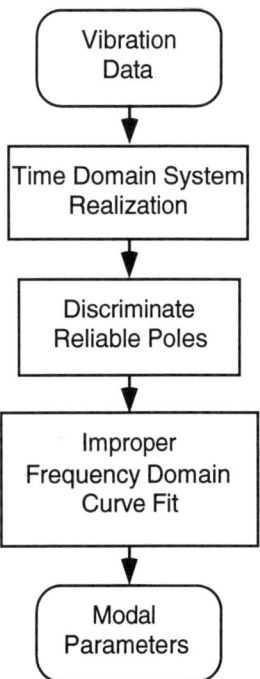

Figure 1. Block Diagram of the Proposed Time and Frequency Domain Procedure

algorithm has three main parts: a time domain system realization procedure, a discrimination of reliable poles from the realized model, and an improper frequency domain curve fit to correct the modal vectors. The time domain part of the procedure uses the very large ERA data analysis enabled by [12] to obtain pole estimates. By careful discrimination of extraneous poles, our procedure provides converged estimates of the poles which will not require refinement using a nonlinear regression. The frequency domain modal vector determination is a derivative of the method developed by Horta and Juang in [11]. We do not, however, include the system poles as free parameters. In addition, building on the experimental observations reported in [13], we include improper terms in the frequency domain linear regression. An "improper" term is an additional pole at infinity in the s-plane added to the state-space z-domain FRF. Such a formulation is missing from the aforementioned references or any other frequency domain curve fit algorithm of which we are aware. As described below, including such a improper term in the frequency domain parameterization results in very high accuracy near the system zeros as well as the poles using only a linear regression.

The paper is organized to present, discuss, and evaluate each step diagrammed in Figure (1). The first two sections review the time domain system realization procedure and the discrimination of reliable poles using modal quality indicators. The third section presents the improper frequency domain curve fit procedure, and is followed by a detailed specification of the steps followed in the procedure. The last section presents and discusses an application to experimental data which illustrates the relative merit of the proposed method. It focuses on the realized improvement in model accuracy provided by each step in the procedure.

TIME DOMAIN SYSTEM REALIZATION PROCEDURE

The response of a structure to a set of forces or inputs $u(t)$ is usually modeled as a spatially discretized second order matrix differential equation of the form:

$$\mathcal{M}\ddot{q} + \mathcal{D}\dot{q} + \mathcal{K}q = \hat{B}u \qquad (1)$$

where $\mathcal{M}$ is the mass matrix, $\mathcal{D}$ is the damping matrix, $\mathcal{K}$ is the stiffness matrix, and $\hat{B}$ is the force influence matrix. The vector $q(t)$ includes all the physical degrees of freedom (DOF) of the model. If we define the n associated normal modes Φ_n of Equation (1) according to:

$$\mathcal{K}\Phi_n = \mathcal{M}\Phi_n \Omega \qquad (2)$$

$$\begin{aligned} \Phi_n^T \mathcal{K} \Phi_n &= \Omega = \{\omega_{ni}, i = 1...n\} \\ \Phi_n^T \mathcal{M} \Phi_n &= I_{n \times n} \\ \Phi_n^T \mathcal{D} \Phi_n &= \Xi \end{aligned} \qquad (3)$$

then the structural model can be placed into the first order modal state-space form:

$$\begin{aligned} \dot{x}_\eta(t) &= A_\eta x_\eta(t) + Bu(t) \\ y(t) &= C_\eta x(t) + Du(t) \end{aligned} \qquad (4)$$

in which y(t) is a vector of measured responses, and the modal state-space matrices are given by:

$$A_\eta = \begin{bmatrix} 0 & I \\ -\Omega & -\Xi \end{bmatrix} \quad B_\eta = \begin{bmatrix} 0 \\ \Phi_n^T \hat{B} \end{bmatrix} \qquad (5)$$

$$C_\eta = [H_d \ 0] + [H_v \ 0]A_\eta + [H_a \ 0]A_\eta^2$$

in which H_d, H_v, and H_a are the output displacement, velocity, and acceleration location influence arrays, respectively.

Because all experimental vibration data is sampled in time, all time domain linear system realization procedures begin from the presumption that a finite order discrete state-space model of the system exists of the form:

$$\begin{aligned} x(k+1) &= Ax(k) + Bu(k) \\ y(k) &= Cx(k) + Du(k) \end{aligned} \qquad (6)$$

in which k is the time sample index. The procedure by which Equation (4) is sampled to lead to Equation (6) must be done carefully to avoid illconditioning due to the transformation from the continuous (s) plane to the discrete (z) plane. Likewise, the transformation from a realized model of the form of Equation (6) back to the continuous representation of Equation (4) requires careful eigenrotation and mass normalization, as described in Reference [14].

When the model of Equation (6) is used as a predictor, the arbitrary response to an input $u(k)$ is given by:

$$y(k) = \sum_{i=1}^{k} M(k-i)u(i) \qquad 1 \le k < \infty \qquad (7)$$

in which the system Markov parameters $M(k)$ are related to the state-space matrices by

$$M(k) = \begin{cases} D & k = 0 \\ CA^{k-1}B & k > 0 \end{cases} \qquad (8)$$

All state-space time domain realization methods attempt to find the state space matrices A, B, C, D from measurements of the sequence $M(k)$. This is the process known as *system realization*.

The essential considerations in system realization are the selection of the model order (it is presumed that the model form is correct) and the determination of the state-space parameters from a minimization of some prediction error. For ERA, the prediction error is defined in terms of a Hankel matrix of the Markov parameters, as defined by:

$$H_{rs}(k-1) = \begin{bmatrix} M(k) & M(k+1) & \dots & M(k+s-1) \\ M(k+1) & M(k+2) & \dots & M(k+s) \\ \dots & \dots & \dots & \dots \\ M(k+r-1) & M(k+r) & \dots & M(k+r+s-2) \end{bmatrix} \qquad (9)$$

The ERA realization finds the linear least squares solution to minimize the error in the shift in the Hankel matrix of the system model and the data according to:

$$H_{rs}(k-1) = V_r A^{k-1} W_s \quad (10)$$

in which

$$V_r = \begin{bmatrix} C \\ CA \\ \ldots \\ CA^{r-1} \end{bmatrix} \quad W_s = \begin{bmatrix} B & AB & \ldots & A^{s-1}B \end{bmatrix} \quad (11)$$

If the Hankel matrix is formed from the data, then the factors V_r and W_s are obtained from a singular value decomposition (SVD) of the 0-th Hankel matrix according to:

$$H_{rs}(0) = \tilde{P}_r S_{rs} \tilde{Q}_s^T$$
$$V_r = \tilde{P}_r S_{rs}^{1/2} \quad W_s = S_{rs}^{1/2} \tilde{Q}_s^T \quad (12)$$

The model order is selected (in principle) by examining the numerical rank of $H_{rs}(0)$. From this, the system realization problem is solved by:

$$A = S_{ro}^{-1/2} \tilde{P}_r^T H_{rs}(1) \tilde{Q}_s S_{rs}^{-1/2}$$
$$B = S_{rs}^{1/2} \tilde{Q}_s^T(1:n_x, 1:n_u)$$
$$C = \tilde{P}_s(1:n_y, 1:n_x)$$
$$D = M(0) \quad (13)$$

Reference [12] discusses the computationally more efficient approach of factoring $H_{rs}(0)H^T_{rs}(0)$ instead of $H_{rs}(0)$ to obtain the factors of Equation (12). In this case, it is more computationally efficient to calculate the factors using a symmetric eigensolver in place of the SVD. By only computing the largest n_x eigenvalues and vectors of the Hankel matrix product, it is possible to determine realizations using very large values of r and s without calculating the entire spectral decomposition.

DISCRIMINATION OF POLES FROM THE TIME DOMAIN CURVE FIT

The experimental study of Reference [13] found two main problems with determining structural poles from an ERA curve fit, even when using large Hankel matrix sizes:

1. Most structural poles converge only after massive order overspecification, and some poles converge faster than others.

2. Poles which have converged can occasionally split into two or more nearly repeated modes as model order overspecification is increased to converge other poles.

In view of these pathologies, our procedure uses a combination of several quantitative modal quality indicators (MQI) to detect convergence and discriminate unwanted or unreliable modes from the A and B matrices which are subsequently used in the frequency domain curve fit. These MQI are defined in the next several sections. We note that none of these indicators is demonstrably superior over the others. In our experience, only a careful, combined examination of all of these MQI leads to reliable discrimination of structural poles.

Extended Modal Amplitude Coherence (EMAC)

The Extended Modal Amplitude Coherence (EMAC) was presented in Reference [15]. It is a measure of the ability of an identified mode to predict its component of the measured data from the initial Markov parameter. The first step is to project the V_r and W_s matrices into each mode. To do this, the A matrix is placed into 2×2 block modal (McMillan) form:[14]

$$A \rightarrow \Phi^{-1} A \Phi \quad (14)$$

Then, the V_r and W_s factors of $H_{rs}(0)$ are also placed into the same coordinate basis by:

$$V_r \rightarrow V_r \Phi^{-1}$$
$$W_s \rightarrow \Phi W_s \quad (15)$$

In this form, every mode has a corresponding pair of columns $\begin{bmatrix} v_r^n & v_r^{n+1} \end{bmatrix}$ of V_r and rows $\begin{bmatrix} w_s^n & w_s^{n+1} \end{bmatrix}^T$ of W_s which are the modal observability and controllability factors of the data matrix. For comparison, the simulated equivalent factors are formed for each mode by simulating the response of the ERA digital model using the block modal form. This leads to estimates $\begin{bmatrix} \hat{v}_r^n & \hat{v}_r^{n+1} \end{bmatrix}$ and $\begin{bmatrix} \hat{w}_s^n & \hat{w}_s^{n+1} \end{bmatrix}^T$. In this notation, then, the EMAC for a particular mode is defined to be the correlation between the first and last block rows of $\begin{bmatrix} \hat{v}_r^n & \hat{v}_r^{n+1} \end{bmatrix}$ and $\begin{bmatrix} v_r^n & v_r^{n+1} \end{bmatrix}$, combined with the correlation between the first and last block columns of $\begin{bmatrix} \hat{w}_s^n & \hat{w}_s^{n+1} \end{bmatrix}^T$ and $\begin{bmatrix} w_s^n & w_s^{n+1} \end{bmatrix}^T$. In this sense, EMAC is a relative measure of the ability of a particular mode in the ERA model to project forward onto the data. Reference [15] provides more details.

Modal Singular Value (MSV)

The Modal Singular Value (MSV) was proposed in Reference [16] as an MQI. Using the same matrices defined above for the EMAC calculation, it computes the MSV as the froebenius norm of the columns of $\begin{bmatrix} v_r^n & v_r^{n+1} \end{bmatrix}$ plus that of the rows of $\begin{bmatrix} w_s^n & w_s^{n+1} \end{bmatrix}^T$. This number is the contribution of the individual modes time history to the Hankel singular value sum. An interpretation of the MSV, then, is that is the relative power contributed by the individual identified mode. To use MSV as an MQI, it is presumed that poorly identified modes contribute little power to the measured data.

Consistent Mode Indicator (CMI)

The Consistent Mode Indicator (CMI) was developed in Reference [15]. It defines the CMI to be the product of the individual modal EMAC with the Modal Phase Colinearity (MPC) of the mode. The MPC value is an indication of the relative complexity of the individual mode shape. In this sense, CMI discriminates individual modes which are not only well-predicting in the data (EMAC) but also closer to the true normal mode shapes of the structure, rather than the damped complex modes.

Mass Scaling Parameter (MSP)

The Mass Scaling Parameter (MSP) evaluates the extent to which the identified mode shape can be normalized by the driving point measurement collocated with the applied forces. This parameter was developed as an outgrowth of our work in Reference [14]. Suppose that a continuous time modal state-space model has been developed in the form of Equation (4). If C_{ij} is the displacement mode shape value for mode i at DOF j, C_{iMAX} is the highest magnitude modal displacement for mode i, and B_{ij} is the continuous time modal participation factor of force for mode i at DOF j, then:

$$\text{MSP} = \left| \frac{C_{iMAX}}{C_{ij}} \right| sgn \frac{B_{ij}}{C_{ij}} \quad (16)$$

Modes with small components at the driving points are poorly excited and poorly normalized. In addition, a negative sign change between the B_{ij} and the C_{ij} indicates that the mode cannot be easily rotated into normal modal coordinates. Therefore, by discriminating modes with low or negative values of MSP, one will exclude modes for which the extracted normalized mode shape is unreliable.

IMPROPER FREQUENCY DOMAIN CURVE FIT

As explained previously, computationally efficient state space realization algorithms allow for a high degree of modal parameter convergence given massive model order overspecification. In addition, a variety of MQI allow discrimination and elimination of the nonphysical modes engendered by this overspecification. Utilizing this approach on the measured acceleration response of a variety of structures, however, we found consistent inaccuracies in the frequency domain reconstructions of the identified models. Thus, as described in the introduction, we examined a variety of state space model-based frequency domain identification methods to improve the frequency domain accuracy of the identified models.

Our goal is a curve fitting algorithm which will match all resonances to an accuracy equal to that of the time domain identification, but further improve the accuracy at points between resonances, particularly the zeros. We initially chose to implement the matrix fraction approach of [11] as a candidate alternative to ERA because of its attractive linear least-squares form. Problems were quickly found, however, with its utility in the identification of large order models and densely spaced modes. This is because the accurate identification of frequencies and damping ratios is highly sensitive to the selection of frequency domain data samples used in the least squares problem. The use of all frequency domain points to reduce this sensitivity was also found to be highly inefficient as compared to time domain algorithms and did not lead to an appreciable improvement in frequency domain accuracy. The first r Markov parameters used in time-domain realization, on the other hand, are rich in the response of all the observable modes and can produce modes with very low frequency and damping variances.

We came to realize that the highly converged modes enabled by the efficient ERA procedure were a strength to be exploited. Therefore, we reformulated the least squares problem to optimize new estimates of the output matrices while fixing the converged pole parameters obtained from the time domain identification. Using this approach, we found it possible to obtain highly accurate frequency domain reconstructions within limited frequency ranges. Unfortunately, when the entire frequency range was optimized, or at least when all resonance peaks were fit, the overall accuracy was similar to that obtained from the ERA modes and mode shapes without the frequency domain adjustment. Furthermore, in examining the highly accurate limited-range curve fit models, it was found that the modes above the curve fit range were significantly overestimated. Although this was in part a result of ill-conditioning in the numerics related to those modes, it was felt more importantly that this overestimation of the higher frequency modes was a consequence of the higher accuracy obtained at the low frequency transmission zeros.

This can be explained by examining the effects of high frequency modes on the frequency-based force-to-acceleration (admittance) response functions at low frequency. In the case of N proportionally damped modes, the FRF component from input location k to output location i can be written as:

$$G_{ik}(s) = \sum_{n=1}^{N} \frac{s^2 \Phi_{in} \Phi_{kn}}{s^2 + 2\zeta_n \omega_i s + \omega_i^2} \quad (17)$$

If we partition the modes into the N_{ID} whose resonances are identified, and those whose resonances are above the measurement bandwidth, we have

$$G_{ik}(s) = \sum_{n=1}^{N_{ID}} \frac{s^2 \Phi_{in} \Phi_{kn}}{s^2 + 2\zeta_n \omega_n s + \omega_n^2} + R_{ik}(s) \quad (18)$$

where the residual of the high frequency poles is:

$$R_{ik}(s) = \sum_{n=N_{ID}+1}^{N} \frac{s^2 \Phi_{in} \Phi_{kn}}{s^2 + 2\zeta_n \omega_n s + \omega_n^2} \quad (19)$$

Next, take the limit of this term at low frequency:

$$\lim_{s/\omega_n \to 0} R_{ik}(s) = \sum_{n=N_{ID}+1}^{N} \lim_{s/\omega_n \to 0} \frac{s^2 \Phi_{in} \Phi_{kn}}{s^2 + 2\zeta_n \omega_n s + \omega_n^2} \quad (20)$$
$$= s^2 F_{ik} + s^4 E_{ik} + \ldots$$

where

$$F_{ik} = \sum_{n=N_{ID}+1}^{N} \frac{\Phi_{in} \Phi_{kn}}{\omega_n^2} \quad E_{ik} = \sum_{n=N_{ID}+1}^{N} \frac{\Phi_{in} \Phi_{kn}}{\omega_n^4} \quad (21)$$

Thus, the modes above the Nyquist contribute to a sum term which biases the inertance FRF transmission zeros without contributing a resonance response within the measurement bandwidth. This term, however, cannot be properly expressed in the discrete state space model form. Therefore, contrasting this analytical result with our experiments in frequency domain model adjustment, we concluded that the overestimated modes outside the curve fitting region were attempting to account for the summed contribution of higher frequency modes toward low frequency biasing of the FRF zeros. This result also suggested that, by explicit inclusion of this improper term, our goal of increasing frequency domain accuracy while maintaining the resonance accuracy of the time domain model could be achieved.

This procedure is outlined as followed. If the data consists of N samples at an interval of Δt, the discrete ERA model's FRF can be sampled at equally spaced points on the unit circle in the z-plane to be equivalent to the continuous FRF at those frequencies. This means:

$$\hat{G}(f_k = \frac{k}{N\Delta t}) = C\left(e^{\frac{j2\pi k}{N}} I - A\right)^{-1} B + D \quad (22)$$

To this, we add the continuous s-plane form of the contributions of the modes outside the bandwidth, since there is no corresponding discrete state-space model for these modes. In the case of acceleration (admittance) data, our representation of the FRF becomes:

$$\hat{G}(f_k) = C\left(e^{\frac{j2\pi k}{N}} I - A\right)^{-1} B + D + \left(\frac{2\pi k}{N\Delta t}\right)^2 F \quad (23)$$

Note that we retain the continuous s-plane form of the modal contributions outside the Nyquist frequency up to the first (F) term in Equation (20).

To maintain the linear least squares form, and to exploit the highly converged pole estimates from the overspecified ERA analysis, we use the A and B matrices from the ERA curve fit, but retain C and F as unknown parameters. These are then estimated by minimizing the Froebenius norm of the error summed over all frequency samples. First, note that Equation (23) can be written as the matrix equation:

$$\hat{G}(f_k) = \begin{bmatrix} C & D & F \end{bmatrix} \begin{bmatrix} \left(e^{\frac{j2\pi k}{N}}I - A\right)^{-1} B \\ I \\ \left(\frac{2\pi k}{N\Delta t}\right)^2 I \end{bmatrix} \quad (24)$$

$$= \begin{bmatrix} C & D & F \end{bmatrix} S(f_k)$$

In this form, the error minimization problem is given by

$$\min_{[C\ D\ F]} \sum_{k \in k_{set}} \left\| G(f_k) - \hat{G}(f_k) \right\|^2 \quad (25)$$

and the solution is given by:

$$\begin{bmatrix} C & D & F \end{bmatrix} = \left\{ \sum_{k \in k_{set}} G(f_k) S(f_k)^* \right\} \left\{ \sum_{k \in k_{set}} S(f_k) S(f_k)^* \right\}^{-1} \quad (26)$$

in which $k \in k_{set}$ denotes the set of frequency samples to be included in the curve fit. We note that a pseudo inverse is generally unnecessary, as rich data typically results in a full rank matrix that can be inverted directly. We also note that frequency weights could be included to proportionately emphasize frequency bands of interest, but we have found that this is usually not necessary.

We should point out that the necessity of the improper term is most pronounced in data sets using admittance data. For other data sets, using displacement or velocity sensors, the error due to modes outside the Nyquist bandwidth is diminished by virtue of the integration.

As a final note, we make the observation that the inclusion of the improper term amounts to a change in the model form assumed in the parameter identification of ERA. In other words, the success of our identifications using this additional term suggests that the fundamental notion in modern system identification theory may be in question: namely, there may not generally exist a finite order discrete domain realization of data sampled from continuous systems.

SUMMARY OF THE PROPOSED PROCEDURE

Having discussed the elements of the proposed procedure in detail, it is possible to enumerate the precise steps in our procedure. This is done in the following list:

1. Estimate the number of modes in the data using the MIMO FRF Singular Value (SV) and the Modal Indicator Function (MIF).[15]

   a. Remember that single inputs cannot detect repeated modes; multiple inputs can.

   b. MIF's < 0.3 together with no apparent SV peak is probably not a mode.

2. Study the convergence of the Hankel singular values.

   a. Set a large initial model order.

   b. Vary number of Hankel block rows up to 2500.

   c. Vary number of Hankel matrix columns up to 4000 to 10000.

   d. Choose Hankel matrix sizes that have converged singular values of the Hankel matrix.

3. Vary model order for given Hankel matrix dimensions.

   a. Minimum model order = 1/2 number of identified peaks.

   b. Maximum model order = 4 times the number of identified peaks.

4. Study modal convergence and apply modal quality indicators (MQI) to eliminate poles as follows:

   a. Eliminate real and unstable poles

   b. Eliminate poles with EMAC < 70%

   c. Eliminate poles with CMI < 50%

   d. Eliminate modes with MSP < 1%

5. Examine split modal vectors and determine if the mode is repeated or single. If single, eliminate one of the repeated, split mode.

6. Reconstruct the MIMO model and compare with the data. Increase model order until all important peaks are captured and converged.

7. Use the system A and B matrices in a frequency domain curve fit via Equation (26) to find C, D, and the improper corrections to the discrete transfer function, F.

8. Change frequency domain bandwidth and weighting as necessary to adjust model accuracy in critical regions.

EXAMPLE APPLICATION

Test Structure and Data Collection Procedure

Figure (2) is a photograph of the testbed used to illustrate our modal identification procedure. It is a cantilevered truss consisting of four half meter bays with a 50 pound steel plate tip mass. FRF data was collected using a pseudo-random force input applied with an electromagnetic shaker in voltage-coupled mode. Auto and cross-spectra were averaged for 50 ensembles consisting of 8192 samples collected at a rate of 500 Hz. A total of 62 accelerometer measurements were collected, including a driving point measurement for mass normalization of the mode shapes. The input signal was band limited to 240 Hz and the sensors were appropriately filtered to avoid aliasing of signals above the Nyquist frequency.

Convergence of the Hankel Matrix Singular Values

The first step is to determine the dimension of the Hankel matrix by varying the number of block rows, r. In this example, we fixed the number of block columns at 4000, and increased r until the singular values were converged. For the subsequent analyses, we selected a value of $r=40$.

Modal Convergence and Discrimination of Unreliable Poles

Having fixed the dimensions of the Hankel matrix, the next step is to vary the model order from a minimum of half the number of detected peaks in the FRF's to 4 times the number of peaks. For this data, there were approximately 50 resonances identified using the MIF, so the convergence was studied for model orders from 50 states to 400 states. Based on the convergence of the value of the dominant natural frequencies, the model was presumed converged at a model order of 400 states, or 200 modes. As a basis for

Figure 2. Four-Bay Cantilevered Truss Used in this Research

comparing the subsequent results, Figure (3) shows the driving point FRF data and its reconstruction using this 200 mode ERA model. Note that this greatly overspecified model captures the resonance peaks, but has degraded accuracy near the structural FRF zeros. Note also the phase lead error in the model from 50 to 200 Hz, an error which could lead to degraded performance were this model used in an active control design.

From this data set, 124 modes were identified using the discrimination criteria set forth above. Table (1) shows the MQI for the first 30 modes identified by the ERA analysis. The far right column indicates the modes which were chosen for subsequent analysis. Figure (4) shows the reconstruction of the driving point data for the resulting 124 mode ERA model. Note that there is little substantial change between this reconstruction and that of the 200 mode ERA

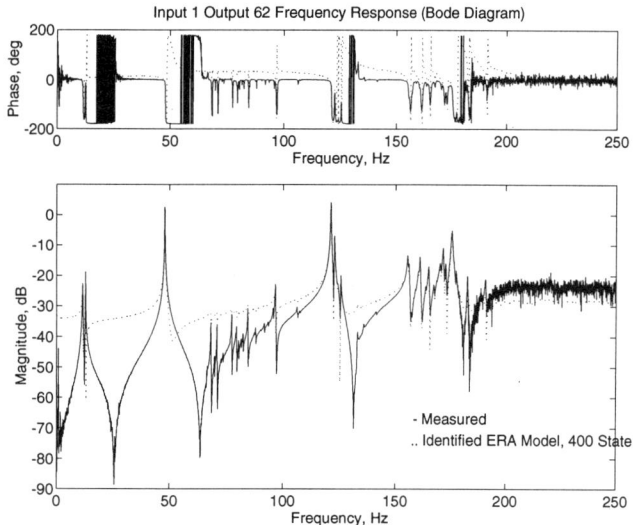

Figure 3. Reconstruction of the Driving Point FRF for the Initial 200-Mode ERA Model

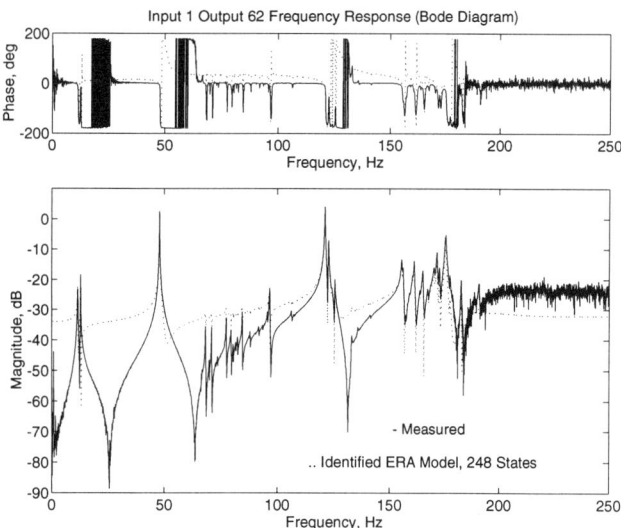

Figure 4. Reconstruction of the Driving Point FRF for the Discriminated 124-Mode ERA Model

model in Figure (3), so the model reduction by simple exclusion of the poorly identified modes is justified.

Modal Splitting

The next step in the procedure is to examine closely spaced modes and determine whether the modes are true repeated modes or are numerically repeated modes. This phenomenon is referred to as "Modal Splitting." Figure (5) illustrates the problem of modal splitting. It shows a restricted region of the plot of model order against modal frequency between 11.5 Hz and 13 Hz. Notice that at lower model orders, the x-y bending mode near 12.8 Hz is extracted first and converges to a stable value near a model order of 100 states. Unfortunately, the mode near 11.5 Hz

Table 1. Modal Quality Indicators for the ERA Identified Modes (124 Mode Model)

ERA Mode	f(Hz)	ζ_i	EMAC	MSV	MPC	MSP	ID'd Mode
1	0.0	86.5%	94.9%	16.5%	96.0%	15.3%	
2	0.9	6.8%	78.3%	3.6%	38.2%	2.8%	
3	11.5	0.8%	98.8%	23.5%	100%	64.0%	1
4	12.8	0.2%	99.1%	20.7%	99.8%	48.9%	2
5	12.9	6.8%	50.7%	4.0%	74.6%	-26.1%	
6	47.9	0.1%	94.9%	71.3%	99.1%	68.4%	3
7	48.0	0.1%	97.2%	47.0%	24.6%	-2.9%	
8	52.9	0.5%	73.0%	6.2%	97.4%	-0.6%	
9	66.6	0.1%	87.8%	7.9%	89.7%	-3.2%	
10	66.8	0.1%	91.6%	8.7%	98.0%	-2.2%	
11	68.3	0.1%	96.9%	14.7%	98.0%	19.9%	4
12	68.9	0.4%	72.2%	5.0%	88.1%	8.6%	5
13	70.0	0.1%	91.7%	9.7%	99.2%	4.3%	6
14	71.0	0.1%	94.8%	18.7%	43.6%	12.1%	
15	71.1	0.2%	90.4%	13.2%	59.3%	7.2%	7
16	73.0	0.2%	83.0%	6.8%	97.1%	2.2%	8
17	73.8	0.1%	87.8%	3.6%	98.3%	2.8%	9
18	74.5	0.3%	68.7%	23.5%	82.3%	64.0%	
19	77.0	0.2%	75.7%	20.7%	95.4%	48.9%	
20	77.6	0.1%	96.5%	4.0%	99.4%	-26.1%	10
21	79.5	0.4%	90.9%	71.3%	78.3%	68.4%	11
22	79.7	0.1%	95.2%	47.0%	79.3%	-2.9%	12
23	80.6	0.1%	93.6%	6.2%	97.1%	-0.6%	13
24	81.7	0.1%	90.5%	7.9%	97.4%	-3.2%	14
25	82.6	0.1%	93.8%	8.7%	98.8%	-2.2%	15
26	84.6	0.1%	96.0%	14.7%	95.2%	19.9%	16
27	85.1	0.1%	83.9%	5.0%	90.8%	8.6%	17
28	85.3	0.6%	63.6%	9.7%	66.3%	4.3%	
29	87.7	0.1%	89.9%	18.7%	99.6%	12.1%	18
30	88.1	0.1%	96.6%	13.2%	98.0%	7.2%	19

does not converge until the model order is extended to approximately 200 states. Other modes require further overspecification of the model order to 400 states, at which order the mode near 12.8 Hz has clearly split into two numerical modes.

Often, the MQI values lead to discrimination of one mode of a split modal pair. An examination of Table (1) shows that this happened for ERA modes 4 and 5, 19 and 20, and 27 and 28. However, modes 14 and 15, near 71 Hz, both passed the MQI tests, so that mode 14 was removed manually. A similar procedure was followed for the remaining modes of the identified set.

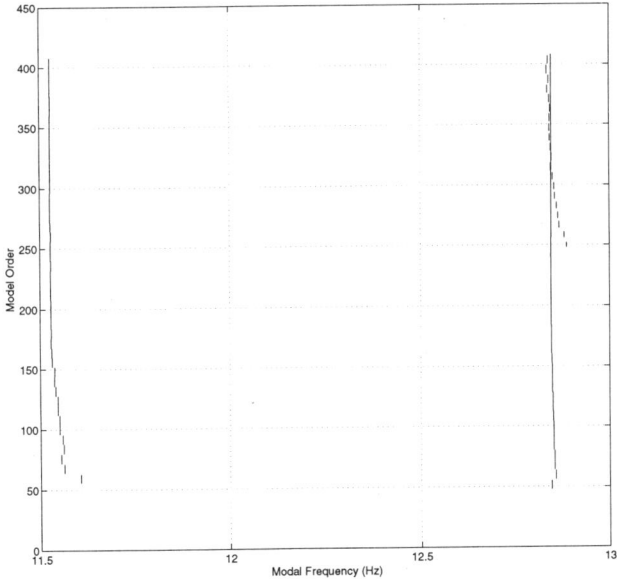

Figure 5. Modal Splitting upon Massive Order Overspecification

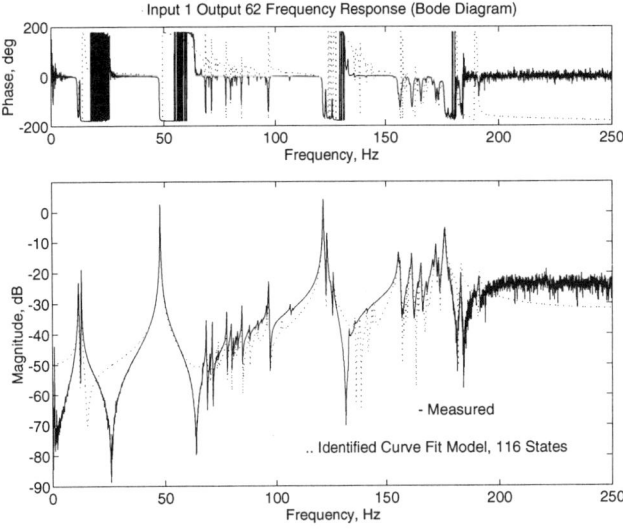

Figure 6. Data Reconstruction after Frequency Domain Adjustment and Reduction to 58 Modes

Effect of the Improper Term in the FRF Curve Fit Adjustment

At this point we should clarify the fact that, after each model reduction or adjustment in the modal vectors, it is important to reapply the MQI criteria. Using this, the initial 124 mode discriminated ERA model was reduced first to 58 modes and finally to 42 modes at the end of this procedure.

To illustrate the benefit of the improper term in the frequency domain curve fit adjustment of the modal vectors, we first present the results of adjusting the modal vectors (in this case the C matrix) by least squares curve fit. The results using the FRF data from all frequencies are illustrated in Figure (6). Note the improvement in the amplitude error, but the persistent error in phase as was observed above for the larger order models.

In contrast, Figure (7) shows the final curve fit including the improper term. Note that this final model has only 42 modes, a number consistent with the original estimate of 50 modal resonance peaks made using the MIF. Compare the accuracy of this 42 mode reconstruction to the 200 mode reconstruction shown in Figure (3) using the original ERA model. Note also that the phase error is also greatly reduced over the entire band.

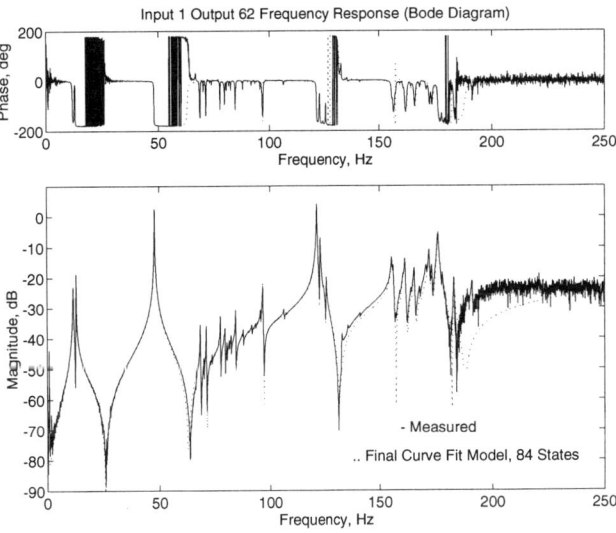

Figure 7. Data Reconstruction After Improper Frequency Domain Curve Fit and Reduction to 42 Modes

CONCLUSIONS

The method reported in this paper obtains a minimal order modal model with both time and frequency domain accuracy using only overdetermined linear regressions. The converged poles from an overdetermined ERA model are carefully discriminated using a combination of modal quality indicators. The model vectors are then determined using a linear least squares curve fit to the FRF data which includes an improper term. The resulting method produces an accurate, minimal order model in a computationally efficient manner.

ACKNOWLEDGMENTS

This research was sponsored by NASA/LaRC under contract NAG-1-1490 and by donation from the Shimizu corporation. The authors would like to specifically acknowledge the contributions of undergraduates lab assistants Stephanie Gow and Nikki Robinson. The authors were also grateful for the encouragement, support and valuable technical insights of Prof. K.C. Park.

REFERENCES

[1] Juang, J.N. and Pappa, R.S., "An Eigensystem Realization Algorithm for Modal Parameter Identification and Model Reduction," *J. Guidance, Control and Dynamics,* Vol. 8, No. 5, pp. 620-627, 1985.

[2] Juang, J.N., Cooper, J.E., and Wright, J.R., "An Eigensystem Realization Algorithm Using Data Correlations (ERA/DC) for Modal Parameter Identification," *Control Theory and Advanced Technology,* Vol. 4, No. 1, pp 5-14, 1988.

[3] Juang, J.N., Phan, M., Horta, L.G., and Longman, R.W., "Identification of Observer/Kalman Filter Markov Parameters: Theory and Experiments," *Proc. of the AIAA Guidance, Navigation, and Control Conference,* August 1991, pp 1195-1207.

[4] King, A.M, Desai, U.B., and Skelton, R.E., "A Generalized Approach to q-Markov Covariance Equivalent Realizations for Discrete Systems," *Automatica,* Vol. 24, No. 4, pp 507-515, 1988.

[5] Peterson, L.D. and Bullock, S.J., "A Comparison of the Eigensystem Realization Algorithm and the Q-Markov COVER," *Proc. 32nd AIAA/ASME/ASCE/AHS/ASC Structures, Structural Dynamics and Materials Conference,* Baltimore, April, 1991, pp 2169-2178.

[6] Vold, H., Kundrat, J., Rocklin, G.T., and Russell, R., "A Multi-Input Modal Estimation Algorithm for Mini-Computers," SAE Paper Number 820194, 1982.

[7] Juang, J.-N., "Mathematical Correlation of Modal Parameter Identification Methods via System Realization Theory," *Int. Journal of Anal. and Expt. Modal Analysis,* Vol. 2, No. 1, 1987, pp. 1-18.

[8] Peterson, L.D., Bullock, S.J., and Doebling, S.W., "Experimental Investigation of the Statistical Variance of Modal Parameters for a Precision Truss" *Proceedings of the 33rd AIAA/ASME/ASCE/AHS/ASC Structures, Structural Dynamics and Materials Conference,* Dallas, April 1992.

[9] Jacques, R. N. and Miller, D. W., "Multivariable Model Identification from Frequency Response Data," submitted to the *32nd IEEE Conf. Decision and Control.*

[10] Mayes, R.L., "A Multi-Degree-of-Freedom Mode Shape Estimation Algorithm Using Quadrature Response," *Proc. of the 11th International Modal Analysis Conference,* February, 1993.

[11] Horta, L. and Juang, J.-N., "Frequency Domain System Identification Methods: Matrix Fraction Description Approach," *Proc. 1993 AIAA Guidance, Navigation and Controls Conf.,* Monterey, CA, Aug. 9-11, 1993, pp. 1236-1242.

[12] Peterson, L.D. "Efficient Computation of the Eigensystem Realization Algorithm" *Proceedings of the 10th International Modal Analysis Conference,* San Diego, February, 1992. Submitted to *J. Guidance, Control and Dynamics.*

[13] Doebling, S.W., Alvin, K.F., and Peterson, L.D., "Limitations of State-Space System Identification Algorithms for Structures with High Modal Density" *Proc. of the 12th International Modal Analysis Conference,* February, 1994.

[14] Alvin, K. F. and Park, K. C., "Second-Order Structural Identification Procedure via State-Space-Based System Identification," *AIAA Journal,* Vol. 32, No. 2, February 1994, pp. 397-406.

[15] Pappa, R. S., Elliot, K. B., and Schenk, A., "A Consistent-Mode Indicator for the Eigensystem Realization Algorithm," AIAA Paper 92-2136-CP, *Proc. of the 1992 AIAA Structures, Structural Dynamics, and Materials Conference,* April, 1992, pp. 556-565.

[16] Longman, R. W., Lew, J.-S., and Juang, J.-N., "Comparison of Candidate Methods to Distinguish Noise Modes from System Modes in Structural Identification," AIAA Paper 92-2518-CP, *Proc. of the 1992 AIAA Structures, Structural Dynamics, and Materials Conference,* April 1992, pp. 2307-2317.

A COMPARISON AND ENHANCEMENT OF FLEXIBLE STRUCTURE SYSTEM IDENTIFICATION METHODS FOR CONTROL[†]

L. L. Hufford[††] and W. D. Robinson[†††]

Sverdrup Technology
620 Discovery Drive
Huntsville, AL 35806

ABSTRACT

The Q-Markov Covariance Equivalent Realization (COVER) and Eigensystem Realization Algorithm (ERA) system identification methodologies for flexible structures are studied in this paper. These methods were selected for study since the models resulting from their use are suitable for applying modern state space control methodologies. A comparison of the methods is performed. As a result of this comparison, the Degree of Modal Purity is developed for the Q-Markov COVER methodology from a similar concept in the ERA. Finally, to demonstrate the concepts discussed in this paper, the two methods are applied to a two degree-of-freedom spring-mass-damper system.

1.0 INTRODUCTION

The development of methods for actively controlling the vibrations in large flexible structures has increased the need for accurate time domain models of the structure which can be used with modern control techniques. This need has led to the development of many system identification techniques. Specifically, the Q-Markov Covariance Equivalent Realization (COVER)[1] and the Eigensystem Realization Algorithm (ERA)[2,3] have received considerable attention. These methods have been applied to large flexible structure testbeds such as the ACES facility at NASA Marshall Space Flight Center[4,5,6,7] and the Mini-Mast facility at NASA Langley Research Center[8]. These methods were developed independently and little work has been done on comparing the two methods. The objective of this paper is to provide a qualitative and quantitative comparison of these two methods of system identification. The emphasis is placed on the practical application of the methods. Also, to illustrate the concepts presented in this paper, the two methods are applied to a two degree-of-freedom spring-mass-damper system.

Section 2 provides an analysis of the ERA and Q-Markov COVER identification methods. Included in this analysis is a comparison of the two methods and the development of the Degree of Modal Purity (DMP) for Q-Markov COVER. The DMP is a measure of the identification accuracy which was originally developed for the ERA[2,3] and was extended to the Q-Markov COVER method. Section 3 presents results of applying these methods to a spring-mass-damper system. Included in this section is a description of the system to be identified, analysis of the parameters to be chosen when applying the two methods, and an illustration of the use of the DMP for Q-Markov COVER.

2.0 METHOD ANALYSIS

This section summarizes the analysis of the methods. Literature on the theory and practical application of the Q-Markov COVER and ERA system identification techniques was studied. A comparison

---

[†] Funded by NASA contract NAS8-37814
[††] Engineer, Member AIAA
[†††] Senior Engineer, Member ASME

Copyright © 1994 by the American Institute of Aeronautics and Astronautics, Inc. All rights reserved.

of the two methods is presented in Section 2.1. As a result of this study, the Degree of Modal Purity was developed for the Q-Markov COVER algorithm from a similar concept in the ERA. Details are given in Section 2.2.

2.1 Method Comparison

It is assumed that the structure to be identified can be represented as a discrete-time, linear, time-invariant, finite dimensional system of the form

$$x(k+1) = Ax(k) + Bu(k)$$
$$y(k) = Cx(k) + Du(k) \quad (1)$$

where $x \in \Re^{n_x}, y \in \Re^{n_y}, u \in \Re^{n_u}$. The Q-Markov COVER and ERA system identification methods[1,2,3,8] are based on the concept of obtaining a discrete-time state space realization of a system from time domain test data. Concepts from system/control theory are used to produce the realizations. Detailed descriptions of the algorithms can be found in the listed references.

When applying the Q-Markov COVER algorithm, one must form the matrix

$$D_q = O_q X O_q^T \quad (2)$$

where T indicates the matrix transpose. In Eq. 2, X, the state covariance matrix, is given by

$$X = AXA^T + BWB^T =$$
$$= [B \ AB \ A^2B \ ...] * \overline{W} * \begin{bmatrix} B^T \\ B^T A^T \\ B^T (A^2)^T \\ \vdots \end{bmatrix} \quad (3)$$

$$\overline{W} = \text{diag}[..., W, ...] \quad (4)$$

If the input to the system is white noise, the matrix W in Eq. 4 is the covariance of that white noise; if the input to the system is a pulse, the matrix W is a diagonal matrix with the square of the pulse heights on the diagonal. Also in Eq. 2, the matrix O_q is given by

$$O_q = \begin{bmatrix} C \\ CA \\ \vdots \\ CA^{q-1} \end{bmatrix} \quad (5)$$

It is possible to form D_q from pulse response data obtained from the system. A pulse is applied to each input channel, one at a time. This input is given by

$$w(j,k) = \begin{cases} w_j \delta_K(k) & \text{input } j \\ 0 & \text{all other inputs} \end{cases} \quad (6)$$

δ_K is the Kronecker delta

Define $y(j,k)$ to be the output at time k to the input $w(j,k)$. The Markov Parameters, H_i, and Covariance Parameters, R_i, can be calculated from the pulse response using the following equations:

$$H_i = Y_i W^{-1/2} = [y(1,i), y(2,i), ..., y(n_u,i)] W^{-1/2}$$
$$R_i = Y_i(d \to \infty, i) = \sum_{j=1}^{n_u} \sum_{k=0}^{d \to \infty} y(j,k+i) y^T(j,k) \quad (7)$$
$$W = \text{diag}[..., w_j^2, ...]$$

The matrix D_q can then be calculated as follows:

$$\overline{H}_q = \begin{bmatrix} H_0 & 0 & \cdots & 0 \\ H_1 & H_0 & \cdots & 0 \\ \vdots & \vdots & \ddots & \vdots \\ H_{q-1} & H_{q-2} & \cdots & H_0 \end{bmatrix} \quad (8)$$

$$\overline{R}_q = \begin{bmatrix} R_0 & R_1^T & \cdots & R_{q-1}^T \\ R_1 & R_0 & \cdots & R_{q-2}^T \\ \vdots & \vdots & \ddots & \vdots \\ R_{q-1} & R_{q-2} & \cdots & R_0 \end{bmatrix} \quad (9)$$

$$D_q = \overline{R}_q - \overline{H}_q W \overline{H}_q^T \quad (10)$$

A singular value decomposition is performed on D_q to determine the principle components. The resulting direction and gain matrices along with another matrix composed of pulse response data (Markov parameters) are used to compute a state space realization of the system. In the absence of noise, the first q Markov parameters and the first q covariance parameters of the identified system are guaranteed to be identical to those of the true system.

When applying the ERA, one must form the block data matrix

$$H_{\xi\eta}(k) = V_{\xi} A^k W_{\eta} \tag{11}$$

where V_{ξ} is a generalized observability matrix, and W_{η} is a generalized controllability matrix given by

$$V_{\xi} = \begin{bmatrix} C \\ C_{J_1} A^{s_1} \\ \vdots \\ C_{J_{\xi}} A^{s_{\xi}} \end{bmatrix} \tag{12}$$

$$W_{\eta} = \begin{bmatrix} B & A^{t_1} B_{K_1} & \cdots & A^{t_{\eta}} B_{K_{\eta}} \end{bmatrix} \tag{13}$$

Note that the block data matrix (Eq. 11) is equivalent to a generalized Hankel matrix with selected rows and columns deleted. The parameters J_j and K_p determine which rows and columns remain.

It is possible to form the matrix $H_{\xi\eta}$ from pulse response data (Markov parameters) obtained from the system. This matrix is formed as follows:

$$H_{\xi\eta}(k-1) =$$

$$\begin{bmatrix} H_k & (H_{k+t_1})_{K_1} & \cdots & (H_{k+t_{v-1}})_{K_v} \\ (H_{s_1+k})_{J_1} & (H_{s_1+k+t_1})_{J_1 K_1} & \cdots & (H_{s_1+k+t_{v-1}})_{J_1 K_v} \\ \vdots & \vdots & \ddots & \vdots \\ (H_{s_{\eta-1}+k})_{J_{\eta}} & (H_{s_{\eta-1}+k+t_1})_{J_{\eta} K_1} & \cdots & (H_{s_{\eta-1}+k+t_{v-1}})_{J_{\eta} K_v} \end{bmatrix} \tag{14}$$

where H_i is the ith Markov parameter from the pulse response data given in Eq. 7. The subscripts $J_j K_p$ indicate which rows (outputs) and columns (inputs) of the Markov parameters to include respectively. These values can be chosen so that data that is known to be questionable a priori can be selectively eliminated from the identification process. For example, the J_j can be chosen to eliminate some data taken from a particularly noisy sensor. The parameters s_i, t_l are used to select data from particular times to include in the block data matrix. A singular value decomposition is performed on the block data matrix with $k = 0$. The resulting direction and gain matrices along with a shifted block data matrix ($k = k_{delayed}$) are used to compute a state space realization of the system. In both algorithms noisy data is handled by truncating the number of singular values used in calculating the state space realization.

As can be discerned from the brief descriptions, several concepts are common to the two methods. For example, both use Markov parameters as structural data and both use a singular value decomposition to break the data into its principle components. One can also see that the concepts of observability and controllability from system/control theory are highly prevalent in both methodologies. As a result of this last similarity, the identified model will contain more accurate information on the state, input, and output relationships of the structure than models obtained from standard modal analysis methods. Since many modern control methods rely on the accuracy of these relationships, the use of these identification algorithms is desirable. Finally, it can be shown that Q-Markov COVER is a special case of ERA with parameters chosen to be $s = (1,...,q-1)$, $t = (1,...,\infty)$, $J_j = (1,...,n_y)$, $K_p = (1,...,n_u)$, and $k_{delayed} = 1$ where $n_y(q-1) \geq n_x$[11]. A proof of this equivalence is given in the reference. This equivalence, however, does not hold when non-zero singular values are truncated to reduce the effect of noise.

When the parameters are not chosen as discussed previously, there are differences between the methods. One major difference between the two algorithms is that ERA does not necessarily incorporate covariance information. The covariance depends on the long term time domain behavior of the system as does damping and low frequency gain; thus, the lack of covariance information results in an increase in damping biases and low frequency gain errors in the resulting identified systems. On the other hand, calculating covariance is computationally expensive and time consuming; thus, there are benefits to not including it. Another major difference between the algorithms is the flexibility that ERA allows in choosing data with the parameters s, t, K, and J. This allows one to eliminate particularly noisy data from the identification process; yet, still identify the system.

Recently, two new identification methods have been developed based on ERA and Q-Markov COVER. The first method, Eigensystem Realization with Data Correlations (ERA/DC)[9], incorporates correlation information into the ERA method. The second method, Observability Range Subspace Extraction (ORSE)[10], is an extension of the Q-Markov COVER algorithm that produces an identified model from colored noise response data. While both of these methods expand the scope of the methods on which they are based, the expense is an increase in computational complexity.

2.2 Development of DMP For Q-Markov COVER

The concept of Degree of Modal Purity (DMP) was initially developed for the Eigensystem Realization Algorithm[2,3]. It is a method for determining the accuracy of each mode in the identified model by calculating the coherence of the extrapolated modal time histories from the identified model and modal components of the time domain pulse response data. This information is used to find and truncate spurious modes produced by noisy data and to provide an accuracy measure for each mode of the identified system. Previously, there was no such tool to provide these functions for the Q-Markov COVER algorithm.

Using the singular value decomposition, it is possible to write[2,3]

$$H_{\xi\eta} = PDQ^T \in \Re^{n_r \times n_c} \quad (15)$$

From Eqs. 11, 12 and 13, $PD^{1/2}$ is simply the observability-like matrix for some state space representation, namely that obtained from ERA.

Likewise, $D^{1/2}Q^T$ is the controllability-like matrix for the same representation. Let $\hat{A}_e$, $\hat{B}_e$, $\hat{C}_e$, $\hat{D}_e$ be the state space representation of the discrete-time identified system obtained directly from the ERA procedure. Let p_i be the eigenvalues of $\hat{A}_e$ and ψ_i be the eigenvectors. Form the modal representation, A_{em}, B_{em}, C_{em}, D_{em}, by using a transformation matrix, V, whose columns are ψ_i:

$$A_{em} = V^{-1}\hat{A}_e V$$
$$B_{em} = V^{-1}\hat{B}_e$$
$$C_{em} = \hat{C}_e V$$

$$A_{em} = \mathrm{diag}(p_1, p_2, \ldots, p_n) \quad (16)$$

$$B_{em} = \begin{bmatrix} b_1 \\ \vdots \\ b_n \end{bmatrix}$$

$$C_{em} = \begin{bmatrix} c_1 & \cdots & c_n \end{bmatrix}$$

For each mode of the identified system, form the extrapolated controllability-like modal time history where the K_p are used as in Eqs. 13 and 14:

$$\bar{f}_i = \begin{bmatrix} b_i & p_i b_{i_{K_1}} & \cdots & p_i^\eta b_{i_{K_\eta}} \end{bmatrix} \quad (17)$$

Project the modal time histories of the data using the corresponding eigenvector:

$$f_i = \chi_i D^{1/2} Q^T$$

$$\text{where } \begin{bmatrix} \chi_1 \\ \vdots \\ \chi_n \end{bmatrix} = \begin{bmatrix} \psi_1 & \cdots & \psi_n \end{bmatrix}^{-1} \quad (18)$$

If the ith mode is a true linear mode of the system, then $\bar{f}_i$ and f_i will be aligned. The value of the coherence, γ_i, given by

$$\gamma_i = \frac{f_i^T \bar{f}_i}{|f_i^T f_i| |\bar{f}_i^T \bar{f}_i|} \quad (19)$$

will be very close to 1 for true linear system modes. Noise modes, inaccurately identified modes, and modes induced by strong nonlinearities will have values of γ_i less than 1. Note that the value of γ_i quantifies the degree to which the ith mode is controlled from the input. A similar quantity can be calculated using the observability-like portion of the decomposition of the block data matrix.

The DMP can be derived for use with Q-Markov COVER. From the singular value decomposition of D_q it is possible to write

$$D_q = P_q P_q^T \qquad (20)$$

From Eqs. 2 and 5, it is clear that P_q is simply the observability-like matrix for a state space representation of the structure with unitary state covariance, namely that obtained from Q-Markov COVER.

Let $\hat{A}_Q$, $\hat{B}_Q$, $\hat{C}_Q$, $\hat{D}_Q$ be the state space representation of the discrete-time identified system obtained directly from the Q-Markov COVER procedure. Form the modal representation, A_{Qm}, B_{Qm}, C_{Qm}, D_{Qm}, of the system using the eigenvalue decomposition of $\hat{A}_Q$ as was done for ERA. The same notation for the modal realization will be used in this development as was used previously for ERA. For each mode of the identified system, form the extrapolated observability-like modal time history

$$\bar{r}_i = \begin{bmatrix} c_i^T & c_i^T p_i & \cdots & c_i^T p_i^{(q-1)} \end{bmatrix}^T \qquad (21)$$

Project the modal time histories of the data using the corresponding eigenvector:

$$r_i = P_q * \psi_i \qquad (22)$$

If the ith mode is a true linear mode of the system, then $\bar{r}_i$ and r_i will be aligned. The value of the coherence, γ_i, given by

$$\gamma_i = \frac{r_i^T \bar{r}_i}{|r_i^T r_i| |\bar{r}_i^T \bar{r}_i|} \qquad (23)$$

will be very close to 1 for true linear system modes.

The interpretation is the same as for the quantity defined for measuring DMP for the ERA. Note that the value of γ_i quantifies the degree to which the ith mode is observed in the output.

3.0 IDENTIFICATION EXAMPLE

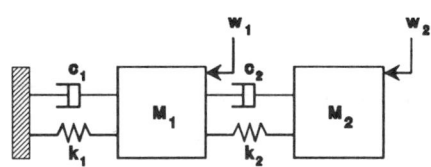

Figure 1. Spring-Mass-Damper System

3.1 System Description

In order to demonstrate some of the method comparisons and the usefulness of the Degree of Modal Purity for Q-Markov COVER, the two methods discussed in this report are applied to a two degree-of-freedom, two-input, two-output, spring-mass-damper system (Figure 1). The inputs are the forces on the two masses, and the outputs are the positions of the two masses. The state space representation of the continuous time system is given by

$$\left[\begin{array}{c|c} A & B \\ \hline C & D \end{array}\right] = \left[\begin{array}{cccc|cc} 0 & 0 & 1 & 0 & 0 & 0 \\ 0 & 0 & 0 & 1 & 0 & 0 \\ -15 & 7.5 & -0.4 & 0.2 & 0 & -0.25 \\ 30 & -30 & 0.8 & -0.8 & -1 & 0 \\ \hline 1 & 0 & 0 & 0 & 0 & 0 \\ 0 & 1 & 0 & 0 & 0 & 0 \end{array}\right] \qquad (24)$$

The continuous-time modes of the system are described in Table I. The system was designed to have one long slow mode (Mode 1) and one short fast mode (Mode 2). The data used for the identification was the output responses to pulses of height 50 sampled at a rate of 10 Hz. Zero mean white noise

Table I. Modal Data

Mode #	Frequency	Damping
1	0.9939 Hz	0.0836
2	0.3808 Hz	0.0319

with variance 0.01 was injected at the sensors, producing a low signal-to-noise ratio. The continuous-time response (without noise) to a pulse (height 50 and width 0.1) and the noisy sampled response to the same input are shown in Figure 2 and Figure 3 for comparison.

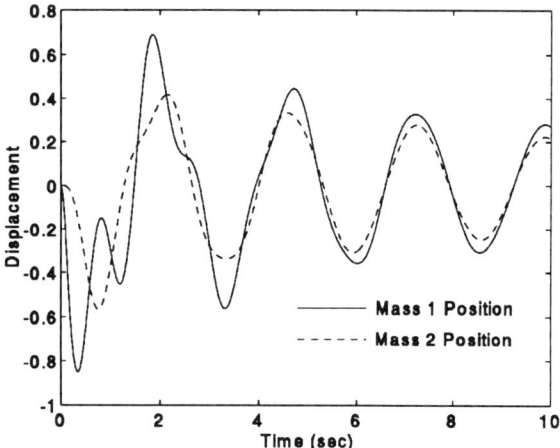

Figure 2. Continuous-Time Response to Pulse on Input 1 (Width 0.1, Height 50)

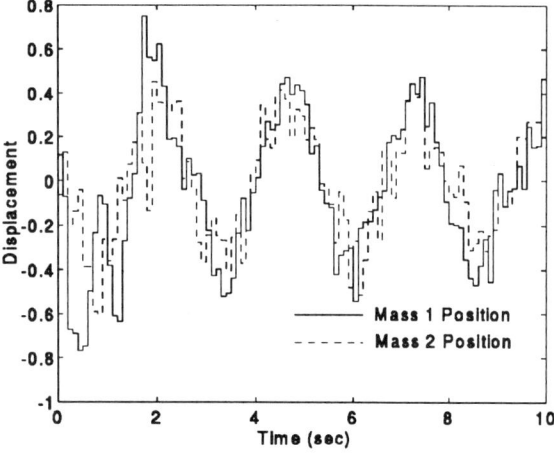

Figure 3. Noisy Sampled Response to Pulse on Input 1 (Width 0.1, Height 50)

3.2 Choice of Parameters

In each method, parameters specifying the amount of data to be used must be chosen. The parameter q in Q-Markov COVER determines the number of Markov and covariance parameters of the identified systems to be matched with the data from the true system. It also determines the dimension of D_q and the maximum order of the identified system. The parameters ξ and η determine the dimensions of the Hankel matrix used in ERA. As discussed in Section 3.1, other parameters (s, t, J, K) are available with ERA to reduce the data in the Hankel matrix by eliminating data from certain times, inputs, or outputs. This reduction of data is not used in this example.

Table II and Table III show the results of varying these parameters on the eigenvalues of the discrete-time identified system. Both methods produced accurate identified models (see highlighted rows). Note that the accuracy of the identified model is much more sensitive to parameter variations for ERA than for Q-Markov COVER. Including covariance information in Q-Markov COVER has an averaging effect which reduces the sensitivity to the parameter q. For both methods, however, it is clear that there is an optimal choice of parameters. When the parameters are chosen too large, noise effects decrease accuracy. Due to the natural damping of the structure, the amplitude of the pulse response of the system will decay while the magnitude of the sensor noise will remain constant, thus decreasing the signal-to-noise ratio with time. When parameters are increased, the additional data has a lower signal-to-noise ratio than previous data. When enough of this noisy data is included, the identification results become corrupted. This effect is especially prevalent in Mode 2 which dies out more quickly than Mode 1. When the parameters are chosen too small, not enough data is included to properly identify the modes. As general guidelines one should choose parameters to include data from a full period of the lowest frequency mode, while not including data long past the point when the signal-to-noise ratio of major modes has significantly degraded. If these guidelines conflict, decisions must be made as to modal priority. Comparing frequency responses of the identified system with the FFT of the test data is also beneficial in fine tuning the choice of parameters.

Table II. Choice of Parameters for ERA Identification

ζ,η	Order	eigenvalues of $\hat{A}_e$ (discrete-time)	% error frequency	% error damping	Pulse Response Error*
10	4	$0.9600 \pm 0.2290i$	2.01	-75.97	4.5753
		$0.7203 \pm 0.5883i$	-9.90	-26.16	
20	4	$0.9537 \pm 0.2410i$	-3.65	-107.47	4.9880
		$0.7658 \pm 0.5578i$	-0.83	-2.26	
29	4	$0.9636 \pm 0.2360i$	-0.39	-3.47	0.8992
		$0.7603 \pm 0.5595i$	-1.66	-8.29	
40	4	$0.9679 \pm 0.2369i$	-0.31	53.42	2.9177
		$0.7589 \pm 0.5581i$	-1.64	-12.28	
50	4	$0.9688 \pm 0.2358i$	0.02	36.51	1.8151
		$0.7496 \pm 0.5556i$	-2.38	-29.32	

Table III. Choice of Parameters for Q-Markov COVER Identification

Q	Order	eigenvalues of $\hat{A}_Q$ (discrete-time)	% error frequency	% error damping	Pulse Response Error *
10	4	$0.9643 \pm 0.2339i$	0.52	-2.22	1.7726
		$0.7676 \pm 0.5399i$	1.67	-23.33	
20	4	$0.9639 \pm 0.2359i$	-0.31	-0.87	0.7754
		$0.7735 \pm 0.5529i$	0.64	2.97	
30	4	$0.9645 \pm 0.2350i$	0.11	4.26	0.7118
		$0.7676 \pm 0.5558i$	-0.38	-2.29	
40	4	$0.9649 \pm 0.2353i$	0.03	10.32	0.8777
		$0.7616 \pm 0.5553i$	-0.97	-11.85	
50	4	$0.9646 \pm 0.2350i$	0.14	5.93	1.0491
		$0.7457 \pm 0.5546i$	-2.71	-36.26	

* The Pulse Response Error was calculated as the sum of the ℓ_2 norm of the difference between the identified and true discrete-time pulse response for 20 seconds.

Another choice which must be made in the identification process is the order of the system. This choice is made in the two methods discussed in this paper by selecting the number of singular values of D_q or $H_{\xi\eta}$ to retain in the design. When the data used in the identification process is noisy, this choice can be difficult. If no noise were present in the system, then the number of nonzero singular values of either D_q or $H_{\xi\eta}$ would be equal to the true system order; however, when the data is corrupted by noise, the noise appears as spurious "modes" of the system. If the signal-to-noise ratio is very high, then the singular values corresponding to the "noise modes" will be very small compared to those corresponding to the true system modes. These small singular values can be truncated. If the signal-to-noise ratio is smaller, there may be no good place to truncate the singular values.[3,8] It is noted in this example that when the "best" choice of parameters ($Q = 30$ or $\xi = \eta = 29$) is made, the separation is larger between singular values corresponding to true system modes and those which are a result of the noise (See Tables IV and V). This makes it easier to identify the true system order. When the parameters are chosen too small, the system modes are not identified accurately due to a lack of information, so there is not a clear break in the singular values. The system modes are not as easily distinguished from noise. On the other hand, when the parameters are chosen too large, the signal-to-noise ratio of the component of the output corresponding to Mode 2 decreases, and this mode appears more as a noise mode. This also explains why there is an increase in separation between the singular values corresponding to Mode 1 and Mode 2 for larger values of the parameters. Mode 1 is clearly identified when the parameters are chosen to be large.

3.3 Use of DMP with Q-Markov COVER

The usefulness of DMP with ERA has been demonstrated[2,3]. It is shown in this report that DMP can provide similar benefits for Q-Markov COVER. Table VII shows the results of the identification using Q-Markov COVER with DMP for different choices of system order. When the model order is made larger than 4, the DMP for the spurious modes decreases, indicating that the mode is not a true system mode.

Note that the value of the DMP for Mode 2 is always lower than for Mode 1. Mode 1 is the dominant mode, so it is reasonable that identification for Mode 2 would be slightly less accurate. This information could be incorporated into norm based control methods such as H_∞ and ℓ_1 as uncertainty via frequency domain weighting. The modal uncertainty could also be incorporated into the Maximum Entropy/Optimal Projection method via parametric uncertainty in the A matrix. Similar results were obtained using ERA with DMP. They are shown in Table VI for comparison.

Table IV. Singular Values for ERA

ζ,η	10	29	50
s_i	5.4624	11.1648	17.7956
	3.2364	9.7650	16.2296
	2.0806	3.2951	3.4087
	1.7837	2.8491	2.9746
	1.0468	1.7957	2.5134
	0.9711	1.7492	2.4557
	0.7802	1.4720	2.0433
	0.6903	1.4424	2.0018
	0.6340	1.3751	1.9676
	0.5860	1.3431	1.9336
	0.2313	1.2833	1.9005
	⋮	⋮	⋮

Table V. Singular Values for Q-Markov COVER

Q	10	30	50
s_i	179.4360	395.7980	556.0506
	95.3819	301.7890	467.2621
	10.4471	15.2426	15.5337
	9.0028	12.5833	13.1546
	6.0392	8.0004	10.7812
	5.9355	7.7511	10.4971
	5.4122	6.6140	8.4022
	5.0880	6.3724	8.2696
	4.8097	6.2038	7.7349
	4.7242	6.0843	7.3781
	4.2333	5.8632	7.2334
	⋮	⋮	⋮

Table VI. Use of DMP for Identification of Example Using ERA

ζ,η	Order	Mode #	eigenvalues (discrete-time)	DMP	% error frequency	% error damping
29	2	1	0.9638 ± 0.2346i	0.9967	0.21	-5.61
29	4	1	0.9636 ± 0.2360i	0.9986	-0.39	-3.47
		2	0.7603 ± 0.5595i	0.9802	-1.66	-8.29
29	6	1	0.9638 ± 0.2358i	0.9986	-.29	-2.54
		2	0.7615 ± 0.5599i	0.9817	-1.57	-5.90
		-	-0.1055 ± 0.9517i	0.9345	N/A	N/A
29	8	1	0.9637 ± 0.2359i	0.9986	-0.35	-2.97
		2	0.7608 ± 0.5623i	0.9755	-1.95	-3.89
		-	-0.1120 ± 0.9428i	0.9391	N/A	N/A
		-	-0.1462 ± 0.7265i	0.7965	N/A	N/A

Table VII. Use of DMP for Identification of Example Using Q-Markov COVER

Q	Order	Mode#	eigenvalues (discrete-time)	DMP	% error frequency	% error damping
30	2	1	0.9646 ± 0.2345i	0.9998	0.30	3.31
30	4	1	0.9639 ± 0.2350i	0.9999	-0.11	4.26
		2	0.7676 ± 0.5558i	0.9831	-0.38	-2.29
30	6	1	0.9645 ± 0.2350i	0.9999	0.11	4.55
		2	0.7746 ± 0.5561i	0.9827	0.36	8.85
		-	-0.3724 ± 0.8914i	0.9137	N/A	N/A
30	8	1	0.9646 ± 0.2348i	0.9999	-0.23	-0.21
		2	0.7801 ± 0.5537i	0.9888	1.15	10.80
		-	-0.3718 ± 0.8914i	0.9136	N/A	N/A
		-	0.4101 ± 0.5429i	0.5088	N/A	N/A

4.0 CONCLUSION

This paper presents the results of an investigation of the Q-Markov COVER and the ERA system identification techniques. Several key results were presented from the preliminary investigation of the methods. Specifically, the similarities and differences between the two methods were discussed. The two methods were found to be equivalent for a specific choice of parameters. Also, the Degree of Modal Purity, which provides a quantitative measure of the accuracy of identified modes, was developed for Q-Markov COVER. The two methods were then applied to a two degree-of-freedom system. Guidelines for the choice of parameters and truncation of singular values to reduce the effect of noise were presented for both methods. Also, the use of the Degree of Modal Purity was illustrated in the example.

REFERENCES

[1] King, A.M., Desai, U.B., and Skelton, R.E., "A Generalized Approach to q-Markov Covariance Equivalent Realizations of Discrete Systems," Automatica, Vol. 24, No. 4, pp. 507-515.

[2] Juang, J.-N., and Pappa, R.S., "An Eigensystem Realization Algorithm (ERA) for Modal Parameter Identification and Model Reduction," Journal of Guidance, Control and Dynamics, Vol. 8, Sept.-Oct. 1985, pp. 620-627.

[3] Juang, J.-N., and Pappa, R.S., "Effects of Noise on Modal Parameters Identified by the Eigensystem Realization Algorithm," Journal of Guidance, Control and Dynamics, Vol. 9, No. 3, May-June 1986, pp. 294-303.

[4] Liu, K., and Skelton, R.E., "Identification and Control of NASA's ACES Structure," Proceedings of the 1991 American Control Conference, Boston, MA, June, 1991.

[5] Liu, K., and Skelton, R.E., "Model Identification and Controller Design for Large Flexible Space Structures -- an Experiment on NASA's ACES Structure," NASA CSI Guest Investigator Program Second Year Report, July, 1991.

[6] Collins, E.G. Jr., Phillips, D.J., and Hyland, D.C., "Robust Decentralized Control Laws for the ACES Structure," *IEEE Control Systems Magazine*, April 1991, pp. 62-70.

[7] Collins, E.G. Jr., Phillips D.J., and Hyland, D.C., "Design and Implementation of Robust Decentralized Control Laws for the ACES Structure and Marshall Space Flight Center," NASA Contractor Report 4310, July 1990.

[8] Hsieh, C., Kim, J.H., Liu, K., Zhu, G., and Skelton, R.E., "Model Identification and Controller Design for Large Flexible Space Structures - An Experiment on NASA's Minimast," CSI/GI Technical Report, July 1990.

[9] Juang, J.-N., Cooper, J. E., and Wright, J. R., "An Eigensystem Realization Algorithm Using Data Correlations (ERA/DC) for Modal Parameter Identification," Control-Theory and Advanced Technology, Vol. 4, No. 1, March, 1988, pp. 5-14.

[10] Liu, K., and Skelton, R. E., "Identification of Linear Systems from their Pulse Responses," Proceedings of the 1992 American Control Conference, June, 1992, pp. 1243-1247.

[11] Robinson, W. D., and Hufford, L. L., "Q-Markov COVER and ERA Identification Methods for the CASES Test Facility", Sverdrup Technical Report #642-001-93-007.

EXPERIMENTAL DYNAMIC CHARACTERIZATION OF A RECONFIGURABLE ADAPTIVE PRECISION TRUSS

J. D. Hinkle[*] and L. D. Peterson[†]

*Center for Aerospace Structures
and Department of Aerospace Engineering Sciences
University of Colorado at Boulder
Boulder, CO 80309-0429*

ABSTRACT

The dynamic behavior of a reconfigurable adaptive truss structure with nonlinear joints is investigated. The objective is to experimentally examine the effects of the local nonlinearities on the global dynamics of the structure. Amplitude changes in the frequency response functions are measured at micron levels of motion. The amplitude and frequency variations of a number of modes indicate a nonlinear Coulomb friction response. Hysteretic bifurcation behavior is also measured at an amplitude approximately equal to the specified freeplay in the joint. Under the 1g preload, however, the nonlinearity was dominantly characteristic of Coulomb friction with little evidence of freeplay stiffening.

INTRODUCTION

Future small spacecraft structures must be designed to be deployed while retaining the structural rigidity necessary for precision instrumentation. In contrast with most erectable spacecraft structures which have been investigated over the past fifteen years, small, compact structures are complex, nonlinear, time-variable dynamical systems. To successfully control the dynamics of these structures, it will be necessary to obtain high fidelity models based on ground test experiments. Unfortunately, there is little experience with system identification methods which are successful for nonlinear, time-varying structures. Furthermore, little is known about the confidence with which these structures can be modeled using existing finite element codes.

Examples of precision deployable structure concepts were assembled by Mikulas and Withnell [1]. Effective design of this type of structure will require a more thorough experimental background in deployable structural dynamics with a particular focus on nonlinear joint behavior.

---
[*] Graduate Research Assistant, Student Member AIAA
[†] Assistant Professor, Senior Member AIAA, Associate Member ASME

Previous research in nonlinear joint effects on space structure dynamics included examinations of modal coupling effects (Sarver and Crawley [2]), and attempts to develop equivalent beam finite element models using an equivalent energy approach (Webster and Velde [3]). Bowden [4] examined the effects of a number of simple nonlinear joint models of on global beam and truss dynamics.

Tests on deployable, erectable, and rotary modules in the Middeck 0-gravity Dynamics Experiment (MODE) provided data on nonlinear variations in frequency, amplitude, and damping as functions of input amplitude, suspension stiffness, and preload [5]. A primary conclusion from this testing was a general rule of one percent variance in the experimental determination of frequency and one-half percent error in that of damping for testing of this type of structure.

A significant difference between the MODE test structures and the batten actuated truss (BAT) being studied here is the shapes of the lowest modes of the structures. The beam-like behavior of the MODE test structures fails to exemplify the unique dynamic displacements which result from the variety of geometries found in more compact deployable structures such as precision reflectors and BAT. Unfortunately, the strong dependence of the dynamics of these structures on local displacements serves to increase the influence of local sources of nonlinearity on the global dynamics.

Initial results from force-state mapping experimentation on the pin-clevis joints used in BAT have revealed extremely complex behavior[6]. As a result, accurate modeling of these elements may require multiple degrees of freedom in addition to nonlinear considerations.

The experiment was organized as follows. Initially, multiple modal surveys of the structure are performed across the structure's deployment range. Then, to obtain piecewise constant models of the structure, the data was modeled by the Eigensystem Realization Algorithm (ERA) [7] as well as the Common Basis-normalized Structural Identification (CBSI) procedure [8]. These results were then compared with the predictions of a NASTRAN finite element model. In addition, a study of phase plane responses of the structure as a function of input amplitude and frequency is performed.

THE BATTEN-ACTUATED TRUSS

Our experiments were performed on the batten actuated deployable truss depicted in Figures 1-4. Six screw jack actuators driven by 0.52 lb motors control the motion from the packaged to fully deployed state. By independently changing the length of the screw jack actuators, the truss can be reconfigured with three degrees of freedom. The bearings and all mechanism components of this structure are preloaded by opposing springs to eliminate backlash and freeplay in the structure. As a result, it is a precision, reconfigurable adaptive truss capable of erectable structure precision. Throughout these tests, the truss was cantilevered from a backstop with two linear bearing connections allowing deployment.

Stainless steel and aluminum pin-clevis hinges with bronze fittings connect the members. These nodes are the primary source of nonlinearity in the structure. The truss members are connected to the hinges with a shoulder bolt mounted to the hinge and a compression nut that clamps the member and shoulder bolt together. It should be noted that these hinge-member connections are identical to those used in other highly linear erectable truss hardware and most likely did not contribute to the observed nonlinear behavior.

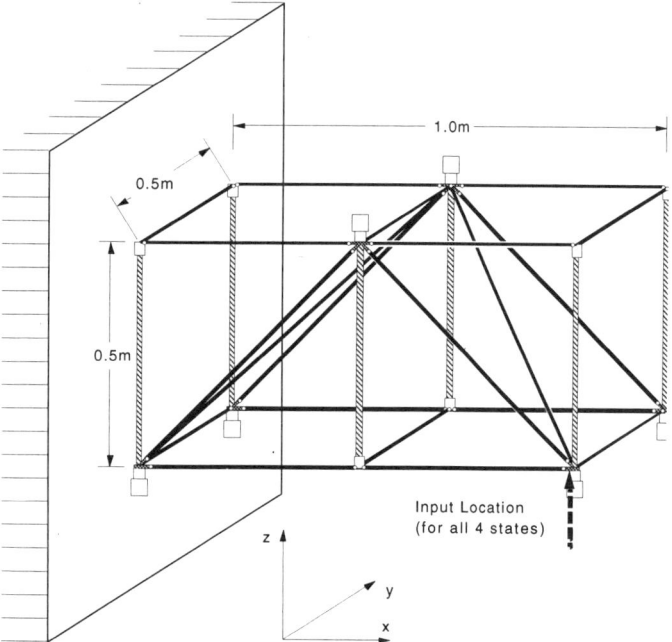

Figure 1: Reconfigurable Truss in Deployed State

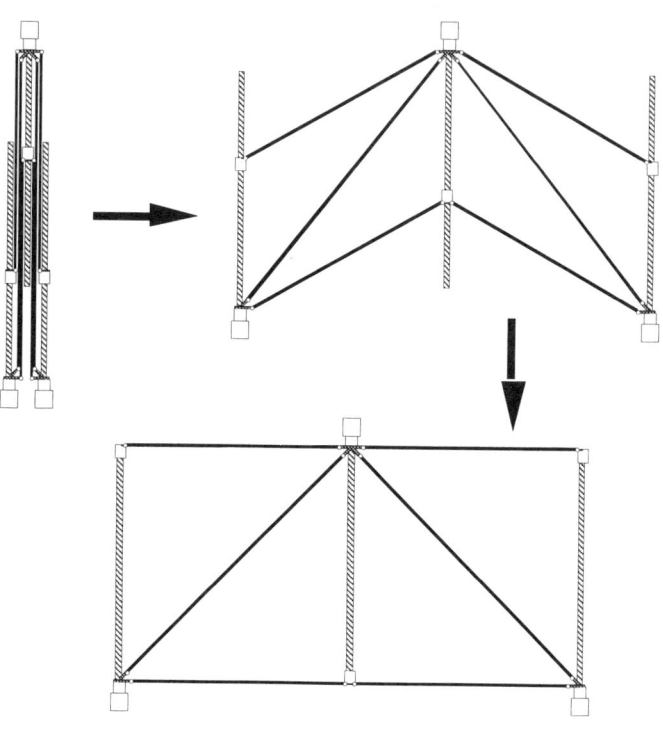

Figure 3: Batten Actuated Deployment

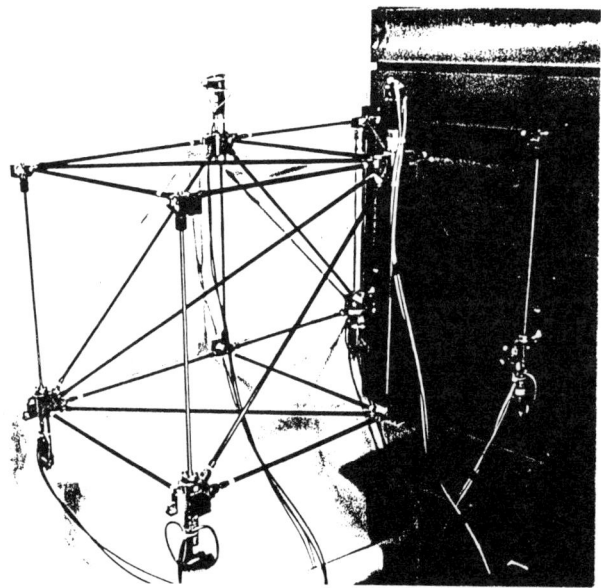

Figure 2: Photograph of Reconfigurable Truss in Deployed State

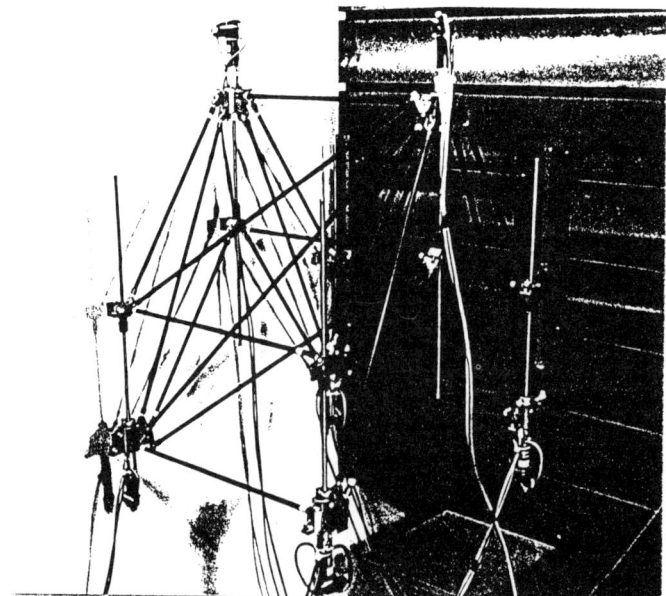

Figure 4: Photograph of Reconfigurable Truss in Half Deployed State

MODELING

Finite element models of the structure were created in MSC/NASTRAN [9] for each of the four tested deployment states. The models consisted solely of linear CBAR, RBAR, and point mass elements assembled to accurately represent the geometry of the structure. Each member was meshed into five CBAR elements and the inertial contributions of the motors were modeled as point masses offset from the appropriate nodes. The pin-clevis joints were modeled as y rotational degrees of freedom connections to the appropriate members. These connections were offset from the node centers by rigid members.

SYSTEM IDENTIFICATION PROCEDURE

Experimental Procedure

An initial bandwidth of interest of 0-200Hz was selected and testing was performed at a 500 Hz sample rate to provide a safe Nyquist value. The high bandwidth is necessary to obtain high fidelity identified models for use in control, but increases the effect of dynamic nonlinearity on the identification. A pseudo burst-random input signal was used across 80% of a sixteen second window. Expecting nonlinear response characteristics, the test case matrix included four approximately logarithmically separated input amplitudes.

Sets of 300 trials were performed for each input amplitude at four equally spaced stages of deployment. The spectral data from these trials was averaged for each case to reduce the standard deviation of the data.

Along with collocated force and acceleration measurements at the input location, tri-axial acceleration measurements were made at each of the nodes. Additional pairs of vertically oriented accelerometers were located on the motorized joints to capture the "rocking" behavior of the joints depicted later. These led to a total of 54 acceleration output channels.

Data Analysis Procedure

Frequency response functions were generated for individual input-output pairs for each input amplitude. This data was then compared between deployment state cases.

The pseudo burst random FRFs were then curve fit using the time domain ERA system identification technique. The ERA Hankel matrix had 40 block rows and 1000 block columns. The choice of the number of block rows meant that the first 0.08 seconds of each impulse response was used, ensuring that at least two full periods of the lowest mode of interest was used in the identification. The number of block columns was as large as feasible given computation time constraints.

These specifications were applied to identification procedures for models of order 300 to 20 to examine the sensitivity of the modal parameters of interest to the truncation order of the model. The results proved to be fairly consistent for models with 80 to 20 degrees of freedom. A statistical analysis of this data was then performed to quantify the amount of error expected from the model and assure that it was sufficiently less then the trends that were being observed. Trends in the identified modal frequencies and damping ratios as input amplitude increased are then examined for a pair of consistently identified modes.

PHASE PLANE RESPONSE INVESTIGATION PROCEDURE

Although pseudo burst-random is a common modal excitation waveform for a full bandwidth survey, it can mask the effects of nonlinearity. Consequently, we also investigated sine dwell resonance tests to examine the nonlinear boundaries of a number of modes.

This portion of the investigation was performed in a manner similar to the modal survey with the input provided by the shaker and outputs by the collocated input and output force and acceleration transducers used in the initial testing. A constant sinusoidal input was provided by a waveform generator. The phase plane response of the collocated input-output transfer function was then examined on an oscilloscope. In this manner, dependence of the phase plane response of the structure was measured as a function of input amplitude at a variety of modal resonance frequencies.

RESULTS

Linear Vibration

Some modal predictions from the finite element model are presented in Table 1. The initially surprising result was that the model predicts a softer response from the fully deployed state of the structure. This is more easily understood once one considers the shapes of the modes developed by the model (see Figures 5 & 6). From these shapes, we notice that the greatest deflections of the structure occur with rotations of the mid-plane nodes allowing a pin-pin deflection shape of the screw-jack members. There is little torsional stiffness in the y direction about these nodes due to the degrees of freedom of all of the pin-clevis joints removing the y direction torsional stiffness contributions of the connected members. This lack of stiffness leads to the shapes of the first few lowest modes with larger deflections developing in the rightmost batten plane for the following modes.

Table 1: Modeled and Measured Modal Frequencies

Mode #	Frequency (Hz)					
	Half Deployment			Full Deployment		
	Modeled	Measured	Error	Modeled	Measured	Error
1	39.0	31.4	+7.6	37.8	42.7	-4.9
2	45.4	42.1	+3.3	39.7	44.8	-5.1
3	52.5	46.7	+5.8	50.2	53.0	-2.8

The result of this is that the shortening of the screw-jack members at half deployment serves to stiffen the softest point in the structure. Therefore, even with the weaker geometry of the half-deployed configuration, the overall stiffness of the structure is increased. These extreme node rotations are examples of the atypical low frequency displacements to be found in compact deployable structures.

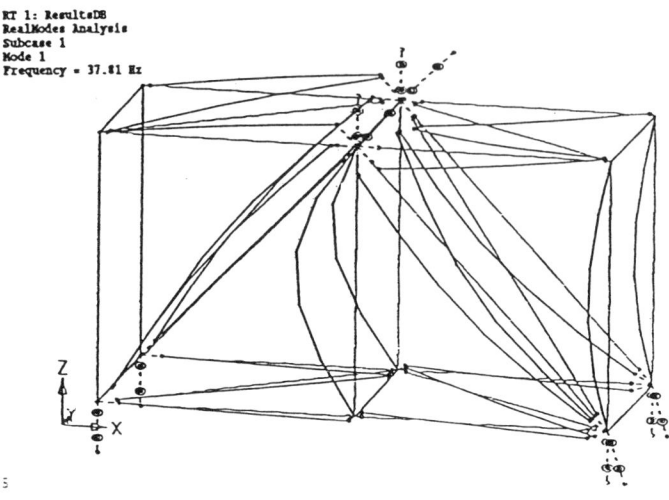

Figure 6: First Mode of Fully Deployed Structure

One of the first areas to be analyzed in the frequency response of the structure was the accuracy of the predicted stiffening of the structure at half deployment. In Table 1, we see that this is not supported by the modal test results. Also, in Figure 7, we observe a general softening in the measured response of the structure from full to half deployment. Evidence for this softening includes decreases in the frequencies of the lowest modes as well as increases in their amplitudes. Possible explanations for the inaccuracy of the models include the complexity of the geometry of the structure as well as failure to include the nonlinear behavior of the joints in the models.

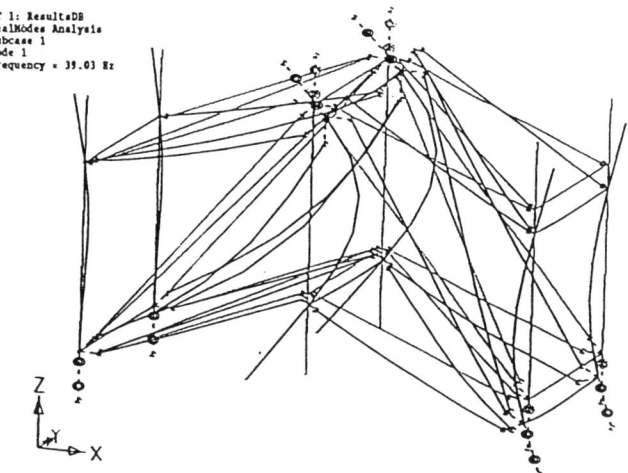

Figure 5: First Mode of Half Deployed Structure

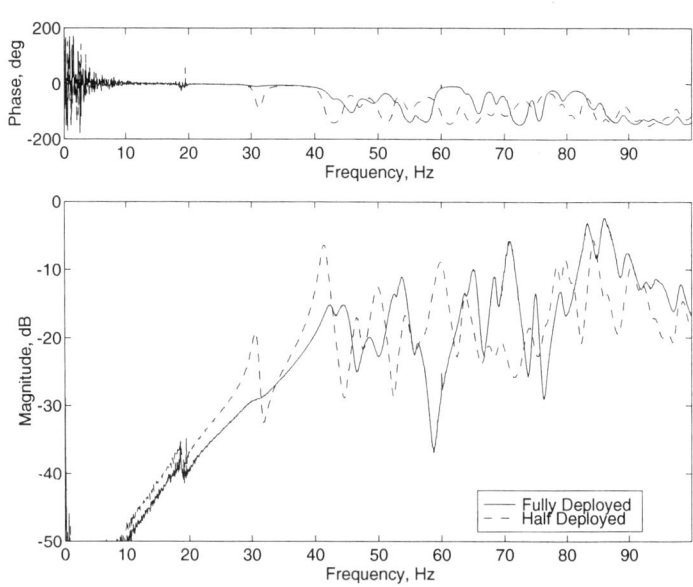

Figure 7: FRFs of Fully Deployed and Half Deployed Structure

Nonlinear Results

Figure 8 displays the frequency responses of the fully deployed structure for all four input amplitudes. Comparison of these responses provides immediate evidence of the high level of nonlinearity in the structure. Trends to be observed include varying amounts of softening (reduced frequency), increased damping, and decreasing response amplitudes as input amplitude is increased. As seen in Bowden's study of fundamental nonlinear joint models [4], decreasing modal frequencies following increasing input amplitude are characteristic of a Coulomb friction nonlinearity. This is in contrast to the nature of joint freeplay dynamics acting to stiffen the dynamic response as input amplitude is increased.

In order to quantify the frequency and damping variations across all of the inputs, it was necessary to focus on modes that were best identified. This was a result of the highly nonlinear response for the highest input amplitude. Even allowing this, the identification algorithm was able to regularly identify only one of the selected modes for the highest input amplitude case. The mean identified modal parameter values for model sizes from 20 to 80 degrees of freedom is presented in Table 2 and in Figures 9 and 10.

The identified modal frequencies decreased by up to 2.7Hz from the lowest to highest input amplitudes for the two selected modes. This variation was a nearly linear function of the input amplitude. The identified frequencies for mode A decreased at a rate of about 0.22 Hz/N, while mode B decreased at about 0.53 Hz/N.

As seen in Figure 10, identified damping ratios for the two modes revealed a slightly less consistent linear dependence on input amplitude. While the identified damping ratio for mode A decreased slightly initially, both modes eventually experienced approximately a 140% increase in damping from the lowest to highest input amplitudes.

In an attempt to better understand the dynamic displacements of the structure, the displacements involved in these modes were simulated by the model identification software. The motion of the measured points resembled global bending modes for modes A and B.

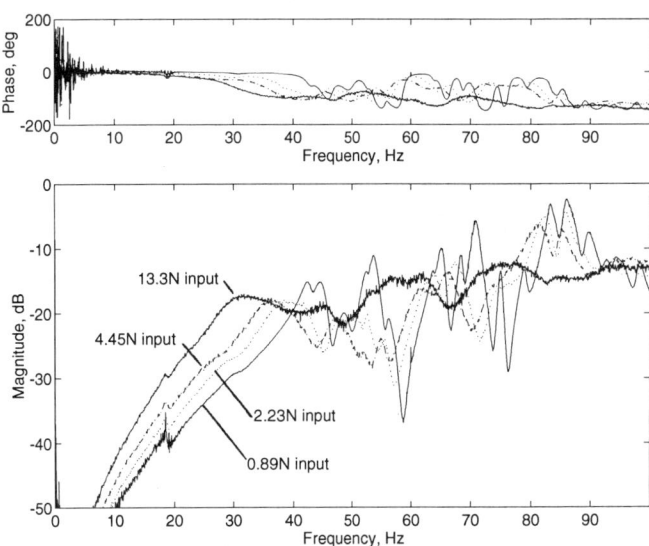

Figure 8: FRFs of Fully Deployed Structure for Multiple Input Amplitudes

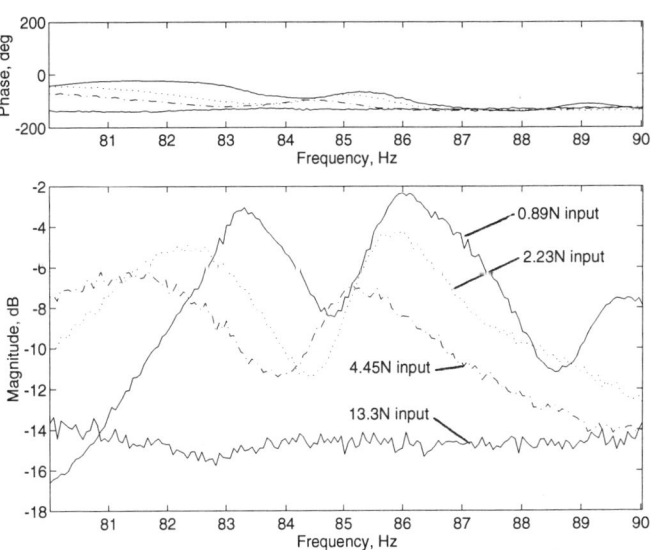

Figure 9: FRFs of Fully Deployed Structure for Multiple Input Amplitudes (Amplification of examined Modes A and B)

Table 2: Trends in Modal Parameters

Input Amplitude(N)	Mode A		Mode B	
	ω (Hz)	ζ (%)	ω (Hz)	ζ (%)
0.89	86.0	1.17	83.7	1.15
2.23	85.7	0.97	82.7	1.75
4.45	84.8	1.11	81.8	2.90
13.3	83.28	2.83	NA	NA

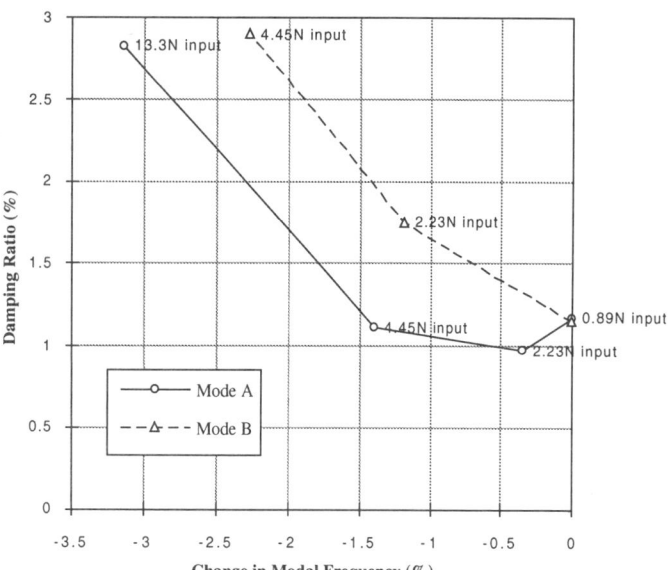

Figure 10: Modal Parameter Variations for Multiple Input Amplitudes

<u>Phase Plane Response Investigation Results</u>

Figure 11 displays a typical series of scans of the acceleration vs. input force phase plots. As can be seen from the equally scaled images, the shift into the obviously nonlinear state occurs over a relatively short increase in input amplitude. This bifurcation of resonance states was observed across the spectrum at varying levels of input amplitude.

Also noted during this part of the study was a hysteretic state path as a function of both input amplitude and frequency. This action was seen as the bifurcation would occur at one input amplitude as the amplitude was increased while the system would return to the previous state at a lower input amplitude as it was decreased. This hysteretic behavior also occurred upon increasing and decreasing the input frequency while the input amplitude was held constant.

During this experimentation, the question arose as to at what amplitudes these bifurcations were occurring at. The specifications for the pin-clevis joint parts suggested that free-play in the joints might occur near 15 μm. Recording the acceleration at which these bifurcations occurred at 5Hz intervals across the spectrum allowed calculation of the maximum displacement of the structure at these points. Figure 12 portrays the results of this nonlinear boundary survey. These values proved to be repeatable to within ±4μm, but no correlation was evident between this boundary and the frequency response of the structure.

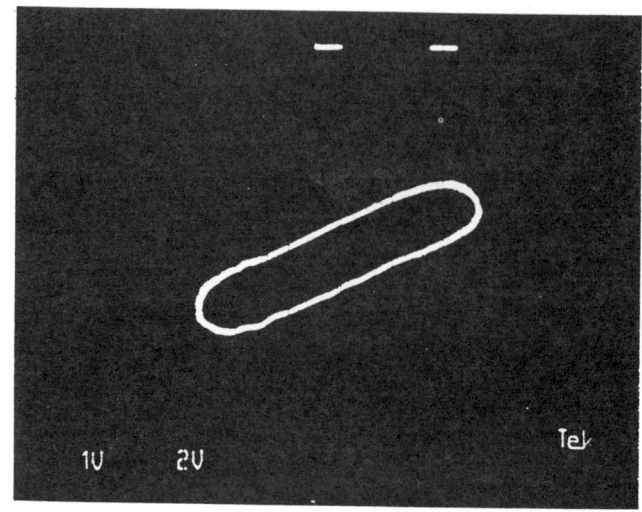
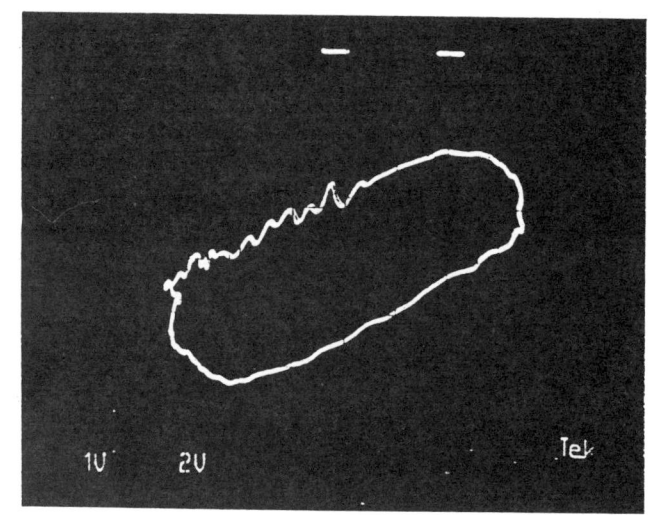
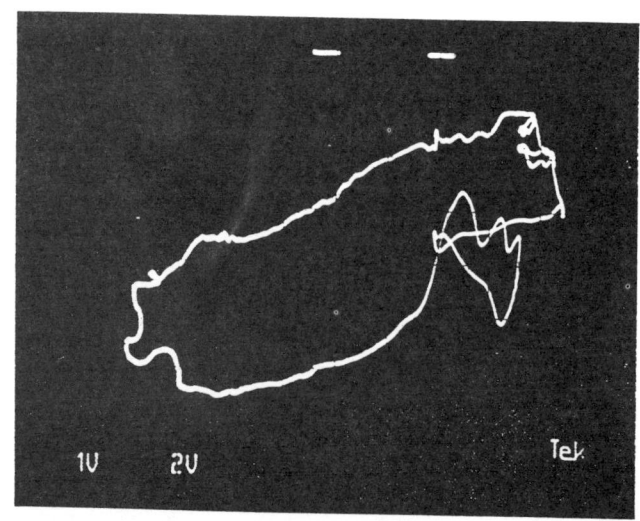

Figure 11: Phase Plane Bifurcation (Input force along the horizontal axes and acceleration along the vertical axes)

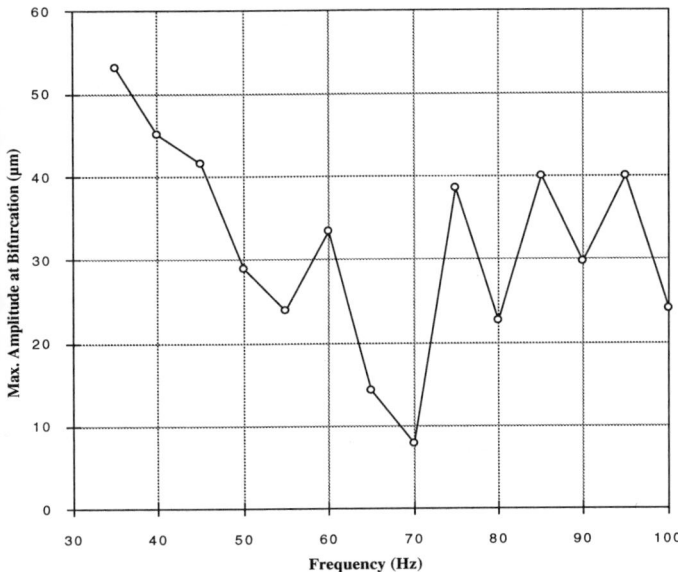

Figure 12: Map of the Nonlinear Boundary at Different Resonances

CONCLUSIONS

Modal displacements of the structure predicted by the finite element model portrayed unique mode shapes that are to be expected from such a compact deployable structure. The dependency of the dynamics of the structure on local displacements increased the influence of the nonlinear joint connections on the global response of the structure. Input amplitude dependent variations of modal frequencies and damping ratios indicated a dominance of nonlinear Coulomb friction. However, an examination of the phase plane response of the structure as input amplitude was increased revealed a hysteretic bifurcation of resonance states at a micron displacement level near the freeplay of the joints. The dominance of the Coulomb friction characteristics is most likely a result of the gravitational preloading of the joints increasing the friction while reducing the freeplay mechanics.

Further study of the structure under a different preloading condition is recommended for a better understanding of its influence. An attempt to model the structure with nonlinear considerations included is also recommended in order to determine the limits of our modeling accuracy.

ACKNOWLEDGMENTS

This work was supported by the Center for Space Construction (CSC) through NASA Grant No. NAGW-1388 and by NASA Langley Research Center (LaRC) through NASA Grant No. NAG-1-1490. Development of the test structure was supported by a grant from the Shimizu Corporation. The authors would like to acknowledge the assistance provided by Trudy Wittrock and Stephanie Gow, undergraduate research assistants, for their aid in the assembly of the finite element model and performance of the modal testing.

REFERENCES

1. Mikulas, M.M., and Withnell, P.R., "Construction Concepts for a Precision Segmented Reflector," Proceeding of the 34rd AIAA Structures, Structural Dynamics and Materials Conference, April, 1993, AIAA-93-1461.

2. Sarver, G.L. and Crawley, E.F., "Energy Transfer and Dissipation in Structures with Discrete Nonlinearities," Ph.D. Thesis, Dept. of Aeronautics and Astronautics, MIT, November, 1987.

3. Webster, M. and Velde, W.V., "Modelling Beam-Like Space Trusses with Nonlinear Joints with Application to Control," Ph.D. Thesis, Dept. of Aeronautics and Astronautics, MIT, June, 1991.

4. Bowden, M.L., "Dynamics of Structures with Nonlinear Joints," Ph.D. Thesis, Department of Aeronautics and Astronautics, MIT, May, 1988.

5. Crawley, E.F., Barlow, M.S., and van Schoor, M.C., "Variation in the Modal Parameters of Space Structures," Proceeding of the 33rd AIAA Structures, Structural Dynamics and Materials Conference, April, 1992, AIAA-92-2209.

6. Bullock, S.J., and Peterson, L.D., "Identification of Nonlinear Joint Mechanics for a Deployable Joint," Proceedings of the 35th AIAA Structures, Structural Dynamics and Materials Conference, April, 1994, AIAA-94-1347.

7. Juang, J.-N. and Pappa, R.S., "An Eigensystem Realization Algorithm for Modal Parameter Identification and Model Reduction," Journal of Guidance, Control and Dynamics, Vol. 8, Sept.-Oct. 1985, pp. 620-627.

8. Alvin, K., Park, K.C. and Belvin, W., "A Second-Order Structural Identification Procedure via State Space-Based System Identification," Proceedings of the 1992 AIAA Guidance, Navigation, and Control Conference, Hilton Head Island, August 1992, AIAA-92-4387.

9. MSC/NASTRAN User's Manual, The MacNeal-Schwendler Corporation, Version 67, August, 1991, ISSN 0741-8019.

10. LabVIEW 2 User Manual, National Instruments Corporation, September 1991 Edition, Austin, TX.

AIAA-94-1735-CP

INTERVAL MODEL IDENTIFICATION FOR FLEXIBLE STRUCTURES WITH UNCERTAIN PARAMETERS

Jiann-Shiun Lew[1]
Center of Excellence in Information Systems
Tennessee State University
Nashville, TN 37203-3401

Troy L. Link[2] and Ephrahim Garcia[3]
Department of Mechanical Engineering
Vanderbilt University
Nashville, TN 37235

Lee H. Keel[4]
Center of Excellence in Information Systems
Tennessee State University
Nashville, TN 37203-3401

ABSTRACT

A novel approach is developed for modeling a structure with uncertain parameters directly from experimental data. In the recent years the study of interval control system has been the subject of much controls and systems research. The interval model of transfer function coefficients is the model structure chosen in this paper. The system identification algorithm proposed in this paper bridges the recently developed interval system techniques to the practical problems. A ten bay truss structure with various added masses is designed to represent a structure with uncertainties. The results of the open-loop and closed-loop experiments of the ten bay truss structure illustrate and verify the algorithm developed.

1. INTRODUCTION

Modeling a dynamic system with uncertain parameters is an important and challenging task for researchers in structural dynamics, system identification, and robust control. For large structures assembled in space, there is the possibility of minor configuration changes and additions to the structure. Any of these changes may drastically change the dynamics of the structure. To maintain structural control, it is necessary to consider these types of system changes.

A structure with uncertainties can be modeled in different ways. In the H_∞ approach[1], the system is described as a nominal model with its corresponding H_∞ error bound transfer function. The minor changes of lightly damping flexible structures may result in

[1] Assistant Professor, Member AIAA.
[2] Graduate student.
[3] Assistant Professor, Member AIAA.
[4] Associate Professor, Member AIAA.

Copyright © 1994 American Institute of Aeronautics and Astronautics, Inc. All rights reserved.

the changes of natural frequencies, damping ratios and mode shapes. For instance, a small change of natural frequency may result in significant H_∞ error bound and the error bound even close to the magnitude of the original model. Clearly a choice of largely over-bounded H_∞ error bound transfer function results in overly conservative H_∞ control design. Using a time invariant system described by a set of differential equations to model a structure with uncertainties provides the physical representation[2]. Each differential equation can be expressed as a transfer function. So this model set can be represented as a set of transfer functions. The transfer function set is the model structure considered in this paper.

Here we address the problem of obtaining a model set to represent a structure with parameter changes so that the analysis and design of the entire system is achieved from those of the model set. In order to take advantage of the well-studied properties found in the recent parametric robust control literature[3,4], we consider an interval system of the transfer functions to represent the model set. The linear interval system, which is the system used in this paper, is given as transfer functions whose numerator and denominator are linear combinations of polynomials that the coefficients of the linear combinations are bounded with known values.

In this paper, an algorithm is developed to obtain an interval model to represent a structure with uncertainties directly from experimental data. First, a least squares technique is used to identify the models for a structure with various changes. Then an algorithm developed by using Singular Value Decomposition(SVD)[5] technique is used to obtain a model structure of the transfer function coefficients to generate an optimal linear interval model. A ten-bay truss structure with various added masses is designed [6] to demonstrate and test the developed algorithm. For the first time, an experimental interval model is constructed and tested. The developed system identification algorithm demonstrates the application of interval control theory to an experimental robust control problem.

2. PROBLEM AND MODEL STRUCTURE

Let us consider the frequency domain experimental data for a structure with various changes

$$\{G_i(w_j), \quad j=1,2,\ldots,N, \quad i=1,\ldots,n\} \quad (1)$$

where G_i is the data for the ith case and each case has N data points. There are n cases for the structure with various changes. There is no single model which can represent all these cases. In this paper, the object is to obtain a model set to represent and cover these n different cases.

The linear time invariant system can be represented as a transfer function. A linear system with uncertainties of physical parameters, such as spring constants and masses, can be represented via bounds of the transfer function coefficients[2]. The model structure chosen in this paper is a linear interval system of the transfer function[7]

$$G(s,p) = \{g(s)|g(s) = \frac{n_0(s) + \sum_{i=1}^{m+1}\alpha_i r_i(s)}{d_0(s) + \sum_{j=1}^{m}\beta_j q_j(s)}$$

$$\alpha_i \in [\alpha_i^-\ \alpha_i^+], \beta_j \in [\beta_j^-\ \beta_j^+]\} \quad (2)$$

where

$$p = [\alpha_1 \ldots \alpha_{m+1}\ \beta_1 \ldots \beta_m]^T \quad (3)$$

$\{r_1, r_2, \ldots, r_{m+1}\}$ is a basis for the polynomials with degree m, and $\{q_1, q_2, \ldots q_m\}$ is a basis for the polynomials with degree $m-1$. The mth order polynomial $n_0(s)$ and monic polynomial $d_0(s)$ are the numerator and denominator of the nominal model. The variables α_i and β_i represent the parameter uncertainty part. This model structure has more degrees of freedom and is relatively less conservative than the most popular interval system[7]

$$G(s,p) = \{g(s)|g(s) = \frac{n_0(s) + \sum_{i=1}^{m+1}\alpha_i s^{m+1-i}}{d_0(s) + \sum_{j=1}^{m}\beta_j s^{m-j}}$$

$$\alpha_i \in [\alpha_i^-,\alpha_i^+], \beta_j \in [\beta_j^-,\beta_j^+]\} \quad (4)$$

Note that the model structure in Eq. (4) is a special case of the model structure in Eq. (2). Thus, the problem is to find the set of basis polynomials $\{r_1(s), \ldots, r_{m+1}(s), q_1(s), \ldots q_m(s)\}$, nominal polynomials $n_0(s), d_0(s)$ and bounds of parameters, α_i^-, α_i^+, β_i^-, β_i^+ so that the interval model $G(s,p)$ represents and covers the previous n sets of data in the frequency domain.

3. INTERVAL MODEL IDENTIFICATION

The algorithm to generate an interval model starts with identifying a model for each case by fitting each data set $\{G_i(jw_k), k=1,2,\ldots,N\}$. To identify the ith model, we use a least squares based system identification technique which is widely used in practice[8].

Using the technique in [8], we first determine the order of the transfer function

$$g_i(s) = \frac{n_i(s)}{d_i(s)} = \frac{n_m^i s^m + n_{m-1}^i s^{m-1} + \ldots + n_0^i}{s^m + d_{m-1}^i s^{m-1} + \ldots + d_0^i} \quad (5)$$

with the parameter vector $p_i = [d_{m-1}^i \ldots d_0^i\ n_m^i \ldots n_0^i]^T$ and let

$$n_i = [n_m^i \ldots n_0^i]^T, \quad d_i = [d_{m-1}^i \ldots d_0^i]^T \quad (6)$$

Then using the least squares technique, we compute the coefficients of the polynomials $n_i(s)$ and $d_i(s)$ that minimize the cost function:

$$J_i = \sum_{k=1}^{N} W(w_k)[(Real(G_i(w_k)d_i(jw_k) - n_i(jw_k)))^2$$

$$+ (Imag(G_i(w_k)d_i(jw_k) - n_i(jw_k)))^2] \quad (7)$$

where $W(w_k)$ is a positive weighting function, $d_i(s)$ and $n_i(s)$ are the denominator and numerator of the transfer function $g_i(s)$, respectively. We repeat this procedure for all the n sets of test data to obtain the transfer functions $g_i(s), i=1,\ldots,n$. After the models for all the cases are obtained, a judicious choice for the nominal model is the average of all the identified models

$$n_0 = \frac{1}{n}\sum_{i=1}^{n} n_i, \quad d_0 = \frac{1}{n}\sum_{i=1}^{n} d_i \quad (8)$$

The uncertainty part in Eq. (2) is contributed from the difference between the nominal model and the identified models. The parameter perturbation vectors of the ith identified model are defined as

$$\Delta n_i = n_i - n_0, \quad \Delta d_i = d_i - d_0 \quad (9)$$

For the model structure in Eq. (4), the minimum interval of each transfer function coefficient is the range of the perturbations of the corresponding coefficients[7]. This model structure may be too conservative. For example, if all the perturbations Δn_i are on a straight line. The dimension of the perturbation space for this model structure is still $m+1$. But the perturbation is actually on a straight line with one dimension. In this paper, a SVD technique is developed to find an optimal linear model structure. To illustrate this SVD technique, we first form a matrix for the uncertainty of denominator as

$$\Delta D = [\Delta d_1\ \Delta d_2\ \ldots\ \Delta d_n] \quad (10)$$

Using SVD to factorize the matrix ΔD yields

$$\Delta D = USV^T, \quad (11)$$

with

$$U = [U_1\ U_2\ \ldots\ U_m], \quad V = [V_1\ V_2\ \ldots\ V_n] \quad (12)$$

where U and V are orthonormal matrices, and S is a rectangular matrix

$$S = [S_m\ 0] \quad (13)$$

with $S_m = diag[s_1, s_2, \ldots, s_m]$ and monotically non-increasing $s_i (i = 1, 2, \ldots, m)$

$$s_1 \geq s_2 \geq \ldots \geq s_m \geq 0 \quad (14)$$

where the number of the non-zero singular values s_i is the same as the rank of the matrix ΔD.

Theorem: If the size of the perturbation matrix ΔD is $m \times n$ with $n \geq m$ and the number of the non-zero singular values is m_k. Then the column vectors Δd_i are inside the space spanned by the first m_k orthonormal vectors of $\{U_1, \ldots, U_m\}$.

Proof:

The proof starts with representing the perturbation vectors in terms of the linear combinations of the orthogonal vectors of the basis $\{U_1, \ldots, U_m\}$. The corresponding coordinate vector of the perturbation Δd_i relative to this basis is

$$\Delta q_i = U^T \Delta d_i \quad (15)$$

So we can obtain

$$[\Delta q_1 \ldots \Delta q_n] = U^T[\Delta d_1 \ldots \Delta d_n] = U^T \Delta D \quad (16)$$

Here the singular value matrix in Eq. (13) can be expressed as

$$S = \begin{pmatrix} S_{m_k} & 0 \\ 0 & 0 \end{pmatrix} = \begin{pmatrix} S^0_{m_k} \\ 0 \end{pmatrix} \quad (17)$$

with $S_{m_k} = diag[s_1, s_2, \ldots, s_{m_k}]$. Substituting Eqs. (11) and (17) into Eq. (16) yields

$$\begin{aligned}[] [\Delta q_1 \Delta q_2 \ldots \Delta q_n] &= U^T U \begin{pmatrix} S^0_{m_k} V^T \\ 0 \end{pmatrix} \\ &= I_m \begin{pmatrix} S^0_{m_k} V^T \\ 0 \end{pmatrix} \\ &= \begin{pmatrix} S^0_{m_k} V^T \\ 0 \end{pmatrix} \end{aligned} \quad (18)$$

where $S^0_{m_k} V^T$ is an $m_k \times n$ matrix. The last $m - m_k$ elements of any Δq_i are zero. So all the uncertainty vectors Δd_i are inside the space spanned by the first m_k orthonormal vectors of $\{U_1, \ldots, U_m\}$.

Q.E.D.

The full rank of $S^0_{m_k} V^T$ shows that the dimension of the uncertainty matrix ΔD is m_k. Notice that the $i^{th} (i \leq m_k)$ row vector of matrix $[\Delta q_1 \ldots \Delta q_n]$ is $s_i V_i^T$ with norm s_i. The jth element of vector $s_i V_i$ represents the perturbation Δd_j in the direction of U_i. Each singular value s_i indicates the magnitude of the perturbations in the direction of U_i.

For structures with low damping, the scalar d^0_0, which is the multiplicative product of the squares of all the natural frequencies, may be many orders larger than the scalar d^0_{m-1}, which is the sum of $2\xi_i w_i$ with ξ_i and w_i as the damping ratio and the natural frequency of the ith mode, respectively. For example, later we will show that the d^0_0 of the first two modes model of the ten-bay structure is around 10^7 times larger than d^0_3 of this model. Due to the level of the perturbation for each coefficient, we compute the weighted perturbation matrix ΔD^W

$$\Delta D^W = W_d^{-1} \Delta D \quad (19)$$

where $W_d = diag[w_d^1\ w_d^2\ \ldots\ w_d^m]$ and w_d^i is the standard deviation of the ith row vector of ΔD. To find the distribution of the weighted uncertainty in space, the SVD is used to factorize the matrix

$$\Delta D^W = U^W S^W (V^W)^T, \quad U^W = [U_1^W \ldots U_m^W] \quad (20)$$

Here the singular value s_i indicates the weighted perturbations distributed in the direction of U_i^W. The corresponding coordinate vector of the uncertainty Δd_i^W relative to the basis $\{U_1^W, \ldots, U_m^W\}$ is

$$\Delta q_i^W = (U^W)^T \Delta d_i^W \quad (21)$$

Since the basis $\{U_1^W, \ldots, U_m^W\}$ is corresponding to the weighted perturbation matrix ΔD^W. We want to find a basis corresponding to the nonweighted perturbation matrix ΔD. From Eqs. (19) and (20), the matrix ΔD can be written as

$$\Delta D = W_d U^W S (V^W)^T \quad (22)$$

The basis for the nonweighted perturbation matrix ΔD is computed as

$$U_d = W_d U^W, \quad U_d = [U_{d1}\ U_{d2}\ \ldots\ U_{dm}] \quad (23)$$

The basis $\{U_{d1}, \ldots, U_{dm}\}$ to the nonweighted perturbation matrix ΔD is equivalent to the basis $\{U_1^W, \ldots, U_m^W\}$ to the weighted perturbation matrix ΔD^W. The corresponding coordinate vector of the uncertainty Δd_i relative to the basis $\{U_{d1}, \ldots, U_{dm}\}$ is

$$\begin{aligned} \Delta q_i &= U_d^{-1} \Delta d_i \\ &= (U^W)^{-1}(W_d)^{-1}(W_d) \Delta d_i^W \\ &= (U^W)^T \Delta d_i^W = \Delta q_i^W \end{aligned} \quad (24)$$

Notice that the nonweighted perturbation coordinate vector Δq_i is the same as the weighted coordinate vector Δq_i^W. Both perturbation matrices share the same singular values to indicate the distribution of perturbation. The fixed polynomials $q_i(s)$ of the interval model in Equation (2) are now composed of the basis vectors of U_d,

$$q_i(s) = \sum_{j=1}^{m} U_{di}(j) s^{m-j} \quad (25)$$

where $U_{di}(j)$ is the jth element of vector U_{di}. Finally, we determine the bounds for the corresponding polynomial $q_j(s)$ as

$$\beta_j^+ = max_{1 \leq i \leq n}(\Delta q_i(j)), \quad j = 1, \ldots, m \quad (26)$$

$$\beta_j^- = min_{1 \leq i \leq n}(\Delta q_i(j)), \quad j = 1, \ldots, m \quad (27)$$

where $\Delta q_i(j)$ is the jth element of vector Δq_i. This interval model represents an optimal linear box of the determined polynomials to cover the perturbation. Similarly, we also apply the SVD technique to the numerator perturbation matrix to obtain $r_i(s)$, α_i^-, and α_i^+.

4. EXPERIMENTAL RESULTS

In this section we will use the experimental results of a ten-bay truss structure[6] as shown in Figure 1 to demonstrate the developed algorithm. This structure is a 8 bay truss with two additional bays forming a T-section at one end. This T-section is designed so that the torsional modes are more pronounced[6]. Each bay is a 0.5 meter cube made of hollow aluminum tubes with threaded fasteners that are connected to nodal components. The truss is cantilevered to a massive monolith structure which is fixed to the ground. The structure weights 31.5 kilograms, is four meters long, one-half meter wide, and one and one-half meters tall. The parameter uncertainty is generated by adding various masses to nodes 17 and 44. In this paper, we consider six different cases with added masses as shown in Table 1 in units of kilograms. A reaction mass actuator(RMA)[6] located between nodes 39 and 40 is used to excite and control the structure's motion. This RMA is similar in design to the one used in [9]. An accelerometer located at node 40 is used to measure the acceleration in the x direction at this position. In this paper, we consider the model for the

Table 1: Masses Added to Structure

Case	1	2	3	4	5	6
Node 17	0	0.5	1.0	1.5	2.0	2.5
Node 44	0	0.5	1.0	1.5	2.0	2.5

modes within the $0 - 40\ HZ$ frequency range. There are two modes[6], one bending mode and one torsional mode, in this frequency range. The experimental frequency domain data are collected using a Tektronix 2642A Fourier Analyzer[6]. The experimental data including actuator dynamics are expressed as

$$G_i(w_k) = G_a(jw_k)G_i^0(w_k), \quad k = 1, 2, \ldots, N \quad (28)$$

where G_i is the ith set of experimental data for the ith case, and G_i^0 is the corresponding experimental data excluding actuator dynamics. The transfer function of the actuator is

$$G_a(s) = \frac{n_a(s)}{d_a(s)} \quad (29)$$
$$= \frac{2.3853 \times 10^{-3}s^3 + 1.6745s^2}{2.3853 \times 10^{-3}s^3 + 1.6745s^2 + 25.781s + 1251.3}$$

Follow the procedure in the last section, we first apply the least squares technique to each set of data G_i^0 to obtain the identified model for each case in Table 1. Figure 2 shows the experimental data and the identified models including actuator dynamics. The identified models well fit the experimental data for all the cases. The two peaks of each set of experimental data represent two structural modes and the natural frequency of each mode decreases when the added weight increases. Also the magnitude of each mode changes little for all the cases. Table 2 shows the structural eigenvalues of these six identified models. After we obtain the iden-

Table 2: Eigenvalues of the identified models

Model No.	First mode	Second mode
1	-1.3864±61.094	-1.4733±123.00
2	-1.3065±59.930	-1.3677±117.76
3	-1.1854±58.892	-1.4871±113.96
4	-1.1081±57.902	-1.4167±110.79
5	-1.0424±56.929	-1.4873±108.03
6	-0.9581±55.724	-1.5050±105.10

tified structural models excluding actuator dynamics, we compute the nominal model as the average of these models

$$G_0(s) = \frac{n_0(s)}{d_0(s)} \quad (30)$$
$$= \frac{-3.463 \times 10^{-3}s^4 - 1.173 \times 10^{-3}s^3 - 24.68s^2}{s^4 + 5.241s^3 + 16255s^2 + 40225s + 4.412 \times 10^7}$$

We now apply the algorithm in the last section to obtain the structural interval model. Tables 3 and 4 show the results of the interval model.

Table 3: Results of denominator part

	q_1	q_2	q_3	q_4
s^3	1.4326e-1	-2.5118e-1	-1.8767e-2	-1.0399e-3
s^2	8.6678e02	5.5011e02	-8.0878e02	1.1301e03
s^1	4.1568e03	1.8545e03	6.8649e03	8.2205e02
s^0	4.0831e06	2.5353e06	-2.4265e06	-6.1023e06
β^+	3.1117e00	2.8036e-1	2.5123e-2	2.5140e-3
β^-	-2.3396e00	-2.1055e-1	-3.3913e-2	-2.5161e-3

Table 4: Results of numerator part

	r_1	r_2	r_3	r_4	r_5
s^4	-8.7360e-4	-3.2529e-5	9.2234e-5	0	0
s^3	-1.3370e-4	3.8897e-4	1.0547e-5	0	0
s^2	-3.3372e00	-1.0513e00	-3.5316e00	0	0
s^1	0	0	0	1	0
s^0	0	0	0	0	1
α^+	2.4339e00	1.0485e00	6.3217e-2	0	0
α^-	-1.8816e00	-1.3450e00	-3.5834e-2	0	0

Table 3 shows that the model uncertainty of the denominator part is dominated by the uncertainty in the direction of the first singular vector. This phenomenon shows that the perturbation of the denominator coefficients is almost linear due to the changes of the added masses. The perturbation distributed in the direction of the fourth singular vector is around 1000 times smaller than that of the first singular vector.

Table 4 shows that the model uncertainty of the numerator part is dominated by the uncertainty in the directions of the first two singular vectors.

Actually the experimental data include actuator dynamics. After we obtain the structural interval model, we can generate the system interval model including actuator dynamics as

$$G(s) = \frac{n_a(a)[n_0(s) + \sum_{i=1}^{m+1} \alpha_i r_i(s)]}{d_a(s)[d_0(s) + \sum_{i=1}^{m} \beta_i q_i(s)]}$$
$$= \frac{n_a(a)n_0(s) + \sum_{i=1}^{m+1} \alpha_i(n_a(s)r_i(s))}{d_a(s)d_0(s) + \sum_{i=1}^{m} \beta_i(d_a(s)q_i(s))} \quad (31)$$

Next, we apply the interval system techniques[4], which can be used to find the boundaries of polar plots and bode plots of the interval model in Eq. (2), to obtain the magnitude envelopes of the system interval model. Figure 3 shows the magnitude envelopes and the magnitude plots of the experimental data. The magnitude envelopes precisely bound the experimental data throughout the frequency domain except that few points in the valley between 12 to 15 HZ are outside and close to the lower envelope. For each case, the minimum magnitude at valley is around 1000 times smaller than the magnitudes at peaks. In general, the data in the valley are much more sensitive to noise, data process and the other unmodelled high frequency modes. The envelopes in the peak ranges precisely cover the experimental data. So the system interval model well represents and covers the considered two modes for all the cases.

To verify the identified interval system by using the closed-loop experiment, we first design a local velocity feedback controller using root-locus techniques for the vibration suppression of the structure[6]. The transfer function of this designed controller is[6]

$$K(s) = \frac{n_k(s)}{d_k(s)} = \frac{-3300s}{s^2 + \pi s + \pi^2}$$

This controller is designed for the structure without added mass. This controller increases the damping of the first mode from 2.2% to 6.6% and the damping of the second mode from 1.2% to 3.9%. The closed-loop interval system with the designed controller can be computed as

$$T(s) = \frac{d_k(s)[n_a(s)n_0(s)+}{d_k(s)[d_a(s)d_0(s) + \sum_{i=1}^{m} \beta_i(d_a(s)q_i(s))]}$$
$$\frac{\sum_{i=1}^{m+1} \alpha_i(n_a(s)r_i(s))]}{+n_k(s)[n_a(a)n_0(s) + \sum_{i=1}^{m+1} \alpha_i(n_a(s)r_i(s))]} \quad (32)$$

And this interval system can be written as the form with fixed parts and perturbation parts

$$T(s) = \frac{\alpha(s) + \sum_{i=1}^{m+1} \alpha_i \alpha_{1i}(s)}{\beta(s) + \sum_{i=1}^{m} \beta_i \beta_{1i}(s) + \sum_{i=1}^{m+1} \alpha_i \beta_{2i}(s)} \quad (33)$$

where the fix part polynomials are

$$\alpha(s) = d_k(s)n_a(s)n_0(s) \quad (34)$$

$$\beta(s) = d_k(s)d_a(s)d_0(s) + n_k(s)n_a(s)n_0(s) \quad (35)$$

and the perturbation part polynomials are

$$\alpha_{1i}(s) = d_k(s)n_a(s)r_i(s) \quad (36)$$

$$\beta_{1i}(s) = d_k(s)d_a(s)q_i(s) \quad (37)$$

$$\beta_{2i}(s) = n_k(s)n_a(s)r_i(s) \quad (38)$$

Then we apply the interval system techniques to obtain the magnitude envelopes for the closed-loop interval system. Figure 4 shows that the magnitude envelopes precisely bound the magnitude of the experimental data except that few points in the valley are outside and close to the lower envelope. The results in Figure 4 verify that the identified interval model well represents and covers all the cases. Also we apply edge theory[10], which can be used to find the boundary of the roots of the linear interval polynomials, to obtain the root clusters of the closed-loop interval model. Figure 5 shows the root clusters of the structural eigenvalues of the closed-loop interval system and the closed-loop experimental structural eigenvalues, which are obtained by applying the previous least squares technique to the closed-loop experimental data. The root clusters bound the experimental eigenvalues with some eigenvalues close to the boundaries. Also the damping of each mode is increased by comparing the boundaries of the root clusters in figure 5 and the eigenvalues in Table 2. The closed-loop performance can be precisely predicted by analyzing the simulation of the identified interval model. In the real experiment, there are many factors that may affect the analysis. For example, the actuator dynamics is not exactly linear[6]. The identified model does not exactly fit the experimental data, which change with the environmental changes. Therefore, the agreement between the analysis and the experimental results is quite satisfactory.

5. CONCLUDING REMARKS

This paper presents an algorithm to obtain a linear interval model of the transfer function coefficients to represent a structure with uncertainty. A singular value decomposition technique is developed to find the perturbation part for the interval model. The singular vectors and singular values of the perturbation matrix give the information of model structure and perturbation distributed in the directions of the singular vectors. We also apply the interval system technique to obtain the magnitude envelopes of the interval model and the results show that the identified interval model precisely covers and represents the experimental data. The results of the closed-loop experiment successfully verify the developed algorithm. The experimental results show that the developed algorithm successfully applies the control theory based on interval systems to a real experimental system.

References

[1] Bayard, D. S., Hadaegh, F. Y., Yam, Y., Scheid, R. E., Mettler, E., and Milman, M. H., "Automated On-Orbit Frequency Domain Identification for Large Space Structures," *Automatica*, vol. 27, pp. 931 – 946, November 1991.

[2] Bhattacharyya, S. P., *Robust Stabilization Against Structured Perturbations*, Lecture Notes in Control and Information Sciences, Springer-Verlag, 1987.

[3] Kharitonov, V. L., "Asymptotic Stability of an Equilibrium Position of a Family of Systems of Linear Differential Equations," *Differential Uravnen*, vol. 14, pp. 2086 – 2088, 1978.

[4] Bhattacharyya, S. P., and Keel, L. H., "Robust Stability and Control of Linear and Multilinear Interval Systems," *Control and Dynamic Systems*, vol. 51, pp. 31 – 78, 1992.

[5] Klema, V. C., and Laub, A. J., "The Singular Value Decomposition: Its Computation and Some Application," *IEEE Transactions on Automatic Control*, vol. AC-25, pp. 164 – 176, 1980.

[6] Link, T. L., "Vibration Suppression of Flexible Structures with Parametric Uncertainty Using Interval Control Systems," M.S. Thesis, Department of Mechanical Engineering, Vanderbilt, Nashville, TN, 1993.

[7] Lew, J. -S., Keel, L. H., and Juang, J. N., "Quantification of Model Error via an Interval Model with Nonparametric Error Bound," *Proceedings of the AIAA Guidance, Navigation, and Control Conference*, Monterey, CA, August 1993.

[8] Adcock, J. L., "Curve Fitter for Pole-Zero Analysis," *Hewlett-Packard Journal*, January 1987.

[9] Garcia, E., Webb, S., and Duke, J., "Passive and Active Control of Complex Flexible Structure Using Reaction Mass Actuator," *Vibrations and Acoustics*, to appear.

[10] Bartlett, A. C., Hollot, C. V., and Lin, H., "Root Location of an Entire Polytope of Polynomials: It Suffices to Check the Edges," *Mathematics of Controls, Signals and Systems*, vol. 1 pp. 61-71, 1988.

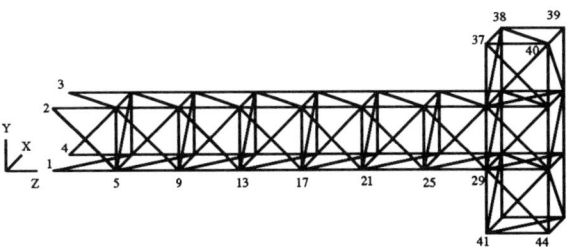

Figure 1: Finite Element Model of 10 Bay Truss Including Node Numbers

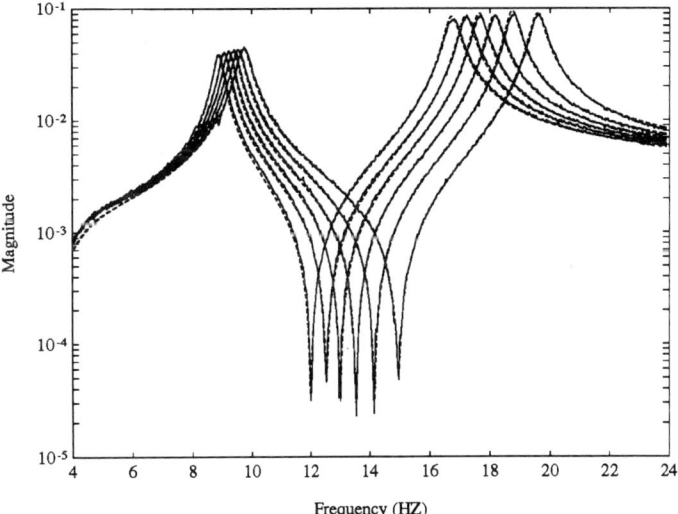

Figure 2: Experimental Data(—) and Identified Models(- -)From Right to Left
$0Kg$, $1Kg$, $2Kg$, $3Kg$, $4Kg$, $5Kg$

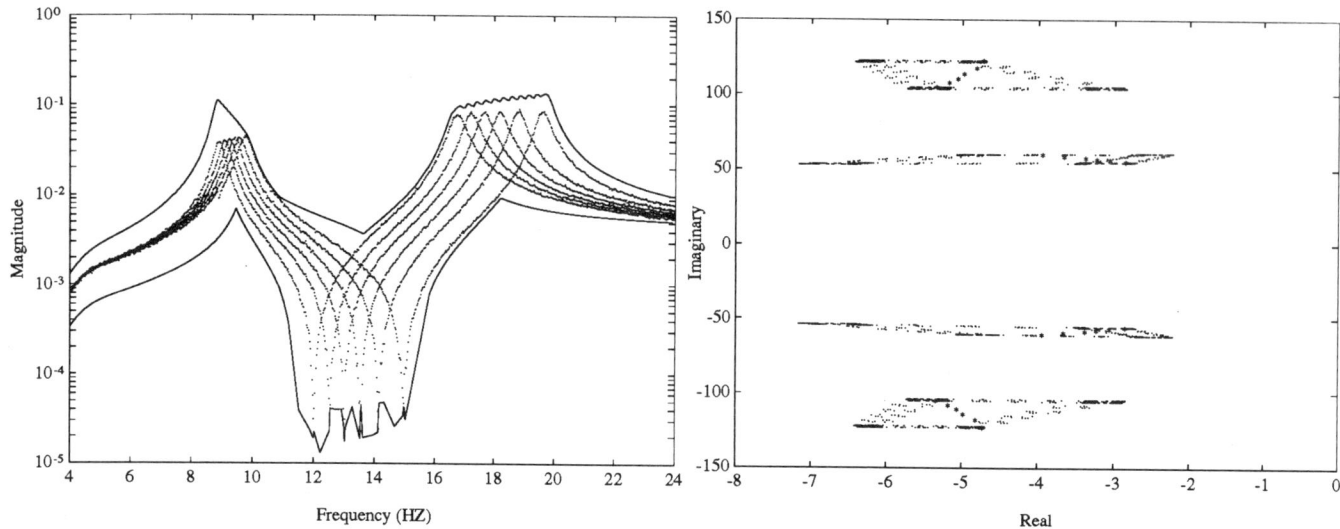

Figure 3: Magnitude Envelopes(—) of Interval Model and Experimental Data(...) for the Open-Loop System

Figure 5: Root Cluster(.) of Closed-Loop Interval Model and Eigenvalues(*) of Closed-Loop Experiment

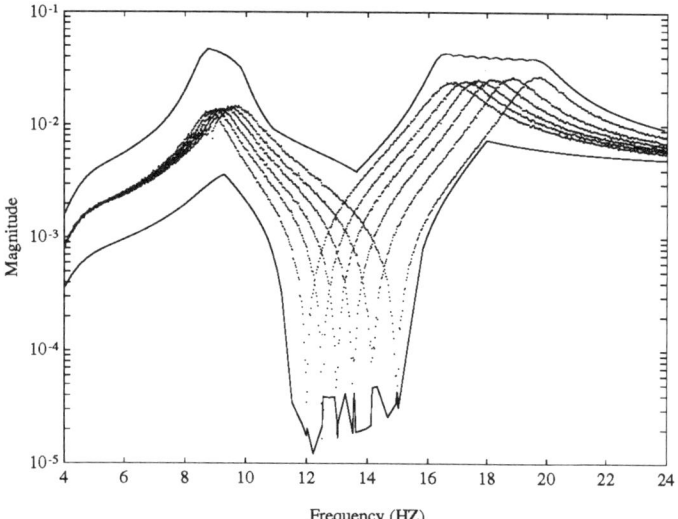

Figure 4: Magnitude Envelopes(—) of Interval Model and Experimental Data(...) for the Closed-Loop System

PIEZOELECTRIC ACTUATOR CAPABLE OF INDUCING TORSIONAL AND BENDING CONTROL LOADS WITH NO STRUCTURAL COUPLING

Grzegorz Kawiecki*
The University of Tennessee
Knoxville, TN

Winthrop P. Smith**
The University of Tennessee
Knoxville, TN

ABSTRACT

The objective of this paper is to show the feasibility of a novel method to achieve complete vibration control of a thin-walled structural member using piezoelectric elements. It is found that the control of combined torsional, bending and longitudinal vibrations of thin-walled members without inducing structural coupling is technically feasible. Numerical, analytical and experimental validation of the concept are presented.

NOTATION

ROMAN LETTERS

A	laminate extensional stiffness matrix, [N/m]
B	laminate coupling stiffness matrix, [N]
D	laminate bending stiffness matrix, [Nm]
E	modulus of elasticity, [N/m^2]
G	shear modulus, [N/m^2]
N	force per unit length, [N/m]
M	bending moment per unit length, [N]
Q	transformed lamina stiffness matrix, [N/m^2]
t	lamina thickness, [m]
z_k	distance from midplane to outer surface of the kth lamina, [m]

GREEK LETTERS

ε	strain
κ	curvature
Λ	induced strain
ν	Poisson's ratio

---

* Assistant Professor, Mechanical and Aerospace Engineering, AIAA Member
** Graduate Student, Mechanical and Aerospace Engineering
Copyright © 1994 by Grzegorz Kawiecki and Winthrop P. Smith. Published by the American Institute of Aeronautics and Astronautics, Inc. with permission.

SUBSCRIPTS

l	laminate
r	resultant
p	piezoelement
x	longitudinal coordinate
y	transverse coordinate

INTRODUCTION

The application of piezoelectric materials as actuating and sensing devices in active control systems has been investigated extensively. Hovewer, a survey of available literature shows that very little effort has been put into the application of piezoelectric elements to modify the torsional and torsional-bending behavior of structures. The most common mode of actuation described in the literature is bending and extension. To the best of the knowledge of the authors, the only studies related to generating twist using piezoelectric materials have been presented in Refs. 1, 2, 3 and 4. Although the results of the studies presented in those papers were very interesting, each of the approaches had some limitations. The study presented in Ref. 1 investigated the application of ring type and shell type piezoelectric transducers to reduce torsional vibrations in circular cross-section cylinders. This method seems to offer no possibility to control bending vibrations and would be difficult to use for non-circular cross-section beams. The approach proposed in Refs. 2 and 3 is similar to the method described in this study. However, the piezoelement distribution layout proposed in Refs. 2 and 3 generates coupling between torsional and longitudinal displacements. The study presented in Ref. 4 describes generating twist in rectangular cross-section members by extension-twist coupling. This is achieved by the warping of component plates of a box-like structural member. A similar effect of twisting thin plates by warping is presented in Ref. 5. This effect was obtained by bonding appropriately polarized layers of PVDF piezopolymer film on a metal plate.

The approach proposed in the present study differs from those described above because it offers a method to control torsional and combined torsional-bending-longitudinal vibrations in closed- and open-section structural members without coupling among torsional, bending or longitudinal control loads or displacements. Extension/compression action of a set of piezoelectric elements laminated to the walls of a structural element is used to generate torsion. The same set of elements is used to control bending as well as longitudinal vibrations.

CONCEPT EXPLANATION

Consider a structural element of rectangular cross-section made of fiber reinforced plastic as shown in Fig. 1. Assume that we can control the extension or contraction of shown fibers using piezoelectric material, that the base EFGH is fixed, and the top ABCD is free. The extension of fibers AH, BE, CF, DG (broken lines) and the contraction of fibers DE, AF, BG and CH (solid lines) will produce a clockwise twist of the top ABCD, or a torque. An additional extension of fibers AH, DE and an additional contraction of fibers BG and CF will produce a bending moment with respect to the plane parallel to planes ADEH and BCFG. In a similar manner we can generate longitudinal forces. We can obtain arbitrary combinations of control loads, provided that the overall voltage supplied to each piezoelement will not exceed the maximum allowable voltage for the given piezomaterial.

CONCEPT VALIDATION

The validation of this concept has been done numerically, analytically and experimentally.

NUMERICAL VALIDATION

The numerical validation has been done in two steps. In the first step a finite element model of a plate subjected to shear by a set of piezoelectric elements has been built and solved for displacements. The plate has been considered to be a projection on a plane of a part of a wall of a cylinder. Therefore, a shear displacement in the plate would indicate piezoactuator capability to induce a torsional displacement in a cylinder. The finite element model included the loads generated by piezolectric elements bonded to both sides of the plate in a manner shown in Fig. 2, where 1 represents piezoelectric elements in -45° direction, 2 identifies piezoelectric elements in +45° direction and 3 denotes the substrate. Edge EH of the plate is fixed and edge AD is free.

In the second step a simple model of a cylinder made of fiber reinforced plastic subjected to the strains generated by a set of distributed piezoelectric actuators has been created and solved for displacements. See Fig. 3.

The finite element package SDRC-IDEAS has been used for the numerical validation.

EXPERIMENTAL VALIDATION

The experimental validation has been done also in two steps.

In the first step a 96 mm x 96 mm x 0.28 mm metal plate with a set of 25.4 mm x 6.4 mm x 0.2 mm piezolectric elements laminated to both sides of the plate was fabricated and tested. The plate material was 3003 aluminum. The aluminum was anodized before mounting the piezoelements. The properties of the anodized plate are shown in Table 1. The piezoelements were bonded to the plate with Eccobond LV-45/LV15 epoxy resin. Sixteen elements were bonded to each side of the plate. The piezoelements used were PZT G1195 from Piezo Systems, Inc.. All piezoelements were directionally attached[2]. The electromechanical properties of the G1195 piezoelements are listed in Table 2. The plate was installed in a specially designed stand. The piezolectric elements were activated and the resulting shear displacement was measured using an analog micrometer.

In the second step a thin-wall, closed cross-section composite beam with laminated piezoelements was fabricated and tested. The beam was manufactured by wrapping one ply of style #120 [0°/90°] fiberglass cloth saturated with plast #88/87 epoxy resin around a foam core. The core was dissolved with acetone after the composite beam was fully cured at room-temperature. Eight pairs of piezoelements were bonded to each of the walls of the beam. Altogether, 64 elements were bonded forming a distributed actuator over a base portion of the beam of 85 mm length. The dimensions of the beam are shown in Fig. 4. The beam was installed in a specially designed stand. Tip displacements were measured using a laser beam reflected off a mirror mounted on the top of the beam at the elastic axis.

ANALYTICAL VALIDATION

The Classical Laminated Plate Theory (CLPT) was used to model the displacements of the plate subjected to strains generated by the distributed piezoelectric actuators. The resultant system of loads acting on a plate or on a single ply of laminate can be represented as[6]

$$\begin{Bmatrix} N_x \\ N_y \\ N_{xy} \\ M_x \\ M_y \\ M_{xy} \end{Bmatrix}_r =$$

$$\begin{bmatrix} A_{11} & A_{12} & A_{16} & B_{11} & B_{12} & B_{16} \\ A_{12} & A_{22} & A_{26} & B_{12} & B_{22} & B_{26} \\ A_{16} & A_{26} & A_{66} & B_{16} & B_{26} & B_{66} \\ B_{11} & B_{12} & B_{16} & D_{11} & D_{12} & D_{16} \\ B_{12} & B_{22} & B_{26} & D_{12} & D_{22} & D_{26} \\ B_{16} & B_{26} & B_{66} & D_{16} & D_{26} & D_{66} \end{bmatrix}_l \begin{Bmatrix} \varepsilon_x \\ \varepsilon_y \\ \varepsilon_{xy} \\ \kappa_x \\ \kappa_y \\ \kappa_{xy} \end{Bmatrix}_l - \begin{Bmatrix} N_x \\ N_y \\ N_{xy} \\ M_x \\ M_y \\ M_{xy} \end{Bmatrix}_p$$

(1)

where

$$A_{ij} = \sum_{k=1}^{N} \left(\overline{Q_{ij}}\right)_k (z_k - z_{k-1}) \quad (2)$$

$$B_{ij} = \frac{1}{2}\sum_{k=1}^{N} \left(\overline{Q_{ij}}\right)_k (z_k^2 - z_{k-1}^2) \quad (3)$$

$$D_{ij} = \frac{1}{3}\sum_{k=1}^{N} \left(\overline{Q_{ij}}\right)_k (z_k^3 - z_{k-1}^3) \quad (4)$$

Assume that the laminate is loaded only by activated piezoelements (no external loads). Also, the coupling stiffnesses can be neglected because the piezoelements are located symmetrically with respect to the neutral plane of the substrate. Therefore, we can rewrite Eq. 1 as

$$\begin{bmatrix} N_p \\ M_p \end{bmatrix} = \begin{bmatrix} A & 0 \\ 0 & D \end{bmatrix} \begin{bmatrix} \varepsilon_l \\ \kappa_l \end{bmatrix} \quad (5)$$

where

$$N_p = \int_{-\frac{t_l}{2}}^{\frac{t_l}{2}} Q \Lambda \, dz \quad (6)$$

$$M_p = \int_{-\frac{t_l}{2}}^{\frac{t_l}{2}} Q \Lambda z \, dz \quad (7)$$

and

$$\Lambda = \begin{bmatrix} \Lambda_x \\ \Lambda_y \\ \Lambda_{xy} \end{bmatrix} \quad (8)$$

The strains and curvatures generated by the piezoelements are calculated using two decoupled equations

$$\{\varepsilon\} = [A]^{-1}\{\Lambda\} \quad (9)$$

and

$$\{\kappa\} = [D]^{-1}\{\Lambda\} \quad (10)$$

The following assumptions have been made:

(a) Piezoelectric elements are isotropic and therefore $\Lambda_x = \Lambda_y = \Lambda$, where Λ is actuation strain.
(b) The actual set of piezolectric elements consists of two components, as shown in Fig 5. Component (a) consists of piezoelements bonded at $\alpha = 45°$ and generating tension. Component (b) consists of elements bonded at $\alpha = -45°$ and generating compression.
(c) Poisson ratios for the layer of piezoelements are equal to Poisson ratios for the substrate: $\nu_{xy_p} = \nu_{yx_p} = \nu_{xy_s} = \nu_{yx_s} = \nu$.
(c) The plate is free on all edges.
(d) The equivalent elastic moduli for the piezoelements: longitudinal E_{x_p} and transverse E_{y_p} have been determined following the approach outlined in Ref. 2.
(e) The bond between piezoelements and the substrate is perfect and of zero thickness.

RESULTS

NUMERICAL RESULTS

Shear Plate FEM Analysis

Two versions of the shear plate were modeled using the I-DEAS package. One of the models had the restraining rigid bars attached to the upper and lower edges, and the other model had the upper edge free and the lower edge supported at four points. The displacements of the upper left corner (A) and the upper right corner (D) of each plate are shown in Table 3. Although only four piezoelements, two in tension and two in contraction, were modeled on each side, the vertical displacements of the free edge are one order of magnitude smaller than the horizontal displacements. In a real structure that would mean very small coupling between induced twist and longitudinal displacements. The difference between horizontal displacements of restrained and unrestrained plate is between 2% and 8%.

Closed Cross-section Beam FEM Analysis

The finite element model represented a cylinder made of composite laminate with a height of 0.065m, a radius of 0.05m and of 0.25 mm thickness. The cylinder was loaded in a similar manner as was the shear plate. When activated, the cylinder underwent a rotational displacement shown by a representative column of elements. See Fig. 3. The maximum angular displacement of the cylinder was $1.192 \cdot 10^{-5}$ m which corresponds to an angle of twist of $\theta = 0.01365°$. See Ref. 7 for more details.

ANALYTICAL RESULTS

Eqs. (9) and (10) have been solved separately for piezoelements in tension (+Λ) bonded at 45° (Fig. 5a) and for piezoelements in compression (-Λ) bonded at -45° (Fig. 5b). The most important results are given below. First, compare the longitudinal strain and shear strain generated by +Λ, +45° and -Λ, -45° piezoelements.

Strains for +Λ, +45° elements (Fig. 5a) can be expressed as

$$\varepsilon_x(+\Lambda, +45°) = \frac{\lambda t_p\left(-4E_{xp}E_{yp}t_p + E_s t_s\left((E_{xp}-E_{yp})v - E_{xp} - E_{yp}\right)\right)}{(2E_{yp}t_p + E_s t_s)(-2E_{xp}t_p - E_s t_s + 2E_{yp}t_p v^2 + E_s t_s v^2)}$$
$$+ \frac{\lambda t_p\left(4E_{yp}^2 t_p v^2 + 2E_s E_{yp} t_s v^2\right)}{(2E_{yp}t_p + E_s t_s)(-2E_{xp}t_p - E_s t_s + 2E_{yp}t_p v^2 + E_s t_s v^2)}$$
(11)

$$\varepsilon_{xy}(+\Lambda, +45°) = \frac{2E_s(-E_{xp}+E_{yp})\lambda t_p t_s (1+v)}{(2E_{yp}t_p + E_s t_s)(-2E_{xp}t_p - E_s t_s + 2E_{yp}t_p v^2 + E_s t_s v^2)}$$
(12)

Similarly, for -Λ and -45° (Fig. 5b), we arrive at

$$\varepsilon_x(-\Lambda, -45°) = \frac{\lambda t_p\left(-4E_{xp}E_{yp}t_p + E_s t_s\left((E_{xp}-E_{yp})v - E_{xp} - E_{yp}\right)\right)}{(2E_{yp}t_p + E_s t_s)(2E_{xp}t_p + E_s t_s - 2E_{yp}t_p v^2 - E_s t_s v^2)}$$
$$+ \frac{\lambda t_p\left(4E_{yp}^2 t_p v^2 + 2E_s E_{yp} t_s v^2\right)}{(2E_{yp}t_p + E_s t_s)(2E_{xp}t_p + E_s t_s - 2E_{yp}t_p v^2 - E_s t_s v^2)}$$
(13)

$$\varepsilon_{xy}(-\Lambda, -45°) = \frac{2E_s(-E_{xp}+E_{yp})\lambda t_p t_s (1+v)}{(2E_{yp}t_p + E_s t_s)(-2E_{xp}t_p - E_s t_s + 2E_{yp}t_p v^2 + E_s t_s v^2)}$$
(14)

We can notice that $\varepsilon_x(+\Lambda, +45°) = -\varepsilon_x(-\Lambda, -45°)$ and $\varepsilon_{xy}(+\Lambda, +45°) = \varepsilon_{xy}(-\Lambda, -45°)$. Therefore, the actuators bonded at an angle of +45° and subjected to tension contribute as much to the shear deflection as actuators bonded at an angle of -45° and subjected to compression. However, the strains generated by those two types of actuators along the longitudinal axis of the laminate cancel out each other. We can make similar observations for transverse strains ε_y.

The conclusion is that the investigated layout of piezoelements eliminates the coupling between extensional and shear strains. This is equivalent to eliminating the coupling between longitudinal and torsional displacements and loads generated in beams.

The results of solving Eq. (10) for curvatures κ, not shown, have been used to validate further the assumptions made in this study against experimental results. The agreement between plate bending displacements computed using κ obtained from Eq. 10 and experimental results is very good (see Fig. 9).

EXPERIMENTAL RESULTS

The shear plate testing set-up is shown in Fig. 6, where 1 represents a specially designed stand, 2 is an analog micrometer and 3 identifies the shear plate. The tested composite beam with a piezoelectric actuator is shown in Fig. 7, where 1 is the composite beam and 2 represents the piezoelectric actuator.

Plate Static Testing Results

Experiments with the shear plate have shown that a voltage of 150 V (750 V/mm) applied to piezoelements generated a 8.0 E-6 m vertical displacement of the free edge of the plate. The displacement estimated using Eq. (9) was 7.3 E-6 m. The agreement is very good.

Only one experimental data point was obtained for shear displacement (although several independent measurements were made to confirm this number) because the repeatability of available micrometer indications was poor for lower activation voltages. This data point is shown against the analytical results in Fig. 8.

Bending displacements are shown against the analytical results in Fig. 9. The agreement is excellent.

Beam Static Testing Results

Experiments with the composite beam showed that the actuator worked as expected. The set of piezoelectric elements laminated to the walls of the beam could induce both torsional and bending moments. Longitudinal load generating capability was not tested due to the lack of appropriate equipment.

The magnitudes of twist and torque generated in the composite beam, versus applied voltage, are shown in Fig. 10. The flexural rigidity of the beam was determined to be about four times higher than the torsional rigidity of the beam. Therefore, the slope change under applied load was difficult to detect. Even at the highest allowable voltage, 750 V/mm, the displacement of the laser dot was very small and the fuzzy character of the edges of the dot made an accurate measurement impossible.

Beam Dynamic Testing Results

The distributed piezoactuator was used to induce the composite beam resonance in both torsional and bending modes. The results are summarized in Table 4. The frequencies listed were measured for the composite beam with a mirror attached to the free end of the beam.

CONCLUSIONS

1. It is possible to fabricate a torsional-bending piezoelectric actuator which generates no structural coupling.

2. Such an actuator can generate sufficient control loads to attenuate torsional and bending vibrations simultaneously.

3. Additional advantages, such as no maintenance requirement and a convenient method of control make this type of actuator particularly attractive.

7 REFERENCES

[1] Sung, C.-C., Varadan, V. V., Bao, X. Q. and Varadan, V. K., 1990, "Active Control of Torsional Vibration Using Piezoceramic Sensors and Actuators," Proceedings of the 31st AIAA/ASME/ASCE/AHS Structures, Structural Dynamics and Materials Conference, Long Beach, CA, April 2-4, 1990, Paper No. AIAA 90-1130.

[2] Barrett, R., Intelligent Rotor Blade and Structures Development Using Directionally Attached Piezoelectric Crystals, Master's Thesis, University of Maryland, College Park, 1990.

[3] Chen, P. and Chopra, I., A Feasibility Study to Build a Smart Rotor: Induced-Strain Actuation of Airfoil Twist with Piezoceramic Elements, Smart-Structures Technology: Innovations & Applications to Rotorcraft Systems, First Annual Review, October 6, 1993, College Park, Maryland.

[4] Lazarus, K.B., Crawley, E.F. and Bohlmann, J.D., Static Aeroelastic Control Using Strain Actuated Adaptive Structures, Proceedings of First Joint U.S./Japan Conference on Adaptive Structures, November 13-15, 1990, Maui, Hawaii, Wada, B.K., Fanson, J.L. and Miura, K., Editors.

[5] Lee, C. - K. and F. C. Moon, June 1989. "Laminated Piezopolymer Plates for Torsion and Bending Sensors and Actuators," J. Acoustical Society of America, Vol. 85, No. 6, June, 1989.

[6] Jones, M. R, Mechanics of Composite Materials, Scripta Book Company, 1975.

[7] Smith, W. P. and Kawiecki, G., "Twist and Torque Generating in Thin-Walled Composite Members Using Piezoelectric Elements", Proceedings of 7th Technical Conference of the American Society for Composites, University Park, Pa, October 13-15, 1992

Table 1 Properties of anodized aluminum plate

Modulus of elasticity	Poisson's ratio	Thickness
63 E9 N/m^2	0.33	2.79 E-4 m

Table 2 Properties of piezoelectric material

Curie Temperature	Max. allowable field	Density	Strain coeff. d_{31}	Modulus of elasticity
360 °C	6 E5 V/m	7.6 E3 kg/m^3	1.9 E-9	63 E9 N/m^2

Table 3 Free edge shear displacement (m)

	Horizontal displ.		Vertical displ.		Transverse displ.	
	Corner A	Corner D	Corner A	Corner D	Corner A	Corner D
Restrained plate	1.97E-5	2.05E-5	1.13E-7	-2.6E-7	-5.64E-8	1.61E-9
Unrestrained plate	2.00E-5	2.24E-5	5.05E-7	-3.26E-6	9.44E-12	9.27E-12

Table 4 Dynamic testing results

Type of displacement	Mode	Frequency	Tip Displacement
Bending	first	38 Hz	4E-4 m
	second	162 Hz	not measured
Torsional	first	190 Hz	0.161°

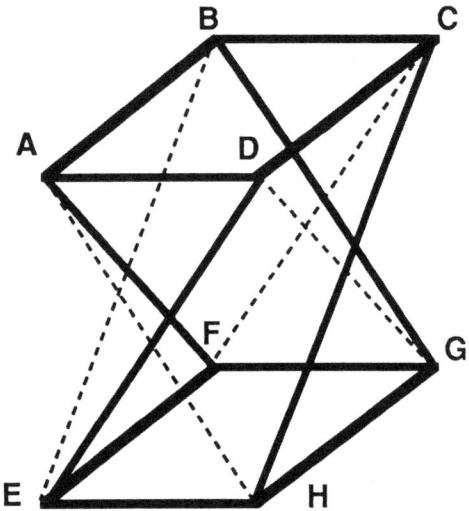

Fig. 1 Schematic of a structural element.

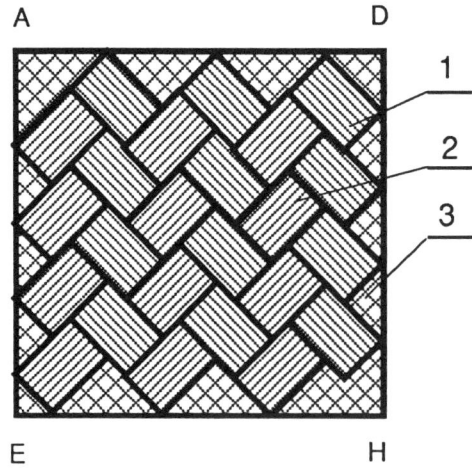

Fig. 2 Shear plate specimen.

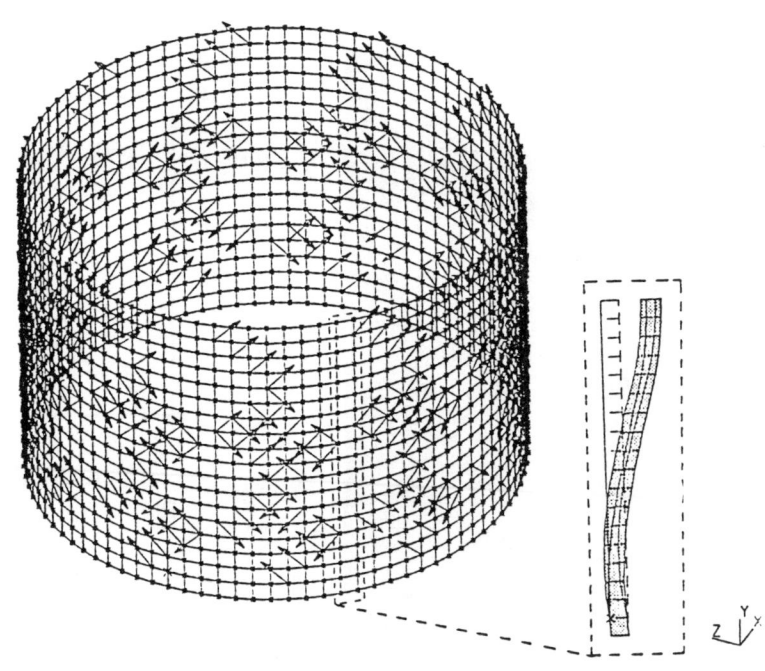

Fig. 3 Actuated cylinder and resulting displacements.

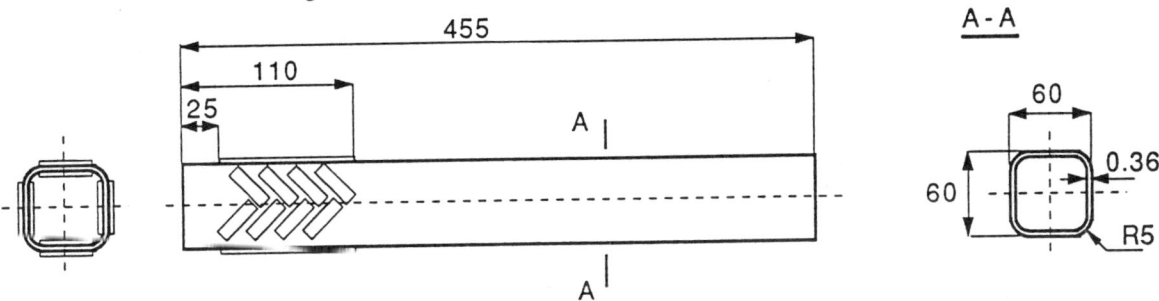

Fig. 4 Smart beam with laminated piezoelectric elements (dimensions in mm).

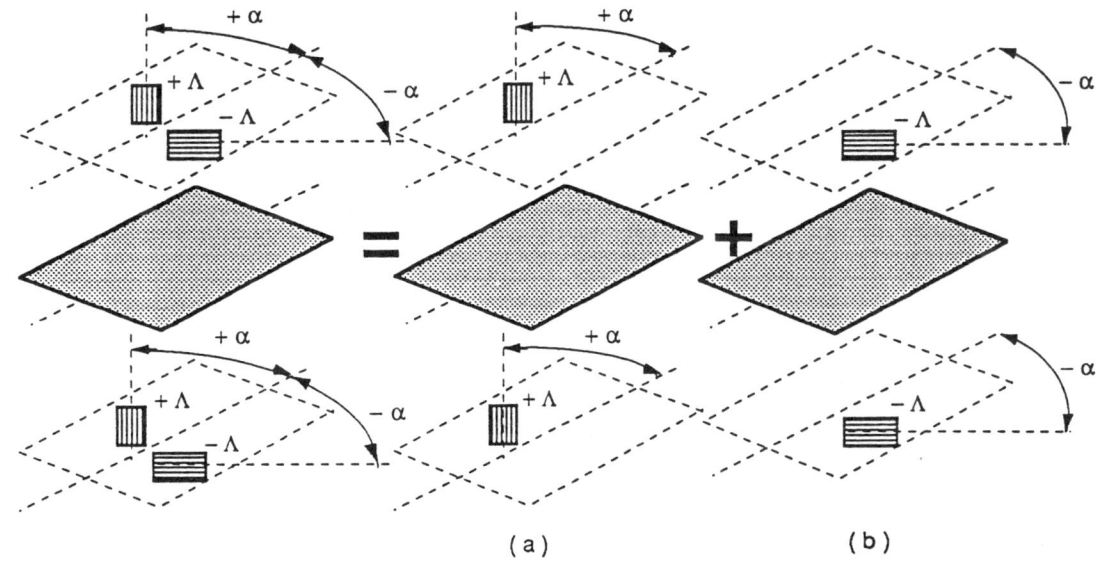

Fig. 5 Resolution of a piezoelectric ply into two components.

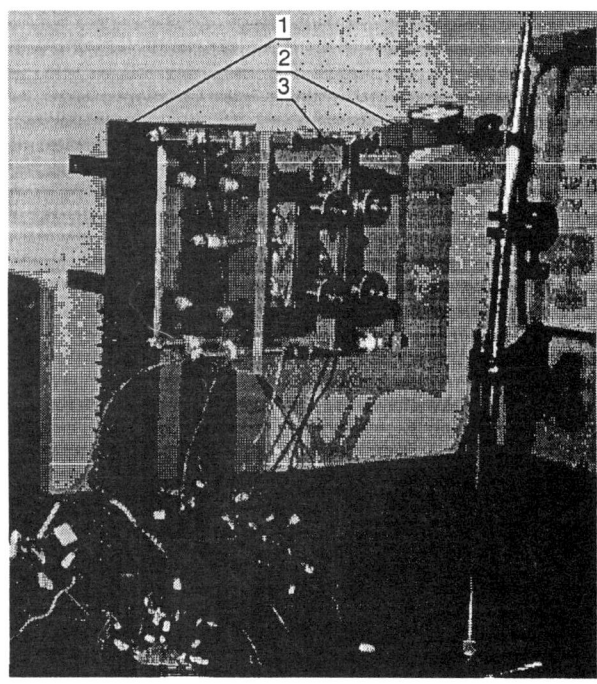

Fig. 6 Experimental set-up for shear plate testing.

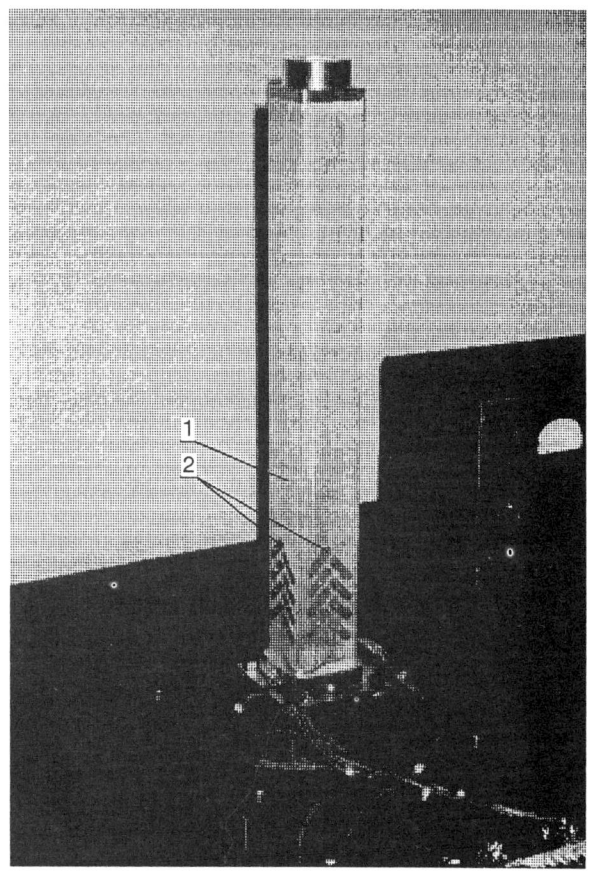

Fig. 7 Composite beam with piezoelectric actuator.

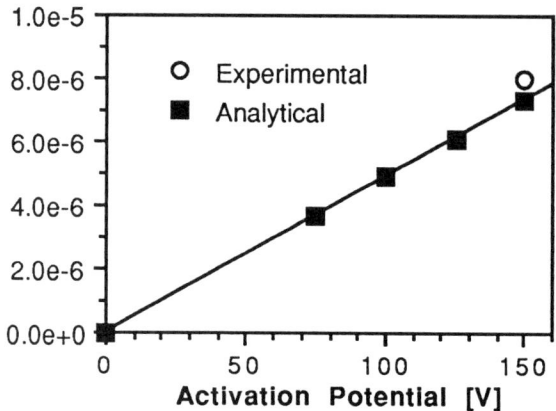

Fig. 8 Plate shear displacement: experimental vs. analytical.

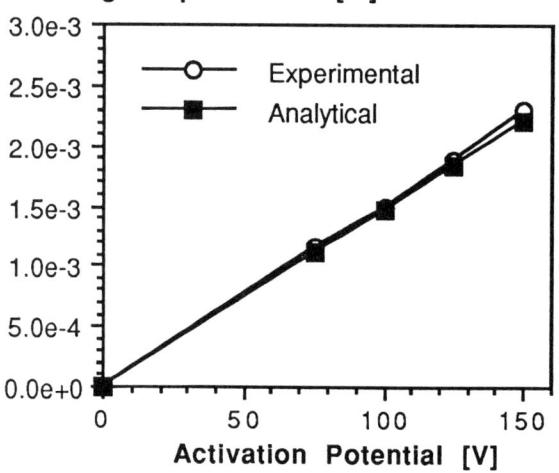

Fig. 9 Plate bending displacement: experimental vs. analytical.

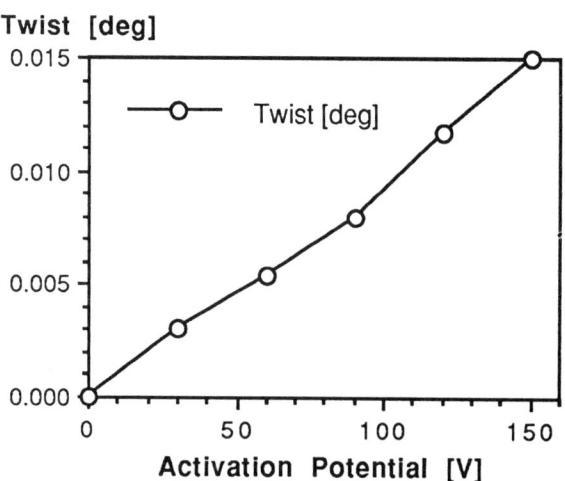

Fig. 10 Induced twist vs. activation potential.

COLLOCATED INDEPENDENT MODAL CONTROL WITH SELF–SENSING ORTHOGONAL PIEZOELECTRIC ACTUATORS
(Theory and Experiment)

H. S. Tzou [1,2][†] and J. J. Hollkamp [2]

[1] Department of Mechanical Engineering
University of Kentucky, Lexington, KY 40506

[2] Wright Laboratory, Flight Dynamics Directorate
WL/FIBG, WPAFB, Ohio 45433

ABSTRACT

Distributed self–sensing piezoelectric actuators provide a perfect collocation of sensors and actuators in closed–loop structural controls. To achieve independent control of various natural modes, spatially distributed self–sensing orthogonal piezoelectric actuators are proposed in this study. A generic spatially shaped orthogonal sensor/actuator theory is derived first, followed by an application to a Bernoulli–Euler beam. Spatially distributed orthogonal sensors/actuators are designed based on the modal strain functions and they are fabricated using a 40μm piezoelectric polymer. A cantilever beam laminated with these self–sensing orthogonal piezoelectric actuators is tested. Collocated independent modal control of the cantilever beam with spatially distributed self–sensing orthogonal actuators is demonstrated and control effectiveness studied.

INTRODUCTION

A perfect sensor/actuator collocation usually provides a stable performance in closed–loop feedback controls. A self–sensing piezoelectric actuator is a single piece of piezoelectric device simultaneously used for both sensing and control. (The sensor signal is separated from the control signal by using a differential amplifier; this signal is then amplified and fed back to induce control actions.) Self–sensing piezoelectric actuators have been proposed in recent years. Dosch, Inman, and Garcia (1992) proposed a self–sensing piezoelectric actuator for collocated control of a cantilever beam. Anderson, Hagood, and Goodliffe (1992) presented an analytical modeling of the self–sensing actuator system, and studied its applications to beam and truss structures. Rectangular–shape piezoelectric devices attached near the fixed end were used in both studies.

It is known that the spatially distributed orthogonal sensors and actuators are sensitive to a mode or a group of natural modes (Tzou, 1993; Lee, 1992). Spatially distributed piezoelectric sensors and actuators were investigated in a number of recent studies, such as beams, plates, rings, shells, etc. (Lee and Moon, 1990; Lee, 1992; Anderson and Crawley, 1991; Collins, Miller, and von Flotow, 1991; Hubbard and Burke, 1992; Tzou and Fu, 1993; Tzou and Tseng, 1990; Tzou, Zhong, and Natori, 1993; Tzou, 1993; Tzou, Zhong, and Hollkamp, 1994). Based on the modal orthogonality, a spatially shaped self–sensing orthogonal modal actuator is effective to only a single mode; consequently, each vibration mode can be independently controlled, *independent modal control*, while the feedback control system is kept simple. This paper is to investigate the sensing and control characteristics of self–sensing orthogonal modal actuators. A generic theory for a spatially distributed self–sensing orthogonal actuator is proposed first, followed by an experimental study of self–sensing orthogonal piezoelectric actuators. Independent modal control with the self–sensing orthogonal actuators are demonstrated. (Note that the emphasis is placed on the experimental aspect.)

THEORY

It is assumed that a spatially distributed piezoelectric layer is laminated on a one–dimensional (1–D) structure, such as arches, rings, beams, rods,

---

† Visiting

Copyright © 1993 American Institute of Aeronautics and Astronautics, Inc. All rights reserved.

etc., Figure 1. Both the piezoelectric layer and the elastic continuum have a constant thickness. It is assumed that the piezoelectric material is hexagonal symmetrical such that the piezoelectric constants $e_{31} = e_{32}$.

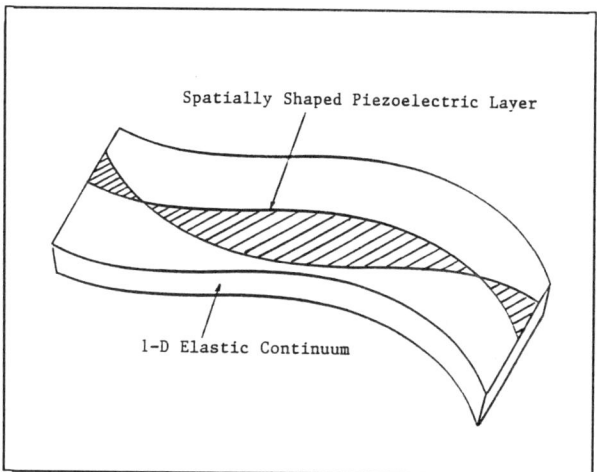

Fig.1 A 1-D spatially distributed orthogonal sensor/actuator.

An open-circuit sensor signal ϕ_3^s from a 1-D spatially distributed orthogonal sensor can be estimated from its strains:

$$\phi_3^s = -\frac{e_{31}}{\epsilon_{33}S^e} \int_{\alpha_1} \text{sgn}[U_3(\alpha_1)] \bigg(h^s(S_{11}^o + S_{22}^o)$$
$$+ 0.5h^s(h+h^s)(k_{11}+k_{22}) \bigg) A_1 A_2 \int_0^{W_s(\alpha_1)} d\alpha_2 d\alpha_1$$
$$= -\frac{e_{31}}{\epsilon_{33}S^e} \int_{\alpha_1} W_s(\alpha_1) \text{sgn}[U_3(\alpha_1)] \bigg[h^s(S_{11}^o + S_{22}^o)$$
$$+ 0.5h^s(h+h^s)(k_{11}+k_{22}) \bigg] A_1 A_2 \, d\alpha_1, \qquad (1)$$

where e_{31} is the piezoelectric constant; ϵ_{33} is the dielectric constant; S^e is the effective electrode area; $W_s(\alpha_1)$ is a 1-D shape function; $U_3(\alpha_1)$ is an orthogonal function; h^s is the thickness of the sensor layer; h is the continuum thickness; S_{ii}^o is the membrane strain; k_{11} denotes the bending strain; A_1 and A_2 are Lamé parameters; and sgn[·] is a signum function which defines the polarity changes of the orthogonal sensor. Note that S_{22}^o and k_{22} are usually neglected since $\partial(\cdot)/\partial\alpha_2 = 0$. In addition, the first term (leading by h^s) denotes the membrane strain contribution to the sensor output, and the second term (leading by $0.5h^s$) the bending strain contribution. The total output signal is contributed by the sum of membrane and bending strain components. In a 1-D elastic continuum with finite radius of curvatures, e.g., arches and rings, both membrane and bending components contribute to the output signal. However, for flat 1-D continua with infinite radius of curvature, e.g., beams, the output signal is contributed either by the membrane component, e.g., rods, or the bending component, e.g., beams (Tzou, 1993).

The distributed velocity (strain–rate) feedback can be derived using the modal expansion method and a spatially distributed modal feedback force (Tzou, Zhong, Hollkamp, 1994). The k-th modal equation can be written as

$$\ddot{\eta}_k + \frac{c}{\rho h}\dot{\eta}_k + \omega_k^2 \eta_k = \frac{1}{\rho h N_k} \sum_{j=1}^{3} \sum_{m=1}^{\infty} \int_{\alpha_1} \int_{\alpha_2}$$
$$\cdot \bigg(\mathcal{G}_{jm}^{vf}(\alpha_1,\alpha_2) \dot{\eta}_m U_{jk}(\alpha_1,\alpha_2) \bigg) A_1 A_2 d\alpha_1 d\alpha_2 . \qquad (2)$$

where $\eta(t)$ is a modal coordinate; c is the damping constant; $\mathcal{G}_{jm}^{vf}(\alpha_1,\alpha_2)$ is the distributed *velocity feedback function*; and $N_k = \int_{\alpha_1}\int_{\alpha_2}[\sum_{j=1}^{\infty}(U_{jk})^2]A_1 A_2 d\alpha_1 d\alpha_2$. Using the modal orthogonality, one can write the distributed **velocity feedback function** as

$$\mathcal{G}_{jm}^{vf}(\alpha_1,\alpha_2) = \mathcal{G}_i^{vf} U_{3n}(\alpha_1,\alpha_2) , \qquad (3)$$

where $\mathcal{G}_i^{vf}$ is a velocity weighting factor (gain constant). All modes other than the n = k mode are filtered out due to their orthogonalities. Considering the transverse oscillation only, one can derive the n-th modal equation (*independent modal control equation*) with the **velocity modal feedback control force** as

$$\ddot{\eta}_n + \frac{1}{\rho h}\bigg(c - \frac{\mathcal{G}_3^{vf}}{N_n}\int_{\alpha_1}\int_{\alpha_2} U_{3n}^2$$
$$\cdot A_1 A_2 d\alpha_1 d\alpha_2\bigg)\dot{\eta}_n + \omega_n^2 \eta_n = 0 . \qquad (4)$$

For 1–D continua, the transverse mode shape U_{3n} is only a function of one coordinate α_1, e.g., the circumferential direction in rings and arches, the longitudinal direction in beams and rods, etc. If electrode areas of the actuators are designed as a 1–D shape function of $W(\alpha_1)$, the modal control force for a 1–D spatially shaped actuator can be rewritten as

$$\ddot{\eta}_n + \frac{1}{\rho h}\left(c - \frac{\mathcal{G}_3^{yf}}{N_n}\int_{\alpha_1} W(\alpha_1) U_{3n}^2 \cdot A_1 A_2 d\alpha_1\right)\dot{\eta}_n + \omega_n^2 \eta_n = 0 . \quad (5)$$

Note that the modal coupling and the spillover from all other natural modes are eliminated. This modal filtering characteristics will be demonstrated in an experimental study on a cantilever beam laminated with orthogonal sensors/actuators presented later.

Orthogonal Sensor/Actuator for a Cantilever Beam

A 1–D cantilever Bernoulli–Euler beam usually exhibits transverse oscillations only. (The in–plane longitudinal oscillation is neglected.) The Lamé parameters for a flat uniform beam are $A_1 = 1$, $A_2 = 1$; the radii are $R_1 = \infty$ and $R_2 = \infty$. In addition, $\partial(\cdot)/\partial\alpha_2 = 0$. Accordingly, the closed–loop equation of motion of a cantilever beam can be derived.

$$\rho h \ddot{u}_3 + YI\frac{\partial^4 u_3}{\partial x^4} - b\frac{\partial^2(M_{11}^a)}{\partial x^2} = bF_3 , \quad (6)$$

where ρ is the mass density; Y is Young's modulus; I is the area–moment of inertia; b is the beam width; M_{11}^a is the induced control moment; and F_3 is the external mechanical force. As discussed previously, the orthogonal modal sensors/actuators are designed based on the *modal function* $U_{3m}(x)$:

$$U_{3m}(x) = \frac{1}{\lambda_m^2}\frac{d^2 U_{3m}(0)}{dx^2}\left[C(\lambda_m x) - \frac{A(\lambda_m L)}{B(\lambda_m L)} D(\lambda_m x)\right] , \quad (7)$$

where

$$A(\lambda_m x) = 0.5[\cosh(\lambda x) + \cos(\lambda x)] , \quad (8a)$$
$$B(\lambda_m x) = 0.5[\sinh(\lambda x) + \sin(\lambda x)] , \quad (8b)$$
$$C(\lambda_m x) = 0.5[\cosh(\lambda x) - \cos(\lambda x)] , \quad (8c)$$
$$D(\lambda_m x) = 0.5[\sinh(\lambda x) - \sin(\lambda x)] , \quad (8d)$$

where x defines the distance measured from the fixed end. The eigenvalue λ_m is determined by its characteristic equation:

$$\cos(\lambda L)\cosh(\lambda L) + 1 = 0 , \quad (9)$$

where $\lambda_1 L = 1.875$; $\lambda_2 L = 4.694$; $\lambda_3 L = 7.855$; $\lambda_4 L = 10.996$; $\lambda_5 L = 14.137$; etc. L is the beam length. The first derivative $\frac{d}{dx}\left[U_{3m}(x)\right]$ is the *modal slope function* and the second derivative $\frac{d^2}{dx^2}\left[U_{3m}(x)\right]$ is the *modal strain function*. The *modal strain function* is used to define the shapes of orthogonal modal sensors/actuators:

$$U_{3m}''(x) = \left[\left[e^{(\lambda L - \lambda x)}[e^{\lambda L} + \cos(\lambda L) + \sin(\lambda L)]/\{2[e^{2\lambda L} + 2e^{\lambda L}\sin(\lambda L) - 1]\}\right]\right.$$
$$+ \left[-e^{\lambda x}\{e^{\lambda L}[0.5\cos(\lambda L) - 0.5\sin(\lambda L)] + 0.5\}\right.$$
$$+ e^{2\lambda L}[0.5\cos(\lambda x) - 0.5\sin(\lambda x)]$$
$$+ e^{\lambda L}[\cos(\lambda x)\sin(\lambda L) - \sin(\lambda x)\cos(\lambda L)]$$
$$\left.\left. - 0.5\cos(\lambda x) - 0.5\sin(\lambda x)\right]\right.$$
$$\left./[e^{2\lambda L} + 2e^{\lambda L}\sin(\lambda L) - 1]\right]/(\lambda L)^2 . \quad (10)$$

Note that each orthogonal modal sensor/actuator has a distinct shape based on its modal strain function and eigenvalue. Detailed layouts of the spatially shaped orthogonal sensors/actuators are presented next.

MODEL FABRICATION AND EXPERIMENTAL SETUP

The shapes of distributed orthogonal sensors/actuators follow the definitions of modal strain functions defined by their eigenvalues. The first four modal function are plotted in Figure 2, and their modal strain functions are plotted in Figure 3.

Note that the effective regions are from zero to one, since they are normalized in the length direction.

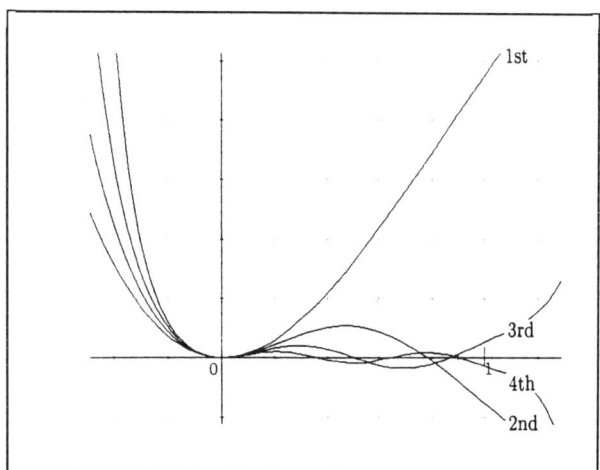

Fig.2 Mode shape functions of the cantilever beam.

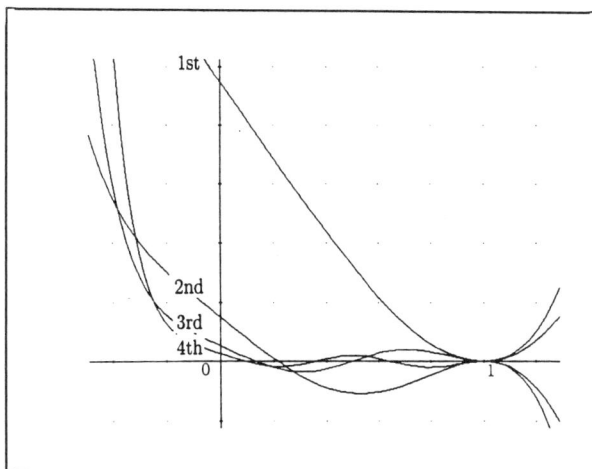

Fig.3 Modal strain functions of the cantilever beam.

Polymeric piezoelectric polyvinylidene fluoride (PVDF) is flexible and easy to cut into various shapes in a laboratory environment. In this study, a 40μm biaxially oriented PVDF sheet is used for the orthogonal sensors/actuators. These sensor/actuator layers are cut according to their strain functions and then glued on a plexiglas beam (15×1×1/8–in). Patterns of surface electrodes are first laid out on the plexiglas beam using a thin–film silver paste to ensure a good electrical conductivity. Individual silver electrodes are connected by either thin silver–paste lines (internal connections) or .5mm Teflon coated surgical wires (external connections). Polarity changes are achieved by reversing the cut PVDF sheets. The finished PVDF/plexiglas beam is shown in Figure 4.

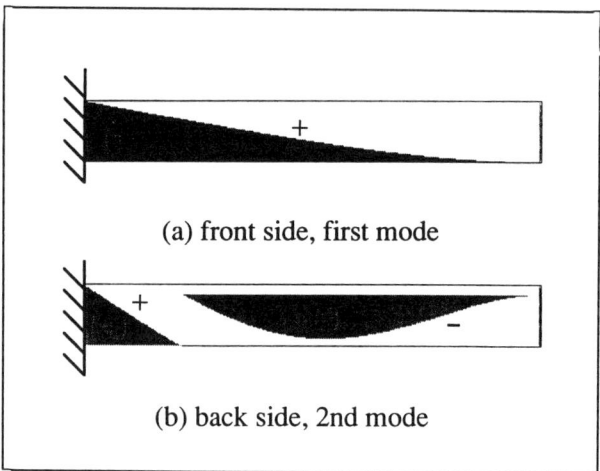

Fig.4 Experimental beam model with orthogonal PVDF sensors/actuators.

Apparatus

A self–sensing feedback control circuit is setup with two current amplifiers and a differential circuit (Anderson, et al., 1992; Dosch, et al., 1992). Three operational amplifiers (AD–711JN), six resistors (24.9kΩ, 8MΩ, and 16MΩ) and a capacitor (14nF) are used to build the circuits for the first and second orthogonal modal sensors/actuators. Figure 5 shows the circuit. The capacitor is used to match the capacitance of the orthogonal piezoelectric sensor/actuator. A power amplifier (BK–1651) supplies a 30V to the operational amplifiers, and an signal amplifier is used to amplify the sensor signal to induce control actions in the piezoelectric layers. A reference accelerometer (Kistler 5205) is mounted at the free end to provide a reference signal. All signals are input into an HP data acquisition system (HP3566A) for signal processing and recording.

Experimental Procedures

There are two sets of experiments carried out in this study. The first set is to test the modal orthogonality of orthogonal modal sensors; the second set is to evaluate the control effectiveness of the self–sensing orthogonal actuators.

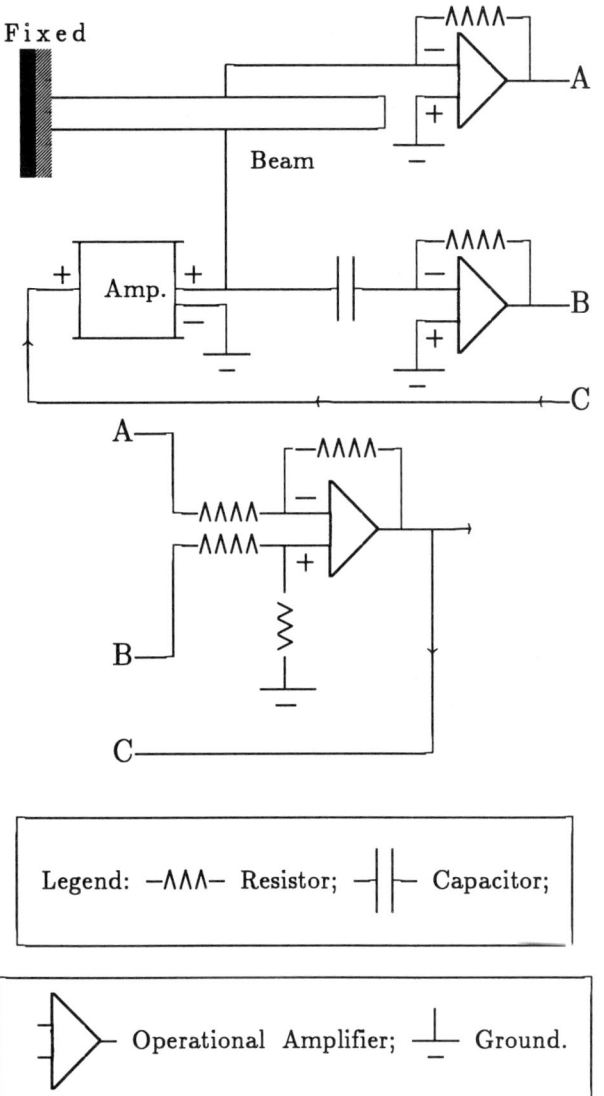

Fig.5 A self-sensing feedback control circuit.

The first set involves two tests: 1) strain signals and 2) strain-rate signals. The strain signal is contributed by elastic bending strains of the cantilever beam; it is ultimately related to the transverse deflection u_3. Thus, the strain signal is often referred as "displacement" signal; the strain-rate can be regarded as the "velocity" signal. The signs of these signals are individually checked to ensure correct feedback signals in the self-sensing feedback control. A self-sensing feedback control circuit, discussed previously, is used in the second set experiments. Controlled time histories from the accelerometer are acquired; modal damping ratios are calculated using the eigensystem realization algorithm (ERA) method (Juang and Pappa, 1985).

RESULTS AND DISCUSSION

There are two sets of experiments carried out in this study. The first set is to test the modal orthogonality of orthogonal modal sensors; the second set is to evaluate the modal control effectiveness of the self-sensing orthogonal piezoelectric actuators.

Modal Sensing

In order to evaluate the sensing effectiveness of orthogonal modal sensors, the spatially shaped piezoelectric layers were subjected to external excitations and their dynamic responses recorded. Strain and strain-rate responses were also tested using a strain-rate circuit (Lee, 1992). Figure 6 shows the spectra of the first modal sensor, the second modal sensor, and the accelerometer. It is observed that the accelerometer senses multiple modes of the cantilever beam, and the modal sensors only respond to their respective modes.

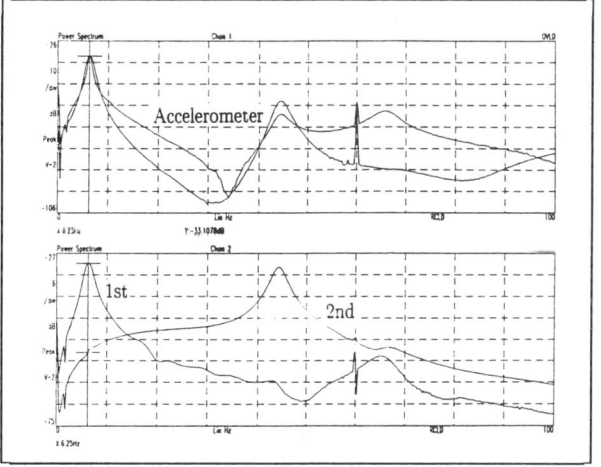

Fig.6 Spectra of the orthogonal modal sensors and the accelerometer.

Self-Sensing and Feedback Control

As discussed previously, a self-sensing piezoelectric actuator provides a perfect collocation of sensor and actuator. In this section, free oscillation and controlled time histories are presented and their respective damping ratios are calculated.

1) Free Oscillations

For the first mode, an initial displacement was applied to the free end and the snap-back response recorded. The free oscillation time histories of strain and strain-rate signals of the first modal sensor/actuator are plotted in Figures 7 and 8, and those of the second sensor/actuator are plotted in Figures 9 and 10, respectively. Note that those time histories of the second sensor/actuator were obtained via impulse excitations.

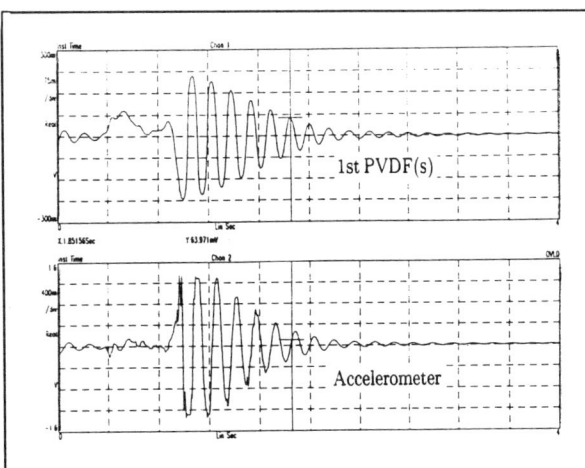

Fig.7 Free oscillation of the plexiglas beam (1st strain).

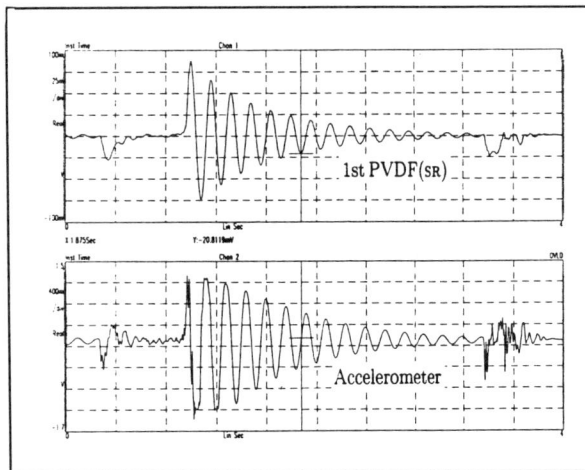

Fig.8 Free oscillation of the plexiglas beam (1st strain—rate).

Damping ratios of the first and second modes were calculated using the free vibration time histories. The first modal damping ratio is 4.0% and the second modal damping ratio is 3.4%. (Note that these data were calculated using more than five sample time histories.) It should be pointed that there was an accelerometer cable taped on the plexiglas beam, which caused a higher damping for the first natural mode.

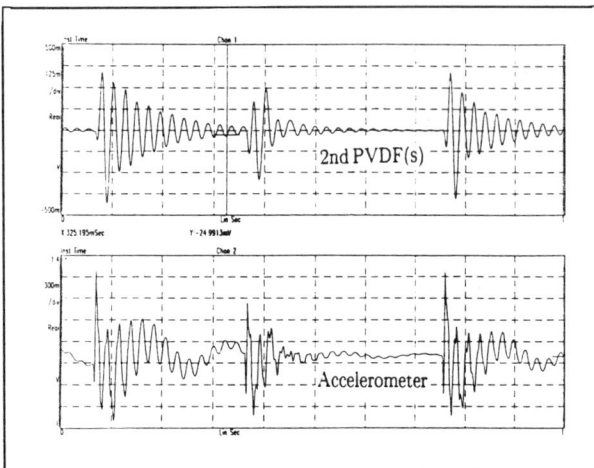

Fig.9 Free oscillation of the plexiglas beam (2nd strain).

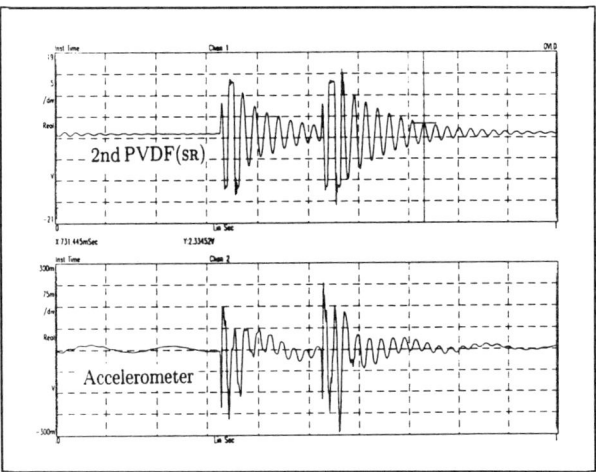

Fig.10 Free oscillation of the plexiglas beam (2nd strain—rate).

2) Self—Sensing Control
— Independent Modal Control

Control effectivenesses of the self—sensing orthogonal actuators were evaluated when the self—sensing control circuit was powered on. The sensing (strain—rate) signal was separated from the actuating signal via the circuit shown in Figure 5. The controlled responses (via the accelerometer signals) of the plexiglas beam were plotted in Figures 11 and 12.

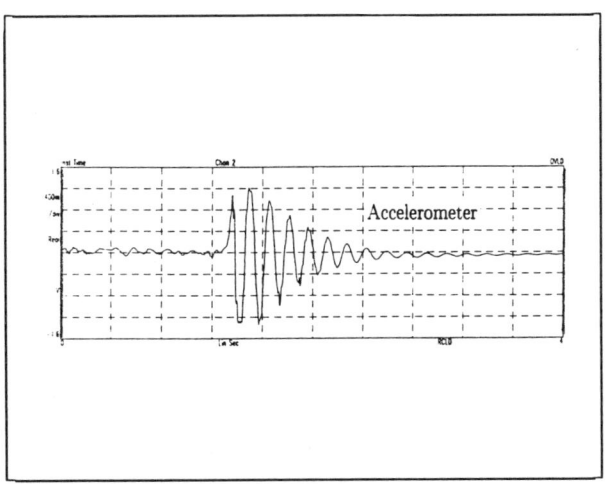

Fig.11 Controlled time history of the plexiglas beam (1st).

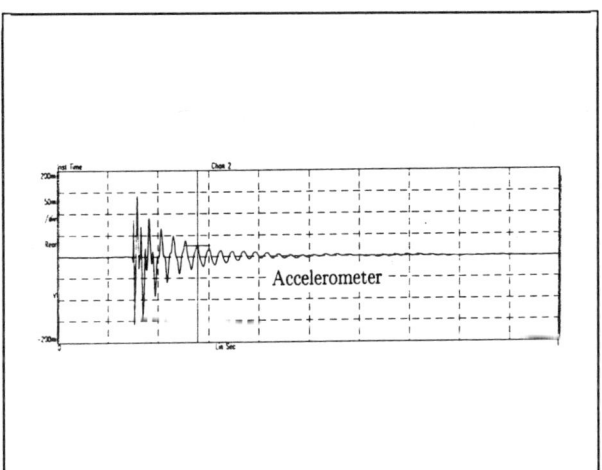

Fig.12 Controlled time history of the plexiglas beam (2nd).

It was noted that the strain–rate signals are rather noisy, which is probably introduced by the electrical line noises, the circuit, the amplifier, etc. Although the strain–rate signals are very noisy, the plexiglas beam is controlled well via the self–sensing orthogonal piezoelectric actuator. The averaged damping ratios of the controlled responses are 7.1% for the first mode and 4.2% for the second mode. (Note that since the second mode was relatively difficult to excite by an impulse excitation, twelve samples were used to obtained the averaged data. Control of the second mode could be improved by changing the resistors in the circuit.) It should be pointed out that the convergence of a modal response is determined by the product of the damping ratio and the modal frequency, i.e., $e^{-\zeta_n \omega_n t}$. Consequently, responses of the higher modes usually converge much faster than those of the lower modes.

SUMMARY AND CONCLUSION

Distributed self–sensing piezoelectric actuators provide a perfect collocation of sensors and actuators in closed–loop structural controls. To separate control actions for different natural modes (independent modal control), spatially distributed self–sensing orthogonal piezoelectric actuators were proposed in this study.

A generic orthogonal sensor/actuator theory was presented first, followed by an application to a Bernoulli–Euler beam. Spatially distributed orthogonal sensors/actuators were designed based on the modal strain functions. A physical model was fabricated and its self–sensing control effectiveness tested. A 40μm polymeric piezoelectric PVDF sheets were cut and laminated on a plexiglas beam. Surface electrodes were connected by either silver pastes or surgical wires. A self–sensing feedback control circuit was setup and tested.

Experimental results showed that the orthogonal modal sensors are sensitive to their respective modes. Free and controlled (via the self–sensing feedback control circuit) time histories were recorded and their modal damping ratios calculated. The calculated results suggested that the modal damping ratios were enhanced by 77.5% for the first mode and by 23.5% for the second mode. The convergence of modal responses is determined by the product of the modal damping and the modal frequency. Thus, the independent modal control of continua can be effectively achieved by using the spatially distributed self–sensing orthogonal piezoelectric actuators.

ACKNOWLEDGEMENT

A fellowship supported by the Wright Laboratory (Flight Dynamics Directorate) under the AFOSR RDL Program is gratefully acknowledged. Contents of the information do not necessarily reflect the position or the policy of the government, nor should official endorsement be inferred.

REFERENCES

Anderson, M.S. and Crawley, E.F., 1991, "Discrete Distributed Strain Sensing of Intelligent Structures," *Proceedings of the Second Joint Japan/USA Conference on Adaptive Structures*, pp.737–754.

Anderson, E.H., Hagood, N.W., and Goodliffe, J.M., 1992, "Self–Sensing Piezoelectric Actuation: Analysis and Application to Controlled Structures," AIAA–92–2465–CP, 33rd SDM Conference.

Colins, S.A., Miller, D.W., von Flotow, A.H., 1991, "Piezoelectric Spatial Filters for Active Vibration Control," *Recent Advances in Active Control of Sound and Vibration*, pp.219–234.

Dosch, J.J., Inman, D., and Garcia, E., 1992, "A Self–Sensing Piezoelectric Actuator for Collocated Control," *J. of Intelligent Material Systems and Structures*, Vol.(3), pp.166–185, Jan. 1992.

Hubbard, J.E. and Burke, S.E., 1992, "Distributed Transducer Design for Intelligent Structural Components," *Intelligent Structural Systems*, Tzou and Anderson (Ed.), Kluwer Academic Publishers, Dordrecht/Boston/London, August 1992, pp.305–324.

Juang, J.N. and Pappa, R.S., 1985, "An Eigensystem Realization Algorithm for Modal Parameter Identification and Model Reduction," *J. of Guidance and Control*, Vol.8, No.5, pp.620–627.

Lee, C.K., 1992, "Piezoelectric Laminates: Theory and Experimentation for Distributed Sensors and Actuators," *Intelligent Structural Systems*, Tzou and Anderson (Ed.), Kluwer Academic Pub., pp.75–167.

Lee, C.K. and Moon, F.C., 1990, "Modal Sensors/Actuators," ASME *Journal of Applied Mechanics*, Vol.(57), pp.434–441.

Tzou, H.S., 1993, *Piezoelectric Shells (Distributed Sensing and Control of Continua)*, Kluwer Academic Publishers, February 1993.

Tzou, H.S. and Fu, H., 1993, "A Study on Segmentation of Distributed Sensors and Actuators, Part–1, Theoretical Analysis, Part–2, Parametric Study and Vibration Controls," *Journal of Sound & Vibration*, Vol.(168), No.19, Nov. 1993. (To appear)

Tzou, H.S. and Tseng, C.I., 1990, "Distributed Piezoelectric Sensor/Actuator Design for Dynamic Measurement/Control of Distributed Parameter Systems: A Finite Element Approach," *Journal of Sound and Vibration*, Vol.(138), No.(1), pp.17–34.

Tzou, H.S., Zhong, J.P., and Natori, M.C., 1993, "Sensor Mechanics of Distributed Shell Convolving Sensors Applied to Flexible Rings," *ASME Journal of Vibration & Acoustics*, Vol.(115), No.1, pp.40–46, January 1993.

Tzou, H.S., Zhong, J.P., and Hollkamp, J.J., 1994, "Spatially Distributed Orthogonal Piezoelectric Shell Actuators (Theory and Applications)," *Journal of Sound & Vibration*, 1994. (To appear)

(RDL–Report93.Wp/Fibg1)(C/SelfAct.Wp/Fibg2)

A COMPARISON OF ACTUATION AND SENSING TECHNIQUES FOR AIRCRAFT CABIN NOISE CONTROL

D. J. Rossetti, M. A. Norris
Thomas Lord Research Center
Lord Corporation
405 Gregson Drive
Cary, North Carolina 27511

Abstract

An experimental comparison of various configurations of actuators and sensors for active noise and vibration control in turboprop aircraft was completed on a deHavilland DASH-7 fuselage section. Noise control performance was experimentally compared for systems using speakers and structural-based actuators. It was found that at low frequencies, structural-based noise control systems considerably outperform speaker-based noise control systems for reducing interior noise in turboprop aircraft. At higher frequencies, the two forms of actuation provided similar noise reduction performance.

In addition, various sensor configurations were investigated for interior noise control. Using a structural-based noise control system, vibration and noise levels are presented for control systems that use accelerometers and microphones as error sensors. It was found that a combination of microphones and accelerometers was necessary to ensure noise and vibration reduction throughout the frequency range studied, and that a frequency dependent sensor weighting can be used to enhance noise control performance. At lower frequencies, the dominant modes of vibration couple directly to the interior acoustics and adding accelerometers to the error sensor array decreased noise reduction performance only slightly. At higher frequencies, adding accelerometers to the error sensor array significantly reduced noise control performance.

Introduction

The principles of active aircraft cabin noise control can be addressed in two fundamental ways. One approach is to use acoustic sources (speakers) imbedded in the aircraft trim which are driven to provide interior noise cancellation. Examples of this approach are techniques that use speakers in the aircraft interior as well as active trim panels with imbedded actuation. The second approach is more fundamental in that actuators attached directly to the cabin-wall are driven to reduce noise and cabin-wall structural vibration.

Airframe manufacturers have traditionally used passive materials and devices for noise reduction. Passive materials are effective in reducing high frequency noise; however, prohibitive size and weight prevent them from being effective for reducing low-frequency noise in aircraft. As an alternative, turboprop manufacturers have begun to develop active technology for reducing low-frequency noise [1-3].

In turboprop applications, speaker-based noise control systems tend to be less practical compared with structural-based noise control systems. In practice, the speaker-based systems incorporate speakers in the trim of the aircraft. Incorporating speakers in the trim is a significant and expensive procedure, considering the small amount of available space and weight allowable in aerospace applications.

Structural-based actuators can be attached to the rib of the aircraft, behind the trim, and are invisible to the passenger. The actuators can be attached to the airframe in the same manner in which a passive dynamic absorber is attached. As an example, Lord uses an actuator that weighs approximately 1 lb and is cylindrical in shape (1.8 inches in height, 2.4 inches in diameter). This actuator can directly replace passive dynamic absorbers used for turboprop aircraft, even using the same attachment hardware. Compared with the speaker-based systems, structural-based systems provide better noise control performance with a system that has lower weight and is smaller. Furthermore, the speaker and structural-based systems can both employ electromagnetic actuation with nearly the same components, and hence can be equally as reliable.

In an active noise control system, secondary sources are used to create a sound field which has the same spatial distribution as the primary sound field but is temporally out-of-phase. Thus, the secondary field cancels the primary field and quieting is accomplished. The effectiveness of an active control system depends on how well the secondary field matches the primary field. At low frequencies, where acoustic wavelengths are long, the secondary field can be made to match the primary field with a relatively low number of actuators. At higher frequencies, the primary field becomes more complex (due to the shorter wavelengths) and more actuators are needed to create an adequate secondary field.

The complexity of an active control system increases with increasing frequency. In addition to the sound field being spatially more complicated, there is also an issue of the time scale growing smaller with increasing frequency. Thus, there is a frequency at which an active system becomes too complex, and a passive system is more practical. The frequency at which this occurs depends on the application.

Cabin noise in turboprop aircraft is primarily airborne, in which the propeller wash excites the cabin wall and creates noise in the cabin. The cabin noise consists mainly of low-frequency airborne noise, where passive treatments are impractical for enhancing passenger comfort.

Low-frequency noise in turboprop aircraft has been reduced using speakers to cancel the interior noise [1-3]. The work done by Bernhard, Elliott, Fuller, Ross, Silcox and the ASANCA team (a European consortium on "Advanced Study for Active Noise Control in Aircraft") has shown significant quieting results in active noise control in 3-D enclosures [1-6].

Bernhard demonstrated a localized (i.e., at a microphone sensor) 10-20 dB broad band noise reduction in an 18 passenger aircraft cabin using a single channel microphone/loudspeaker system [2].

Elliott and Ross have studied noise problems in a rectangular box, a sports car, and a fuselage cabin. In a B.Ae. 748 turboprop aircraft, their system obtained a spatially averaged noise reduction of 13 dB at the first blade passage frequency (BPF), a 12 dB noise reduction at the second and third BPF, and an overall noise reduction of 7 dBA [1]. The system used 16 loudspeakers and 32 microphones.

In a flight test on a Dornier 228 turboprop aircraft, the ASANCA team demonstrated a spatially averaged quieting of 11, 12, and 6 dB at the first, second, and third BPF, respectively. The system used 32 loudspeakers and 48 microphones [3,4].

Work using structural actuators for noise control in cabin interiors has been performed by Fuller, Silcox and co-workers [5,6] and **Lord Corporation** [7]. Demonstration of the structural-based active noise control technology has been performed on full-scale test articles [5,6,7].

Active control of interior noise of turboprop aircraft is quickly approaching commercialization [8,9]. However, a number of important research areas must be investigated. Much of the current research in the field involves the development and comparison of different kinds of actuation and sensing.

An analytical comparison between speaker-based and structural-based noise control systems was completed in Refs. [10,11]. The comparison concluded that large global reductions in interior acoustic energy can be obtained by the use of a few structural actuators (as secondary force inputs) compared with acoustic sources. In addition, the work concludes that at the second BPF "...the benefits of using secondary force inputs are even more significant (compared with the first BPF), as they allow global reductions in sound pressure level to be obtained where the use of acoustic secondary sources had only allowed localized reductions in the sound pressure level". This paper completes an experimental comparison of speaker-based and structural-based noise control system performance in reducing turboprop cabin noise.

In addition, the paper compares noise control system performance using two forms of sensing. Typically, microphones and accelerometers are used as sensors for controlling noise and vibration, respectively. The sensor comparison addresses the following question:

In a turboprop environment where noise *and* vibration are problems for passenger comfort, what kind of sensors (microphones, accelerometers, or a combination) should be used as part of an active control system, for greatest reduction of both noise and vibration?

Note that structural-based actuators can control both noise and vibration, whereas speaker-based systems cannot be practically used to provide vibration reduction for turboprop applications. Hence, the sensor comparison is only relevant to structural-based noise control.

The underlying physical phenomena which govern the above question relate to the modal dynamics of both the turboprop structure and its interior acoustic space. If the modes that dominate vibration in the structure at a particular frequency couple well to the acoustics, then the overall acceleration of the structure and the interior noise level will be directly related. In this case, controlling accelerometers will reduce interior noise levels; likewise, controlling microphones will reduce structural vibrations.

On the other hand, when modes that dominate the structural vibration do not couple well to the acoustic space, then microphones and accelerometers will not be directly related. In this case, the structural modes that drive the acoustics may represent only a small portion of the structural vibration. When this occurs, controlling accelerometers will not, in general, reduce noise levels. In fact, noise levels may be significantly increased [10]. Similarly, the use of microphones as sensors may increase vibration levels [11].

The study discussed in this paper was completed on a deHavilland DASH-7 fuselage section at the Thomas Lord Research Center. The testing consisted of measuring frequency response functions between the actuators and sensors and using these to predict performance of an active noise and vibration control system. For the actuator comparisons, frequency response functions were obtained using four actuators (either speakers or structural-based actuators) and sixteen microphones, and the predicted performance of structural-based and speaker-based systems was compared.

For the sensor comparisons, results using eight accelerometer and eight microphone sensors were compared, based on a system using four structural-based actuators. Control performance was predicted using the entire sensor array as well as sensor array subsets.

Experimental Apparatus and Approach

The DASH-7 turboprop fuselage section used in this study is a 12 foot long mock-up of the production aircraft. The test article contains no trim, seats, etc. Thus, its structural and acoustic dynamics may be significantly different from those of the actual aircraft, although the trends in the results should be similar to those found in a production aircraft.

Figure 1 shows a sketch of the test article revealing the placement of actuators (speakers and structural actuators) and sensors (microphones and accelerometers). Figure 1 also shows the location of exterior speakers, used to emulate the turboprop propeller wash.

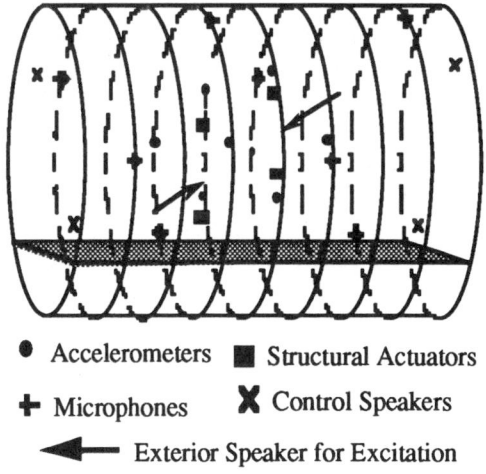

Figure 1. DASH-7 test article, experimental setup.

System Equations

The system steady-state (frequency domain) equations can be expressed as follows:

$$y_m(\omega) = \mathbf{P_m}(\omega) u(\omega) - d_m(\omega) \quad (1)$$

$$y_a(\omega) = \mathbf{P_a}(\omega) u(\omega) - d_a(\omega) \quad (2)$$

where y corresponds to the error sensor signal, d is the disturbance measured at each of the error sensors, u is the control input created by the structural actuators, and $\mathbf{P_m}$, $\mathbf{P_a}$ are the frequency response function matrices, containing the plant dynamics, relating the actuator signals to the sensor signals. The subscript m refers to measurements based on the microphone sensors and the subscript a denotes measurements based on the accelerometer sensors. From equations (1) and (2), the error signals can be computed for any control input u.

For this study, three cost functions J were defined as follows:

$$J_m(\omega) = y_m^H y_m \quad (3)$$

$$J_a(\omega) = y_a^H y_a \quad (4)$$

$$J_c(\omega) = J_m + \alpha(\omega)^2 J_a. \quad (5)$$

The superscript H denotes the Hermitian operator. The subscript c in (5) refers to a combination of accelerometers and microphones. The variable α in (5) is a weighting factor for the accelerometers, and can be expressed as a function of frequency. For convenience, the independent variable ω is dropped in future expressions, as it is understood that the system has been described in the frequency domain.

By minimizing these three cost functions, three possible control inputs can be computed. It can be shown that these will take the form:

$$u_m = \mathbf{P_m}^+ d_m \quad (6)$$

$$u_a = \mathbf{P_a}^+ d_a \quad (7)$$

$$u_c = \mathbf{P_c}^+ d_c \quad (8)$$

where u_m is the control inputs necessary to minimize J_m, u_a is the control inputs that minimize J_a, and u_c are those that minimize J_c. The superscript $+$ denotes the pseudo inverse. The matrix $\mathbf{P_c}$ in Eq. (8) is the combined frequency response function matrix containing both the microphone and accelerometer frequency response functions and is given by:

$$\mathbf{P_c} = \begin{bmatrix} \mathbf{P_m} \\ \alpha \mathbf{P_a} \end{bmatrix} \quad (9)$$

and, d_c is the combined disturbance vector which has the following form:

$$d_c = \begin{Bmatrix} d_m \\ \alpha d_a \end{Bmatrix} \quad (10)$$

The variable α can be set arbitrarily to weight the accelerometers with respect to the microphones. Depending on the application and the airframe manufacturers preference, the variable α can be chosen to vary with frequency to provide a comfortable environment in the aircraft cabin. In this application, the control system would balance the noise and vibration control performance. Indeed, at higher frequencies, vibration becomes increasingly imperceptible and would not be required to be reduced. In this case, α would approach zero. At lower frequencies where vibration is perceptible, α would increase in order to enhance the vibration reduction.

In order to illustrate a rather simple "equal" weighting between each of the microphones and accelerometers at every frequency, α was computed at each spectral line as follows:

$$\alpha(\omega) = \mathrm{norm}(\mathbf{P_m}(\omega)) / \mathrm{norm}(\mathbf{P_a}(\omega)) \quad (11)$$

In the next two sections, the results were generated by measuring the frequency response function matrices $\mathbf{P_m}$ and $\mathbf{P_a}$ on the DASH-7 fuselage section. The actuator comparison was completed using Eq. (6) for both structural-based and speaker-based systems. The performance predictions plot the uncontrolled (corresponding to the disturbance) average microphone level in various frequency ranges, as well as the average predicted noise levels with control using speakers and structural actuators.

Results: Actuator Comparison

Figures 2 through 4 reveal the noise control performance predictions using structural-based and speaker-based noise control systems, for the frequency ranges from 60 to 80 Hz, 130 to 150 Hz and 200 to 220 Hz, respectively. The spectral line resolution was 0.5 Hz. The figures also provide the uncontrolled average sound pressure levels created by the disturbance at the sixteen microphones.

Figure 2 shows that below about 70 Hz, the structural-based system outperforms the speaker system by at least 10 dB. Above 70 Hz, however, the performance of the two systems begins to converge. This behavior supports the hypothesis that at low frequencies, the spatial coupling between the structure and the acoustics dominates the acoustic response. In other words, the modes that dominate the acoustic response are those that spatially couple well to the structure. Naturally, the structural-based actuators control these modes well and provide superior performance when compared with the speaker system.

Above 70 Hz, the acoustic dynamics dominate the acoustic response. Speakers can control the acoustic modes as well as structural-based actuators, and the performance generated by the two systems are comparable. Figures 3 and 4 demonstrate this trend. Note that Figs. 2 - 4 have different vertical scales.

Results: Sensor Comparison

Figures 5 - 10 show acceleration levels and microphone levels for the various sensor configurations in frequency ranges from 60 to 220 Hz. Each figure provides the noise or vibration control result using microphones (Eq. (6)), accelerometers (Eq. (7)), and the accelerometer/microphone combination (Eq. (8)), as well as the uncontrolled result. Average acceleration (shown in Figures 2, 4, and 6) and average sound pressure levels (Figures 3, 5 and 7) were measured using the eight control accelerometers and eight microphones, respectively. Each graph has a spectral line resolution of 0.5 Hz.

Note that the results presented in Figs. 5 - 10 indicate that the best average vibration (acceleration) reduction is produced by the system that uses only the accelerometers as sensors. This corresponds to a very large accelerometer weighting ($\alpha = \infty$) in Eq. (8). Analogously, the system that provides the best

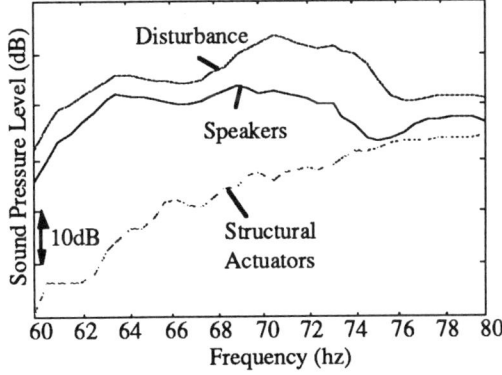

Figure 2. Average sound pressure level at 16 microphone error sensors for speakers and structural-based actuators for the 60 to 80 Hz frequency range.

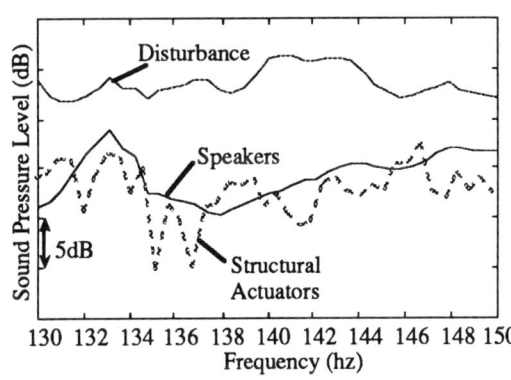

Figure 3. Average sound pressure level at 16 microphone error sensors for speakers and structural-based actuators for the 130 to 150 Hz frequency range.

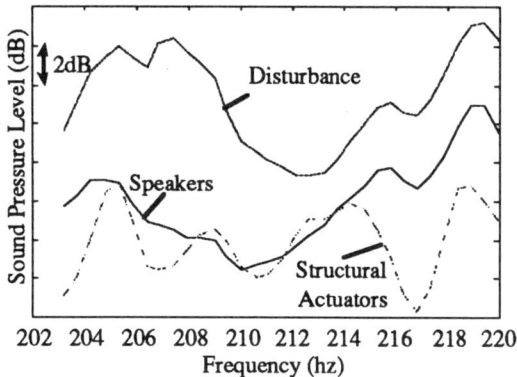

Figure 4. Average sound pressure level at 16 microphone error sensors for speakers and structural-based actuators for the 200 to 220 Hz frequency range.

average sound pressure reduction is provided by one that uses only microphones as error sensors, where $\alpha = 0$ in Eq. (8). The results that use both accelerometers and microphones in Figs. 5 - 10 were generated using Eq. (8), with α given by Eq. (11). Of course, these results are dependent on the weighting α, which varies as a function of frequency.

Figures 5 and 6 reveal interesting behavior of the system at low frequencies. Below about 68 Hz, it appears that the vibration of the structure is directly coupled to the acoustic field, due to the fact that at low frequencies there are very few acoustic modes. Accelerometers or microphones (or combinations of both) both provide good reduction of the noise and vibration at low frequencies.

At higher frequencies, the coupling between the vibration levels and interior acoustics changes. At 72 Hz, for example, the mode(s) that dominate the structural vibration do not couple well to the acoustics, due to an acoustic resonance which dominates the response. Thus, the levels of vibration and noise follow different trends depending on which kinds of sensors are used. Figure 5 shows that vibration levels actually increase at 72 Hz when microphones are used as control sensors. Similarly, the noise levels increase when accelerometers are used. Figures 5 and 6 reveal that when both kinds of sensors are used, good reductions can be attained in both noise and vibration.

Figures 7 and 8 shows that there is no direct coupling between the dominant modes of structural vibration and the acoustic modes in the frequency range 130 to 150 Hz. Indeed, Figure 8 shows that controlling accelerometers results in little or no reduction (and possibly an increase) in the noise levels at the microphones. Also, controlling microphones has the result of increasing acceleration levels. Once again, when a combination of accelerometers and microphones is used, relatively good reductions are attained at the microphones. At the accelerometers, however, a combination of sensors can result in a slight increase in vibration levels. This increase is much smaller than that caused by just using microphones.

Compared with Figures 7 and 8, Figures 9 and 10 show similar trends in the 200 to 220 Hz range. At these frequencies, the microphones and accelerometers produce significantly different solutions as evidenced by the large increases in acceleration levels when microphones are used for control. However, as Figures 9 and 10 indicate, there are frequencies (above 216 Hz) where using both accelerometers and microphones will increase average noise levels compared with the uncontrolled case. In a practical application, these increases are unacceptable and will require different weighting between accelerometers and microphones at higher frequencies to ensure that this does not occur.

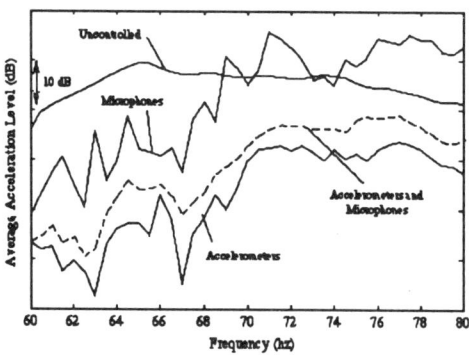

Figure 5. Average acceleration levels for various control sensor configurations.

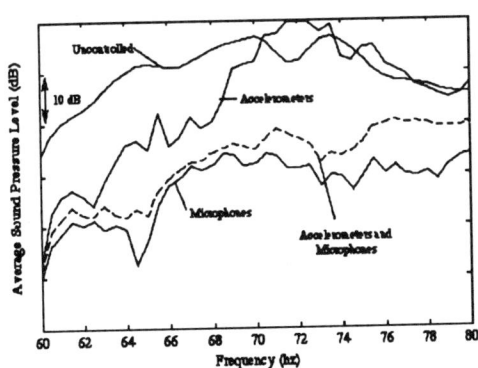

Figure 6. Average sound pressure levels for various control sensor configurations.

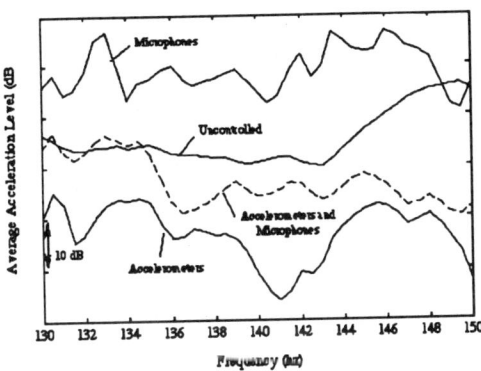

Figure 7. Average acceleration levels for various control sensor configurations.

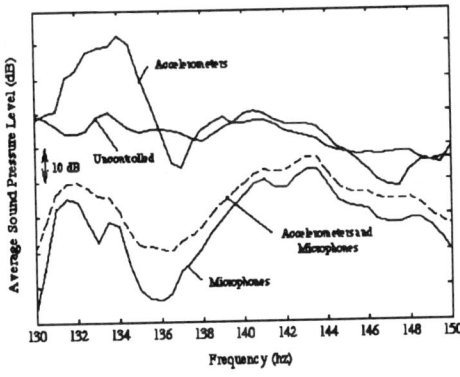

Figure 8. Average sound pressure levels for various control sensor configurations.

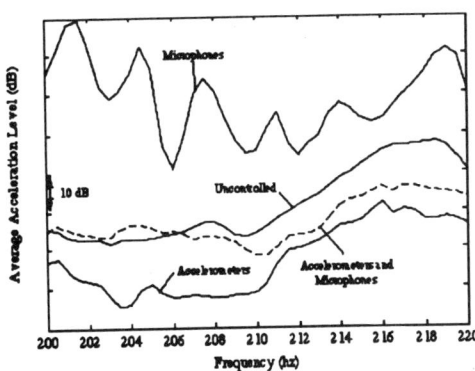

Figure 9. Average acceleration levels for various control sensor configurations.

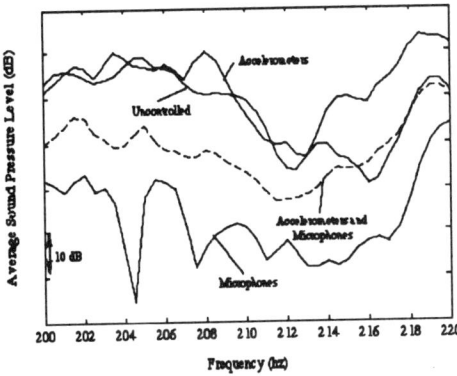

Figure 10. Average sound pressure levels for various control sensor configurations.

Conclusions

Noise control performance was compared for systems using speakers and structural-based actuators, on a deHavilland DASH-7 turboprop fuselage. It was found that at low-frequencies, where low-order structural motion drives the interior noise, structural-based systems out perform speaker-based noise control systems considerably. At higher frequencies, the two forms of actuation provided similar noise reduction performance.

Vibration and noise levels were predicted for structural-based noise control systems that use accelerometers and microphones as error sensors. It was found that a combination of microphones and accelerometers was necessary to ensure noise and vibration reduction throughout the frequency range studied. At lower frequencies, the dominant modes of fuselage vibration couple directly to the interior acoustics. In this range, when accelerometers were added to the error sensor array, noise control performance deteriorated only slightly. At higher frequencies, adding accelerometers to the error sensor array significantly reduced noise control performance. Depending on the turboprop customer's preference, frequency dependent sensor weighting can be used to enhance passenger comfort.

References

1. Elliott, S. J., P. A. Nelson, I. M. Stothers and C. C. Boucher, 1990. "In-flight experiments on the active control of propeller-induced cabin noise," Journal of Sound and Vibration, 140(2):219-238.

2. Warner, J. V. and R. J. Bernhard, 1990. "Digital control of local sound fields in an aircraft passenger compartment," AIAA Journal, 28(2):284-289.

3. Borchers, I. U. and e. al., 1992. "Advanced study for active noise control in aircraft (ASANCA)," Proceedings of the DGLR/AIAA 14th Aeroacoustics Conference, pp. 1-11.

4. Lefebvre, S., G. Billoud, H. Ribet and J. Paillard, "In-flight validation of an active noise control system for propeller aircraft," Proceedings of the 124th Meeting of the Acoustical Society of America, 1992.

5. Silcox, R. J., H. C. Lester and S. B. Abler, 1989. "An evaluation of active noise control in a cylindrical shell," Journal of Vibration, Stress, and Reliability in Design, 111, pp. 337-342.

6. Simpson, M. A., T. M. Luong, C. R. Fuller and J. D. Jones, "Full-scale demonstration tests of cabin noise reduction using active vibration control," Proceedings of the 12th AIAA Aeroacoustics Conference, 1989.

7. Rossetti, D. J., M. A. Norris, S. C. Southward, and J. Q. Sun, 1993. "A Comparison of Speakers and Structural-Based Actuators for Aircraft Cabin Noise Control.", Second Conference on Recent Advances in Active Control of Sound and Vibration, p S1.

8. Emborg, U and C. F. Ross, 1993. "Active Control in the Saab 340.", Second Conference on Recent Advances in Active Control of Sound and Vibration, p S67.

9. Paxton, M. et. al., 1993. "Active Control of Sound in a MD-80.", Second Conference on Recent Advances in Active Control of Sound and Vibration, p S100.

10. Thomas, D. R., P. A. Nelson and S. J. Elliott, 1993. "Active control of the transmission of sound through a thin cylindrical shell, Part I: The minimisation of vibrational energy," Journal of Sound and Vibration, 167(1), 91-111.

11. Thomas, D. R., P. A. Nelson and S. J. Elliott, 1993. "Active control of the transmission of sound through a thin cylindrical shell, Part II: The minimisation of acoustic potential energy," Journal of Sound and Vibration, 167(1), 113-128.

12. Nelson, P. A. and S. J. Elliott. 1992. Active Control of Sound, San Diego, CA: Academic Press Limited, pp. 381-382.

13. Golub, G. H. 1988. Matrix Computations, Baltimore, MD: The Johns Hopkins University Press, pp. 139.

DETERMINATION OF THE POWER CONSUMPTION OF STACKED PZT ACTUATOR-DRIVEN UNDERWATER PLATES FOR ACTIVE STRUCTURAL ACOUSTIC CONTROL

C. Liang, F. P. Sun and C.A. Rogers
Center for Intelligent Materials Systems and Structures
Virginia Polytechnic Institute and State University
Blacksburg, VA 24061-0261

Yi Gu
NCT Inc.
Linthicum, MD 21090

Abstract

This paper presents an analysis technique to determine various energy dissipation in a PZT actuator-driven underwater plate based on an electro-mechanical impedance model. The entire actuator/structure/fluid system is modelled as an electrical network which may be represented by a coupled electro-mechanical admittance or impedance. Various dissipation, including the radiated acoustic power, may be determined from the electro-mechanical admittance. This paper also discusses the transduction characteristics of stacked PZT actuators and the determination of the acousto-mechanical impedance of a simply-supported plate considering fluid interaction. The method outlined in this paper may be used to determine the structural response, acoustic radiation, and power flow of active structures with integrated induced strain actuators. The acoustic radiation actuator power factor discussed in this paper may also be utilized as an efficiency index in the design of active structural acoustic control systems.

1. Introduction

Active structural acoustic control (ASAC) using induced strain actuators has recently gained attention as a possible means of controlling low frequency noise (< 1000 Hz). It has been demonstrated experimentally for composite beams with embedded shape memory alloy (SMA) actuators (Rogers, 1989; Saunders et al., 1990), beams with bonded piezoelectric (PZT) actuators (Fuller et al., 1990) and plates with bonded PZT actuators (Clark and Fuller, 1992). Other research has focused on optimization of multiple PZT actuators for ASAC of plates (Wang, 1990) and ASAC of cylindrical shells using PZT actuators (Lester and Lefebvre, 1991; Sonti et al., 1991). These efforts have been useful in identifying the significant design parameters for induced strain actuator utilized in an ASAC system (i.e. size, location, thickness, phasing, etc.).

In most, if not all, of the research related to ASAC with induced strain actuators, the investigation of energy consumption and requirement has been omitted. For laboratory scale structures, such as thin beams, plates, and shells, the energy consumption may not be a concern. However, for realistic and large scale complex structures, such as marine vessels which may require a great number of relatively large actuators, the power consumed by the actuators in the active control scenario may be very significant. Large actuator power consumption signifies large power supplies which are expensive and massive. Reducing the power consumption of active control structures, or increasing their energy efficiency can have a significant impart in the development of active control technologies.

In ASAC with PZT actuators, the sound pressure is linearly related to the activation forces and moments. Increasing the sound pressure level in dB is therefore limited by the amount of induced force/moment an actuator can deliver. The voltage level and power consumption, however, can be significantly reduced by increasing the force/moment output of the actuator for a set electric field. As an example, consider an actuator configured in such a way as to obtain a 100% increase in the activation moment or force. This increase in activation increases the sound pressure level by only about 3 dB. However, the voltage level needed to maintain the original sound pressure level will drop by 50% and the power consumption can be reduced by up to 75%. Thus, power consumption of actuators should be included as a primary variable in the objective function for optimization of intelligent material systems for active control.

Copyright © 1994 by CIMSS
Published by the AIAA, Inc. with Permission

The first step to determine the energy consumption and requirement of an active control structure is to determine how the energy is transferred from the integrated actuator to the host structure to create intended motions for active control. In this paper, the actuator is the only excitation source. The objective is to quantify the energy conversion between the electrical source and an underwater plate driven by a stacked PZT actuator.

A similar work on the system power consumption has been presented by Stein, Liang, and Rogers (1993) considering a beam structure. This paper considers a plate structure and provides a more in-depth and comprehensive discussion on various energy dissipation in an actuator-driven fluid-loaded structure. The acoustic radiation calculation program (fluid loaded) used in this paper was developed by Gu (1992) and has been extensively verified.

The method used in this analysis to determine the electrical to mechanical and acoustic energy conversion is an electro-mechanical impedance modeling approach (Liang et al, 1993a; 1993b), and will be briefly introduced first.

2. Introduction to the Electro-Mechanical Impedance Model

The electro-mechanical impedance model considers an electrically powered active structure, including the actuator, host structure, and acoustic medium to be an electrical network, as shown in Fig. 1. The actuator can generally be described by its transduction relations (Kinsler et al., 1982; Liang et al., 1993c) which are represented by a two-port electrical system, as shown in Fig. 1. The mechanical interaction between the actuator and the host structure, including fluid loading, may be represented by an acousto-mechanical impedance, Z, as shown in Fig. 1. The transduction relations of an actuator may be expressed as:

$$\begin{cases} u = F/Z_{ms} + \phi V \\ I = \phi F + V/Z_{EF} \end{cases} \quad (1)$$

where Z_{EF} is the free electrical impedance of the actuator, Z_{ms} the short-circuit mechanical impedance of the actuator, ϕ is the transduction coefficient, F is the force, u is velocity, V is the electrical voltage, and I is the electrical current.

The mechanical interaction force, F, and the coupled electro-mechanical admittance, Y, may be determined by substituting F=-Zu to Eq. (1), yielding:

$$\begin{cases} F = -\dfrac{Z_{ms}Z}{Z_{ms}+Z}\phi V \\ Y = \dfrac{I}{V} = \dfrac{1}{Z_{EF}} - \dfrac{\phi^2 Z_{ms}Z}{Z_{ms}+Z} \end{cases} \quad (2)$$

where Z is a generalized mechanical impedance which includes the fluid interaction in this paper.

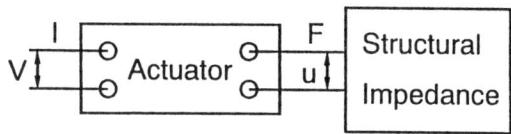

Figure 1. A generic electro-mechanical representation of active structures with an integrated induced strain actuator described by a two-port electro-mechanical system.

The important components in the electro-mechanical impedance modeling technique are the electro-mechanic transduction relations of the actuator and the dynamics of the host structures, represented in this case by the acousto-mechanical impedance. The key parameters to determine for an electro-mechanical system are the mechanical interaction force and the electro-mechanical admittance as given by Eq. (2).

A stacked PZT actuator is used to excite the plate in this paper. In general, a commercially available stacked PZT actuator is very sophisticated. It may consist of a rigid casing, pre-stress fixture, output rod, and possibly a hysteresis compensator. The modeling of the electrodynamic transduction relations can be very complicated (Flint et al., 1994). Since the objective here is not to model a stacked PZT actuator, the stacked PZT actuator considered in this paper is assumed to be constructed by simply stacking many PZT disks in their thickness direction. The effect of the glue and electrodes on the dynamics of the stacked actuator is not considered.

For a stacked PZT actuator, the PZT layers are mechanically in series and electrically in parallel. The electrical field is generally applied in the z or 3-3 direction and d_{33} effect is used to achieve larger stroke output (d_{33} is generally twice as much as d_{31} or d_{32}). Based on the electro-mechanical modeling technique presented in Liang et al. (1993a and 1993b), the interaction force between a stacked PZT actuator and its host structure may be derived as:

$$\bar{F} = -\frac{Z}{(Z_A+Z)} d_{33} \bar{E} \bar{Y}_{33}^E A \qquad (3)$$

where A is the cross-section area of the PZT actuator, d_{33} the piezoelectric coefficient, $\bar{E}$ the applied electric field, $\bar{Y}_{33}^E = Y_{33}^E(1+i\eta_A)$ is the complex elastic modulus of the PZT at a constant electrical field in the 3-3 direction, η_A the mechanical loss factor for the PZT, and Z the driving point mechanical impedance of the structure. The short circuit actuator mechanical impedance, Z_A, is:

$$Z_A = \frac{K_A}{i\omega} \frac{k_A n h_A}{\tan(k_A n h_A)} = \frac{\bar{Y}_{33}^E A}{n h_A i\omega} \frac{k_A n h_A}{\tan(k_A h h_A)}, \qquad (4)$$

where K_A is the complex stiffness of the PZT actuator, n is the number of layers for the stacked actuator, h_A is the thickness of each PZT layer, and the complex wave number k_A is given by $k_A^2 = \omega^2 \rho_A / \bar{Y}_{33}^E$ (ρ_A is the density of the PZT actuator).

The coupled electro-mechanical admittance, which reflects the electrical behavior of the entire system may be obtained as:

$$Y = in\omega \frac{A}{h_A} \left[\bar{\epsilon}_{33}^T - \frac{Z}{Z_A+Z} d_{33}^2 \bar{Y}_{33}^E \right], \qquad (5)$$

where $\bar{\epsilon}_{33}^T$ is the complex dielectric constant of the PZT material at a constant stress given by $\epsilon_{33}^T(1 - i\delta)$. δ is the dielectric loss factor.

The influence of the mechanical interaction between the structure and the actuator (through impedance) is clear in Eq. (5). The electrical power is related to the electro-mechanical admittance through the following:

the apparent power, W_A:

$$W_A = \frac{V^2 |Y|}{2} \qquad (6)$$

the dissipative power, W_D:

$$W_D = \frac{V^2 \text{Re}(Y)}{2} \qquad (7)$$

and the reactive power, W_R:

$$W_R = \frac{V^2 \text{Im}(Y)}{2}. \qquad (8)$$

where V is the amplitude of the applied electrical voltage.

For a mechanical system vibrating in vacuo, the power consumed includes three parts: the first two result from the mechanical and dielectric losses in the PZT actuator and the third results from the damping in the mechanical system. For a system with fluid loading, the additional interaction between the structure and the fluid medium causes an increase in power consumption of the actuator in the form of acoustic radiation.

3. Determination of the Acousto-Mechanical Impedance

The structural impedance or admittance of fluid-loaded elastic plates has been extensively investigated by Nayak (1970), Crighton (1972, 1977 and 1980), and Smith (1978). The coupled acousto-mechanical impedance of a fluid-loaded structure contains information on the acoustic radiation and near field fluid/structure interaction. For example, the real part of the acousto-mechanical admittance of a lossless structure is directly related to the radiated acoustic power. The research reported in this paper, to a certain degree, is an extension of the work conducted by Crighton et al. The electro-mechanics of the dynamic transducers (PZT actuator in this case) which excites fluid-loaded structures are considered in this analysis. The electro-mechanical admittance of a transducer/structure/fluid system contains the same amount of information as the acousto-mechanical admittance obtained by Crighton and other researchers mentioned above, but they may be directly measured and easily converted into structural admittance (Sun et al., 1994).

The plate investigated herein is simply-supported. One end of the stacked PZT actuator is fixed at (x_0, y_0) on the plate. The other end of the actuator is fixed on the ground. It is assumed that the actuator creates a point loading. The driving point mechanical (acousto-mechanical) impedance at (x_0, y_0) corresponding to a point force (perpendicular to the plate) is determined first. The acousto-mechanical impedance is then substituted into Eqs. (3) and (5) to determine the interaction force between the PZT actuator and the plate and the electro-mechanical (or electro-acousto-mechanical) admittance. The determination of the acousto-mechanical impedance here is based on the modeling technique (fluid loaded plate) developed by Sandman (1977). Detailed derivation of the entire modeling procedure of Sandman's approach may also be found in Gu (1992). A brief derivation is provided below.

Consider a simply-supported plate within an infinite baffle. One side of the plate/baffle is loaded with fluid (air or water) and the other side is vacuum. The equation of motion for the plate is:

$$D\nabla^4 w + \rho_p h \frac{\partial^2 w}{\partial t^2} = q(x,y,t) - p_0(x,y,t), \quad (9)$$

where D is the complex flexural rigidity of the plate, h is the thickness of the plate, ρ_p is the density of the plate, $q(x,y,t)$ is the dynamic loading on the plate, and $p_0(x,y,t)$ is the acoustic pressure acting on the plate. The mechanical loss factor for the plate material is assumed to be η_p. In this case study, the dynamic loading is a point force at (x_0, y_0). Since the analysis is conducted in the frequency domain, the time variable may be separated from the spatial variables. To determine the acousto-mechanical impedance, the excitation force may be assumed to be unity and represented by a Delta function as:

$$\bar{q}(x,y,t) = \delta(x_0, y_0)\exp(i\omega t), \quad (10)$$

where ω is the angular velocity of the excitation force. The acoustic radiation and plate response subsequently calculated will be the frequency response function of the excitation force, q, and will be denoted with an overhead bar.

The unit excitation force may also be expressed in the following form:

$$\bar{q} = \sum_{m=1}^{\infty} \sum_{n=1}^{\infty} \bar{a}_{mn} X_{mn}(x,y)\exp(i\omega t), \quad (11)$$

where $X_{mn}(x,y)$ is the mn^{th} mode shape. For a simply-supported plate, $X_{mn}(x,y) = 2/(lw)^{1/2} \sin(m\pi x/l)\sin(n\pi y/w)$, where 'l' and 'w' are the length and width of the plate, respectively.

The modal force coefficient $\bar{a}_{mn}$ in Eq. (11) can then be solved as:

$$\bar{a}_{mn} = \frac{2}{\sqrt{wl}} \sin\left(\frac{m\pi x_0}{l}\right)\sin\left(\frac{n\pi y_0}{w}\right). \quad (12)$$

The displacement response of the plate under the unit excitation force and resulting acoustic pressure, $\bar{p}_0$, may be solved to have the following form:

$$\bar{w} = \sum_{n=1}^{\infty} \sum_{m=1}^{\infty} \bar{\Delta}_{mn} X_{mn}(x,y)\exp(i\omega t). \quad (13)$$

The spatial distribution of the acoustic pressure may then be solved (Sandman, 1977) as:

$$\bar{P}(x,y,z) = -\frac{\rho_f k_f \omega}{2\pi} \sum_{n=1}^{\infty} \sum_{m=1}^{\infty} \Delta_{mn} \int_0^l \int_0^w X_{mn}(x_1,y_1) \times \frac{\exp(-ik_f R)}{R} dx_1 dy_1, \quad (14)$$

where ρ_f is the density of the fluid, k_f is the wave number (acoustic medium) given by ω/c_f (c_f is the speed of the sound in the fluid medium), and R is given by:

$$R = \sqrt{(x-x_1)^2 + (y-y_1)^2 + z^2}. \quad (15)$$

The acoustic pressure acting on the plate, $\bar{p}_0(x,y)$, may be determined to be $\bar{P}(x,y,z=0)$. After substituting Eqs. (11), (12), (13), and the acoustic pressure on the plate, $\bar{p}_0(x,y)$, into Eq. (9), the displacement modal amplitude, $\bar{\Delta}_{mn}$, may be determined from the following equations:

$$A_{mn}\bar{\Delta}_{mn} + B\sum_{r=1}^{\infty}\sum_{t=1}^{\infty} Z_{mnst}\bar{\Delta}_{st} = \bar{a}_{mn}, \quad (16)$$

where

$$A_{mn} = D\left[\left(\frac{m\pi}{l}\right)^2 + \left(\frac{n\pi}{w}\right)^2\right]^2 - \rho_p h \omega^2 \quad (17)$$

and

$$Z_{mnst} = \int_0^l \int_0^w [X_{mn}(x,y) \int_0^l \int_0^w X_{st}(x_1,y_1)\frac{\exp(-ik_f R)}{R} dx_1 dy_1] dx dy \quad (18)$$

and $B = \rho_f k_f \omega / 2\pi$.

In the numerical analysis, the infinite modal expansion used to approximate the plate displacement response, excitation force, and acoustic radiation are truncated to finite MxN terms. In this analysis, M=N=10, which provides a converged solution in the frequency range of interest for the plate.

4. Determination of System Dissipation

Once the displacement modal amplitude, $\bar{\Delta}_{mn}$, is solved from Eq. (16), the displacement response $\bar{w}$, acoustic pressure, $\bar{P}(x,y,z)$ (frequency response functions of the excitation force q) can be determined from Eqs. (13) and

(14), respectively. The radiated acoustic power under the unit excitation force (frequency response function, as well) may be determined by integrating the radiated acoustic intensity over the infinite semi-sphere on the fluid side as:

$$\bar{\Pi} = \int_{\phi=0}^{2\pi} \int_{\theta=0}^{\pi/2} \frac{1}{2\rho_f c_f} |\bar{P}(r,\theta,\phi)|^2 r^2 \sin(\theta) d\theta d\phi , \quad (19)$$

where $\bar{P}(r,\theta,\phi)$ is the same as radiated acoustic pressure (under unit excitation), as given by Eq. (14), except it has been converted into a spherical coordinate system. θ is the polar angle, ϕ is the azimuthal angle, and r is the radius.

The acousto-mechanical impedance, Z_{AM} which interacts with the actuator, may be determined as:

$$Z_{AM} = 1/i\omega \bar{w}(x_0, y_0) . \quad (20)$$

Substituting the acousto-mechanical impedance, Z_{AM}, into Eqs. (3) and (5), the interaction force between the actuator and plate and the electro-mechanical admittance may be determined. Once the interaction force (F), which is frequency dependent, is determined, the actual structural response, w, may be determined as:

$$w(x,y) = F \times \bar{w}(x,y) . \quad (21)$$

The actual radiated acoustic pressure:

$$P(x,y,z) = F \times \bar{P}(x,y,z) . \quad (22)$$

The actual radiated acoustic power:

$$\Pi = \bar{\Pi} |F|^2 . \quad (23)$$

The damping of the system is incorporated by using complex stiffness for the PZT and the elastic plate, and complex dielectric constant for the PZT actuator. The parameters involved in the above derivation are complex variables.

The four basic sources of energy dissipation for a PZT actuator-driven underwater plate are:
- Mechanical loss of the PZT actuator which is accounted by its complex modulus,
- mechanical loss of the plate which is accounted by its complex modulus,
- electrical loss of the PZT actuator which is accounted by its complex dielectric constant, and
- radiated acoustic energy, which is included in the real part of the acousto-mechanical impedance.

The dissipation of the acoustic medium is not considered in this paper. It may also be included by using complex sound wave speed.

Every dissipated power listed above may be calculated using Eq. (7). The dissipative power from the dielectric loss of the PZT actuator may be determined as:

$$W_{DE} = \frac{V^2 \text{Re}(Y|_{\eta_A=0, Re(Z_{AM})=0})}{2} , \quad (24)$$

where $Y|_{\eta_A=0, Re(Z_{AM})=0}$ is obtained from Eq. (5) while assuming no mechanical loss for the PZT actuator and no plate dissipation and radiated acoustic energy, which is achieved by assuming the real part of the acousto-mechanical impedance to be zero ($Re(Z_{AM})=0$).

The dissipative power resulting from the mechanical loss of the PZT actuator may be approximated as:

$$W_{DMA} \approx \frac{V^2 \text{Re}(Y|_{\delta=0, Re(Z_{AM})=0})}{2} , \quad (25)$$

where $Y|_{\delta=0, Re(Z_{AM})=0}$ is obtained from Eq. (5) while assuming no electrical loss for the PZT actuator and no plate dissipation and radiated acoustic energy ($Re(Z_{AM})=0$). The power dissipation resulting from the mechanical loss of the PZT actuator is much smaller than the other dissipation because of the relatively small volume of the actuator compared with the host structures.

The dissipative power due to the mechanical loss within the plate may be approximated as:

$$W_{DMP} \approx \frac{V^2 \text{Re}(Y|_{\eta_A=\delta=0, Im(Z_{mnst})=0})}{2} , \quad (26)$$

where $Y|_{\eta_A=\delta=0, Im(Z_{mnst})=0}$ is obtained from Eq. (5) while assuming no mechanical and electrical loss for the PZT actuator and no radiated acoustic energy which may be modeled by assuming the imaginary part of Z_{mnst} (dissipative aspect of fluid interaction) to be zero. The real part of Z_{mnst} represents the equivalent modal mass and stiffness of the acoustic medium to the plate and needs to be maintained in the system dynamic analysis.

The radiated acoustic power can also be approximated as:

$$W_{AR} \approx \frac{V^2 \text{Re}(Y|_{\eta_A=\delta=\eta_p=0})}{2} , \quad (27)$$

where $Y|_{\eta_A=\delta=\eta_p=0}$ is obtained from Eq. (5) while assuming no mechanical and electrical loss for the PZT actuator and no mechanical loss for the plate. The radiated acoustic power obtained from Eq. (23) is very close to the radiated acoustic power predicted using Eq. (27).

The total dissipative power given by Eq. (7) and individual dissipation are related by:

$$W_D = W_{DE} + W_{DMA} + W_{DMP} + W_{AR}. \quad (28)$$

5. Numerical Case Studies

A simply-supported aluminum plate (0.80 meter by 0.55 meter and 8 mm thick) is used in the numerical case study. The Young's modulus of the aluminum is 65 GPa, possion's ratio is 0.3, density is 2700 Kg/m^3, and the mechanical loss factor is 0.005. The plate is driven by a PZT stack actuator which may be simplified as point force excitation. The piezoelectric coefficient of the PZT material, d_{33}, is 300x10^{-12} m/m/volt, its modulus, Y_{33}^E is 4.5x10^{10} Pa, its density is 7650 kg/m^3, its dielectric constant, ϵ_{33}^T, is 1.5x10^{-8} Farads/m, its dielectric loss factor, δ, is 1.5%, and the mechanical loss factor, η, is 1%.

The cross-sectional area of the stacked PZT actuator is 5x5 mm^2. The thickness of each layer is 0.5 mm. The stacked actuator has 100 layers. The voltage applied to each layer is 100 volts.

Figure 2 shows the driving point mechanical impedance of the plate in vacuum, air, and water. The short-circuit mechanical impedance of the PZT stack actuator is also plotted in Fig. 2 (dotted line). The driving point mechanical impedance for plate in vacuum (solid line) is almost the same as that in air (dashed line). However, considering air loading or not does make a difference, in particular, the dissipative aspect (real part) of the mechanical impedance, which will be discussed further in this section. The driving point mechanical impedance of the plate with water loading is plotted as the dash-dotted line. As expected, the fluid interaction, to a great degree, creates a mass loading effect on the plate, which increases the modal density in the frequency range of interest. The short-circuit mechanical impedance of the actuator is generally higher than the driving point acousto-mechanical impedance of the plate, which causes a significant stiffening to the plate/fluid system. The resonant frequency of the actuator/plate/fluid system may be determined from the intersection points of the actuator and structural impedance curves. If the actuator impedance curve is above the structural impedance curve, the resonant frequency of the actuator/plate/fluid system will be the frequency corresponding to the peak of the structural impedance curve (anti-resonant frequency of the plate/fluid system). If the actuator impedance curve intersects two points around a peak of the structural impedance curve, only the first one, which corresponds to the complex conjugate of the actuator impedance, is the resonant frequency of the actuator/plate/fluid system.

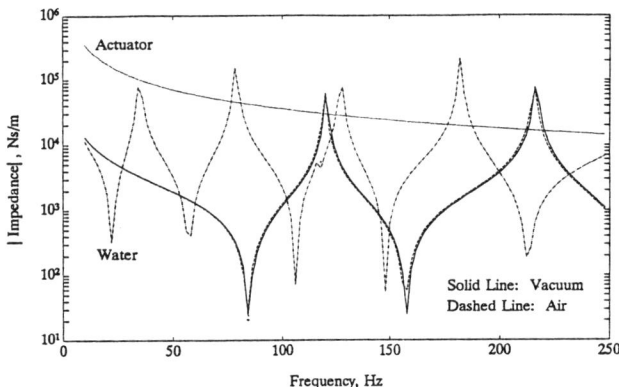

Figure 2. Driving point acousto-mechanical impedance of a baffled simply-supported plate in vacuum (solid line), in air (dashed line), and in water (dash-dotted line). The dotted line is the short-circuit mechanical impedance of the stacked PZT actuator.

Various energy dissipation in a PZT actuator-driven plate/water system is shown in Fig. 3. The solid line is the total dissipative power of the actuator/plate/water (APW) system, which includes actuator dissipation (dotted line), dissipation due to damping within the plate (dash-dotted line) and radiated sound power (dashed line). The actuator dissipative power is calculated from Eq. (24) and (25). The radiated sound power is calculated using Eq. (27), which is almost the same (not exactly the same) as what obtained from Eq. (23), because assuming a zero plate damping in Eq. (27) would result in slight variations in the plate response and radiated acoustic radiation, especially around the system resonant frequencies. At off-resonance, the difference between the prediction from Eqs. (23) and (27) is much less than one percent. At on-resonance, the difference may reach 5% of the actual radiated power (from Eq. (23)). The energy dissipation due to the mechanical damping within the plate (dash-dotted line) is calculated using Eq. (28).

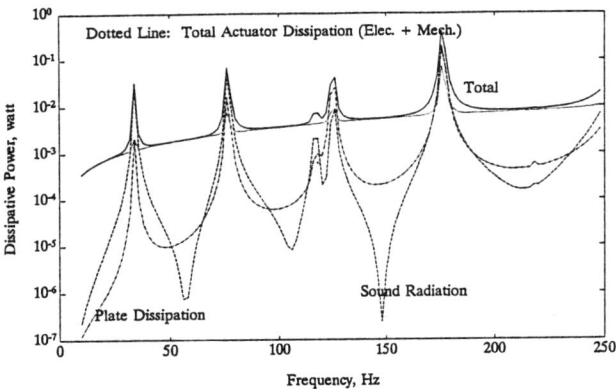

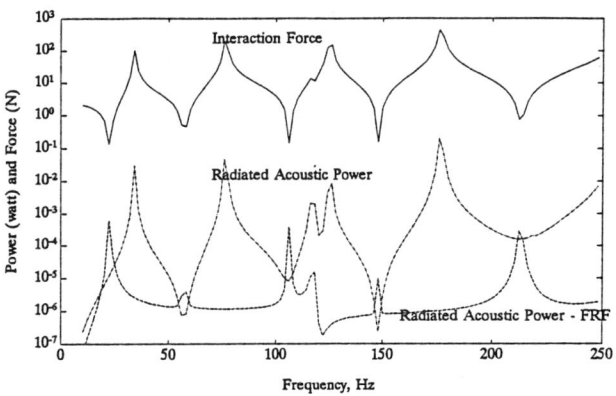

Figure 3. The power consumption in underwater acoustic radiation. The solid line is the total system power consumption, which includes the dissipative power due to the dielectric and mechanical loss of the PZT actuator (dotted line), structural damping of the plate (dash-dotted line), and radiated sound power (dashed line).

Figure 4. The determination of the actual radiated sound power using Eq. (23). The solid line is the interaction force (absolute value) between the PZT actuator and the plate/water system. The dashed line is the frequency response function of the radiated acoustic power. The dash-dotted line is the actual radiated power determined using Eq. (23).

Figure 4 illustrates the process of using Eq. (23) to determine the radiated acoustic power. The solid line in Fig. 4 is the absolute value of the interaction force between the actuator and plate with water loading. The interaction force is frequency dependent and is at its minimum at the resonant frequency of the plate/water system and maximum at the resonant frequency of the APW system (see Liang et al. (1993b) for more discussion on this subject). The radiated power under a unit excitation force (power frequency response function) is plotted as the dashed line. The actual radiated sound power obtained using Eq. (23) is plotted as the dash-dotted line. The difference between the profile of the power frequency response function and the profile of the actual radiated power is significant in this case study because of the stiffening from the actuator. It is apparent that modeling of the actuator interaction force as a frequency in-dependent quantity in the prediction of sound radiation (Clark and Fuller, 1992; Lester and Levebvre, 1991; Sonti et al., 1991) may cause some errors even for the case of surface bonded PZT actuators.

Figure 5 shows the total power consumption and radiated sound power of a plate/air system. The electrical loss and mechanical loss of the PZT actuator is not considered in this case study. The ratio of the radiated sound power to total power consumption is plotted in Fig. 6. It is very

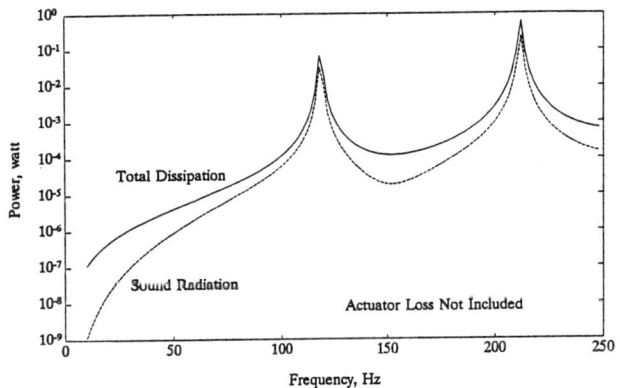

Figure 5. The power consumption of the plate to radiate in air. Actuator loss is not included. Solid line is the total dissipative power. The dashed line is the radiated acoustic power calculated using Eq. (27). The radiated acoustic power calculated from Eq. (23) is almost the same as the one determined using Eq. (27).

important to note that the acoustic radiation can be more than 50% of the total dissipation at some frequencies. This indicates that using in-air dynamic testing to determine structural damping may yield completely misrepresenting results if the dissipation resulting from the acoustic radiation is not properly considered. The results shown in Fig. 6 also represent the efficiency of a <u>structure</u> to radiate sound. A structural radiation efficiency provided by Fig. 6 may have more practical utilities than modal radiation efficiency.

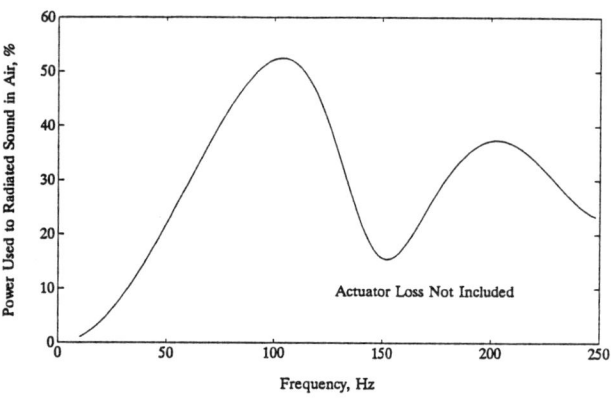

Figure 6. Percentage of power used to radiated sound in air, which represents the structural radiation efficiency. Actuator loss is not considered.

Figures 7 and 8 basically repeat the same plots as shown in Figs. 5 and 6, except the results are for a plate/water system. The radiated acoustic power can be more than 90% of the total power consumption in this case study.

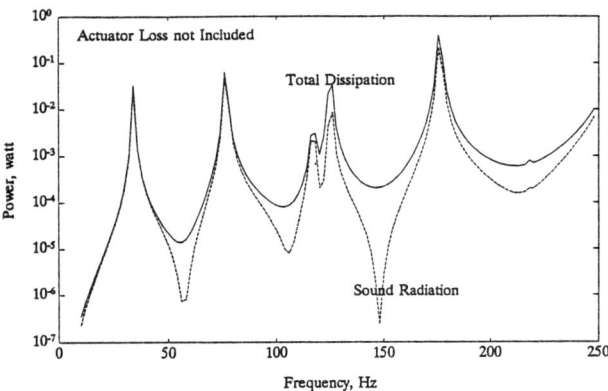

Figure 7. The power consumption of the plate to radiate in water. Actuator loss is not included. Solid line is the total dissipative power. The dashed line is the radiated acoustic power calculated using Eq. (27). The radiated acoustic power calculated from Eq. (23) is almost the same as the ones determined using Eq. (27).

Figure 9 plots the actuator power factor (Liang et al., 1993d) for sound radiation. The first actuator power factor (APF-1), defined as the ratio of the sum of the radiated acoustic power and the plate dissipative power to the supplied electrical power, represents the effectiveness of an actuator to create structural vibration and consequently to radiate sound. The second actuator power factor (APF-2), defined as ratio of the radiated acoustic power to supplied electrical power, represents the effectiveness of the actuator to drive the plate to radiate sound. Both actuator power factors may be used to assist in locating the effective actuator positions for active structural acoustic control. APF-1 is a measurable parameter (Liang et al., 1994). APF-2 directly reflects the sound radiation, but cannot be directly measured. In general, APF-1 agrees with APF-2 except at the anti-resonant frequencies, which are not important to most active structural acoustic control problems. The relatively large difference between APF-1 and APF-2 at the fourth peak indicates that this vibration mode has a lower modal radiation efficiency, which is also implied in the results shown in Fig. 8.

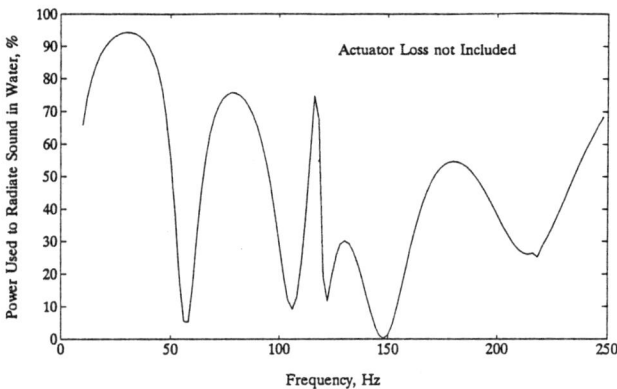

Figure 8. Percentage of power used to radiated sound in water, which represents the structural radiation efficiency. Actuator loss is not considered. One of the future work of this research is to determine this radiation efficiency based on structural dynamic and transducer power consumption measurement.

6. Summary

This paper has presented an analysis technique to determine the power consumption for PZT actuator-driven underwater plate. This is the first step to determine the power consumption and requirement for active underwater structural acoustic control. The principle outlined in the paper should be used in the acoustic analysis of active structures with integrated actuators to more accurately determine the structural response and radiated acoustic field.

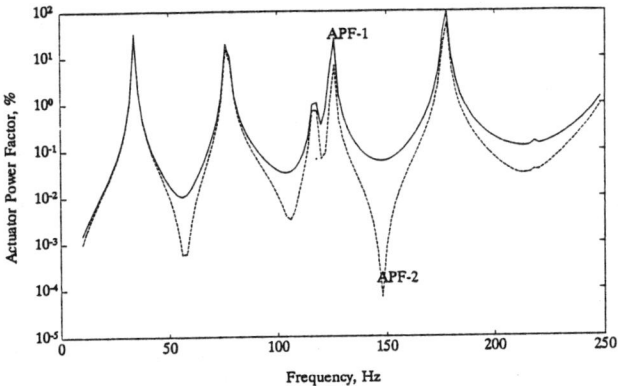

Figure 9. Actuator power factor for underwater acoustic radiation. APF-1 is defined as the ratio of the sum of the plate dissipation and the acoustic radiation to the total supplied electric power. APF-1 is defined as ratio of the acoustic radiation to the total supplied electrical power. APF-1 may be directly measured.

The results shown in Fig. 9 indicate that the power consumption (dissipative power) is generally much smaller than the reactive power in a PZT actuator driven-mechanical/fluid system, except at the system resonant frequencies. However, the investigation of the power consumption will assist in the understanding of the mechanics of active structures for underwater acoustic control and the design of more efficient active structures.

This paper has described and theoretically quantified various dissipative power involved in an actuator-driven plate with fluid loading. One interesting problem that will be further investigated is the determination of radiated acoustic power by measuring the actuator power consumption without using microphones. The total actuator power consumption can be easily measured. The actuator (electrical) dissipation is also a measurable quantity (assuming the mechanical loss of the actuator can be ignored because of its much smaller volume). If the energy dissipation within the structure can be determined, the radiated acoustic power may be easily calculated according to Eq. (28). Being able to measure radiated acoustic power without using microphones can have many important utilities, such as the acoustic radiation control of underwater subjects.

7. Acknowledgements

The authors gratefully acknowledge support from the Office of Naval Research, Grant ONR-00014-92-J-1170, Dr. Kam Ng, the Technical Monitor.

8. References

Clark, R. L, and Fuller, C. R., 1992, "Experiments on Active Control of Structurally Radiated Sound Using Multiple Piezoceramic Actuators", *Journal of Acoustical Society of America*, **91**(6), 3313-3320.

Crighton, D. G., 1972, "Force and Moment Admittance of Plates Under Arbitrary Fluid Loading", *Journal of Sound and Vibration,* Vol. 20, No. 2, pp. 209-218, 1972.

Crighton, D. G., 1977, "Point Admittance of an Infinite Thin Elastic Plate Under Fluid Loading", *Journal of Sound and Vibration,* Vol. 54, pp. 389-391, 1977.

Crighton, D. G., 1980, "Approximations to the Admittances and Free Wavenumbers of Fluid-Loaded Panels", *Journal of Sound and Vibration,* Vol. 68, No. 1, pp. 15-33, 1980.

Flint, E., Liang, C., and Rogers, C. A., "Design and Analysis of an PZT Stack Strut Member -An Impedance Based Approach", Proceedings, the Second International Conference on Intelligent Materials, June, 1994, Williamsburg, Virginia; in press.

Fuller, C. R., Gibbs, G. P., and Silcox, R. J., 1990, "Simultaneous Active Control of Flexural and Extensional Waves in Beams", *Journal of Intelligent Material Systems and Structures,* 1(2).

Gu, Y., 1991, "Active Control of Sound Radiation from Fluid Loaded Plates", PhD Dissertation, Department of Mechanical Engineering, Virginia Tech, Blacksburg, VA.

Lester, H. C., and Lefebvre, S., 1991, "Piezoelectric Actuator Models for Active Sound and Vibration Control of Cylinders", Proceedings, Recent Advances in Active Control of Sound and Vibration, pp. 3-26, April 15-17 1991, Technomic Publishing Co. Inc., Lancaster, PA.

Liang, C., Sun, F. P., and Rogers, C. A., 1993a, "Coupled Electric-Mechanical Analysis of Piezoelectric Ceramic Actuator Driven Systems - Determination of the Actuator Power Consumption and System Energy Transfer", Proceedings, SPIE Conference on Smart Structures and Materials '93, 31 Jan - 4 Feb, Albuquerque, NM, pp. 286--298. Also in *Journal of Intelligent Material Systems and Structures,* Vol. 5, No. 1, pp. 12-20, 1994.

Liang, C., Sun, F. P. and Rogers, C. A., 1993b, "An Impedance Method for Dynamic Analysis of Active Material Systems", Proceedings, 34th SDM Conference, Lajolla, CA, April, 19-21, pp. 3587-3599. Also in ASME *Journal of Vibration and Acoustics,* Vol. 116, No. 1, pp. 120-128, January, 1994.

Liang, C., Sun, F. P., and Rogers, C. A., 1993c, "Dynamic Output Characteristics of PZT Actuators", Proceedings, SPIE Conference on Smart Structures and Materials'93, 31 Jan - 4 Feb, Albuquerque, NM.

Liang, C., Sun, F. P., and Rogers, C. A., 1993d, "Determination of the Optimal Actuator Locations and Configurations Based on Actuator Power Factor," Proceedings, Fourth International Conference on Adaptive Structures, Cologne, Germany, Nov. 2-4, 1993; in press.

Nayak, P. R., 1970, "Line Admittance of Infinite Isotropic Fluid-Loaded Plate", *Journal of Acoustic Society of America,* Vol. 47, pp. 191-201, 1970.

Rogers, C. A., 1990, "Active Vibration and Structural Acoustic Control of Shape Memory Alloy Hybrid Composites : Experimental Results", *Journal of Acoustical Society of America,* **90**(1), 2803-2811.

Saundman, B. E., 1977, "Fluid-loaded Vibration of an Elastic Plate Carrying a Concentrated Mass", *Journal of Acoustical Society of America,* **61**(6), 1503-1510, (1977).

Saunders, W. R., Robertshaw, H. H., and Rogers, C. A., 1990, "Experimental Studies of Structural Acoustic Control for a Shape Memory Alloy Composite Beam", AIAA-90-1090-CP, Proceedings, 31st SDM Conference, Washington D.C., (1990).

Smith, P. W., 1978, "The Imaginary Part of Input Admittance: a Physical Explanation of Fluid-Loading Effects on Plates", *Journal of Sound and Vibration,* Vol. 60, pp. 213-216, 1978.

Sonti, V. R., Jones, J. D., and Herrick, R. W., 1991, "Active Vibration Control of Thin Cylindrical Shells Using Piezo-Electric Actuators", Proceedings, Recent Advances in Active Control of Sound and Vibration, pp. 27-38, April 15-17 1991, Technomic Publishing Co. Inc., Lancaster, PA.

Stein, S., Liang, C. and Rogers, C. A., 1993, "Power Consumption of Piezoelectric Actuators in Underwater Active Structural Acoustic Control", Proceedings, The Second Conferene on Recent Advances in Active Control of Sound and Vibration, pp. 19-203, Technomic Publishing Co. Lancaster, PA.

Sun, F. P., Liang, C., and Rogers, C. A., 1994, "Modal Analysis Using Collocated PZT Actuator/Sensor - an Electro-Mechanical Approach", Proceedings, SPIE 1994 North American Conference on Smart Structures and Materials, Feb. 13-18, 1994; in press.

Wang, B. T., 1991, "Active Control of Sound Transmission/Radiation from Elastic Plates Using Multiple Piezoelectric Actuators", PhD Dissertation, Dept. of Mech. Eng., Virginia Polytechnic Institute and State University, June (1991).

Wang, B. T., Dimitriadis, E. K., and Fuller, C. R., 1990, "Active Control of Structurally Radiated Noise Using Multiple Piezoelectric Actuators", Proceedings, AIAA SDM Conference, AIAA 90-1172-CP, Long Beach, CA (1990).

MODAL ANALYSIS USING PIEZOELECTRIC SENSORS*

J. Callahan[†] and H. Baruh[‡]
Department of Mechanical and Aerospace Engineering
Rutgers University
P.O. Box 909, Piscataway, NJ 08855-0909

Abstract

This paper is concerned with the use of piezoelectric film sensors for the modal analysis of structures. We make use of the relation between the charge generated by the piezoelectric sensor and the modal coordinates, the asymptotic behavior of the eigensolution of structures, and the concept of modal observers to extract the modal coordinates and velocities from the sensor output. We also conduct sensitivity studies in an effort to determine the optimal size and shape of the piezo film sensor, and we investigate the effects of noise, unmodelled dynamics, and parameter errors on the accuracy of the modal coordinate and velocity extraction. We analyze implementation of piezo elements on beams as well as on plates.

1 Nomenclature

$u_0(x,y,t)$:	Midplane displacement in x-direction
$v_0(x,y,t)$:	Midplane displacement in y-direction
$w(x,y,t)$:	Midplane displacement in z-direction
h_k:	Thickness of k_{th} lamina
h:	Thickness of overall laminate
$\bar{z}_k$:	Midplane-to-centroid distance of k_{th} lamina
ρ:	Equivalent density of overall laminate
a:	Length of composite plate (x-direction)
b:	Width of composite plate (y-direction)
$e_{31,k}, e_{32,k}, e_{36,k}$:	Piezoelectric stress/charge material constants of k_{th} lamina
$\epsilon_{33,k}$:	Permittivity of the k_{th} lamina
$E_{33,k}(x,y,t)$:	Electric field applied to k_{th} lamina
$S_k^{(12)}$:	Area of the effective surface electrode of k_{th} lamina
$F_k(x,y)$:	Spatial pattern of effective surface electrode of k_{th} lamina
$P_{0,k}(x,y)$:	Polarization profile of k_{th} lamina
p:	Coverage parameter ($0 \leq p \leq 1$)

*Supported by the Federal Aviation Administration
[†]Graduate Assistant
[‡]Associate Professor, Member AIAA

Copyright © 1994 by Haim Baruh. Published by the American Institute of Aeronautics and Astronautics, Inc. with permission.

2 Introduction

In the past few years, the introduction of piezoelectric film components for motion sensing and control has changed the nature of vibration measurement and analysis tremendously. Hardware issues that have plagued experimentalists in the past, such as cost, size, weight, number and reliability of sensors have almost disappeared with the introduction of piezo film components. The potential that such components have is evident from the interest these materials have generated (e.g., [4, 5, 6, 7, 9, 13]).

Piezosensors in a sense can be treated as piecewise continuous or continuous sensors, with the exception that a single electrical charge comes out of each sensor. For 1-D structures, such as beams, the concept in [9] is to use film elements running throughout the beam which are cut into shapes proportional to the second derivative of the eigenfunctions. By virtue of the orthogonality with respect to the stiffness of the structure, the piezosensor acts as an orthogonal filter, giving a charge signal directly proportional to the amplitude of the associated mode. The advantage of this approach is that it gives the modal coordinate directly, with no computational effort and excellent accuracy [4, 9]. The disadvantage is that the procedure requires one layer of film which has to run throughout the entire length of the beam for each observed mode, making it difficult to extract several modal amplitudes. Also, for 2-D structures, the approach becomes more complicated.

In this paper we analyze the application of piezoelectric film sensors to modal analysis and specifically to the extraction of several modal coordinates and velocities from the system output. In addition to the approach in [9], the process of obtaining modal information from traditional sensors has been in the form of modal filters or full-order or reduced-order observers [1, 11].

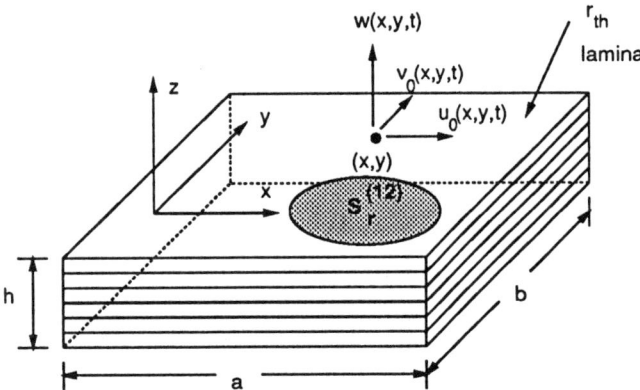

Figure 1: Composite plate with surface electrode.

3 General charge equation

We begin with the general charge-deformation relation formulated by Lee [8]. For a thin laminated rectangular composite plate with in-plane mechanical isotropy, the charge generated by the r_{th} lamina due to plate deformation and applied electric field is

$$q_r(t) = \iint_{S_r^{(12)}} \left[e_{31}\frac{\partial u_0}{\partial x} + e_{32}\frac{\partial v_0}{\partial y} + \epsilon_{33}E_{33} \right.$$
$$\left. + e_{36}\left(\frac{\partial u_0}{\partial y} + \frac{\partial v_0}{\partial x}\right) \right]_r dxdy \quad (1)$$
$$- \bar{z}_r \iint_{S_r^{(12)}} \left(e_{31}\frac{\partial^2 w}{\partial x^2} + e_{32}\frac{\partial^2 w}{\partial y^2} + 2e_{36}\frac{\partial^2 w}{\partial x \partial y} \right)_r dxdy$$

Figure 1 shows the coordinate axes and dimensions of the plate.

4 2-D Sensor Equation

If the in-plane displacements (u_0, v_0) are negligible compared to the transverse deflection (w) and the output signal is generated by deformation only, (1) reduces to the general 2-D sensor equation:

$$q_r(t) = -\bar{z}_r \int_0^a \int_0^b \left\{ F(x,y)P_0(x,y) \left(e_{31}^0 \frac{\partial^2 w}{\partial x^2} \right. \right.$$
$$\left. \left. + e_{32}^0 \frac{\partial^2 w}{\partial y^2} + 2e_{36}^0 \frac{\partial^2 w}{\partial x \partial y} \right) \right\}_r dxdy \quad (2)$$

$$F_r(x,y) = \begin{cases} 1, & \text{if } (x,y) \text{ is within sensor coverage} \\ 0, & \text{otherwise} \end{cases}$$

Here we have modelled the piezoelectric properties as the product of a constant part and a polarization profile to account for spatial variation.

$$\begin{bmatrix} e_{31} \\ e_{32} \\ e_{36} \end{bmatrix} = \begin{bmatrix} e_{31}^0 P_0(x,y) \\ e_{32}^0 P_0(x,y) \\ e_{36}^0 P_0(x,y) \end{bmatrix}$$

5 1-D Case: Composite beam

If we further assume that the midplane displacement has no y-dependence, then we can simplify (2) to the **1-D sensor equation**:

$$q_r(t) = -e_{31,r}^0 \bar{z}_r \int_0^a \mathcal{F}_r(x) \frac{\partial^2 w(x)}{\partial x^2} dx \quad (3)$$

where,

$$\mathcal{F}_r(x) = \int_{-\frac{b}{2}}^{\frac{b}{2}} F_r(x,y) P_{0,r}(x,y) dy$$

From linear beam theory, we can express the transverse displacement of the beam as a linear combination of the natural modes.

$$w(x,t) = \sum_{s=1}^{\infty} w_s(x) \eta_s(t) \quad (4)$$

After substitution of (4) into (3), we can express the sensor equation in a convenient matrix form.

$$\{q(t)\} = [B]\{\eta(t)\}$$

where,

$$B_{rs} = -e_{31,r}^0 \bar{z}_r \int_0^a \mathcal{F}_r(x) \frac{d^2 w_s(x)}{dx^2} dx \quad (5)$$

The entries of $[B]$ need to be calculated only once, before the response of the structure is measured. Note that for an exact modal expansion, the matrix $[B]$ has dimension $N \times \infty$, with N as the number of lamina. For practical modelling purposes, we used the first $n = 10$ modes for our beam simulation. The next step is to choose the variation of the spatial electrode pattern $\mathcal{F}_r(x)$ for each layer.

5.1 $\mathcal{F}_r(x)$ proportional to modal strain

Lee and Moon [9] have shown that if the spatial electrode pattern of the r_{th} lamina is of the form:

$$\mathcal{F}_r(x) = \mu_r \frac{d^2 w_r(x)}{dx^2} \quad (6)$$

we obtain a true modal sensor based on the orthogonality of the eigensolution with respect to the stiffness.

$$B_{rs} = -\frac{\mu_r e_{31,r}^0 \bar{z}_r (\beta_r a)^4}{\rho h a^4} \delta_{rs} \quad (7)$$

where $\beta_r a$ is the r_{th} eigenvalue. Note that the lamina must span the entire length of the beam (integral is from

0 to a). This can be done for beams, since the laminae can be put side by side along the beam width or a few of them can be stacked, but we also need to cut the sensors into specific patterns due to the requirement of $\mathcal{F}_r(x)$. However, this approach may not be feasible for large bodies or complicated structures and is the main reason for investigating the following approach.

5.2 $\mathcal{F}_r(x)$ constant

Rather than etch the lamina into complicated full-length patterns, we place segmented strips at various locations on the beam. The spatial pattern of the r_{th} strip has the following form:

$$\mathcal{F}_r(x) = \mu_r[u(x - x_r) - u(x - x_r - a_r)] \quad (8)$$

where x_r is the left-hand x-coordinate of the r_{th} strip and a_r is its length. Putting (8) into (5), we obtain

$$B_{rs} = -e^0_{31,r}\overline{z}_r\mu_r\left[\left.\frac{dw_s(x)}{dx}\right|_{x_r+a_r} - \left.\frac{dw_s(x)}{dx}\right|_{x_r}\right] \quad (9)$$

We now have a fully populated $[B]$ matrix.

We chose a simply supported beam for our test case, with the following eigensolution:

$$w_s(x) = \sqrt{\frac{2}{\rho h a}}\sin\left(\frac{s\pi x}{a}\right), \quad (s = 1, 2, \ldots)$$

$$\omega_s = (s\pi)^2\sqrt{\frac{D_{11}}{\rho h a^4}}$$

For mathematical simplicity, set

$$\mu_r = \frac{1}{e^0_{31,r}\overline{z}_r}\sqrt{\frac{\rho h a}{2}}\left(\frac{a}{\pi}\right)$$

Thus, for our model,

$$B_{rs} = s\left[\cos\left(\frac{s\pi x_r}{a}\right) - \cos\left(\frac{s\pi(x_r + a_r)}{a}\right)\right] \quad (10)$$

6   2-D Case: Composite plate

We consider here plates subjected to boundary conditions which permit orthogonalization. A simply supported plate is one such case. We can decompose the linear vibration of plates into their constituent modes.

$$w(x, y, t) = \sum_{l=1}^{\infty}\sum_{m=1}^{\infty} w_{lm}(x, y)\eta_{lm}(t) \quad (11)$$

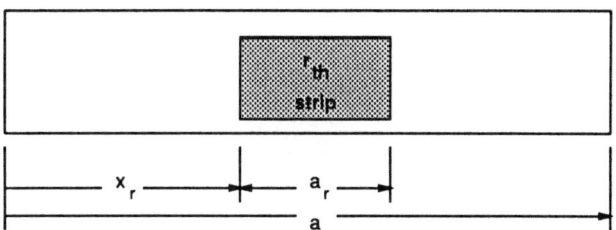

Figure 2: Piezoelectric strip dimensions for beam.

Substituting (11) into (2) we obtain

$$q_r(t) = -\overline{z}_r\sum_{l=1}^{\infty}\sum_{m=1}^{\infty}\left\{\int_0^a\int_0^b (FP_0)_r\left(e^0_{31,r}\frac{\partial^2 w_{lm}}{\partial x^2}\right.\right.$$
$$\left.\left.+ e^0_{32,r}\frac{\partial^2 w_{lm}}{\partial y^2} + 2e^0_{36,r}\frac{\partial^2 w_{lm}}{\partial x\partial y}\right)dydx\right\}\eta_{lm}(t) \quad (12)$$

At this point in the 1-D case, we could express the sensor equation in a convenient matrix form. We cannot, however, express the double summation in (12) as the product of two matrices. Rather, arrange the modal coordinates in a column vector according to

$$\begin{bmatrix} \eta_1 & \eta_2 & \eta_5 & \cdots \\ \eta_3 & \eta_4 & \eta_7 & \cdots \\ \eta_6 & \eta_8 & \eta_9 & \cdots \\ \vdots & \vdots & \vdots & \ddots \end{bmatrix} = \begin{bmatrix} \eta_{11} & \eta_{12} & \eta_{13} & \cdots \\ \eta_{21} & \eta_{22} & \eta_{23} & \cdots \\ \eta_{31} & \eta_{32} & \eta_{33} & \cdots \\ \vdots & \vdots & \vdots & \ddots \end{bmatrix}$$

The logic behind this ordering sequence is to read the upper left square of the $[\eta_{lm}(t)]$ matrix at all times. For example, if we have only four strips, we want to sense the first four modal coordinates, η_{11}, η_{12}, η_{21}, and η_{22}. Note that this index assignment applies to $w_s(x, y)$ as well. Thus, equation (12) becomes:

$$q_r(t) = -\overline{z}_r\sum_{s=1}^{\infty}\left\{\int_0^a\int_0^b FP_{0,r}\left(e^0_{31,r}\frac{\partial^2 w_s}{\partial x^2} + e^0_{32,r}\frac{\partial^2 w_s}{\partial y^2}\right.\right.$$
$$\left.\left.+ 2e^0_{36,r}\frac{\partial^2 w_s}{\partial x\partial y}\right)dydx\right\}\eta_s(t) \quad (13)$$

Or in matrix notation,

$$\{q(t)\} = [B]\{\eta(t)\}$$

where,

$$B_{rs} = -\overline{z}_r\int_0^a\int_0^b FP_{0,r}\left(e^0_{31,r}\frac{\partial^2 w_s}{\partial x^2} + e^0_{32,r}\frac{\partial^2 w_s}{\partial y^2}\right.$$
$$\left.+ 2e^0_{36,r}\frac{\partial^2 w_s}{\partial x\partial y}\right)dydx \quad (14)$$

We examine a simply supported plate, whose normalized modes and frequencies are given by:

$$w_s(x,y) = \frac{2}{\sqrt{\rho h a b}} \sin\frac{l_s \pi x}{a} \sin\frac{m_s \pi y}{b}, \quad (s=1,2,\ldots)$$

$$\omega_s = \pi^2 \left[\left(\frac{l_s}{a}\right)^2 + \left(\frac{m_s}{b}\right)^2\right]\sqrt{\frac{D_{11}}{\rho h}}$$

The indices are given the subscript s to denote that their values now depend on their location in the column vector. Thus, for the simply supported plate,

$$B_{rs} = \frac{2\bar{z}_r}{\sqrt{\rho h a b}} \int_0^a \int_0^b FP_{0,r} \left\{ \left[e^0_{31,r}\left(\frac{l_s\pi}{a}\right)^2 + e^0_{32,r}\left(\frac{m_s\pi}{b}\right)^2\right] \right.$$
$$\times \sin\frac{l_s\pi x}{a}\sin\frac{m_s\pi y}{b} \quad (15)$$
$$\left. - 2e^0_{36,r}\left(\frac{l_s\pi}{a}\right)\left(\frac{m_s\pi}{b}\right)\cos\frac{l_s\pi x}{a}\cos\frac{m_s\pi y}{b}\right\}dy\,dx$$

6.1 $(FP_0)_r$ proportional to mode shapes

One possible choice for the spatial electrode pattern is [9]:

$$(FP_0)_r(x,y) = \mu_r\left[\sin\left(\frac{i_r\pi x}{a}\right)\sin\left(\frac{j_r\pi y}{b}\right)\right] \quad (16)$$

where the indices ij correspond to r just as lm to s. Putting (16) into (15) yields

$$B_{rs} = \frac{2\pi^2 \mu_r \bar{z}_r}{\sqrt{\rho h a b}}\left(e^0_{31,r} i_r^2 \frac{b}{a} + e^0_{32,r} j_r^2 \frac{a}{b}\right)\delta_{rs} \quad (17)$$

which is indeed a modal sensor based on linear plate theory. For this approach, not only do the lamina have to be cut, but to sense more than one mode, they must be stacked. This increases the overall thickness of the plate.

6.2 $(FP_0)_r$ constant

Analogous to the 1-D case, try rectangular segmented strips of the form

$$(FP_0)_r = \mu_r[u(x-x_r) - u(x-x_r-a_r)]$$
$$\times [u(y-y_r) - u(y-y_r-b_r)] \quad (18)$$

In the y-direction, r_{th} the strip begins at y_r and has length b_r, with x_r and a_r defined earlier. Upon substitution of (18) into (15), we obtain

$$B_{rs} = \frac{2\mu_r\bar{z}_r}{\sqrt{\rho h a b}}\left\{\left[e^0_{31,r}\left(\frac{l_s b}{m_s a}\right) + e^0_{32,r}\left(\frac{m_s a}{l_s b}\right)\right]\right.$$
$$\times \cos\frac{l_s\pi x}{a}\Big|_{x_r}^{x_r+a_r}\cos\frac{m_s\pi y}{b}\Big|_{y_r}^{y_r+b_r}$$
$$\left. - 2e^0_{36,r}\sin\frac{l_s\pi x}{a}\Big|_{x_r}^{x_r+a_r}\sin\frac{m_s\pi y}{b}\Big|_{y_r}^{y_r+a_r}\right\} \quad (19)$$

As for the beam, $[B]$ is fully populated. For our plate simulation, we used the first $10\times 10 = 100$ natural modes. For bending sensors, it is common to set $e_{36} = 0$, since for commercially available films, it is introduced through a coordinate transformation [8] of the plate axes. Since we want to sense only bending modes, we will do the same. To facilitate the analysis, we set $e^0_{31,r} = e^0_{32,r}$ and the scaling constants as

$$\mu_r = \frac{\sqrt{\rho h a b}}{2e^0_{31,r}\bar{z}_r}$$

This leads to

$$B_{rs} = \left[\frac{l_s b}{m_s a} + \frac{m_s a}{l_s b}\right]\cos\frac{l_s\pi x}{a}\Big|_{x_r}^{x_r+a_r}\cos\frac{m_s\pi y}{b}\Big|_{y_r}^{y_r+b_r} \quad (20)$$

7 Higher Mode Effects

We have formulated the previous plate analysis so that we can apply the same techniques to *both* the beam and plate. Now partition $[B]$ into N (N^2 for the plate) modelled and $n-N$ (n^2-N^2) residual portions

$$\{q(t)\} = [\,B_m \quad B_R\,]\left\{\begin{array}{c}\eta_m(t)\\ \eta_R(t)\end{array}\right\} \quad (21)$$

Since there are only N (N^2) strips, we can sense only N (N^2) modal coordinates at one time. We accomplish this by inverting $[B_m]$ and using

$$\{\hat{\eta}(t)\} = [B_m]^{-1}\{q(t)\} \quad (22)$$

Substituting (22) into (21) yields

$$\{\hat{\eta}(t)\} = \{\eta_m(t)\} + [B_m]^{-1}[B_R]\{\eta_R(t)\} \quad (23)$$

The measured modal coordinate $\hat{\eta}_r(t)$ is the true coordinate plus residual mode contamination. In the example below, we used equal length sensors and defined a length parameter p, where $0 \leq p \leq 1$. The segmented strips then have the form

$$x_r = \left[r - \frac{1}{2}(1+p)\right]\frac{a}{N}, \qquad a_r = p\frac{a}{N}$$

For our $n=10$ mode simulation, we tried to sense the first $N=3$ coordinates. With continuous coverage,

$$[B_m]^{-1}[B_R] = $$

$$\begin{bmatrix} 0 & 5 & 0 & 7 & 0 & 0 & 0 \\ 2 & 0 & 0 & 0 & 4 & 0 & 5 \\ 0 & 0 & 0 & 0 & 0 & 3 & 0 \end{bmatrix}, \quad p=1\ [\mathcal{O}(s)] \quad (24)$$

As coverage vanishes,

$$[B_m]^{-1}[B_R] = \begin{bmatrix} 0 & 25 & 0 & -49 & 0 & 0 & 0 \\ 4 & 0 & 0 & 0 & -16 & 0 & -25 \\ 0 & 0 & 0 & 0 & 0 & -9 & 0 \end{bmatrix}, \quad \begin{matrix} p \to 0 \\ [\mathcal{O}(s^2)] \end{matrix}$$

For true point displacement sensors located where coverage goes to zero, we have

$$[B_m]^{-1}[B_R] = \begin{bmatrix} 0 & 1 & 0 & -1 & 0 & 0 & 0 \\ 1 & 0 & 0 & 0 & -1 & 0 & -1 \\ 0 & 0 & 0 & 0 & 0 & -1 & 0 \end{bmatrix}, [\mathcal{O}(1)]$$

So, as $p \to 0$, the strips act as point sensors, with residual mode effects proportional to s^2 (as opposed to s for $p = 1$). From this observation, for equal distribution among the strips, we see that $p = 1$ leads to the least amount of contamination by higher modes.

At this point, we would like to compare the order of magnitude of the contamination caused by residual dynamics. It is shown in [12] that the asymptotic behavior of a one-dimensional structure is governed by

$$w_s \sim c_s s^h$$
$$w_s(x) \sim d_s \sin(s\pi x) + e_s \cos(s\pi x), \quad (s = 1, 2, \ldots)$$

where $2h$ is the highest order spatial derivative in the equation of motion. For example, for a beam $2h = 4$ and for a string $2h = 2$. Here, c_s, d_s, e_s are coefficients.

The above results for the simply supported beam apply exaclty the same way for any type of beam, regardless of boundary conditions and mass and stiffness distributions. We can now classify the different types of sensors and examine the orders of $[B_m]^{-1}[B_R]$. Note that for this analysis to be valid for the general case, we will assume that the sensors are relatively evenly distributed and that elements $[B_m]$ and $[B_m]^{-1}$ are of order $\mathcal{O}(1)$. We then analyze the entries of $[B_R]$.

Point displacement sensors (displacement or velocity):	$\to \mathcal{O}(1)$
Point slope sensors (slope or velocity of slope):	$\to \mathcal{O}(s)$
Strain gages:	$\to \mathcal{O}(s^2)$
Full coverage piezoelectric lamina:	$\to \mathcal{O}(s)$
Small coverage piezoelectric lamina:	$\to \mathcal{O}(s^2)$

For strain gages, ϵ is proportional to the second derivative of deformation for beams. Actually, we see that piezosensors with extremely small surface area perform just like strain gages. This is to be expected, as equation (1) indicates that the charge is proportional to the integral of the strain. These orders have to be considered together with the amplitudes of the modal coordinates.

It follows from the above argument that to minimize residual dynamic effects, one is best served by using point displacement sensors. However, this is not possible under many realistic conditions because the displacement sensors need to be attached to a fixed reference point. As we shall show in the numerical examples, piezosensors with a relatively significant amount of coverage do a good job, especially after their signal is processed by a modal observer. Figure 3 shows the measurements obtained from a three strip (full coverage) beam along with the true modal coordinates for the following loading and initial conditions:

$$q(x,t) = 20\sin(3\pi^2 t)\frac{x}{a}\left[1 - \left(\frac{x}{a}\right)^2\right]$$

$$w_0(x) = \frac{x}{a}\left[1 - \frac{x}{a}\right], \qquad \dot{w}_0(x) = \frac{x}{a}$$

Similar to the beam, we examined rectangular strips for the plate model, except we now measure N^2 modes for symmetry. Using the same coverage parameter p, we then have

$$x_r = \left[l_r - \frac{1}{2}(1+p)\right]\frac{a}{N}, \qquad a_r = p\frac{a}{N}$$
$$y_r = \left[m_r - \frac{1}{2}(1+p)\right]\frac{b}{N}, \qquad b_r = p\frac{b}{N}$$

For our 2-D model, we used $N^2 = 3 \times 3 = 9$ sensors with a 100 mode simulation. For this case, $[B_m]^{-1}[B_R]$ is a 9×91 matrix, so we will not include it here, but the same principles apply, as the asymptotic behavior of the eigensolution of a plate is similar to that of a beam.

8 Modal Observers

We have extracted a modal coordinate measurement, but we may still need a modal velocity. One can take the time derivative of (23) to obtain modal velocity estimates, but using modal observers has proven to be more reliable [3]. For a linear system with time invariant coefficients of the form:

$$\begin{aligned} \dot{x}(t) &= Ax(t) + Bu(t) \\ y(t) &= Cx(t) \end{aligned} \qquad (25)$$

we can construct a full-order state estimator (observer) of the form [2]

$$\dot{z}(t) = Az(t) + Bu(t) + K[y(t) - Cz(t)] \qquad (26)$$

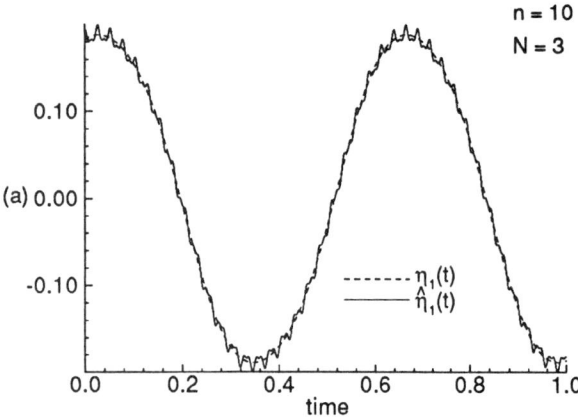

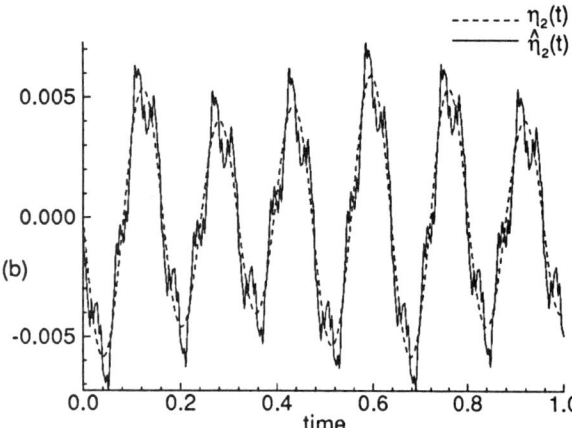

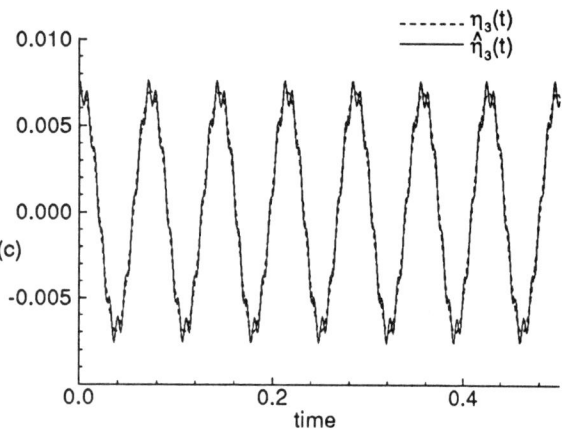

Figure 3: Measurements and coordinates for (a) $\eta_1(t)$, (b) $\eta_2(t)$, and (c) $\eta_3(t)$.

where, $z(t)$ are the estimates of $x(t)$ and K is a gain matrix to be determined such that the observer will converge. We could have constructed a reduced order observer, since one of the state variables is the output itself, but we shall see the benefit of reconstructing the modal coordinate $\eta_r(t)$ as well as $\dot{\eta}_r(t)$. Since the equations for each mode are uncoupled, we can design observers for each mode independently. These uncoupled (undamped) equations in the form of (25) are:

$$\dot{x}_r(t) = \begin{bmatrix} 0 & 1 \\ -\omega_r^2 & 0 \end{bmatrix} x_r(t) + \begin{bmatrix} 0 \\ 1 \end{bmatrix} u_r(t)$$
$$y_r(t) = \begin{bmatrix} 1 & 0 \end{bmatrix} x_r(t)$$

where,

$$x_r(t) = \begin{bmatrix} \eta_r(t) \\ \dot{\eta}_r(t) \end{bmatrix}, \qquad u_r(t) = F_r(t), \qquad y_r(t) = \eta_r(t)$$

Each modal observer will then have the form

$$\dot{z}_r(t) = \begin{bmatrix} -k_{r,1} & 1 \\ -\omega_r^2 - k_{r,2} & 0 \end{bmatrix} z_r(t) + \begin{bmatrix} k_{r,1} \\ k_{r,2} \end{bmatrix} \eta_r(t) + \begin{bmatrix} 0 \\ 1 \end{bmatrix} F_r(t)$$

Closed form expressions can be found for the K matrix:

$$\begin{bmatrix} k_{r,1} \\ k_{r,2} \end{bmatrix} = \begin{bmatrix} -(\lambda_1 + \lambda_2) \\ \lambda_1 \lambda_2 - \omega_r^2 \end{bmatrix} \qquad (27)$$

If entries of K are chosen such that eigenvalues λ_1, λ_2 of $[A - KC]$ have negative real parts, then

$$z(t) - x(t) \to 0 \quad \text{as} \quad t \to \infty$$

This is true for $y_r(t) = \eta_r(t)$, but what happens when we use $\hat{\eta}_r(t)$ as the output? We next analyze the effects that need to be considered in an actual real time observation. All of these effects couple the modal observer equations with each other and should result in deterioration of the performance of the modal observers.

9 Noise Effects

In an attempt to model noise, assume that the charge is affected by some small quantity

$$\hat{q}_r(t) = q_r(t) + \epsilon_r \qquad (28)$$
$$\epsilon_r = \delta_r q_r(t)$$

in which δ_r is a uniform random number between $\pm \delta$. Using the previous approach, we obtain for the measurements

$$\{\hat{\eta}(t)\} = \{\eta_m(t)\} + [B_m]^{-1}[\delta][B_m]\{\eta_m(t)\}$$
$$+ [B_m]^{-1}[I + \delta][B_R]\{\eta_R(t)\} \qquad (29)$$

$[\delta]$ is a diagonal matrix with random elements between $\pm \delta$. We see that noise modelled in this fashion further corrupts the modal coordinate measurements by coupling the observed modes.

10 Damping effects

In all real physical systems, there is some damping present. Not including a modal damping term or including an incorrect damping term in the observer equations makes the model unrealistic, but the modal observers inherently account for this by choice of the observer gains. In the presence of damping, the modal equations become

$$\dot{x}_r(t) = \begin{bmatrix} 0 & 1 \\ -\omega_r^2 & -2\zeta_r\omega_r \end{bmatrix} x_r(t) + \begin{bmatrix} 0 \\ 1 \end{bmatrix} u_r(t)$$
$$y_r(t) = \begin{bmatrix} 1 & 0 \end{bmatrix} x_r(t)$$

For a desired set of eigenvalues λ_1, λ_2, the gain matrix becomes

$$\begin{bmatrix} k_{r,1} \\ k_{r,2} \end{bmatrix} = \begin{bmatrix} -(\lambda_1 + \lambda_2 + 2\zeta_r\omega_r) \\ \lambda_1\lambda_2 - \omega_r^2 + 2\zeta_r\omega_r(\lambda_1 + \lambda_2 + 2\zeta_r\omega_r) \end{bmatrix}$$

In comparison to (27), by including damping in the observer model, we require less gain for the same pole placement.

11 Discretization Effects

For very simple plate and beam geometries, exact solutions of the eigenvalue problem exist. In general, however, complicated geometries and boundary conditions do not permit such results. To analyze the effects of spatial discretization, we use the *assumed modes method* [10]. For this approach we choose a finite series of n twice-differentiable linearly independent *admissible* functions $\psi_j(x)$ which satisfy the geometric boundary conditions without overconstraining the problem. The mode shape approximations $\phi_i(x)$ are then related to these functions by

$$\phi_i(x) = \sum_{j=1}^{n} U_{ji} \psi_j(x) \tag{30}$$

where the columns of the matrix U_{ij} are the normalized eigenvectors of the discrete eigenvalue problem resulting from the application of the assumed modes method. For the beam case, we used the following set of linearly independent polynomials:

$$\psi_j\left(\frac{x}{a}\right) = \frac{x}{a} \prod_{i=1}^{j} \left(1 - \frac{j}{i}\frac{x}{a}\right) \tag{31}$$

The natural frequencies of the discretized system (for order of discretization $n = 5$) are

$$\{^5\omega\} = \{9.870, 39.65, 90.34, 276.0, 508.4\} \sqrt{\frac{D_{11}}{\rho h a^4}}$$

From the exact solution to the eigenvalue problem,

$$\{\omega\} = \{9.870, 39.48, 88.83\} \sqrt{\frac{D_{11}}{\rho h a^4}}$$

we see that five comparison functions give good approximations to the first three.

Discretization can also be done for the simply supported plate model, except some care must be used when integrating the biharmonic stiffness operator by parts. Our set of 2-D trial functions has a similar form:

$$\psi_{rs}(x,y) = \frac{x}{a}\frac{y}{b} \prod_{i=1}^{r}\left(1 - \frac{r}{i}\frac{x}{a}\right) \prod_{j=1}^{s}\left(1 - \frac{s}{j}\frac{y}{b}\right) \tag{32}$$

One must use caution when numerically computing the discrete eigenvectors. Depending on method of solution, they may not be properly normalized and the order of magnitude of the natural frequencies may not coincide with the index arrangement defined earlier.

Continuing with the beam, now denote $[\hat{B}]$ as the charge/modal coordinate matrix using the trial functions. Thus, (23) becomes

$$\{\hat{\eta}(t)\} = [\hat{B}_m]^{-1}[I + \delta][B_m]\{\eta_m(t)\}$$
$$+ [\hat{B}_m]^{-1}[I + \delta][B_R]\{\eta_R(t)\} \tag{33}$$

with,

$$\hat{B}_{rs} = -e_{31,r}^0 \bar{z}_r \mu_r \left[\left.\frac{d\phi_s(x)}{dx}\right|_{x_r+a_r} - \left.\frac{d\phi_s(x)}{dx}\right|_{x_r} \right]$$

Using the set of trial functions (31), we obtain for the three sensor beam case

$$[\hat{B}_m]^{-1}[B_m] = \begin{bmatrix} 1 & 0 & -0.214 \\ 0 & 0.981 & 0 \\ 0 & 0 & 1.024 \end{bmatrix} \tag{34}$$

$$[\hat{B}_m]^{-1}[B_R] =$$

$$\begin{bmatrix} 0 & 5 & 0 & 7 & 0 & -0.641 & 0 \\ 1.963 & 0 & 0 & 0 & 3.926 & 0 & 4.907 \\ 0 & 0 & 0 & 0 & 0 & 3.073 & 0 \end{bmatrix} \tag{35}$$

$[\hat{B}_m]^{-1}[B_m]$ now has off-diagonal entries. Comparing the higher mode contribution to (24), we see that it is slightly different, due to the approximate mode shapes.

For the plate discretization, the computed eigenvectors must be arranged according to the previous index assignment before computing $[\hat{B}_m]$.

12 Observer Performance

12.1 Beam results

We will first show the reconstructed modal velocities and coordinates for the beam, using a three strip model with full coverage. We used the approximate mode shapes derived earlier and included measurement noise with $\delta = 0.05$. Thus, the natural frequencies that we use in the observers will be those computed from the discretization analysis. We included system damping $\zeta_r = 0.01$ in the solution but modelled no observer damping to illustrate that it is not necessary. After experimenting with several different observer pole configurations, we found that complex conjugate poles of the form

$$\left.\begin{array}{c} \lambda_1 \\ \lambda_2 \end{array}\right\}_r = Re \pm j\omega_r \qquad (36)$$

gave the best results. This is not as evident in the beam analysis as it was in the observer plate response. The imaginary part of the pair gives the oscillating frequency of the observer response, so we set it equal to the corresponding natural frequency. This is why we need accurate estimates for ω_r. Figure 4 gives the true modal velocities $\dot{\eta}_r(t), (r = 1, 2, 3)$ plotted with the observer results for $Re = -5$ for all modal observers.

We next checked the observer quantities corresponding to the modal displacements which are given in Figure 5. Comparing Figure 5 with Figure 3 (note that Figure 3 is for an undamped model), the observer quantity corresponding to modal displacement was better than the modal measurement $\hat{\eta}_r(t)$ itself. This is the main reason we prefer to use a full-order observer rather than a reduced-order observer. Further, we used the same beam loading here as in Figure 3

$$q(x, t) = 20\sin(3\pi^2 t)\frac{x}{a}\left[1 - \left(\frac{x}{a}\right)^2\right]$$

We also ran cases where beam loading was present but not modelled in the observer. The results indicated that the observers work very well even in the presence of known as well as unknown external excitation. Also note that for the third mode, there is a well-defined lag between the true modal quantities and their observer counterparts. The modal observers are constantly trying to keep up with the system motion.

We next reduce the lengths of the strips by decreasing the coverage parameter p. The results for the second mode are shown in Figure 6. Note that two distinct real poles were used for the modal observers; we found that this best illustrates the effect of decreasing coverage. As can be expected, the accuracy of the modal velocity extraction deteriorates as the coverage becomes

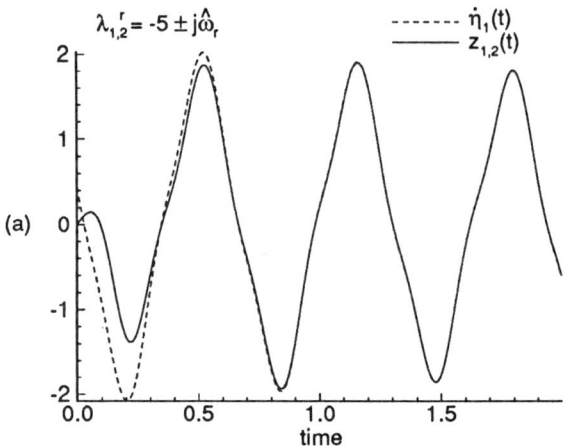
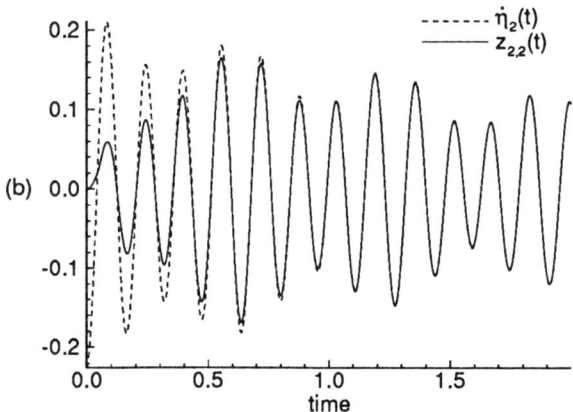
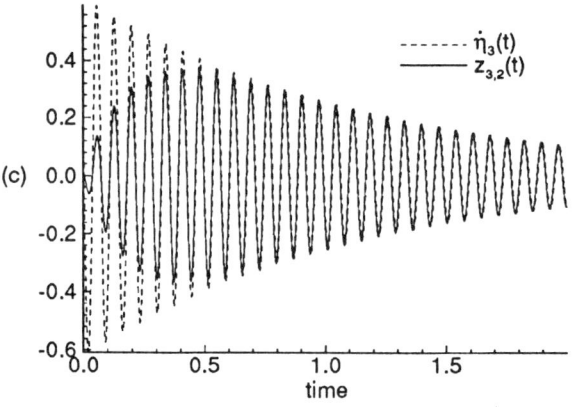

Figure 4: Estimated modal velocities for (a) $\dot{\eta}_1(t)$, (b) $\dot{\eta}_2(t)$, and (c) $\dot{\eta}_3(t)$.

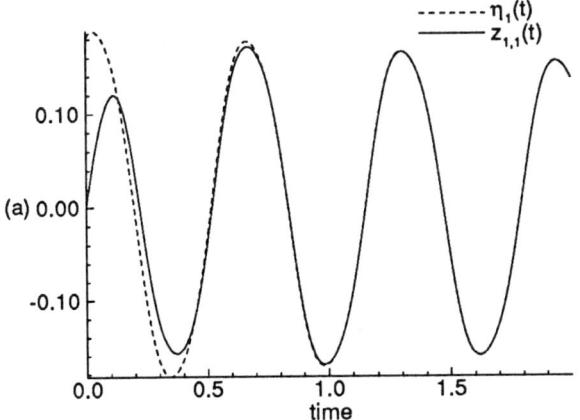

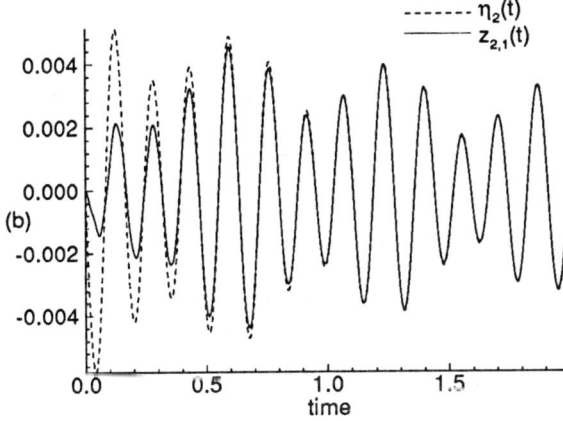

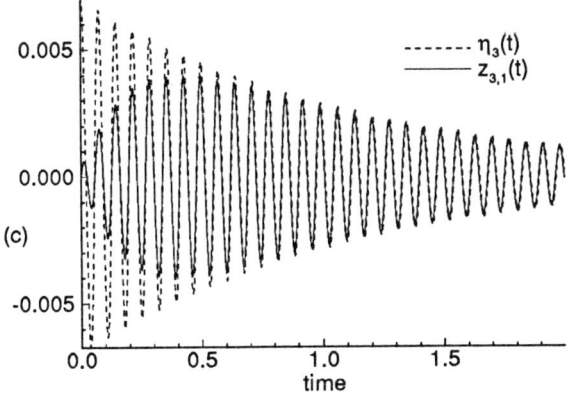

Figure 5: Reconstructed modal coordinates for (a) $\eta_1(t)$, (b) $\eta_2(t)$, and (c) $\eta_3(t)$.

smaller. This deterioration, however, is not very large. Higher mode effects manifest themselves as increasing oscillations within a given peak. The observer output gets more choppy as the strip coverage is decreased, but still within an acceptable range of accuracy. A similar pattern was observed for the modal displacements. It should be noted that all of the simulations described above were also conducted for cases where each piezosensor was of different length or the distribution of the sensors on the beam was not even. There was no discernable difference in the results.

We next introduce parameter uncertainties to the modal coordinate extraction process. These deteriorate the extraction results by making the $[\hat{B}_m]$ matrix even more different than $[B_m]$. As an illustration, consider a point mass added to the beam, modelled as:

$$M = 0.1\rho ha \quad \text{at} \quad \frac{x}{a} = 0.25 \qquad (37)$$

This gives the first three natural frequencies

$$\{\omega\} = \{9.405, 36.45, 87.56\}\sqrt{\frac{D_{11}}{\rho ha}}$$

We treat the point mass as a mass distribution error by including it in the discretization calculation of $[\hat{B}_m]$, but not in the true system dynamics. For $\delta = 0$,

$$[\hat{B}_m]^{-1}[B_m] = \begin{bmatrix} 1.049 & 0.152 & -0.095 \\ -0.009 & 1.069 & 0.151 \\ -0.001 & -0.023 & 1.059 \end{bmatrix}$$

$$[\hat{B}_m]^{-1}[B_R] =$$

$$\begin{bmatrix} 0.303 & 5.246 & 0 & 7.345 & 0.607 & -0.284 & 0.758 \\ 2.138 & -0.046 & 0 & -0.065 & 4.277 & 0.453 & 5.346 \\ -0.046 & 0.004 & 0 & -0.006 & -0.091 & 3.176 & -0.114 \end{bmatrix}$$

Even in the absence of measurement noise, we now have full higher mode contamination and observer coupling. The observer response for the second modal velocity with $\delta = 0.05$ and a concentrated mass prescribed by (37) is shown in Figure 7. Damping effects were omitted for clarity. Although the amplitude is smaller than the true value, the modal observer gives accurate frequency response with a constant phase lag. The modal observers converge to the proper frequency, even with a significant error in the mass distribution.

For a true modal sensor [9], it is clear that $\hat{\eta}_r(t) = \eta_r(t)$. For our case, this was not true, but $z_r(t) \to \begin{bmatrix} \eta_r(t) & \dot{\eta}_r(t) \end{bmatrix}^T$ for no noise, and came very close with a conservative noise model and approximate mode shapes. The modal observer acts to filter out residual dynamics and smooth out noise contamination, as well as give a velocity estimate.

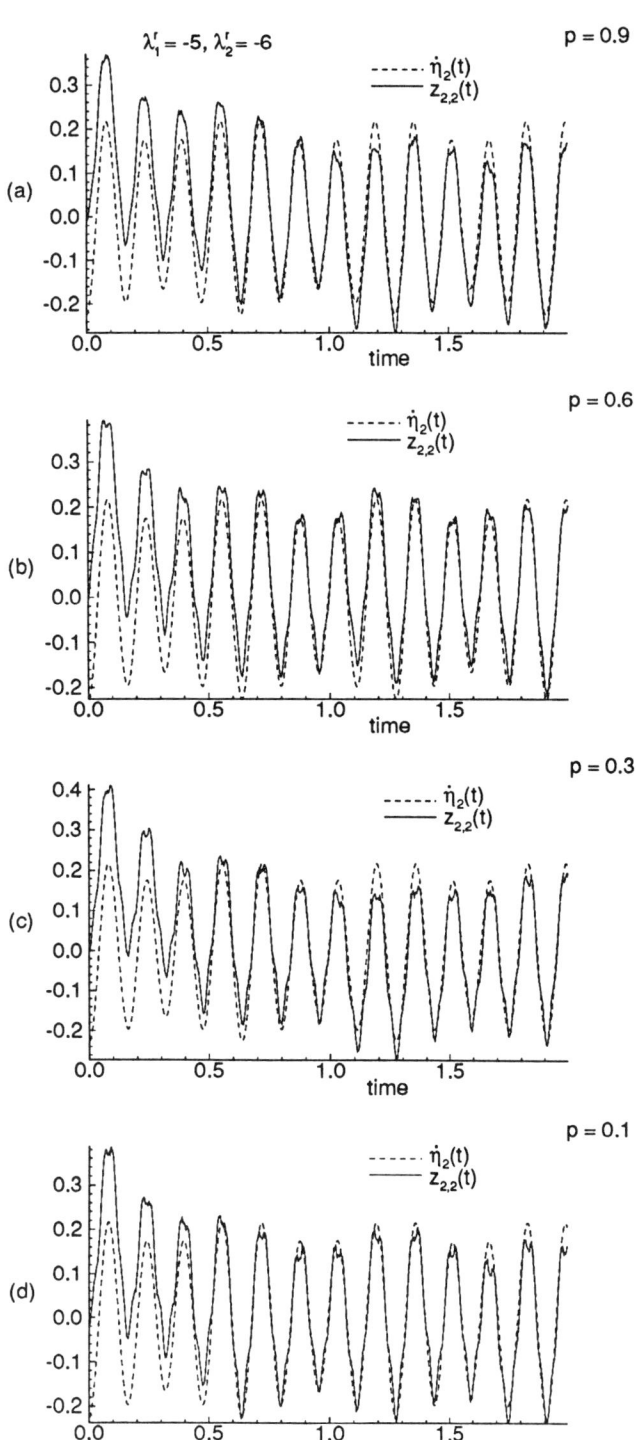

Figure 6: Modal observer for $\dot{\eta}_2(t)$ with (a) $p = 0.9$, (b) $p = 0.6$, (c) $p = 0.3$, and (d) $p = 0.1$.

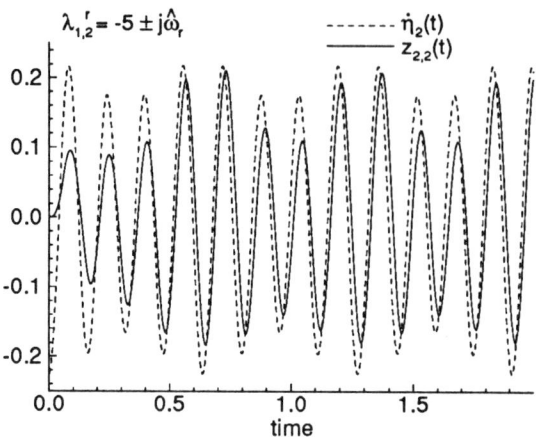

Figure 7: Modal observer response for $\dot{\eta}_2(t)$ with a point mass error.

12.2 Plate results

We simulated a rectangular plate with $b = 1.5a$ and full 3×3 sensor coverage subject to an impulse loading:

$$\hat{F}_0 = 10, \text{ at } \frac{x}{a} = 0.4, \frac{y}{b} = 0.7$$

Measurement noise with $\delta = 0.05$ was also included; approximate mode shapes and observer frequencies were found using a 5×5 discretization with the 2-D trial functions given by (32). Damping was excluded to help show the discretization effects. Figures 8 and 9 show modal observer results with $Re = -8$ for several velocities and coordinates, respectively.

In Figure 9(a), a slight discretization error in ω_{11} caused the observer response to occasionally deviate from the true coordinate $\eta_{11}(t)$. When we select the imaginary part of the observer poles significantly different from the modal frequencies, we have observed a beating phenomenon in the response, due to interference between the two frequencies. As long as the imaginary part of the observer poles is set very close to the frequency used by the modal observers, we have found that convergence was satisfactory.

For the higher modes, the amplitudes of both the reconstructed coordinates and velocities were smaller than the true modal values. The frequencies for these modes were simply too fast for the observers to keep up. This can be remedied by making the real part of the modal observer poles more negative, speeding up the convergence rate but also increasing sensitivity to sources of error. All modal observers showed excellent agreement with the true vibration frequencies.

From previous discussion, we have selected $e_{36} = 0$ for all laminae in these plots. We initially included a nonzero

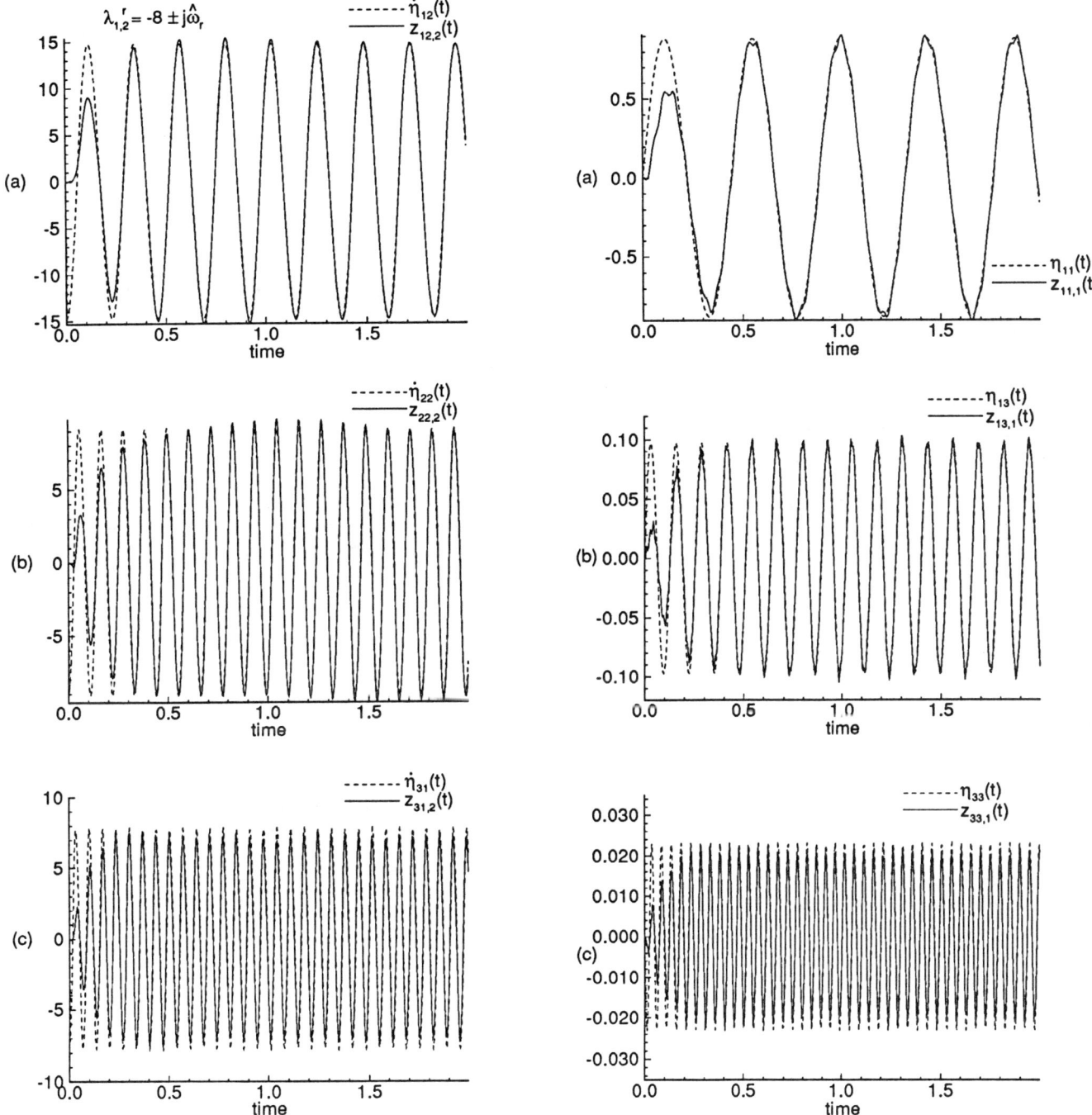

Figure 8: Reconstructed modal velocities for (a) $\dot{\eta}_{12}(t)$, (b) $\dot{\eta}_{22}(t)$, and (c) $\dot{\eta}_{31}(t)$.

Figure 9: Reconstructed modal coordinates for (a) $\eta_{11}(t)$, (b) $\eta_{13}(t)$, and (c) $\eta_{33}(t)$.

value in the simulation, but we found that it caused substantial steady-state error in the higher mode observers. This material constant is generally introduced for torsion sensors [8], which may confuse the modal observer in the filtering process as to which bending frequency to converge upon. Regardless, since it may be set to zero in the manufacturing process, we have done so in the model.

13 Conclusions

We have found that convergence of the modal observer is directly dependent on reliable estimates of the natural frequencies ω_r. In our model, measurement noise and residual modes prevent absolute convergence of the reconstructed modal quantities, but the asymptotic stability of the observers keep them close to the true values.

We have presented a method to sense structural vibration by combining modern piezoelectric film technology and modal analysis. When the geometry does not permit accurate knowledge of the vibration modes, we may not be able to cut the lamina into their required shapes. We can, however, cut the lamina into several strips without stacking them. This approach does not need any spatial variation in the polarization profile of the lamina. Feeding the measurements through modal observers gives very good results, even with noise, discretization, and modelling errors present.

The results obtained in the one- and two- dimensional analyses indicate that piezo elements are very promising for the measurement of the vibration of elastic bodies. They are lighter in weight, less expensive, and more versatile. Their location on the structure is not a critical issue, as long as there is some evenness in placing them. The shape of the film is not a critical issue at all, so that one can cut the sensors to any desired shape to fit with the geometry of the elastic body. The size of the individual elements and the percent area of the structure covered by the piezo films is a more critical issue. However, once the sensor signals are processed through the modal observers, this issue ceases to be a critical one, as long as about a quarter of the structure is covered. In conclusion, piezo elements as sensors to analyze modal quantities present themselves as viable alternatives to existing measurement approaches.

References

[1] Baruh, H. and Choe, K., "Sensor Placement is Structural Control", *J. Guid. Cont. & Dyn.*, Vol. 13, No. 3, 1990, pp. 524-533.

[2] Chen, Chi-Tsong, *Linear System Theory and Design*, Holt, Rinehart and Winston, Inc., New York, 1984.

[3] Choe, K. and Baruh, H., "Sensor Failure Detection in Flexible Structures Using Modal Observers", *J. Dyn. Syst. Meas. & Cont.*, Vol. 115, No. 3, 1993, pp. 411-418.

[4] Collins, S. A., Padilla, C. E., Notestine, R. J., von Flotow, A. H., Schmitz, E., and Ramey, M., "Design, Manufacture, and Application to Space Robotics of Distributed Piezoelectric Sensors", *J. Guid. Cont. & Dyn.*, Vol. 15, No. 2, 1992, pp. 396-403.

[5] Crawley, E. F. and de Luis, J., "Experimental Verification of Distributed Piezoelectric Actuators for Use in Precision Space Structures", , *Proc. of the 27th AIAA Struct. Dyn. Conf.*, 1986, pp. 116-124.

[6] Hubbard, J. E. Jr., "Distributed Transducers for Smart Structural Components", *Proc. of the 6th Int. IMACS Conf.*, Feb. 1-5, 1988, Orlando, FL.

[7] Kepler, R. G. and Anderson, R. A., "Ferroelectric Polymers", *Advances in Physics*, Vol. 41, No. 1, 1992, pp. 1-57.

[8] Lee, C.K., "Theory of Laminated Piezoelectric plates for the design of Distributed Sensors/Actuators. Part I: Governing Equations and Reciprocal Relationships", *J. Acoust. Soc. Am.*, Vol. 87, No. 3, March 1990, pp. 1144-1158.

[9] —— and Moon, F. C., "Modal Sensors/Actuators", *J. Appl. Mech.*, Vol. 57, June 1990, pp. 434-441.

[10] Meirovitch, Leonard, *Analytical Methods in Vibrations*, Macmillan Publishing Company, New York, 1967.

[11] —— and Baruh, H., "Implementation of Modal Filters for Control of Structures", *J. Guid. Cont. & Dyn.*, Vol. 8, No. 6, 1985, pp. 707-716.

[12] Tadikonda, S. S. K. and Baruh, H., "Gibbs Phenomenon in Structural Mechanics", *AIAA J.*, Vol. 29, No. 9, 1991, pp. 1488-1497.

[13] Tzou, H.-S., "Lightweight Transducer Development using Piezoelectric Polymers", *Proc. of the 6th Int. IMACS Conf.*, Feb. 1-5, 1988, Orlando, FL.

[14] —— and Fu, H. Q., "A Study on Segmentation of Distributed Piezoelectric Sensors and Actuators: Part 1 - Theoretical Analysis", *Active Control of Noise and Vibration*, Vol. 38, 1992, pp. 239-246.

[15] ——, "A Study on Segmentation of Distributed Piezoelectric Sensors and Actuators: Part 2 - Parametric Study and Active Vibration Controls", *Active Control of Noise and Vibration*, Vol. 38, 1992, pp. 247-253.

MODAL FILTERING USING LINEAL SENSORS

Eric J. Lang*
Tsu-Wei Chou**

Center for Composite Materials, and
Department of Mechanical Engineering
University of Delaware
Newark, DE 19716

Abstract

The use of lineal sensors as modal sensors is discussed. Modal filtering is achieved without reliance on the orthogonality of eigenfunctions or the weighting of sensor sensitivity. Instead, the sensors exhibit sensitivity to specific modes of deformation and to remain insensitivity to other modes by taking appropriate paths through the structure. The mathematical problem of determining the required paths and practical methods of solution are discussed.

Introduction

The process of estimating the deformed state of a structure is becoming increasingly more important as the desire for lighter and higher performance structures leads to more flexible designs. There are several methods of achieving an estimate of the deformed state of a structure. One method involves observing the deformation at a small number of points over a period of time. Then, in conjunction with a dynamic model of the structure, the state is calculated. Another approach is to observe the deformation at several points and to use some sort of interpolation to arrive at the state of the structure. Both of these methods require a estimator capable of complex and rapid calculations. While modern computers have vast computing power, the capability of even the fastest computers is still finite and as system complexity increases, there are real limits to these approaches. Even without computational concerns, the cost and complexity of these systems can be prohibitive in many applications.

An alternative to discrete sensors is to use modal sensors which give useful information about the state of the structure directly. Several researchers have investigated the use of distributed piezoelectrics as modal sensors and/or modal actuators.[1,2] The modal sensors rely on the ability to vary the sensitivity of a piezoelectric sensor and thus produce a spacially weighted sensor. The appropriate weighting distributions are derived from the orthogonality of eigenfunctions. Once the modal coordinates are known, it is relatively easy to synthesize an effective control law.[3,4]

In this paper, a different class of distributed sensors is investigated. Specifically, the sensors are *lineal* sensors, that is, they are sensors which measure along a line. We restrict ourselves to uniformly weighted sensors. By uniform weighting we mean that the output of the sensor to a unit deformation is the same at each point along its length. The restriction to uniform sensitivity makes the task of manufacture easier and less expensive. However, the task of making a modal sensor is more difficult.

The output of a piezoelectric sensor is the cumulative effect over the entire area of the sensor. Similarly, the signal from a lineal sensor is the integral of the sensor output per unit length over the entire length. In general, the lineal sensor could be one of many different types of sensors – for example fiber optic, resistive wires, or wires with a piezoelectric coating.

The lineal sensors analyzed in this article are strain wires approximately 0.005″ in diameter which have been incorporated into a twisted fiberglass yarn. The yarn is woven into a multi-layered fiberglass preform which is then made into a textile composite by the resin transfer molding (RTM) process. See Figure 1. The path of the sensor yarn is controlled using a special weaving loom which is able to control the placement

---

*Research Assistant
**Professor
Copyright ©1994 by the American Institute of Aeronautics and Astronautics, Inc. All rights reserved.

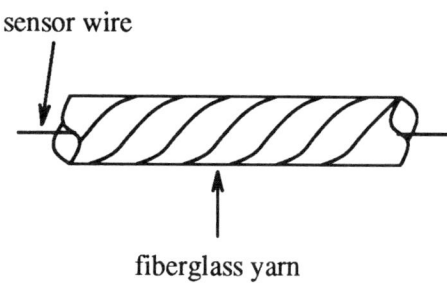

Figure 1: Sensor Wire and Twisted Yarn

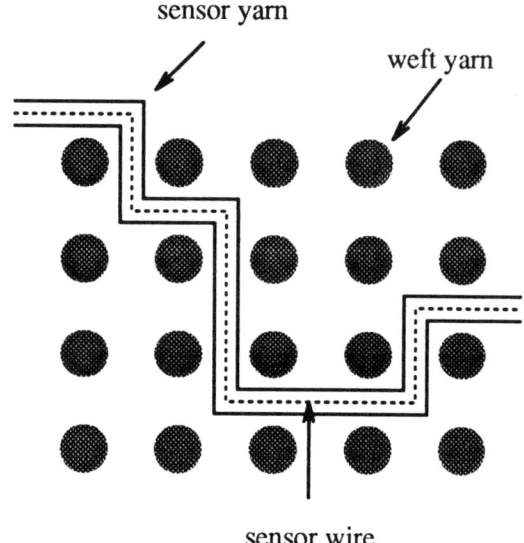

Figure 2: Multi-layer textile preform with sensor

of individual warp yarns. See Figure 2. The purpose of each lineal sensor is to give the magnitude of a specific mode of deformation while remaining insensitive to the presence of other concurrent modes of deformation. The sensor is able to achieve this specific sensitivity by following an appropriate path through the structure.

This paper is significant because it introduces the idea of lineal sensors for the measurement of modes of deformation in adaptive structures. This is in contrast to many other methods which use discrete [5] or areal sensors.[1] Also, the lineal sensors achieve modal filtering [4] by following appropriate paths through the structure. The mathematical problem of determining the required path of the sensors is formulated and a method of solution is presented and evaluated.

Formulation

This section contains a fairly general formulation of the lineal sensor path problem. A more specific example is given for an isotropic beam in bending in a later section.

Consider some closed domain Ω (which represents a body such as a beam) such that

$$\Omega \subset \mathcal{R}^n \qquad n = 2, 3 \qquad (1)$$

where the value of n depends on the problem. A value of 2 for n means the sensor path must lie in a plane and a value of 3 for n means that the sensor path must be contained in a volume.

A point in the domain is given by the vector x

$$x \in \Omega \qquad (2)$$

The individual components are represented by subscripts.

$$x = (x_1, x_2, \ldots, x_n) \qquad (3)$$

Choose the x_1 direction to be the domain of a function $y(x_1)$ which gives the sensor path. The range of y is the remaining dimensions of Ω, that is, x_2 for $n = 2$ and (x_2, x_3) for $n = 3$.

We assume the deformation of the body can be represented with sufficient accuracy by a finite number of modes of deformation and that the displacements are such that the total deformation is the superposition of each mode. Or,

$$U(x) = \sum_{i=1}^{N} q_i u_i(x) \qquad (4)$$

where $U(x)$ is the total displacement at a point x, $u_i(x)$ is the displacement due to the ith mode, and q_i is the modal coordinate for the ith mode. The modal coordinates give the magnitude of each mode.

Choose some reference state in which all modes of deformation are zero. The strain in the reference state is also zero by definition. We denote the strain in the

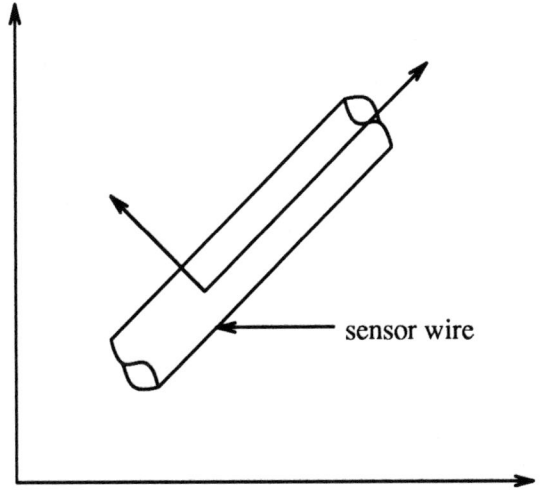

Figure 3: Local and Global Coordinate Systems

global coordinate system by ϵ^G and use ϵ^L to denote the global strain after it has been transformed from the global coordinate system into the local coordinate system coincident with the sensor. See Figure 3. The signal from the sensor is dependent on the strain in the sensor. Let ϵ^S be the strain in the sensor due to the local strain ϵ^L. If the material properties of the sensor are different from the material properties of the surrounding body, the strains ϵ^S and ϵ^L will be different.

In order to discover the relationship between the local strain and the sensor strain, either a micromechanical analysis or experiments must be performed.

Let ρ be defined as the sensor output per unit length along the sensor. The total output of the sensor is

$$s = \int_0^L \rho \, dl \quad (5)$$

When the body is in the reference state, the sensor output is zero.

Now let us consider a set of N sensors. The sensor outputs can be represented by a vector $\{S\}$

$$\{S\} = \left\{ \begin{array}{c} s_1 \\ \cdots \\ s_N \end{array} \right\} \quad s_i = \text{output from sensor } i \quad (6)$$

The modal coordinates can be represented by a vector $\{Q\}$.

$$\{Q\} = \left\{ \begin{array}{c} q_1 \\ \cdots \\ q_N \end{array} \right\} \quad q_i = \text{coefficient for mode } i \quad (7)$$

Let use define a matrix $\{A\}$ whose elements, a_{ij} are the output of the ith sensor due to a unit deformation of the jth mode. The sensor output is the combined effect of all modes, or

$$s_i[y_i] = \sum_j a_{ij} q_j \quad (8)$$

The subscript on y_i is to remind us that each sensor has a different path.

By definition

$$\{S\} = [A]\{Q\} \quad (9)$$

If $\{A\}$ is invertible, the modal coordinates can be determined from the sensor output via

$$\{Q\} = [A]^{-1}\{S\} \quad (10)$$

In general, there will be several modes of deformation taking place simultaneously. In order to have each sensor sensitive to one and only one mode, the matrix $\{A\}$ must be diagonal. Additionally, in order to have the greatest sensitivity, we want the diagonal elements of $\{A\}$ to have the largest magnitude possible. Thus, we want

$$\{A\} = \left[\begin{array}{ccc} a_{11} & 0 & 0 \\ 0 & a_{22} & 0 \\ 0 & 0 & a_{33} \end{array} \right] \quad \|a_{ii}\| = \max \quad (11)$$

We refer to the case when $\{A\}$ is diagonal as uncoupled sensing of the modes. The task now is to determine the paths y_i such that $\{A\}$ is diagonal and extremized.

Solution Techniques

In general, it is difficult to find an analytical solution for the optimal path of a sensor through a structure that meets the requirements of uncoupled sensing and maximum sensitivity. Thus, approximate methods are investigated. Let us consider a path made of piecewise linear segments. In other words, the path of the sensor is made of a series of straight line segments. The unknowns in the problem are then the endpoints of each line segment. See Figure 4. Let $\{Y_i\}$ be the vector of endpoints of the line segments for the i_{th} sensor. The problem of finding the path for the ith sensor is now the problem of finding the vector $\{Y_i\}$

Several methods of solving the sensor path problem have been investigated however, the method of multipliers [6] has proven to be the most useful, efficient, and reliable for the determination of the sensor path.

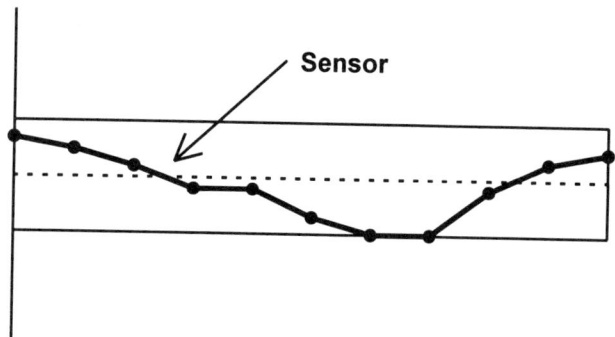

Figure 4: Sensor Path made of Line Segments

Method of Multipliers

The method of multipliers is a method of optimization in which a single minimization subject to nonlinear constraints is converted into a sequence of unconstrained minimizations. For example, a classical nonlinear programming problem is

$$\text{minimize} \quad f(x) \quad (12)$$
$$\text{subject to} \quad h(x) = 0 \quad (13)$$

where $f : R^n \mapsto R$ and $h : R^n \mapsto R^m$ are given functions. Here, we have n unknown x values and m nonlinear constraints.

Fiacco and McCormick [7] introduced a method of sequentially unconstrained minimizations sometimes referred to as penalty function methods which take the form

$$\text{minimize} \quad f(x) + \tfrac{1}{2} c_k \|h(x)\|^2 \quad (14)$$
$$\text{subject to} \quad x \in R^n, \quad (15)$$

where the c_k's are a sequence of positive numbers with $c_k < c_{k+1}$ and $\|h(x)\|$ is the magnitude of the constraint vector. The method of multipliers is an extension of this method which includes multipliers as in the method of Lagrange multipliers. First, we form the modified Lagrangian

$$L_{c_K}(x, \lambda_k) = f(x) + \lambda_k h(x) + \frac{1}{2} c_k \|h(x)\|^2 \quad (16)$$

where λ_k is a vector of multipliers of length m. Then, a sequence of minimizations of the form

$$\text{minimize} \quad L_{c_k}(x, \lambda_k) \quad (17)$$
$$\text{subject to} \quad x \in R^n, \quad (18)$$

is performed where c_k is again a sequence of increasing positive numbers. The multiplier sequence λ_k is generated by the iteration

$$\lambda_{k+1} = \lambda_k + c_k h(x_k), \quad (19)$$

where x_k is the solution to Eq. (17) on the kth iteration.

In order find the path for the ith sensor which is optimized to measure the ith mode, we must first form the modified Lagrangian

$$L_{c_K}(Y_i, \lambda_k) = -\|a_{ii}(Y_i)\| + \lambda_k g(Y_i) + \frac{1}{2} c_k \|g(Y_i)\|^2 \quad (20)$$

where Y_i is the vector of endpoints of the sensor path and $g(Y_i)$ is a vector of length $N-1$ (one less than the total number of modes) containing the off-diagonal terms in the matrix $\{A\}$. After a sequence of minimizations

$$\text{minimize} \quad L_{c_k}(Y_i, \lambda_k) \quad (21)$$
$$\text{subject to} \quad Y_i \in \Omega, \quad (22)$$

the values of Y_i converge to give the endpoints of the optimized piecewise linear sensor path.

The method of multipliers has been successful in solving the sensor path problem while other standard minimization methods have often failed due to the highly nonlinear constraints.

A Cantilever Beam: An Example

In the proceeding section, no assumptions were made about the strain distributions or the relationship between ϵ^L and ϵ^S. We will now consider the case of an isotropic beam and an isotropic sensor with the same elastic properties. In this case ϵ^L is equal to ϵ^S. Also, for the case of a beam the relationship between the strains and the deformation is known. Thus all the relations needed to formulate the sensor path problem for the beam are known.

Figure 5 shows the first three mode shapes of a uniform cantilever beam. As a hypothetical example, we determined the sensor paths required to sense each of these modes while remaining insensitive to the other two modes. The sensor paths are shown in Figure 6-8

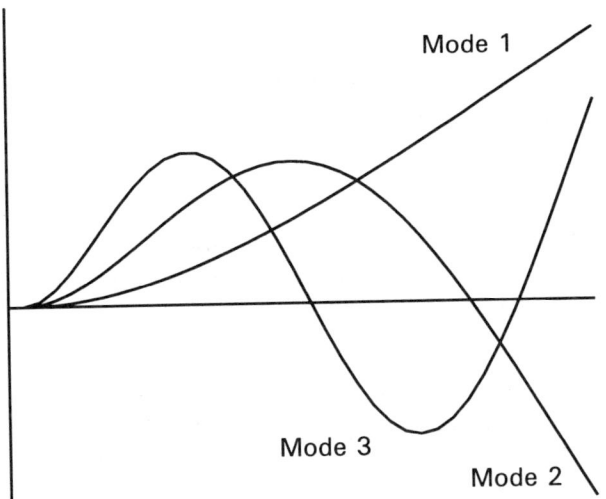

Figure 5: Modes of Vibration of a Cantilevered Beam

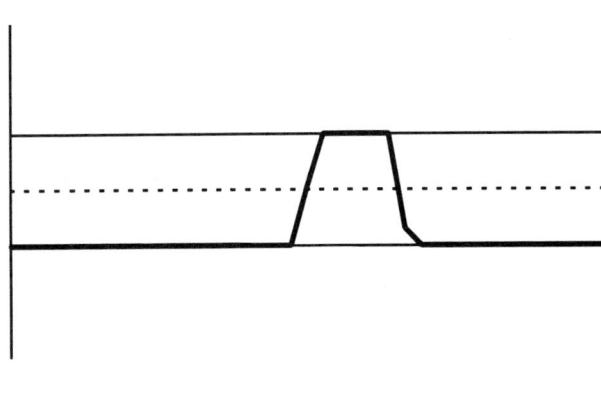

Figure 6: Sensor Path for Mode 1. Sensor insensitive to Mode 2 and Mode 3

Since the sensor paths are calculated without reliance on the orthogonality of eigenfunctions, we can calculate static mode sensor paths which are orthogonal to dynamic modes and other static modes. Figure 9 shows the sensor path which is optimized for Mode 1 while insensitive to Mode 2 and to an applied end load. A sensor like this would be useful for an application such as the vibration control of a wind turbine. As the wind speed increases, the applied static load increases. It is desirable to uncouple the vibration (which is to be controlled) from the static deformation.

Figure 10 shows the sensor path which is optimized for Mode 1 while insensitive to Mode 2 and to an applied end moment.

Experiments must be conducted to determine whether the details of the complex microstructure of a three dimensional textile composite affect the average response of a sensor. If the details of the microstructure do affect the output, then either the response of the sensors must be experimentally determined or a more detailed micromechanical analysis must be conducted.

Conclusion

Lineal sensors offer a novel way of providing modal sensing. They may be used as dynamic mode sensors or static mode sensors. The method of multipliers allows efficient solution of the required sensor path. The uncoupled nature of the sensor output may prove to

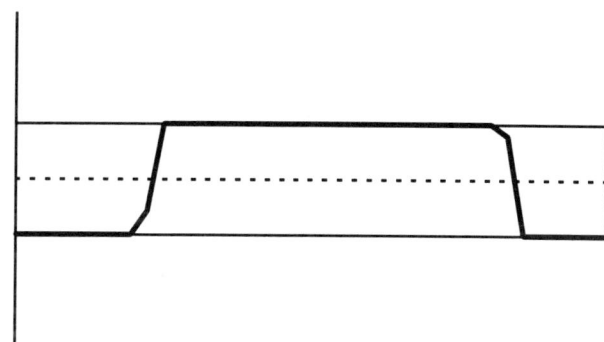

Figure 7: Sensor Path for Mode 2. Sensor insensitive to Mode 1 and Mode 3

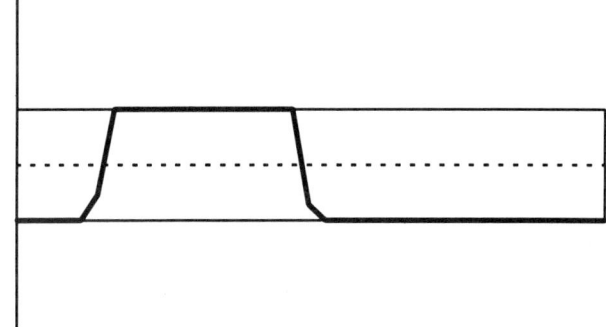

Figure 8: Sensor Path for Mode 3. Sensor insensitive to Mode 1 and Mode2

be useful in certain design situations when a complex controller is not available to decouple the signal set.

References

1. LeeMoon C.-K Lee and F.C. Moon. Modal sensors/actuators. *Journal of Applied Mechanics*, 57:434–441, June 1990.

2. Lee C.K. Lee et al. Piezoelectric modal sensors and actuators achieving critical active damping on a cantilever plate. In *AIAA/ASME/ASCE/AHS/ASC 30st Structures, Structural Dynamics and Materials Conference*, pages 2018–2026, September 1989.

3. Baz2 A. Baz and S. Poh. Experimental implementation of the modified independent modal space control method. *Journal of Sound and Vibration*, 139(1):133–149, 1990.

4. Meirovitch1 L. Meirovitch et al. The implementation of modal filters for control of structures. *Journal of Guidance and Control*, 8(6):707–716, 1985.

5. Andersson Mark S. Andersson and Edward F Crawley. Discrete distributed strain sensing of intelligent structures. In Yuji Matsuzaki and Ben K. Wada, editors, *Second Joint Japan/U.S. Conference on Adaptive Structures*, pages 737–754, November 1991.

6. Bertsekas Dimitri P. Bertsekas. *Constrained Optimization and Lagrange Multiplier Methods*. Computer Science and Applied Mathematics. Academic Press, New York, 1982.

7. Fiacco Anthony V. Fiacco. *Nonlinear programming; sequential unconstrained minimization techniques*. Wiley, New York, 1968.

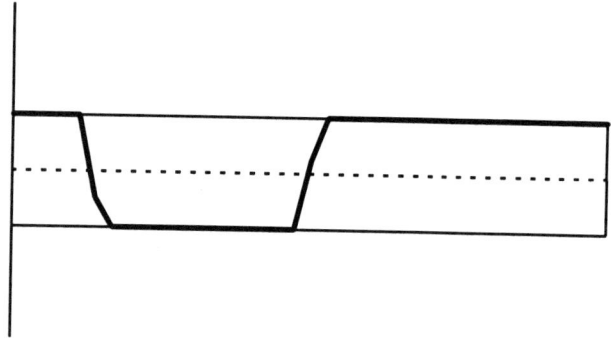

Figure 9: Sensor Path for Mode 1. Sensor insensitive to Mode 2 and End Load

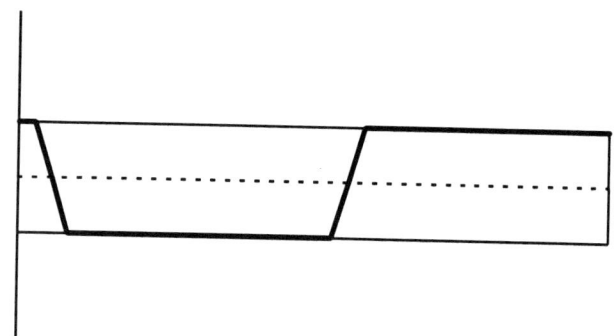

Figure 10: Sensor Path for Mode 1. Sensor insensitive to Mode 2 and End Moment

DEVELOPMENT OF A LINEAR PIEZOELECTRIC MOTOR

David Newton[*], Ephrahim Garcia[†]

Smart Structures Laboratory
Mechanical Engineering Department
Vanderbilt University
Nashville, TN 37235

and

Garnett C. Horner[‡]
Spacecraft Dynamics Branch
NASA Langley Research Center
Hampton, VA 23665

Abstract

The development of a linear hybrid transducer type piezoelectric motor is shown. This design has the advantages of light weight, macro and micro positioning, and large force/velocity output. The motor consists of a longitudinal actuator that provides output displacement and force, and two alternating clamping actuators which provide the holding force. The advantages of this design over other types of piezoelectric motors are shown. Significant design considerations for the development of a piezoelectric motor (including material choices, shear loads on piezoelectric-materials and contact surface interaction) are discussed. A simulation, including a finite element analysis of the system, is used to predict the motor response to various input sets. A prototype motor is built and its characteristics are compared to theoretical predictions.

Introduction

This new variation of linear motor has numerous applications. Since the Vanderbilt piezoelectric-motor has few moving parts, is light-weight and is bearingless it is ideal for space applications where problems of bearing seizure, reliability and weight are key concerns. Articulation of solar arrays and deployable trusses are two space applications. A group of six such actuators can be incorporated into a Steward's platform which allows positioning in all six degrees of freedom. Applications on earth are also abundant; in the field of micro-biology there is a need for long travel fast and powerful precise actuators. The current technology of mounting a micro-positioner onto a lead screw can be eliminated by the use of the Vanderbilt motor as a macro/micro positioner.

Numerous variations of piezoelectric motor designs have been investigated,[1,2,3,4,5,6,7,8] of which the two most reoccurring configurations are the traveling wave type and the hybrid transducer (inchworm) type. Three variations of piezoelectric motors are shown in Figures 1,2, and 3.

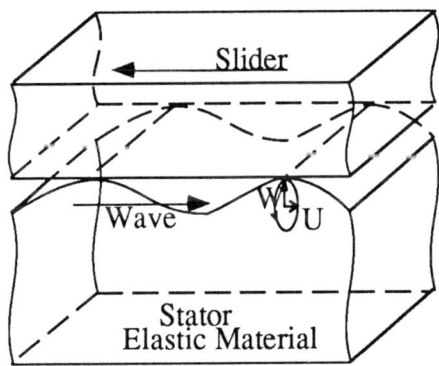

Figure 1. Traveling Wave Type Motor

The traveling wave type ultrasonic motor, Figure 1, has a vibration system that requires high power electronics[1]. With this type of motor higher efficiencies are difficult to achieve because of the power consumption of the vibration system. The inchworm type motor, Figure 2, has the advantage of decoupling the vibration system to allow greater efficiency over the traveling wave type. An inchworm type motor such as the one proposed consists of a longitudinal vibrator to provide extension-contraction motion and output force, and two clamping actuators that provide alternating holding forces. Burleigh Instruments Inc. of Fishers NY patented and manufactures their Inchworm® motor for use as a low speed micropositioner.[9,10] A third type of piezoelectric-motor is the inertial mass

---

[*] Graduate student
[†] Assistant Professor, Mechanical Engineering
   Member AIAA
[‡] Aerospace technologist

type[8]. The inertial mass type motor in Figure 3 operates via a stick-slip motion of two unequal masses. This type of motor is unstable at low speeds and has a low output force.

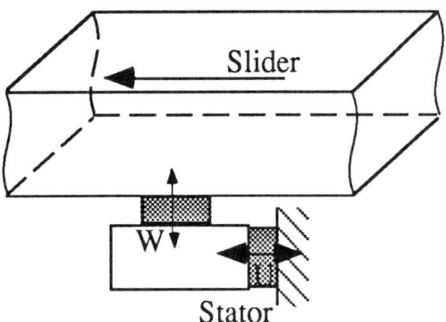

Figure 2. Hybrid Transducer Type Motor

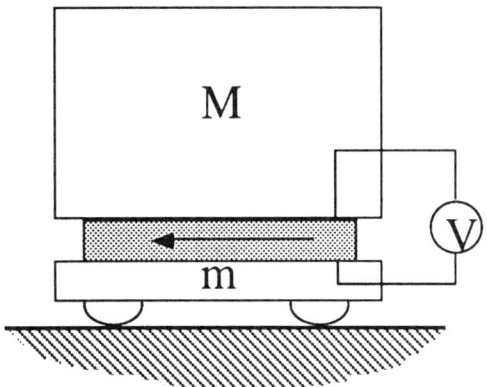

Figure 3. Inertial Mass Type Motor

Operation of a hybrid transducer type motor

Piezoelectrics have the ability to provide large forces at high speed but with very small strain. Using these properties an inchworm motor attempts to take multiple steps at high frequencies to realize large displacement and velocity. Figure 4 shows a schematic of a typical inchworm motor. The motor consists of the left and right clamping actuators (LCA and RCA) which when extended grip onto a shaft to provide a positive holding force. Between the two clamping actuators is the longitudinal piezoelectric vibrator (LPV) providing the extension and contraction motion for the system. Referring to Figure 4, the operating cycle is as follows:

1) The LCA is extended to clamp onto the shaft.
2) The LPV extends to provide a positive step.
3) With the LPV extended, the RCA extends to clamp onto the shaft.
4) The LCA then releases the shaft.
5) The LPV contracts to bring the rear of the motor into its advanced position.
6) The LCA then extends to provide a positive holding force.
7) The RCA contracts to complete the cycle.

Continuous motion is realized through repetition of the cycle. A positive holding force is achieved through applying a positive DC voltage to both of the clamping actuators. (similar to positions 3 and 6 in Figure 4.) The motor is free to slide when a negative DC voltage is applied to both clamping actuators. Because of the capacitive nature of piezoelectrics these hold and free states can be achieved with very little power consumption.

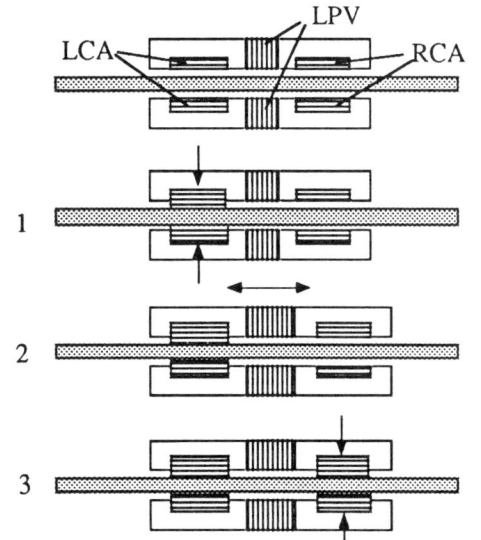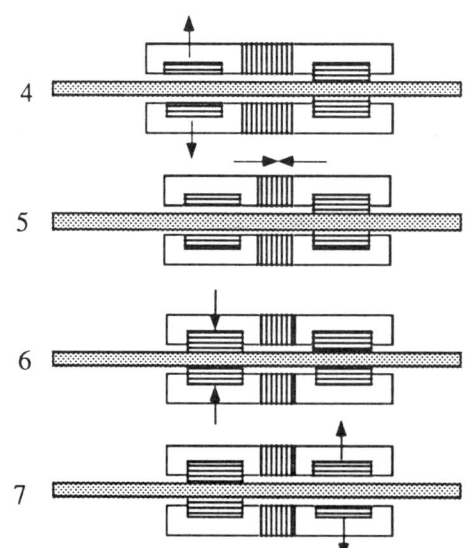

Figure 4. Schematic operation diagram of an inchworm motor

There are two ways to control the velocity of the motor. One method is to vary the step size per cycle by adjusting the voltage across the LPV. The second method is to vary the steps per second by adjusting the frequency of operation. For large motions the operating cycle described in Figure 4 is used. The direction of the motor is reversed by switching the left and right clamping actuator signals. The motor can be operated in a "push rod" mode where the motor is fixed and the shaft is actuated, or in a stationary shaft mode where the motor moves over the fixed shaft.

For precise microscopic motion, one clamping actuator is clamped while the other, load bearing, side is free (as in position 2 of Figure 4). Voltage across the LPV can then be used for precise motion about the datum determined by the clamped actuator. Internal position sensing can be added to the motor to increase the resolution of the micro-motion.

Current design and design considerations

The goal of the current design is to develop an inchworm motor with increased speed and strength characteristics. The Burleigh motor operates by exciting a single, cylindrical piezoelectric element at high voltages (1000 V). By incorporating piezoelectric stacks the current design will be able to operate at lower voltages with relatively large displacements. The Burleigh motor operates well as a micropositioner with a maximum speed of 2 mm/sec and a maximum force output of 1.5 kgf[10]. Speed improvements are realized by the ability to drive lower voltages at higher frequencies. The Kurosawa motor, which incorporates piezoelectric stacks, can be driven with sinusoidal voltages at high frequencies to achieve a reported maximum speed of 50 cm/sec, and a maximum force output of 0.5 kgf[1]. The force output of both of these designs is limited by a piezoceramic's inability to withstand shear forces.

The use of piezoelectric stacks in a novel clamping design will allow the Vanderbilt motor to avoid problems of shear failure. The clamping design incorporated in the Vanderbilt motor utilizes a mechanical flexure that provides an alternate shear path. Shear forces are transmitted into the flexure rather than into the piezoelectric, thus avoiding shear failure. The flexure also includes a tilt bearing to further maintain purely compressive loads on the stack. Mechanical pre-load on the clamping stacks is achieved though elastic deformation of the flexure.

In order to operate the stacks in the most efficient manner, a mechanical pre-load will be applied to both the clamping actuators and the longitudinal vibrator. The total deflection of a piezoelectric stack is virtually unchanged by a near constant pre-load[10]. The pre-load applies a mechanical bias on the stack that helps avoid failure due to the ceramic's low tensile strength. For the LPV, the inverse piezoelectric effect provides the extension force of the motor and the pre-load provides the return force. Stiff springs are used to apply a pre-load on the LPV. The pre-load on the clamping stacks is provided by the elastic spring constant of the leg flexure. Finite element analysis of the clamping actuators has been used to determine a shape that will insure the proper pre-load on the clamping stack. A stress analysis of the clamping flexures allows a fatigue calculation that shows that the maximum and cyclic stress levels are well below the endurance limit of the chosen material.

In order to avoid excessive heat generation by the motor, a hard PZT material (such as industry type PZT-4) was chosen over the soft PZT material (such as industry type PZT-5). Hard PZT allows the motor to be driven at or near resonance frequency for extended duration without the dielectric loss heating characteristic of a soft PZT. The disadvantages of a hard ceramic are that it produces less strain per voltage and is difficult to manufacture in stack layers below 150 microns. These disadvantages are outweighed, however, by the hard PZT's ability to drive a mechanical load more efficiently and by its higher Q_m (mechanical resonator quality factor). This higher Q_m manifests itself in that the displacement amplification at resonance of a typical hard PZT is on the order of 10 times, where as the displacement amplification at resonance for a typical soft PZT is only on the order of 2 times.

The motor design above was modeled for modal and harmonic analysis using ANSYS code from Swanson Analysis Systems, Inc. ANSYS allows the inclusion of piezoelectric elements. A modal analysis was performed and the modal response was determined well beyond the intended driving frequencies. The analysis showed groups of three mode shapes corresponding to the tilting of one half mass with respect to the other and torsional modes caused by bending of the LPV stacks. A harmonic analysis of the motor showed that operating below the first structural resonance or between two structural resonances yielded the ideal longitudinal response with minimal modal interaction. For example when operating between a tilting mode and a torsional mode, purely longitudinal motion was achieved with insignificant tilting or torsional motion. This FEM allows confident operation over a wide frequency range with regard to structural resonances.

Simulated performance and discussion

A dynamic simulation of the motor system is used to investigate some of the design parameters such as input signal wave form and the effects of phase lags between the three driving signals. The motor was modeled using SIMULAB[11] from Math Works. The stack output was modeled using a simple one dimensional assumption of an equivalent spring and damper. The equivalent spring and damping constants were determined from an Eigenvalue Realization Algorithm[12] analysis of one of the stack to be used in the prototype motor. Further details of the modeling are given in Jones, Newton, and Garcia[13].

The simulation shows that the motor velocity will increase with driving frequency until a maximum is reached near resonance. Above resonance the motor velocity decreases as a result of the inertial effects between the left and right half masses. From this it is determined that the optimal frequency at which to operate is resonance, which is to be expected because at resonance the stack achieves its maximum extension and contraction for a set of cyclic voltages. Also, this result shows that the ideal method of controlling velocity is by varying voltage while setting frequency to the resonance of the system.

The simulation was run with varying loads on the motor. As is typical with other linear motors the velocity decreases with increasing loading. Investigating the motor response over a small time and distance reveals that this decrease in speed is a result of smaller steps. An increase in the pre-load spring on the longitudinal vibrators allows larger loads to be carried.

The phase relationship between the clamping signals and the vibrator signal was perturbed. A delay was put in line with one or both of the clamping signals in order to investigate the sensitivity of the motor performance with regard to some phase shift caused by the amplifier dynamics and stack dynamics. The simulation shows that the motor output is not significantly effected by phase differences of up to ±15 degrees. For phase shifts over 20 degrees the motor output degenerates.

From the ideal parameters considered in the dynamic simulation the maximum velocity predicted is 17 cm/sec, driving at 22 kHz, the maximum force output predicted is 15 kgf, and a maximum holding force predicted is 30 kgf. These predictions are highly dependent on the choice of frictional contact material for the clamping actuators and the assumption of a simple spring-mass-damper for the piezoelectric stacks.

Experimental motor construction

An experimental prototype motor was built with an aluminum body and steel clamping flexures. The piezoelectric actuators were hand assembled stacks consisting of 40 active ceramic layers for the clamping actuators and 60 active ceramic layers for the longitudinal vibrators. Individual layers are 0.25 mm thick hard PZT wafers supplied by Aura Ceramics. These prototype stacks were tested and confirmed prior to incorporation in the motor assembly. The longitudinal vibrator stacks showed an amplification of displacement of 12 times at resonance. Quasi-static displacement of the clamping actuators and longitudinal vibrators is .015 µm/V and 0.023 µm/V respectively.

The assembled motor runs in a "push rod mode"; one half mass is rigidly fixed to ground, the other is free to move, and a 1/2 inch diameter stainless steal hardened precision shaft is driven linearly. The shaft/motor interface is bronze on steel. Prototype electronics were designed with an Apex PB58A power booster to provide the high driving current needed for the highly reactive loads. Front end analog circuitry was built to shape the wave forms and insure the proper phase difference between signals. For the prototype electronics and the given load, the operation bandwidth is limited to approximately 4 kHz at 200 V_{pp} for the clamping actuators and approximately 5 kHz at 50 Vpp for the longitudinal exciters.

The clamping flexures provided a pre-load on the claming actuators of 3000 kPa. Stiff compression springs provide a mechanical pre-load of 500 kPa on the longitudinal exciters.

Motor performance and discussion

The holding force of the motor was tested for DC voltages fed to both clamping actuators. The results are shown in Figure 5. As can be seen the holding force increases dramatically with increased voltage. The maximum holding force achieved was 5.1 kgf at 400 V DC. The holding force is governed by the static friction between the clamping contact material and the shaft and the nominal displacement of the clamping actuator. The material chosen as a contact material needs to be hard in order to avoid deflection; this will allow the normal force provided by the clamping stack to more closely approach its blocked or maximum force.

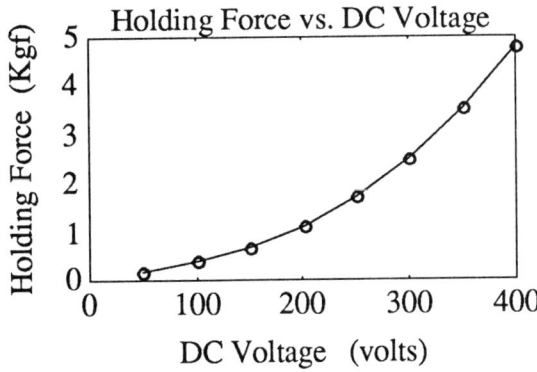

Figure 5. Static holding Force Test

The dynamic test conditions were 200 V_{pp} with a 100 V offset for the clamping actuators, and 50 V_{pp} with a 25 V offset for the LPV. Both resonant and non resonant operations were tested showing trends in speed that matched the simulation model; i.e. a maximum speed at resonance. The prototype electromechanical motor system resonated longitudinally at about 3.3 kHz with an electrical impedance of 12 ohms. The maximum speed of the motor was 1.2 cm/sec operating at 3.5 kHz. Figure 6 shows the motor velocity as a function of input frequency. One important issue of voltage control over a range of frequencies is that of a phase lag between command voltage and stack displacement. The voltage-displacement phase relationship for the longitudinal stacks is different from that of the clamping stacks, therefore for any given frequency the phase difference between the clamping command voltage and the LPV command voltage must be tuned to give the desired 90 degree phase shift in displacement (LPV vs. RCA).

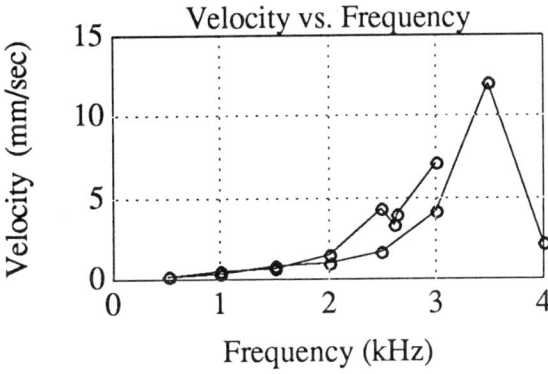

Figure 6. Frequency Response Test

The motor's response to loading was tested using a 500 Hz input frequency. Figure 7 shows the results. As is typical with other motors, and as the simulation predicted, the speed decreases with increasing load. The 500 Hz stall speed was 250 gf. The load characteristics of the motor were also tested at the 3 kHz operating frequency and a stall load of 700 gf was found. The load carrying capability can be increased by increasing the pre-load on the LPV's. Future test will use stiffer springs to pre-load the LPV stacks.

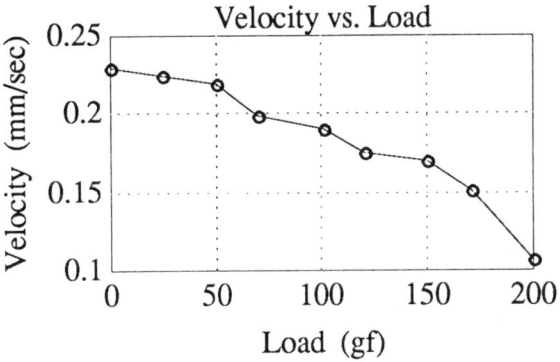

Figure 7. Dynamic Loading Test

The performance of the motor degrades over time because of the wear of the contact material. This degradation makes repeatable tests difficult. The motor will be made stronger and more reliable be the incorporation of a harder more durable contact material.

Conclusions

Design of a linear piezoelectric inchworm type motor was shown. The characteristics of a prototype were enumerated and discussed and were found to match those predicted by our simulation. The study showed that in order to increase the force and velocity output of a piezoelectric motor, shear loads on piezoelectric stacks must be avoided, and a hard wear resistant wear material must be used for a contact material.

Acknowledgment

The authors wish to acknowledge the support for this research provided by NASA under grant number NAG-1-1484.

References

1. M. Kurosawa, H. Yamada, and S. Ueha, "Hybrid transducer type ultrasonic linear motor," *Japanese Journal of Applied Physics*, Vol. 28 Supplement 28-1, pp. 158-160, 1988.
2. G. Schadegrodt and B. Salomon, "The piezo traveling wave motor--a new element in actuation," *Control Engineering*, pp. 10-18, May 1990.

3. M. Kurosawa and S. Ueha, "Hybrid transducer type ultrasonic motor," *IEEE Transactions on Ultrasonics, Ferroelectrics, and Frequency Control*, Vol. 38 No. 2, pp. 89-92, 1991.

4. M. Fleischer, D. Stein, and H. Meixner, "New type of piezoelectric ultrasonic motor," *IEEE Transactions on Ultrasonics, Ferroelectrics, and Frequency Control*, Vol. 36 No. 6, pp. 614-619, 1989.

5. H. Goto and T. Sasaoka, "Vertical micro positioning system using PZT actuators," *Bulletin of Japanese Society of Precision Engineering*, Vol. 22 No. 4, pp. 277-282, 1988.

6. A. Kumada, "A piezoelectric ultrasonic motor," *Japanese Journal of Applied Physics*, Vol. 24 Supplement 24-2, pp. 739-741, 1985.

7. O. Ohnishi, O. Myohga, T. Uchikawa, M. Tamegai, T. Inoue, and S. Takahashi, "Piezoelectric ultrasonic motor using longitudinal torsional composite vibration of a cylindrical resonator," *Proceedings from IEEE Ultrasonic Symposium*, pp. 739-743, 1989.

8. Ph. Niedermann, R. Emch, and P. Descouts, "Simple Piezoelectric Translation Device," Rev. Sci. Instrum., Vol. 59 (2), pp. 368-369, 1988.

9. Burleigh Instruments Inc., "Piezoelectric electromechanical translation apparatus," U.S. Patent 3,902,084, 1975.

10. Burleigh micropositioning systems brochure, Burleigh Instruments Inc., Fishers, NY, U.S.A.

11. SIMULAB User's Guide, © Copyright 1990 by The Math Works, Inc., Natick, Mass, U.S.A.

12. J.-N. Juang and R. S. Pappa, "An Eigensystem Realization Algorithm for Modal Parameter Identification and Model Reduction," AIAA Jounal of Guidance, Control, and Dynamics, Vol. 8, No. 5, pp. 620-627, 1985.

13. L. Jones, D. Newton, E. Garcia, "Adaptive Devices for Precise Position Control," SPIE Proceedings: Conference on Smart Structures and Materials, 1993, Alb. N.M., Vol. 1917, pp. 648-659.

Nonlinear Flutter Control of Composite Panels with Embedded Piezoelectric Materials Using Finite Element Method

R.C. Zhou*, Min-Hung Hsiao*, David Y. Xue**, Zhihong Lai*,
Chuh Mei† and Jen-Kuang Huang††
School of Mechanical and Aerospace Engineering
Old Dominion University, Norfolk, VA 23529-0247

Abstract

The nonlinear finite element equations of motion are derived for composite panels with embedded piezoelectric layers subjected to aerodynamic and thermal loads. The quasi-steady first-order piston theory is utilized for the aerodynamic load in panel flutter at supersonic flow. The nonlinear dynamic equations of motion in the modal coordinates are obtained through a model reduction. These modal equations are then employed for time domain numerical integration and control design. An optimal controller is developed based on a linear control theory. Numerical simulations show that the critical dynamic pressure can be increased by the piezoelectric actuation so that the vehicles are able to fly faster without flutter. The panel flutter large amplitude limit-cycle motions are shown to be able completely suppressed within a certain dynamic pressure. For completely embedded piezoelectric actuators within top and bottom of the composite laminates, two-set actuator design performs better than one-set actuator design. An optimal shape and location of small-size or patched piezoelectric actuators are also investigated by using the norms of feedback control gain matrix(NFCG). The results show that the embedded piezoelectric actuators are effective in nonlinear panel flutter suppression.

Introduction

Panel flutter is a self-excited, dynamic instability of thin plate or shell-like structural components of flight vehicles. This aeroelastic phenomenon is often encountered in the operation of aircraft and missiles at supersonic speed. Prevention of panel flutter is one of the many problems that designers must face. The earliest reported structural failures due to panel flutter are the failure of the 60 to 70 early German V-2 rockets during World War II.[1] A most recent panel flutter (also sonic fatigue) failure was reported at the 1992 AIAA Dynamics Specialist Conference.[2] After the flight tests of the F-117A Stealth Fighter, cracks due to panel flutter were found in about half of the laminated composite skin panels. Those panels were then redesigned and stiffened and thus a tremendous weight penalty was paid.

Panel flutter becomes one of the important issues to be considered in the development of high speed flight vehicles such as the High Speed Civil Transport (HSCT), the National Aerospace Plane (NASP) and the YF-22 advanced Tractical Fighter (ATF). When a flight vehicle travels at high speeds, not only will it experience flutter due to the dynamic pressure, but it also will be affected by great amounts of heating owing to the friction from the surrounding air. The presence of high temperature load results in a flutter motion at lower dynamic pressure. In addition, the temperature rise may also cause large aerodynamic-thermal deflections of the skin panels, thus affects flutter response. Due to severe thermal environment induced by the aerodynamic heating, the composite materials, such as carbon-carbon, will be considered when designing those vehicles. Gray and Mei[3] gave a complete survey on various theoretical considerations and analytical methods for the investigation of nonlinear panel flutter up to 1991. The limit-cycle theoretical results obtained from the finite element[3] and classical[4] analyses agree very well with the results obtained experimentally. The classical analytical method employs, in general, the Galerkin's approach in the spatial domain and the panel deflection is expressed in terms of the linear normal modes; the direct numerical integration is then applied in the time domain. Dowell[4] determined that four or six linear modes are required for obtaining a converged limit-cycle amplitude and frequency of a simply-supported isotropic plate. Dixon and Mei[5] studied the nonlinear flutter of rectangular composite panels under a uniform thermal load by the finite element frequency domain method. The panels considered were anisotropic composite laminates and were modeled as thin rectangular plate finite elements. The aerodynamic pressure load was modeled through the use of the first-order piston theory.

During the past few years, a significant amount of research has been reported in the field of control of flexible structures by the use of smart sensors and actuators.[6] Piezoelectric materials can be used as such sensors and actuators. Piezoelectric phenomena were first observed by Curies' brothers in 1880.[7] In general it can develop an electrical charge when subjected to a mechanical strain. This is referred to as the direct piezoelectric effect which forms a basis of using such materials as sensors. The converse piezoelectric effect, the development of mechanical strain when subjected to an electrical field, can be utilized to actuate a structure. Thus, actuation of a structure may be accomplished at the material level. Piezoelectric materials in current use include polyvinylidene fluoride (PVDF), a semicrystalline polymer film, and lead zirconate titanate (PZT), a piezoelectric ceramic material. PZT generally is stiffer and has large electromechanical coupling coefficients and is thus better suited for actuator applications. The film product (PVDF), on the other hand, has higher voltage limit with lower stiffness and coupling coefficients and is thus better for sensor applications. Piezoelectric laminates bonded to the surface of, or manufactured and embedded into flexible structure members can act as either control actuators or sensors. The effectiveness of using passive or active control of smart systems has been demonstrated by many

---

* Ph.D. Student. Student Member AIAA.
** Research Associate, Member AIAA.
† Professor, Associate Fellow AIAA.
†† Associate Professor, Member AIAA.

Copyright © 1993 American Institute of Aeronautics and Astronautics, Inc. All rights reserved.

researchers. However, most of the control designs have been applied to the beam-like and truss structures. Only a few research papers have been reported in the area of control of panel flutter response using adaptive materials. Scott and Weisshaar[8] were the first to study controlling the linear panel flutter using piezoelectric materials. The piezoelectric materials were fully covered on the surface of the panel. They used an assumed mode approach and the Ritz method. The panel was modeled as a simply-supported isotropic plate. Only four modes in the x-direction (one in y-direction) were retained in the study. Linear optimal control theory was applied in the simulation. Lai et al.[9] extended to control the nonlinear flutter of a simply-supported isotropic plate by using piezoelectric materials as actuators. The piezoelectric materials were also completely covered over the plate. The Galerkin's approach was adopted in obtaining the nonlinear modal equations of motion. The optimal control theory and numerical integration were used in the simulation. They concluded that the bending moment was more effective in flutter suppression than the in-plane force. Xue and Mei[10] recently studied the feasibility of the application of SMA in panel flutter control. The finite element equations of motion were developed for panel flutter including thermal and SMA effects. They demonstrated that the SMA was effective in passive control of panel flutter at high temperatures. To the authors' knowledge, the present paper will be the first attempt in studying of control of nonlinear flutter response of composite panels by using piezoelectric materials and the finite element approach.

In this paper, an optimal control method is employed to suppress nonlinear flutter limit-cycle motions of laminated composite panels at elevated temperatures ($\Delta T/\Delta T_{cr} > 1$) using piezoelectric actuators. The nonlinear finite element equations of motion, based on the von Karman's large-deflection theory, are developed for composite panels with piezoelectric layers subjected to aerodynamic and thermal loads. A modal transformation and reduction is then performed to obtain a set of equations of motion in the modal coordinates. By using the optimal control theory, control actions can be determined based on the linearized modal equations of motion and the dynamic system will be stable with small perturbations. Due to the limitation of the maximum operating (one half of depolarization) electric field of the piezoelectric actuator, the success of flutter suppression will depend on how large the dynamic pressure is. The number of modes retained to achieve a convergent solution are also determined. Numerical simulations based on the nonlinear modal equations are performed to demonstrate the effectiveness of panel flutter suppression under the piezoelectric actuation. The location and shape of the embedded piezoelectric actuators are also investigated by the use of norms of feedback control gain matrix (NFCG). The increased dynamic pressure due to the piezoelectric actuation is investigated and presented.

Finite Element Equations of Motion

As disclosed in all the panel flutter surveys, the aerodynamic theory employed for most panel flutter studies at supersonic flow ($M_\infty > \sqrt{2}$) is the quasi-steady first-order piston theory.[1,3,4] The aerodynamic load can be expressed as

$$p_a = -\frac{2q_a}{\beta}\left(\frac{\partial w}{\partial x} + \left(\frac{M_\infty^2 - 2}{M_\infty^2 - 1}\right)\frac{1}{V_\infty}\frac{\partial w}{\partial t}\right)$$

$$= -\left(\lambda \frac{D_{110}}{a^3}\frac{\partial w}{\partial x} + \frac{g_a}{\omega_0}\frac{D_{110}}{a^4}\frac{\partial w}{\partial t}\right) \quad (1)$$

where $q_a = \rho_a V_\infty^2 / 2$ is the dynamic pressure, ρ_a is the air mass density, V_∞ is the free stream airflow speed, M_∞ is the Mach number, $\beta = \sqrt{M_\infty^2 - 1}$ and w is the transverse displacement of panel. The nondimensional dynamic pressure and aerodynamic damping are given by

$$\lambda = \frac{2q_a a^3}{\beta D_{110}}, \qquad g_a = \frac{\rho_a V_\infty (M_\infty^2 - 2)}{\rho h \omega_0 \beta^3} \quad (2)$$

where a is the panel length, ρ is the average mass density of the panel, D_{110} is the first entry of the laminate bending stiffness matrix calculated when all the fibers of the composite layers are aligned in the direction of the airflow (x-direction), and $\omega_0 = (D_{110}/\rho h a^4)^{1/2}$ is a convenient reference frequency. Other commonly used non-dimensional parameters are the air-panel mass ratio and aerodynamic damping coefficient[4] which are defined as

$$\mu = \frac{\rho_a a}{\rho h}, \qquad c_a = \left(\frac{M_\infty^2 - 2}{M_\infty^2 - 1}\right)^2 \frac{\mu}{\beta} \quad (3)$$

From Eqs. (2) and (3), $g_a = \sqrt{\lambda c_a}$ and $c_a \approx \mu/M_\infty$ for $M_\infty \gg 1$ as used in reference 4.

In the derivation of equations of motion, it is assumed that the panel is thin, i.e. the ratio of length or width over thickness of the panel is greater than 20. The rotary inertia and transverse shear deformation effects are thus negligible. The von Karman's nonlinear strain-displacement relationships are given by

$$\{\epsilon\} = \begin{Bmatrix} u_{,x} \\ v_{,y} \\ u_{,y} + v_{,x} \end{Bmatrix} + \frac{1}{2}\begin{Bmatrix} w_{,x}^2 \\ w_{,y}^2 \\ 2w_{,x}w_{,y} \end{Bmatrix} - z\begin{Bmatrix} w_{,xx} \\ w_{,yy} \\ 2w_{,xy} \end{Bmatrix}$$

$$= \{\epsilon_m^0\} + \{\epsilon_\theta^0\} + z\{\kappa\} \quad (4)$$

where u, v are the inplane displacements. The membrane strains $\{\epsilon_m^0\}$ and $\{\epsilon_\theta^0\}$ are due to inplane displacements and large deflections respectively, and $\{\kappa\}$ is the curvature.

For a laminated aircraft panel consists of fiber-reinforced composite and piezoelectric layers subjected to a temperature change of $\Delta T(x,y,z)$, the stress-strain relationships of an orthotropic composite layer are given by

$$\begin{Bmatrix} \sigma_1 \\ \sigma_2 \\ \tau_{12} \end{Bmatrix}_s = \begin{bmatrix} Q_{11} & Q_{12} & 0 \\ Q_{12} & Q_{22} & 0 \\ 0 & 0 & Q_{66} \end{bmatrix}_s \left(\begin{Bmatrix} \epsilon_1 \\ \epsilon_2 \\ \gamma_{12} \end{Bmatrix} - \Delta T \begin{Bmatrix} \alpha_1 \\ \alpha_2 \\ 0 \end{Bmatrix}\right)_s \quad (5)$$

where α_1 and α_2 are the major and minor coefficients of thermal expansion respectively, and $[Q]$ is the reduced lamina stiffness matrix. The stress-strain relationships of an orthotropic piezoelectric layer can be written as

$$\begin{Bmatrix} \sigma_1 \\ \sigma_2 \\ \tau_{12} \end{Bmatrix}_p = \begin{bmatrix} Q_{11} & Q_{12} & 0 \\ Q_{12} & Q_{22} & 0 \\ 0 & 0 & Q_{66} \end{bmatrix}_p \left(\begin{Bmatrix} \epsilon_1 \\ \epsilon_2 \\ \gamma_{12} \end{Bmatrix} - \Delta T \begin{Bmatrix} \alpha_1 \\ \alpha_2 \\ 0 \end{Bmatrix}_p - e_3 \begin{Bmatrix} d_{31} \\ d_{32} \\ 0 \end{Bmatrix} \right) \quad (6)$$

where e_3 is the electric field applied, and d_{31} and d_{32} are the electro-mechanical coefficients. The pyroelectricity effect has been neglected in equation (6) and e_{3max} is equal to one half of depolarizing electric field. Thus, for a general k-th layer, either the fiber-reinforced composite (set $e_{3k}=0$) or the piezoelectric layer, the stress-strain relations can be expressed as

$$\begin{Bmatrix} \sigma_x \\ \sigma_y \\ \tau_{xy} \end{Bmatrix}_k = \begin{bmatrix} \overline{Q}_{11} & \overline{Q}_{12} & \overline{Q}_{16} \\ \overline{Q}_{12} & \overline{Q}_{22} & \overline{Q}_{26} \\ \overline{Q}_{16} & \overline{Q}_{26} & \overline{Q}_{66} \end{bmatrix}_k \left(\begin{Bmatrix} \epsilon_x \\ \epsilon_y \\ \gamma_{xy} \end{Bmatrix} - \Delta T \begin{Bmatrix} \alpha_x \\ \alpha_y \\ \alpha_{xy} \end{Bmatrix}_k - e_{3k} \begin{Bmatrix} d_x \\ d_y \\ d_{xy} \end{Bmatrix}_k \right)$$

or

$$\{\sigma\}_k = [\overline{Q}]_k (\{\epsilon\} - \Delta T\{\alpha\}_k - e_{3k}\{d\}_k) \quad (7)$$

where $[\overline{Q}]_k$ is the transformed reduced lamina stiffness matrix.

The stress resultants, per unit length, are defined as

$$(\{N\},\{M\}) = \int_{-h/2}^{h/2} \{\sigma\}_k (1,z) dz \quad (8)$$

which lead to the constitutive relations for a laminated panel as

$$\begin{Bmatrix} N \\ M \end{Bmatrix} = \begin{bmatrix} A & B \\ B & D \end{bmatrix} \begin{Bmatrix} \epsilon^0 \\ \kappa \end{Bmatrix} - \begin{Bmatrix} N_{\Delta T} \\ M_{\Delta T} \end{Bmatrix} - \begin{Bmatrix} N_e \\ M_e \end{Bmatrix} \quad (9)$$

where $\{\epsilon^0\}$ ($= \{\epsilon^0_m\} + \{\epsilon^0_\theta\}$) is the membrane strain, and the laminate stiffness matrices are given by

$$([A],[B],[D]) = \int_{-h/2}^{h/2} [\overline{Q}]_k (1,z,z^2) dz \quad (10)$$

The thermal force and moment, per unit length, are

$$(\{N_{\Delta T}\},\{M_{\Delta T}\}) = \int_{-h/2}^{h/2} [\overline{Q}]_k \{\alpha\}_k \Delta T(1,z) dz \quad (11)$$

and the piezoelectric force and moment, per unit length, are

$$(\{N_e\},\{M_e\}) = \int_{-h/2}^{h/2} [\overline{Q}]_k \{d\}_k e_{3k}(1,z) dz \quad (12)$$

The element equations of motion and matrices are derived using the principle of virtual work

$$\delta W = \delta W_{int} - \delta W_{ext} = 0 \quad (13)$$

The internal virtual work, considered first, is in general

$$\delta W_{int} = \int_A \left[\{\delta\epsilon^0\}^T \{N\} + \{\delta\kappa\}^T \{M\} \right] dA \quad (14)$$

where A is the element area.

The virtual work of the external forces, due to the inertia force and the aerodynamic pressure, is given by

$$\delta W_{ext} = \int_A \left[\delta w \left(p_a - \rho h \left\{ \frac{\partial^2 u}{\partial t^2} + \frac{\partial^2 v}{\partial t^2} + \frac{\partial^2 w}{\partial t^2} \right\} \right) \right] dA \quad (15)$$

The plate bending element employed in the present study is the C^1 continuous rectangular element. The 16 element bending nodal displacements at the four vertices are

$$\{w_b\}^T = \{w_1, w_2, w_3, w_4, w_{x1}, w_{x2}, w_{x3}, w_{x4},$$
$$w_{y1}, w_{y2}, w_{y3}, w_{y4}, w_{xy1}, w_{xy2}, w_{xy3}, w_{xy4}\} \quad (16)$$

The eight element in-plane nodal displacements are defined as

$$\{w_m\}^T = \{u_1, u_2, u_3, u_4, v_1, v_2, v_3, v_4\} \quad (17)$$

and the element membrane displacement functions, u and v, are both linear function in x and y.

With the application of virtual work, Eqs. (13), (14) and (15), and the use of finite element expressions, the equations of motion for a rectangular plate element subjected to a dynamic pressure, temperature change and electric field with aerodynamic damping are obtained and can be expressed as

$$\frac{1}{\omega_o^2} \begin{bmatrix} [m_b] & 0 \\ 0 & [m_m] \end{bmatrix} \begin{Bmatrix} \ddot{w}_b \\ \ddot{w}_m \end{Bmatrix} + \frac{g_a}{\omega_o} \begin{bmatrix} [g] & 0 \\ 0 & 0 \end{bmatrix} \begin{Bmatrix} \dot{w}_b \\ \dot{w}_m \end{Bmatrix}$$
$$+ \left(\lambda \begin{bmatrix} [a_a] & 0 \\ 0 & 0 \end{bmatrix} + \begin{bmatrix} [k_b] & [k_B]_{bm} \\ [k_B]_{mb} & [k_m] \end{bmatrix} - \begin{bmatrix} [k_{N\Delta T}] & 0 \\ 0 & 0 \end{bmatrix} \right.$$
$$- \begin{bmatrix} [k_{Ne}] & 0 \\ 0 & 0 \end{bmatrix} + \begin{bmatrix} [k1_{Nm}]_b + [k1_{NB}]_b & [k1]_{bm} \\ [k1]_{mb} & 0 \end{bmatrix}$$
$$\left. + \begin{bmatrix} [k2] & 0 \\ 0 & 0 \end{bmatrix} \right) \begin{Bmatrix} w_b \\ w_m \end{Bmatrix} = \begin{Bmatrix} p_{b\Delta T} \\ p_{m\Delta T} \end{Bmatrix} + \begin{Bmatrix} p_{be} \\ p_{me} \end{Bmatrix} + \{f\} \quad (18)$$

where $[m]$, $[g]$, $[a_a]$ and $[k]$ are the element mass, aerodynamic damping, aerodynamic influence and linear stiffness matrices respectively; $[k1]$ and $[k2]$ are the first and second-order nonlinear stiffness matrices which depend linearly and quadratically on the element displacements respectively; $\{p\}$ and $\{f\}$ are the element load and internal equilibrium force vectors respectively. The subscripts b and m denote bending and membrane components respectively, and the subscripts $B, N\Delta T, Ne, Nm, NB, \Delta T$ and e denote that the corresponding stiffness matrix or load vector is due to the laminate stiffness matrix $[B]$, thermal membrane force $\{N_{\Delta T}\}$, piezoelectric membrane force $\{N_e\}$, membrane forces $\{N_m\}$ ($= [A]\{\epsilon^0_m\}$) and

$\{N_B\}$ ($=[B]\{\kappa\}$), thermal ΔT and piezoelectric e_3 respectively. All the element matrices are symmetrical except the aerodynamic influence matrix $[a_a]$ which is skew-symmetric. The detailed derivation and expressions of element mass, thermal and geometric stiffness, nonlinear stiffness matrices and load vectors are given in references 11 and 12. These matrices and load vectors due to piezoelectricity can be found in Appendix A.

By summing up the contributions from all the elements and taking account of the kinematic boundary conditions, the system equations of motion for a given rectangular panel can be written as

$$\frac{1}{\omega_o^2}\begin{bmatrix}[M_b] & 0 \\ 0 & [M_m]\end{bmatrix}\begin{Bmatrix}\ddot{W}_b \\ \ddot{W}_m\end{Bmatrix}+\frac{g_a}{\omega_o}\begin{bmatrix}[G] & 0 \\ 0 & 0\end{bmatrix}\begin{Bmatrix}\dot{W}_b \\ \dot{W}_m\end{Bmatrix}$$
$$+\left(\lambda\begin{bmatrix}[A_a] & 0 \\ 0 & 0\end{bmatrix}+\begin{bmatrix}[K_b] & [K_B]_{bm} \\ [K_B]_{mb} & [K_m]\end{bmatrix}-\begin{bmatrix}[K_{N\Delta T}] & 0 \\ 0 & 0\end{bmatrix}\right.$$
$$-\begin{bmatrix}[K_{Ne}] & 0 \\ 0 & 0\end{bmatrix}+\begin{bmatrix}[K1_{Nm}]_b+[K1_{NB}]_b & [K1]_{bm} \\ [K1]_{mb} & 0\end{bmatrix}$$
$$+\left.\begin{bmatrix}[K2] & 0 \\ 0 & 0\end{bmatrix}\right)\begin{Bmatrix}W_b \\ W_m\end{Bmatrix}=\begin{Bmatrix}P_{b\Delta T} \\ P_{m\Delta T}\end{Bmatrix}+\begin{Bmatrix}P_{be} \\ P_{me}\end{Bmatrix}$$

or simply

$$\frac{1}{\omega_0^2}[M]\{\ddot{W}\}+\frac{g_a}{\omega_0}[G]\{\dot{W}\}+(\lambda[A_a]+[K]-[K_{N\Delta T}]$$
$$-[K_{Ne}]+[K1]+[K2])\{W\}=\{P_{\Delta T}\}+\{P_e\} \quad (19)$$

Model Reduction and Actuators

Unsymmetric Panel

The system equation of motion presented in Eq. (19) is not suitable for control design in two shortcomings: (a) the order of degrees-of-freedom of the system displacement vector is too large, and (b) the nonlinear stiffness matrices are functions of system displacement vector. Therefore, this equation has to be transferred into a set of properly chosen modal coordinates with much smaller and manageable degrees-of-freedom (say smaller than 10). This is accomplished by a modal transformation

$$\{W\}=\sum_{r=1}^{n}\sum_{s=1}^{m}q_{rs}(t)\phi_{rs} \quad (20)$$

where ϕ_{rs} corresponds to the normal mode (r,s) from the linear vibration problem of a panel

$$\left(\frac{\omega_{rs}}{\omega_0}\right)^2[M]\phi_{rs}=[K]\phi_{rs} \quad (21)$$

For a rectangular panel with air flow along its length, the proper modal in y-direction is only the first mode (m=1), Eq. (20) becomes

$$\{W\}=\sum_{r=1}^{n}q_r(t)\phi_r=\Phi\mathbf{q} \quad (22)$$

where q is the modal coordinate vector, and n is generally less than 10.

The nonlinear stiffness matrices [K1] and [K2] can be expressed in terms of the modal coordinates as

$$[K1]=\sum_{r=1}^{n}q_r[K1]^{(r)} \quad (23)$$

and

$$[K2]=\sum_{r=1}^{n}\sum_{s=1}^{n}q_rq_s[K2]^{(rs)} \quad (24)$$

The nonlinear modal stiffness matrices $[K1]^{(r)}$ and $[K2]^{(rs)}$ are evaluated with the corresponding element components (see Eq. (18)) obtained from the system modes ϕ_r and ϕ_s as

$$[K1]^{(r)}=\sum_{\substack{all \\ elements}}\begin{bmatrix}[k1_{Nm}]_b^{(r)}+[k1_{NB}]_b^{(r)} & [k1]_{bm}^{(r)} \\ [k1]_{mb}^{(r)} & 0\end{bmatrix} \quad (25)$$

and

$$[K2]^{(rs)}=\sum_{\substack{all \\ elements}}\begin{bmatrix}[k2]^{(rs)} & 0 \\ 0 & 0\end{bmatrix} \quad (26)$$

Thus, all entries of the nonlinear modal stiffness matrices $[K1]^{(r)}$ and $[K2]^{(rs)}$ are known quantities.

With the modal transformation of Eq. (22) and the nonlinear modal stiffness matrices in Eqs. (23) and (24), the system equation of motion, Eq. (19), is thus transferred into the forced Duffing equation in reduced modal coordinates as

$$\mathbf{M}\frac{d^2\mathbf{q}}{d\tau^2}+\mathbf{C}\frac{d\mathbf{q}}{d\tau}+(\mathbf{K}+\mathbf{K}_e+\mathbf{K}_q+\mathbf{K}_{qq})\mathbf{q}=\mathbf{F} \quad (27)$$

where $\tau=\omega_0 t$ is the nondimensional time, and the modal mass $\mathbf{M}$ and aerodynamic damping $\mathbf{C}$ matrices both are diagonal, $\mathbf{K}$ and $\mathbf{K}_e$ are the linear modal stiffness matrices, $\mathbf{K}_q$ and $\mathbf{K}_{qq}$ are linear and quadratic in terms of the modal coordinate $\mathbf{q}$ respectively. The transformation of the equations of motion from the system displacements, Eq. (19), to the model coordinates $\mathbf{q}$, Eq. (27), is given in Appendix B.

Symmetric Panel

Equation (19) can be simplified if we design the piezoelectric laminated composite panel with certain rules. Figure 1 shows a typical laminated composite panel containing the embedded piezoelectric layers of equal thickness placed symmetrically with respect to the midplane of the panel. For a symmetrically laminated panel configuration under a steady-state uniform temperature distribution ΔT, the stiffness matrices due to the laminate stiffness $[B]$ and the thermal bending load vector are all null, that is

$$[K_B]_{bm} = [K_B]_{mb} = [K1_{NB}]_b = \{P_{b\Delta T}\} = 0 \quad (28)$$

Then Eq. (19) can be modified by neglecting the inplane inertia and then substituting the inplane displacement in terms of transverse displacement. The dynamic equations of motion can be expressed, by collecting the terms, as

$$\frac{1}{\omega_o^2}[M_b]\{\ddot{W}_b\} + \frac{g_a}{\omega_o}[G]\{\dot{W}_b\} + ([\overline{K}] + [\overline{K1}] + [\overline{K2}])\{W_b\} = \{P_{be}\} \quad (29)$$

where $[M_b]$ and $[G]$ are the panel mass and aerodynamic damping matrices, $\{P_{be}\}$ is the piezoelectric load vector respectively. The detailed expressions of the matrices $[\overline{K}]$, $[\overline{K1}]$ and $[\overline{K2}]$ are given in Appendix C.

Since the bending ϕ_{rb} and the membrane ϕ_{rm} mode shapes in Eq. (22) are uncoupled, a properly chosen modal transformation instead of Eq. (22) will have the form

$$\{W_b\} = \sum_{r=1}^{n} q_r(t)\phi_{rb} = \Phi q \quad (30)$$

where ϕ_{rb} is the normal mode (m=1) from

$$\left(\frac{\omega_{rs}}{\omega_0}\right)^2 [M_b]\phi_{rb} = [K_b]\phi_{rb} \quad (31)$$

And the system equation of motion in this symmetric case reduces to the Duffing equation as

$$M\frac{d^2q}{d\tau^2} + C\frac{dq}{d\tau} + (K + K_e + K_{qq})q = F \quad (32)$$

The detailed expressions for the modal matrices and vector in above equation can be also found in Appendix C.

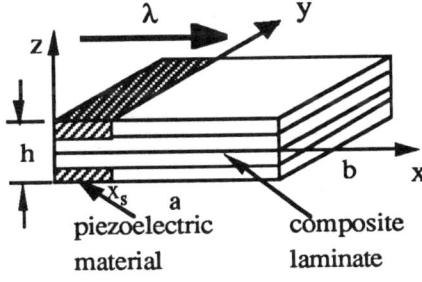

Figure 1 Panel geometry with piezoelectric layers

The piezoelectric layers (top and bottom) of the present design in Figure 1 can be divided to several separate actuation sets. Each actuator set can be stimulated such that one layer contracts, another expands, to create bending moment, and inplane force if the strains or stresses induced by electric fields in both layers are not balanced. The bending moment has been proved to be more effective to suppress the nonlinear flutter motions than the inplane force,[9] therefore only bending moment is considered in the present analysis. The piezoelectric stiffness matrix K_e and the piezoelectric inplane load vector $\{P_{me}\}$ in this case are null. The load vector F is a function of the electric field e_b only, or ($U_b = e_b/e_{3\max}$), and $e_{3\max}$ is the maximum operating electric field of the piezoelectric material. Then the reduced modal equations of motion, Eq. (32), can be expressed as

$$M\frac{d^2q}{d\tau^2} + C\frac{dq}{d\tau} + (K + K_{qq})q = GU_b \quad (33)$$

where G is the control influence matrix which is related with piezoelectric layers. The detailed form of G is given in Appendix A.

Optimal Control Design

For suppression of panel flutter limit-cycle motions, an optimal control approach based on the linear optimal control theory is proposed. The linear model is obtained by ignoring the nonlinear terms from the equations of motion as

$$M\frac{d^2q}{d\tau^2} + C\frac{dq}{d\tau} + Kq = GU_b \quad (34)$$

where U_b is control variable which induces bending moment only. The equation of motion can be cast into a state space form as

$$\frac{dX}{d\tau} = AX + BU \quad (35)$$

where

$$X = \begin{bmatrix} q \\ \dfrac{dq}{d\tau} \end{bmatrix}, \quad A = \begin{bmatrix} 0 & I \\ -M^{-1}K & -M^{-1}C \end{bmatrix}$$

$$B = \begin{bmatrix} 0 \\ M^{-1}G \end{bmatrix}, \text{ and } U = U_b$$

The linear quadratic performance index for optimal control can be formulated as

$$J = \frac{1}{2}\int_0^\infty (X^T Q X + U^T R U)d\tau \quad (36)$$

where Q is a positive semi-definite state penalty matrix and R is a positive definite control penalty matrix. From the optimal control theory, the optimal controller for this linear quadratic regulator problem is

$$U = -R^{-1}B^T P X \quad (37)$$

where P is a positive definite matrix obtained from the following Riccati equation[13]

$$A^T P + PA - PBR^{-1}B^T P = -Q \quad (38)$$

The weighting matrix R is chosen as an identity matrix

times a positive constant. The weighting matrix **Q** can be chosen in many ways as long as it is a positive definite matrix. In this study the form is selected as the energy weighting, i.e. kinetic and potential energy of the system.

$$Q = \begin{bmatrix} K & 0 \\ 0 & M \end{bmatrix} \quad (39)$$

Due to the limitation of the maximum electric field which can be applied to the piezoelectric material, the constraint on the normalized control variable should be

$$|U_{bi}| \leq 1, \quad i=1,2,\ldots,N \quad (40)$$

where N is the number of piezoelectric actuator sets.

Nonlinear equations of motion are used for all the numerical simulations. Optimal control design is performed for increasing values of dynamic pressure. When the controller is activated, the composite panel deflection will be completely suppressed within a certain dynamic pressure which is referred to the maximum flutter-free dynamic pressure λ_{max}. This dynamic pressure is normally increased two to four times of the critical dynamic pressure λ_{cr} by using the properly designed feedback control gain.

To determine the shape of the small piezoelectric actuators, a norm of the feedback control gain (NFCG) is defined. The feedback control gain matrix can be determined from Eq. (37) as

$$k = R^{-1}B^T P \quad (41)$$

The norm of the feedback control gain for each set can thus be written as

$$\text{NFCG} = \sqrt{\sum_{j=1}^{2n} k_{ij}^2}, \quad (i=1,2,\ldots,N) \quad (42)$$

where 2n is the total number of state variables defined in Eq. (35).

Numerical Simulations

The numerical simulations are based on the panel shown in Figure 1, i.e. a simply supported square symmetric composite panel with two embedded piezoelectric layers which can be any size and the same thickness as that of the lamina. The piezoelectric layer is assumed to behave isotropically. The C^1 conforming rectangular element is used in the finite element model and the panel is modeled as 10x6 (10 in x- and 6 in y-direction, total 60 elements) mesh.

The inplane displacements at the edges are considered to be immovable, i.e. $u = v = 0$ for all boundaries. A simply supported square [0/45/-45/90]$_s$ graphite-epoxy composite panel (referred as type I panel) is considered in the example. The dimensions and material properties of the panel are:

$a = b = 12$ in. (30.5 cm)
$h = 0.05$ in. (0.127 cm)

$E_1 = 22.5$ Msi (155 GPa)
$E_2 = 1.17$ Msi (8.07 GPa)
$G_{12} = 0.66$ Msi (4.55 GPa)
$\nu_{12} = 0.22$
$\alpha_1 = -0.04 \times 10^{-6}$ /°F (-0.07×10^{-6}/°C)
$\alpha_2 = 16.7 \times 10^{-6}$ /°F (30.1×10^{-6}/°C)
$\rho = 0.1458 \times 10^{-3}$ lb-sec^2/in.^{4} (1550 Kg/m^3)
$c_a = 0.1$

The properties for the piezoelectric material are given as follows:

$E_p = 9.0$ Msi (62.1 GPa)
$\alpha_p = 3 \times 10^{-6}$ /°F (5.41×10^{-6}/°C)
$\nu_p = 0.3$
$\rho_p = 0.7101 \times 10^{-3}$ lb-sec^2/in.^{4} (7549 Kg/m^3)
$d_{31} = -7.51 \times 10^{-9}$ in./v (-1.9×10^{-10} m/v)
$e_{3max} = 1.52 \times 10^4$ v/in. (6.0×10^5 v/m)

The inplane forces induced by piezoelectric actuators are set to zero in this study since the bending moment is more effective in flutter suppression.[9] Figure 2 studies the convergence of limit-cycle results by retaining different numbers of modes (n) in the modal transformation equation, Eq. (30). It can be seen that a six-mode model will give converged results. Thus, six modes are used for all the simulations. Figure 3 shows the flutter flight envelop at elevated temperatures. The data in this graph is generated by a finite element frequency domain method.[11] It provides us some general information of the region where the flutter motion would occur with a combination of temperature rise and dynamic pressure.

The Runge-Kutta method is used for numerical integration. Nonlinear modal equations of motion are used for all the numerical simulations. Optimal control design is performed for increasing dynamic pressure. The piezoelectric materials are embedded within the composite panel at top and bottom layers which have the lamination angle 0°. When more and more piezoelectric materials are added in, the panel becomes less stiff, and the critical dynamic pressure λ_{cr} will be decreased. As the piezoelectric materials increase to replace the top and bottom layers completely, the composite panel under study becomes [Piezo/45/-45/90]$_s$ (referred as type II panel). The D_{110} for both types of panels are the same for comparison purpose. The λ_{cr} for type I and II panels are found to be 298 and 227, respectively.

Table 1 illustrates λ_{max} for the one-set actuator design of different sizes of piezoelectric layers (see Fig. 1) and two constants for **R** matrix. The piezoelectric actuator is fully extended in y-direction while it is varied in x-direction with a length of x_s from the leading edge. For the small-size actuator design ($x_s/a < 0.3$), λ_{max} is low due to the limited moments induced by the piezoelectric material. When piezoelectric layers completely replace the top and bottom layers of the composite laminate ($x_s/a = 1.0$), λ_{max} drops since the panel reduces to type II with smaller critical dynamic pressure. It can be conclude that piezoelectric materials required to suppress the flutter motion can be neither too little nor too much. The λ_{max} can be increased slightly by using larger **R**. The time to suppress the flutter motions, however, is longer for this case.

Table 1. Comparison of Maximum Flutter-Free Dynamic Pressure λ_{max} for One-Set Actuator Design

x_s/a	λ_{max} (R=500xI)	λ_{max} (R=1000xI)
0.0	λ_{cr}=298	
0.1	600	616
0.2	819	878
0.3	877	975
0.4	1157	1262
0.5	1100	1171
0.6	1037	1150
1.0	597	596
1.0	λ_{cr}=227	

For the type II [Piezo/45/-45/90]$_s$ panel, the piezoelectric materials are completely embedded within the composite laminate. Using two-set piezoelectric actuator normally would lead to a better performance than one-set actuator design due to the flexibility of two control variables. One-set actuator design, in this case, indicates that only the part of the completely embedded piezoelectric layers from the leading edge of the panel actuates. If normalized separating position x_s/a is set to 1, then the two-set design will reduce to the one-set design. The λ_{max} varying with the normalized separating position of one-set and two-set actuator designs for this type of panel is shown in Fig. 4, where the solid line indicates the two-set actuator design and dotted line represents the one-set actuator design. The result clearly shows the better performance obtained by the two-set actuator design. The constant for weighting matrix **R** is chosen to be 500. It is also shown that the λ_{max} reaches to its maximum at about x_s/a at half of the panel. This separating position for the maximum λ_{max}, however, will be changed for different constants chosen for the matrix **R**. The two-set actuator design will increase the critical dynamic pressure about three times for this type of composite panel. Figure 5 shows the time history of the panel deflection and control effort of the type II panel with two-set actuator design. The normalized separating position x_s/a is taken to be 0.4. The dynamic pressure is 600, and the inplane force is set to zero. The control weighting is also a positive constant 500 in this case. The time response of the panel deflection shows a limit-cycle motion. After the controller is activated, the panel deflection is quickly suppressed in about three cycles. It can be seen from the figure that the two voltages U_{b1} and U_{b2} which are applied on two different sets of actuators are approximately 180° out of phase. This means that the ups and downs of the panel deflection are controlled by these voltages accordingly. The numerical simulations show that the flutter can be suppressed completely within the maximum flutter-free dynamic pressure λ_{max} by using the constant feedback control gain properly designed at the λ_{max}.

To use as little as possible the piezoelectric material and reduce the required electric power, one needs to determine the shape of the small patched actuator and its possible location. In this paper, the norms of the feedback control gains (NFCG) are proposed to achieve the optimal location and shape of the piezoelectric actuators. The type II panel with fully embedded piezoelectric material is first divided into 60 (the same as the number of elements) small sets. The norms are then calculated for each set using Eq. (42) at λ=1000 and **R**=1000x**I**. The NFCG values are shown in Fig. 6. The higher the values, the more important in that corresponding area for the flutter control, which means that the piezoelectric actuators should be placed there. With this in mind, the location and the approximate shape of the actuators can be determined. The precise shape can be obtained by refining the element mesh or using the triangular finite elements. Figure 7 illustrates a few configurations of one-set design based on the NFCG. Even some of the piezoelectric actuator configurations will be modified for practical usage. The purpose is to demonstrate the advantages of using the NFCG for actuator design. The panel is considered as the type I. The size of piezoelectric material is limited to 18 elements because the panel would become very soft as the size increases. Table 2 compares the λ_{max} and the ratio of λ_{max} to λ_{cr} for three different sizes of actuators (6,12 and 18 element-size) based on two types of designs. One is based on Fig. 1 where the actuators are placed at the leading edge (LE). Another is based on Fig. 7 where the shape and location of the actuators are obtained by using the concept of the NFCG. The piezoelectric materials are embedded within the composite laminate and are limited only to the small area (maximum 18 elements) for the sake of keeping the panel close to the type I. The critical dynamic pressure λ_{cr} for the type I panel without piezoelectric materials is calculated to be 298. It can be seen from the table that the maximum flutter-free dynamic pressure can be increased a lot for the same size (or area) of piezoelectric material by properly designed one-set piezoelectric actuator using the NFCG. For example, the λ_{max} is increased from 616 of actuator x_s/a=0.1 at the leading edge to 870 of actuator using the NFCG for one-set actuator containing one piece of 6 element-size piezoelectric material. On the other hand, to have the same λ_{max}, the use of the NFCG will result in smaller size of piezoelectric actuator. Furthermore, the λ_{cr} can be increased up to about four times as compared with that of the panel without piezoelectric actuator. The optimal shape and location for maximizing the dynamic pressure are thus obtained based upon the NFCG. This, of cause, is still limiting to the size of the piezoelectric materials that have been used (cf. Table 1).

Table 2. Comparison of λ_{max} of Two Different Designs

	Actuator at the LE		Actuator by NFCG		
x_s/a	λ_{max}	$\lambda_{max}/\lambda_{cr}$	Size	λ_{max}	$\lambda_{max}/\lambda_{cr}$
0.1	616	2.1	6	870	2.9
0.2	878	2.9	12	987	3.3
0.3	975	3.3	18	1170	3.9

Conclusions

The presented set of reduced nonlinear modal equations of motion is efficient and successful for the flutter analysis. The limit-cycle motions of composite panels can be suppressed by using piezoelectric actuators. Only the bending moment induced by piezoelectric materials is used to control the panel flutter since the in-plane force from the materials is insignificant. Once the optimal control gain is properly determined by linear control theory at the maximum flutter-free dynamic pressure, it can be used for any lower dynamic pressure to completely suppress the flutter limit-cycle motions. The norms of feedback control gain matrix are used as a basis to determine the location and shape of the embedded piezoelectric actuators. For the same size of piezoelectric materials, the optimal designed one-set actuator based on the norms of the feedback gain (NFCG) is shown to have a better performance than the actuators placed at the leading edge. The use of NFCG will also result in smaller size of piezoelectric actuators to achieve similar performance (same λ_{max}) as compared with the actuator design without considering the NFCG. The completely embedded piezoelectric layers ($x_s/a=1.0$) will not yield the better performance for the flutter suppression as one usually expects since the panel (type II) becomes less stiff. With optimal feedback control through one-set of small sized piezoelectric actuator design, the critical dynamic pressure can be increased about four times. Relatively larger sized one-set actuator can increase the critical dynamic pressure even more but more materials and power supply will be needed. The proposed concept of the norms of the feedback control gain matrix has been utilized to determine the shape and location of the piezoelectric actuators successfully. And more studies are planned and needed.

Acknowledgments

This work is partially supported by AFOSR under SBIR contract #F49620-93-C-0021. This support is gratefully acknowledged.

References

[1] Bisplinghoff, R. L. and Ashley, H., *Principles of Aeroelasticity*, John Wiley, 1962, pp. 419.

[2] Baker, R., "F-117A Structures and Dynamics Design Considerations," Plenary Session 8, *AIAA Dynamics Specialist Conference*, Dallas, TX. April 1992.

[3] Gray, C. E., Jr. and Mei, C., "Large Amplitude Finite Element Flutter Analysis of Composite Panels in Hypersonic Flow," *Proceedings of the AIAA Dynamics Specialist Conference*, Dallas, TX, April 16-17, 1992, pp. 492-512, (Accepted for publication in *AIAA Journal*).

[4] Dowell, E. H., "Nonlinear Oscillations of a Fluttering Plate," *AIAA Journal*, Vol. 4, No. 7, 1966, pp. 1267-1275.

[5] Dixon, I. R. and Mei, C., "Nonlinear Flutter of Rectangular Composite Panels Under Uniform Temperature Using Finite Elements," In *Nonlinear Vibrations*, DE-Vol. 50/AMD-Vol. 144, ASME Winter Annual Meeting, Anaheim, CA, November 1992, pp. 123-132.

[6] Dehart, D. and Griffin, S., "Astronautical Laboratory Smart Structure/Skins Overview," *Proceedings of the First Joint U.S./Japan Conference on Adaptive Structures*, Maui, Hawaii. Nov. 13-15, 1990, pp. 3-10.

[7] Curie, J. and Curie, P., Acad. Science (Paris), Vol. 91, 1880, pp. 294 and 383.

[8] Scott, R. C., Weisshaar, T. A., "Controlling Panel Flutter Using Adaptive Materials," *Proceedings of the AIAA/ASME/ASCE/AHS/ASC 32nd Structures, Structural Dynamics, and Materials Conference*, Baltimore, MD, April 8-10, 1991, pp. 2218-2229.

[9] Lai, Z., Xue, D.Y., Huang, J.-K. and Mei, C., "Nonlinear Panel Flutter Suppression with Piezoelectric Actuation," *Proceedings of the Second Conference on Recent Advances in Active Control of Sound and Vibration*, Blacksburg, VA, April 28-30, 1993. (Submitted for publication in Journal of Intelligent Material Systems & Structures).

[10] Xue, D.Y. and Mei, C., "A Study of the Application of Shape Memory Alloy in Panel Flutter Control," Accepted for presentation at the 5th International Conference on Recent Advances in Structural Dynamics, Institute of Sound and Vibration Research, University of Southampton, Southampton, UK, July 1994.

[11] Xue, D.Y., "Finite Element Frequency Domain Solution of Nonlinear Panel Flutter with Temperature Effects and Fatigue Life Analysis," Ph.D. Dissertation, Department of Mechanical Engineering and Mechanics, Old Dominion University, Norfolk, VA, 1991. Also *AIAA Journal*, Vol. 31, No. 1, 1993, pp. 154-162.

[12] Dixon, I. R., "Finite Element Analysis of Nonlinear Panel Flutter of Rectangular Composite Plates under a Uniform Thermal Load," M.S. Thesis, Department of Mechanical Engineering and Mechanics, Old Dominion University, Norfolk, VA, 1991. (Accepted for publication in *AIAA Journal*).

[13] Lewis, F. L., *Optimal Control*, Wiley-Interscience Publication, 1986.

Appendix A

The piezoelectric force and moment, per unit length, are given by

$$\{N_e\} = \int_{-h/2}^{h/2} [\overline{Q}]_k \{d\}_k (e_3)_k dz \quad (A1)$$

$$\{M_e\} = \int_{-h/2}^{h/2} [\overline{Q}]_k \{d\}_k (e_3)_k z \, dz \quad (A2)$$

The element load vector due to the electric field e_3 is defined as

$$\{p_{me}\} = [T_m]^T \int_A [C_m]^T \{N_e\} dA \quad (A3)$$

$$\{p_{be}\} = [T_b]^T \int_A [C_b]^T \{M_e\} dA \quad (A4)$$

The piezoelectric stiffness matrix is given by

$$[k_{N_e}] = [T_b]^T \int_A [C_\theta]^T [N_e][C_\theta] dA [T_b] \quad (A5)$$

The matrices $[T_b]$, $[C_b]$, $[T_m]$, $[C_m]$ and $[C_\theta]$ can be found in reference 12.

The control influence matrix **G** in Eq. (33) can be determined as follows:

$$\mathbf{G}\mathbf{U}_b = \Phi^T \{P_{be}\} = \Phi^T \mathbf{P}_e \mathbf{U}_b \quad (A6)$$

where the control variable vector can be expressed as

$$\mathbf{U}_b = \{U_{b1} \quad U_{b2} \quad \cdots \quad U_{bN}\}^T \quad (A7)$$

and the matrix related to piezoelectric layers is given by

$$\mathbf{P}_e = [\mathbf{P}_1 \quad \mathbf{P}_2 \quad \cdots \quad \mathbf{P}_N] \quad (A8)$$

where N in above expressions represents the total number of piezoelectric actuator sets (N=1 shown in Fig. 1); $\mathbf{P}_i$ is a vector assembled from piezoelectric element force vector (A4), which can be written as

$$P_i = (e_{3\max})_i \sum_{\substack{all \\ elements}} (\{p_{be}\}\delta_i) \quad i=1,2,\ldots,N \quad (A9)$$

and

$$\delta_i = \begin{cases} 0, & \text{when the element does not belong to set i} \\ 1, & \text{when the element belongs to set i} \end{cases} \quad (A10)$$

Appendix B

The diagonal modal mass and aerodynamic damping matrices in Eq. (27) are

$$(\mathbf{M},\mathbf{C}) = \Phi^T([M], g_a[G])\Phi \quad (B1)$$

the linear terms in $\mathbf{q}$ are

$$\mathbf{Kq} = \Phi^T(\lambda[A_a] + [K] - [K_{N\Delta T}])\Phi\mathbf{q} \quad (B2)$$

$$\mathbf{K_e q} = -\Phi^T[K_{Ne}]\Phi\mathbf{q} \quad (B3)$$

the quadratic and cubic terms in $\mathbf{q}$ are

$$\mathbf{K_q q} = \Phi^T\left(\sum_{r=1}^n q_r [K1]^{(r)}\right)\Phi\mathbf{q} \quad (B4)$$

$$\mathbf{K_{qq} q} = \Phi^T\left(\sum_{r=1}^n \sum_{s=1}^n q_r q_s [K2]^{(rs)}\right)\Phi\mathbf{q} \quad (B5)$$

and the modal force vector is

$$\mathbf{F} = \Phi^T(\{P_{\Delta T}\} + \{P_e\}) \quad (B6)$$

Appendix C

Linear Stiffness Terms in $\{W_b\}$

$$[\overline{K}]\{W_b\} = (\lambda[A_a] + [K_b] - [K_{N\Delta T}] - [K_{Ne}])\{W_b\} + [K1]_{bm}[K_m]^{-1}(\{P_{m\Delta T}\} + \{P_{me}\}) \quad (C1)$$

First-order Nonlinear Stiffness in $\{W_m\}$

$$[\overline{K1}] = [K1_{Nm}]_b \quad (C2)$$

Second-order Nonlinear Stiffness in $\{W_b\}$

$$[\overline{K2}] = [K2] - [K1]_{bm}[K_m]^{-1}[K1]_{mb} \quad (C3)$$

where all the system matrices in equations (C1)-(C3) have been defined in equation (19).

The nonlinear stiffness matrices $[K1]_{bm}$ and $[K2]$ can be expressed in terms of the modal coordinates as

$$[K1]_{bm} = \sum_{r=1}^n q_r [K1]^{(r)}_{bm} \quad (C4)$$

and

$$[K2] = \sum_{r=1}^n \sum_{s=1}^n q_r q_s [K2]^{(rs)} \quad (C5)$$

where the nonlinear modal stiffness matrices $[K1]^{(r)}_{bm}$ and $[K2]^{(rs)}$ are evaluated with the normal modes ϕ_{rb} and ϕ_{sb}.

The nonlinear stiffness matrix $[K1_{Nm}]_b$ is linearly dependent upon the inplane displacement $\{W_m\}$ which can be expressed in term of the modal coordinates as

$$\{W_m\} = [K_m]^{-1}(\{P_{m\Delta T}\} + \{P_{me}\} - [K1]_{mb}\{W_b\})$$
$$= \{W_m\}_0 + \sum_{r=1}^n \sum_{s=1}^n q_r q_s \{W_m\}_{rs} \quad (C6)$$

where

$$\{W_m\}_0 = [K_m]^{-1}(\{P_{m\Delta T}\} + \{P_{me}\}) \quad (C7)$$

$$\{W_m\}_{rs} = -[K_m]^{-1}[K1]^{(r)}_{mb}\phi_{sb} \quad (C8)$$

This leads to that the matrix $[K1_{Nm}]_b$ is the sum of two matrices: the first matrix $[K_{Nm}]$ is evaluated with $\{W_m\}_0$ and the second matrix $[K2_{Nm}]^{(rs)}$ is evaluated with $\{W_m\}_{rs}$ as

$$[K1_{Nm}]_b = [K_{Nm}] + \sum_{r=1}^n \sum_{s=1}^n q_r q_s [K2_{Nm}]^{(rs)} \quad (C9)$$

With the modal transformation Eq. (30) and Eqs (C4)-(C9), Eq. (29) becomes

$$\mathbf{M}\frac{d^2\mathbf{q}}{d\tau^2} + \mathbf{C}\frac{d\mathbf{q}}{d\tau} + \mathbf{Kq} + \mathbf{K_e q} + \mathbf{K_{qq} q} = \mathbf{F} \quad (C10)$$

where the diagonal modal mass and aerodynamic damping matrices are

$$(\mathbf{M},\mathbf{C}) = \Phi^T([M_b], g_a[G])\Phi \quad (C11)$$

the linear terms in $\mathbf{q}$ are

$$\mathbf{Kq} = \Phi^T(\lambda[A_a] + [K_b] - [K_{N\Delta T}] + [K_{Nm}])\Phi\mathbf{q} + \Phi^T\left(\sum_{r=1}^n q_r [K1]^{(r)}_{bm}\{W_m\}_0\right) \quad (C12)$$

$$\mathbf{K_e q} = -\Phi^T[K_{Ne}]\Phi\mathbf{q} \quad (C13)$$

the cubic term in $\mathbf{q}$ is

$$\mathbf{K_{qq} q} = \Phi^T\left\{\sum_{r=1}^n \sum_{s=1}^n q_r q_s [K2_{Nm}]^{(rs)} + \sum_{r=1}^n \sum_{s=1}^n q_r q_s [K2]^{(rs)} - \sum_{r=1}^n \sum_{s=1}^n q_r q_s [K1]^{(r)}_{bm}[K_m]^{-1}[K1]^{(s)}_{mb}\right\}\Phi\mathbf{q} \quad (C14)$$

and the modal force vector is

$$\mathbf{F} = \Phi^T\{P_{be}\} \quad (C15)$$

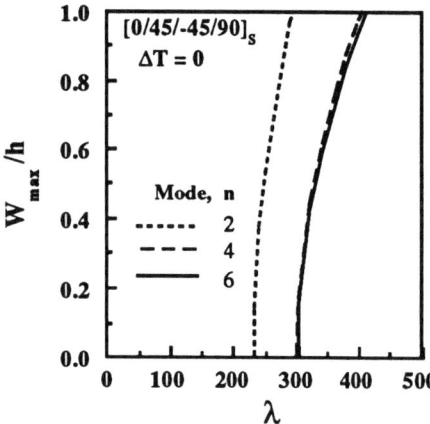

Figure 2 Limit-cycle amplitudes vs. dynamic pressure with different number of linear modes

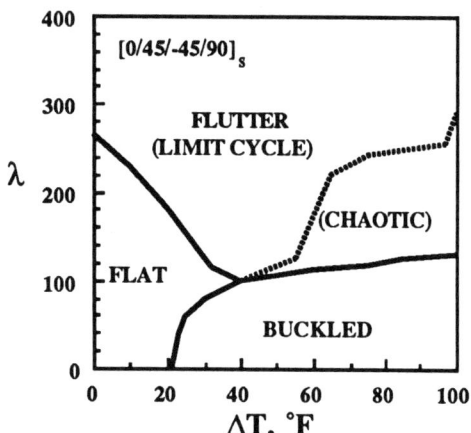

Figure 3 Stability boundaries of a simply-supported square composite $[0/45/-45/90]_s$ panel with uniform temperatures

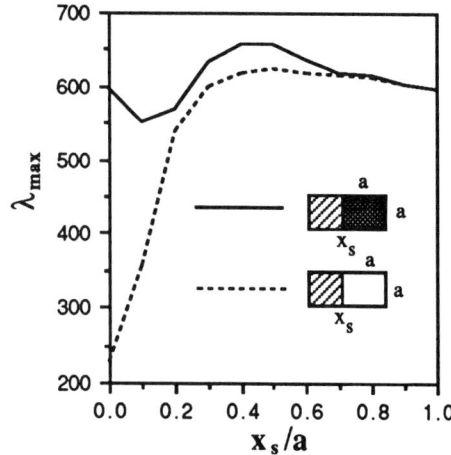

Figure 4 Maximum flutter-free dynamic pressure vs. normalized separating position

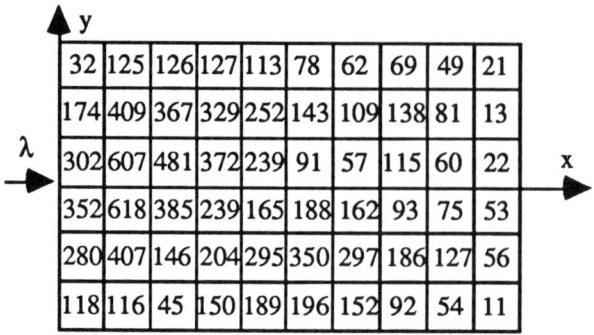

Figure 6 The norms of feedback control gain matrix

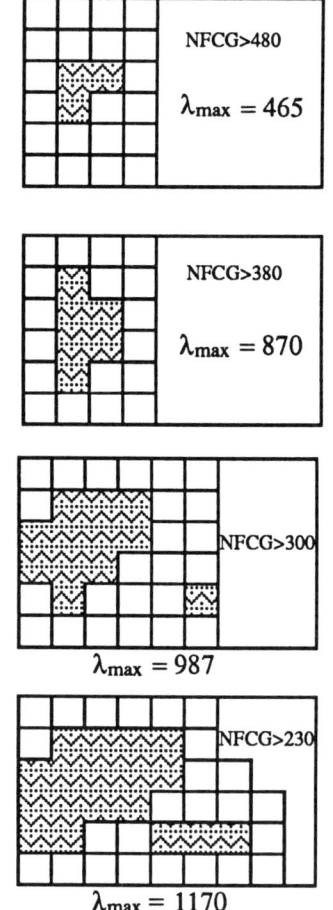

Figure 7 The shape and location of embedded one-set small patched piezoelectric actuators and maximum flutter-free dynamic pressures

116

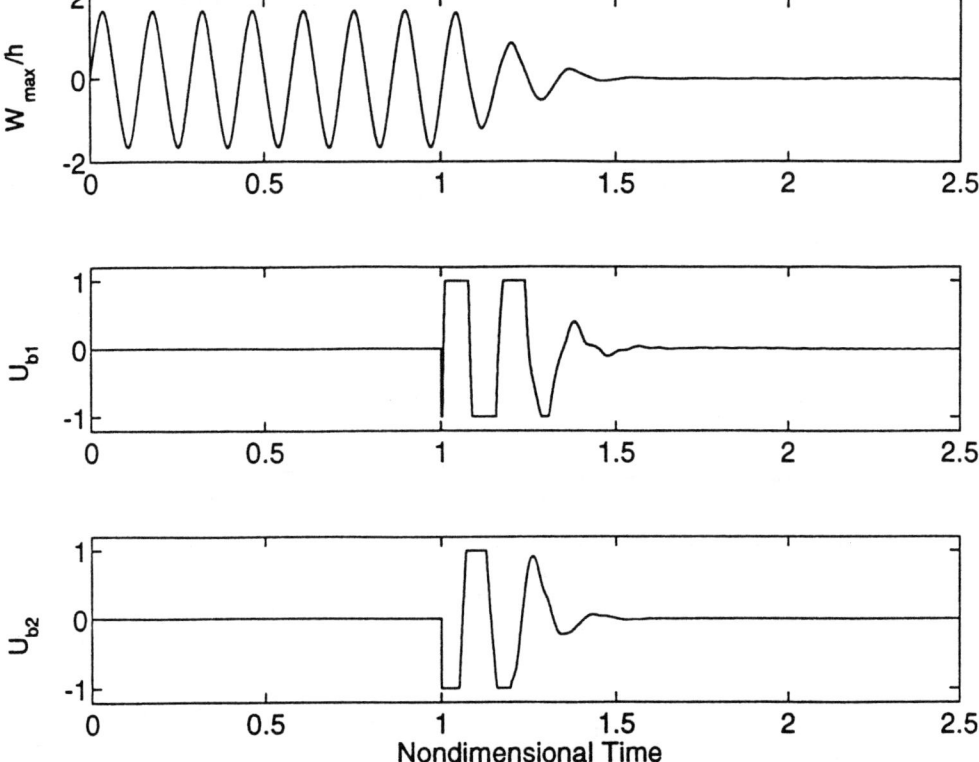

Figure 5 Time history of limit-cycle amplitudes and control efforts for two-set actuator design at $\lambda=600$ and $x_s/a=0.4$

FLUTTER CONTROL OF AN ADAPTIVE COMPOSITE PANEL

A. Suleman* and V.B. Venkayya**

Wright-Patterson AFB
Wright Laboratory, WL/FIBAD
Ohio 45433-6553

Abstract

A finite element formulation for flutter control of an adaptive composite panel with piezoelectric sensors and actuators is presented. Classical laminate theory with electromechanical induced actuation and variational principles are used to formulate the equations of motion. First order piston theory is used to model the supersonic flow. The equations of motion are discretized with four-node, 24 degree of freedom quadrilateral shell elements with one electrical degree of freedom per piezoelectric layer. Performance of the coupled electromechanical composite plate finite element formulation developed here has been tested and compared to the experimental and analytical results documented in the literature.

1. Introduction

In the past decade, technological developments in materials and computer sciences have evolved to the point where their synergistic combination have culminated in a new field of multi-disciplinary research in adaptation. The advances in material sciences have provided a comprehensive and theoretical framework for implementing multifunctionality into materials, and the development of high speed digital computers has permitted the transformation of that framework into methodologies for practical design and production. The concept is elementary: a highly integrated sensor system provides data on the structures environment to a processing and control system which in turn signals integrated actuators to modify the structural properties in an appropriate fashion.

---

* NRC Research Associate, Member
** Principal Scientist, Member
Copyright ©1994 by the American Institute of Aeronautics and Astronautics, Inc. All rights reserved.

Numerous investigators have recently demonstrated the feasibility of the integrated concept. The use of lead zirconate titanate (PZT) ceramics and polyvinylidene fluoride (PVDF) thin film as actuators and sensors for beam and plate vibration has been studied and elastic models have been developed. Some of the analyses are based on analytical approaches[1-4]. Other studies have used finite element methods for beams and plates with integrated piezoelectric sensors and actuators[5,6], where the host structure and the sensor/actuator system are modelled by stacking isoparametric solid elements. This approach makes the problem large and requires some special techniques such as Guyan reduction to reduce the total degrees of freedom. Other problems associated with the isoparametric solid element in thin plate analysis are the excessive shear strain energies and the higher stiffness coefficients in the thickness direction. To overcome these problems, three internal degrees of freedom were added in the formulation, which makes the problem large and complex.

The purpose of this study is to develop a more efficient finite element formulation for vibration control of a laminated composite plate with piezoelectric sensors and actuators. By modelling the plate and the sensor/actuator system with the four noded bilinear Mindlin plate element, the problems associated with the solid element are eliminated and the problem size is considerably reduced.

The applicability and effectiveness of the electromechanical composite plate to panel flutter control problem is proposed, given its resurgent interest in the development of high speed flight vehicles such as the HSCT. Panel futter is a dynamic instability that results from the interaction between the motion of the panel and the aerodynamic forces created by this motion, which results in failure due to fatigue. In order to extend the life of the panel, this type of motion must be suppressed.

Performance of the adaptive composite plate is to be characterized by the degree of control that is involved when compared with more conventional flutter suppression techniques such as thickening of panels, aeroelastically tailoring the panel, adding stiffeners, using new materials, and reshaping the panel.

2. Electromechanical Plate Model

Consider a laminated composite plate containing distributed piezoelectric layers that can be either bonded to the surface or embedded within the structure as shown in Figure 1(a).

To derive the equations of motion for the laminated composite plate, in an aerodynamic field with piezoelectrically coupled electromechanical properties, we use the generalized form of Hamilton's principle

$$\delta \int_{t_1}^{t_2} [T - \Pi + W_e + W_a] dt = 0 \quad (1)$$

where T is the kinetic energy, Π is the potential energy, W_e is the work done by the electrical field and W_a is the work done by the aerodynamic forces. The kinetic and potential energies can be written in the form

$$T = \int_V \frac{1}{2} \rho \dot{\bar{u}}^T \dot{\bar{u}} dV; \quad \Pi = \int_V \frac{1}{2} \bar{S}^{c^T} \bar{T}^c dV$$

where $\bar{S}^c$ and $\bar{T}^c$ are the generalized elastic strain and stress vectors. The work done by the electrical forces can be written as

$$W_e = \int_{V_p} \frac{1}{2} \bar{S}^{e^T} \bar{T}^e dV_p$$

where $\bar{S}^e$ is a vector of electrical field (volts/meter) in the piezoelectric material, and $\bar{T}^e$ is a vector of electrical displacements (charge/area).

2.1 Electromechanical Constitutive Relations

For piezoelectrics the properties are defined relative to the local poling direction. Available piezoelectric materials have the direction of poling associated with the 3 direction (Figure 1b) and the material is approximately isotropic in the other two directions. In matrix form the equations governing these material properties can be written as

$$\bar{T}^e = \mathbf{e}^T \bar{S}^c + \epsilon \bar{S}^e$$

$$\bar{T}^c = \mathbf{c} \, \bar{S}^c - \mathbf{e} \, \bar{S}^e$$

where $\bar{T}^e$ is the electric displacement vector; $\mathbf{e}$ is the dielectric permittivity matrix; $\bar{S}^c$ is the elastic strain vector; ϵ is the dielectric matrix at constant mechanical strain; $\bar{S}^e$ is the electric field vector; $\bar{T}^c$ is the elastic stress vector and $\mathbf{c}$ is the matrix of elastic coefficients at constant electric field strength.

2.2 Strain-Displacement Relations

The state of deformation for a Mindlin plate with electromechanical properties is described by eight generalized strains and one electrical field parameter per lamina. Thus, the augmented generalized strain vector takes the form

$$\bar{S} = \{\bar{S}^m \quad \bar{S}^b \quad \bar{S}^{ts} \quad \bar{S}^e\}$$
$$= \{S_x^m \quad S_y^m \quad S_{xy}^m \quad S_x^b \quad S_y^b \quad S_{xy}^b$$
$$S_{xz}^{ts} \quad S_{yz}^{ts} \quad -E_1 \quad \ldots -E_{n_p}\}$$

where n_p is the number of piezoelectric layers in the element. For the bilinear finite element with four nodal points, we use the shape functions:

$$\mathbf{N} = \frac{1}{4}(\bar{C} + \bar{\xi}\xi + \bar{\eta}\eta + \bar{H}\xi\eta)$$

in which

$$\bar{C}^T = [\;1,\;\;1,\;\;1,\;\;1];$$
$$\bar{\xi}^T = [-1,\;1,\;\;1,-1];$$
$$\bar{\eta}^T = [-1,-1,\;1,\;\;1];$$
$$\bar{H}^T = [\;1,-1,\;1,-1].$$

There are six degrees of displacement degrees of freedom at each node for the elastic behaviour, and there is one potential degree of freedom per layer for the piezoelectric effect. Thus

$$\bar{q}_i^c = \{u \quad v \quad w \quad \theta_x \quad \theta_y \quad \theta_z\}_i;\; i=1,...,4$$
$$\bar{q}^e = \{\phi_1 \quad \ldots \phi_{n_p}\}$$

The strain-displacement relations for the bilinear element are based on the Mindlin first order shear deformation theory and the electric field-potential relations $\bar{S}^e = -\nabla.\phi$. The potential degrees of freedom are constant throughout the plane of the piezoelectric layer and they are assumed to vary linearly through the thickness. Thus the matrix relating the generalized strains to the nodal displacements and electric potentials can be written as follows:

$$\bar{S} = \left\{ \begin{array}{c} \bar{S}^c \\ \bar{S}^e \end{array} \right\} = \begin{bmatrix} \mathbf{b}^c & 0 \\ 0 & \mathbf{b}^e \end{bmatrix} \left\{ \begin{array}{c} \bar{q}^c \\ \bar{q}^e \end{array} \right\}$$

where

$$\mathbf{b}_i^c = \begin{bmatrix} \frac{\partial N_i}{\partial x} & 0 & 0 & 0 & 0 & 0 \\ 0 & \frac{\partial N_i}{\partial y} & 0 & 0 & 0 & 0 \\ \frac{\partial N_i}{\partial y} & \frac{\partial N_i}{\partial y} & 0 & 0 & 0 & 0 \\ 0 & 0 & 0 & 0 & z\frac{\partial N_i}{\partial x} & 0 \\ 0 & 0 & 0 & -z\frac{\partial N_i}{\partial y} & 0 & 0 \\ 0 & 0 & 0 & -z\frac{\partial N_i}{\partial x} & z\frac{\partial N_i}{\partial y} & 0 \\ 0 & 0 & \frac{\partial N_i}{\partial x} & 0 & N_i & 0 \\ 0 & 0 & \frac{\partial N_i}{\partial y} & -N_i & 0 & 0 \end{bmatrix}$$

$$\mathbf{b}^e = \begin{bmatrix} \frac{1}{t_1} & \cdots & 0 \\ \vdots & \ddots & \vdots \\ 0 & \cdots & \frac{1}{t_{n_p}} \end{bmatrix}$$

2.3 Stress-Strain Relations

The composite laminate plate is persumed to consist of perfectly bonded laminae. Moreover, the bonds are presumed to be infinitesimally thin as well as non-shear-deformable.

Thus, following the classical lamination theory[7], the state of stress in the element is given by

$$\bar{T} = \{\bar{T}^m \quad \bar{T}^b \quad \bar{T}^{ts} \quad \bar{T}^e\}$$
$$= \{T_x^m \quad T_y^m \quad T_{xy}^m \quad T_x^b \quad T_y^b \quad T_{xy}^b$$
$$T_{xz}^{ts} \quad T_{yz}^{ts} \quad D_1 \quad \ldots \quad D_{n_p}\}$$

and the stress-strain relationship takes the form

$$\bar{T} = \left\{ \begin{array}{c} \bar{T}^c \\ \bar{T}^e \end{array} \right\} = \begin{bmatrix} \mathbf{c} & \mathbf{c} & 0 & \mathbf{e} \\ \mathbf{c} & \mathbf{c} & 0 & \mathbf{e} \\ 0 & 0 & \mathbf{g} & 0 \\ \mathbf{e}^T & \mathbf{e}^T & 0 & \epsilon \end{bmatrix} \left\{ \begin{array}{c} \bar{S}^m \\ \bar{S}^b \\ \bar{S}^{ts} \\ \bar{S}^e \end{array} \right\}$$

where $\mathbf{c}$ is the transformed moduli matrix for each lamina including the piezoelectric layers. The transverse shear stiffness matrix $\mathbf{g}$ is defined in terms of the transverse strain energy through the thickness. Substituting for the generalized stress and strain expressions into Eq. (1), we obtain the mass, elastic stiffness and piezoelectric stiffness matrices:

$$\mathbf{M}^j_{cc} = \int_{V_j} \rho \mathbf{N}^T \mathbf{N} dV_j, \quad j = 1, \ldots, n_{el};$$

$$\mathbf{K}^j_{cc} = \int_{V_j} \mathbf{b}^{cT} \mathbf{c} \mathbf{b}^c dV_j, \quad j = 1, \ldots, n_{el};$$

$$\mathbf{K}^j_{ce} = \int_{V_j} \mathbf{b}^{cT} \mathbf{e} \mathbf{b}^e dV_j, \quad j = 1, \ldots, n_{el};$$

$$\mathbf{K}^j_{ee} = \int_{V_j} \mathbf{b}^{eT} \mathbf{e} \mathbf{b}^e dV_j, \quad j = 1, \ldots, n_{el}.$$

2.4 Aerodynamic Loads

Using the first order high Mach number approximation to the linear potential flow theory[8], the work done by the external aerodynamic forces is given by

$$W_a = -\int_A \frac{2Q}{V\beta} \left(V \frac{\partial w}{\partial x} + \frac{M^2-2}{M^2-1} \frac{\partial w}{\partial t} \right) w dA$$

where $Q = \rho V^2/2$ is the free stream dynamic pressure, $\beta = \sqrt{M^2-1}$, V is the free stream velocity, M is the free stream Mach number, and ρ is the air density.

Substituting into Eq. (1), we obtain the following aerodynamic damping and stiffness matrices for each element:

$$\mathbf{A}_1^j = \lambda_1 \int_A \mathbf{N}^T \mathbf{N} dA, \quad j = 1, \ldots, n_{el};$$

$$\mathbf{A}_2^j = \lambda_2 \int_A \mathbf{N}^T \mathbf{N}_{,x} dA, \quad j = 1, \ldots, n_{el};$$

and

$$\lambda_1 = \frac{2Q(M^2-2)}{V(M^2-1)^{\frac{3}{2}}}; \quad \lambda_2 = \frac{2Q}{(M^2-1)^{\frac{1}{2}}}$$

where λ_1 is the aerodynamic damping constant and λ_2 is the dynamic presure parameter. The aerodynamic damping matrix is proportional to the mass matrix and reads

$$\mathbf{A}_1^j = \frac{\mathbf{M}_{cc}^j}{\rho}, \quad j = 1, \ldots, n_{el}$$

where ρ is the material mass density. The aerodynamic stiffness matrix $\mathbf{A}_2^j$ is non-symmetric, due to the nonconservative nature of the aerodynamic loading.

2.5 Equations of Motion

For the entire structure, using the standard assembly technique for the finite element method and applying the appropriate boundary conditions, we obtain the complete equations of motion for a piezoelectrically coupled electromechanical composite panel in a flow field

$$\overbrace{\begin{bmatrix} \mathbf{M}_{cc} & 0 \\ 0 & 0 \end{bmatrix}}^{Inertia} \begin{Bmatrix} \ddot{\bar{U}}^c \\ \ddot{\bar{U}}^e \end{Bmatrix} + \overbrace{\begin{bmatrix} \mathbf{K}_{cc} & 0 \\ 0 & 0 \end{bmatrix}}^{Elastic\,Stiffness} \begin{Bmatrix} \bar{U}^c \\ \bar{U}^e \end{Bmatrix} +$$

$$\overbrace{\begin{bmatrix} 0 & \mathbf{K}_{ce} \\ \mathbf{K}_{ec} & \mathbf{K}_{ee} \end{bmatrix}}^{Piezo\,Stiffness} \begin{Bmatrix} \bar{U}^c \\ \bar{U}^e \end{Bmatrix} +$$

$$\overbrace{\begin{bmatrix} \mathbf{A}_1 & 0 \\ 0 & 0 \end{bmatrix}}^{Aero\,Damping} \begin{Bmatrix} \dot{\bar{U}}^c \\ \dot{\bar{U}}^e \end{Bmatrix} + \overbrace{\begin{bmatrix} \mathbf{A}_2 & 0 \\ 0 & 0 \end{bmatrix}}^{Aero\,Stiffness} \begin{Bmatrix} \bar{U}^c \\ \bar{U}^e \end{Bmatrix} = 0$$

3. Performance Results

The proposed piezoelectric composite plate finite element formulation was incorporated into PROTEC[9]. To demonstrate the usefulness of the derived finite element scheme, several case studies are presented in this section. The numerical results were compared to experiments and finite element simulations performed by Ha, Keilers and Chang[5], Tzou and Tseng[6], and Crawley and Lazarus[10].

3.1 Static Actuation and Sensing

The first validation test case was based on an experiment conducted by Tzou and Tseng. The experimental apparatus consists of a cantilivered piezoelectric bimorph beam with two PVDF layers bonded together and polarized in opposite directions (Figure 2a). The finite element model was divided into five equal elements, each with two piezoelectric layers bonded together. This produced a model with 53 total degrees of freedom (Figure 2b).

First, the actuation mechanism, governed by $\bar{U}^c = -\mathbf{K}_{cc}^{-1}\mathbf{K}_{ce}\bar{U}^e$, is investigated. The top and bottom surface of the beam were subjected to an electric potential across the thickness of the beam and the resulting displacements were determined. A unit voltage (1 V) produces a tip deflection of $3.45 \times 10^{-7} m$ as shown by the results tabulated in Figure 2(c). It is observed that there is no difference between the results evaluated by the composite finite element model and the theoretical results. The slightly lower tip deflection observed in the experiment could be caused by non-perfect bonding, voltage leakage, energy dissipation, etc. The total number of degrees of freedom used in this analysis (63 - 53 structural and 10 electrical) is considerably lower than the model studied by Tzou and Tseng (144 - 108 structural and 36 electrical), resulting is lower computational memory and time requirements.

The bimorph beam is also studied for sensing voltage distribution for a given static deflection. This is the sensing mechanism, governed by $\bar{U}^e = -\mathbf{K}_{ee}^{-1}\mathbf{K}_{ec}\bar{U}^c$. When external tip loads are applied to produce a given deflection pattern, the electrical degrees of freedom output a sensing voltage. The results in Figure 2(d) show that a voltage of 290 V is produced for an imposed tip deflection of 1 cm. The results are in good agreement with the solid finite element solution. However, since the finite plate element used in this study guarantees the continuity of strains due to bending, i.e. rotation at nodes, the accuracy of sensing may be higher than the brick element that guarantees only displacement continuity at nodes. Furthermore, it can be observed that while the results for Tzou and Tseng are given in terms of nodal voltages, the present theory produces elemental voltages, constant over each piezoelectric layer.

The second validation case was based on the experiments conducted by Crawley and Lazarus. The experimental apparatus consists of a cantilevered laminated composite graphite/epoxy plate with distributed G-1195 piezoceramic (PZT) actuators bonded to the top and bottom surfaces (Figure 3a). The finite element model consists of 160 elements with a total of 880 degrees of freedom (Figure 3b).

During actuation, a constant voltage with an opposite sign was applied to the actuators on each side of the plate. The deflections of the center line and both edges were measured by proximity sensors. Figure 3(c) shows the comparison of the deflection due to longitudinal bending for a $[0/\pm 45]_s$ layup between the present plate formulation, the solid brick element finite element model and the experimental results. All the solutions are in close agreement, with lower deflection observed for the solid element formulation due to shear locking effects associated with such finite element models. The discrepancy observed in the experimental results may be attributed to shear losses in the bonding layers.

Next the sensing mechanism was tested. The comparisons are performed against the numerical simulations performed by Ha, Keilers and Chang. During this simulation, the center row of piezoceramics were considered sensors while the outer two rows were used as actuators. A constant voltage of 100 V was applied to one row of actuators with a positive sign on the top surface and a negative sign on the bottom surface. The same voltage was applied to the other row but in this instance the polarity was reversed, thus inducing a twisting motion to the plate. Furthermore, a constant mechanical load of 0.2 N was applied at the tip of the plate. Thus, the output sensor voltages were numerically determined for the combination of electrical and mechanical loads. Figure 3(d) shows good agreement between the solid brick finite element model and the composite plate model formulations.

Finally, consider the same cantilever plate with only two pairs of piezoelectric patches located near the clamped boundary. It was of interest to determine the amount of voltage required by the piezoelectrics for keeping a zero tip deflection of the plate, when subjected to a concentrated load at the tip. Figure 4(a) shows the deformation of the plate under both the mechanical loading and the calculated electrical loading, with the piezoelectric layers in actuation. The voltage required to preserve the zero tip deflection as a function of the applied mechanical loading is presented in Figure 4(b). It is observed that a linear relationship between the voltage and the applied load was obtained.

3.2 Frequency Analysis

The experimental apparatus used in this experiment consists of an aluminum cantilever beam with six pairs of lead zirconate titanate (PZT) tiles attached to the locations shown in Figure 5(a). This experiment has been set up by Hollkamp[11] in the Wright Laboratory. A measurement/excitation spectrum analyzer is connected to the structure to carry out control studies in the frequency domain. The natural frequencies and damping factors of the first three bending modes have been estimated using experimental data. The natural frequencies are 9.2, 57.5 and 160.5 Hz and the estimated damping factors are 1.1%, 0.3%, and 0.4% respectively. In this study, the finite element model used to simulate this system is comprised of 26 elements, 195 structural degrees of freedom, and 24 electrical degrees of freedom (Figure 5b). The frequencies estimated by the finite element model are 9.1, 58.2, and 168.3 Hz (Figure 5c). Further work is currently under way to implement passive and active control design strategies into the finite element code and these are to be used for active panel flutter control problems.

4. Concluding Remarks

An analytical investigation has been performed for analyzing the performance of an electromechanical laminated composite finite element plate model containing piezoelectric actuators and sensors. Based on this study, the following remarks can be made:

(i) The numerical results generated by the electromechanical finite element plate model simulations agree well with experimental data and solid element formulations reported in the literature;

(ii) the finite element model based on the one-point integration Mindlin plate formulation, and with one electrical degree of freedom per piezoelectric layer is much simpler to formulate and more computationally efficient than models based on solid element formulations, where the number of degrees of freedom used to model the problem is significantly larger.

(iii) Dynamic effects and appropriate control algorithms are being considered for application to the panel flutter control problem.

5. References

1. Bayley, T., and Hubbard, J.E., Jr., "Distributed Piezoelectric Polymer Active Vibration Control of a Cantilever Beam", *Journal of Guidance, Control, and Dynamics*, Vol. 8, No. 5, 1985, pp. 606-610.

2. Crawley, E.F., and de Luis, J., "Use of Piezoelectric Actuators as Elements of Intelligent Structures", *AIAA Journal*, Vol. 25, No. 10, 1987, pp. 1373-1385.

3. Crawley, E.F., and Anderson E.H., "Detailed Models of Piezoceramic Actuation of Beams", *J. of Intell. Mater. Syst. and Struct.*, Vol. 1, January 1990, pp. 4-25.

4. Lee, C.K., "Theory of Laminated Piezoelectric Plates for the Design of Distributed Sensors/Actuators. Part I: Governing Equations and Reciprocal Relationships", *J. Acoust. Soc. Am.*, Vol. 87, No. 3, March 1990, pp. 1144-1158.

5. Ha, S.k., Keilers, C., and Chang, F.-K., "Finite Element Analysis of Composite Structures Containing Distributed Piezoceramic Sensors and Actuators" *AIAA Journal*, Vol. 30, No. 3, March 1992, pp. 772-780.

6. Tzou, H.S., and Tseng, C.I., "Distributed Piezoelectric Sensor/Actuator Design for Dynamic Measurement/Control of Distributed Parameter Systems: a Piezoelectric Finite Element Approach", *Journal of Sound and Vibration*, Vol. 138, No. 1, 1990, pp. 17-34.

7. Jones, R.M., *Mechanics of Composite Materials*, Washington, D.C., 1975, pp. 147-187.

8. Nasr-Bismarck, M.N., "Finite Element Analysis of Aeroelasticity of Plates and Shells", *Appl. Mech. REv.*, Vol. 45, No. 12, Part 1, December 1992, pp. 461-482.

9. Brockman, R.A., Lung, F.Y., and Braisted, W.R., "Probabilistic Finite Element Analysis of Dynamic Structural Response", Air Force Wright Aeronautical Laboratories, Report No. AFWAL-TR-88-2149.

10. Crawley, E.F., and Lazarus, K.B., "Induced Strain Actuation of Isotropic and Anisotropic Plates", *AIAA Journal*, Vol. 29, No. 6, 1991, pp. 944-951.

11. Hollkamp, J.J., "Multimodal Passive Vibration Suppression with Piezoelectrics", *AIAA/ASME/ASCE/AHS/ACS Structures, Structural Dynamics and Materials Conference*, April 19-22, 1993, La Jolla, California, Paper No. AIAA-93-1683.

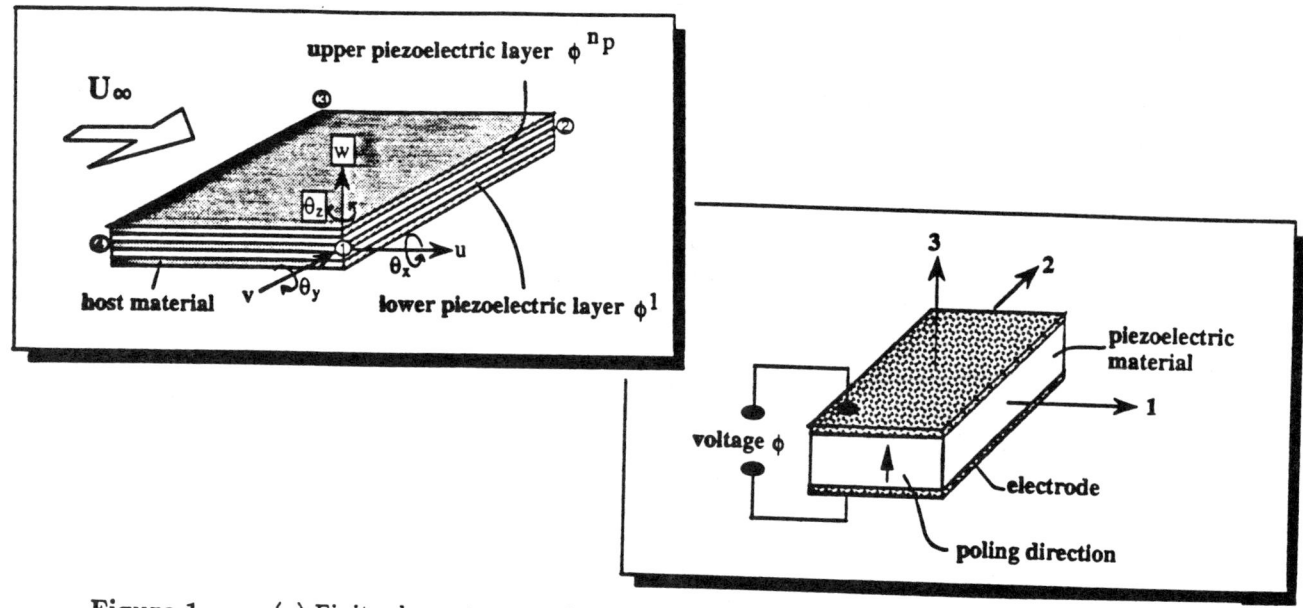

Figure 1   (a) Finite element composite plate element in a flow field showing the displacement degrees of freedom for the elastic properties and the electrical degrees of freedom for the piezoelectric properties; and (b) Piezoelectric layer showing the poling direction and electrode arrangement.

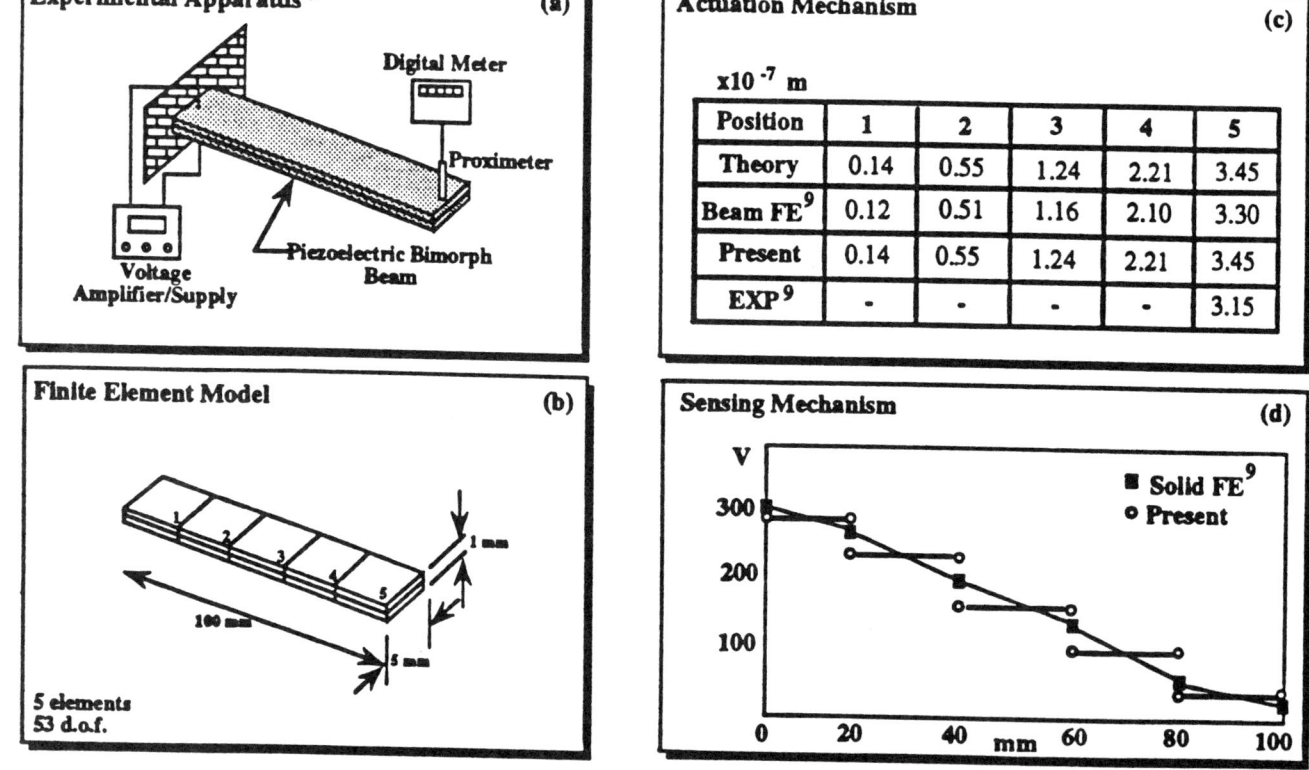

Figure 2   Test study between the present formulation with Tzou and Tseng[6] showing: (a) the experimental apparatus for the piezoelectric bimorph plate; (b) the finite element model with 53 degrees of freedom; (c) the comparison of results for the static actuation mechanism; and (d) static sensing mechanism.

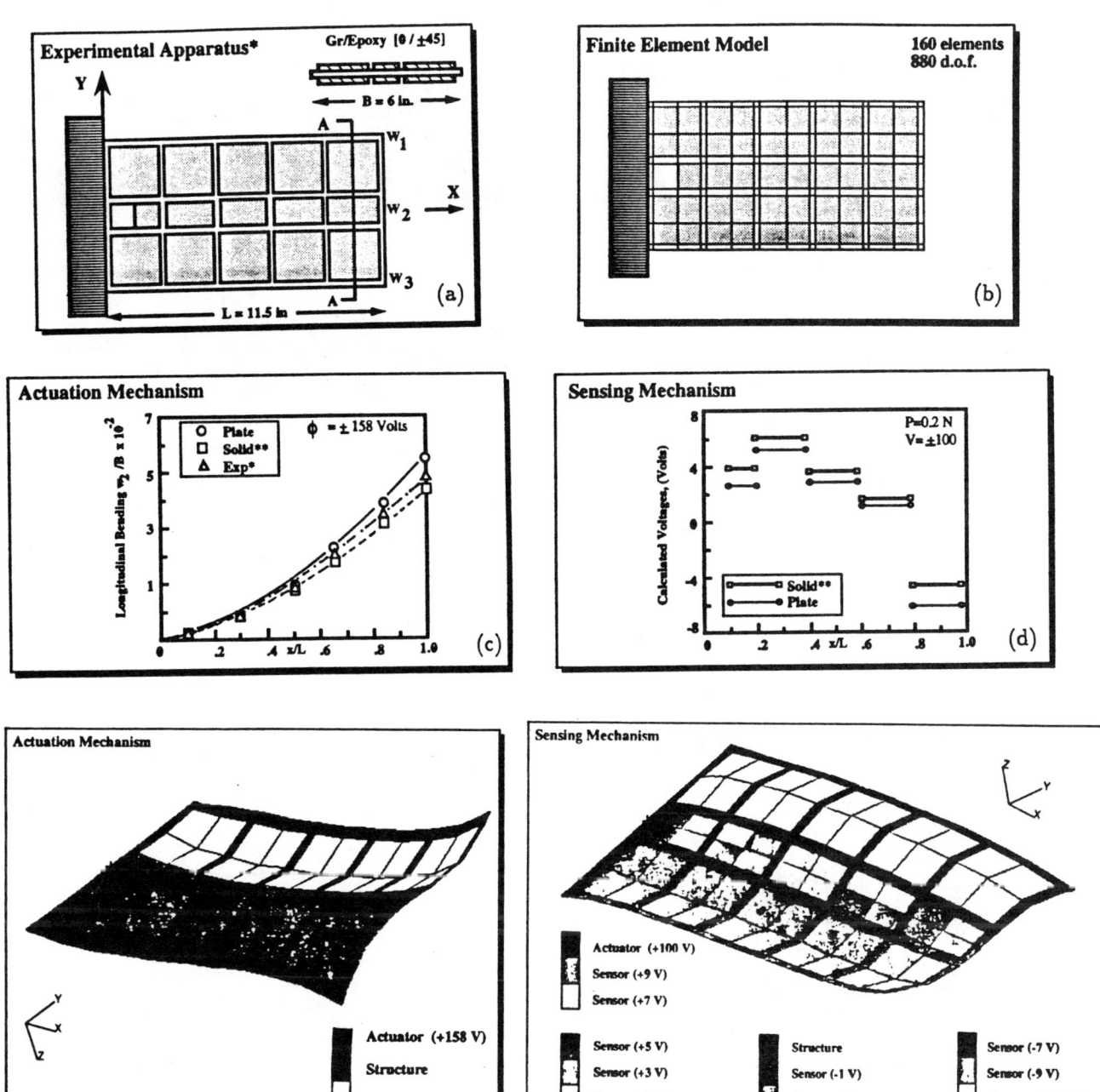

Figure 3 Case study showing the comparison between the prsent formulation and Crawley and Lazarus[10]: (a) the experimental apparatus for the Gr/Epoxy [0/±45] cantilevered plate; (b) the finite element modal with 880 degrees of freedom; (c) the actuation mechanism and the resultant deformed shape; and (d) the sensing mechanism with the sensor voltages shown with deformed plate configuration.

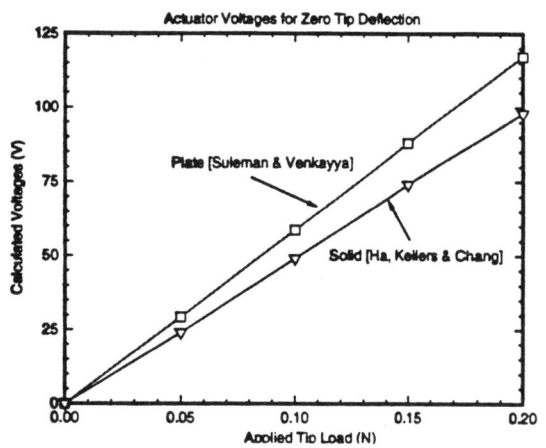 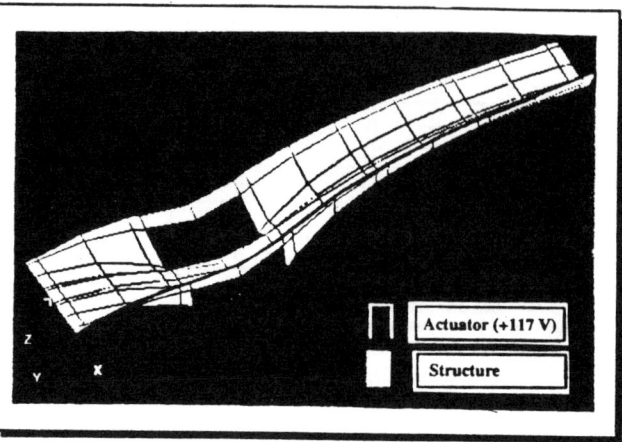

Figure 4   Comparison between the present formulation and Ha, Keilers and Chang[5] showing: (a) the relationship between the required actuation voltages and the various tip loading conditions; and (b) the deformed configuration of the cantilevered plate under the action of the mechanical tip load and electric actuation.

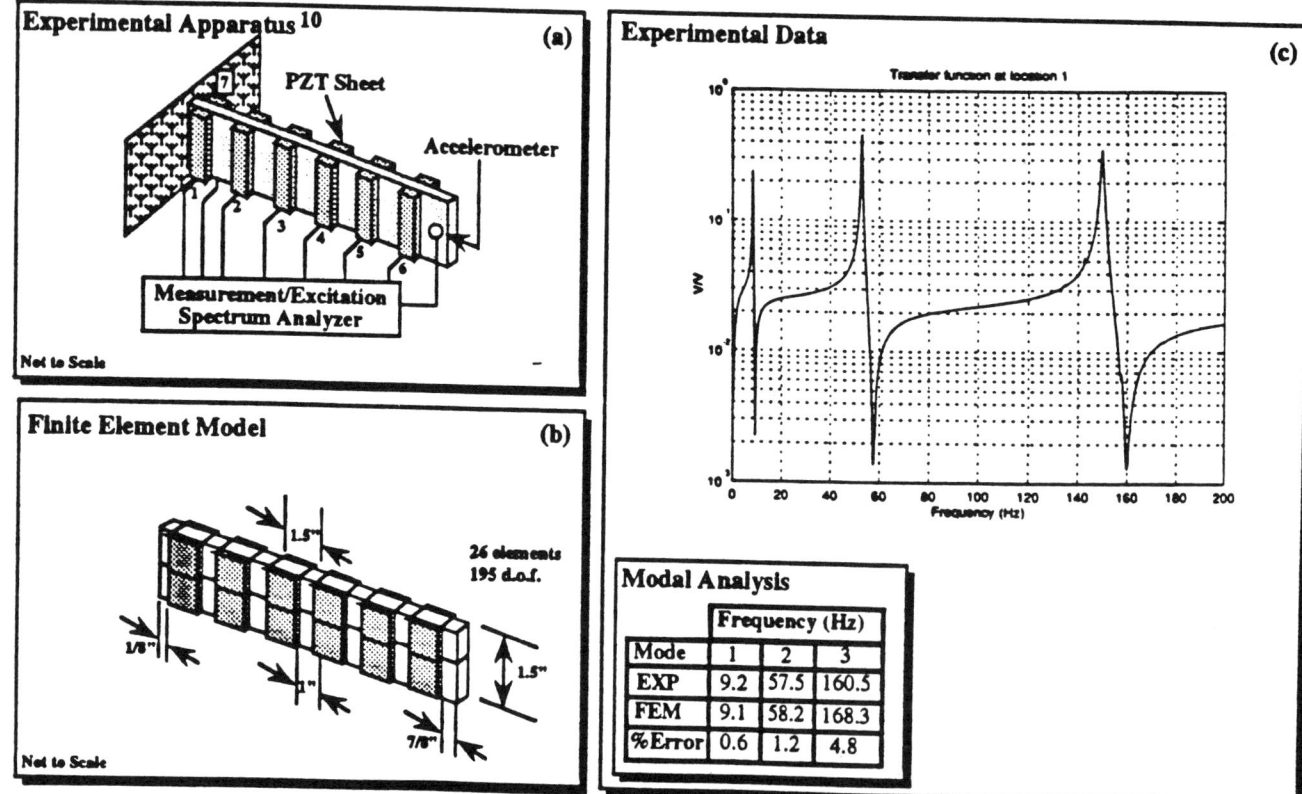

Figure 5   Dynamic test comparison between the present formulation and Hollkamp[11] showing: (a) the experimental apparatus for the aluminum cantilever beam with 6 pairs of PZT tiles; (b) the finite element model with 195 degrees of freedom; and (c) the comparison between the experimentally obtained abd numerical scheme solutions.

ACTIVE FLUTTER SUPPRESSION OF COMPOSITE PLATE WITH PIEZOELECTRIC ACTUATORS

Changho Nam[*]
Hankuk Aviation Univ., Seoul, Korea

Wiedae Kim[†]
Seoul National University, Seoul, Korea

Seungmin Oh[‡]
Agency for Defense Development, Taejeon, Korea

ABSTRACT

This paper describes the use of piezoelectric actuators for the design of active flutter suppression system of a laminated composite plate-wing. The Rayleigh-Ritz method is used to develop the equations of motion of a laminated plate-wing model with piezo actuators. State space aeroservoelastic mathematical model by Rational Function Approximation(RFA) of the unsteady aerodynamic forces is derived. The Minimum State method combined with the optimization technique is adapted for RFA. The linear quadratic regulator theory is employed in active control of the system. The effects of flutter suppression on locations of piezo actuators are examined. The optimal placement of piezoelectric actuators for flutter suppression subject to minimize the performance index is determined analytically by using the optimization technique. The results show the capability of piezo actuators for the control of wing flutter. Numerical simulations of a model with the optimal actuators placement show a substantial saving in control effort compared with the initial model.

Introduction

In order to prevent the static and dynamic instability in a composite wing, we can passively suppress the instability by a proper choice of ply orientation, or actively control by the control surface force of servo actuator. The application of piezoelectric materials provides a new dimension in wing design to eliminate the instability by changing wing configuration to cause lift distribution variation. There has been a considerable amount of research activity to use piezo sensor/actuators for the vibration control of flexible structure. Since the piezoelectric materials exhibit elastic deformation in proportion to the magnitude of an applied electric field and transfer forces to the structure, piezoelectric materials that are bonded at the proper location of a mother structure can be used as actuators.

An analytic model for induced strain actuator coupled with beams and plates has been developed by many authors[1-5]. Recently, the static and dynamic aeroelastic behavior of wing structure with piezo actuators have been studied. Ehlers and Weisshaar[6,7] showed the effects of the piezoelectric actuators on the static aeroelastic behavior such as lift effectiveness, divergence, roll effectiveness. Lazarus, Crawley and Bohlmann[8] used the typical section to represent wing model and performed the analytic study to find the optimal placement thickness of piezo actuator for camber or twist control. They also used TSO code to evaluate the static aeroelastic response of the adaptive wing. Lazarus, Crawley and Lin[9] showed the ability of both articulated control surfaces and piezoelectric actuators to control the dynamic aeroelastic systems. Scott and Weisshaar[10] also examined the capability of panel flutter suppression with piezoelectric actuators and shape memory alloy actuators. Hajela and Glowasky[11] conducted parameter study to suppress the supersonic panel flutter with piezoelectric actuators. They used optimization technique to find the best configuration of panels for both structural weight reduction and maximum flutter speed. Heeg[12] conducted the experimental study for flutter suppression of beam model and demonstrated the application of piezoelectric materials for flutter suppression, which is affixed to the flexible mount system.

The purpose of this paper is to design the active control system for flutter suppression of composite plate-wing with segmented piezoelectric actuators. The optimal placement of piezoelectric actuators on the composite wing structure for the flutter suppression is investigated. The analysis for laminated composite wing with segmented piezoelectric actuators is conducted by Rayleigh-Ritz method. The active control system design for flutter suppression requires the equation of motion to be expressed in a linear time-invariant state-space form. Therefore, it is necessary to approximate the unsteady aerodynamic forces in terms of rational functions of the Laplace variable. Doublet lattice method is used to compute unsteady aerodynamic forces at Mach number 0.7, which are approximated as the transfer functions of the Laplace variable by Minimum State method combined with optimization technique. The optimal placement of piezo actuators is determined by optimization technique referred to as the sequential linear programming method in ADS system. The objective function is the control system performance index associated with the linear quadratic regulator optimal control problem, design variables are the locations of piezoelectric actuators.

Modeling of Adaptive Wing Structures

Forces and moments acting on the laminated com-

---

[*]Member of AIAA, Dept. of Aeronautical & Mechanical Eng.
[†]Associate member of AIAA
[‡]Associate member of AIAA

Copyright ©American Institute of Aeronautics and Astronautics, Inc., 1994. All rights reserved.

posite plate with segmented piezo actuators are derived by the classical laminate plate theory. The stress-strain relations of a thin piezoelectric layer are

$$\left\{\begin{array}{c}\sigma_1\\\sigma_2\\\sigma_{12}\end{array}\right\}_p = \left[\begin{array}{ccc}Q_{p11} & Q_{p12} & 0\\Q_{p12} & Q_{p22} & 0\\0 & 0 & Q_{p66}\end{array}\right]\left\{\left\{\begin{array}{c}\epsilon_1\\\epsilon_2\\\gamma_{12}\end{array}\right\} - \left\{\begin{array}{c}d_{31}\\d_{32}\\0\end{array}\right\}e_3\right\}$$

or

$$\{\sigma\}_p = [Q_p]\{\{\epsilon\} - \{\Lambda\}\} \quad (1)$$

$[Q_p]$ is stiffness matrix of piezoelectric layer, d_{ij} is the piezoelectric strain coefficient, and e_3 is the applied electric field. This equation is similar to stress-strain equation with thermal effect considering the fact that piezoelectric strain Λ has the same form with thermal strain $\{\alpha\}\Delta T$. Inplane forces and moments of laminated plate including the loads of the piezo actuators are obtained by integrating stresses over the ply thickness and expressed as follows[13,14],

$$\{N\} = \int \{\sigma\} dz = [A]\{\epsilon\} + [B]\{\kappa\} - \{N_\Lambda\} \quad (2)$$

$$\{M\} = \int \{\sigma\} z \, dz = [B]\{\epsilon\} + [D]\{\kappa\} - \{M_\Lambda\} \quad (3)$$

where $\{\epsilon\}$, $\{\kappa\}$ are the midplane strain and curvature, respectively. $[A]$, $[D]$, and $[B]$ are extension, bending, and extension/bending coupling stiffness matrices, respectively. These matrices are influenced by not only ply orientation of the laminate but also geometry of the piezo actuators. Inplane forces and moments due to actuator strain are given by

$$\{N_\Lambda\} = \int_{z_p} [Q_p]\{\Lambda\} \, dz \quad (4.a)$$

$$\{M_\Lambda\} = \int_{z_p} [Q_p]\{\Lambda\} z \, dz \quad (4.b)$$

where z_p is the coordinate through thickness in the piezoelectric material.

Strain energy of plate with piezoelectric material can be expressed as

$$U = \frac{1}{2}\int\int_{A,A_p}\lfloor\epsilon \; \kappa\rfloor\left[\begin{array}{cc}A & B\\B & D\end{array}\right]\left\{\begin{array}{c}\epsilon\\\kappa\end{array}\right\}dx\,dy$$
$$- \int\int_{A_p}\lfloor N_\Lambda \; M_\Lambda\rfloor\left\{\begin{array}{c}\epsilon\\\kappa\end{array}\right\}dx\,dy \quad (5)$$

The kinetic energy is

$$T = \frac{1}{2}\int\int_{A,A_p}\rho(\dot{u}^2 + \dot{v}^2 + \dot{w}^2)dx\,dy \quad (6)$$

where A, A_p represent area of composite plate and piezoelectric material, respectively. u, v, w are displacements in x, y, z direction. In Eqs. (5) and (6), strain energy and kinetic energy include effects of both the composite laminate and the piezoelectric materials. Therefore we must carefully apply different integral region as A or A_p and different variable values according to the material.

It is impossible to obtain the closed form solutions to the laminated plate due to the complexity of equation, arbitrary boundary conditions, and external forces such as unsteady aerodynamic forces. In this study, the Rayleigh-Ritz method which is faster without loss of accuracy than finite element method is adopted for the structural analysis. To apply the Rayleigh-Ritz method, we introduce displacement functions in generalized coordinate system to represent displacements u, v, w

$$u(x,y,t) = \sum_{i=1}^{l} X_i(x,y)q_i(t) \quad (7.a)$$

$$v(x,y,t) = \sum_{j=l+1}^{l+m} Y_j(x,y)q_j(t) \quad (7.b)$$

$$w(x,y,t) = \sum_{k=l+m+1}^{l+m+n} Z_k(x,y)q_k(t) \quad (7.c)$$

where $X(x,y)$, $Y(x,y)$, $Z(x,y)$ are shape functions which satisfy boundary conditions of the structures. Using these displacement expressions, equations for strain energy and kinetic energy are written in matrix form,

$$U = \frac{1}{2}\{q\}^T[K_s]\{q\} - \lfloor Q_\Lambda\rfloor\{q\} \quad (8)$$

$$T = \frac{1}{2}\{\dot{q}\}^T[M_s]\{\dot{q}\} \quad (9)$$

For the design of suppression system of a symmetric laminated plate model with piezoelectric actuators, a model is assumed to have the surface bonded piezoelectric actuators on opposite side of plate at the same location. It is also assumed that the same magnitude but opposite direction of the electric field is applied to the actuator so as to create a pure bending moment for flutter suppression as shown in Fig. 1(a). With these assumptions, the displacements in the x and y directions can be neglected. In order to express the displacement in the z direction, we used the free-free beam vibration modes in the x direction (chordwise direction) and cantilever beam vibration modes in the y direction (spanwise direction) as follows;

$$Z_k(x,y) = \Phi_i(x)\Psi_j(y) \quad (10)$$
$$\Phi_i(x) : 1, \; i=1$$
$$: \sqrt{3}(1-2\frac{x}{c}), \; i=2$$
$$: (cos\,\alpha_i\frac{x}{c} + cosh\,\alpha_i\frac{x}{c})$$
$$-\beta_i(sin\,\alpha_i\frac{x}{c} + sinh\,\alpha_i\frac{x}{c}),$$
$$i = 3,4,5...$$
$$\Psi_j(y) : (-cos\,\bar{\alpha}_j\frac{y}{l} + cosh\,\bar{\alpha}_j\frac{y}{l})$$
$$+\bar{\beta}_j(-sin\,\bar{\alpha}_j\frac{y}{l} + sinh\,\bar{\alpha}_j\frac{y}{l}),$$
$$j = 1,2,3...$$

In Eq. (8), if we express $\{Q_\Lambda\}$ in terms of the applied voltage $\{u\}$ and the others $[F_p]$, it becomes as follows;

$$\{Q_\Lambda\} = 2 \sum_{i=1}^{np} \frac{1}{t_p} \left\{ \left(\frac{T}{2} + t_p\right)^2 - \left(\frac{T}{2}\right)^2 \right\}$$

$$\iint_{A_{p_i}} \left\{ [Q_p] \left\{ \begin{array}{c} d_{31} \\ d_{32} \\ 0 \end{array} \right\} \right\}^T \{Z''\} dx dy \, V(i)$$

$$= [F_p]\{u\} \qquad (11)$$

where

$$\{Z''\}^T = -\lfloor \frac{\partial^2 Z}{\partial x^2}, \frac{\partial^2 Z}{\partial y^2}, 2\frac{\partial^2 Z}{\partial x \partial y} \rfloor \qquad (12.a)$$

$$\{u\}^T = \lfloor V(1), V(2), .., V(i), ..V(np) \rfloor \qquad (12.b)$$

T, t_p are the thicknesses of the laminate and piezoelectric layer. $[F_p]$ is the control force per unit voltage, np is the number of bonded piezoelectric materials, $V(i)$ is the applied voltage to the i-th piezoelectric actuator.

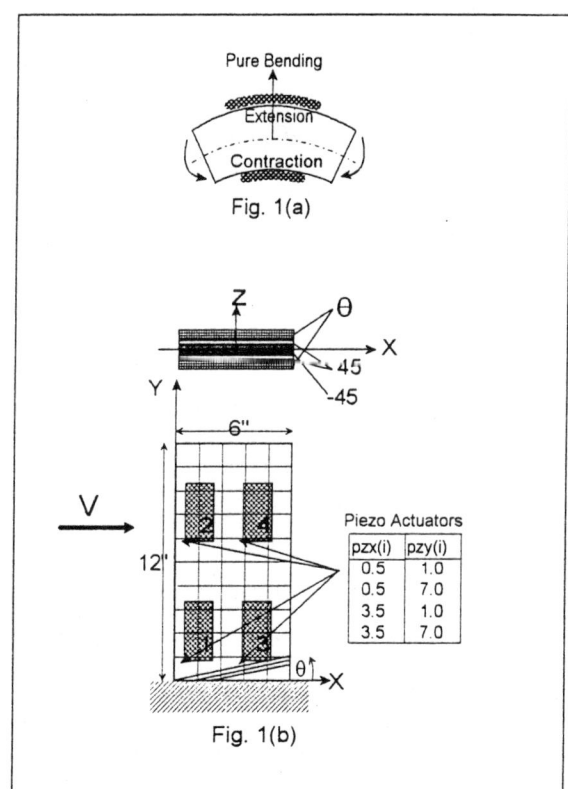

Fig. 1. (a) Pure bending control by piezo actuators, (b) Composite plate wing model with piezo actuators.

Lagrange's equation results in a set of ordinary differential equations of motion as follows;

$$[M_s]\{\ddot{q}\} + [K_s]\{q\} = [F_p]\{u\} + \bar{q}[A]\{q\} \qquad (13)$$

where $\bar{q}$ is the dynamic pressure, $[A]$ is the unsteady aerodynamic force matrix. The aerodynamic force matrix is calculated using doublet-lattice method[15] for Mach 0.7 and 12 reduced frequencies ranging from 0 to 1.2. For calculation of the pressure distribution on an oscillating plate-wing undergoing simple harmonic motion, the plate is divided into total 50 panels arranged in 10 spanwise direction and 5 chordwise direction. $[M_s]$ and $[K_s]$ are respectively the generalized mass and stiffness matrices including the effect of the piezoelectric actuators placement, $[F_p]$ is the control force matrix due to unit applied electric field. After the vibration analysis, a model reduction is performed to obtain a set of equations of motion in the modal coordinates using first six vibration modes.

Rational Function Approximation

The classical aeroelastic analysis such as $V-g$ method and $p-k$ method is performed in frequency domain using unsteady aerodynamic forces computed for simple harmonic motion. However, for the aeroservoelastic analysis and design, it is necessary to transform the equation of motion into the state space form. This requires to approximate the unsteady aerodynamic forces in terms of rational functions of the Laplace variable. There are several methods for the rational function approximation(RFA) such as Roger's method, Matrix Pade method, and Minimum State method[16]. However, when the RFA is conducted, it causes an increase in the total number of states due to the number of augmented aerodynamic states to represent unsteady aerodynamic forces accurately. If the optimization technique is applied, the RFA can be improved without increasing the additional augmented aerodynamic states. In this paper, Minimum State method[17] combined with optimization technique is adopted for the rational function approximation, since the increase in the size of the augmented aerodynamic state is smaller than any other methods.

<u>Minimum State Method</u>

Minimum State method[17] approximates the aerodynamic force matrix by

$$[A_{ap}(\bar{s})] = [A_0] + [A_1]\bar{s} + [A_2]\bar{s}^2 + [D'][\bar{s}I - R']^{-1}[E]\bar{s} \qquad (14)$$

The components of diagonalized matrix $[R']$ are negative constants which are selected arbitrarily. For given $[R']$ matrix, $[A_0], [A_1], [A_2], [D']$, and $[E]$ are determined by using repeated least-square fit. $[R']$ terms are chosen using optimization technique which minimizes the least-square errors without an increase in additional aerodynamic lag terms. Davidon-Fletcher-Powell method[18] is applied to minimize the overall least-square errors with components of the diagonalized matrix $[R']$ as design variables.

Using Eq. (14) for the RFA and the state vector $\{x\}^T = \lfloor q \ \dot{q} \ q_a \rfloor$, the linear time-invariant state space equations of motion which include the effects of piezoelectric control forces are expressed as follows

$$\{\dot{x}\} = [F]\{x\} + [G]\{u\} \qquad (15)$$

where

$$[F] = \begin{bmatrix} [0] & [I] & [0] \\ -[M]^{-1}[K] & -[M]^{-1}[B] & [M]^{-1}[D] \\ [0] & [E] & [R] \end{bmatrix} \qquad (16)$$

$$[G] = \begin{bmatrix} [0] \\ [M]^{-1}[F_p] \\ [0] \end{bmatrix} \quad (17)$$

$$[M] = [M_s] - \frac{1}{2}\rho b^2 [A_2] \quad (18)$$

$$[B] = -\frac{1}{2}\rho b V [A_1] \quad (19)$$

$$[K] = [K_s] - \frac{1}{2}\rho V^2 [A_0] \quad (20)$$

$$[D] = \frac{1}{2}\rho V^2 [D'] \quad (21)$$

$$[R] = \frac{V}{b}[R'] \quad (22)$$

If $[R']$ is set to be a $m \times m$ matrix, the total number of states is $12 + m$.

Suppression System Design

The linear quadratic regulator(LQR) theory is applied for the design of flutter suppression system. The LQR theory determines the optimal control $\{u\}$ to minimize the performance index expressed as follows[19],

$$J = \int_0^\infty [\{y\}^T[Q]\{y\} + \{u\}^T[R]\{u\}] \, dt$$

$$subject\ to\ \{\dot{x}\} = [F]\{x\} + [G]\{u\} \quad (23)$$
$$\{y\} = [C]\{x\}$$

$[Q]$ and $[R]$ are weighting matrices and chosen as identity matrices. The output matrix $[C]$ is specified such that the output vector $\{y\}$ be a vector where the aerodynamic states are deleted from the state $\{x\}$. The corresponding optimal control is given by

$$\{u\} = -[K_G]\{x\} = -[R]^{-1}[G]^T[P]\{x\} \quad (24)$$

where $[K_G]$ is optimal feedback gain matrix, which can be obtained from the Riccati matrix $[P]$. The Riccati matrix $[P]$ is determined by the solution of the following Steady State Riccati equation:

$$[P][F] + [F]^T[P] + [C]^T[Q][C] - [P][G][R]^{-1}[G]^T[P] = 0 \quad (25)$$

It is well known that the performance index J is

$$J = \{x_0\}^T[P]\{x_0\} \quad (26)$$

for a given specified initial condition $\{x_0\}$. The closed loop state equation can be written as

$$\{\dot{x}\} = [F^+]\{x\} = [[F] - [G][K_G]]\{x\} \quad (27)$$

The design speed V_{Design} which is usually higher than the open loop flutter speed is specified to obtain the control gain matrix $[K_G]$ through the minimization of the performance index.

Since the flutter characteristics are dependent on both ply orientation of the laminate and piezo actuators placement which affects the mass, stiffness, and control force matrices, it is necessary to devise some strategy for the design of flutter suppression system.

Optimization Problem Formulation

The optimization technique is applied to determine the best piezoelectric actuators placement for the design of efficient flutter suppression system. The objective function is the performance index of control system defined as Eq. (26). The formulation of this optimization problem may be stated as follows;

$$Find\ actuators\ locations\ \vec{X}$$
$$to\ minimize\ J(\vec{X}) = \{x_0\}^T[P]\{x_0\}$$
$$subject\ to\ G_j(\vec{X}) \leq 0 \quad (28)$$
$$and\ \vec{X}_l \leq \vec{X} \leq \vec{X}_u$$

$\vec{X}$ is the design variables vector, $\vec{X}_l$ and $\vec{X}_u$ are the lower and upper limit of design variables, respectively. G_j is the inequality constraint. Total eight design variables $\vec{X}$ are considered in this study. These are x,y coordinates of the four piezo actuators, $pzx(i), pzy(i)$ shown in Fig. 1(b). 11 constraints are considered and defined as the followings.

- 5 constraints on the closed loop damping ratios;

$$\xi_i = -\frac{\bar{\sigma}_i}{\sqrt{\bar{\sigma}_i^2 + \bar{\omega}_i^2}} \geq \xi_{0i}, \quad i = 1, 2, ..., 5 \quad (29)$$

where $\bar{\sigma}_i, \bar{\omega}_i$ are real and imaginary part of the closed loop eigenvalues. The first five eigenvalues of the closed loop are considered as the constraints.

- 6 constraints on placement of four piezo actuators without overlapping;

$$pzx(i) - pzx(j) \geq 1.5$$
$$or\ pzy(i) - pzy(j) \geq 3.0, \ i,j = 1,2,3,4 \quad (30)$$

There are also side constraints on design variables for the actuators to be placed within plate boundary. The sequential linear programming method[18] in the general purpose optimization program ADS is used to solve this problem.

Numerical Example and Results

The plate model containing four sets of piezoelectric actuators is used as a model for the control system design(see Fig. 1(b)). It is assumed that a set of actuators is bonded on both top and bottom surfaces of the laminated plate in order to generate a pure bending force. The size of each piezo actuator is $1.5 \times 3.0\ inch$. The material properties used in this study are given in Table 1. The laminate has six symmetric layers, $[\theta/\pm 45]_s$. Each layer has uniform thickness. The thicknesses of the $\theta, \pm 45$ layer are $0.03, 0.015\ inch$, respectively. The initial coordinates of the piezo actuators are $pzx(i) = 0.5, 0.5, 3.5, 3.5$ and $pzy(i) = 1.0, 7.0, 1.0, 7.0$ as in Fig. 1(b). The open loop flutter speed is calculated by $V-g$ method and plotted against ply orientation in Fig. 2. It is seen that the flutter speed is low around $10°$ and $100°$ ply angles. $[105/\pm 45]_s$ laminate model is taken as the base model for the suppression system design. In $[105/\pm 45]_s$ model, the flutter occurs at around 620 $feet\ per\ second(fps)$ without control system.

The open loop flutter analysis of $[105/\pm 45]_s$ laminate is conducted with the unsteady aerodynamics transformed into Laplace domain through Minimum State method. The result is shown in Fig. 3. For the construction of Eq. (15), total six components of $[R']$ matrix are used for the RFA. Therefore, the dimension of the state vector is 18. As in Fig. 3, flutter occurs at $620 fps$ by the first torsion mode and flutter frequency is about $59 Hz$.

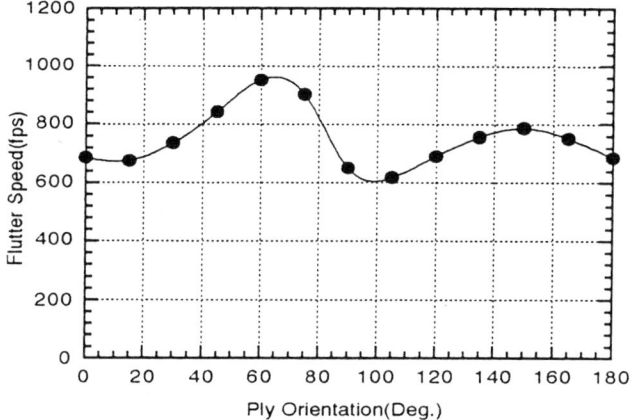

Fig. 2. Flutter speed variation with ply orientations of composite wing.

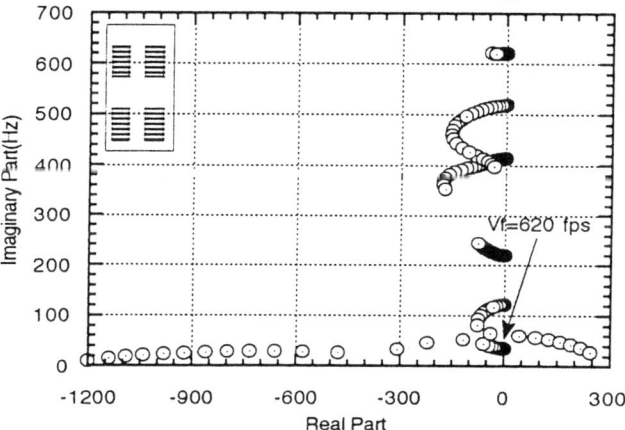

Fig. 3. Open loop root loci for $[105/\pm 45]_s$ initial model.

For the active control design, the design speed V_{Design} is set to be $800 fps$, which is chosen arbitrary. At the design speed the open loop system is unstable in the first torsion mode, yielding unstable eigenvalues at $159.3 \pm j304.1$. The open loop equation of matrix is supplemented with control term. The control input $\{u\}$ is determined by LQR theory subject to minimize the performance index at the design speed. The loci of the closed loop aeroelastic roots with increasing airspeed are plotted in Fig. 4. Above the design speed, the control law provides flutter mode control. The model is stable until a divergence occurs about $890 fps$. The performance index defined in Eq. (26) for the active control system is 154.7.

The optimization technique is applied to find the best placement of piezoelectric actuators for flutter suppression through the minimization of the performance index. Fig. 5 shows the variations in the actuators placement during the optimization process. Two actuators approach to the root region, the other two are attached together and approach to the tip. The final placements of the piezo actuators are $pzx(i) = 0.21, 1.24, 1.72, 2.61$ and $pzy(i) = 0.0, 9.0, 2.25, 9.0$. The optimal value of the objective function which is performance index is 39.4. There is a reduction in performance index about 75% of the inital value. Fig. 6 shows the closed loop root loci about the optimized geometry. The system is stable until a divergence occurs about $900 fps$. The results of the root loci about the optimized geometry are very similar to those of the initial geometry.

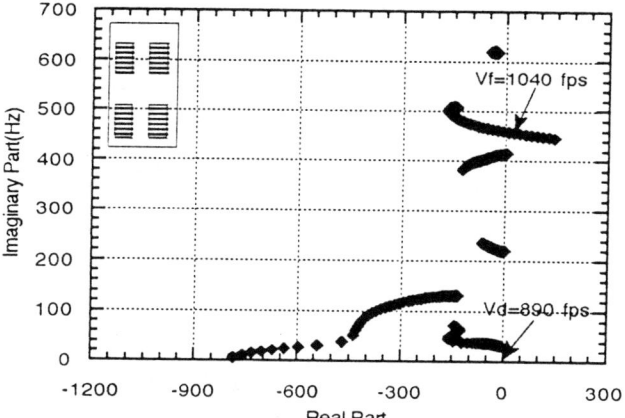

Fig. 4. Closed loop root loci for $[105/\pm 45]_s$ initial model.

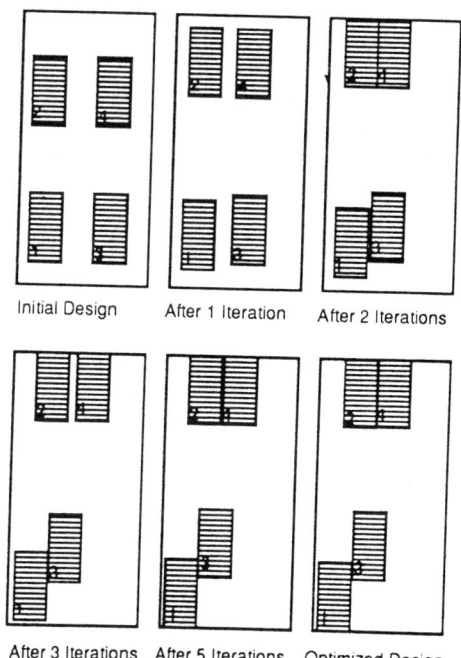

Fig. 5. Variation in piezo actuators placement during the optimization process.

In order to compare the performance of the two control systems, the responses and the control voltages applied to the piezo actuators at the design speed are simulated for each geometry. Fig.7 and Fig. 8 show the time histories of tip displacement and tip twist at the midchord in the model with initial and optimal actuators placement, respectively. The initial conditions $\{x_0\}$ are chosen arbitrary. Although the difference is hardly noticeable, tip displacement of the optimized geometry is smaller than that of the initial geometry. However, the tip twist of the optimized geometry is greater than that of the inital geometry.

Fig. 9 and Fig. 10 show the control voltages applied to the piezo actuators for each geometry. Figures show that the optimized geometry is more efficient for the system control. There is also a substantial saving in the control effort through the optimization process.

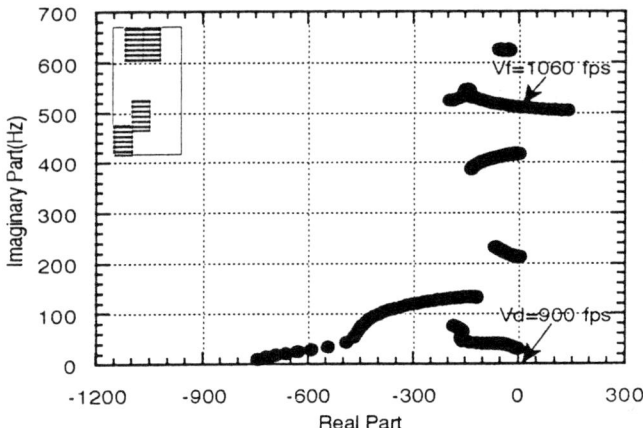

Fig. 6. Closed loop root loci for $[105/\pm 45]_s$ optimized model.

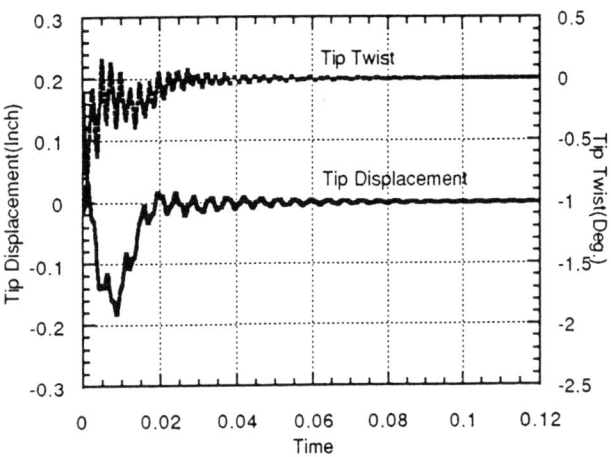

Fig. 7. Time histories of the model with initial piezo actuators placement.

Conclusions

This paper describes the use of piezoelectric actuators for the design of an active flutter suppression system of a composite wing. The flutter analysis of a composite lifting surface with segmented piezoelectric actuators is presented. The control system design is performed for the active flutter suppression. Considering the fact that piezo actuators placement affects both the structural paramater and the control parameter, the optimization technique is applied to find the best actuators placement for flutter suppression. The numerical example demonstrates the feasibility of application of the piezoelectric actuators for flutter suppression. The results show the application of piezo actuators to the control of wing flutter, and about 43% increase in flutter speed compared with the open loop configuration. Numerical result shows a substantial saving in control effort compared with the initial model.

Acknowledgement

The authors would like to thank Dr. Y. Kim of Seoul National University for his helpful discussions.

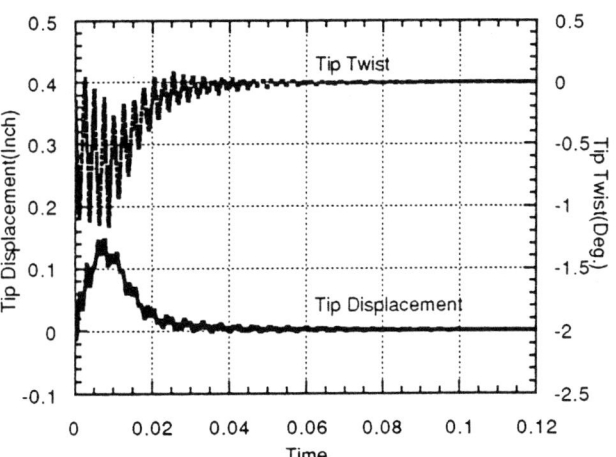

Fig. 8. Time histories of the model with optimal piezo actuators placement.

Table 1. Material properties

Composite Materials
$E_1 = 14.21 \times 10^6\ psi$
$E_2 = 1.146 \times 10^6\ psi$
$G_{12} = 0.8122 \times 10^6\ psi$
$\bar{\rho} = 0.05491\ lb/in^3$
$\nu_{12} = 0.28$

Piezoelectric Materials
$E_p = 9.137 \times 10^6\ psi$
$\rho_p = 0.28\ lb/in^3$
$\nu_p = 0.3$
$d_{31}, d_{32} = 6.5 \times 10^{-9}\ in/V$

References

[1] Crawley, E.F., deLuis, J., "Use of Piezo-Ceramics as Distributed Actuators in Large Space Structures," AIAA Paper No. 85-0626, *Proceedings of the*

26th Structures, Structural Dynamics and Material Conference, NY, April 1985.

[2] Crawley, E.F., deLuis, J., "Use of Piezoelectric Actuators as Elements of Intelligent Structures," AIAA Journal, Vol.25, No.10, Oct. 1987, pp.1373-1385.

[3] Crawley, E.F., Lazarus, K.B., "Induced Strain Actuation of Isotropic and Anisotropic Plates," AIAA Paper No. 89-1326, Proceedings of the 30th Structures, Structural Dynamics and Material Conference, AL, April 1989.

[4] Crawley, E.F., Anderson, E.H., "Detailed Models of Piezoceramic Actuction of Beams," AIAA Paper No. 89-1388, Proceedings of the 30th Structures, Structural Dynamics and Material Conference, AL, April 1989.

[5] Wang, B., Rogers, C.A., "Modelling of Finite-Length Spatially-Distributed Induced Strain Actuators for Laminate Beams and Plates," AIAA Paper no. 91-1258.

[6] Ehlers, S.M., Weisshaar, T.A., "Static Aeroelastic Behavior of an Adaptive Laminated Piezoelectric Composite Wing," AIAA Paper No. 90-1078, Proceedings of the 31th Structures, Structural Dynamics and Material Conference, CA, May 1990.

[7] Ehlers, S.M., Weisshaar, T.A., "Effects of Adaptive Material Properties on Static Aeroelastic Control," AIAA Paper No. 92-2526, Proceedings of the 33th Structures, Structural Dynamics and Material Conference, April 1992.

[8] Crawley, E.F., Lazarus, K.B., Bohlmann, J.D., "Static Aeroelastic Control Using Strain Actuated Adaptive Structures," Proceedings of the First Joint U.S./Japan Conference on Adaptive Structures, Hawaii, Oct. 1990.

[9] Lazarus, K.B., Crawley, E.F., Lin, C.Y., "Fundamental Mechanisms of Aeroelastic Control with Control Surface and Strain Actuation," AIAA Paper No. 91-0985, Proceedings of the 32th Structures, Structural Dynamics and Material Conference, April 1991.

[10] Scott, R.C., Weisshaar, T.A., "Controlling Panel Flutter Using Adaptive Materials," AIAA Paper No. 91-1067, Proceedings of the 32th Structures, Structural Dynamics and Material Conference, AL, April 1991.

[11] Hajela, P., Glowasky, R., "Application of Piezoelectric Elements in Supersonic Panel Flutter Suppession," AIAA Paper No. 91-3191, AIAA Aircraft Design Systems and Operations Meeting, Sep. 1991.

[12] Heeg, J., " An Analytical and Experimental Investigation of Flutter Suppression Via Piezoelectric Actuators," AIAA Paper No. 92-2106, Proceedings of the AIAA Dynamics Specialist Conference, April 1992.

[13] Jones, R.M. , Mechanics of Composite Materials, McGraw-Hill Book Co., 1975.

[14] Whitney, J.M., Structural Analysis of Laminated Anisotropic Plates, Technomic Publishing Co. INC. 1987.

[15] Albano, E., Rodden, W.P., "A Doublet-Lattice Method for Calculating Lift Distributions on Oscillating Surfaces in Subsonic Flows," AIAA Journal, Vol. 7, No. 2, Feb. 1969, pp. 279-285.

[16] Karpel, M., "Design for Active Flutter Suppression and Gust Alleviation Using State-Space Aeroelastic Modeling," Journal of Aircraft, Vol.19, No.3, March 1982, pp.221-227.

[17] Hoadley, S.T., Karpel, M., "Application of Aeroservoelastic Modeling Using Minimum-state Unsteady Aerodynamic Approximations," Journal of Guidance, Control, and Dynamics, Vol.14, No.6, November-December 1991, pp. 1267-1276.

[18] Vanderplaats, G.N., Numerical Optimization Techniques for Engineering Design with Application, McGraw-Hill, 1984.

[19] Meirovitch, L., Dynamics and Control of Structures, Wiley-Interscience, New York 1990.

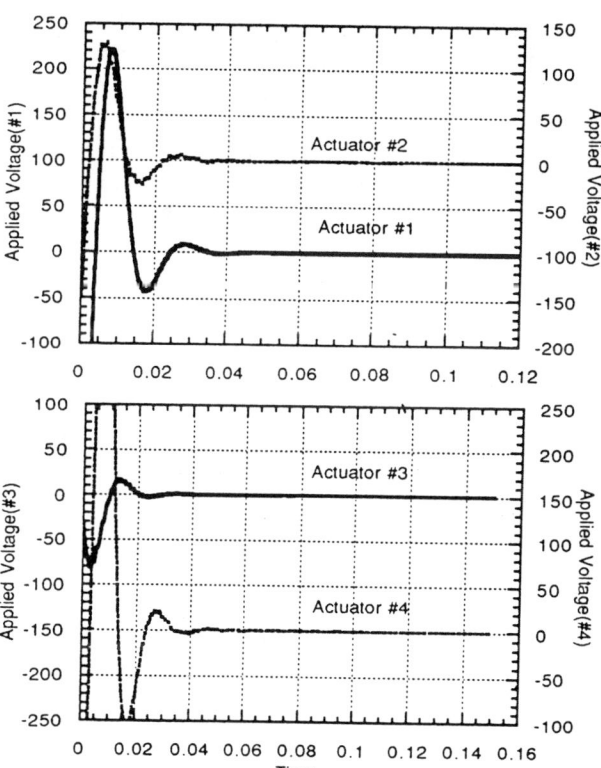

Fig. 9. The control voltages applied to the piezoelectric actuators of the initial design.

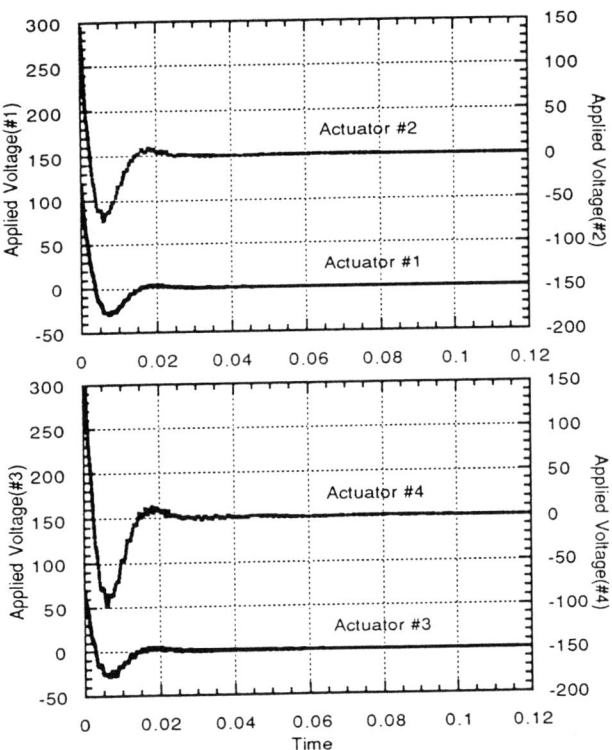

Fig. 10. The control voltages applied to the piezoelectric actuators of the optimized design.

STRUCTURE-CONTROL INTERACTION AND THE DESIGN OF PIEZOCERAMIC ACTUATED ADAPTIVE AIRFOILS

D.L. Steadman[*], S.F. Griffin[*], and S. V. Hanagud[**]
School of Aerospace Engineering
Georgia Institute of Technology
Atlanta, Georgia

ABSTRACT

Structure-control interaction has been used beneficially to develop an adaptive airfoil that can be used in the cyclic and vibration control of a helicopter. The camber of the adaptive airfoil can be changed by using piezoceramic actuators. The deformations produced by the piezoceramic actuators are magnified using an elastic beam that can undergo large deformations with small strains. The needed camber change can be realized over a selected band of frequencies by appropriately tuning the level of structure-control interaction. In this paper, we discuss the use of two actuating surfaces, a piezoceramic sensor, and an active controller to develop a control surface that can produce cyclic camber changes and can be part of an adaptive airfoil.

INTRODUCTION

During the past several years, researchers have discussed many possible applications of smart or adaptive structures. The helicopter system can benefit significantly from the concept of smart structures. One such application, discussed in this paper, concerns the development of a smart rotor system. Here the basic helicopter control is derived from individual blade control rather than the conventional swashplate actuator.

In the conventional control of a helicopter, the vertical climb as well as the longitudinal and lateral control are controlled through the actuation of the swashplate. The swashplate is moved up and down in response to collective control commands to change the pitch of all rotor blades. Similarly, the swashplate is tilted to change the pitch of the rotor blades cyclically to achieve longitudinal and lateral control. Our objective is to remove the swashplate from the helicopter control system and replace it with camber changes of the individual rotor blades. We would like to execute these camber changes by designing adaptive airfoils whose cambers are changed by actuators designed using the principles of smart structures.

In reference (1), we discuss the development of an adaptive airfoil whose camber can be changed by the use of shape memory alloy. These shape memory alloy actuators are ideal for changing the camber of the airfoil and holding the camber change at a given configuration. Thus, the shape memory alloy induced camber changes can be used to achieve a desired collective control of the helicopter. In reference (1), we demonstrate the feasibility of employing the shape memory alloy based adaptive airfoil for helicopter hover control. We have demonstrated the feasibility of achieving a desired hover control by means of flight tests on a remotely piloted helicopter that had been modified to operate with the shape memory alloy based camber actuator.

For individual blade cyclic control, however, we need camber changes at a frequency corresponding to the rotor rotational frequency. For vibration control, we need camber changes at higher frequencies than one per rev. For example, in a four bladed rotor system, we need camber changes at four per rev. Shape memory alloy actuators, which depend on a thermal actuation system, cannot provide camber changes at these frequencies. Most helicopters operate at 3 to 5 Hertz. The remotely piloted helicopter operates in the range 20 to 22 Hertz. Thus, we need an actuator other than shape memory actuators that can respond at these frequencies, provide camber changes corresponding to flap deflections of 3 to 10 degrees, and provide the forces needed to achieve these camber changes.

Piezoceramic actuators can provide the needed frequency response and the forces required

[*] Graduate Student
[**] Professor, Member AIAA

Copyright © 1994 by the American Institute of Aeronautics and Astronautics, Inc. All rights reserved for cyclic and vibration control. However, the deformations realized by piezoceramic actuators are small. Many attempts have been made to multiply the deformations through mechanical lever arrangements. A weakness of such an approach is the loss of the small deformations of the piezoceramic transducers in the tolerances of the mechanical lever system. Thus, we seek an amplification system that can amplify the small strains realized by the piezoceramic actuators.

A beam structure is ideally suited for such an amplification system. An elastic beam in flexure can undergo large deformations with small strains. To achieve cyclic camber changes suitable for the cyclic or vibration control of a helicopter, we need to deflect a beam in its first mode. Because large deflections can only be realized by exciting the beam at its resonant frequency, the actuation will be restricted to one frequency. Therefore, we must be able to change the natural frequency of the beam over a narrow band of frequencies. We have been able to achieve this goal by using an active control mechanism and the beneficial use of structure-control interaction.

Structure-control interaction is usually observed as an undesired feature in the active vibration control of a flexible structure. However, in reference (2), we successfully exploit these interactions to damp the first mode and create a new structure-control mode within a narrow band of frequencies. This beam can be excited by the controller to produce a cyclic or higher harmonic camber change at the desired frequency. In reference (2), we discuss only the feasibility of such an application. In this paper, we discuss our next step in translating the use of structure-control interaction to develop an adaptive airfoil with cyclic camber changes.

ADAPTIVE AIRFOIL

In reference (2), we discussed the use of a cantilever beam as a means of multiplying the piezoceramic actuator and the beneficial use of structure-control interaction to achieve a desired deflection of the cantilever beam over a narrow band of frequencies. The test article used in reference (2) is shown in Figure 1.

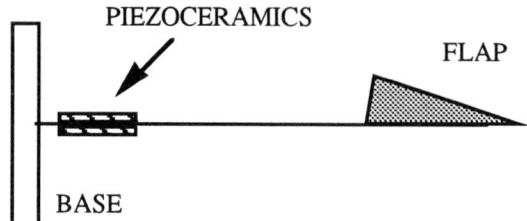

Fig. 1 Test Article 1

In this paper, we are considering a second test article that is much closer to the integrated control surface of an adaptive airfoil. The cantilever of the first article is replaced with two strips of .025 in. aluminum sheet. In actual operation this sheet can be the aircraft's outer skin. Each strip of the airfoil is pinned at one end, and attached to the fairing section at the other end. This test article has stiffness and frequency response characteristics comparable to the fixed-free flexure, but it produces camber change of the airfoil instead of flap tip deflections. The airfoil in its first mode deformed shape is shown in Figure 4. Since the camber line is located between the two strips, its deflected shape depends on the deflected shape of the upper and lower surfaces.

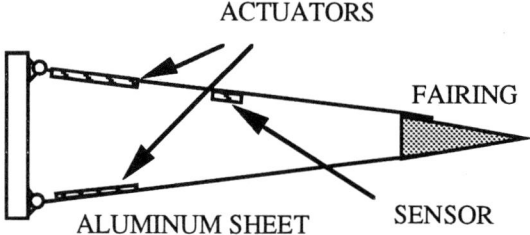

Fig. 2 Test Article 2

Piezoceramic actuators are bonded to the inner surface of each strip. In reference (2), we use an eddy current probe to sense the flap deflection of the cantilever beam and implemented an active controller to change the frequency of the beam. An eddy current probe supplies voltage as a linear function of target displacement. In a practical airfoil, it is difficult to use an eddy probe as a sensor. Therefore, an additional piezoceramic transducer acts as the sensor. The sensor is located on the inside of the skin strips, and is shown in Figure 2. This sensor has two distinct advantages over the eddy current probe. First, the sensor in this structure is designed to measure camber changes, while the objective in reference (2) is to measure tip deflections. This location is an area of high curvature in the desired vibration mode. Voltage from the sensor is directly related to curvature of the skin, and therefore related to camber of the airfoil. Second, all actuators, sensors, and wiring will be located inside the airfoil surface, as is desired in a practical airfoil. The block diagram of the structure system is shown below in Figure 3.

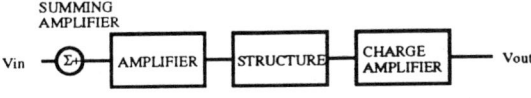

Fig. 3 Structure Loop

The amplifier and charge amplifier are required when using the piezoelectric transducers. The value of V_{out} from the charge amplifier corresponds to camber change in the structure.

CONTROL OBJECTIVES

The first mode shape is the desired shape, for it produces camber over the entire flapped portion of the airfoil. A depiction of this mode shape is shown in Figure 4. The second mode cannot be allowed to participate in the motion. Not only does the second mode's contribution counteracts camber changes over part of the flap, but more importantly, the increased curvature of this mode shape can destroy the actuators. Piezoceramic transducers can only tolerate a small amount of curvature, and any significant amount of second mode motion results in cracks and short-circuits in the actuators. Therefore, our objective is to create a first mode at the desired frequency, while insuring that the dynamic response of second mode is reduced to near zero.

Fig. 4 Airfoil in Deflected Position

CONTROL LOOP

The resonant behavior modification is accomplished by using a local control loop. This local controller actively modifies the structure's frequency response, for it is seen as part of the control surface by the global controller. In the frequency domain, the local controller transforms the lightly damped first natural frequency into a highly damped, relatively flat frequency response function while creating a relatively high gain frequency response function at the frequency desired for cyclic control. We can actively tune to this given frequency by tuning the controller. This transformation is achieved through a local feedback loop. Input for this feedback loop is the global control surface deflection and control forces for this actuation is through the piezoceramic actuators. The structure-control resonant frequency can be tailored through adjustment of the compensator in the feedback loop. The block diagram for the complete structure-control system is shown below in Figure 5.

The first step in control loop design is to construct a pole-zero model of the structure from experimental measurements. The transfer function of the system is shown in Figure 6.

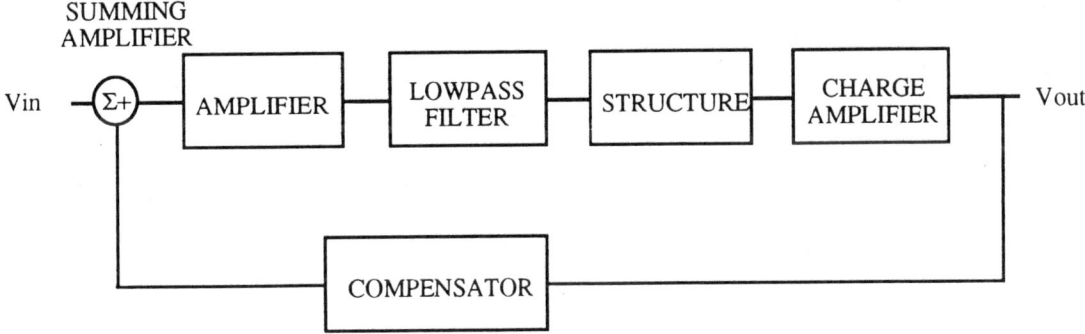

Fig. 5 Structure-Control Loop

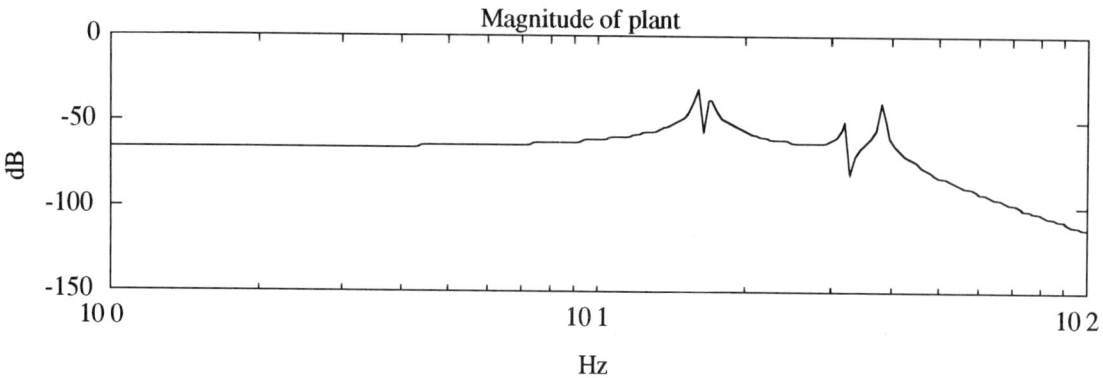

Fig. 6 Magnitude of Structure

The transfer function shows two resonant groups, or modes; one centered at 16 Hz. and a second centered at 36 Hz. Within each group are two peaks. Each peak within a group corresponds to the natural frequency of one skin strip of the control surface. That is, the first group of peaks corresponds to the first natural frequency of both strips, and the second group corresponds to the second natural frequency for both strips. This effect is the result of the slight differences in structural characteristics between the top and bottom surfaces of the flap, and should be anticipated in any structure of this type. Also, because of the location of the sensor, the first and second modes are 180 degrees out of phase.

There are two important features in this control loop. The first is a low-pass filter connected with the amplifier and the structure. The filter is tuned so that there is a 180 degree phase difference between the phase shift at the first mode and the phase shift at the second mode. The low-pass cutoff frequency is in between the two modes. This filter performs two important functions. First, it lowers the gain admitted to the unwanted second mode. Second, it eliminates the 180 degree phase shift between the modes. Without this filter, any attempt to damp one mode will inevitably remove damping from the other. Using this filter, damping can be added to both modes simultaneously.

The second feature is the feedback compensator loop. The compensator implemented to perform this transformation is a simple second order band-pass filter known as a Strain Rate Filter, or SRF[3]. This compensator has been used in the past to provide high levels of active damping to space structure test-beds[4]. The compensator is a bandpass filter, designed in this application with a very narrow pass band. A resistor within the filter is adjustable, so that the location of the gain peak can be moved as required. The effect of this compensator loop is to add damping to the two structural modes, while removing damping at the desired frequency. For the greatest altered resonance response, the compensator is tuned so that the maximum feedback gain is available A bode plot of a SRF is shown in Figure 7.

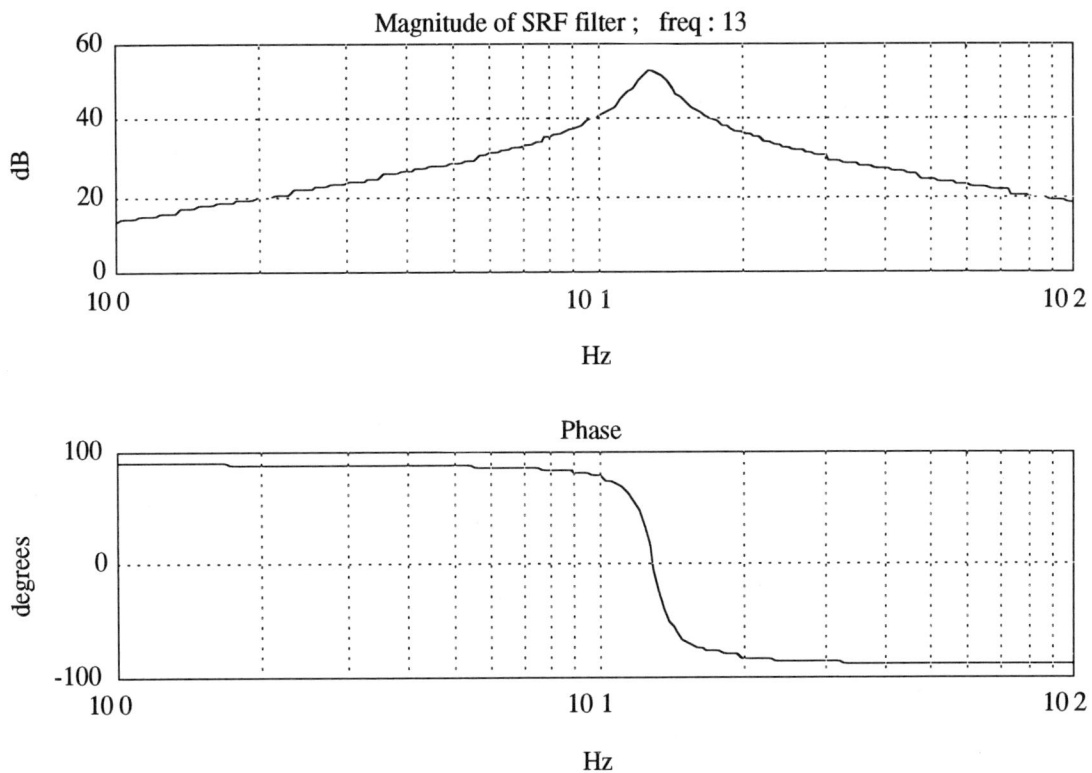

Fig. 7 Bode Plot of SRF

INSTRUMENTATION

PZT 5A piezoceramic transducers are used as sensors and actuators. The thickness of these transducers is .005 inches, and they are poled through the thickness. The high voltage amplifier is a Krohn-Hite 50 watt amplifier, while other amplifiers in the circuit are based on the 741 OP Amp. All experimental frequency response measurements were taken with a GENRAD dynamic signal analyzer, with random noise as the input.

RESULTS

The compensator bandpass frequency was set to values used in simulation. Closed loop frequency response measurements were taken as the feedback gain was adjusted. The closed loop frequency response showed the development of a structure-control mode just before the feedback gain reached an unstable level, while the original structural modes were damped. These results qualitatively matched those predicted in simulation, as shown in Figure 8.

The real test of the interaction is the systems motion when driven to produce camber changes. For example, with the filter frequency response tuned as shown in Figure 7, the structural open loop first mode occurred at 16 Hz. With the loop closed, a structure-control mode existed near 13 Hz. At this mode, camber changes were produced at the desired frequency. Although the system was adjusted to be just short of unstable, there was no evidence of a second mode instability.

ANALYTICAL EXTRAPOLATION

The second objective is to evaluate the smart control surface's ability to operate in a flight environment. When the control surface structure is discussed above, the first structural resonant peak is treated as being of constant location and magnitude. However, in the aircraft operating environment, the aeroelastic effects of flight may modify the effective resonant frequencies and damping of the flap structure. The feedback compensator is required to adapt to this changing frequency response as the stiffness and damping effects change due to flight condition. As an example, we have considered the aeroelastic effects on its operation.

SCHEMATIC REPRESENTATION

In order to conduct the flight environment simulation of the control surface system, pole-zero models of the flap structure and the local controller are extrapolated from the

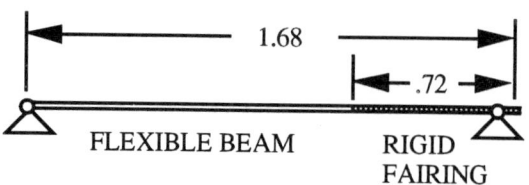

Fig. 9 Idealized Structure

The beam under aeroelastic loading takes the place of the structure in the control block diagram. In our simplified model, the surrounding atmosphere is modeled as finite number of negative, linear springs to be considered as part of the structure. Thin airfoil theory is used to calculate the stiffness values of these springs. Aerodynamic damping is not considered.

We assume that the aerodynamic loads are small, and that they do not alter the mode

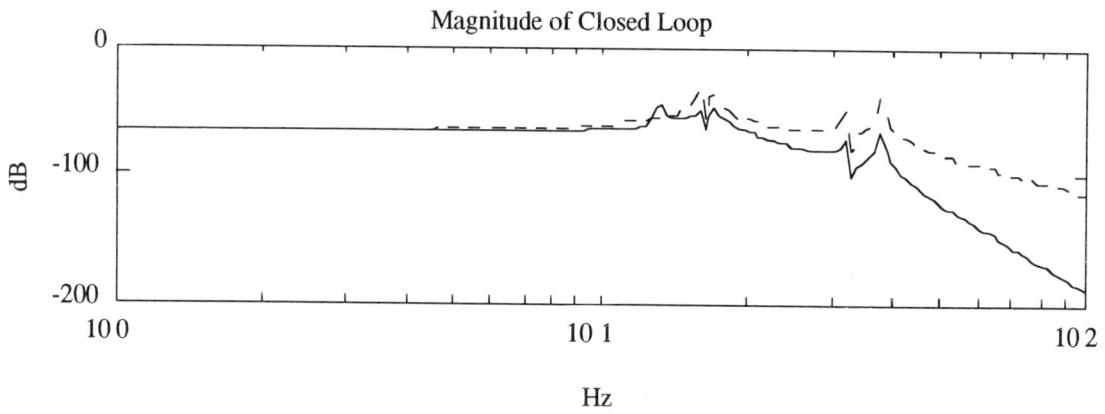

Fig. 8 Magnitude of Structure/Control Loop

experimental data. The analysis assumes a NACA 0012 airfoil with the integral trailing edge flap operating on 30 percent of the chord. The chord length is assumed to be 2.4 feet, but the proportions of the integral flap remain constant. Induced lift is then calculated using thin airfoil theory. Consistent with the thin airfoil theory, which considers the mean camber line as representative of the entire airfoil, the structure is modeled by considering only the vibrations of the mean camber line. This beam also has a first natural frequency at 16 Hz. A schematic of this representation is shown below in Figure 9.

shapes of the beam. If this assumption is correct, then only the natural frequency of the beam will be affected. We can verify this assumption at the end of the analysis by comparing strain energy of the deflected beam to work done by the atmosphere.

AIRLOAD CALCULATION

In this example, the airfoil is moving forward at 300 ft/s at sea level. The initial angle of attack is zero, although the change in camber also causes some change in angle of attack. The values below are for a fairing rotation of six degrees.

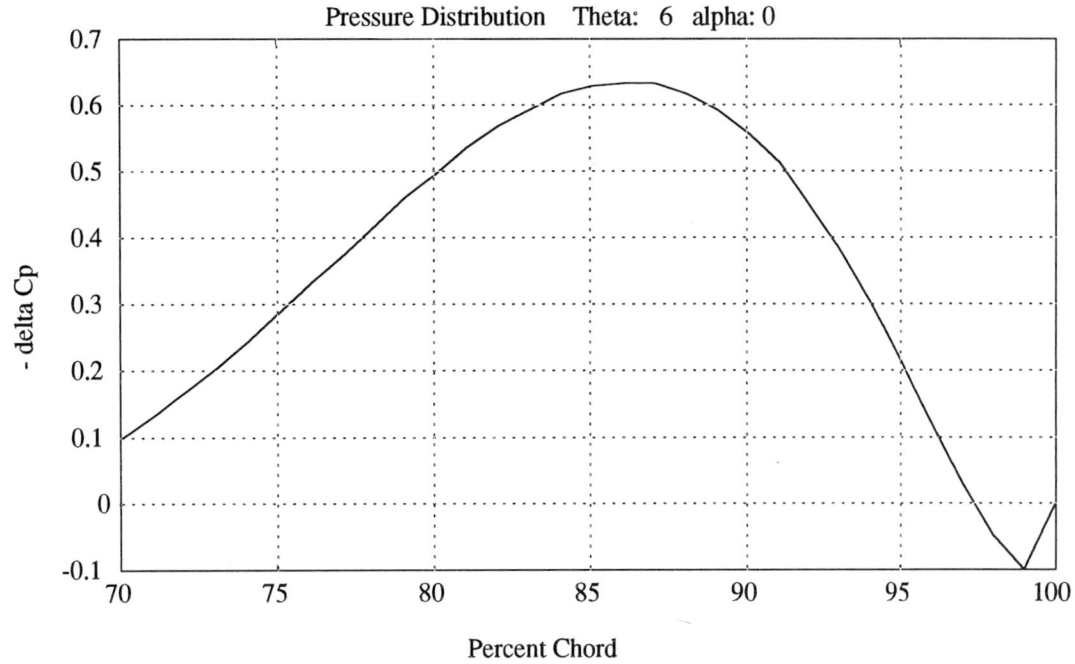

Fig. 10 Coefficient of Pressure over Airfoil

The coefficient of pressure as calculated over the integral flap is shown in Figure 10.

These pressures, combined with beam displacement, are used to calculate the effective stiffness of the beam under aerodynamic loading. For this flight condition, the new first natural frequency is calculated as 15.47 Hz. The transfer function for the idealized smart airfoil in flight is shown in Figure 11.

As in the experimental case, a new structure-control resonant peak for the idealized

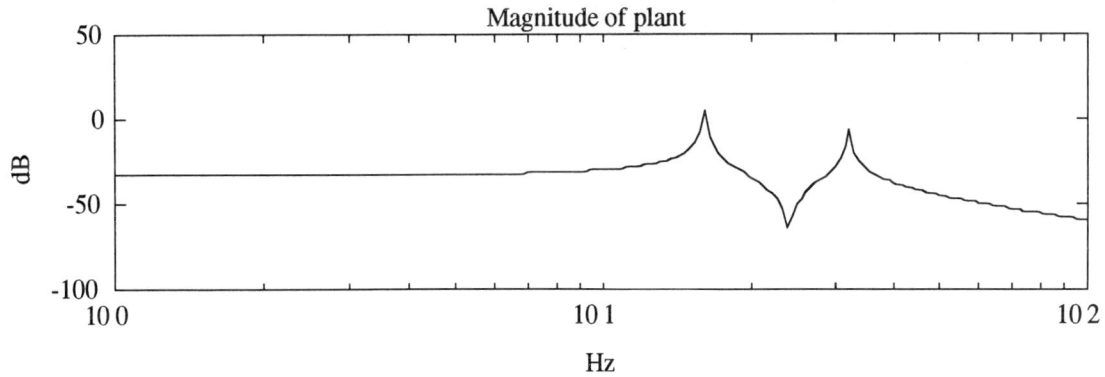

Fig. 11 Magnitude of Idealized Structure

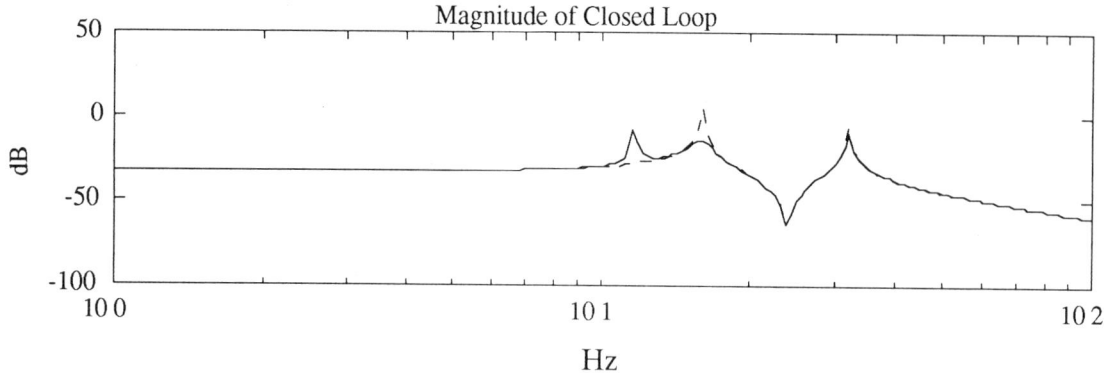

Fig. 12 Magnitude of Idealized Structure/Control Loop

system can be created through a correctly adjusted feedback compensator, as shown in Figure 12.

RESULTS

While the addition of aerodynamic loading did lower the natural frequency, it did not affect the structure so much that our control is ineffective. In our analysis, the frequency never shifted more than 1 Hertz. Also, energy analysis shows the strain energy of the deflected beam to be an order of magnitude higher than work done by the aerodynamic forces, validating our original assumption.

While the second-order filters constructed for this experimentation are effective for our simple model, a more sophisticated control algorithm would be required for an environment involving unsteady dynamic pressure and aerodynamic damping.

CONCLUSION

The result of this work is the development of a adaptive airfoil based on the modified resonant elastic vibrations of the lifting surface. This concept has been demonstrated experimentally and evaluated in flight analytically. This airfoil can potentially be used to provide cyclic control at any target frequency within a selected range.

References:

1. Roglin, R.L., Hanagud, S.V., and Kondor, S., "Adaptive Airfoils for Helicopters", 35th Structures, Structural Dynamics, and Materials Conference, AIAA 1994.

2. Griffin, S.F., and Hanagud, S.V., "Smart Aerodynamic Control Surface Actuator," 1994 North American Conference on Smart Structures and Materials, SPIE 1994.

3. Huelsman, L. P., "Active and Passive Analog Filter Design," Mcgraw-Hill inc. 1993.

4. Bronowicki, A.J et al., "ACESA Active Member Damping Performance," SPIE Proceedings Series on Smart Structures and Materials 1993, Vol. 1917, pp.836-847, Feb. 1993.

5. Lazarus, K.B., Crawley, E.F., and Bohlmann, J.D., "Static Aeroelastic Control Using Strain Actuated Adaptive Structures," Journal of Intelligent Materials, Systems, and Structures, Vol. 2, pp386-410, July 1991.

6. Giurgiutiu, V., Chaudhry, Z., and Rogers, C. A., " Engineering Feasibility of Induced Strain Actuators for Rotor Blade Active Vibration Control," 1994 North American Conference on Smart Structures and Materials, SPIE 1994.

Seismic Vibration Suppression by the Axial-Force-Rated Actuators

Vikas Chawla*, Murat Şener†, Senol Utku‡
Duke University, Durham, NC 27708-0288, USA

Ben K. Wada§
Jet Propulsion Laboratory, Pasadena, CA 91109, USA

Abstract

This work proposes a method of preventing critical damage to civil structures due to earthquakes by suppression of damaging vibrations. The axial-force-rated (AFR) actuators enable vibration suppression without the need for external power which is unavailable during earthquakes. The design of the control system is based on the energy to be dissipated. A method for estimation of energy to be dissipated is outlined. In statically indeterminate structures complete dissipation of vibrational energy with AFR actuators is not possible due to the presence of redundant members. The permanent strains in the members with AFR actuators may create elastic strains in all the members. This paper focuses on vibration control in statically indeterminate structures with AFR actuators.

1 Introduction

Seismic design of civil structures for very strong earthquakes is expensive, considering the average life of the structures. Furthermore, due to the variable soil conditions and lack of test data, large amplitude behavior of civil structures, like tall buildings, under very strong earthquakes is difficult to predict. Vibration suppression with active members provides a feasible solution to

---
*Graduate Student
†Graduate Student
‡Professor of Civil Engineering, Professor of Computer Science
§Deputy Manager, Applied Mechanics Technologies Section, AIAA Fellow

the problem of inadequate seismic resistance of civil structures.

In the case of elasto-plastic material behavior, a part of the strain energy is dissipated in producing plastic strains. Normally, this phenomenon occurs in civil structures during strong earthquakes in the form of cracking of the structural elements which, without control, can lead to collapse of the structure. If the structural behavior can be controlled by allowing local non-structural failures a considerable amount of the strain energy can be dissipated without jeopardizing the safety of the structure. The energy absorption in this manner can far exceed the natural damping of the vibrations with elastic strains.

Based on the idea of energy dissipation by introducing plastic strains, a vibration control strategy was proposed in [1] with actuators, henceforth referred to as the axial-force-rated (AFR) actuators, that limit the maximum axial forces in the active members. The use of the AFR actuators in civil structures can help in dissipating the strain energy and thus suppress the undesired seismic vibrations. For simplicity, the proposed actuator simulates *linear elastic-perfectly plastic-linear elastic* behavior without *Bauschinger effect*.

This paper focuses on the problem of vibration suppression in statically indeterminate structures. In order to simplify the theoretical analysis, truss structures are considered here. The statically determinate case was discussed in [1, 2]. The main areas of investigation can be identified as the estimation of the energy to be dissipated per unit time, vibration control in statically indeterminate trusses with the AFR actuators, estimation of the number of the AFR actuators required and the location of the AFR actuators.

2 Linear Excitation-Response Relations for Discrete Structures

The basic equations of statics for discrete structures are used to derive the governing expressions for the dynamical system. For simplicity the theory is developed for the truss structure. Since truss is multiple degree of freedom structure, the equations are written in matrix form.

Consider M as the number of elements in a truss, N as the number of nodes, e as the number of deflection degrees of freedom per node, a as the number of element forces per element, b as the number of deflection constraints, and f as the number of internal force constraints, then

$$n = Ne - b \quad (1)$$
$$m = Ma - f \quad (2)$$

give the number of independent nodal displacement components and independent element force components, respectively. In trusses $a=1$ and $f=0$. For statically determinate trusses,

$$m = n \quad (3)$$

and for statically indeterminate trusses,

$$m > n \quad (4)$$

2.1 Equations of Statics for Discrete Structures

The linear relations between loads and responses, ignoring the induced inertial forces and the effects of initial stresses on stiffness in the analysis, may be displayed as [3]:

$$\begin{bmatrix} diag(k^j) & -\mathbf{I} & 0 \\ -\mathbf{I} & 0 & \mathbf{B}^T \\ 0 & \mathbf{B} & 0 \end{bmatrix} \begin{Bmatrix} \mathbf{v}_e \\ \mathbf{s} \\ \boldsymbol{\xi} \end{Bmatrix} = \begin{Bmatrix} \mathbf{o} \\ \mathbf{v}_0 \\ \mathbf{p} \end{Bmatrix} \quad (5)$$

where $\mathbf{v}_0$ and $\mathbf{p}$, representing prescribed element elongations and nodal loads, are the excitation quantities, and $\mathbf{v}$, $\mathbf{s}$ and $\boldsymbol{\xi}$ representing element deformations, element forces, and nodal displacements, are the response quantities. $\mathbf{B}$ is the sparse coefficient matrix of element forces in nodal equilibrium equations, and $diag(k^j)$ is the diagonal matrix of element stiffnesses.

2.2 Equations of Dynamics for Discrete Structures

Assuming linear behavior, for a freely vibrating damped truss, i.e., $\mathbf{p}=\mathbf{o}$, Eqn. (5) can be written as

$$\begin{bmatrix} diag(k^j) & -\mathbf{I} & 0 \\ -\mathbf{I} & 0 & \mathbf{B}^T \\ 0 & \mathbf{B} & 0 \end{bmatrix} \begin{bmatrix} \mathbf{v}_e \\ \mathbf{s} \\ \boldsymbol{\xi} \end{bmatrix} = \begin{bmatrix} \mathbf{o} \\ \mathbf{v}_0 \\ -\mathbf{M}\ddot{\boldsymbol{\xi}} - \mathbf{D}\dot{\boldsymbol{\xi}} \end{bmatrix} \quad (6)$$

where $\mathbf{M}$ is the mass matrix, and $\mathbf{D}$ is the damping matrix of the statically indeterminate truss. In the case of statically indeterminate structures $\mathbf{B}$, $\mathbf{v}_e, \mathbf{v}_0$ and $\mathbf{s}$ can be partitioned as $[\mathbf{B}_d, \mathbf{B}_r]$, $(\mathbf{v}_{ed}{}^T, \mathbf{v}_{er}{}^T)^T$, $(\mathbf{v}_{0d}{}^T, \mathbf{v}_{0r}{}^T)^T$ and $(\mathbf{s}_d{}^T, \mathbf{s}_r{}^T)^T$, where indices d and r indicate statically determinate and redundant partitions. Now Eqn. (6) can be manipulated to give

$$\mathbf{M}\ddot{\boldsymbol{\xi}} + \mathbf{D}\dot{\boldsymbol{\xi}} + \mathbf{K}\boldsymbol{\xi} = \mathbf{B}_d diag((k^j)_d)\mathbf{v}_{0d}$$
$$+ \mathbf{B}_r diag((k^j)_r)\mathbf{v}_{0r} \quad (7)$$

where
$\mathbf{K} = \mathbf{K}_d + \mathbf{K}_r$
$\mathbf{K}_d = \mathbf{B}_d diag((k^j)_d)\mathbf{B}_d^T$,
$\mathbf{K}_r = \mathbf{B}_r diag((k^j)_r)\mathbf{B}_r^T$

The mathematical law governing the working of the AFR actuator is discussed in [1, 2].

The second-order system in Eqn. (7) can be converted to first-order system for the purposes of step-by-step integration using the Runge-Kutta method.

3 Estimation of the Energy to be Dissipated

The design of control system for seismic vibrations has to be based on the amount of energy

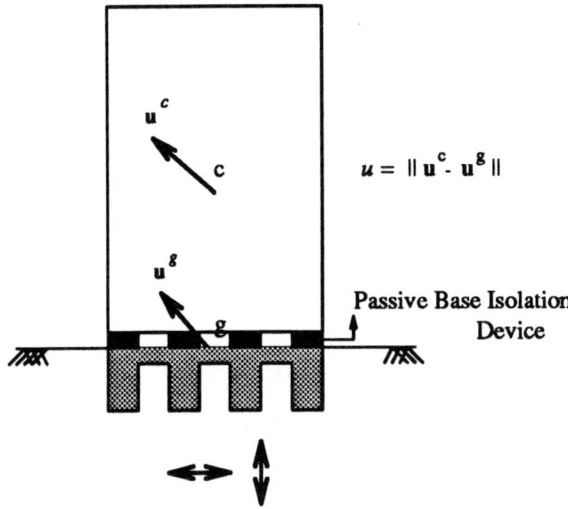

Figure 1: A typical building under earthquake

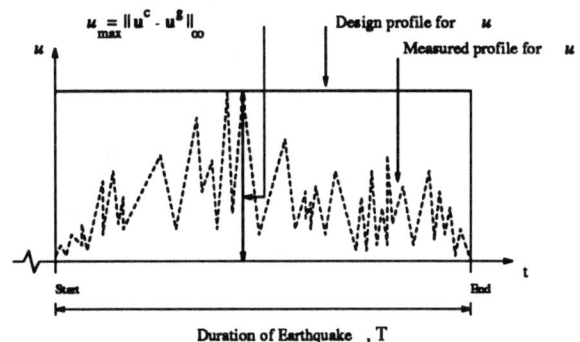

Figure 2: Time history of relative velocity u of the building under earthquake

to be dissipated per unit time. Since earthquake is a natural phenomenon it is difficult to predict *a priori* the power input to the structure. Instead, it is possible to estimate a maxima for the energy that the building might be subjected to. This can be obtained from data collected from past earthquakes by instrumentation of similar structures. Using the measured values of velocities at the ground level, $\mathbf{u}^g$, and the center of mass, $\mathbf{u}^c$, the magnitude of relative velocity of the structure, u, can be obtained. A plot of u with time gives us the time history as shown in Fig. 2. The maximum value of u over the duration of the earthquake is defined as the maximum velocity, u_{max}. Using this velocity and the available value of the total mass of the structure, the maximum instantaneous kinetic energy of the building subjected to seismic vibrations can be evaluated as

$$E_{max} = \frac{1}{2} M u_{max}^2 \qquad (8)$$

A conservative design of the control system may be such that the energy dissipated per unit time, $\dot{E}_d(t)$, satisfies $\int_0^T \dot{E}_d(t) dt \geq E_{max}$.

4 Vibration Control in Statically Indeterminate Structures

4.1 Effect of Statical Indeterminacy

Civil structures, like tall buildings, are multiple degree of freedom structures with high degree of statical indeterminacy. The use of the AFR actuators for vibration suppression introduces plastic strains into the structure. In case of statically indeterminate truss structures, the plastic strains caused by the AFR actuators can cause elastic strains throughout the structure. This results in storage of a part of the vibrational energy as elastic strain energy. These strains may remain in the structure as prestressing as long as the plastic strains remain or they may be released by eliminating the plastic strains with appropriate movements in the redundant members.

4.2 Control Strategy

Vibration control involves dissipation of vibrational energy of the structure to minimize the damage to the structure. The AFR actuators are useful in dissipating the vibration energy by consuming it as work against friction. In the case of statically indeterminate truss structures, however, part of the vibrational energy may be frozen as strain energy due to the presence of redundant members.

By using the force method for the solution of Eqn. (5) the vector of element forces, **s** can be defined as

$$\mathbf{s} = \mathbf{Cx} + \bar{\mathbf{C}}\mathbf{p} \quad (9)$$

where **x** is the vector containing forces in redundant members and **C** and $\bar{\mathbf{C}}$ are defined such that

$$\mathbf{BC} = \mathbf{0} \quad (10)$$

and

$$\mathbf{B}\bar{\mathbf{C}} = \mathbf{I} \quad (11)$$

Using these equations, the response quantities, when the vibrations have ceased, can be expressed as

$$\left\{ \begin{array}{c} \mathbf{v} \\ \mathbf{s} \\ \boldsymbol{\xi} \end{array} \right\} = \left[\begin{array}{c} \mathbf{FK}_c \\ \mathbf{K}_c \\ -\bar{\mathbf{C}}^T(\mathbf{I} - \mathbf{FK}_c) \end{array} \right] \mathbf{v}_0 \quad (12)$$

where $\mathbf{F}=diag(1/k^j)$ and $\mathbf{K}_c=\mathbf{C}(\mathbf{C}^T\mathbf{FC})^{-1}\mathbf{C}^T$.

In case of statically determinate truss structure $\mathbf{K}_c = \mathbf{0}$ so that **v**=**o** and **s**=**o** for any value of $\mathbf{v}_0$. So no permanent elastic strains are created inspite of $\mathbf{v}_0$. In case of statically indeterminate truss structure, due to nonzero entries in $\mathbf{K}_c$, plastic strains in the active members in the determinate base structure give nonzero entries in **v** and **s**. Thus, statically indeterminate truss structure would have permanent elastic strains. If these elastic strains are sufficiently small, i.e., they do not cause excessive stresses in the structural members, they may be tolerated. However, if the stresses are in excess of the failure limits for the members then special actuators, called *slave* actuators [4] may be used in appropriate redundant members to eliminate the stresses. These actuators allow sufficient movement in these members so as to prevent the development of plastic strains induced elastic strain in the structure.

The AFR actuators dissipate energy by cyclic movement of the friction device as explained in [2]. This requires alternate tensile and compressive forces in axial members containing the AFR actuators. In the case of earth-bound structures the AFR actuators can be used only on members with no static loads to prevent compressive preloading of actuators.

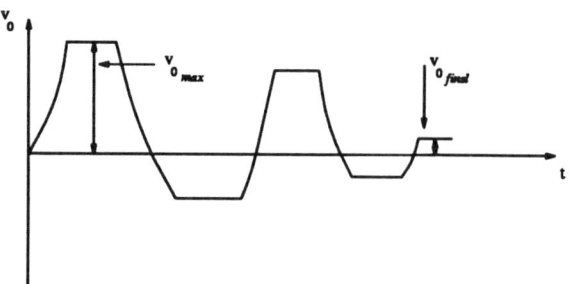

Figure 3: Variation of plastic deformation in a typical active member

4.2.1 Control without Extra Actuators

In statically indeterminate truss structure, the plastic deformations due to the AFR actuators cause elastic strains in the entire structure. Due to cyclic movement of the AFR actuators there is variation of plastic deformations as shown in Fig. 3 and hence the elastic strains with time. If the truss structure is designed to withstand the elastic strains caused by the maximum plastic deformations in the AFR actuators during the control process, it will be safe against the permanent elastic strains. Thus, when the stresses in the structure are low, the structure may be allowed to remain in a permanently strained state.

4.2.2 Control with Synchronized Actuators

In truss structures, it is possible to achieve vibration control by introducing minimal additional stresses in the members of the structure. Additional actuators, referred to as *slave* actuators, are placed on some if not all of the redundant members of the structure. These actuators can compensate for the additional stresses created as a result of the deformations in the vibration controlling actuators. Thus no additional stresses are created in the members of the structure due to the actuation in the vibration controlling actuators. The control strategy for stress-free geometry control has been discussed in [5], and applies to dynamic control. The complete list of actuator elongations can be computed from the plastic deformations in the AFR actuators. If the deformations in the AFR actuators are represented by $\mathbf{v}_{0d}$, the complete list of actuator elongations, $\mathbf{v}_0$, can be written as,

$$\mathbf{v}_0(t+\Delta t) = \mathbf{B}^T \mathbf{B}_d^{-T} \mathbf{v}_{0d}(t) \quad (13)$$

The elongations in the synchronized actuators on the redundant members are then computed as

$$\mathbf{v}_{0r}(t+\Delta t) = \mathbf{B}_r^T \mathbf{B}_d^{-T} \mathbf{v}_{0d}(t) \quad (14)$$

A detailed derivation of the expressions is given in [5]. The measured plastic deformations in the AFR actuators are used to calculate the necessary elongations in the synchronized actuators. The actuation procedure has to be synchronized, i.e., the time-lapse (delay), Δt, between the measurement of the plastic deformations in the AFR actuators and the activation of the synchronized actuators has to be minimized.

An important consideration for using synchronized actuators is that additional energy may be required to operate the synchronized actuators on the redundant members.

5 Energy Dissipation in Statically Indeterminate Structure

In order to prove the feasibility of the proposed control strategies for statically indeterminate structures, energy balance is studied to show dissipation of vibrational energy by the AFR actuator. The total energy, $E(t)$, at any time is given as sum of the kinetic energy and the potential energy of the system

$$E(t) = \frac{1}{2}\dot{\boldsymbol{\xi}}^T \mathbf{M} \dot{\boldsymbol{\xi}} + \frac{1}{2}\boldsymbol{\xi}^T \mathbf{K} \boldsymbol{\xi} \quad (15)$$

Then the rate of dissipation of energy, $\dot{E}_d$, can be given as negative of the rate of the total energy at any instant t

$$dE_d(t)/dt = -dE(t)/dt = -\dot{\boldsymbol{\xi}}^T(\mathbf{M}\ddot{\boldsymbol{\xi}} + \mathbf{K}\boldsymbol{\xi}) \quad (16)$$

Substituting the equations of motion (neglecting the natural damping) into the expression for the rate of energy dissipation it can be seen that the energy is dissipated due to the plastic deformations in the AFR actuators in the primary structure and slave actuators in the redundant members

$$\dot{E}_d(t) = -\dot{\boldsymbol{\xi}}^T (\mathbf{B}_d diag(k^j)_d \mathbf{v}_{0d} + \mathbf{B}_r diag(k^j)_r \mathbf{v}_{0r}) \quad (17)$$

Using Eqn. (14) and assuming Δt to be negligibly small, Eqn. (17) can be rewritten as

$$\dot{E}_d(t) = -\dot{\boldsymbol{\xi}}^T (\mathbf{B}_d diag(k^j)_d + \mathbf{B}_r diag(k^j)_r \mathbf{B}_r^T \mathbf{B}_d^{-T}) \mathbf{v}_{0d} \quad (18)$$

From design considerations the rate of energy dissipation should at least be equal to the maximum instantaneous energy input rate into the structure.

In the case of control without extra actuators in redundant members there is no energy dissipation in redundant members so there is no contribution from the last term on the right hand side of Eqn. (17) and for control with synchronized actuators extra energy may be required to move some of the actuators in redundant members due to the different directions of forces and deformations.

6 Actuator Placement

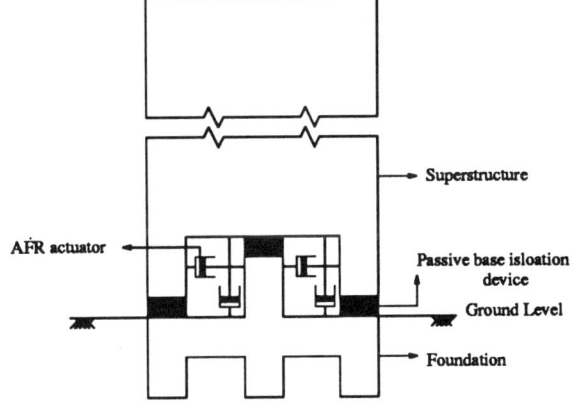

Figure 4: AFR actuators used for base isolation

The AFR actuator is based on linear elastic member deformation with small magnitude of actuator stroke. Due to these limitations, the control parameters can be varied over only a small range with limited energy dissipation. Assuming full actuator stroke over a complete cycle of the fundamental vibration mode, an estimation of energy dissipation per cycle for an AFR actuator can be obtained. Using this value and an estimate of the maximum seismic energy, an approximate number of actuators for vibration suppression in real time can be computed.

Optimization of the control process requires maximum energy dissipation over minimum time. The AFR actuators when located at points of the maximum strain energy undergo large *total deformation* [1]. This allows maximum actuator movement and thus maximum energy dissipation [6]. In the case of bending, torsion or shear modes, the maximum strain energy occurs near the base of the structure, so the AFR actuators can be located in the base columns of the structure.

One possible application of the AFR actuator technology for seismic vibration control is in the area of base isolation. Due to their working mechanism AFR actuators cannot be used in preloaded members, such as axially loaded columns. They may be placed on additional, redundant, members of the structure upon the completion of the construction. It is feasible to use them as energy dissipators between the foundation and the superstructure, as shown in Fig. 4.

7 Numerical Example

In order to test the proposed method of vibration control a plane truss structure is considered. Numerical simulations are performed to control the desired vibration mode and to demonstrate the control schemes.

7.1 Description of the Plane Truss Structure

A statically indeterminate plane truss as shown in Fig. 5, is used for the purpose of analysis. The truss consists of six (6) members and four (4)

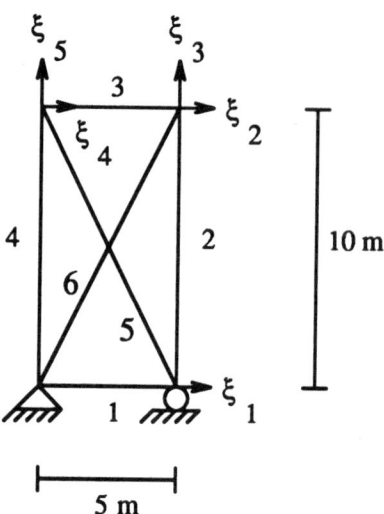

Figure 5: A 2-dimensional example truss

Natural Frequencies (rad/sec)	First Mode Shape ξ_1
16.13	0.0344
63.77	0.6903
68.08	-0.1526
79.66	0.6912
106.50	0.1452

Table 1: Dynamic characteristics of the example truss

nodes and is supported on a hinge and a roller. Using the notation defined previously, the truss has $e = 2$, $N = 4$, $M = 6$ and $b = 3$. From Eqn. (1) the number of independent degrees of freedom, $n = 5$ and from Eqn. (2) the number of independent element forces, $m = 6$. The material properties of the truss are chosen as: E (Young's Modulus) $= 7 \times 10^8$ Pa, and ρ (Density) $= 2000$ kg/m^3. The cross-section area for each bar is uniform and is equal to 1.492×10^{-3} m^2 (pipe section of 0.1 m outer diameter and 0.09 m inner diameter).

In order to obtain the dynamic characteristics of the truss, the nodal mass lumping scheme is used to get the mass matrix. The mass in each

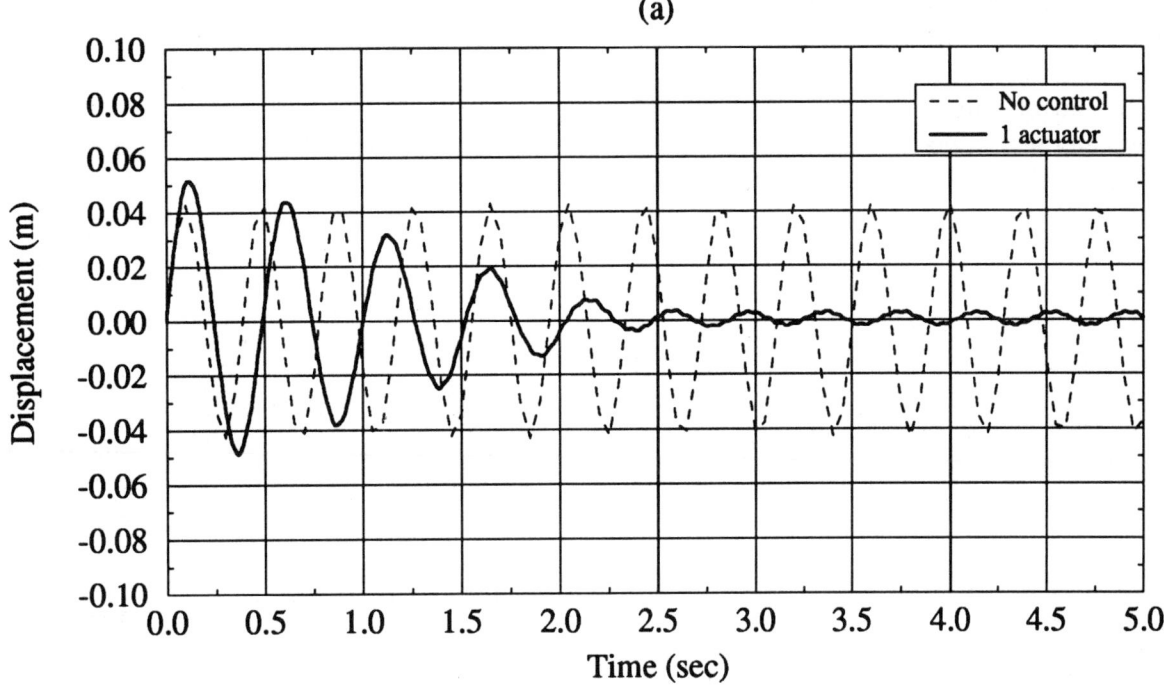

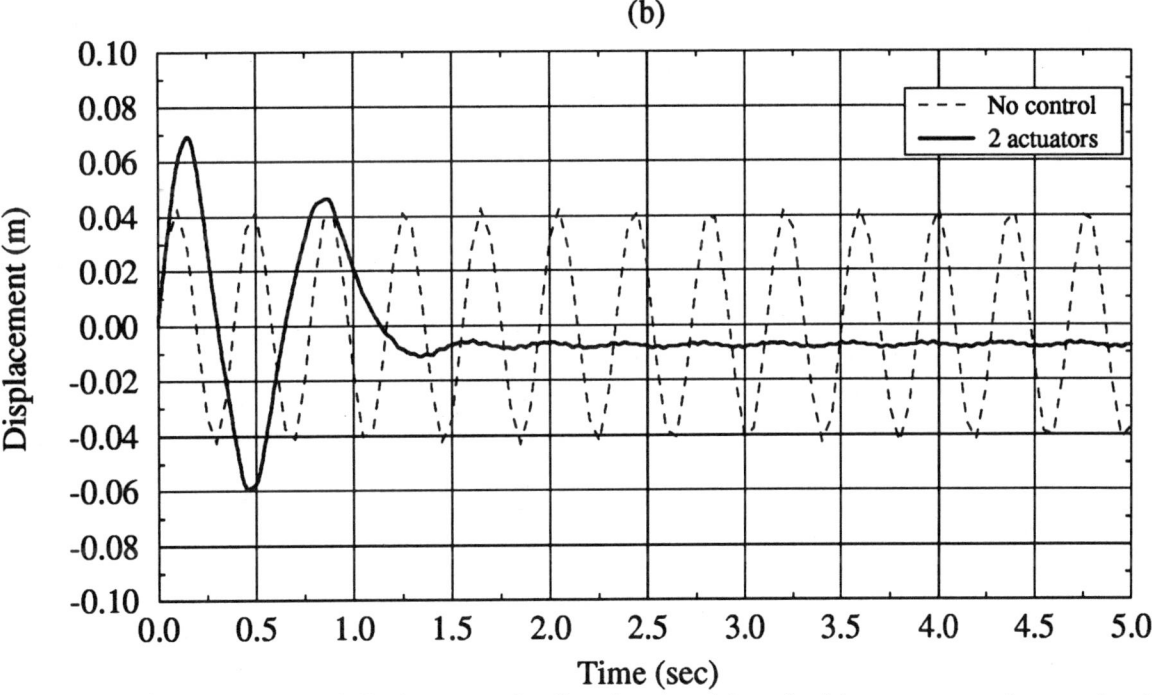

Figure 6: (a) Comparison of displacement in direction ξ_2 with and without actuator in member 5, (b) Comparison of displacement in direction ξ_2 with and without actuator in members 5 and 6 (refer Fig. 5 for direction ξ_2 and member identification).

bar is assumed to be equally distributed between the two end points so that the mass matrix is a diagonal matrix. The numerical simulation for vibration control is aimed at studying the control of a statically determinate truss. For simplicity, the damping in the truss is assumed to be non-existent for all modes of vibration. The natural frequencies and the mode shapes are shown in Table 1.

To test the proposed scheme of vibration control, the truss is disturbed with an initial velocity corresponding to the first vibration mode shape. Since the mode shapes are normalized quantities, an amplification factor α is chosen and fixed at $\alpha=2.0$.

7.2 Discussion of Results

Numerical simulations are performed on the statically indeterminate truss for three different cases: no control, control with one actuator on member 5 and control with two actuators on members 5 and 6. The no control case is used to get reference response of the truss. Control without extra actuator is demonstrated by use of one actuator on member 5 and control with synchronized actuators is demonstrated by two actuators on members 5 and 6.

Figs. 6(a) and (b) shows the variation of the horizontal displacement of node 3, ξ_2, with time for the cases with and without control. Fig. 6(a) demonstrates that the first mode vibrations are damped satisfactorily by utilizing single actuator in member 5. The damping rate is faster if two actuators are used simultaneously, in members 5 and 6, as shown in Fig. 6(b). The steady state displacement, ξ_2, in the controlled case in Fig. 6(b) is the result of permanent deformations in the two AFR actuators.

8 Conclusions

It is possible to control vibrations in statically indeterminate structures by dissipating vibrational energy with the AFR actuators. The energy dissipation rate must be greater than the maximum input power at any time. Vibration control without using extra actuators introduces permanent elastic strains in the structural members. By utilizing synchronized actuators, the additional stresses can be minimized. Since the AFR actuators are not designed to work in preloaded members, base isolation provides a feasible application of the control strategies for seismic vibration suppression.

References

[1] V. Chawla, S. Utku, and B. K. Wada, "Vibration control by limiting the maximum axial forces in space trusses," in *Proceedings of the AIAA/ASME/ASCE/AHS/ASC 34th Structures, Structural Dynamics and Material Conference*, no. AIAA-93-1690, pp. 3279-3288, April 19-22 1993.

[2] V. Chawla, "Vibration control by limiting the maximum axial forces," Master's thesis, Duke University, 1993 (to be published).

[3] S. Utku, C. H. Norris and J. B. Wilbur, *Elementary Structural Analysis*, Fourth Edition, pages 737-810, McGraw-Hill, New York, 1991.

[4] C. Baycan, S. Utku, and B. K. Wada, "Vibration control in statically indeterminate adaptive truss structures," in *Proceedings of the AIAA/ASME/ASCE/AHS/ASC 34th Structures, Structural Dynamics and Material Conference*, no. AIAA-93-1691, pp. 3289-3296, April 19-22 1993.

[5] M. Şener, S. Utku, and B. K. Wada, "Geometry Control in Prestressed Adaptive Space Trusses," to appear in *Journal of Smart Materials and Structures*.

[6] B. K. Wada and S. Das, "Application of Adaptive Structures Concepts to Civil Structures", presented at *US-Italy-Japan Workshop/Seminar on Intelligent Systems*, June 27-29, 1991, Perugia, Italy, 1991.

AIAA-94-1751-CP

AN ARTIFICIAL NEURAL NETWORK APPROACH TO STRUCTURAL DAMAGE DETECTION USING FREQUENCY RESPONSE FUNCTIONS

Clinton R. Povich* and Tae W. Lim**
University of Kansas, Lawrence, Kansas 66045

Abstract

Structural damage detection in a twenty-bay planar truss was accomplished using an artificial neural network. Instead of using natural frequencies and mode shapes, the frequency response functions (FRF's) experimentally obtained from accelerometers at two locations on the truss were directly used to distinguish among damage cases and to train the network. Unlike conventional approaches based on system identification techniques, the neural network approach does not require an analytical model of the structure. The direct use of the FRF's eliminates the need for modal parameter identification. Out of the 60 damage cases considered in this study, the neural network was able to identify uniquely 21 damage cases and narrowed 38 others down to two possible damaged struts.

Introduction

To maintain the performance and safe operation of aerospace vehicles, structural integrity must be monitored periodically. Numerous studies (e.g., Smith and Hendricks, 1988; Lim, 1991; Zimmerman and Kaouk, 1992; Lim and Kashangaki, 1994) have been conducted to detect structural damage using measured mode data and system identification techniques. A survey on the subject is available [Kashangaki, 1991]. Recently, attention was given to the application of the artificial neural network to structural damage detection (Tsou and Shen, 1993; Yen and Kwak, 1993). The natural frequencies and mode shapes were used to train the network and to conduct damage detection. Unlike the conventional approaches based on system identification techniques, the neural network approach does not require a refined model of the undamaged structure. The fact that damage detection can be performed using the measured mode data alone without an analytical model relieves the burden of conducting analytical model refinement to match measured mode data, which is typically costly and time consuming.

This paper investigates an effective use of the artificial neural network to conduct structural damage detection. In a typical modal test, the modal parameters are identified using the frequency response functions (FRF's). Instead of using mode shapes and frequencies, the neural network can be trained directly using the FRF's. Some of the advantages using the FRF's directly include: (1) the modal parameter identification process is not required and thus the error and cost associated with it are eliminated, (2) nonlinear behavior exhibited in FRF's can be incorporated in the damage detection process, and (3) the number of transducers required to perform damage detection can be significantly reduced as will be shown in the twenty-bay truss structure example.

Test Apparatus and Procedure

The focus of this study is damage detection in a twenty-bay two-dimensional truss as shown in Fig. 1. The 7.07 m long truss fixed at one end and free at the other lies flat on a table. The truss consists of hollow aluminum tubes and steel node balls. Underneath each batten, a steel bar is bolted to the node balls. Each steel bar rests on two steel ball bearings that allow the truss to move in the horizontal plane. For detailed information on the truss, refer to Hallauer and Lamberson (1989). At the end of the truss, a shaker is attached to provide force excitation. Accelerometers are located at the numbered positions as shown in Fig. 1. Since the minimum number of accelerometers needed was unknown, eleven accelerometers were installed to provide sufficient modal test data. The minimum number of accelerometers and their locations are determined by inspecting the FRF's and evaluating the damage detection performance of the trained neural network. Consequently, only the FRF's from accelerometers 5 and 11 are selected to train the neural network and to conduct damaged detection. Accelerometer 11 is chosen because the FRF's at this point displayed the greatest diversity. The other accelerometer is used to provide uniqueness to the few damage cases that exhibit similar FRF's at accelerometer 11.

A different strut is removed for each of the 60 damage cases investigated to simulate complete damage in that strut. In each bay, only the diagonals and longerons are removed. It is difficult to remove the battens because of the steel bars connecting the node balls underneath each batten. Thus, they are eliminated from damage consideration in this study. The FRF's are then found using a Fourier analyzer. The Fourier analyzer controls the random force input to the shaker and records the FRF's over a range of 0 to 50 Hz, which were obtained using 20 averagings. The FRF's obtained for all damaged and undamaged cases contain at least four bending modes below 50 Hz. Figures 2 and

---

*Graduate Student, Department of Aerospace Engineering, Member, AIAA.
**Assistant Professor, Department of Aerospace Engineering, Member, AIAA.

Copyright © 1994 by the American Institute of Aeronautics and Astronautics, Inc. All rights reserved.

3 show the FRF's measured at accelerometers 5 and 11, respectively. The figures show the changes in FRF's due to damage in upper longerons, diagonals, and lower longerons. These changes were used to train the neural network and to detect damage. Each figure has 21 FRF's corresponding to the undamaged case and the damage cases in bay 1 through 20. Thus, the FRF corresponding to damage case number 10 for the upper longeron case in Fig. 2 indicates the FRF measured at accelerometer 5 with upper longeron removed in bay 9. Figure 4 shows in detail the FRF of a typical damage case plotted with the FRF of the undamaged case at accelerometer 11.

The sample rate employed to measure the FRF's was 2.56 times the bandwidth (50 Hz) and 512 samples were taken for each test. Since only data up to 50 Hz is desired, each FRF consists of 200 frames (201 data points). Of these 201 points, the first four are not included in the neural network training due to the large 'spike' in the data at these points in nearly every FRF of every damage case. The resulting FRF's then have a range of 1.25 to 50 Hz (197 data points). Finally, the FRF data is converted to a Matlab file to conduct neural network training.

Artificial Neural Network

The Matlab Neural Network Toolbox (Demuth and Beale, 1992) was used to construct and train the neural network. As shown in Fig. 5, the type of network used in this study was a three-layer neural network using log-sigmoid neurons. The backpropagation method (Rumelhart et al., 1986) was used to train the network.

A neural network is composed of one or more layers and each layer contains a series of processing units (neurons) that act in parallel within the layer. Input data arranged in a vector (or a matrix of vectors) is transferred through the layers. Each data value within an input vector is multiplied by the weight associated with each neuron in a layer and the results are summed at each neuron input creating a new vector with a size equal to the number of neurons in that layer. Then, the transfer function at each neuron converts its input, which is a sum of weighted input and bias, into a neuron layer output. For hidden layers, the neuron layer output vector is used as input to the next layer. For the output layer, the output vector represents the final output of the entire neural network. The transfer function used in this study is a log-sigmoid neuron which converts any real number to a value between 0 and 1. In the target output for the network, a '1' signifies no damage in a strut while a '0' indicates complete damage (missing).

Initially, the weights and biases for all of the neurons are randomly chosen. After one pass (epoch) through the network, the output vectors are compared to the target output vectors. The neural network computes the delta error vectors (derivatives of error) which are backpropagated through the network until delta vectors are found for each layer. The learning rate is then used to change the weights and biases in the direction of steepest decent with respect to the delta vectors. For example, the change in a weight, delW, and the change in a bias, delB, in a layer, are found as

$$delW(i,j) = lr*D(i)*P(j) \quad (1)$$
$$delB(i) = lr*D(i) \quad (2)$$

where

lr = learning rate
D = delta vector of the layer
P = input vector to the layer
i = ith neuron in the layer
j = jth element in input vector to the layer

Once the weights and biases for each layer are updated, the inputs are sent through the network layers again and the process is repeated. This process is referred to as training. Training stops when the sum squared error, which is a sum of squared values of network errors, is less than the error goal specified by the user. In this study, the error goal is selected as 1e-10 to confine individual errors within about ±1e-6.

An adaptive learning rate (Vogl et al., 1988) allows the network to be trained faster by comparing the new error vector after each epoch with the previous (old) error vector. If the new error vector is larger than the old (by 1.04), the new weights and biases are discarded and the learning rate decreases by 0.7. If the new error vector is less than the old, the learning rate is increased by 1.05 and the new weights and biases are kept. Also, momentum is used such that the network avoids getting stuck in a local minimum of the overall error surface. Basically, momentum allows the network to ignore small changes and respond to the general trend in the error surface. A momentum constant is chosen before training with a value between 0 and 1. The quantity 0 signifies a weight change based solely on the gradient while 1 sets the new weight change equal to the last weight change and the gradient is ignored, i.e.,

$$delW(i,j) = mc*delW(i,j) + (1 - mc)*lr*D(i)*P(j) \quad (3)$$

where the quantity mc is a momentum constant. A typical value of mc is 0.95.

For the network in this study, 61 input vectors are used to train the network: 60 damage cases and 1 undamaged case. Each input vector contains the FRF's with the FRF from accelerometer 5 listed first than that of accelerometer 11. Thus, the size of the input matrix becomes 394 by 61. Two hidden layers of 125 and 40 neurons, respectively, are included in the network to achieve the desired error goal with a reasonable amount of training time. The target output vector corresponding to each input vector was 60 element vector consisting of 1's (denoting struts with no damage) and a 0 (damaged strut). The undamaged case contains all 1's. Therefore, the size of the target matrix is 60 by 61. Each element in the target vector corresponds to a truss strut with following pattern

Element		Strut description
1		upper longeron
2	Bay 1	diagonal
3		lower longeron
4		upper longeron
5	Bay 2	diagonal
6		lower longeron
...		...
58		upper longeron
59	Bay 20	diagonal
60		lower longeron

The planar truss structure that is used to perform damage detection exhibits interesting dynamic characteristics mainly due to the lacing pattern of the diagonals. Each longeron does not produce unique changes in the FRF's. The two longerons located at the open end of each 'V' created by two diagonals in adjacent bays produced nearly identical FRF's when either was removed as shown in Fig. 6. This resulted in 19 pairs of longeron damage cases with indistinguishable dynamic characteristics. Therefore, the best that can be achieved for longeron damage cases was to reduce from 60 damage candidates to two adjacent longerons. Also, when the lower longeron in the 20th bay is removed the resulting FRF is similar to that for the undamaged case; thus making it difficult to detect damage in that strut as indicated in Fig. 7. To accommodate these characteristics of the truss structure, the target output vectors are modified as follows: for the 19 longeron pairs, each vector contains two 0's representing the two possible damage cases and for the 20th lower longeron, no 0's are present. This eliminates the latter from consideration as a damage case. All the diagonals produced dynamic characteristics that are uniquely identifiable.

The number of epochs taken for the network described above to be trained is between 900 and 1000 epochs. This range is considered reasonable and is obtained by adjusting factors such as initial learning rate, number of layers and number of neurons in each layer. There are approximately 57,000 variables (weights and biases) in the network trained successfully compared to the size of the input matrix (24,034) and output matrix (3,660). Reducing the number of neurons in the layers and/or the number of layers to a smaller number would be beneficial since it would result in less computing time when presenting independent test cases to the trained network. However, convergence is not always assured. Many smaller sizes of the hidden layers were tested in this network and none were close to converging in the amount of time this configuration took to train (~2 hours) even after 1 or 2 days of training. The summed squared error in the failed configurations reached a value which is significantly larger than 1e-10 and remained there. Poor damage detection performance is expected when the networks with larger summed squared errors are used.

Figure 8 shows the error distribution between the target output and actual output computed using the final weights and biases of the trained network. Most of the individual errors are confined within ±1e-6 thanks to the small summed squared error goal of 1e-10. To evaluate the performance of the trained network, the FRF's of several damage cases which were not used to train the network, were presented to the network. Figure 9 shows the plot of the resulting output vectors of several damage cases. To clarify the location and magnitude of damaged elements, each element of the output vector was subtracted by 1. Thus, the quantities 0 and -1 represent no damage and complete damage, respectively, in the figure. The location and the magnitude of damage is clearly indicated in the plot. In nearly all damage cases, the location and magnitude of damage was clearly identified. For longeron damage cases, two struts are identified because of the reasons discussed previously.

Conclusions

Successful damage detection was accomplished using the artificial neural network trained with experimentally obtained FRF's from just two accelerometers on a twenty-bay planar truss. The neural network successfully located a missing strut in 21 cases out of the 60 damage cases investigated. The 21 cases include all diagonal damage cases and the lower longeron damage in the first bay. In 38 of the 60 damage cases, the network reduced the 60 candidates to two adjacent longerons. For one case, which corresponds to the lower longeron in the 20th bay, it was not successful in locating the missing strut. The success of damage detection was directly tied to the uniqueness of the FRF that was produced by a certain damage case. For instance, due to the lacing pattern of the diagonals of the truss, 19 longeron pairs produced almost identical FRF's. Thus, pinpointing the damaged member was impossible.

A neural network trained using natural frequencies and/or mode shapes may have produced the same results but the number of accelerometers required would have been at least four to be able to distinguish four bending modes below 50 Hz. The direct use of FRF's investigated in this report eliminates the need for the modal parameter identification process and helps avoid the cost and error associated with it. Also, a refined analytical model is not required unlike the conventional approaches based on system identification techniques. Thus, the burden of performing modal refinement to obtain high fidelity analytical model is removed.

Further study should include the response of the trained network to partially damaged struts and/or combinations of damaged struts and the need to include these in training. Additional FRF's from other accelerometers may also be needed for these cases. The FRF of an axially mounted accelerometer may help separate the indistinguishable pairs of damage cases. A similar study on a three-dimensional truss might prove more successful since the number of damage cases that produce almost identical FRF's are expected to be fewer.

Acknowledgments

The first author of the paper was supported by the Air Force Office of Scientific Research Summer Graduate Student Research Program and the research was conducted at the Frank J. Seiler Laboratory at the United States Air Force Academy in Colorado Springs, Colorado. The authors are grateful for the help from the personnel in the Frank J. Seiler Laboratory.

References

Demuth, H. and Beale, M., 1992, *Neural Network Toolbox User's Guide*, MathWorks, Natick, Massachusetts.

Hallauer, W. L. and Lamberson, S. E., 1989, "A Laboratory Planar Truss for Structural Dynamics Testing," *Experimental Techniques*, Vol. 13, No. 9, pp. 24-27.

Kashangaki, T., 1991, "On-Orbit Damage Detection and Health Monitoring of Large Space Trusses - Status and Critical Issues," *Proceedings of the 32nd Structures, Structural Dynamics and Materials Conference*, AIAA, Washington, D.C., pp. 2947-2958.

Lim, T. W., 1991, "Structural Damage Detection Using Modal Test Data," *AIAA Journal*, Vol. 29, No. 12, pp. 2271-227.

Lim, T. W. and Kashangaki, T. A.-L., 1994, "Structural Damage Detection of Space Truss Structures Using Best Achievable Eigenvectors," *AIAA Journal*, accepted for publication.

Rumelhart, D. E, Hinton, G. E, and Williams, R. J., 1986, "Learning Internal Representations by Error Propagation," *Parallel Data Processing*, Rumelhart, D. and McClelland, J., editors, MIT Press, Cambridge, MA, Vol. 1, Chap. 8, pp. 318-362.

Smith, S. W. and Hendricks, S. L., 1988, "Damage Detection and Location in Large Space Trusses," *AIAA SDM Issues of the International Space Station, A collection of Technical Papers*, AIAA, Washington, D.C., pp. 56-63.

Tsou, P. and Shen, M.-H. H., 1993, "Structural Damage Detection and Identification Using Neural Network," *Proceedings of the 34th Structures, Structural Dynamics and Materials Conference*, AIAA, Washington, D.C., pp. 3552-3560.

Vogl, T. P., Mangis, J. K., Rigler, A. K., Zink, D. L., and Alkon, D. L., 1988, "Accelerating the Convergence of the Backpropagation Method," *Biological Cybernetics*, Vol. 59, pp. 257-263.

Yen, G. G. and Kwak, M. K., 1993, "Neural Network Approach of the Damage Detection of Structures," *Proceedings of the 34th Structures, Structural Dynamics and Materials Conference*, AIAA, Washington, D.C., pp. 1549-1555.

Zimmerman, D. C. and Kaouk, M., 1992, "Structural Damage Detection Using a Subspace Rotation Algorithm," *Proceedings of the 33rd Structures, Structural Dynamics and Materials Conference*, AIAA, Washington, D.C., pp. 2341-2350.

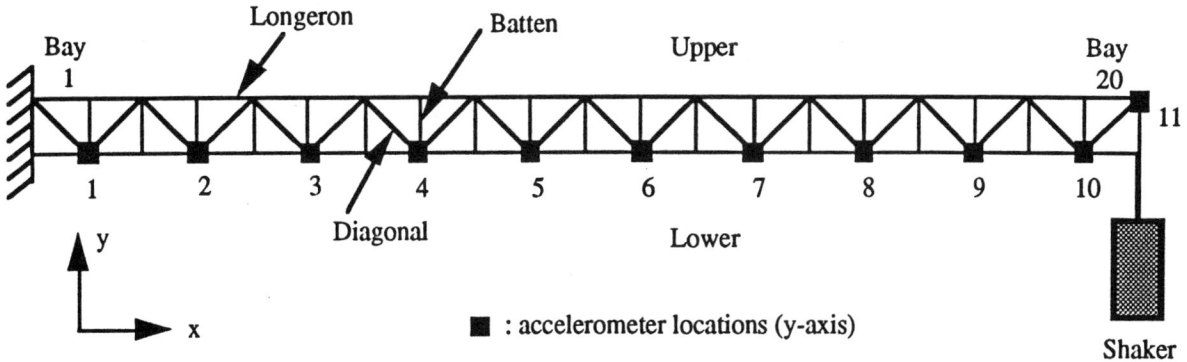

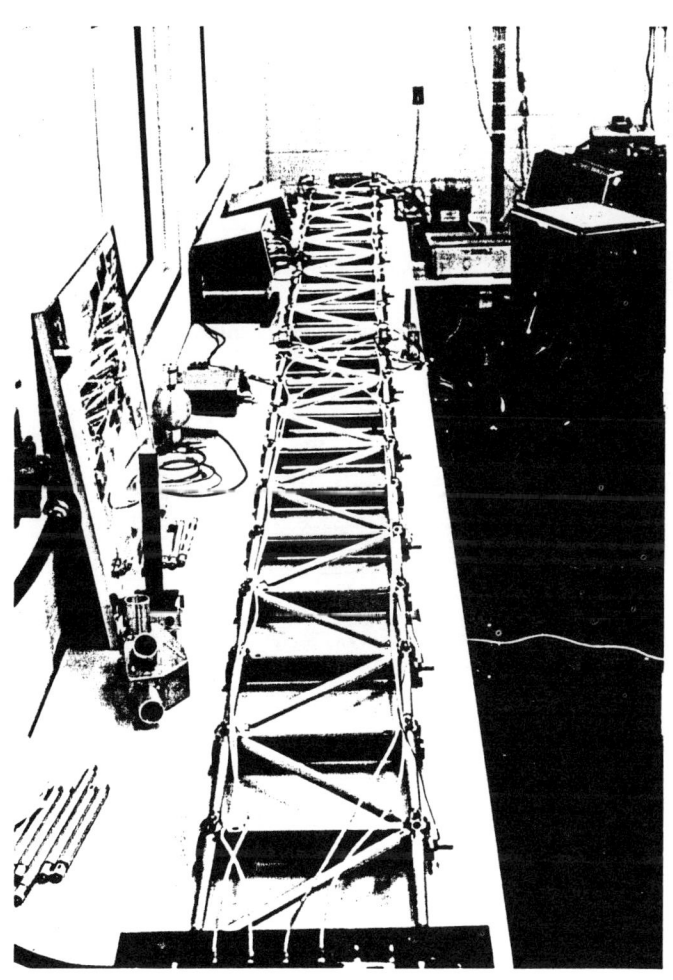

Fig. 1 A twenty-bay planar truss structure used for damage detection

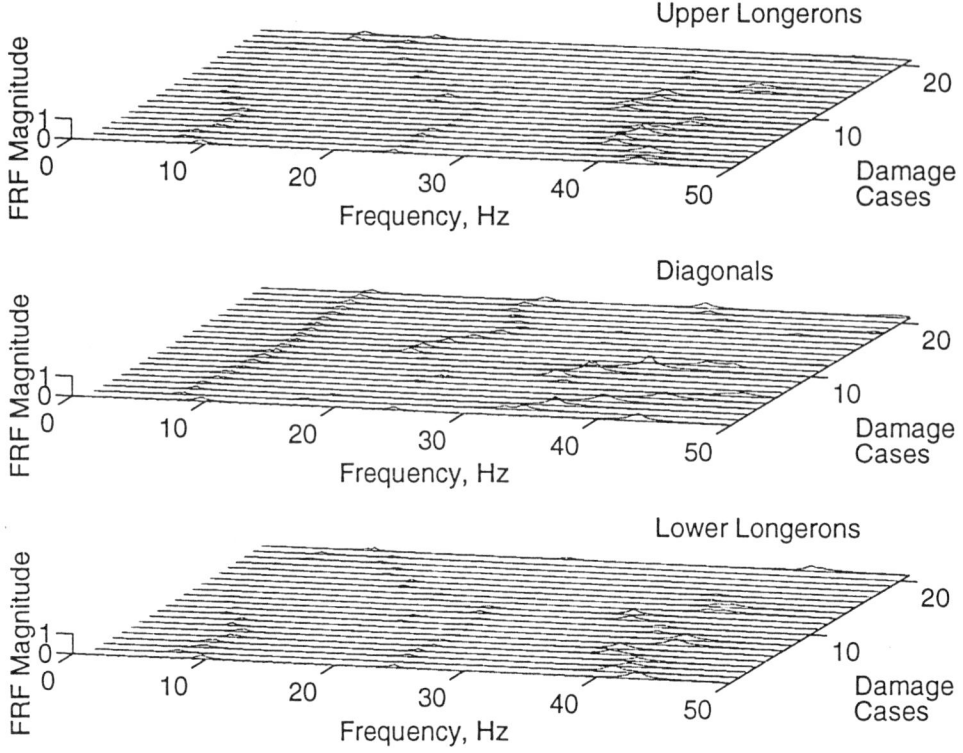

Fig. 2 FRF's measured at accelerometer 5

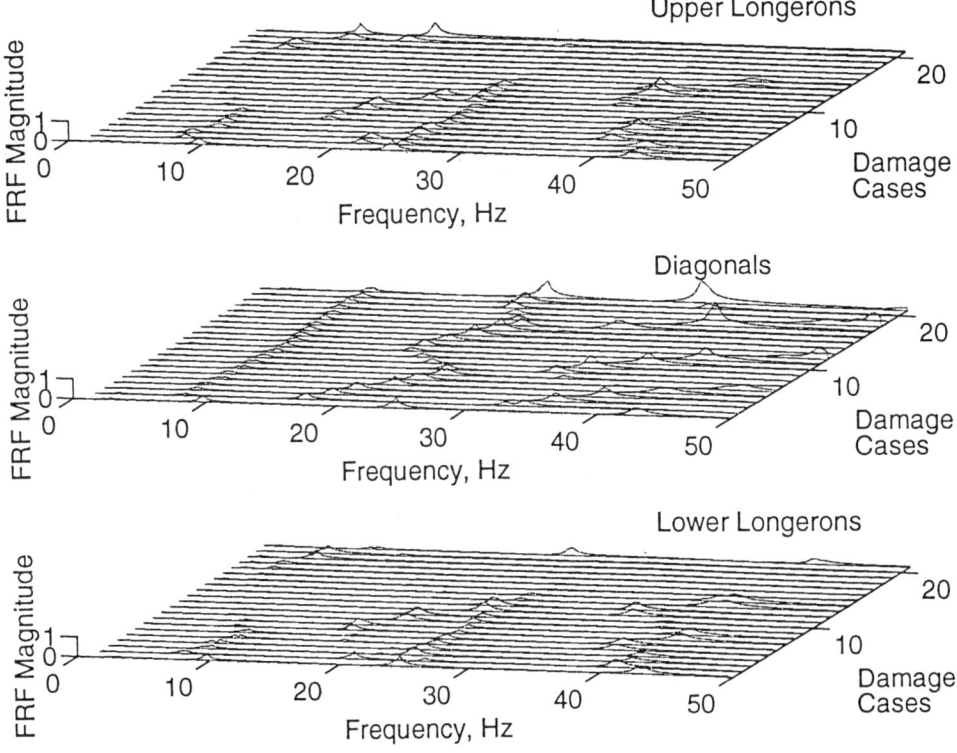

Fig. 3 FRF's measured at accelerometer 11

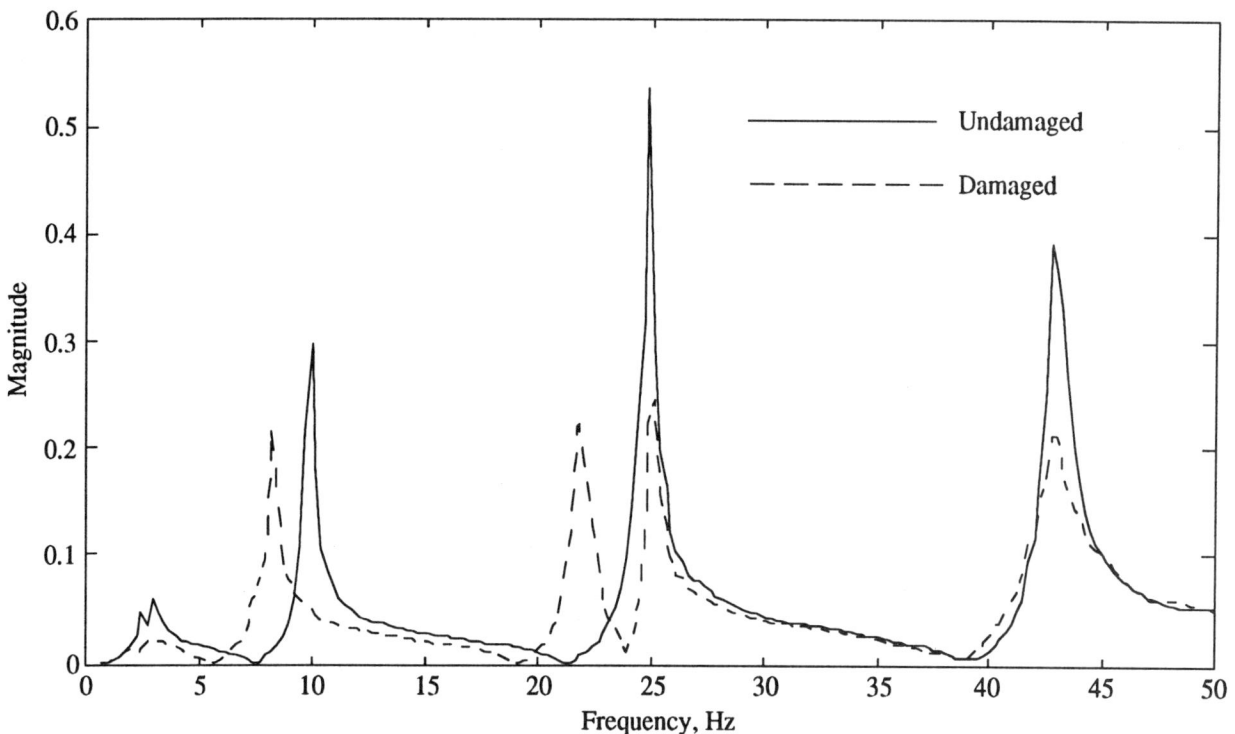

Fig. 4 Comparison of the undamaged and a typical damaged truss FRF's at accelerometer 11

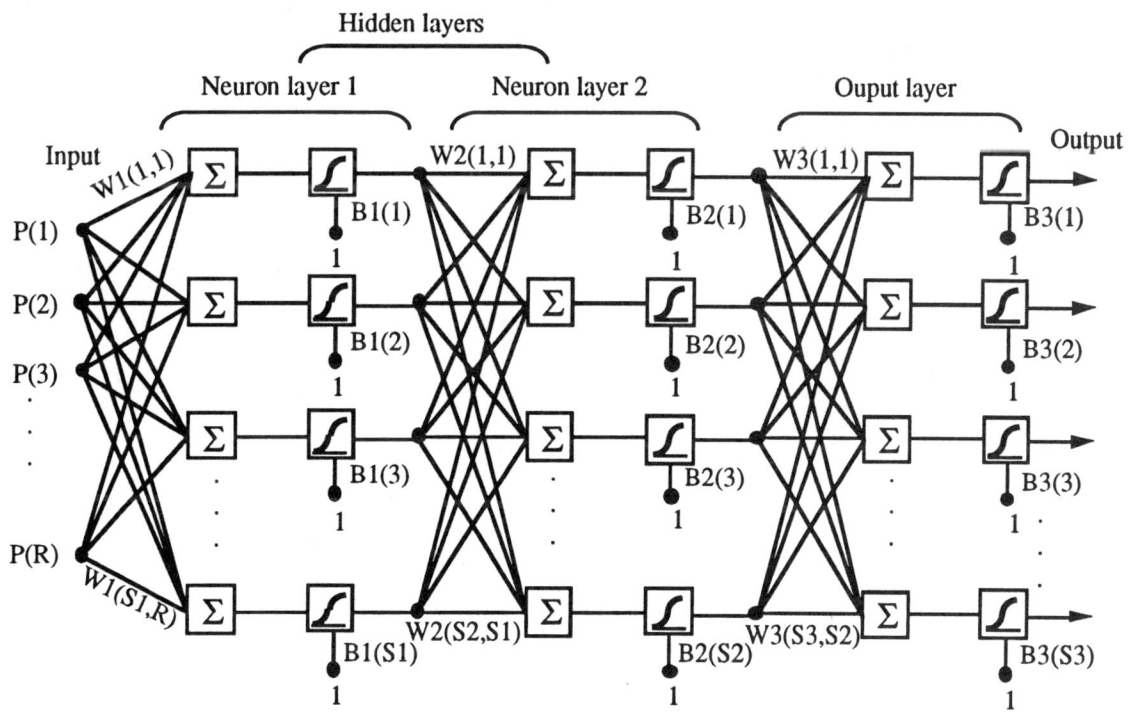

P: input vector
W1, W2, and W3 : weight matrices
B1, B2, and B3: bias vectors

R: number of inputs
S1, S2: number of neurons in the hidden layers
S3 : number of neurons in the output layer

Fig. 5 Schematic of three layer sigmoid backpropagation network designed for damage detection

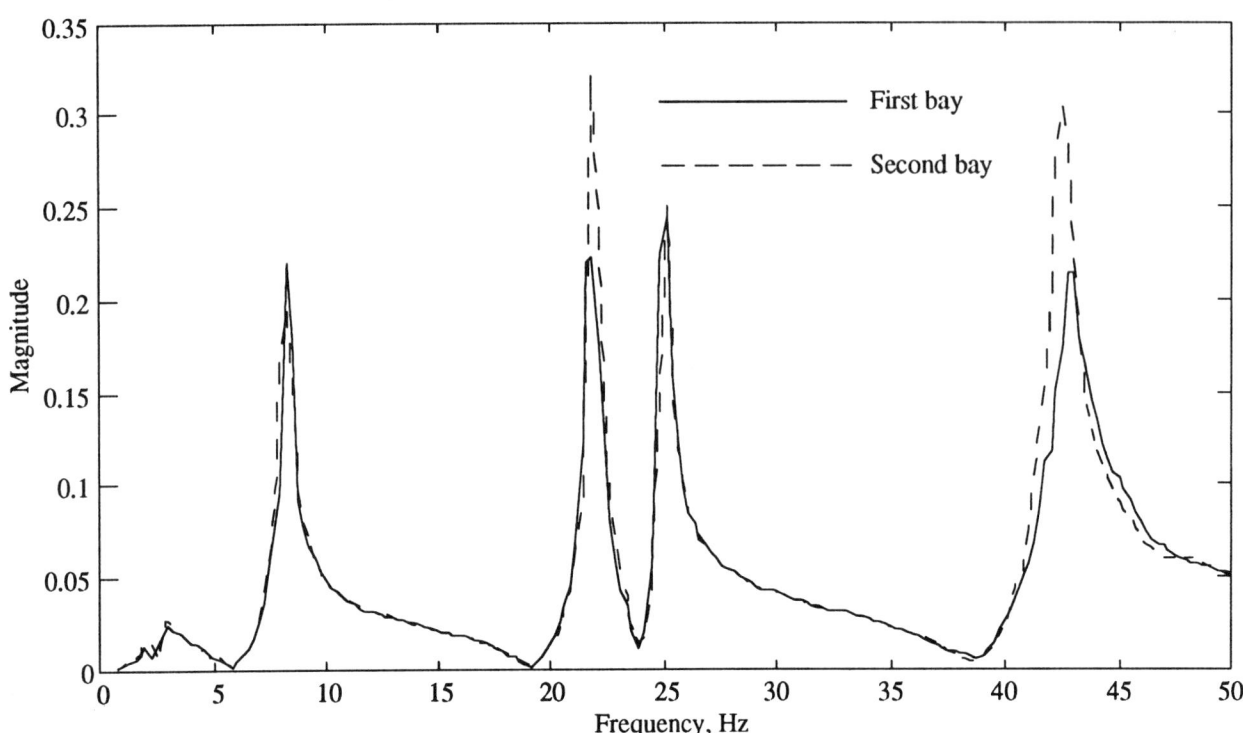

Fig. 6 FRF's at accelerometer 11 for missing upper longerons in the 1st and 2nd bays

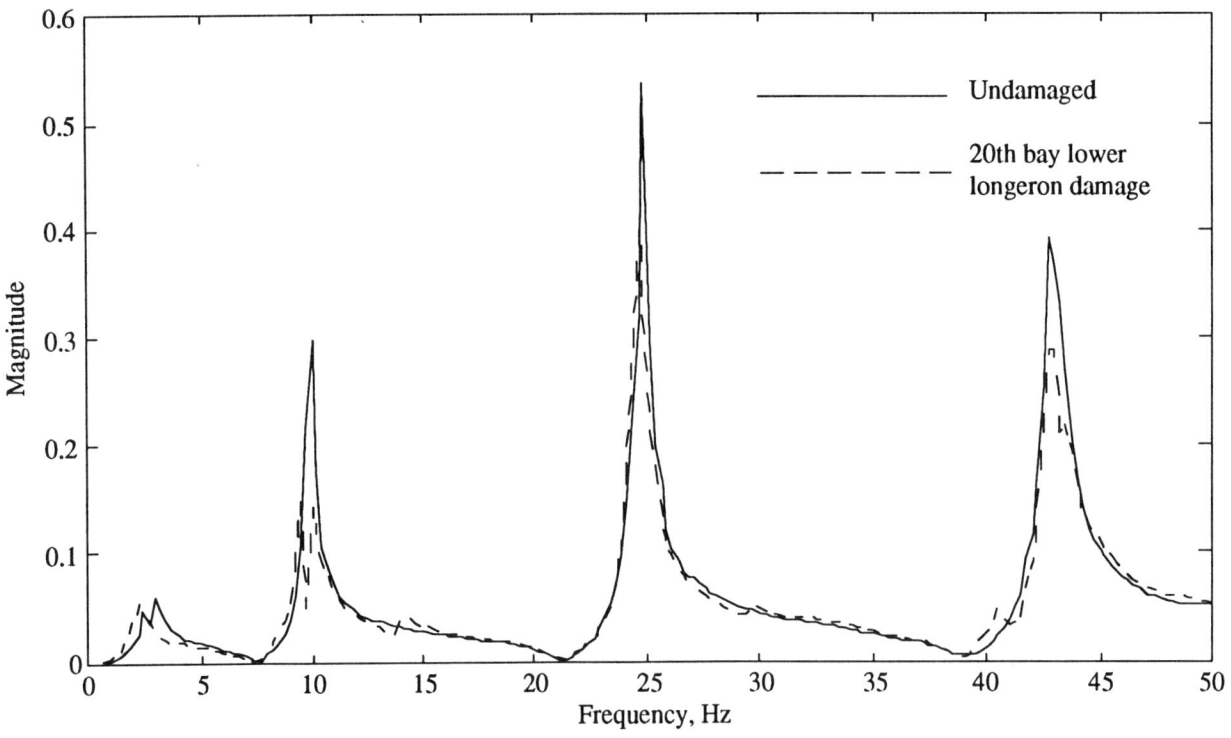

Fig. 7 FRF's at accelerometer 11 for the undamaged truss and missing lower longeron in the 20th bay

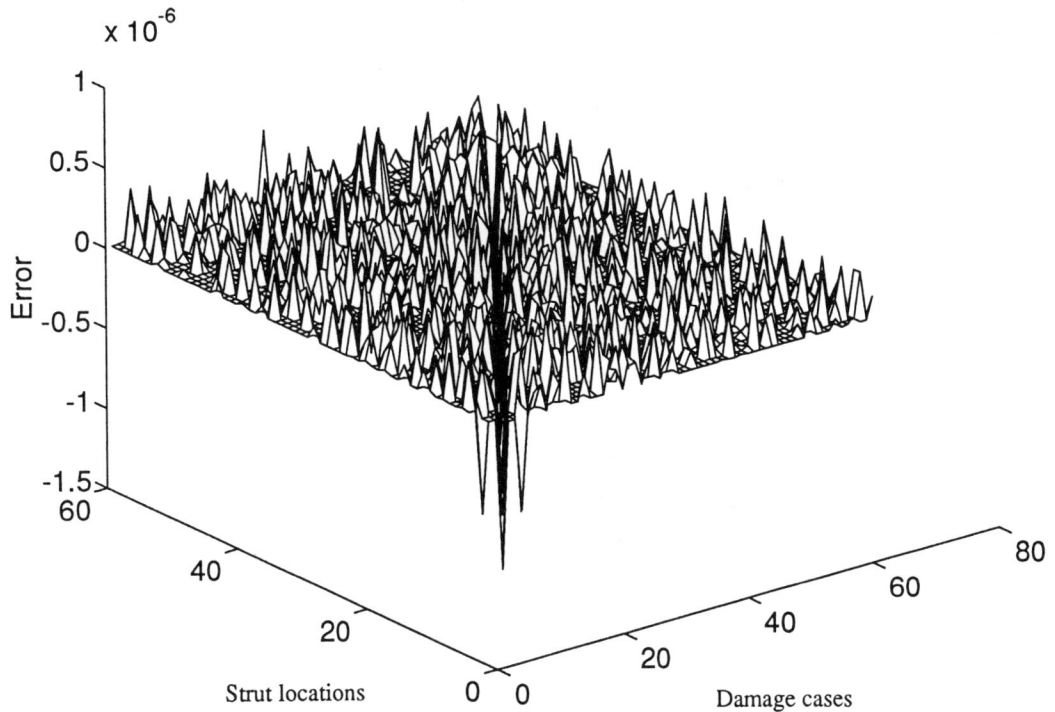

Fig. 8 Error distribution of the trained network

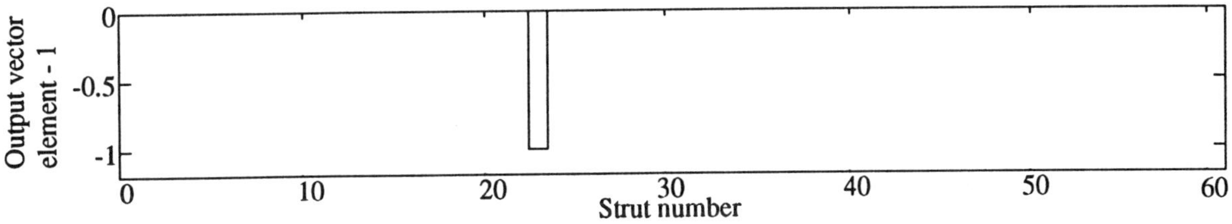

(a) Diagonal in the 8th bay damage case

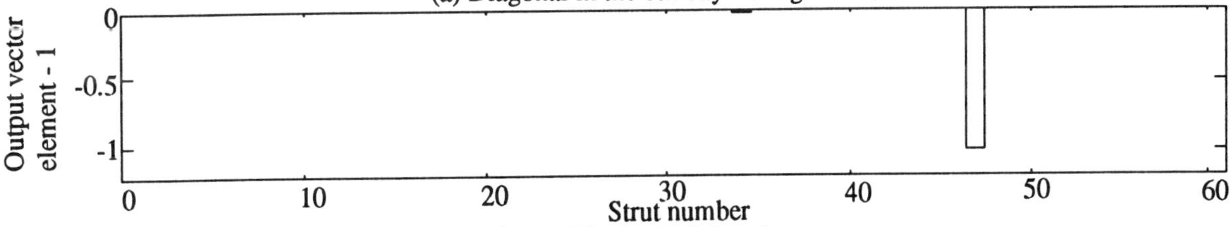

(b) Diagonal in the 16th bay damage case

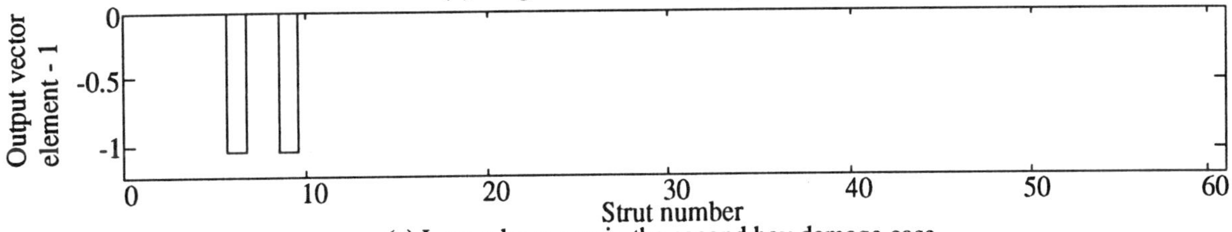

(c) Lower longeron in the second bay damage case

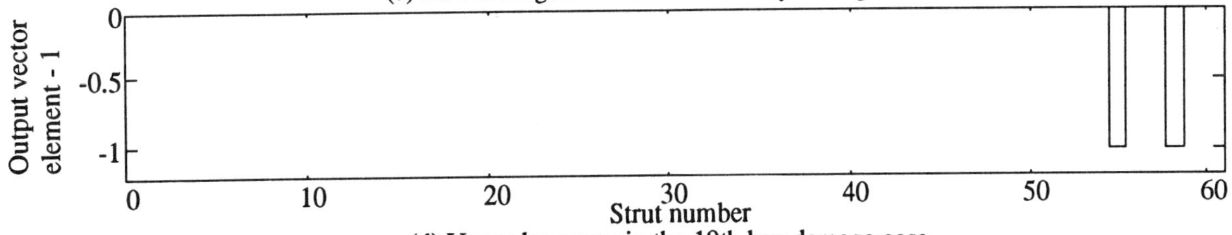

(d) Upper longeron in the 19th bay damage case

Fig. 9 Damage detection results using the trained neural network

AIAA-94-1752-CP

Damage Detection in Adaptive Structures Using Neural Networks

R. A. Manning*
TRW Space and Electronics Group

Abstract

A method is presented for determining the location and amount of damage in a structure based on active member transfer function characteristics and artificial neural networks. The method relies on using the active members, which are already present for structural control, to detect and locate damage in the structure. The neural network is trained for a number of known damage cases where the poles and zeros in the active member transfer functions are used as input training data. Two sample problems are given which demonstrate the feasibility of the method. Various simulated damage cases were run; some where the damage is within the domain of the training data and some where the damage is outside the bounds of the training data. In either case, the neural network is able to locate the damaged members and give a good estimate as to the amount of damage in the member.

Introduction

The detection of damage in structures is a topic that has considerable interest in many fields. Detecting damage in space structures subjected to the harsh environment of space could allow the repair of the structure to occur before the damage threatens the mission objectives. Offshore oil platforms constantly have problems with potential member failure in the corrosive sea environment. Buildings and bridges, where structural failure proves catastrophic, would also benefit from a reliable method of detecting and pinpointing structural damage.

In the past many methods for detecting damage in structures have relied on finite element model refinement methods [1-4]. Hajela and Soeiro [1] determined the damage present in a structure by updating the finite element model to match the static and dynamic characteristics of the damaged structure. Their method was an outgrowth of those presented in References 2 and 3 where undamaged members' section properties changed during the model update process, thus smearing the damage over a wide portion of the structure and making specific damage location difficult. Hajela and Soreiro also extended their damage detection techniques to composite structures [4] where a similar gradient-based optimization scheme was used to update the finite element model.

Other methods of detecting damage in structures rely strictly on measured data. Cawley and Adams [5] used only natural frequency data, Pandey et al [6] used mode shape curvature data, and Swamidas and Chen [7] used strain, displacement, and acceleration data to monitor and detect changes and damages in various structures. These methods require comparing measurements of the structure in the nominal (undamaged) state with those at a later date where some damage is potentially present in the structure. These methods have the drawback that they can only identify that the structure has changed; they cannot identify the location and extent of the damage.

---

* Staff Engineer,
  AIAA Member, ASME Member

For space structures in particular, stringent pointing and/or slew/settle performance requirements will require some form of structural control. The latest methods of structural control utilize active members, composite members with embedded sensors and actuators [8], for controlling flexible modes. With the sensors and actuators already present for the control task, these structures are ready for the development of damage detection algorithms that rely solely on the measurements available from the active members. But the difficulty remains in processing the measurements generated by the active members and in correlating the data with actual locations and levels of damage.

Neural networks have the unique ability to be trained to recognize known patterns and classify data based on these patterns. Neural networks have been used with success for structural design tasks [9] and for classification of experimental data such as sonar target classification [10]. With proper training neural networks should be able to process the transfer function measurements from the active members, classify the data, and provide a tool for determining the location and level of damage present in a structure.

This paper presents a structural damage methodology in which only active member transfer function data is used in conjunction with an artificial neural network to detect damage in structures. Specifically, the method relies on training a neural network using active member transfer function pole/zero information to classify damaged structure measurements and to predict the degree of damage in a structure. The method differs from many of the past damage detection algorithms in that no attempt is made to update a finite element model or to match measured data with new finite element analyses of the structure in a damaged state.

Damage Detection Methodology Overview

Transfer functions taken of structures before and after some form of damage has been introduced show changes in the pole/zero spacing and, perhaps, pole/zero patterns. It is easy to see these changes when reviewing them, but it is difficult to classify them. For example, Figure 1 shows two transfer functions taken of a structure with and without damage. The differences in pole/zero spacing are small, yet detectable, to the naked eye. However, there is no convenient way to correlate the pole/zero spacing and the location and amount of damage present in the structure. Furthermore, given the transfer function of the damaged structure, no adequate method exists for locating which structural members are damaged and how much damage is present.

The method presented in this paper utilizes finite element data to simulate damage in a structure, with the resulting active member transfer functions used as input training data in an artificial neural network. The method assumes that a reasonable finite element model of the structure in the nominal configuration (i.e., without damage) is available and yields transfer functions that properly characterize the structure.

A flow diagram, outlining the details of the damage detection methodology, is shown in Figure 2. A set of members that are assumed to be at most risk within the structure are identified. These members, which may be a subset or the complete set of members within the structure, will subsequently be used to generate training data for the neural network. Each of the selected "at risk" members' cross sectional areas are varied and the resulting pole/zero information within the active member transfer functions saved. Using the pole/zero information as inputs to the neural network and the corresponding member cross sectional areas as outputs, the neural network is batch trained until a suitable level of error bound is achieved.

(Achieving this error bound is most likely an iterative process involving the number of neurons in the hidden layer, the learning rate, and the number of iterations used to train the network.) The resulting neural network weights and biases represent a mapping from pole/zero information to structural member cross sectional areas. Given a measured set of pole/zero data on a potentially damaged structure, the neural network output provides the location of the damaged members and an estimate of the cross sectional area of the damaged members.

Active Member Description

The active members used in this work are similar to those described in References 8 and 11. Figure 3 shows a schematic of the active portion of the members. Each active member consists of a host material, either graphite composite or a metallic material, with piezoceramic sensors and actuators resident with the host material. In the case of a graphite composite host material, the sensors and actuators are usually embedded within the layup of the composite for enhanced sensing and actuation and for added protection from hostile environmental conditions. In the case of a metallic host material, the sensors and actuators can be bonded to the external surface of the host member. For the case of a truss member, where only axial sensing and actuation are required, the sensors and actuators on all four sides of the active member are tied together to cancel any imperfections in the alignment and layup of the sensors and actuators and to produce (or sense) only axial motions. On each face of the active member are two sensors; one colocated with the actuator and one nearly-colocated with the actuator. Averaging the two sensors together can give a transfer function that is advantageous for control purposes [12]. This is accomplished by varying the pole/zero spacing and pattern within the active member transfer function by changing the relative weights between the colocated and nearly-colocated sensors.

As a structure changes, the transfer functions between the actuators and the colocated and nearly-colocated sensors change. By monitoring these changes, specifically the pole/zero pattern and spacing within the transfer functions, damage to the structure can be detected.

Neural Network Description

A neural network consists of many simple elements operating in parallel. The elements were originally conceived to simulate the processes of biological systems where many processes occur in parallel. The function of the neural network is determined by the connectivity of the network and the weights assigned to the neurons. Neural networks have been used in the areas of speech interpretation, pattern recognition, and process control. One of the main features of neural networks is their ability to be trained to recognize known patterns and classify data. Once trained, the neural nets can be used to predict future outcomes or classify data when given a new set of input data.

Shown in Figure 4 is the schematic of a typical neural network. This network has a set of inputs, a single hidden layer of neurons, and a set of outputs. In general, multiple hidden layers of neurons can be used, but this point will be discussed later regarding the structural damage detection problem. The output from each neuron in the hidden layer is given by the tangent sigmoidal function

$$f(\beta) = \tanh(\beta)$$

where the input to the neuron is

$$\beta = \sum w_{ij} x_{ij} - \theta_j$$

For the tangent sigmoidal function, input values between +∞ and −∞ are mapped to output values between +1 and -1. Outputs from the hidden layer were linearly combined to produce the outputs of the neural network.

For the work reported on herein, the inputs consisted of the imaginary parts of the transfer function poles and zeros and the outputs consisted of the cross sectional areas of the truss members. For the generic structure with n active members, the input training data consists of 2n sets of zeros (i.e., a set of zeros for the colocated sensor transfer function and a set of zeros for the nearly-colocated transfer function), a single set of structural poles, and the feedforward voltage produced by the sensors when operating the actuators well below the dynamics of the system. This methodology has assumed that local "surge" modes of the active member are beyond the frequency band of interest.

Example Problems

Ten Bar Truss

The example structure on which the previously outlined damage detection methodology will be demonstrated is the ubiquitous ten bar truss structure shown in Figure 5. This structure has been used for many structural optimization methodology demonstrations including one utilizing neural networks [9]. The nominal design for the structure without active members typically consists of all ten aluminum members having a cross sectional area of 1.0 in^2. Active members were substituted for element number 1 (the bottom root longeron) and for element number 8 (the upwardly pointing root diagonal). The piezoceramic sensors and actuators were designed to have matched stiffness to the local region of placement. This involved cutting the aluminum portion of the active truss members so that the overall stiffness characteristics of the active member approximately match those of the inert aluminum members. The ten bar truss structure with this baseline design has the natural frequencies of 13.6, 39.0, 40.2, 75.6, 82.3, 93.0, and 94.0 Hz.

Transfer functions between the active member actuators and sensors were generated. A typical set of transfer functions for the two active members is shown in Figure 6. Note that the colocated sensor in either case has a relatively large feedforward term when compared with the nearly-colocated sensor. This feedforward term gives an indication of the stiffness of the active member relative to the remainder of the structure. Thus it can be used as an indicator of the health of the active member itself. In addition, the location of the poles and zeros gives an indication of the health of the remainder of the structure.

Input training data for the neural network consisted of the level of feedforward at the four sensors as well as the imaginary parts of the transfer function poles and zeros. Output training data for the neural network consisted of the cross sectional areas of each of the ten bars in the truss. Additional training sets were obtained by decreasing the stiffness of a member of the truss by a known amount and presenting the resulting input and output training data, as described above, to the neural network.

All results presented below were obtained using a neural network with a single hidden layer of 9 tangent sigmoidal neurons. Additional configurations of neural networks were trained and used to locate and predict the damage in the ten bar truss, but did not achieve better results than the single layer, 9 neuron network. Two networks that achieved approximately equivalent results were a double layer network (with 5 and 4 tangent sigmoidal neurons) and a single layer network with 17 log sigmoidal neurons.

Table 1 contains a list of the simulated damage cases that were run on the ten bar truss structure. The resulting neural network predictions of the member cross sectional areas are also given in Table 1 and presented pictorially in Figures 7 through 9. Test Case 1 represents a condition where a single member was damaged (i.e., member number 4). This type of damage is within the domain of the training data and gives an indication of the adequateness of the training of the neural network. The damage assessment from the neural network indicates that member 4 is damaged and the predicted level of damage, $A_4 = 0.74$, compares well with the actual level of damage used to generate the damaged structure transfer functions (see Figure 7). The network also predicts slight damage to members 2 and 9 that is a result of the static indeterminacy in the ten bar truss. Test Cases 2 and 3 represent multiple member damage conditions where 2 and 3 members are damaged simultaneously, respectively. These types of damage are outside the domain of the training data of the neural network. Nonetheless, the neural network pinpoints the damage very well for both cases (see Figures 8 and 9). In addition, the level of damage is predicted within a few percent for Test Case 2 and within approximately 8% for Test Case 3.

Twenty Five Bar Transmission Tower

A second example structure for demonstrating the damage detection methodology is the twenty five bar transmission tower shown in Figure 10. This structure has been used in a number of design optimization studies and has behavior closer to realistic structures that would benefit from the damage detection algorithm than the ten bar truss structure.

Initially the twenty five bars in the truss were linked to produce four "design variables". The lateral batten across the top of the tower was designated as an independent design variable, the eight upper diagonals were linked to a second independent design variable, the four mid-tower battens were linked to a third independent design variable, and the lower ten inert diagonals were linked to a fourth independent design variable. The cross sections of the two active members, located in the lower diagonals as shown in Figure 10, were held fixed in generating the training data.

A baseline set of transfer functions between the actuators in the active members and both the colocated and nearly-colocated sensors were obtained. A typical baseline transfer function is shown in Figure 11. Each design variable was then perturbed and the resulting structural poles and active member transfer function zeros recorded for use in training the neural network. A typical perturbed transfer function is also shown in Figure 11. As in the case with the ten bar truss, small, but distinguishable, differences can be seen in the transfer functions. Without the neural networks, no existing methodology can take these differences and determine the location and amount of damage present in the structure.

A neural network with a set of input neurons, two layers of hidden neurons, and a single layer of output neurons was batch trained with the baseline and perturbed structure data. The input layer contained forty neurons corresponding to 8 poles and 32 zeros. The two hidden layers contained 7 and 5 neurons each, while the output layer contained four neurons corresponding to the cross sectional area of each independent design variable. The neural network was batch trained until the network had approximately converged to its "best" solution (i.e., 8000 epochs of training). Examining the trained neural network indicated that predictions of damage in design variables 2 through 4 should be relatively good. Predictions for damage in design variable 1 would produce some inaccuracy because the neural network has not converged to the weights that yield high quality estimates. This failure to converge is due to the fact that design variable 1 does not have a great influence on the first 8 modes of the structure (i.e., the modes that were used to train the neural network). Using additional modes for training the neural network could alleviate this difficulty.

Three cases of simulated damage were run on the truss and compared with the predictions made by the trained neural network. These three cases are given in Table 2 and correspond to: Case 1) 25% reduction in stiffness in design variable 2 only; Case 2) 5% reduction in design variable 2 and a 25% reduction in design variable 4; and, Case 3) 5% reduction in the active member stiffness. The first case corresponds to an extrapolation of the training data, the second corresponds to a combination and extrapolation of training sets, and the third corresponds to the presence of damage in the structure that was outside the domain of the training data.

The results of these three case are given in Table 2 and shown pictorially in Figures 12 through 14. For Case 1, the neural network is able to locate the damaged member and give a reasonable estimate of the damage. The network predicts a 35% stiffness reduction in design variable 2 as compared with the actual reduction of 25%. However, the neural network tended to smear the damage across design variables 3 and 4. This smearing effect is due to the inadequacy of the training data in representing damage to design variable 1. (The neural network predicts some increase in stiffness for variable 1 and counteracts this with a smeared reduction in the remaining design variables.) For Case 2, a reasonable prediction of the damage to design variables 2 and 4 is given, along with the same difficulty associated with design variable 1 as seen in Case 1. In Case 3, the neural network recognizes that the damaged structure does not include damage to any of the 4 design variables. Though damage is present in the structure, with a moderate movement of the transfer function poles and zeros, the neural network successfully predicts that the four design variables that it can accurately classify are undamaged. In this case, the neural network does not try to smear the damage across the design variables, preferring instead to correctly recognize design variables 1 through 4 as undamaged.

Conclusions

A methodology for detecting damage in structural systems has been described. The method utilizes the active members that are already present for a controlled structure in conjunction with a trained artificial neural network. Two numerical examples demonstrated the feasibility of the method by pinpointing the damaged members and by giving a very good estimate regarding the level of damage present for each member. Better estimates of the levels of damage could be obtained if training data that encompasses the majority of most likely damage scenarios is used.

The damage detection methodology presented herein is potentially applicable to a wide range of structures where sensor/actuator transfer function pole and zero information is available. Though demonstrated only on simple truss structures, the method could be applied to bending active members or to active plate and shell structures. The keys to making the problem tractable for larger problems are adequately identifying the areas of the structure at high risk for potential damage and including enough pole/zero information in the training of the neural network.

The numerical results from both the 10 bar truss and the 25 bar transmission tower demonstrate the feasibility of the method for detecting and locating damage within structures. The neural network was able to locate patterns of damage even for cases where the damage was outside the domain of the training data.

References

1. Hajela, P. and Soeiro, F.J., "Structural Damage Detection Based on Static and Modal Analysis", *Proceedings of the 30th Structures, Structural Dynamics and Materials Conference*, Mobile, AL, April 3-5, 1989, pp. 1172-1182.
2. Chen, J-C. and Garba, J.A., "On Orbit Damage Assessment for Large Space Structures", *AIAA Journal*, Vol. 26, No. 9, 1988.
3. Smith, S.W. and Hendricks, S.L., "Evaluation of Two Identification Methods for Damage Detection in Large Space Structures", *Proceedings of the 6th VPI&SU/AIAA Symposium on Dynamics and Control of Large Structures*, 1987.
4. Soeiro, F.J. and Hajela, P., "Damage Detection in Composite Materials Using Identification Techniques", *Proceedings of the 31st Structures, Structural Dynamics and Materials Conference*, Long Beach, CA, April 2-4, 1990, pp. 950-960.
5. Cawley, P. and Adams, R.D., "The Localization of Defects in Structures from Measurements of Natural Frequencies", *Journal of Strain Analysis*, Vol. 14, No. 2, 1979.
6. Pandey, A.K., Biswas, M., and Samman, M.M., "Damage Detection from Changes in Curvature Mode Shapes", *Journal of Sound and Vibration*, Vol. 145, No. 2, March 8, 1991, pp. 321-332.
7. Swamidas, A.S.J. and Chen, Y., "Damage Detection in a Tripod Tower Platform Using Modal Analysis", *Proceedings of the 11th International Conference on Offshore Mechanics and Arctic Engineering*, Vol. 1, Part B, June 7-12, 1992.
8. Bronowicki, A.J., Mendenhall, T.L., Betros, R.S., Wyse, R.E., and Innis, J.W., "ACESA Structural Control System Design", presented at the First Joint U.S./Japan Conference on Adaptive Structures, Maui, HI, November 13-15, 1990.
9. Swift, R.A. and Batill, S.M., "Application of Neural Networks to Preliminary Structural Design", *Proceedings of the 32nd Structures, Structural Dynamics and Materials Conference*, Baltimore, MD, April 8-10, 1991, pp. 335-343.
10. Gorman, R. and Sejnowski, T., "Learned Classification of Sonar Targets Using a Massively Parallel Network", *IEEE Transactions on Acoustics, Speech, and Signal Processing*, Vol. 36, No. 7, July 1988.
11. Manning, R.A., "Optimum Design of Intelligent Truss Structures", *Proceedings of the 32nd Structures, Structural Dynamics and Materials Conference*, Baltimore, MD, April 8-10, 1991, pp. 528-533.
12. Bronowicki, A.J. and Mendenhall, T.L., "Locally Compensated Deformation Sensor", U.S. Patent No. 5022272, Issued June 11, 1991.

Table 1. Simulated Damage Test Cases - Ten Bar Truss

Member Number	Test Case 1		Test Case 2		Test Case 3	
	Actual Area	NN Area	Actual Area	NN Area	Actual Area	NN Area
1	1.00	1.00	1.00	0.98	1.00	1.05
2	1.00	0.92	1.00	1.00	1.00	0.96
3	1.00	0.99	0.80	0.82	0.80	0.88
4	0.75	0.74	1.00	0.99	1.00	1.05
5	1.00	1.02	1.00	0.99	1.00	1.00
6	1.00	0.98	1.00	1.02	0.80	0.85
7	1.00	0.99	0.95	0.95	1.00	0.92
8	1.00	1.07	1.00	0.99	1.00	1.11
9	1.00	0.98	1.00	1.00	1.00	0.95
10	1.00	1.02	1.00	1.00	0.70	0.76

Table 2. Simulated Damage Test Cases - Twenty Five Bar Tower

Member Number	Test Case 1		Test Case 2		Test Case 3	
	Actual Area	NN Area	Actual Area	NN Area	Actual Area	NN Area
1	1.00	1.04	1.00	0.91	1.00	0.97
2	0.75	0.65	0.95	0.93	1.00	1.00
3	1.00	0.95	1.00	0.96	1.00	1.00
4	1.00	0.92	0.75	0.67	1.00	1.00

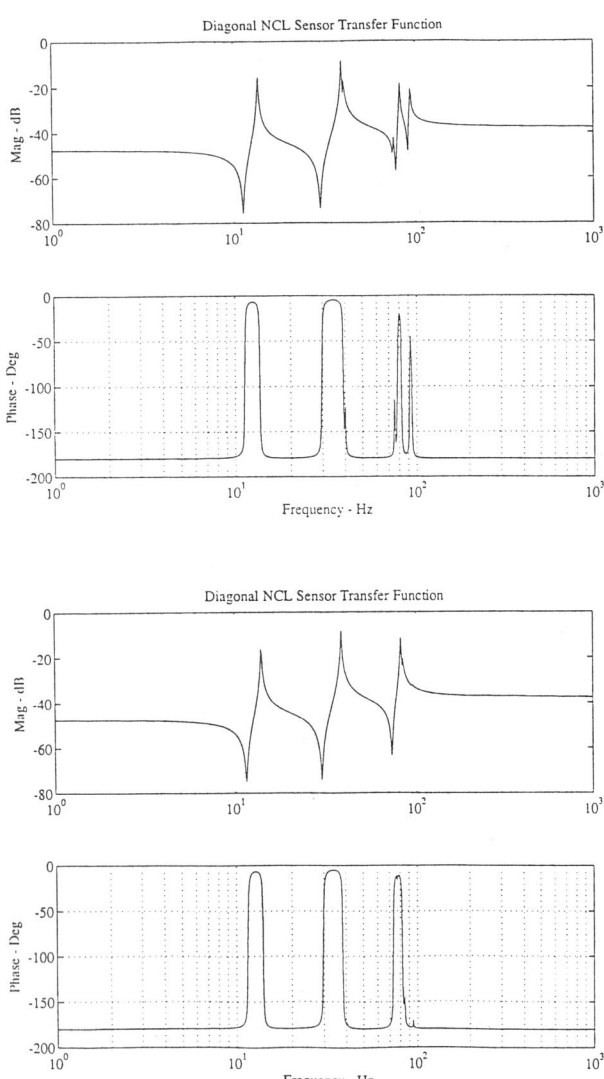

Figure 1. Typical Undamaged and Damaged Structure Transfer Functions

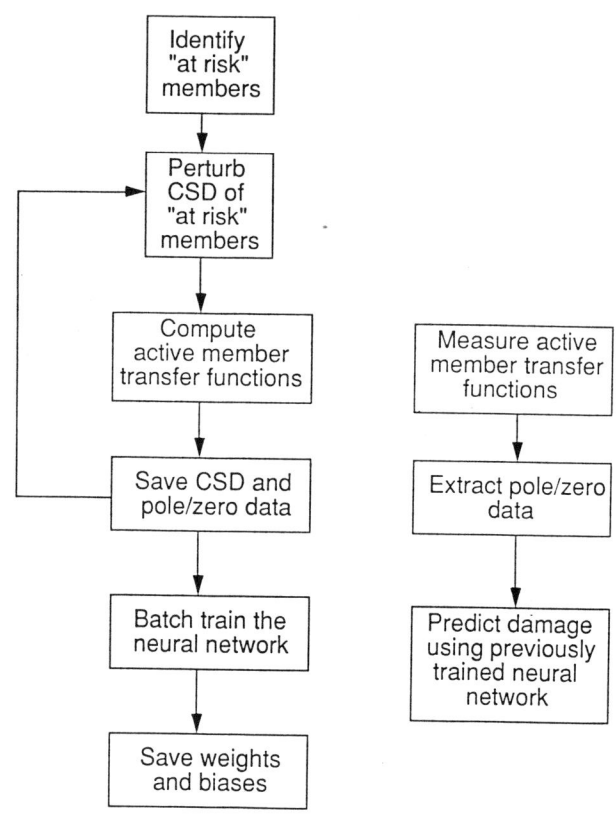

Figure 2. Structural Damage Detection Flow Diagram

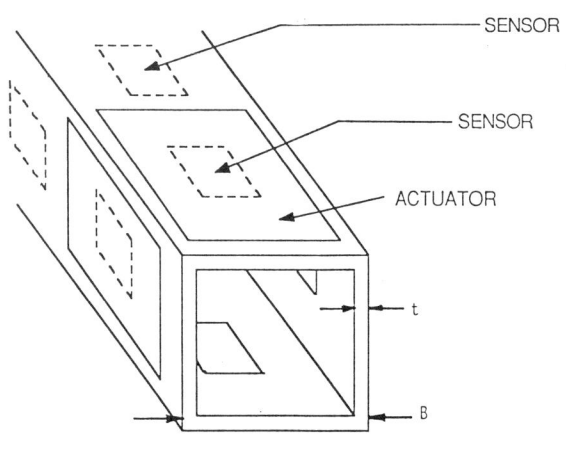

Figure 3. Typical Active Member Schematic

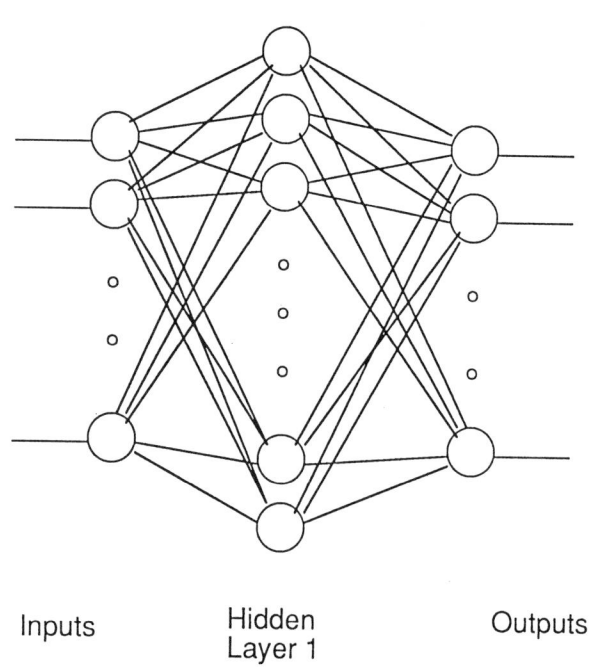

Figure 4. Generic Neural Network Layout

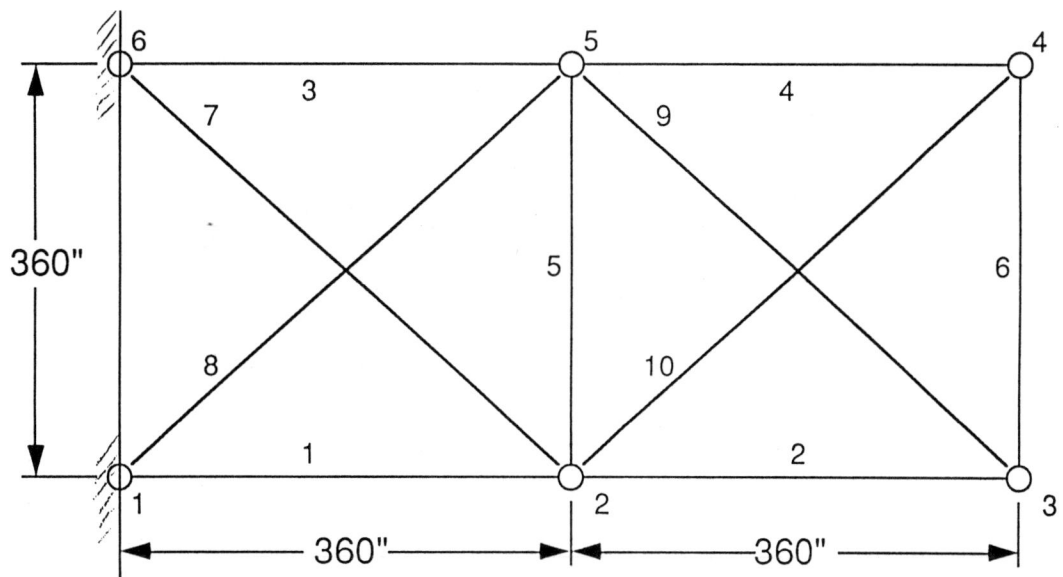

Figure 5. Ten Bar Truss Example Structure

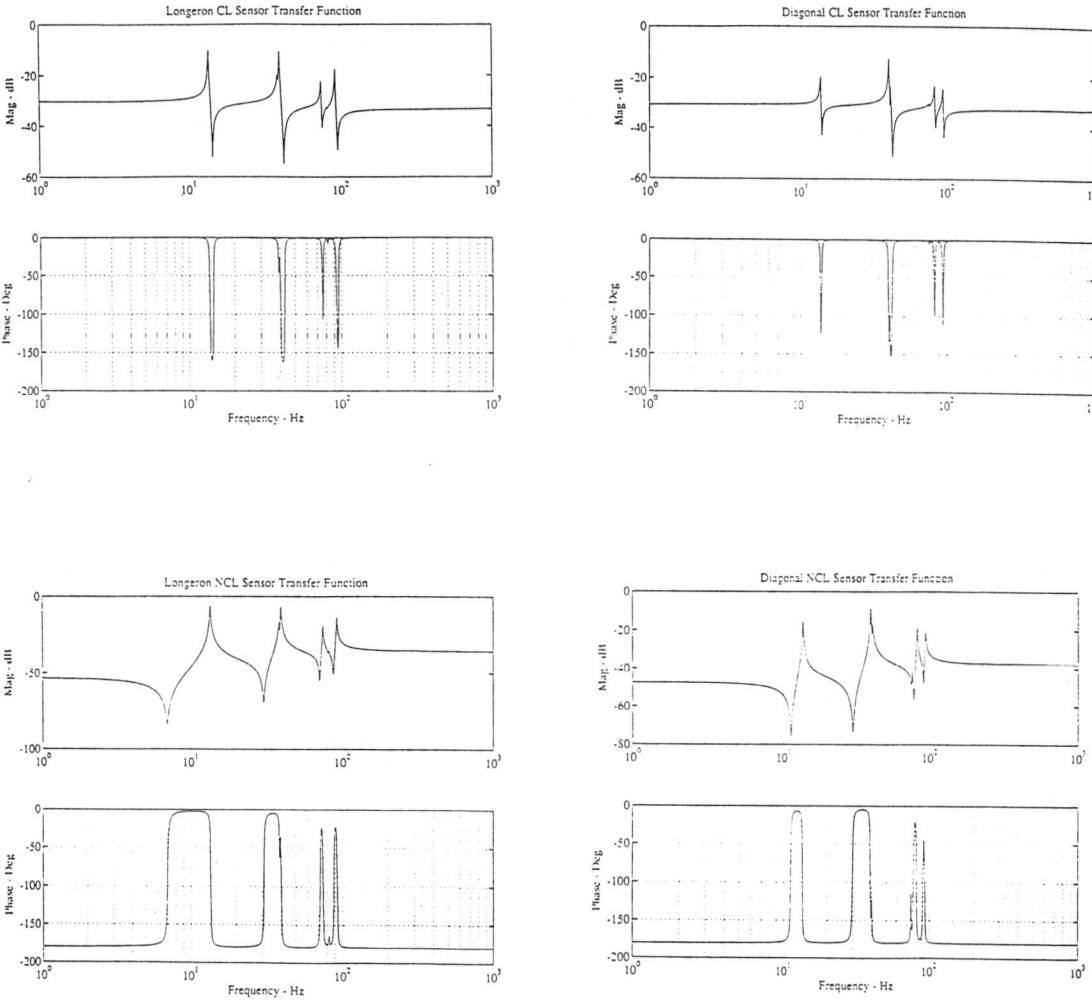

Figure 6. Ten Bar Truss Active Member Transfer Functions

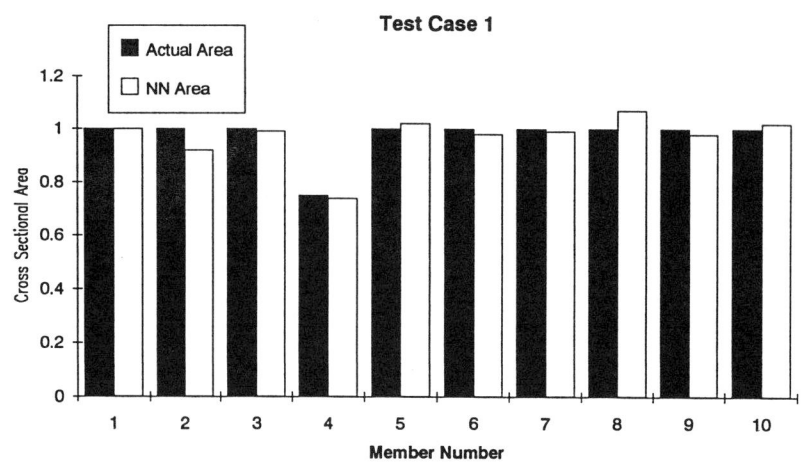

Figure 7. Actual and Predicted Cross Sectional Areas - Test Case 1

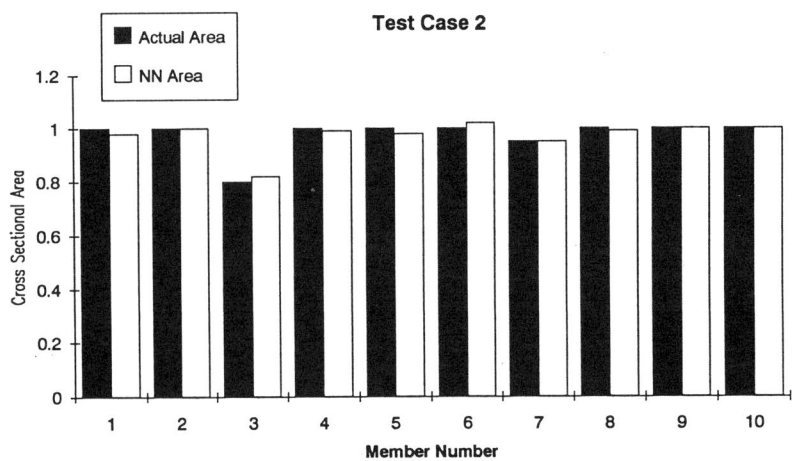

Figure 8. Actual and Predicted Cross Sectional Areas - Test Case 2

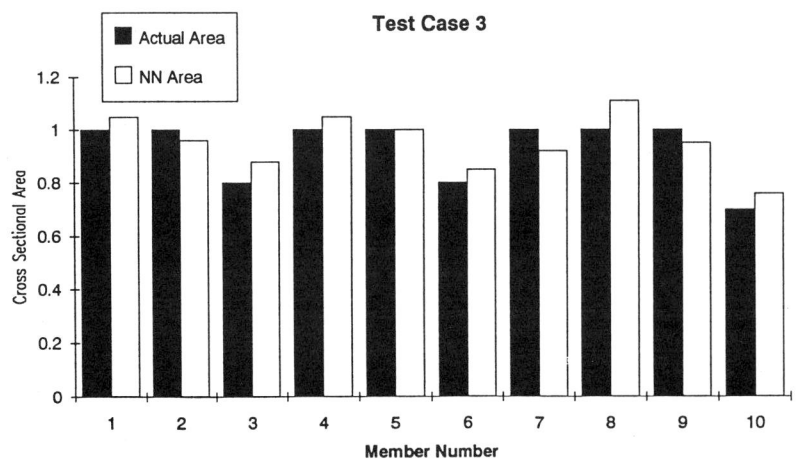

Figure 9. Actual and Predicted Cross Sectional Areas - Test Case 3

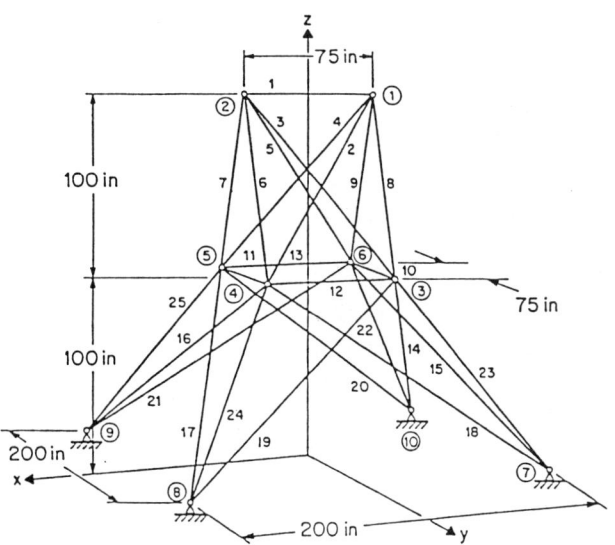

Figure 10. Twenty Five Bar Transmission Tower

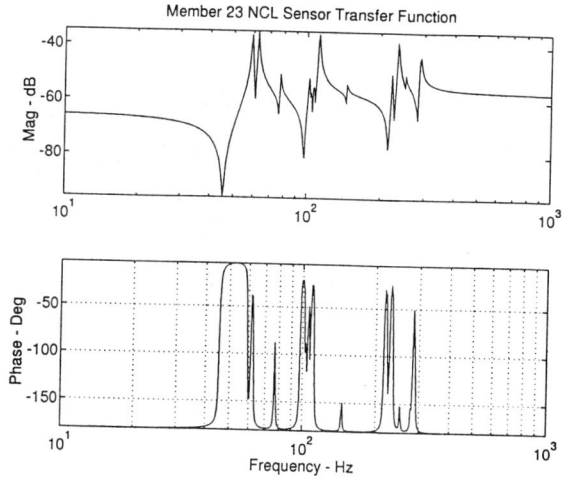

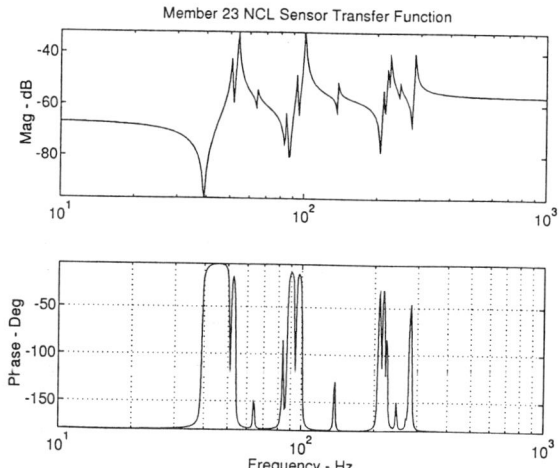

Figure 11. Baseline and Perturbed Active Member Transfer Functions

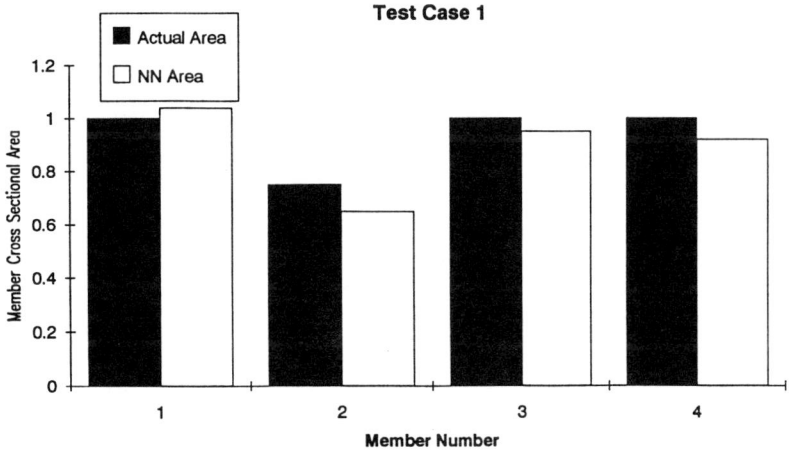

Figure 12. Actual and Predicted Cross Sectional Areas - Test Case 1

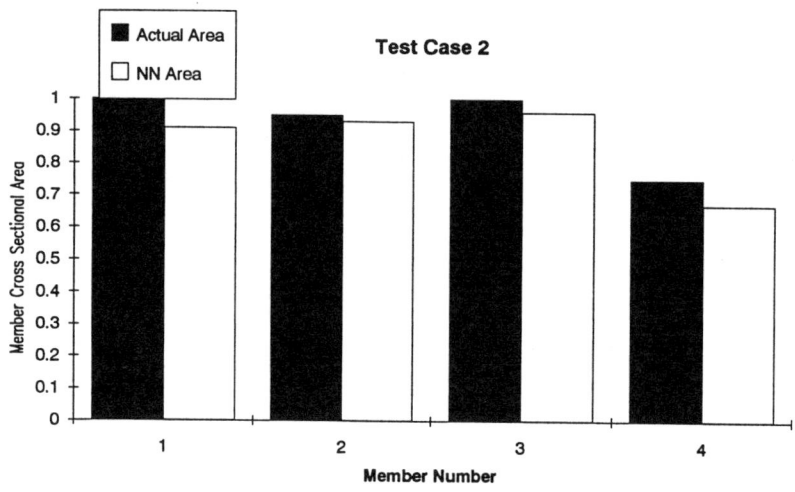

Figure 13. Actual and Predicted Cross Sectional Areas - Test Case 2

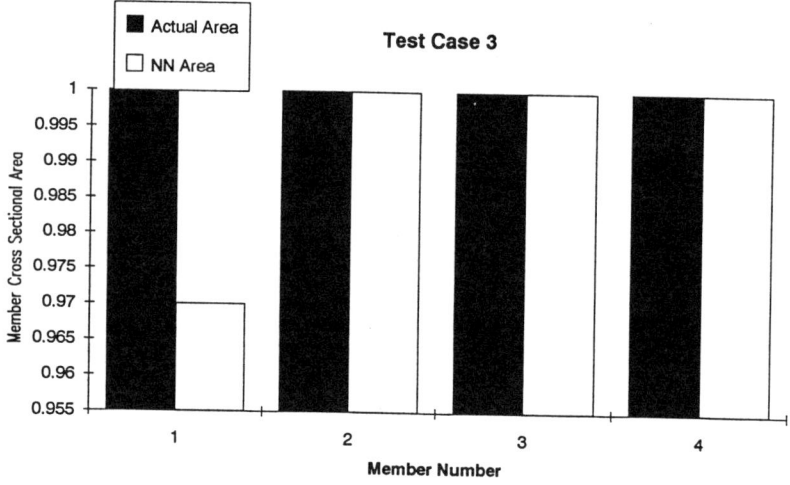

Figure 14. Actual and Predicted Cross Sectional Areas - Test Case 3

AIAA-94-1753-CP

A Neural Network Approach for Damage Detection and Identification of Structures

Jaewook Rhim[1] and Sung W. Lee[2]
Department of Aerospace Engineering
University of Maryland at College Park
College Park, MD 20742

Abstract

This study examines the feasibility of using artificial neural network in conjunction with system identification techniques to detect the existence and to identify the characteristics of damage in composite structures. The methodology proposed here includes a training phase and a recognition phase. In the training phase, candidate models for structures with various types of damage are designated as the patterns. These patterns are organized into pattern classes according to the location and the severity of the damage. Then system identifications are performed to extract the transfer functions as the features of the structural systems. These transfer functions are fed into a multi-layer perceptron (MLP) as the input patterns for training. The MLP serves as a nearest neighborhood classifier. In the pattern recognition phase, a structure with unforeseen damage is classified within the closest class in the training set and the damage in the structure is identified as that of the class. Numerical tests are conducted to demonstrate the feasibility of the proposed method.

Introduction

Recently on-line damage detection and identification (DDI) of flexible structures has been a subject of intensive investigations[1-7]. Among these is a class of damage detection methods that resemble the algorithms for finite element model update or test-analysis correlation. Most of these works begin with measured dynamic response and modal parameters (i.e., natural frequencies and mode shapes) to find the differences, such as the changes in the stiffness and mass, between the undamaged and damaged

---
[1] Graduate Research Assistant
[2] Professor, Associate Fellow

Copyright © 1994 by Jaewook Rhim and Sung W. Lee. Published by the American Institute of Aeronautics and Astronautics, Inc. with permission.

system. There are two major disadvantages of these methods. First, they need a large number of sensors to measure mode shapes accurately, while the number of sensors is limited in practical applications. Second, the comparison of the new model with the old one requires heavy computational efforts to assess the damage extent and severity. Also it has been shown that these methods can generate ambiguous results unless the modes that are high in strain energy are used[6,7].

A potentially powerful structural health monitoring system is that of using a neural network approach[8-10]. Because of their ability to learn from past experiences and to memorize the patterns in the form of an associative memory, special attention has been drawn to pattern recognition and classification (for an excellent review, see Lippman[8]). Neural networks are attractive for detecting damage-caused signals because they are capable of automatically discovering patterns of interrelated features which serve to define the corresponding class of an unforeseen signal. In addition, they provide fast answer once they are trained. Accordingly, the objective of the present study is to investigate the feasibility of approaching the DDI of structures from a pattern classification perspective.

Some investigators have examined the feasibility of neural networks for DDI[11,12]. Barga et al.[11] used the time history of acoustic emission waveforms as their inputs to a neural network in order to classify two types of acoustic emission events, crack and fretting. Tsou and Shen[12] proposed the use of modal parameters to identify the location and the severity of damage in spring-mass-damper systems. In the present study, transfer functions of auto-regressive model with exogenous input (ARX)[13,14] serve as input patterns for an MLP. Transfer function is used as the system feature since it represents the complete and compact information on a dynamic system from a given input-output data. An advantage of this method is that the number of sensors can be kept as small as possible. The performance of the proposed DDI scheme is evaluated using cantilever composite beams with delaminations as the test beds.

System identification in auto-regressive model with exogenous input (ARX)

The ARX model of a single-input single-output dynamic system is described by the following mathematical expression[13,14]:

$$A(q^{-1})y(t) = q^{-nk}B(q^{-1})u(t) + e(t) \quad (1)$$

where

$$A(q^{-1}) = 1 + a_1 q^{-1} + \cdots + a_{na} q^{-na}$$
$$B(q^{-1}) = b_0 + b_1 q^{-1} + \cdots + b_{nb} q^{-nb} \quad (2)$$

In eq. (1), $y(t)$ and $u(t)$ are the output and input of the system respectively, nk is the time-delay, $e(t)$ is a zero-mean white noise process representing measurement noise and modeling error, and q is the time-shifting operator, i.e.,

$$q^{-1}y(t) = y(t-1) \quad (3)$$

In eq. (2), $a_1, \ldots, a_{na}$ and $b_0, \ldots, b_{nb}$ are the coefficients of $A(q^{-1})$ and $B(q^{-1})$ respectively. Because the inherent nature of sampling is discrete, such a representation of a dynamic system has been widely used in systems and control literature to establish input-output relationship. The ARX model is a special case of general input-output model, so called ARMAX model (auto-regressive moving average model with exogenous input)

$$A(q^{-1})y(t) = B(q^{-1})u(t) + C(q^{-1})e(t) \quad (4)$$

with

$$C(q^{-1}) = c_0 + c_1 q^{-1} + \cdots + c_{nc} q^{-nc} \quad (5)$$

where nc is the order of $C(q^{-1})$ and $c_0, \ldots, c_{nc}$ are the coefficients. As seen in eqs. (4) and (5), the ARMAX model has an advantage over the ARX model owing to its ability to handle the noise by another filter $C(q^{-1})$. However, since the major goal of this study is to examine a DDI scheme based on the neural network combined with a system identification, $C(q^{-1})$ is assumed to be 1 and consequently, noise is treated

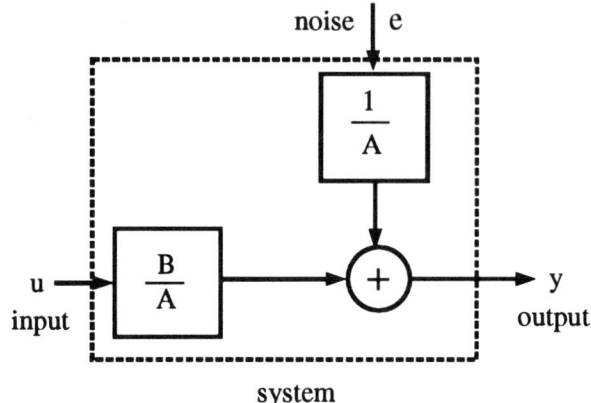

Figure 1 Structure of ARX model

as a perfect white noise. Figure 1 depicts the structure of ARX model.

Ignoring $e(t)$ in eq. (1), we can write

$$y(t) = G(q)u(t) \quad (6)$$

where

$$G(q) = \frac{q^{-nk}B(q^{-1})}{A(q^{-1})} \quad (7)$$

is defined as the transfer function. Therefore $G(q)$ is an effective estimation of the discrete transfer function of a system in noisy environment. The transfer function contains the complete information on the behavior of a linear system and therefore represents the feature of the system.

The coefficients in eq. (2) are called *system parameters* and define the transfer function $G(q)$. In this regard, estimating the transfer function is equivalent to estimating the system parameters. Estimation of parameters is normally done by least squares algorithm given a set of input-output data[13,14].

It should be mentioned that using transfer function as input pattern is different from using modal parameters (natural frequencies and damping ratios) although they are closely related. Once the sampling frequency is set, the frequency range of observation is also limited to approximately a half of it. If natural frequencies are used as input pattern, the inconsistency of input elements can cause confusion since the number of natural frequencies in the range varies when damage exists in the structure. This inconsistency can be avoided by using transfer function or characteristic polynomial because the coefficients for lower power in q are actually zero even though the order of the polynomial changes.

Multi-layer perceptron (MLP)

Multi-layer perceptron (MLP) is a class of neural networks that can be used mainly as a nearest neighborhood classifier and sometimes as an associative memory. An MLP can be completely described by its interconnecting topology, neuronic characteristics, and learning rule. An individual unit (neuron or node) performs calculation of the weighted sum and activation of nonlinear mapping on the sum. The activation function should be differentiable and nonlinear and should have a threshold type of behavior. The conceptual structure of a neuron is diagrammed in figure 2. First, the weighted sum of input components are calculated as

$$S_j^k = \sum_{i=0}^{N^{k-1}-1} w_{ij}^k O_i^{k-1} \quad (8)$$

where S_j^k is the weighted sum of the j'th neuron in the k'th layer, N^{k-1} is the number of neurons in the $k-1$'th layer, w_{ij}^k is the weight between the i'th neuron in the $k-1$'th layer and the j'th neuron in the k'th layer, and O_i^{k-1} is the output of the i'th neuron in the $k-1$'th layer. Once the weighted sum S_j^k is calculated, output for the j'th neuron in the k'th layer O_j^k is computed with a sigmoid function as follows:

$$O_j^k = f(S_j^k) = \frac{1}{1 + e^{-(S_j^k - p_j^k)}} \quad (9)$$

where p is the node offset which is helpful for fast convergence of learning. The sigmoid nonlinearity activates in every layer except the input layer. The topology of an MLP is shown in figure 3. Many researchers proved that three layer MLP can perform arbitrarily complex classification while the complexity is dependent on the number of neurons in the hidden layers[8].

A three layer MLP can be trained with the conventional back-propagation algorithm. The back-propagation algorithm uses a gradient search technique to minimize a cost function which is equal to the mean square difference between the desired and the actual network outputs. It is essentially a generalization of the least mean square (LMS) algorithm also used in the system identification phase. The output nodes corresponds to the pattern class. For example, if the current input pattern is from class α, only the output node corresponding to the class α is selected by the MLP. The desired output of all nodes is typically "low" (nearly 0) except the output node corresponding to class α in which case it is "high" (nearly 1). The network is trained by initially selecting small random weights and internal offsets and then presenting all training data repeatedly. Weights are adjusted after every trial using the side information specifying the correct class until weights converge and the cost function is reduced to satisfy an acceptable criterion. An essential component of the algorithm is the iterative method that propagates the error terms required to adapt weights back from the nodes in the output layer to the nodes in the lower layers.

Usually, the criterion for convergence can be determined in terms of the maximum mean square error between the desired outputs and the actual outputs. The mean square error for the i'th training input-output pair

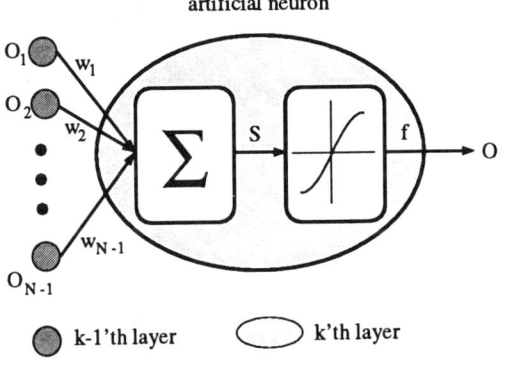

Figure 2 Fundamental structure of a single neuron

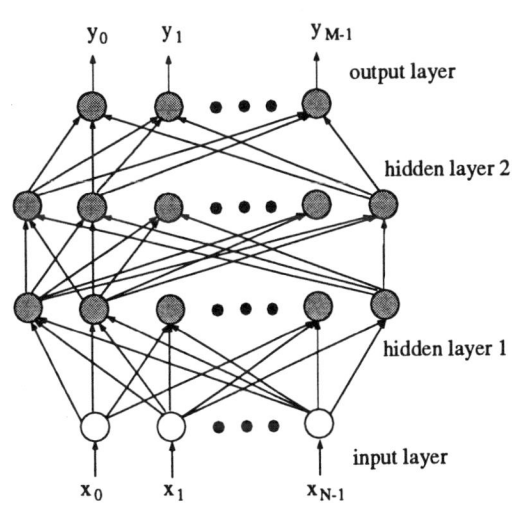

Figure 3 Topology of a typical multi-layer perceptron

is expressed as

$$M.S.E.(i) = \frac{1}{M} \sum_{j=0}^{M-1} (y_j^i - d_j^i)^2 \quad (10)$$

where M is the number of output neurons (i.e., classes), y_j^i and d_j^i are the actual output and the desired output of the j'th neuron in output layer, respectively. Letting the total number of training pairs be NT, the cost function is defined as

$$J(t, \mathbf{W}, \mathbf{P}) = \max_{k=0}^{NT-1} \{M.S.E.(k)\} \quad (11)$$

where $\mathbf{W}$ and $\mathbf{P}$ are the sets of weights and offsets respectively. The iteration stops if the following condition holds:

$$J(t, \mathbf{W}, \mathbf{P}) < \varepsilon \quad (12)$$

This condition ensures the maximum mean square error of the MLP under the prescribed convergence criterion ε.

Numerical test

A model problem of cantilever $[0/90]_{4s}$ T300/934 graphite/epoxy laminate beams with various delaminations is used to illustrate the DDI method proposed in this study and test the performance. It should be pointed out that this is a hypothetical example and does not necessarily represent a realistic problem. All beams are of the same length, thickness and width. The material properties and the beam geometry are summarized in table 1. These beams are organized into several classes according to the coherence of their delaminations as depicted in figure 4. For example, the beams without damage or with negligible delamination are organized into the class 0 and the beams with near-root delaminations of significant lengths are organized into the class 1. Table 2 lists the location and the length of delaminations and the pattern classes used to train the MLP. For simplicity, the delaminations are assumed to be on the symmetric plane.

The geometry of a delaminated beam is shown in figure 5. The delamination is assumed to be uniform through the width of the beam and located arbitrarily, as defined by the parameters d_i, d_f, and h. For convenience of modeling, the beam is divided into two parts: one is above the delamination plane and the other is below the delamination plane. Each part is modeled separately by beam elements based on the Timoshenko theory and then assembled into a global finite element model.

A delaminated beam exhibits the coupling between longitudinal motion and bending motion[15]. By taking into account this coupling effect, a finite element model was developed (figure 6). In the undamaged region, all nodal degrees of freedom of the upper part and the lower part are constrained to be the same. At the points of delamination boundary, say at d_i and d_f, the upper and lower part share common u_o and w, while θ is left to be independent. Within

Longditudinal Young's modulus (E_{11})	19.5×10^6 psi
Transverse Young's modulus (E_{22})	1.5×10^6 psi
Inplane shear modulus (G_{12})	0.725×10^6 psi
Major Poisson's ratio (ν_{12})	0.33
Mass density (ρ)	1.3821×10^{-4} lb-s^2/in^4
Beam length (L)	10 in
Width (b)	1 in
ply thickness (t)	0.005 in

Table 1 Material properties and geometry

Model number	Delamination (in.)	Pattern class
0	No damage	Class 0
1	4.5 - 5.5	Class 1
2	4.0 - 6.0	
3	3.5 - 6.5	
4	2.0 - 3.0	Class 2
5	1.5 - 3.5	
6	1.0 - 4.0	
7	7.0 - 8.0	Class 3
8	6.5 - 8.5	
9	6.0 - 9.0	

Table 2 Training set and pattern classes

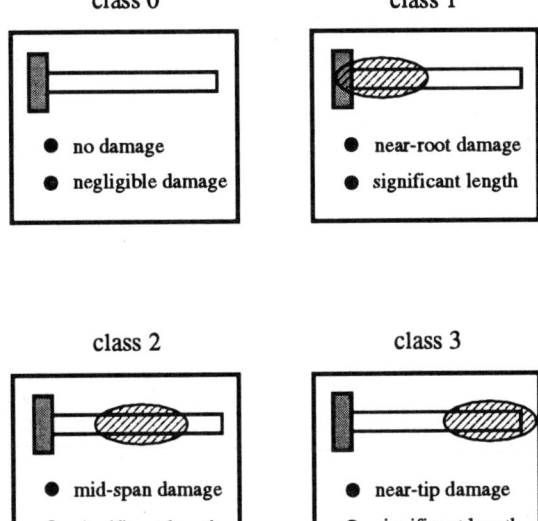

Figure 4 Organizing pattern classes of delaminated beams

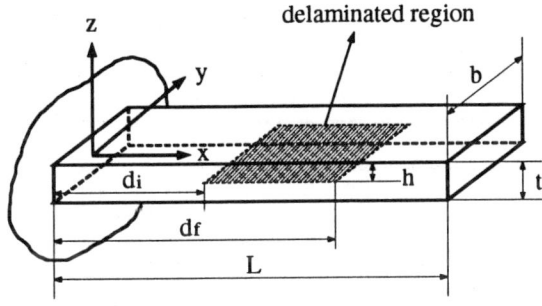

Figure 5 Beam geometry

the delaminated region, all degrees of freedom are independent.

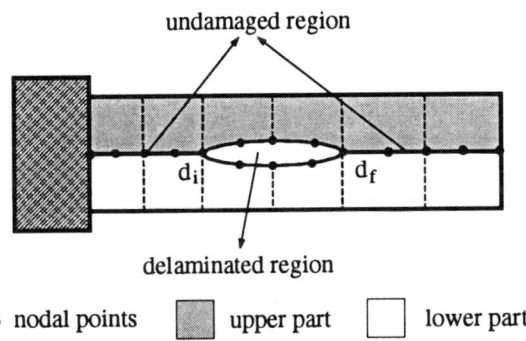

Figure 6 Finite element model of a delaminated beam

A three-node nine degree of freedom isoparametric beam element with quadratic shape functions is used in this study. To avoid the effect of transverse shear locking, two-point reduced integration is chosen for the calculation of element stiffness matrices.

Forced responses of the undamaged beam and the delaminated beams are simulated by FEM. The input $u(t)$ is the concentrated force at the beam tip and the output signal $y(t)$ is acceleration also at the tip. The input $u(t)$ is a pseudo random binary sequence, which is a nearly zero-mean white noise and thus is rich in frequency content. Such an input sequence is very important for good system identification results[16]. ARX system identification is then performed to determine the transfer function and thus the system parameters corresponding to every beam from the input-output sequences obtained from FEM. The output is artificially corrupted by random noise. The system parameters are fed into the MLP as input patterns. The MLP is then trained with the input patterns and classes defined in table 2. The complete DDI procedures are schematically shown in figure 7.

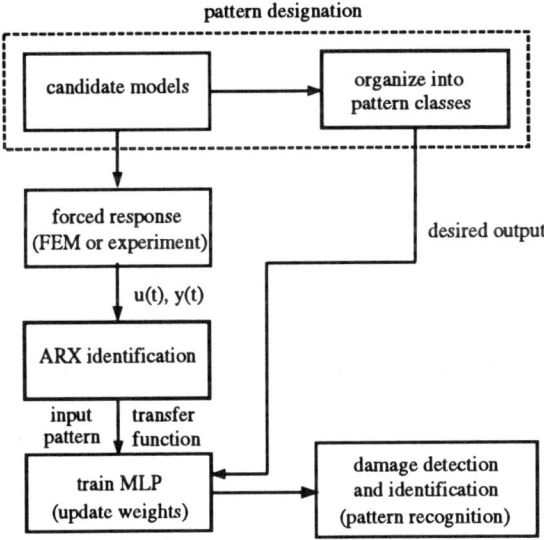

Figure 7 Schematic diagram of DDI procedure

The sampling frequency is 4,000 Hz. and the signal is prefiltered to avoid aliasing. Therefore frequency responses under 2,000 Hz. are extracted by the ARX system identification algorithm. The system identification results are summarized in Table 3. It is noteworthy that the order of $A(q^{-1})$ of delaminated beams generally tends to increase as delamination length grows. The identified transfer functions for the candidate models in the class 2, for example, are presented in figure 8 through figure 10 in the form of Bode plots. From the figures, one can see that the number of peaks in the given frequency bandwidth increases when the length of delamination grows. Also downward shifting of fundamental frequencies is observed (softening effect).

Instead of feeding all parameters into MLP as the input patterns, an alternate way is chosen. Because $A(q^{-1})$ is the characteristic polynomial and has information on natural frequencies and damping ratios in the form of poles, only $a_0, a_1, \ldots, a_{na}$ are selected to be the input patterns. This is also because $B(q^{-1})$ is sensitive to noises and less accurate than $A(q^{-1})$. An MLP is trained with a gain factor of $\eta = 0.1$. Since the largest order of $A(q^{-1})$ of the candidate models is 12, the number of input neurons is set to 13. The number of neurons in each hidden layer is set to 30 and the number of output neurons is set to 4, equal to the number of the pattern classes.

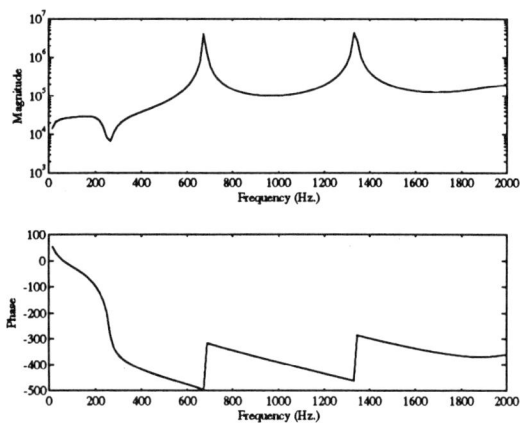

Figure 8 Bode plot for model 4 in class 2

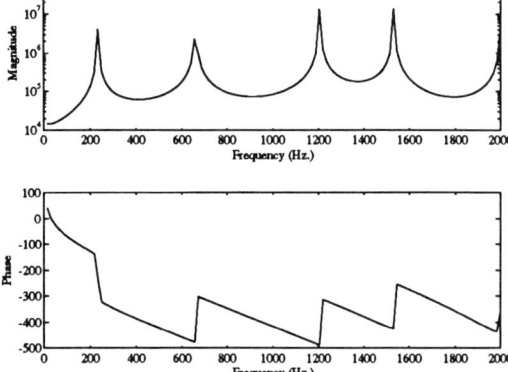

Figure 9 Bode plot for model 5 in class 2

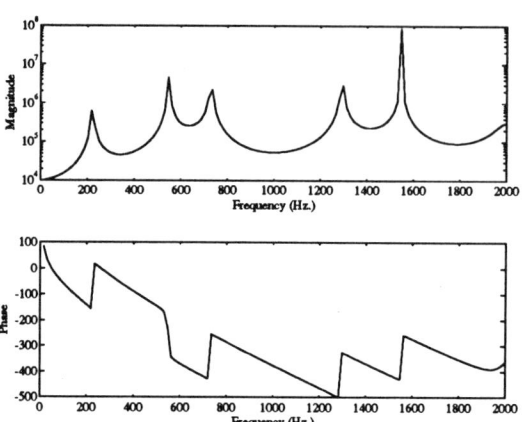

Figure 10 Bode plot for model 6 in class 2

Model number	Delamination (in.)	na	nb	nk
0	No damage	8	4	5
1	4.5 - 5.5	8	4	5
2	4.0 - 6.0	12	5	8
3	3.5 - 6.5	10	4	7
4	2.0 - 3.0	8	4	5
5	1.5 - 3.5	10	4	7
6	1.0 - 4.0	12	4	9
7	7.0 - 8.0	8	4	5
8	6.5 - 8.5	12	4	9
9	6.0 - 9.0	12	4	9

Table 3 Identification summary for candidate models

DDI results

The MLP is successfully trained (i.e., the weights converged in a stable manner) in 393,000 iterations. The cost function defined in eq. (11) is calculated at every 1,000 iterations and plotted in figure 11.

The criterion of convergence ε in eq. (12) is 0.002. The performance of the network is examined through three unforeseen delaminated beams none of which is identical to any of the training pairs defined in table 2. The identified transfer functions are summarized in table 4 and figure 12 through figure 14. One can easily determine by inspection to which class each of them belongs. The ability of the MLP to recognize the delaminations is then examined by comparing the actual output of the MLP with the expected output. This comparison is shown in table 5 along with the configurations of three delaminated beams. As seen in

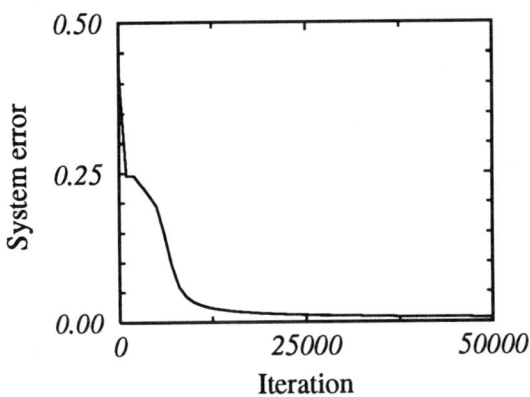

Figure 11 Cost function versus iteration number

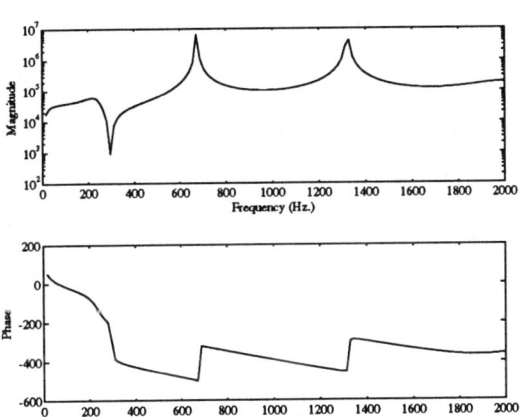

Figure 12 Bode plot for test example 1

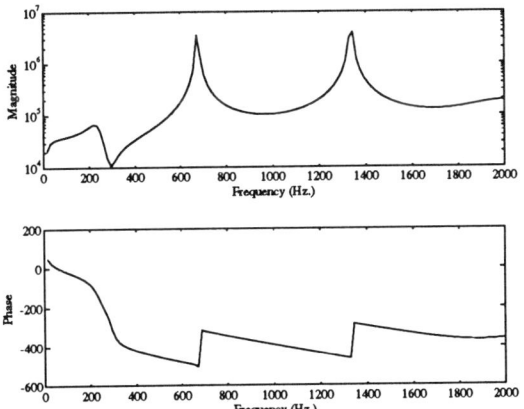

Figure 13 Bode plot for test example 2

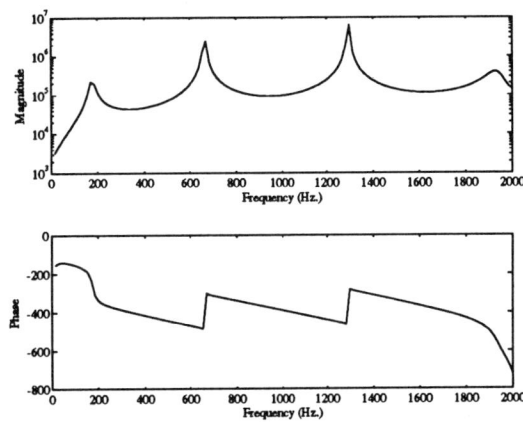

Figure 14 Bode plot for test example 3

Test number	Delamination (in.)	na	nb	nk
1	4.0 - 4.4	8	4	5
2	7.1 - 8.3	8	4	5
3	1.6 - 3.3	10	5	6

Table 4 Identification summary for test models

Test number	Actual delamination (in.)	Expected class	Estimated class by MLP
1	4.0 - 4.4	Class 0 or Class 1	Class 0
2	7.1 - 8.3	Class 3	Class 3
3	1.6 - 3.3	Class 2	Class 2

Table 5 Damage detection and identification results

table 5, all of the three cases are identified correctly. Since numerically obtained acceleration signals which are the system output are corrupted by noise and the three example models are not identical to any of the candidate models in the training set, the results demonstrate that the proposed DDI method is robust to measurement noise and distortion of the input patterns.

Conclusions

An approach for damage detection and identification based on a neural net and an ARX system identification technique is proposed and examined in this study. The results of the numerical simulation indicate that the proposed DDI method has the potential for practical application since it is robust to measurement noise and distortion of input patterns. An advantage of this method is that the number of sensors required can be kept as small as possible.

The finite element models used to generate input patterns (i.e., system parameters) can be replaced by experiments because the proposed DDI method does not need the mass and stiffness matrices. Excessively small number of training pairs (i.e., candidate models) might cause failure of pattern recognition when an ambiguous pattern is given to the MLP. Although increasing the number of training pairs and the number of pattern classes can avoid this problem and strengthen the capability of the MLP, it also raises the cost. Further studies should investigate the methods of reducing the number of training pairs without degrading the performance of the MLP.

References

1. Pandey, A. K., "Damage detection from changes in curvature mode shapes", *Journal of Sound and Vibration*, Vol. 145, No. 2, 1991, pp. 321–332.
2. Wolff, T. and Richardson, M., "Fault detection in structures from changes in their modal parameters", *Proceedings of the 7th International Modal Analysis Conference*, 1989, pp. 87–94.
3. Zimmerman, D. C. and Kaouk, M., "Structural damage detection using a subspace rotation algorithm", *Proceedings of the 33rd AIAA/ASME/AHS/ASC Structures, Structural Dynamics, and Materials Conference*, 1992, pp. 2341–2350.
4. Ricles, J. M. and Kosmatka J. B., "Damage detection in elastic structures using vibratory residual forces and weighted sensitivity", *AIAA Journal*, Vol. 30, No. 9, 1992, pp. 2310–2316.
5. Smith, S. W. and McGowan, P. E., "Locating damaged members in a truss structure using modal data: A demonstration experiment", *NASA Technical Memorandum 101595*, 1989
6. Peterson, L. D., Alvin, K. F., Doebling, S. W., and Park, K. C., "Damage detection using experimentally measures mass and stiffness matrices", *Proceedings of the 34th AIAA/ASME/AHS/ASC Structures, Structural Dynamics, and Materials Conference*, 1993, pp. 1518–1528.
7. Doebling, S. W., Hemez, F., Barlow, M. S., Peterson, L. D., and Farhat, C., "Selection of experimental modal data sets for damage detection via model update", *Proceedings of the 34th AIAA/ASME/AHS/ASC Structures, Structural Dynamics, and Materials Conference*, 1993, pp. 1506–1517.
8. Lippmann, R. P., "An introduction to computing with neural nets", *IEEE ASSP Magazine*, Vol. 4, 1987, pp. 4–22.
9. Pao, Y. H., "Adaptive pattern recognition and neural networks", Massachusetts, 1989, Addison-Wesley.
10. Wasserman, P. D., "Neural computing: Theory and practice", New York, 1989, Van Nostrand Reinhold.
11. Barga., R. S., Friesel., M. A., and Melton, R. B., "Classification of acoustic emission waveforms for nondestructive evaluation using neural networks", *SPIE: Applications of Artificial Neural Networks*, Vol. 1294, 1990, pp. 545–556.
12. Tsou, P. and Shen, M. H. H., "Structural damage detection and identification using neural network", *Proceedings of the 34th AIAA/ASME/AHS/ASC Structures, Structural Dynamics, and Materials Conference*, 1993, pp.3551–3560.
13. Ljung, L., "System identification: Theory for the user", New Jersey, 1987, Prentice-Hall.
14. Middleton, R. H. and Goodwin, G. C., "Digital control and estimation: A unified approach", New-Jersey, 1990, Prentice-Hall.
15. Shen, M. H. H. and Grady, J. E., "Free vibrations of delaminated beams", *AIAA Journal*, Vol. 30, No. 5, 1992, pp. 1361–1370.
16. Davies, W. D. T., "System identification for adaptive-self control", London, New York, 1970, Wiley-Interscience.

AIAA-94-1754-CP

DETECTION OF DELAMINATIONS IN COMPOSITE BEAMS USING PIEZOELECTRIC SENSORS

Dimitris A. Saravanos[1], Victor B. Birman[2] and Dale A. Hopkins
Structural Mechanics Branch
NASA Lewis Research Center, MS 49-8
Cleveland, Ohio

Abstract

This paper investigates the feasibility of a proposed technique for detecting delaminations using piezoelectric layers or patches embedded or bonded to a composite structure. Variations in the voltage of the piezoelectric layers indicates the presence and location of delamination while the structure is excited either externally or via piezoelectric actuators. The theoretical foundations of a method for predicting the dynamic response of delaminated composite beams with piezoelectric layers are described. The governing equations are presented for the case of external vibroacoustic excitation, as well as, for the case of locally induced vibrations by some of the embedded piezoelectric elements. An exact solution is developed within the limits of linear laminate theory. Applications illustrate the feasibility of delamination detection in cantilever beams. The results illustrate that the proposed technique may provide accurate detection of the presence, size and location of a delamination.

Introduction

Since the advent of composite materials, the development of delamination cracks have represented

[1]Senior Research Associate, Ohio Aerospace Institute

[2]Associate Professor, University of Missouri-Rolla, Engineering Education Center, 8001 Natural Bridge Road, St. Louis, MO 63121

"Copyright © 1994 by the American Institute of Aeronautics and Astronautics, Inc. No copyright is asserted in the United States Under Title 17, U.S. Code. The U.S Government has a royalty-free license to exercise all rights under the copyright claimed herein for Government purposes. All other rights are reserved by the copyright owner."

one of the principal concerns to their engineering applications. Delamination cracks may result to catastrophic failure, but also may easily remain undetected. Consequently, numerous theoretical and experimental studies have been conducted to elucidate the effects of delaminations on the static and dynamic response of composite structures. Additional research has addressed the nondestructive detection of delaminations.

The problem of delaminations as a fracture mechanics problem has received considerable attention and the recent papers of Chatterjee[1], Pindera[2], Nilsson and Storakers[3] and Thangjitham and Choi[4] are mentioned. The vibration and buckling of delaminated composite structures has been also investigated. Simitses[5] provided a comprehensive bibliography on the subject of buckling of composite plates and beams and a typical formulation of a delaminated beam problem based on a subdivision of the beam into four regions. Anastasiadis and Simitses[6] reported also analytical treatments for the case of weak bonding between the delaminated plies. Lee et al.[7] applied a layer-wise theory for the buckling of a composite beam and considered multiple delaminations through the thickness of the beam using a finite element formulation.

A large number of nondestructive, vibration based, techniques for detection of delaminations have been also developed. A review of vibration methods used for nondestructive evaluations (NDE) of laminated structures was published by Cawley and Adams[8]. Cawley[9] also reviewed a number of methods suitable for low-frequency NDE. Tracy and Pardoen[10] presented theoretical and experimental vibration studies on the effects of delamination on the natural frequencies of specialty composite beams. They subdivided the beam in four regions, similar to the approach employed in the buckling studies, and used traditional Euler beam theory which neglects extension-flexure coupling effects. Effects of the delamination on the higher natural frequencies were observed, depending on the location and size of delamination. Nagesh Babu and Hanagud[11] and Paolozzi and Peroni[12] used finite element models to study the same problem. Their results confirmed the conclusions of Tracy and Pardoen (1989) regarding a reduction in natural frequencies as a result of delamination. Additional finite element solutions have been published by Tenek et al[13] (1993). They concluded that the effect of delaminations increases for higher natural frequencies. Shen and Grady[14] developed equations of motion for a delaminated beam using the Hu-Washizu variational principle and Timoshenko beam theory, and also reported experimental results and correlations. Saravanos[15] developed an exact analytical procedure for the analysis of delamination effects on the natural frequencies, mode shapes and modal damping of composite beams. The last approach, that represents a version of a layer-wise theory, allows tracing a number of delaminations through the thickness of the laminate.

"High-frequency" nondestructive methods have been also proposed for delamination detection and are briefly reviewed here for the shake of completeness. Velocity measurements of surface waves have been used to nondestructively detect various defects, including delaminations[16]. Buynak et al[17] used an ultrasonic technique to produce images of delaminations in graphite-epoxy laminates. Acousto-graphic imaging techniques[18], and a photothermal method was discussed[19]. Acoustic emission measurements were proposed[20,21] to monitor propagation of delaminations. Thermo-acoustic emission was considered[22] and shearography techniques have been also proposed[23]. Moire interferometry[24] was also used to detect strain concentrations associated with the presence of delaminations in graphite-epoxy laminates.

A relatively new area of non-destructive delamination monitoring is associated with adaptive or smart structures. These structures can incorporate piezoelectric materials, shape memory alloys, electro-rheological fluids, electrostrictive, and magnetostrictive materials. Nagesh Babu[11] and Hanagud et al[25] recommended the use of piezoelectric sensors to detect delaminations in composite beams. An analytical model for automatic detection and control of delaminations growth using piezoelectric sensors and actuators was discussed[26]. They used a four-region model of a delaminated beam, using Timoshenko beam theory for each region, and solutions were obtained by the finite element method. Teboub and Hajela[27] also used a four region model of a delaminated beam with piezoelectric sensors for delamination detection under static loads using Kirchhoff type assumptions in each region. An experimental and numerical investigation on applications of piezoelectric sensors to detect delaminations in composite laminates was carried out[28]. Kim et al[29] proposed the use of strain gages bonded to the surface or embedded within the laminate.

The present paper, proposes a detection technique based on monitoring the spatial voltage distributions at distributed piezoelectric sensors. Mechanics for composite piezoelectric beams with multiple delaminations and piezoelectric patches bonded to the top and bottom surfaces are developed. An exact analytical solution is presented, within the limits of the assumptions of linear theory of elastic composite beams. The solution is further concentrated on the application of piezoelectric sensors to detect delaminations. Evaluations of the method on cantilever beams illustrate the merits and effectiveness of the proposed technique.

Theoretical Background

This section presents the theoretical formulation for modelling the dynamic response of delaminated composite beams with embedded piezoelectric sensors and actuators. A number of issues can be addressed in the framework of this formulation including, the detection of delaminations and the control of the dynamic or static response of delaminated beams using some of the piezoelectric elements as sensors and the remaining piezoelectric elements as actuators.

Consider a multilayered beam formed from composite plies and piezoelectric layers, as shown in Fig. 1a, with delamination cracks between the composite plies. The delaminations may be at any axial location and may overlap each other. The piezoelectric patches may be bonded to the surfaces or embedded between the composite plies of the beam (Fig. 1a). Although the theory can handle embedded layers, additional consideration may be required in this case due to local stresses in the vicinity of piezoelectric patches[30-32].

For analytical purposes, the beam may be subdivided into four typical segments, A, B, C, and D respectively, as shown in Fig. 1b. Type A corresponds to an "ordinary" beam without delaminations and patches. Segments B correspond to the portions of the beam with piezoelectric patches but without delaminations. The segments of the beam with delaminations but without patches are denoted as type C. Finally, the segments D include both delaminations and piezoelectric layers. Note that because of the delamination, the segments C and D include two additional subregions separated by a delamination. According to the approach used in this paper, equations of motion are formulated for each subregion within the segments C and D and for each segment (region) A and B independently. The coupling between the admissible solutions is introduced by the continuity conditions for displacements, forces and moments at the boundaries of each region.

The three-dimensional constitutive relations for a piezoceramic material with a macroscopic polar axis aligned with the z-axis are[33]:

$$\begin{Bmatrix} e_x \\ e_y \\ e_z \\ e_{yz} \\ e_{xz} \\ e_{xy} \end{Bmatrix} = \begin{bmatrix} s^E_{11} & s^E_{12} & s^E_{13} & 0 & 0 & s^E_{16} \\ & s^E_{11} & s^E_{13} & 0 & 0 & s^E_{26} \\ & & s^E_{33} & 0 & 0 & s^E_{36} \\ & & & s^E_{44} & s^E_{45} & 0 \\ & & & & s^E_{55} & 0 \\ sym & & & & & s^E_{66} \end{bmatrix} \begin{Bmatrix} \sigma_x \\ \sigma_y \\ \sigma_z \\ \sigma_{yz} \\ \sigma_{xz} \\ \sigma_{xy} \end{Bmatrix} + \begin{Bmatrix} d_{31}E_z \\ d_{31}E_z \\ d_{33}E_z \\ d_{15}E_y \\ d_{15}E_x \\ 0 \end{Bmatrix}$$

(1.1)

$$D_x = d_{15}\sigma_{xz} + \varepsilon^T_{11}E_x \quad (1.2)$$

$$D_y = d_{15}\sigma_{yz} + \varepsilon^T_{11}E_y \quad (1.3)$$

$$D_z = d_{31}(\sigma_x + \sigma_y) + d_{33}\sigma_z + \varepsilon^T_{33}E_z \quad (1.4)$$

where e and σ indicate engineering strains and stresses respectively. The matrices s^E_{mn}, d_{mn} and ε^T_{mm} are the elastic compliances, piezoelectric constants, and dielectric permittivity respectively. Superscripts E and T indicate that the corresponding properties should be measured in constant electric field and constant stress conditions respectively. E_i and D_i are the components of the electric field and electric displacements. In the case of a beam problem it may be assumed that $\sigma_y = \sigma_z = \sigma_{ij} = 0$, which simplifies the constitutive equations as follows,

$$\sigma_x = c^*_{11}(e_x - d_{31}E_z) \quad (2.1)$$

$$D_z = d_{31}\sigma_x + \varepsilon^T_{33}E_z \quad (2.2)$$

where $c^*_{11} = 1/S_{11}$ is the equivalent axial modulus of the material.

It is assumed that the problem is geometrically linear, moreover, the transverse shear deformations and rotary inertia are considered negligible. It is further assumed that $u^k = u^k(x)$ and $w^k = w^k(x)$ are the axial and through-the-thickness displacements at the mid-plane of the k-th region (or subregion), while the transverse displacement is assumed negligible ($v^k(x) = 0$). Then the strain-displacement relationship for thin sections is:

$$e_x = u^k_{,x} - z_k w^k_{,xx} \quad (3)$$

where z_k is the distance from the mid-plane axis of each region.

Integration of the previous equation through the thickness of the beam and combination with eqs. (2) yields the stress and moment resultants N_x^k and M_x^k respectively, in the k-th region of the beam,

$$\begin{Bmatrix} N_x^k \\ M_x^k \end{Bmatrix} = \begin{bmatrix} A_k & B_k \\ B_k & D_k \end{bmatrix} \begin{Bmatrix} u^k_{,x} \\ -w^k_{,xx} \end{Bmatrix} - \begin{Bmatrix} F_k \\ G_k \end{Bmatrix} \quad (4)$$

where A, B, and D are the extensional, coupling and bending stiffness of the beam,

$$\{A_k, B_k, D_k\} = \sum_{m=1}^{n_{lk}} \int_{z_m}^{z_{m+1}} c^*_{11}\{1, z, z^2\} dz \quad (5)$$

and F, G are the piezoelectric force and moment resultants respectively,

$$\{F_k, G_k\} = \sum_{m=1}^{n_{lk}} \int_{z_m}^{z_{m+1}} c_{11}^* d_{31} \{1, z\} E_z dz \qquad (6)$$

n_{lk} indicates the number of plies across the k-th region (or subregion) of the beam, In segment A (see Fig. 1b), n_{lk} is the number of composite plies through the depth h of the beam; in segment(s) B n_{lk} must be modified to incorporate the piezoelectric patches; in segments C and D, the limits of integration must be chosen to include the corresponding delaminated sublaminates.

The equations of motion in the k-th region of the beam, assuming that the axial inertia effects are negligible, are:

$$N_{x,x}^k = 0$$
$$M_{x,xx}^k = \rho_k \ddot{w}^k - q(x,t) \qquad (7)$$

where q(x,t) is an external load, and

$$\rho_k = \int_{z_k}^{z_{k+1}} \rho(z) dz \qquad (8)$$

where ρ is the linear density of the beam.

Substitution of eqs. (4) into equations (7) yields the equation of motion for the kth region:

$$A_k u_{,xx}^k - B_k w_{,xxx}^k - F_k(E_{z,x}, t) = 0 \qquad (9.1)$$

$$B_k u_{,xxx}^k - D_k w_{,xxxx}^k - G_k(E_{z,xx}, t) - \rho_k \ddot{w}^k = q(x,t) \qquad (9.2)$$

Eqs. (9) can be used to analyze the motion of the beam when it is excited by external loads (sensory beam), as well as, when it is excited by some of the piezoelectric layers (sensory/active beam). In the sensory case, an external load q(x,t) is present, while the imposed electric field E_z in the piezoelectric layers is zero. In a sensory/active beam, electric fields E_z may be imposed in some of the piezoelectric layers to induce additional vibrations, while external mechanical loads q(x,t) may be present.

The solution must satisfy the boundary conditions at the supports. For example, the free-end conditions are $N_x = M_x = V_x = 0$, where V_x is the shear force, and the clamped conditions are $u = w = w_{,x} = 0$. Additional kinematic continuity conditions between the k-th and (k+1)-th regions at $x = x_r$ are imposed

$$u^k(x_r) = u^{k+1}(x_r), \quad w^k(x_r) = w^{k+1}(x_r),$$
$$w_{,x}^k(x_r) = w_{,x}^{k+1}(x_r) \qquad (11)$$

Continuity in the balance of axial forces, bending moments and shear forces is also required

$$\sum_k N_x^k(x_r^-) = \sum_j N_x^j(x_r^+) \qquad (12.1)$$

$$\sum_k [M_x^k(x_r^-) + N_x^k(x_r^-)\bar{z}_k] = \sum_j [M_x^j(x_r^+) + N_x^j(x_r^+)\bar{z}_j] \qquad (12.2)$$

$$\sum_k V_x^k(x_r^-) = \sum_j V_x^j(x_r^+) \qquad (12.3)$$

where the indices k and j identify the number of subregions to the left (x_r^-) and to the right (x_r^+), respectively, of a cross-section at $x = x_r$. The coordinates $\bar{z}_k$ reflect the distance of the elastic axes of the corresponding subregions to the mid-plane of the beam. The shear forces in eq. (12.3) are given by:

$$V_x^r = M_{x,x}^r + N_x^r w_{,x}^r \qquad (13)$$

where (r=k or j). In the absence of external axial forces, the last term in the right side of eq. (13) may be neglected and eq. (12.3) becomes:

$$\sum_k M_{x,x}^k(x_r) = \sum_j M_{x,x}^j(x_r) \qquad (14)$$

The governing equations (9) together with the boundary conditions and continuity conditions (12) formulated in this section represent the necessary analytical relations for the solution of problems of active and passive control of composite beams with delaminations. In the next section applications of the theory for the detection of delaminations are presented.

Detection of Delaminations

This section further extends the method for the detection of delaminations under dynamic excitation. In the proposed scheme, distributed piezoelectric sensors are used to detect the presence, size and location of delaminations, when the beam is excited either passively or actively by some of the piezoelectric layers. In both cases, the voltage generated across the piezoelectric sensors in an intact (pristine) beam, are different than those in the delaminated counterpart.

Consider the more general case where the beam is excited by either an external harmonic load $q(x)e^{i\omega t}$ or an applied harmonic electric field $\bar{E}_z(x)e^{i\omega t}$ at some of the piezoelectric layers (actuators). It is customary to neglect the effect of a self-generated voltage on the motion[34], however, techniques for incorporating these effects have been reported by Heyliger and Saravanos[32] and may be included in future work. The displacements are assumed to have the form

$$\{u^k, w^k\} = \{U^k, W^k\}e^{sx}e^{i\omega t} \quad (15)$$

and the equations of motion (9) are reduced to:

$$A_k U^k_{,x} - B_k W^k_{,xx} = \bar{N}_k \quad (16.1)$$

$$B_k U^k_{,xxx} - D_k W^k_{,xxxx} = -\rho_k \omega^2 W^k - q_k(x) \quad (16.2)$$

where $\bar{N}_k$ is a constant and q_k is the amplitude of the external load acting on the region.

The solution of equations (16) is:

$$W^k = C_1^k \sin s_k x + C_2^k \cos s_k x + C_3^k e^{s\underline{k}x} + C_4^k e^{-s\underline{k}x} + Q_k + R_k \quad (17.1)$$

$$U^k = C_5^k + \frac{\bar{N}_k}{A_k}x + \frac{B_k s_k}{A_k}(C_1^k \cos s_k x - C_2^k \sin s_k x + C_3^k e^{s\underline{k}x} - C_4^k e^{-s\underline{k}x}) + P_k \quad (17.2)$$

where C_1^k through C_5^k are coefficients of integration and

$$s_k = \frac{\rho_k \omega^2}{D_k - B_k^2/A_k} \quad (18.1)$$

$$P_k = \frac{1}{A_k}\int F_k(\bar{E}_z)dz, \quad R_k = G_k(\bar{E}_{z,xx})/\rho_k \omega^2 \quad (18.2)$$

$$Q_k = -q_k/\rho_k \omega^2 \quad (18.3)$$

Note that integration of the system of equations (16) results in 5 constants of integration for each beam region including $\bar{N}_k$. The number of boundary and continuity conditions is proved to be always equal to the number of coefficients of integration. Hence, a linear system of algebraic equations results by substituting eq. (17) into the boundary and continuity conditions, of the form

$$[a_{ij}]\{C_j\} = \{f_i\} \quad (19)$$

where $[a_{ij}]$ is a square matrix, $[C_j]$ is the vector of unknown coefficients in eqs. (17), and $[f_i]$ is a vector of loading terms. The elements of the matrix [a] are independent to the method of excitation (passive or active). Once the constants of integration are determined, the displacements, strains and stresses can be evaluated.

The voltage difference $\delta\phi$ generated across the sensory piezoelectric layers or patches can be calculated as

$$\delta\phi(t) = -\int_{h_p} E_z(t)dz \quad (20)$$

where h_p is a thickness of the corresponding piezoelectric layer. Combining eqs. (2) the electric field can be obtained from

$$E_z = \frac{1}{c_{11}^* d_{31}^2 - \varepsilon_{33}^T}\left(c_{11}^* d_{31} e_x - D_z\right) \quad (21)$$

In a sensory piezoelectric layer the electric circuit remains practically open, hence, the total charge D_z over the length l_p of the electrodes remains constant and can be assumed equal to zero,

$$\int_{l_p} D_z dx = 0 \quad (22)$$

Combining eqs. (21) and (22), and considering that the terminals impose a constant electric potential condition along its length l_p, then the voltage output across the electrodes of a sensor is:

$$\delta\phi(t) = -\frac{c_{11}d_{31}}{\left(c_{11}^* d_{31}^2 - \varepsilon_{33}^T\right)l_p}\int_{l_p}\int_{h_p} e_x(t)\,dz\,dx \quad (23)$$

Passive Excitation. In the case of external mechanical excitation, the terms in eq. (18.2) vanish, hence, the right hand side term $\{f_i\}$ in eq. (19) contains only terms related to the mechanical load.

Piezoelectric Excitation. In addition to external

mechanical excitation, voltage may be applied across some of the piezoelectric layers/patches to actively excite the beam. This inclusion of piezoelectric excitation requires only modification of the loading vector $\{f_i\}$ in eq. (19). If the motion of the beam is excited by a uniform sinusoidal voltage applied to a continuous actuator-layer, the solution of the equations of motion is given by equations (17) (where $F_k=0$ may or may not be zero) and U^k is complemented by the following additional term:

$$P_k = \frac{\bar{E}_z x}{A_k} \int_{z_k}^{z_{k+1}} c_{11}^k d_{31}^k dz \qquad (24)$$

Applications and Discussion

Materials and Assumptions

Evaluations of the previously described model were performed for sensory composite beams. The composite plies were T300/934 graphite/epoxy of 0.60 fiber volume ratio. The piezoceramic PZT5A was selected as the material for the piezoelectric layer(s) with properties provided by the manufacturer[35]. This material is recommended for sensor applications by the manufacturer.

Eight (8) ply symmetrically laminated beams with a single delamination and 2 piezoceramic layers, each bonded to the top and bottom surface of the composite beam, were considered in the numerical examples. Unless otherwise stated, a cross-ply layup $[p/0/90/0/90]_s$ was mostly considered, where "p" denotes a piezoceramic layer. Laminations of the type $[p/0/90/45/-45]_s$ and $[p/0/45/-45/0]_s$ were also investigated. The length of the beam was equal to 5 in, its width was 1 in, and the thickness of each layer was 0.01 in. Results for a cantilever beam clamped at the left.

The beams were assumed to be excited by a uniformly distributed load of 1 lb/in amplitude applied on the upper surface of the beam at a frequency of 80 Hz which is lower than the fundamental frequency. This low frequency was selected to avoid excitation of the higher modes that could complicate the interpretation of the results. The terminals were assumed closely spaced having infinitesimal length, such that they yield continuous voltage readings along the free surfaces.

Effect on Dynamic Deflections

The first phase of the analysis evaluates the effect of delamination on the transverse dynamic deflections of the sensory beam. In the case of a cantilever beam these effects are quite significant. As follows from Figure 2, increases in the delamination length resulted in significant increases in the amplitude of vibration. This is associated with the reduction of stiffness in the delaminated beam. The predicted influence of local vibration modes in the delaminated regions on the global mode shape appears negligible both in the present and in the following examples.

The location of a delamination through the depth of the cantilever beam had a minimal effect on the motion (see Fig. 3). The two cases compared in Fig. 4 include: a delamination located at the mid-plane of the beam (mid-plane case); and a delamination located near the upper surface (off-plane case) such that $[p/0/90/0/90/90/0//90/0/p]$, where // indicates the presence of the crack. Considering various axial locations of delaminations in Fig. 4, the axial location of a delamination has a relatively larger effect. As expected, the displacement amplitudes increase as the delamination approaches the clamped end of the beam, indicating that such delaminations result in more pronounced reductions of the stiffness.

Sensory Response

The following paragraphs present the predicted voltage output across the piezoelectric sensors. The outputs from the cantilever beam with centrally located delaminations of various lengths are shown in Figs. 5-8. In these Figures, as well as in the following Figures, the two curves represent the amplitudes of the resultant AC voltage across the lower and upper piezoelectric layer (referred to as "lower" and "upper" layers) respectively. Apparently, the presence of the delamination can be easily detected from discontinuities in voltage output along the axis (see Figs. 5-8). Moreover, the edges of the delamination are sharply outlined. In practice such an accurate detection of a delamination will be limited by the finite number and size of sensors/electrodes, however, the spacing of the sensors will be a design problem that should be solved in conjunction the allowable minimum length of a detectable delamination.

The predicted effect of an axially shifted delamination on the voltage outputs of a cantilever beam is shown in Figs. 9 and 10. In both cases, voltage readings clearly indicate the presence, axial location and size of the delamination. The resultant voltage change is lower when the delamination is near the free-end, as opposed to a delamination near the root. This happens because the axial strain in the piezoelectric layers also decreases towards the free end.

The voltage output of the piezoelectric sensors is also sensitive to the through-the-thickness location of delamination (see Figs. 11 and 12). Fig. 11 corresponds to a delamination located near the top surface of the beam, while Fig. 12 voltages to a delamination near the bottom surface. A difference between the Figures (compare also to Fig. 10 where the delamination lies on the mid-plane) indicates that the proposed technique may reveal the lateral location of delaminations, in addition to detecting the presence and axial location.

The laminate layup did not affect significantly the conclusions obtained in the previous examples. This was verified for a central delamination in cantilever beams with laminations $[p/0/90/45/-45]_s$ and $[p/0/45/-45/0]_s$ respectively, however, the results are not shown for space limitations.

Summary and Conclusions

A technique for detecting delaminations in vibrating composite beams was proposed, and the associated mechanics were developed. The proposed scheme is based on monitoring the changes in the voltage output signature of distributed piezoelectric sensors bonded to or embedded into the vibrating structure. The vibrations may be induced externally (mechanically or acoustically) or by using piezoelectric actuators.

The method was applied to detect delaminations in composite beams. Based on the obtained predicted results, the following conclusions are summarized. The predicted response of the piezoelectric sensors was very sensitive to the presence of a delamination. It was observed that in the case of continuous piezoelectric layers with infinitesimally spaced terminals bonded to the upper and lower surfaces, abrupt axial discontinuities in the AC voltage output across the sensors indicate clearly the crack tips. Hence, it seems that the proposed concept can detect the presence of a delamination and predict its size and location both axially and through-the-thickness of a composite structure. The effectiveness of the detection was not affected by the type of boundary conditions. This was illustrated by analyzing simply supported and cantilever beams.

In closing, the reported research has demonstrated the feasibility of the proposed technique. The method described herein can be successfully extended to detect delaminations in composite plates and shells. Future work will address the effectiveness of the method when limited numbers of sensors are implemented, and experimental verification of the predicted results.

Acknowledgement

This research was conducted during the ASEE summer faculty fellowship of Dr. V. Birman to the NASA Lewis Research Center.

References

1. Chatterjee, S. N., 1987, "Three and Two-Dimensional Stress Fields Near Delaminations in Laminated Composite Plates", International Journal of Solids and Structures, Vol. 23, pp. 1535-1549.

2. Pindera, M.-J., 1991, "Local/Global Stiffness Matrix Formulation for Composite Materials and Structures", Composites Engineering, Vol. 1, pp. 69-83.

3. Nilsson, K.-F. and Storakers, B., 1992, "On Interface Crack Growth in Composite Plates", Journal of Applied Mechanics, Vol. 59, pp. 530-538.

4. Thangjitham, S. and Choi, H. J., 1993, "Interlaminar Crack Problems of a Laminated Anisotropic Medium", International Journal of Solids and Structures, Vol. 30, pp. 963-980.

5. Simitses, G. J., 1993, "Delamination Buckling of Flat Laminates", Buckling and Postbuckling of Composite Plates, Eds., Turvey, G. J. and Marshall, I., Elsevier, In press.

6. Anastasiadis, J. S. and Simitses, G. J., 1991, "Spring

Simulated Delamination of Axially-Loaded Flat Laminates", Composite Structures, Vol. 17, pp. 67-85.

7. Lee, J., Gurdal, Z. and Griffin, O. H., Jr., 1992, "A Layer-Wise Approach for the Bifurcation Problem in Laminated Composites with Delaminations", AIAA Paper 92-2224-CP.

8. Cawley, P., 1990, "Low Frequency NDT Techniques for the Detection of Disbonds and Delaminations", British Journal of Non-Destructive Testing, Vol. 32, pp. 454-461.

9. Cawley, P. and Adams, R. D., 1987, "Vibration Techniques (of NDT)", Non-Destructive Testing of Fibre-Reinforced Plastics Composites, Vol. 1, Ed., Summerscales, J., Elsevier, London, pp. 151-200.

10. Tracy, J. J. and Pardoen, G. C., 1989, "Effect of Delamination on the Natural Frequencies of Composite Laminates", Journal of Composite Materials, Vol. 23, pp. 1200-1215.

11. Nagesh Babu, G. L. and Hanagud, S., 1990, "Delaminations in Smart Composite Structures: A Parametric Study on Vibrations", AIAA Paper 90-1173-CP, 31st AIAA/ASME/ASCE/AHS/ASC SDM Conference, Part 4, pp. 2417-2426.

12. Paolozzi, A. and Peroni, I., 1990, "Detection of Debonding Damage in a Composite Plate Through Natural Frequency Variations", Journal of Reinforced Plastics and Composites, Vol. 9, pp. 369-389.

13. Tenek, L. H., Henneke, E. G. II and Gunzburger, M. D., 1993, "Vibration of Delaminated Composite Plates and Some Applications to Non-Destructive Testing", Composite Structures, Vol. 23, 253-262.

14. Shen, M. H. H. and Grady, J. E., 1992, "Free Vibrations of Delaminated Beams", NASA TM 105582.

15. Saravanos, D. A., 1993, "Mechanics for the Effects of Delaminations on the Dynamic Characteristics of Composite Laminates", Proceedings of the 1993 ASME Winter Annual Meeting, New Orleans, LA. In press.

16. Rose, J. L., Pilarski, A. and Huang, Y., 1990, "Surface Wave Utility in Composite Material Characterization", Research in Nondestructive Evaluation, Vol. 1, pp. 247-265.

17. Buynak, C. F., Moran, T. J. and Martin, R. W., 1989, "Delamination and Crack Imaging in Graphite-Epoxy Composites", Materials Evaluation, Vol. 47, pp. 438-441.

18. Sandhu, J. S., 1988, "Acoustography: A New Imaging Technique and its Applications to Nondestructive Evaluation", Materials Evaluation, Vol. 46, pp. 608-613.

19. Balageas, D. L., Deom, A. A. and Boscher, D. M., 1987, "Characterization and Nondestructive Testing of Carbon-Epoxy Composites by a Pulsed Photothermal Method", Materials Evaluation, Vol. 45, pp. 461-465.

20. Garg, A. and Ishai, O., 1985, "Characterization of Damage Initiation and Propagation in Graphite/Epoxy Laminates by Acoustic Emission", Engineering Fracture Mechanics, Vol. 22, pp. 595-608.

21. Kexing, L., Davis, A., Ohn, M.M., Byung, P., Measures, R. M., 1992, "Embedded Optical Fiber Sensors for Damage Detection and Cure Monitoring", Proceedings of the ADPA/AIAA/ASME/SPIE Conference on Active Materials and Adaptive Structures, IOP Publishing, Bristol, pp. 395-398.

22. Sato, N., Kurauchi, T. and Kamigaito, O., 1986, "Thermo-Acoustic Emission from Damaged Composite", Proceedings, 31st International SAMPE Symposium: Materials Sciences for the Future, Eds., Bauer, J. L. and Dunaetz, R., Soc. Adv. Mater. & Process Engineering, Covina, CA, pp. 342-351.

23. Hung, Y. Y., 1989, "Shearography: A Novel and Practical Approach for Nondestructive Testing," J. of Nondestructive Evaluation, Vol. 8, No. 1 pp. 55-67.

24. Wood, J. D., 1985, "Detection of Delamination Onset in a Composite Laminate using Moire Interferometry", Journal of Composites Technology and Research, Vol. 7, pp. 121-128.

25. Hanagud, S., Nagesh Babu, G. L. and Won, C. C., 1990, "Delaminations in Smart Composite Structures", Proceedings, The 1990 SEM Spring Conference on Experimental Mechanics, Bethel, CT, Society for

Experimental Mechanics, Inc., pp. 776-781.

26. Hanagud, S., Nagesh Babu, G. L., Roglin, R. L. and Savanur, S. G., 1992, "Active Control of Delaminations in Composite Structures", Proceedings, 33rd AIAA/ASME/ASCE/AHS/ASC SDM Conference, Dallas, TX, pp. 1819-1829 (AIAA Paper 92-2387-CP).

27. Teboub, Y. and Hajela, P., 1992, "A Neural Network Based Damage Analysis of Smart Composite Beams", AIAA Paper 92-4685, Fourth AIAA/USAF/NASA/OAI Symposium on Multidisciplinary Analysis and Optimization, September 21-23, 1992, Cleveland, OH.

28. Keilers, C. H., Jr., and Chang, F-K., 1993, "Damage Detection and Diagnosis of Composites Using Built-In Piezoceramics", Proceedings of the 1993, North American Conference on Smart Structures and Materials, Albuquerque, NM, In press.

29. Kim, K-S, Segall, A. and Springer, G. S., 1993, "The Use of Strain Measurements for Detecting Delaminations in Composite Laminates", Composite Structures, Vol. 23, pp. 75-84.

30. Joshi, S. P. and Chan, W. S., 1992, "Damage-Survivable and Damage-Tolerant Laminated Composites with Optimally Placed Piezoelectric Layers", Final Report No.1, U.S. Army Research Office, Grant No. DAAL03-89-G-0090, University of Texas at Arlington, Arlington, Texas.

31. Warkentin, D. J., Crawley, E. F. and Senturia, S. D., 1992, "The Feasibility of Embedded Electronics for Intelligent Structures", Journal for Intelligent Material Systems and Structures, Vol. 3, pp. 462-482.

32. Heyliger P.R. and Saravanos D.A., 1993, "On Discrete-Layer Mechanics for Health Monitoring Application In Smart Composite Structures", ASME Winter Annual Meeting, New Orleans, Louisiana, Nov. 28-Dec. 3.

33. Smith, W. A., 1992, "The Key Design Principle for Piezoelectric Ceramic/Polymer Composites", Recent Advances in Adaptive and Sensory Materials and their Applications, Eds., Rogers, C. A. and Rogers R. C., Technomic, Lancaster, pp. 825-838.

34. Tzou, H. S., 1993, "Piezoelectric Shells", Kluwer Academic Publishers, Dordrecht.

35. "Guide to Modern Piezoelectric Ceramics", 1993, Morgan Matroc, Inc., Electro Ceramics Division, Bedford, Ohio 44145.

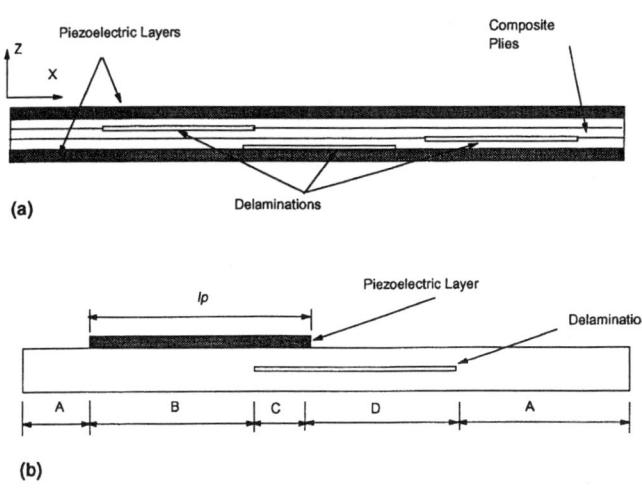

Fig. 1. Delaminated Sensory beam. (a) Typical configuration; (b) Typical beam segments.

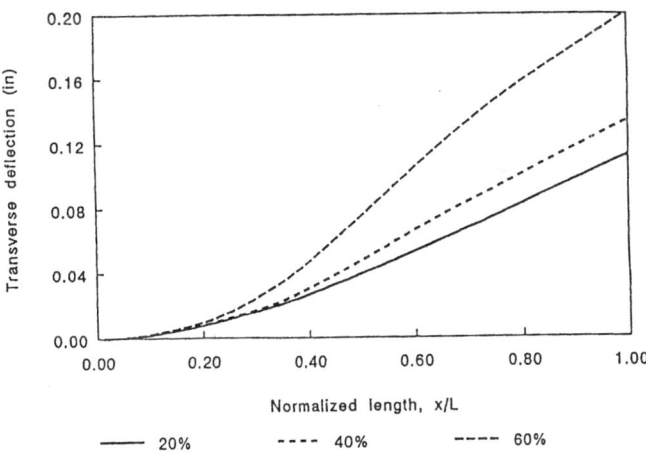

Fig. 2. Effect of delamination length on the dynamic displacement of a cantilever beam. Centrally located delamination.

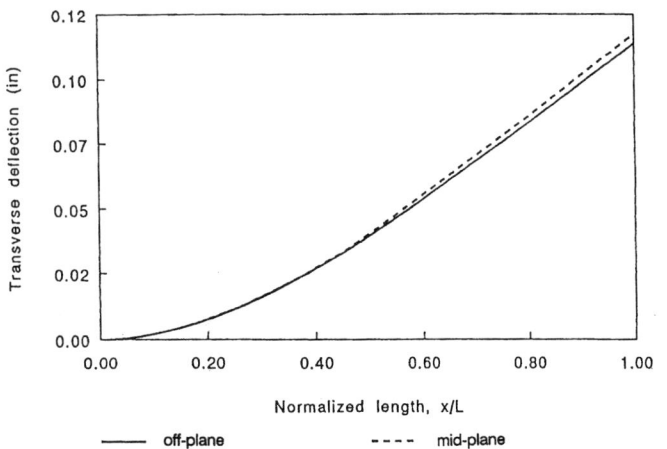

Fig. 3. Effect of the through-the-thickness location of a delamination on the dynamic displacement. Delamination length = 20%; lamination: $[p/(0/90)_2/90/0//90/0/p]$.

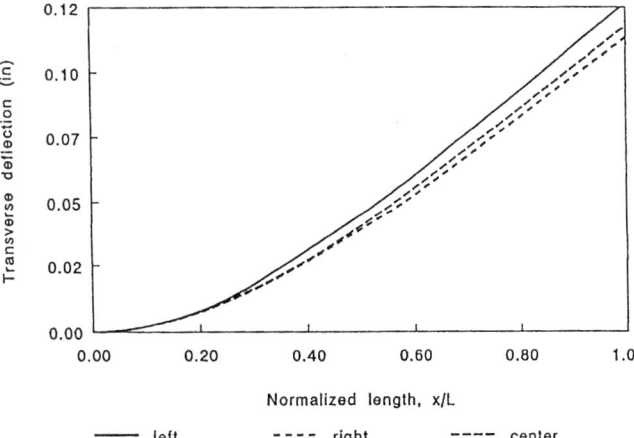

Fig. 4. Effect of the axial location of a delamination on the dynamic displacement. Delamination length = 20%. Coordinate of delamination center is x_c/L. Left: x_c/L = 0.3; Right: x_c/L = 0.7; Center: x_c/L = 0.5.

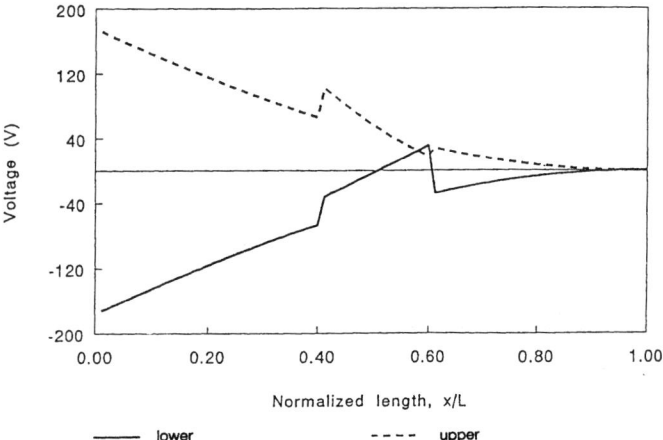

Fig. 5. Detection of a centrally located delamination in a cantilever beam. Delamination length = 20%.

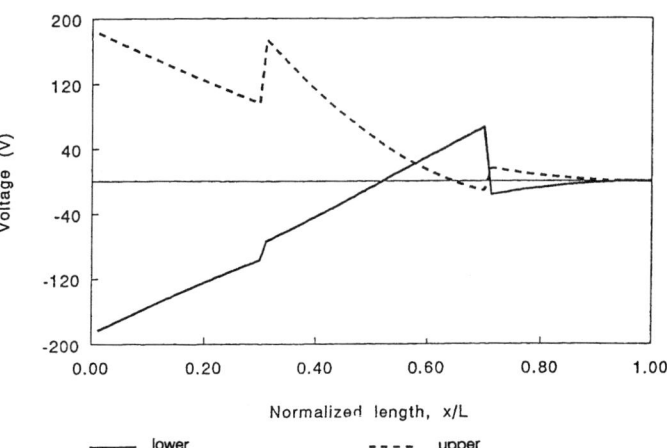

Fig. 6. Detection of a centrally located delamination in a cantilever beam. Delamination length = 40%.

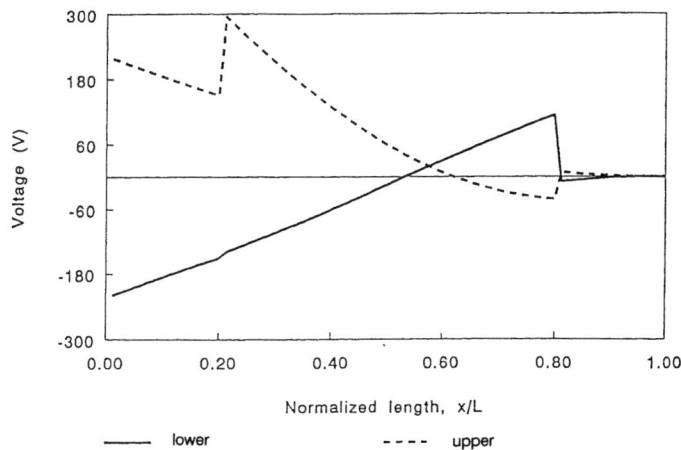

Fig. 7. Detection of a centrally located delamination in a cantilever beam. Delamination length = 60%.

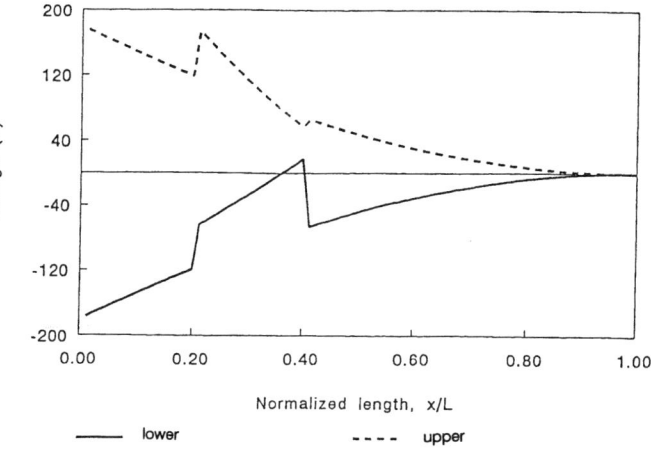

Fig. 8. Detection of an axially shifted delamination in a cantilever beam. Delamination length = 20%. Delamination center is located at 30% span from the clamped end.

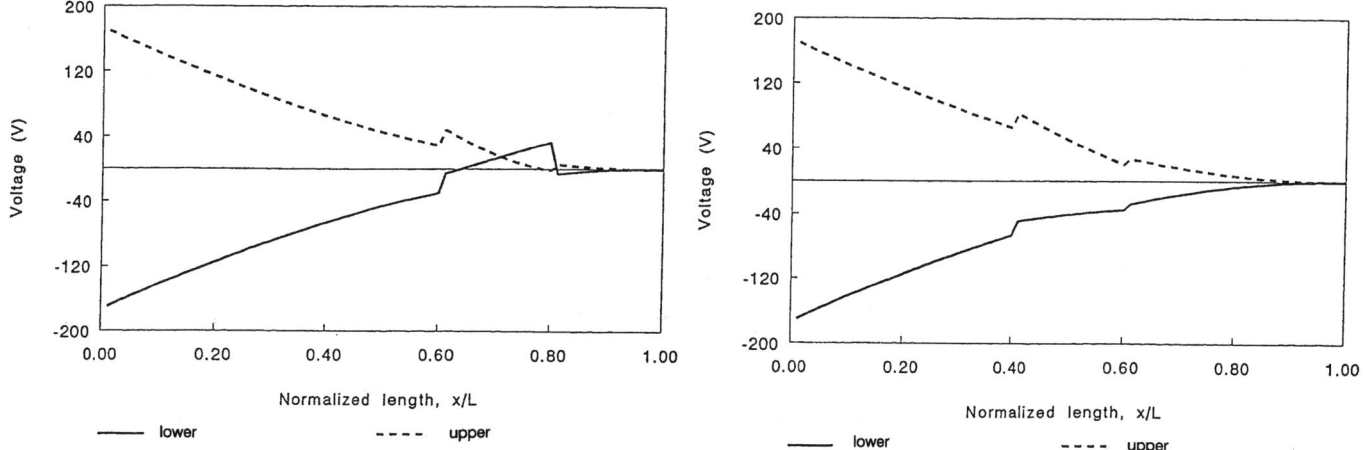

Fig. 9. Detection of an asymmetric delamination in a cantilever beam. Delamination length = 20%. Delamination center is located at 70% span from the clamped end.

Fig. 11. Detection of a delamination located near the bottom surface of a cantilever beam. Delamination length is 20%; off-plane delamination: [p/0/90//0/90/(90/0)$_2$/p].

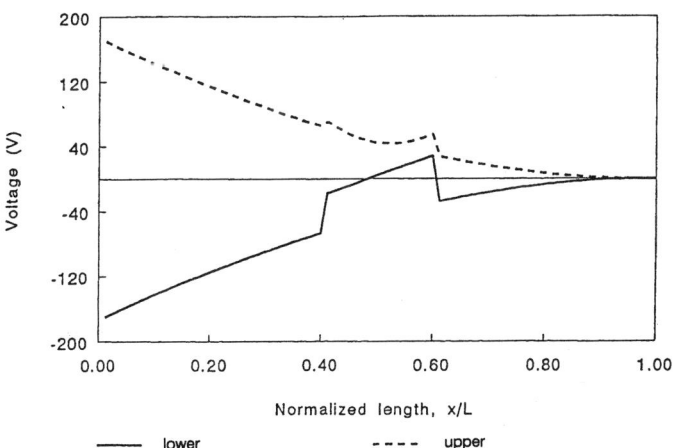

Fig. 10. Detection of a delamination located near the top surface of a cantilever beam. Delamination length is 20%; off-plane delamination: [p/(0/90)$_2$/90/0//90/0/p].

AIAA-94-1755-CP

LOCATION AND ESTIMATION OF DAMAGE IN A BEAM USING IDENTIFICATION ALGORITHMS

Douglas K. Lindner*
Virginia Tech
Blacksburg, Virginia

George Kirby**
Naval Research Laboratory
Washington, DC

Abstract

We describe an algorithm for locating and estimating damage in a beam used in connection with an identification algorithm. The damage location and estimation is based on a finite element model of the beam. The concept is to start with a model of the beam which is known to accurately model the healthy beam. This model also models the damage that is expected in the beam. Using dynamic response data, an a estimated damage coefficient for each beam element is computed for each identified modeshape and modal frequency. Damage in each element is determined by comparing estimated damage coefficients for each element across all modes. The performance of this algorithm is illustrated on a cantilever beam through simulation and experiment.

Introduction

In this paper we consider the location and estimation of damage in a beam using parameter identification techniques.[1] The concept is to start with a model of the healthy beam which is known to accurately model the beam before damage has occurred. After damage has occurred a damage model of the beam is constructed from dynamic response data using an identification algorithm. By comparing the parameters in the truth model with the parameters of the damage model, the damage is detected, located, and the extent of the damage accessed.

This approach to damage location and estimation has been proposed based on model refinement techniques.[2-11] The identified modeshapes are used to update the healthy model of the beam. Damage is detected by comparing the elements of the healthy and damaged stiffness matrices.

A second approach to damage location and estimation is based on computing a perturbation of the healthy stiffness matrices from the modeshapes and frequencies of the damaged model. From this perturbation the algorithm produces an estimate the damage and its location. One approach to computing this perturbation is to use a model refinement algorithm.[12-14] In a variation on this approach, the connectivity information is included into the perturbation term.[15] A second approach to computing the perturbation is based on sensitivity methods.[16-18] The perturbation of the healthy model using a Taylor series expansion is computed using an unconstrained optimization procedure. A change in the parameter of a particular element will indicate damage to that element.

In this paper we extend a damage location and estimation algorithm for trusses to lightly damped beams.[19] Both changes in the mass and stiffness are considered. The algorithm follows the general approach of computing a perturbation of the undamaged beam model. The perturbation is derived directly from the finite element model. In this way we are able to concentrate the model of the damage into a few parameters. The damage location and estimation is accomplished by computing a estimate of the damage for each beam finite element using each mode. For each beam element, damage to that element is located and estimated by comparing the damage estimates from each mode. That is, multiple estimates of the damage are used to check the consistency of the damage estimates with respect to the dynamic response data.

The calculations for this algorithm are a few matrix multiplications on the order of the size of the stiffness and mass matrices. No constrained or unconstrained optimization is required. The performance of this algorithm is illustrated on a cantilever beam through simulation and experiment.

Model Development

Beam Model

In this paper we will consider beams in bending such as a cantilevered beam. We assume a finite element model of the beam constructed using Euler-Bernoulli beam elements. First, the beam is divided into n_e elements of length s_i. The stiffness matrix for the ith element is

---

* Associate Professor, Bradley Department of Electrical Engineering
** Mechanics of Materials Branch

Copyright © 1993 American Institute of Aeronautics and Astronautics, Inc. All rights reserved.

$$\frac{E_i I_i}{s_i^3} \begin{bmatrix} 12 & -6s_i & -12 & -6s_i \\ -6s_i & 4s_i^2 & 6s_i & 2s_i^2 \\ -12 & 6s_i & 12 & 6s_i \\ -6s_i & 2s_i^2 & 6s_i & 4s_i^2 \end{bmatrix} = \left[\frac{E_i I_i}{s_i^3}\right] \tilde{K}_i \quad (1)$$

where E_i is Young's modulus, I_i is the second moment of inertia, and s_i is the length of the element. The mass matrix is

$$\frac{\rho_i A_i s_i}{420} \begin{bmatrix} 156 & 22s_i & 54 & -13s_i \\ 22s_i & 4s_i^2 & 13s_i & -3s_i^2 \\ 54 & 13s_i & 156 & -22s_i \\ -3s_i & -3s_i^2 & -22s_i & 4s_i^2 \end{bmatrix} \quad (2)$$
$$= \left[\frac{\rho_i A_i s_i}{420}\right] \tilde{M}$$

where A_i is the cross-sectional area and ρ_i is the mass density. To assemble the stiffness matrix in global coordinates assume that the ith element is attached at the ith and (i+1)th node respectively. According to the numbering of the nodes, the displacements at the ends of the element contribute to the 2i-1, 2i, 2i+1, and 2i+2 dofs. Define the matrix

$$B_i = [e_{2i-1} \quad e_{2i} \quad e_{2i+1} \quad e_{2i+2}] \quad (3)$$

where e_j is the jth unit vector; an vector of zeros except for a 1 in the jth position. The expanded stiffness matrix in global coordinates is

$$\left(\frac{E_i I_i}{s_i}\right) B_i \tilde{K}_i B_i^T = \left(\frac{E_i I_i}{s_i}\right) Q_i = K_{hi}. \quad (4)$$

Summing the individual stiffness matrices in global coordinates we have

$$K_h = \sum_{i=1}^{n_e} \left(\frac{E_i I_i}{s_i}\right) Q_i = \sum_{i=1}^{n_e} K_{hi}. \quad (5)$$

Similarly, the mass matrix for each element is given by

$$\left(\frac{\rho_i A_i s_i}{420}\right) B_i \tilde{M}_i B_i^T = \left(\frac{\rho_i A_i s_i}{420}\right) P_i = M_{hi}. \quad (6)$$

Then mass matrix of the finite element model is

$$M_h = \sum_{i=1}^{n_e} \left(\frac{\rho_i A_i s_i}{420}\right) P_i = \sum_{i=1}^{n_e} M_{hi}. \quad (7)$$

The beam model is given by

$$M_h \ddot{z}(t) + K_h z(t) = 0 \quad (8)$$

where M_h is the n_h x n_h mass matrix, K_h is the n_h x n_h stiffness matrix of the beam.

Damaged Beam Model

When the dynamic model of the beam is accurately represented by (8), we say the beam is healthy. Hence, the subscript "h" in (8). The subscript "d" will denote quantities from the damage model. We assume that the damage results in a change in the flexural rigidity "$E_i I_i$" of the beam but leaves the mass of the beam unaltered. (We extend this problem formulation to changes in mass in Section 5.) To model this damage, we write the scalar factor in (4) as

$$(1-d_i)\left[\frac{E_i I_i}{s_i}\right] \quad \text{for} \quad 0 \leq d_i \leq 1. \quad (9)$$

We call d_i the <u>stiffness damage coefficient of the ith element of the beam</u>. Suppose that the ith element of the beam has sustained damage. Using (9) in (4), the contribution of this damaged element to the global stiffness matrix is

$$\left[\frac{(1-d_i)E_i I_i}{s_i}\right] Q_i = K_i - d_i K_i. \quad (10)$$

Substituting (10) into (5) the stiffness matrix of the damaged beam is

$$K_d = K_h - \sum_{i=1}^{n_e} d_i K_i. \quad (11)$$

We assume that the dynamic behavior of the beam after it sustains damage is modeled by the <u>damaged</u> beam model

$$M_d \ddot{x}(t) + K_d x(t) = 0. \quad (12)$$

Define the <u>set of element stiffness damage coefficients</u>

$$\mathcal{D} = \{d_m, \quad m = 1, \cdots, n_e\}. \quad (13)$$

Locating and estimating damage to a structure is equivalent to determining the set of damage coefficients. If the damage coefficient for the ith beam element is zero, $d_i = 0$, then that beam element is healthy. If the damage coefficient is positive, $d_i > 0$, then that element of the beam has suffered a d_i x 100% reduction in stiffness.

The Damage Location And Estimation Algorithm

Modal Damage Coefficients

In this section we develop an algorithm for damage location and estimation based of the models of the healthy and damaged beam in Section 2. From the damaged structure we extract n_r pairs of modeshapes and mode frequencies using an identification algorithm. The modeshape/mode frequency pair $(\phi_{dj}, \omega_{dj}^2)$ satisfies

$$M_h \omega_{dj}^2 \phi_{dj} = K_d \phi_{dj} = \left(K_h - \sum_{i=1}^{n_e} d_i K_i\right) \phi_{dj}. \quad (14)$$

First, we assume that the only one element is damaged, the ith element. With this assumption, (14) reduces to

$$d_i K_i \phi_{dj} = \left(K_h - M_h \omega_{di}^2\right) \phi_{dj}. \quad (15)$$

We define the vector

$$F_{ij} = K_i \phi_j \quad (16)$$

as the corresponding force vector. Define the righthand side of (15) as

$$F_j = \left(K_h - \omega_{dj}^2 M_h\right) \phi_{dj}. \quad (17)$$

Substituting (17) into (15) we have

$$d_i F_{ij} = F_j. \quad (18)$$

Furthermore, (18) must hold for all modeshapes/mode frequency pairs.

The equations in (18) are a set of over determined equations. A least squares solution is given by

$$d_i = \left[F_{ij}^T F_{ij}\right]^{-1} F_{ij}^T F_j. \quad (19)$$

Based on this analysis we make two important observations for the damage DLE algorithm proposed below.

1. Damage to the ith element is parameterized as a scalar change to the system stiffness matrix. This damage parameter is directly related to the physical parameters subject to damage.

2. The contribution of the ith element to the global stiffness matrix is given by

$$(1-d_i)K_i = (1-d_i)\left[\frac{E_i I_i}{s_i}\right] Q_i. \quad (20)$$

This element is located by the ith element location matrix, Q_i. This matrix, Q_i, is not subject to change when the beam sustains damage, because this matrix is solely determined by the geometry of the beam and the location of the ith element in the beam. Hence, if the damage coefficient is nonzero, the element location matrix locates the beam element where damage has been sustained. Also the damage to the ith element is determined solely by the ith element damage coefficient.

Based on these observations and the models of the healthy beam and the damaged beam above, we can reformulate the damage detection, location, and estimation problem more carefully.

Estimated Modal Damage Coefficients

The location and estimation algorithm is derived form the calculation of the estimated modal damage coefficient in (19). We start with the healthy beam stiffness matrix, K_h, and the modal frequencies, ω_{dj}, and the modeshapes, ϕ_{dj}, from the damage beam. We do not know if there is damage, and, if there is damage, where the damage has occurred and what is the extent of the damage. We postulate that the mth element is damaged. Inserting the known data into (19), we obtain

$$\hat{d}_{mj} = \frac{\Delta F_{mj}^T F_j}{\left\|\Delta F_{mj}\right\|^2}. \quad (21)$$

We call the number, $\hat{d}_{mj}$, in (21) the estimated modal stiffness damage coefficient. This coefficient is indexed to the mth beam element and the jth modal frequency/modeshape pair. Considering all elements and all modes, we attain an array of numbers we call the estimated modal stiffness damage coefficients into the estimated modal stiffness damage coefficient array, KDA, where $KDA(m,j) = \hat{d}_{mj}$. The algorithm for calculating these coefficients is shown in Fig. 1.

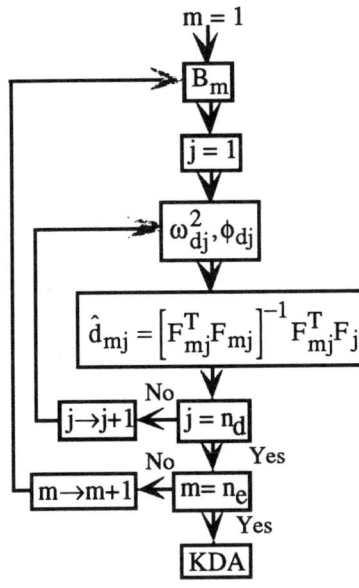

Fig. 1. Calculation of the Estimated Damage Coefficient Array.

Even if the mth element is healthy, (21) will return a number for $\hat{d}_{mj}$. In fact, the algorithm is trying to fit an undamaged model to damaged data. The algorithm proceeds by identifying which coefficients are erroneous and which coefficients represent real damage. To that end we will investigate the structure of the update algorithm to distinguish between true and misleading coefficients.

(A) For the damaged element, the *same* estimated damage coefficient is returned for every pair of damage modal frequencies and damage modeshapes, i.e.

$$\hat{d}_{mj} = \text{constant} \quad j = 1, \cdots, n_e. \tag{22}$$

If the beam element is not damaged, then the returned estimated modal damage coefficients for that element will not be the same for all considered modes. If only one beam element is damaged, the damage model is unique for the damage model above. Hence, all modeshapes and modal frequencies can't be assigned with the same estimated damage coefficient.[19]

(B) We assumed that the damage to the beam is a reduction in stiffness. In that case the damage coefficient of the damaged element satisfies $0 \leq d_i \leq 1$. If the estimated modal damage coefficient is not in that range, then that number is not providing any information and it can be eliminated form the KDA array by setting that particular entry in the array to zero.

After the estimated damage coefficient array is constructed including the filtering (B), the columns of the array are searched to see if the entries corresponding to the ith element, say, satisfy criteria (A) above. If the estimated damage coefficients of the ith column do satisfy criteria (A), the damage is detected, the damaged beam element is located (the ith beam element), and the estimated damage coefficient in (22) gives an estimate of the extent of the damage.

Cantilevered Beam

Consider the cantilevered beam with parameters in Table 1.

TABLE 2.1 BEAM PARAMETERS

Young's Modulus	E	10^7 psi
Mass Density	ρA	12.14×10^{-6} sl/in^3
Length	L_b	22 in
Width	h_b	1.5 in
Thickness	t_b	1/32 in

After reconstructing the modeshapes and extracting the modal frequencies, an estimated modal stiffness damage coefficient array, KDA, is computed for each beam element. We display the results using a mesh plot as shown in Fig. 2.

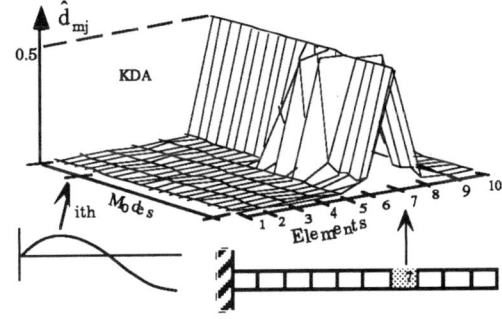

Fig. 2. A Beam with a 5% Reduction in Stiffness to Element 7.

On one edge of the array is indexed with the beam elements numbers. The second index is the modes as determined in an arbitrary way from the output of the identification algorithm. Each mesh intersection is an element from the KDA which has been filtered. That is the KDA element is set to zero if it is outside the interval $0 \leq \hat{d}_{mj} \leq 1$. The damaged beam element is located by comparing estimated modal damage coefficients for one beam element across several modes. If these coefficients satisfy (22) then they identify the damaged beam element. The display of the estimated modal damage coefficient array in Fig. 2 makes this selection criteria obvious. The actual damage estimate is obtained from the entries of the

KDA. The random spikes seen in Fig. 2 are the algorithm trying to fit data to the wrong model.

As a second example, Fig. 3 shows the KDA when the beam has damage in elements 3 and 7.

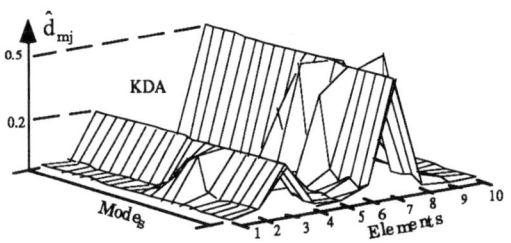

Fig. 3. A Beam with a 20% Reduction in Stiffness to Element 3 and a 50% Reduction in Stiffness to Element 7.

This example shows that multiple damage locations can be detected if these damage sites are geographically separated. Fig. 4 shows the KDA when the beam has damage in elements 7 and 8.

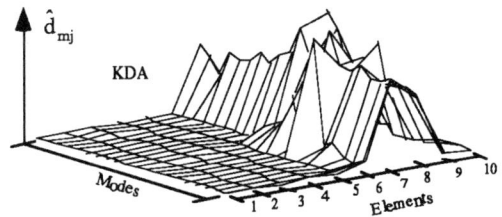

Fig. 4. A Beam with a 50% Reduction in Stiffness to Element 7 and a 40% Reduction in Stiffness to Element 8.

When the damage to the beam occurs in beam elements that are adjacent, the algorithm will indicate damage roughly in the right area but may not select the correct element.

<u>Changes In The Mass Distribution</u>

In this section we develop an algorithm for detecting changes to the mass distribution along the beam. The approach here follows the development of the algorithm for detection of changes to the stiffness matrix. First, we consider damage in the form of a reduction in mass to a single beam element, say the ith element. To model this damage, we write the scalar in (6) as

$$(1+c_i)\left(\frac{\rho_i A_i h_i}{420}\right). \tag{23}$$

Using (23) in (7), the contribution of this damaged element to the global mass matrix is

$$(1+c_i)\left[\frac{\rho_i A_i h_i}{420}\right]P_i = M_i + c_i M_i. \tag{24}$$

We call c_i the <u>mass damage coefficient of the ith element of the beam</u>. The ith mass damage coefficient is negative, then this element experiences an decrease in mass. If $c_i = -1$, then the mass of the corresponding finite element is entirely removed. Obviously, this is impossible if the beam is to continue to have a dynamic response similar to the dynamic response before the damage event. Define the <u>set of element mass damage coefficients</u>

$$C = \{c_i, \quad i = 1, \cdots, n_e\}. \tag{25}$$

Locating and estimating damage to a structure is equivalent to determining the set of damage coefficients.

The damage location and estimation algorithm proceeds as along the same lines as the damage location and estimation algorithm for the stiffness coefficients. The modeshape/mode frequency pair $(\phi_{dj}, \omega_{dj}^2)$ satisfies

$$\omega_{dj}^2 M_d \phi_{dj} = K_h \phi_{dj}$$
$$= \omega_{dj}^2 \left(M_h + \sum_{i=1}^{n_e} c_i M_i\right) \phi_{dj}. \tag{26}$$

First, we assume that the ith element is damaged. With this assumption, (26) reduces to

$$c_i \omega_{di}^2 M_i \phi_{dj} = \left(K_h - \omega_{di}^2 M_h\right) \phi_{dj}. \tag{27}$$

We can define

$$c_i \omega_{di}^2 M_i \phi_{dj} = G_{ij}. \tag{28}$$

We also define

$$F_j = \left(K_h - \omega_{dj}^2 M_h\right) \phi_{dj}. \tag{29}$$

Note that the this vector is the same as the vector used in the algorithm used to find the stiffness damage coefficient. Substituting (29) into (27) we have

$$c_i G_{ij} = F_j. \tag{30}$$

Then the mass damage coefficient is given by

$$c_i = \left[G_{ij}^T F_j\right] \left\|G_{ij}\right\|^{-2} \tag{31}$$

If we don't know the location of the damage, for each modeshape/mode frequency pair and beam element we can still compute a number

$$\hat{c}_{mj} = \|G_{mj}\|^{-2} \left[G_{mj}^T F_j \right]. \quad (32)$$

We call the estimated modal damage coefficients into the <u>estimated mass modal damage coefficient array</u> MDA, where MDA(m,j) = $\hat{c}_{mj}$. The algorithm for calculating these coefficients follows the algorithm for computing the stiffness estimated damage coefficients. The damage is located and estimated by using the criteria

$$\hat{c}_{mj} = \text{constant} \quad \text{for} \quad j = 1, \cdots, n_e. \quad (33)$$

We consider again the cantilevered beam discussed in Section 4. Here we assume that the mass of the 7th finite element is increased by 20%; i.e. $c_7 = .2$. The estimated mass modal damage coefficient array, MDA, is shown in Fig. 5.

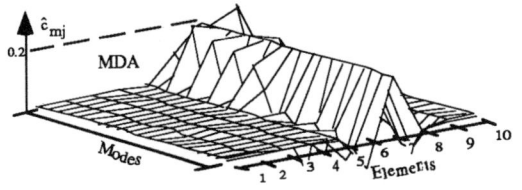

Fig. 5. A 20% increase in the Mass of the 7th Element of a Cantilevered Beam.

The performance of the algorithm detecting changes in mass is similar to the its ability to detect changes in stiffness.

Conclusions

In this paper we have described a new algorithm for detecting damage in beams. A finite element model of the beam is used to parameterize the damage to the beam in terms of individual elements of the beam. This parameterization of the damage allows the algorithm to narrowly focus on the perturbation of the stiffness matrix caused by the damage. Because these two factors were taken into account, the damage location and estimation algorithm proposed here was able to identify damage in individual struts of the beam. Furthermore, the algorithm produces an estimated of the extent of the damage. The proposed algorithm also is computationally less intensive than other algorithms proposed thus far. It is clear that this algorithm is closely related to other reported algorithms. Because of space limitations, these relationships will be reported elsewhere.

Acknowledgments

This research was conducted, in part, while the author was with the Mechanics of Materials Branch at the Naval Research laboratory as a Navy/ASEE Faculty Fellow.

References

1. Zimmerman, D. C. and S. W. Smith, "Model Refinement and Damage Location for Intelligent Structures," to appear in *Intelligent Structural Systems*, H. S. Tzou, ed., 1991.

2. Baruch, M. and I. Y. Bar Izhack, "Optimum Weighted Orthogonalization of Measured Modes, *AIAA Journal*, Vol. 16, 1978, pp. 346-351.

3. Berman, A. and E. J. Nagy, "Improvements of a Large Analytical Model Using Test Data," *AIAA Journal*, Vol. 9, 1971, pp. 1481-1487.

4. Hajela, P. and F. J. Soeiro, "Structural Damage Detection Based on Static and Modal Analysis," AIAA Journal, Vol. 28, No 6, June 1990, pp. 1110 - 1115.

5. Kabe, A. M., "Stiffness Matrix Adjustment Using Mode Data," *AIAA Journal*, Vol. 23, 1985, pp. 1431-1436.

6. Kashangaki, T. A-L, S. W. Smith, and T. W. Lim, "Underlying Modal Data Issues For Detecting Damage in Truss Structures," *Proceedings of the 33rd AIAA/ASME/AHS/ASC Structures, Structural Dynamics and Materials Conference*, Dallas, TX, 1992, pp. 1437 1446.

7. Lim, T. W., "Structural Damage Detection Using Modal Test Data," *AIAA Journal*, Vol. 29, December 1991, pp. 1431-1436.

8. Lim, T. W., "A Submatrix Approach to Stiffness Matrix Correction Using Modal Test Data," *AIAA Journal*, Vol. 28, June 1990, pp. 1123-1126.

9. Peterson, L., K. Alvin, S. Doebling, and K. Park, "Damage Detection Using Experimentally Measured Mass and Stiffness Matrices," *Proceedings of the AIAA/ASME/ASCE/AHS/ASC 34th Structures, Structural Dynamics, and Materials Conference*, La Jolla, CA, April 19 - 21, 1993, pp 1518 - 1528.

10. Smith, S. W. and C. A. Beattie, "Secant Method Adjustment for Structural Models," *AIAA Journal*, Vol. 29, 1990, pp. 119-126.

11. White, C. W. and B. D. Maytum, "Eigensolution Sensitivity to Parametric Modal Perturbation," *Shock and*

Vibration Bulletin, Bulletin 46, August 1976, pp. 193 - 198.

12. Zimmerman, D. C. and M. Kaouk, "An Inverse Problem Approach for Structural Damage Detection - Finite Element Model Refinement," to appear in the Proceedings of the Eighth VPI&SU Symposium on Dynamics and Control For Large Structures, Blacksburg, VA, 1991.

13. Zimmerman, D. C. and M. Kaouk, "Eigenstructure Assignment Approach for Structural Damage Detection," *AIAA Journal*, Vol. 29, July 1992, pp. 1848 - 1885.

14. Kaouk, M. and D. Zimmermann, "Structural Damage Assessment Using Minimum Rank Perturbation Theory," *Proceedings of the AIAA/ASME/ASCE/AHS/ASC 34th Structures, Structural Dynamics, and Materials Conference*, La Jolla, CA, April 19 - 21, 1993, pp 1529 - 1538.

15. Chen, J. Y. and J. A. Garba, "On-Orbit Damage Assessment for Large Space Structures," *AIAA Journal*, Vol. 26, September 1988, pp. 1119-1126.

16. Hemex, F. and C. Farhat, "Locating and Identifying Structural Damage Using A Sensitivity-Based Model Updating Methodology," *Proceedings of the AIAA/ASME/ASCE/AHS/ASC 34th Structures, Structural Dynamics, and Materials Conference*, La Jolla, CA, April 19 - 21, 1993, pp 264 - 2653.

17. Ricles, J. M. and J. B. Kosmatka, "Damage Detection in Elastic Structures Using Vibratory Residual Forces and Weighted Sensitivity," *AIAA Journal*, Vol. 30, September 1992, pp. 2310-2316.

18. Tavares, R., J. Kosmatka, J. Ricles, and A. Wicks, "Using Experimental Modal Data to Detect Damage in a Space Truss," *Proceedings of the AIAA/ASME/ASCE/AHS/ASC 34th Structures, Structural Dynamics, and Materials Conference*, La Jolla, CA, April 19 - 21, 1993, pp 1556 - 1564.

19. Lindner, Douglas, K., G. Twitty, and R. Goff, "Damage Detection, Location, and Estimation For Large Truss Structures," *Proceedings of the AIAA/ASME/ASCE/AHS/ASC 34th Structures, Structural Dynamics, and Materials Conference*, La Jolla, CA, April 19 - 21, 1993, pp 1539 - 1548.

AIAA-94-1756-CP

MECHANICS FOR THE COUPLED DYNAMIC RESPONSE OF ACTIVE/SENSORY COMPOSITE STRUCTURES

D. A. Saravanos[1], P. R. Heyliger[2] and D. A. Hopkins
Structural Mechanics Branch
NASA Lewis Research Center
Mail Stop 49-8
21000 Brookpark Rd.
Cleveland, Ohio 44135

Abstract

Semi-analytical and finite-element models based on discrete-layer theories are presented for the coupled-field dynamic analysis of "smart" laminated composite structures containing piezoelectric layers as sensors and actuators. The theories implement layerwise representations of the displacements and electrostatic potential, hence, they can model both global and local dynamic response. Applications on the free vibration response of composite beams and simply supported plates with added piezoelectric sensors/actuator demonstrate the advantages and versatility of the method to analyze the dynamic characteristics and simulate both sensory and active dynamic response.

Introduction

The development of a new class of "smart" composite materials and adaptive structures with sensory/active capabilities will further improve the performance and reliability of aeronautical propulsion systems.

---

[1] *Senior Research Associate, Ohio Aerospace Institute; member AIAA*

[2] *Associate Professor, Civil Engineering Dept., Colorado State University, Fort Collins, Colorado 80523*

"Copyright © 1994 by the American Institute of Aeronautics and Astronautics, Inc. No copyright is asserted in the United States Under Title 17, U.S. Code. The U.S Government has a royalty-free license to exercise all rights under the copyright claimed herein for Government purposes. All other rights are reserved by the copyright owner."

Such materials will combine the superior mechanical properties of composite materials, as well as, incorporate the additional inherent capability to sense and adapt their static and vibro-acoustic response (adaptive composite structures), or continuously monitor the type, location, and extent of eminent damage (health monitoring). However, this approach requires, among other things, the development of admissible mechanics entailing capabilities to model the unified electromechanical response of sensory/active structures including the coupling between sensors and actuators. Additionally, the mechanics should address local through-the-thickness effects, such as the evolution of complicated stress-strain fields in sensory/active composites, interfacial phenomena between the embedded micro-devices and passive composite plies, and the evolution of critical damage modes in a smart laminate. The present paper presents such a methodology for analyzing the dynamic response of smart composite structures with embedded piezoelectric sensors and actuators.

There have been many approximate models proposed for the analysis of laminated composite plates containing active or passive piezoelectric layers[1-8]. Most of these theories attempt to replicate the induced strain or electric fields generated by a piezoelectric solid under an external electric field or applied load. Exact solutions[9-10] for piezoelectric laminates have shown that the electric and elastic field distributions are often poorly modeled using simplified theories.

Variational methods and finite element models for piezoelectric solids have also seen extensive development[11-12]. Related work for infinite piezoelectric laminates has been reported by Pauley[13]. The work in this paper builds on previous work on discrete layer theories[13-15], as well as, on that of the authors[16-18] to consider the complete dynamic electromechanical response of smart piezoelectric plate structures under external mechanical or electrical loading.

The paper describes discrete-layer laminate theories for composite structures containing embedded piezoelectric layers as sensors/actuators. Each layer is modeled implementing independent approximations for the in-plane displacement components and the electrostatic potential using unified representations as provided by the linear theory of piezoelectricity. The theories assume either constant or variable transverse displacements through the laminate thickness. Structural solutions for the dynamic response of smart structures are further developed, using additional two-dimensional in-plane approximations. Applications demonstrate the advantages and versatility of the mechanics to analyze the dynamic characteristics and simulate both sensory and active dynamic response of composite beams and plates with added piezoelectric layers.

Piezoelectric Laminates

This section describes admissible mechanics for representing the global/local response of composite laminates with embedded piezoelectric sensors/actuators. The coupled material equations for each ply are first presented in a unified way which may represent either piezoelectric or passive composite layers. Discrete-layer approximations are subsequently defined and the resultant laminate mechanics are presented. A standard laminate coordinate system is assumed, such that x and y are the in-plane axes, while the thickness dimension of the laminate coincides with the z-direction. The general problem considered in this study is to determine the behavior of the elastic and electric field components throughout the laminate under an applied mechanical or electrical loading. Each layer of the laminate can be composed of a purely elastic, piezoelectric, or conducting material. The forcing function is introduced through either an applied surface displacement, traction, potential, or electric charge.

Material Representation

The constitutive equations of a piezoelectric material are given by[19],

$$\sigma_i = C_{ij}^E S_j - e_{ji} E_j$$
$$D_i = e_{ij} S_j + \epsilon_{ij}^S E_j \quad (1)$$

where σ_i are the components of the stress tensor, C_{ij}^E are the elastic stiffness components measured under constant voltage, S_j are the components of engineering strain, e_{ij} is the matrix of piezoelectric coefficients, E_i are the components of the electric field, D_i are the components of the electric displacement and ϵ_{ij}^S are the dielectric constants measured in constant strain conditions. The electric field vector E_i is the gradient of the electrostatic potential ϕ

$$E_i = -\partial \phi / \partial x_i \quad (2)$$

The poling direction of the material is assumed coincident with the z axis. Descriptions of the stiffness, piezoelectric and dielectric permittivity matrices are provided in the Appendix. The assumed elastic stiffness matrix is the one of an orthotropic material rotated about the z axis, hence eq. (1) encompasses also the behavior of passive off-axis composite plies ([e]=0).

The energy stored (electric enthalpy) in the piezoelectric layer includes the components of elastic strain energy, piezoelectric energy and electric energy[19]

$$H_l = \frac{1}{2}(C_{ijkl} S_{ij} S_{kl} - 2 e_{ijk} E_i S_{jk} - e_{ij} E_i E_j) \quad (3)$$

Discrete-Layer Laminate Theory

A new discrete-layer laminate theory is proposed in this paper using piecewise continuous approximations along the z-axis for both displacement and electric potential fields. Previous work has demonstrated the advantages of discrete-layer approaches in capturing local intralaminar and interlaminar effects in elastic[14] and elastodynamic problems[16,17] of composite laminates. Consequently, the suitability of discrete-layer theories to represent the additional heterogeneity induced by the presence of embedded piezoelectric sensors/actuators in composite laminates can not be understated.

The electric potential is included in the state variables and piecewise continuous representations are assumed (see Fig. 1) of the following form:

$$u(x,y,z,t) = \sum_{j=1}^{N} u^j(x,y,t)\Psi^j(z)$$
$$v(x,y,z,t) = \sum_{j=1}^{N} v^j(x,y,t)\Psi^j(z)$$
$$w(x,y,z,t) = \sum_{j=1}^{N} w^j(x,y,t)\Psi^{wj}(z) \quad (4)$$
$$\phi(x,y,z,t) = \sum_{j=1}^{N} \phi^j(x,y,t)\Psi^{\phi j}(z)$$

Two unique advantages of the method are obvious: (1) the complete electromechanical state of the smart laminate is represented; and (2) the formulation entails the inherent option to select the detail of representation in both electric and displacement fields. At the lower limit (N=2) the method reduces to "single-layer" type of assumptions, and for linear $\Psi(z)$ it may further reduce to the Kirchoff-Love assumptions (CLPT) or first order shear theory.

Moreover, the through the thickness displacement can be assumed constant ($\Psi^w(z)=1$) or variable, which establishes the difference between the two laminate theories considered here. For the case of constant transverse displacement through the thickness, this function is equal to 1. For a variable-w approximation, this function can be any Lagrangian interpolation polynomial. It is also possible to use functions in the z-direction which are non-zero only over specified regions.

In the context of eqs. (4), the engineering strain and electric field components become respectively,

$$S_i(x,y,z,t) = \sum_{j=1}^{N} S_i^j(x,y,t)\Psi^j(z) \quad i=1,2,6$$
$$S_3(x,y,z,t) = \sum_{j=1}^{N} S_3^j(x,y,t)\Psi^j_{w,z}(z)$$
$$S_i(x,y,z,t) = \sum_{j=1}^{N} S_i^j(x,y,t)\Psi^j_{,z}(z) + \bar{S}_i^j(x,y,t)\Psi_w^j(z) \quad i=4,5$$
$$E_i(x,y,z,t) = \sum_{j=1}^{N} E_i^j(x,y,t)\Psi_\phi^j(z) \quad i=1,2$$
$$E_3(x,y,z,t) = \sum_{j=1}^{N} E_3^j(x,y,t)\Psi_{\phi,z}^j(z)$$
(5)

where the $\{S^j\}$ and $\{E^j\}$ are respectively the generalized strain and electric potential vectors defined as follows:

$$S_1^j = U_{,x}^j \quad S_2^j = V_{,y}^j \quad S_6^j = U_{,y}^j + V_{,x}^j \quad S_3^j = W^j$$
$$S_4^j = V^j \quad S_5^j = U^j$$
$$\bar{S}_4^j = W_{,y}^j \quad \bar{S}_5^j = W_{,x}^j \quad (6)$$
$$E_1^j = \Phi_{,x}^j \quad E_2^j = \Phi_{,y}^j \quad E_3^j = \Phi^j$$

The electric enthalpy of the laminate is by definition

$$H_L = \int_{-h/2}^{h/2} H_l \, dz \quad (7)$$

Combination eqs. (3,5,7) and integration through-the thickness provides the energy stored in a piezoelectric laminate as a quadratic expression of the generalized strain/electric field and the generalized laminate stiffness, piezoelectric and dielectric permittivity matrices,

$$H_L = 1/2 \sum_{m=1}^{M} \sum_{n=1}^{N} [S_i^m A_{ij}^{mn} S_j^n + S_i^m B_{ij}^{mn} \bar{S}_j^n + \bar{S}_i^m D_{ij}^{mn} \bar{S}_j^n - 2(E_i^m E_{ij}^{mn} S_j^m + E_i^m \bar{E}_{ij}^{mn} \bar{S}_j^m + E_3^m \hat{E}_{ij}^{mn} S_3^m) - (E_i^m G_{ij}^{mn} E_j^m + E_3^m \hat{G}_{33}^{mn} E_3^m)] \quad (8)$$

In the above equation, A^{mn}, B^{mn}, and D^{mn} are the generalized laminate stiffness matrices,

$$A_{ij}^{mn} = \sum_{l=1}^{L} \int_{z_l}^{z_{l+1}} C_{ij} \Psi^m(z) \Psi^n(z) dz \quad i,j=1,2,6$$
$$A_{ij}^{mn} = \sum_{l=1}^{L} \int_{z_l}^{z_{l+1}} C_{ij} \Psi^m(z) \Psi_{,z}^n(z) dz \quad i=1,2,6, \ j=3$$
$$A_{ij}^{mn} = \sum_{l=1}^{L} \int_{z_l}^{z_{l+1}} C_{ij} \Psi_{,z}^m(z) \Psi_{,z}^n(z) dz \quad i,j=4,5 \quad (9)$$
$$B_{ij}^{mn} = \sum_{l=1}^{L} \int_{z_l}^{z_{l+1}} C_{ij} \Psi_{,z}^m(z) \Psi^n(z) dz \quad i,j=4,5$$
$$D_{ij}^{mn} = \sum_{l=1}^{L} \int_{z_l}^{z_{l+1}} C_{ij} \Psi^m(z) \Psi^n(z) dz \quad i,j=4,5$$

E^{mn} are the equivalent piezoelectric laminate matrices,

$$E_{ij}^{mn} = \sum_{l=1}^{L} \int_{z_l}^{z_{l+1}} e_{ij} \Psi^m(z) \Psi_{,z}^n(z) dz \quad i=1,2 \ j=4,5$$
$$\text{and} \quad i=3 \ j=1,2,6 \quad (10)$$
$$\bar{E}_{ij}^{mn} = \sum_{l=1}^{L} \int_{z_l}^{z_{l+1}} e_{ij} \Psi^m(z) \Psi^n(z) dz \quad i=1,2 \ j=4,5$$
$$\hat{E}_{ij}^{mn} = \sum_{l=1}^{L} \int_{z_l}^{z_{l+1}} e_{ij} \Psi_{,z}^m(z) \Psi_{,z}^n(z) dz \quad i,j=3$$

and G^{mn} are the equivalent laminate matrices of dielectric permittivity,

$$G_{ij}^{mn} = \sum_{l=1}^{L} \int_{z_l}^{z_{l+1}} \varepsilon_{ij} \Psi^m(z) \Psi^n(z) dz \qquad i=j=1,2 \quad (11)$$

$$G_{ij}^{mn} = \sum_{l=1}^{L} \int_{z_l}^{z_{l+1}} \varepsilon_{ij} \Psi^m_{,z}(z) \Psi^n_{,z}(z) dz \qquad i=j=3$$

where L is the number of plies including the piezoelectric layers.

Dynamic Response of Piezoelectric Structures

The starting point for the equations of motion of a composite structure with embedded active/sensory piezoelectric layers in a variational form is the Hamilton's principle for a piezoelectric medium[19]

$$\delta \int_{t_0}^{t} dt \int_{V} \left[\frac{1}{2} \rho \dot{u}_j \dot{u}_j - H_l(S_{kl}, E_k) \right] dV + \int_{t_0}^{t} dt \int_{S} (\bar{\sigma}_k \delta u_k - \bar{q} \delta \phi) dS = 0 \quad (12)$$

Here t is time, V and S are the volume and surface occupied by and bounding the solid, bar σ and bar q are the specified surface tractions and surface charge, respectively, δ is the variational operator, the dot superscript represents differentiation with respect to time, and H_l represents the local electric enthalpy given by eq. (3)

Integration through-the-thickness yields the Hamilton's principle (equations of motion) in terms of the generalized laminate quantities defined in the previous section

$$\delta \int_{t_0}^{t} dt \int_{A} \left(\frac{1}{2} \sum_{i=1}^{N} \sum_{j=1}^{N} P^{ij} \dot{U}^i \dot{U}^j - H_L \right) dA + \int_{t_0}^{t} dt \int_{S} (\bar{\sigma}_k \delta u_k - \bar{q} \delta \phi) dS \quad (13)$$

where, A is the mid-plane area occupied by the laminated structure, H_L is the electric enthalpy of the laminate given by eq. (8), and P^{ij} are the generalized densities (per unit area) of the laminate given by

$$P^{ij} = \sum_{l=1}^{L} \int_{z_l}^{z_{l+1}} \rho_l \Psi_i(z) \Psi_j(z) dz \quad (14)$$

To develop structural solutions, in-plane approximations of the generalized electromechanical state (displacements and electric potential) in eq. (4) are proposed of the following type.

$$U^j(x,y,t) = \sum_{i=1}^{M} U^{ji}(t) \bar{\Psi}_i^u(x,y)$$

$$V^j(x,y,t) = \sum_{i=1}^{M} V^{ji}(t) \bar{\Psi}_i^v(x,y)$$

$$W^j(x,y,t) = \sum_{i=1}^{M} W^{ji}(t) \bar{\Psi}_i^w(x,y) \quad (15)$$

$$\Phi^j(x,y,t) = \sum_{i=1}^{M} \Phi^{ji}(t) \bar{\Psi}_i^\phi(x,y)$$

where U^{ji} is the value of the generalized displacement component U^j corresponding to the i-th in-plane approximation function, and so forth. This formulation allows for either global (such as Fourier series, Legendre polynomials, etc.), or local approximations (such as finite element techniques) to be used. A semi-analytical method for specialty smart composite plates, as well as, a finite element based technique developed within this framework are subsequently described.

Semi-Analytical Solution

There are several cases of dynamic problems for which semi-analytical solutions may result, that is exact in-plane, yet, approximate through-the-thickness. An example is simply-supported orthotropic composite plates with embedded piezoelectric continuous layers in the following boundary conditions:

$$w(x,0,z) = w(x,L_y,z) = w(0,y,z) = w(L_x,y,z) = 0$$
$$\phi(x,0,z) = \phi(x,L_y,z) = \phi(0,y,z) = \phi(L_x,y,z) = 0 \quad (16)$$
$$u(x,0,z) = u(x,L_y,z) = v(0,y,z) = v(L_x,y,z) = 0$$

where L_x, L_y are length and width of the plate along the x and y axes, respectively. The engineering strain and electric field components The boundary conditions (16) and the governing equations (13) are satisfied exactly with the Fourier functions

$$U^j(x,y,t) = U^j_{pq}(t) \cos(px) \sin(qy)$$
$$U^j(x,y,t) = V^j_{pq}(t) \sin(px) \cos(qy)$$
$$V^j(x,y,t) = W^j_{pq}(t) \sin(px) \sin(qy)$$
$$\phi^j(x,y,t) = \Phi^j_{pq}(t) \sin(px) \sin(qy)$$

Incorporation into the equations of motion (13) yields finally a system of dynamic equations (equal to the number of generalized displacements and electric potentials) for each in-plane mode (p,q). The formulation, structure and solution of this dynamic system are similar to the ones described in the following section, hence are omitted here.

Finite Element Approximation

Local in-plane approximations may be also used in eqs. (15). Substituting these approximations into the equations of motion (13) and collecting the coefficients allows the governing dynamic equations of the smart composite structure to be expressed in a discrete matrix form as

$$\begin{bmatrix} [M_{11}] & [0] & [0] & [0] \\ [0] & [M_{22}] & [0] & [0] \\ [0] & [0] & [M_{33}] & [0] \\ [0] & [0] & [0] & [0] \end{bmatrix} \begin{Bmatrix} \{\ddot{U}\} \\ \{\ddot{V}\} \\ \{\ddot{W}\} \\ \{\ddot{\Phi}\} \end{Bmatrix} +$$

$$+ \begin{bmatrix} [K_{11}] & [K_{12}] & [K_{13}] & [K_{14}] \\ [K_{21}] & [K_{22}] & [K_{23}] & [K_{24}] \\ [K_{31}] & [K_{32}] & [K_{33}] & [K_{34}] \\ [K_{41}] & [K_{42}] & [K_{43}] & [K_{44}] \end{bmatrix} \begin{Bmatrix} \{U\} \\ \{V\} \\ \{W\} \\ \{\Phi\} \end{Bmatrix} = \begin{Bmatrix} \{F_1(t)\} \\ \{F_2(t)\} \\ \{F_3(t)\} \\ \{Q(t)\} \end{Bmatrix} \quad (18)$$

The elements of these matrices are calculated from the generalized laminate matrices defined in eqs. (9-11) as determined by the variational statement. The nature of the submatrices depends also on the approximation used for w. For variable w, the structure of $[K^{j3}]$ is similar to those of the other matrices. For constant w, the submatrices within $[K^{13}]$, $[K^{23}]$, and $[K^{43}]$ are column vectors and those in $[K^{33}]$ become scalars. In general, the submatrices within each $[K^{ij}]$ are each of order (N+1), while the $[K^{ij}]$ themselves depend on the order of in-plane approximation. The final representation of the coupled dynamic system can be expressed in fairly compact form:

$$\begin{bmatrix} [M_{uu}] & 0 \\ 0 & 0 \end{bmatrix} \begin{Bmatrix} \{\ddot{U}\} \\ \{\ddot{\phi}\} \end{Bmatrix} + \begin{bmatrix} [K_{uu}] & [K_{u\phi}] \\ [K_{\phi u}] & [K_{\phi\phi}] \end{bmatrix} \begin{Bmatrix} \{U\} \\ \{\phi\} \end{Bmatrix} = \begin{Bmatrix} \{F(t)\} \\ \{Q(t)\} \end{Bmatrix} \quad (19)$$

Assuming that both sensory and active piezoelectric layers are embedded into the structure, the electric potential vector is subdivided in a sensory component ϕ^S representing the voltage output at the sensors, and an active component ϕ^A representing the voltage imposed on the active piezoelectric layers, such that $\{\phi\} = \{\phi^S ; \phi^A\}$. Separating the active and sensory potential components in eq. (19), the dynamic equations take the following form

$$\begin{bmatrix} [M_{uu}] & 0 \\ 0 & 0 \end{bmatrix} \begin{Bmatrix} \{\ddot{u}\} \\ \{\ddot{\phi}^S\} \end{Bmatrix} + \begin{bmatrix} [K_{uu}] & [K_{u\phi}^{SS}] \\ [K_{\phi u}^{SS}] & [K_{\phi\phi}^{SS}] \end{bmatrix} \begin{Bmatrix} \{u\} \\ \{\phi^S\} \end{Bmatrix} = $$
$$= \begin{Bmatrix} \{F(t)\} - [K_{u\phi}^{SA}]\{\phi^A\} \\ \{Q^S(t)\} - [K_{\phi\phi}^{SA}]\{\phi^A\} \end{Bmatrix} \quad (20)$$

where superscripts S and A indicate the partitioned submatrices in accordance with the selected sensory/active configuration. The left-hand side includes the unknown electromechanical response of the structure $\{u, \phi^S\}$, that is, the resultant displacements and voltage at the sensors. The right-hand includes the excitation of the structure in terms of mechanical loads and applied voltages on the actuators. The electric charge at the sensors $Q^S(t)$ remains constant with time (practically open-circuit conditions) and is assumed equal to zero.

Among the obvious advantages is the capability of the mechanics to model the response of the smart structure either: in "active" mode, that is, with specified voltages $\Delta\phi^A$ applied across the piezoelectric layers to induce a desirable deflection/strain state; or in "sensory" mode where displacements or mechanical loads are applied to the structure and the resultant voltage or charge is monitored; or in combined "active/sensory" mode.

Additional manipulation of eqs. (20) results in the following uncoupled dynamic equations for the structural displacements and sensory voltages respectively

$$[M_{uu}]\{\ddot{u}\} + ([K_{uu}] - [K_{u\phi}^{SS}][K_{\phi\phi}^{SS}]^{-1}[K_{\phi u}^{SS}])\{u\} = $$
$$\{F(t)\} + ([K_{u\phi}^{SS}][K_{\phi\phi}^{SS}]^{-1}[K_{\phi\phi}^{SA}] - [K_{u\phi}^{SA}])\{\phi^A\} \quad (21)$$

$$\{\phi^S\} = -[K_{\phi\phi}^{SS}]^{-1}([K_{\phi u}^{SS}]\{u\} - [K_{\phi\phi}^{SA}]\{\phi^A\}) \quad (22)$$

Further inspection of the above system reveals that "induced-strain" approaches are only a subcase of the above system ($\phi^S=0$), and in the presence of sensors neglect the coupling effects on both stiffness and induced piezoelectric force.

The above dynamic system are solved to obtain, the modal characteristics (free-vibration), the frequency response or the transient response of the smart structure.

Evaluations and Discussion

A number of application cases are considered in this section. First, the modal characteristics of (natural frequencies and corresponding modal state) of a uniaxial composite beam with piezoelectric layers on the top and bottom surface are presented using finite-element approximations developed in the concept of this formulation. Subsequently, a semi-analytical solution for the free-vibration of simply supported cross-ply hybrid composite plates is investigated.

Sensory/Active Beams. Predicted natural frequencies of a cantilever aluminum beam with an adhered piezoelectric layer on the top surface (Fig. 2a) are shown in Table 1. The length of the beam is 152.4 mm (6 in) and the thicknesses of the aluminum, adhesive, and piezoelectric layers are 15.24 mm (0.6 in), 0.254mm (0.01in), and 1.524mm (0.06in). The mechanical properties of the materials are those reported in Ref. 8. The dielectric permittivity of PZT-4, as provided by the manufacturer (Morgan Matroc[20]), was used. Results for open-circuit (free electric potential on top surface of piezoelectric layer) and closed-circuit conditions (top surface of piezoelectric layer grounded) are shown. The results compare favorable, yet, the differences are attributed to the inclusion of the new terms into the formulation.

The resultant unified mode shapes for a cantilever composite beam, consisting of uniaxial T300/Epoxy plies 3.05mm (0.12in) thick and 2 piezoceramic layers (PZT-4[20]) on the top and bottom (Fig. 2b) each 5.09mm (0.020in) thick are also shown in Fig. 3. Each mode consists of the characteristic displacements and the associated electric potential distribution on the free surfaces. The significance of these unified modes should not be understated, as they indicate the type of the characteristic sensory response that uniquely identifies the corresponding characteristic shape. Conversely, the results indicate the active voltage required to be applied to induce the corresponding dynamic deflections. Hence, the results may provide valuable information for the placement of sensors and actuators.

Sensory/Active Square Plate. A 5-ply laminated simply-supported square plate incorporating piezoelectric layers and composite plies was studied to assess the accuracy of the discrete-layer theory. In this case the transverse displacement w is also assumed to vary through the thickness. The laminate configuration consists of a [0/90/0] cross-ply orthotropic composite sublaminate with ply stiffness components C11=134.9, C22=14.35, C33=14.35, C12=5.156, C13=5.156, C23=7.133, C44=3.606, C55=5.654, C66= 5.654 (all in GPa). The three composite plies have equal thickness of 0.267h each, where h is the plate thickness. Two layers of the piezoceramic material PZT-4 of thickness 0.1h were also bonded to the upper and lower surfaces of the laminate.

Two sets of electric boundary conditions were considered in this analysis. In the first, the top and bottom free surfaces of the piezoelectric layer are grounded. In the second, the terminals in the top and bottom free surfaces of the plate remain open (zero electric displacements). These cases represent closed-circuit and open-circuit conditions respectively, and are termed accordingly with (C) and (O).

The predicted natural frequencies of the first six modes are shown in Table 2 for various numbers of degrees of freedom (N) through the thickness. Two thickness aspect ratios are considered corresponding to a thin plate ($L_x/h=50$) and a thick one ($L_x/h=4$). Comparisons with an exact through-the-thickness solution[10] indicate that the predicted frequencies converge to the values of the exact solution, yet, the consideration of minimal DOFs may provide reasonable accuracy. The normalized through-the-thickness distributions of the modal displacements and modal electric potential of the fundamental mode are shown in Figs. 4 and 5 for the two thickness aspect ratios. Closed-circuit conditions were assumed. Clearly many assumptions of simplified theories are only valid for low thicknesses (Fig. 4), while significant deviations from linear variations are observed in thick plates (Fig. 5).

Summary

Admissible mechanics and the corresponding electromechanical models for the dynamic analysis of smart composite structures with embedded piezoelectric layers were developed. The described theories entail the capability to simulate both sensory and active dynamic response of smart composite structures either at the global structural or the local laminate levels. Initial evaluations of the method on smart composite beams and smart composite plates illustrated the advantages of the mechanics, the capability to represent both active and sensory configurations, and the merits of the approach in representing local dynamic response.

References

1. C.-K. Lee and F. C. Moon, "Laminated Piezopolymer Plates for Torsion and Bending Sensors and Actuators", Journal of the Acoustical Society of America, Vol. 85, 2432-2439, 1989.

2. C.-K. Lee, "Theory of Laminated Piezoelectric Plates for the Design of Distributed Sensors/Actuators. Part I: Governing Equations and Reciprocal Relationships", Journal of the Acoustical Society of America, Vol. 87, 1144-1158, 1990.

3. C.-K. Lee and F. C. Moon, "Modal Sensors/Actuators", ASME Journal of Applied Mechanics, Vol. 57, 434-441, 1990.

4. B.-T. Wang and C. A. Rogers, "Laminate Plate Theory for Spatially Distributed Induced Strain Actuators", Journal of Composite Materials, Vol. 25, 433-452, 1991.

5. K. B. Lazarus and E. F. Crawley, "Induced Strain Actuation of Composite Plates", GTL Report No. 197, Massachusetts Institute of Technology, Cambridge, Massachusetts, 1989.

6. H. S. Tzou and C. I. Tseng, "Distributed Piezoelectric Sensor/Actuator Design for Dynamic Measurement/Control of Distributed Parametric Systems: A Piezoelectric Finite Element Approach", Journal of Sound and Vibration, Vol. 138, 17-34, 1990.

7. R. Lammering, "The Application of a Finite Shell Element for Composites Containing Piezoelectric Polymers in Vibration Control", Computers and Structures, Vol. 41, 1101-1109, 1991.

8. D. H. Robbins and J. N. Reddy, "Analysis of Piezoelectrically Actuated Beams Using a Layer-Wise Displacement Theory", Computers and Structures, Vol. 41, 265-279, 1991.

9. P. R. Heyliger and S. P. Brooks, "Exact Solutions for Piezoelectric Laminates in Cylindrical Bending", Journal of Applied Mechanics (in review).

10. P. R. Heyliger, "Exact Solutions for Simply-Supported Laminated Piezoelectric Plates", Journal of Applied Mechanics (in review).

11. H. Allik and T. J. R. Hughes, "Finite Element for Piezoelectric Vibration", International Journal for Numerical Methods in Engineering, Vol. 2, 151-157, 1970.

12. M. Naillon, R. H. Coursant, and F. Besnier, "Analysis of Piezoelectric Structures by a Finite Element Method", Acta Electronica, Vol. 25, 341-362, 1983.

13. K. P. Pauley, "Analysis of Plane Waves in Infinite, Laminated, Piezoelectric Plates", Ph.D. dissertation, University of California at Los Angeles, 1974.

14. J. N. Reddy, "A Generalization of Displacement-Based Laminate Theories", Communications in Applied Numerical Methods, Vol. 3, 173-181, 1987.

15. D. H. Robbins Jr. and J. N. Reddy, "Modelling of Thick Composites Using a Layerwise Laminate Theory", International Journal for Numerical Methods in Engineering, Vol. 36, 655-677, 1993.

16. D. A. Saravanos, and J.M. Pereira, "Effects of Interply Damping Layers on the Dynamic response of Composite Plates," AIAA Journal, Vol. 30, No. 12, Dec. 1992, pp. 2906-2913.

17. D. A. Saravanos, "Analysis of Passive Damping in Thick Composite Structures," AIAA Journal, Vol. 31, No. 8, Aug. 1993, pp. 1503-1510.

18. P. R. Heyliger and D. A. Saravanos, "On Discrete-Layer Mechanics for Health-Monitoring Applications in Sensory/Active Composite Laminates", Proceedings, ASME Winter Annual Meeting, New Orleans, LA, Nov. 28-Dec. 3, 1993.

19. Tiersten, H. F., Linear Piezoelectric Plate Vibrations, Plenum Press, New York, 1969.

20. "Guide to Modern Piezoelectric Ceramics", 1993, Morgan Matroc, Inc., Electro Ceramics Division, Bedford, Ohio 44145.

Table 1 Predicted natural frequencies.
Aluminum with single PZT-4 layer

	Modal Frequency (Hz)					
Mode	Present(w=const)		Robbins[8]	Present(w=w(z))		Robbins[8]
	(open)	(closed)		(open)	(closed)	
1	544	551	538	545	555	538
2	3238	3275	3196	3232	3286	3181
3	7540	7577	7576	7530	7586	7566
4	8464	8550	8340	8466	8589	8201
5	15275	15416	15017	15309	15501	14434
6	22490	22628	22462	22476	22674	21371

Table 2. Natural Frequencies of Simply Supported Plate.
(C)- closed circuit; (O)- open circuit

	Mode					
a/h=4 (C)	1	2	3	4	5	6
N=5	57.2531	194.840	255.648	282.168	368.461	389.525
N=10	57.1249	192.190	252.024	276.853	364.218	383.352
N=64	57.0773	191.313	250.786	274.969	362.516	381.072
Exact	57.0745	191.301	250.769	274.941	362.522	381.049
a/h=4 (O)						
N=5	57.2707	194.843	255.648	282.168	368.505	389.534
N=10	57.1403	192.192	252.025	276.853	364.252	383.364
N=64	57.0921	191.316	250.786	274.969	362.548	381.084
Exact	57.0893	191.304	250.770	274.941	362.522	381.049
a/h=50 (C)						
N=5	0.619025	15.6835	21.4947	212.811	214.690	384.953
N=10	0.618348	15.6820	21.4933	210.561	211.596	379.943
N=64	0.618127	15.6816	21.4928	209.718	210.538	378.132
Exact	0.618118	15.6816	21.4928	209.704	210.522	378.104
a/h=50 (O)						
N=5	0.619038	15.6835	21.4949	212.827	214.736	384.953
N=10	0.618351	15.6821	21.4935	210.568	211.645	379.944
N=64	0.618141	15.6816	21.4930	209.721	210.589	378.133
Exact	0.618120	15.6816	21.4930	209.707	210.573	378.105

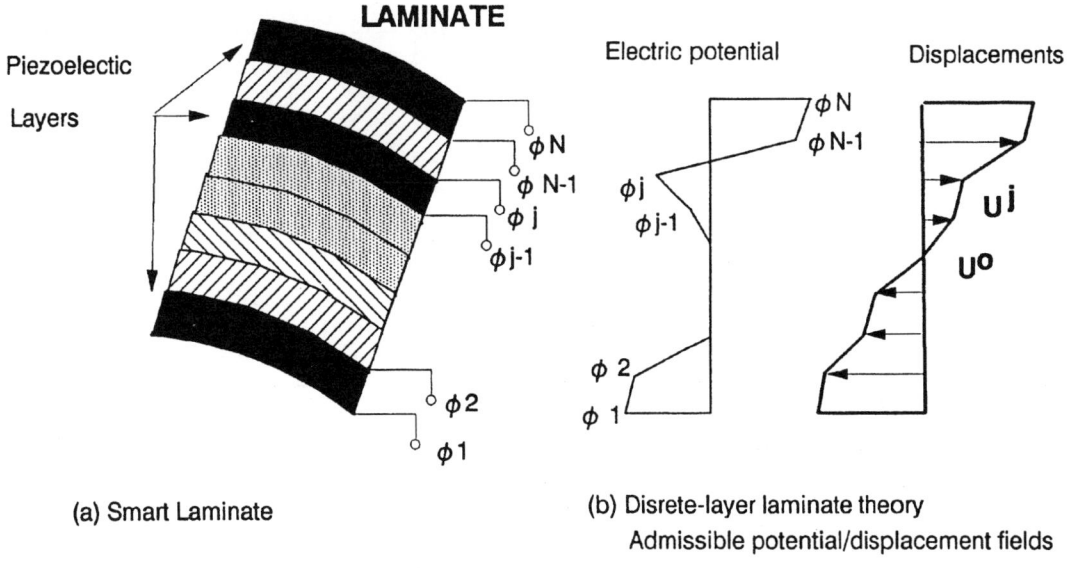

Fig. 1 Typical sensory/active laminate configuration. (a) Concept; (b) Assumed displacement and electric potential fields through the thickness of the laminate

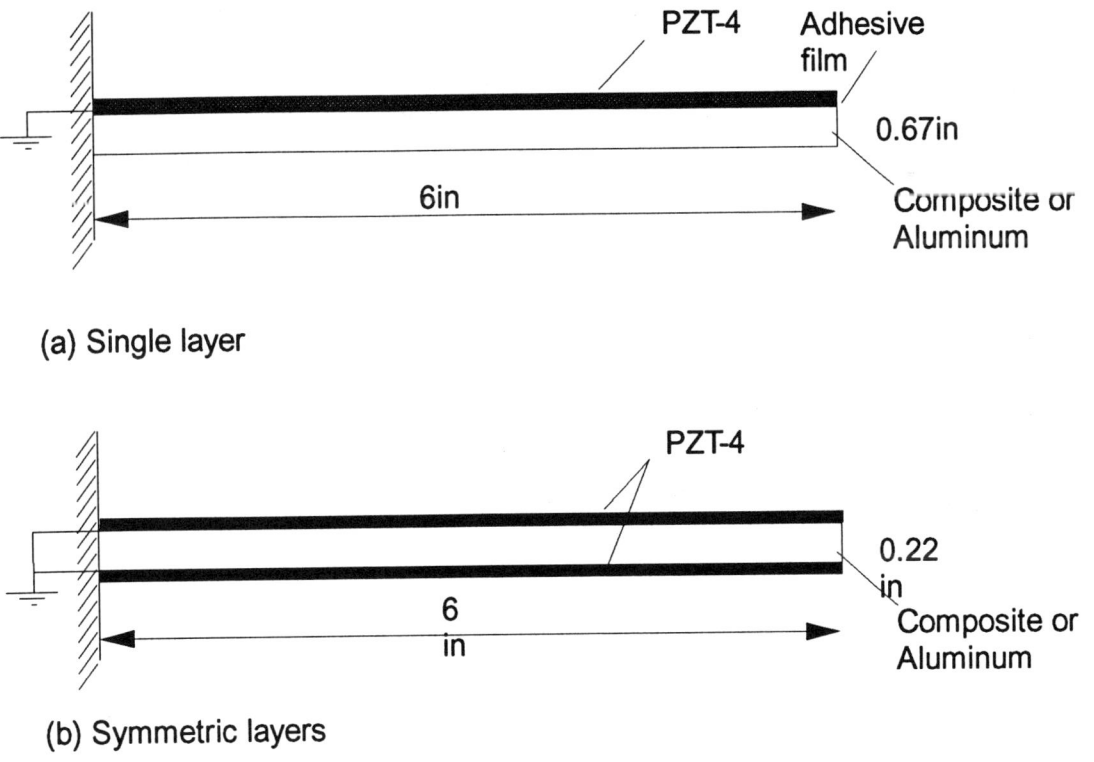

Fig. 2 Beam configurations. (a) Single layer; (b) symmetric layer

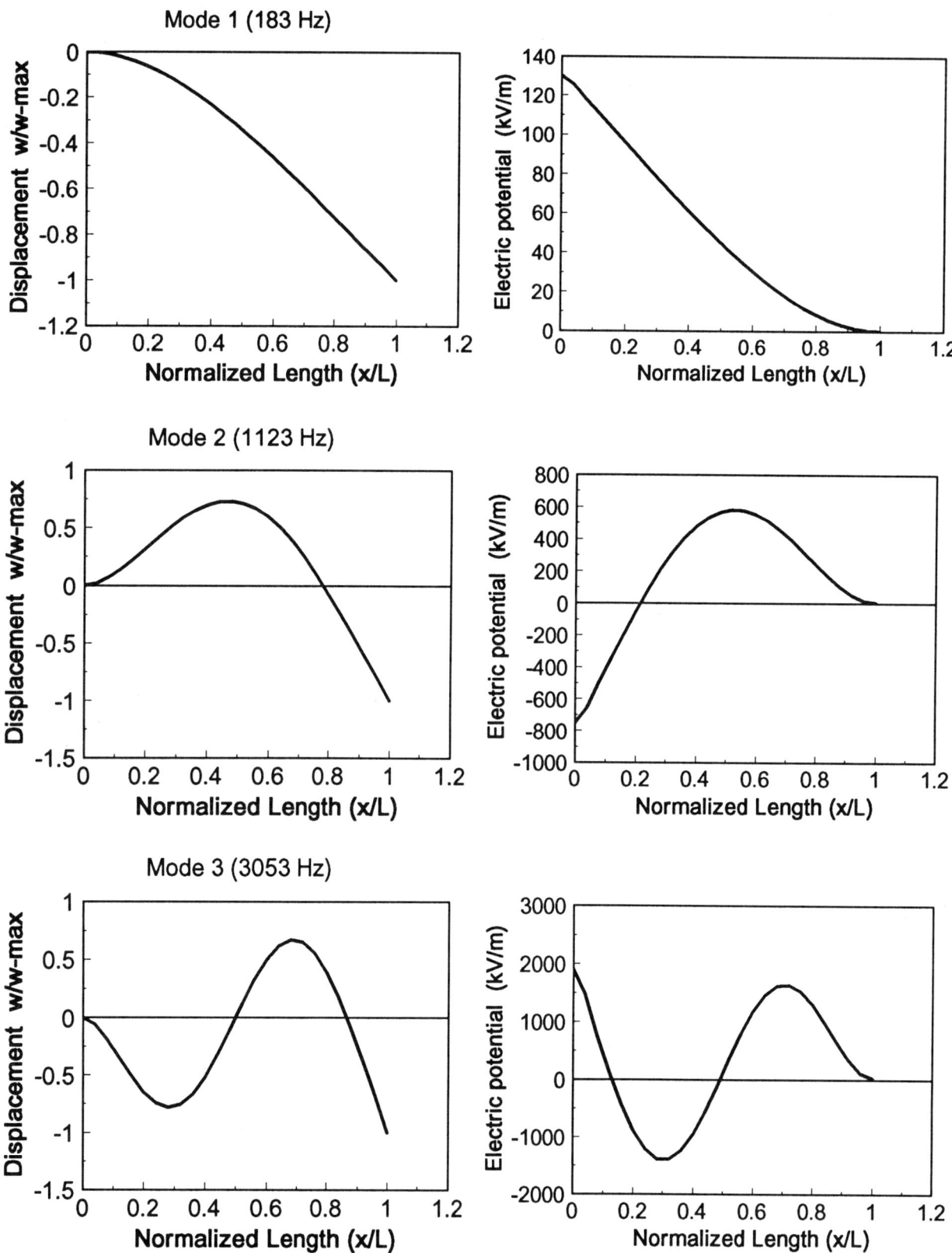

Fig. 3   Predicted electromechanical modes of the smart composite beam.

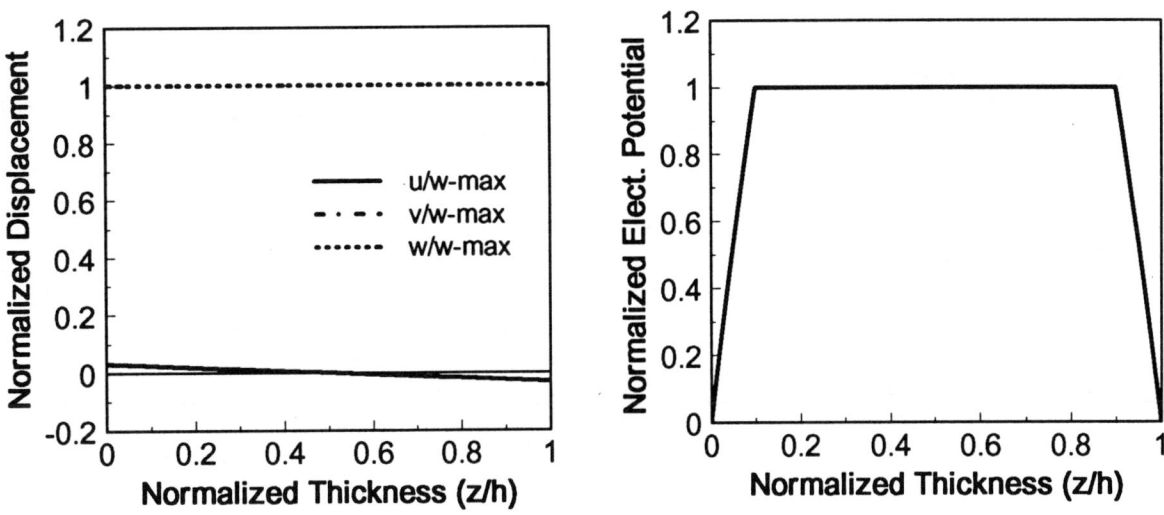

Fig. 4 Through the thickness modal displacments and electric potential fields of the fundamental mode (1,1). Thin Composite Plate (L/h=50)

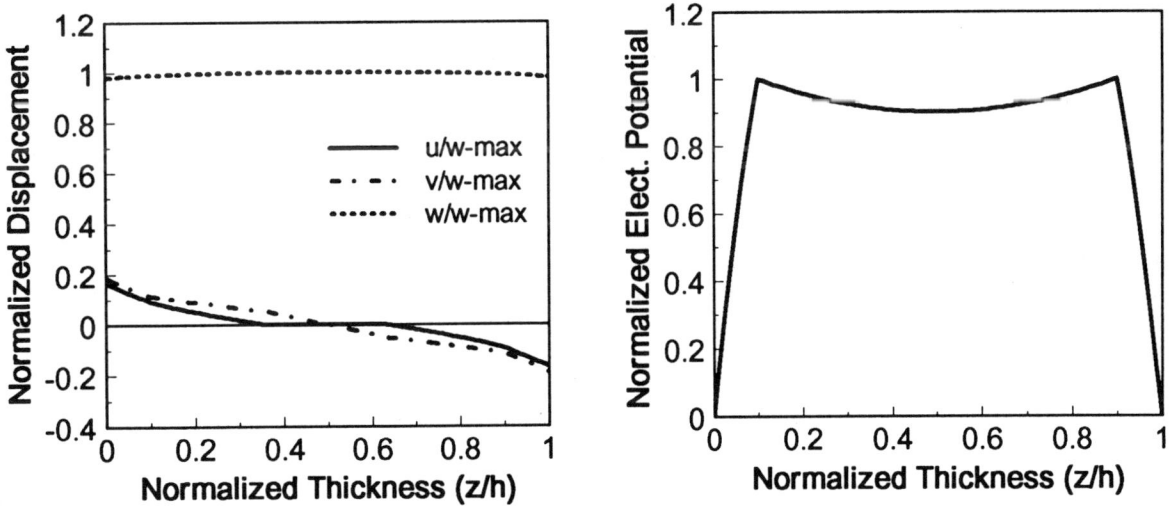

Fig. 5 Through the thickness modal displacments and electric potential fields of the fundamental mode (1,1). Thick Composite Plate (L/h=4)

THE DESIGN OF EXTENDED BANDWIDTH SHAPE MEMORY ALLOY ACTUATORS

Jason B. Ditman[*], Lawrence A. Bergman[†], and Tsu-Chin Tsao[‡]
University of Illinois at Urbana-Champaign
Urbana, Ilinois

Abstract

Extending the usable bandwidth of shape memory alloy actuators for vibration control is investigated, both analytically and experimentally. The differential actuator, which uses SMA wires in agonist-antagonist pairs, is a potential solution. Simulation shows that only modest power requirements are necessary to achieve bandwidths between 10 and 20 Hz.

Introduction

As space structures become more lightweight and flexible, control of response due to both external and internal forces becomes increasingly important. A small maneuver by a space structure with flexible appendages will induce low frequency vibration which will require many cycles to attenuate by passive means. Traditionally, such response has been attenuated by applying passive control systems such as tuned mass dampers and viscoelastic damping materials. These can be effective, but in extreme cases may provide insufficient improvement in performance. For example, an antenna on a communications satellite may be required to slew rapidly, stop, and send or receive information in a fraction of a second. Such performance objectives are inconsistent with the capabilities of passive vibration control systems and lead the designer to active strategies.

Recently, a large body of research has been devoted to developing applications for the unique properties of shape memory alloys. Shape memory materials possess the ability to remember their initial configuration. After deformation, these materials return to their initial shape when heated through a transformation region. In addition, the material properties of shape memory alloys change throughout the phase transformation. It is these unique characteristics that make shape memory alloys potentially valuable components in smart material systems.

In terms of application to structural control, Rogers and coworkers[10-12,16-22] have made significant contributions, both analytically and experimentally. Rogers, Liang and Jia[17] and Rogers and Barker[21] have used active strain energy tuning (ASET) and active properties tuning (APT) to alter the vibrational characteristics of composite plates containing embedded shape memory alloy wires. In both cases, increasing the effective stiffness of the structure caused an upward shift in the natural frequencies of vibration. Selective activation of the embedded wires was also shown to be useful in altering mode shapes. Baz and Ro[3] developed a thermal finite element model of shape memory alloy reinforced composite beams and investigated the effect of selective activation on response. In other applications, Chaudry and Rogers[5,6] used the forces developed during restrained recovery to control buckling of cantilevered beams. Wires embedded along the neutral axis of a beam generate recovery forces that oppose buckling. The inplane tensile loads add to the overall structural stiffness thereby increasing the critical buckling load.

The above systems can be considered to be passive in nature because they do not require that the shape memory alloy wires be thermally cycled. The use of shape memory materials as active components in vibration control systems has been studied by Baz and coworkers[1,2] and Rogers and coworkers.[16-19,21,22] Baz, Imam and McCoy[1] used the shape recovery effect to actively control the vibration of a cantilevered beam. Wires mounted to the surface of the beam were used to generate control forces opposing low frequency vibration. The bandwidth of such a control system, however, is limited by the ability to cool the nitinol wires for the next cycle. Wilson, Anderson, Rempt and Ikegami[25] extended the bandwidth of such a system by using multiple shape memory alloy wires attached to the surface of a cantilevered beam. The multiple wire configuration allowed a higher actuation frequency to be achieved by heating a wire every other cycle rather than on every cycle. This provided more cooling time for each wire. Liang and Rogers[10] extended the capabilities of SMA-based actuator systems by developing design criteria for actuators utilizing shape memory alloy wires. The performance of these actuators is highly dependent on the material properties of the shape memory alloy wires and the means by which the wires are cooled. Baz, Imam and McCoy[3] showed experimentally that the maximum attainable bandwidth of SMA-based actuators is highly dependent on the means of cooling the wires. In both cases, only relatively low bandwidths were achieved. The objective of the work reported herein is to demonstrate a concept by which the usable bandwidth of SMA-based actuators can be extended.

---

[*] Staff Engineer, General Motors Power Train Group, Ypsilanti, Michigan. Formerly, Research Assistant, Department of Aeronautical and Astronautical Engineering, Member AIAA

[†] Associate Professor, Department of Aeronautical and Astronautical Engineering, Associate Fellow AIAA

[‡] Assistant Professor, Department of Mechanical and Industrial Engineering, Member ASME

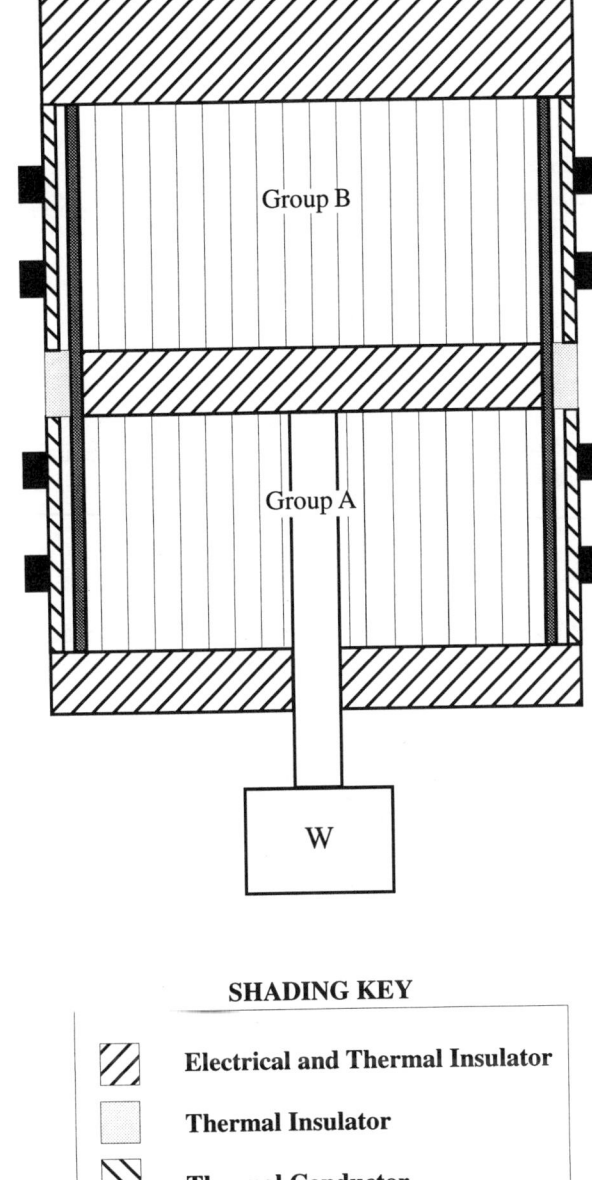

order to simplify the notation, both groups of wires are assumed to have the same material properties and transformation temperatures.

The temperature of each group of wires is to be controlled independently. As a result, the chambers housing the two groups will be thermally insulated from each other. Two thermocouples, one for each group, will be used to provide temperature feedback. All of the wires in a group are connected in series to allow for resistive heating, and thermoelectric devices will be used to maintain predetermined ambient temperatures. A set of thermoelectric devices will be used for each chamber to allow different ambient temperatures to be maintained.

The typical operation of this actuator system is illustrated in Figure 2. First, wire group B is heated while the temperature of group A is held constant. The group B wires are prestrained, and heating causes the deformed martensite to revert to austenite. The transformation process recovers the prestrain and produces a force against the group A wires. The group A wires serve as a nonlinear spring with a high initial stiffness and a lower stiffness during the transformation. If the maximum force produced by the group B wires cannot get the group A wires past the first yield strength, the group A wires will be deforming within the linear elastic range. In this case the wires in group A function as a bias spring.[12] The advantage of using shape memory alloy wires over a coil spring to provide the restoring force becomes evident when the stress in the wires surpasses the yield stress. Once the stress in the group A wires is above the first yield strength, the wires undergo a stress induced martensitic transformation. This process stores energy that will later be used to restore the actuator to its initial position. During the restoration process, the group A wires are heated while the group B wires cool. The group A wires generate a much greater recovery force, due to the energy stored in the martensitic structure, that pulls the group B wires back to the initial position. The external stress on the group B wires will increase the martensitic transformation temperature, resulting in an earlier martensitic transformation and shorter cycle times. The actuator can be heated for the next cycle once the group B wires have returned to the initial state.

Figure 1. Differential force actuator configuration.

The Differential Force Actuator

Configuration

A diagram of the differential force actuator is shown in Figure 1.[15] The actuator consists of two groups (A and B), each with N shape memory alloy wires. The groups form an agonist-antagonist pair with one group providing the deformation force and the other providing the restoring force. The group A wires are of length L^A and diameter d^A, while the group B wires are of length L^B and diameter d^B. In

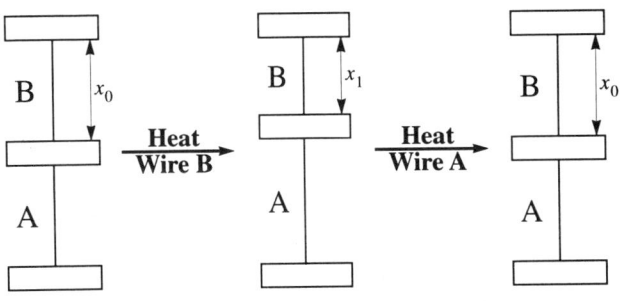

Figure 2. Operating cycle of the differential force actuator.

Differential Force Relation

The rate form of the constitutive relations for wire groups A and B can be expressed as[12,23]

$$\left. \begin{array}{l} \dot{\sigma}^{A_i} = D\dot{\varepsilon}^{A_i} + \Omega\dot{\xi}^{A_i} + \theta\dot{T}^{A_i} \\ \dot{\sigma}^{B_i} = D\dot{\varepsilon}^{B_i} + \Omega\dot{\xi}^{B_i} + \theta\dot{T}^{B_i} \end{array} \right\} i = 1, 2, \ldots, N \quad (1)$$

where N is the total number of wires in each of groups A and B. The subscript, i, is used to refer to the properties of an individual wire, and the overdot indicates the derivative with respect to time. The equilibrium condition of the system is given by

$$\Sigma \sigma^{A_i} = \Sigma \left(\sigma^{B_i} - \sigma_0^{B_i} \right) \quad (2)$$

It is assumed that the total force being exerted by each group is evenly distributed among all of the wires which results in the relations

$$\begin{array}{l} \sigma^{A_1} = \sigma^{A_2} = \ldots = \sigma^{A_{n_A}} \\ \sigma^{B_1} = \sigma^{B_2} = \ldots = \sigma^{B_{n_B}} \end{array} \quad (3)$$

where n_A and n_B refer to the number of active wires in each group. Substituting (3) into (2) yields the generalized equilibrium equation,

$$n_A \sigma^A = n_B \left(\sigma^B - \sigma_0^B \right) \quad (4)$$

The effect of activating only a small number of the wires in a group will be discussed in a later section.

The geometric compatibility between groups A and B can be written as

$$-\dot{\varepsilon}^{A_i} = \dot{\varepsilon}^{B_i} \quad (5)$$

The rate form of the constitutive law for the actuator can be expressed as[11,12]

$$L_A \dot{\sigma}^A - L_A \Omega \dot{\xi}^A - L_A \theta \dot{T}^A = -L_B \dot{\sigma}^B + L_B \Omega \dot{\xi}^B + L_B \theta \dot{T}^B \quad (6)$$

The subscripts in (1) have been omitted in order to simplify the notation. Equation (6), however, is still the rate form of the constitutive law for a single wire pair. Moving the terms involving the group B wire to the left hand side and factoring out the wire lengths yields

$$L_A (\dot{\sigma}^A - \Omega \dot{\xi}^A - \theta \dot{T}^A) + L_B (\dot{\sigma}^B - \Omega \dot{\xi}^B - \theta \dot{T}^B) = 0 \quad (7)$$

Liang and Rogers[12] showed that this equation could be integrated resulting in

$$L_A (\sigma^A - \Omega \xi^A - \theta T^A) + L_B (\sigma^B - \Omega \xi^B - \theta T^B) = C_1 \quad (8)$$

where the constant C_1 depends on the initial conditions of the actuator. At this point, it will be assumed with no loss of generality that $L^A = L^B = L$ and $d^A = d^B = d$. Substituting the equilibrium equation into (8) yields

$$\frac{n_A}{n_B}\left(\sigma^B - \sigma_0^B \right) - \Omega \xi^A - \theta T^A + \sigma^B - \Omega \xi^B - \theta T^B = C_1, \quad (9)$$

the generalized differential actuator relation. To solve for the unknown constant, let $\sigma^B = \sigma_0^B$ and substitute into (9) to obtain

$$C_1 = -\Omega \xi_0^A - \theta T_0^A + \sigma_0^B - \Omega \xi_0^B - \theta T_0^B \quad (10)$$

Substituting (10) into (9) and solving for σ^B yields the generalized stress-temperature relation for the differential actuator

$$\sigma^B = \sigma_0^B - \frac{\left[\Omega\left(\xi^A - \xi_0^A \right) + \theta\left(T^A - T_0^A \right) \\ + \Omega\left(\xi^B - \xi_0^B \right) + \theta\left(T^B - T_0^B \right) \right]}{1 + n_B/n_A} \quad (11)$$

Defining the quantities θ'' and Ω'' such that

$$\theta'' = \frac{\theta}{1 + n_B/n_A} \quad (12)$$

$$\Omega'' = \frac{\Omega}{1 + n_B/n_A} \quad (13)$$

the generalized stress-temperature relation becomes

$$\sigma^B = \sigma_0^B - \Omega''\left(\xi^A - \xi_0^A \right) - \theta''\left(T^A - T_0^A \right) \\ - \Omega''\left(\xi^B - \xi_0^B \right) - \theta''\left(T^B - T_0^B \right) \quad (14)$$

The actuator displacement is directly proportional to the strain in the group A wires and can be expressed as,

$$x = L\varepsilon^A \quad (15)$$

The strain in the group A wires can be derived from the one dimensional thermomechanical relations. Using the equilibrium condition (5) to relate the stress in the two wire groups, the thermomechanical relations for the two groups can be equated resulting in

$$D(\varepsilon^B - \varepsilon^B) + \Omega\left(\xi^B - \xi_0^B\right) + \theta\left(T^B - T_0^B\right) =$$
$$\frac{n_A}{n_B}\left[D\varepsilon^A + \theta\left(T^A - T_0^A\right) + \Omega\left(\xi^A - \xi_0^A\right)\right] \quad (16)$$

Using the compatibility relation (4), the strain in the group A wires can be expressed as

$$\varepsilon^B = \frac{\left[\begin{array}{c}\Omega\left(\xi^B - \xi_0^B\right) + \theta\left(T^B - T_0^B\right) \\ -\frac{n_A}{n_B}\{\Omega\left(\xi^A - \xi_0^A\right) + \theta\left(T^A - T_0^A\right)\}\end{array}\right]}{(1 + n_B/n_A)D} \quad (17)$$

Transient Response of SMA Wires

The transient thermal analysis of the wire is carried out using a lumped capacitance model. On heating from an initial temperature, T_0, and assuming convective boundary conditions, the temperature response can be expressed as

$$T - T_0 = T_f(1 - e^{-t/\tau}) \quad (18)$$

For resistive heating, the final stable temperature, T_f, and thermal time constant, τ, are given by

$$T_f = \frac{I^2 R}{\pi d h}, \quad \tau = \frac{d\rho c}{4h} \quad (19)$$

where I is the input current, R is the electrical resistance of the SMA wire, h is the heat transfer coefficient, ρ is the density, and c is the specific heat. On cooling, the response can be expressed as

$$T - T_0 = (T_i - T_0)e^{-t/\tau} \quad (20)$$

where T_i is the temperature at the beginning of cooling cycle.

For the case of laminar flow around a rod of diameter d and subjected to a temperature gradient ΔT, the heat transfer coefficient can be approximated by[8]

$$h = 1.3\left(\frac{\Delta T}{d}\right)^{0.25} \quad (21)$$

Substituting (21) into (19) yields

$$T_f = \frac{I^2 R}{1.3\pi d}\left(\frac{d}{\Delta T}\right)^{1/4}, \quad \tau = \frac{\rho c}{4(1.32)}\left(\frac{d^{5/4}}{\Delta T^{1/4}}\right) \quad (22)$$

Temperature Control

Assume now that each of the groups are divided into subgroups containing n wires. The wires in the individual subgroups are connected in series and contain a thermocouple for temperature feedback. In this configuration, subgroups from chambers A and B form N/n agonist-antagonist pairs. The governing relations for each subgroup pair are the same as those given in the previous section but with $n_A = n_B = n$. The subgroup configuration has two benefits. First, it reduces the actuator reset problems encountered with previous actuator designs[2,10] by requiring that each subgroup be used only once out of every N/n cycles. Once one subgroup pair has completed its cycle, another pair will activate for the next. This eases the ambient temperature requirements, because each pair will now have N/n times longer to cool. If the actuator contained only one large group in each chamber, they would have to be kept at extremely low temperatures in order to achieve reasonable actuation frequencies. With the subgroup configuration, the actuator can complete a cycle by using one subgroup combination and begin another cycle with a different combination.

Since multiple subgroups are to be used, the ambient temperatures in two chambers only need to be low enough to assure that the wires will reset before their next activation cycle. The temperature to which the wire must be cooled in order to achieve reset will be called T_{reset} which can be calculated from the thermomechanical relations. Thermoelectric devices will be used to maintain this temperature in the chambers. The only requirement on the actuator is that it contain sufficient subgroups to allow reset to occur before the next activation cycle.

A thermocouple will be used to monitor the temperature of each subgroup. During the heating portion of each subgroup's cycle, the thermocouple will activate an electrical switch. Once the wires in the subgroup reach the austenite finish temperature, the input signal will be turned off, and the input for the other subgroup in the pair turned on. This control scheme allows for high input voltages which result in shorter cycle times.

The maximum temperature will also be used to control the restoration process. Since the stress in wire group A is less than the stress in group B, the austenitic transformation temperatures of wire group A do not change as much as the martensitic transformation temperatures of group B. As a result, the transformation in the group A wires starts before the group B wires. Similarly, if the maximum temperature is high enough, the group A wires can be heated through their transformation before the group B wires begin the reverse transformation. This simplifies the analysis significantly.

Design Case Studies

In this section, the governing relations derived above will be used to simulate the response of the differential actuator. Liang and Rogers[10] investigated single wire pair actuators and showed that they could be cycled more quickly than bias spring actuators. The case studies that follow will investigate the effect of the actuator physical parameters on actuation time, wire stress and actuator stroke. The parameters under investigation are the maximum wire temperature, the initial martensitic fractions of subgroups A and B, and the wire diameter. The results of these studies will be used to determine the effectiveness of this actuator at extended bandwidths. The case studies are carried out with reference to a baseline actuator system whose parameters are shown in Table 1.

Parameter	Value
M_s	-27°C
M_f	-34°C
A_s	-25°C
A_f	-14°C
D	7000 MPa
L	2 in
θ	0.1 MPa/°C
Ω	-70 MPa
T_f	100°C
T_0^A	-25°C
T_0^B	-25°C
d	0.025 in
ξ_0^A	0
ξ_0^B	0.6
$n = N$	1
W	1 lb

Table 1. Parameters for the baseline differential actuator.

Maximum Wire Temperature

Figure 3 shows the effect of the maximum attainable temperature on the heating cycle time for groups A and B. This figure indicates that for equal maximum temperatures, the group B heating cycle is shorter than the group A cycle. The difference is due to the fact that the stress in the group A

wires at the start of the restoration process is higher than the stress in the group B wires at the start of the deformation process. Higher stress results in higher austenitic transformation temperatures for the group A wires which causes longer actuation times. In order to maintain equal cycle times, the maximum attainable temperatures must be different. Using the governing relations it is possible to predict what the two maximum temperatures need to be in order to maintain equal cycle times.

Figure 3 also indicates the existence of a hyperbolic relationship between the maximum temperature and the heating cycle time. The figure shows that small increases in the maximum temperature can result in large decreases in the heating cycle time. For the group B wires, a change in the maximum temperature from 100°C to 200°C results in nearly a factor of six decrease in the heating cycle time. For the group A wires, a change in the maximum temperature from 200°C to 400°C results in a similar decrease.

Wire Diameter

The wire diameter has a significant effect on the thermal time constant. Figure 4 shows the effect of the wire diameter on the cycle time of wire groups A and B. Once again the difference in the cycle times for the two groups is due to the stress at the start of the respective transformations. The figure indicates that extremely small wire diameters can result in a small heating cycle time for both groups. As a result an actuator with a large number of small diameter wires can achieve higher actuation frequencies than one with a small number of large diameter wires.

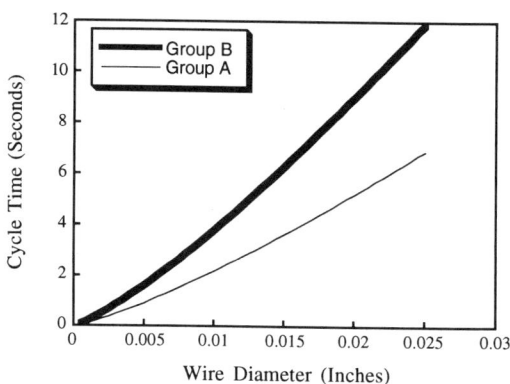

Figure 4. Effect of wire diameter on the group A and B cycle times.

Initial Martensitic Fraction of Wire B

The initial martensitic fraction of the group B wires has a significant effect on the response of the actuator. The initial martensitic fraction determines the maximum actuator stroke. Figure 5 shows the effect of the initial martensitic fraction of the group B wires on the actuator stroke. The larger the martensitic fraction of the group B wires, the larger the maximum actuator stroke. Though the increased martensitic fraction increases the stroke, it also decreases the

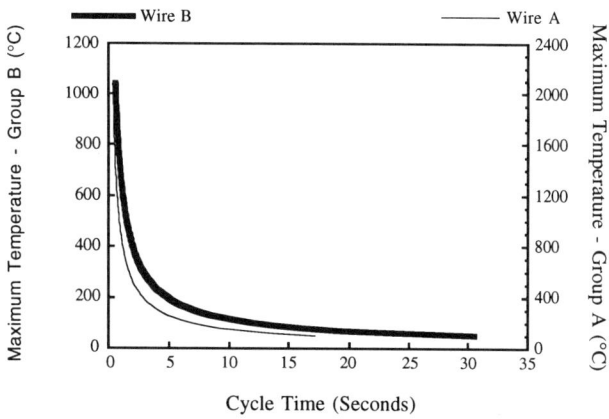

Figure 3. Effect of the maximum wire temperature on the group A and B cycle times.

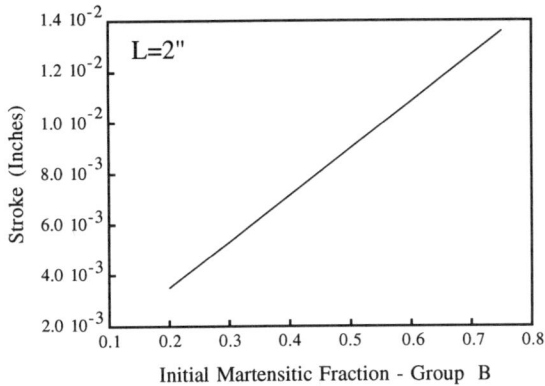

Figure 5. Effect of the initial martensitic fraction of the group B wires on the actuator stroke.

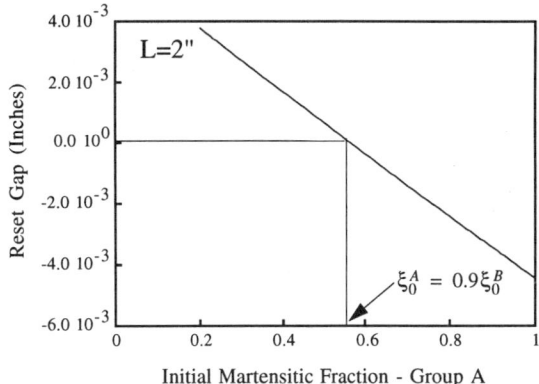

Figure 7. Effect of the initial martensitic fraction of the group A wires on the actuator reset gap.

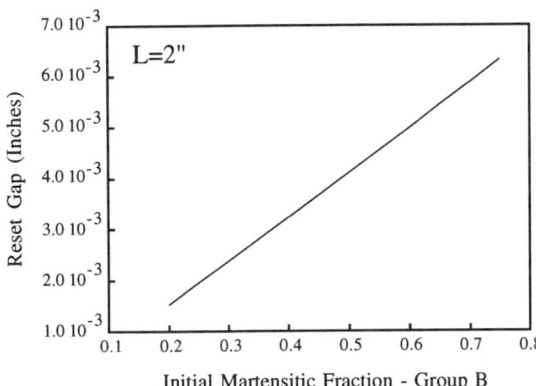

Figure 6. Effect of the initial martensitic fraction of the group B wires on the actuator reset gap.

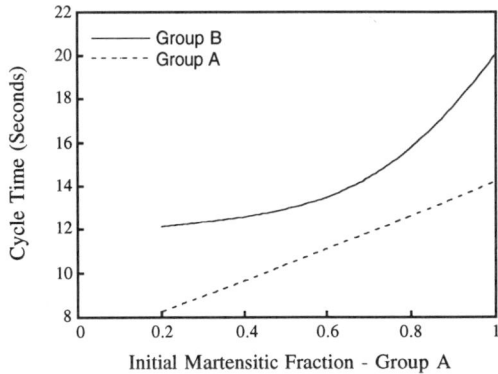

Figure 8. Effect of initial martensitic fraction of the group A wires on the cycle times.

ability of the group A wires to return the actuator to the initial position. Figure 6 shows the reset gap versus the initial martensitic fraction of the group B wires. The reset gap is the difference between the initial actuator position and the position at the end of the group A transformation. The figure indicates that if the group A wires have no initial martensitic fraction, they can only recover approximately half the deformation caused by the group B wires. As a result, in order to minimize the reset gap, the group A wires must also have an initial non-zero martensitic fraction. Figure 7 shows the martensitic fraction of the group A wires versus the reset gap. The figure shows that a martensitic fraction of 0.55 or above is required to achieve a complete cycle. From this result, a critical ratio between the two martensitic fractions can be derived. Simulations indicate the ratio of the initial martensitic fraction of the group A wires to the initial martensitic fraction of the group B wires must be 0.9 or larger in order to achieve complete reset.

Initial Martensitic Fraction of Wire A
The initial martensitic fraction of the group A wires has a larger effect on the cycle time than the group B wires. Figure 8 shows the cycle time plotted against the initial martensitic fraction of the group A wires. The cycle times increase because higher martensitic fractions result in higher stress. The maximum wire stress in the group A wires increases linearly with the initial martensitic fraction as shown in Figure 9. Increasing the initial martensitic fraction of the group A wires from 0.2 to 1.0 causes the wire stress to double. The increased stress increases the transformation temperatures resulting in longer cycle times.

The initial martensitic fraction of the group A wires also has an effect on the actuator stroke. Though increasing the martensitic fraction of the group A wires decreases the reset gap, it has the reverse influence on the stroke. Figure 10 shows the effect of the initial martensitic fraction of the group A wires on the actuator stroke. Increasing the martensitic fraction results in almost a 40% decrease in the actuator stroke. As a result, the martensitic fraction of the group A wires should be limited to the value that preserves reset.

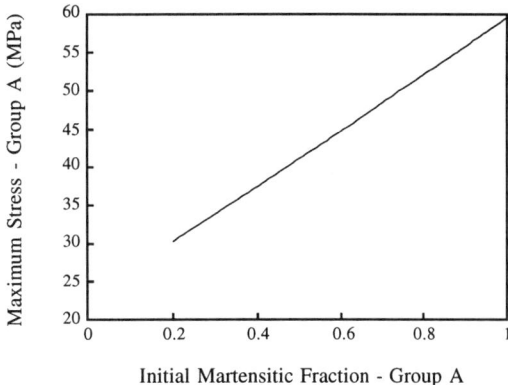

Figure 9. Effect of the initial martensitic fraction of the group A wires on the maximum wire stress.

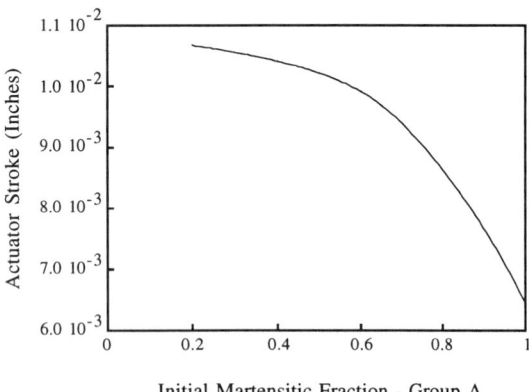

Figure 10. Effect of the initial martensitic fraction of the group A wires on the actuator stroke.

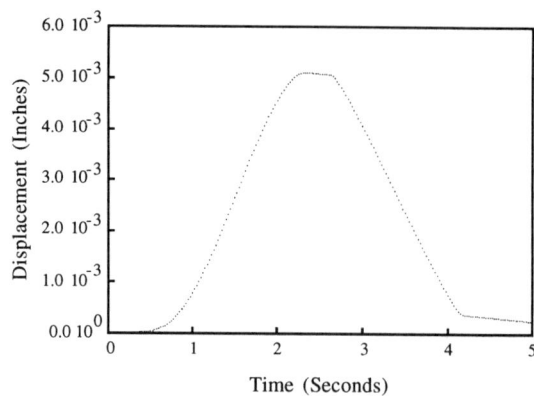

Figure 11. Simulated differential actuator response with $T_{max} = 100°C$.

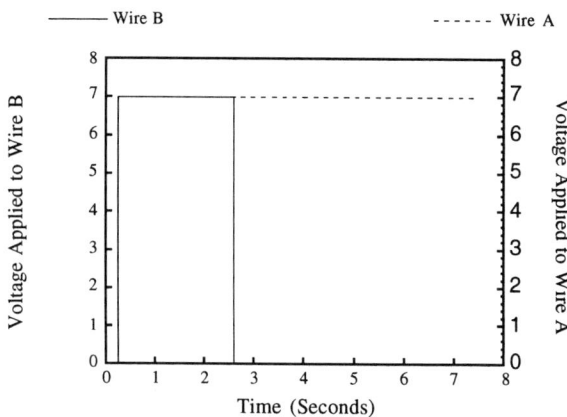

Figure 12. Control signal for simulated differential actuator.

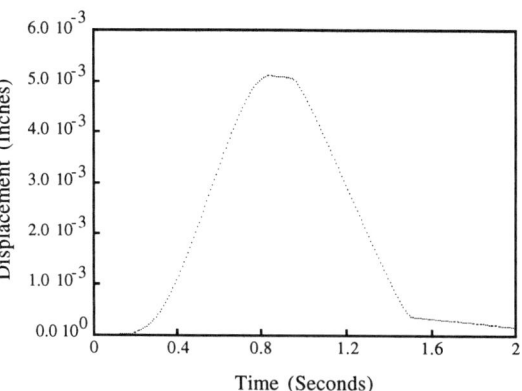

Figure 13. Simulated differential actuator response with $T_{max} = 200°C$.

Simulated Sinusoidal Response

The simulated response of the differential actuator and its associated control signals are shown in Figures 11 and 12 respectively. The physical parameters used for the simulation are the same as those in Table 1 except that $\xi_0^A = 0.9\xi_0^B$ and $d = 0.005$ inches. The frequency of oscillation depends on the maximum attainable wire temperature. For the maximum temperature given in Table 1, the response is nearly sinusoidal with a frequency of 0.3 Hz. As the maximum temperature is increased, the frequency of the response will increase as shown in Figure 13. Using a diameter of 5 millinches and a maximum temperature of 200°C above the austenite finish temperature, the maximum frequency of oscillation is 0.8 Hz (Figure 13). This is an improvement of 167 percent over previous designs[2,10] with similar material properties.

Since cooling is not used for either part of the actuation cycle, the maximum frequency is no longer limited by

Frequency Bandwidth	10 Hertz	20 Hertz
T_{max}	300 °C	500 °C

Table 2. Maximum wire temperatures required to maintain extended bandwidths.

natural convection. Providing that small diameter wires are used in the subgroups, and that sufficient amplification is available to provide the input power, the differential force actuator is capable of operating in the range from 10 to 20 Hz. Table 2 gives the maximum wire temperatures required to maintain a 10 and 20 Hz bandwidth for a wire diameter of 5 millinches and indicates that relatively low maximum wire temperatures are required. As a result, differential force actuators are feasible for use as extended bandwidth actuators.

Experimental Response of the Differential Actuator

Setup and Procedure

A photograph of the apparatus used for the differential force actuator tests is shown in Figure 14. The actuator consists of a single pair of wires, each 2 inches in length. The top and bottom sections of the actuator are affixed to the stand while the middle section is free to displace. The actuator displacement is monitored with an LVDT that is connected to the middle section. Set screws are used to attach the wires to the sections. A set of leads for resistive heating are connected to each wire, and thermocouples are used to monitor wire temperature. The thermocouples provide temperature feedback that allow the heating processes to be coordinated. An automated data acquisition system equipped with a digital signal processor is utilized to record temperature and displacement as well as to provide control signals.

Parameter	Value
M_s	0°C
M_f	-24.3°C
A_s	25.7°C
A_f	34.4°C
D	7000 MPa
L	2 in
θ	0.1 MPa/°C
Ω	-70 MPa
T_f	170, 500 °C
T_0^A	25°C
T_0^B	25°C
d	0.025, 0.0075 in
ξ_0^A	0, 0.72
ξ_0^B	0.8
$n = N$	1
W	0.5 lb

Table 3. Parameters for the experimental differential actuator.

Table 3 lists the physical parameters of the wires used for the tests. Two types of wires were used in the tests. For the initial martensitic fraction tests, wires with a diameter of 25.0 millinches and an austenite start temperature of 27°C were used. For the actuator frequency tests, wires with a diameter of 7.5 millinches and an austenite start temperature of 27°C were also used. In preparation for the tests, the wires were heated in an oven to a temperature of 100°C and then placed in a liquid nitrogen bath. This process set all of the wires to pure martensite, which was used as the reference state.

The group B wires were given initial martensitic strains by straining them while in the martensitic phase. For all of the tests, a martensitic strain of 8% was used. Straining while in this phase causes twin boundary movement which is later recovered by heating. The initial martensitic fractions in the A wire were obtained by heating it part, or all, of the way through the austenitic transformation. This process drives away part of the martensite that formed while in the liquid nitrogen bath. The temperature to which the A wire was heated depended on the desired initial martensitic fraction and will be discussed on the next section. Once both groups were set to their initial martensitic fractions, the wires were loaded into the fixture. Care was taken not to deform the wires during the loading process as this would change the initial martensitic fraction.

Figure 14. Differential actuator testing apparatus.

Results

Initial Martensitic Fraction Ratio

Figures 15, 16, and 17 show the results of the initial martensitic fraction studies. Figure 15 gives the response of an actuator in which no stress induced martensite is formed in the A wire. The A wire for this test was heated completely through the austenitic transformation which left it with no initial martensite. In this case, the B wire recovered its initial strain but could not get the A wire past its first yield point. No stress induced martensite formed, and as a result the A wire had no strain martensitic structure to recover and could not return the actuator to its initial position. The small amount of recovery shown in Figure 15 is due to normal thermal contraction of wire B as it cools.

The A wire for the actuator whose response is shown in Figure 16 also had no initial martensitic fraction. In this test though, the stress generated by the B wire forced the A wire past the first yield point. As a result some martensite was induced, but not enough to allow the A wire to return the actuator to its initial position. The force generated by the B wire only induced enough martensite to restore half of the original deformation.

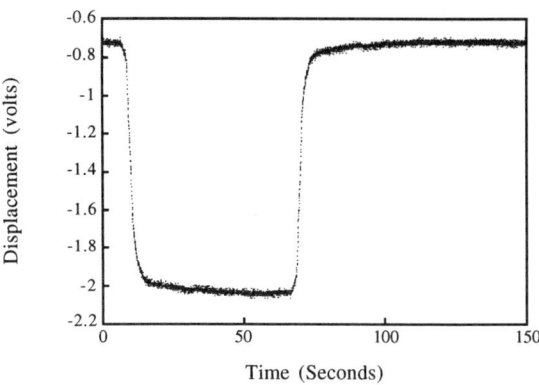

Figure 17. Complete differential actuator cycle.

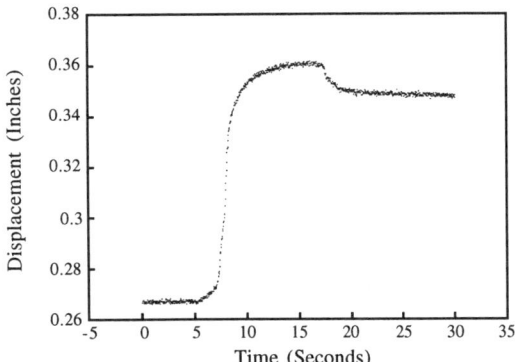

Figure 15. Actuator cycle with no stress induced martensite.

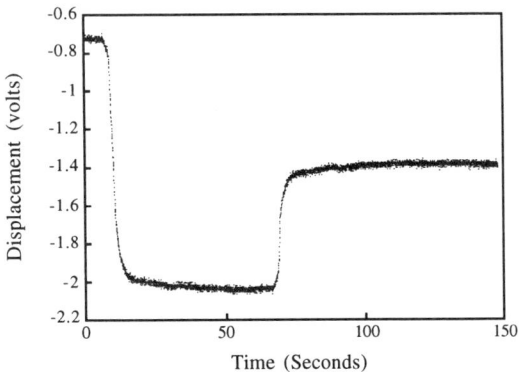

Figure 16. Actuator cycle with no initial martensitic fraction in A wire.

Figure 17 shows a complete actuation cycle. In this test, the A wire was given an initial martensitic fraction equal to 0.9 times the initial martensitic fraction in wire B. The initial martensitic fraction of the B wire was 0.8. Therefore, the initial martensitic fraction of the A wire required to achieve complete recovery was 0.72. The temperature that the A wire needed to be heated to in order to reach a martensitic fraction of 0.72 was found to be 28°C. Preparing the A wire in this manner allows it to be able to fully recover the deformation caused by the B wire.

The results shown in Figures 15, 16, and 17 support the existence of a critical martensitic fraction ratio. The numerical simulations indicate that the critical ratio of the martensitic fraction of the A wire to the martensitic fraction of the B wire is 0.9. The results shown in Figure 17 verify this value.

Maximum Actuator Frequency

Figures 18 and 19 show single actuation cycles for actuators containing two different wire diameters. A 3 amp current was used to heat the wires. The wire diameter affects the maximum attainable wire temperature and the thermal time constant, both of which reduce the actuation time. Figure 18 shows a cycle for an actuator with physical parameters equal to the those in Table 3, a wire diameter of 25 millinches, and an initial martensitic fraction in the A wire of 0.72. Measurements of the time required to complete the cycle indicate that the actuator operates at a frequency of 0.1 Hz. The maximum frequency predicted by numerical simulation is 0.212 Hz. Figure 19 shows the response of the same actuator but with 7.5 millinch diameter wires. The oscillation frequency for this case is 1.667 Hz, while the frequency predicted in the numerical simulation is 1.853 Hz. In both cases the experimental value is lower than the predicted value. The discrepancy is likely due to the non-

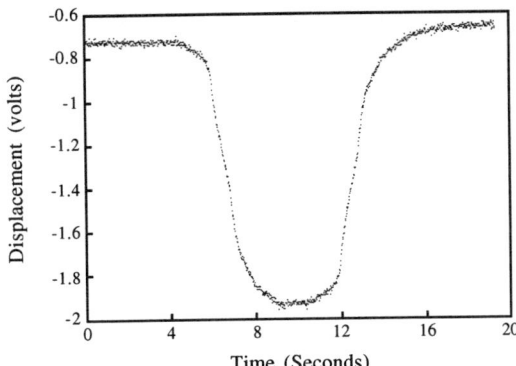

Figure 18. Actuation cycle with d=25.0 millinches.

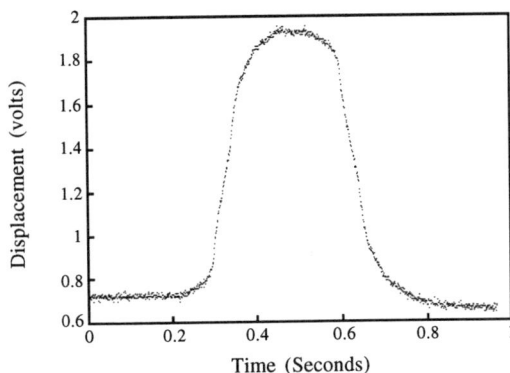

Figure 19. Actuation cycle with d=7.5 millinches.

ideal conditions under which the experiments were performed, which included excessive handling of the wires during setup, slippage in the gripping system during operation, and fluctuating ambient temperatures.

The results shown in Figures 18 and 19 indicate that the actuation frequency is very sensitive to the wire diameter. Simulations indicate that further reductions in the wire to 5 millinches would allow the actuator to achieve a 5-10 hertz bandwidth. In order to optimize the actuation frequency, small diameter wires and high driving power should be used.

Conclusions

The purpose of this study was to investigate a method of the extending the usable bandwidth of shape memory alloy actuators for vibration control. The differential actuator, which uses SMA wires in agonist-antagonist pairs, appears to be a solution despite the fact that it does not rely on cooling for the return cycle. Simulation shows that only modest power requirements are necessary to achieve bandwidths between 10 and 20 Hz. The results from the simulations and experiments also indicate that the ratio of the initial martensitic fraction of the group A wires to the group B wires must be greater than 0.9 in order to achieve a complete actuation cycle. If the ratio is less than this value, the group A wires will not be able to return the actuator to the initial position.

References

1. Baz, A., Imam, K., and McCoy, J., "Active Vibration Control of Flexible Beams Using Shape Memory Actuators," *Journal of Sound and Vibration*, **140**(30), 1990, pp.437-456.

2. Baz, A., Imam, K., and McCoy, J., "The Dynamics of Helical Shape Memory Actuators," *Journal of Intelligent Material Systems and Structures*, **1**, 1990, pp.105-133.

3. Baz, A., and Ro, J., "Thermo-Dynamic Characteristics of Nitinol-Reinforced Composite Beams," *Composites Engineering*, **2**(5-7), 1992, pp. 527-542.

4. Buehler, W. J., and Wiley, C. R., "Nickel Based Alloys, U.S. Patent 3,147,851, Naval Ordinance Laboratory, 1965.

5. Chaudhry, Z., and Rogers, C. A., "Bending and Shape Control of Beams Using SMA Actuators," *Journal of Intelligent Material Systems and Structures*, **2**, 1991, pp. 581-602.

6. Chaudhry, Z., and Rogers, C. A., "Response of Composite Beams to an Internal Actuator Force," *Proceedings of the 32nd Structures, Structural Dynamics and Materials Conference*, AIAA-91-1166-CP, 1991, pp. 186-193.

7. Cross, W. B, Kariotis, A. H., and Stimler, F. J., "Nitinol Characterization Study," No. GER-14188 (NASA-CR-1433), Goodyear Aerospace Corporation, Akron, Ohio, 1970.

8. Holman, J. P., *Heat Transfer*, McGraw-Hill Book Company, New York, 1981, pp. 240-250.

9. Jackson, C. M., Wagner, A. H., and Wasilweski, R. J., "55-Nitinol-The Alloy with a Memory: It's Physical Metallurgy, Properties and Applications," NASA-SP-5110, Battelle Memorial Institute, 1972.

10. Liang, C., and Rogers, C. A., "Design of Shape Memory Alloy Actuators," *Journal of Mechanical Design*, **114**, 1992, pp. 223-230.

11. Liang, C., and Rogers, C. A., "One-Dimensional Thermomechanical Constitutive Relations for Shape Memory Materials," *Journal of Intelligent Material Systems and Structures*, **1**, 1990, pp. 207-234.

12. Liang, C., and Rogers, C. A., "The Multi-Dimensional Constitutive Relations of Shape Memory Alloys," *Proceedings of the 32nd Structures, Structural Dynamics and Materials Conference*, AIAA-91-1165-CP, 1991, pp. 178-185.

13. Materials Electronic Products Corp., *Thermoelectric Heat Pumps*, Materials Electronic Products Corp, New Jersey, 1990.

14. Paine, J. S., and Rogers, C. A., "The Effect of Thermoplastic Composite Processing on the Performance of Embedded Nitinol Actuators," *Proceedings of the 32nd Structures, Structural Dynamics and Materials Conference*, AIAA-91-1167-CP, 1991, pp. 194-204.

15. Quattrone, R. A., *Two-Way Force Actuators*, U.S. Patent Applicaiton 07/984619, 1992.

16. Rogers, C. A., Fuller, C. R., and Liang, C., "Active Control of Sound Radiation from Panels Using Embedded Shape Memory Alloy Fibers," *Journal of Sound and Vibration*, **136**(1), 1990, pp. 164-170.

17. Rogers, C. A., Liang, C., and Jia, J., "Behavior of Shape Memory Alloy Reinforced Composite Plates Part I: Model Formulations and Control Concepts," *Proceedings of the 30th Structures, Structural Dynamics and Materials Conference*, AIAA-89-1389-CP, 1989, pp. 2011-2017.

18. Rogers, C. A., Liang, C., and Jia, J., "Behavior of Shape Memory Alloy Reinforced Composite Plates Part II: Results," *Proceedings of the 30th Structures, Structural Dynamics and Materials Conference*, AIAA-89-1331-CP, 1989, pp. 1504-1513.

19. Rogers, C. A., and Robertshaw, H. H., "Development of a Novel Smart Material," *ASME 1988 Winter Annual Meeting*, 88-wa/DE-9, Chicago, IL, November 27-December 2, 1988.

20. Rogers, C. A., Barker, D. K., "Experimental Studies of Active Strain Energy Tuning of Adaptive Composites," *Proceedings of the 31st Structures, Structural Dynamics and Materials Conference*, AIAA-91-1086-CP, 1990, pp. 2234-2241.

21. Rogers, C. A., Liang, C., and Jia, J., "Structural Modification of Simply Supported Laminated Plates Using Embedded Shape Memory Alloy Fibers," *Computers and Structures*, **38**(5/6), 1991, pp. 569-580.

22. Rogers, C. A., and Liang, C., "Design of Shape Memory Alloy Springs with Applications in Vibration Control," *Journal of Vibration and Acoustics*, **115**, 1993, pp. 129-135.

23. Tanaka, K., Hayashi, T., Itoh, Y., and Tobushi, H., "Analysis of Thermomechanical Behavior of Shape Memory Alloys," *Mechanics of Materials*, **13**, 1992, pp. 207-215.

24. Wayman, C. M., and Duerig, T. W., "An Introduction to Martensite and Shape Memory Alloys," *Engineering Aspects of Shape Memory Alloys*, Butterworth-Heinemann, Boston, 1990, pp. 3-20.

25. Wilson, D. G., Anderson, J. R., Rempt, R. D., and Ikegami, R., "Shape Memory Alloys and Fiber Optics for Flexible Structure Control," *Proceeding of the SPIE: Fiber Optic Smart Structures and Skins III*, 1370, 1990, pp. 286-295.

AIAA-94-1758-CP

SHAPE MEMORY CERAMIC ACTUATION OF ADAPTIVE STRUCTURES

Kamyar Ghandi* and Nesbitt W. Hagood**
MIT Department of Aeronautics and Astronautics
77 Massachusetts Ave. Cambridge MA 02139

Abstract

Field induced phase transitions in piezoceramics with the advantages of large strain, and shape memory, are investigated as a mechanism for structural actuation. Phase transition and shape memory behavior were found to depend heavily on temperature. The effects of stress and frequency on the response of the material, which are relevant for structural actuation, have also been examined. Data is presented for Lanthanum and Niobium doped piezoelectric compositions. A prototype adaptive structure using Lanthanum based shape memory ceramics has been constructed and tested.

Introduction

Piezoceramics have been utilized in structural actuation for some time. Continuing research into materials possessing electro-mechanical coupling holds the promise of higher actuation authority as well as features such as shape memory. An area of particular interest is field induced phase transitions in electroceramics.

Field induced phase transitions have been investigated by several authors[1,2,3]. Large strains have been observed to occur during the antiferroelectric (AFE) to ferroelectric (FE) phase transition. Furthermore, certain classes of material, due to metastability of both phases, are capable of maintaining a residual strain even after the electric field has been switched off. Two families of materials displaying such behavior are the Lead Lanthanum Zirconate Stannate Titanate (PLZST) family of ceramics studied by L. E. Cross[1], and the Lead Niobium Zirconate Stannate Titanate (PNZST) family of ceramics studied by K. Uchino[2].

The feasibility of using the strain associated with phase transition in actuating adaptive structures is the primary focus of the present work. Consideration has been given to factors such as temperature and structural stress, which affect the response of the material.

The study is motivated by two factors. First, strain levels as high as 0.9% have been reported to occur during the phase transition. Second, the use of shape memory in shape control has clear advantages. Utilizing these materials in actual applications requires the capability of predicting the nonlinear behavior of the material. A phenomenological model for the phase transition, similar to the Chen-Montgomery model[3], is presented for antiferroelectric-ferroelectric switching.

---

* Graduate Research Assistant, Room 37-356, Tel: (617) 253-5488.
** Assistant Professor, AIAA member, Room 33-313, Tel: (617) 253-2738.

Copyright © 1994 by the American Institute of Aeronautics and Astronautics, Inc. All rights reserved

The material has been used in the construction of small adaptive mirror, and both theoretical and experimental data on the response of the mirrors are presented.

Material Behavior

In ferroelectrics an electric field can cause mechanical strain due to the polarization of the material (Figure 1). Unlike ferroelectrics, antiferroelectrics do not display any macroscopic polarization. Rather, they contain two sub-lattices which are spontaneously poled in opposite directions. This leads to zero net polarization. As a result very little strain is achieved at low electric fields. If the composition of an antiferroelectric material is close to the ferroelectric phase boundary (Figures 2,3), a phase transition from AFE to FE can be induced by the application of a large electric field. This phase transition, which is caused by the switching of the domain orientations, is accompanied by a lattice distortion leading to a net volume expansion in the material[3]. The isotropic expansion manifests itself as a sharp jump in the strain level of the material (Figure 1). Longitudinal strain levels ranging from 0.08% to 0.87% have been reported for various compositions[1,2]. The compositions studied in the present work have shown strain levels of ~0.3%. Following this phase transition, the ferroelectric material displays anisotropic strains due to the rotation of the ferroelectric domains.

As the electric field is reduced, the material may either return to its original state, or remain in the ferroelectric state (exhibiting shape memory), depending on the exact composition of the material. The ceramics lacking shape memory are only stable in the antiferroelectric state at zero field, while those with shape memory appear to be metastable in either of the ferroelectric and the antiferroelectric states at zero field. In the latter case, a field in the opposite direction must be applied to remove the residual strain.

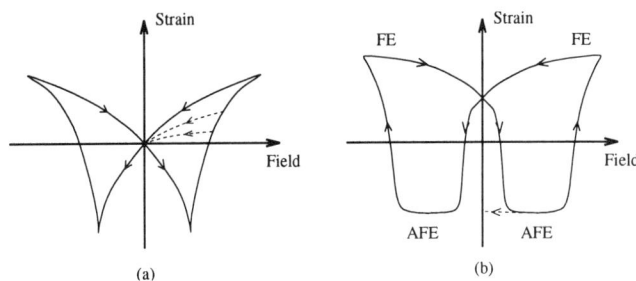

Figure 1- Comparison of the longitudinal strains for ferroelectric and antiferroelectric materials. (a) is ferroelectric. (b) undergoes field induced phase transition from antiferroelectric to ferroelectric and vice versa.

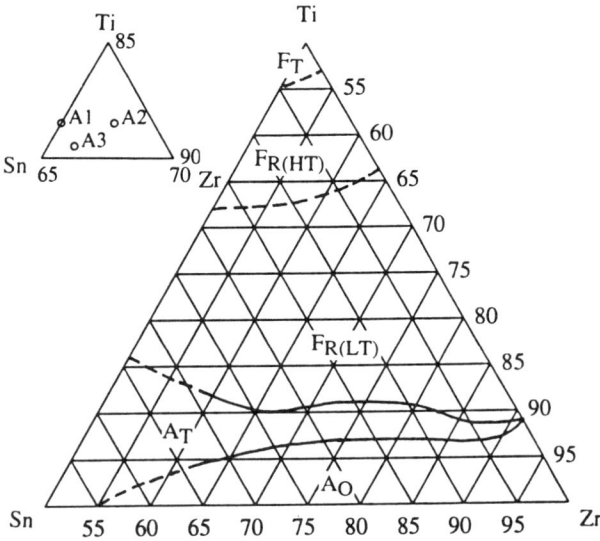

Figure 2- Phase diagram for $Pb_{0.97}La_{0.02}(ZrTiSn)O_3$. The inset shows the compositions manufactured for this study.[4]

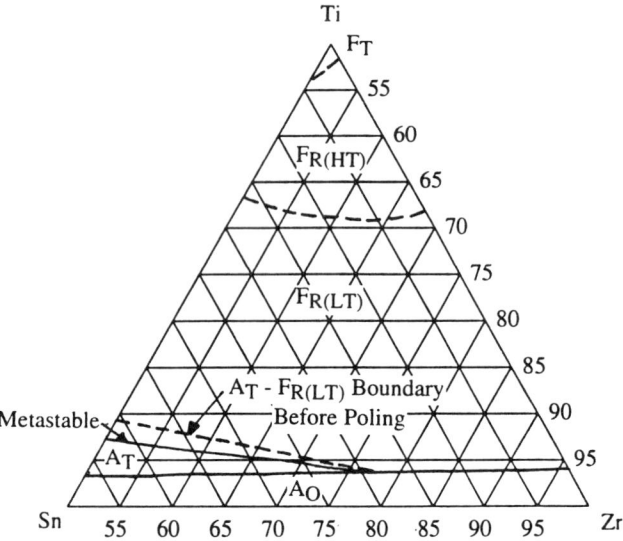

Figure 3- Phase diagram for $Pb_{0.99}Nb_{0.02}(ZrTiSn)O_3$.[4]

Material Manufacturing

In order to evaluate the response of the various material compositions, and to obtain sufficient data for modeling the material, several piezoceramic compositions have been manufactured. A list of the materials presented in this paper are shown in Table 1. Standard ceramics processing techniques were used in the manufacturing. Reagent grade oxides were mixed and ballmilled for 24 hours. The powder was then sieved through a 45μm mesh and calcined at 800°C for 10 hours. The ballmilling and calcination steps were then repeated in order to ensure complete reaction of the powder and uniformity of composition. The powder was then pressed into 1 inch diameter cylinders and isopressed to 40000 psi. Each cylinder was sintered at 1350°C in a PbO rich atmosphere for 5 hours. The cylinders were then sliced into wafers using an ID saw at Staveley Sensors Inc[*]. Each wafer was electroded using thermal evaporation of Aluminum.

Table 1 - Compositions of the electroceramics for which data is presented.

A1	$Pb_{0.97}La_{0.02}(Zr_{0.65}Ti_{0.115}Sn_{0.235})\,O_3$
A2	$Pb_{0.97}La_{0.02}(Zr_{0.67}Ti_{0.115}Sn_{0.215})\,O_3$
A3	$Pb_{0.97}La_{0.02}(Zr_{0.66}Ti_{0.105}Sn_{0.235})\,O_3$
B1[†]	$Pb_{0.99}Nb_{0.02}(Zr_{0.551}Ti_{0.0617}Sn_{0.3673})\,O_3$

† was supplied by Kenji Uchino's group at Penn. State University.

Material Characterization

Free Strain Tests

A series of tests were performed on each sample. The first of these was the measurement of transverse, and longitudinal strains in the ceramic wafers, under an applied electric field. The experimental apparatus for obtaining simultaneous transverse and longitudinal measurements is shown in Figure 4. An interferometric system was utilized to obtain longitudinal displacement of the sample, while a strain gage provided transverse strain measurements. In addition, the current supplied to the sample was monitored and integrated numerically to provide a measure of the electric displacement or polarization of the sample. Characteristic field-strain curves for three of the compositions studied are shown in Figures 5, 6, and 7. These curves correspond to an electric field with frequency of 0.05Hz, and a triangular waveform. The first two compositions (A1 and A2), display purely ferroelectric response (at 23°C). As expected for a ferroelectric material, the transverse strain curve is similar in shape to the longitudinal strain curve, but with opposite sign. A3 and B1 demonstrate antiferroelectric-ferroelectric switching, with longitudinal strains of approximately 2200με and 2700με respectively. Note that for both of these samples the strain associated with the phase transition has the same sign in the longitudinal and transverse directions.

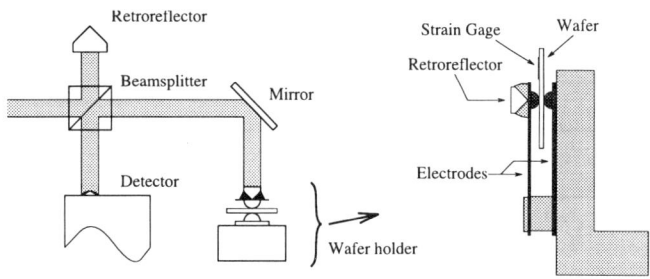

Figure 4- Experimental setup used for simultaneous measurement of transverse and longitudinal strain. An interferometer provides longitudinal measurements, while a strain gage provides transverse measurements.

---

[*] East Hartford, CT, U.S.A., Tel: (203) 289-5428

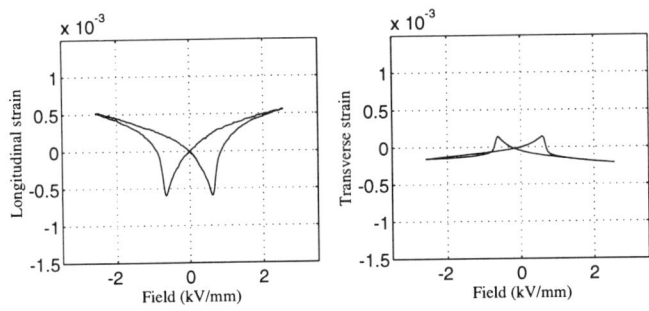

Figure 5-Transverse and longitudinal field-strain curves for A2 ceramics. A1 material, showed similar behavior.

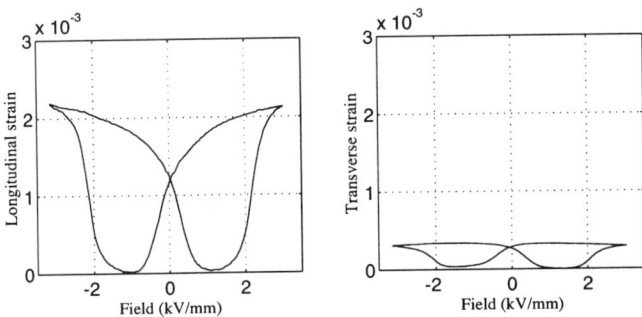

Figure 6- Transverse and longitudinal field-strain curves for A3 ceramics.

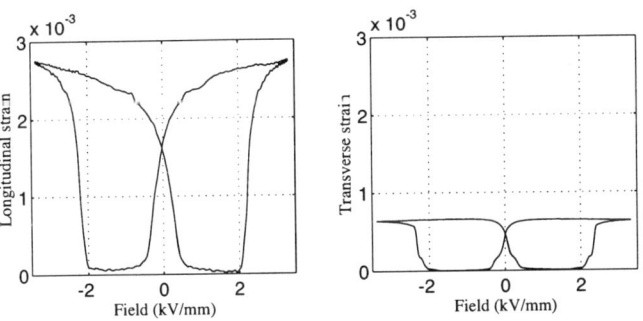

Figure 7- Transverse and longitudinal field-strain curves for B1 ceramics.

An interesting observation made during experiments was that a virgin antiferroelectric sample required much higher electric field for the first phase transition than subsequent phase transitions. For example, the A3 sample did not undergo a phase transition until the electric field exceeded 2.7kV/mm. During subsequent cycles however, phase transition occurred at approximately 2kV/mm. This phenomenon is most likely caused by the initially random polarization of the various domains.

It can be seen from the residual strain at zero field, that both compositions A3 and B1 possess some shape memory. It should be noted that the transverse strain at zero field was as high as 75% of the peak strain. The longitudinal strain at zero field however, was much lower than the peak value. This results from the combination of the anisotropic ferroelectric strain and the isotropic strain associated with the phase transition. This may be understood by considering the strains as the field is reduced from its maximum amplitude to zero. In the longitudinal direction, the strain due to the phase transition and the strain due to the ferroelectric coupling between field and strain contribute to each other. As the field is reduced, the drop in the ferroelectric portion of the strain causes a large decrease in the overall strain. In the transverse direction, the ferroelectric portion of the strain has the opposite sign. Thus as the field is reduced, the strain actually increases (this is more evident in Figure 8). In either direction there is some drop in the strain which can be attributed to a fraction of the domains switching back to the antiferroelectric phase.

In a shape memory application, one is interested in having the actuator in either the AFE or FE phase at zero field. If the ceramic is used for longitudinal actuation one has to contend with the large overshoot in strain when the phase transition to FE phase is induced. In the transverse direction however, there is much lower overshoot. Hence transverse actuation is more suitable than longitudinal actuation if use of the shape memory effect is intended.

Temperature Tests

A series of tests were performed to characterize the behavior of the materials at different temperatures. The phase of the material is a function of its state. Along with the applied electric field, it is expected that parameters such as temperature and pressure will be among the state variables. One can think of the phase boundary shifting due to a temperature change. Since the phase transition depends on the position of the specific material composition relative to the phase boundary, the behavior of the material depends on the operating temperature. This dependence has been investigated by conducting strain measurements at various temperatures in the range 0°C-25°C. Figure 8 shows the field-strain response for material B1. Several effects were observed. As the temperature was decreased, the onset of the reverse transition (i.e. from FE to AFE) was delayed. This would be an advantage in terms of shape memory. Unfortunately however, this was accompanied by a drop in the overall strain associated with the phase transition. At low enough temperatures, the phase transition was inhibited all together, and the material appeared to behave as a ferroelectric. Similar tests on A3 samples show a somewhat different behavior. As before, the strain associated with phase transition decreased as the temperature was lowered. However, a phase transition was still observed even at the lower end of the temperature range studied. The higher sensitivity of the Niobium doped composition to temperature is possibly related to the fact that the material is antiferroelectric for a smaller range of compositions (See Figures 2 and 3).

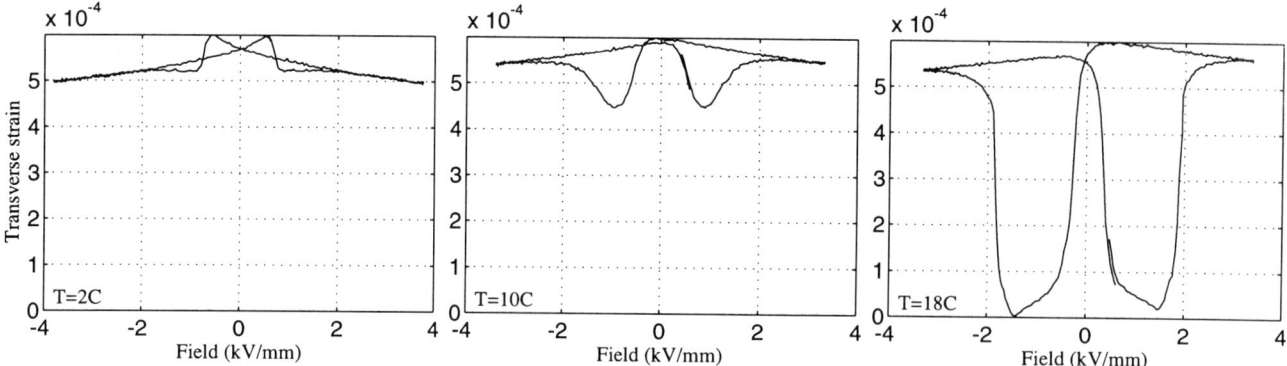

Figure 8- Field strain curves for the B1 shape memory ceramics at different temperatures.

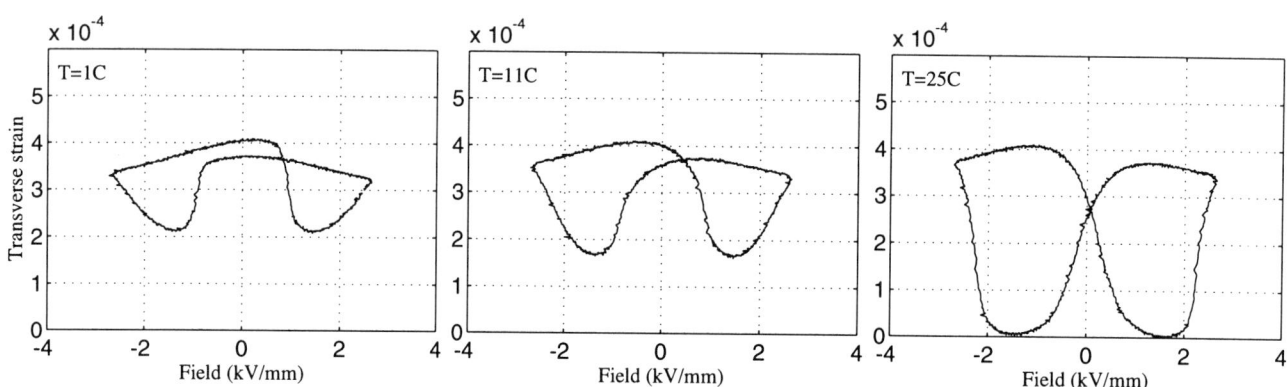

Figure 9- Field strain curves for the A3 shape memory ceramics at different temperatures.

The shape memory of the sample is seen more clearly in time domain tests. Figure 10b shows the strains in the sample when applying an electric field with the time history shown in Figure 10a. The magnitude of the positive voltage pulse was selected to force transition from AFE to FE. The negative pulse was made just large enough to cause the reverse transition. Of interest are the strain levels while the electric field is switched off. Several curves are shown corresponding to different temperatures. It is immediately obvious that over the temperature range considered, the peak transverse strain increases with temperature. However at higher temperatures the strain decays more rapidly. Figure 11 shows these effects in more detail. The top curve demonstrates the increase in peak strain with temperature. The bottom curve shows the residual strain after the field has been off for 4.5 secs. This residual strain peaks around 18°C. The rapid decay at higher temperatures can be attributed to an increase in the thermal vibrations of the lattice, which reduces the stability of the FE phase at zero field. Clearly, as far as the memory in the ceramic is concerned there is an optimal operating temperature where the magnitude of the residual strain is maximized. This optimal temperature can possibly be tailored for specific applications by making slight variations in the composition of the material.

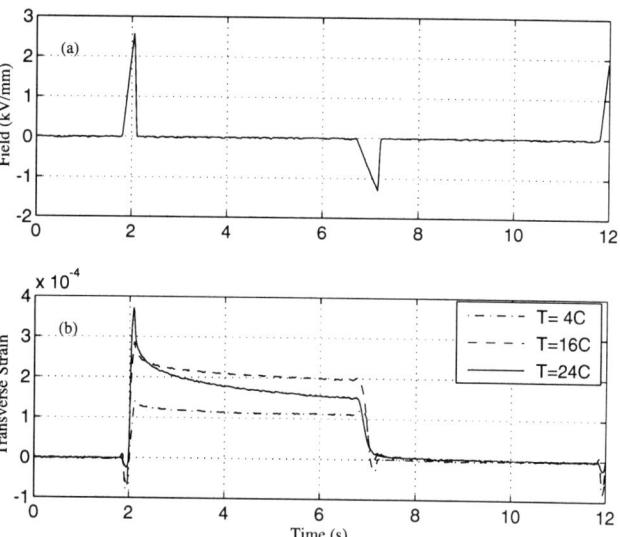

Figure 10- The transverse strains arising from the application of the specified electric field at different temperatures. The increasing strains correspond to increasing temperature.

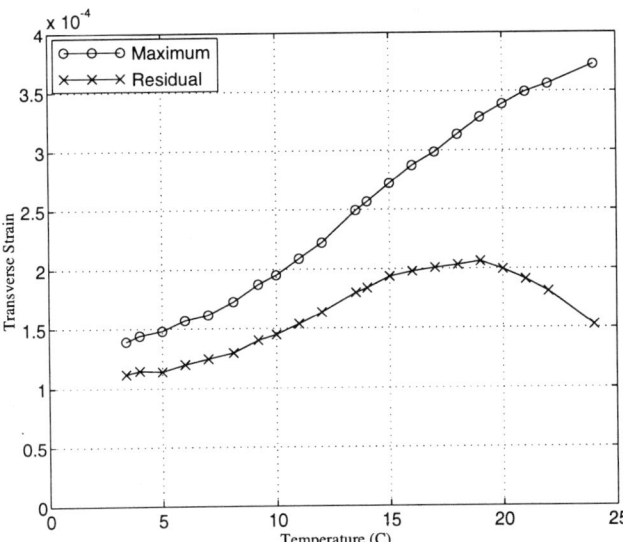

Figure 11- Transverse strains of A3 sample during application of the electric field in Figure 12a. The two curves correspond to the peak strain, and the residual strain after 4.5 seconds.

Stress Tests

Another series of tests were aimed at determining the effect of a backstress on the sample. The effect of stress on the response of the material is of great concern when the material is used in structural applications. For example, it is necessary to ensure that structural stresses do not impede the phase transition in the actuator. An 11mil thick B1 wafer was partially clamped by bonding a 40mil thick Aluminum to each side. Figure 12 shows the transverse strain of the partially clamped sample compared to one tested under free conditions. Note that due to the resulting stress in the partially clamped sample, the phase transition was not accompanied by the large strain observed under free conditions. Also, no shape memory is observed.

Figure 13 shows the observed transverse strains for A3 wafers sandwiched between metal shims of various stiffness. Curve (a) corresponds to a free wafer. Curves (b-d) are the transverse strains produced by a 15mil wafer sandwiched between metal shims with the properties listed in Table 2. An immediate observation is that under the stiffness ratios examined, phase transition was not impeded. However, Figures 12 and 13 are not directly comparable since a much higher stiffness ratio was used in clamping the B1 sample.

Table 2 - Properties of the material used in performing partially clamped wafer tests

	Material	E (GPa)	ν	t (mil)	ψ*
b	Brass	105	0.35	2.5	0.23
c	Steel	190	0.30	4	0.62
d	Aluminum	70	0.35	20	1.24

* ψ values based on actuator stiffness c^a=116GPa which was calculated as described below.

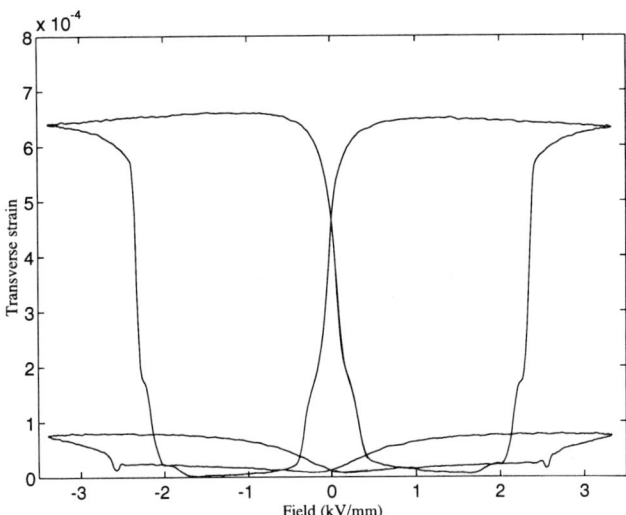

Figure 12 -Field-strain curves for wafers of sample B1 under free, and partially clamped conditions.

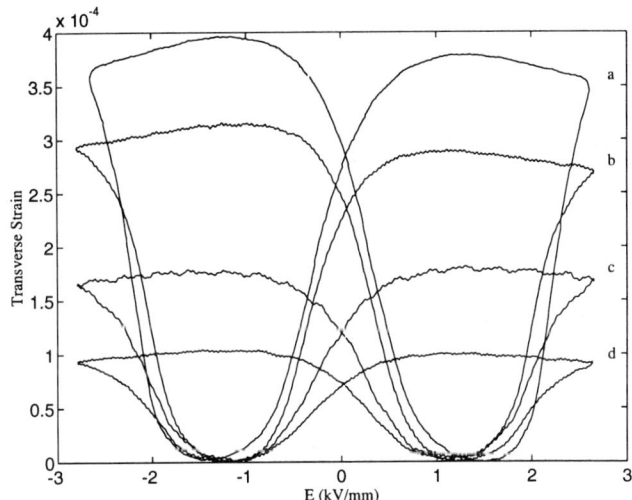

Figure 13- Field-strain curves for partially clamped wafers: (a) free wafer. (b-d) wafer sandwiched between metal shims of increasing stiffness. (See Table 2)

The data from these clamped tests can be used to approximate the stiffness of the ceramic while it is in the ferroelectric phase. The geometry for the problems under consideration are shown in Figure 14. We start with the Bernoulli-Euler assumption for a plate problem,

$$\varepsilon = \varepsilon^o + \kappa z \qquad (1)$$

Placing this into the stress-strain relation,

$$\mathbf{T} = \mathbf{c}(\varepsilon^o + \kappa z) \qquad (2)$$

A plane stress assumption is used. Thus **c** corresponds to the 3x3 stiffness matrix appropriate for plane stress. Proceeding with standard plate theory, we calculate the resultant loads and

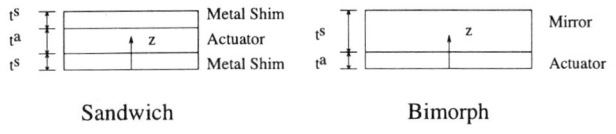

Figure 14- Cross section of sandwich and bimorph test samples. The bimorph corresponds to the adaptive mirror discussed in the section below.

moments in the plate, by integrating the stress through the thickness of the test article,

$$\mathbf{N} = \int \mathbf{T} dz = \left(\int \mathbf{c} dz\right) \varepsilon^o + \left(\int \mathbf{c} z dz\right) \kappa \qquad (3)$$

$$\mathbf{M} = \int \mathbf{T} z dz = \left(\int \mathbf{c} z dz\right) \varepsilon^o + \left(\int \mathbf{c} z^2 dz\right) \kappa \qquad (4)$$

These must balance the induced loads and moments caused by the actuator which are given by

$$\mathbf{N_e} = \int_a \mathbf{c} dz \Lambda \qquad \mathbf{M_e} = \int_a \mathbf{c} z dz \Lambda \qquad (5)$$

Here, Λ is the vector of the ceramic free strains, and the integrals are evaluated over the thickness of the ceramic only. Evaluating the integrals in equations (3), (4), and (5), which are function of geometry only, and equating the structural and induced loads and moments, we have

$$\begin{bmatrix} \mathbf{A} & \mathbf{B} \\ \mathbf{B} & \mathbf{D} \end{bmatrix} \begin{bmatrix} \mathbf{N} \\ \mathbf{M} \end{bmatrix} = \begin{bmatrix} \mathbf{N_e} \\ \mathbf{M_e} \end{bmatrix} \qquad (6)$$

The problem under consideration is symmetric in the x and y directions. This implies that the strains are symmetric. Hence,

$$\kappa = \begin{bmatrix} \kappa_x \\ \kappa_y \\ \kappa_{xy} \end{bmatrix} = \begin{bmatrix} \kappa \\ \kappa \\ 0 \end{bmatrix} \quad \varepsilon^o = \begin{bmatrix} \varepsilon^o_x \\ \varepsilon^o_y \\ \varepsilon^o_{xy} \end{bmatrix} = \begin{bmatrix} \varepsilon^o \\ \varepsilon^o \\ 0 \end{bmatrix} \quad \Lambda = \begin{bmatrix} \Lambda_x \\ \Lambda_y \\ \Lambda_{xy} \end{bmatrix} = \begin{bmatrix} \Lambda \\ \Lambda \\ 0 \end{bmatrix} \qquad (7)$$

This reduces the problem to a one dimensional problem with the effective stiffnesses for the structure and the actuator given by

$$c^s = c^s_{11} + c^s_{12} = E/(1-\nu)$$
$$c^a = c^a_{11} + c^a_{12} \qquad (8)$$

The final expressions will be simplified by defining the non-dimensional stiffness and thickness ratios as

$$\psi = t^s c^s / t^a c^a \qquad \tau = t^s / t^a \qquad (9)$$

Using the above definitions, (6) can be solved for the strains and curvatures. The equations for the sandwich problem yield

$$\kappa = 0 \qquad \varepsilon^o = \frac{\Lambda}{1+2\psi} \qquad (10)$$

This can be rearranged and written as:

$$\Lambda/\varepsilon^o = \left(2t^s c^s / t^a\right)\frac{1}{c^a} + 1 \qquad (11)$$

In this form a, least squares fit can be performed to obtain c^a. To decoupling the c_{11} and c_{12} values one would need to assume a value for ν. However, since only the sum is important in structural actuation, this is not necessary. Using the strains at E=1kV/mm from Figure 13, a value of 116GPa was calculated for the stiffness ($c_{11}+c_{12}$) of A3 samples.

Frequency Tests

The final series of tests performed on the sample involved obtaining strain data, while changing the frequency of the applied electric field. A triangular waveform was used. These tests were performed in silicon oil to prevent arcing across the electrodes at higher frequencies. Power supply current limitations restricted the maximum frequency of the tests to 10Hz. Figure 15 shows some of the field-strain curves obtained from these tests. The asymmetry of the 10Hz curve is caused by current saturation. The peak strains attained at each frequency are plotted in Figure 16. The peak strain appears to decrease almost linearly with log of frequency.

It should be noted that the hysteresis of supplied current with voltage, implies a power dissipation in the sample. This causes a rise in the sample temperature at higher frequencies, as seen in Figure 17. It would be informative to determine how much of the strain decrease observed in Figure 16 is caused by the frequency response of the material, and how much is due to the dependence of the material behavior on temperature. To accomplish this, one could compare field-strain curves obtained at different frequencies but fixed temperatures. Unfortunately, this data is not yet available.

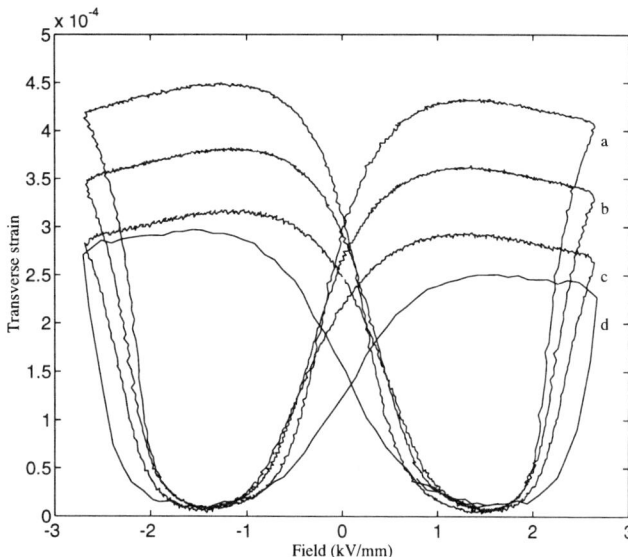

Figure 15- Field-strain curves produced by a triangular electric field of amplitude 2.6kV/mm at the frequencies: a) 0.01Hz, b) 0.1Hz, c) 1Hz d) 10Hz.

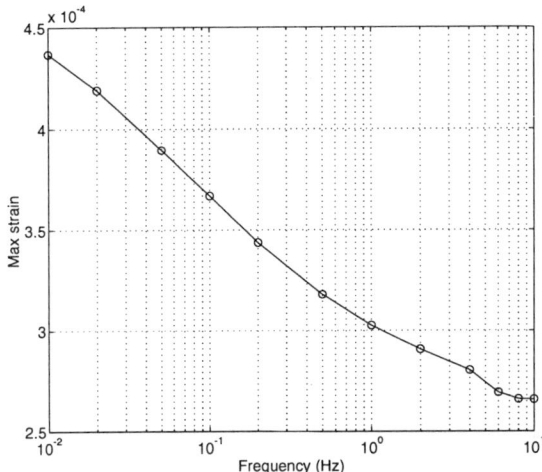

Figure 16- Peak transverse strain in the ceramic wafer produced by a triangular electric field of amplitude 2.6kV/mm at different frequencies.

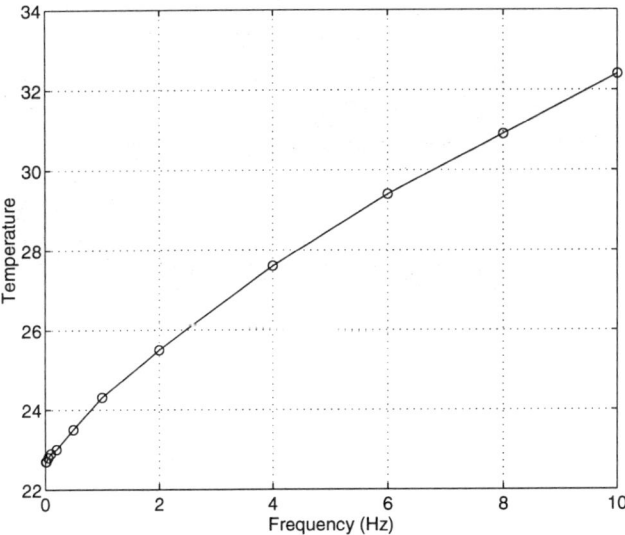

Figure 17- Increase in the temperature of the ceramic wafer caused by the power dissipation at high actuation frequencies.

Phenomenological Model

Over the years, numerous models have been proposed for describing the behavior of shape memory alloys[5,6,7,8]. These theories include completely phenomenological models as well as more theoretical models based on non-equilibrium thermodynamics. Various models have also been proposed for describing the phase transition in electroceramics[9,10]. These models take a thermodynamics approach and describe the phase transition by obtaining a phase diagram for the material. While these models are capable of predicting the phase transition, they do not deal with the rate of the transition or the relation of the transition to stress and strain in the material. A model for the behavior of ferroelectrics, accounting for the dependence of the transition on the applied field and incorporating the constitutive relationship for the material, has been developed by Chen and Montgomery[3].

The Chen-Montgomery model employs an internal variable representing the degree of polarization in the ferroelectric. This internal variable (N) is incorporated into the constitutive relation involving the traditional state variables of stress (T), strain (S), and electric field (E):

$$T = cS - eNE + a|N| \qquad (12)$$

where c, e, and a are constants. N varies between -1 and 1. It is assumed to consist of two components.

$$N = N_p + N_r \qquad (13)$$

where N_p denotes the dipoles which are permanently switchable, and N_r refers to dipoles which return to their original states after removal of the field. The evolution of each component, given an applied electric field is governed by a rate law. The rate law takes the form:

$$\frac{d}{dt}N_p = 0 \quad \text{if} \quad \begin{cases} |N_p| \geq |p\beta(E)/\alpha(E)| \\ \& \\ \text{sgn}(N_p) = \text{sgn}(p\beta(E)/\alpha(E)) \end{cases}$$

$$\frac{d}{dt}N_p + \alpha(E)N_p = p\beta(E) \quad \text{otherwise}$$

$$\frac{d}{dt}N_r + \alpha_r N_r = 0 \quad \text{if} \quad \begin{cases} |N_r| > |r\beta(E)/\alpha(E)| \\ \& \\ \text{sgn}(N_r) = \text{sgn}(r\beta(E)/\alpha(E)) \end{cases}$$

$$\frac{d}{dt}N_r/q + \alpha(E)N_r = r\beta(E) \quad \text{otherwise}$$

$$(14)$$

The curve $\beta(E)/\alpha(E)$ is effectively the steady state value of N, and $\alpha(E)$ is the rate parameter. α_r, p, q, and r are constants. The reader is referred to [3] for further discussion of these quantities. The conditions imposed on the rate law are required in order to capture the hysteretic behavior of the material. $\alpha(E)$ and $\beta(E)$, as well as the various constants in the rate law, are parameters which can be adjusted to produce experimentally observed behavior.

In order to model the phase transition in antiferroelectrics, we employ a similar approach. However, two internal variables are used, representing the polarization of each of the two sub-lattices separately (N_1, N_2). Thus N_1 and N_2 having the same sign would be representative of the ferroelectric phase. Opposite signs would be representative of the antiferroelectric phase. With this interpretation, the quantity a in Equation 12 describes the strain due to phase transition. It should be noted that this is an extreme simplification of reality. In an actual sample, various crystal domains will be either ferroelectric or antiferroelectric at any given time. The macroscopic behavior of the material represents an average over crystal domains with different orientations. Proceeding with the simplified view of the material, the net polarization of the sample can be written as

$$N = \tfrac{1}{2}(N_1 + N_2) = \tfrac{1}{2}(N_{1p} + N_{1r} + N_{2p} + N_{2r}) \qquad (15)$$

Each of the two internal variables, then follows a rate law as in the previous model but with modified α(E) and β(E) functions. The α(E) and β(E) functions used by Chen and Montgomery were even and odd functions respectively. The functions used to determine the evolution of N_1 and N_2 however are offset with respect to zero field as shown in Figure 18. If there is no stress on the sample, the strain can be calculated from (12) as,

$$S = (eNE - a|N|) / c \qquad (16)$$

Figure 19 shows an example of the result produced by the model, with the various parameters adjusted to reproduce the transverse strain of the A3 sample. The parameters used are listed below, and the $\alpha_i(E)$ and $\beta_i(E)$ used are those shown in Figure 18.

$\alpha_r=0.1 \qquad q=0.15 \qquad p=0.9 \qquad r=1-p$
$c=116 \times 10^9 \qquad e=-2.04 \qquad a=46.6 \times 10^6$

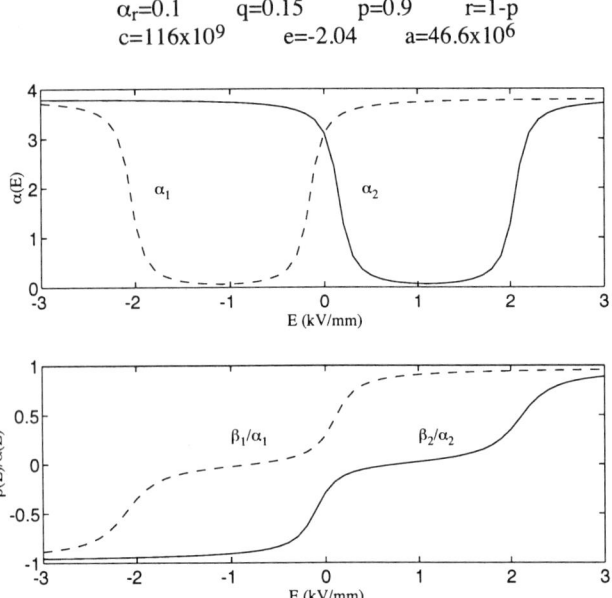

Figure 18- $\alpha_i(E)$ and $\beta_i(E)/\alpha_i(E)$ curves used in modeling phase transition.

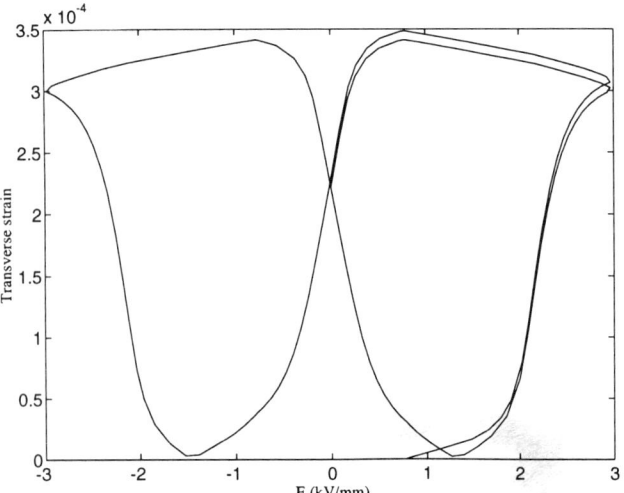

Figure 19- Example of the transverse strain generated by the model.

Adaptive Mirror Development

The knowledge gained from the material tests has been put into use in the design of an adaptive optical mirror. An array of piezoceramics bonded to the rear surface of the mirror may be used to correct deformations in the mirror (Figure 20). While a great deal of work has been done in adaptive optics, the demonstration of the benefits of using shape memory ceramics in such applications is intended. There are several advantages associated with utilizing the strain associated with phase transition. The first is the large actuation capability attained from the high strain levels. The second is the residual strain at zero field. Rather than using a large array of amplifiers to actuate each actuator individually, a single amplifier can be used to program the actuators sequentially. This simplifies the requirements on the power supply and the driving circuitry.

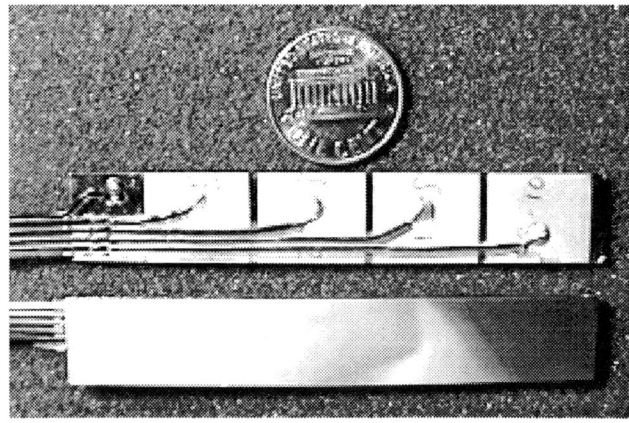

Figure 20- The front and back views of the adaptive mirror.

An even greater advantage can be gained by utilizing the nonlinearity of the antiferroelectric actuators in future applications. With the material in its antiferroelectric phase, no response is observed until the electric field exceeds the minimum required for switching. This suggests the use of a row/column addressing technique for large arrays of actuators. Figure 21 shows a mirror with a large array of actuators. Columns of actuators share the top electrode, while rows of actuators share the bottom electrodes. To activate a specific actuator, the voltage on the corresponding row and column electrodes can be set to slightly larger than +Vs/2 and -Vs/2 volts respectively, while the remaining electrodes are grounded. The actuator in question will experience a large enough field to undergo phase transition. Other actuators on the same row and column however, will only experience half of the required switching field, and will maintain their previous state. This type of row column addressing would be a major advantage when constructing adaptive systems with thousands of actuators, as required for example in adaptive telescope mirrors.

Two prototype mirrors have been constructed, each with four piezoceramic wafers mounted on the rear (Figure 20). The first mirror is actuated with PZT-5H (12.5mil thick) piezoceramic wafers, and the second with A3 (15mil thick) wafers. A 0.5 mil thick copper coated kapton sheet was bonded to the rear surface of a 7.5x1.25x0.1cm plane glass mirror, to provide a ground terminal for the ceramic wafers which were bonded on top. The ceramic wafers are 1.5x1.25cm in dimension and are separated by 1mm.

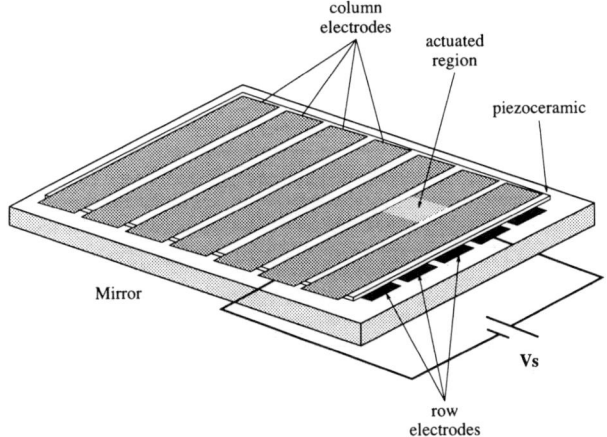

Figure 21- Schematic of row/column actuator addressing.

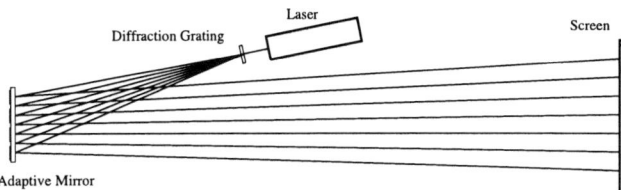

Figure 22- Experimental setup used to measure the deflections of the adaptive mirror.

The same analysis used to calculate the stiffness of the ceramic from partial clamping data, can be used to obtain a prediction of the mirror deflection. Solving equation (6) for the geometry of bimorph actuation (Figure 14) yields

$$\kappa = -6\frac{\psi(1+\tau)\Lambda}{t^a(1+4\psi+6\psi\tau+4\psi\tau^2+\psi^2\tau^2)}$$
$$\varepsilon^o = \frac{(4\psi\tau^2+9\psi\tau+6\psi+1)\Lambda}{(1+4\psi+6\psi\tau+4\psi\tau^2+\psi^2\tau^2)} \quad (17)$$

Thus the curvature induced by the actuator can be calculated based on the stiffness ratio of the mirror and actuator, their thicknesses, and the free strain of the actuator. Making the simplifying assumption that the curvature of the mirror is κ directly above the actuators, and zero in between actuators, it can be integrated to obtain the profile of the mirror.

Published stiffness values were used in the calculation of plane stress c_{11} and c_{12}, and thus c^a, for the PZT-5H wafers. The experimentally derived value was used for the A3 wafers. The relevant data for the model are listed in Table 3.

The deformations were measured by directing multiple laser beams to points on the surface of the adaptive mirror (Figure 22). The reflected beams were observed on a screen placed 3m from the mirror. Deformation of the mirror causes movement of the image points on the screen, which were recorded. Since the deflections of the mirror are quite small, only the slope of the mirror surface causes movement of the laser beams. Using a small angle approximation, it is a simple task to calculate the slope of the surface from the movement of the laser beams. The slope can then be integrated numerically to obtain the profile of the mirror. Figure 23 shows a comparison of the deflection of the PZT-5H mirror along its centerline, as determined from the model and measured experimentally. The curves correspond to the actuation of all four wafers on the mirror with an applied electric field of 0.6kV/mm. These curves serve to validate the calculations based on the experimental setup as well as the model.

Figure 24 shows the tip deflection of the A3 mirror at various applied field levels. A 30 second settling time was allowed before each measurement. The presence of shape memory is clearly observed in this application. While the PZT mirror would require a sustained electric field to maintain its deflection, the A3 mirror remains deformed even after the field has been switched off. Figure 25 shows the centerline deflection of the mirror at three different field levels. The model prediction for the maximum deflection is shown for comparison. Discrepancy between experiment and model can be attributed to the inaccuracy in material stiffnesses.

Summary & Conclusions

Lanthanum and Niobium doped piezoceramics, exhibiting antiferroelectric to ferroelectric phase transition, have been investigated for the purpose of utilizing their shape memory in structural actuation. A Lanthanum doped composition was selected which exhibited the desired shape memory. While Niobium doped compositions also provide shape memory, preliminary tests indicate that they are much more sensitive to changes in temperature.

Numerous issues important in the use of such a material in applications have been examined. The residual strain as compared with the peak strain was found to be higher in the transverse direction than in the longitudinal direction. This makes transverse actuation more suitable for shape memory based switching applications.

It was found that the observed strains are strongly temperature dependent. There is a specific range of temperatures where the material is metastable in both the ferroelectric and antiferroelectric phases at zero field. In this case, field induced phase transitions provide a mechanism for actuation. If only the ferroelectric phase is stable, the material behaves similarly to a conventional piezoelectric. This was the case with B1 material at temperatures slightly below room temperature. If only the antiferroelectric phase is stable at zero field, no shape memory

Table 3- Data used in mirror deflection predictions.

	PZT-5H	A3		Glass
t^a (mm)	0.317	0.381	t^s (mm)	1.0
c_{11} (GPa)	66.2	--	E (GPa)	65.0
c_{12} (GPa)	19.2	--	ν	0.16
c^a (GPa)	85.4	116	c^s (GPa)	77.4
Λ (µε)	220[†]	410[††]		

[†] Corresponding to E=0.6kV/mm (85% of coercive field)
[††] Corresponding to E=1.0 kV/mm

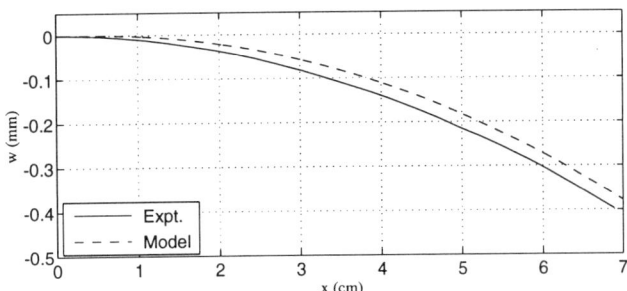

Figure 23- Centerline deflection of the PZT-5H actuated adaptive mirror, with an electric field of 0.6kV/mm applied to all four wafers.

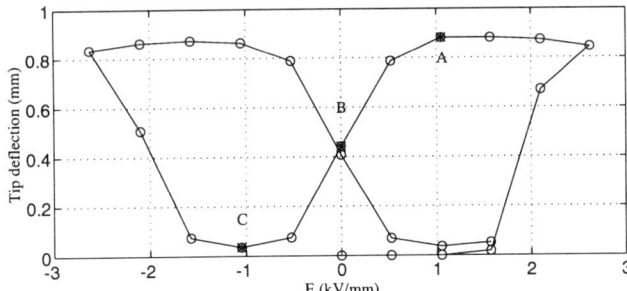

Figure 24- Experimentally determined tip deflection of the shape memory ceramic actuated adaptive mirror at different applied field levels.

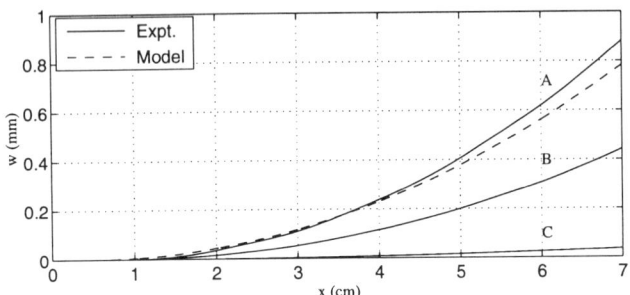

Figure 25- Experimentally determined centerline deflection of the shape memory ceramic actuated mirror. The three curves correspond to the three points labeled in Figure 24.

will be observed. It was also observed that at higher temperatures the strain decays more quickly.

The importance of investigating strain under partially clamped conditions was presented. It is necessary to ensure that structural backstress will not inhibit phase transition in the material as seen in the PNZST. In addition, clamped tests provide a means of calculating the stiffness of the ceramic. The brittle nature of ceramic wafers makes it difficult to use conventional tensile testing machines for this purpose. Another approach for stiffness measurements which is being considered is dynamic resonant testing of the ceramic wafers bonded to a beam.

A preliminary model intended to capture the antiferroelectric to ferroelectric phase transition has been presented. The model can reproduce the overall shape of the field-strain curves observed in experiments. However, attempts at reproducing the frequency dependence was found to be very difficult. Further work on the model, with explicit consideration of individual domains of differing orientation, is in progress.

The results of the present study have demonstrated the feasibility of structural actuation using shape memory ceramics. The large strains associated with phase transition provide a great deal of actuation authority. The use of the shape memory effect is possible even under structural loads.

Numerous factors need to be considered in the design of such applications. Operating temperatures, and the stiffness of the controlled structure have a significant effect on the phase transition. Furthermore, the response of the material decreases with frequency. The effect of each of these factors needs to be fully characterized to allow a successful application of these materials in adaptive structures.

Nomenclature

A,B,D stiffness weighted 0th, 1st, and 2nd moments.
c [3x3] plane stress stiffness matrix.
E Young's modulus.
E Electric field.
N internal polarization variable.
N_p permanently switchable dipoles.
N_r nonpermanently switchable dipoles.
N Resultant loads.
N_e Field induced loads.
M Resultant moments.
M_e Field induced moments.
T Plane stress vector in engineering notation.
t^s, t^a Structure and actuator thicknesses.
z Through thickness coordinate.
ε^o Plate strains along z=0.
κ Plate curvatures.
Λ Free strains in engineering notation.
τ nondimensional thickness ratio.
ψ nondimensional stiffness ratio.
ν Poisson's ratio.

References

1. L. E. Cross, "Polarization Controlled Ferroelectric High Strain Actuators", *J. of Intell. Mater. Syst. and Struct.*, **2** (Jul 1991) pp. 241-260.

2. K. Oh, A. Furuta, K. Uchino, "Field Induced Strains In Antiferroelectrics", *IEEE Ultrasonics Symposium* (1990) pp. 743-746.

3. P. J. Chen, S.T. Montgomery, "A Macroscopic Theory for the existence of the Hysteresis and Butterfly Loops In Ferroelectricity", *Ferroelectrics*, **23** (1980) pp. 199-208.

4. D. Berlincourt, "Transducers Using Forced Transitions Between Ferroelectric and Antiferroelectric States" *IEEE Trans. Sonics Ultrason.* **SU-13**, 116 (1966).

5. C. Liang, C.A. Rogers, "One-Dimensional Thermomechanical Constitutive Relations for Shape Memory Materials", *J. of Intell. Mater. Syst. and Struct.*, **1** (Apr 1990) pp. 207-234.

6. K.H. Hoffmann, "Mathematical Models of Dynamical Martensitic Transformations in Shape Memory Alloys", {em *J. of Intell. Mater. Syst. and Struct.*, **1** (Jul 1990) pp. 355-374.

7. J.S.Cory, J.L. McNichols, Jr. "Nonequilibrium Thermostatics", *J. Appl. Phys.*, **58** (9), 1 Nov 1985, pp. 3282-3294.

8. J.L. McNichols, Jr., J.S. Cory, "Thermodynamics of Nitinol", *J. Appl. Phys.*, **61** (3), 1 Feb 1987, pp. 972-984.

9. L.E. Cross, "Antiferroelectric-Ferroelectric Switching in a Simple "Kittel" Antiferroelectric", *J. of the Physical Soc. of Japan*, **23** (1) Jul 1967, pp. 77-82.

10. K. Okada, "Phenomenological Theory of Antiferroelectric Transition. III. Phase Diagram and Bias Effects of First-Order Transition", *J. of the Physical Soc. of Japan*, **37** (5) Nov 1974, pp. 1226-1232.

STRUCTURAL DAMPING AND SELF-SENSING ACTUATION IN TERFENOL-D MAGNETOSTRICTIVE MATERIALS

Ralph C. Fenn*
Michael J. Gerver
SatCon Technology Corporation
161 First Street, Cambridge, MA 02142

Abstract

Terfenol-D magnetostrictive material has been used for actuation but rarely for "reverse" transduction from mechanical to electrical energy. A Terfenol-D transducer produces intense magnetic fields when compressed which are converted into electrical energy by a surrounding coil. This electrical energy can be dissipated in a coil shunt resistor to increase structural damping. During motion, the transducer coil produces a voltage that is proportional to velocity. High Terfenol-D energy density produces high modal damping, as well as high velocity sensor resolution. Terfenol-D has several benefits including low mass, stable properties, lack of fatigue, low operating voltages, and low temperature sensitivity. Models of transducer damping and velocity sensing performance are developed from Terfenol-D properties and transducer geometry. The models are validated by experiments showing modal loss factors up to 0.22, and velocity scale factors of 183 V/(m/s). Velocity sensing is demonstrated simultaneously with actuation, and during damping operation. The feasibility of Terfenol-D transducers for passive damping and self-sensing actuation is supported.

1. Introduction

This work studies structural damping and sensing using Terfenol-D magnetostrictive material. Terfenol has been used for actuation, however, "reverse" transduction from mechanical to electrical energy has not been explored with few exceptions[1]. Terfenol-D can rapidly damp structural vibration because of its large energy density. It can be used as a primary load bearing member so that large amounts of strain energy can be concentrated in the Terfenol and dissipated, resulting in high modal damping. Terfenol can transduce fully 50 percent of strain energy into magnetic field energy because of its large coupling coefficient. When a force is applied to the Terfenol, the magnetic domains in the materials rotate and generate a magnetic field. This field cuts through the surrounding coil, generating a voltage and current. This configuration is directly analogous to the "back EMF" produced in electrical motors that can also be used as generators. Structural damping results when this electrical energy is dissipated in a resistive shunt in series with the coil. High passive damping is possible even when temperatures vary between cryogenic and room temperature because of the low temperature sensitivity. Terfenol transducers are also high resolution velocity sensors which produce large coil voltages during Terfenol straining. Co-location of the sensor with the actuation element is another advantage.

Previous work has shown promising results using piezoelectric materials for damping[2] and sensing[3]. However Terfenol has several advantages over these materials. It has higher practical energy density and is therefore lighter. Terfenol is less prone to cracking and has more stable properties. If actuation is required in addition to sensing then low voltage requirements increase reliability and safety, and reduce power supply weight.

Another conventional method of structural damping uses viscoelastic materials. This approach has other disadvantages, especially the large temperature dependence of the material loss factor. Also difficult is forming a geometry where the viscoelastic dissipation is optimized for high modal loss factors.

This work studies damping and sensing in the magnetostrictive material Terfenol through theory and experiments. Four tasks are performed: 1. Modeling of the transducer to relate Terfenol coupling coefficient, coil resistance and inductance to the capacity for damping; 2. Modeling which relates velocity sensor scale factor to the number of coil turns, Terfenol area and length, and Terfenol magnetostrictive properties; 3. Experiments to determine the material and modal loss factor and validation of the task 1 model; 4. Experiments to determine the velocity sensor scale factor and validation of the task 2 model.

Transduction from mechanical to electrical domains occurs with equal efficiency as actuation. When a force is applied to the Terfenol, the domains rotate and generate a magnetic field. This field cuts through the surrounding coil, generating a voltage and current (Fig. 1). This electrical energy that originated as strain energy can then be dissipated to provide damping or can be measured to deduce strain rate or velocity. This configuration is directly analogous to the "back EMF" produced in electrical motors that can also be used as generators.

---

*Professional Member

Copyright c 1994 by the American Institute of Aeronautics and Astronautics, Inc. All rights reserved.

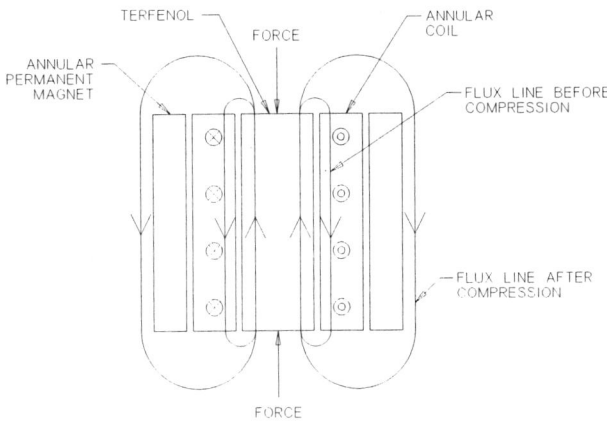

Figure 1. A magnetostrictive strut showing the expanding field caused by compression.

2. Introduction to Magnetostrictive Actuators

Magnetostriction describes the property of materials that causes them to change shape (strain) when in the presence of a magnetic field. The magnetostrictive effect was first discovered in nickel by James Joule in 1840. These strains are typically limited to approximately 50 parts per million (ppm).

Scientists at the Ames Laboratory discovered that the rare earth element terbium exhibited much larger magnetostrictive strains, greater than 1000 ppm. This element exhibits these "giant" magnetostrictive strains, however, only at cryogenic temperatures and requires very large magnetizing fields. Dr. Clark and fellow researchers at the Naval Surface Weapons Center combined the highly magnetostrictive lanthanides terbium and dysprosium with magnetic transition metals such as nickel, cobalt, and iron and achieved giant magnetostrictive strains at room temperatures. Further studies conducted in collaboration with Dale McMasters of the Ames Laboratory yielded the alloy family $Tb_xDy_{1-x}Fe_2$ (Terfenol-D) that produce giant magnetostrictive strains without the requirement of large magnetizing fields.

The important magnetostrictive properties of Terfenol-D can be seen in Figure 2. Shown are curves of magnetostriction, measured in parts per million, versus applied magnetizing field H, measured in oersteds, for various compressive stress levels in the material. The magnetostriction in Terfenol-D causes the material to increase in length when the magnetizing field is applied parallel to the material drive axis. It is evident that strains of over one part in a thousand are possible. Another important property is that magnetostrictive performance improves if the material is under compressive stress. As can be seen by the symmetry of the curves, the magnetostrictive strain depends only on the magnitude of the applied magnetizing field, not its sign. For transducer applications, the material is usually magnetically biased, for bidirectional operation and improved linearity and gain.

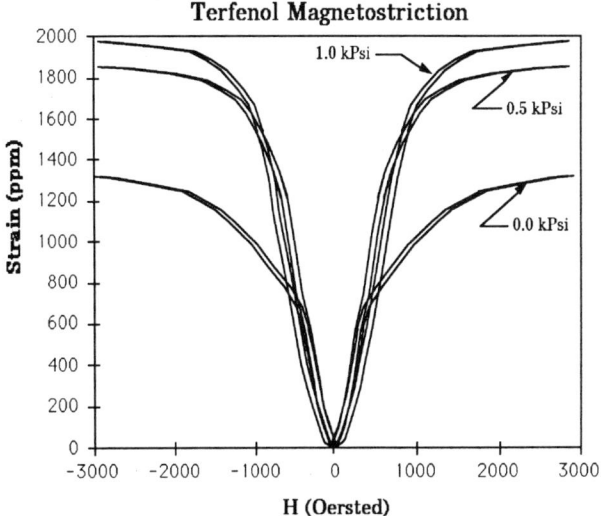

Figure 2 Terfenol magnetostriction.

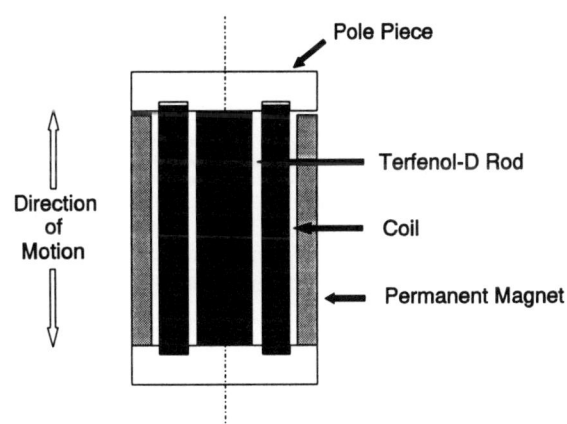

Figure 3 Typical permanent magnet biased Terfenol actuator configuration.

A typical actuator design is shown in Figure 3. The axially symmetric design features a Terfenol-D rod running down the center. The bias magnetic field is supplied by the annular permanent magnet, which is axially magnetized. Surrounding the rod is the coil, which provides the changing magnetizing field used to actively control the magnetostrictive strain. The Terfenol-D rod is normally placed under axially compressive stress by mechanical preloading not shown in Figure 3.

3. Modeling Magnetostrictive Transducers as Passive Dampers

If the coil in a magnetostrictive actuator is shorted out, with finite resistance, then it becomes a passive damper, converting mechanical energy into electrical energy which is dissipated in the coil. This occurs because the permanent magnets produce flux Φ in the Terfenol-D even without current, and this flux varies with the externally applied stress σ, because the permeability of the Terfenol-D varies with stress. The change in flux induces an emf in the coil, which drives current.

The time-averaged dissipated mechanical power may be expressed as

$$P_{mech} = \frac{\omega}{2\pi} \ell_{Terf} A_{Terf} \oint d\varepsilon \, \sigma \qquad (1)$$

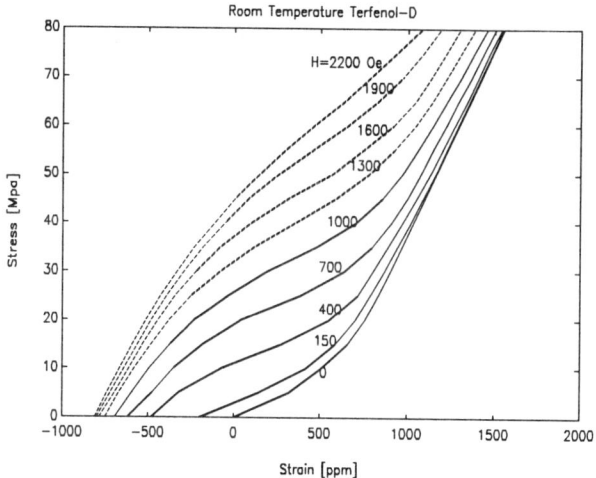

Figure 4. Anhysteretic stress σ vs. strain ε at various values of H, for Terfenol-D based on the data given in Fig. 4 of Moffett et al.[4]

where ε is the strain, ℓ_{Terf} and A_{Terf} are the length and cross-sectional area of the Terfenol-D, and ω is the frequency at which the stress is being varied. This must be equal to the time-averaged dissipated electric power

$$P_{elec} = \frac{\omega}{2\pi} \int_0^{2\pi} dt \, I^2 R \qquad (2)$$

where I is current in the coil, and R is the coil resistance. To evaluate the dissipated power for a given applied variation in stress, we write H, B, and σ as sums of equilibrium parts $H^{(0)}$, $B^{(0)}$, $\sigma^{(0)}$, and perturbed parts $H^{(1)}e^{-j\omega t}$, $B^{(1)}e^{-j\omega t}$, $\sigma^{(1)}e^{-j\omega t}$. Since there is no equilibrium current in the coil, $I = I^{(1)}e^{-j\omega t}$.

The equilibrium quantities must satisfy the zero-order equations

$$\ell_{Terf} H^{(0)} = A_{mag} B_r \mathfrak{R}_{ext} - A_{Terf} \mathfrak{R}_{ext} B^{(0)} \qquad (3)$$

$$B^{(0)} = B_{Terf}(H^{(0)}, \sigma^{(0)}) \qquad (4)$$

where $\mathfrak{R}_{ext} \approx 4/\pi \mu_0 \ell_{tot}$ is the external reluctance, and ℓ_{tot} is the actuator length, $B_{Terf}(H,\sigma)$ depends on the properties of the Terfenol-D, and may be found from the data in Fig. 4 and Fig. 5. The first order equations are

$$RI^{(1)} = -NA_{Terf} \frac{dB^{(1)}}{dt} = j\omega NA_{Terf} B^{(1)} \qquad (5)$$

$$H^{(1)} = \sigma^{(1)} \left(\frac{\partial H}{\partial \sigma}\right)_{I=0} + I^{(1)} \left(\frac{\partial H}{\partial I}\right)_{\sigma} \qquad (6)$$

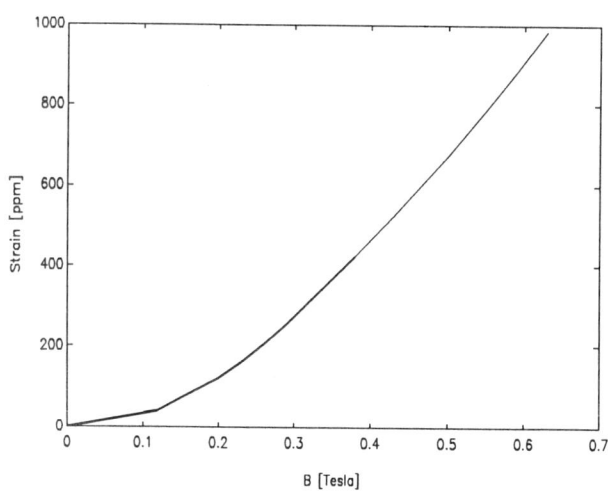

Figure 5 Strain ε vs. B, for Terfenol-D, for any stress σ > 1 ksi.

$$B^{(1)} = \sigma^{(1)} \left(\frac{\partial B}{\partial \sigma}\right)_{I=0} + I^{(1)} \left(\frac{\partial B}{\partial I}\right)_{\sigma} \qquad (7)$$

where

$$\left(\frac{\partial H}{\partial I}\right)_{\sigma} = \frac{-A_{Terf} \mathfrak{R}_{ext}}{\ell_{Terf}} \left(\frac{\partial B}{\partial I}\right)_{\sigma} - \frac{N}{\ell_{Terf}} \qquad (8)$$

$$\left(\frac{\partial B}{\partial I}\right)_\sigma = \left(\frac{\partial B_{Terf}}{\partial H}\right)_\sigma \left(\frac{\partial H}{\partial I}\right)_\sigma \equiv \mu_T \left(\frac{\partial H}{\partial I}\right)_\sigma \quad (9)$$

The Terfenol-D permeability at constant stress, μ_T, is either the initial permeability if the perturbation is small compared to hysteresis (i.e. $H^{(1)} \ll H_c$), or the anhysteretic permeability if the perturbation is large, $H^{(1)} \gg H_c$. If $H^{(1)}$ is comparable to H_c, then μ_T will be intermediate between these limits, and will have a significant imaginary part, due to hysteresis. It will also have a significant imaginary part if the frequency is high enough for skin effects to be important. Hysteresis and skin effects can both cause damping in addition to the damping caused by the resistance of the coil. We will neglect hysteresis and skin effects in this analysis, but their contribution to damping can be derived from our equations by assigning an appropriate imaginary part to μ_T. The partial derivatives $(\partial H/\partial \sigma)_{I=0}$ and $(\partial B/\partial \sigma)_{I=0}$ can be found from the data in Fig. 5, together with the data in Fig. 4 and Eq. (3).

The dissipated power is

$$P = \omega \, \text{Im}\{H^{(1)*}B^{(1)}\} A_{Terf} \ell_{Terf} \quad (10)$$

where the asterisk indicates the complex conjugate. If we assume that μ_T is real (i.e. we neglect damping due to hysteresis and eddy currents), then we find, after some algebraic manipulation,

$$P = \frac{\omega^2 F^2}{(1 + \mathcal{R}_{ext}/\mathcal{R}_{Terf})} Re\left\{\frac{N^2}{R+j\omega L}\right\}\left[\left(\frac{\partial B}{\partial \sigma}\right)_{I=0} - \mu_T \left(\frac{\partial H}{\partial \sigma}\right)_{I=0}\right] \quad (11)$$

where $F = \sigma^{(1)} A_{Terf}$ is the perturbed force, $\mathcal{R}_{Terf} = \ell_{Terf}/\mu_T A_{Terf}$ is the reluctance of the Terfenol at constant stress, and

$$L = \frac{N^2 A_{Terf} \mu_T}{\ell_{Terf} + A_{Terf} \mathcal{R}_{ext} \mu_T} \quad (12)$$

is the inductance of the coil.

The damping coefficient η is defined as $P/\omega W$ where W is the energy of the oscillation,

$$W = \frac{F^2 \ell_{Terf}}{A_{Terf}} Re\left\{\left(\frac{\partial \varepsilon}{\partial \sigma}\right)_{V=0}\right\} \quad (13)$$

We wish to express η in terms of the magnetomechanical coupling constant k, defined by

$$k^2 = d^2 \left(\frac{\partial B}{\partial H}\right)_\sigma^{-1} \left(\frac{\partial \varepsilon}{\partial \sigma}\right)_H^{-1} \quad (14)$$

where $d = (\partial \varepsilon/\partial H)_\sigma = (\partial B/\partial \sigma)_H$. To do this we must express the partial derivatives at $I = 0$ found in Eq. (11) in terms of partial derivatives at constant H and σ. We find,

$$\eta = Re\left\{\frac{\omega L}{R+j\omega L}\right\} k^2 \left(\frac{\partial \varepsilon}{\partial \sigma}\right)_H \left[Re\left(\frac{\partial \varepsilon}{\partial \sigma}\right)_{V=0}\right]^{-1} \left(1 + \frac{\mathcal{R}_{ext}}{\mathcal{R}_{Terf}}\right)^{-1} \quad (15)$$

When $\mathcal{R}_{Terf} \ll \mathcal{R}_{ext}$, then almost all of the permanent magnet flux passes through the Terfenol, and B in the Terfenol will be $B_r A_{mag}/A_{Terf}$ independent of the stress. In this circumstance there will be very little magnetomechanical coupling, even if the coupling constant k is large. In the opposite limit, the permanent magnet will deliver a constant H to the Terfenol, and B will be sensitive to stress, so there will be substantial magnetomechanical coupling if k is large. In the limit of small k, at least, when $(\partial \varepsilon/\partial \sigma)_{V=0} \approx (\partial \varepsilon/\partial \sigma)_H$, the damping is proportional to k^2. To determine the behavior when k is comparable to 1, we find a general expression for $(\partial \varepsilon/\partial \sigma)_{V=0}$.

$$\left(\frac{\partial \varepsilon}{\partial \sigma}\right)_{V=0} = \left(\frac{\partial \varepsilon}{\partial \sigma}\right)_H + \left(\frac{\partial \varepsilon}{\partial H}\right)_\sigma \left(\frac{\partial H}{\partial \sigma}\right)_{V=0}$$

$$= \left(\frac{\partial \varepsilon}{\partial \sigma}\right)_H \left[1 - k^2\left(1 + \frac{\mathcal{R}_{Terf}}{\mathcal{R}_{ext}(1 + jR/\omega L)}\right)\left(1 + \frac{\mathcal{R}_{Terf}}{\mathcal{R}_{ext}}\right)^{-1}\right] \quad (16)$$

If there is finite compressibility in the rest of the structure, in series with the Terfenol, then this compressibility can be added to the right hand side of Eq. (16). This will reduce the damping coefficient, since it appears in the denominator of Eq. (15). Taking the real part of Eq. (16) and putting it into Eq. (15), and taking the limit that $\mathcal{R}_{Terf} \gg \mathcal{R}_{ext}$, we obtain

$$\eta = \frac{k^2 R/\omega L}{1 - k^2 + R^2/\omega^2 L^2} \quad (17)$$

This is identical to the expression found by Hagood and von Flotow[2] for piezoelectrics. For a given k, it is greatest at $R/\omega L = 1 - k^2$, at which point

$$\eta = \frac{k^2}{2(1-k^2)^{1/2}} \quad (18)$$

For Terfenol-D, which has k as high as 0.7, η can be as high as 0.35, so a large fraction of the oscillation energy can be damped in one cycle.

4. Modeling Magnetostrictive Transducers as Velocity Sensors

In this Section a first-principles model is developed to predict the velocity sensing signal produced by a magnetostictive transducer. This model will be needed for future application related design of self-sensing actuators and dampers. This analysis predicts a velocity sensor scale factor of 125 V/(m/s) for the room temperature magnetostrictive strut. The observed velocity scale factor for large amplitudes was 183 V/(m/s). The similarity between prediction and observation validates the model, which can be used for design of self-sensing actuators in future applications.

An expression is derived relating the coil voltage to the strain rate and consequently velocity of the magnetostrictive material. This analysis focuses on the open coil case. However, the transducer can also be used as a velocity sensor with a shunted coil during damping, as observed in tests.

In the case of an open coil the current is zero. Therefore the voltage V across the coil is entirely due to the rate of change in flux through it, associated with the rate of change of strain of the Terfenol,

$$V = N\left(A_{Terf}\frac{dB}{dt} + \frac{\ell_{Terf}}{\Re_{coil}}\frac{dH}{dt}\right) \quad (19)$$

where B and H refer to the Terfenol. The second term involving the reluctance $\Re_{coil}$ due to the finite coil thickness, is fairly small, of order 10% or 20%, and we neglect it. The time derivative of B is related to the velocity v (the time derivative of strain ε) by

$$\frac{dB}{dt} = \frac{v}{\ell_{Terf}}\left(\frac{\partial B}{\partial \varepsilon}\right)_{I=0} \quad (20)$$

where

$$\left(\frac{\partial B}{\partial \varepsilon}\right)_{I=0} = \left(\frac{\partial B}{\partial \varepsilon}\right)_\sigma - \left(\frac{\partial B}{\partial \sigma}\right)_\varepsilon \left(\frac{\partial \sigma}{\partial \varepsilon}\right)_{I=0}$$

$$= \left(\frac{\partial B}{\partial \varepsilon}\right)_\sigma \left[1 - \left(\frac{\partial \sigma}{\partial \varepsilon}\right)_{I=0}\left(\frac{\partial \sigma}{\partial \varepsilon}\right)_{H=0}^{-1}\right] \quad (21)$$

We have used the fact that ε(B) - ε(B=0) depends only on B, to good approximation, not on σ, at least for σ in the normal operating range, as shown in Fig. 4.

Here $(\partial B/\partial \varepsilon)_\sigma$ is independent of σ, depending only on B, and may be found from Figure 4. The elastic modulus $\partial \sigma/\partial \varepsilon$ at H=0 may be found from Figure 5. The elastic modulus at I=0 is given by

$$\left(\frac{\partial \sigma}{\partial \varepsilon}\right)_{I=0} = \left(\frac{\partial \sigma}{\partial \varepsilon}\right)_H - \left(\frac{\partial \sigma}{\partial H}\right)_\varepsilon \left(\frac{\partial H}{\partial \varepsilon}\right)_{I=0} \quad (22)$$

Using Eq. (3) to find $(\partial H/\partial \varepsilon)_{I=0}$ in terms of $(\partial B/\partial \varepsilon)_{I=0}$, we find

$$\left(\frac{\partial B}{\partial \varepsilon}\right)_{I=0}\left[1 + \left(\frac{\partial B}{\partial \varepsilon}\right)_\sigma \left(\frac{\partial \sigma}{\partial H}\right)_\varepsilon \left(\frac{\partial \sigma}{\partial \varepsilon}\right)_{H=0}^{-1} \frac{4A_{mag}}{\pi \mu_0 \ell_{Terf} \ell_{tot}}\right]$$

$$= \left(\frac{\partial B}{\partial \varepsilon}\right)_\sigma \left[1 - \left(\frac{\partial \sigma}{\partial \varepsilon}\right)_H \left(\frac{\partial \sigma}{\partial \varepsilon}\right)_{H=0}^{-1}\right] \quad (23)$$

Here $(\partial \sigma/\partial H)_\varepsilon$, the clamped derivative of stress with respect to H, and $(\partial \sigma/\partial \varepsilon)_H$, the elastic modulus at constant H, may be read off the plot in Figure 5, at the equilibrium values of σ and H, given by solving Eqs. (23) and (24). The elastic modulus at H=0, $(\partial \sigma/\partial \varepsilon)_{H=0}$, is to be evaluated at the equilibrium stress σ, but at H=0 rather than at the equilibrium H. Using Eqs. (19) (neglecting the second term of the right-hand side), (20), and (23), the voltage V across the coil is related to the velocity v by

$$V = \frac{NA_{Terf}\, v\left[1 - \left(\frac{\partial \sigma}{\partial \varepsilon}\right)_H \left(\frac{\partial \sigma}{\partial \varepsilon}\right)_{H=0}^{-1}\right]}{\ell_{Terf}\left(\frac{\partial \varepsilon}{\partial B}\right)_\sigma + \frac{4A_{mag}}{\pi \mu_0 \ell_{tot}}\left(\frac{\partial \sigma}{\partial H}\right)_\varepsilon \left(\frac{\partial \sigma}{\partial \varepsilon}\right)_{H=0}^{-1}} \quad (24)$$

where: N=coil turns; A_{Terf}=Terfenol area; v=velocity; σ=Terfenol stress; ε=Terfenol strain; H=magnetic field in Terfenol; B=flux density in Terfenol; A_{mag}=permanent magnet area; μ_0=permeability of free space;

ℓ_{Terf}=Terfenol length; ℓ_{tot}=actuator length (including Terfenol and pole pieces).

Use Terfenol dimensions and material properties:
$A_{Terf} = 7.13 \times 10^{-5} m^2$ $(\partial\varepsilon/\partial B)_a \approx 2 \times 10^{-3} Tesla^{-1}$
$\ell_{tot} \approx 7 \times 10^{-2} m$ $(\partial\sigma/\partial\varepsilon)_H \approx \frac{1}{2}(\partial\sigma/\partial\varepsilon)_{H=0}$ $A_{mag} = 1.27 \times 10^{-4} m^2$ $(\partial\sigma/\partial H)_\varepsilon = 3.5 \times 10^2 Nm^{-1}A^{-1}$
$\ell_{Terf} = 5.72 \times 10^{-2} m$ $(\partial\sigma/\partial\varepsilon)_{H=0} = 4 \times 10^{10} Pa$

Substituting these parameters into Eq. (24):

$$\frac{V}{Nv} = 0.156 \; volt \cdot sec \cdot m^{-1} turns^{-1} \qquad (25)$$

For 800 turns Eq. (25) predicts 125 V/(m/s), with probably 30% uncertainty due to the Terfenol properties, which vary somewhat with bias field and stress.

The **measured** open lead voltage/velocity scale factor is 170 V/(m/s).

5. Magnetostrictive Transducer Experiments

A series of experiments were performed to determine the degree of passive damping produced by a Terfenol magnetostrictive transducer. Damping was found to be optimized by using particular coil shunt resistances as predicted by theory. Tests were also performed to measure the sensitivity of the Terfenol transducer to velocity. These tests also showed large velocity voltages even during simultaneous optimal damping shunt use. Velocity sensing was also effective during actuation.

Damping and sensing tests were performed using two magnetostrictive struts designed for space structure actuation. One used Terfenol-D optimized for room temperature use. The second strut used Terfenol-D optimized for cryogenic use at 77°K but was also tested at room temperature.

Test Hardware

Resonant Testbed

The first testbed is shown in Figure 6. This testbed concentrates the strain energy of a resonant structure in the Terfenol material where it can then be transformed into electrical energy for damping or sensing. When the transducer is mounted between a granite surface plate and a 15 pound mass, a resonant system is formed. The rate of decay of resonance in this system indicates the efficiency of energy transformation of the Terfenol transducer. While extracting structural vibration energy, the transducer also produces a voltage in the coil or shunt resistor that is directly proportional to velocity of the strut extension and compression, or strain rate.

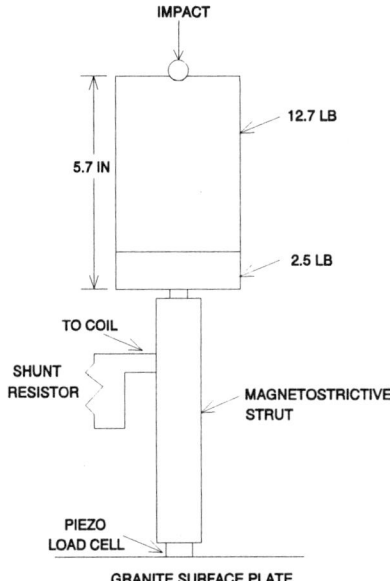

Figure 6. Apparatus used to measure preliminary strut damping and sensing properties.

Boundary Impedance Testbed

The requirements of this testbed are control of transducer velocity during actuation. Two velocity conditions were desired; zero velocity and unloaded velocity. The testbed was designed to be very stiff to enforce the zero velocity condition even when actuated. This was done by clamping the transducer ends to a milling machine bed. The second boundary condition, free expansion or unloaded velocity was created by simply releasing the clamp. Any changes in coil voltage when velocity conditions were switched are proportional to transducer velocity.

Simultaneous Passive Damping and Sensing Tests

Tests using the resonant system testbed are discussed in this section. These tests include measurements of passive damping and velocity sensing during transducer impulse responses. Both the room temperature and cryogenic Terfenol-D materials were tested for damping and sensing performance.

Damping Experiment Data

Measurements of passive damping modal loss factors were performed for three perturbations. These variations were Terfenol type (room temperature or cryogenic), shunt resistance value (open circuit or optimal resistance), and oscillation amplitude. The results of the tests using the impulse response data are summarized in Table 1. Material loss factors up to 0.23 were achieved for high amplitude vibrations of 80 μm

Table 1. Summary of damping and velocity sensing for the two magnetostrictive struts during impulse response tests.

Conditions	Loss Fctr (η)	Rel. Damp. (ζ)	Vel. Scale Fctr (V/(m/s))
Room Temp, 70 μm P-P, 5 ohm shunt	0.22	0.11	
Room Temp, 8 μm P-P, 5 ohm shunt	0.13	0.07	83
Cryo. Terf., 80 μm P-P, 8 ohm shunt	0.23	0.12	67
Cryo. Terf., 8 μm P-P, 16 ohm shunt	0.09	0.04	
Room Temp, 80 μm P-P, open coil	0.13	0.07	
Room Temp, 8 μm P-P, open coil	0.06	0.03	183
Cryo. Terf., 80 μm P-P, open coil	0.14	0.07	155
Cryo. Terf., 8 μm P-P, open coil	0.06	0.03	

peak-to-peak or 1400 microstrain peak-to-peak. Damping in the room temperature material and cryogenic material tested at room temperature were almost identical for large amplitudes.

Damping of large amplitude strains in the Terfenol formulated for optimal performance at 77°K is shown in the top plot of Figure 7. Here the material loss factor is also 0.22. The bottom plot of Figure 7 shows the coil voltage across an 8 ohm shunt resistance, which maximizes damping. The plot shows the high level voltage produced, which is proportional to strut velocity.

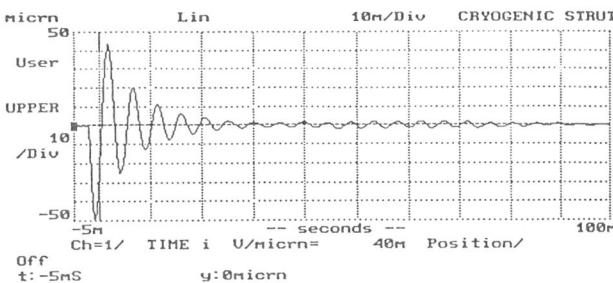

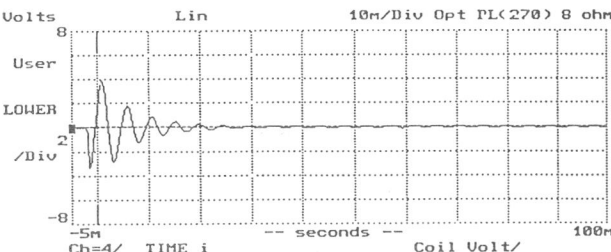

Figure 7. Impulse response of a cryogenic Terfenol strut at room temperature using the optimal shunt resistance. Material loss factor = 0.22.

It is important to note that the *modal loss factor* is equal to the *material loss factor* in these tests. Much previous work has shown material loss factors of piezoelectric appliques to have similar high material loss factors. However, the maximum damping of specific structural modes is significantly lower because the applique is not a primary structural load bearing member. Because the applique does not contain the majority of the modal strain energy it is less effective for modal damping compared to the magnetostrictive material studied here.

Lower damping at lower vibration amplitudes indicates two possible mechanisms. First, the stiffness of the joints in the resonating structure may be low for small amplitudes due to nonlinear stiffness. Such soft joints would limit energy in the Terfenol for transduction and dissipation. Another mechanism for changing loss factors could be amplitude dependent variations in the effect of hysteresis damping.

Velocity Sensor Experiment Data

During Terfenol transducer resonance tests various strain rates were produced in the material. This transducer produces a voltage proportional to strain rate and ultimately velocity. Two types of impulse response tests were performed; open coil tests and optimally shunted tests. In each type of test the voltage was simply measured. During open coil tests the coil lead voltage was measured. During shunted tests, the voltage drop across the resistor was measured. The measured voltages were then divided by the velocity which was derived from the internal position measurement sensor signal. The result was the velocity sensor scale factor which is theoretically predicted to be a constant for equivalent for equivalent shunt conditions. The values of the sensor scale factor are presented in Table 1 above with the damping data.

Measurement of Coupling Coefficient Figure of Merit from Damping

One of the most important figures of merit for transducer materials is its coupling coefficient. This number describes the efficiency of energy conversion from the mechanical domain to the electrical domain and vice versa. The coupling coefficient affects both the degree of passive damping as well as the velocity sensitivity. A value of one implies perfect conversion and complete energy domain coupling. A value of zero implies no energy conversion.

The material loss factor is determined by the coupling coefficient, therefore the measured damping can be used to estimate the coupling coefficient for comparison to expected values. The equation below relates the measured loss factor η to the magnetomechanical coupling coefficient k.

$$\eta_{max} = \frac{k^2}{2(1-k^2)^{0.5}}$$

Solving this equation with measured material loss factor = 0.22 produces an effective coupling coefficient of 0.59. This is 84 percent of the typical magnetomechanical coupling coefficient k of 0.70 for high quality, optimally biased Terfenol without mounting effects. This measurement shows that the observed damping levels are predicted by theory, and that the strut and testbed are operating properly.

Simultaneous Actuation and Sensing Tests

In the above section, passive damping and velocity sensing during damping were discussed. In this section tests of velocity sensing *during actuation* are addressed. Being able to sense the velocity of the transducer at the same time as actuation should be very useful for simplified, low mass and volume active damping systems. The transducer tested is capable of sensing its velocity, using proper circuitry, and responding to the undesirable vibration using only one pair of leads and one coil. Because the relationship between voltage and current is known, the part of the coil voltage due to the actuation current can be determined. Any added voltage that is measured is due to the transducer velocity, which arises as a "back EMF" voltage as in many other motors.

Figure 8 shows the coil voltage during both clamped and free operation. The clamped signal is the dashed plot with lower amplitude. Because the velocity is zero while clamped, there is no back EMF added to the actuation voltage. The second, larger trace shows the coil voltage when the ends are allowed to move. The back EMF, which is proportional to velocity, is added to the voltage required to maintain the desired coil current.

The difference in voltage between the two traces in Figure 8 is the voltage due to the velocity of the strut during free actuation. The difference is plotted in Figure 9. The amplitude of velocity signal is 0.52 V. The measured value of the velocity sensor scale factor during actuation was 200 V/(m/s). This value is very similar to

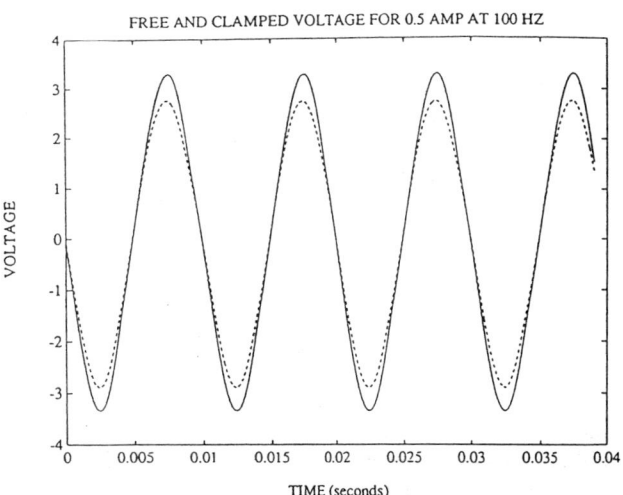

Figure 8. Coil voltage traces showing the effect of strut velocity. The larger trace is free, the smaller trace is clamped.

the measured value of 183 V/(m/s) produced during impact tests with open coil leads. Open coil leads prevents the back EMF from changing the current in the coil, just as the current controller used in the actuation tests held current constant. These results show that a signal proportional to velocity can be derived from the coil lead voltage during actuation.

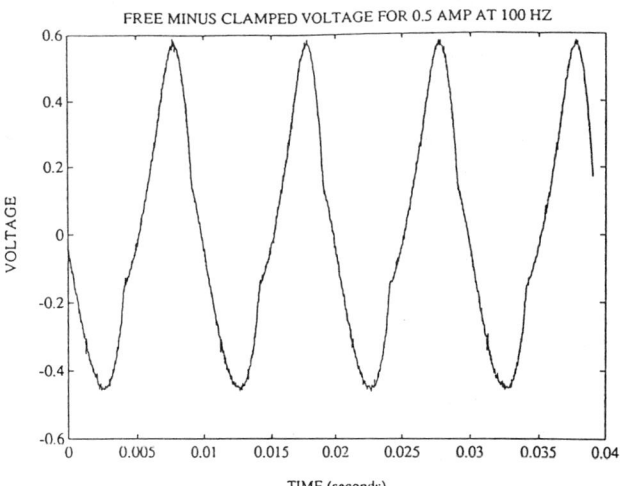

Figure 9. The difference between the coil voltage during free and clamped conditions is proportional to strut velocity.

6. Conclusions

This paper explores the use of magnetostrictive transducers for structural vibration damping and velocity sensing. High modal damping rates are predicted by theory and observed in experiments. Large velocity resolution is predicted and also observed in tests. Self-sensing of velocity during actuation was observed. Magnetostrictive transducers using Terfenol-D show promise as low mass, high reliability devices for damping and sensing in smart structures.

7. Acknowledgements

This work was conducted by SatCon Technology Corporation under Contract No. NAS8-39828 with National Aeronautics and Space Administration Marshall Space Flight Center. On behalf of SatCon, the authors would like to acknowledge the guidance provided by Mr. Bill Walker and Dr. Henry Waites.

The magnetostrictive "motors" in these struts were designed, built, and tested for the Jet Propulsion Laboratory. The loan of this equipment for tests at SatCon is gratefully acknowledged.

Any opinions, findings, and conclusions or recommendations expressed in this publication are those of the authors and do not necessarily reflect the views of NASA Marshall Space Flight Center.

8. References

1. Hall, D.L. and Flatau A.B., "Nonlinearities, harmonics, and trends in dynamic applications of Terfenol-D," *Proc. 1993 North American Congerence Son Smart Structures and Materials*, Albuquerque, NM.

2. Hagood, N.W. and von Flotow, A., "Damping of Structural Vibration with Piezoelectric Materials and Passive Electrical Networks," J. Sound and Vib., 146(2), 243-268.

3. Anderson, E.H., Hagood, N.W., and Goodliffe, J.M., "Self-sensing piexolelectric actuation: Analysis and application to controlled structures," AIAA-92-2645, Proc. 33rd AIAA SDM Conf., April, 1992.

4. Moffett, et al: "Theory of the Magnetization Process in Ferromagnets and its Application to the Magnetomechanical Effect," J. Acoust. Soc. Am. 89, 1448-1455 (1991).

AIAA-94-1760-CP

TORSIONAL ACTUATION WITH EXTENSION-TORSION COMPOSITE COUPLING AND MAGNETOSTRICTIVE ACTUATORS

Christopher M. Bothwell*, Ramesh Chandra**, and Inderjit Chopra***

Center for Rotorcraft Education and Research
Department of Aerospace Engineering
University of Maryland
College Park, MD 20742

Abstract

This paper presents an analytical cum experimental study of using magnetostrictive actuators in conjunction with an extension-torsion coupled composite tube to actuate a rotor blade trailing edge flap to actively control helicopter vibration. Thin-walled beam analysis based on Vlasov theory is used to predict the axial force-induced twist and extension in the composite tube. $[20/-70]_{2s}$ graphite-epoxy as well as $[20/-70]_S$ and $[11]_2$ kevlar-epoxy tubes were fabricated using an autoclave molding technique. The tubes were tested under static mechanical loads, and tip twist and axial extension were measured by means of a laser optical system and strain gages respectively. The tubes showed good correlation between theory and experiment for the external load case. The magnetostrictive actuator/composite tube systems were then assembled and tested. The $[20/-70]_{2s}$ graphite-epoxy tube system exhibited 0.03 degrees of tip twist in both tension and compression, while the $[20/-70]_S$ kevlar-epoxy tube resulted in a tip twist of 0.089 degrees in tension and 0.102 degrees in compression. The $[11]_2$ kevlar-epoxy tube system generated the most twist, 0.19 degrees in tension and 0.20 degrees in compression. The kevlar-epoxy systems showed good correlation between measured and predicted twist values. Further parametric studies were then performed to determine the important design variables that would result in maximum induced twist and actuator force.

Nomenclature

D = composite tube inner diameter
F = axial force applied to composite tube/ actuator force
F_{block} = actuator block force
K_{ij} = tube stiffness matrix
l = length of composite tube
M_z, M_{zs} = moment resultants referring to tube plate segment
N = axial force referring to tube
N_z, N_{zs} = stress resultants referring to tube plate segment
n, s, z = plate segment coordinate system
T_s = torsion moment referring to tube
W = axial displacement of tube
$W_{actuator}$ = actuator stroke
W_{free} = actuator free displacement
W_{tube} = axial displacement of tube
x, y, z = tube coordinate system
$\varepsilon_s, \varepsilon_z, \varepsilon_{sz}$ = strains in tube plate segment
$\varepsilon_{xz}, \varepsilon_{yz}$ = transverse shear strains for tube in xz and yz planes, respectively
ϕ_x, ϕ_y, ϕ_z = rotations about x, y, z axes
$\kappa_s, \kappa_z, \kappa_{sz}$ = bending curvatures referring to tube plate segment
$\sigma_s, \sigma_z, \sigma_{sz}$ = stress field referring to tube plate segment
φ = warping function

Introduction

Recently, active control of helicopter vibrations has received considerable attention from researchers. These vibrations are caused by unsteady aerodynamic forces occurring at a dominant frequency of $N\Omega$, where N is the

---

Presented at the 35th Structures, Structural Dynamics and Materials Conference and Adaptive Structures Forum, April 18-21, 1994.
* Graduate Research Assistant, Member AIAA
** Assistant Research Scientist, Senior Member AIAA
*** Professor and Director, Fellow AIAA, Member AHS

Copyright © 1993 American Institute of Aeronautics and Astronautics, Inc. All rights reserved.

number of blades and Ω is the rotational speed. The concept of Higher Harmonic Control (HHC) has been investigated as a possible method of actively suppressing helicopter vibration[1]. HHC reduces helicopter vibration through excitation of the blade pitch at higher harmonics of the rotational speed. The changes in blade pitch generate new unsteady airloads which act to cancel out the blade loads that caused the original vibration. A common method of HHC relies on multicyclic blade pitch control by means of swash plate actuation. However, the swash plate actuation is restricted to frequencies that are integral multiples of the rotor blades. Therefore, with an HHC system, it is not possible to actuate other frequencies that are important for rotor performance and blade life. The frequency limitation of the swash plate can be overcome with Individual Blade Control (IBC). The IBC concept, developed by Ham et al[2], employs actuators on individual blades in the rotating frame with independent control of each rotor blade, thus the actuation frequency is not limited to integer multiples of $N\Omega$. However, there is a mechanical complexity of hydraulic sliprings with an IBC system.

Straub and Robinson[3] presented an analytical study of an IBC system using a trailing edge flap concept to actively control rotor noise caused by blade vortex interaction (BVI). The study modeled the dynamics of both the rotor and the individual flap actuation. Results showed that nonharmonic flap inputs induced significant elastic torsion responses at 1/rev, 5/rev, and 6/rev. These responses can substantially increase oscillatory blade section loads and therefore may be used to reduce helicopter vibration and/or noise. Currently, a 1/4 model scale AH-64 rotor is being developed for a proof of concept test.

At the University of Maryland, the use of piezoceramic technology for vibration suppression in IBC systems is being investigated. Two piezoelectric actuation applications currently being explored are blade twist induced by embedded piezoceramics and trailing edge flap actuation by means of piezoelectric bimorphs. Chen and Chopra[4-5] presented an experimental study of a dynamically-scaled rotor blade embedded with piezoceramic elements for the purpose of vibration control. Banks of piezoelectric crystals at +/- 45 degree angles were embedded into the top and bottom surfaces of the blade. Equal and in-phase potentials were applied to the top and bottom piezoelectric elements inducing a twist distribution along the blade. Testing of a 6 ft diameter Froude-scaled rotor model on the hover test facility showed that the magnitude of twist achieved was approximately 20 times less than the twist needed for significant vibration control of the rotor. Samak and Chopra[6] performed a feasibility study on the development of a Froude-scaled rotor model in which vibrations were suppressed by means of trailing edge flap actuation. Here the flap was actuated by piezoceramic "bimorphs" (two piezo-elements with a brass shim in the middle). If opposite potentials are applied to each element, a pure bending motion is caused. Through the use of a hinge and leverage arrangement, the bending motion generated by three bimorphs lined-up side by side induced amplified deflection of the trailing edge flap. Dynamic testing of the rotor on the hover stand showed flap deflections of the order of 2 degrees at a tip speed of 258 ft/sec (RPM = 900), approximately one quarter of the deflection necessary for significant vibration suppression (for a full-scale rotor blade, a flap deflection of the order of ± 8 degrees is needed to effectively reduce vibration). Currently, improved bimorphs are being incorporated into this model to increase the flap deflection[7]. Major limitations with both these approaches are low actuator force and small induced displacement.

The use of magnetostrictive elements in a blade trailing edge flap design was investigated by Fenn et al[8]. In this analytical study, pairs of magnetostrictive terfenol-D rods were mechanically linked to a control rod in a "Y" configuration where tension or compression induced in the terfenol-D caused linear motion of the control rod. Since the control rod was attached to a trailing edge flap, the strain in the magnetostrictive elements induced actuation of the flap. This study was based on the UH-60A helicopter and the theory predicted that three pairs of magnetostrictive rods driving four flaps per blade would provide the ± 2 degree total blade deflection necessary to effectively reduce helicopter vibration. However, an experimental validation of this study has yet to be carried out.

This paper examines the concept of using magnetostrictive actuators in conjunction with an extension-torsion coupled tube to actuate the twisting motion of a trailing edge flap. Figure 1 shows a schematic drawing of this concept. The axial force generated by a magnetostrictive actuator induces elastic twist in the coupled composite tube. This concept is especially suited for full-scale applications as the twist induced in the tube increases with length. This study focuses on accurately modeling the twist induced in the composite tube. Beam theory developed in Ref. [9] is used to predict the axial force-induced twist in a thin-walled coupled composite tube. Then, composite tubes are fabricated, followed by testing under mechanical loads as well as magnetostrictive-induced actuation to validate both the analysis and actuation concept.

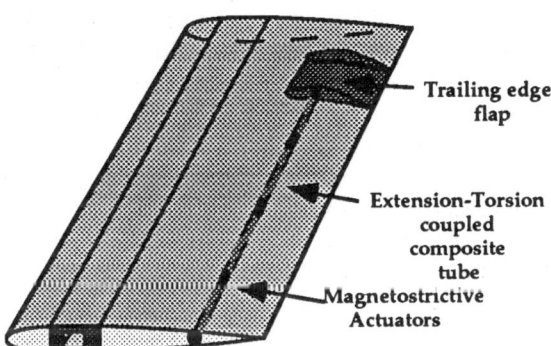

Figure 1: Schematic Drawing of Proposed Trailing Edge Flap Actuation Concept

Extension-Torsion Coupled Beam Analysis

In this paper, the beam analysis presented in Ref. [9] is specialized to predict the axial displacement and induced twist of an extension-torsion coupled composite circular cylindrical tube due to an axial force. The thin-walled analysis takes into account effects of transverse shear, section warping (Saint Venant torsion) and constrained warping at the edges (Vlasov torsion). The essence of this analysis is that two-dimensional stress and displacement fields associated with any shell segment of the tube are reduced to the generalized one-dimensional beam displacements and forces (Vlasov theory). Moreover, the development of an extension-twist coupling relationship is the key component for this study.

The elements making up the composite tube are idealized as thin laminated composite plates. The coordinate systems used to describe the tube are shown in Figure 2. A Cartesian coordinate system (x, y, z) is used for the tube, while an orthogonal coordinate system (n, s, z) describes a general plate segment where the n axis is normal to the mid surface of any plate segment, the s axis is tangential to the mid surface, and the z axis is along the longitudinal axis of the composite tube. Further, in-plane warping of the cross-section is neglected and both the surface normal strain ε_s and normal stress σ_s are assumed small compared to the axial strain ε_z and axial stress σ_z respectively. Thus, the non zero strains and bending curvature for the shell segment are ε_z, ε_{sz}, κ_z, and κ_{sz}.

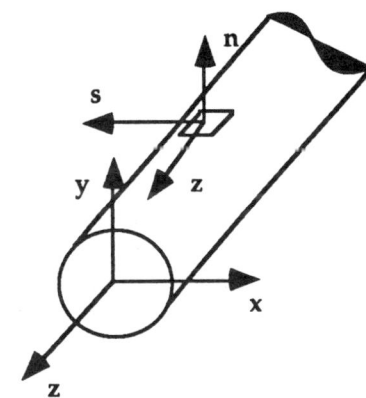

Figure 2: Cartesian coordinates in the composite tube

The shear strain ε_{sz} consists of two components; one due to transverse shear effects and the other due to torsion. Hence, ε_{sz} is given by:

$$\varepsilon_{sz} = \varepsilon_{xz} \cos\theta + \varepsilon_{yz} \sin\theta + \varepsilon_{sz}^{(s)} \quad (1)$$

Using relations for displacement from Ref. [9], ε_z is obtained as:

$$\varepsilon_z = W' + x\phi'_x + y\phi'_y - \varphi\phi''_z \quad (2)$$

Similarly, the laminate curvatures κ_z and κ_{zs} are found to be:

$$\kappa_z = -\phi'_x \sin\theta + \phi'_y \cos\theta - \phi''_z q + \varepsilon'_{xz} \sin\theta - \varepsilon'_{yz} \cos\theta \quad (3)$$

$$\kappa_{zs} = -2\phi'_z \quad (4)$$

Using classical laminated plate theory, the stress resultants and moment resultants are:

$$\begin{aligned} N_z &= A_{11}\varepsilon_z + A_{16}\varepsilon_{zs} + B_{11}\kappa_z + B_{16}\kappa_{zs} \\ N_{zs} &= A_{16}\varepsilon_z + A_{66}\varepsilon_{zs} + B_{16}\kappa_z + B_{66}\kappa_{zs} \\ M_z &= B_{11}\varepsilon_z + B_{16}\varepsilon_{zs} + D_{11}\kappa_z + D_{16}\kappa_{zs} \\ M_{zs} &= B_{16}\varepsilon_z + B_{66}\varepsilon_{zs} + D_{16}\kappa_z + D_{66}\kappa_{zs} \end{aligned} \quad (5)$$

The generalized beam forces are obtained from the plate forces. The force-displacement relations for an extension-torsion coupled tube are:

$$\begin{Bmatrix} N \\ T_s \end{Bmatrix} = \begin{bmatrix} K_{11} & K_{15} \\ K_{15} & K_{55} \end{bmatrix} \begin{Bmatrix} W' \\ \phi'_z \end{Bmatrix} \quad (6)$$

where K_{11}, K_{15}, and K_{55} are respectively the extension, extension-torsion, and torsion stiffness coefficients as defined in Ref. [9]. Thus, the axial displacement and tip induced twist in an extension-torsion coupled uniform tube under axial force are obtained as:

$$W = \frac{K_{55}}{K_{11}K_{55} - K_{15}^2} Fl \quad (7)$$

$$\phi_z = \frac{K_{15}}{K_{11}K_{55} - K_{15}^2} Fl \quad (8)$$

where F is the applied force and l is the length of the composite tube.

Testing of Coupled Tubes Under Mechanical Loading

Fabrication of Composite Tubes

The kevlar-epoxy and graphite epoxy composite tubes are fabricated using an autoclave molding technique. Unidirectional prepregs are laid-up on the split metal mold shown in Figure 3. For the first round of testing, thermally stable lay-ups of [20/-70]$_S$ kevlar-epoxy and [20/-70]$_{2S}$ graphite-epoxy were chosen in order to prevent the tube from constricting around and adhering to the mold during the curing process. Later, a [11]$_2$ kevlar-epoxy tube was attempted and successfully removed from the mold indicating that temperature-induced stresses were less of a problem than anticipated for this particular tube lay-up. The plies are sealed into a vacuum bag and a pump is used for further compacting of the layers. After laminating the desired number of plies, peel ply is wrapped to provide surface finish to the tube. Bleeder, breather and barrier plies are added to control epoxy flow during the cure process. The lay-up is again sealed in a vacuum bag and cured in a microprocessor-controlled autoclave. The cure cycle and pressure given by the prepeg manufacturer are used. At the end of the cure, the lay-up is removed from the autoclave. Finally, the vacuum bag is removed and the tube is released from the mold.

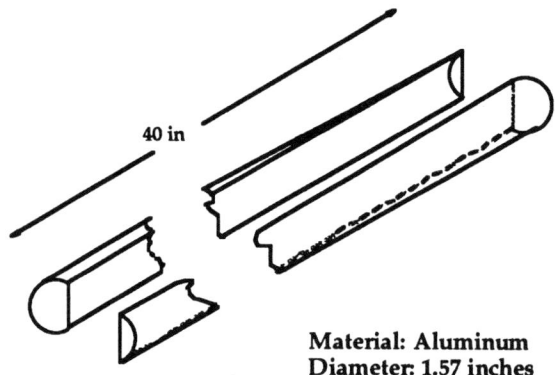

Material: Aluminum
Diameter: 1.57 inches

Figure 3: Split mold for composite tube fabrication

Experimental Setup

Two end pieces are attached to the composite tube using a casting metal alloy which liquefies at approximately 200°F and re-solidifies at room temperature. The end pieces are necessary to mechanically link the tube to the actuator assembly as well as to fix one end of the tube for external load tests.

The twist of the tube is measured using a laser optical system. One tube end is bolted to rectangular aluminum bar which is secured in a vise. External force is applied to the free end of the tube by means of a pulley and dead weight. The laser beam is reflected off a small mirror attached to the free end of the tube approximately 30 feet onto a wall. Displacement of the beam on the wall is converted to twist using trigonometric relations. In this particular setup, 1.0 inch of displacement on the wall corresponds to 0.083 degrees of induced twist in the tube.

The axial extension of the composite tube is measured using a pair of Micro-Measurements precision strain gages aligned along the longitudinal axis of the tube. The composite tube is allowed to hang free and dead weight is again used to apply an external axial load. The strain gages are placed diametrically opposite on the tube to eliminate any bending-induced strain in the response. An electronic strain indicator gives the gage output in microstrain which is converted to axial displacement.

The induced tip twist of each of the two thermally stable tubes are measured for external loads of up to 100 lb in increments of 10 lb. In the extension tests, dead weight is applied in 20 lb increments to 100 lb. The extension test is repeated four times for the kevlar-epoxy tube and six times for the graphite-epoxy tube to ensure data accuracy. The results of the induced twist and axial displacement tests are presented in Figures 4-7.

Results

Figures 4 and 5 show tip twist versus external load for the kevlar-epoxy and graphite-epoxy cantilevered tubes respectively. Predictions for both tubes correlate well with measured values. A small error in the kevlar-epoxy tube plot at higher loads may be due to wrinkles on the tube which occurred during the fabrication process. Tube wrinkles can be eliminated with an outer-mold, but for this study the correlation between theory and experiment appears adequate.

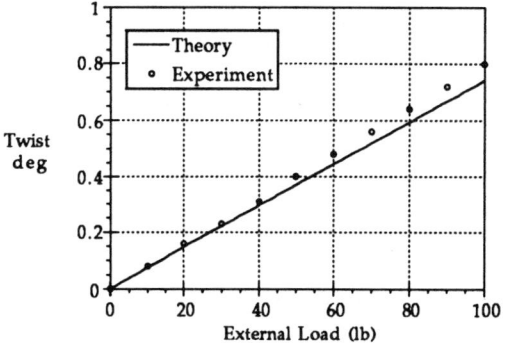

Figure 4: Tip twist vs. external load for [20/-70]$_S$ kevlar-epoxy tube (L = 34.375 in, D = 1.57 in)

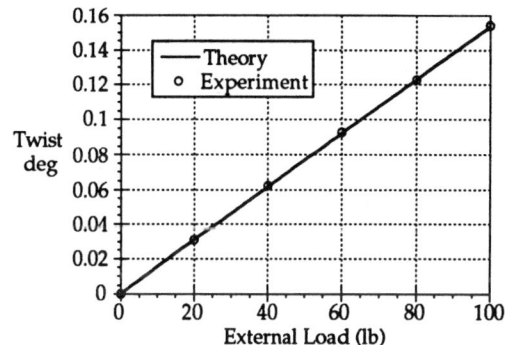

Figure 5: Tip twist vs. external load for [20/-70]$_{2S}$ graphite-epoxy tube (L = 20.0 in, D = 1.57 in)

Figures 6 and 7 show axial extension versus external load for the kevlar-epoxy and graphite-epoxy tubes respectively. Once again there is good agreement with theory. Predicted longitudinal extensions for the graphite-epoxy tube are within 10% of measured values, while for the kevlar-epoxy tube, theoretical axial displacements are within 20% of the experimental measurements. Again, the wrinkles in the kevlar tube may be the cause for most of the difference between theory and experiment.

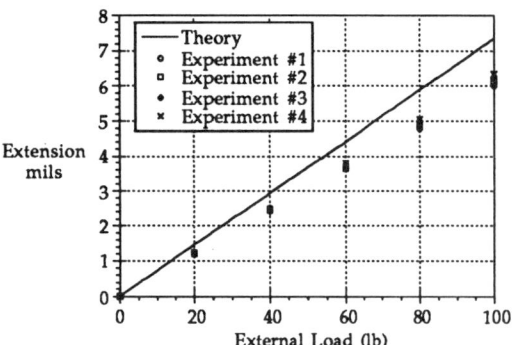

Figure 6: Extension vs. external load for [20/-70]$_S$ kevlar-epoxy tube (L = 34.375 in, D = 1.57 in)

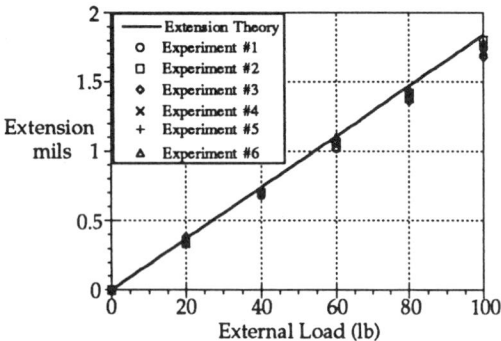

Figure 7: Extension vs. external load for [20/-70]$_{2S}$ graphite-epoxy tube (L = 20.0 in, D = 1.57 in)

Analysis: Extension-Torsion Coupled Tube with Actuator

The force (F) and displacement (W) of the magnetostrictive actuator for a known excitation voltage are represented in linear form as:

$$F = F_{block}(1 - \frac{W}{W_{free}}) \qquad (9)$$

where F_{block} is the block force (zero displacement condition) and W_{free} is the free displacement of the actuator.

Actuator Force-Displacement Curve

The values of F_{block} and W_{free} were determined experimentally for a known voltage field using the setup in Figure 8. A lever arm is used to amplify the force and axial displacement of the actuator. A known load is applied by means of dead weight and a Kaman KDM-7200 proximity sensor measures deflection at the end of the lever arm taking advantage of the amplification of the actuator motion. Since the leverage ratios are known the force and displacement of the actuator can be calculated from the weight and deflection at the end of the lever arm.

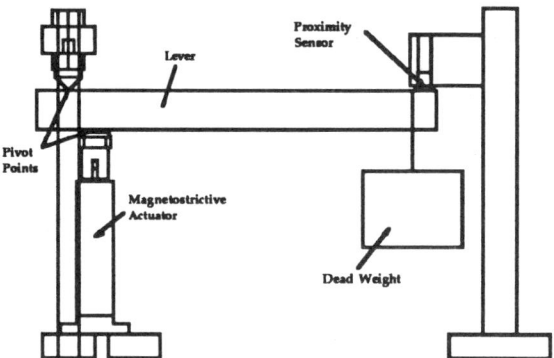

Figure 8: Setup for determining actuator force-displacement curve

Figure 9 shows actuator force versus displacement for electric currents of 1.0 amp (4.0 Volts) and 1.5 amps (6.0 Volts). The magnetostrictive actuator used in the study is the 100/6-MP from Etrema Products Inc., and 1.5 amps is the maximum input current recommended by the manufacturer.

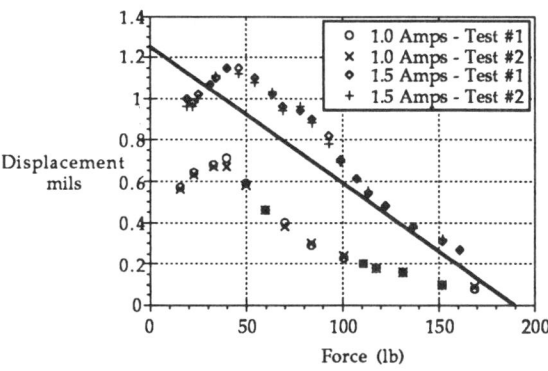

Figure 9: Experimentally determined force vs. displacement curve for 100-MP magnetostrictive actuator

A line of best fit is drawn to the data to determine F_{block} and W_{free} for the 1.5 amp case, and the tests are repeated to ensure consistency. From Figure 9, the actuator block force is estimated to be about 190 lb and the free displacement is approximately 1.25 mils for a current of 1.5 amps.

Actuator/Tube System

In order to introduce the actuator characteristics into the analysis of the combined actuator/tube system, it is assumed that the axial displacement of the actuator equals the corresponding axial extension in the composite tube, i.e.

$$W_{actuator} = W_{tube} = W \quad (10)$$

By combining equations (7), (9), and (10) the axial extension of the composite tube system is:

$$W = \frac{K_{55}F_{block}W_{free}l}{(K_{11}K_{55} - K_{15}^2)W_{free} + K_{55}F_{block}l} \quad (11)$$

Once the axial displacement is known, the induced twist can be rewritten in terms of equation (11) as:

$$\phi_z = \frac{K_{15}}{K_{55}} W \quad (12)$$

Testing of Coupled Tube/Actuator System

Experimental Setup

In order to utilize the extension-torsion coupling properties of the composite tube, the magnetostrictive actuator must be attached to the tube in such a manner that both twist and axial strain are allowed. Figure 10 shows the schematic drawing of the composite tube and actuator assembly. In this design, two end pieces are integrally connected to the ends of the composite tube by means of metal casting. The dummy rod is used to lengthen the actuator for composite tubes longer than 6 inches (in this study the graphite-epoxy tube is 20.0 inches long and the kevlar-epoxy tube is 34.375 inches long). On the tube assembly's fixed end, the base is bolted to the tube end piece, while a separate bolt secures the dummy rod to the base piece, keeping the end rigidly constrained. At the tip of the magnetostrictive actuator (i.e. the push rod) an adapter with an end nut lengthens the push rod allowing for the addition of two thrust bearings and a plug. These bearings are essential to the design as they allow the plug to rotate freely when the magnetostrictive actuator applies an axial force to the system. Since the plug is bolted to the tube end piece, the composite tube will twist as the plug rotates.

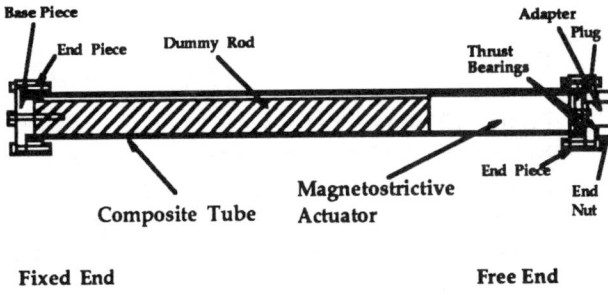

Figure 10: Schematic drawing of composite tube and actuator assembly

One end of the tube/actuator assembly is bolted to rectangular aluminum bar which is secured in a vise. The twist and extension of the tube are again determined using the laser optical system and strain gages respectively.

Each composite tube assembly is tested at an electric current level of 1.5 amps to produce the largest induced twist and axial extension possible. The 100/6-MP actuator has the capability of bi-directional motion. As shown in Figure 11, the actuator is made up of a terfenol-D rod surrounded by a coil of wire. An electric current in the coil induces a magnetic field which causes extension of the terfenol-D. The permanent magnet introduces an initial strain to the rod allowing the actuator to compress when a negative DC current is applied. Tests are performed for the actuator in tension (positive DC current) and compression (negative DC current).

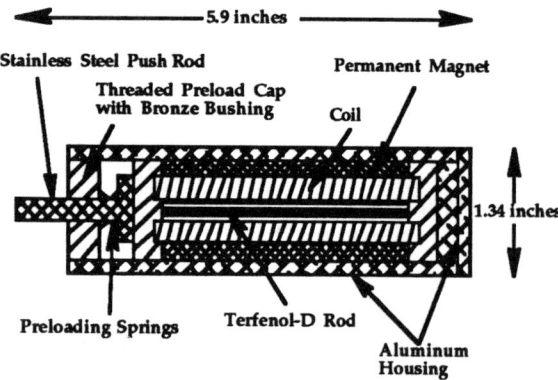

Figure 11: Schematic drawing of 100/6-MP magnetostrictive actuator from Etrema Products Inc.

Results

Figure 12 shows the results from the induced twist tests. The graphite-epoxy tube system measured twists of 0.030 degrees and 0.032 degrees in tension and compression respectively, about 40% of the theoretical value for both cases. This represents an axial force of about 20 lb. The kevlar-epoxy tube system measured twist values of 0.089 degrees in tension and 0.102 degrees in compression. Both values are close to the predicted 0.111 degrees, within 20% and 10% respectively. This translates into an axial force of 11 lb in tension and 12 lb in compression.

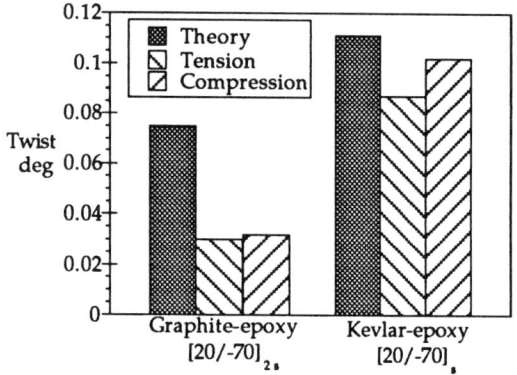

Figure 12: Twist comparison for graphite and kevlar tube/actuator assemblies

Figure 13 shows the axial extension test results. The graphite-epoxy tube system measured displacements of 0.33 mils in tension and -0.31 mils in compression, again about 40% of the theoretical value for both cases. The kevlar-epoxy tube had extension values of 0.76 mils and -0.79 mils in tension and compression respectively. Both values are about 70% of the theoretical prediction of 1.11 mils.

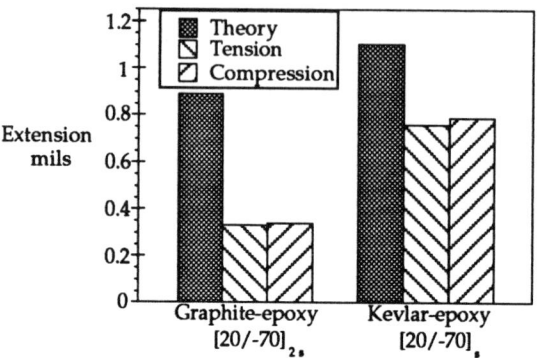

Figure 13: Extension comparison for graphite and kevlar tube/actuator assemblies

For both composite tube actuator systems, the experimental values are less than the theoretical values because of losses in the assembly. Because the axial extension predicted is on the order of one mil or less, sloppiness of the system and a slight misalignment of the actuator or dummy rod can be quite appreciable and cause error, especially as the actuator force increases. Therefore the difference between theory and experiment of the graphite-epoxy tube system is more than that of the kevlar-epoxy tube, since the predicted actuator force for the graphite-tube is about 50 lb, while the predicted force in the kevlar-epoxy tube is only 15 lb.

Tube Design Optimization

Up to this point, the experiment and analysis are carried out on a baseline thermally stable tube design, that is $[20/-70]_S$ for kevlar-epoxy and $[20/-70]_{2s}$ for graphite-epoxy. Since there is a reasonable correlation between theory and experiment (at least for the kevlar-epoxy tube/actuator system), the tube design can be optimized in order to maximize induced twist for the given actuator characteristics.

Composite Tube Ply Angle Studies

Figures 14 and 15 show tip twist versus ply angle (all plies have the same ply angle) for kevlar-epoxy and graphite-epoxy tube/actuator systems respectively. The number of plies is varied from 2 to 8 in increments of 2, and the tube length is 34.375 inches, same as the length of the above kevlar-epoxy tube. It is clear that a kevlar-epoxy tube system, made up of two 11 degree plies will generate a twist magnitude of 0.20 degrees, over 80% more than the thermally stable kevlar-epoxy lay-up. The graphite-epoxy tube system will generate a maximum twist of 0.17 degrees with 2 plies, about 15% lower than the maximum twist of kevlar-epoxy tube.

Figures 16 and 17 illustrate the corresponding variation of force with ply angle and number of plies, again for a tube 34.375 inches long. In an actual trailing edge flap application, in addition to a high induced twist, a large force in the tube is desired to counteract the aerodynamic forces exerted on the flap. From the results, it is clear that large displacement and therefore twist is achieved at the expense of actuation force. It is interesting to note that in changing from 2 ply to 8 ply kevlar-epoxy (ply angle = 11 degrees) the twist decreases less than 25 % while at the same time the force increases from 15 lb to almost 60 lb. This suggests that for full-scale applications increasing the number of plies may be advantageous.

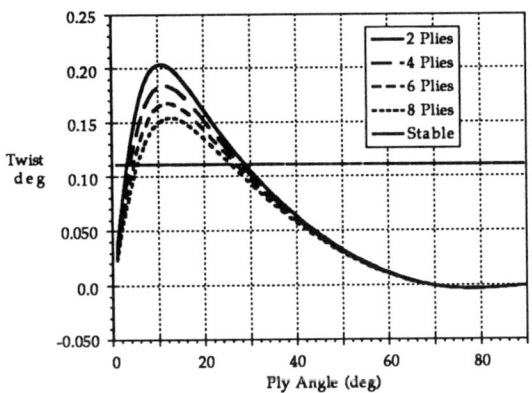

Figure 14: Tip twist with ply angle and number of plies for kevlar-epoxy tube system (l = 34.375 in, D = 1.57 in)

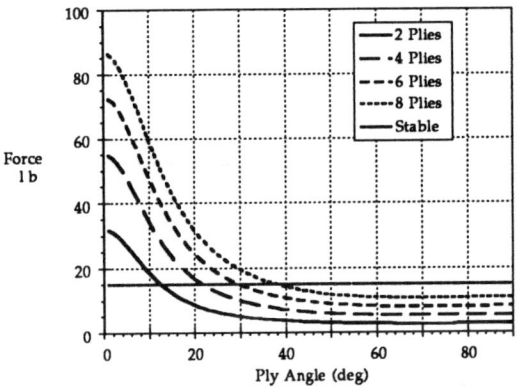

Figure 16: Axial force vs. ply angle vs. number of plies for kevlar-epoxy tube system (l = 34.375 in, D = 1.57 in)

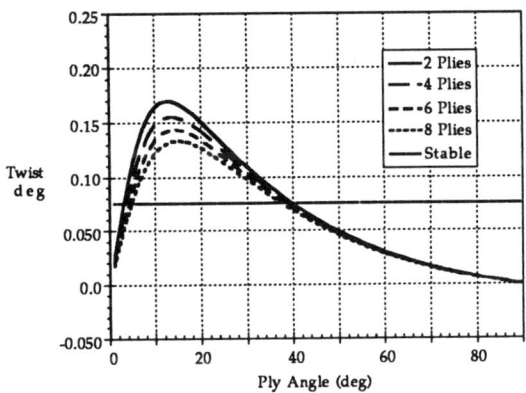

Figure 15: Tip twist with ply angle and number of plies for graphite-epoxy tube system (l = 34.375 in, D = 1.57 in)

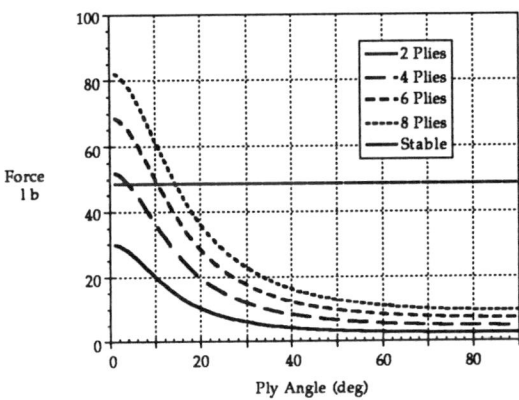

Figure 17: Axial force vs. ply angle vs. number of plies for graphite-epoxy tube system (l = 34.375 in, D = 1.57 in)

Composite Tube Length Studies

Figures 18 and 19 show predicted twist and force respectively versus tube length. Data are plotted for two ply, four ply, and the thermally stable kevlar-epoxy tube systems. As the tube length increases, there is a large increase in induced twist until the length becomes about 15 inches, after that the change is small. The actuator force decreases rapidly with length of the tube. Therefore, for a full-scale application where the total tube length required is approximately 200 inches, the key may be to use many smaller tube sections (say of 15-20 inch lengths) instead of a few longer sections to keep the force level above 50 lb.

Composite Tube Diameter Studies

Figures 20 and 21 present induced twist and axial force versus tube inner diameter. Once again the two ply, four ply, and thermally stable kevlar-epoxy tube cases are shown. The inner diameter is limited to 1.4 inches because the outer diameter of the magnetostrictive actuator is 1.34 inches. As the tube diameter becomes larger, the tip twist decreases while the axial force increases. The slopes of both curves are of opposite sign, indicating that in varying the diameter there is a trade-off between twist and force.

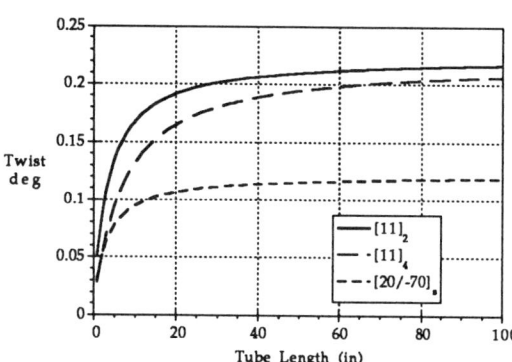

Figure 18: Tip twist vs. tube length for various kevlar-epoxy tube systems (D = 1.57 in)

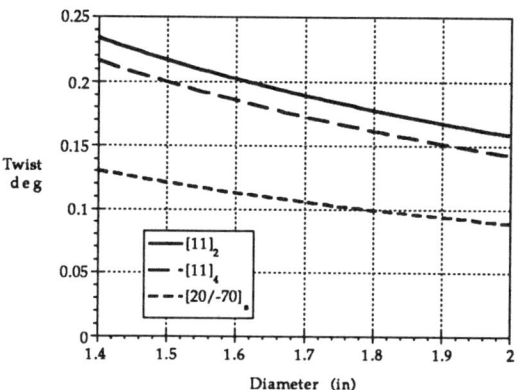

Figure 20: Tip twist vs. tube inner diameter for various kevlar-epoxy tube systems (l = 34.375 in)

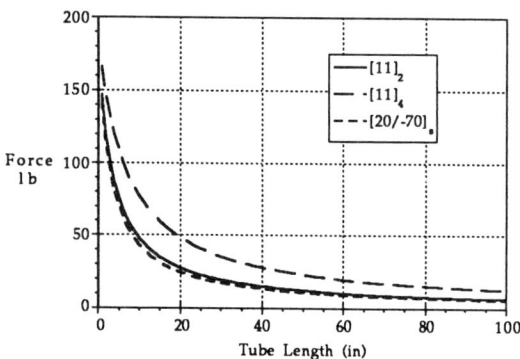

Figure 19: Axial force vs. tube length for various kevlar-epoxy tube systems (D = 1.57 in)

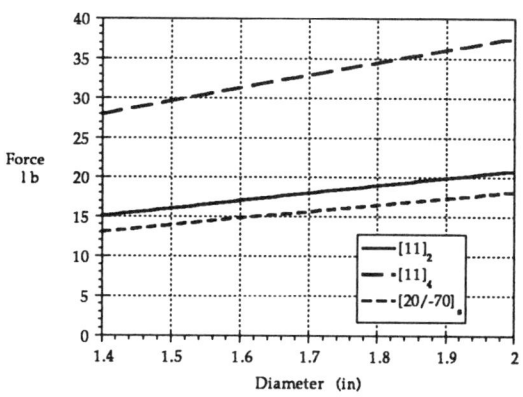

Figure 21: Axial force vs. tube inner diameter for various kevlar-epoxy tube systems (l = 34.375 in)

Experiment

A [11]$_2$ kevlar-epoxy tube is fabricated and tested with the actuator in both tension and compression, where the twist and extension are again measured using the laser optical system and strain gages.

Results

Figure 22 shows the results from the induced twist tests. In both the tension and compression case, correlation with theory is quite good with induced twists of 0.19 degrees in tension and 0.20 degrees in compression. For the [11]$_2$ kevlar-epoxy tube/actuator system 0.20 degrees translates into about 16 lb of force.

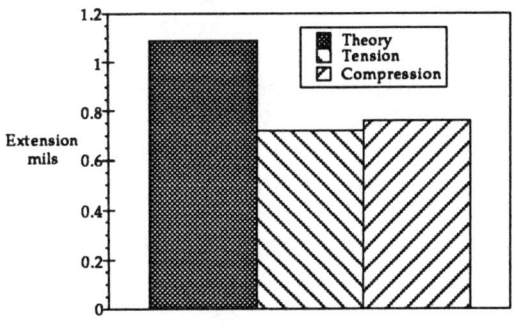

Figure 23: Axial extension experiment results for [11]$_2$ kevlar-epoxy tube system (l = 34.375 in, D = 1.57 in)

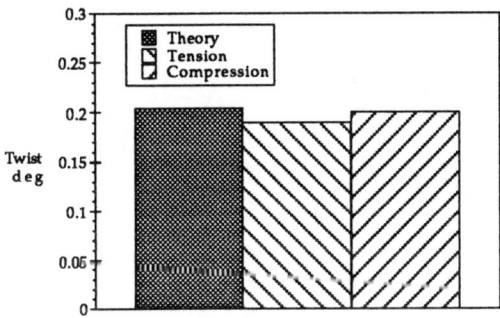

Figure 22: Twist experiment results for [11]$_2$ kevlar-epoxy tube system (l = 34.375 in, D = 1.57 in)

Figure 23 shows the axial extension test results. The measured displacement of the tube in tension is 0.72 mils in tension and -0.76 mils in compression. Both values are approximately 70% of the value predicted by theory. This general under-prediction of extension is similar to earlier test results of the [20/-70]$_s$ kevlar-epoxy tube shown in Figure 13.

Conclusions

Graphite-epoxy and Kevlar-epoxy tubes were tested under mechanical loading as well as magnetostrictive-induced actuation. The analytical model correctly predicts the induced twist and axial displacement in the graphite-epoxy tube for the case of mechanical loading. The twist and extension for the mechanically loaded kevlar-epoxy tube are predicted adequately, with a slight difference at higher loads. When the actuator's force-displacement characteristics are introduced, the model appears satisfactory in predicting the induced twist of the kevlar-epoxy tube/actuator systems (within 20% of measured values), but is quite off in the prediction of twist for the graphite-epoxy tube system (over-prediction of more than twice the measured values). There is a considerable difference between predicted and measured values of extension for both tubes (within 30% of measured values for the kevlar-epoxy tubes and over two times the measured values for the graphite-epoxy tube). The correlation with theory improves for cases where the actuator force is low, suggesting that sloppiness and misalignment of the assembly may be the cause of this large difference. Improving the mechanical linkages in the actuator/tube assembly may improve accuracy for higher actuator force designs.

Tube design studies were performed to show the effect of varying ply angle, length, number of plies, and diameter on predicted tube twist and axial force. A 20 inch long kevlar-epoxy tube system made up of four 11 degree plies resulted in the best balance of induced twist and axial force (approximately 0.17 degrees of twist and 48 lb of axial force respectively) of the configurations considered.

For a full scale flap application, the length of the system can be as large as 200 inches, the axial force in the system must be high enough to counteract aerodynamic loads, and the flap deflection should be of the order of ± 8 degrees in order to effectively reduce vibration. Based on the current best tube/actuator system design, ten 20 inch $[11]_4$ kevlar-epoxy tube/actuator systems in series would generate approximately ± 1.7 degrees of flap deflection and 48 lb of axial loading. Future research will focus on optimizing the tube/actuator design based on full-scale deflections and aerodynamic forces.

Acknowledgments

This work was supported by the Army Research Office under the grant DAAL 03-92-G-0121 with Dr. Gary Anderson as Technical Monitor. The authors would also like to thank Mr. Kwok Yu and Mr. Jon Velapoldi for their assistance in manufacturing and testing of composite tubes.

References

[1] Chopra, I. and McCloud III, J. C., "A Numerical Simulation Study of Open-Loop, Closed-Loop and Adaptive Multicycle Control Systems," *Journal of the American Helicopter Society*, Vol. 28, (1), Jan. 1983, pp. 311-325.

[2] Ham, N., Behal, B. and McKillip, R., Jr., "Helicopter Rotor Lag Damping Augmentation Through Individual Blade Control," *Vertica*, Vol. 7, No. 4, 1983, pp. 361-371.

[3] Straub, Friedrich K., and Robinson, Lawson H., "Dynamics of a Rotor with Nonharmonic Control," *Proceedings of the 49th Annual American Helicopter Society Forum*, May 19-21, 1993. St. Louis, Missouri.

[4] Chen, Peter C. and Chopra, I., "A Feasibility Study to Build a Smart Rotor: Induced-Strain Actuation of Airfoil Twisting Using Piezoceramic Crystals," *SPIE's North American Conference on Smart Structures and Materials*, February 1993, Albuquerque, New Mexico.

[5] Chen, Peter C. and Chopra, I., "Induced Strain Actuation of Composite Beams and Rotor Blades with Embedded Piezoceramic Elements," *SPIE's North American Conference on Smart Structures and Materials*, February 1994, Orlando, Florida.

[6] Samak, Dhananjay K. and Chopra, I., "A Feasibility Study to Build a Smart Rotor: Trailing Edge Flap Actuation," *SPIE's North American Conference on Smart Structures and Materials*, February 1993, Albuquerque, New Mexico.

[7] Walz, Curtis and Chopra, I., "Design and Testing of a Helicopter Rotor Model with Smart Trailing Edge Flaps," *Proceedings of the 35th AIAA/ASME/ASCE/AHS/ASC Structures, Structural Dynamics and Materials Conference*, Paper #94-1767, Hilton Head, South Carolina, April 18-21, 1994.

[8] Fenn, Ralph C., Downer, James R., Bushko, Dariusz A., Gondhalekar, Vijay., Ham, Norman D., "Terfenol-D Driven Flaps for Helicopter Vibration Reduction," *SPIE's North American Conference on Smart Structures and Materials*, February 1993, Albuquerque, New Mexico.

[9] Chandra, R. and Chopra, I., "Structural Behavior of Two-Cell Composite Rotor Blades with Elastic Couplings," *AIAA Journal*, Vol. 30, No. 12, 1992, pp. 2914-2921.

ACTIVE DAMPING OF A FLEXIBLE BEAM USING ER FLUID ACTUATORS *

Norman M. Wereley [†]
Department of Aerospace Engineering
University of Maryland, College Park, Maryland 20742

Abstract

Advanced rotor systems including hingeless and bearingless rotors have air and ground resonance instabilities due to coalescence of low frequency rotor modes with landing gear and fuselage modes, respectively. The rotor blade is connected to the rotor hub via a flexure or flexbeam. The modal coalescence is of difficulty due to the lack of a clear hinge in these advanced rotor systems. We are currently exploring the mitigation of this modal coalescence through the use of active damping techniques and electro-rheological fluid technology. An emulation of a flexbeam was fabricated from a glass epoxy composite, and filled with ER fluid, to determine feasibility of using ER fluid for active damping of the flexbeam.

Introduction

Advanced rotor systems such as soft hingeless and bearingless rotors are mechanically less complex than articulated rotors. However, this is accomplished by eliminating various hinges that have been traditonally used to reduce vibratory hub loads. The rotor blade is connected to the rotor hub via a flexure or *flexbeam*, instead of a mechanical hinge. These rotors are designed as soft in-plane rotors (lag frequency is less than rotational frequency), so that rigid body airframe modes can interact with rotor lag modes. As a result, air and ground resonance instabilities have emerged as major problems in these advanced rotors [1]. Ground resonance is characterized by the coalescence of landing gear or support modes, and a low frequency lag mode, while the helicopter is on the ground. Air resonance is characterized by the coupling of low frequency flap and lag modes with rigid body airframe modes while the helicopter is airborne. Although damping levels have been shown to be quite low (ranging from 1 to 2 percent of critical depending on trim condition [2]), damping has been shown to be one of the major stabilizing influences, so that all helicopters have mechanical lag dampers. At low rotor speeds, the lag mode damping tends to be the lowest, so that any added lag mode damping would provide benefits when the rotor is spooling up to its operational RPM. However, active damping may provide improved control over this instability, especially in the case of soft hingeless and bearingless rotors.

Electro-rheological (ER) fluids exhibit dramatic reversible changes in viscosity when an electric field is applied [3]. This behavior is similar to that of elastomeric dampers where the damping varies nonlinearly as a function of frequency and amplitude. By modulating the electric field that is applied across two electrodes sandwiching an electro-rheological fluid, the viscosity of the ER fluid between the electrodes can be increased, and the level of damping may be controlled. Thus, actuators based on ER fluids are a candidate for active damping and stability augmentation of the lag mode for both ground and air resonance.

This paper describes work in progress at the University of Maryland to improve lag mode damping of a rotor through application of electro-rheological smart material technology to elasto-dynamically tailor the damping in a flexbeam.

Bearingless Rotors

In order to decrease the mechanical complexity of the rotor due to the mechanical swashplate and pitch, lag, and flap hinges, advanced rotor concepts based on hingeless and bearingless rotor concepts will be used in future helicopters. In place of the lag hinge will be a flexure, or flexbeam, that is intended to permit motion of the rotor blade in the lag direction in much the same way that current lag hinges permit lag motion. However, the flexbeam will sustain large twist deflections.

Typical cross-sections that are currently being considered for application as a flexbeam include the rectangular, I and cruciform cross-sections as illustrated

---

*Copyright ©1994 by the American Institute of Aeronautics and Astronautics, Inc. All Rights Reserved. Presented at the AIAA/ASME Adaptive Structures Forum, 21-22 April 1994, Hilton Head, SC.

[†]Senior Member AIAA

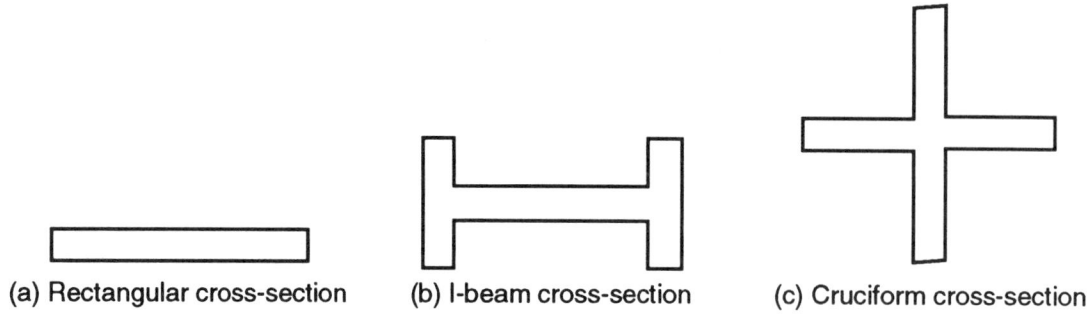

Figure 1: Typical rotor flexbeam cross-sections.

in Figure 1.

ER Fluid Approach

ER fluids are colloidal suspensions that undergo changes in properties when subjected to an electric field. They exhibit a significant increase in viscosity when a field strength of the order of 3 kV/mm is applied. The particles suspended in the fluid tend to fibrillate and form chain–like structures between the electrodes along the direction of the electric field. These structures inhibit the free flow of ER fluids. By varying the applied electric field, the viscosity can be controlled. The behavior of ER fluids under the influence of an electric field is complex. ER fluids behave like Newtonian fluids under zero field conditions. With the field applied, ER fluids behave similarly to Bingham plastics, exhibiting a finite yield stress with the shear stress dependent on the shear rate [4]. Thus, for field strengths on the order of 3 kV/mm, ER fluids 'solidify' with static and dynamic yield stresses as high as 10 kPa and 5 kPa respectively [5]. This reversible liquid to solid transition has been shown to be useful in many engineering applications. Moreover, these changes could be reversed in time intervals on the order of a few milliseconds, thus, offering excellent actuation capabilities.

The properties mentioned above make ER fluids an attractive choice for use in a wide range of engineering devices [6] such as clutches, valves and engine mounts. ER fluids have excellent potential for use in active vibration control. An ER fluid's controllable viscosity has lead to its use in the construction of various kinds of dampers [7, 8, 9], typically used as discrete dampers in a vibration control strategy. These damping applications have typically explored the substitution of ER fluids for hydraulic fluids. Successful attempts have been made in distributed vibration control by using ER fluid as a viscoelastic layer [11, 10, 12, 13, 14].

Gandhi *et al* [13, 14] constructed beams with ER fluids sandwiched between two composite layers. Graphite epoxy (AS4/3501-6) served both as a structural member and the electrode needed to apply the electric field. Four smart beam specimens were constructed with varying properties, as follows: (1) ER fluid used, (2) volume fraction of ER fluid used, and (3) lay–up of the composite layers. The beams were mounted in the cantilevered condition and static tests performed to determine the load-deflection characteristics. From the data presented, it appears that the flexural rigidity (EI) of the beam specimens in the zero-field condition varied between 0.06 N-m^2 and 0.3 N-m^2. The transient response characteristics of the beam were then studied with varying electric field strength. These results showed increments in the damping ratio and natural frequency of up to 45 and 25 percent, respectively.

Coulter *et al* [12] have also explored the use of ER-fluids to tailor the modal parameters of a fluid-filled beam. Experiments with simply supported aluminum beams were performed and the results compared with the theoretical predictions based on the Ross-Kerwin-Ungar (RKU) model [15]. Their experimental results showed an appreciable increase in modal frequencies and loss factors as the electric field applied to the ER fluid sandwiched by the aluminum beams is increased.

This approach of using ER fluid filled flexbeams to vary the frequency and the damping characteristics may be appropriate for use in rotor flexbeams. Thus, by varying the applied electric field, the coalescence of modes could be averted and the existing vibrations damped out. An important caveat is that rotor flexbeams have flexural rigidities of a much higher order of magnitude than that of the beams used in the work mentioned above. Hence, the need to perform experiments with beams of much higher stiffnesses is paramount. This would give us a realistic estimate of the damping and frequency shift that could be achieved with the existing ER fluid technology as applied to the flexbeam problem.

Fluid-filled Flexbeam

The initial experiments reported in this paper attempt to characterize how ER fluid can augment damping in an emulation of a rotor flexbeam. Essentially, a composite cantilevered beam is filled with ER fluid, and the natural frequency and damping ratios of the fundamental mode were experimentally measured as the electric field was varied. The logarithmic decrement was used to calculate the damping ratio based on a measured time response to an intial deflection of the free end of the beam. The test configuration is shown in Figure 2.

The structural beam elements are fabricated from a glass epoxy composite. The copper electrode was deposited on the glass-epoxy layer, and was exposed to the ER fluid. Two types of beams were tested. The first class of beams were designated as HERB (Hollow Electro-Rheological Box-beam), and the second class of beams were designated as SIBER (SIlicone-spaced Beam with Electro-Rheological fluid). A DC electric field was supplied by a TREK 609-A 10kV 2mA amplifierin in of the all cases reported in this paper. Four beams were tested:

HERB-2-DC-364-MC: A box beam was formed from two plates, each of nominal 3/64 inch thickness, with a nominal 2 mm spacer of glass epoxy composite. This beam was filled with a mixture of mineral oil and corn starch (50 percent each by volume). For this beam, $EI = 7.97\text{N-m}^2$ was measured. The volume fraction of ER fluid was approximately 40 percent.

SIBER-2-DC-364-MC: A box beam was formed from two plates, each of nominal 3/64 inch thickness, with a nominal 2 mm spacer of silicone. This beam was filled with a mixture of mineral oil and corn starch (50 percent each by volume). For this beam, $EI = 1.50\text{N-m}^2$ was measured. The volume fraction of ER fluid was approximately 40 percent.

HERB-1-DC-264-VF: The second box beam was formed from two plates, each of 1/32 inch thickness, with a nominal 1 mm spacer of glass epoxy composite. This beam was filled with commercially available Lord Corporation Versa-Flo 100 ER fluid. For this beam, $EI = 1.62\text{N-m}^2$ was measured. The volume fraction of ER fluid was approximately 40 percent.

SIBER-1-DC-264-VF: A box beam was formed from two plates, each of 1/32 inch thickness, with a nominal 1 mm spacer of silicone. This beam was filled with commercially available Lord Corporation Versa-Flo 100 ER fluid. For this beam, $EI = 0.475\text{N-m}^2$ was measured. The volume fraction of ER fluid was approximately 40 percent.

The properties of each beam are summarized in Table 1.

The HERB class of box beam, exhibited very low passive damping (i.e. for the zero-field condition). In contrast, the addition of the silicone spacer in the SIBER class of beam tended to increase the passive damping, much as would be expected from a constrained visco-elastic layer. The addition of the ER fluid, for any applied electric field did not cause any measurable change in natural frequency for these beams. The damping ratio measured for each of these beams is shown in Figure 3, as the applied electric field is varied from 0 to as high as 2500 V/mm. The damping measured in these tests did not change for the HERB class of beams. For the SIBER class of beams, a slight increase in damping is manifested, but is believed to be an experimental artifact.

At high voltages (typically where the damping ratio curves terminate in Figure 3), tremendous amounts of arcing resulted in subsstantial noise in the accelerometer signal, and beam damage. This arcing occurred in spite of taking great care to etch about 1/4 inch of copper electrode from the sides of each plate, so that no copper layer was exposed on the outside of the beam.

Although these results appear to be discouraging for a fluid filled flexbeam, it should be noted that the beams tested by Gandhi *et al* were orders of magnitude softer than those tested in this paper. We estimate that $EI = 0.06\text{N-m}^2$ was a nominal value of flexural rigidity for the graphite epoxy beam tested in [13, 14]. The lag mode of a typical flexbeam can be hundreds of times stiffer [2]. The soft beams used by Gandhi *et al* allow a larger ratio of strain energy to be concentrated in the ER fluid layer. The volume fraction of ER fluid used for the beams tested in this paper was nominally 40 percent, which is comparable to the volume fraction suggested by Gandhi *et al.*

It appears that for the SIBER class of beams, the silicone spacer tends to dominate the damping characteristic, so that the damping added to the beam by the ER fluid is difficult to measure. However, the damping ratio was incremented on the order of tens of percent in the work by Gandhi *et al*, which is a large change that was not observed in the beams tested in this paper.

These experimental results tend to suggest that much stronger ER materials need to be developed for structural applications such as the augmentation of damping in a helicopter rotor using an ER fluid-filled flexbeam. Lord Corporation Versaflo 201 will be evaluated as a candidate ER fluid, requiring an AC excitation, since it has much better damping potential. A General Radio Company high power frequency generator is currently being retrofitted with a transformer to provide the required voltage and current levels.

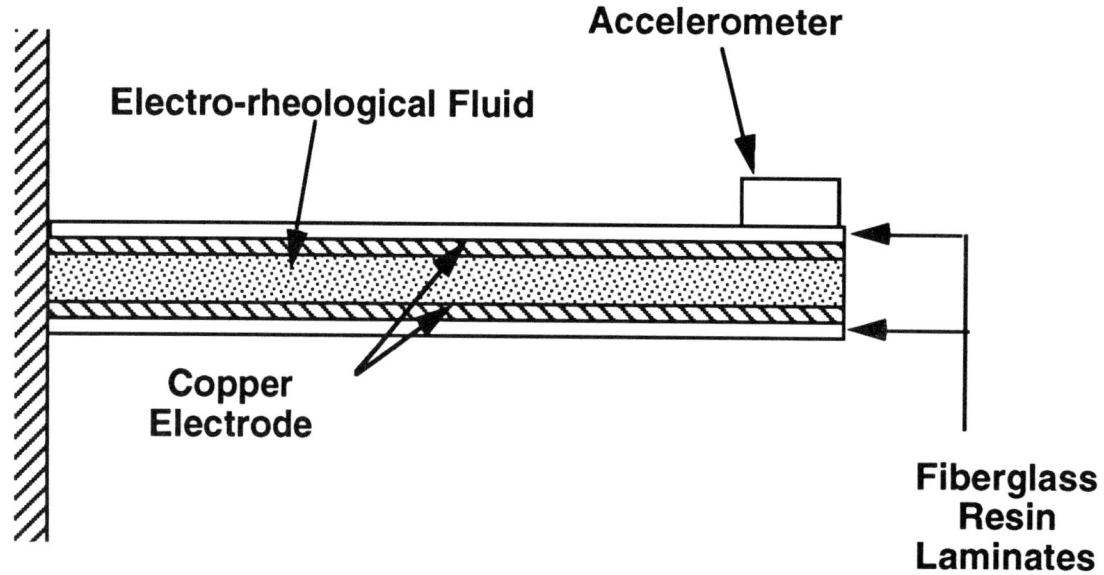

Figure 2: Testing of an ER fluid filled cantilevered composite beam.

Table 1: Summary of beam properties.

Beam	EI [N-m^2]	Nominal Gap [mm]	Plate Width [in]	Spacer Material	ER Fluid
HERB-2-DC-364	7.97	2	3/64	Glass-Epoxy	Mineral Oil / Corn Starch
SIBER-2-DC-364	1.50	2	3/64	Silicone	50 percent each by volume
HERB-1-DC-264	1.62	1	1/32	Glass-Epoxy	VersaFlo
SIBER-1-DC-264	0.475	1	1/32	Silicone	100

Work is continuing to model the ER effects, and softer beams are under construction to verify the ER effects and models. The objective is to develop a model that will permit the systematic design of smart structures with components based on ER fluids.

Conclusions

Results of testing the increased damping supplied by an electro-rheological fluid were reported for a cantilevered composite beam with approximately 40 percent volume fraction of ER fluid. The experimental results tend to suggest that much stronger ER materials need to be developed for structural applications such as a helicopter fluid-filled flexbeam.

Acknowledgements

This work is supported by the Army Research Office under grant DAAL 03-92-G-0121 with Dr. Gary Anderson as Technical Monitor. Graduate Research Assistants, Mr. Gopal Kamath and Mr. Clifford Smith, performed the preliminary measurements described in this paper.

References

[1] I. Chopra. "Perspectives in aeromechanical stability of helicopter rotors." *Vertica*, Vol. 14 No. 4, pp. 457–458, 1990.

[2] C. J. Niggemeier and I. Chopra. "Bearingless rotor aeromechanical stability testing." Technical Report UM-AERO 93-15, Center for Rotorcraft Education and Research, Dept. of Aerospace Engineering, University of Maryland at College Park, May, 1993.

[3] M. V. Gandhi, B. S. Thompson, S. R. Kasiviswanathan, S. B. Choi, B. Hansknecht, M. Soomar, X. Huang, C. Chmielewski, and C. Foiles. "An innovative class of smart materials and structures incorporating hybrid actuator and sensing systems." *Active Materials and Adaptive Structures, Proceedings of the ADPA / AIAA /*

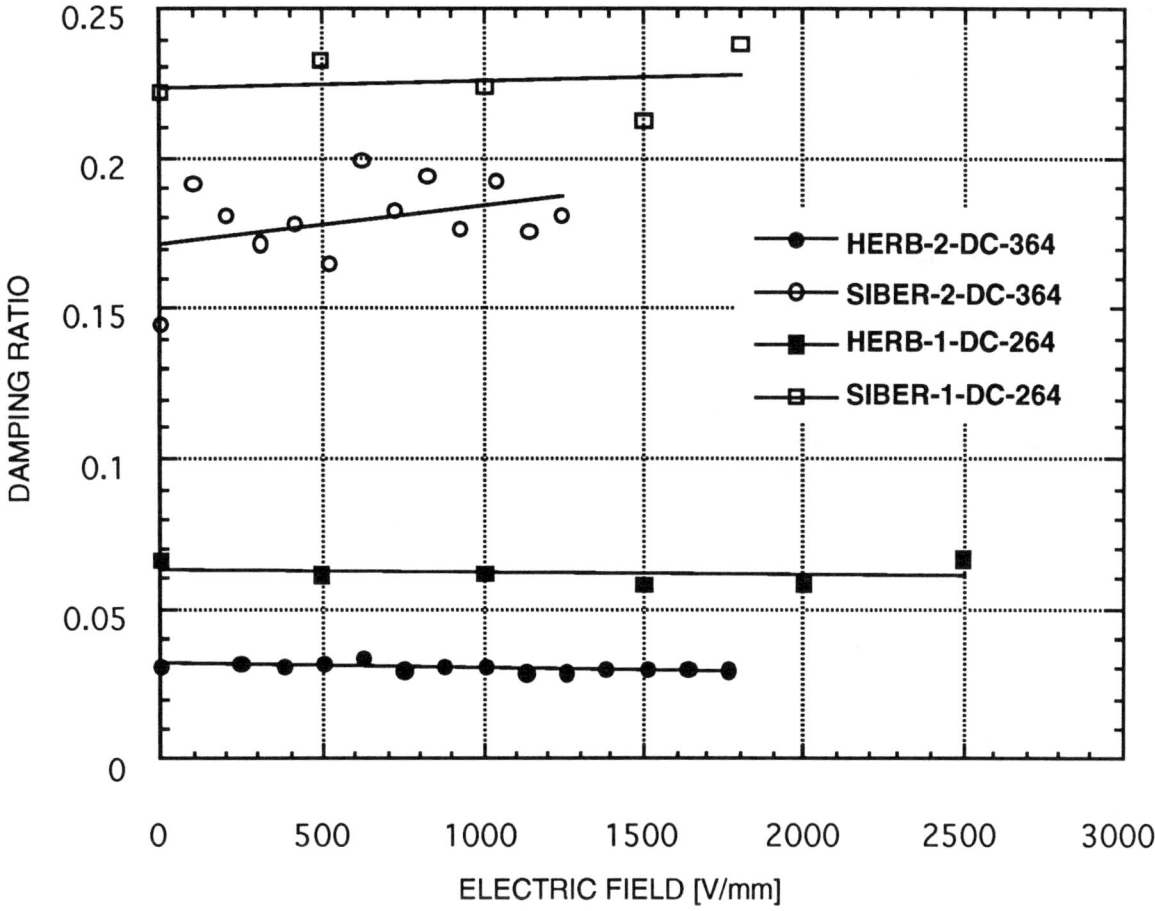

Figure 3: Damping ratio changes for an ER-fluid filled cantilevered beam.

ASME / SPIE Conference on Active Materials and Adaptive Structures, pp. 751–756, Alexandria, VA, 1991.

[4] D. L. Klass and T. W. Martinek. "Electroviscous fluids. Part I. Rheological properties." *Journal of Applied Physics*, Vol. 38, No. 1, pp. 67–74, January 1967.

[5] H. Blockand J. P. Kelly. "Electro-rheology." *Journal of Physics D:Applied. Physics*, Vol. 21, pp. 1661–1667, 1988.

[6] T. G. Duclos. "Design of devices using electro-rheological fluids." *Automotive Engineering*, pp. 2.532–2.536, 1988.

[7] S. Morishita and Y. Kuroda. "Controllable dynamic damper as an application of electro-rheological fluid." *Pressure Vessels and Piping Conference - Active and Passive Damping*, pp. 1–6, San Diego, CA, 1991.

[8] J. R. Salois. "Concept verification of an electro-rheological torsional steering system damper." *Active Materials and Adaptive Structures, Proceedings of the ADPA / AIAA / ASME / SPIE Conference on Active Materials and Adaptive Structures*, pp. 745–750, Alexandria, VA, 1991.

[9] N. G. Stevens, J. L. Sproston, and R. Stanway. "Experimental evaluation of a simple electroviscous damper." *Journal of Electrostatics*, Vol. 15, pp. 275–283, 1984.

[10] S. B. Choi, M. V. Gandhi, and B. S. Thompson. "Control of smart flexible structures incorporating electro-rheological fluids: a proof of concept investigation." *1989 Automatic Control Conference*, Pittsburg, PA, June 1989.

[11] S. B. Choi, B. S. Thompson, and M. V. Gandhi. "An experimental investigation on the active-damping characteristics of a class of ultra-advanced intelligent composite materials featuring electro-rheological fluids." *Damping 89*, Vol.

CAC, pp. 1–14, West Palm Beach, FL, February 1989.

[12] J. P. Coulter and T. G. Duclos. "Applications of electrorheological materials in vibration control." *Proceedings of the Second International Conference on ER Fluids*, pp. 300–325, Raleigh, NC, 1989.

[13] M. V. Gandhi and B. S. Thompson. "Dynamically-tunable smart composites featuring electro-rheological fluids." *SPIE Conference*, Boston, MA, 1989.

[14] M. V. Gandhi, B. S. Thompson, and S. B. Choi. "A new generation of innovative ultra-advanced intelligent composite materials featuring electro-rheological fluids: an experimental investigation." *Journal of Composite Materials*, Vol. 23, pp. 1232–1255, 1989.

[15] A. D. Nashif, D. I. G. Jones, and J. P. Henderson. *Vibration Damping*. John Wiley and Sons, 1985.

AIAA-94-1762-CP

COUPLED ELECTRO-MECHANICAL IMPEDANCE MODELING TO PREDICT POWER REQUIREMENT AND ENERGY EFFICIENCY OF PIEZOELECTRIC ACTUATORS INTEGRATED WITH PLATE-LIKE STRUCTURES

Suwei Zhou[*], C. Liang[†], and C. A. Rogers[+]
Center for Intelligent Material Systems and Structures
Virginia Polytechnic Institute and State University
Blacksburg, VA 24061-0261

Abstract

In a piezoelectric (PZT) actuator-driven adaptive structure, the energy conversion and power consumption of the system are dominated by the complex electro-mechanical impedance of the system. The entire actuator/substrate system can essentially be represented by a coupled electro-mechanical impedance model. This paper presents such an electro-mechanical impedance model to quantitatively determine how the energy is consumed and how much of the energy is required in an adaptive system. The current work was expanded from the mechanical impedance model developed by the authors, to include the electrical parameters of the PZT actuator. First, the formulation of a coupled electro-mechanical admittance for two-dimensional PZT actuator-driven adaptive structures was developed. Then, the model was used to predict the power factor, the power dissipation, and the power requirement of the system. As a numerical example, the modeling approach was applied to a simply-supported thin plate that is excited by a pair of PZT patch actuators in pure bending mode. A parametric study was performed to examine how the electrical power is consumed by different dissipators in the system. The effects of the location and the thickness of the PZT actuator on the power consumption and the energy conversion efficiency were also investigated. An experiment was conducted to directly measure the complex electro-mechanical admittance so that the theoretical model has been verified.

---

[*]Graduate Research Assistant, Student Member AIAA
[†]Research Scientist, Member AIAA
[+]Professor, Mechanical Engineering, Member AIAA, ASME

Copyright © 1994 by
Center for Intelligent Material Systems and Structures.
Published by the AIAA, Inc. with permission

Introduction

Piezoelectric (PZT) materials have been widely used as actuators and sensors in active structural vibration and acoustic control because compact PZT actuators and sensors allow more flexible design of control systems. A great deal of effort has focused on the analysis and optimization of the mechanical performance of PZT actuators and sensors.[2,3,4,11,12,13] Investigations of power consumption and electro-dynamics of the piezoelectric actuator-driven adaptive structures have been limited. In fact, adaptive structures are a complex electro-mechanical coupling systems in which electrical energy is converted into mechanical energy and vice-versa. The ability and efficiency of the energy conversion of the actuator is always of concern in the design and application of intelligent structures, especially for space applications. Minimizing power consumption and enhancing energy efficiency of actuators will result in reductions in the cost and mass of the system, two of major objectives in designing intelligent structures[9]. Therefore, it is important in the design stage to develop a theoretical model for the prediction of system power requirement and the optimization of design parameters in intelligent material systems.

Liang et al.[7] suggested a coupled electro-mechanical analysis for a piezoelectric actuator-driven spring-mass-damper system and discussed the concept of actuator power factor and energy transfer of the system. Stein et al.[10] investigated the power consumption of piezoelectric actuators for underwater acoustic control. The effect of the fluid-structure interaction on the actuator power requirements was included in the modeling. In a recent paper[8] the concept of the actuator power factor was applied to optimize actuator location and configuration. However, the research done to date on energy conversion efficiency and power consumption has been limited to generic one-dimensional structures. An impedance model applicable to two-dimensional structures, developed by the authors,[12,13] now needs to be extended to include the electrical parameters of the PZT actuator for a coupled electro-mechanical energy analysis.

In this paper, the formulation of a coupled electro-mechanical admittance for two-dimensional PZT actuator-driven plate-like structures will first be derived. Then, the model will be used to predict and analyze the power consumption, power requirement, and energy conversion efficiency of the integrated PZT/substrate system. A parametric study will be presented to examine the effects of different elements in the system on power dissipation and energy transfer, including the effects of different damping elements and geometric parameters in the system. An experiment on a simply-supported thin plate using an impedance analyzer will be conducted. The experimental data will be compared with the analytical results from the coupled electro-mechanical impedance model so that the theoretical model can be verified.

A Coupled Electro-Mechanical Impedance Model

In this section, a brief derivation for the mechanical impedance model[13] is given in order to clearly demonstrate the procedure for obtaining the coupled electro-mechanical admittance of a generic two-dimensional piezoelectric actuator-driven system.

A schematic representation of a two-dimensional PZT/substrate coupling system is shown in Fig. 1. When a voltage is applied to the piezoelectric patches along the polarization direction (3), the dynamic strains are induced in the PZT actuator in both the x (1) and y (2) directions. The equation of motion of the PZT actuator may be expressed by:

$$\rho_p \frac{\partial^2 u}{\partial t^2} = S_p^* \frac{\partial^2 u}{\partial x^2} \tag{1a}$$

$$\rho_p \frac{\partial^2 v}{\partial t^2} = S_p^* \frac{\partial^2 v}{\partial y^2}, \tag{1b}$$

where the subscript p refers to the parameters of the PZT patch and the superscript * indicates a complex parameter. ρ is the mass density and S_p^* is the complex Young's modulus at a constant electrical field:

$$S_p^* = S_p(1 + i\eta_p), \tag{2}$$

in which η_p is the structural loss factor. The in-plane displacement response of the PZT actuator is then described by:

$$u = \left[A \sin(k_{p11}x) + B\cos(k_{p11}x)\right] e^{i\omega t} \tag{3a}$$

$$v = \left[C \sin(k_{p22}y) + D\cos(k_{p22}y)\right] e^{i\omega t}, \tag{3b}$$

where A, B, C, and D are unknowns and can be determined by the boundary conditions. Assuming the isotropy

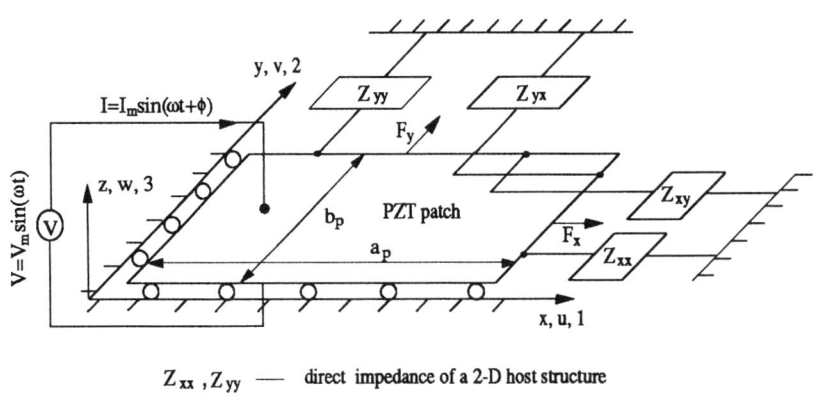

Z_{xx}, Z_{yy} — direct impedance of a 2-D host structure
Z_{xy}, Z_{yx} — cross impedance of a 2-D host structure

Figure 1: A schematic representation of a coupled electro-mechanical impedance model of an integrated PZT/Substrate system.

of the PZT material gives the wave number:

$$k_p = k_{p11} = k_{p22} = \omega \sqrt{\frac{\rho_p}{S_p^*}}, \quad (4)$$

where ω is the input angular frequency. Applying the equivalent displacement boundary condition, $u_{x=0}=0$ and $v_{y=0}=0$, as shown in Fig. 1, to Eq. (3), the response of the PZT actuator is reduced to:

$$u = A \sin(k_p x) e^{j\omega t} \quad (5a)$$

$$v = C \sin(k_p y) e^{j\omega t}. \quad (5b)$$

The unknowns, A and C, may be evaluated from the constitutive equation of the PZT material at $x=a_p$ and $y=b_p$:

$$\begin{pmatrix} \varepsilon_x \\ \varepsilon_y \end{pmatrix} = \begin{pmatrix} \frac{\partial u}{\partial x} \\ \frac{\partial v}{\partial y} \end{pmatrix} = \begin{pmatrix} \frac{1}{S_p^*} & -\frac{\nu_p}{S_p^*} \\ -\frac{\nu_p}{S_p^*} & \frac{1}{S_p^*} \end{pmatrix} \begin{pmatrix} \sigma_x \\ \sigma_y \end{pmatrix} + \begin{pmatrix} d_{31} \\ d_{32} \end{pmatrix} E, \quad (6)$$

where ε is the dynamic strain, d_{31} and d_{32} are the piezoelectric constants of the PZT material in the *1* and *2* directions, respectively, and E is the electric field. The Poisson's ratio of the PZT material, ν_p, is introduced such that the mechanical coupling of the in-plane motion in the x and y directions of the PZT actuator can be included in the modeling. The induced stress at the edge of the PZT patch, $\sigma_{x(y)}$, is equal and opposite to the host structural stress, and may be expressed using the mechanical impedance of the host structure:

$$\begin{pmatrix} \sigma_x \\ \sigma_y \end{pmatrix} = -\begin{pmatrix} 1/(b_p h_p) & 0 \\ 0 & 1/(a_p h_p) \end{pmatrix} \begin{pmatrix} Z_{xx} & Z_{xy} \\ Z_{yx} & Z_{yy} \end{pmatrix} \begin{pmatrix} \dot{u} \\ \dot{v} \end{pmatrix}, \quad (7)$$

where a_p, b_p, and h_p are the length, width, and thickness of the PZT patch, respectively. Z_{xx} and Z_{yy} are the direct structural impedances in the x and y directions, respectively. Z_{xy} and Z_{yx} are the cross structural impedances. Substituting Eqs. (5) and (7) into Eq. (6) and taking the algebraic operation to rearrange A and C yields:

$$\begin{pmatrix} k_p \cos(k_p a_p)(1-\nu_p \frac{b_p}{a_p}\frac{Z_{xy}}{Z_{pxx}} + \frac{Z_{xx}}{Z_{pxx}}) & k_p \cos(k_p b_p)(\frac{a_p}{b_p}\frac{Z_{yx}}{Z_{pyy}} - \nu_p \frac{Z_{yy}}{Z_{pyy}}) \\ k_p \cos(k_p a_p)(\frac{b_p}{a_p}\frac{Z_{xy}}{Z_{pxx}} - \nu_p \frac{Z_{xx}}{Z_{pxx}}) & k_p \cos(k_p b_p)(1-\nu_p \frac{a_p}{b_p}\frac{Z_{yx}}{Z_{pyy}} + \frac{Z_{yy}}{Z_{pyy}}) \end{pmatrix} \begin{pmatrix} A \\ C \end{pmatrix} = \begin{pmatrix} d_{31} \\ d_{32} \end{pmatrix} E, \quad (8)$$

where Z_{pxx} and Z_{pyy} are the input mechanical impedance of the PZT actuator in the x and y directions, defined as:

$$Z_{pxx} = -i \frac{K_{px}}{\omega} \frac{k_p a_p}{\tan(k_p a_p)} \quad (9a)$$

$$Z_{pyy} = -i \frac{K_{py}}{\omega} \frac{k_p b_p}{\tan(k_p b_p)}, \quad (9b)$$

with the static extension stiffness of the PZT actuator in the x and y directions, K_{px} and K_{py}:

$$K_{px} = \frac{S_p^* b_p h_p}{a_p} \quad (10a)$$

$$K_{py} = \frac{S_p^* a_p h_p}{b_p}. \quad (10b)$$

Solving the unknowns A and C from Eq. (8) leads to:

$$\begin{pmatrix} A \\ C \end{pmatrix} = [M]^{-1} \begin{pmatrix} d_{31} \\ d_{32} \end{pmatrix} E, \quad (11)$$

where $[M]$ is the coefficient matrix of A and C in Eq. (8). Substituting Eqs. (5) and (11) into Eq. (7), the dynamic stress output of the PZT actuator is then determined by:

$$\begin{pmatrix} \sigma_x \\ \sigma_y \end{pmatrix} = \frac{S_p^* E}{1-\nu_p^2} \begin{pmatrix} 1 & \nu_p \\ \nu_p & 1 \end{pmatrix} \left[k_p \begin{pmatrix} \cos(k_p x) & 0 \\ 0 & \cos(k_p y) \end{pmatrix} [M]^{-1} \begin{pmatrix} d_{31} \\ d_{32} \end{pmatrix} - \begin{pmatrix} d_{31} \\ d_{32} \end{pmatrix} \right]. \quad (12)$$

The constitutive equation of the PZT material is again revoked in terms of the electrical displacement field[6], D_3, in the z (3) direction:

$$D_3 = \varepsilon_{33}^* E + d_{31}\sigma_x + d_{32}\sigma_y, \quad (13)$$

where

$$\varepsilon_{33}^* = \varepsilon_{33}(1-\delta_p i), \quad (14)$$

is the complex dielectric constant at a constant stress and δ_p is the dielectric loss factor. Substituting Eq. (12) into Eq. (13) and letting $d_{32}=d_{31}$ yields the electrical displacement field:

$$D_3 = E\left\{\varepsilon_{33}^* + \frac{d_{31}^2 S_p^*}{1-\nu_p}\left(k_p[\cos(k_p x) \quad \cos(k_p y)][M]^{-1}\begin{pmatrix}1\\1\end{pmatrix} - 2\right)\right\}, \quad (15)$$

where the applied electric field E can be expressed in terms of the applied voltage V:

$$E = \frac{V}{h_p} = \frac{V_0}{h_p} e^{i\omega t}, \quad (16)$$

in which the subscript 0 of the V denotes the magnitude of an electrical parameter. The charge in the PZT patch can be obtained by integrating the displacement in Eq. (15) with respect to x and y:

$$q = \int_0^{a_p}\int_0^{b_p} D_3 \, dx \, dy. \quad (17)$$

The current is thus given by:

$$I = I_0 e^{i\omega t} = \dot{q}$$
$$= iV_0 \omega \frac{a_p b_p}{h_p}\left\{\varepsilon_{33}^* - \frac{2d_{31}^2 S_p^*}{1-\nu_p} + \frac{d_{31}^2 S_p^*}{1-\nu_p}\begin{bmatrix}\frac{s_1}{a_p} & \frac{s_2}{b_p}\end{bmatrix}[M]^{-1}\begin{pmatrix}1\\1\end{pmatrix}\right\} e^{i\omega t}, \quad (18)$$

where $s_1 = \sin(k_p a_p)$ and $s_2 = \sin(k_p b_p)$. When the PZT actuator is driven by an active voltage V, the current in the circuit, I, is related to the driven voltage through the coupled electro-mechanical admittance, $Y^* = I/V$. Accordingly, rewriting Eq. (18) produces the coupled electro-mechanical admittance:

$$Y^* = i\omega \frac{a_p b_p}{h_p}\left\{\varepsilon_{33}^* - \frac{2d_{31}^2 S_p^*}{1-\nu_p} + \frac{d_{31}^2 S_p^*}{1-\nu_p}\begin{bmatrix}\frac{s_1}{a_p} & \frac{s_2}{b_p}\end{bmatrix}[M]^{-1}\begin{pmatrix}1\\1\end{pmatrix}\right\}. \quad (19)$$

It is noted that the admittance, Y^*, represents the integrated electro-mechanical characteristics of a piezoelectric actuator-driven system, and it is frequency dependent. Apparently, the complex admittance, Y^*, contains all of the parameters concerning the system electro-dynamics performance, including mass, stiffness, damping, material and physical properties, electrical parameters, and boundary conditions. Once these parameters are selected, the admittance will be determined and the power consumption of the system can be predicted. The input structural impedance of the PZT actuator is given by Eq. (9). The mechanical impedance matrix of the host structure in Eq. (7) is determined by:

$$Z = \begin{pmatrix}Z_{xx} & Z_{xy}\\Z_{yx} & Z_{yy}\end{pmatrix} = \begin{pmatrix}H_{xx} & H_{yx}\\H_{xy} & H_{yy}\end{pmatrix}^{-1}, \quad (20)$$

where H_{xx} and H_{yy} are the direct mechanical admittance of the host structure at the edge of the PZT patches, and H_{xy} and H_{yx} are the cross mechanical admittance. As an example, we consider a simply-supported thin plate, as shown in Fig. 2.

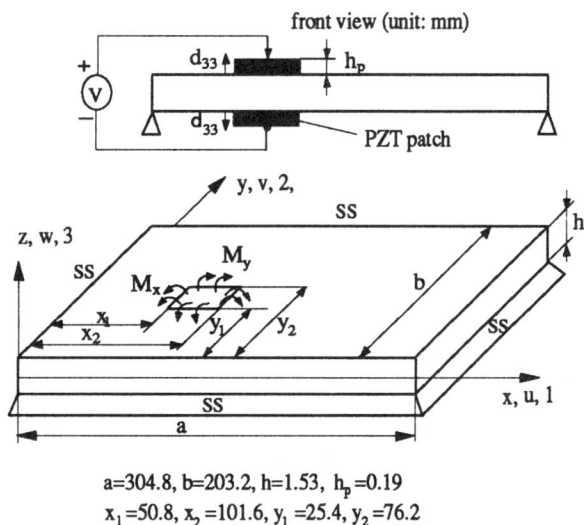

$a=304.8, b=203.2, h=1.53, h_p=0.19$
$x_1=50.8, x_2=101.6, y_1=25.4, y_2=76.2$

Figure 2: A geometric configuration of a surface-bonded PZT actuator-driven simply-supported (ss) aluminum plate.

Two PZT patch actuators are assumed to perfectly bonded on the top and bottom surfaces of the plate. When an active voltage is applied to the PZT patches along the polarization direction (3). The two PZT actuators can be actuated out-of-phase. A pure bending moment excitation is thus created at the boundaries of the PZT patch. Under the actuation of the distributed moments, an analytical solution of the admittance matrix of a simply-supported plate at the mid-point of the edge of the PZT patch was derived by the authors. The detailed derivation can be found in a separate reference.[13] The results are directly given here:

$$H_{xx} = i\frac{2\pi(h+h_p)^2\omega}{\rho h a^3 b_p}\sum_{m=1}^{\infty}\sum_{n=1}^{\infty}\left(\frac{m^2(\cos\frac{m\pi x_1}{a}-\cos\frac{m\pi x_2}{a})^2(\cos\frac{n\pi y_1}{b}-\cos\frac{n\pi y_2}{b})\sin\frac{n\pi(y_1+y_2)}{2b}}{n(\omega_{mn}^2-\omega^2)}\right), \quad (21)$$

$$H_{yy} = i\frac{2\pi(h+h_p)^2\omega}{\rho h b^3 a_p}\sum_{m=1}^{\infty}\sum_{n=1}^{\infty}\left(\frac{n^2(\cos\frac{m\pi x_1}{a}-\cos\frac{m\pi x_2}{a})(\cos\frac{n\pi y_1}{b}-\cos\frac{n\pi y_2}{b})^2\sin\frac{m\pi(x_1+x_2)}{2a}}{m(\omega_{mn}^2-\omega^2)}\right), \quad (22)$$

and

$$H_{xy} = H_{yx} = i\frac{2\pi(h+h_p)^2\omega}{\rho h a^2 b b_p}\sum_{m=1}^{\infty}\sum_{n=1}^{\infty}\left(\frac{m(\cos\frac{m\pi x_1}{a}-\cos\frac{m\pi x_2}{a})(\cos\frac{n\pi y_1}{b}-\cos\frac{n\pi y_2}{b})^2\sin\frac{m\pi(x_1+x_2)}{2a}}{\omega_{mn}^2-\omega^2}\right), \quad (23)$$

where x_1, x_2, y_1, and y_2 are the location coordinates of the edges of the PZT patches on the plate, as illustrated in Fig. 2 and a, b, and h are the length, width, and thickness of the plate, respectively. ω_{mn} is the resonant frequency of the plate determined from the homogenous equation of the transverse motion of the plate:

$$\omega_{mn} = \pi^2\sqrt{\frac{h^3 S^*}{12\rho h(1-\nu^2)}}\left(\left(\frac{m}{a}\right)^2+\left(\frac{n}{b}\right)^2\right), \quad (24)$$

in which m and n are the modal indices in the x and y directions, respectively.

Power Consumption and Energy Conversion Efficiency

Since a PZT actuator acts as a plate capacitor, when a voltage, $V=V_0 e^{i\omega t}$, is applied to it to activate a host structure, the current in the circuit is expressed by:

$$I = I_0 e^{i(\omega t+\phi)}, \quad (25)$$

where ϕ is the phase between the current and voltage[1,5]. The electrical power supplied to the PZT actuator is actually decomposed into two components: one is the real power,

$$P = \frac{I_0 V_0}{2}\cos\phi = \frac{V_0^2}{2}Re(Y^*), \quad (26)$$

where $\cos\phi$ is called the power factor of the electrical system. The other is the reactive power,

$$Q = \frac{I_0 V_0}{2}\sin\phi = \frac{V_0^2}{2}Im(Y^*). \quad (27)$$

Physically, the real power is the power being transmitted to the PZT actuator and is dissipated in driving the host structure. The real power is also called the dissipative power. While the reactive power circulates within the system. The total complex power is then expressed as:

$$W^* = P + iQ. \quad (28)$$

The magnitude of the complex power is defined as the apparent power:

$$W_0 = \sqrt{P^2+Q^2} = \frac{I_0 V_0}{2} = \frac{V_0^2}{2}Y_0, \quad (29)$$

where Y_0 is the magnitude of the admittance Y^*. The apparent power reflects the power requirement of the system. Obviously, the reactive power exerts an important influence on the power requirement. Substituting Eq. (29) into Eq. (26), the power factor is rewritten as:

$$FP = \cos\phi = \frac{Re(Y^*)}{Y_0} = \frac{P}{W_0}. \quad (30)$$

The power factor, FP is the ratio of the dissipative power to the apparent power and represents the energy conversion efficiency of the system. It is noted that the dissipative power consumption, the power requirement, and the energy conversion efficiency of the system, is strongly related to the coupled electro-mechanical admittance.

The dissipative (real) power in an integrated

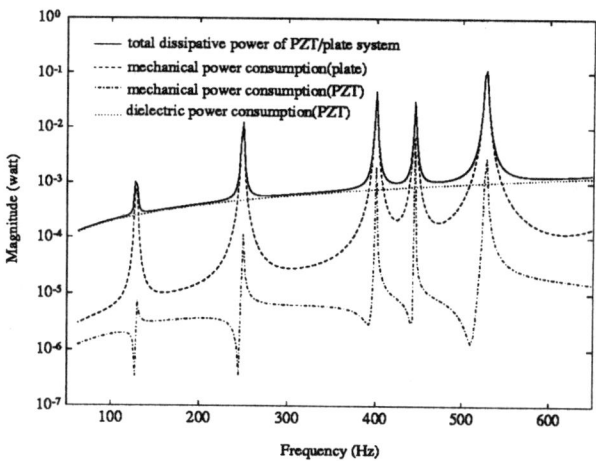

Figure 3: The constitution of the dissipative power of the integrated PZT/Substrate system.

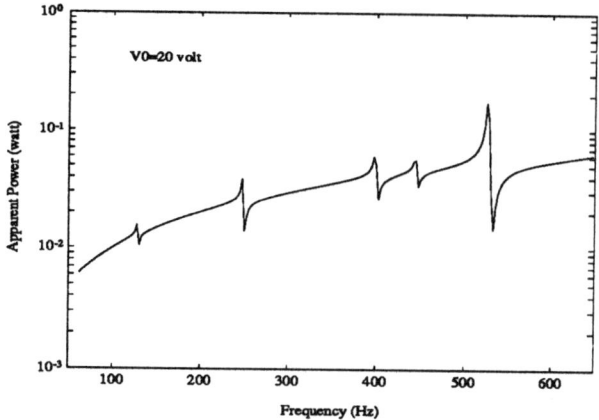

Figure 4: The apparent power of the integrated PZT/plate system.

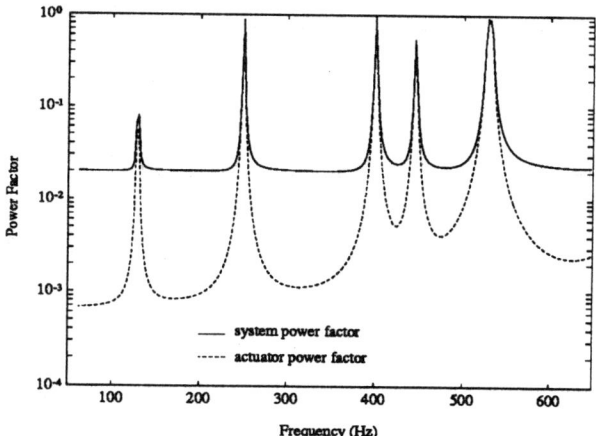

Figure 5: A comparison of the system power factor and the actuator power factor.

dashed line is the power consumption only due to the mechanical loss of the plate, obtained from Eq. (26) assuming $\delta_p=\eta_p=0$. The dash-dotted line is the power consumption only from the mechanical loss of the PZT actuator, found by setting $\delta_p=\eta=0$ in Eq. (26). The dotted line is the power consumed only by the dielectric loss of the PZT actuator, found by assuming $\eta=\eta_p=0$ in Eq. (26). It is clearly seen in Fig. 3 that at the resonant frequencies, the dissipative power is primarily consumed by the structural damping of the plate, while at off-resonance, it is significantly affected by the dielectric loss of the PZT itself. However, the mechanical damping of the PZT has a slight influence on the real power consumption because of its small size.

Figure 4 shows the characteristics of apparent power of the system. The peak power at the 5th mode can be used to estimate the power requirement of the system. Obviously, the power consumption significantly goes up when the excitation frequency increases. Once the interesting frequency band is determined, the power requirement of the system can be numerically predicted.

Figure 5 demonstrates the difference of the power factor predicted from Eq. (30) and Eq. (31), respectively. At off-resonance, the actuator power factor is much smaller than the system power factor. The reason is that the dissipative power is dominated by the dielectric loss behavior of the PZT actuator at off-resonance. Assuming a zero dielectric loss gives a very low power consumption, resulting in a low power factor. At resonant frequencies, the power factor is maximized because of the minimum resistance to the structural vibration. Thus, both the system power factor and the actuator power factor gives the same prediction.

When the damping value in the system changes, the basic relations of the power consumption as shown in Fig. 3 are still applicable. Figure 6 illustrates that if the loss factor of the plate increases from 0.001 to 0.005, the

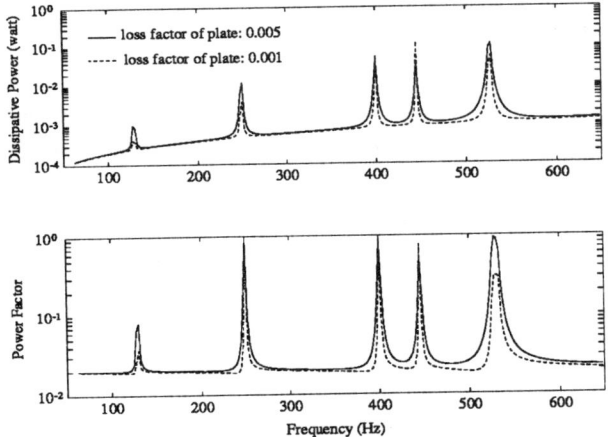

Figure 6: The influence of the loss factor of the plate on the dissipative power and power factor of the integrated PZT/plate system.

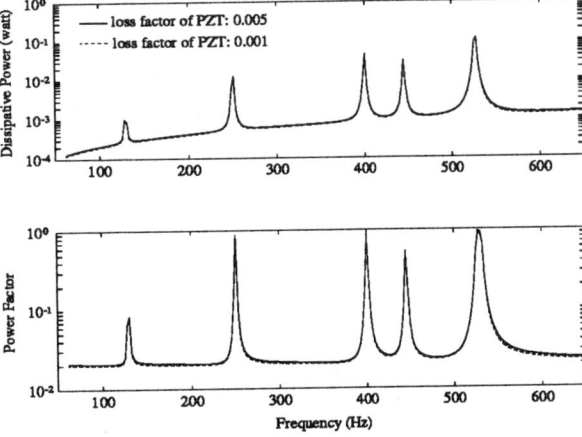

Figure 8: The influence of the loss factor of the PZT actuator on dissipative power and power factor of the integrated PZT/plate system.

dissipative power goes up at resonant frequencies and increases slightly at off-resonance. In contrast, if the dielectric loss factor of the PZT actuator doubles from the 0.015 to 0.03, the dissipative power and the power factor increases by 50% at off-resonant frequencies and remains the same at the resonant frequencies, as displayed in Fig. 7. The structural loss factor of the PZT actuator does not have much impact on the power consumption of the system, as shown in Fig. 8, although the loss factor increases by four times from 0.005 to 0.025.

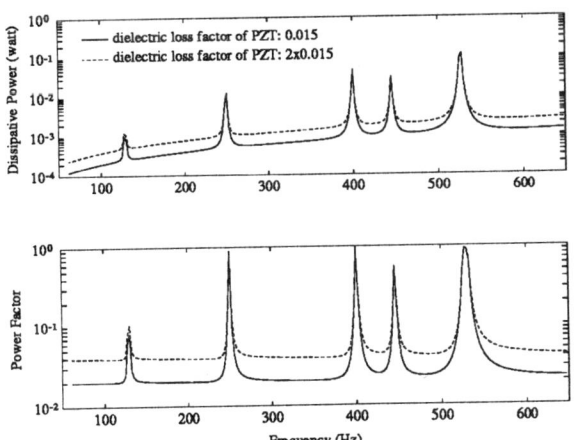

Figure 7: The influence of the dielectric loss factor of the PZT actuator on the dissipative power and power factor of the integrated PZT/plate system.

The geometric parameters of the PZT actuator, such as the thickness and the location, have significant influence on the dissipative power and power factor because the mechanical impedance of the system strongly varies with these geometric parameters.[12,13] Under the assumption of the constant magnitude of the applied voltage (v_0=20 volt), when the thickness of the PZT patch increases from 0.19 mm to 2 × 0.19 mm and 4 × 0.19 mm, the real power consumption decreases on the whole frequency band, as displayed in Fig. 9. The power factor also decreases at off-resonance. However, the power factor has a complicated variation at the resonant frequencies. This observation indicates that when the thickness of the PZT actuator varies, the maximum power factor will depend on the individual mode. Another important observation in Fig. 9 is that the resonant frequencies of the system apparently shift to higher values when the thickness of the PZT actuator increases. It may be explained by the stiffening effect of the PZT actuator.[12,13]

Figure 10 shows that when the location of the PZT actuator on the structure varies, the mechanical impedance changes and so do the resonant frequencies of the system. Note that when the center of the PZT actuator locates at the node line position in the x direction (location #2), the corresponding modes [(2,1) and (2,2)] will be tailored off since little mechanical vibration energy is supplied to these

PZT/substrate system includes three parts:

(1) the power dissipated by the structural damping of the host structure, which is related to the structural loss factor η in the complex Young's modulus. This power is proportional to the mechanical vibration power of the host structure.
(2) the power consumed by the internal damping of the PZT actuator, which is associated with the mechanical loss factor η_p of the PZT actuator.
(3) the power consumption caused by the dielectric loss of the PZT actuator, which is related to the dielectric loss factor δ_p in the complex dielectric constant.

The consumed power is eventually converted into heat in the system, resulting a internal heat generation and inducing a thermal stress in the PZT actuator.[14] The first two parts of the system power consumption mentioned above are directly used in driving the host structure. They may be considered to represent the total mechanical power dissipation in the system. A concept of actuator power factor, suggested by Liang et al.[7], is defined as:

$$PF_p = \frac{dissipative\ mechanical\ power}{apparent\ power} = \frac{Re(Y_p^*)}{Y_0}, \quad (31)$$

in which Y_p^* is calculated by assuming the dielectric loss factor (δ_p) to be zero in Eq. (19). Apparently, the difference between the system power factor [Eq. (30)] and the actuator power factor [Eq. (31)] is that the former includes the dielectric power consumption of the PZT actuator in the dissipative power consumption. The actuator power factor defined in Eq. (31) has a clear physical meaning. However, when the power factor is experimentally determined, the electrical system power factor should be used because the dielectric loss of the PZT actuator is not zero in a real system. In this paper, the system power factor is used in the following numerical examples. A comparison between the system power factor and the actuator power factor is also performed.

Parametric Study and Discussion

A parametric study is conducted in this section to quantitatively examine how the dissipative power is consumed by the different dissipators in the system and how much power is required to drive the system. The influence of the thickness and the location of the PZT actuator is also discussed.

A thin plate made of aluminum is used in the numerical example. The PZT material is G1195. The geometric parameters are shown in Fig. 2, the basic material properties are listed in Table 1.

It is assumed in the numerical case that the magnitude of the voltage applied to the PZT actuator is 20 volt. Figure 3 illustrates the constitution and distribution of the dissipative (real) power in the frequency domain. The solid line represents the total dissipative power. The

Table 1: Material Properties of the PZT* and Aluminum

	Young's modulus (N/m²)	mass density (kg/m³)	Poisson ratio	piezoelectric constants (m/volt)	dielectric constant (Farads/m)	dielectric loss factor	loss factor
PZT	6.3x10¹⁰	7650	0.3	-1.66x10⁻¹⁰	1.5 x 10⁻⁸	0.015	0.005
Aluminum	6.9x10¹⁰	2700	0.33	N/A	N/A	N/A	0.005

*From Piezo System, Inc.

vibrational modes.

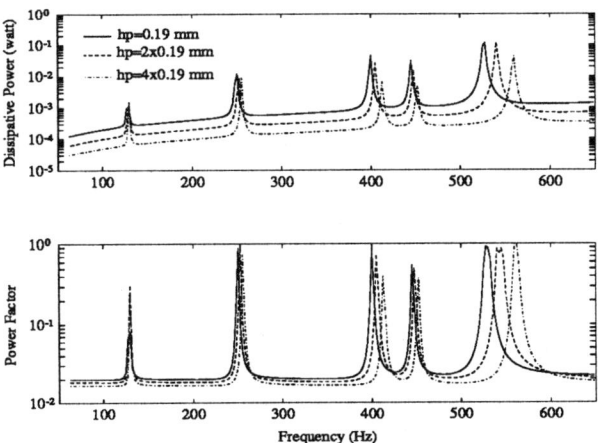

Figure 9: The influence of the thickness of the PZT actuator on dissipative power and power factor of the integrated PZT/plate system.

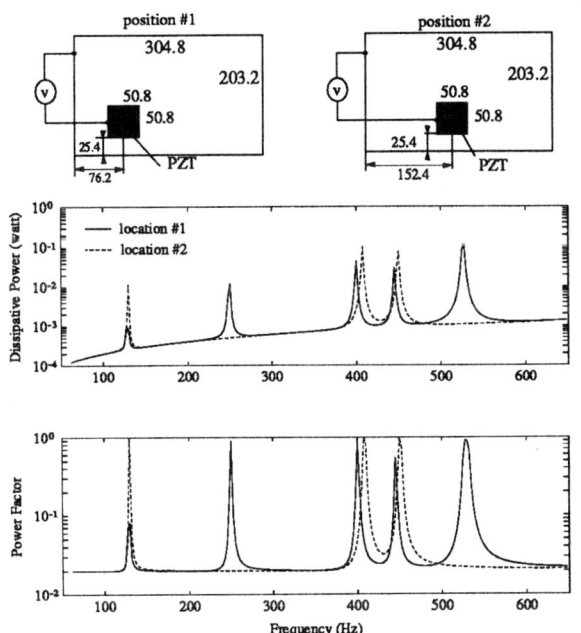

Figure 10: The influence of the location of the PZT actuator on dissipative power and power factor of the integrated PZT/plate system.

Experimental Verification

A simply-supported thin plate integrated with surface-bonded PZT patches was built and tested to validate the coupled electro-mechanical impedance model. The size and physical properties of the plate and the PZT material are the same as those used in the numerical calculation in the previous section, as listed in Fig. 2 and Table 1. An HP 4194A Impedance/Gain-Phase Analyzer was used to directly measure the coupled electro-mechanical admittance of the piezoelectric actuator-driven plate. Then, the comparison between the theoretical model and the experimental results was performed.

Figure 11 illustrates the measured and predicted complex admittance of the system in terms of the real part and the imaginary part, respectively. The first eight modes are excited. The corresponding power factor, calculated from the experimental data using Eq. (30), is displayed in Fig. 11. In both figures, the theoretical prediction based upon the complex impedance model (dashed line) agrees well with the experimental data (solid line). The coupled electro-mechanical impedance model has provided a reasonably accurate prediction of the power consumption and the energy conversion efficiency of the PZT actuator-driven systems. It should be noted that the maximum difference between the theoretical model and the experimental results appears at the 5th mode, i.e., (2,2) mode. It may be explained that when the geometric center of the PZT actuator locates on the anti-node position of the host plate (x=76.2 mm, y=50.8 mm), the excitation of that mode is maximized. The inertial effect caused by the added mass loading of the PZT patch is then intensified. The measured resonant frequency and the response of this mode [the 5th mode (2,2)] are thus smaller than those predicted by the theoretical model. Another possible factor causing this difference is the real boundary condition of the simply-supported plate.

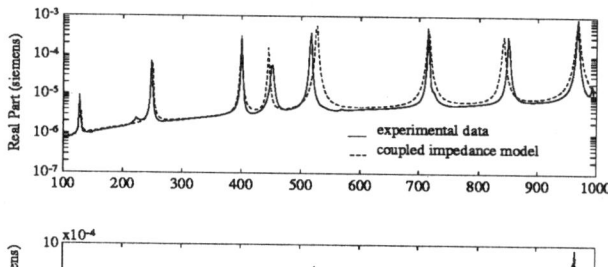
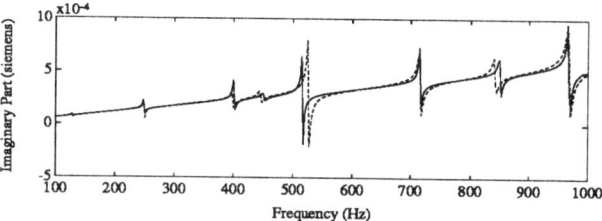

Figure 11: The measured and predicted complex electro-mechanical admittance of the integrated PZT/plate system.

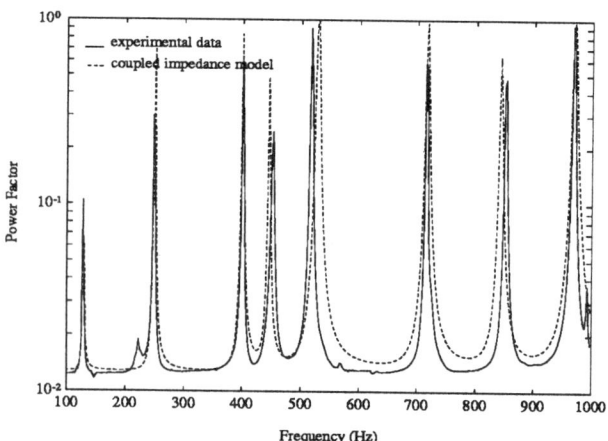

Figure 12: The measured and predicted power factor of the integrated PZT/plate system.

Summary

- A coupled electro-mechanical impedance model for a two-dimensional piezoelectric actuator-driven structure has been developed to predict the power consumption and the energy efficiency of the system. This modeling approach is helpful in designing energy-efficient intelligent structures and induced strain actuators, and their associated power electronics.

- The coupled electro-mechanical impedance model provides an effective tool to estimate the rate of internal heat generation in the PZT actuator. The thermal stress, thus, can be numerically predicted.

- The coupled electro-mechanical impedance model has been experimentally verified.

- The parametric study has demonstrated that the dissipative power supplied to the PZT actuator is primarily consumed by the mechanical damping of the host structure at the resonant frequencies and is dissipated by the dielectric loss of the PZT itself at the off-resonance.

- The thickness and location of the PZT actuator have an impact on the dissipative power consumption and the energy efficiency of the system because these geometric parameters can cause significant changes in the mechanical impedance of the system.

Acknowledgment

The authors gratefully acknowledge the support of the Air Force Office of Scientific Research for this work under AFOSR Grant No. F49620-93-1-0166; Dr. Jim Chang, Program Manager.

Reference

1. Carlson, A. B. and D. G. Gisser, 1990, Electrical Engineering: Concepts and applications, Addison-Wesley, California.

2. Crawley, E. F. and J. de Luist, 1987, "Use of Piezoelectric Actuators as Elements of Intelligent Structures", AIAA Journal, pp. 1373-1385.

3. Crawley, E. F. and K. B. Lazarus, 1991, "Induced Strain Actuation of Isotropic and Anisotropic Plates", AIAA Journal, pp. 945-951.

4. Dimitriadis, E. K., C. R. Fuller, and C. A. Rogers,

1989, "Piezoelectric Actuators for Distributed Noise and Vibration Excitation of Thin Plates", ASME Failure Prevention and Reliability, DE-Vol. 16, pp. 223-233.

5. Elgerd, O. I., 1981, Electric Energy Systems Theory, McGraw-Hill Inc., New York.

6. Ikeda, T., 1990, Fundamentals of Piezoelectricity, Oxford University Press, New York.

7. Liang, C., F. P. Sun, and C. A. Rogers, 1993a, " Coupled Electro-Mechanical Analysis of Piezoelectric Ceramic Actuator-Driven Systems—Determination of the Actuator Power Consumption and System Energy Transfer", Proceedings of Smart Structures and Materials '93, SPIE, Albuquerque, NM.

8. Liang, C., F. P. Sun, and C. A. Rogers, 1993b, "Design of Optimal Actuator Location and Configuration Based on Actuator Power Factor", Proceedings of the Fourth International Conference on Adaptive Structure, Cologne, Germany.

9. Rogers, C. A., 1993, Intelligent Material Systems: The Dawn of a New Materials Age", Journal of Intelligent Material Systems and Structures, Vol. 4, No. 1, pp. 4-12.

10. Stein, S. C., C. Liang, and C. A. Rogers, 1993, "Power Consumption of Piezoelectric Actuators in Underwater Active Structural Acoustic Control", The Second Conference on Recent Advances in Active Control of Sound and Vibration, Blacksburg, VA.

11. Wang, B. T., 1991, "Active Control of Sound Transmission/Radiation from Elastic Plates Using Multiple Piezoelectric Actuators", Ph.D. Dissertation, Department of Mechanical Engineering, Virginia Polytechnic Institute and State University, Blacksburg, VA.

12. Zhou, S. W., C. Liang, and C. A. Rogers, 1993, "Impedance Modeling of Two-Dimensional Piezoelectric Actuators Bonded on a Cylinder", Proceeding of the Adaptive Structures and Material Systems, ASME Winter Annual Meeting, New Orleans, pp. 247-256.

13. Zhou, S. W., C. Liang, and C. A. Rogers, 1994, "A Dynamic Model of a Piezoelectric Actuator-Driven Thin Plate", Proceedings of Smart Structures and Materials '94, SPIE, Orlando, FL, in press.

14. Zhou, S. W., C. Liang, and C. A. Rogers, 1994, "Dynamic Design Method and Stress Characteristics Analysis of Integrated Piezoelectric Patch Actuators", Proceedings of 1994 International Conference on Intelligent Materials, Williamsburg, VA, in press.

THE EFFECTS OF SHAPED PIEZOCERAMIC ACTUATORS ON THE EXCITATION OF BEAMS

Gregory W. Diehl[1]
Harley H. Cudney[2]

Adaptive Structures Laboratory
Virginia Polytechnic Institute and State University
Blacksburg, Virginia 24061-0238

Abstract

The effect of the shape of piezoceramic actuators on the vibration response of a simply supported beam is investigated. An equation is derived to convert between the shape of the piezoceramic actuator and the resulting moment distribution caused on the structure. A beam simulation program is then created to model the vibrations caused by various shaped moment distributions exciting a simply supported beam. Another equation is then derived to explain the resulting minimums and maximums of the modal amplitudes from the beam simulation. The equation is shown to be a useful tool in designing shapes to meet specific control criteria. An example is given showing how the shape of the actuator can be designed to give superior performance for specific control criteria than a traditional rectangular shape. Two possible actuator shapes are shown for the situation. One shape is optimized for the given control criteria by causing the maximum response for the critical mode. The results from the beam simulation for both shapes are shown. The shape of the actuator may now be used as a variable in the cost function for control optimization.

Nomenclature

b_a actuator width
d_{31} voltage to strain piezoelectric constant
E_a Young's modulus of the actuator
E_b Young's modulus of the beam
$f(x,t)$ external force on beam
L length of beam
M bending moment
M_{eq} equivalent bending moment
M_i generalized mass
$m(x)$ mass per unit length
M moment distribution
q_i generalized coordinate
t time
t_a thickness of piezoceramic actuator
t_b thickness of beam
t_s thickness of shear bonding layer
$V(t)$ voltage applied to actuator
x,y,z Cartesian coordinates
ϕ_i mode shape
Λ actuation strain
ω_i natural frequency
Ψ non-dimensionalized property relating the thickness and stiffness of the beam and actuator

Subscript
a   actuator
b   beam
eq   equivalent
i   mode number
s   shear bonding layer

Introduction

Piezoceramic materials have been widely investigated for use in both vibration and noise control because they strain when excited by an electric field. When piezoceramic actuators are bonded to or embedded in a structure, the strain is induced in the structure which induces forces onto the structure. This fact has led to research in applications involving piezoelectric actuators, including active vibration control of large space structures,[1,2] and active acoustic control of plates and beams.[3] Other likely applications involve controlling acoustical noise on a cylinder in an underwater environment[4] and noise reduction in jet engines, pumps, turbines, and automobiles.[5]

---

[1] Graduate Student, Mechanical Engineering
[2] Assistant Professor, Mechanical Engineering
© 1994 by Harley H. Cudney. Published by American Institute of Aeronautics and Astronautics, Inc. with permission

Traditionally, when controlling structures by bonding piezoceramic actuators to the surface or embedding them within the structural members, only rectangular shaped actuators have been used, while little work has been done exploring other shapes. Shaped actuators potentially have several advantages over rectangular actuators. They may lower stress concentrations on the surface to which they are bonded. This can be very important in controlling composite structures where a high stress concentration may cause delaminations. Shaped piezoceramic actuators may also be able to control certain modes of vibration more effectively, especially when actuator location and size may be physically limited. Finally, there may be less control spillover when using shaped piezoceramic actuators, allowing certain modes to be excited while other modes are not.

In the past, research has focused on rectangular actuators, and more generally shaped actuators that span the entire length of the beam. Burke and Hubbard applied a spatially shaped distributed actuator for vibration control of a simply-supported beam and analyzed it experimentally and analytically.[6] Lee and Moon developed equations for modal sensors when the sensor covers the length of the beam and showed the reciprocal relationship between sensors and actuators.[7] Collins, et al. investigated piezoelectric modal sensors and showed they perform as well as typical estimators for the flexible states and require less computational overhead.[8] Burke, et al. extend the concept of colocation to distributed piezoelectric sensors and actuators.[9] Kim, et al. did experimental studies of various shaped actuators applied to plates, but did not go into detail exactly how the shape of the actuator affects the modal response and hence the controllability of the structure.[10]

This paper presents an equation which is useful in predicting the response of shaped piezoceramic actuators attached to structures. It describes how the shape of the actuator may be modelled as a moment distribution on the beam. A computer simulation is used to model a structure excited by various shaped piezoceramic actuators. The equation is shown to be useful as a tool for design of specific actuator shapes for specific control situations and shows that the shape may be used as a variable in the optimization problem.

Theory

In this section, the theory used for shaped actuators is be developed. The assumptions used for modelling the actuator as a moment distribution are presented as well as the equations involved. An equation is then derived to explain the beam simulation response.

While the shape of the actuator spreads the moment distribution across the width of the beam, the moment distribution in this research is modelled as acting along the centerline of the beam as shown in Figure 1.

This can be done since the moments can be moved anywhere across the width of the beam without affecting the transverse vibrations of the beam. The result is a simplified one-dimensional moment distribution which is easy to implement within the beam vibration program and is also used in the beam vibration equation to be derived.

The assumption of a perfectly bonded actuator is used which allows the moments from the actuator attached to the beam to be modelled as concentrated moments at the endlines of the actuator. The fact that the moments act at the endlines of the actuator is important when finding the moment distribution from different shaped actuators. It was also important to establish that a two-dimensional moment distribution can be converted to one-dimension.

Conversion between the actuator shape and the resulting moment distribution relies on the fact that the shape can be modelled as a series of rectangles. Crawley and de Luis showed that the magnitude of the moment at the endlines of a rectangular actuator is proportional to its width, b_a.[2] The resulting equation is shown below.

$$M_{eq} = \frac{t_b^2 E_b}{6+\Psi} b_a \Lambda, \qquad (1)$$

Here the subscripts a and b refer to the actuator and beam respectively, t is thickness, b_a is the actuator width, E_b is Young's modulus of the beam, and Ψ is a non-dimensionalized property relating the thickness and stiffness of the beam and actuator, where

$$\Psi = \frac{t_b E_b}{t_a E_a}. \qquad (2)$$

The actuation strain, Λ, is

$$\Lambda = \frac{V(t)d_{31}}{t_a}, \quad (3)$$

where $V(t)$ is the voltage applied to each actuator, and d_{31} is the piezoelectric constant relating strain to voltage.

Because the total moment produced is directly proportional to the width of the actuator, the change in width of the shape at a point is proportional to the moment at that point. By transforming Equation (1) to the continuous domain, the equation for the moment distribution becomes

$$m(x,t) = CV(t)\frac{db(x)}{dx}, \quad (4)$$

where V(t) is the voltage input, b(x) is the function describing the actuator shape, and x is the position along the beam. The constant term, C, is derived from combining Equations (1) and (3) and is given here as

$$C = \frac{t_b^2 E_b d_{31}}{(6+\Psi)t_a}, \quad (5)$$

which has units of force per length.

The total moment produced must equal zero and is the integration of the moment distribution over the length of the beam, given by

$$M(t) = \int_0^L m(x,t)dx = 0, \quad (6)$$

which has the units of force multiplied by length. This expression may also be used to find the total moment acting on any portion of the beam by changing the integration limits.

An example using this equation is given as follows. The shape analyzed is shown in Figure 2.

The function that describes the shape is given by

$$b(x) = \frac{b\max}{(x_2 - x_1)}[(x - x_1) \cdot H(x - x_1)$$
$$-(x - x_2) \cdot H(x - x_2)$$
$$-(x_2 - x_1) \cdot H(x - x_2)], \quad (7)$$

where bmax is the maximum width of the shape, x_1 and x_2 are the positions of the edges of the shape, and H is the Heaviside function defined by

$$H(a) = 0 \ \forall \ a < 0$$
$$= 1 \ \forall \ a \geq 0, \quad (8)$$

where H is unitless. Taking the derivative with respect to x of Equation (7) gives

$$\frac{db(x)}{dx} = \frac{b\max}{(x_2 - x_1)}[H(x - x_1)$$
$$-H(x - x_2) - (x_2 - x_1)\delta(x - x_1)$$
$$-(x - x_1)\delta(x - x_1)$$
$$-(x - x_2)\delta(x - x_2)]. \quad (9)$$

Here δ represents the Dirac delta function defined by

$$\delta(a) = 0 \ \forall \ a \neq 0$$
$$= 1 \ \forall \ a = 0, \quad (10)$$

and has the units of 1/length. Here it is important to note that the last two Dirac delta functions in Equation (9) will not affect the results since the terms they are multiplied with will always be zero when the Dirac delta function is one, and for all other cases the Dirac delta function will be zero, causing the expressions to go to zero for all cases of x. Substituting Equation (9) into Equation (4) and simplifying yields

$$m(x,t) = \frac{CV(t)b\max}{(x_2 - x_1)}[H(x - x_1)$$
$$-H(x - x_2) - (x_2 - x_1)\delta(x - x_1)$$
$$-(x - x_1)\delta(x - x_1)$$
$$-(x - x_2)\delta(x - x_2)]. \quad (11)$$

The resulting moment distribution from Equation (11) is shown in Fig. 3.

Figure 4 shows examples of various shapes and the resulting moment distributions arrived at using Equation (4). The equivalent non-symmetric shapes are also given. Due to the modelling of the moment distribution along the centerline of the beam, actuator shapes are not unique to a particular moment distribution.

An equation is derived in the following paragraphs which is used extensively to explain the results from the simulations. The resulting equation is especially useful in predicting the modal amplitudes produced by specific moment distributions.

The governing equation for modal vibrations in beams is derived in a text by Thompson,[11] and is given here as

$$\ddot{q}_i(t) + \omega_i^2 q_i(t) = \frac{1}{M_i} * \left[\int f(x,t)\phi_i(x)dx + \int m(x,t)\phi_i'(x)dx \right] \quad (12)$$

where the subscript i is the mode number, q_i is the generalized coordinate, $m(x,t)$ is the moment distribution over the length of the beam as a function of time, ω_i is the natural frequency of the mode, and the terms ϕ_i, and ϕ_i' are the mode shape and its first spatial derivative respectively. M_i is the generalized mass and satisfies the equation

$$M_i = \int_0^l \phi_i^2(x) m(x) dx, \quad (13)$$

where $m(x)$ is the mass of the beam per unit length as a function of location, and l is the length of the beam. M_i is a constant for each particular mode. Since external forces are not being considered here, the term $f(x,t)$ goes to zero, which results in,

$$\ddot{q}_i(t) + \omega_i^2 q_i(t) = \frac{1}{M_i} \int m(x,t)\phi_i'(x)dx \quad (14)$$

Here, the amplitude of q_i is

$$q_{i\max} = \frac{1}{\omega_i^2 M_i} \int m(x,t)\phi_i'(x)dx. \quad (15)$$

The equation that governs the how each point on the beam vibrates with respect to time, $y(x,t)$, is given by

$$y(x,t) = \sum_i q_i(t)\phi_i(x)dx. \quad (16)$$

The modal amplitude will be zero when q_i is zero and the amplitude will be maximum when q_i is a maximum. Thus, looking at Eq. (14), it can be seen that the following applies: for zero actuation of beam mode i, the relationship

$$\int m(x,t)\phi_i'(x)dx = 0 \quad (17)$$

must hold, and for maximum modal actuation the equation

$$\int m(x,t)\phi_i'(x)dx = Maximum \quad (18)$$

must be true. In general, the modal amplitude at a mode peak is

$$Modal\ Amplitude = A \int m(x,t)\phi'(x)dx, \quad (19)$$

where A is a constant given by

$$A = q_{\max}\phi_i. \quad (20)$$

Equation (19) is used extensively to explain the maximum and minimum modal amplitudes from the beam program. It will also be used to establish the capabilities of shaped actuators.

<u>Beam Simulation Model</u>
A thin simply-supported beam is considered in this work. The mode shapes of this Euler-Bernoulli beam are sinusoidal, which simplifies calculations and gives a better qualitative understanding of the relationship between the moment distribution and modal amplitudes. The piezoceramic actuator is assumed to be thin compared to the beam and does not affect the mode shapes. It should be noted that the theory used here can be adapted to a beam with any boundary conditions provided the mode shapes are known.

The beam program simulates a shaped actuator, perfectly bonded to the beam, excited by an ideal impulse voltage. The impulse voltage ensures that a large bandwidth of frequencies is excited. The beam is nondimensional. The width and thickness of the beam are not important in this work since they will not affect the relative amplitude differences between the modes. The sensor location is chosen near the end of the beam so that it will not be on a node of any of the modes examined. Modal amplitudes are then scaled to their maximum values along the beam.

Results

The modal information from the beam simulations is presented as follows. The maximum amplitude of each mode when excited by the moment distribution is obtained. The length of the moment distribution is iterated from zero to the length of the beam. The length is then plotted versus each modal amplitude to show trends in the modal response and to show the length of the distribution that causes minimum and maximum excitation of each mode. Figure 5 graphically shows how the length of the distribution is varied during the simulation. Notice that the amplitude of the distribution increases as the length of the distribution decreases. The total moment produced has to be the same regardless of the actuator length since the actuator width is assumed to be constant as the actuator decreases in length.

The beam simulation was run using a cosine and sine moment distributions where the area under the moment distribution was held constant as the length of the distributions varied from zero to the length of the beam. Figures 6 and 7 show the modal amplitude results from the program. As can be seen, the even modes are not actuated. Figure 8 shows graphically why the even modes are not excited for this case.

From Equation (19), the integration between ϕ' for an even mode and the cosine moment distribution centered on the beam will always be zero, regardless of the length of the distribution. This applies to any symmetrical distribution centered on the beam.

The points of insensitivities of the modes occur when Equation (19) goes to zero. Correspondingly, the points of maximum actuation occur when Equation (19) is at a maximum. All peaks and insensitivities were verified by a numerical integration.

Non-Symmetric Moment Distributions

Symmetrically shaped actuators exhibit limited performance because their ability to selectively excite (or sense) modes is limited. There are more possible shapes of non-symmetric actuators, giving them more flexibility in selective modal excitation. Non-symmetric actuators were implemented within the beam simulation program to compare their responses to that of symmetric shaped actuators.

Figure 9 shows how the non-symmetric moment distributions are modelled by the beam simulation.

In this example, the right edge is held at 1, and the left edge is varied from zero to the midpoint of the beam, giving a range of modal amplitude results that can be plotted.

Non-symmetric cosine and sine shaped moment distributions were implemented in the beam simulation for the same reasons as the symmetric case. Figures 10 and 11 show the simulation results from non-symmetric distributions with the right half held constant and the left half varied. The maximums and minimums on the graph can be explained once again by Equation (19). The maximums occur where the integration is at a peak and the insensitivities occur where the integration goes to zero.

Actuator Design

An example will now be given to show how shaped actuators may be useful. This also goes through the process of using Equation (19) to choose a suitable shape that will meet the given criteria.

Suppose the 4th mode of a beam must be excited while not affecting the 2nd mode. This presents problems if a single rectangular actuator is being used. A rectangular actuator cannot excite the 4th mode without also exciting the 2nd mode. Figure 12 shows the situation.

The rectangular actuators producing the moment pairs shown that cause zero actuation of the 2nd mode must also cause zero actuation of the 4th mode. Shaped actuators allow the 4th mode to be excited, while the 2nd mode is not affected. Figure 13 shows the moment distribution from an actuator that will excite the 4th mode while not exciting the 2nd mode.

Looking first at the 2nd mode, it is seen that there will be no actuation. The moment on the left is at a zero of ϕ', causing no actuation of the mode. The moment distribution on the right is symmetric about a zero point of ϕ', with negative values to the right and positive values to the left. Here Equation (19) goes to zero causing no actuation of the mode. Looking at mode 4, the left moment will cause the maximum actuation possible since it is a concentrated moment at a peak of ϕ'. The moment distribution on the right will tend to lessen the actuation, since the result of Equation (19) will be opposite in sign of that for the left half. However, since the moments are not concentrated at a peak, the right half will not fully cancel the left half. The result will be actuation of mode 4 without actuating mode 2, although the amplitude of mode 4 will be small compared to the following solution. A solution which will give a better actuation of mode 4 is shown in Figure 14.

Mode 2 will not be actuated since the left moment is on a zero of ϕ' and the response from the other moments cancel due to ϕ' being equal and opposite for each moment. Using Equation (19) shows that there will be zero actuation of mode 2. The 4th mode will be excited more than in the previous example. All of the moments are placed on peaks of ϕ', causing the maximum actuation of mode 4. Both these situations were implemented in the beam program and the results are shown in Figures 15 and 16. As predicted, the amplitude of the 4th mode is much greater for the second case. The 2nd mode is not actuated in either case. Also notice that the other modes shown vary widely in both examples. Depending on the needed criteria, the output from shaped actuators is predictable and adjustable.

When optimization of control is needed, Equation (19) can be used to create a cost function. In this case, the cost function would be the modal amplitudes desired. The optimization loop would include the shape, location, and length of the piezoceramic actuator as variables and would minimize the cost function. This allows more flexibility and lower cost functions than previously possible.

Conclusions

This paper studied the effects of shape, size, and location of piezoceramic actuators on the modal response of beams, and develops tools used to design actuator shapes to meet specific control criteria. An equation was derived to convert between the actuator shape and the corresponding moment distribution. Then, a beam simulation was created which models a simply supported beam excited by moment distributions caused by shaped actuators, and predicts the resulting modal amplitudes. Different moment distributions were simulated, including the moment distributions from traditional rectangular shaped actuators, cosine moment distributions, and sine moment distributions. Both symmetric and non-symmetric shaped actuators were simulated. The modal responses were obtained and another equation was derived to explain the maximums and minimums for each mode. This equation allows specific modal amplitudes to be extracted for a specific moment distribution without resorting to a full scale simulation which saves computational time. It is extremely useful in designing the shape and size, as well as the placement of the actuator to meet control criteria.

Shaped actuators can produce modal responses in beams not possible using rectangular actuators. An example was given that shows how the shapes of actuators were chosen to actuate the 4th mode while not actuating the 2nd mode, which cannot be done using a rectangular actuator. The moment distribution from a shaped actuator can be distributed over a portion of a structure. This gives it more flexibility in meeting control criteria than a rectangular actuator with moments concentrated only at the edges. Using this theory, the shape of the actuator may now be made part of the design space for optimizing actuator configurations.

References

[1]Bailey, T. and J. E. Hubbard, 1985. "Distributed Piezoelectric-Polymer Active Vibration Control of a Cantilever Beam," *AIAA Journal of Guidance and Control*, **8**(5), pp. 605-611.

[2]Crawley, E. F., and J. de Luis, 1987. "Use of Piezoelectric Actuators as Elements of Intelligent Structures," *AIAA Journal*, **25**(10), pp. 1373-1385.

[3]Clark, R. L., C. R. Fuller, and R. A. Burdisso, 1991. "Design Approaches for Shaping Polyvinylidene Fluoride Sensors in Active Structural Acoustic Control (ASAC)," *Journal of Intelligent Material Systems and Structures*, **4**(3), pp. 354-365.

[4]Sumali, H., H. Cudney, J. Vipperman, 1992. "Vibration Control of Cylinders Using Piezoelectric

Sensors and Actuators," *ASPA/AIAA/ASME/SPIE Conference on Active Materials and Adaptive Structures*, Alexandria, Virginia, pp. 467-472.

[5]Wang, B. T., 1991. "Active Control of Sound Transmission/Radiation From Elastic Plates Using Multiple Piezoelectric Actuators," Doctoral Dissertation, Virginia Polytechnic Institute and State University, Blacksburg, Virginia.

[6]Burke, S. E., and J. E. Hubbard, Jr., 1987. "Active Vibration Control of a Simply Supported Beam Using a Spatially Distributed Actuator," *IEEE Control Systems Magazine*, 7(4), pp. 25-30.

[7]Lee, C. K., and F. C. Moon, 1990. "Modal Sensors/Actuators," *Journal of Applied Mechanics*, 57(6), pp 434-441.

[8]Collins, S. A., C. E. Padilla, R. J. Notestine, and A. H. von Flotow, 1992. "Design, Manufacture, and Application to Space Robotics of Distributed Piezoelectric Film Sensors," *Journal of Guidance, Control, and Dynamics*, **15**(2), pp. 396-403.

[9]Burke, S. E., J. E. Hubbard, Jr., and J. E. Meyer, 1993. "Distributed Transducers and Colocation," *Mechanical Systems and Signal Processing*, 7(4), pp. 349-361.

[10]Kim, S. J., V. R. Sonti, and J. D. Jones, "Equivalent Forces and Wavenumber Spectra of Shaped Piezoelectric Actuators," *2nd Conference on Recent Advances in Active Control of Sound and Vibration*, Blacksburg, Virginia, April, 1993, pp. 216-227.

[11]Thomson, W. T., 1988. <u>Theory of Vibration with Applications</u>, 3rd Ed., Prentice Hall Publishing Company, Englewood Cliffs, New Jersey.

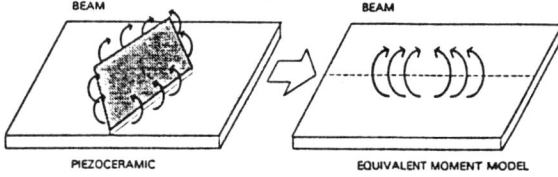

Figure 1. Moment Shown Modelled Along Beam Centerline

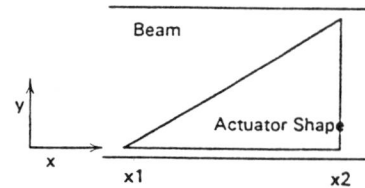

Figure 2. Actuator Shape with Finite and Infinite Slopes

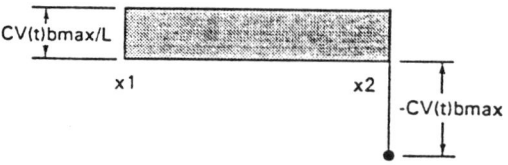

Figure 3. Moment Distribution for Finite and Infinite Slopes

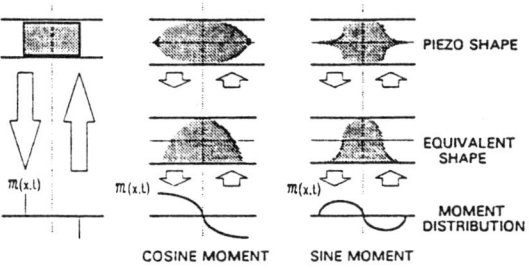

Figure 4. How Actuator Shapes Affect Moment Distribution

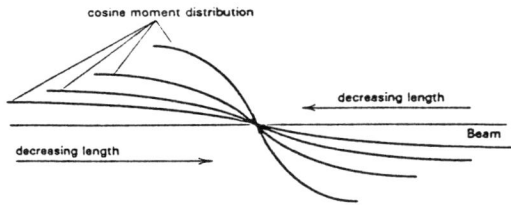

Figure 5. Length of Cosine Distribution as Varied during Simulation

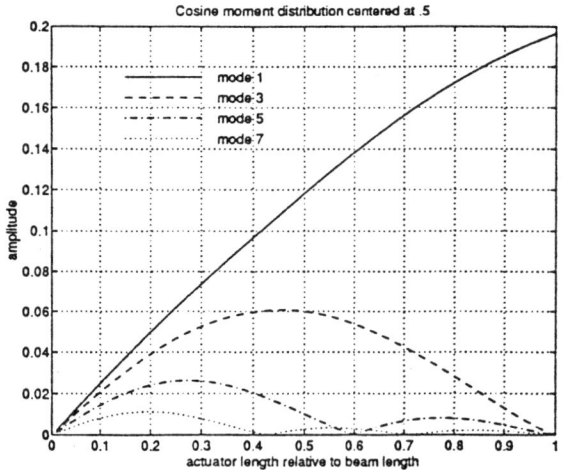

Figure 6. Cosine Moment Distribution Actuation

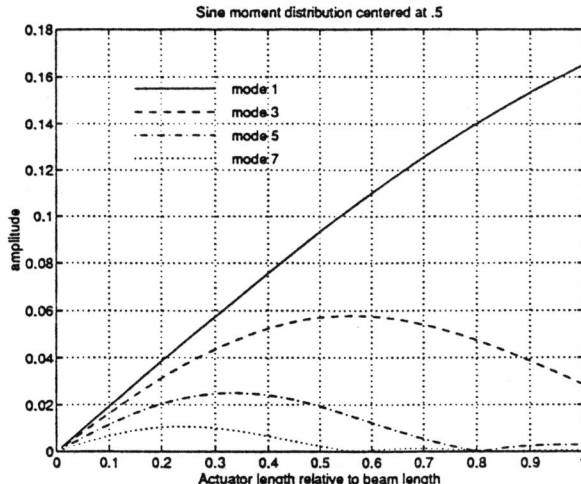

Figure 7. Sine Moment Distribution Actuation

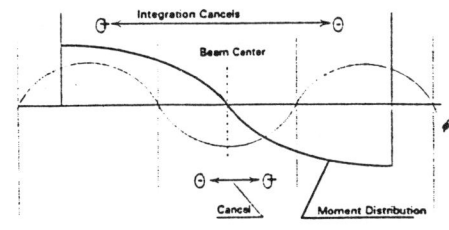

Figure 8. Cosine Moment at Center of Beam cannot Excite Even Modes

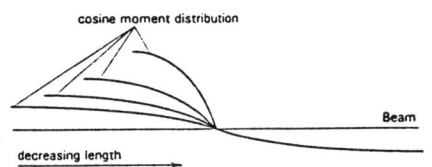

Figure 9. Non-Symmetric Distribution Simulation Model

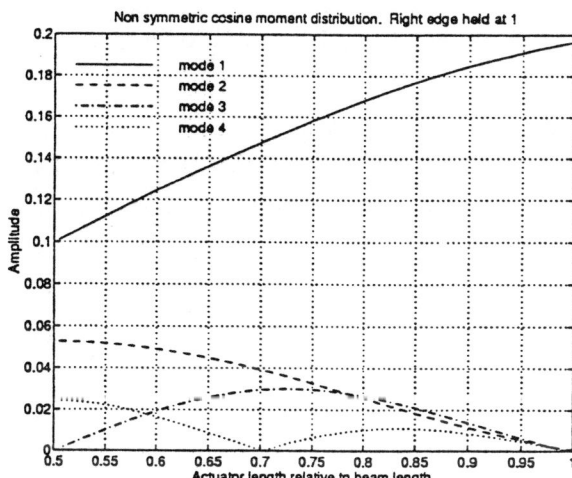

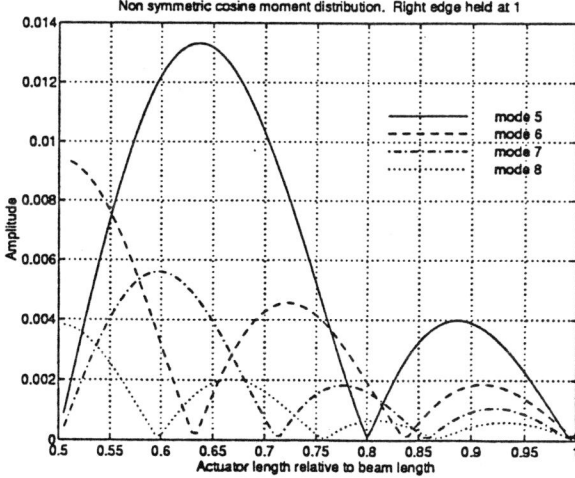

Figure 10. Cosine Moment Distribution Actuation
a. Modes 1-4
b. Modes 5-8

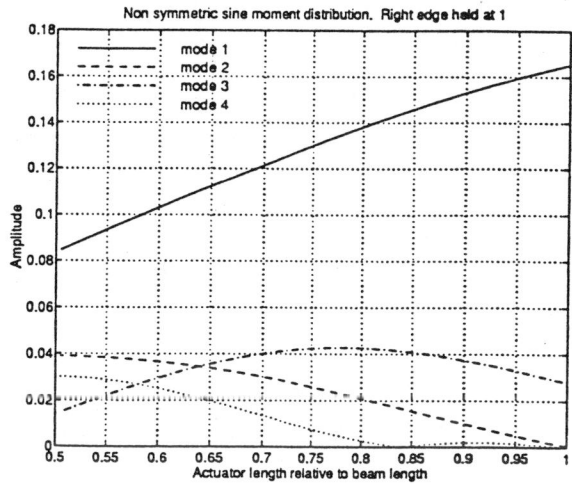

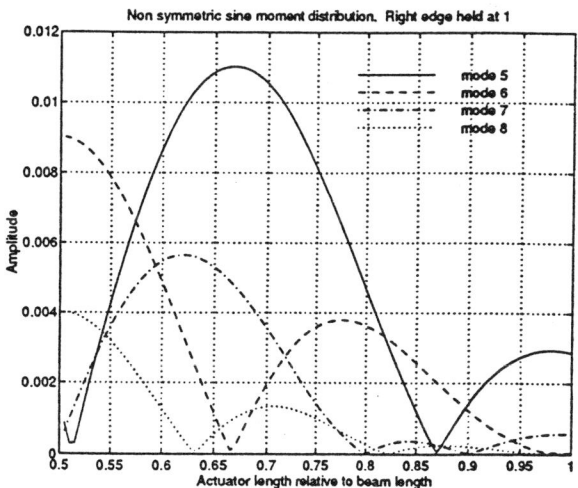

Figure 11. Sine Moment Distribution Actuation
a. Modes 1-4
b. Modes 5-8

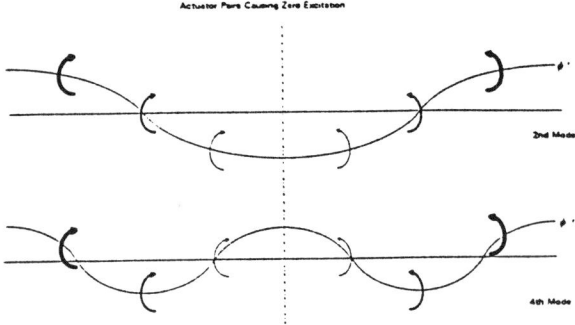

Figure 12. Rectangular Actuator Excites 2nd and 4th Mode

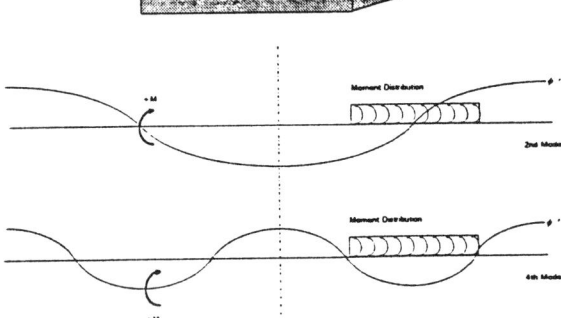

Figure 13. Actuating Mode 4 without Actuating Mode 2

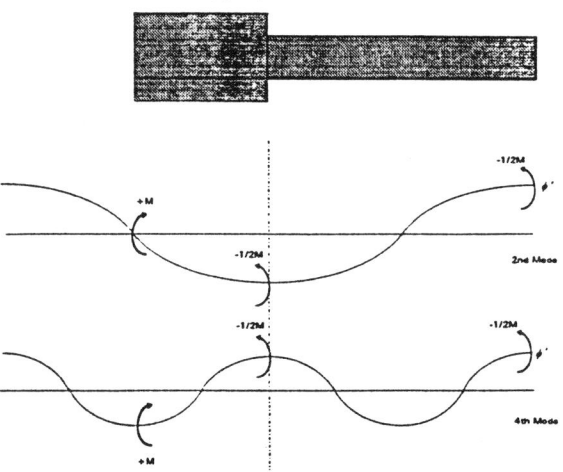

Figure 14. Mode 4 Actuated without Actuating Mode 2

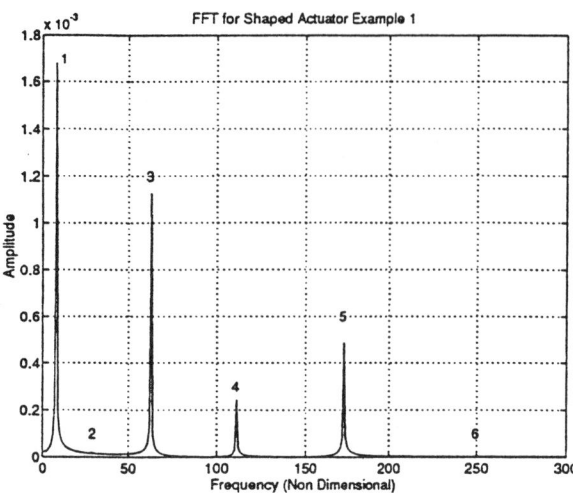

Figure 15. Peaks of FFT for Example 1

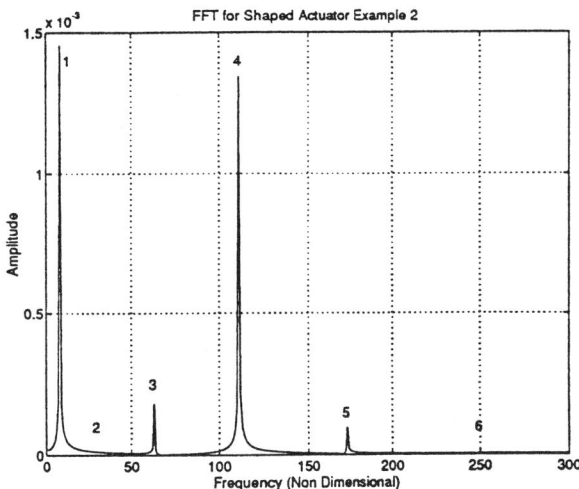

Figure 16. Peaks of FFT for Example 2

ADAPTIVE AIRFOILS FOR HELICOPTERS

by

R.L.Roglin[*], S. Kondor and S.Hanagud[**]

School of Aerospace Engineering
Georgia Institute of Technology
Atlanta, Georgia, U.S.A.

ABSTRACT

The development of an adaptive airfoil is discussed in this paper. The adaptive airfoil uses a shape memory alloy actuator mechanism to actively change the camber change of an airfoil. The lift resulting from this camber change is then the primary source of the helicopter thrust replacing the conventional collective control mechanism used by a helicopter. The paper describes the development of a helicopter test platform and the ground and flight tests performed to study the application of this type thrust generation. The testing provided many interesting pieces of data but the most interesting was the successful employment of the shape memory alloy actuator to lift a remotely piloted helicopter that was used as a test vehicle.

INTRODUCTION

The ability to change the shape of an airfoil cross section in real time operation has been the dream of many engineers. Such a real time change of the airfoil cross section has the potential of providing numerous benefits in the field of helicopters. Some possible applications include vibration reduction, aeroelastic tailoring, flight controls and minimization of performance losses due to nonuniform inflow and induced power requirements.

Through the use of smart or adaptive materials, real time computational capability, and the development of light weight controllers the dream of real time change of airfoil shape is now possible. Currently, piezoelectric materials, shape memory alloys, and electrorheological fluids are being used in these smart, adaptive or intelligent structures. In this paper, we propose to use a shape memory alloy to change the camber of a rotor blade and develop collective control.

SHAPE MEMORY ALLOYS

In 1965 at the U.S. Naval Ordnance Laboratory Beuhler and Wiley received a United States Patent for a series of engineering alloys that possessed a unique mechanical "memory". Generically named 55-Nitinol due to the fact that the material contains approximately 55 weight percent nickel in chemical composition. The Naval Ordnance Laboratory now called the Naval Surface Warfare Center has been very active in characterization of this material. Other laboratories including Battelle-Memorial Institute and National Aeronautics and Space Administration have also made major contributions toward understanding the mechanism associated with shape memory alloys (SMA) [R1 - R11].

This phenomenon of shape memory or the shape memory effect (SME) is illustrated in Figure 1. Let us consider a wire made of SMA (Step 1). Let us also assume that this wire is stretched and deformed inelastically at low temperature (Step 2). The wire is then heated to a temperature above a certain critical temperature (Step 3) to determine the shape that is to be remembered and then cooled. A mechanical stress is then applied to deform the cold wire (STEP 4). Then if the wire is heated it returns to its memorized geometric shape (Step 5).

Other researchers [R12 - R14] have also been attempting to characterize these materials and still have not completely come to a consensus as to how the SME occurs. These alloys have the mechanical ability to accommodate plastic strains of typically six to eight percent and when restrained from regaining their original shape they can generate stresses in the range of 100,000 psi [R15]. These force and displacement capabilities make NITINOL an attractive material for electromechanical actuators.

The temperature deformation relationship is illustrated in Figure 2. In this figure there is a transition temperature where the alloy undergoes a material phase transformation from austenite to martensite. It is this phase transformation which is responsible for the shape memory effect. The effect of the end conditions is also illustrated in Figure 2. As the end condition stiffness increases from free to fixed the deformation of the wire decreases and internal wire stress increases. This can lead to a phenomenon known as stress induced martensitic transformation of the alloy.

The fabrication of SMA hybrid composites are an example of a material with "Intelligence at the most primitive levels". These composite materials contain a lamina of SMA fibers. There is still much to be learned about the interaction of stress and temperature in relation to the extent, reliability and fatigue life of shape memory alloy dynamic actuators [R16,R17]. Composite materials which incorporate shape memory fibers have tremendous potential for creating structures which can adapt to changes in their loading environment [R18,R19]. When these composite materials are heated they can demonstrate stiffness increases of as much as three times that of the unheated material. There are many applications for such an actuator. These include

---

[*] Research Engineer, GTRI Aerospace Laboratory, AIAA Member
[**] Professor, Georgia Institute of Technology Aeospace Engineering, AIAA Member
Copyright © 1994 by Georgia Institute of Technology.
Published by the American Institute of Aeronautics and Astronautics, Inc. with permission

vibration control through the processes of active stiffness tuning or active strain energy tuning.
characteristics, frequencies and mode shapes, of a structure or mechanical component can be altered in a controlled fashion by heating the embedded or bonded shape memory alloy fibers associated with the structure. As the fibers heat, the alloy experiences a phase change when it transforms from the martensitic state to an austenitic state. Associated with this phase transformation is an increase in Young's modulus by a factor of three and an increase in yield strength by a factor of ten [R1]. Since the natural frequencies of any structure are proportional to the Young's modulus these increases will result in a stiffer structure with higher natural frequencies.

Active strain energy tuning [R17] is a process in which a shape memory lamina is located and oriented in such a way that upon heating it will not deform the structure but instead will impose a residual state of strain. The resulting stored strain energy (compression or tension) changes the total energy distribution and modifies the modal response of the structure.

The application to structural mechanics problems of these hybrid shape memory alloy composites is far greater than just the control of vibrations. The ability of the material to change modulus of elasticity would allow discretely placed shape memory alloy fibers to increase the critical buckling load [R8] of flexible structures or alter the critical speed of a flexible drive shaft [R20]. Another area where shape memory alloys will excel is in shape control of flexible structures [R26 - R23]. This application will include both motion and shape control. To accomplish this a controller must simultaneously employ force actuators and stiffness actuators for transition to occur from one predetermined shape to another similar to the way the human muscular system works.

HELICOPTER TEST PLATFORM

This project started with a generic radio controlled helicopter. A three view of the helicopter and the characteristic data associated with the helicopter are shown in Figure 3. The recommended glow fuel motor to be used with a helicopter of this size has an output of 1.8 horsepower and can generate 17,000 rpm under load.

Control of this helicopter is achieved with five small servo actuators. These function as follows:

- Longitudinal Cyclic
- Lateral Cyclic
- Collective Main Rotor
- Collective Tail Rotor
- Throttle

There is mechanical coupling between the main rotor collective and lateral cyclic and electronic coupling between the main rotor collective, tail rotor collective and throttle.

Active stiffness tuning [R16] is a method of steady state vibration control. The vibration

Helicopter Modifications

This baseline helicopter was then modified to remove the glow fuel power plant and incorporate an electric power plant. The reason for this modification was to improve the reliability and repeatability of helicopter flights. The gasoline motors are sensitive to many factors such as fuel content and carburetor adjustments. This sensitivity results in many hours of downtime spent diagnosing and correcting power plant problems. This sensitivity also leads to variations in the power available and the associated performance of the helicopter which adversely effect the repeatability of the experimental results. There is also the problem of testing the helicopter indoors with a internal combustion power plant. Since volatile liquid fuels are used there is a real possibility of fire or explosion and the problem of venting the exhaust.

The first step was to determine the required input and output of an electric motor to be used as the helicopter power plant. An analysis of the rotor system lift and drag was performed using a combination of the momentum and blade element theories [R24,R25]. There are two basic equations associated with the approach. These equations are

$$dT = b \frac{1}{2} \rho (\Omega r)^2 a (\theta - \phi) c \, dr \qquad (1)$$

$$dT = (2 \pi r \, dr \, \rho \, u) \, 2v \qquad (2)$$

Equation 1 is the thrust developed according to the blade element theory and equation 2 is the thrust developed according to the momentum theory. These two are then set equal to each other and the result is a formula for the induced velocity.

$$v = \left(\frac{V_v}{2} + \frac{bca\Omega}{16\pi} \right) * \left(-1 + \sqrt{1 + \frac{2\Omega r (\theta - \frac{V_v}{\Omega r})}{\frac{4\pi V_v^2}{bca\Omega} + V_v + \frac{bca\Omega}{16\pi}}} \right) \qquad (3)$$

Once the induced velocity is known the inflow angle can be determined

$$\phi = \frac{v + V_v}{\Omega r} \qquad (4)$$

and then substitution of equation 4 into equation 1 and integration from the blade root to the effective blade tip results in the thrust being developed by the rotor system. Using these equations a parametric study was performed where the parameters were motor rpm (1400 rpm to 1600 rpm) and collective setting (1° to 5°) to determine if an electric motor could be found for this application. The results of this parametric study are tabluated in Table 1.

Collective Setting (Degrees)	1400 rpm Thrust	1400 rpm Power	1450 rpm Thrust	1450 rpm Power	1500 rpm Thrust	1500 rpm Power	1550 rpm Thrust	1550 rpm Power	1600 rpm Thrust	1600 rpm Power
1.0	0.43	0.32	0.47	0.36	0.52	0.41	0.57	0.46	0.62	0.52
1.5	1.07	0.79	1.16	0.89	1.26	1.00	1.36	1.11	1.46	1.24
2.0	1.85	1.37	2.00	1.53	2.16	1.71	2.33	1.90	2.50	2.11
2.5	2.74	2.03	2.96	2.27	3.19	2.52	3.42	2.80	3.67	3.10
3.0	3.71	2.74	4.01	3.07	4.31	3.41	4.62	3.78	4.94	4.18
3.5	4.75	3.51	5.12	3.92	5.50	4.36	5.89	4.82	6.30	5.32
4.0	5.84	4.32	6.29	4.81	6.75	5.35	7.23	5.92	7.73	6.53
4.5	6.98	5.16	7.51	5.75	8.06	6.38	8.63	7.06	9.22	7.79
5.0	8.15	6.03	8.77	6.71	9.41	7.45	10.07	8.24	10.76	9.08

Units: Thrust (lbs) Power (watts/100)

Table 1 Thrust & Power Output
(Hover & OGE)

Based on these results and the power rating of the currently used gasoline motor an electric motor with an output of 1200 watts (1.6 Horsepower) was selected. Although from Table 1 it would appear that this motor is very conservative it was selected because an actual value for the losses associated with the drive train and the power required for the tail rotor are estimated as a 25% reduction in available power. In addition to the estimation of power loss due to the tail rotor the effect of the vertical drag associated with the airframe has been neglected. The helicopter airframe had to be modified and a mounting block had to be designed and fabricated to integrate the electric motor into the airframe.

The motor mount was a rectangular block of aluminum with two orthogonal set screws to hold the electric motor and four mounting screws to mate with the mounting holes on the airframe. To achieve the proper alignment of the motor and the drive train the mounting holes located on the helicopter were elongated to allow longitudinal adjustment of the mounting block. In addition to the motor mount a coupler was also fabricated which allowed the motor to mate with the existing centrifugal clutch which is the first component of the drive train. These two components and the modification of the airframe mounting holes allowed the electric motor to be mated to the airframe.

Unlike the internal combustion motor which uses a throttle the electric motor requires an electronic speed controller. This device regulates the motor speed by adding or subtracting resistance from the input circuit which then decreases or increases voltage to the electric motor. The problem with this device is the removal of the heat generated by the resistors. This was not a trivial problem and during early ground tests of the helicopter a speed controller caught on fire and was destroyed. This problem was solved by locating the speed controller in the down wash of the main rotor. The device was moved and then a series of tests were performed where the speed setting was varied and using a temperature probe the controller temperature was measured after each test. This helped to locate the speed controller in the downwash where it would get sufficient cooling to run at any speed indefinitely. In addition to the new location a heat sink was added with fins to remove heat from the controller. The speed controller also was subject to overload once because of a voltage spike due to inductance in the power tether. The voltage was a result of the long (60 foot) tether power cord used to provide power to the electric motor. This problem was solved by including a large capacitor (2200 µf \ 50 volt) in the input power line to drain off the AC voltage before it could get to the speed controller. There also was a thermal problem with the electric motor. This problem is not present when these motors are used in an airplane application because the downwash from the propeller is used to cool the motor but in our application the downwash is obstructed from cooling the motor. This problem was solved by attaching a cooling fan to the bottom output shaft of the motor and sucking air through the motor to provide sufficient cooling to keep the motor core below 300° F.

ROTOR BLADES

A NACA 0012 is a standard main rotor blade cross section for this size helicopter. The blade span is 28 inches (71.12 cm) and the blade chordwise dimension is 2.5 inches (6.35 cm) resulting in a solidity of 0.0568. The design goal was to match the thrust generated at a collective setting greater than that required to hover the helicopter by changing the camber of a portion of the rotor blade. The tethered helicopter has a weight of 8 pounds (35.58 NT) therefore the cambered portion of the blade must generate thrust greater than 8 pounds. Blade element and momentum theory aerodynamics in the form of equation 1, 2, 3, and 4 were used to determine the required collective setting (4.1°) to hover at 8 pounds. Conservation of momentum neglecting any download effect due to the small size of the helicopter fuselage was then used to determine the climb velocity V_n. This is illustrated in Figure 4.

Once the climb velocity had been established, size and location of the control surface had to be determined. The design methodology was to set the collective such that the helicopter would generate a thrust equal to the weight of the vehicle. The thrust generated by the control surface would then be employed for vertical climb or descent. Based on this methodology the collective was set at 4.1°. This collective setting would generate the 8 pounds of thrust required for the helicopter to hover. The control surface was located at 75% of the span of the rotor blade [R26]. The dimensions of the control surface are 8 inches in the spanwise direction (28% of the total span) and 1 inch in the chordwise direction (40% of the total chord).

A 2D panel method airfoil analysis was performed to determine upper and lower airfoil section pressure distributions as well as sectional lift and drag coefficients. These coefficients were used to determine the incremental lift associated with rotor blade segments and the resultant pressure distribution on the airfoil. The resultant pressure distribution is illustrated in Figure 5. From the pressure distribution illustrated in Figure 5 the effect of the deflected control surface can been seen as an increase in the pressure differential at approximately 55% of the chord.

The area under this portion of the curve is the airload on the control surface and is used to calculate the required moment for deflecting the surface. Equation 5 is the integral for calculating the area and is equal to the lift coefficient associated with just the control surface portion of the airfoil. This calculation assumes the airfoil to be torsional rigid with no load redistribution which would occur if the blade twisted as center of pressure moved aft.

$$c_l = \frac{1}{c} \int_0^c Cp_{lo} - Cp_{up}\, dx \implies 0.17 \quad (5)$$

In Table 2 the incremental airloads on the control surface from the inside edge radius 1.5 feet to the outside edge at radius 2.08 feet are tabulated. The airload on the control surface is then the sum of the incremental airloads. This total lift then was assumed to act at the chordwise station of 1.83 inches from the leading edge or 0.33 inches aft of the hinge line. Based on these numbers the moment due to airloads to be overcome is .34 inch-pounds or 5.44 inch-ounces. Once the airload has been overcome the control surface must be deflected 1.5° for the effective angle of attack to be the required 4.7°. The geometric relationship between control surface angle and effective angle of attack is illustrated in Figure 6.

Radius (ft)	Velocity (ft/sec)	Dynamic Pressure (lbs/sqft)	Lift (lbs)
1.50	250.50	74.67	0.09
1.58	264.42	83.20	0.10
1.67	278.33	92.19	0.11
1.75	292.25	101.64	0.12
1.83	306.17	111.55	0.13
1.92	320.08	121.92	0.14
2.00	334.00	132.75	0.16
2.08	347.92	144.04	0.17

Total Lift => 1.02

Table 2 Control Surface Airload

Again estimating the lift by the combined blade element and momentum theories the total lift for this rotor system would be 9.02 pounds. Where the blades are set to a collective pitch of 4.1° and the control surface described above is deflected by 1.5°. The actuation element used to manifest this deflection is a Shape Memory Alloy.

The next issue to resolve was the power requirement for the SMA actuator element. In Figure 7 the power required to activate a 2 inch long strand of 0.01 inch diameter SMA wire is shown. It can been seen from this figure that the shape memory effect can be used to deflect the control surface with a reasonable amount of electric power. Even though the time responses that one encounters when shape memory alloys are used for shape changes (1 hertz) are not as fast as the time responses of piezoelectric transducers, a shape memory alloy can maintain the changed shape for a prescribed time duration. This makes the shape memory alloy an ideal smart or adaptive material for collective control.

In Figure 8 the design configuration of the control surface and the shape memory alloy actuation element are illustrated [R27 - R29]. Based on this figure when the SMA wire is heated it will contract pulling free end point "A" of the actuator lever arm forward. As the free end of the actuator lever arm moves forward the portion between points "B" and "C" will rotate clockwise and the embedded end point "D" of the control surface lever arm will move downward. The length of the actuator lever arm was based on the amount of strain recovery and force associated with this work. The amount of strain allowed in an application is a function of the number of cycles the material will experience this strain. The manufacturers recommended strain for applications such as this is 4% and the available force is 6.5 ounces for SMA wire with a diameter of 0.010 inches. Since the required moment to be overcome is 5.44 inch-ounces the actuator lever arm (point "A" to "B") must be 0.84 inches in length. The 4% strain and a right angle pull orientation were then used to calculate the amount of SMA wire required and the voltage required. Figure 9 illustrates the layout of the SMA wire.

The 0.1" dimension shown in this figure represents the actual magnitude of the displacement at point "A" (free end of the actuator lever arm). The required displacement from the calculations is 0.022" (0.84*Tan1.5°) which is a 0.9% strain.

R = r*L => 0.5W/in * 2.24in = 1.12W
V = R*i => 1.12W*1A = 1.12V
P = V*i => 1.12V*1A = 1.12W

The above calculation based on 4% strain being developed indicates that 1.12 watts of power are required per blade. Since, only 0.9% strain is required and a 0.9% strain corresponds to a temperature of 60°F the curve in Figure 7 indicates that only 0.6 amps are required to power the actuator mechanism.

FLIGHT AND GROUND TESTS

After the design of the test platform a number of ground and flight tests were performed to demonstrate the application of a shape memory alloy actuated control surface. Table 3 lists all test which were performed.

TEST	DESCRIPTION	OBJECTIVE
1	Ground spin up of rotor system	Determine electric power requirements
2	Flight test of electric powered helicopter	Determine flight characteristics of electric powered helicopter
3	Wind tunnel test of modified rotor blade	Investigate structural integrity of modified rotor blade
4	Ground spin up of modified rotor blade	Investigate structural integrity of modified rotor blade
5	Ground spin up of modified rotor blade	Track and balance modified main rotor blades
6	Flight test with modified blades	Determine flight characteristics with modified blades
7	Ground test of activation circuit	Investigate activation circuit under no external load condition
8	Flight test using control surface	Demonstrate the application of a SMA actuated control surface

Table 3 Test Program

The first series of ground tests were performed on the basic system in order to determine electrical power requirements and limitations on the electric motor and motor speed control unit; these tests also were used to setup and make preliminary adjustments to the radio control unit. All tests were performed using an unmodified blade (NACA 0012 with a 2.5 inch chord). The next series of tests were flight tests of the helicopter system with unmodified blades. The purpose of these tests were to provide the pilot with flight time on an electric powered helicopter and to refine the radio control unit setup. In addition to this many mechanical problems were encountered which were not present during the ground spin tests. An example of one problem was the tendency for the gear teeth on the tail rotor drive gear to strip during flights when the electric motor rpm would be slightly above the rpm developed by a glow motor.

Results from these tests showed that an electric helicopter of this size can generate sufficient lift with the unmodified blades, but care must be taken not to overheat the electric motor or the motor speed control unit. The main rotor speed during spinup on the ground was 1500 rpm and the hover rpm was 1400 rpm. The required battery voltage was 36 volts to overcome a 10 volt drop in the 60 foot tether line between the batteries and the helicopter. The operating temperature of by the electric motor was 150° F and the motor speed control was 120° F; these temperatures are slightly below the upper operating limits for both these components.

The next test was a wind tunnel test to insure the structural integrity of the modified helicopter rotor blade. The standard blade was first modified by cutting a control surface into the blade. The control surface was located 4 inches from the tip of the blade, the spanwise length of the control surface was 8 inches and the chordwise length of the control surface was 40 percent of the airfoil chord or 1 inch. A plastic hinge was attached to the top surface of both the control surface and the blade to attach the two pieces. This configuration was then placed in the wind tunnel in a vertical orientation. The blade was subjected to a series of dynamic pressures minimum of 0 psf maximum of 50 psf the pressure was increased to this maximum in 10 psf increments and was subjected to each dynamic pressure increment for approximately 10 minutes.

The results of this test were the modified blade demonstrated sufficient structural integrity for dynamic pressures from 0 to 50 psf and the control surface did not flutter.

The next test was to subject the modified rotor blade to a ground spin test to evaluated the structural integrity to centrifugal load. For this test, the helicopter was weighted with approximately 12 pounds of lead. The rotor was spun to 1600 rpm with zero collective input and was kept at this speed for approximately 5 minutes. Examination of the blade during the test showed no sign of vibration or flutter even though the blades did not track very well. After the test examination of the blades showed no structural integrity problems. The rotor was then spun up again and collective was applied in a controlled manner till the helicopter actually lifted off with the 12 pounds of ballast.

The results of this test were the modified blade demonstrated sufficient structural integrity for dynamic pressures from 0 to 50 psf and centrifugal loads associated with a rotational speed of 1600 rpm. Also under these conditions the control surface did not flutter.

The next tests were flight tests using the modified blade and the conventional control system. In this test the aircraft was hovered at 6 to 8 feet off the ground for approximately 5 minutes. This test was performed 8 times and in all flights the helicopter experienced only mechanical failures in the drive system. The major failure being the tendency for the gear teeth on the tail rotor drive gear to strip and there was one tail rotor blade strike on a hard landing during a gust. The purpose of these tests were also to provide the pilot with flight time on the electric powered helicopter with modified rotor blades.

The results of these test flights again emphasized the sensitivity of the drive system to overspeed problems. But the major outcome was the modified blades are able to lift the helicopter and can be controlled with conventional control inputs.

Ground tests of the helicopter with the modified blade activated to a camber change were the next set of tests performed. The objective of these was to demonstrate that a camber change could be maintained using the shape memory alloy actuation system. These tests gave the pilot some practice at activating and deactivating the control surface. The conditions for these tests were similar to the other ground tests where the rpm was varied from 0 to 1600 rpm in increments of 100 rpm and each increment was maintained for 5 minutes. Since, the actuator wire was not directly exposed to the airstream there was minimal aerodynamic cooling of the wire actuator and power requirements were low, 1 amp at 3 volts.

The results of this test demonstrated the structural integrity of the shape memory alloy and the control surface actuation system.

The final test was a flight test of the helicopter where the shape memory actuator was activated and resulted in a vertical climb of the helicopter. This test was performed in the following sequence: (1) the helicopter was hovered at 6 feet off the ground, (2) once a stable hover was achieved the actuator was activated and the helicopter climbed vertically with the pilot maintaining stability through cyclic and increased tail rotor pitch to overcome the additional torque placed on the main rotor system, (3) after a climb of 3 feet the pilot deactivated the actuator and varied the collective to land the helicopter. This test was performed by the tethered helicopter with a tether line which had 6 inch increments marked on it for maintaining hover altitude and measurement of the vertical climb.

The results of this test demonstrated the ability of the shape memory alloy actuator system to change blade camber. This camber change generated enough thrust to result in a vertical climb of the 8 pound helicopter. Therefore a shape memory alloy actuator system has the ability to effect the magnitude of camber changes required to replace the current collective control system on the helicopter.

CONCLUSIONS

In conclusion, this paper documents the development of a flight test platform and the subsequent testing done using this platform. These tests were performed to investigate the capability of a shape memory alloy actuated control surface. The platform developed was an electric powered remotely controlled scaled helicopter. Power for this helicopter was provided by a tethered power cord or by an onboard NiCad battery pack. The reason an electric motor was selected was to allow indoor flying of the helicopter and to improve the repeatability of experimental results. Both ground and flight tests were performed with the electric helicopter. The ground testing was done primarily to insure structural integrity of the modified helicopter blades and the flight tests were done to demonstrate the application of a shape memory alloy actuated control surface.

REFERENCES

1 Cross,W.B.,Kariotis,A.H. and Stimler,F.J.. "Nitinol Characterization Study." In *NASA CR-1433.* , 1969.

2 Jackson,C.M.,Wagner,H.J. and Wasilewski,R.J.. "55-Nitinol The Alloy with a Memory: Its Physical Metallurgy, Properties and Applications." In *NASA-SP-5110.* , 1972.

3 Jai,J. and Rogers,C.A.. "Formulation of a Mechanical Model for Composites with Embedded SMA Actuators." In *ASME Failure Preventions and Reliability* . , 1989.

4 Laing,C.,Jia,J. and Rogers,C.A.. "Behavior of Shape Memory Alloy Reinforced Composite Plates, Part1: Model Formulation & Control Concept and Part2 : Results." In *30st AIAA Structures, Structural Dunamics and Materials Conference.* , 1989.

5 Liang,C. and Rogers,C.A.. "A One-Dimensional Thermomechanical Constitutive Relation of Shape Memory Materials." In *31st AIAA Structures, Structural Dunamics and Materials Conference.* , 1990.

6 Stoeckel,D. and Simpson,J.. "Actustion and Control with Ni-Ti Shape Memory Alloys." In *Active Materials & Adaptive Structures*. Philadelphia, Pennsylvania: IOP Publishing Ltd., 1991.

7 Wayman,C.M.. "Shape Memory and Related Effects." In *Active Materials & Adaptive Structures*. Philadelphia, Pennsylvania: IOP Publishing Ltd., 1991.

8 Baz,A.,Ro,J.,Mutua,M. and Gilheany,J.. "Active Buckling Control of Nitinol Reinforced Composite Beams." In *Active Materials & Adaptive Structures*. Philadelphia, Pennsylvania: IOP Publishing Ltd., 1991.

9 Wang,F.E.. "Recent Advances in Nitinol Technology." In *Active Materials & Adaptive Structures*. Philadelphia, Pennsylvania: IOP Publishing Ltd., 1991.

10 Perkins,J.. *Shape Memory Effect in Alloys*. New York, New York: Plenum Press, 1975.

11 Schetky,L.. "Shape Memory Alloys." *Scientific American*, 1 November 1979

12 Cliff,E.M.. "Time Scale Effects in Shape Memory Alloys." In *Active Materials & Adaptive Structures*. Philadelphia, Pennsylvania: IOP Publishing Ltd., 1991.

13 Burns,J.A. and Spies,R.D.. "Finite Element Approximation of a Shape Memory Alloy." In *Active Materials & Adaptive Structures*. Philadelphia, Pennsylvania: IOP Publishing Ltd., 1991.

14 Negahban,M.. "Constitutive Modeling of Phase Transition in Smart Materials." In *Active Materials & Adaptive Structures*. Philadelphia, Pennsylvania: IOP Publishing Ltd., 1991.

15 Brown,W.. *Technical Characteristics of FLEXINOL*. Irvine California: Dynalloy,Inc.

16 Rogers,C.A.. "Active Vibration and Structural Acoustic Control of Shape Memory Alloy Hybrid Composites: Experimental Results." In *International Conference on Recent*

17. Rogers,C.A. and Barker,D.K.. "Experimental Studies of Active Strain Energy Tuning of Adaptive Composites." In *31st AIAA Structures, Structural Dynamics and Materials Conference.* , 1990.

   Developments in Air and Structural Borne Sound and Vibration. , 1990.

18. Rogers,C.A.. "Novel Design Concepts Utilizing Shape Memory Alloy Reinforced Composites." In *American Society of Composites 3rd Technical Conference on Composite Materials.* Lancaster, Pennsylvania: Technomic, 1988.

19. Rogers,C.A. and Robertshaw,H.H.. "Shape Memory Alloy Reinforced Composites." *Engineering Science Reprints 25, ESP25.88027*, (1988).

20. Nagaya,K., Takeda,S., Tsukui,Y. and Knmaido,Y.. "Active Control Method for Passing Through Critical Speeds of Rotating Shafts by Changing Stiffness of the Supports with Use of Memory Metals." *Journal of Sound & Vibration*, 113 (1987).

21. Baz,A., Ro,J., Poh,S. and Gilheany,J.. "Control of Smart Traversing Beam." In *Active Materials & Adaptive Structures.* Philadelphia, Pennsylvania: IOP Publishing Ltd., 1991.

22. Lagoudas,D.C. and Tadjbakhsh,I.G.. "Active Flexible Rods with Embedded SMA Fibers." In *Active Materials & Adaptive Structures.* Philadelphia, Pennsylvania: IOP Publishing Ltd., 1991.

23. Baz,A.,Inman,K. and McCoy,J. "Active Control of Flexible Beams using Shape Memory Actuators." *Journal of Sound & Vibration*, 140 (1990).

24. Johnson, W., *Helicopter Theory*, Princeton University Press ,New Jersey, 1980.

25. Gessow, A. and Myers, G.C., Jr., *Aerodynamics of the Helicopter 8th Edition*, pp 29, Fredrick Ungar Publishing Co.,New York, 1985.

26. Personal communication with Cliff Gunsallus, Kaman Aerospace Corporation, 1993

27. Roglin, R.L. and Hanagud, S.V.,1992,Shape Memory Alloy Camber Control,Patent Issued.

28. Hanagud, S.V. and Roglin, R.L., "Adaptive Airfoils", *Souteastern Confrence of Theoretical and Applied Mechanics*, 1992.

29. Hanagud, S.V., Roglin, R.L. and Nagesh Babu, G.L., "Smart Airfoils for Helicopter Control", *Eighteenth European Rotorcraft Form*, 1992.

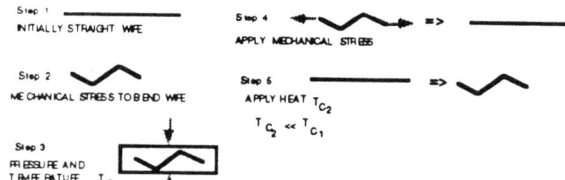

FIGURE 1 Shape Memory Effect

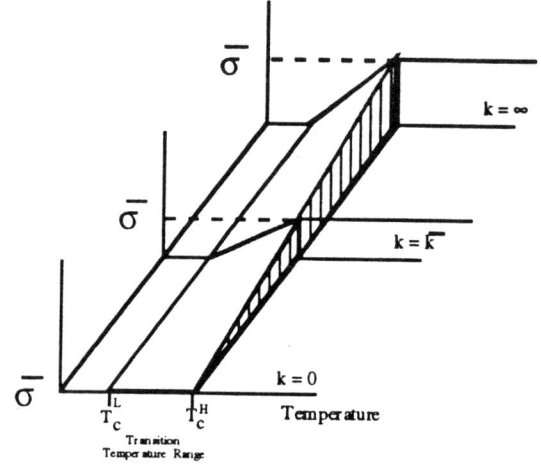

Temperature Relationship

$$\sigma = \bar{\sigma} \left(\frac{T - T_C^L}{T_C^H - T_C^L} \right)$$

Deformation Relationship

$$\delta = 0 => k = \infty : \bar{\sigma} = \sigma_{max}$$

$$\delta = \delta_{max} => k = 0 : \bar{\sigma} = 0$$

$$\bar{\sigma} = \sigma_{max} \left(\frac{\delta_{max} - \delta}{\delta_{max}} \right)$$

FIGURE 2 Temperature Deformation Relationship

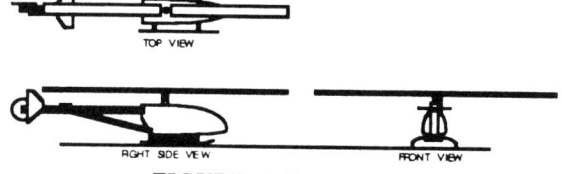

FIGURE 3 Three View

Fuselage Length	35 in
Fuselage Width	8 in
Helicopter Weight	108 oz
Power Plant Weight	20 oz

ROTOR PARAMETERS	MAIN ROTOR	TAIL ROTOR
Radius	28 in	6 in
Chord	2.5 in	0.75 in
Solidity	0.0568	0.0796
No. of Blades	2	2
Tip Speed	4693 in/sec	4023 in/sec
Airfoil Section	NACA 0012	NACA 0012
Collective Range	+8° -> -8°	+8° -> -8°
Longitudinal Cyclic Range	+5° -> -5°	N/A
Lateral Cyclic Range	+5° -> -5°	N/A

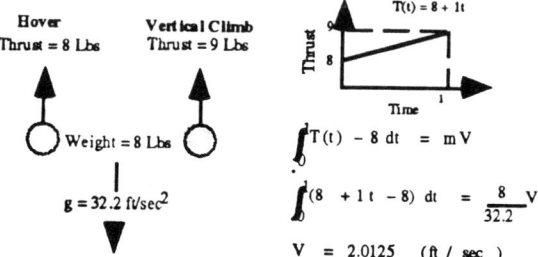

FIGURE 4 Vertical Climb Velocity

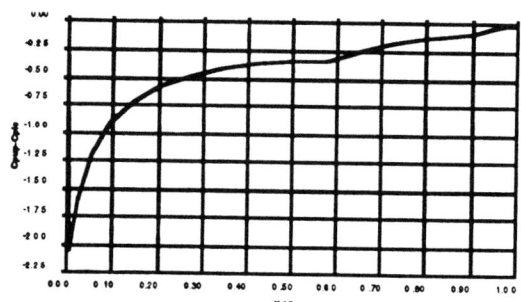

FIGURE 5 Airfoil Pressure Distribution

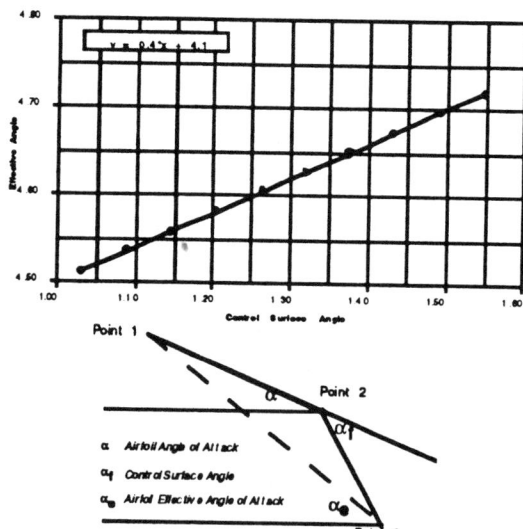

FIGURE 6 Effective Angle of Attack

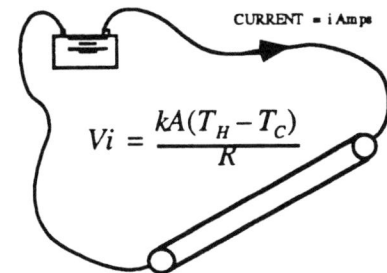

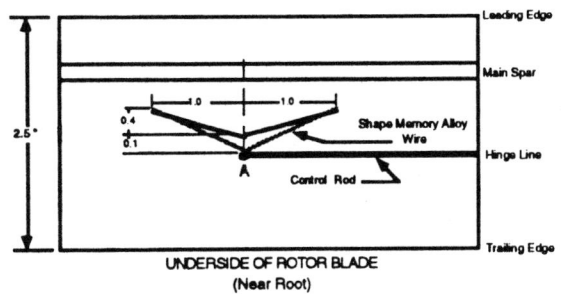

FIGURE 9 Actuator Wire Layout

A = Surface Area of the SMA Wire
k = Thermal Conductivity of the SMA Wire
R = Radius of the SMA Wire
T_H = Temperature of the SMA Wire Center
T_C = Temperature of the SMA Wire Surface

$$(6) i = \frac{(0.05)(0.4054)(T_H - 20)}{(0.0127)}(4.186)$$

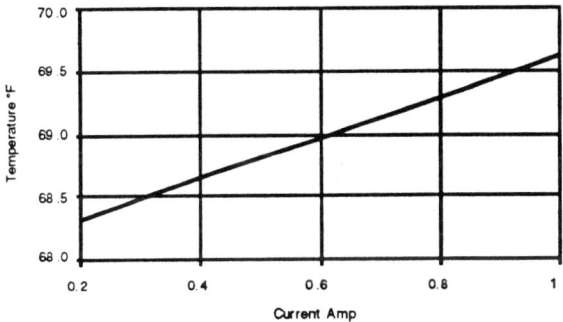

FIGURE 7 SMA Power vs. Temperature Relationship

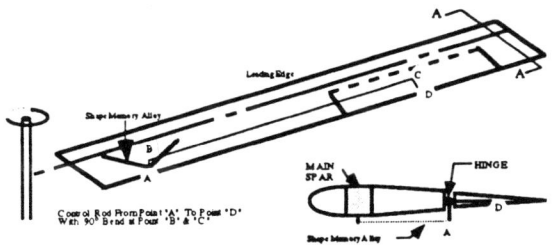

FIGURE 8 Adaptive Airfoil with Shape Memory Alloy

ACTIVE CONTROL OF HELICOPTER ROTOR BLADES
WITH INDUCED STRAIN ACTUATORS

AIAA-94-1765-CP

Victor Giurgiutiu[*], Zaffir Chaudhry[**], Craig A. Rogers[†]

Virginia Polytechnic Institute and State University,
Blacksburg, VA 24061-0261, U.S.A.

Abstract

Rotor blade vibration reduction based on higher harmonic control - individual blade control (HHC-IBC) principles is presented as a possible area of application of induced strain actuators (ISA). Recent theoretical and experimental work on achieving HHC-IBC through conventional and ISA means is reviewed. Although the force-displacement and power-energy estimates vary significantly, some common-base values are identified. Hence, a benchmark specification for a tentative HHC-IBC device based on the aerodynamic servo-flap principle using ISA is developed. Values for the invariant quantities of energy, power and force-displacement product are identified, along with actual displacement and force values of practical interest. The implementation feasibility of this specification into an actual ISA device is then discussed. It is shown that direct actuation is not feasible due to the large required length of the ISA device, resulting in excessive compressibility effects (displacement loss and parasitic strain energy). Indirect actuation through a displacement amplifier was found to be more feasible, since this arrangement allows the matching of internal and external stiffness. A closed form formula was developed for finding the optimal amplification gain for each required value of the closed-loop amplification ratio. Preliminary studies based on force, stroke, energy and output power requirements show that available ISA stacks coupled with an optimally designed displacement amplifier might meet the benchmark specifications.

1 Introduction

1.1 Rotor Blade Active Vibration Control through HHC and IBC

A traditional way of controlling the pitch feathering motion of helicopter rotor blades is the **swash-plate mechanism**. This low-frequency device is designed for collective (quasi-steady) and cyclic (1/rev) control of the pitch motion $\theta(t) = \theta_0 - a_1 \cos\Omega t - b_1 \sin\Omega t$, where Ω is the rotor angular speed. **Higher harmonic control (HHC)** superposes an oscillatory modulation of frequency ω on the basic pitch control of frequency Ω :

$$\theta(t)\sin\omega t = (\theta_0 - a_1\cos\Omega t - b_1\sin\Omega t)\sin\omega t$$
$$= \theta_0\sin\omega t - \frac{a_1}{2}[\sin(\omega-\Omega)t + \sin(\omega+\Omega)t]$$
$$- \frac{b_1}{2}[\cos(\omega-\Omega)t + \cos(\omega+\Omega)t]$$

Since helicopter vibrations appear at multiples of rotor speed, HHC frequencies of the form $\omega = N\Omega$ are usually used, thus resulting in pitch modulations not only at $N\Omega$ but also at $(N-1)\Omega$ and $(N+1)\Omega$. Extensive theoretical[1,2] and experimental studies[3] have shown that HHC can be an effective means of rotor blade vibration control, and reductions as large as 90% have been reported.[3,4] However, two major issues limit the usefulness of conventional HHC: (a) the swash plate is an inherently low-frequency device and hence acts as a low-pass filter on the HHC input, and (b) HHC affects all the blades equally, while real rotor blades may differ noticeably in their detailed aerodynamic and inertial characteristics.

Individual Blade Control (IBC) denotes a method by which the blades of a rotor system have their pitch motion controlled individually. Thus, many additional issues may be addressed with this control concept, such as: blade tracking, and lift improvement through a 2/rev modulation. As a means of vibration control, the IBC implementation of the HHC concept can address the vibrations due to manufacturing variability between the blades, and also is not restricted by the N+1, N-1 phenomenon specific to swash plate control. An effective HHC-IBC device should also have a good frequency response characteristic, and thus produce the least possible filtering of the HHC signal.

Various engineering options are possible for implementing the HHC-IBC concept. One is to transform the conventional push-rods, connecting the swash-plate with the blade root, into active devices, e.g., hydraulic actuators.[5] However, significant values of the aerodynamic pitch moment (450 - 780 Nm) must be overcome.[6] A more efficient way is to utilize the principle of servo-aerodynamic

---

[*]Visiting Professor, Engineering Science and Mechanics Department, Senior Member AIAA
[**]Research Scientist, Center for Intelligent Materials Systems and Structures, Member AIAA, ASME
[†]Professor and Director, Center for Intelligent Materials Systems and Structures, Member AIAA, ASME

Copyright 1994 by
Center for Intelligent material Systems and Structures
Published by the AIAA, Inc. with permission

that takes energy from the airstream to modify the rotor blade pitch. One application of this principle is the servo-flap concept. A limited span, trailing edge flap is used to produce aerodynamic pitching moments that in turn modify the pitch setting of the flexibly-restrained blade. The Kaman SH-2G Seasprite helicopter uses a mechanically operated servo-flap for primary (collective and cyclic) rotor blade control[7]. However, the SH-2G servo-flaps are operated by a traditional swash-plate through push rods inside the blade and are not suited for HHC-IBC. Efficient vibration control requires frequency characteristics higher than can be offered by such mechanical systems. Remote electrical operation is an attractive alternative, and solid state ISA devices present numerous opportunities in this direction.

1.2 Induced Strain Actuation Principles

In certain materials, electro-mechanical coupling occurs in the form of an electrical charge being generated by an externally applied stress and, *vice-versa*, a strain being generated by an applied electric field. For active control applications, the interest is directed towards materials displaying mainly the latter effect, i.e., the **induced strain actuation (ISA)** effect. Such ISA materials can respond either linearly (e.g., **piezoelectric**) or quadratically (e.g., **electrostrictive**) to the applied electric field. An associate behavior is displayed by the **magnetostrictive** materials where an induced strain is produced by an applied magnetic field. The general constitutive equations for an ISA material are

$$S_{ij} = s^E_{ijkl}T_{kl} + d_{kij}E_k + M_{klij}E_kE_l$$
$$D_i = d_{ijk}T_{kl} + \epsilon^T_{ik}E_k ,$$
(1.1)

where S_{ij} is the strain, T_{kl} the stress, E_k (or E_l) the electric field, s_{ijkl} the compliance, d_{ijk} the linear ISA coefficients, and M_{ijkl} the quadratic ISA coefficients. For piezoelectric materials, the linear coefficients d_{ijk} are dominant, whereas for an electrostrictive material, the quadratic terms M_{ijkl} are dominant. Magnetostrictive materials have a behavior similar to that of electrostrictive materials (i.e., dominantly quadratic). Widely developed in the last decade, ISA materials have shown remarkable performance, and positive induced strains in excess of 0.075% have been reported. The variation of induced strain with electric (or magnetic) field is comparatively presented in Figure 1 for commercially available ISA materials:[8,21] PZT (a piezoelectric ceramic); PMN (an electrostrictive ceramic); and TERFENOL (a magnetostrictive alloy consisting of iron and two rare earth elements). The maximum strain is about 0.075% for PZT, 0.075% for PMN, and 0.160% for TERFENOL. (Higher values have also been occasionally reported.) Due to their quadratic characteristic, PMN and TERFENOL materials can only produce positive (i.e., expansion) strain, and hence compressive loads. The linear characteristic of the PZT material suggests a tension-compression capability, but its strength under tension loading is much less than under compression. Hence, it is reasonable to say that direct ISA applications are limited to compression loads. For bilateral, tension-compression, cycles, either a biased (offset) operation, or a counter-acting actuator pair, are normally used.

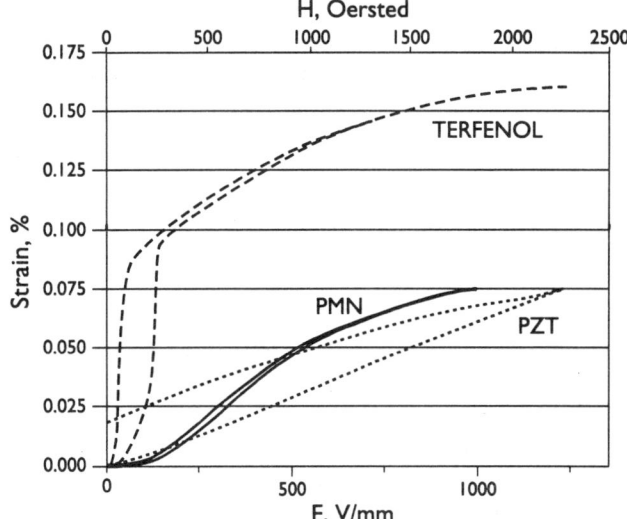

Figure 1 Induced strain *vs* applied field for several ISA materials[8,21].

Though equation (1.1) represents a tensorially fully coupled behavior, most actuator applications of ISA materials make use of the ISA effect in the polarization (Ox_3) direction only. A common example is that of an ISA stack (resembling the early voltaic piles of the last century) in which thin washers of the ISA material are intercalated with metallic electrodes alternatively charged. Thus, a high electric field can be applied throughout the ISA material, ensuring good performance. Commercially available ISA stacks of typically 150 mm length display free ISA displacement (x_A) in excess of 0.100 mm. However, due to internal compressibility, only a portion of the ISA displacement can be delivered externally. Denoting the quasi-static internal stiffness by k_i, one computes the compressibility loss as $x_i = F/k_i$, and hence the externally delivered displacement is $x_e = x_A - x_i$. The higher the external load, the lower the externally deliverable displacement. For the fully constrained case, the actuator is blocked and all the ISA displacement is consumed internally. Since work is the product of force and displacement, it is apparent that the work done externally by an ISA actuator is zero both at zero load and at stalling load. Its variation is parabolic, and a maximum value is attained when the internal and external components of the ISA displacement are equal. A similar conclusion is obtained if we reason in terms of energy. Assuming a quasi-steady external stiffness k_e, one writes the internal and external strain energies, and the corresponding ISA energy, as:

$$E_i = \frac{1}{2} k_i x_i^2, \quad E_e = \frac{1}{2} k_e x_e^2,$$
$$E_{ISA} = E_i + E_e = \frac{1}{2} k_i x_i^2 + \frac{1}{2} k_e x_e^2, \quad (1.2)$$

Liang, Sun and Rogers,[9] performed a comprehensive analysis of energy and power aspects using an electro-mechanically coupled model. Plots of E_i, E_e, and E_{ISA} against the external stiffness parameter $r = k_e / k_i$ are given in Figure 2. It can be seen that while the parasitic internal energy steadily increases, the externally delivered energy reaches a maximum, after which it starts to decrease.

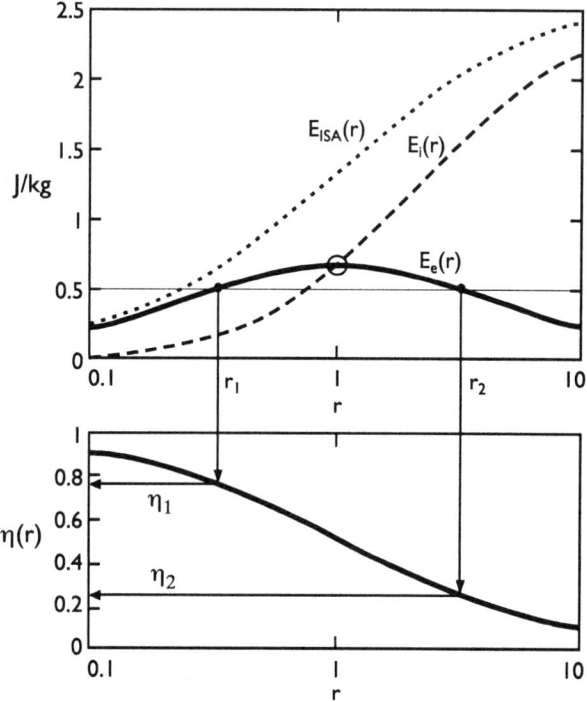

Figure 2 Variation of internal, external, and ISA energies with stiffness ratio r. Maximum external energy is attained at $r = 1$.

Hence an optimum condition exists ($r = 1$, i.e., $k_e = k_i$) for which the delivered energy, and hence the energy per unit volume, and per unit weight, are maximum. Typical values (Table 1) are 4.078 kJ/m^3 and 0.666 J/kg, respectively. Another aspect to be addressed is that of energy transmission efficiency, i.e. what percentage of the ISA energy gets delivered externally. Since part of the energy is stored internally due to compressibility, we define the energy transmission efficiency as

$$\eta_{Energy}(r) = \frac{E_e(r)}{E_{ISA}(r)} \quad (1.3)$$

Figure 2 shows that a given external energy demand can be satisfied by two different values of the stiffness ratio, r_1 and r_2. The lower value, r_1 yields a better energy transmission efficiency, since $\eta(r_1) > \eta(r_2)$. Hence, it can be concluded that optimal designs are obtained by matching, as much as practically possible, the external and internal stiffness. This matching must be done from below, i.e. by keeping the apparent external stiffness of the application lower than, or equal to, the internal stiffness of the ISA actuator.

Table 1 Typical Parameters for ISA Stacks (PI-270-70 Actuator[12])

Maximum displacement	0.120 mm
Length	144 mm
Internal stiffness	370 kN/mm
Volume	163 10^{-6} m^3
Weight	1 kg
Maximum free strain	0.083 %
Maximum deliverable energy per unit volume	4.078 kJ/m^3
Maximum deliverable energy per unit mass	0.666 J/kg

2 Review of Existing Proposals for Individual Blade Control

Several proposals have recently been developed for achieving individual blade control (IBC) through conventional and unconventional means. These proposals range from the direct distributed control of twist along the blade span, to the use of aerodynamic control surfaces discretely placed at the blade outer stations. Application of ISA technology in rotary wing IBC follows similar developments taking place in the ISA control of conventional wings,[10] and a significant degree of technological cross-fertilization is taking place.

2.1 Induced Twist Concept

ISA control of bending and twisting seems obviously the most direct and potentially effective way of aeroelastic control. Theoretical studies on ISA adaptive aeroelastic control were first done on classical wings,[10,11] and then extended to rotor blades.[20]

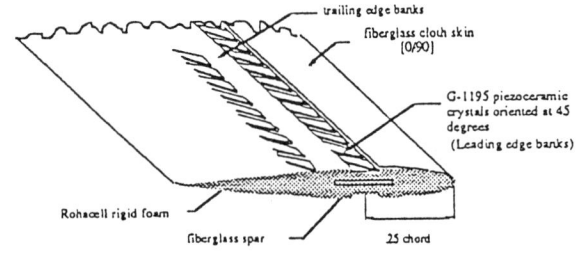

Figure 3 Diagonal PZT twist actuation concept.[14]

Theoretical studies of ISA twist control of rotor blades conducted at DLR in Germany[12,13] showed the benefits of spanwise twist distribution over discrete control surface movement, especially when wide-band reduction of rotor systems vibration is sought. However, numerical simulations with existing PZT technology have shown that only the vibration modes with very low aerodynamic damping will benefit. Hence, it was concluded that the practical implementation of the method relies on "future development of a new generation of PZT materials having higher strain/volt capabilities".[13]

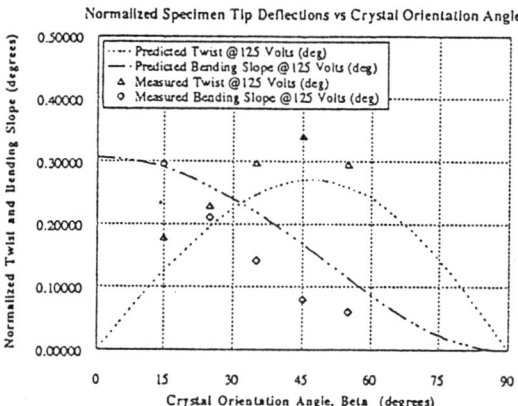

Figure 4: Twist and bending slopes vs PZT angle.[14]

A 1/8 Froude scale composite rotor was constructed at the University of Maryland.[14] It used diagonally oriented PZT crystals embedded at 45° in the fiberglass skin (Fig. 3) to achieve induced twist when an electric field was applied. A value of 1° tip rotation was targeted in these trials. Extensive theoretical and experimental studies were conducted. Good prediction of tip twist and bending slopes was reported[14] (Fig. 4). The largest recorded values, in agreement with predicted results, did not exceed 0.35°. Dynamic tests were also performed in non-rotating and rotating conditions. Significant twist response was measured when excitation was close to resonant bending or torsion frequencies (50 Hz and 95 Hz, respectively). Maximum tip twist values at these frequencies were 0.35° and 1.1°, respectively. Outside resonance, a very small response was observed. These experimental results seem to confirm the theoretical conclusion[13] that the practical implementation of induced twist actuation through embedded PZT technology is difficult.

2.2 Servo-flap Concept

The servo-flap concept has received considerably more attention due to the inherent advantages of extracting additional power and energy from the airstream through the servo-aerodynamic effect. This concept was first pioneered by Kaman Aerospace Corp.[8]. The helicopters SH-2G Seasprite and K-Max are flying proof of its feasibility. However, these two helicopters still employ a conventional swashplate to achieve control of the servo-flap through levers and control rods running through the blade. The purpose of recent studies was to replace these mechanical controls with other means (electrical, electro-hydraulic, etc.).

2.2.1 Advanced Rotor Control System (ARCS) Studies

Conventional electrical motors and electro-hydraulic actuators were considered in an extensive industry study of the advanced rotor control system (ARCS) to be implemented on existing helicopters such as the McDonnell Douglas Helicopter Co. (MDHC) AH-64A Apache[15], and the Helicopter Textron Inc. (BHTI) AH-1W Super Cobra[16]. Extensive studies ranging from aerodynamic predictions to detailed engineering design were performed. The results showed consistently that if the servo-flap concept were to be used for achieving both basic flying controls (collective and cyclic pitch) as well as HHC-IBC, then extensive flap travel and hinge moment capabilities would be required. The MDHC study[15] used previous experience[3] and the CAMRAD-JA prediction code to evaluate a 42.5% chord, 17.5% span (0.65R - 0.825R) flap design extending 32.5% beyond the trailing edge of the blade. The peak deflection and hinge moment values resulting from this study were placed around +/-19.2°, and 2445 lb-in, with an associated electrical power of 15 Kw per blade. The BHTI study[16] used the COPTER aerodynamic code to evaluate a servo-flap design of slightly different dimensions: 20% chord, 20% span centered about the 0.70 spanwise blade station. This study[16] yielded peak values of ±18.9° and 1200 lb-in, with 1.95 Kw per blade. The nearly 8 times difference in power values between the two predictions points out the sensitivity of this kind of analysis on modeling methods, engineering solutions, and design selections.

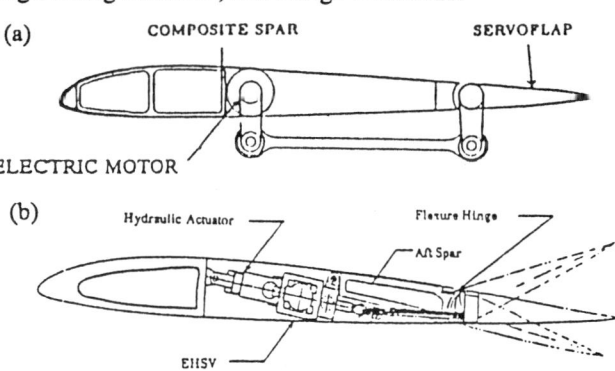

Figure 5 Advanced rotor control system (ARCS) servo-flap actuation concepts: (a) electric motor (McDonnell Douglas Helicopter Co.[15]); (b) hydraulic cylinder (Bell Helicopter Textron Inc.[16]).

However, both studies pointed out that achieving both primary (collective and cyclic) and vibration control through the same actuation system does not lead to an efficient design. Hence, the alternative of separating primary control from vibration control was contemplated, and either collective, or both collective and cyclic controls were achieved through conventional root actuation. Thus, the requirements imposed on the servo-flap system could be brought within more feasible bounds. The HHC-IBC values estimated by these studies as: ±0.90°, with 512 lb-in at 24.1 Hz for MDHC (flap reduced to 35% chord, 15% span), requiring 512 W per blade; and ±3.8°, with 320 lb-in at 20 Hz for BHTI (same flap as before), requiring 388 W per blade.

2.2.2 ISA Servo-flap Studies

2.2.2.1 Theoretical studies at UCLA

Extensive theoretical studies performed at UCLA[17] showed the benefits of using the servo-flap concept for active control of helicopter rotor blades. Both spring-restrained rigid blade and fully elastic blade models were used. Geometrical non-linearities were included. Advanced unsteady aerodynamic 2-D models were employed, together with a flap efficiency coefficient of 60% to compensate for previously reported discrepancies between theory and experiments. A vibration reduction controller was connected with the aeroelastic model, and HHC input to the flap was assumed. Substantial vibration reduction was demonstrated at various helicopter airspeeds corresponding to advance ratios $\mu = 0.0 - 0.4$. The study provides estimates of the required flap travel and hinge moment. The average power consumption was also calculated using the formula[17]:

$$P_{cs} = \sum_{k=1}^{N_b=4} \frac{1}{2\pi} \int_0^{2\pi} M_H(\psi_k) \dot{\delta}(\psi_k) \, d\psi_k , \qquad (2.1)$$

where P_{cs} signifies control surface power, N_b is the number of blades, $M_H(\psi)$ is the control surface hinge moment, $\dot{\delta}(\psi)$ is the time derivative of control surface deflection, and ψ_k is the azimuthal position of the k-th blade. Similar calculations were also obtained from conventional IBC blade root actuation, and comparisons of the servo-flap and conventional HHC-IBC were performed. Though the values varied with the blade modeling complexity, the conclusion could be drawn that substantially less power is required to implement HHC-IBC through the servo-flap concept than through the conventional root actuation. The numerical values required to achieve satisfactory HHC-IBC were found to vary with the rotational and torsional stiffness of the blade. The softer (lower frequency) blades were easier to control through the servo-flap system. In this case ($\omega_T = 3$/rev), it was found[17] that the peak deflections of the flap should be +3°/-4°. No values were given for the corresponding hinge moment. The corresponding power requirement was found to be around 0.25% of the rotor power.[17] The flap deflection and power requirements for the torsionally stiffer blade ($\omega_T = 5$/rev) were considerably higher.

2.2.2.2 Theoretical and experimental studies at MIT

Extensive theoretical and experimental studies of ISA controlled flaps were performed at MIT. In one investigation[18], bimorph arrangements of PZT material were used to produce a lateral displacement that was transformed into flap rotation through a hinge-and-lever mechanism (Fig. 6). The studies were targeted at the Boeing CH-47D tandem helicopter having a large (R = 30 ft) 3-blade rotor rotating at 225 rpm (3.75 Hz). HHC input at 3/rev (11.25 Hz) were considered in the theoretical study, and extensive scaling and modeling work was performed in order to obtain experimental data. A 1/5 scale wind tunnel test was set up using a stationary blade section equipped with a 10% chord flap. A 1/10 velocity scale and 1/2 frequency scale was employed in the model design. Experiments were successfully conducted at various airspeeds between zero and 78 ft/sec, and at frequencies up to 100 Hz. The flap deflection capability, as well as the resulting lift and pitch moment coefficients created by this deflection, were observed and discussed.

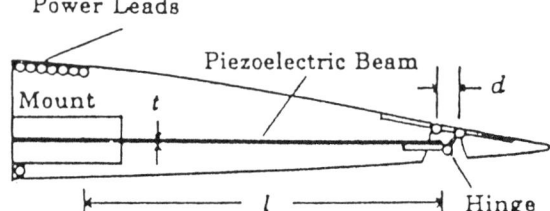

Figure 6 Bimorph PZT flap actuator.[18]

As a general trend, values significantly below the theoretical predictions were reported. This discrepancy was attributed to certain inconsistencies in Reynolds' number (and hence boundary layer thickness), as well as to high mechanical losses due to friction in the hinges, the low stiffness of the trailing edge design, and the spanwise bending response of the model.

Another study was independently conducted by SatCon Technology Co. in cooperation with MIT[6] to design an ISA rotor blade flap actuator using the magnetostrictive TERFENOL-D material for the Sikorsky UH-60A Black Hawk helicopter. Analytical predictions of flap angle, hinge moments and control power requirements, as well as ISA actuator sizing based on the peak energy needs were reported. The requirement of ±2° for conventional (root actuation) HHC-IBC was taken as basis, and then translated into equivalent flap deflection based on flap chord ratio and spanwise dimension and location. It was found that, for a full span flap, the required deflection varies from ±5° with a 10% chord, to ±3° with 40% chord. Adjustment for the partial span flap was done using the percentage positions of the inner and outer ends of the flap (ϵ_{in} and ϵ_{out}, respectively) in the formula:

$$\delta_{ps} = \frac{1}{\epsilon_{out}^3 - \epsilon_{in}^3} \delta_f , \qquad (2.2)$$

where ps and f signify "partial span" and "full span", respectively. Eventually, a 17.5% chord, 46% span (0.52R - 0.98R) flap was selected, requiring a travel range of ±5.7°. The corresponding aerodynamic hinge moment was estimated at ±30 ft-lb, with an additional constant contribution of 11 ft-lb being added to the positive side to account for steady state maneuver loads. Next, the peak energy required for achieving these deflection-moment values was calculated, and it was found to be 3.5 J for the positive side, and 2.0 J for the negative side of the cycle. With a 50% margin of error assumption, a final design based on 6 TERFENOL-D actuators per blade was achieved. The estimated electrical power consumption was 7.2 Kw (i.e., 1.8 Kw per blade). The combined weight of all the actuators was 43 kg (i.e., 10.75 kg per blade). These values were shown to be well within the accepted power and weight budgets of the complete actuation system taken as 1% of helicopter power and weight (11.2 W and 81 kg, respectively).

2.2.2.3 Experimental and theoretical studies at University of Maryland

Experimental and theoretical studies with the bimorph ISA principle for rotor blade flap actuation (Fig. 7) were conducted at the Center for Rotorcraft Education and Research at the University of Maryland[19].

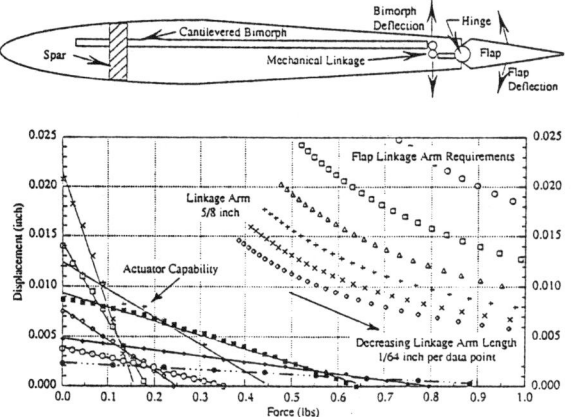

Figure 7 Parameter study of bimorph effectiveness: (a) device schematic; (b) linkage arm and actuator capability plots.[19]

Trailing edge flaps of 20% chord, 12% span (0.85R - 0.97R) were built into a 36-in radius, 3-in chord composite blade model. Initial tests were performed with the stationary blade placed in a conventional wind tunnel operated at speeds up to 111 ft/sec, and at two airfoil set angles (4° and 8°). Excitation at frequencies up to 15 Hz showed good frequency response, but a very significant decrease in amplitude with increasing airspeed. Further experiments were conducted with the blade installed in a rotating rig operated at up to 900 rpm (15 Hz). Excitation frequencies of 1, 2, 3, and 4/rev were investigated. The trailing edge flap response was to be rather constant with frequency, but strongly decreasing with rpm. The blade flapping response varied with excitation frequency and rotor speed due to interdependence between flapping resonant frequency and rotor rpm. To counteract the decrease in trailing edge flap response with airspeed, further investigation was conducted with 2-layer and 4-layer bimorph PZT designs[19]. The use of the 2-layer actuator generated an increase of about 21% in the flap angle at the higher speed value. The 4-layer actuator "showed much improved force capability, but lower displacement"[19]. The goal of these investigations was initially set at achieving 2° of flap deflection at 258 ft/sec airspeed. Subsequently, the goal was redefined as "5% flap authority (additional steady lift due to a flap deflection divided by total steady lift with zero flap deflection) at 8° collective blade pitch".[19] A theoretical parameter study was performed[19] to determine the required linkage arm length that will meet this specification (Fig. 7). At present, a second experimental rotor with a larger span and multi-layered actuators is being built and will be tested on the hover stand for a range of rotor speeds and collective settings.

3 Analysis of Rotor Blade Servo-flap Vibration Control with ISA Devices

Various attempts reported in the literature regarding the study and experimentation of Individual Blade Control (IBC) through conventional (electro-mechanical, hydraulic) and solid state (electro-mechanical ISA) means were reviewed in the previous section. Two major directions are apparent: the direct twist of the rotor blade through tension-torsion-bending coupling, and indirect modification of the aerodynamic lift and pitch moment of the blade using the servo-flap and servo-tab principles. The former concept is more direct and allows for properly addressing modal control, since continuous variation of induced strain effects along the blade length can be achieved. The latter concept has the shortcoming of an additional complication in the rotor construction, and uses a rather rigid device (flap or tab) with only limited span-wise extent; however, it is much more efficient in achieving a palpable result due to the amplification properties of servo-aerodynamics. For this latter reason, and considering the present day state-of-the-art in ISA technology, most engineering efforts reported in the literature have been **generally directed towards the servo-flap concept**, and hence, we shall restrict our investigation in subsequent chapters towards this option. However, one should recognize that as yet there is *nolo contendere* in passing judgement on the feasibility of the **induced twist concept**.

3.1 Definition of an ISA Servo-flap for Rotor Blade Vibration Control

Within the servo-flap concept, one observes that the requirements for rotor blade primary steady state and dynamic control (collective and cyclic pitch) cannot be simultaneously met by the same ISA device. Both collective and cyclic pitch controls require large servo-flap deflection authority ($\approx \pm 20°$) and are too demanding for existing ISA technology. Hence, our attention regarding the use of dynamic ISA devices for helicopter IBC will be restricted to vibration control, i.e. HHC applications.

Table 2 Proposed bench-mark design parameters for rotor blade vibration control ISA servo-flap system

Parameter	Value and units	Notes
Flap deflection	$\pm 1/30$ rad ($\approx \pm 2°$)	
Hinge moment	± 75 Nm	simultaneous with flap deflection
Frequency	25 - 30 Hz	
Maximum instantaneous energy transmitted to the airstream	1.25 J / blade	(zero steady force is assumed)
Maximum instantaneous power transmitted to the airstream	0.5 kW	
ISA actuator weight budget	10 kg / blade	$\approx 10\%$ of typical blade weight
ISA system weight budget (including power supply, lead wires, controls, etc.)	80 kg / 4-blade helicopter	$\approx 1\%$ of typical helicopter weight
Overall power consumption budget	10-12 kW / 4-blade helicopter	$\approx 1\%$ of typical rotor power
Specific transmitted energy	0.125 J/kg	
Specific transmitted power	25 W/kg	
Specific power consumption	≤ 140 W/kg	including all losses

In a previous study,[22] we used the data reviewed in section 2 to identify baseline values for an ISA servo-flap system to be used for rotor blade vibration control. Though large variations were noticed, a benchmark set of values for flap deflection, hinge moment, frequency, mass, energy, and power, as well as energy and power densities could be estimated (Table 2). The ISA devices reviewed in Table 1, section 1.2, can deliver 0.666 J/kg and hence the 0.125 J/kg specific energy requirement of Table 2 can be met. (The weight of the power supply, lead wires, control electronics, etc. was not included in Table 1, but even with their contribution the overall trend will be the same.)

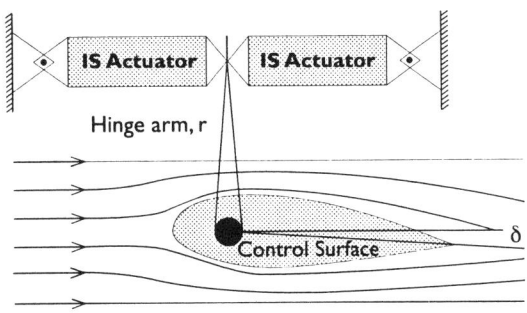

Figure 8 ISA servo-flap for rotor blade vibration control

Commercially available ISA stacks could be used to meet the specifications laid down in Table 2. The output energy requirement can be met[22] by a double pair of ISA stacks, each pair 300 mm long, 40 mm diameter, and 1 kg mass, with free travel of ≈ 0.125 mm. Each pair is L = 300 mm long, and has a total free travel $x_A \approx 0.250$ mm. Hence 4 kg of ISA mass is required for each blade. Adequate weight allowance for system implementation must be made.

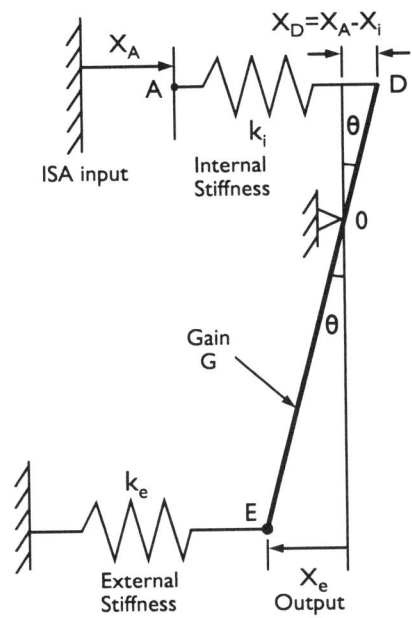

Figure 9 Schematic drawing of displacement amplification principles.

We propose to allow 10% of the blade weight (i.e., 10 kg / blade) for the ISA actuator, servo-flap linkages, and local reinforcement of the blade section.

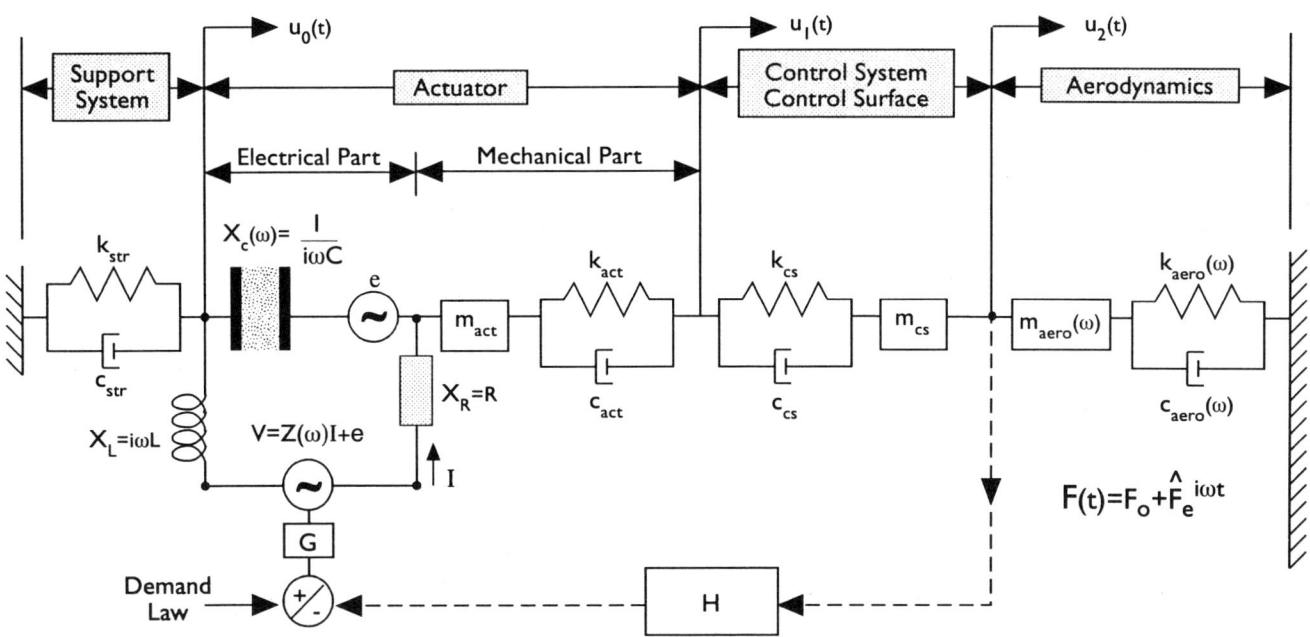

Figure 10 Proposed electro-aero-mechanical model for ISA servo-flap vibration control

For the complete system, including power supply, control electronics, etc., a larger weight allowance can be made, since part of it will be accommodated in the fuselage. Hence we propose that the complete system does not exceed 1% of the helicopter weight (i.e., 80 kg / 4-blade helicopter).

Since the ISA travel is too small for practical applications, a displacement amplifier is affixed to the end of the ISA stack. (The ISA stack and the displacement amplifier make up the actuator.) A practical actuator output of 1 mm is desired since this value leads to a hinge arm r = 30 mm to produce the required flap deflection δ = 1/30 rad. Suitable values for the amplification gain G can be selected to achieve the η = 4 amplification required to increase the 0.250 mm ISA displacement to the required 1 mm actuator output. A formula can be derive for finding the optimal gain values using feedback principles:[22]

$$G(\eta) = \frac{1}{2\eta} \sqrt{\frac{k_i}{k_e}} \left(\sqrt{\frac{k_i}{k_e}} - \sqrt{\frac{k_i}{k_e} - 4\eta^2} \right). \quad (3.1)$$

This formula ensures that optimum use is made of the ISA capabilities for a given pair of internal and external stiffness values, k_i and k_e. In this particular application, the ± 1 mm output target means a 4:1 displacement ratio and yields the value G = 5.85 for the optimal gain.

3.2 Aero-electro-mechanical Modeling of the ISA Servo-flap System

Figure 10 presents a conceptual electro-aero-mechanical model that can be used to study the actuation scheme shown in Figure 8. It contains several distinct parts: the electrical impedance associated with the ISA device; the mechanical impedance of the support system; the mechanical impedance of the control system and control surface; and the equivalent mechanical impedance simulating the reaction of the airstream. A feedback sensor is positioned conveniently to pick up the servo-flap response. Its signal is fed into the control loop of basic gain G and feedback gain H. The predominant electric parameter of the ISA stack is its capacitance C. The resistance R models the electrical losses inside the ISA material, while the inductance L is designed to compensate the capacitive reactance and hence reduce the reactive power requirements. For a system operating at around a given frequency ω_0, the inductance values can be estimated such as to minimize the reactance

$$X(\omega_0) = \omega_0 L - \frac{1}{\omega_0 C} \quad (3.2)$$

Application of this concept (also known as power factor correction) to piezoelectric actuators was discussed.[24] The counter-electro-motive force e represents the piezoelectric effect which appears as an electric reaction when the actuator is operating under mechanical load. (A similar effect takes place in conventional electric motors.)

The actuator mechanical impedance can be calculated using its dynamic stiffness, $\bar{k}_{act} = (1 + i \eta_{act}) k_{act}$. For modeling purposes, it is sometimes convenient to replace the internal loss coefficient η_{act} with an equivalent viscous coefficient c_{act}. The actuator impedance depicted in this model also includes the effects due to the displacement amplification device incorporated into the actuator. The control circuit and control surface impedance incorporates the stiffness, damping and mass effects of both the servo-flap and its hinge arm and supports.

The aerodynamic impedance incorporates equivalent aerodynamic mass, damping and stiffness associated with the oscillatory movement of the servo-flap. Their estimation is not straightforward, and includes non-steady aerodynamics, transonic effects on the advancing blade, and stall and reverse flow on the retreating blade. For first analysis, Theodorsen's linear theory may be used.[17,18] Hence, the servo-flap hinge moment takes the form:

$$M_\delta(\omega) = [-\omega^2 m_{aero}(k) + i\omega c_{aero}(k) + k_{aero}(k)] \delta , \quad (3.3)$$

where $k = \omega b/U$ is the reduced frequency. The aerodynamic stiffness, mass and damping depend on the reduced frequency k, and hence their values vary depending on the flap geometry, local airspeed, and frequency.

4 Discussion

The work presented herein is far from being complete. Though a benchmark set of values and an electro-aero-mechanical model have been proposed, several avenues are still open for further research. **First**, a more precise definition the base-line requirements for achieving HHC-IBC vibration control of helicopter rotors should be investigated. This is a formidable analytical effort, since load prediction codes for rotor blade unsteady aerodynamics still show large variation in results. However, a concerted effort of industry and academia could yield better estimates for the HHC-IBC actuator than the "rough-and-ready" approximations proposed herein. **Second**, the present analysis should be extended to non-ideal displacement amplifier incorporating transmission losses due to their own compliance and damping. **Third**, the more comprehensive subject of identifying and numerically defining the power dissipation mechanisms involved in the three stages of an ISA system (ISA material, power supply, and mechanical and electrical transmissions) should be addressed in more detail. **Fourth**, no discussion was made here of the effects of D.C. bias field, and of the associated steady load, usually present in all ISA devices (This offset is necessary for at least two reasons: (a) to avoid putting the ISA material into tension; (b) to compensate for the asymmetric tension-compression behavior of the ISA material). Finally, a major concern with any rotor blade device operating in the outer blade span is the effect of very high g loading (a point at 5 m radius rotating at 30 rad/sec experiences $R\Omega^2 = 5 \times 30^2 = 4500$ m/s^2 $= 450 g$!). This was a major concern with the ARCS studies[15,16] considering conventional electro-mechanical and hydraulic devices. Regarding the effect of g loading on ISA devices, very little theoretical and experimental research has been reported so far. Though the solid-state nature of an ISA device justifies optimism in this respect, properly conducted studies are still required.

5 Conclusions

A review of the existing literature produced baseline requirements for implementing active HHC-IBC vibration alleviation using ISA devices. Typical values of $\pm 2°$ servo-flap displacement, and ± 75 Nm of hinge moment at 25-30 Hz were identified and proposed as benchmark values for future investigations. Maximum instantaneous energy of 1.25 J and maximum instantaneous power output of 0.5 kW, per blade, were also identified. Weight budget allowance of 10 kg/blade for the actuators and linkages, and 80 kg / 4-blade helicopter for the complete ISA system are proposed, along with corresponding power budget values. The resulting energy density requirement of 0.125 J/kg was found within the capabilities of commercially available ISA stacks. However, displacement amplification devices are necessary at the ISA stack level. Using an optimal design for the displacement amplifier, realistic values for actuator output (1 mm) and servo-flap hinge arm (30 mm) could be considered.

An electro-aero-mechanical model is proposed for the analysis of an ISA servo-flap for rotor blade vibration control. Opportunities for further research are identified. The experimentation of an actual ISA device coupled with an optimally designed displacement amplifier would substantiate the main conclusions of this study and identify practical issues associated with its real-time implementation. The baseline specification developed in this and previous studies[22] is also proposed as a benchmark for testing other possible ISA solutions.

6 Acknowledgments

The authors gratefully acknowledge the support of the Army Research Office - University Research Initiative Program, Grant No. DAAL03-92-G-0181, Dr. Gary Anderson, Program Manager.

7 References

1. Robinson, L., and Friedmann, P.P., "A Study of Fundamental Issues in Higher Harmonic Control Using Aeroelastic Simulation", *Journal American Helicopter Society*, Vol. 36, No. 2, April 1991, pp. 32-43.
2. Nguyen, K, and Chopra, I., "Application of Higher Harmonic Control (HHC) to Rotors Operating at High Speed and Thrust", *Journal American Helicopter Society*, Vol. 35, No. 3, July 1990, pp. 78-89.
3. Straub, F.K., and Byrns, E.V. Jr., "Application of Higher Harmonic Blade Feathering on the OH-6A Helicopter for Vibration Reduction", NASA CR-4013, December 1986.
4. Ham, N.D., "Helicopter Individual Blade Control Research at MIT 1977-1985", *Vertica*, Vol. 11, No. 1/2, 1987, pp. 109-122.
5. Jacklin, S.A., Leyland, J.A., and Blaas, A., "Full-

Scale Wind Tunnel Investigation of a Helicopter Individual Blade Control System", *Proceedings of the AIAA/ASME/ASCE/AHS/ASC 34th Structures, Structural Dynamics, and Materials Conference,* LaJolla, CA, April 1993, paper AIAA-93-1361-CP, pp. 576-586.

6. Fenn, R.C., Downer, J.R., Bushko, D.A., Gondhalekar, V., and Ham, N.D., "Terfenol-D Driven Flaps for Helicopter Vibration Reduction", *Proceedings of SPIE 1993 North American Conference on Smart Structures and Materials,* Albuquerque, 1-4 Feb., 1993, SPIE Vol. 1917 *Smart Structures and Intelligent Systems (1993),* pp. 407-418.

7. Lemnios and Smith, "An Analytical Evaluation of the Controllable Twist Rotor Performance and Dynamic Behavior", Report R-794, Kaman Aerospace Corp., 1972.

8. Galvagni, J., Rawal, B., "A Comparison of Piezoelectric and Electrostrictive Actuator Stacks", Paper 1543-50, SPIE Vol. 1543 *Active and Adaptive Optical Components",* 1991, pp 296-300.

9. Liang, C., Sun, F.P., Rogers, C.A., "Coupled Electro-Mechanical Analysis of Piezoelectric Ceramic Actuator Driven Systems - Determination of Actuator Power Consumption and System Energy Transfer", *Proceedings of SPIE 1993 North American Conference on Smart Structures and Materials,* Albuquerque, 1-4 Feb., 1993, SPIE Vol. 1917 *Smart Structures and Intelligent Systems (1993),* pp. 286-298

10. Lazarus, K.B., Ceawley, E.F., Bohlmann, J.D., "Static Aeroelastic Control Using Strain Actuated Adaptive Structures", *First Joint U.S./Japan Conference on Adaptive Structures,* Nov. 13-15, 1990, Maui, Hawaii, Technomic Pub. Co., Inc., pp.197-224.

11. Song, O., Librescu, L., and Rogers, C.A., "Application of Adaptive Technology to Static Aeroelastic Control of Wing Structures", *AIAA Journal,* Vol.30, No. 12, Dec. 1993, pp. 2882-2889.

12. Nitzsche, F., and Breitbach, E., "Study of Feasibility of Using Adaptive Structures in the Attenuation of Vibration Characteristics of Rotary Wings", AIAA paper No. AIAA-92-2452-CP, 1992.

13. Nitzsche, F., "Modal Sensors and Actuators for Individual Blade Control", *Proceedings of the AIAA/ASME/ASCE/AHS/ASC 34th Structures, Structural Dynamics, and Materials Conference,* LaJolla, CA, April 1993, paper AIAA-93-1703-CP, pp. 3507-3516.

14. Chen, P.C., and Chopra, I., "Induced-Strain Actuation of Composite Beams and Rotor Blades with Embedded Piezoceramic Elements", *1st Workshop on Smart Structures,* University of Texas, Arlington, TX, Sept. 22-24, 1993 (in press)

15. Straub, F.K., AND Charles, B.D., "Preliminary Assessment of Advanced Rotor/Control System Concepts (ARCS)", McDonnell Douglas Helicopter Co., Aviation Applied Technology Directorate, US Army Aviation Systems Command, Fort Eustis, VA, USAAVSCOM TR 90-D-3, August 1990.

16. Phillips, N.B., and Murphy, M.R., "Preliminary Assessment of Advanced Rotor/Control System Concepts", Bell Helicopter Textron Inc., Aviation Applied Technology Directorate, US Army Aviation Systems Command, Fort Eustis, VA, USAAVSCOM TR 90-D-18, August 1990.

17. Millott, T.A., and Friedmann, P.P., "The Practical Implementation of an Actively Controlled Flap to Reduce Vibrations in Helicopter Rotors", *Proceeding of the American Helicopter Society 49th Annual Forum,* St. Louis, Missouri, May 19-21, 1993, p.p. 1079-1092.

18. Spangler, R.L., Jr., and Hall, S.R., "Piezoelectric Actuators for Helicopter Rotor Control", *Proceedings of the AIAA/ASME/ASCE/AHS/ASC 31st Structures, Structural Dynamics, and Materials Conference,* Long Beach, CA, April 1990, Paper No. AIAA-90-1076-CP, pp. 1589-1599.

19. Walz, C., and Chopra, I., "Design, Fabrication, and Testing of a Scaled Rotor with Smart Trailing Edge Flaps", *1st Workshop on Smart Structures,* University of Texas, Arlington, TX, Sept. 22-24, 1993 (in press).

20. Song, O., Librescu, L., "Vibration Behavior of Rotating Helicopter Blades Incorporating Adaptive Capabilities", *Proceedings of SPIE 1993 North American Conference on Smart Structures and Materials,* Albuquerque, 1-4 Feb., 1993, pp. 354-365

21. Clark, A.E., "High Power Rare Earth Magnetostrictive Materials", in *Proceedings of the Conference on Recent Advances in Adaptive and Sensory Materials,* Technomic Publishing, 1992, pp. 387-397.

22. Giurgiutiu, V., Chaudhry, Z., Rogers, C., "Engineering Feasibility of Induced-Strain Actuators for Rotor Blade Active Vibration Control", *1994 North American Conference on Smart Structures and Materials,* Orlando, FL, 13-18 February 1994 (in press).

23. Hanagud, S., Roglin, R.L., NageshBabu, G.L., "Adaptive Airfoils for Helicopters", SPIE Vol. 1917, *Smart Structures and Intelligent Systems (1993),* pp. 217-224.

24. Hagood, N.W., Chung, W.H., and Flotow, A., "Modelling of Piezoelectric Actuator Dynamics for Active Structural Control", *Proceedings of the AIAA/ASME/ASCE/AHS/ASC 31st Structures, Structural Dynamics, and Materials Conference,* Long Beach, CA, April 1990, paper AIAA-90-1087-CP, pp. 2242-2256

Designing Efficient Helicopter Individual Blade Controllers Using Smart Structures

F. Nitzsche*
DLR - Institute of Aeroelasticity
Bunsenstr. 10, D-37073 Goettingen, Germany

Abstract

The individual blade control of bearingless composite helicopter rotors aiming at vibration attenuation using adaptive structures is studied in this paper. A periodic model-following regulator is developed in the frequency domain to approximate the aeroelastic response of a single blade in forward flight to its hover characteristics. Two control strategies are investigated: a more robust one using distributed control, and a more cost-efficient one using lumped control. Selective linearization of the periodic system at determined frequencies is demonstrated in both cases.

Nomenclature

b	blade semichord
c	blade wing box width
c_θ, c_φ	blade-root flexibility coefficients
D_{ij}^*	laminate compliance coefficients
d_a	mean distance between actuation layers in the upper and lower skins
EI_R	cross-section reference stiffness
E^*	saturation electric field
E_3	applied electric field
e_{3x}, e_{3s}	piezoelectric material stress per charge coefficients in the blade radial direction (normal, shear)
H	cross-section shear resultant
I_b	blade total mass moment of inertia in flapping
k_θ	cross-section radius of gyration
M	cross-section bending moment
m	blade running mass, number of rationed modes
n	number of discretizing points for the integrating-matrix method
r	local radius
R	rotor radius
T	cross-section tension
t	time
w	cross-section bending displacement
x_a	aerodynamic center offset lying aft the blade pitch axis
z_s	mean distance between sensing layers and the blade neutral plane
β	blade flap angle (positive upwards)
γ	Lock number, $\rho a(2b)R^4/I_b$
ι	sensor's output current
θ	cross-section torsion angle
θ_a, θ_s	torsion-related shape functions (actuator, sensor)
Λ	matrix of complex eigenvalues
λ^s, λ^u	Fundamental Transition Matrix eigenvalues (stable, unstable)
μ	rotor advance ratio
ν	$m\Omega^2 R^4/EI_R$
ξ	modal damping in units of *rev*
τ	cross-section torque
Φ	transition matrices
φ	cross-section bending slope
φ_a, φ_s	bending-related shape functions (actuator, sensor)
Ψ	Fundamental Transition Matrix eigenvectors
ψ	azimuth angle, Ωt
Ω	rotor frequency
ω	modal frequency in units of *rev*
$\mathbf{1}$	unit matrix

Superscripts

H	Hermitian transpose
$\cdot$	time derivative
$'$	space derivative
$\check{}$	diagonal matrix
$-$	non-dimensional parameter

Introduction

In recent years, with the advent of adaptive materials technology, the development of the so-called "smart rotor" has become a matter of remarkable interest.[1] A recent paper by Chopra presents a review of the most important activities related to the field.[2] In summary, two strategies have been investigated in connection with the individual blade control (IBC) philosophy. The first calls for embedding of

---

* Scientist, Member AIAA.
Copyright © 1994 by the American Institute of Aeronautics and Astronautics, Inc. All rights reserved.

the adaptive material in the blade laminate in order to induce strain deformations that can be controlled externally. It not only has the advantage of being cleaner under the aerodynamic point of view, but it also allows the possibility of exploiting some excellent features related to the distributed control theory, such as its inherent robustness, and superior authority over the system's degrees of freedom. However, it has been recognized now that this solution leads to an overestimation of the adaptive material's capabilities to deliver strain deformations in the rotor blade laminate under the typical loads experienced in forward flight. Therefore, since the introductory study by Spangler and Hall, the use of lumped control configurations has gained particular interest.[3] The most accepted solution has been the introduction of a "smart flap" at the blade's airfoil trailing edge.[3-7] An interesting alternative to the flap has been proposed by Nitzsche, Lammering and Breitbach.[8] According to this solution, actuation is performed at the blade root, locally inducing a torsion displacement. Other attempts to actively control rotary wing problems such as dynamic stall, blade vortex interaction, and shock have been made by deforming the airfoil shape.[9-11] These attempts also can be classified as lumped control techniques, because the control is localized at determined positions along the blade span. Most of the aforementioned studies, however, do not try to elaborate on the controller, although it represents a fundamental aspect of "smart" structures. In particular, for rotary wings the controller must be either fully-adaptive or periodic.[8,12] The present work intends to address this point, taking one example from each class of the candidate control configurations: distributed and lumped.

Control Rationale

Basically, IBC, as introduced by Kratz and Ham, will be proposed now.[13,14] Under this philosophy, two points are fundamental: 1) IBC involves not just the control of each blade individually, but presents feedback loops for each blade in the rotating frame; and 2) IBC is more effective if comprised of several subsystems, each controlling a specific structural blade mode. In fact, *experimental* results indicated that overall reduction on the level of helicopter vibration is observed if *determined* blade modes are controlled.[14] This observation results from the well-known "filtering" characteristics of the rotor hub at integer multiples of the number of blades. It is also well established that it is possible to achieve reduction on the dynamic amplification of blade modes, for example reducing blade root fatigue, by "aeroelastic detuning".[15] Since the aeroelastic modes of the rotating blade retain all the information about the system's dynamics, they should be precisely the elements to be tuned by active control. Although the aeroelastic modes for a rotating blade in forward flight change with the azimuth angle, it will be demonstrated throughout the present investigation that some degree of *independent modal control* is still possible. For this, a periodic model-following regulator is studied in the reduced-order subspace spanned by the aeroelastic modes associated with the blade at its hover (*and invariant*) condition.

Actuation Model

The aeroelastic equations for a single blade modelled as a rotating beam were derived in a previous work.[16] Two degrees of freedom were considered in the simplified linear model: flatwise bending and torsion. The possibility of actuation through an external strain due to piezoelectricity is allowed by including the last terms in Eq. 1c and Eq. 1f (distributed control). In the state vector form the complete equations read:

$$M' = H + \nu T\varphi$$
$$H' = \nu m\ddot{w} - F_w$$
$$\varphi' = D^*_{11}M + D^*_{13}\tau + \Gamma_\varphi E_3$$
$$w' = -\varphi$$
$$\tau' = \nu m k_\theta^2 (\theta + \ddot{\theta}) - M_\theta$$
$$\theta' = D^*_{13}M + D^*_{33}\tau + \Gamma_\theta E_3.$$

(1a-f)

The boundary conditions of the above six first-order differential equations in the space variable are the ones associated with the bearingless rotor. They also include the possibility of actuation by commanding a local torsion displacement at the blade's root (lumped control). Hence,

$$M(0) = c_\varphi \varphi(0)$$
$$M(R) = 0$$
$$H(R) = 0$$
$$w(0) = 0$$
$$\tau(R) = 0$$
$$\tau(0) = c_\theta(\theta(0) - \theta_c).$$

(2a-f)

In the case of distributed control, the actuation coefficients

$$\Gamma_\varphi = \varphi_a\left(D^*_{11}e_{3x} + D^*_{13}e_{3s}\right)t_a d_a$$
$$\Gamma_\theta = \theta_a\left(D^*_{13}e_{3x} + D^*_{33}e_{3s}\right)t_a d_a$$
(3a-b)

are dependent upon the piezoelectric material's (PZT) stress coefficient per unit of charge (in C/m² or N/V.m). The latter coefficients are obtained in the blade global axis considering the clamped-clamped situation.[17] The actuation mechanism assumes that the piezoelectric layers are constructed in pairs symmetrically distributed with respect to the blade neutral plane (Fig. 1). The electrodes are wired to generate forces in opposite directions, respectively in the upper and lower skins of the laminate, when a voltage is applied. At least four layers of adaptive material are necessary to provide independent actuation in both bending and torsion when using this method. In Eqs. 3a-b the shape functions, which define the effective width of the adaptive material embedded in the blade (Fig. 1), are optimized to achieve *independent modal control* by actuating selectively in one of the aeroelastic modes in the rotating frame. For the bending-torsion degree-of-freedom model, only two shape functions are necessary. In a more complete formulation, including the lead-lag degree-of-freedom, a third shape function would be required. It is feasible to construct the so-called *modal filters* that allow the blade *aeroelastic tuning* to be performed actively by adjusting the shape functions according to a closed-form solution.[17]

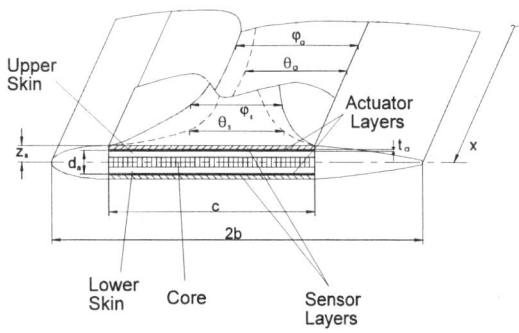

Fig.1 Distributed control solution. The layers of adaptive material with effective width φ_a, θ_a (actuator), and φ_s, θ_s (sensor) are embedded in the wing box.

The perturbation loads assumed in this work are introduced in Eq. 1b and Eq. 1e, and are attributed to the blade running lift and pitch moment, respectively. Ignoring the unsteady terms, a simplified aerodynamics for the forward flight condition may be derived using the blade element theory:[18]

$$\begin{bmatrix} F_w \\ M_\theta \end{bmatrix} = \gamma I_b \begin{bmatrix} 1 \\ x_a \end{bmatrix}\left(\begin{bmatrix} F_{\dot\beta} & 0 \end{bmatrix}\dot{\vec q} + \begin{bmatrix} F_\beta & F_\theta \end{bmatrix}\vec q\right)$$
(4)

where

$$F_{\dot\beta} = -r^2(1 + r\mu\sin\psi)/2$$
$$F_\beta = -\mu r\cos\psi/2 - \mu^2\sin 2\psi/4$$
$$F_\theta = -r^2/2 - \mu r\sin\psi + \mu^2(\cos 2\psi - 1)/4$$
(5a-c)

are dependent on both the rotor advance ratio and the blade azimuth position. It is worthwhile to point out that if the advance ratio is zero (hover condition), the aerodynamic loads do not depend on the azimuth angle. In the present formulation, dynamic stall is also neglected. Although the aerodynamic model may be considered over-simplified, it is clear that the procedure described in the following analyses is general. It is straightforward to extend it when dealing with more sophisticated blade aerodynamics. In Eq. 4, the dependent variables are collected in

$$\vec q = \begin{bmatrix} \varphi & \theta \end{bmatrix}^T$$
(6)

The governing equations, Eqs. 1a-f, and their boundary conditions, Eqs. 2a-f, are solved in the space domain using the integrating matrix method.[8,16] Summarizing, Eqs. 1a-f must be normalized in the interval *[0,1]*, discretized along *n* points in the blade spanwise direction, and pre-multiplied by an *integrating matrix operator* **L** of order *n*.[8,16,19] Each one of the 6 differential equations is transformed into a set of *n* algebraic equations in the state vector components defining the local properties of the blade. The boundary conditions are used to determine the integration constants (constant vectors of dimension *n*) which are generated in the process of integration. Next, the resulting *6n* algebraic equations are reduced to *2n* by analytically solving the system for the "fundamental" dependent variables, chosen to be the flatwise bending slope and the torsion angle. Thus,

$$\mathbf{F\dot x + Gx = Hu}$$
(7)

where

$$\mathbf{x} = \begin{bmatrix} \vec q_{1\times n} & \dot{\vec q}_{1\times n} \end{bmatrix}^T$$
(8)

and

$$\mathbf{F} = \mathbf{F}_0 + \mathbf{F}_1 \sin\psi$$

$$\mathbf{G} = \mathbf{G}_0 + \sum_{k=1}^{2} \mathbf{G}_k (\cos k\psi + \sin k\psi)$$

$$\mathbf{H} = \begin{bmatrix} \mathbf{H}_1 \\ \mathbf{0} \end{bmatrix}.$$

(9a-c)

In the case of distributed control, one has

$$\mathbf{H}_1 = -\begin{bmatrix} \mathbf{L} & \mathbf{0} \\ \mathbf{0} & \mathbf{L} \end{bmatrix} \begin{bmatrix} \overline{\Gamma}_{\varphi n \times 1} \\ \overline{\Gamma}_{\theta n \times 1} \end{bmatrix}$$

$$\mathbf{u} = \overline{E}_3 = (c/R)^3 E_3 / E^*,$$ (10a-b)

whereas for the lumped control case,

$$\mathbf{H}_1 = 1/\nu \begin{bmatrix} \mathbf{0}_{n \times 1} \\ \theta_c^* \mathbf{1}_{n \times 1} \end{bmatrix}$$

$$\mathbf{u} = \theta_c / \theta_c^*.$$ (11a-b)

In both cases the problem admits a single input, either the normalized electric field applied across the piezoelectric elements or the normalized torsion angle applied at the blade root. In the case of distributed control, the integrating matrix operator appears explicitly because the local contributions from each layer of adaptive material must be integrated along the blade.

Since the aim of the present work is to make aeroelastic tuning by actuating on determined modes of the blade in the rotating frame, the modal approach seems to be attractive. However, the aeroelastic modes of a single blade change from one azimuth position to another due to the presence of the periodic terms that multiply the powers of the advance ratio. Nonetheless, at the hover condition the aeroelastic modes are invariant, which makes them an ideal basis for the modal decomposition. Therefore, considering the subspace spanned by the hover modes, a set of generalized coordinates is defined:

$$\mathbf{x} = \mathbf{U}\vec{\eta}$$

$$\vec{\eta} = \mathbf{V}\mathbf{x},$$ (12a-b)

where **U** and **V** are the modal matrices collecting the right and left eigenvectors of the homogeneous system associated with Eq. 7 *at the hover condition* (the periodic terms of Eqs. 9a-b are cancelled out). Both the right and left eigenvectors are necessary to decouple the homogeneous system because the problem is non-conservative due to the presence of the aerodynamic terms. Truncating the modal expansion after the first *2m* complex-conjugate modes, Eq. 7's decomposition yields:

$$\tilde{\mathbf{F}} = \mathbf{VFU}$$

$$\tilde{\mathbf{G}} = \mathbf{VGU}$$

$$\tilde{\mathbf{H}} = \mathbf{VH}.$$ (13a-c)

Hence, the problem is reduced to *2m* differential equations in time:

$$\dot{\vec{\eta}} = \tilde{\mathbf{A}}(\psi)\vec{\eta} + \tilde{\mathbf{B}}(\psi)\mathbf{u}$$ (14)

where

$$\tilde{\mathbf{A}} = \tilde{\mathbf{F}}^{-1}\tilde{\mathbf{G}}$$

$$\tilde{\mathbf{B}} = \tilde{\mathbf{F}}^{-1}\tilde{\mathbf{H}}$$ (15a-b)

Provided that the dependence on the azimuth angle in Eqs. 9a-b is retained explicitly, as suggested in the same equations, the inversion required by Eqs. 15a-b is performed in the reduced-order system. Formally, the aeroelasticity of the rotating blade in the modal subspace is governed by Eq. 14, where the complex coefficients require a numerical evaluation of Eqs. 15a-b at each azimuth angle. Obviously, at the hover condition a constant coefficient system where

$$\tilde{\mathbf{A}} = \text{`}\Lambda$$ (16)

is obtained.

Sensing Model

At least two membranes of soft piezoelectric material (PVDF) must be embedded in the upper skin of the laminate to independently measure both bending and torsion deformations of the blade. The effective width of the sensing elements are, as in the case of the actuation, defined by two shape functions. The latter also may be optimized to achieve *independent modal control*. In general, the same arrangement must be repeated in the lower skin of the laminate in order to account for the sign changes normally encountered in the shape functions associated with sensing.[17]

The sensor equations are derived by adapting Lee and Moon's formula for the electric charge developed across the electrodes of a plate made of adaptive material.[20] If current is measured instead of charge, and if all electrodes have the same length (the blade span), one has:

$$\iota = -z_s e_{3x} R \int_0^R \left(\varphi_s \dot{\varphi}' - 2\frac{e_{3s}}{e_{3x}} \theta_s \dot{\theta}' \right) dr \quad (17)$$

Equation 17 can be normalized in the interval [0,1], discretized at n points along the blade, and transformed into modal coordinates, yielding

$$\vec{\iota} = \tilde{\mathbf{C}}\vec{\eta} = \mathbf{C}\mathbf{U}\vec{\eta} \quad (18)$$

where

$$\mathbf{C} = \begin{bmatrix} \mathbf{0} & \mathbf{C}_2 \end{bmatrix} \quad (19)$$

$$\mathbf{C}_2 = -\overline{z_s e_{3x}} \begin{bmatrix} \overline{\varphi}_{s\,1\times n} & -2\frac{\overline{e_{3s}}}{e_{3x}}\overline{\theta}_{s\,1\times n} \end{bmatrix} \begin{bmatrix} \mathbf{D} & \mathbf{0} \\ \mathbf{0} & \mathbf{D} \end{bmatrix}$$
(20)

In Eq. 20, the *differentiating matrix operator* **D** is introduced to replace the space derivatives involved in Eq. 17.[8,16,19]

Periodic Model-Following Controller

It is well known that the hover condition is characterized by low levels of rotor vibration. Provided that the *formal difference* between the hover and the forward flight conditions is the periodic nature of the aeroelastic system given by Eq. 14 and Eq. 18, it is worthwhile to examine the feasibility of eliminating (or at least minimizing) the sinusoidal terms of Eqs. 9a-b using feedback control. For this, a model-following regulator which penalizes the difference between the time derivative of the state vector (in modal coordinates) and a *model* that generates the time derivative of same state vector *if the hover condition were verified* (Eq. 16), is constructed. The process may be called *linearization of the blade's aeroelasticity*. Thus, the following cost function is formulated:

$$J = \frac{1}{2}\int_0^\infty \left\{ \left(\dot{\vec{\eta}} - \dot{\vec{\eta}}^*\right)^H \mathbf{Q}\left(\dot{\vec{\eta}} - \dot{\vec{\eta}}^*\right) + \mathbf{u}^H \mathbf{R}\mathbf{u} \right\} dt,$$
(21)

where the model is

$$\dot{\vec{\eta}}^* = \Lambda \vec{\eta}$$
$$\vec{\iota}^* = \tilde{\mathbf{C}}\vec{\eta}^* \quad (22a\text{-}b)$$

The model-following technique was extended to periodic systems by Nishimura,[21] and particularly applied to rotary wings by McKillip,[22,23] who called the present scheme *implicit* because the model is constructed using the actual system's states. It has the important advantage of not requiring the internal generation of the model's states at each instant of time, as the explicit formulation does. As a result, the closed-loop system may use a direct feedback based on the sensor output. In a previous investigation, the same technique was generalized to complex-coefficient systems, as represented by the equations governing the aeroelastic behavior of a rotating blade in the modal domain.[12] For the sake of completeness, the derivation is summarized next.

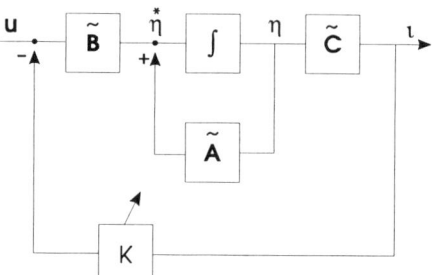

Fig.2 Control block diagram for a typical IBC subsystem.

For the two configurations studied, discrete and lumped control, a single input is considered. Hence, the relative cost of the control is expressed by a scalar quantity:

$$\mathbf{R} = \rho \quad (23)$$

Two control strategies will be examined: *full-state* and *output* feedback. In the former, **Q** in Eq. 21 is a diagonal matrix of dimension *2m* whose elements are chosen to attribute a relative weight to the state vector components in the modal domain. Therefore, the system can be *linearized within any specific frequency range*. The latter proposition seems particularly attractive if one recalls that the rotor hub naturally filters out the vibration associated with frequencies (in *1/rev* units) different from an integer multiple of the number of blades. In the output feedback situation (Fig. 2), **Q** is constructed using the sensor matrix as the weighting factor:

$$\mathbf{Q} = \tilde{\mathbf{C}}^H \tilde{\mathbf{C}} \quad (24)$$

The proposed "linearization" of the system in specific ranges of frequencies also can be achieved with the output feedback if a modal filter is employed.[17] In this case, the output matrix in Eq. 24 is obtained using the shape functions optimized for a particular aeroelastic mode.

The optimum control problem involves the solution of the time-varying Riccati equation:

$$-\dot{\mathbf{P}} = \mathbf{PS} + \mathbf{S}^H\mathbf{P} - \mathbf{PEP} + \mathbf{N}, \qquad (25)$$

where

$$\mathbf{S} = \tilde{\mathbf{A}} - \tilde{\mathbf{B}}\mathbf{W}_{uu}^{-1}\mathbf{W}_{\eta\mu}^{H}$$
$$\mathbf{E} = \tilde{\mathbf{B}}\mathbf{W}_{uu}^{-1}\tilde{\mathbf{B}}^{H}$$
$$\mathbf{N} = \mathbf{W}_{\eta\eta} - \mathbf{W}_{\eta\mu}\mathbf{W}_{uu}^{-1}\mathbf{W}_{\eta\mu}^{H} \qquad \text{(26a-c)}$$

and

$$\mathbf{W}_{\eta\eta} = (\tilde{\mathbf{A}} - \grave{}\Lambda)^{H}\mathbf{Q}(\tilde{\mathbf{A}} - \grave{}\Lambda)$$
$$\mathbf{W}_{uu} = \mathbf{R} + \tilde{\mathbf{B}}^{H}\mathbf{Q}\tilde{\mathbf{B}}$$
$$\mathbf{W}_{\eta\mu} = (\tilde{\mathbf{A}} - \grave{}\Lambda)^{H}\mathbf{Q}\tilde{\mathbf{B}}. \qquad \text{(27a-c)}$$

Nishimura determined that the periodic solution of Eq. 25 is obtained by the so-called *spectral factorization*.[21] First, it is necessary to compute the transition matrices along one complete period (the integrating matrix scheme offers an interesting alternative solution to the traditional step-by-step integration methods based on Range-Kutta).[12] The transition matrices are obtained from:[24]

$$\dot{\Phi} = \mathbf{T}\Phi(t,t_0)$$
$$\Phi(t,t) = \mathbf{1}$$
$$2\pi/\Omega \geq t \geq t_0, \qquad \text{(28a-c)}$$

with

$$\mathbf{T} = \begin{bmatrix} \mathbf{S} & -\mathbf{E} \\ -\mathbf{N} & -\mathbf{S}^{H} \end{bmatrix} \qquad (29)$$

In particular, the fundamental transition matrix (FTM) is recovered if $t=2\pi/\Omega$ and $t_0=0$. Next, the FTM is factored into its stable and unstable modes:

$$\Phi^{*} = \begin{bmatrix} \Psi_{11} & \Psi_{12} \\ \Psi_{21} & \Psi_{22} \end{bmatrix} \begin{bmatrix} \grave{}\lambda^{s} & 0 \\ 0 & \grave{}\lambda^{u} \end{bmatrix} \begin{bmatrix} \Psi_{11} & \Psi_{12} \\ \Psi_{21} & \Psi_{22} \end{bmatrix}^{-1} \qquad (30)$$

Finally, the periodic solution of Eq. 25 is found at any instant using (1) the partitions of the intermediate transition matrices between t and $t_0=0$ (which can be saved during the process of finding the FTM) and (2) the FTM's eigenvectors factored according to Eq. 30:

$$\mathbf{P} = [\Phi_{11}\Psi_{11} + \Phi_{12}\Psi_{21}][\Phi_{21}\Psi_{11} + \Phi_{22}\Psi_{21}]^{-1}. \qquad (31)$$

The closed-loop gain associated with this solution is now obtained. It is a matrix with *2m* columns whose elements represent periodic functions of time:

$$\mathbf{K}(t) = \mathbf{W}_{uu}^{-1}(\mathbf{W}_{\eta\mu}^{H} + \tilde{\mathbf{B}}^{H}\mathbf{P}) \qquad (32)$$

The regulator's input is then simply

$$\mathbf{u}(t) = -\mathbf{K}\vec{\eta} \qquad (33)$$

Substitution of Eq. 33 into Eq. 14 yields the closed-loop system's equation:

$$\dot{\vec{\eta}} = (\tilde{\mathbf{A}} - \tilde{\mathbf{B}}\mathbf{K})\vec{\eta} \qquad (34)$$

The complex eigenvalues associated with Eq. 34 describe the aeroelastic behavior of the controlled system. Ideally, if a perfect linearized situation is achieved, they are constant with time.

Results

The typical bearingless rotor presented in previous works is studied.[8,12,16,17] The main properties are defined in Table 1.

Table 1: Typical Rotor Parameters

parameter	value	units
c/R	0.0555	-
c_φ	∞	-
c_θ	∞	-
$D^*_{11}EI_R/R$	1.0	-
$D^*_{13}EI_R/R$	0.0	-
$D^*_{33}EI_R/R$	1.756	-
d_a/c	0.09	-
E^*	381.[(1)]	V/m
E^*	6,350.[(2)]	V/m
EI_R	6.89×10^4	N.m²
$e_{3x}E^*/m\Omega^2$	0.4272[(1)]	-
$e_{3s}E^*/m\Omega^2$	0.1617[(1)]	-
$e_{3x}E^*/m\Omega^2$	0.03163[(2)]	-
$e_{3s}E^*/m\Omega^2$	0.008785[(2)]	-
I_b/mR^3	0.333	-
k_θ/R	0.03	-
m	5.5	kg/m
R	4.926	m
t_a/c	0.015	-
x_a/c	0.1	-
z_s/c	0.06	-
Ω	44.4	rad/s
θ_c^*	1.0	deg
γ	5.65	-
μ	0.32	-
ν	92.7	-

(1) *PZT (actuator)*
(2) *PVDF (sensor)*

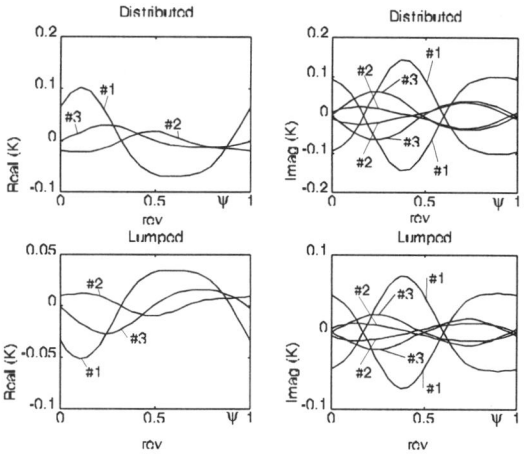

Fig.3 One-period evolution of the real and imaginary components of the control gain for the distributed and lumped control configurations.

Periodic Model-Following Controller

Plots from the real and imaginary parts of the control gain are depicted in Fig. 3 to illustrate the typical results that can be obtained with a periodic model-following controller designed in the complex domain. The method described in the previous section is followed. The modal expansion is truncated after the first six complex-conjugate modes ($m=3$). These are related to the first bending (#1), second bending (#2), and first torsion (#3) of the non-rotating blade, respectively. The integrating-matrix method is employed to obtain the transition matrices.[12] It was observed that a minimum of 20 discretizing points are necessary to get numerically stable solutions, which demands the inversion of a linear system with complex coefficients of order 240 (**T** in Eq. 29 is a $4m \times 4m$ matrix). However, a comparative solution using a third-order Range-Kutta step-by-step integration scheme was slower by a factor of three.

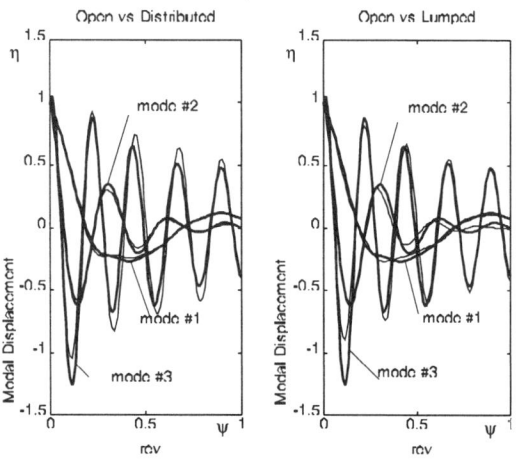

Fig.4 Periodic model-following controller: time response of the generalized coordinates to unitary initial conditions (distributed and lumped control configurations). The thicker lines represent the open-loop; the thiner lines the closed-loop cases. Full-state feedback, **Q** = 1, ρ = 0.

The results for both the distributed and lumped control configurations are obtained. It is observed that in both cases **K** is a perfect periodic complex-conjugate quantity for each one of the modes. Output feedback is assumed (Eq. 24). The shape functions, not optimized for a particular mode, are constants equal to $0.8c$.

Figure 4 shows comparative plots between the open-loop and closed-loop time-responses for the two control configurations studied: lumped and distributed (arbitrary unity initial conditions are assigned for all modes). Here, full-state feedback is assumed (**Q**=1). In Fig. 2, the feedback loop is closed right after the plant dynamics, excluding the sensing block from the main path. It is clear that in both cases the modal damping associated with each mode is not significantly affected by closing the loop. In fact, the controller *is not* designed to increase modal damping. However, a closer observation of the third mode's decaying characteristics indicates that in a closed-loop situation it behaves like a linear one-degree-of-freedom oscillator with viscous damping. This suggests that at least for the third mode a *linearization* of the periodic system was achieved, which is confirmed in Fig. 5. For the same cases presented in Fig. 4, the evolution of the modal parameters along one period is plotted. The third aeroelastic mode in both closed-loop cases presents an almost linear behavior: the frequency is virtually constant, whereas the damping ratio varies just a small fraction from the original open-loop characteristics. The extreme situation of no control cost is assumed ($\rho=0$). Although the same relative weight is given to all modes, the first and second modes are only marginally affected by control, indicating that there is a limit beyond which the implicit model proposed in this investigation is saturated. Furthermore, the lumped and distributed control configurations presented similar performance. Nevertheless, recalling the filtering properties of the system at the frequencies multiple of the number of blades, such a "linearization" of the torsion mode at *4/rev* would be beneficial regarding the overall vibration reduction of a four-bladed bearingless rotor.

Distributed Control Configuration

According to the distributed control configuration, the actuation is performed by layers of PZT embedded in the structure. One half of the layers are specialized to produce structural bending, the other half torsion. An electric field applied by a pair of electrodes across the adaptive material's thickness generates internal strain that deforms the structure in both bending and torsion. Optimum distributions of the electrodes' effective width are able to induce the "right" amount of bending and torsion that yields superior authority over a given aeroelastic mode (which is originated from the superposition of pure bending and torsion natural modes). An *aeroelastic modal filter* is created. Considering the lack of power inherent to the available PZT material, it is expected that by concentrating the energy in one particular aeroelastic mode a more efficient actuation mechanism is created. In the following, the results are presented for the situations with and without modal filters.

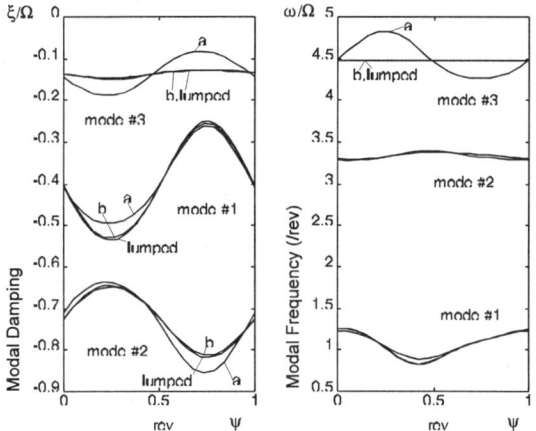

Fig.5 One-period evolution of the modal parameters. Full-state feedback, **Q** = **1**, ρ = 0. (a) open-loop; (b) distributed; and lumped control.

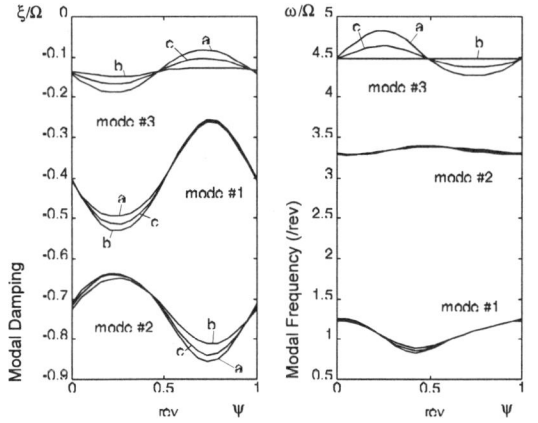

Fig.6 One-period evolution of the modal parameters (distributed control). (a) open-loop; (b) full-state feedback, **Q** = **1**, ρ = 0; and (c) output feedback, $\mathbf{Q} = 10^4 \tilde{\mathbf{C}}^H \tilde{\mathbf{C}}$ (no modal filter), ρ = 0.001.

Figure 6 presents the comparative evolution of the modal parameters for the first three modes in the cases of open-loop, full-state feedback (**Q=1**, repeated from Fig. 5) and output feedback without modal filters (the shape functions have a constant width of *0.8c*). Therefore, actuation on all modes is verified in both controlled cases. The overall performance of the output feedback configuration is inferior to its full-state counterpart, even though some linearization of the third mode is observed with the latter approach. This is not an unexpected result, since here the problem was reduced to one of a single input and single output without the introduction of a state estimator.

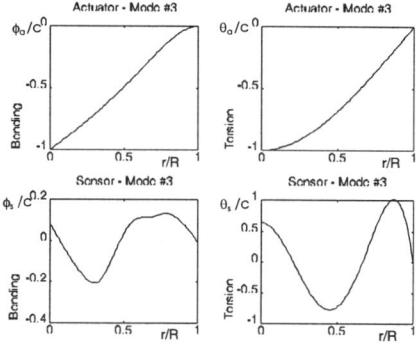

Fig.7 Modal filter: normalized actuator (upper), and sensor (lower) shape functions for Mode #3 at hover condition. Flatwise bending components (left) and torsion components (right).

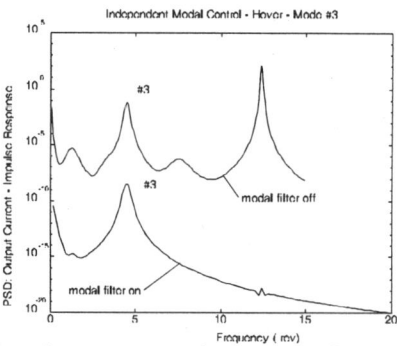

Fig.8 Impulse response for *m = 5* (five aeroelastic modes retained). Without modal filter (upper), and with the modal filter of Fig.7 (lower).

The shape functions that optimize the actuator and sensor width distributions along the blade are depicted in Fig. 7 for the third mode *at the hover condition* (where they become independent of the azimuth angle). They are calculated in closed, semi-analytical form using the method presented in a former investigation.[17] The output feedback strategy requires four distinct shape functions to perform independent modal control in the present non-self-adjoint problem (unlike in self-adjoint problems, where the same shape functions define both the actuator and sensor distributions). The changes in sign observed on a given shape function may be realized in actual applications by duplicating the number of layers on opposite sides of the neutral plane.[20] The performance of the aeroelastic modal filter defined in Fig. 7 is analyzed in Fig. 8, where the system's response to an impulse is plotted for the situations with and without the filter ($m=5$ was taken in this analysis). It is clear that only mode number three (at about 4/rev) is dominating when the filter is present.

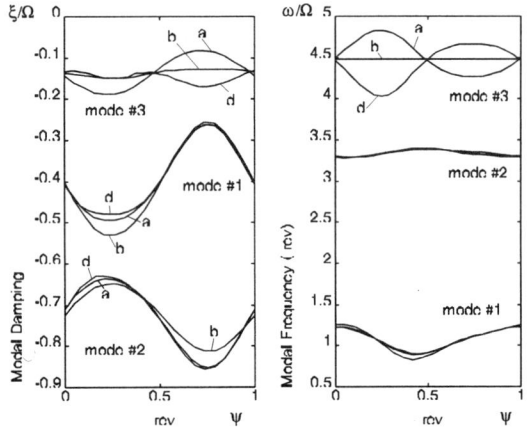

Fig.9 One-period evolution of the modal parameters (distributed control). (a) open-loop; (b) full-state feedback, $\mathbf{Q} = diag[0,0,0,0,1,1]$, $\rho = 0$, and $\tilde{\mathbf{B}}$ from the modal filter of Fig.7; and (d) output feedback, $\mathbf{Q} = 10^{12} \tilde{\mathbf{C}}^H \tilde{\mathbf{C}}$ (with modal filter), $\rho = 100$.

The results shown in Fig. 9 are for the open loop, the full-state feedback - with 1) the actuator width distribution optimized for the third mode, and 2) the **Q** matrix selecting only mode number three and its complex conjugate ($\mathbf{Q}_{ij}=0$; $\mathbf{Q}_{55}=\mathbf{Q}_{66}=1$), - and the output feedback considering the complete modal filter (sensors and actuators). Although in the case of full-state feedback only mode three was aimed to be controlled, the modal damping plots indicate that the same undesired actuation is performed on both the first and second modes. The performance evaluated in terms of linearization of the third mode is, however, still good. It seems that an "extra filtering step" provided by the sensing device is necessary to achieve perfect independent modal control. In fact, a superior rejection to modes one and two is observed when the complete modal filter is used. Unfortunately, in the latter case, linearization of mode number three failed completely. The most probable reason for such behavior is a mismatch between the gain needs and the sensor characteristics (whose shape functions were designed to perform at the hover condition) at different azimuth angles. It was verified that the sensor shape functions, which are derived from the system's left eigenvectors, are extremely sensitive to non-zero advance ratios.[17] Therefore, the most feasible design in real rotor applications (adhering to the distributed control configuration) leads to a compromise of the optimization only to the actuator elements, keeping the sensor distribution constant.

Figure 10 presents the sensor output plots for the situations studied above. As expected, the sensor output in the presence of the complete modal filter has the cleaner, single-frequency behavior associated with the third mode. The plot on the left-hand side of Fig. 10 includes the output of a closed-loop system using the lumped control configuration, which will be focused on next.

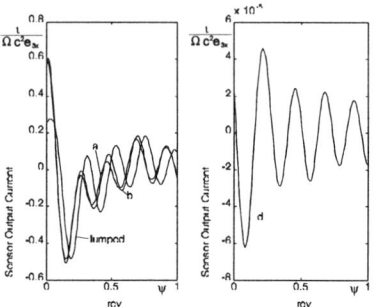

Fig.10 Sensor output $\bar{\iota} = \iota/(\Omega c^2 e_{3x})$ versus azimuth angle. On the left: (a) open-loop; (b) taken from Fig.9; and a lumped control case (output feedback, $\mathbf{Q} = 10^2 \tilde{\mathbf{C}}^H \tilde{\mathbf{C}}$, $\rho = 1$). On the right: (d) taken from Fig.9.

<u>Lumped Control Configuration</u>

The most important drawback of the distributed control configuration is the substantial lack of power of the available PZT to work against the loads developed by a typical rotor in forward flight. Among the virtues of the distributed control, the most interesting is the possibility of performing independent modal control, avoiding spillover, and other "bad" closed-loop effects. At the present stage of the materials technology, however, the lumped control approach seems to be the most promising. As lumped control is concerned, aside from the flap solution mentioned in the introductory section, an alternative approach deserves special attention. Torsion deformations can be induced at the blade root in order to control the aeroelastic response. Several methods may be devised to create the desired actuation (with or without adaptive materials). These methods are not discussed in the present work; only the assumption is made that the actuation is bounded by a maximum angle of one degree.

As in the previous section, the periodic model-following controller is synthesized considering two cases: full-state and output feedback. Of course, in the latter no attempt will be made to find an optimum shape for the sensor width distribution because, conceptually, the control itself modifies the aeroelastic modes. Hence, the sensor shape functions are again assumed constant, and equal to *0.8c*. A case of full-state feedback (**Q=1**) was already studied in Fig. 4 and Fig. 5, indicating that for $\rho=0$ the performances of both the lumped and distributed control configurations are indistinguishable.

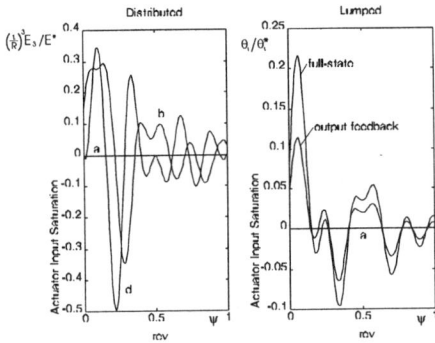

Fig.11 Normalized actuator input **u** versus azimuth angle. On the left, distributed control cases: (a) open-loop; and (b) and (d) taken from Fig.9. On the right, lumped control: (a) open-loop; full-state feedback with **Q** = 1 and ρ = 0; and output feedback with $\mathbf{Q} = 10^2 \tilde{\mathbf{C}}^H \tilde{\mathbf{C}}$, ρ= 1.

Figure 11 demonstrates the most remarkable advantage of the lumped over the distributed control configuration. In this figure, plots from the normalized input variable **u** versus azimuth angle along one period are presented. Therefore, a non-saturated situation (corresponding to a feasible solution regarding the actuator's ability to work against the external loads) is characterized by values of **u** lying in the interval *[-1,1]*. The cases of full-state feedback are, for both control configurations, based on **Q=1**. In the left-hand side figure, where the distributed configuration is studied, the input electric field is normalized by the PZT's saturation field and $(c/R)^3$, a very small value in the order of 10^{-3}. The result recovers the well-known limitations of the distributed configuration. However, in the right-hand side plot, where the lumped configuration is analyzed, the control angle is normalized only by its maximum acceptable value. Hence, in contrast with the distributed control, a perfectly balanced situation is achieved. Actuation at the blade root seems to be a much more efficient way of changing the torsion mode characteristics. Obviously, similar success shall not be expected with the aeroelastic modes whose torsion components are not the dominant ones. Finally, the time-history presented in Fig. 10 for one lumped control configuration corresponds to the case of output feedback in Fig. 11.

Conclusions

The main conclusions that can be drawn from the present investigation are: 1) The implicit model-following technique may be extended to the complex modal domain. 2) Integrating matrices are a powerful tool to obtain state transition matrices useful in the solution of periodic systems. 3) An implicit periodic model-following regulator may be used to linearize the open-loop system at determined frequencies, approximating the aeroelastic response characteristics of a single blade in forward flight to its hover conditions. 4) Aeroelastic modal filters constructed with adaptive materials are useful to perform independent modal control, even in time-varying systems such as rotary wings. 5) In the distributed control configuration, attempts to optimize the sensor's distribution aiming at independent modal control should be taken cautiously. A better solution is to optimize only the actuator's distribution. 6) Available piezoceramic material lacks enough power to provide suitable actuation in the distributed control case. 7) Presently, lumped control by inducing torsion at the blade root seems to be the most feasible solution for the individual blade control of rotary wings using adaptive materials.

References

[1] Ormiston, R.A., "Can Today's Smart Structures Make Helicopters Better?" *Fourth Workshop on Dynamics and Aeroelastic Stability Modeling of Rotorcraft Systems*, The University of Maryland, November 19-22, 1991.

[2] Chopra, I., "Development of a Smart Rotor," *Proceedings: 19. European Rotorcraft Forum*, Associazione Italiana di Aeronautica ed Astronautica, Cernobbio, Italy, 1993, pp. N6.1-N6.18.

[3] Spangler, R.L., Jr. and Hall, S.R., "Piezoelectric Actuators for Helicopter Rotor Control," *Proceedings: AIAA/ASME/ASCE/AHS/ ASC 31. Structures, Structural Dynamics and Materials Conference,* AIAA, Washington, DC, 1990, Part 3, pp. 1589-1599.

[4] Millott, T.A. and Friedmann, P.P., "Vibration Reduction in Helicopter Rotors Using an Active Control Surface Located on the Blade," *Proceedings: AIAA/ASME/ASCE/AHS/ASC 33.*

Structures, Structural Dynamics and Materials Conference, AIAA, Washington, DC, 1992, Part 4, pp. 1975-1988.

[5] Samak, D.K. and Chopra, I., "A Feasibility Study to Build a Smart Rotor: Trailing Edge Flap Actuation," Proceedings: Smart Structures and Intelligent Systems 1993, SPIE - The International Society for Optical Engineering, Washington, DC, 1993, Vol. 1917, Part 1, pp. 225-237.

[6] Strehlow, H. and Rapp, H., "Smart Materials for Helicopter Rotor Active Control," Paper No. 5, AGARD/SMP Specialists' Meeting on Smart Structures for Aircraft and Spacecraft, Lindau, Germany, October 5-7, 1992.

[7] Narkiewicz, J. and Roguz, M., "Smart Flap for Helicopter Rotor Blade Performance Improvement," Proceedings: 19. European Rotorcraft Forum, Associazione Italiana di Aeronautica ed Astronautica, Cernobbio, Italy, 1993, pp. G9.1-G9.11.

[8] Nitzsche, F., Lammering, R. and Breitbach, E., "Can Smart Materials Modify Blade Root Boundary Conditions to Attenuate Helicopter Vibration?" Fourth International Conference on Adaptive Structures, DGLR Paper No. 93-04-12, Cologne, Germany, November 2-4, 1993 (to be published in the Proceedings).

[9] Geissler, W. and Raffel, M., "Dynamic Stall Control by Airfoil Deformation," Proceedings: 19. European Rotorcraft Forum, Associazione Italiana di Aeronautica ed Astronautica, Cernobbio, Italy, 1993, pp. C2.1-C2.13.

[10] Hanagud, S., Prasad, J.V.R., Bowles, T. and Nagesh-Babu, G.L., "Smart Structures in the Active Control of Blade Vortex Interaction," Paper No. 75, 17. European Rotorcraft Forum, Berlin, Germany, September 24-27, 1991.

[11] Prasad, J.V.R., Sankar, L.N. and Park, W.G., "Active Control of Rotor Noise," Proceedings: Smart Structures and Intelligent Systems 1993, SPIE- The International Society for Optical Engineering, Washington, DC, 1993, Vol. 1917, Part 1, pp. 255-266.

[12] Nitzsche, F., "Periodic Model-Following Controller for the Independent Modal Control of Individual Blades using Smart Structures," Proceedings: 19. European Rotorcraft Forum, Associazione Italiana di Aeronautica ed Astronautica, Cernobbio, Italy, 1993, pp. G6.1-G6.12.

[13] Kretz, M., "Research in Multicyclic and Active Control of Rotary Wings," Vertica, Vol. 1, No. 1/2, 1976, pp. 95-105.

[14] Ham, N.D., "Helicopter Individual Blade Control at MIT 1977-1985," Vertica, Vol. 11, No. 1/2, 1987, pp. 109-122.

[15] Bielawa, R.L., Rotary Wing Structural Dynamics and Aeroelasticity, AIAA, Washington DC, 1992, pp. 211-212.

[16] Nitzsche, F. and Breitbach, E., "A Study on the Feasibility of Using Adaptive Structures in the Attenuation of the Vibration Characteristics of Rotary Wings," Proceedings: AIAA/ASME/ASCE/AHS/ASC 33. Structures, Structural Dynamics and Materials Conference, AIAA, Washington, DC, 1992, Part 3, pp. 1391-1402.

[17] Nitzsche, F., "Modal Sensors and Actuators for Individual Blade Control," Proceedings: AIAA/ASME/ASCE/AHS/ASC 34. Structures, Structural Dynamics and Materials Conference, AIAA, Washington, DC, 1993, Part 6, pp. 3507-3516.

[18] Johnson, W., Helicopter Theory, Princeton University Press, Princeton, New Jersey, 1980, pp. 548-600.

[19] Lehman, L.L, "Hybrid State Vector Methods for Structural Dynamic and Aeroelastic Boundary Value Problems," NASA CR-3591, August 1982.

[20] Lee, C.-K. and Moon, F.C., "Modal Sensors/Actuators," Journal of Applied Mechanics, Vol. 57, June 1990, pp. 434-441.

[21] Nishimura, T., "Spectral Factorization in Periodically Time-Varying Systems and Application to Navigation Problems," Journal Spacecraft, Vol. 9, July 1972, pp. 540-546.

[22] McKillip, R.M., Jr., "Periodic Control of the Individual Blade Control Helicopter Rotor," Vertica, Vol. 9, No. 2, 1985, pp. 199-225.

[23] McKillip, R.M., Jr., "Periodic Model-Following for the Control-Configured Helicopter," Journal of the American Helicopter Society, July 1991, pp. 4-12.

[24] Bryson, A.E. and Ho, Y.-C., Applied Optimal Control, Hemisphere Publishing Corporation, Washington, DC, 1975, pp. 148-176, 451.

AIAA-94-1767-CP

Design and Testing of a Helicopter Rotor Model with Smart Trailing Edge Flaps

Curtis Walz[*] and Inderjit Chopra[**]

Center for Rotorcraft Education and Research
Department of Aerospace Engineering
University of Maryland, College Park, MD 20742

Abstract

This paper discusses the development of a Froude scale helicopter rotor model with trailing edge flap to be used for vibration suppression. A previous feasibility study verified that a trailing edge flap could be actuated using bimorph actuators mounted inside the rotor blade. However, the flap deflections obtained were too small to be used effectively for vibration suppression. The objective of the present research was to obtain larger flap deflections by improving the bimorph flap actuation system. An actuator model was developed using experimental data to characterize the force and displacement output of the bimorph actuator. Higher force bimorphs were developed and constructed in-house. Using quasi-steady aerodynamic characteristics of a flap, parametric studies are used to match the flap requirements with the capability of the actuators. Modifications to the mechanical leverage arrangement between actuator and flap are incorporated in the design and construction of an improved rotor model. Construction methods and some preliminary results of the improved rotor model are given.

Nomenclature

l, L	lift per unit span, total lift
h, H	hinge moment per unit span, total hinge moment
X	flap authority
δ	flap deflection
c	blade chord
c_f	flap chord
C_h	hinge moment coefficient
C_T	thrust coefficient
C_l	lift coefficient
C_{l_α}	lift curve slope
α_o	zero lift angle of attack
θ_o	blade collective pitch
V	velocity
U_P	normal velocity
U_T	inplane velocity
r	radius
R_{root}, R_{tip}	blade root radius, blade tip radius
R_1, R_2	inboard flap radius, outboard flap radius
Ω	rotation speed
σ	rotor solidity
λ	rotor inflow
N_b	number of blades
k_p	tip loss factor
ρ	density
1 mil	=.001 inch
subscripts	
1	rotor without trailing edge flap
2	rotor with trailing edge flap

Presented at the 35th Structures, Structural Dynamics and Materials Conference, Adaptive Structures Forum, April 1994.
[*] Graduate Research Assistant, Member AIAA
[**] Professor and Director, Fellow AIAA, Member AHS
Copyright © 1994 by American Institute of Aeronautics and Astronautics, Inc. All rights reserved.

Introduction

Helicopters suffer in performance due to high vibration levels, fatigue loads, excessive noise, and deficiencies in handling qualities. The focus of this research is active vibration suppression using smart structures technology. Because of the unsteady aerodynamic environment at the rotor disk, the blades experience high levels of vibratory forces, which are transmitted to the helicopter fuselage. Currently, passive isolators and absorbers are routinely used to reduce vibration. However, these devices have weight penalties and they rapidly degrade in performance away from the tuned flight condition. In a conventional helicopter, primary control of the rotor is achieved by cyclic variation of blade pitch through a mechanical swashplate. Higher harmonic control (HHC) systems have been shown to reduce helicopter vibration by shaking the swashplate at N/rev frequency with servo-actuators where N is the number of blades.[1] Oscillation of the blade pitch at higher harmonics generates additional unsteady airloads which can surpress vibration when properly controlled. This type of HHC system again has a weight penalty and is limited to N/rev excitation frequencies.

The purpose of the trailing edge flap in this research is to provide an active vibration reduction device that cancels the vibration at its source, by directly altering the airloads on the blades. It is envisioned that using a trailing edge flap driven by smart-materials actuators embedded in the blade with the necessary force and displacement characteristics can provide a lightweight, compact, vibration suppression system. One of the key benefits of this scheme is that the blade can be excited at any frequency and, therefore, can act as an individual blade control (IBC) device.[2]

At the University of Maryland, an experimental six foot diameter Froude scale rotor model was built and tested in hover using bimorphs as flap actuators.[3] The small bending deflection at the tip of the bimorph was amplified using a mechanical leverage arrangement to actuate the flap.

Results from this initial rotor test showed that the bimorph mechanism worked and was able to deflect the flap at low dynamic pressures, but a decrease in flap deflection occurred with increased dynamic pressure or rotor RPM. At 900 RPM, a flap rotation of ±2° was obtained. In order to have sufficient vibration suppression capability it is estimated that a flap deflection of ±8° is required.

The objective of this research is to redesign and improve the flap actuation system by increasing the force output of the bimorph actuators, developing a simple aerodynamic model to properly size the flap, and modifying the mechanical leverage arrangement between the actuator and flap.

Bimorph Flap Actuation System

A sheet of piezoceramic material (PZT) undergoes contraction or extension when a positive or negative electric field is applied. Typical piezoceramic sheets available today develop between 180 to 220 microstrain with an applied field of 15 volt/mil.[4] With these values of strain, the resulting deflections of moderate size PZT sheets are too small to effectively drive a trailing edge flap. By bonding a piezoceramic sheet to the upper and lower surfaces of a thin beam (shim), a much larger bending deflection can be obtained. An opposite voltage field is applied to the upper and lower sheets causing a bending deformation as shown in Figure 1. This type of device is referred to as a bimorph bender element.

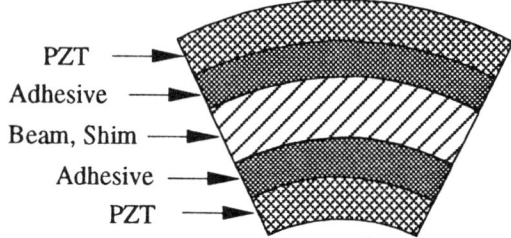

Figure 1. Bimorph Cross-Section.

The bimorph can be used as an actuator by cantilevering one end and making use of the tip displacement at the free end as shown in Figure 2. A typical 20 mil thick 1 inch long bimorph cantilevered at one end can develop as much as 20 mils displacement at the free end under an applied field of 13.3 volt/mil to each of its sheets.

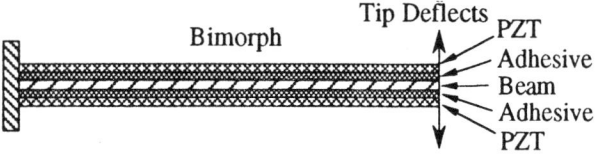

Figure 2. Cantilevered Bimorph.

The bimorph bender element can be used to actuate a flap by cantilevering one end of the bimorph to the blade spar and connecting the free end to a mechanical linkage arm which rotates the flap around a hinge. The proposed bimorph flap actuation system is shown in Figure 3.

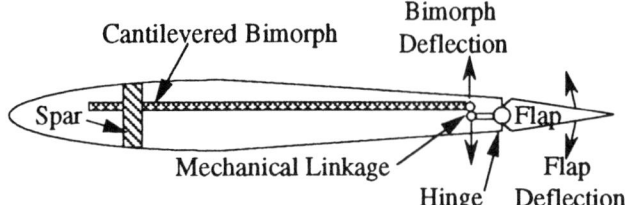

Figure 3. Bimorph Flap Actuation System.

Since the power source is placed in the fixed frame, signals to the bimorphs in the rotating blades are passed through a slip ring unit. Such a system that is able to generate significant flap deflections to effectively alter the loads on each blade at any desired frequency will provide individual blade pitch control, and potentially a vibration suppression system.

Results from Initial Rotor Test

At the University of Maryland, a two bladed six foot diameter rotor with three inch chord, NACA 0012 airfoil was tested on a hover stand to determine the feasibility of a trailing edge flap actuated by piezoceramic bimorphs. The trailing edge flap system consisted of a 20% chord, 4.5 inch span, trailing edge flap located near the blade tip from 85.4% to 97.9% span. The flap was actuated by three twenty mil thick G-1195 commercial bimorphs. The 3.0 inch chord blades were built in a mold using rigid foam covered with fiberglass prepreg fabric. The arrangement with the three bimorphs cantilevered at the blade spar in shown in Figure 4.

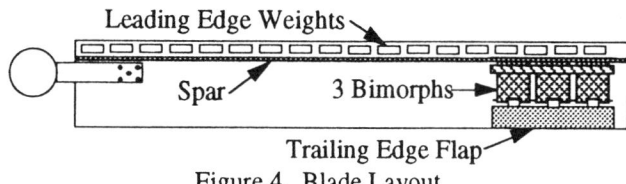

Figure 4. Blade Layout.

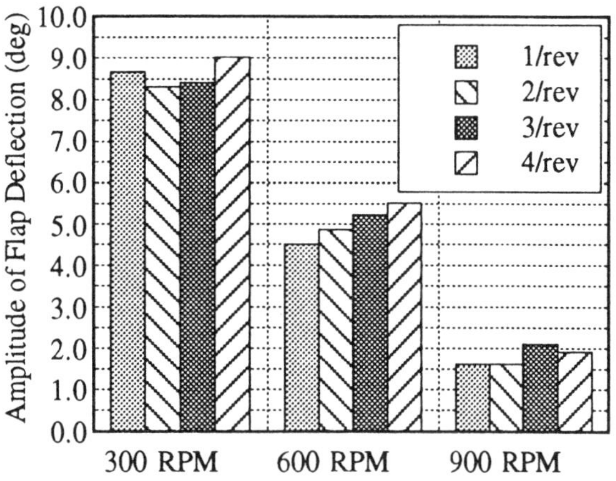

Figure 5. Results of Initial Hover Test.

As the rotor rpm increased to the operating condition of 900 RPM, the trailing edge flap actuation decreased to about ±2° degrees as shown in Figure 5. These results were obtained for a blade collective pitch of 8° and an applied oscillatory field of 14.5 volts RMS/mil to each sheet of the bimorph at frequencies of 1, 2, 3, and 4 per rev. More results are available in Ref. 3.

In order to have sufficient vibration suppression capability for this rotor model, it is estimated that a flap deflection of about ±8° is needed. Thus, to increase the flap authority it is essential to develop higher force actuators.

Development of Higher Force Bimorph Actuators

To increase the force capability of the actuators, thicker bimorphs with higher bending stiffness were built by gluing two additional 7.5 mil G-1195 piezoceramic layers to the commercially available two layer bimorph using conducting adhesive. Figure 6 shows the cross-section of a two layer 20 mil commercial bimorph used in the initial rotor model and an in-house built four layer 40 mil bimorph. In the case of the four layer actuator, both sheets on either side of the brass shim are electrically driven the same way so they effectively act as a regular bimorph with thicker piezoceramic sheets (15 mil, rather then 7.5 mil).

The displacement and force characteristics of the two and four layer bimorph actuators with one inch width and various cantilever lengths were measured statically by measuring the displacements at the free end subjected to various hanging weights. Figures 7 and 8 show the results with an applied field of 13.3 volt/mil dc applied to each sheet. The data show a linear decrease in the tip displacement as the applied force is increased. Increasing the cantilever length increases the free displacement (zero force condition), but decreases the block force (zero displacement condition). The four layer actuator showed a decrease in free displacement, but a substantial increase in block force compared to the two layer actuator. A slight non-linearity in the data is due to creeping of PZT material under dc voltages. The potential for the four layer actuator is its higher actuation force capability, however, its lower actuation displacements require further magnification of the mechanical leverage mechanism.

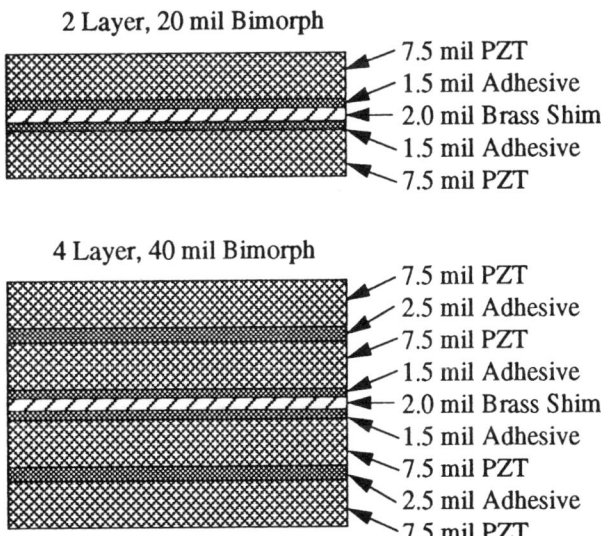

Figure 6. Cross-Section of Bimorphs.

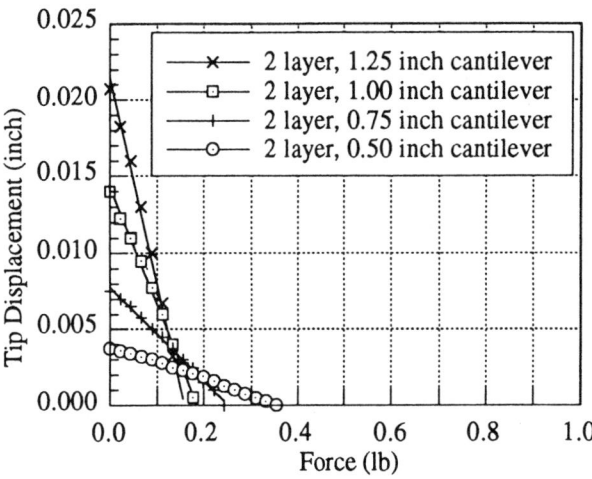

Figure 7. Displacement vs. Force, Two Layer Bimorph.

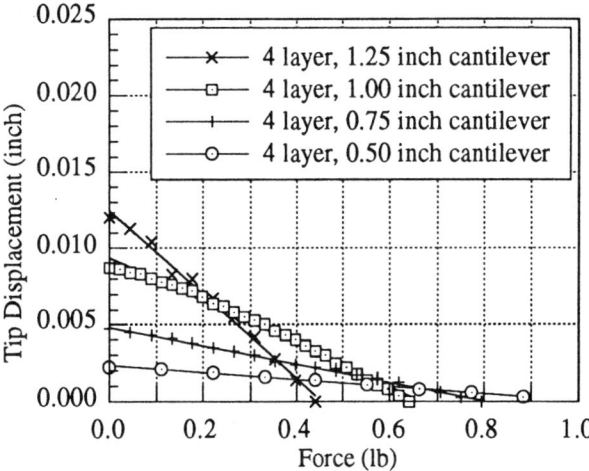

Figure 8. Displacement vs. Force, Four Layer Bimorph.

Prediction of Aerodynamic Hinge Moments

The next step was to develop a simple aerodynamic model that could be used to properly size the dimensions of the flap. The following quasi-steady, incompressible aerodynamic calculations are carried out for a rotor blade to predict the flap hinge moment for a desired flap authority. These calculations are intended to obtain an estimate of the magnitude of the flap hinge moment such that the flap's linkage arm displacement and force requirements can be achieved.

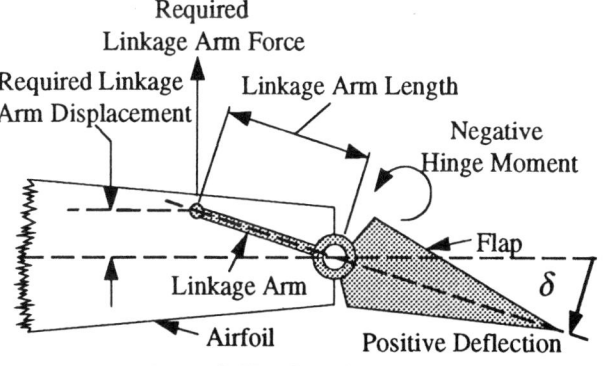

Figure 9. Flap Requirements.

Due to the aerodynamic forces, a positive flap deflection causes a negative hinge moment. Figure 9 shows for a particular hinge moment and linkage arm length the tip of the linkage arm needs to be reacted by an actuation force to maintain the flap deflection.

Calculation of flap deflection and hinge moment as a function of flap authority is developed using blade element/momentum theory for hover assuming uniform inflow and symmetric, untwisted blades. First, equations for a rotor without a trailing edge flap are given.

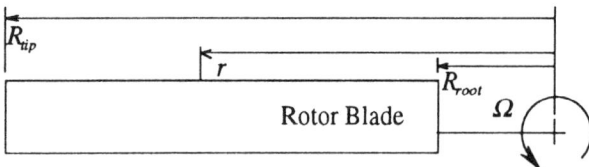

Figure 10. Rotor Blade.

$$l_1 = \frac{1}{2}\rho V^2 c C_{l_\alpha} \alpha_{eff} = \frac{1}{2}\rho V^2 c C_{l_\alpha}\left(\theta(r) - \alpha_o - \frac{U_p}{U_t}\right) \quad (1)$$

Equation (1) is the lift per unit span along the blade.

$$U_p = \lambda_1 \Omega R_{tip} \quad (2)$$

where $\lambda_1 \equiv k_p\sqrt{\frac{C_{T_1}}{2}} \quad (3)$

$$U_t = \Omega r \approx V \quad (4)$$

$\theta(r) = \theta_o$ for an untwisted blade $\quad (5)$

$\alpha_o = 0$ for a symmetric airfoil $\quad (6)$

Substituting equations (2), (4), (5), and (6) into equation (1) and integrating along the span from R_{root} to R_{tip} gives the total blade lift.

$$L_1 = \int_{R_{root}}^{R_{tip}} l_1 dr = \int_{R_{root}}^{R_{tip}} \frac{1}{2}\rho\Omega^2 c C_{l_\alpha}\left(\theta_o r^2 - \alpha_o r^2 - \lambda_1 R_{tip} r\right) dr \quad (7)$$

$$L_1 = \frac{1}{2}\rho\Omega^2 c C_{l_\alpha}\left[\theta_o\left(\frac{R_{tip}^3}{3} - \frac{R_{root}^3}{3}\right) - \lambda_1 R_{tip}\left(\frac{R_{tip}^2}{2} - \frac{R_{root}^2}{2}\right)\right] \quad (8)$$

$$T_1 = N_b L_1 \quad (9)$$

The total rotor thrust is obtained for N_b blades. The thrust coefficient is defined as

$$C_{T_1} \equiv \frac{T_1}{\rho \pi R_{tip}^2 \Omega^2 R_{tip}^2} \quad (10)$$

$$C_{T_1} = \frac{\sigma C_{l_\alpha}}{2}\left[\frac{\theta_o}{3}\frac{\left(R_{tip}^3 - R_{root}^3\right)}{R_{tip}^3} - \frac{\lambda_1}{2}\frac{\left(R_{tip}^2 - R_{root}^2\right)}{R_{tip}^2}\right] \quad (11)$$

where $\sigma = \frac{N_b c R_{tip}}{\pi R_{tip}^2} = \frac{N_b c}{\pi R_{tip}} \quad (12)$

The roots of the resulting equation can be found from the quadratic equation. One of the roots is the physical parameter for rotor inflow given as

$$\lambda_1 = \frac{\sigma C_{l_\alpha} k_p^2}{16}\left[\left\{\frac{64\left(R_{tip}^3 - R_{root}^3\right)}{3\sigma C_{l_\alpha} k_p^2 R_{tip}^3}(\theta_o) + \frac{\left(R_{tip}^2 - R_{root}^2\right)^2}{R_{tip}^4}\right\}^{\frac{1}{2}}\right.$$

$$\left. - \frac{\left(R_{tip}^2 - R_{root}^2\right)}{R_{tip}^2}\right] \quad (13)$$

The equations for a rotor with a trailing edge flap can be derived in a similar manner. The equations contain an additional term $\Delta\alpha_o/\Delta\delta$, the change in effective angle of attack due to a flap deflection over the portion of the blade with the flap.

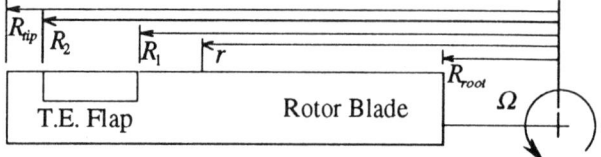

Figure 11. Rotor Blade with Trailing Edge Flap.

$$l_2 = \frac{1}{2}\rho V^2 c C_{l_\alpha}\left(\theta(r) - \alpha_o + \frac{\Delta\alpha_o}{\Delta\delta}\delta - \frac{U_p}{U_t}\right) \quad (14)$$

$$U_p = \lambda_2 \Omega R_{tip} \quad (15)$$

where $\lambda_2 \equiv k_p\sqrt{\frac{C_{T_2}}{2}} \quad (16)$

$$U_t = \Omega r \approx V \quad (17)$$

The lift per unit span from equation (14) can be integrated along the entire blade span to give the total lift.

$$L_2 = \int_{R_{root}}^{R_{tip}} \frac{1}{2}\rho\Omega^2 c C_{l_\alpha}\left(\theta_o r^2 - \alpha_o r^2 + \frac{\Delta\alpha_o}{\Delta\delta}\delta r^2 - \lambda_2 R_{tip} r\right) dr \quad (18)$$

$$L_2 = \frac{1}{2}\rho\Omega^2 c C_{l_\alpha}\left[\theta_o\left(\frac{R_{tip}^3}{3} - \frac{R_{root}^3}{3}\right) + \frac{\Delta\alpha_o}{\Delta\delta}\delta\left(\frac{R_2^3}{3} - \frac{R_1^3}{3}\right)\right.$$

$$\left. -\lambda_2 R_{tip}\left(\frac{R_{tip}^2}{2} - \frac{R_{root}^2}{2}\right)\right] \quad (19)$$

The thrust coefficient is given as

$$C_{T_2} = \frac{\sigma C_{l_\alpha}}{2}\left[\frac{\theta_o}{3}\frac{\left(R_{tip}^3 - R_{root}^3\right)}{R_{tip}^3} + \frac{\Delta\alpha_o}{\Delta\delta}\frac{\delta}{3}\frac{\left(R_2^3 - R_1^3\right)}{R_{tip}^3}\right.$$

$$\left. -\frac{\lambda_2}{2}\frac{\left(R_{tip}^2 - R_{root}^2\right)}{R_{tip}^2}\right] \quad (20)$$

The inflow is given as

$$\lambda_2 = \frac{\sigma C_{l_\alpha} k_p^2}{16}\left[\left\{\frac{64\left(R_{tip}^3 - R_{root}^3\right)}{3\sigma C_{l_\alpha} k_p^2 R_{tip}^3}\left(\theta_o + \frac{\Delta\alpha_o}{\Delta\delta}\delta\frac{\left(R_2^3 - R_1^3\right)}{\left(R_{tip}^3 - R_{root}^3\right)}\right)\right.\right.$$

$$\left.\left. + \frac{\left(R_{tip}^2 - R_{root}^2\right)^2}{R_{tip}^4}\right\}^{\frac{1}{2}} - \frac{\left(R_{tip}^2 - R_{root}^2\right)}{R_{tip}^2}\right] \quad (21)$$

The change in effective angle of attack due to a flap deflection, $\Delta\alpha_o/\Delta\delta$, is taken from Ref. 5 for a symmetric airfoil with a plain flap and is shown in Figure 12. A polynomial curve fit through the data is given by

$$\frac{\Delta\alpha_o}{\Delta\delta}\left\{\frac{c_f}{c}\right\} = -.002192 + 2.669\left(\frac{c_f}{c}\right) - 2.323\left(\frac{c_f}{c}\right)^2 \quad (22)$$

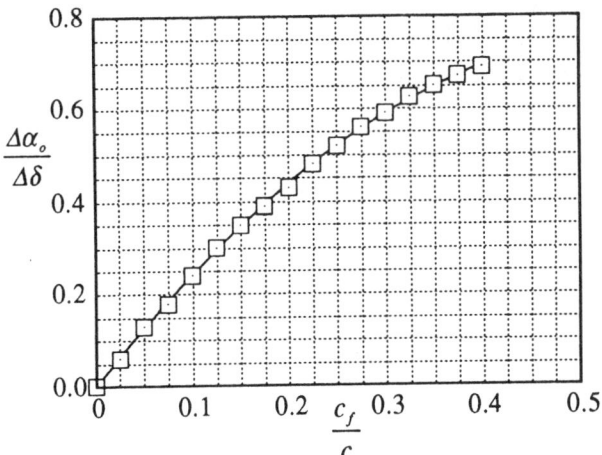

Figure 12. Section Flap Effectiveness.

A steady flap authority X can be defined as the ratio of additional rotor thrust generated by a flap deflection to the rotor thrust without a flap deflection.

$$X = \frac{C_{T_2} - C_{T_1}}{C_{T_1}} \quad (23)$$

$$\delta = \frac{X\theta_o\left(R_{tip}^3 - R_{root}^3\right) + \frac{3}{2}R_{tip}\left(R_{tip}^2 - R_{root}^2\right)(\lambda_1 - X\lambda_1 - \lambda_2)}{\frac{\Delta\alpha_o}{\Delta\delta}\left(R_2^3 - R_1^3\right)} \quad (24)$$

Equation (24) is obtained by substituting the expressions for thrust coefficient into equation (23) and solving for the flap deflection, δ. It represents the flap deflection necessary to obtain a flap authority X. Once the flap deflection is calculated, the resulting hinge moment coefficient can be determined using

$$C_h = C_l\left(\frac{dC_h}{dC_l}\right) + \delta\left(\frac{dC_h}{d\delta}\right) \quad (25)$$

where

$$\frac{dC_h}{dC_l}\left\{\frac{c_f}{c}\right\} = -.01018 - .5494\left(\frac{c_f}{c}\right) + 1.028\left(\frac{c_f}{c}\right)^2 - .9934\left(\frac{c_f}{c}\right)^3 + .2770\left(\frac{c_f}{c}\right)^4 \quad (26)$$

$$\frac{dC_h}{d\delta}\left\{\frac{c_f}{c}\right\} = -.8469 + .9833\left(\frac{c_f}{c}\right) - .07663\left(\frac{c_f}{c}\right)^2 + .2567\left(\frac{c_f}{c}\right)^3 - .3205\left(\frac{c_f}{c}\right)^4 \quad (27)$$

These theoretical values used from Ref. 5 are plotted in Figure 13 for different flap chords. The flap hinge moment coefficient is defined as

$$C_h \equiv \frac{h}{\frac{1}{2}\rho V^2 c_f^2} \quad (28)$$

The hinge moment per unit span is given as

$$h = \frac{1}{2}\rho(\Omega r)^2 c_f^2\left[C_{l_\alpha}\left(\theta_o - \frac{\lambda_2 R_{tip}}{r}\right)\frac{dC_h}{dC_l}\left\{\frac{c_f}{c}\right\}\right.$$

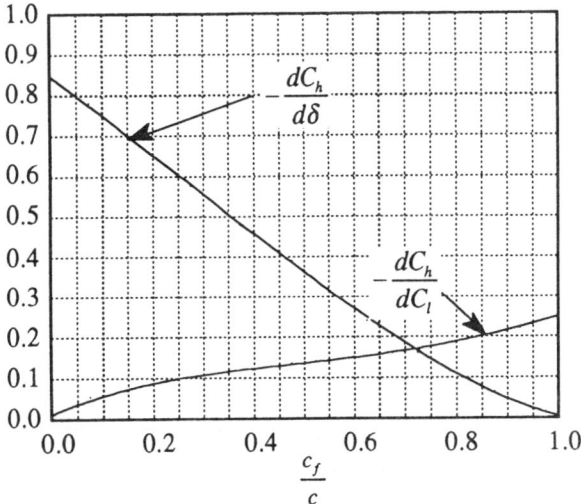

Figure 13. Hinge Moment Characteristics.

$$\left. + \frac{dC_h}{d\delta}\left\{\frac{c_f}{c}\right\}\delta\right] \quad (29)$$

The hinge moment per unit span can be can be integrated along the flap span to give the total hinge moment.

$$H = \frac{1}{2}\rho\Omega^2 c_f^2\left[\left(\frac{R_2^3}{3} - \frac{R_1^3}{3}\right)\left(C_{l_\alpha}\theta_o\frac{dC_h}{dC_l}\left\{\frac{c_f}{c}\right\} + \frac{dC_h}{d\delta}\left\{\frac{c_f}{c}\right\}\delta\right)\right.$$
$$\left. - C_{l_\alpha}\lambda_1 R_{tip}\left(\frac{R_2^2}{2} - \frac{R_1^2}{2}\right)\frac{dC_h}{dC_l}\left\{\frac{c_f}{c}\right\}\right] \quad (30)$$

Referring to Figure 9, once the flap deflection and hinge moment is determined, combinations of the required displacement and force can be determined as a function of the flap linkage arm length.

Linkage Arm Displacement $= (\text{Linkage Arm Length})(\delta)$ (31)

$$\text{Linkage Arm Force} = \frac{(H)}{(\text{Linkage Arm Length})} \quad (32)$$

A small angle assumption has been used in equations (31) and (32). This completes the equations necessary to find the flap requirements at the tip of the flap's linkage arm. It is important to note Equation (23) represents the steady flap deflection necessary to generate a steady percentage change X in rotor thrust. Obviously when using the flap for vibration suppression it will be deflected in a complicated time varying manner that will generate the appropriate time varying changes in blade lift. In this case, the steady desired flap authority is used to determine a flap deflection and corresponding hinge moment that is representative of the amplitude or peak value of any time varying flap schedule within that range of flap authority. In other words, the flap actuation system is designed to be capable of generating steady changes in blade lift, which gives an indication of the peak changes in blade lift attainable during a time varying flap oscillation.

The following analytic results are shown for a four bladed rotor at a collective pitch of 8° and a rotational speed of 900 RPM. The rotor radius is 36 inches and the root cutout is 9.5 inches. The blades consist of NACA 0012 airfoils with 3.0 inch chord. A tip loss factor of 1.15 is used. The flap is located .75 inch inboard from the blade tip. Figure 14 shows flap deflection versus flap chord to generate 10% flap authority for different flap spans. A larger flap chord requires a smaller flap deflection to generate the same change in blade lift. Also, increasing the flap span reduces the required flap deflection.

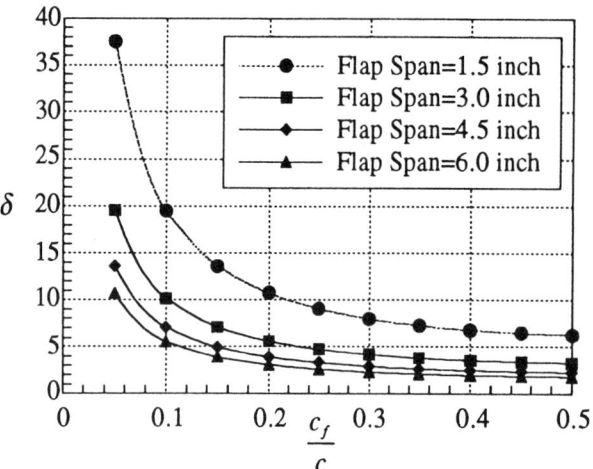

Figure 14. Flap Deflection for 10% Flap Authority.

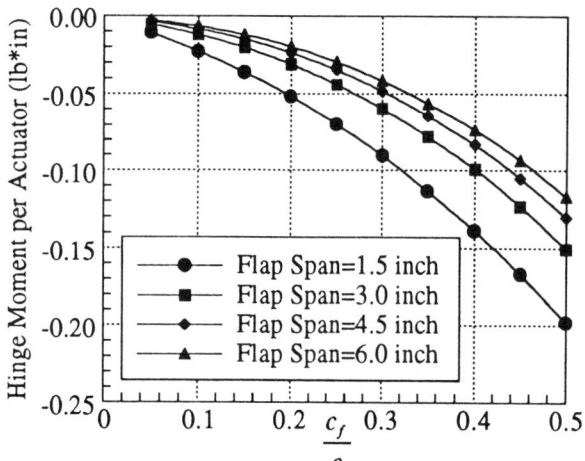

Figure 15. Hinge Moment for 10% Flap Authority.

Although increasing the flap chord shows a decrease in the required flap deflection, the corresponding hinge moment increases substantially. In the present design configuration, an actuator can be placed every 1.5 inch of flap span. The hinge moment per actuator decreases as the flap span is increased. For example, changing the configuration from 1.5 inch flap span with one actuator to 4.5 inch flap span with three actuators significantly reduces the hinge moment per actuator as shown in Figure 15. Further increasing the flap span and number of actuators reduces the hinge moment, but substantially increases the blade weight. Thus, the moderate flap span of 4.5 inches was chosen for the improved rotor model design.

The flap deflection and hinge moment results shown for the 4.5 inch flap span in the two previous plots are used to determine the required displacement and force at the tip of the flap linkage arm for different linkage arm lengths. Figure 16 shows the results for several flap chords.

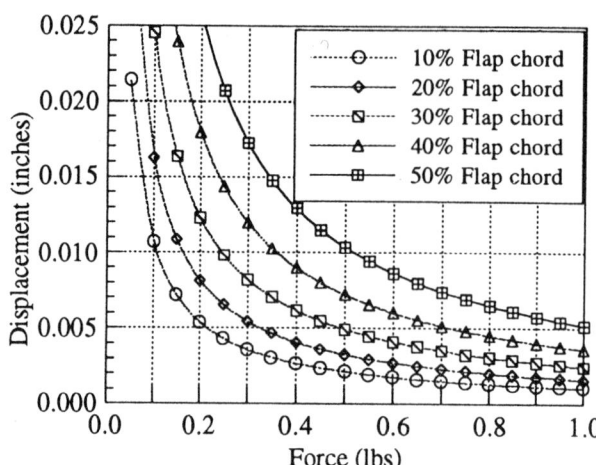

Figure 16. Flap Displacement and Force Requirements for 10% Flap Authority, 4.5 inch Flap Span.

The plot shows that it is desirable to use the smallest flap chord possible. However, there are limitations during blade construction. For less than about 20% flap chord, the thickness of the airfoil's trailing edge becomes too small to accommodate the bimorph deflections, flap hinge mechanism, and Hall-effect transducer used to sense flap position. It was therefore decided to use a 20% flap chord.

The displacement and force requirements curve for a 4.5 inch span and 20% flap chord is compared to the actuator displacement and force capability respectively for the two and four layer bimorph actuators in Figures 17 and 18. There is no overlap between actuator capability and flap requirements for the two layer actuator. The initial rotor model had 4.5 inch span, 20% chord flap and used 1.0 inch cantilevered two layer actuators. Although the steady flap authority was not measured in the initial hover test, this plot reveals that it was probably much less than 10% for applied field of 9.33 volts RMS/mil.

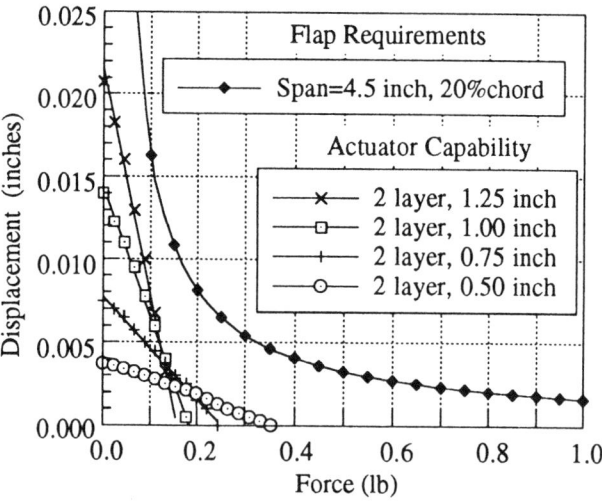

Figure 17. Comparison of Two Layer Actuator Capability and Flap Requirements for 10% Flap Authority.

Figure 18 shows an overlap or intersection between actuator capability and flap requirements for the four layer actuator. The plot indicates that a 10% flap authority can be obtained with these actuators using the appropriate linkage arm length. The design point with the actuators operating at .3 lb force and 5.4 mils displacement is circled on the plot.

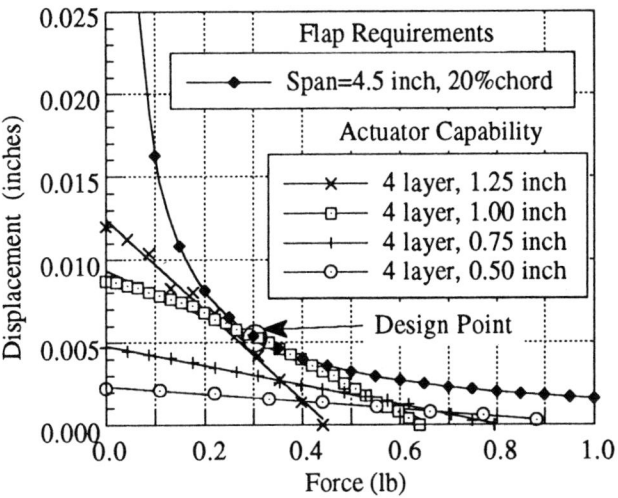

Figure 18. Comparison of Four Layer Actuator Capability and Flap Requirements for 10% Authority.

Figure 19 shows the needed linkage arm length for the 4.5 inch span, 20% chord flap corresponding to the appropriate displacement and force values as shown in Figure 16. For the actuators to operate at .3 lb force, the linkage arm length needs to be 80 mils.

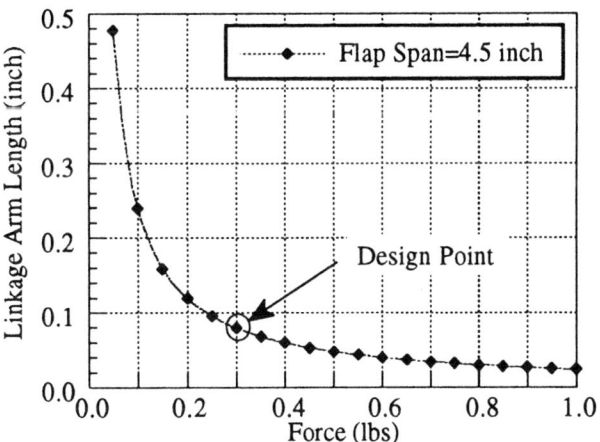

Figure 19. Corresponding Linkage Arm Length for 10% Flap Authority, 20% chord.

Thus, the improved rotor model design using three actuators capable of operating at .3 lb force at 5.4 mils displacement connected with an 80 mil linkage arm to a 4.5 inch span, 20% chord flap would be capable of generating 10% flap authority. From Figure 14, the improved design would be able to achieve a flap deflection of 3.9°. This would nearly double the flap deflection obtained from the initial rotor model. These calculations show the flap actuation system can be made more effective using the higher force four layer actuators. Thus, it was decided to build and test the improved rotor model design.

Mechanical System Improvements

Figure 20 shows the flap and bimorph tip movement for several possible flap deflections. The distance between the tip of the bimorph and tip of the flap linkage increases as the flap rotates, so the tip of the bimorph can not be directly pinned to the linkage mechanism. In the initial rotor model, this was accounted for by a vertical offset the bimorph from the chord line which allowed an additional rigid linkage to be pinned from the tip of the bimorph to the flap linkage arm as shown in Figure 21. As the bimorph deflects, it pulls on the additional linkage and deflects the flap. The disadvantage of this system with the bimorph off center was the deflection exceeded the boundaries of thin airfoil on one side. In the initial rotor model there was a small cutout to allow for this which led to aerodynamically unclean surface.

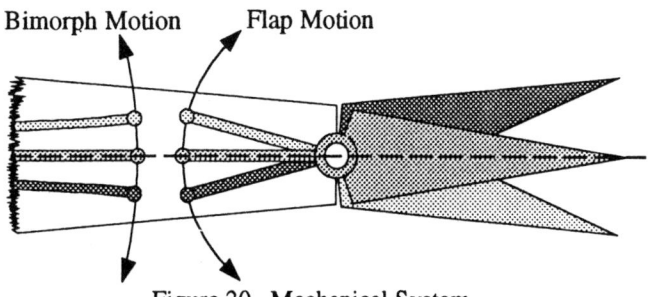

Figure 20. Mechanical System.

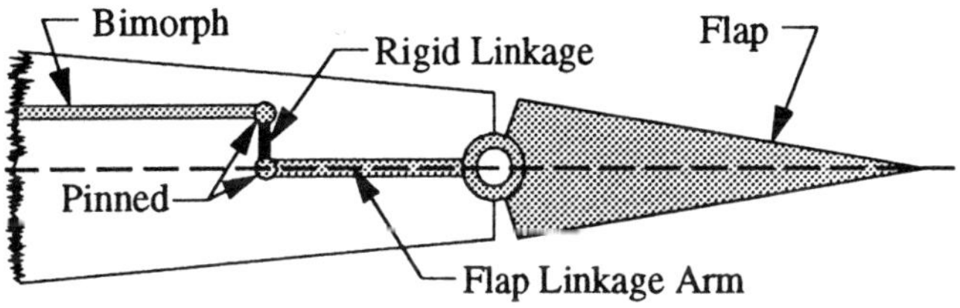

Figure 21. Initial Rotor Model Design.

An improved mechanical system was designed as shown in Figure 22. The bimorph has a round molded rod on its tip which fits inside a tiny machined cusp. The cusp is an integral part of the flap. As the bimorph moves up and down and rotates the flap, the molded tip moves slightly in and out of the cusp. The advantage here is the actuator is centered and there is no cutout required in the skin for the range of deflections desired. Also, the design is simple and has a minimal number of parts. It was decided to incorporate this type of mechanical system in the improved rotor model. In this design, the distance between the flap hinge and contact point of the bimorph tip and cusp acts as the effective linkage arm length.

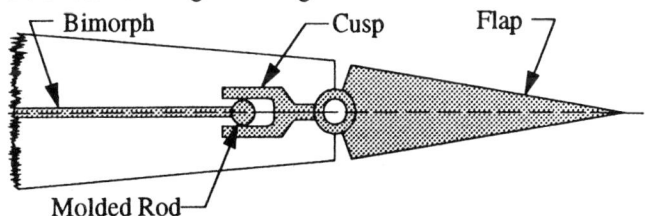

Figure 22. Cusp Design.

Rotor Blade Construction

At the University of Maryland, a four bladed bearingless rotor hub mounted on a 15 foot hover tower is used to test rotor models in hover. To save time, it was decided to construct two blades with the improved flap design and test the rotor model in the two bladed configuration before building a set of four blades.

The first step was to build three four layer actuators for each blade. The actuators were made by bonding additional G-1195 piezoceramic sheets to commercial bimorphs using EccoBond 64C conducting adhesive. As shown in Figure 23, the additional sheets were cut short at one end for electrical connections to be made to each layer. The actuators were completed by molding a 1/16 inch diameter circular rod on the tip using a steel tube and several plies prepreg fiberglass cloth. This was done in a specially built mold to give a smooth finish. The molded tip is designed to fit inside a cusp attached to the trailing edge flap.

The next step was to build the flap, cusp, and hinge as a single integral part. A tiny aluminum cusp, with a cross-section as shown in Figure 24, was carefully machined using a ball endmil. The 1/16 inch opening in the cusp was designed to match the rod molded on the actuator's tip. A .042 inch outside diameter, .027 inside diameter steel hinge tube was wrapped with film adhesive and bonded in the aft section of the cusp opening. The tube is used in conjunction with a .026 inch diameter steel pin to hinge the flap to the rotor blade. A mold was used to keep the film adhesive from completely filling the opening in the cusp during the cure cycle. The airfoil mold was then used to mold a piece of foam and a section from the trailing edge was removed and bonded to the cusp to form the trailing edge flap assembly as shown in Figure 24, except for the outer skin which is added later.

A 1/16 inch thick aluminum anchor plate and clamp was built to cantilever the actuators as shown in Figure 25. Provisions were made for electrical connections and different clamp positions so the actuator cantilever length could be varied. Steel hinge tubes were bonded to the aft edge of the anchor plate. Portions of the cusp and hinge tube on the trailing edge flap assembly were cut so the trailing edge flap assembly and anchor plate could be interlocked similar to a door hinge. Figure 25 shows the two parts unhinged and separated. A magnet is attached to one end of the flap along the hinge axis. When the blade is completed the magnet rotates with the flap and a Hall-effect transducer bonded to the airfoil's skin in the proximity of the magnet senses changes in magnetic flux. The transducer is calibrated to sense flap deflection.

A .026 inch diameter steel pin was used to hinge the trailing edge flap to the anchor plate. Figure 26 shows the flap hinged to the anchor plate. This figure also shows the fiberglass spar designed to withstand the centrifugal loads of the anchor plate and flap assembly. The cruciform cross section spar was attached to the anchor plate at one end and a steel ring at the other end using several plies of fiberglass. The steel ring is used to bolt the blade to the rotor rig. A piece of mahogany wood carefully shaped to the airfoil's cross-section is notched to fit around the steel ring and a portion of the spar to strengthen the root of the blade. Figure 26 shows the basic skeletal structure of the flap actuation system before it is surrounded with foam and fiberglass and molded into an aerodynamically conformal rotor blade.

Using half of the airfoil mold and a flat plate, pieces of foam for the upper and lower half of the airfoil were shaped. The airfoil foam was made in two pieces so the trailing edge flap assembly, anchor plate, spar, and root section could be sandwiched between the upper an lower pieces with the anchor plate effectively along the airfoil chord line. The foam pieces were milled and cut to fit properly around the flap assembly. The skeletal assembly and foam pieces were bonded together with film adhesive and cured in the airfoil mold. After removing from the mold, this cured structure was effectively a rotor blade without its skin. The foam blade with flap assembly inside was balanced by inserting weights in the leading edge. The foam blade was then wrapped with two five mil plies of fiberglass and cured in the airfoil mold. The blade was completed by cutting a removable access hatch and removing a small amount of foam and fiberglass skin between the flap and the main portion of the rotor blade to provide clearance for the flap to deflect. During the molding process, the actuators were replaced with aluminum plates to prevent damage to the actuators. The access hatch was used to remove the plates and reinsert the actuators. The mahogany root with its surface area bonded to a portion of the skin, also provides a load path for centrifugal loads in the airfoil's skin to be transferred through the wood to the steel ring. Two identical blades were constructed using these methods. The finished blades are shown in Figure 27.

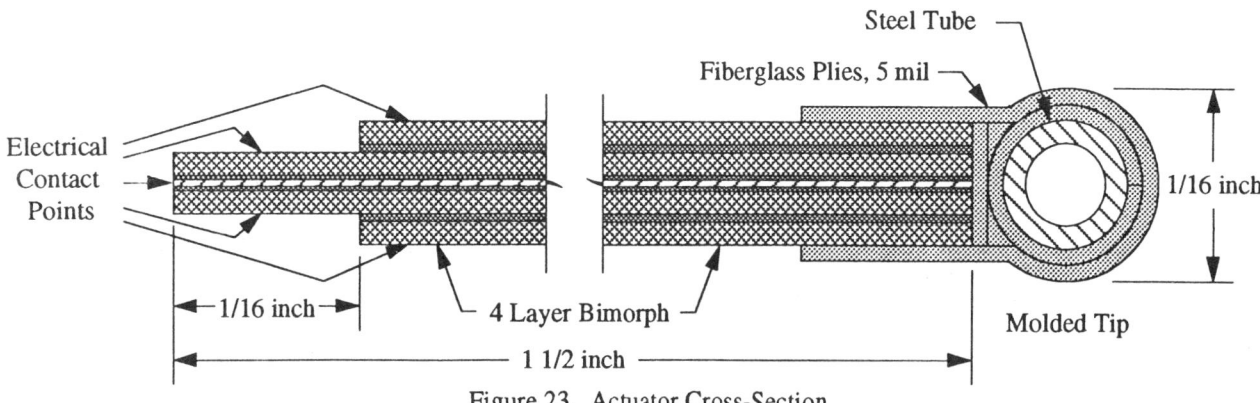

Figure 23. Actuator Cross-Section.

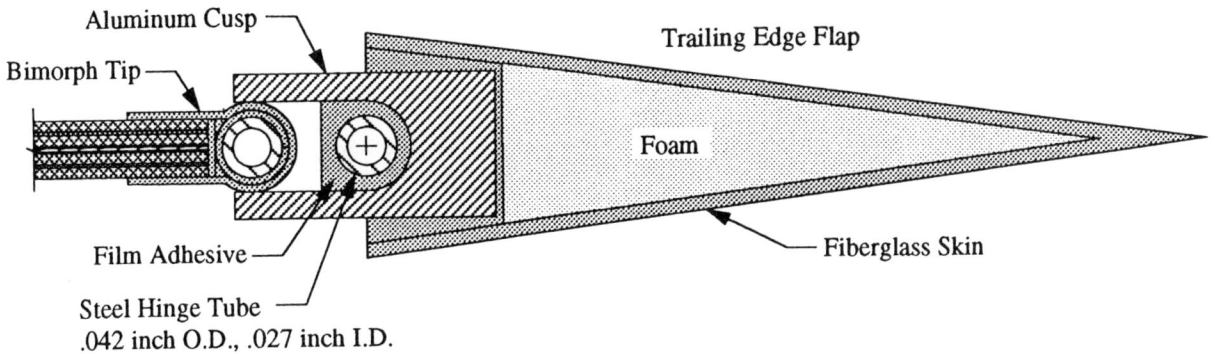

Figure 24. Cross-Section of Actuator Tip and Trailing Edge Flap Assembly.

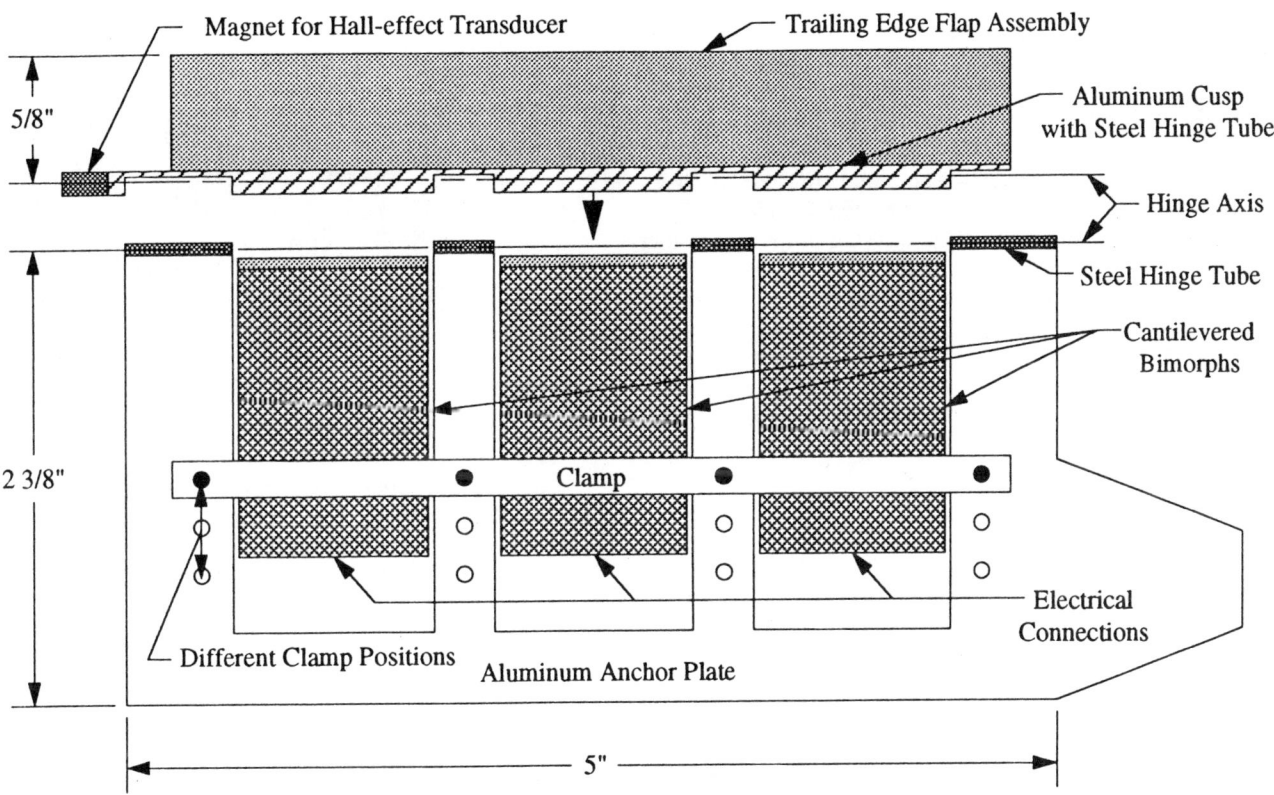

Figure 25. Top View of Anchor Plate and Trailing Edge Flap Assembly.

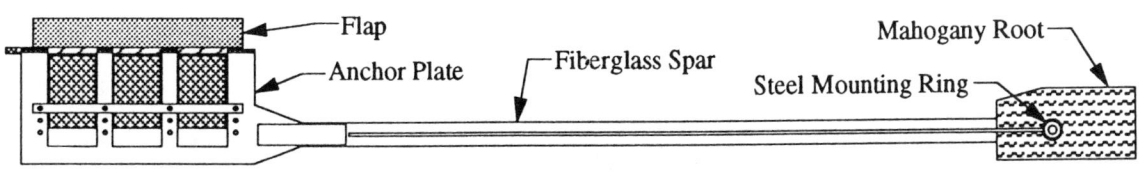

Figure 26. Flap Assembly, Anchor Plate, Spar, and Blade Root.

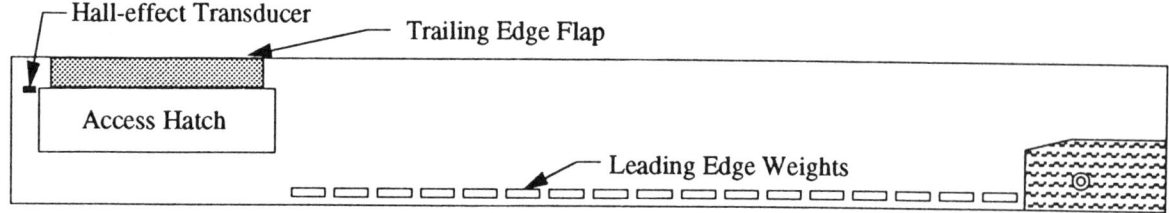

Figure 27. Finished Blade with Foam Core and Fiberglass Skin.

Nonrotating Results.

The finished blades were clamped in a vise and tested in the nonrotating environment. The Hall-effect transducer in each blade was calibrated to sense flap deflection. An excitation sine wave of 9.3 volts/mil RMS (13.3 volts/mil amplitude) was applied to each layer in the actuators. Figure 28 shows a deflection of ±5.5° is obtained at low frequency and a steady increase to about ±10.5° occurs as the frequency is increased to 60 Hz. Using the free displacement data for the 1.0 inch cantilevered, 4 layer actuator, under an applied 13.3 volt/mil and the known 80 mil linkage arm length, the estimated maximum static flap deflection with an ideal mechanical system would be about ±6.5°. The low frequency amplitude of ±5.5° is close to the expected static rotation of an ideal system. This shows that the flap mechanism is working properly and the one degree difference may be due to friction, changes in the effective linkage arm length as the flap rotates, and small amounts of play in the cusp and flap hinge tube. The zero load flap deflections are lower then the initial rotor model, however, the improved rotor model with higher force actuators is expected to perform better at the operating speed of 900 RPM by retaining higher flap deflections under aerodynamic loading.

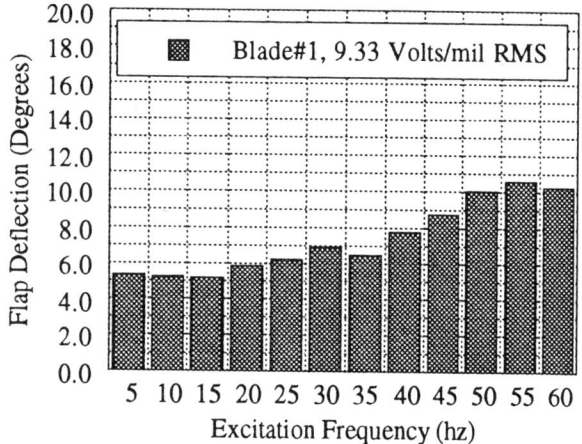

Figure 28. Flap Rotation vs. Frequency.

Piezoceramics can be damaged at high negative voltages. To apply higher voltages to the actuators without damaging them, a circuit was built to modify the sine wave in such a way that larger positive voltages could be applied. The circuit generated a wave with the negative portion of the wave (1/2 cycle) at one amplitude and the positive portion of the wave at twice the amplitude of the negative portion. The type of excitation allows the voltage to be increased without damaging the actuators. An example of the benefits of using a higher voltage is shown in Figure 29. At low frequency, the flap deflection increased from ±5.5° to ±7.0° and at high frequency, say 60 Hz, the flap deflection increased from ±10.5° to ±17.0°.

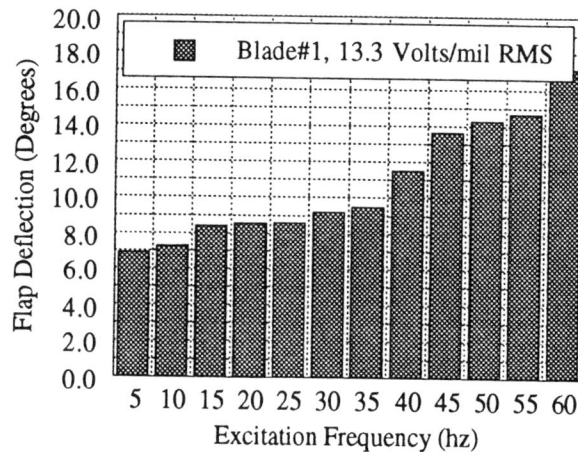

Figure 29. Flap Rotation vs. Frequency.

Hover testing to confirm the design improvements is presently underway. The improved rotor model will also make use of this circuit during hover testing.

Conclusions

This research involved the development of a scaled rotor model with trailing edge flap actuated with piezoceramic bimorphs for active vibration control. Fabrication and testing of current and earlier rotor models demonstrated that the bimorph actuators can provide a simple, lightweight, compact flap actuation system. The system can be controlled individually on each blade and operate over a range of frequencies. In the earlier model, the maximum ±2° flap deflection at the operating speed of 900 RPM in hover was not large enough to generate significant changes in blade lift to be used for effective vibration suppression. To increase flap authority, actuators with higher actuation force were built in-house by doubling the bimorph thickness from 20 mils to 40 mils. The 40 mil actuators produced about three times as much block force at the expense of reduction of one third of free displacement. A simple aerodynamic model of a flap was developed to properly match the improved actuator characteristics to the flap requirements. The lower displacements were compensated in the improved rotor model design by shortening the flap linkage arm length and making use of the higher block force. Although the flap deflections of the improved design at zero load condition were lower than the initial rotor model, it is expected to achieve larger flap deflections at the operating rotational speed. The expected flap deflection of ±3.9° at 900 RPM is about twice what was achieved with the earlier model in hover.

Construction of the improved rotor model showed that by altering the mechanical arrangement from a pinned linkage system to a simple cusp design, the flap actuation system could be completely enclosed inside the thin rotor blade. This design provided an aerodynamically clean surface.

Nonrotating test on the new model showed an expected flap deflection of ±5.5° at low frequency and a higher flap deflection of ±10.5° as the frequency was increased to 60 Hz. An electronic circuit was used to increase the output at low frequency to ±7.0° and at higher frequency to ±17.0°. Future work will explore a design using six layer actuators or using a pairs of 4 layer actuators working in parallel to obtain the goal of ±8° flap deflection. To examine the performance of these actuators in the rotating environment, testing of the rotor model on the hover stand is in progress.

Acknowledgments

This work was supported in part by the Army Research Office under Grant DAAL 03-92-G-0121 with Dr. Gary Anderson. The authors would like also like to thank Mr. Kwok Yu for his assistance in blade construction.

References

1. Nguyen, K. and Chopra I., "Application of Higher Harmonic Control to Rotors Operating at High Speed and Thrust," *Journal of the American Helicopter Society*, Vol. 35, (3), July 1990.

2. Ham, N. D., " A Simple System for Helicopter Individual-Blade-Control Using Modal Decomposition," Vertica, Vol. 4, (1), 1980.

3. Samak, D.K. and Chopra, I., "A Feasibility Study to Build a Smart Rotor: Trailing Edge Flap Actuation,", SPIE Conference, Albuquerque, NM, 1993.

4. Park, C., Walz, C. and Chopra, I., "Bending and Torsion Models of Beams with Induced Strain Actuators", SPIE Conference, Albuquerque, NM, 1993.

5. Abbott, Ira H., and Von Doenhoff, Albert E., Theory of Wing Sections, Dover Publications, Inc., New York, 1959.

ACTIVE CLOSED CELL BEAM SHAPE CONTROL

Steven M. Ehlers [*]
McDonnell Douglas Technologies, Inc.
San Diego, California

Abstract

Displaced shape control of an anisotropic closed cell beam with integral active materials is examined. An analytical model of an arbitrary cross-section uniform closed-cell anisotropic beam, including the effects of out-of-plane warping, is developed. Examples are used to study various methods of bending and torsional actuation. The use of actuators in bending pairs is shown to be the most effective method of bending actuation. Best torsional actuation is achieved through the use of piezoelectric fiber composite skins that induce shear in the cell walls. A new method of torsional actuation that employs warping-torsion coupling is introduced. Axially aligned actuators are used to generate out-of-plane warping forces that result in twisting of the section. The twist distribution created by this method is quite different from the other torsional actuation methods. For a beam free to warp at both ends the twist displacement can be made nearly uniform over a large portion of the beam.

Introduction

Adaptive structures can be used to actively tailor and control structural response to enhance performance and provide new capabilities. Such structures combine active materials and conventional structural materials to control static shape, vibration and internal loads. Adaptive structures have application in a wide range of areas including robotics, large space structures, flight vehicles, ground transportation systems and civil structures.

An adaptive structure is composed of embedded sensors and active materials, a control system and parent structure. The embedded active materials and parent structure form the active structure which is capable of inducing loads and deformations in response to an external stimulus. The active structure is a critical element of an adaptive structural system. Acting as both muscle and skeleton it must combine active and conventional materials in a structurally efficient manner to produce an optimum combination of stiffness, strength, damage tolerance, structural stability and control.

In flight vehicles adaptive structures can be used to control a wide variety of aeroelastic phenomena such as flexible lift and rolling moment, flutter and divergence, buffet, vibration and gust loads. The adaptive structure provides the ability to control static and dynamic shape of the flight vehicle structure in the presence of displacement dependent aerodynamic loads. As a result the control of displaced shape can be very useful.

In many cases the structural configuration of lifting surface or other flight vehicle components can be modeled as a closed cell beam. Closed cell beams are efficient structures that provide high bending and torsional stiffness at a relatively low weight. They also allow internal volume for such things as fuel, payload and propulsion systems. Examples of flight vehicle closed cell beam structures include wing torqueboxes, rotor blades, fuselages, tail booms and engine pylons.

In the case of moderate to high aspect ratio swept wings the control of bending in the vertical plane and twist is of interest. These displacements can be controlled by an active closed cell wing box structure. The active structure is created by embedding active materials in the cell wall. Active materials are positioned to create the desired distribution of induced extensional and shear forces. Beam force resultants are determined by the distribution of induced stresses and cross-section geometry. The resulting active structure can be structurally efficient and provide a high degree of control.

Previous studies of active and adaptive flight vehicle structures have employed plate[1] and laminated beam[2] models to represent bending and torsional stiffness, elastic coupling and induced strain actuation. Bending actuation of a closed cell beam for vibration control has also been studied[3]. An initial investigation of warping and differential bending to create torsional

---

[*] Mgr. of Advanced Concepts, Senior Member AIAA

Copyright © American Institute of Aeronautics and Astronautics, Inc., 1994. All rights reserved.

deformation of a solid laminated beam using a finite element analysis has also been conducted[4].

The purpose of this paper is to examine and compare approaches to closed cell beam bending and twist shape control that might be applied to lifting surface aeroelastic control. In particular, a method which uses warping-torsion coupling to generate twist deformations through the use of actuators aligned with the beam axis, will be introduced.

A closed cell beam model based on common displacement assumptions and membrane material behavior is developed. The beam is considered to be uniform in the spanwise direction with arbitrary variation of material properties along the cell perimeter. Beam piezoelectric force resultants are found by assuming that the same electric field is applied to all active material segments. The piezoelectric force resultants and beam section properties are used to develop expressions for the piezoelectric strain actuation coefficients. Examples are used to illustrate various methods of bending and torsion actuation of a closed cell beam. Deformed shapes are determined by integrating the governing differential equations. Effectiveness of each approach is evaluated based on actuation deformation per unit induced active material strain and beam weight per unit length.

Active Closed Cell Beam Model

The active closed cell beam is idealized as a thin-walled uniform beam as shown in Figure 1. Active materials are embedded in the cross-section at selected positions to create the desired bending and torsion actuation. In this study the active material is assumed to be a linear piezoelectric material.

Active Cell Wall

The cell wall is treated as a membrane in plane stress with negligible transverse loads ($N_{ss}=0$). The resulting general form of the constituitive relation for an active wall segment is:

$$\begin{Bmatrix} N_{xx} \\ N_{xs} \end{Bmatrix} = \begin{bmatrix} \tilde{A}_{11} & \tilde{A}_{16} \\ \tilde{A}_{16} & \tilde{A}_{66} \end{bmatrix} \begin{Bmatrix} \varepsilon_{xx} \\ \gamma_{xs} \end{Bmatrix} - \begin{Bmatrix} \tilde{f}_{3x} \\ \tilde{f}_{3s} \end{Bmatrix} E_3 \quad (1)$$

where the reduced elastic constants are:

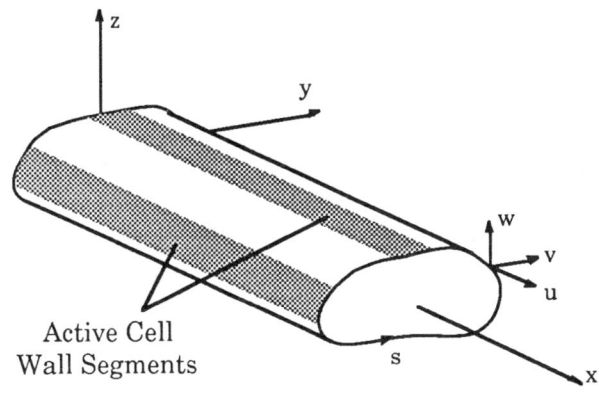

Figure 1.   Uniform closed cell beam.

$$\tilde{A}_{ij} = A_{ij} - \frac{A_{i2}A_{j2}}{A_{22}} \quad ; \quad i,j=1,6 \quad (2)$$

and the reduced piezoelectric stress resultant coefficients are:

$$\tilde{f}_{3x} = f_{xx} - \frac{A_{12}}{A_{22}} f_{yy} \quad (3a)$$

$$\tilde{f}_{3s} = f_{xy} - \frac{A_{26}}{A_{22}} f_{yy} \quad (3b)$$

The above relation is sufficiently general to represent the membrane behavior of monoclinic piezoelectric and passive materials. The stiffness terms, A_{ij}, and piezoelectric stress resultants (f_{xx}, f_{yy} and f_{xy}) can be found using laminated plate theory[5] where the piezoelectric stress resultant terms are analogous to stresses induced by thermal expansion. Passive structural materials are represented by setting the applied field to zero. It should also be emphasized that the stiffness coefficients include the contributions of both the active and passive materials that comprise the closed cell cross-section.

Beam Constituitive Relation

The total potential energy of an active closed cell beam under distributed transverse loads (p_y and p_z) and torque (m_x) is given by:

$$U = \frac{1}{2}\int_0^L \oint_C \left(N_{xx}(\varepsilon_{xx} - d_{3x}E_3) + N_{xs}(\gamma_{xs} - d_{3s}E_3)\right) ds\, dx$$
$$- \int_0^L (p_y V + p_z W + m_x \theta)\, dx \quad (4)$$

where d_{3x} and d_{3s} are the piezoelectric strain coefficients that give the induced axial and shear strain, respectively, per unit applied transverse electric field.

Local displacements of the cell wall in the x, y and z directions are denoted by u, v and w, respectively. The beam cross-section is assumed to remain rigid. In-plane displacements of points on the beam cross-section are given by:

$$v(x,s) = V(x) - z(s)\theta(x) \quad (5a)$$

$$w(x,s) = W(x) + y(s)\theta(x) \quad (5b)$$

where V and W are the y and z deflections of the reference axis and θ is the rotation of the section about the x-axis. Axial deformation of the cross-section is assumed to be of the form:

$$u(x,s) = U(x) - y(s)\beta_z(x) - z(s)\beta_y(x)$$
$$+ \psi(s)\theta'(x) \quad (6)$$

where U is the x displacement of the cross-section and β_y and β_z are rotations of the section about the y and z axes, respectively. The final term in Eqn. 6 is the warping displacement which is assumed to be proportional to the twist rate. The warping function, $\psi(s)$, used in this study was developed by Rehfield and Atiligan[6] and is given by:

$$\psi(s) = \int_0^s \left(\frac{2A_c}{l\bar{a}_{66}}a_{66} - r_n\right) d\xi \quad (7)$$

where A_c is the cross-sectional area enclosed by the cell wall midplane, l is the perimeter of the section and r_n is the normal distance from the wall tangent to the structural reference axis. The wall shear compliance and average shear compliance are defined as:

$$a_{66} = \frac{\tilde{A}_{11}}{\tilde{A}_{11}\tilde{A}_{66} - \tilde{A}_{16}^2} \quad (8)$$

and,

$$\bar{a}_{66} = \frac{1}{l}\oint a_{66}\, ds \quad (9)$$

The warping function in Equation 7 has been shown to be in good agreement with test data and a higher order beam theory for a variety of cross-section types[7].

The axial and shear strains in the cross-section wall are:

$$\varepsilon_{xx} = \frac{\partial u}{\partial x} \quad (10a)$$

$$\gamma_{xs} = \frac{\partial v_t}{\partial x} + \frac{\partial u}{\partial s} \quad (10b)$$

where v_t is the tangential displacement of the wall.

The total potential energy of the beam can be expressed in terms of the six displacement variables by combining the assumed displacement field, strain-displacement relation and cell wall constituitive relation with Equation 4. In general, the applied electric field, E_3, can vary along the cell perimeter. In this paper the electric field will be assumed to be constant within each active material segment. The beam constuitive relation is derived from the resulting total potential energy expression and can be expressed in the form:

$$\{F\} = [C]\{\varepsilon\} - [f]\{E_3\} \quad (11)$$

where the beam forces are:

$$\{F\} = \lfloor P \quad Q_y \quad Q_z \quad T \quad M_y \quad M_z \quad Q_w \rfloor^T \quad (12)$$

The beam forces include the axial force resultant, P, the transverse shear forces, Q_y and Q_z, torque, T, and the bending moments, M_y and M_z. The warping force is denoted by Q_w. The vector of beam strains is:

$$\{\varepsilon\} = \lfloor U' \quad \gamma_{xy}^o \quad \gamma_{xz}^o \quad \theta' \quad \beta_y' \quad \beta_z' \quad \theta'' \rfloor^T \quad (13)$$

where the beam transverse shear strains are:

$$\gamma_{xy}^o = V' - \beta_z \quad (14a)$$

$$\gamma_{xz}^o = W' - \beta_y \tag{14b}$$

In the beam constituitive relation the piezoelectric force coefficients, $[f]$, form a 7xN matrix where N is the number of independent active material segments in the closed cell. Correspondingly, $\{E_3\}$, is a vector of length N containing the electric field applied to each of the active material segments. In this study we shall further restrict the applied electric field to being the same for all active material segments. This reduces E_3 to a single scalar quantity. In this case [f] is a column vector of the form:

$$\{f\} = \lfloor f_1 \; f_2 \; f_3 \; f_4 \; f_5 \; f_6 \; f_7 \rfloor^T \tag{15}$$

The above vector of piezoelectric force resultants represents the ability to induce extensional (f_1), transverse shear (f_2,f_3), torsion (f_4) and bending (f_5,f_6) resultants. The last element, f_7, is unique as it represents the ability to create twist deformation through the generation of axial warping forces. This effect will be studied in greater detail in the examples. The resulting form of the constituitive relation is:

$$\{F\} = [C]\{\varepsilon\} - \{f\}E_3 \tag{16}$$

The stiffness coefficients, C_{ij}, are a function of the cell wall elastic properties and cross-section geometry. They represent the stiffness contributions of both the active and conventional materials. The wall extensional stiffness, $\tilde{A}_{11}$, contributes to extensional, bending and bending-extension coupling terms which are given by:

$$C_{11}, C_{15}, C_{16}, C_{55}, C_{66}, C_{56} =$$
$$\oint_C \tilde{A}_{11} [1, z, y, z^2, y^2, yz] ds \tag{17}$$

Wall shear stiffness, $\tilde{A}_{66}$, contributes to the section transverse shear and torsional stiffness, and related coupling terms. The transverse shear stiffness and coupling terms are:

$$C_{22}, C_{33}, C_{23} = \oint_C \tilde{A}_{66} \left[\left(\frac{dy}{ds}\right)^2, \left(\frac{dz}{ds}\right)^2, \frac{dy}{ds}\frac{dz}{ds}\right] ds \tag{18}$$

The section torsional stiffness and torsion-transverse shear coupling terms are given by:

$$C_{24}, C_{34}, C_{44} =$$
$$\oint_C \tilde{A}_{66}\left(r_n + \frac{d\psi}{ds}\right)\left[\frac{dy}{ds}, \frac{dz}{ds}, \left(r_n + \frac{d\psi}{ds}\right)\right] ds \tag{19}$$

Coupling between extension and bending, and transverse shear and bending arises due to extension-shear coupling, $\tilde{A}_{16}$, in the cell wall. This can be a result of the use of anisotropic materials or skewed stiffeners. The corresponding stiffness coefficients are:

$$C_{12}, C_{13}, C_{25}, C_{26}, C_{35}, C_{36} =$$
$$\oint_C \tilde{A}_{16}\left[\frac{dy}{ds}, \frac{dz}{ds}, z\frac{dy}{ds}, y\frac{dy}{ds}, z\frac{dz}{ds}, y\frac{dz}{ds}\right] ds \tag{20}$$

Stiffness terms representing the coupling between extension and bending, and torsion are:

$$C_{14}, C_{45}, C_{46} = \oint_C \tilde{A}_{16}\left(r_n + \frac{d\psi}{ds}\right)[1, z, y] ds \tag{21}$$

Warping related terms are determined by the shape and magnitude of the warping function. Two groups of such terms exist. The first group is governed by the cell wall extensional stiffness distribution. The warping stiffness, warping-extension coupling and warping-bending coupling terms are:

$$C_{17}, C_{57}, C_{67}, C_{77} = \oint_C \tilde{A}_{11}\psi[1, z, y, \psi] ds \tag{22}$$

The second group of terms are nonzero when extension-shear coupling is present in the cell wall. This results in coupling between shear in the cell wall and out-of-plane warping. The warping-shear coupling coefficients are:

$$C_{27}, C_{37}, C_{47} = \oint_C \tilde{A}_{16}\psi\left[\frac{dy}{ds}, \frac{dz}{ds}, \left(r_n + \frac{d\psi}{ds}\right)\right] ds \tag{23}$$

The piezoelectric force resultants depend on the properties of the active material and parent structure, location of the active materials and shape of the cross-section. The piezoelectric force coefficients, $\{f\}$, are found by integrating the extensional and shear stresses due to a unit applied electric field. The generation of extensional actuation forces in the cell wall can be use to create extensional, bending and warping force resultants. The corresponding force coefficients are:

$$f_1, f_5, f_6, f_7 = \oint_C \tilde{f}_{3x}[1, z, y, \psi] ds \qquad (24)$$

Inducing in-plane shear forces in the cell wall can result in transverse shear and torque resultants. The transverse shear and torsional force actuation coefficients are:

$$f_2, f_3, f_4 = \oint_C \tilde{f}_{3s}\left[\frac{dy}{ds}, \frac{dz}{ds}, \left(r_n + \frac{d\psi}{ds}\right)\right] ds \qquad (25)$$

The collection of seven coefficients defined by Equations 24 and 25 give the active beam force resultants per unit applied electric field.

Strain Actuation Coefficients

The ability of the embedded active materials to control the displaced shape of an unloaded beam is revealed when the force coefficients are transformed into strain coefficients. The resulting form of the beam constituitive relation is:

$$\{F\} = [C](\{\varepsilon\} - \{\Gamma\}E_3) \qquad (26)$$

where the vector of strain actuation coefficients is:

$$\{\Gamma\} = \lfloor \Gamma_1 \ \Gamma_2 \ \Gamma_3 \ \Gamma_4 \ \Gamma_5 \ \Gamma_6 \ \Gamma_7 \rfloor^T \qquad (27)$$

The piezoelectric strain coefficients are related to the force coefficients by:

$$\{f\} = [C](\{\Gamma\}) \qquad (28)$$

In general, the above relation cannot be inverted because [C] may be singular. This situation occurs in sections that do not exhibit coupling between torsion and out-of-plane warping. The simplest example of this is a uniform isotropic circular cross-section. However, it is also possible for the warping function to be zero for noncircular sections. By properly tailoring the section shape and cell wall elastic properties a noncircular section can be made "warpless." To account for this possibility the stiffness matrix is partitioned into warping, warping coupling and non-warping related terms:

$$[C] = \left[\begin{array}{c|c} \overline{C} & c \\ \hline c^T & C_{77} \end{array}\right] \qquad (29)$$

where the warping coupling terms are:

$$\{c\}^T = \lfloor C_{17} \ C_{27} \ C_{37} \ C_{47} \ C_{57} \ C_{67} \rfloor \qquad (30)$$

and $[\overline{C}]$ is the remaining 6x6 partition of the stiffness matrix.

The piezoelectric strain coefficients are given by:

$$\{\overline{\Gamma}\} = [\overline{S}](\{\overline{f}\} - \Gamma_7\{c\}) \qquad (31a)$$

$$\Gamma_7 = \frac{f_7 - \{c\}^T[\overline{S}]\{\overline{f}\}}{C_{77} + \{c\}^T[\overline{S}]\{c\}} \qquad (31b)$$

where $\{\overline{\Gamma}\}$ and $\{\overline{f}\}$ contain the first six elements of $\{\Gamma\}$ and $\{f\}$, respectively, and:

$$[\overline{S}] = [\overline{C}]^{-1} \qquad (32)$$

For warpless sections Γ_7 is zero and Equation 31a is used to compute the remaining actuation coefficients.

Governing Equations

The differential equations that describe the displaced shape of the active closed cell beam are found by minimizing the total potential energy expression with respect each of the six displacement variables. The accompanying boundary conditions are obtained from the displacement constraints at the root and the natural boundary conditions. The differential equations form a fourteenth order system which can be expressed in the following form:

$$\left[\begin{array}{c|c} 0 & 0 \\ \hline 0 & L_{bb} \end{array}\right] \frac{d^4}{dx^4} \left\{\begin{array}{c} \delta_a \\ \delta_b \end{array}\right\} + \left[\begin{array}{c|c} 0 & M_{ab} \\ \hline -M_{ab}^T & M_{bb} \end{array}\right] \frac{d^3}{dx^3} \left\{\begin{array}{c} \delta_a \\ \delta_b \end{array}\right\}$$

$$+\begin{bmatrix} N_{aa} & N_{ab} \\ N_{ab}^T & N_{bb} \end{bmatrix}\frac{d^2}{dx^2}\begin{Bmatrix}\delta_a\\\delta_b\end{Bmatrix} + \begin{bmatrix} 0 & P_{ab} \\ -P_{ab}^T & P_{bb} \end{bmatrix}\frac{d}{dx}\begin{Bmatrix}\delta_a\\\delta_b\end{Bmatrix}$$

$$+\begin{bmatrix} 0 & 0 \\ 0 & R_{bb} \end{bmatrix}\begin{Bmatrix}\delta_a\\\delta_b\end{Bmatrix} = \begin{Bmatrix}d_a\\d_b\end{Bmatrix} \quad (33)$$

where the vector of beam displacements is partitioned into translational and rotational components:

$$\begin{Bmatrix}\delta_a\\\delta_b\end{Bmatrix} = \lfloor U \quad V \quad W \mid \theta \quad \beta_y \quad \beta_z \rfloor^T \quad (34)$$

and the right hand side of Equation 33 is given by:

$$\begin{Bmatrix}d_a\\d_b\end{Bmatrix} = \lfloor 0 \quad p_y \quad p_z \mid m_x \quad -f_3E_3 \quad -f_2E_2 \rfloor^T \quad (35)$$

The coefficient matrices in Eqn. 33, which are detailed in the Appendix, are expressed in terms of the the beam stiffness coefficients. In this study external loads are not present ($p_x=p_y=m_x=0$) and transverse shear actuation will not be considered ($f_2=f_3=0$). As a result only homogenous solutions of the governing equations are used.

At the tip the beam is assumed to be free. The resulting boundary conditions are:

$$P(L) = Q_y(L) = Q_z(L) = M_y(L) = M_z(L) = 0 \quad (36a)$$

$$T(L) - Q_w'(L) = 0 \quad (36b)$$

$$Q_w(L) = 0 \quad (36c)$$

At the root the beam is fixed such that:

$$U(0) = V(0) = W(0) = \theta(0) = \beta_y(0) = \beta_z(0) = 0 \quad (37)$$

The final boundary condition concerns warping deformation of the section at the root of the beam. In most cases the root is constrained from warping:

$$\theta'(0) = 0 \quad (38a)$$

An alternate case, where the root is free to warp, will also be considered:

$$Q_w(0) = 0 \quad (38b)$$

The displaced shape of an active closed cell beam under the action of the active materials is found by solving the sytem of differential equations and boundary conditions using a finite difference solution procedure.

The above method has been implemented in a computer code that generates section properties and finds the deformed shape of a uniform active closed cell anisotropic beam. Either the warping restrained or warping free root boundary conditions can be invoked The following examples were analyzed using the code.

Examples

Controlling the shape of a closed cell beam in bending and torsion can be accomplished by a number of methods. To examine and compare these different approaches a basic rectangular cross-section boxbeam will be used. The section, shown in Figure 2, is comprised of the principal elements of a closed cell structure including the upper and lower skins, and forward and aft spars. Each spar possesses an upper and lower flange and a vertical shear web. Active materials may be present in any of the structural elements. In addition, the parent structural material can be a general anisotropic laminated composite. The structural reference axis of the beam is fixed at the geometric center of the section.

The example section is 2.0 in. wide by 0.25 in. high with 0.1 in. thick ±45° glass/epoxy fabric vertical shear webs. Each of the four spar flanges is 0.5 in. wide. Material properties for the example section structural materials and active

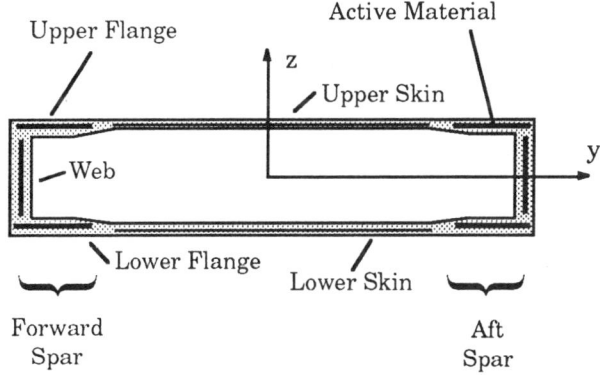

Figure 2.   Rectangular boxbeam.

Table 1

Structural And Active Material Properties

Material Property	Glass/Epoxy Fabric	Glass/Epoxy UD Tape	PZT	PZT Fiber/Epoxy UD Tape
E_{11} (msi)	4.0	8.0	9.1	6.1
E_{22} (msi)	4.0	2.0	9.1	2.2
ν_{12}	0.15	0.25	0.3	.25
G_{12} (msi)	1.0	1.0	3.3	1.0
d_{31} (in/Volt x 10^{-9})	0	0	-6.54	-4.33
d_{32} (in/Volt x 10^{-9})	0	0	-6.54	-0.79
ρ (lb/in^3)	.073	.073	.287	.197

materials are presented in Table 1. Cross-sectional shape of the beam is assumed to be rigid. This assumption represents the effect of closely spaced ribs that are rigid in in-plane bending and shear yet possess negligible out-of-plane stiffness.

Bending and twisting actuation of the rectangular boxbeam are addressed separately. Alternate methods of actuation for each type of deformation are examined. Only bending in the xz-plane and twist about the x-axis will be considered.

Bending Actuation

Three methods of bending actuation are examined:

- Bending pair actuation
- Bending-extension coupling actuation
- Asymmetric actuation

The approaches are compared by determining the deformed shape of a section with fixed geometry and effective bending rigidity. In this paper the effective bending rigidity is defined as the ratio of bending moment to bending curvature (M_y/β_y') with all other beam forces and the applied field zero. Each of the sections is designed to a bending rigidity of 5170 lb.in.^{2}.

The sections are each characterized by the bending curvature per unit induced strain (Γ_5/ε) and weight per unit length (ρA). A figure of merit for effectiveness of the active structure is obtained by dividing the bending curvature per unit induced strain by the section distributed weight ($\Gamma_5/\rho A\varepsilon$).

Bending Pair

Inducing tensile and compressive strains in the spar upper and lower flanges causes the section to deform in pure bending. In this example the upper and lower skins are .009 in. 0° glass/epoxy fabric. At each spar flange position is bonded a 0.5 in. wide by .0075 in. thick PZT wafer. Adhesive layers and bond lines are neglected in this analysis. The resulting deformed shape is shown in Figure 3 where the vertical displacement has been normalized by the beam length and piezoelectric induced strain.

Extension-Bending Coupling

Extensional forces induced by axially aligned actuators can be used with elastic bending-

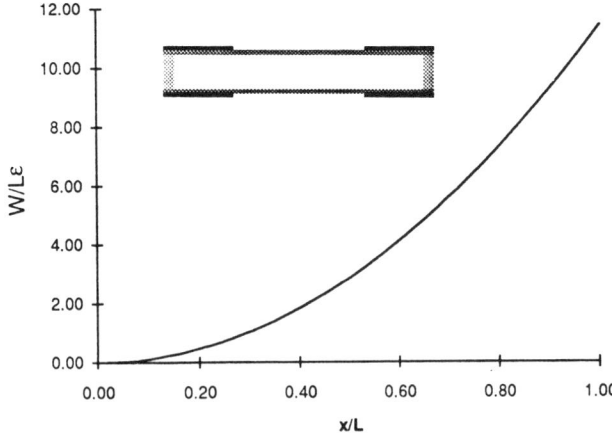

Figure 3. Beam deformation due to bending pair actuation.

extension coupling to produce bending deformation. In this case only extensional forces are created by the active materials ($f_1 \neq 0$). Bending-extension coupling ($C_{15} \neq 0$) is designed into the section to generate bending deformation.

In Figure 4 the extensional and bending deformation of an example beam is plotted. The section shown is comprised of a .015 in. upper and .005 in. lower 0° glass/epoxy fabric skins. The active material elements are in the same configuration as the bending pair case.

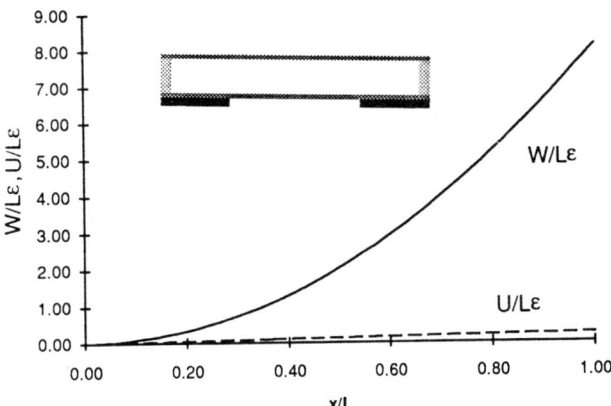

Figure 5. Beam deformation due to asymmetric actuation.

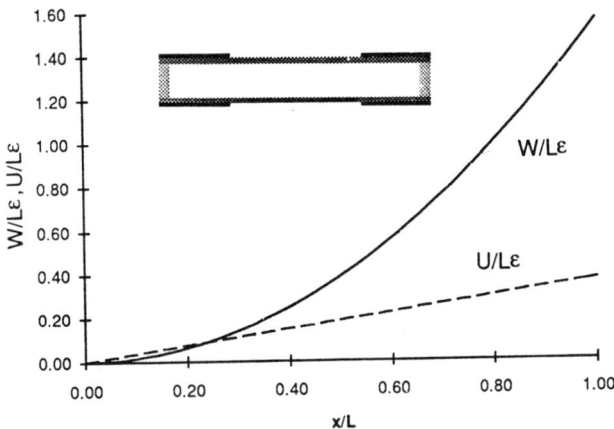

Figure 4. Beam deformation due to elastic bending-extension coupling actuation.

Asymmetric Actuation

In the case of asymmetric actuation both extensional (f_1) and bending (f_5) actuation force resultants are present. In general, some amount of bending-extension coupling ($C_{15} \neq 0$) also exists. For this example the actuators are located only in the lower flanges, with two 0.5 in. by .0075 in. PZT actuators in each flange. The upper skin and lower skin are .0125 and .009 in. glass/epoxy fabric, respectively. Axial and vertical deflection of the beam are presented in Figure 5.

Comparison

The bending curvature per unit induced strain, section weight per unit length and figure of merit for each of the three bending actuation methods are presented in Table 2. Bending pair actuation yields the highest curvature per unit induced strain. The poorest actuation effectiveness is provided by the bending-extension coupling method. Weight per unit length is similar for each of the three cases. This is primarily due to the dominant effect of the active material mass, which is the same in each case. As a result, the figure of merit values show the same relative performance as the bending curvature per unit induced strain.

While asymmetric actuation is an effective method it also has some advantages due to placement of the actuators on one surface of the beam. Actuators on lower surfaces are better protected from low velocity impact damage due to falling objects. Alternatively, placing active materials on the upper surfaces can significantly reduce exposure to tensile strains in structures such as wings.

Torsional Actuation

Torsion actuation of a closed cell beam can be achieved by a number of different approaches.

Table 2

Bending Actuation Comparison

	Γ_5/ε (in.$^{-1}$)	ρA (lb./in.)	$\Gamma_5/\rho A\varepsilon$ (lb.$^{-1}$)
Bending Pairs	3.81	.0102	374
Bending-Extension	0.52	.0105	49.7
Asymmetric Actuation	2.72	.0107	255

Methods compared in this study are:

- Cell wall shear actuation
- Twist-extension coupling actuation
- Warping-torsion coupling actuation

Section geometry is the same as in the bending study. Effective torsional rigidity, the ratio of torque to twist rate (T/θ') with all other forces and the applied field set to zero, is 3520 lb. in.^{2}.

Torsional actuation methods are characterized by the average twist angle per induced strain (θ_{avg}/ε) and section weight per unit length (ρA). The torsional actuation figure of merit is defined as the ratio of the average twist angle per unit strain to the weight per unit length ($\theta_{avg}/\rho A\varepsilon$).

Cell Wall Shear

Deforming a closed section by inducing shear in the cell walls requires either a shear deforming active material or an anisotropic active material. Although some materials do exhibit shear strain-electric field coupling they are not in forms useful for thin-walled structures. There are currently three approaches to achieving induced strain anisotropy:

- Orthotropic piezoelectric materials[8]
- Specially attached isotropic piezoelectrics[9]
- Unidirectional piezoelectric fiber composites[10]

Unidirectional piezoelectric fiber composite, which combine active fibers in a matrix to produce an orthotropic active material will be used in this example. Such active composites are especially useful because they possess significant stiffness and strain anisotropy. Acting in ±45° pairs they can be used to induce shear in structurally efficient thin-walled configurations.

In Figure 6 the twist distribution for a beam consisting of .008 in. upper and lower skins made from a ±45° PZT/epoxy composite laminate is shown. This configuration permits the piezocomposite to serve as the active material and structural material without the need for additional materials to carry and distribute loads.

Extension-Twist Coupling

Using conventional isotropic active materials aligned longitudinally ($f_1 \neq 0$) with an anisotropic

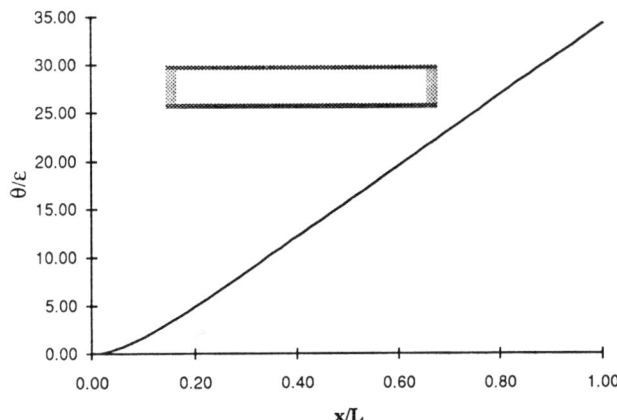

Figure 6. Beam twist deformation due to cell wall shear actuation.

parent structure permits twist actuation through the use of extension-twist coupling ($C_{14} \neq 0$). The axial and twist deformations of a rectangular beam with extension-shear coupling in the upper and lower skins is shown in Figure 7. The upper and lower skins in this example are .006 in. unidirectional glass/epoxy tape oriented 45° with respect to the local x-axis.

Warping-Torsion Coupling

Coupling between out-of-plane warping and torsion in noncircular cross-sections can also be exploited to twist a closed cell beam. Axially aligned actuators are used to generate warping forces ($f_7 \neq 0$) which, due to internal equilibrium, result in a twist deflection. This approach yields a

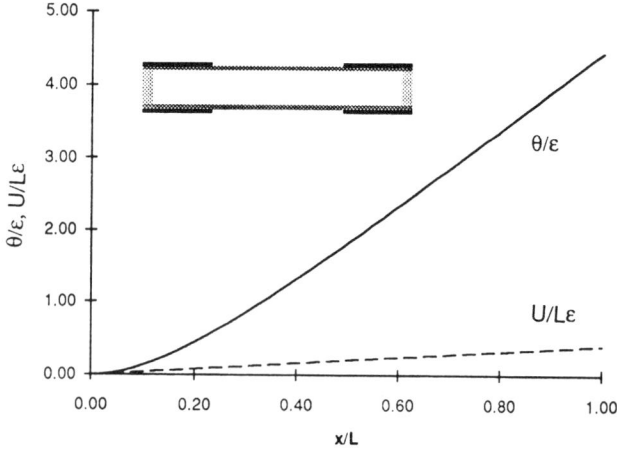

Figure 7. Twist deformation due to extension-twist coupling actuation.

twist distribution much different from the previous cases. Deformations are concentrated near the the free end in a boundary layer[11] with a decay length that is a function of the ratio of the torsional to warping stiffness of the section.

The boundary layer effect is pronounced for large values of μL, where $\mu = \sqrt{C_{44}/C_{77}}$. Warping induced twist deformations are concentrated in a region near the free end where the depth of this region, x^*, is measured from the free end. This decay length can be defined as the point where the displacement has been reduced to 1/e of the maximum value. The expression for the decay length in nondimensional coordinates is:

$$\frac{x^*}{L} = \frac{1}{\mu L} \qquad (39)$$

As the beam is lengthened, with the section properties held constant, the percentage of beam length in the boundary layer decreases.

The internal shear flows that result from warping tend to distort the cross-section shape. Maintaining section shape requires closely spaced ribs or bulkheads. The current analysis method is consistent with this assumption. However, it should be noted that the weight of this structure is not included in the distributed weight calculations.

The cross-section in this example is identical to the bending pairs actuation example. Warping deformation is created by expanding the upper left and lower right actuators and contracting the lower left and upper right. The resulting out-of-plane deformation approximates the section warping mode.

The warping function (Eqn. 7) for the example closed cell beam is plotted in Figure 8. For comparison the first warping mode using Bauchau's anisotropic beam theory[12] and a simple boxbeam warping model are also shown. The boxbeam model is based on a method that assumes all extensional stiffness is lumped at the corner flanges[13] with the skins and webs acting only in shear. The plots show generally good agreement. Greatest differences occur in the displacement of the spar flanges. Warping deformation of the vertical shear webs is nearly identical for all three theories.

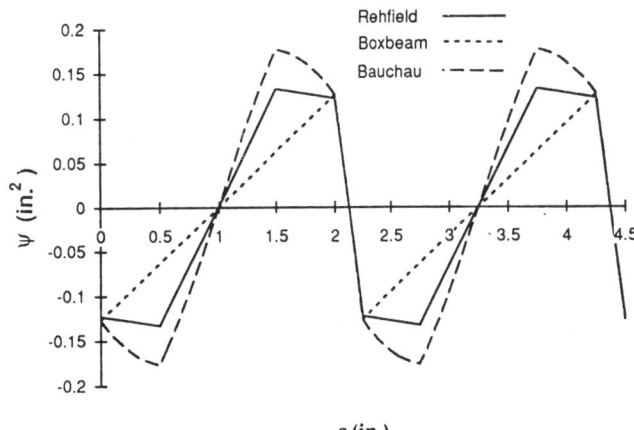

Figure 8. Comparison of warping functions for the baseline rectangular boxbeam.

The twist distribution for a beam constrained from warping at the root and actuated by warping-torsion coupling is shown in Figure 9. In this case much of the beam is essentially undeformed with the majority of the twist displacement occurring in a region near the tip.

However, a beam that is free to warp at both ends and constrained from rotating at the root presents a potentially more useful twist distribution. The twist distribution for the example cross-section with the root free to warp but not twist is shown in Figure 10.

Ordinarily, warping does not affect the deformed shape of such a beam. But in an active beam

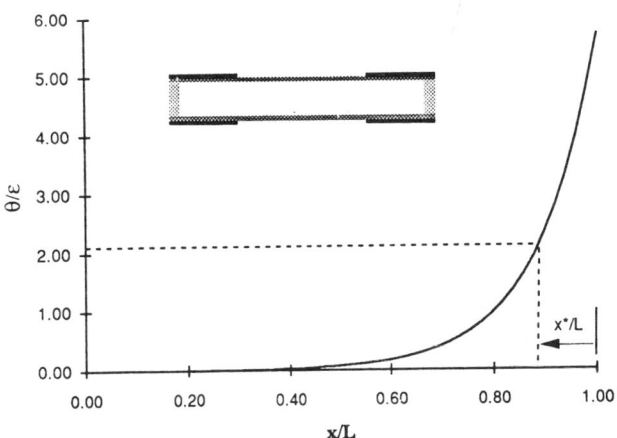

Figure 9. Warping-torsion coupling induced twist distribution with the root constrained from warping.

internally generated warping forces will cause the section to twist. Near the tip and root, in the boundary layer regions, the twist deformation goes to zero. Again, the decay length or depth of the boundary layer is determined by the ratio of the torsion to warping stiffness. For a beam that is properly tailored and of sufficient length the twist displacement can be made nearly constant over a large percentage of the span.

It can also be shown that for large values of μL the maximum twist angle of a beam in torsion is:

$$\theta_{max} = \frac{\Gamma_7 E_3}{\mu^2} \qquad (40)$$

The magnitude of Γ_7 in the above equation is determined by the modulus and strain inducing capabilities of the active materials and their ability to drive the section warping mode. Equation 40 can be used to show how tailoring of the active structure might be accomplished to affect the twist distribution. For instance, adding longitudinal active materials, such as piezo fibers, has virtually no effect on C_{44}. However, this can increase both C_{77} and Γ_7. The resulting decrease in μ (which diminishes the boundary layer effect) is accompanied by an increase in θ_{max}. This creates a tradeoff between maximum twist angle and uniformity of the twist distribution. It is also an example of the potential usefulness of piezoelectric composites in tailoring active structures.

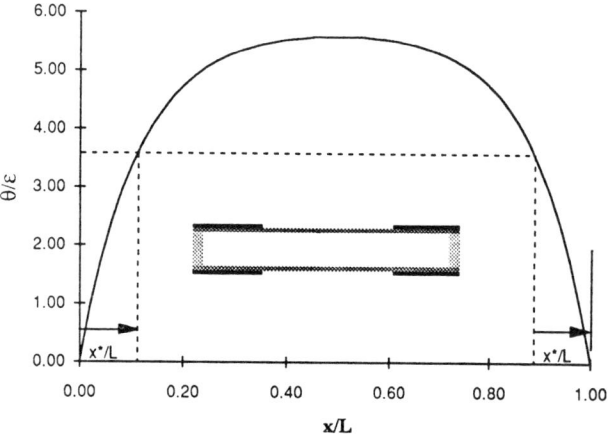

Figure 10. Warping-torsion coupling induced twist distribution with the root free to warp.

Comparison

The average twist angle per unit strain, weight per unit length and figure of merit for each of the torsional actuation schemes is presented in Table 3. Best performance is provided by the cell wall shear approach. As might be expected warping-torsion coupling actuation with the root constrained from warping yields the lowest average twist per unit induced strain. However, freeing the root to warp significantly increases the performance of this method. In these examples the warping-torsion coupling method with the root free to warp provides more than twice the effectiveness of the twist-extension coupling approach.

Table 3

Torsional Actuation Comparision

	θ_{avg}/ε (in.$^{-1}$)	ρA (lb./in.)	$\theta_{avg}/\rho A \varepsilon$ (lb.$^{-1}$)
Cell Wall Shear	15.4	.0098	1570
Twist-Extension	1.92	.0094	205
Warping (Root Fixed)	0.65	.0102	63.8
Warping (Root Free)	4.42	.0102	438

Conclusion

Multiple schemes are available for the bending and torsional deformation control of thin-walled closed cell structures. Most importantly, the ability to deform a structure through the use of coupling with out-of-plane deformation modes has been shown. The use of warping-torsion coupling permits axially aligned actuators to control twist of the closed cell section. The shape of the twist distribution is unique and may be useful in some applications.

The usefulness of piezo fiber composites in creating efficient active structures is also apparent. High specific stiffness and induced strain anisotropy are characteristics that may be exploited in the tailoring of active structures. Such active composites can be integrated into spar caps, shear webs or skins to provide directional stiffness and actuation tailoring. They can also be used to induce in-plane shear strain under a transverse electric field.

Combined actuation that provides simultaneous and independent bending and torsion control is also important. Some of the methods examined can be combined to yield this capability. For instance, using bending pair actuators with extension-twist coupling allows the same set of four flange actuators to be used for both bending and torsion control. This can also be accomplished with bending pair and warping-torsion coupling actuation, without the need for elastic coupling. A third approach, combining bending pair with piezocomposite skins, is also possible.

Further work with more advanced beam configurations will look at multicell sections, taper and nonuniform spanwise segmented actuators. The ability to utilize other higher order deformation modes to control displaced shape will also be investigated.

References

1. Lazarus, K. B., Crawley, E. F. and Bohlmann, J. D., "Static Aeroelastic Control Using Strain Actuated Adaptive Structures," Proceedings of the First Joint U.S./Japan Conference on Adaptive Structures, Maui, Hawaii, October 1990.

2. Ehlers, S. M. and Weisshaar, T. A., " Static Aeroelastic Behavior of an Adaptive Lifting Surface," *Journal of Aircraft,* Vol. 30, No. 4, July-August 1993, pp. 534-540.

3. Song, O., Librescu, L. and Rogers, C. A., "Vibrational Behavior of Adaptive Aircraft Wing Structures Modelled as Composite Thin-Walled Beams," Proceedings of the Ninth DoD/NASA/FAA Conference on Fibrous Composites in Structural Design, Lake Tahoe, Nevada, November 1991, pp. 361-381.

4. Abdul-Wahed, M. N., " Finite Element Modeling of Piezoelectric Structure," AAE Aeroelasticity Report 92-5, School of Aeronautics and Astronautics, Purdue University, West Lafayette, Indiana, August 1992.

5. Jones, R. M., *Mechanics of Composite Materials,* Scripta, Washington, D.C., 1975.

6. Rehfield, L. W. and Atiligan, A. R., "Shear Center and Elastic Axis and Their Usefulness for Composite Thin-Walled Beams," Proceedings of the American Society For Composites Fourth Technical Conference, Blacksburg, Virginia, October 1989, pp. 179-188.

7. Berdichevsky, V., Armanios, E. and Badir, A., "Theory of Anisotropic Thin-Walled Closed-Cross-Section Beams," Composites Engineering, Vol. 2, Nos. 5-7, 1992, pp. 411-432.

8. Lee, C. K., Piezoelectric Laminates for Torsional and Bending Modal Control: Theory and Experiment." PhD Thesis, Cornell University, May 1987.

9. Barret, R., "Intelligent Rotor Blade Actuation Through Directionally Attached Piezoelectric Crystals," presented at the American Helicopter Society National Forum, Washington, D.C., May 1990.

10. Hagood, N. W. and Bent, A. A., " Development of Piezoelectric Fiber Composites for Structural Actuation," Proceedings of the 34th AIAA Structures, Structural Dynamics and Materials Conference, San Diego, Calif., April 1993, pp. 3625-3638.

11. Rehfield, L. W., Atiligan, A. R., and Hodges, D. H., "Nonclassical Behavior of Thin-Walled Composite Beams with Closed Cross Sections," Journal of the American Helicopter Society, May 1990, pp. 42-50.

12. Bauchau, O. A., "A Beam Theory for Anisotropic Materials," Journal of Applied Mechanics, Vol. 52, June 1985, pp. 416-422.

13. Kuhn, P., *Stresses in Aircraft and Shell Structures,* McGraw-Hill, New York, N.Y., 1956, pp. 181-202.

Appendix

The differential equation coefficient matrices are:

$$[L_{bb}] = \begin{bmatrix} C_{77} & 0 & 0 \\ 0 & 0 & 0 \\ 0 & 0 & 0 \end{bmatrix} \quad (A1)$$

$$[M_{ab}] = \begin{bmatrix} -C_{17} & 0 & 0 \\ -C_{27} & 0 & 0 \\ -C_{37} & 0 & 0 \end{bmatrix} \quad (A2)$$

$$[M_{bb}] = \begin{bmatrix} 0 & C_{57} & C_{67} \\ -C_{57} & 0 & 0 \\ -C_{67} & 0 & 0 \end{bmatrix} \quad (A3)$$

$$[N_{aa}] = \begin{bmatrix} -C_{11} & -C_{12} & -C_{13} \\ -C_{12} & -C_{22} & -C_{23} \\ -C_{13} & -C_{23} & -C_{33} \end{bmatrix} \quad (A4)$$

$$[N_{ab}] = \begin{bmatrix} -C_{14} & -C_{15} & -C_{16} \\ -C_{24} & -C_{25} & -C_{26} \\ -C_{34} & -C_{35} & -C_{36} \end{bmatrix} \quad (A5)$$

$$[N_{bb}] = \begin{bmatrix} -C_{44} & -C_{37}-C_{45} & -C_{27}-C_{46} \\ -C_{37}-C_{45} & -C_{55} & -C_{56} \\ -C_{27}-C_{46} & -C_{56} & -C_{66} \end{bmatrix} \quad (A6)$$

$$[P_{ab}] = \begin{bmatrix} 0 & C_{13} & C_{12} \\ 0 & C_{23} & C_{22} \\ 0 & C_{33} & C_{23} \end{bmatrix} \quad (A7)$$

$$[P_{bb}] = \begin{bmatrix} 0 & C_{34} & C_{24} \\ -C_{34} & 0 & C_{25}-C_{36} \\ -C_{24} & C_{36}-C_{25} & 0 \end{bmatrix} \quad (A8)$$

$$[R_{bb}] = \begin{bmatrix} 0 & 0 & 0 \\ 0 & C_{33} & C_{23} \\ 0 & C_{23} & C_{22} \end{bmatrix} \quad (A9)$$

ADVANCED CONTROL LAW FOR VIBRATION SUPPRESSION WITH A TYPE-II VARIABLE STIFFNESS MEMBER

Kenji Minesugi* & Junjiro Onoda†
The Institute of Space & Astronautical Science
3-1-1 Yoshinodai, Sagamihara, Kanagawa, JAPAN

Abstract

An advanced semi-active control law for vibration suppression of large space structures is proposed. This law uses an active member named TYPE-II which is able to vary its axial stiffness, and makes more effective use of the features of TYPE-II member than the existing laws. Numerical simulations are performed on beam like truss model to verify and compare the effectiveness of the law with the existing laws.

1. Introduction

Large Space Structures (LSS) will be very flexible so that some of the expected disturbances can easily excite a low frequency vibration on these structures. Additionally, the vibration excited may remain for a while because LSS are expected to have very low natural damping. Such vibration will prevent the management of missions which require to keep a specific shape or attitude with necessary accuracy. Therefore, vibration suppression will be a difficult and important problem.

Many papers have been published on this problem, and various types of actuators have been proposed, for examples, reaction jet, gyrodamper, proof-mass actuator[1] and so on. An active member is also one of actuators which has been designed to be installed in LSS like structural members. Piezoelectric actuator[2] and moving coil type actuator[3] are the truss member type of active members which can generate axial force. These active members are called Variable Axial Force member (VAF member). Active control with these actuators are very powerful for the vibration suppression. However, if a control system is not adequate or a mathematical model is poor, these actuators may conversely give the energy to the structure and subsequently the system may become unstable. It is inevitable as far as the active control is used for the vibration suppression.

Variable stiffness member is a kind of active members which can vary its stiffness. In particularly, a truss member type of active member which can vary its axial stiffness is called Variable Axial Stiffness member (VAS member). A VAS member which can vary its stiffness continuously at any moment as shown in Fig.1 is called TYPE-I. This member is able to dissipate energy due to the negative work done by its induced axial force because of the axial stiffness variation. From this point of view, this member is the same as VAF member such as piezoelectric actuator. Therefore, the active control with TYPE-I member may induce the system instability. Moreover, TYPE-I member is not able to generate the axial force however its axial stiffness varies at the moment when the strain of this member is zero. This difference from VAF member restricts the degree of freedom for designing the control system using TYPE-I member.

In order to avoid the system instability, a VAS member which never gives the energy to a structure has been designed. It is called TYPE-II as shown in Fig.2. It is not able to absorb the energy of vibration from the structure by itself either. It can enhance the effect of the passive damping which the structure holds naturally. The control using this type of actuator is called semi-active control. Onoda et al.[4] designed it by using a piezoelectric material. They succeeded to damp the vibration excited on the structure with it. The control law used is only to decide the timing of the axial stiffness variation. This law takes advantage of energy dissipation due to strain energy release at the time of the axial stiffness decrease of TYPE-II member.

On the other hand, Minesugi et al.[5] have proposed another method for vibration suppression with TYPE-II member. This method takes advantage of the energy movement between the elastic modes of the structure due to the axial stiffness variation

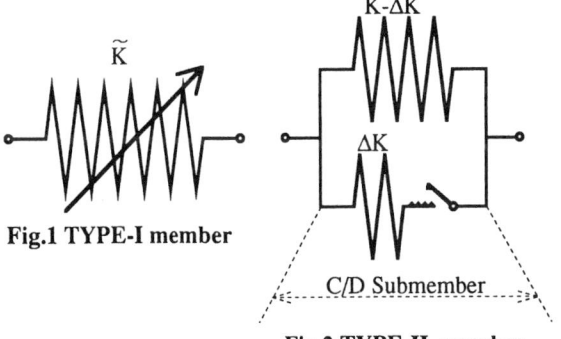

Fig.1 TYPE-I member

Fig.2 TYPE-II member

*Research Associate, AIAA member
†Professor, AIAA member

Copyright © 1994 by the American Institute of Aeronautics and Astronautics, Inc. All rights reserved.

of TYPE-II member. It has been shown that the control law based on the method can move the energy from the critical modes for vibration suppression to the residual modes, while keeping the total energy of the system. It has also been confirmed that the structural damping plays an important role for the semi-active control. However, the control laws proposed in Ref.5 dose not completely make use of the characteristics of TYPE-II member for the comparison with the law based on the method in Ref.4.

In this study, we propose an advanced semi-active control law using TYPE-II member. This law takes advantage of the energy movement between the elastic modes of the structure as well as the laws in Ref.5. It is, however, different from these laws because it accounts for one of the important characteristics of TYPE-II member that its axial stiffness can be recovered at any moment. Numerical simulations are performed on a truss structure to verify and compare the effectiveness of the law with the existing laws. In the next chapter, TYPE-II member is described. In Chapter 3, the equation of motion is obtained. In Chapter 4, the control methods are introduced. In Chapter 5, control laws are described. In Chapter 6, the results of numerical simulations are shown. The last chapter is concluding remarks.

2. TYPE-II Variable Stiffness Member

A variable stiffness member studied in the paper is called TYPE-II member. It is composed of two submembers which is able to bear an axial force as shown in Fig.2. In the figure, $K-\Delta K$ and ΔK denote the axial stiffness of the submembers, respectively. One of the submembers is able to be connected/disconnected arbitrarily so that the total axial stiffness of TYPE-II member is able to vary. This connectable submember is shown as C/D submember in Fig.2. When the C/D submember is connected, the axial stiffness of TYPE-II member is K, and when disconnected, it is $K-\Delta K$.

The C/D submember is able to be connected at any position. Even if it is connected at a little bit different position from the target due to the dead time of the controller and so on, it cannot correct the position because it only has the connect/disconnect mechanism. If it has an additional mechanism generating the axial force to correct the position, the mechanism may induce the system instability because it is able to give the energy to the system. TYPE-II member doesn't have such mechanism so that its operation never induce the system instability. It is one of the most important features of TYPE-II member.

The connecting position of the C/D submember where the axial stiffness of TYPE-II member is recovered at the moment when no strain energy is stored in it is called home position. In case that the C/D submember is connected at the different position from the home position, the static equilibrium shape of a structure with TYPE-II member is varied and subsequently the structure always keeps strain energy even after vibration will have been completely suppressed. It is another important characteristic of TYPE-II member. In this paper, we propose an advanced semi-active control law taking advantage of this characteristic of TYPE-II member for vibration suppression of truss structures.

3. Basic Formulations

In this chapter, the equation of motion of a truss structure into which n_A TYPE-II members are installed are considered with the application of FEM. Now, we define stiffness state number m as follows.

$$m = \sum_{i=1}^{n_A} s_{mi} 2^{i-1} \quad (1)$$

This number m represents the stiffness state of TYPE-II members. If the axial stiffness of the i-th TYPE-II member is in stiffness recovery, s_{mi} is given 0. Otherwise, s_{mi} is given 1. Then, in case that all TYPE-II members are in stiffness recovery, $m=0$, and in case that they are in stiffness decrease, $m=2^{n_A}-1$. When $m=\alpha$, the total stiffness matrix of the structure is represented as

$$K_\alpha = K_s + \sum_{i=1}^{n_A} (K_{Ai} - s_{\alpha i} \Delta K_{Ai}) \quad (2)$$

where K_s and K_{Ai} are the stiffness matrices constructed by normal structural members and the i-th TYPE-II member, respectively. The matrix ΔK_{Ai} is the stiffness variation matrix of the i-th TYPE-II member, for instance, the stiffness matrix constructed by its C/D submember. The relation between the nodal displacement vector x and the nodal force F is written as

$$F = K_\alpha x - \sum_{i=1}^{n_A} (1 - s_{\alpha i}) \Delta K_{Ai} x_{Ai} \quad (3)$$

where x_{Ai} is the displacement vector at the moment when the stiffness of the i-th TYPE-II member has been recovered lately. It is obvious from Eq.(3) that the static equilibrium shape vector is not $x=0$. This equilibrium displacement vector defined as $x_{E\alpha}$ is calculated from the equation below.

$$K_\alpha x_{E\alpha} = \sum_{i=1}^{n_A} (1 - s_{\alpha i}) \Delta K_{Ai} x_{Ai} \quad (4)$$

From these equations, the equation of motion of the structure is represented as follows.

$$M\ddot{x} + K_\alpha x = K_\alpha x_{E\alpha} \quad (5)$$

Where M is the mass matrix and $(\cdot)$ means the differentiation by time in this paper. The variation of mass distribution due to the variation of the axial stiffness is assumed to be negligible so that the matrix M keeps constant. By using the displacement vector which is newly defined as

$$x_\alpha \equiv x - x_{E\alpha} \tag{6}$$

Eq. (5) is rewritten as follows.

$$M\ddot{x}_\alpha + K_\alpha x_\alpha = 0 \tag{7}$$

It means that the center of the vibration is not $x=0$ but $x=x_{E\alpha}$. It is apparent from these equations that the center of the vibration shifts according to both the stiffness state and the history of the axial stiffness variations of TYPE-II members.

The displacement vector x_α is expanded by the modal matrix Φ_α and the modal displacement vector q_α as follows.

$$x_\alpha = \Phi_\alpha q_\alpha \tag{8}$$

The modal matrix Φ_α satisfies the next orthogonal relations.

$$\Phi_\alpha^T M \Phi_\alpha = I \tag{9}$$

$$\Phi_\alpha^T K_\alpha \Phi_\alpha = \Lambda_\alpha \tag{10}$$

The i-th diagonal element of the diagonal matrix Λ_α is the square of the i-th natural frequency $\omega_{\alpha i}$. From Eqs.(8)-(10), Eq.(7) is rewritten with the modal damping as follows.

$$\ddot{q}_\alpha + \eta \Lambda_\alpha^{1/2} \dot{q}_\alpha + \Lambda_\alpha q_\alpha = 0 \tag{11}$$

In this equation, each modal damping is assumed to be proportional to its natural frequency. Then, it can be written with the structural damping coefficient η as $\eta\omega_{\alpha i}$.

4. Control Methods

For vibration suppression, two control methods are introduced to the system using TYPE-II member in Ref.5 and are called Method-I and Method-II, respectively. Firstly, let us describe these two methods.

Method-I

If the axial stiffness of TYPE-II member is decreased by disconnecting the C/D submember when the structure is vibrating, the strain energy stored in the C/D submember is released. Most of this energy moves up to the energy of higher modes including local modes such as longitudinal vibration modes of the C/D submember itself. On the assumption that each modal damping is proportional to its natural frequency, this released energy is dissipated rapidly.

This method intends to decrease the total energy of the system. If no structural damping exists, the method is not available for the vibration suppression. It is not available when no strain energy is stored in the C/D submember either. Therefore, it is of no use at the moment of the axial stiffness recovery. These are important features of the method.

Method-II

Due to the axial stiffness variation of TYPE-II member, the modal coordinate transformation is occurred between the modal coordinates for the axial stiffness state just before the variation and the state just after it. Each modal energy for one modal coordinate is distributed to several modes for the other modal coordinate due to this transformation. Therefore, it is able to move the energy from the critical modes for the vibration suppression to the residual modes by means of adequately determining the timing of the axial stiffness variation of TYPE-II member. This is Method-II.

This method intends to decrease the sum of the energy of several modes selected as control modes. It is able to be carried out even without the structural damping by moving the energy from the control modes to the residual modes while keeping the total energy of the system. Furthermore, the method is of use at the moment of not only the axial stiffness decrease of TYPE-II member but also its recovery. These are important and different features of Method-II from Method-I.

Let us consider the case that the stiffness state number m is changed from α to β. The total stiffness matrix K_β in $m=\beta$ is written from Eq.(2) as follows.

$$K_\beta = K_s + \sum_{i=1}^{n_A} (K_{Ai} - s_{\beta i} \Delta K_{Ai}) \tag{12}$$

The static equilibrium shape vector $x_{E\beta}$ is calculated from the equation below.

$$K_\beta x_{E\beta} = \sum_{i=1}^{n_A} (1 - s_{\beta i}) \Delta K_{Ai} x_{Ai} \tag{13}$$

As well as the case of $m=\alpha$, the equation of motion is represented as

$$M\ddot{x}_\beta + K_\beta x_\beta = 0 \tag{14}$$

where x_β is defined as

$$x_\beta \equiv x - x_{E\beta} \tag{15}$$

The vector x_β is expanded by the modal matrix Φ_β and the modal displacement vector q_β as follows.

$$x_\beta = \Phi_\beta q_\beta \tag{16}$$

The modal matrix Φ_β has the similar orthogonality to the mass matrix as shown in Eqs.(9) and (10). From Eqs.(6), (8), (15), (16), the modal coordinate transformation is represented as follows.

$$q_\beta = \Phi_\beta^T M \Phi_\alpha q_\alpha + \Phi_\beta^T M (x_{E\alpha} - x_{E\beta}) \tag{17}$$

From this equation, it is apparent that each modal energy for one modal coordinate is distributed to several modes for the other modal coordinate due to this transformation.

For the description of Method II, two degree-of-freedom spring-mass model shown in Fig.3 is introduced. Right spring is a TYPE-II member of which spring constant is K at the recovery of the axial stiffness or $K-\Delta K$ at its decrease. In Method-II, if the first mode is selected as the control mode on the system, the axial stiffness of the TYPE-II member is varied at the moment when the modal energy of the first mode is decreased due to this variation.

Figure 4 explains one effect of the modal coordinate transformation on this system. For the separation from another effect described after, the axial stiffness of the TYPE-II member is intentionally recovered at the moment when its C/D submember is connected at the home position. In this figure, x_1-x_2 denotes the actual displacement coordinate, q_{R1}-q_{R2} denotes the modal displacement coordinate at the recovery of the axial stiffness and q_{D1}-q_{D2} denotes the modal displacement coordinate at its decrease. The unit vector along each axis of these modal coordinates corresponds to the eigenvector for each mode.

Let us vary the axial stiffness of the TYPE-II member when the state variables of the system take Point A in the figure. The displacement vector is resolved into the solid vectors for q_{R1}-q_{R2} coordinate or the outlined vectors for q_{D1}-q_{D2} coordinate as shown in Fig.4. The length of each resultant vector represents the corresponding modal displacement. At the moment of the axial stiffness variation, one set of the vectors is changed to the other set according to the modal coordinate transformation. This is one effect due to the stiffness variation of TYPE-II member.

Next, let us consider the case that the axial stiffness of TYPE-II member is recovered by connecting its C/D submember at the different position from the home position. As described before, the static equilibrium shape vector of the system is changed from $x=0$ and subsequently the system continues to vibrate around the new equilibrium position, while each modal shape and natural frequency are the same as those in the case that the C/D submember is connected at the home position. It results that the modal coordinate moves parallel from q_{R1}-q_{R2} coordinate to $\tilde{q}_{R1}$-$\tilde{q}_{R2}$ coordinate through the vector x_E as shown in Fig.5. The vector x_E is the static equilibrium shape vector derived from the connecting position of the C/D submember. Therefore, the modal displacement vector is different according to the connecting position of the C/D submember, even if the axial stiffness of TYPE-II member is recovered at the same state variables. It is obvious in Fig.5 by making a comparison of the outlined and the solid vectors into which the same displacement vector is resolved for $\tilde{q}_{R1}$-$\tilde{q}_{R2}$ coordinate and q_{R1}-q_{R2} coordinate, respectively. This is another effect of the modal transformation and is not applied to the control laws proposed in Ref.5.

In this paper, the proposed control law takes accounts of these two composite effect of the modal coordinate rotation and translation due to the stiffness variation.

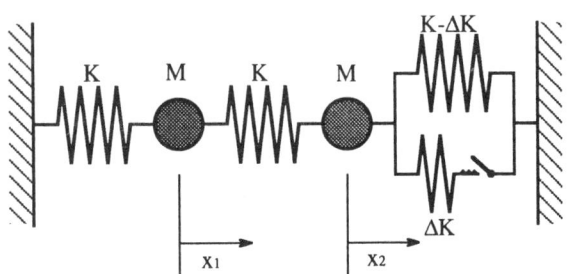

Fig.3 Two DOF Spring-Mass System

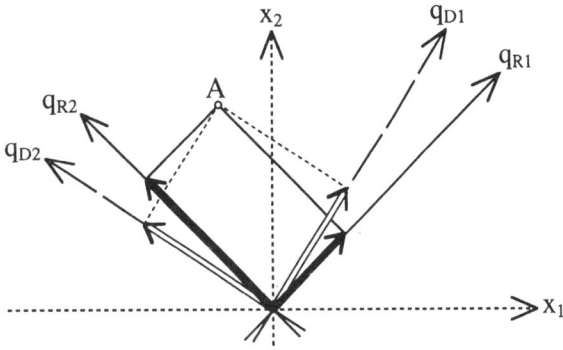

Fig.4 Modal Coordinate Transformation
(Rotation)

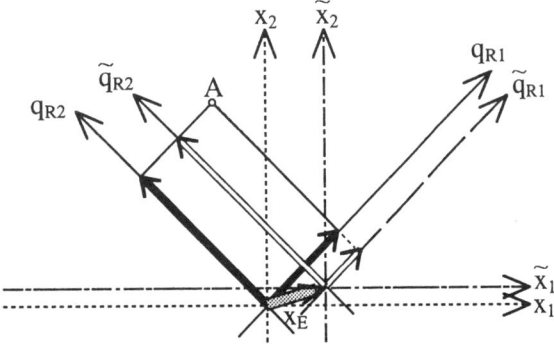

Fig.5 Modal Coordinate Transformation
(Translation)

5. Control Laws

In this chapter, an advanced control law named Law-IV is proposed. This law is based on Method-I and Method-II. Firstly, let us introduce the control laws proposed in Ref.5 for the comparison. Three control laws named Law-I, Law-II and Law-III have been proposed in Ref.5. Law-I is based on Method-I. In this law, the axial stiffness of TYPE-II member is decreased at the moment when the released strain energy becomes maximal and is recovered at the moment when the strain of TYPE-II member is zero. Law-II is based on Method-II, however, without the effect of the modal coordinate translation. In this law, the axial stiffness of TYPE-II member is varied at the moment when the total energy of the control modes, which are selected because they are critical for vibration suppression, is decreased due to this variation.

The energy of the system using Law I is dissipated faster than that using Law II, while the Law I varies the axial stiffness of TYPE-II member much more frequently than Law II. Furthermore, it dose not take account of the energy movement between modes due to the modal coordinate transformation so that the energy once having been moved from the control modes to the residual modes may return to the control modes. On the other hand, such phenomenon never occurs in the system using Law II. However, the condition for the axial stiffness variation may be rarely satisfied if the sum of the energy of the residual modes next to the control modes is considerably larger than the total energy of the control modes.

In order to compensate for these defects, Law III has been proposed. This law makes use of both Method-I and Method-II and its logic is as follows. At first, the several modes are selected as the control modes like Law II. The axial stiffness of TYPE-II member is decreased on the condition that the total energy of the control modes is decreased due to the axial stiffness decrease at the moment when the strain energy release becomes maximal or is decreasing, and it is recovered on the condition that the energy of the control modes is decreased due to its recovery at the moment when the connecting position of the C/D submember comes to the home position. Therefore, Law III dose not take advantage of the effect of the modal coordinate translation.

Next, let us introduce an advanced law named Law IV.
Law IV
Law IV is, of course, based on Method-I and Method-II and takes advantage of the modal coordinate translation. The logic of this law is as follows. At first, the several modes are selected as the control modes like Law III. The axial stiffness of TYPE-II member is varied at the moment when the total energy of the control modes is decreased due to this variation. Different from Law III, Law IV dose not take account of the strain energy release. Let the total energy of the control modes at the stiffness state number $m=\alpha$ denote as $E_{C\alpha}$ and the energy at $m=\beta$ denote as $E_{C\beta}$. These are represented as

$$E_{C\alpha} = \sum_i^N \left(\frac{1}{2}\omega_{\alpha i}^2 q_{\alpha i}^2 + \frac{1}{2}\dot{q}_{\alpha i}^2 \right) \quad (18)$$

$$E_{C\beta} = \sum_i^N \left(\frac{1}{2}\omega_{\beta i}^2 q_{\beta i}^2 + \frac{1}{2}\dot{q}_{\beta i}^2 \right) \quad (19)$$

where N is the number of the control modes. The difference $\Delta E_{\alpha\beta}$ between $E_{C\alpha}$ and $E_{C\beta}$ is defined as follows.

$$\Delta E_{\alpha\beta} = E_{C\beta} - E_{C\alpha} \quad (20)$$

With this notation, the logic is represented as

$$\Delta E_{\alpha\beta} < 0 \text{ and } \Delta \dot{E}_{\alpha\beta} = 0 \Rightarrow \text{varing} \quad (21)$$

in case that the stiffness state number is changed from $m=\alpha$ to $m=\beta$.

6. Numerical Simulations

Numerical simulations are performed in order to verify and compare Law IV with Law III. The structural damping coefficient η is set to zero. With regard to Method-I, the released strain energy of TYPE-II member due to the axial stiffness decrease is assumed to be dissipated rapidly even though the structural damping coefficient η is zero.

The control laws are applied to the beam like two dimensional truss as shown in Fig.6. All members of the truss are identical in terms of material, cross section and length except the length of the diagonal members. A TYPE-II member is

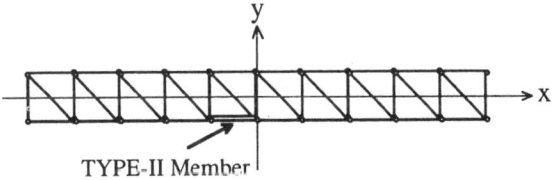

Fig.6 Two Dimensional Truss

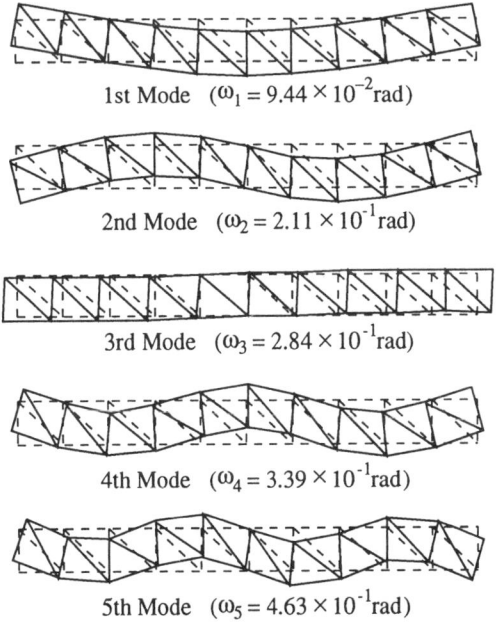

1st Mode ($\omega_1 = 9.44 \times 10^{-2}$ rad)

2nd Mode ($\omega_2 = 2.11 \times 10^{-1}$ rad)

3rd Mode ($\omega_3 = 2.84 \times 10^{-1}$ rad)

4th Mode ($\omega_4 = 3.39 \times 10^{-1}$ rad)

5th Mode ($\omega_5 = 4.63 \times 10^{-1}$ rad)

Fig.7 Modal Shape of the Control Modes

installed near the center of the truss. The axial stiffness of the TYPE-II member at the recovery state are equal to that of the horizontal and the vertical members of the truss. Its axial stiffness at the decrease state is 0.1 times as that at the recovery state.

This truss has 3 rigid modes and 41 elastic vibration modes. The control using TYPE-II member is a kind of the internal force control so that it makes no influence on the motion of the rigid modes. Then, only elastic modes are taken into account in the simulations. The first five modes are selected as the control modes. Figure 7 shows the mode shapes of the control modes at the axial stiffness recovery of the TYPE-II member. Mass, length and time are normalized by the mass, the length and the axial stiffness of the vertical and the horizontal members of the truss. For the initial condition, the modal velocity of each control mode is given 0.1 and the rest of the state variables are set to zero .

The result of the numerical simulations using Law IV is shown in Fig.8. The horizontal axes of Figs.(a) ~ (c) indicate the time from the beginning of the control and are identical. Figure (a) indicates the time history of the energy ratio to the initial total energy of the system. The n-th broken line from the horizontal axis indicates the summation of the energy ratio for the first n mode. Then, the top bold broken line shows the ratio of the total energy of the control modes. Figure (b) shows the axial stiffness variation of the TYPE-II member. Figure (c) shows the displacement of the C/D submember from the home position. It is apparent that Law IV is available for vibration suppression.

The comparison of the control efficiency between Law IV and Law III is shown in Fig.9. This figure indicates each energy ratio to the initial energy obtained by the numerical simulations under the same initial condition. It is apparent that Law IV is more effective for decreasing both the total energy of the system and the energy of the control modes than Law III. The

(a) Energy Ratio

(b) Axial Stiffness Variation

(c) Displacement from Home Position

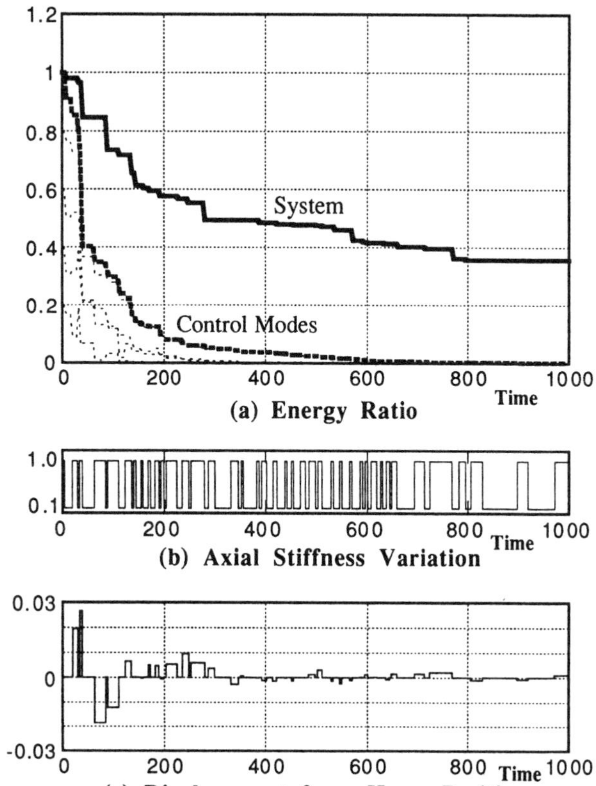

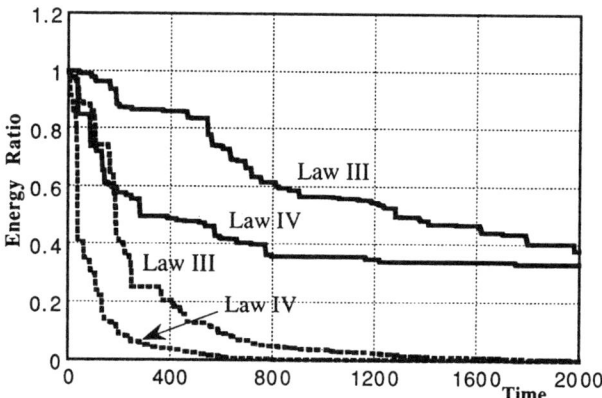

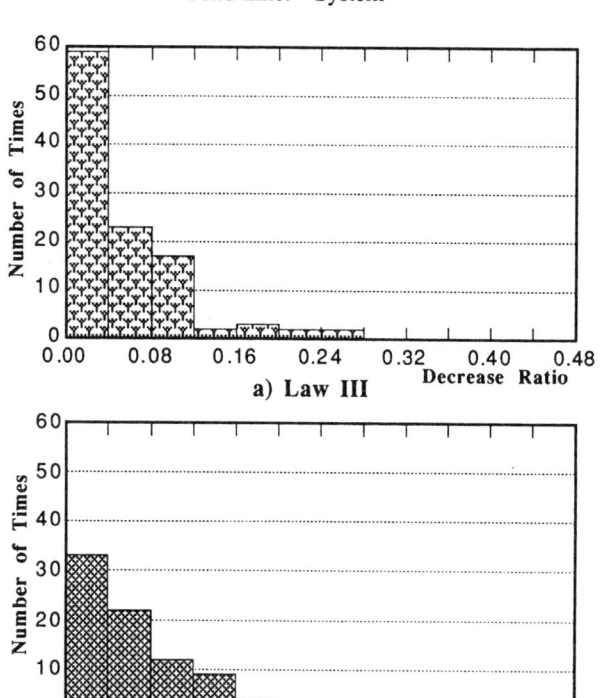

Fig.8 Simulation Result (Law IV)

Fig.9 Comparison of Law IV and Law III
Broken Line: Control Modes
Solid Line: System

Fig.10 Histograms of Energy Decrease Ratio

histograms in Fig.10 indicate the distribution of the energy decrease ratio of the control modes. This ratio means the ratio of the energy decrease of the control modes to the energy of the control modes before the stiffness variation by one stiffness variation. With the notation in Eqs.(18)~(20), the decrease ratio R_E is calculated by

$$R_E = \frac{\Delta E_{\alpha\beta}}{E_{C\alpha}} \qquad (22)$$

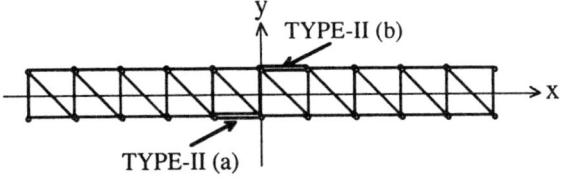

Fig.11 Two Dimensional Truss

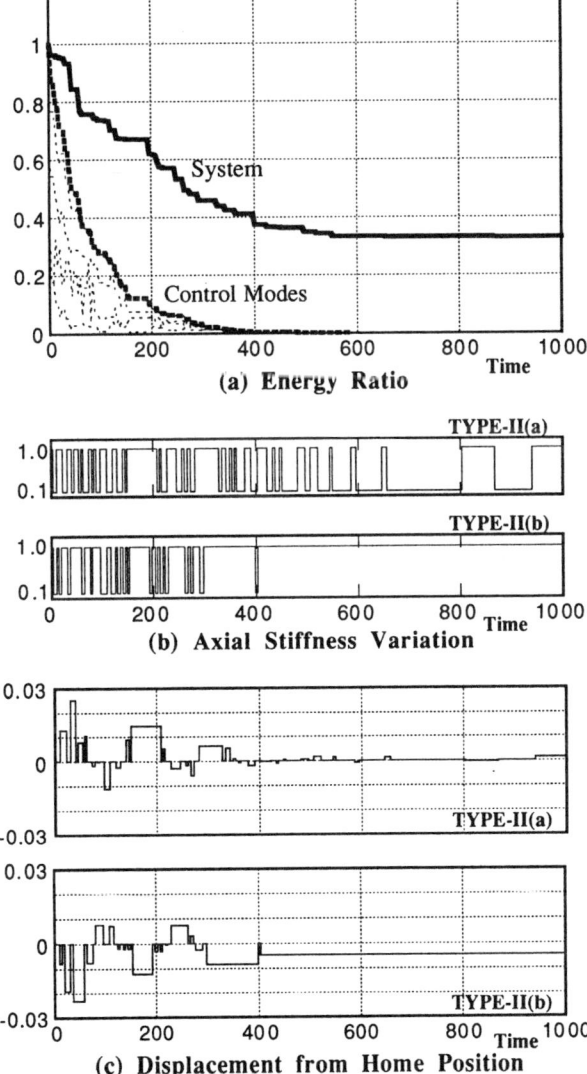

Fig.12 Simulation Result (Law IV) (Two TYPE-II Members)

From Fig.10, it is obvious that the number of the stiffness variation of high ratio over 12% in Law IV is larger than that in Law III. Therefore, Law IV suppress the vibration faster with less number of the stiffness variation compared to Law III. With these results, it is concluded that Law IV is more effective than Law III for vibration suppression.

Next, the same TYPE-II member is added to the truss as shown in FIg.11. Simulation result under the same initial condition is shown in Fig.12. From the comparison of results between Fig.8 and Fig.12, the control efficiency by two TYPE-II members is slightly better than that by one member. The improvement of the control efficiency depends on the position of the added TYPE-II members.

8. Concluding Remarks

An advanced semi-active control law for the vibration suppression of large space structures is proposed. This law uses an active member named TYPE-II which is able to vary its axial stiffness, and takes advantage of the composite effect of the modal coordinate translation and rotation induced by its axial stiffness variation. Numerical simulations are performed on a beam like truss model in order to verify and compare the control efficiency of the advanced law with that of the existing laws. From the results of the numerical simulations, it is shown that the advanced law is more effective than the existing laws.

References

[1] Zimmerman, D.C., Horner, G.C. and Inman, D.J., "Microprocessor Controlled Force Actuator," *J.Guidance, Control and Dynamics*, Vol.11, No.3, 1988.

[2] Forward, R.L., "Electric Damping of Orthogonal Bending Modes in a Cylindrical Mast --- Experiment," *J.Spacecraft and Rockets*, Vol.18, No.1, Jan.-Feb., 1981.

[3] Murohashi, S., "Vibration Suppression of Truss by Axial Force Actuator," Master Thesis, University of Tokyo, 1990 (in Japanese).

[4] Onoda, J., Endo, T., Tamaoki, H., and Watanabe, N., "Vibration Suppression by Variable-Stiffness Members," *AIAA Journal*, Vol.29, No.6, June 1991, pp.977-983.

[5] Minesugi, K. and Kondo, K., "Semi-Active Vibration Suppression of Large Space Structures with a Variable Axial Stiffness Member," 34th SDM Conference, April 1992, LaJolla, California, USA.

AIAA-94-1770-CP

SEMI-ACTIVE VIBRATION SUPPRESSION BY VARIABLE-DAMPING MEMBERS

Junjiro Onoda* and Kenji Minesugi†
The Institute of Space and Astronautical Science
Sagamihara, Kanagawa 229, Japan

ABSTRACT

A semi-active vibration suppression approach is proposed and investigated. It is to control the damping of viscous dampers installed in structures. By using a single-degree-of-freedom (SDOF) example first, vibration damping is shown to be much enhanced by a suitable variation of the damping of the damper according to the phase of vibration. Next, several types of control logic for this approach are proposed, which are applicable to multiple-degree-of-freedom (MDOF) structures with multiple variable-dampers. The performances of these types of control logic are investigated by using a SDOF example and an MDOF truss structure example with three variable-damping truss members excited by impulsive and random forces. Numerical investigations demonstrate that the vibration suppression capability of most of them are much higher than that of optimally tuned passive system, keeping the robustness of passive systems.

1 Introduction

To suppress the vibration of space structure, active vibration suppression is very attractive and powerful[1]. Numerous works have been published on the subject, with various types of actuators. However the possibility of instability, e.g., due to the spill over[2], is still an annoying problem, even though many works have been published on robust control logic. On the other hand, the passive systems, which have no artificial controls and whose vibration energy is dissipated by the structural damping, viscosity, friction, etc., are always stable. This robustness is their great advantage. In many cases, however, passive systems have disadvantages in their damping performance. A possible attractive approach to reduce this disadvantage, keeping the advantages of passive systems, may be the semi-active vibration suppression, which controls the state of the systems such that their inherent damping performances are enhanced.

With the semi-active approach, the vibration is suppressed by passive energy dissipation mechanisms. Therefore, the system is always stable even when the control logic is improper due to, e.g., lack of exact information about the dynamic characteristics of the structures. In the cases of space structures which are deployed or constructed in the orbits, exact estimation of the dynamic characteristics is very difficult. In addition, very high reliability is required for the space structural systems.

---

* Professor, Member AIAA
† Research Associate, Member AIAA
Copyright © 1994 by the American Institute of Aeronautics and Astronautics, Inc. 1994. All rights reserved.

Therefore, the robust semi-active approaches seem to be suitable for space structures.

Several types of semi-active vibration suppression have been proposed and studied. For space truss structure, Onoda et.al.[3] proposed and investigated a vibration suppression by stiffness control using hysteretic variable-stiffness member which is called type-II in his paper. He also proposed to control the friction for semi-active vibration suppression of truss structures[4] and tension-stabilized structures[5]. To suppress the vibration of mechanical system, Karnop et.al.[6] proposed to vary the damping of dash-pot type damper according to the phase of vibration. This type of semi-active approach has been studied mainly for ground vehicles[7,8] and buildings[9,10]. In these studies, the variable-damping dampers are directly connected to lumped masses. To apply the approach to flexible space structures, it may be better to study based on more suitable mathematical model, which may require other control logic.

In this paper, a semi-active vibration suppression is proposed and investigated, which is to control the damping of dashpot-type passive dampers in the structures. Based on a Single-Degree-of-Freedom (SDOF) system example which simulates flexible system, it is first shown that the vibration damping can be much enhanced by varying the damping of the damper according to the phase of vibration. Next, several types of control logic are proposed, which are practically applicable to multiple-degree-of-freedom (MDOF) structures with multiple variable-dampers. The performances of these control laws are investigated and compared with each other by using a SDOF system example and an MDOF truss beam example. Numerical simulations demonstrate that most of them are much more effective than optimally tuned passive dampers for the truss structure excited by impulsive and random forces.

Some may be concerned about the practicality of the application of dash-pot type variable dampers in space, which contain fluid. However, D-strut[11] is an example which seems to be applicable to the space structures. Variable-damping dampers may be a combination of this type of damper and a variable-area orifice. Electro-Rheological (ER) fluid may be used instead of the variable-area orifice.

2. Vibration Suppression Enhancement by a Variable-Damping Element

Let us first investigate the SDOF system shown in Fig. 1. Because of spring k_2, this system represents the flexible structures with variable-dampers more suitably than those studied in Refs. 3-7 in which the variable-damper is directly connected to the mass and the base. When the external force is absent, the equations of motion of this system are

$$m\ddot{x} + k_1 x + k_2(x-e) = 0 \quad (1)$$
$$\dot{e} = (x-e)k_2/c \quad (2)$$

where, x is the displacement of the mass, e is the elongation of the dashpot element, and c is its damping. We define the effective damping rate as

$$\zeta = -\ln(a_{n+1}/a_n)/[T\sqrt{(k_1+k_2)/m}] \quad (3)$$

where, a_n is the amplitude of the nth cycle, and T is the duration of a cycle. Because the cycle duration is a function of c which will be varied in the subsequent section, we use ζ defined by Eq. (3) which represents the damping per unit time rather than the damping per cycle. When $k_2/k_1 = 1.0$ and 0.1 for examples, the effective damping rates can be numerically calculated as a function of β as plotted in Figs. 2 and 3 noted as "constant c", respectively, where

$$\beta \equiv k_2/c \quad (4)$$

Next, let us assume that the value of c is variable in the range

$$c_L \leq c \leq c_U \quad (5)$$

To investigate if higher damping rate is available by varying the damping c in this range, let us minimize c when

$$(x+\varphi_1\dot{x})(\dot{x}-\varphi_2 x) < 0 \quad (6)$$

holds, and keep c maximum during the rest of the cycle, where φ_1 and φ_2 are constants representing the timings of the reduction and recovery of c. After optimizing the values of φ_1 and φ_2, the damping rate is obtained as a function of β^* as plotted in Figs. 2 and 3 noted as "OT" (Optimal Timing) where

$$\beta^* \equiv k_2/c_U \quad (7)$$

and c_U/c_L is fixed to 100 as an example. Figure 2 shows the case where $k_2/k_1=1$, whereas $k_2/k_1=0.1$ in Fig. 3. These values of damping rate are obtained from the numerically simulated time history of free decay vibration after enough number of cycles.

The figures show that the damping rate largely depends on the value of k_2/k_1 in both "constant c" and "OT" cases. However, regardless of the value of k_2/k_1, much higher damping rate has been obtained by varying c with optimal timing than that of passive system. Roughly maximum damping rate is obtained when $\varphi_1=7.0$, $\varphi_2=1.8$ and $\beta^*=0.15$ in Fig. 2, and when $\varphi_1=4.0$, $\varphi_2=0.26$ and $\beta^*=0.10$ in Fig. 3. The optimal timings of the reduction and recovery (i.e., φ_1 and φ_2) are functions of β^* and c_U/c_L. But it may not be straightforward to obtain the functions explicitly. Furthermore, the extension of this control logic to MDOF systems with multiple variable-damping elements does not seem to be easy. Therefore, it is necessary to find more convenient approaches.

Figures 4 and 5 show time histories of the SDOF system whose k_2/k_1 is 0.1, where p is the load on the damper. Figure 4 is the case of semi-active control with the

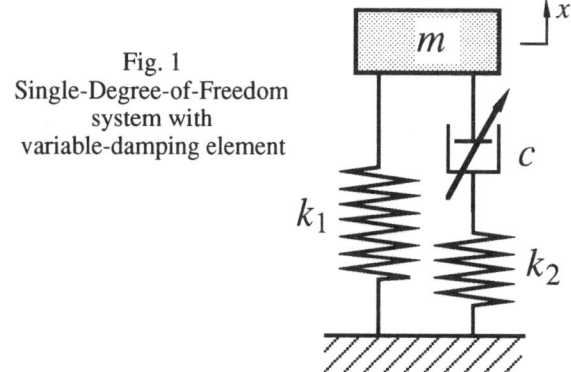

Fig. 1 Single-Degree-of-Freedom system with variable-damping element

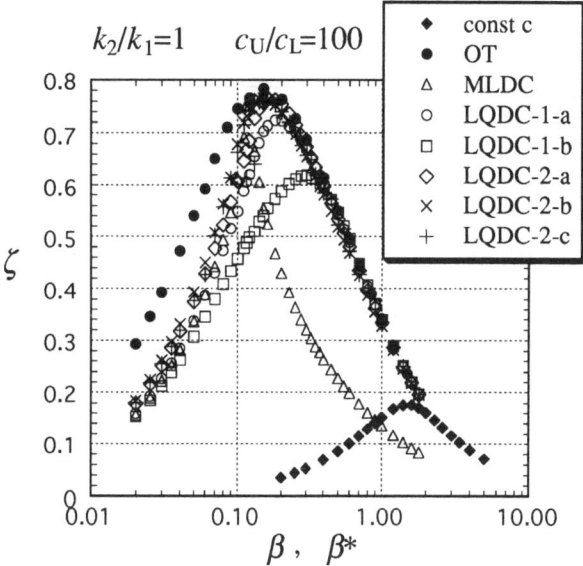

Fig. 2 Damping rates of SDOF system from various approaches ($k_2/k_1=1.0$)

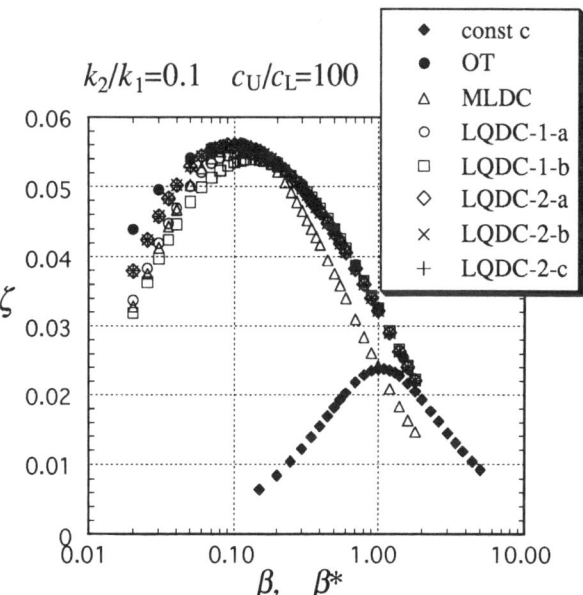

Fig. 3 Damping rates of SDOF system from various control laws ($k_2/k_1=0.1$)

optimal value of β^*, which give the maximum ζ in the "OT" plots in Fig. 3. Figure 5 shows the time histories of optimal passive system, whose value of β gives the maximum ζ in the "constant c" plots in Fig. 3. In Fig. 4, the values of e and p varies stepwise according to the control of β, whereas p varies sinusoidally in Fig. 5. Figure 6 shows the time history of $-p\dot{x}$ of the two cases shown in Figs. 4 and 5, which is the work done by the damper to the mass. As shown in the figure, the work done by the damper is never positive in the case of semi-actively controlled system, indicating that the damper always remove the energy form the mass. However, in the case of passive system, it is positive for short periods. Furthermore, because of the large amplitude of p, the negative work done by the controlled damper is larger than that of passive damper in the initial phase. These are the reason why the vibration damping can be enhanced by the control of β, i.e., c.

3. Equations of Motion of MDOF Systems with Variable-Damping Elements

To investigate the control logic which is applicable to MDOF systems with multiple variable-damping elements, let us first derive their equations of motion. As an example of MDOF system with multiple variable-damping elements, let us investigate a truss structure which has n_a variable-damping members whose axial characteristics can be modeled as Fig. 7. When the external force is absent and the structural damping is negligible, the equations of motion of this system are written in the physical coordinate as

$$M\ddot{x}+Kx+HDe = 0 \qquad (8)$$

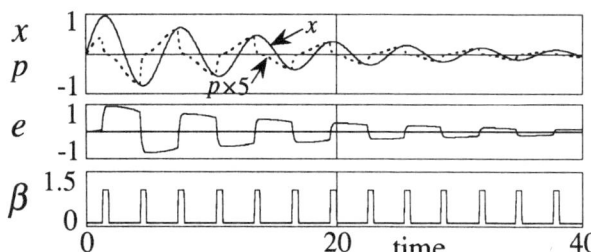

Fig. 4 An example of time history of the response of SDOF system controlled by OT control law (k_2/k_1=0.1, β^*=0.1)

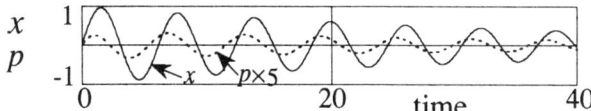

Fig. 5 An example of time history of the response of optimally tuned SDOF system (k_2/k_1=0.1, β=1.0)

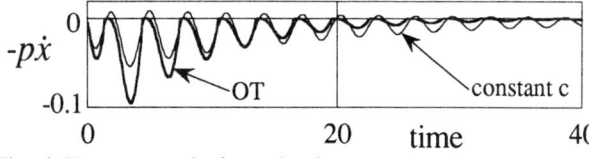

Fig. 6 Energy supply from the damper

$$\dot{e} = -CD(H^{-1}x+e) \qquad (9)$$

where

$$D = \text{diagonal}[k_{2i}] \qquad (10)$$
$$C = \text{diagonal}[1/c_i] \qquad (11)$$
$$e = \lfloor e_1, e_2,..., e_{n_a} \rfloor \qquad (12)$$

and x is the displacement vector, M is the mass matrix, K is the stiffness matrix which includes both k_{1i} and k_{2i} of all the variable-damping members, H is a matrix composed of directional cosines of variable-damping truss members, k_{1i} and k_{2i} are the spring constants shown in Fig. 4, c_i is the value of c, and e_i is the elongation of the dashpot element of the ith active member, respectively.

In the modal coordinate, the equations of motion are written as

$$\ddot{q}+\Omega q+\Phi^{-1}HDe = 0 \qquad (13)$$
$$\dot{e} = -CD(H^T\Phi q+e) \qquad (14)$$

where

$$\Omega = \text{diagonal}(\omega_1^2,..., \omega_m^2) \qquad (15)$$
$$\Phi = (\phi_1,..., \phi_m) \qquad (16)$$

and q is the displacement vector in the modal coordinates, ω_i and ϕ_i are the angular frequency and mode shape of the ith mode obtained from M and K matrices of Eq. (8) (i.e. those of the structure whose values of c are all infinitely large). In this paper, the value of c_i is assumed to be limited in the range

$$c_{Li} \leq c_i \leq c_{Ui} \qquad (17)$$

4. Control Strategies for Variable-Damping Structures

<u>Maximum Load Damping Control (MLDC)</u>
When the values of c_{Li} are small and those of c_{Ui} are large enough, the present system becomes similar to the type-II variable-stiffness system investigated in Ref. 3. Therefore, from the analogy with "control logic C" of the reference, the following on-off approach seems to be effective;
- minimize c_i when $|p_i|$ is at a maximum
- maximize c_i when p_i becomes zero or when $|p_i|$ starts to increase before reaching zero.

where p_i is the load transferred through the ith variable damper. The time history of Fig. 4 qualitatively supports this control law. This control law is simple, and can be easily applied to the MDOF systems with multiple variable-damping elements. In addition, it is suitable for distributed control because only the locally available information is required for the control of each variable damper. In this

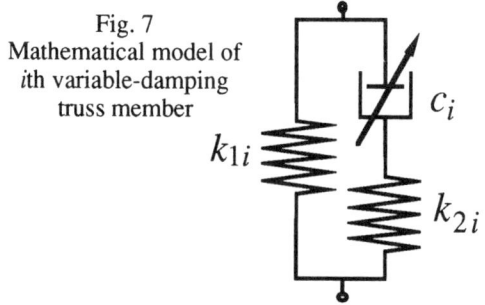

Fig. 7 Mathematical model of ith variable-damping truss member

paper, this control law is referred to as Maximum Load Damping Control (MLDC).

Linear Quadratic Damping Control - 1 (LQDC-1)

Equation (13) can be rewritten as
$$\dot{z}_1 = A_1 z_1 + B_1 e \quad (18)$$
where
$$z_1 = \lfloor q^T, \dot{q}^T \rfloor^T \quad (19)$$
$$A_1 = \begin{bmatrix} 0 & I \\ -\Omega & 0 \end{bmatrix} \quad (20)$$
$$B_1 = \begin{bmatrix} 0 \\ -\Phi^T H D \end{bmatrix} \quad (21)$$

and I is a unit matrix. Equation (18) shows that we could apply the well-developed linear control theories if we could directly control e of this system. As the Linear Quadratic (LQ) control theory, it is known[12] that the optimal linear control which minimizes

$$J \equiv \int_0^\infty \left(z_1^T R_1 z_1 + e^T R_2 e \right) dt \quad (22)$$

is to control e as
$$e = e_T \equiv -F_1 z_1 \quad (23)$$
where
$$F_1 = R_2^{-1} B_1^T P_1 \quad (24)$$
and P_1 is the positive-definite solution of the following Riccati equation;
$$P_1 B_1^T R_2^{-1} B_1 P_1 - A_1^T P_1 - P_1 A_1 - R_1 = 0 \quad (25)$$

In the present system, we cannot directly control e. However, the above investigation suggests that a promising approach for vibration suppression of the present system is to control c_i such that e traces e_T defined by Eq. (23) as exactly as possible. Since lower value of c_i promotes the variation of e_i as shown in Eqs. (11) and (14), a possible strategy is to reduce c_i as long as e_i is varying toward e_{Ti} in order to promote the variation. And otherwise, it is to maximize c_i in order to keep e_i close to e_{Ti}, where e_{Ti} is the ith element of e_T. This strategy can be implemented by, e.g., the following control law.

$$c_i = c_{Li} \quad \text{when } \dot{e}_i(e_{Ti} - e_i) > 0$$
$$c_i = c_{Ui} \quad \text{when } \dot{e}_i(e_{Ti} - e_i) < 0 \quad (26)$$

In this paper, this control law is called as Linear Quadratic Damping Control -1-a (LQDC-1-a).

In the LQ control, a large value of e results in a large penalty as shown by the term R_1 in Eq. (22). However, in the present system, large value of e does not require large energy, and it is not necessary to penalize it. Therefore, better performance may be obtained by controlling c_i such that the absolute value of e_i becomes maximum when the sign of e_i is the same as that of e_{Ti}, and otherwise, such that it becomes minimum. This strategy can be implemented by the following control law.

$$c_i = c_{Li} \quad \text{when } \dot{e}_i e_{Ti} > 0$$
$$c_i = c_{Ui} \quad \text{when } \dot{e}_i e_{Ti} < 0 \quad (27)$$

We can see that this control law promotes the variation of e_i when e_i is moving towards the polarity of e_{Ti}, and otherwise depress its motion. In this paper, this control law is called as LQDC-1-b.

Linear Quadratic Damping Control - 2 (LQDC-2)

Equations (13) and (14) can be combined and rewritten as
$$\dot{z}_2 = A_2 z_2 + B_2 v \quad (28)$$
where,
$$z_2 = \lfloor q^T, \dot{q}^T, e^T \rfloor^T \quad (29)$$
$$A_2 = \begin{bmatrix} 0 & I & 0 \\ -\Omega & 0 & \Phi^T H D \\ 0 & 0 & 0 \end{bmatrix} \quad (30)$$
$$B_2 = \begin{bmatrix} 0 \\ 0 \\ I \end{bmatrix} \quad (31)$$
$$v = -CD(H^T \Phi q + e) \quad (32)$$

If we could control the value of v without any limitation, the LQ control logic tells us that v should be controlled such that
$$v = v_T \equiv -F_2 z_2 \quad (33)$$
where F_2 is the gain matrix given by the LQ control theory similarly as Eqs. (24) and (25).

In the present case, as shown by Eqs. (11) and (32), the ith element of v, v_i, is inversely proportional to c_i which can be directly controlled. However, the value of c_i is limited in the range Eq. (17). Therefore, similarly as LQDC-1, a possible strategy is to control the value of c_i such that the value of v_i becomes as close to v_{Ti} as possible, where v_{Ti} is the ith element of vector v. This strategy can be implemented by

$$c_i = p_i/v_{Li} \quad \text{when } c_{Li} < p_i/v_{Ti} < c_{Ui}$$
$$c_i = c_{Li} \quad \text{when } p_i/v_{Ti} < c_{Li} \quad (34)$$
$$c_i = c_{Ui} \quad \text{when } c_{Ui} < p_i/v_{Ti}$$

where p_i is the ith element of p defined as
$$p = -D(H^T \Phi q + e) \quad (35)$$
In this paper, this control law is called as Linear Quadratic Damping Control -2-c (LQDC-2-c)

To apply LQDC-2-c, the damping c of each variable damping members has to be continuously variable. Such a continuously-variable-damping member may be heavier, more complex and more expensive than an on-off type variable-damping member whose c_i is either c_{Li} or c_{Ui}. A control strategy for the systems with such on-off type variable-damping members is to select the value of c_i which gives closer value of v_i to v_{Ti} than the other value does. This strategy can be implemented by

$$c_i = c_{Ui} \quad \text{when } (v_{Ti} - v_{Ai})(v_{Ui} - v_{Ai}) > 0$$
$$c_i = c_{Li} \quad \text{when } (v_{Ti} - v_{Ai})(v_{Ui} - v_{Ai}) < 0 \quad (36)$$
where,
$$v_{Li} = p_i/c_{Li}$$
$$v_{Ui} = p_i/c_{Ui} \quad (37)$$

and v_{Ai} is an average of v_{Ui} and v_{Li}. This control law is called as LQDC-2-a in this paper.

Similarly as LQDC-1, large values of v_i does not cost in the present system. Therefore, another possible approach is to make the absolute value of v_i maximum when the polarities of v_i and v_{Ti} are identical with each other, and otherwise, to make the absolute value of v_i minimum. This strategy can be implemented by the following control law;

$c_i = c_{Li}$ when $v_i v_{Ti} > 0$

$c_i = c_{Ui}$ when $v_i v_{Ti} < 0$ (38)

This control law is called as LQDC-2-b in this paper.

5 Numerical Examples of SDOF System

MLDC, LQDC-1 and LQDC-2 approaches are applied to the SDOF system shown in Fig. 1. The value of c_U/c_L is assumed to be 100. In LQDC-1 and LQDC-2, all the elements of R_1 are set to be zero except for $R_1(1,1)$ and $R_1(2,2)$, where $R_1(i,j)$ is the ith row jth column element of matrix R_1. $R_1(1,1)$ is set as $(k_1+k_2)/m$, and the values of $R_1(2,2)$ and R_2 are roughly optimized by trial and error so that the value of ζ becomes maximum. These values are shown in Table 1. In LQDC-2-c, the geometrical mean is used for v_{Ai} of Eq. (36) because it gives better results in this example than the arithmetic mean. Figures 2 and 3 show the resulting damping rates from these approaches as functions of β^*, which are obtained by numerical simulations.

The figures show that these approaches result in much higher damping rate than that of optimally tuned "constant" c, demonstrating their effectiveness. When $k_2/k_1=1$, the performance of MLDC is excellent only when the value of β^* is optimal as shown in Fig. 2. The performances of LQDC-2 approaches are also excellent. Furthermore, they are relatively insensitive to the variation of β^*. At the optimal value of β^*, the performances of LQDC-2 approaches are almost the same as that of OT, which is the maximum damping rate achieved by an on-off variation of c per cycle. The performances of LQDC-1 are also relatively insensitive to the variation of β^*, although the damping rates are slightly less than those of LQDC-2 approaches.

When $k_2/k_1=0.1$ also, the damping rates are substantially increased by the control of c as shown in Fig. 3, although the damping rates are lower than those of Fig. 2 in all the cases. In this case, the difference in the performances of the approaches are relatively small, although the damping rates from LQDC-1 approaches are slightly lower than those from the others. Unlike the previous case, the performance of MLDC is relatively insensitive to the variation of β^* in this case.

6 Numerical Examples of MDOF System

To investigate the effectiveness of the above-mentioned vibration suppression strategies in MDOF systems with multiple variable-damping elements, vibration suppression of the truss shown in Fig. 8 is studied. The truss has three variable-damping members, which are modeled as Fig. 7. The axial stiffness (EA) of each passive member is unity. The mass per unit length of each member is unity including the variable-damping members. The length of the members (l) is also unity except for the diagonal members. In this examples, k_{1i} and k_{2i} (shown in Fig. 7) are assumed to be

$k_{1i} = k_{2i} = EA/(2l_i)$ $(i = 1, 2, 3)$ (39)

where l_i is the length of the ith variable-damping member. Further more, it is also assumed that

$c_{Ui}/c_{Li} = 100$ $(i = 1, 2, 3)$ (40)

and the values of k_{2i}/c_{Li} of variable-damping members are identical with each other as

$k_{2i}/c_{Li} = \beta^*$ $(i = 1, 2, 3)$ (41)

A unit impulsive load is applied as shown in Fig. 8, and the subsequent vibration suppression by each approach is numerically simulated. In the simulation, all the 60 modes are included in the mathematical model without any modal truncation. To measure the performance of each approaches, the value of

$$I_1 = \int_0^{16\pi/\omega_1} \delta_{rms} dt \quad (42)$$

is calculated, where δ_{rms} is the root-mean-square (rms) displacement of the truss nodes. In LQDC-1 approaches, matrix R_1 is set as

$R_1 = \text{diagon}(1, 1, ..., 1, r_1/\omega_1^2, r_1/\omega_2^2, ..., r_1/\omega_{n_c}^2)$ (43)

and it is set as

$R_1 = \text{diagon}(1, 1, ..., 1, r_1/\omega_1^2, r_1/\omega_2^2, ..., r_1/\omega_{n_c}^2, 0)$ (44)

for LQDC-2 approaches, where r_1 is a constant, and n_c is the number of controlled modes. In both LQDC-1 and 2 approaches, the lowest 5 modes are controlled (i.e., $n_c = 5$), and R_2 is set as

$R_2 = r_2 I$ (45)

for both LQDC-1 and 2 approaches, where r_2 is a constant. The values of r_1 and r_2 are roughly optimized by trial and error so that the value of I_1 becomes minimum. The roughly optimal values, which are used for the following simulation examples, are listed in table 2.

The values of I_1 obtained by numerical simulations by using the various control laws are plotted in Fig. 9 as

Table 1 Roughly optimal parameter values used for numerical investigation of SDOF system

	$k_2/k_1 = 1$		$k_2/k_1 = 0.1$	
	$R_1(2,2)$	$R_2/2$	$R_1(2,2)$	$R_2/2$
LQDC-1-a	5×10^2	$2k_1/m$	1×10^3	$5k_1/m$
LQDC-1-b	1×10^3	$1 \times 10^3 k_1/m$	1	$1 \times 10^6 k_1/m$
LQDC-2-a	1×10^3	1×10^{-2}	1×10^5	2×10^{-3}
LQDC-2-b	1×10^3	2×10^{-2}	1×10^5	2×10^{-3}
LQDC-2-c	5×10^2	1×10^{-2}	1×10^5	2×10^{-3}

functions of β^*. The values of I_1 of passive system, whose c of each dashpot element are constant, are also plotted as a function of β, where the values of k_{2i}/c_i of the damping members are assumed to be identical with each other as

$$\beta_i \equiv k_{2i}/c_i = \beta \quad (i = 1, 2, 3) \tag{46}$$

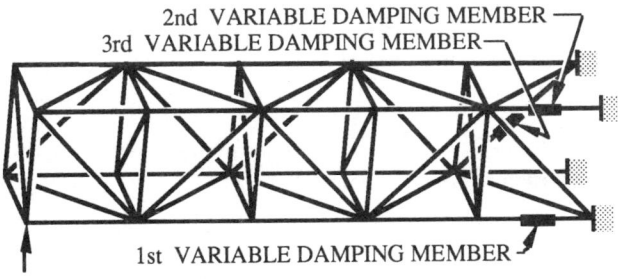

Fig. 8 Truss structure with variable-damping truss members

Table 2 Roughly optimal parameter values used for numerical investigation of MDOF system

	r_1	r_2
LQDC-1-a	100	10
LQDC-1-b	0.1	1×10^4
LQDC-2-a	0.1	100
LQDC-2-b	0.5	5
LQDC-2-c	1	100

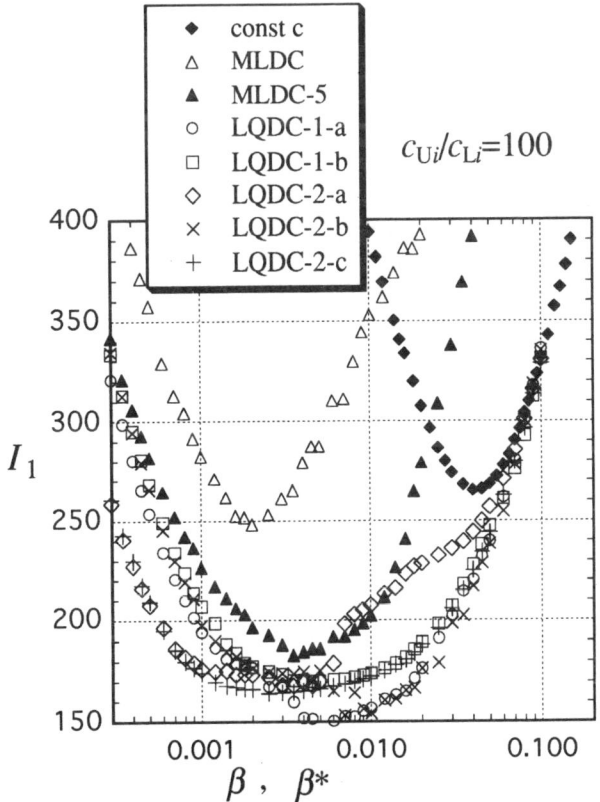

Fig. 9 The values of I_1 resulting from various approaches applied to a truss structure

Figure 9 shows that all the approaches are effective in comparison with the passive (i.e., "constant c") system except for MLDC. Unlike the previous case of SDOF system, MLDC is not very effective. This is because of too frequent variation of c resulting from the high frequency component in the load signal, which MLDC control law monitors. The values of I_1 noted as MLDC-5 in the figure are also obtained from a MLDC control law. However, in this case, the value of c is controlled based on the load caused by the lowest 5 modes instead of the actual load. By using the lower modes component only, the excessive variation of c has been remedied, and the damping performance has been increased. However, this approach require the modal data which is not necessary for the original MLDC approach.

When the value of β^* is optimal, Fig. 9 shows that LQDC-1-a is the best, and LQDC-2-b is the second in their performance. Generally speaking, the sensitivities of the performance of the semi-active approaches to the relative variation of β^* are less than that of the passive system. Especially, the performance of LQDC-2-c is excellent in most wide range of β^*. The best values of I_1 of each control law in Fig. 9 are listed in Table 3 in comparison with the value of I_1 of the structure whose modal damping is 0.5%. The table indicates that the value of I_1 has been substantially reduced by the vibration suppression approaches, especially by the semi-active approaches other than MLDC.

Figure 10 shows an example of time history of the response of the truss with LQDC-1-a approach with optimal value of β^*. The figure shows that the vibration, especially the lower mode vibration, is suppressed nicely. We can see that the values of e's trace their target values e_T's (which are shown by broken lines) when it is possible. However, e_i can be increased or decreased only when the load on the damper is tensile or compressive, respectively. The values of β_1 and β_3 are increased (i.e., c_1 and c_3 are reduced) mainly at the peaks of $|q_3|$ and $|q_1|$, respectively, similarly as Fig. 4. However, β_2 looks to respond various modes. The time history of δ_{rms} suggests that the higher mode vibrations, which are not shown in the figure, are not so large as to significantly contribute to δ_{rms} at $t\omega_1=50$.

Table 3 Minimum Values of I_1 and I_2 from various approaches

	I_1	I_2	I_1/I_2	$\sqrt{I_1}/I_2$
0.5% damping	645.6	64348	0.0100	3.95E-4
constant c	265.1	36916	0.00718	4.41E-4
MLDC	248.1	32361	0.00766	4.87E-4
MLDC-5	182.9	27716	0.00660	4.88E-4
LQDC-1-a	149.8	26209	0.00572	4.67E-4
LQDC-1-b	170.3	27503	0.00619	4.74E-4
LQDC-2-a	167.9	27503	0.00610	4.71E-4
LQDC-2-b	153.2	27947	0.00548	4.43E-4
LQDC-2-c	164.1	27368	0.00600	4.68E-4

Next, random force is applied to the truss structure, and the above-mentioned semi-active and passive approaches are applied to reduce the response of the truss. The direction and the attach point of the random force is the same as those of the impulsive force shown in Fig. 8. As shown in Fig. 11 denoted as f, a random force time history is generated and applied to the simulation, whose rms value is unity. The power spectral density (PSD) per unit angular frequency of the random force is flat in the range from $0.7\omega_1$ to $4.1\omega_1$, and zero in the other range of frequency. This frequency region of non-zero PSD covers ω_1 to ω_4. The parameter values shown in Table 2, which are roughly optimal for the impulsive excitation, are used for this numerical simulations. To measure the performances of the approaches, the value of

$$I_2 = \int_{8\pi/\omega_1}^{16\pi/\omega_1} \delta_{rms} dt \tag{47}$$

is calculated. The value of I_2 obtained from various control laws are plotted in Fig. 12 as functions of β^*. The value of I_2 of passive system is also plotted as a function of β. In Fig. 12, the performance of LQDC-1-a is again the best. The difference between the minimum values from MLDC-5, LQDC-1-b, and LQDC-2 control laws are smaller than that of Fig. 9. The value of β^* which gives the best performance of MLDC is slightly lower than that of Fig. 9. However, we can see that Fig. 12 is similar to Fig. 9. This fact suggests that the semi-active control investigated here is effective for the random excitation similarly as the impulsive excitation.

Figure 11 shows an example of time history of the truss structure which is controlled by LQDC-2-c. The value of v's traces their target values v_T's only when it is possible. As shown in the figure, the values of β are varied continuously unlike Fig. 10. Similarly as Fig. 10, β_1 and β_3 are increased mainly at the peaks of $|q_3|$ and $|q_1|$, respectively.

The best values of I_2 of each control law in Fig. 12 are

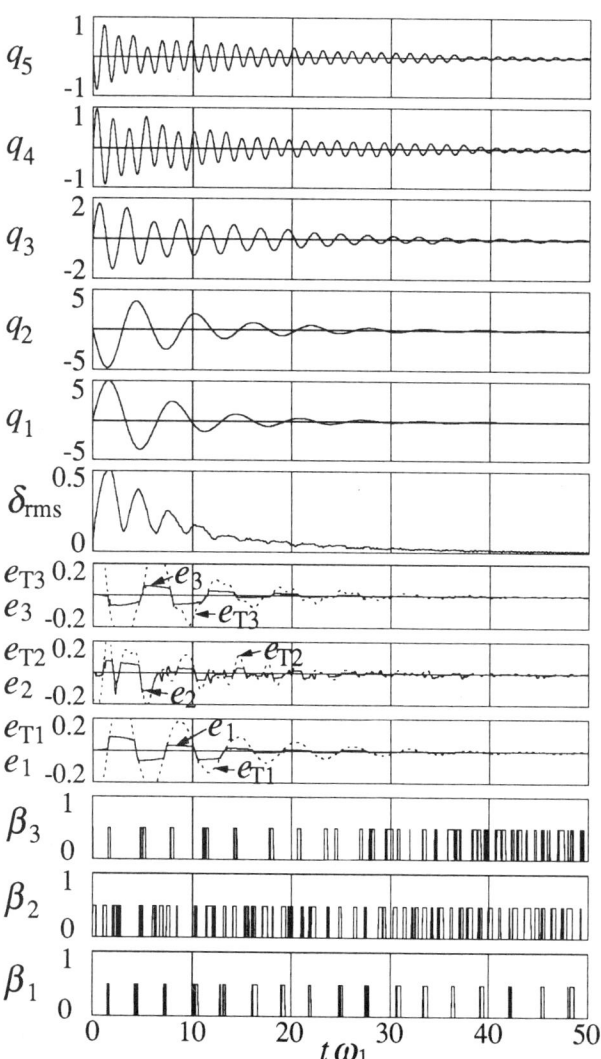

Fig. 10 An example of time history of vibration suppression by LQDC-1-a control law. (Impulsive excitation)

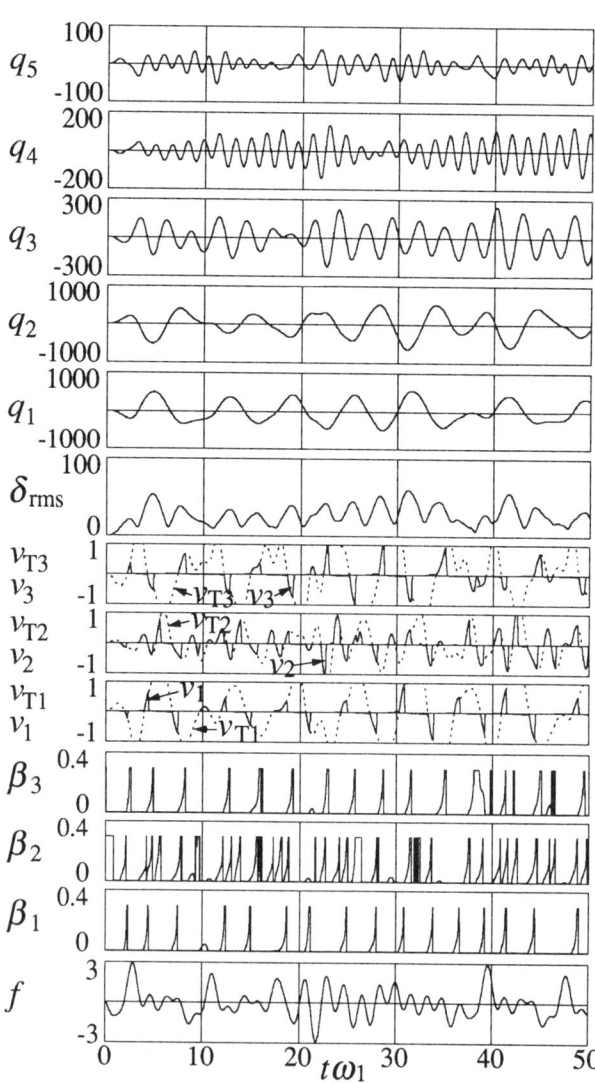

Fig. 11 An example of time history of vibration suppression by LQDC-2-c control law. (Random excitation)

listed in Table 3. In the case of random force excitation also, the value of I_2 has been substantially reduced by the passive and semi-active approaches compared with the case of 0.5% modal damping. We can see that the control law which gives smaller value of I_1 for impulsive excitation gives smaller value of I_2 for the random excitation also, demonstrating its effectiveness for the random excitation. It is reasonable that the value of I_2 obtained from each control law is roughly proportional to $\sqrt{I_1}$ rather than I_1.

7. Concluding Remarks

A semi-active vibration suppression has been proposed and studied. The approach is to control the damping of dampers installed in structures. Several control laws are proposed, which are applicable to MDOF structures with multiple variable-dampers. To study the performance of these control laws, numerical simulations have been performed by using a SDOF example and a MDOF truss example. The numerical simulation results have demonstrated that all of them results in substantially higher vibration damping rate than that of the optimally tuned passive system in the case of SDOF system. In the case of MDOF truss with multiple variable-damper also, the results have demonstrated that all these control laws except for one reduce the response of structure to the impulsive and random excitation to substantially lower level than that of optimally tuned passive system, demonstrating their effectiveness.

With semi-active vibration suppression, the energy is dissipated by passive mechanisms. Therefore, unlike with the active vibration suppression, the system is always stable even when the control system is improperly designed. Furthermore, the performance of semi-active vibration suppression is much higher than passive vibration suppression. Therefore, semi-active vibration suppression is an attractive approach for space structures which will be deployed or constructed in the orbit, whose control system design is difficult because of the lack of exact information about its dynamic characteristics. However, to apply this approach to actual space structure, much more study is required.

References

1. Balas,M.J., Trends in Large Space Structure Control Theory: Fondest Hopes, Wildest Dreams, *IEEE Transactions on Automatic Control*, Vol. AC-27, No. 3, 1982, pp.522-535.
2. Balas,M.J., Active Control of Flexible Systems, *Journal of Optimization Theory and Applications*, Vol.25, No.3, July 1978, pp.415-436.
3. Onoda,J., Endo,T., Tamaoki,H., Watanabe,N., Vibration Suppression by Variable-Stiffness Members, *AIAA Journal*, Vol. 29, No. 6, June 1991, pp.977-983.
4. Onoda,J., Minesugi,K., Semiactive Vibration Suppression of Truss Structures by Coulomb Friction, to appear in *Journal of Spacecraft and Rockets*
5. Onoda,J., Sano,T., Kamiyama,K., Active, Passive and Semi-Active Vibration Suppression by Stiffness Variation, *AIAA Journal*, Vol.30, No.12, 1992, pp.2922-2929.
6. Karnopp,D., Crosby,M.J., Harwood,R.A., Vibration Control Using Semi-Active Force Generator, *Trans. ASME, Journal of Engineering for Industry*, Vol.96, May 1974, pp.619-626.
7. Rakheja,S., Sankar,S., Vibration and Shock Isolation Performance of a Semi-Active "On-Off" Damper, *Trans. ASME, Journal of Vibration, Acoustics, Stress, and Reliability in Design*, Vol.107, 1985, pp.398-403.
8. Hrovat,D., Margolis,D.L., Hubbard,M., An Approach Toward the Optimal Semi-Active Suspension, *Trans. ASME, Journal of Dynamic Systems, Measurement, and Control*, Vol.110, Sept. 1988, pp.288-296.
9. Kobori,T., Takahashi,M., Niwa,N., Kurata,N., Research on Active Seismic Response Control System with Variable Structure Characteristics - Feedback Control with Variable Stiffness and Damping Mechanism - Vol. 37B, March 1991, pp.193--202. (in Japanese)
10. Kobori,T., Takahashi,M., Nasu,T., Niwa,N., Ogasawara,K., Kurata,N., Mizuno,T., Research on Active Seismic Response Control System with Variable Structure Characteristics - Basic Property of Variable Stiffness and Damping Mechanism and Fundamental Experiment by Shaking Table - Vol. 37B, March 1991, pp.193--202. (in Japanese)
11. Davis,L.P., Workman,B.J., Chu,C.C., Anderson,E.H., Design of a D-Strut and Its Application Results in the JPL, MIT and LARC Test Bed, AIAA-92-2274-CP, 33rd AIAA/ASME/ASCE/AHS/ASC Structures, Structural Dynamics and Materials Conference, Dallas, TX, USA, April 1992.
12. Kwakernaak,H. and Sivan,R., *Linear Optimal Control System*, Wiley-Interscience, New York, 1972.

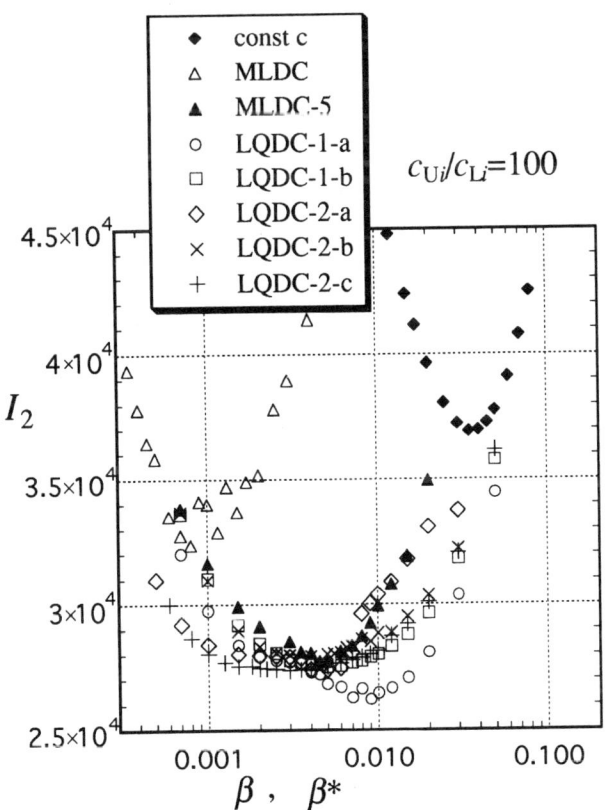

Fig. 12 The value of I_2 resulting from various approaches applied to a truss structure

ON-ORBIT SHAPE CORRECTION OF INFLATABLE STRUCTURES*

M. Salama, C.P. Kuo, J. Garba, B. Wada

Jet Propulsion Laboratory
California Institute of Technology
Pasadena, CA. 91109

M. Thomas

L'Garde, Inc.
Tustin, CA. 92680

ABSTRACT

Piezo induced deformations are proposed as a means of achieving a higher degree of on-orbit surface accuracy of inflatable antennas. A series of tests were performed on an inflated circular membrane and an inflated tube, both made of off-the-shelf piezo film. The same configurations were also analyzed to validate the tests. The concept was demonstrated qualitatively in spite of a discrepancy between the analysis and test results, most likely due to inaccurate knowledge of the piezoelectric coefficients of the film.

Numerical simulations are also reported herein, which demonstrate the piezocontrol of an antenna design concept. A parametric study is carried out to determine the most effective distribution of actuators and their optimal gains required to correct given aberrations in the surface of the antenna. The proposed approach is generally suitable for making small local adjustments in the shape of the antenna.

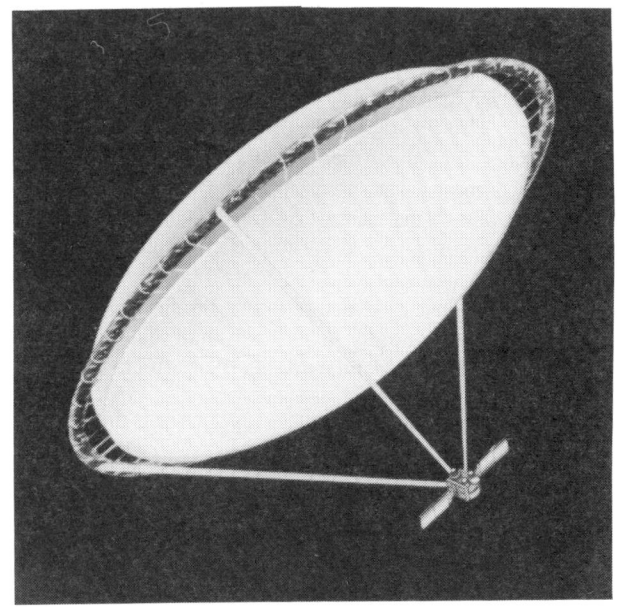

Figure 1: An Inflatable Antenna Concept

I. Introduction:

By comparison to mechanically erectable systems, inflatable structures have been shown [1] to have the advantage of a much lower cost, weight, and packaging volume, as well as more favorable thermal gradients and damping characteristics. The use of inflatable concepts for emergency slides in aircrafts or airbags in cars is an evidence of their deployment reliability. In space applications, an airbag system is currently under consideration as a means of attenuating the landing impact of NASA's MESUR pathfinder spacecraft on the surface of Mars. The use of such technology in the construction of large, lightweight, yet precisely shaped antennas and reflectors can make many important space missions practical.

The precision inflatable antenna in Fig. (1) is an example of a large lightweight antenna concept. Depending on the application, inflatable antennas or reflectors can be made to work at frequencies as high as 3 GHz (radiometry), or 100 GHz (communications). Reference [2] shows that during ground tests, surface deviations from the desired shape of such reflectors are routinely about 1 mm rms. To go to higher frequencies will require a step increase in the surface accuracy of inflatable reflectors. Once attained, the required level

---

* The research described in this paper was carried out in part by the Jet Propulsion Laboratory, California Institute of Technology under contract with the National Aeronautics and Space Adminstration.

Copyright © 1993 by the American Institute of Aeronautics and Astronautics, Inc. No copyright is asserted in the United States under Title 17, U.S. Code. The U.S. Government has a royalty-free license to exercise all rights under the copyright claimed herein for Governmental purposes. All other rights are reserved by the copyright owner.

of accuracy may be maintained by making periodic on-orbit shape adjustments as the antenna is exposed to thermal changes and possible aging of the polymeric film.

The on-orbit adjustment of an inflatable space antenna can be made by several techniques. Varying the inflation pressure can enable substantial adjustment of the focal length by making global shape changes [2]. Unfortunately, this will do little to correct unsymmetric distortions. The use of electrostatic charge to change the shape of a deployed reflecting membrane has also been examined [3]. The idea was to pull on sections of the deployed membrane by applying a static electric charge to certain regions of an adjacent parallel membrane. The technique was partially successful and was not sufficiently lightweight.

The concept explored in the present paper offers a potentially lightweight and simple technique to correct local aberrations in the shape of an inflated membrane by shrinking or stretching sections of the membrane or support structure. This can be done by employing piezoelectric elements in the construction of the membrane or support structure. The piezoelectric element may be a thin polymeric film, much similar to that normally used for the inflatable membrane itself. A positive or negative electric field applied to the piezofilm creates a mechanical strain proportional to that field. This provides the ability to induce controlled on-orbit dimensional changes in gores of the antenna itself or its attachments in order to optimize the reflector performance. The technique is suitable, not for making large focal length changes, but for making small symmetric or unsymmetric local adjustments. As such it could complement the variable pressure approach of Ref. [2]. Some theoretical aspects of the present technique has been examined in an earlier paper [4]. Its feasibility is further explored in the present paper by a combination of laboratory experiments and numerical simulations using simple configurations, as well as numerical simulations and trade studies using the more realistic design concept of Fig. (1).

II. Actuated Inflatables:

The ability to make small shape corrections in an inflated structure is considered here by integrating the actuation mechanism in the inflated membrane itself or in its support structure. Since in most cases the support structure is likely to be also an inflatable membrane, this section will deal with distributed piezoelectric actuation of membranes.

1. Distributed Piezoelectric Actuation: The phenomenon of piezoelectricity couples the electrical and mechanical properties of certain materials. Polymers, for example, are comprised of many randomly oriented positive and negative dipoles. Their piezo property can be induced through a one-time application of a sufficiently high voltage at high temperature, which results in permanent polarization of the dipoles. The process, called "poling", locks in the piezoelectric effect anisotropically, along and normal to the poling direction. Once poled, the strains will depend not only on the applied stress, but also on the applied electric field and dielectric displacement. The coupled electromechanical constitutive relations then take the general form [5]:

$$D_k = d_{k\ell m}^T \sigma_{\ell m} + \eta_{kn}^{\sigma}{}^T E_n$$
$$\epsilon_{ij} = s_{ij\ell m}^E{}^T \sigma_{\ell m} + d_{ijn}^T E_n \quad (1)$$

The first equation describes the "direct" piezoelectric effect and states that the dielectric displacement vector D depends upon the stress tensor σ and electric field E through the piezoelectric coefficient matrix d and the permittivities η, respectively. On the other hand, the second equation describes the "converse" piezoelectric effect and states that in presence of an electric field E, the strain tensor ϵ depends upon the applied stress tensor σ as well as the electric field through the elastic compliance s and the piezoelectric coefficients d, respectively. In both equations, the superscript T indicates transpose. Other superscripts indicate that the coefficient in question is evaluated at a constant value of the superscript.

In polymeric films, piezoelectricity is introduced by poling in the "3"-direction. This is the direction of the normal to the plane of the film. By constructing an inflatable component from this piezoelectric film, a subsequent application of voltage to the film will induce a distributed in-plane actuation strain field in addition to the existing strains due to inflation pressure. From the second of Eq. (1), the components of strain at a point in the film are then obtained from:

$$\epsilon_{11} = s_{11}\sigma_{11} + s_{12}\sigma_{22} + d_{31}E_3$$
$$\epsilon_{22} = s_{22}\sigma_{22} + s_{12}\sigma_{11} + d_{32}E_3$$
$$\epsilon_{12} = s_{44}\sigma_{12} + d_{14}E_2 \quad (2)$$

Thus, the in-plane piezoelectric constants d_{31}, d_{32} are exploited in inflated membranes as a means of inducing a desired state of in-plane membrane strains, sufficient to correct a prescribed amount of out-of-plane deformations. The d_{31}, d_{32} constants describe the strains in the in-plane directions "1" and "2", respectively, when an electrical field is applied in the thickness direction "3". In the presnt application, $E_2 = 0$.

Piezoelectric films are now commercially available off-the-shelf. They are typically employed as sensors in devices for a wide variety of applications, e.g. pressure sensors. Their use in distributed sensing and control has been investigated only recently [6]. The piezo film used here is a polyvinylidene fluoride semicrystalline resin (a.k.a. PVDF), known by the trade name Kynar. The film comes in various thicknesses, with a very thin metal layer deposited on each side to provide electrical connection.

2. Experiments: The purpose of the laboratory experiments that follow is to explore the advantages and limitations of these films as inflatables with built-in distributed actuation ability, suitable for correcting shape aberrations. The test configurations are kept simple to minimize costs and test variables. Thus instead of the relatively complex inflatable structure of Fig. (1), the actuation ability of piezoelectric films was explored experimentally on the inflated circular membrane and closed tube configurations in Fig. (2).

Figure (2a) shows two circular membranes, initially flat and rigidly clamped along their 10. *in* diameter boundary. Inflation pressure is introduced in the air space between the two membranes. Because of the rigid boundary conditions, each membrane deforms under pressure independent from the other. As a check on the test/analysis results (to be discussed later), one membrane was chosen to be 0.001 *in* mylar (isotropic), and the other a 0.002 *in* Kynar membrane (anisotropic), comprised of approximately 0.0015 *in* basic PVDF film, on which two layers (.0005 *in* total thickness) of silver ink metallization are deposited (one layer on each side for electric contact). As the inflation pressure is changed, the deformed shape of each membrane relative to the flat undeformed configuration is monitored at the center and four other points on each membrane. Up to 0.5 *psi* internal pressure was applied, and the corresponding lateral deformations (up to 0.5 *in*) were measured using dial gage type sensors. When up to 300 volts were applied to the pressurized Kynar membrane, additional lateral membrane deformations of the order of 0.001 *in* were recorded using eddy current sensors.

The tube in Fig. (2b) is made of the same 0.002 *in* Kynar membrane with the "1"-axis parallel to the tube axis (i.e. d_{31} is along the longitudinal direction). Separate strips of conducting piezo are created by chem-

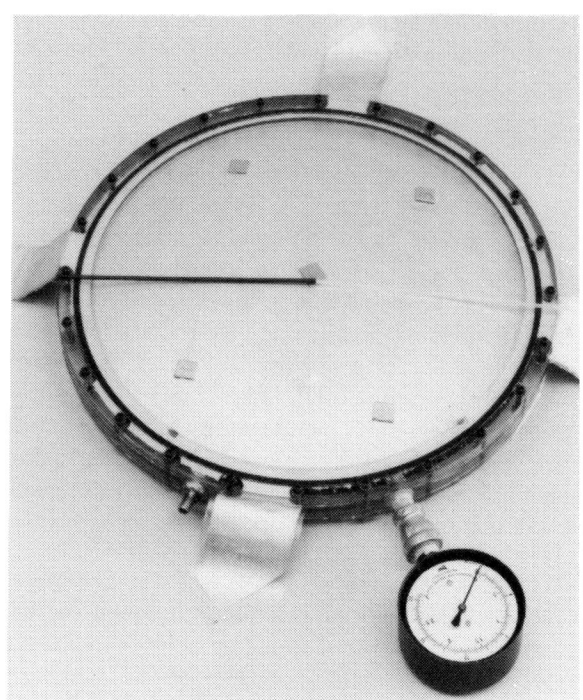

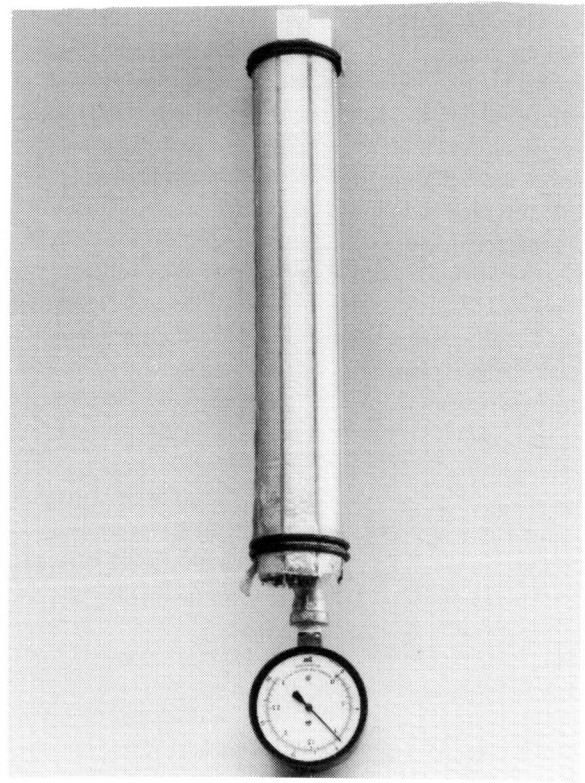

Figure 2: Piezoelectric Actuated Test Components (a) Circular Membrane, (b) Tube

ically etching thin strips of the metallization on both sides of the film. The tube is closed at each end with a 1/2 in plastic disk. During the experiment, the bottom end of the tube is rigidly fixed and the deformations at the top free end are monitored in three directions as the inflation pressure (up to 1.0 psi) is applied, separately or in addition to electric fields applied individually to the metallized strips.

III. Analysis/Test Results:

Analysis was also performed to corroborate the results of tests described above, and to examine the applicability of the concept to more practical configurations. Consistent with the fact that the stiffness of inflatable structures is largely dependent upon the internal pressure, the analysis method assumes large elastic deformations with small membrane strains. For this purpose, the nonlinear finite element analysis capability in NASTRAN was found to be sufficiently accurate.

1. Inflation Pressure Only: In Fig. (3), the test center deformation on the mylar side of the circular membrane (Fig. 2a) is compared to two independent analysis methods; NASTRAN's nonlinear analysis, and the series solution of Ref. [7]. Both methods of analysis give almost identical results, which tend to be approximately 10% stiffer than the test results. Inaccuracy in the mylar's elastic modulus value used in both analyses (0.5375×10^6 psi), is the most likely reason for the discrepancy.

2. Piezoelectric Actuation: In the piezo actuated tests, the Kynar membrane in Fig. (2a) is first inflated then subjected to an electric field through the leads. The applied voltage was varied from zero to 300 volts, and from zero to -300 volts. A series of such tests was performed at different inflation pressures ($p_i = 0.1, 0.2, 0.3, 0.4, and\ 0.5$ psi). The deformed membrane surface Δ_{pi} at pressure p_i is used as a reference, from which additional deformations δ^v_{pi} due to the applied voltage are measured. Figures (4a, 4b) together show typical results for inflation pressure $p_i = 0.2$ psi, in the form of applied voltage and corresponding deformations at the center and four other locations indicated in Fig. (2a). As seen from Fig. (4), the hysteresis is elastic and recoverable. Here, positive voltage induces negative strain (shrinkage), and negative voltage induces positive strain (stretching).

As a measure of the rate of deformation per volt for the present test configuration, the average slope of the hysteretic curve for the center deformation was computed. For inflation pressures ranging from 0.1 psi

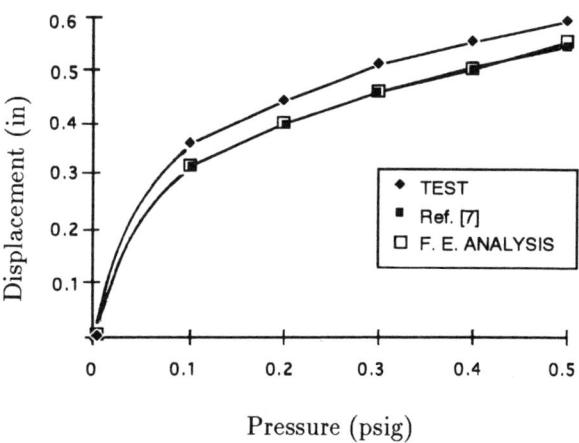

Figure 3: Analysis/Test Comparison of Inflated Circular Membrane

to 0.5 psi, the average rate was found to range from 4.1×10^{-6} to 3.6×10^{-6} in/volt. Larger rate corresponds to lower inflation pressure. These test values are compared bellow to the analysis results.

Referring to Eq. (2), one observes that the piezoelectric strains, $d_{31}E_3$ and $d_{32}E_3$, behave numerically analogous to thermal strains of the form, $-\alpha_{31}\Delta T$ and $-\alpha_{32}\Delta T$, where ΔT is the change in temperature, and α_{31}, α_{32} are anisotropic expansion coefficients in directions "1" and "2", respectively. This analogy was exploited in the numerical simulation of the piezo actuated inflated membrane. Thus in NASTRAN's nonlinear analysis simulations, large membrane deformations are computed iteratively under a simultaneously applied inflation pressure p_i and temperature change ΔT, the magnitude of which is chosen such that;

$$d_{31}E_3 = \alpha_{31}\Delta T, \quad and \quad d_{32}E_3 = \alpha_{32}\Delta T \qquad (3)$$

The following values were used in Eq. (3) and the analysis: $d_{31} = 0.9 \times 10^{-9}$ in/in/volt/in, $d_{32} = 0.117 \times 10^{-9}$ in/in/volt/in, Elastic modulus$= 0.29 \times 10^6$ psi, $\alpha_{31} = 28.0 \times 10^{-6}$ in/in/F^o, and $\alpha_{32} = 4.0 \times 10^{-6}$ in/in/F^o. Note that both d_{31}, d_{32} should be divided by the membrane thickness before their use in Eq. (3). For comparison with the test results in Fig (4), the rate of center deformation per volt was computed from the analysis and were found to range from 13.9×10^{-6} in/volt to 8.2×10^{-6} in/volt for inflation pressures ranging from 0.1 psi to 0.5 psi. Considering the good analysis/test correlation in Fig. (3), the discrepancy of about a factor of three may be due to

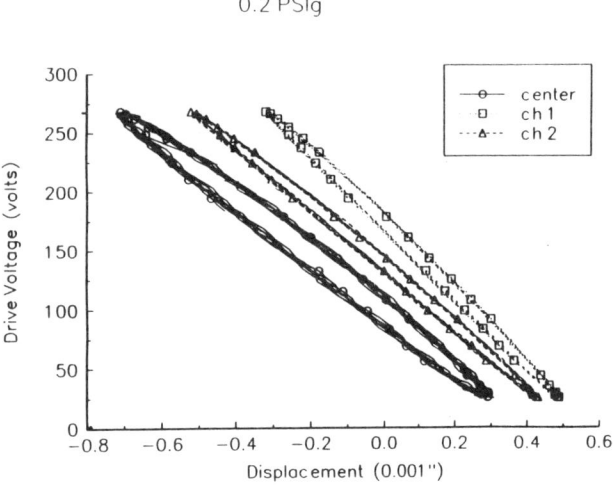

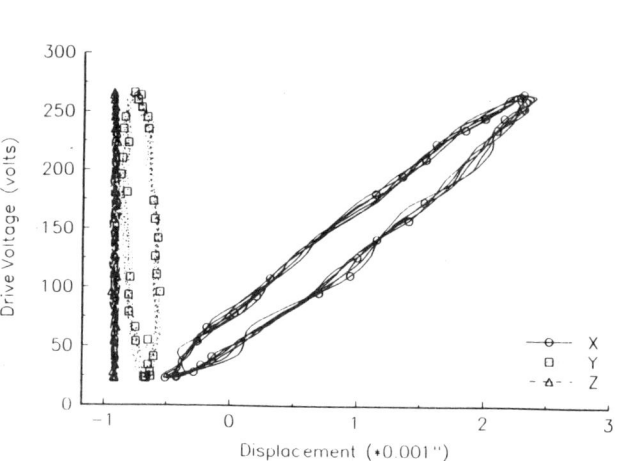

Figure 5: Piezoactuated Deformation At Free End of Inflated Tube

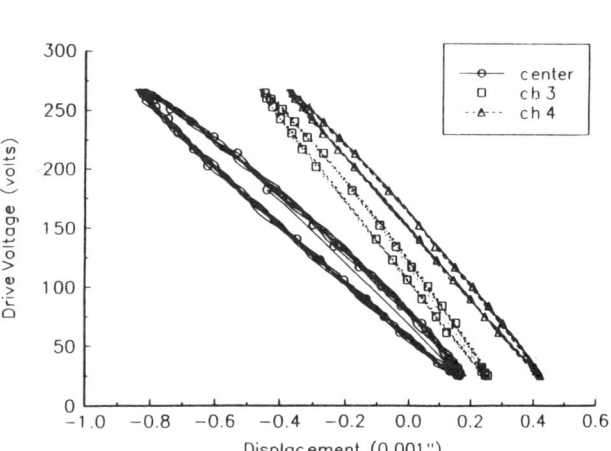

Figure 4: Piezoactuated Deformation of Inflated Circular Membrane At Center and Points Marked in Fig. 2a.

inaccuracies in values of the material constants listed above. According to the manufacturer, measurement of d_{33} can be made much more accurately than for d_{31}, d_{32}.

As in the circular membrane, piezo actuation tests of the tube in Fig. (2b) were performed by applying voltage to the leads of the already inflated tube, with the bottom end rigidly fixed. To simulate line-of-sight adjustments of the free end of the tube relative to the fixed end, an equal and opposite voltage was applied to the leads of only two strips on opposite sides of the tube. The deformations δ_{pi}^v due to the applied voltage alone are then measured at the three degrees of freedom at the free end. Figure (5) is an example of these measurements at pressure $p_i = 1.0 psi$ as the voltage is varied. Here again the deformation rate per volt was computed from the test data at the lateral degree-of-freedom for $p_i = 0.5, ,0.6, 0.7, 0.8, 0.9, and 1.0\ psi$ and was found to range from 12.0×10^{-6} to 11.1×10^{-6} in/volt. The analysis followed the same procedure described above for the circular membrane, and discrepancies of the same order (factor of 3) was also encountered for the tube configuration. Figure (6) is an example of the analysis simulation results of the piezo actuated line-of-sight adjustment.

IV. Piezocontrol of Antenna System:

The antenna system in Fig.(7) is a finite element idealization of Fig. (1) without the tripod. The axisymmetric parabolic antenna membrane has a diameter $D = 120.0 in$, $F/D = 0.5$, thickness $= 0.00025\ in$, and is inflated and taut at the outer edge by 16 tierods to a thin-walled torus. The nominal dimensions of the tie-rods are: length $= 6.0in$, area $= 0.03in^2$, and those of the torus are: radius of centerline $= 72.0in$, tube radius $d_t = 6.0in$, and thickness $t_t = 0.02in$. The surface accuracy of the inflated parabolic antenna depends upon several parameters such as the initial uninflated shape, inflation pressure, edge conditions at the torus interface, elastic interaction between the parabolic membrane and the supporting torus, imperfections in the fabrication process of any of the above components, and the operating conditions in space. On-orbit correction of deviations from the desired shape can be implemented by several piezo actuation methods. These are discussed next.

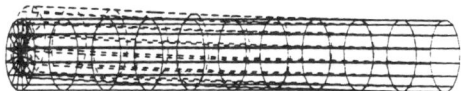

Figure 6: Simulation of Piezoactuated Line-of-sight

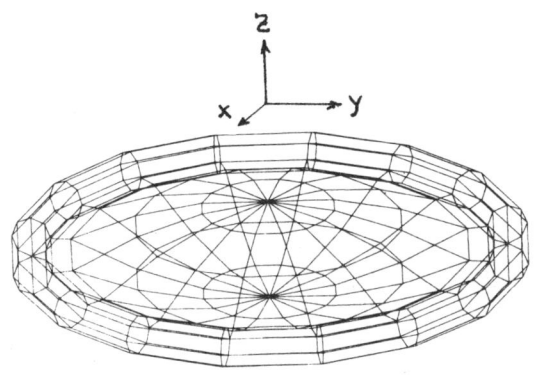

Figure 7: Finite Element Idealization of Inflated Parabolic Antenna/Torus System

1. Simulations: In the following simulation scenario, we assume that aberrations in the antenna surface are the result of out-of-plane circumferential sinusoidal distortion of the torus centerline from the XY-plane in the form: $d_z = d_o sin3\theta$, with zero values at the three interface points to the supporting tripod. Such imperfection in the torus will give rise to a deformed state, u^d, of the inflated antenna which differ from the deformed state, u^o, had the torus been perfect. Then the vector of aberrations we wish to correct is taken to be: $u^* = (u^d - u^o)$. Only deformation components in the Z-direction (parallel to axis of symmetry) at the nodes on the parabolic membrane are included in u^*.

One approach to correct the aberrations u^* on-orbit is by introducing controlled amounts of piezo actuated strains along axes of the tie-rods. To achieve the best correction, the actuation gain (magnitude of voltage or displacement induced in each tie-rod) must be optimally distributed among the tie-rods.

Let n_c = number of degrees of freedom in the controlled set u^*, and n_{ap} = number of possible actuators (here 16 tie-rods). Also let:

$\tilde{S} = (n_c \times n_{ap})$ matrix of influence coefficients, each column of which represents the deformation state at the n_c control d.o.f due to a unit actuation gain at one actuator location,

$\Delta_o = (n_{ap} \times 1)$ vector of unknown actuator gains,

Then, the deformations u at the same n_c controlled d.o.f. due to applying a combination of actuation gains Δ_o at the tie-rods is:

$$u = \tilde{S}\Delta_o \qquad (4)$$

Now the unknown gains Δ_o may be determined optimally in a least square sense by requiring, $u^* - u = 0$, so that;

$$\Delta_o = (\tilde{S}^T \tilde{S})^{-1} \tilde{S}^T u^* \qquad (5)$$

In general, it may be desirable to use fewer actuators n_a than the total number of possible actuator locations n_{ap}. In this case, the actuator locations should be also selected optimally. Since the actuator locations are discrete variables, a discrete optimization technique such as the simulated annealing algorithm [8] may be used for selecting the optimal actuator locations and gains. The algorithm is iterative. It does not require gradient information, and is designed to avoid getting trapped in local minima. During a typical kth iteration, the algorithm selects n_a out of all possible n_{ap} actuator locations. Let,

$J_k^*(j)$ = the integer vector of n_a location indices $j = 1, ..., n_a$, such that $J_k^*(j) \in n_{ap}$ specifies the n_a locations selected during the current simulated annealing iteration, and

$B_k = (n_{ap} \times n_a)$ location selection matrix, whose b_{ij} elements are either "one" or "zero" depending upon whether or not $i = J_k^*(j)$ for any j and for $i = 1, ..., n_{ap}$.

Then, analogous to Eqs. (4,5);

$$u_k = S_k \Delta_{ok}$$
$$S_k = \tilde{S} B_k$$
$$\Delta_{ok} = (S_k^T S_k)^{-1} S_k^T u^* \qquad (6)$$

As a measure of the proximity of u_k to u^*, the algorithm seeks the best actuator locations $J_k^*(j)$ with optimal gains $\Delta_{ok}(j)$, $j = 1, ..., n_a$ which minimize the normalized rms error:

$$e_{min} \equiv e_k = \Big[\frac{(u^* - S_k\Delta_{ok})^T(u^* - S_k\Delta_{ok})}{u^{*T}u^*}\Big]^{1/2} \quad (7)$$

2. Analysis Results: Results of the numerical implementation of the methodology described above are given here for different cases that differ in the number of d.o.f. included in the control set u^*, and in the number of actuators used to perform the shape adjustment. For all cases, the rms of the control d.o.f. in u^* is 1.28×10^{-3} in. Table 1 summarizes the optimization results when the number of control d.o.f in the u^* set are $n_c = 17, 25$, and 49. The specific choice of members of these sets is shown in Fig. (8), and was made to represent progressively finer granularity in the number of sensors one may use in monitoring the shape of the antenna. Notice that the intentional choice of $n_c(17) \in n_c(25) \in n_c(49)$, **does not** guarantee that the resulting optimal locations (see Table 1) satisfy the same relation, $n_a(4) \in n_a(8) \in n_a(16)$.

The effect of increasing the number of control d.o.f. n_c to be monitored and the number of actuators n_a on the ability to adjust the antenna errors is shown in Table 1 and Fig. (9). As one may expect, more accurate adjustment is possible as more actuators are brought to share in the process, and as one tries to enforce the adjustment at a fewer number of d.o.f. Conversely, the accuracy deteriorates as one tries to use the same number of actuators to enforce adjustment of shape at a larger number of d.o.f. The ideal situation is when the number of actuators and number of control d.o.f. are equal, $n_a = n_c$, in which case exact adjustment of the shape is assured. It should be realized, however, that achieving higher degree of adjustment accuracy at the d.o.f. included in u^* **does not** guarantee equally high accuracy at d.o.f. excluded from u^*. The latter may become even worse. It is important, therefore, that the control d.o.f. be selected judiciously.

Rather than volts, the optimal actuator gains in Table 1 represent displacements (or travel) which the actuators must provide. The use of piezo films to provide this level of displacement in the tie-rods would require their redesign. However, off-the-shelf PZT ceramic actuators are usually capable of up to 4×10^{-3} in travel. Thus, except for the case when, $n_a = 16$, $n_c = 17$, one could use existing PZT ceramic actuators. Otherwise, the actuator design must be modified for specific applications. Of course, design limitations will have to be imposed on their energy consumption and weight.

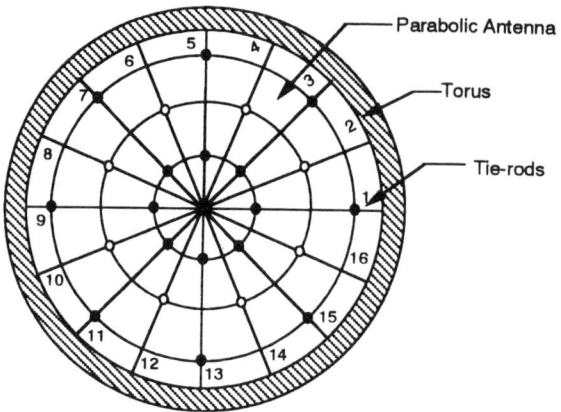

Figure 8: Control d.o.f. and Tie-rod Locations
$n_c = 17$: "●"
$n_c = 25$: "●+o"
$n_c = 49$: All Z-d.o.f. on Antenna

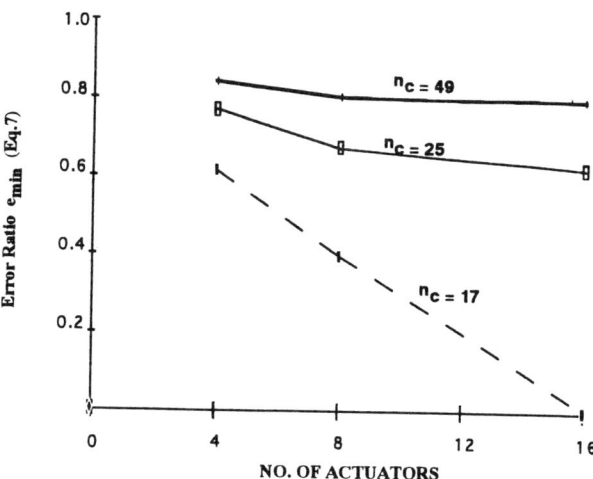

Figure 9: Error Ratio e_{min} As Function of n_a, n_c

3. Other Actuation schemes: Rather than inducing piezo strains in the tie-rods to correct the aberrations in the inflated membrane, another approach may be to induce the piezo strain directly in sectors (gores) of the parabolic membrane. This would be similar the experiments discussed in Sec. II. Electric separation of the different sectors or regions is easily achieved by chemical etching of the metallization. This approach has the advantage of increasing the number of possible actuation regions, n_a, thereby reducing the required voltage.

A third approach involves application of piezoelectric moments locally at different locations along the circumference of the torus. Such moments would correct distortions in the torus directly. This scheme was attempted on the system of Fig. (7), but was abandoned because it required too high voltage to deform the relatively very stiff torus.

V. Concluding Remarks:

The concept of using induced piezo strains to control on-orbit dimensional changes in elements of inflatable structures was investigated here analytically and experimentally. Off-the-shelf material was used to construct relatively simple laboratory experiments. Feasibility of the approach was demonstrated qualitatively. However, due to inaccurate knowledge of the anisotropic piezoelectric coefficients of the piezo film, a factor of three discrepancy was found between the analysis and test results. More accurate characterization of these constants is required.

For a 10.0 in diameter, 0.002 in thick membrane, out-of-plane center deformation of about 0.0012 in was obtained during the experiment when about 300 volts were applied. One would expect this deformation to scale directly with diameter. Thus, the demonstrated piezo induced deformations would be of the order 0.05 in for 14-meter diameter F/1 system, such as the inflatable reflector to be flown on the Inflatable Antenna Experiment of Ref. [9]. Manufacturing tolerances have been shown to cause focal length changes on the order of several inches for a 14-meter diameter antenna. So, the demonstrated piezo approach would not be appropriate for this level adjustments. On the other hand, small adjustments in the gore lengths or tie-rods may be sufficient to correct symmetric or unsymmetric local irregularities in the membrane surface arising from material imperfections, modulus variations, or other causes of small dimensional instabilities. The combination of a pressure control to adjust the focal length, and piezoelectric control to reduce deviations from the desired shape, could provide the added increase in surface accuracy needed to extend the use of inflatable antennas into the high frequency domain.

Table 1: Optimized Performance of Antenna System As Function of n_a, n_c. (Gain $\times 10^{-3} in$)

No. of Actuators	$n_c = 17$			$n_c = 25$			$n_c = 49$		
	e_{min}	Actuator No.	Gain	e_{min}	Actuator No.	Gain	e_{min}	Actuator No.	Gain
$n_a = 4$	0.612	1	-4.4	0.766	6	-1.3	0.837	6	-1.1
		5	2.7		9	-1.0		9	-2.1
		7	4.2		12	-2.6		11	-0.9
		12	-3.8		14	-2.4		12	-1.2
$n_a = 8$	0.395	4	-2.6	0.671	1	-0.5	0.800	1	-0.7
		5	2.6		6	-2.6		3	-0.5
		7	3.9		7	1.3		6	-1.2
		8	-2.4		9	-1.6		7	1.1
		10	-1.8		10	1.1		9	-2.2
		11	3.0		11	0.6		10	0.5
		13	2.9		12	-2.6		11	0.9
		14	-3.3		14	-2.1		12	-1.2
$n_a = 16$	~0.0	1	-16.1	0.619	1	-0.5	0.792	1	-0.7
		2	4.9		2	1.5		2	0.3
		3	-6.6		3	-0.3		3	-0.6
		4	-0.7		4	-1.0		4	-0.1
		5	8.5		5	0.4		5	0.1
		6	11.8		6	-3.0		6	-1.2
		7	12.1		7	0.8		7	0.9
		8	-6.8		8	1.5		8	0.5
		9	0.9		9	-1.5		9	-2.1
		10	-7.4		10	1.6		10	0.5
		11	13.2		11	0.8		11	0.9
		12	20.2		12	-3.1		12	-1.3
		13	6.3		13	0.6		13	0.0
		14	-20.3		14	-2.4		14	-0.3
		15	-18.4		15	0.3		15	-0.1
		16	-5.8		16	1.1		16	0.2

References:

1. Thomas, M. and G.F. Friese, "Pressurized Antennas for Space Radars," AIAA Sensor Systems for the 80's Conference, CP807, December 1980.

2. Thomas, M. and Veal, G.; "Scaling Characteristics of Inflatable Paraboloid Concentrators," ASME Solar Engineering, Ed.: Mancini, T., Watanabe, K., and Klett, D., 1991, pp. 353-358.

3. Goslee, J.W., "Progress Report on the Electrostatic Membrane Antenna Concept Testing," - Large Space Systems Technology Conference, 1981, NASA Conference Publication 2215, part 2, 1981, pp. 681-687.

4. Utku, S., Kuo, C.P., Garba, J., Salama, M., and Wada, B., "Adaptive Inflatable Space Structures: Shape Control of Reflector Surface," 4th International Conference on Adaptive Structures, Maternushaus, Cologne, Germany, November 1993.

5. Jaffe, B., Cook, W., and Jaffe, H., **Piezoelectric Ceramics**, Academic Press Limited, 1971.

6. Tzou, H.S., **Piezoelectric Shells: Distributed Sensing and Control of Continua**, Kluwer Academic Publishers, 1993.

7. Hencky, H., "Uber den Spannungszustand in kreisrunder Platten mit verschwindender Biegungssteifigkeit," Zeits. Math. Phys., Vol. 63, 1915, pp. 311-316.

8. Chen, G-S., Bruno, R., Salama, M., "Optimal Placement of Active / Passive Members in Truss Structures Using Simulated Annealing," AIAA J., Vol. 29, 1991, pp. 1327-1334.

9. Freeland, R.E., Bilyeu, G.D., Veal, G.R., "Validation of a Unique Concept for a Low Cost Lightweight Space Deployable Antenna Structure," 44th Congress of the International Astronautical Federation, October 1993, paper No. IAF-93-I.1.201.

AIAA-94-1772-CP

STATIC SHAPE CONTROL OF SPACE TRUSSES WITH PARTIAL MEASUREMENTS

Hiroshi Furuya*
Nagoya University
Chikusa, Nagoya 464-01, JAPAN
and
Raphael T. Haftka**
Virginia Polytechnic Institute and State University
Blacksburg, Virginia 24061-0203, USA

Abstract

The paper addresses the problem of predicting the statistics of shape distortion of space truss structures when measurements are limited to the distortions in a subset of the members of the truss. Expressions for the average values of a quadratic measure of the distortion in terms of mean values and covariances of member length errors are developed. An example of a 150-member antenna truss is used to assess the performance of static shape control with active members used as collocated sensors and actuators. Actuator locations are selected based on genetic algorithm optimization. It is found that poor performance is achieved even with large numbers of sensor-actuator pairs. This indicates that it may not be practical to rely on incomplete measurements of member distortions to achieve reasonable levels of shape control.

Introduction

Space antennas often have to maintain extreme surface accuracies. This requirement translates to similar shape accuracy requirements for the truss structures that support these antennas. Member length errors are the prime contributors to shape distortion, and these length errors are mostly due to manufacturing errors and thermal expansion. Because of the random nature of both sources of length errors, the treatment of the shape distortion is often probabilistic in nature, with the expected value of some measure of surface distortion used as a measure of performance [1].

To achieve the high accuracy we usually use actuators that respond to shape distortion by corrective action. Most of the work on shape control of space trusses assumed perfect knowledge of the distortion. However, this assumption may not be realistic. Kuwao, Chen and Wada [2] have explored shape estimation and control based on partial measurements. They found that when the distortions of some members are measured, interpolation of unmeasured member distortions may be an effective tool for estimating overall shape error and controlling it. Bruno, Toomarian, and Salama [3] have estimated the effects of incomplete measurements by Neural Network approach. The objective of the present paper is to extend these works so as to include probabilistic treatment of the uncertainties associated with the distortion errors.

We assume that the major source of member length errors is thermal expansion. Manufacturing length errors are also important, but because they are constant, they may be compensated for electronically. That is, constant errors are easier to deal with than variable errors. Because temperature fields are continuous, we expect the temperatures of adjacent members to be correlated, and therefore their distortions to be likewise correlated.

This paper has two parts. In the first part we develop expressions for a quadratic shape distortion measure based on assumed knowledge of the statistics of the member length errors. We first consider the case of no actuators and the case of using actuators with full knowledge of the distortion. Next we consider the case of partial measurements and show that for this case the optimal control strategy starts by estimating the unmeasured quantities and then treats them the same as measured distortions. Finally, we obtain expressions for the loss of performance associated with partial measurements and use a simple example for illustration.

The second part of the paper deals with the expected performance of shape correction based on partial measurements through an example of a 150-member truss antenna structure. We first propose a simple model for generating statistical parameters for such a structure, and then we consider the performance of shape correction as the number of sensors and actuators (assumed to be collocated) increases. To insure that good locations are chosen for the actuators, we employ a genetic algorithm we have previously developed [4].

---
* Assistant Professor, Dept. of Aerospace Eng., Member AIAA
** Christopher Kraft Professor, Dept. of Aerospace and Ocean Eng., Associate Fellow AIAA
Copyright © 1994 by the American Institute of Aeronautics and Astronautics, Inc.

Statistics of Shape Distortion

Shape Errors with Complete Measurements

Consider first the problem of estimating the shape errors when we know only the statistical distribution of the member length errors. The shape error is measured by a quadratic measure, such as the mean square error

$$f = u^T B u, \quad (1)$$

where u is the displacement field, and B is a positive semidefinite weighting matrix. We assume a vector ϵ of member length errors is responsible for the displacement u, and the behavior is linear so that u may be obtained from ϵ via linear analysis

$$u = G\epsilon, \quad (2)$$

where G is a matrix obtained (for example) by finite element analysis. Using (1) and (2) we get

$$f = \epsilon^T A \epsilon, \quad (3)$$

where

$$A = G^T B G. \quad (4)$$

We assume that ϵ is a vector of normally distributed random variables with means given by the vector $\mu = E(\epsilon)$, and covariance given as $\Sigma_0 = \Sigma(\epsilon)$. To calculate the expected value of the error f, we start by defining

$$\epsilon' = L^T (\epsilon - \mu), \quad (5)$$

where L is any factor of A (e.g., the Cholesky factor) that satisfies

$$A = LL^T. \quad (6)$$

Then ϵ' is a vector of normally distributed random variables with zero mean and covariance

$$\Sigma(\epsilon') = L^T \Sigma_0 L. \quad (7)$$

Using (3) and (5) we can write

$$\begin{aligned} f &= (L^T \epsilon)^T L^T \epsilon \\ &= \epsilon'^T \epsilon' + 2\mu^T L \epsilon' + \mu^T A \mu. \end{aligned} \quad (8)$$

Finally, the expected value of f is given as

$$E(f) = trace\left(L^T \Sigma_0 L\right) + \mu^T A \mu. \quad (9)$$

Shape Errors with Partial Measurements

Next assume that we have measured some components of the vector ϵ. We partition the vector into its unmeasured part ϵ_1, and its measured part ϵ_2, and denote the partially measured vector as ϵ^p.

$$\epsilon^p = \left[\epsilon_1^T, \epsilon_2^T\right]^T. \quad (10)$$

We denote the expected values of the two parts as μ_1 and μ_2, respectively

$$\mu_1 = E(\epsilon_1), \quad \mu_2 = E(\epsilon_2). \quad (11)$$

We similarly partition the matrices A and Σ_0

$$A = \begin{bmatrix} A_{11} & A_{12} \\ A_{21} & A_{22} \end{bmatrix}, \quad \Sigma_0 = \begin{bmatrix} \Sigma_{11} & \Sigma_{12} \\ \Sigma_{21} & \Sigma_{22} \end{bmatrix}. \quad (12)$$

Using this notation we have

$$f = \epsilon_1^T A_{11} \epsilon_1 + 2\epsilon_1^T A_{12} \epsilon_{2m} + \epsilon_{2m}^T A_{22} \epsilon_{2m}. \quad (13)$$

We denote the conditional expected values and variances, given $\epsilon_2 = \epsilon_{2m}$ by subscript m. Then from (13)

$$\begin{aligned} E_m(f) &= E(f|(\epsilon_2 = \epsilon_{2m})) \\ &= E_m(f_1) + 2\mu_{1m}^T A_{12} \epsilon_{2m} + \epsilon_{2m}^T A_{22} \epsilon_{2m}, \end{aligned} \quad (14)$$

where

$$f_1 = \epsilon_1^T A_{11} \epsilon_1, \quad (15)$$

and μ_{1m}, the conditional expected value, and Σ_{11m}, the conditional covariance matrix of ϵ_1 are (e.g., Ref. [5] page 92, see also Appendix)

$$\begin{aligned} \mu_{1m} &= E_m(\epsilon_1) \\ &= \mu_1 + \Sigma_{12} \Sigma_{22}^{-1} (\epsilon_{2m} - \mu_2), \end{aligned} \quad (16)$$

and

$$\begin{aligned} \Sigma_{11m} &= \Sigma_m(\epsilon_1) \\ &= \Sigma_{11} - \Sigma_{12} \Sigma_{22}^{-1} \Sigma_{12}^T. \end{aligned} \quad (17)$$

To evaluate $E_m(f_1)$ we can use the results of the previous section with ϵ_1 replacing ϵ, A_{11} replacing A, and the conditional expected values and covariances replacing their unconditional counterparts.

Finally we obtain the following expression.

$$\begin{aligned} E_m(f) = &\ trace\left(L_{11}^T \Sigma_{11m} L_{11}\right) + \mu_{1m}^T A_{11} \mu_{1m} \\ &+ 2\mu_{1m}^T A_{12} \epsilon_{2m} + \epsilon_{2m}^T A_{22} \epsilon_{2m}. \end{aligned} \quad (18)$$

Shape Control with Complete Knowledge

We assume that we have a vector θ of control commands that produce a vector of corrective displacements u_c which depends linearly on the control

$$u_c = T\theta. \quad (19)$$

The total displacement field u_T is given as

$$u_T = G\epsilon + T\theta. \quad (20)$$

Assuming that we have complete knowledge of the effect of the errors, that is, we know $G\epsilon$, we can calculate the best corrective action θ by minimizing f

$$f = (G\epsilon + T\theta)^T B (G\epsilon + T\theta). \qquad (21)$$

Differentiating f with respect to θ and setting to zero, we find that

$$\theta = -(T^T BT)^{-1} T^T BG\epsilon. \qquad (22)$$

Note that when the columns of BT are linearly dependent, or when one of the columns is zero (which happens when an actuator does not have any effect on any of the components of u that have non-zero B entries), then $T^T BT$ is singular. In this case we understand the inverse to mean the generalized inverse (which we have implemented using the singular value decomposition). Substituting θ into (21), we get the corrected value of f

$$f_c = \epsilon^T (A - S)\epsilon, \qquad (23)$$

where

$$S = G^T BT(T^T BT)^{-1} T^T BG = L_s L_s^T. \qquad (24)$$

We can calculate the expected value of $\epsilon^T S\epsilon$ using a procedure similar to the one that led to (9) with A replaced by S and ϵ' defined on the basis of the Cholesky factor L_s of S instead of the Cholesky factor L of A. Altogether we get

$$E(f_c) = trace(L^T \Sigma_0 L) + \mu^T A\mu \\ - trace(L_s^T \Sigma_0 L_s) - \mu^T S\mu. \qquad (25)$$

This expression can be further minimized by choosing optimal locations of sensors and actuators.

Shape Correction with Partial Measurements

Next we consider the problem of applying corrections on the basis of incomplete data. We assume that we select θ so as to minimize the conditional expected value of f. Taking the conditional expected value of f with respect to the measured strain from (21) and using (4) we have

$$E_m(f) = E_m(\epsilon^{pT} A\epsilon^p) \\ + 2E_m(\epsilon^p)^T G^T BT^T \theta + \theta^T T^T BT\theta, \qquad (26)$$

where

$$E_m(\epsilon^p) = \begin{bmatrix} \mu_{1m}^T, & \epsilon_{2m}^T \end{bmatrix}^T. \qquad (27)$$

Differentiating with respect to θ and setting to zero we get that

$$\theta = -(T^T BT)^{-1} T^T BG E_m(\epsilon^p). \qquad (28)$$

Equation (28) is similar to (22) and indicates that in calculating the optimal actuator action we use the expected value of the strains for unmeasured elements. This means that with partial measurements of the strains, we will first estimate the unmeasured strains from (16) and then use these strains to calculate the actuator response as if these strains were measured. We then obtain the corrected value of f by substituting from (28) back into (21) to get,

$$f_{cm} = \epsilon^{pT} A\epsilon^p - \epsilon^{pT} S E_m(\epsilon^p) - E_m(\epsilon^p)^T S\epsilon^p \\ + E_m(\epsilon^p)^T S E_m(\epsilon^p). \qquad (29)$$

Therefore, the expected value of f_{cm} is

$$E(f_{cm}) = E(\epsilon^{pT} A\epsilon^p) - E\left(\epsilon^{pT} L_s L_s^T E_m(\epsilon^p)\right. \\ \left. + E_m(\epsilon^p)^T L_s L_s^T \epsilon^p - E_m(\epsilon^p)^T L_s L_s^T E_m(\epsilon^p)\right). \qquad (30)$$

For simplifying (30) we need the following relationship for arbitrary x and y,

$$\begin{aligned} E(x^T L_s L_s^T y) &= E\left(trace(L_s^T y x^T L_s)\right) \\ &= trace\left(E(L_s^T y x^T L_s)\right) \\ &= trace\left(L_s^T E(y x^T) L_s\right). \end{aligned} \qquad (31)$$

With the help of (31), we transform (30) into

$$\begin{aligned} E(f_{cm}) &= E(\epsilon^{pT} A\epsilon^p) \\ &- trace\left\{L_s^T E\left(\epsilon^p \epsilon^{pT} - \epsilon^p \epsilon^{pT}\right.\right. \\ &\left. + E_m(\epsilon^p)\epsilon^{pT} + \epsilon^p E_m(\epsilon^p)^T \right. \\ &\left. - E_m(\epsilon^p) E_m(\epsilon^p)^T \right) L_s \big\}. \end{aligned} \qquad (32)$$

Finally, one additional application of (31) yields

$$E(f_{cm}) = E\left(\epsilon^{pT}(A-S)\epsilon^p\right) \\ + E\left\{\left(\epsilon^p - E_m(\epsilon^p)\right)^T S\left(\epsilon^p - E_m(\epsilon^p)\right)\right\}. \qquad (33)$$

The second term of the right-hand side of (33) is

$$\begin{aligned} &E\left\{\left(\epsilon^p - E_m(\epsilon^p)\right)^T S\left(\epsilon^p - E_m(\epsilon^p)\right)\right\} \\ &= E\left\{trace\left(L_s^T \left(\epsilon^p - E_m(\epsilon^p)\right)\left(\epsilon - E_m(\epsilon^p)\right)^T L_s^T\right)\right\} \\ &= trace\left(L_s^T E\left((\epsilon_1^T - \mu_{1m}^T, 0^T)^T (\epsilon_1^T - \mu_{1m}^T, 0^T)\right) L_s\right) \\ &= trace\left(L_s^T \begin{bmatrix} \Sigma_{11m} & 0 \\ 0 & 0 \end{bmatrix} L_s\right) \geq 0. \end{aligned} \qquad (34)$$

Also, clearly $E(\epsilon^{pT}(A-S)\epsilon^p) = E(\epsilon^T(A-S)\epsilon)$, and so using (23) we have

$$E(f_{cm}) = E(f_c) + trace\left(L_s^T \begin{bmatrix} \Sigma_{11m} & 0 \\ 0 & 0 \end{bmatrix} L_s\right). \qquad (35)$$

The difference between the expected value with complete and partial will depend on the number of sensors and their locations.

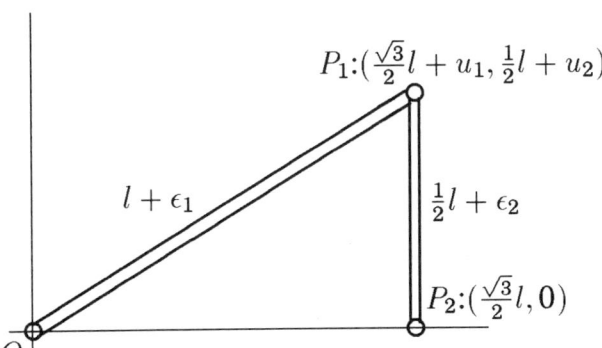

Figure 1: Two-bar truss structure model

Two-Bar Truss Example

Consider the two-bar truss structure in Figure 1. The relationships between the member length errors and the displacements are

$$(l + \epsilon_1)^2 = (\frac{\sqrt{3}}{2}l + u_1)^2 + (\frac{1}{2}l + u_2)^2,$$

$$(\frac{1}{2}l + \epsilon_2)^2 = u_1^2 + (\frac{1}{2}l + u_2)^2.$$

By linearizing we get

$$\left\{ \begin{array}{c} \epsilon_1 \\ \epsilon_2 \end{array} \right\} = \left[\begin{array}{cc} \frac{\sqrt{3}}{2} & \frac{1}{2} \\ 0 & 1 \end{array} \right] \left\{ \begin{array}{c} u_1 \\ u_2 \end{array} \right\} = G^{-1} u,$$

where

$$G = \left[\begin{array}{cc} \frac{2\sqrt{3}}{3} & -\frac{\sqrt{3}}{3} \\ 0 & 1 \end{array} \right].$$

We take the weighting matrix B as the unit matrix

$$B = \left[\begin{array}{cc} 1 & 0 \\ 0 & 1 \end{array} \right],$$

and then

$$L = G^T,$$

and

$$A = G^T G = \left[\begin{array}{cc} \frac{4}{3} & -\frac{2}{3} \\ -\frac{2}{3} & \frac{4}{3} \end{array} \right].$$

The expected values and correlation matrix of the member length errors are

$$\mu = [\mu_1, \mu_2]^T,$$

and

$$\Sigma_0 = \left[\begin{array}{cc} \sigma_1^2 & \rho\sigma_1\sigma_2 \\ \rho\sigma_1\sigma_2 & \sigma_2^2 \end{array} \right].$$

Also

$$\mu^T A \mu = \frac{4}{3}(\mu_1^2 - \mu_1\mu_2 + \mu_2^2)$$

and

$$\Sigma(\epsilon') = \left[\begin{array}{cc} \frac{1}{3}(4\sigma_1^2 - 4\rho\sigma_1\sigma_2 + \sigma_2^2) & \frac{\sqrt{3}}{3}(2\rho\sigma_1\sigma_2 - \sigma_2^2) \\ \frac{\sqrt{3}}{3}(2\rho\sigma_1\sigma_2 - \sigma_2^2) & \sigma_2^2 \end{array} \right],$$

Thus, the expected value of f without measurement is

$$E(f) = \frac{4}{3}(\sigma_1^2 - \rho\sigma_1\sigma_2 + \sigma_2^2 + \mu_1^2 - \mu_1\mu_2 + \mu_2^2).$$

We assume element No. 2 is an active member, so that T is the second column of G,

$$T = (-\frac{\sqrt{3}}{3}, 1)^T,$$

thus,

$$S = G^T BT (T^T BT)^{-1} T^T BG = \left[\begin{array}{cc} \frac{1}{3} & -\frac{2}{3} \\ -\frac{2}{3} & \frac{4}{3} \end{array} \right].$$

If we put the sensor on element No.2, the conditional covariance matrix is

$$\begin{array}{rl} \Sigma_{11m} = & \Sigma_{11} - \Sigma_{12}\Sigma_{22}^{-1}\Sigma_{12}^T \\ = & (1 - \rho^2)\sigma_1^2. \end{array}$$

Next we calculate the following expected value

$$E(\epsilon^T S \epsilon) = trace(L_s^T \Sigma_0 L_s) + \mu^T S \mu,$$

where

$$L_s^T \Sigma_0 L_s = \frac{1}{3}(\sigma_1^2 - 4\rho\sigma_1\sigma_2 + 4\sigma_2^2),$$

and

$$\mu^T S \mu = \frac{1}{3}(\mu_1^2 - 4\mu_1\mu_2 + 4\mu_2^2).$$

Using these values, we finally obtain the expected values with complete measurements and with partial measurements, respectively

$$E(f_c) = \sigma_1^2 + \mu_1^2,$$

and

$$E(f_{cm}) = \sigma_1^2 + \mu_1^2 + \frac{1}{3}(1-\rho^2)\sigma_1^2.$$

If instead of using element 2 as the active element we use element 1, we get the expected symmetrical results

$$E(f_c) = \sigma_2^2 + \mu_2^2,$$

and

$$E(f_{cm}) = \sigma_2^2 + \mu_2^2 + \frac{1}{3}(1-\rho^2)\sigma_2^2.$$

Comparing these results to $E(f)$ we note that beside eliminating the effect of the error in the active member (i.e., no contribution from σ_2 or μ_2 for the case that the element 2 is the active element.), we are able to reduce the effect of the error in the other member as well. With complete measurements this reduction is by one quarter (from 4/3 to 1). With partial measurement there is an extra penalty, unless the correlation coefficient, ρ, is equal to 1. Indeed when the correlation is perfect, knowledge of one component implies knowledge of the other one as well.

Performance of Shape Correction with Partial Measurements

The algebraic experessions developed in the previous section do not tell us how well we can expect to correct shape distortion with partial measurements. To get an answer to this question we turn our attention to a more realistic example than the one analyzed in the previous section, the 150-member truss shown in Figure 2. In order to make use of this example we need to come up with some reasonable statistics of member length errors and a strategy for optimally placing actuators on the truss.

Simple Statistical Model

To generate statistical data for our problem we assume that the vector of length errors ϵ is composed of two uncorrelated contributions

$$\epsilon = \epsilon^0 + \Delta\epsilon, \quad (36)$$

where ϵ^0 represents the well-correlated part of the distortion field, due to continuous thermal fields, and $\Delta\epsilon$ represents random, uncorrelated fluctuations. We denote the mean and covariance matrix of ϵ^0 by μ^0 and Σ^0, respectively, and the corresponding quantities for $\Delta\epsilon$ as $\Delta\mu$ and $\Delta\Sigma$, respectively. Here we assume that $\Delta\mu = 0$, and that the correlation between all the components of ϵ^0 is one, so that Σ^0 is completetly defined in terms of the standard deviations σ_i of its components

$$\Sigma^0_{ij} = \sigma_i \sigma_j. \quad (37)$$

The mean μ and covariance Σ of ϵ are then given as

$$\begin{aligned}
\mu &= E(\epsilon) = E(\epsilon^0 + \Delta\epsilon) = \mu^0, \quad (38)\\
\Sigma &= E[(\epsilon - \mu)(\epsilon - \mu)^T] \\
&= E[(\epsilon^0 - \mu^0)(\epsilon^0 - \mu^0)^T] \\
&\quad + E[(\Delta\epsilon - \Delta\mu)(\Delta\epsilon - \Delta\mu)^T] \\
&= \Sigma^0 + \Delta\Sigma, \quad (39)
\end{aligned}$$

where, we used the assumption that ϵ^0 and $\Delta\epsilon$ are uncorrelated.

With this simple model of member length errors, the relative magnitude of the standard deviations of ϵ^0 and $\Delta\epsilon$ determine the correlation between the components of ϵ as illustrated by the following two member example.

Consider a case where member 1 and member 2 are subjected to thermal loads with the temperature differential (from some ground state) for member 2 expected to be twice as large as the differential for member 1. This would lead to length errors of $(\epsilon_1, 2\epsilon_1)$, with a covariance matrix of

$$\Sigma^0 = \begin{bmatrix} \sigma_1^2 & 2\sigma_1^2 \\ 2\sigma_1^2 & 4\sigma_1^2 \end{bmatrix}. \quad (40)$$

We also have some uncorrelated noise for both members, and if it is of the same magnitude for both, then

$$\Delta\Sigma = \begin{bmatrix} \Delta\sigma^2 & 0 \\ 0 & \Delta\sigma^2 \end{bmatrix}, \quad (41)$$

and

$$\Sigma = \Sigma^0 + \Delta\Sigma = \begin{bmatrix} \sigma_1^2 + \Delta\sigma^2 & 2\sigma_1^2 \\ 2\sigma_1^2 & 4\sigma_1^2 + \Delta\sigma^2 \end{bmatrix}. \quad (42)$$

The correlation coefficient ρ between ϵ_1 and ϵ_2 is

$$\rho = \frac{\Sigma_{12}}{\sqrt{\Sigma_{11}\Sigma_{22}}} = \frac{2\sigma_1^2}{\sqrt{(\sigma_1^2 + \Delta\sigma^2)(4\sigma_1^2 + \Delta\sigma^2)}}. \quad (43)$$

Note that if $\Delta\epsilon$ is small compared to ϵ_1 then $\Delta\sigma$ will be small compared to σ, and then the correlation coefficient will be close to 1. If, on the other hand, $\Delta\epsilon$ dominates, $\Delta\sigma$ will be large and the correlation coefficient will be close to zero.

Thus, if we have some idea about the fraction of ϵ_1 that can be determined from interpolation and the part which is random, we can estimate σ and $\Delta\sigma$, and we can calculate the covariance matrix of these two member-length errors.

The expressions developed in this paper for the expected value of the shape distortion do not depend on the details of the statistical distributions but only on their mean values and covariances. However, for the examples we have also performed simulations by randomly generating shape errors. These simulations employed normally distributed member-error vectors.

Selection of Actuator Locations by Genetic Algorithm

The locations of actuators are selected so as to minimize the expected value of the shape distortion. We furthermore assume that active members are used as both actuators and sensors, so that the actuators and sensors are colocated. Using Eqs. (25) and (35) we get for the expected value of the distortion

$$E(f_{cm}) = trace(L^T \Sigma_0 L) + \mu^T A \mu \\ - trace(L_s^T \Sigma_0 L_s) - \mu^T S \mu \\ + trace\left(L_s^T \begin{bmatrix} \Sigma_{11m} & 0 \\ 0 & 0 \end{bmatrix} L_s\right). \quad (44)$$

As the first and the second terms of the right-hand side of the equation are independent to the location of actuators, the problem reduces to the maximization of

$$trace(L_s^T \Sigma_0 L_s) + \mu^T S \mu - trace\left(L_s^T \begin{bmatrix} \Sigma_{11m} & 0 \\ 0 & 0 \end{bmatrix} L_s\right). \quad (45)$$

The genetic algorithm used for obtaining actuator locations is an integer coding uniform crossover algorithm [4] described briefly in the following. For n_a actuators the locations are coded as a string of n_a integers specifying the active members. Genetic algorithms work simultaneously on a population of n_s designs (strings) subjecting them to operators that mimic the biological processes of evolution and natural selection. The procedure goes through the following steps

1. Generate randomly an initial population of n_s designs.

2. Evaluate the objective function for each member of the population and rank the members.

3. Select a pair of parents from the population by simulating a roulette wheel with each design getting a portion of the wheel proportional to n_s minus its rank.

4. Generate a child design by uniform crossover. That is, for each string position take a location from one parent or the other parent at random.

5. Mutate randomly (but with low probability) some of the actuator locations in the child design to random new locations.

6. Repeat steps 2–5 n_s times to generate a complete new generation.

7. Replace old generation by new generation.

This process is repeated for a specified number of generations.

JPL Truss Antenna Model

The Jet Propulsion Laboratory (JPL) truss antenna model shown in Figure 2 was used to evaluate the performance of shape correction based on partial measurements. The truss has 150 members and 45 nodes.

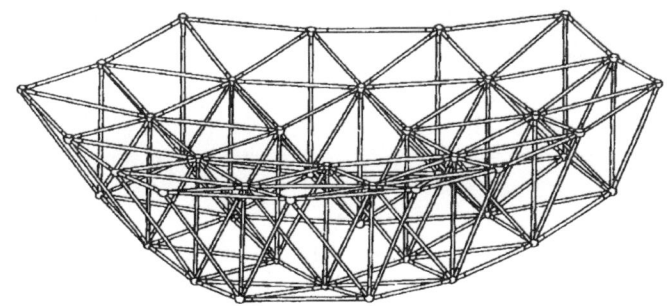

Figure 2: JPL truss antenna model

Of these nodes, 27 support the reflecting surface, and we assumed that only the z–displacement (axsymmetric direction of the parabola) of these nodes affects shape distortion. Accordingly, we used a diagonal matrix B with zeroes everywhere except for the 27 locations corresponding to the z–displacements of the nodes of the reflecting surface.

For testing performance, we have assumed uniform heating of the entire truss. Thus the perfectly correlated part of the member errors has all members with the same mean value and standard deviation of the error. The uncorrelated part of the error has zero mean with all members having the same standard deviations. We considered three ratios between the variance (square of the standard deviation) of the correlated part and the total variance (the sum of the two parts): zero, one half and one. These assumptions also make for a uniform coefficient of correlation between all members equal to the ratio of the correlated variance and total variance. This case appears to be fairly benign in terms of the distribution of errors, but as the results show, it is not easy to control the shape distortion.

Reflector Surface Distortion Results

The effectiveness of the shape correction is measured in terms of ratios of corrected shape distortion to original shape distortions. From the expressions derived in the paper we can calculate the ratio g^2 of the means of the two quantities

$$g^2 = \frac{E(f_{cm})}{E(f)}, \quad (46)$$

which is called the distortion ratio. From Monte Carlo simulations we can calculate beside g^2 also several other quantities

$$E\left(\frac{f_{cm}}{f}\right), \quad \frac{\sigma(f_{cm})}{E(f)}, \quad \sigma\left(\frac{f_{cm}}{f}\right). \quad (47)$$

The first is the average of the distortion-correction ratio, the second is a normalized standard deviation of the corrected distortion, and the third is the standard deviation of the distortion correction ratio.

Distortion ratios with complete and partial measurements were obtained by the analytical formulae presented in the paper and by 1000 random (Monte Carlo) simulations. Results for surface error of the truss reflector antenna are shown in Tables 1. In the tables, the asterisk (*) indicates the results by random simulation. The agreement between the analytical and simulation results is very good. It may appear that with the possiblility of performing such simulations there is no need for analytical results. However, the genetic optimization of the actuator locations would have been prohibitively expensive if we needed to rely on simulations instead of analytical expressions.

As can be seen from Table 1, when the correlation coefficient is very close to one, the distortion ratios with complete and partial measurement are very close. This is due to the fact that with all members highly correlated, the partial measurements provide excellent estimates of the unmeasured distortions. Table 1-(1) also indicates that we can get excellent performance with a small number of actuators. For example, four actuators suffice to reduce the expected distortion to six percent of its original value. This excellent performance reflects the uniformity of the distortion field and the relatively small number (below 27) of displacement components that define the distortion measure.

As can be seen from Tables 1-(2) and -(3), without such perfect correlation the number of actuators required to reduce the distortion is substantially larger even with complete measurements. For example, sixteen actuators reduce the expected distortion by just over an order of magnitude, while for the perfect correlation case they reduce it by almost three orders of magnitude. Note that the numbers in the first column of Table 1-(2) are slightly higher than the corresponding numbers in Table 1-(3), which gives the impression that the case with a correlation of 0.5 is more difficult than the case with almost zero correlation. However, $E(f)$ is almost twice as high in Table 1-(3), so that both the uncorrected and corrected distortions are higher for the low-correlation case.

The results with partial measurements are quite discouraging, with 40 pairs of sensors and actuators not being sufficient for reducing the expected value of the distortion by a factor of 3. The random simulation also gives the standard deviations associated with the distortion ratio. The large standard deviation reinforces the bleakness of the results for correlations of 0.001 and 0.5. These results indicate that when the errors have a substantial uncorrelated random component, partial measurement may not be a viable strategy.

Table 2 shows the distortion-correction ratio obtained by random simulations. The mean values of the distortion-correction ratio are somewhat larger than the corresponding values of the distortion ratio in Table 1. This difference may reflect the influence of cases where f is small but not easily correctible by the actuator configuration.

When the number of displacement components of interest is larger than the 27 we have for this example we may expect to require a larger number of actuators to achieve the same level of shape correction. To investigate this possibility we included all 135 degrees of freedom in the distortion measure by using the unit matrix for the weighting matrix, B.

Table 3 shows the distortion ratios for this case. As expected, the number of actuators required to achieve a given value of distortion ratio increases much even for the full measurement case. For example, with 27 distortion degrees of freedom, 8 actuators were sufficient for a distortion ratio of about 0.25 (see Tables 1-(2) and -(3). Now we need more than about 20 actuators for similar performance. As in Table 1 the distortion ratio improves with the reduction in correlation coefficient, but including the effect of the increase in $E(f)$ the corrected distortion actually increases.

The performance with partial measurements is still quite dismal, although, since the performance with complete measurements is not as good, it does not look as bad in comparison. Still, with 40 pairs of sensors and actuators we can achieve only a distortion ratio of about 0.4.

Table 4 shows the distribution of actuator locations obtained by the genetic algorithm for the fully correlated and uncorrelated cases, respectively. For the completely correlated model, the distribution of the best location of actuators does not appear to be so consistent. For the uncorrelated case the best locations follow a pattern as the number of actuators increase, with the locations used for a low number of actuators retained and neighbouring locations added. This pattern does not hold for the fully correlated case.

Concluding Remarks

Expressions for the statistics of shape distortion of space truss structures when measurements are limited to the distortions in a subset of the members of the truss were obtained for both the uncorrected and corrected cases. It was shown that the mean values and covariance matrices of the errors can be used to estimate the distortions in the rest of the truss members. Furthermore, the best strategy for cotrolling the distortion was shown to be the treatement of estimated member distortions as if they were measured.

The performance of shape correction was investigated

for a 150-member antenna truss. A genetic algorithm was employed for placing pairs of sensors and actuators so as to minimize the expected value of the distortion. Results were obtained for two cases, one with a distortion measure including only axial displacements on the reflector face of the truss, and the other including all degrees of freedom. For the first case it was found that with complete measurement the distortion of the truss could be reduced effectively with a small number of actuators. With partial measurements even a large number of actuators did not provide substantial reduction in shape distortion. For the case when all degrees of freedom were involved, a large number of actuators was required even with complete measurement. The performance was still dismal with partial measurements, but the gap in performance due to having only partial measurements was not so wide.

References

1. Burdisso, R. A., and Haftka, R. T.,"Statistical Analysis of Static Shape Control in Large Space Structures," *AIAA Journal*, Vol. 28, No. 8, pp. 1504–1508, Nov., 1990.

2. Kuwao, F., Chen, G.-S., and Wada, B.K., "Quasi-Static Shape Estimation and Control of Adaptive Truss Structures," AIAA Paper 91-1160-CP, *32nd SDM Conference*, pp. 544–552, April, 1991.

3. Bruno, R., Toomarian, N., and Salama, M., "Shape Estimation from Incomplete Measurements: a Neural Net Approach," *3rd. Int. Conf. on Adaptive Structures*, San Diego, Technomic Publishing Co., pp. 130–141, Nov., 1992.

4. Furuya, H., and Haftka, R. T., "Genetic Algorithms for Placing Actuators on Space Structures," *5th International Conference on Genetic Algorithms*, Urbana, Illinois, Morgan Kaufmann Publishers, pp. 536–542, July, 1993.

5. Morrison, Donald F., Multivariate Statistical Methods, 2nd Edition, McGraw Hill.

Appendix - Conditional Probabilities

The derivations of the conditional expected value and correlation is summarized from Ref. [5], as follows:

The conditional density function of ϵ_1 is given by fixing the elements of ϵ_2 as

$$r(\epsilon_1|\epsilon_2) = \frac{e(\epsilon_1, \epsilon_2)}{h(\epsilon_2)}, \quad (A.1)$$

where e is the joint density of the complete set of ϵ ($= [\epsilon_1^T, \epsilon_2^T]^T$), and h is the joint density of the ϵ_2.

Assuming that the ϵ is a normally distributed random variable,

$$h(\epsilon_2) = \frac{1}{(2\pi)^{q/2}|\Sigma_{22}|^{1/2}}$$
$$\times exp\left[-\frac{1}{2}(\epsilon_2 - \mu_2)^T \Sigma_{22}^{-1}(\epsilon_2 - \mu_2)\right], \quad (A.2)$$

where μ_1 and μ_2 are expected values of ϵ_1 and ϵ_2, respectively.

Considering

$$e(\epsilon) = \frac{1}{(2\pi)^{(p+q)/2}|\Sigma|^{1/2}} exp\left[-\frac{1}{2}(\epsilon - \mu)^T \Sigma^{-1}(\epsilon - \mu)\right], \quad (A.3)$$

where

$$|\Sigma| = |\Sigma_{22}||\Sigma_{11} - \Sigma_{12}\Sigma_{22}^{-1}\Sigma_{21}|, \quad (A.4)$$

and

$$\Sigma^{-1} = \begin{bmatrix} (\Sigma_{11} - \Sigma_{12}\Sigma_{22}^{-1}\Sigma_{21})^{-1} & -(\Sigma_{11} - \Sigma_{12}\Sigma_{22}^{-1}\Sigma_{21})^{-1}\Sigma_{12}\Sigma_{22}^{-1} \\ -\Sigma_{22}^{-1}\Sigma_{21}(\Sigma_{11} - \Sigma_{12}\Sigma_{22}^{-1}\Sigma_{21})^{-1} & \Sigma_{22}^{-1} + \Sigma_{22}^{-1}\Sigma_{21}(\Sigma_{11} - \Sigma_{12}\Sigma_{22}^{-1}\Sigma_{21})^{-1}\Sigma_{12}\Sigma_{22}^{-1} \end{bmatrix}. \quad (A.5)$$

Thus, the conditional density function r is obtained as

$$r(\epsilon_1|\epsilon_2) = \frac{1}{(2\pi)^{p/2}|\Sigma_{11m}|^{1/2}}$$
$$\times exp\left[-\frac{1}{2}(\epsilon_1 - \mu_{1m})^T \Sigma_{11m}^{-1}(\epsilon_1 - \mu_{1m})\right], \quad (A.6)$$

where

$$\begin{aligned} \mu_{1m} &= E_m(\epsilon_1) \\ &= \mu_1 + \Sigma_{12}\Sigma_{22}^{-1}(\epsilon_{2m} - \mu_2), \end{aligned} \quad (A.7)$$

and

$$\begin{aligned} \Sigma_{11m} &= \Sigma_m(\epsilon_1) \\ &= \Sigma_{11} - \Sigma_{12}\Sigma_{22}^{-1}\Sigma_{21} \\ &= \Sigma_{11} - \Sigma_{12}\Sigma_{22}^{-1}\Sigma_{12}^T. \end{aligned} \quad (A.8)$$

Table 1: Distortion ratios with complete and partial measurement: Analytical results and random simulations for reflector surface z–displacements.

(1) $\rho = 0.99999$, $\mu = 0.001$, $\sigma = 0.001$.

number of actuators	distortion ratio					
	with complete measurement			with partial measurement		
	analytical value	random simulation		analytical value	random simulation	
	$\frac{E(f_c)}{E(f)}$	$\frac{E(f_c^*)}{E(f^*)}$	$\frac{\sigma(f_c^*)}{E(f^*)}$	$\frac{E(f_{cm})}{E(f)}$	$\frac{E(f_{cm}^*)}{E(f^*)}$	$\frac{\sigma(f_{cm}^*)}{E(f^*)}$
2	0.4653	0.4653	0.5418	0.4653	0.4653	0.5418
4	0.0604	0.0604	0.0708	0.0605	0.0605	0.0708
6	0.0338	0.0338	0.0421	0.0338	0.0339	0.0421
8	0.0189	0.0189	0.0231	0.0189	0.0189	0.0231
16	0.0017	0.0017	0.0021	0.0018	0.0018	0.0021
24	0.0000	0.0000	0.0000	0.0001	0.0001	0.0001
32	0.0000	0.0000	0.0000	0.0001	0.0001	0.0000
40	0.0000	0.0000	0.0000	0.0001	0.0001	0.0000

$E(f) = 0.1794 \times 10^{-4}$, $\mathrm{E}(f^*) = 0.1827 \times 10^{-4}$, $\sigma(f^*) = 0.2213 \times 10^{-4}$.

(2) $\rho = 0.5$, $\mu = 0.001$, $\sigma = 0.001$.

number of actuators	distortion ratio					
	with complete measurement			with partial measurement		
	analytical value	random simulation		analytical value	random simulation	
	$\frac{E(f_c)}{E(f)}$	$\frac{E(f_c^*)}{E(f^*)}$	$\frac{\sigma(f_c^*)}{E(f^*)}$	$\frac{E(f_{cm})}{E(f)}$	$\frac{E(f_{cm}^*)}{E(f^*)}$	$\frac{\sigma(f_{cm}^*)}{E(f^*)}$
2	0.5957	0.5918	0.3367	0.8854	0.8741	0.5032
4	0.3265	0.3317	0.1426	0.7523	0.7364	0.4417
6	0.2828	0.2797	0.1275	0.6695	0.6745	0.3837
8	0.2420	0.2497	0.1155	0.6000	0.6239	0.3450
16	0.0881	0.0886	0.0453	0.4916	0.4953	0.2511
24	0.0179	0.0179	0.0162	0.4433	0.4416	0.2245
32	0.0000	0.0000	0.0000	0.3980	0.4092	0.2160
40	0.0000	0.0000	0.0000	0.3834	0.3878	0.2455

$E(f) = 0.1325 \times 10^{-3}$, $\mathrm{E}(f^*) = 0.1303 \times 10^{-3}$, $\sigma(f^*) = 0.7897 \times 10^{-4}$.

(3) $\rho = 0.0$, $\mu = 0.001$, $\sigma = 0.001$.

number of actuators	distortion ratio					
	with complete measurement			with partial measurement		
	analytical value	random simulation		analytical value	random simulation	
	$\frac{E(f_c)}{E(f)}$	$\frac{E(f_c^*)}{E(f^*)}$	$\frac{\sigma(f_c^*)}{E(f^*)}$	$\frac{E(f_{cm})}{E(f)}$	$\frac{E(f_{cm}^*)}{E(f^*)}$	$\frac{\sigma(f_{cm}^*)}{E(f^*)}$
2	0.5891	0.5936	0.3787	0.8998	0.8881	0.5528
4	0.3427	0.3458	0.1446	0.7899	0.8001	0.4806
6	0.2998	0.3000	0.1306	0.7132	0.6992	0.3692
8	0.2577	0.2591	0.1218	0.6376	0.6352	0.3299
16	0.0791	0.0800	0.0443	0.5229	0.5283	0.2595
24	0.0130	0.0124	0.0104	0.5131	0.5085	0.2840
32	0.0000	0.0000	0.0000	0.4758	0.4783	0.2585
40	0.0000	0.0000	0.0000	0.3998	0.4048	0.1957

$E(f) = 0.2471 \times 10^{-3}$, $\mathrm{E}(f^*) = 0.2459 \times 10^{-3}$, $\sigma(f^*) = 0.1525 \times 10^{-3}$.

Table 2: Distortion-correction ratios with complete and partial measurement: Random simulations for reflector surface z–displacements.

(1) $\rho = 0.99999$, $\mu = 0.001$, $\sigma = 0.001$.

number of actuators	distortion-correction ratio			
	with complete measurement		with partial measurement	
	$E(\frac{f_c^*}{f^*})$	$\sigma(\frac{f_c^*}{f^*})$	$E(\frac{f_{cm}^*}{f^*})$	$\sigma(\frac{f_{cm}^*}{f^*})$
2	0.4659	0.0239	0.4691	0.0305
4	0.0629	0.0183	0.0671	0.0451
6	0.0376	0.0289	0.0417	0.0523
8	0.0220	0.0221	0.0284	0.0612
16	0.0030	0.0081	0.0129	0.0786
24	0.0002	0.0020	0.0089	0.0587
32	0.0000	0.0000	0.0068	0.0537
40	0.0000	0.0000	0.0058	0.0381

$E(f^*) = 0.1827 \times 10^{-4}$, $\sigma(f^*) = 0.2213 \times 10^{-4}$.

(2) $\rho = 0.5$, $\mu = 0.001$, $\sigma = 0.001$.

number of actuators	distortion-correction ratio			
	with complete measurement		with partial measurement	
	$E(\frac{f_c^*}{f^*})$	$\sigma(\frac{f_c^*}{f^*})$	$E(\frac{f_{cm}^*}{f^*})$	$\sigma(\frac{f_{cm}^*}{f^*})$
2	0.6516	0.2170	0.9202	0.2624
4	0.4035	0.1925	0.8112	0.3461
6	0.3528	0.1904	0.7813	0.4089
8	0.3119	0.1707	0.7305	0.3920
16	0.1167	0.0843	0.6085	0.3490
24	0.0233	0.0250	0.5537	0.3668
32	0.0000	0.0000	0.5135	0.3457
40	0.0000	0.0000	0.4940	0.3283

$E(f^*) = 0.1303 \times 10^{-3}$, $\sigma(f^*) = 0.7897 \times 10^{-4}$.

(3) $\rho = 0.0$, $\mu = 0.001$, $\sigma = 0.001$.

number of actuators	distortion-correction ratio			
	with complete measurement		with partial measurement	
	$E(\frac{f_c^*}{f^*})$	$\sigma(\frac{f_c^*}{f^*})$	$E(\frac{f_{cm}^*}{f^*})$	$\sigma(\frac{f_{cm}^*}{f^*})$
2	0.6485	0.2169	0.9334	0.2744
4	0.4272	0.2018	0.8902	0.4004
6	0.3786	0.2007	0.8090	0.3964
8	0.3248	0.1747	0.7545	0.3994
16	0.1052	0.0773	0.6474	0.3811
24	0.0169	0.0186	0.6260	0.4008
32	0.0000	0.0000	0.5912	0.3675
40	0.0000	0.0000	0.5146	0.3213

$E(f^*) = 0.2459 \times 10^{-3}$, $\sigma(f^*) = 0.1525 \times 10^{-3}$.

Table 3: Distortion ratios with complete and partial measurement: Analytical results and random simulations for all displacements.

(1) $\rho = 0.99999$, $\mu = 0.001$, $\sigma = 0.001$.

number of actuators	distortion ratio					
	with complete measurement			with partial measurement		
	analytical value	random simulation		analytical value	random simulation	
	$\frac{E(f_c)}{E(f)}$	$\frac{E(f_c^*)}{E(f^*)}$	$\frac{\sigma(f_c^*)}{E(f^*)}$	$\frac{E(f_{cm})}{E(f)}$	$\frac{E(f_{cm}^*)}{E(f^*)}$	$\frac{\sigma(f_{cm}^*)}{E(f^*)}$
2	0.9285	0.9285	1.1902	0.9285	0.9285	1.1902
4	0.8395	0.8395	0.9716	0.8395	0.8395	0.9716
6	0.7538	0.7537	0.8588	0.7538	0.7538	0.8588
8	0.6888	0.6888	0.8255	0.6888	0.6888	0.8255
16	0.4694	0.4694	0.5680	0.4694	0.4694	0.5680
24	0.4134	0.4134	0.4850	0.4134	0.4134	0.4850
32	0.3393	0.3393	0.4023	0.3393	0.3393	0.4023
40	0.2798	0.2798	0.3284	0.2798	0.2798	0.3284

$E(f) = 0.3117 \times 10^{-3}$, $E(f^*) = 0.3128 \times 10^{-3}$, $\sigma(f^*) = 0.3727 \times 10^{-3}$.

(2) $\rho = 0.5$, $\mu = 0.001$, $\sigma = 0.001$.

number of actuators	distortion ratio					
	with complete measurement			with partial measurement		
	analytical value	random simulation		analytical value	random simulation	
	$\frac{E(f_c)}{E(f)}$	$\frac{E(f_c^*)}{E(f^*)}$	$\frac{\sigma(f_c^*)}{E(f^*)}$	$\frac{E(f_{cm})}{E(f)}$	$\frac{E(f_{cm}^*)}{E(f^*)}$	$\frac{\sigma(f_{cm}^*)}{E(f^*)}$
2	0.8031	0.8059	0.5064	0.9399	0.9380	0.5462
4	0.6807	0.6767	0.4896	0.8785	0.8724	0.5412
6	0.5987	0.5980	0.3930	0.8163	0.8167	0.4591
8	0.4959	0.5006	0.3668	0.7466	0.7541	0.4175
16	0.3874	0.3952	0.3026	0.5907	0.5942	0.3241
24	0.2665	0.2677	0.1883	0.5240	0.5231	0.2528
32	0.2108	0.2116	0.1492	0.4168	0.4200	0.1868
40	0.1717	0.1716	0.1283	0.3791	0.3831	0.1729

$E(f) = 0.5367 \times 10^{-3}$, $E(f^*) = 0.5340 \times 10^{-3}$, $\sigma(f^*) = 0.3055 \times 10^{-3}$.

(3) $\rho = 0.0$, $\mu = 0.001$, $\sigma = 0.001$.

number of actuators	distortion ratio					
	with complete measurement			with partial measurement		
	analytical value	random simulation		analytical value	random simulation	
	$\frac{E(f_c)}{E(f)}$	$\frac{E(f_c^*)}{E(f^*)}$	$\frac{\sigma(f_c^*)}{E(f^*)}$	$\frac{E(f_{cm})}{E(f)}$	$\frac{E(f_{cm}^*)}{E(f^*)}$	$\frac{\sigma(f_{cm}^*)}{E(f^*)}$
2	0.7285	0.7187	0.2607	0.9213	0.9214	0.4008
4	0.5634	0.5677	0.1471	0.8419	0.8395	0.3096
6	0.4689	0.4655	0.1024	0.7641	0.7603	0.2759
8	0.4014	0.4056	0.0859	0.7202	0.7298	0.2662
16	0.2904	0.2897	0.0537	0.5841	0.5809	0.2030
24	0.2151	0.2163	0.0382	0.5123	0.5161	0.1835
32	0.1960	0.1968	0.0334	0.4580	0.4596	0.1406
40	0.1633	0.1646	0.0286	0.4178	0.4142	0.1340

$E(f) = 0.7618 \times 10^{-3}$, $E(f^*) = 0.7557 \times 10^{-3}$, $\sigma(f^*) = 0.3395 \times 10^{-3}$.

Table 4: Actuator locations for reflector surface z-displacements case.

location of actuators	number of actuators $\rho = 0.99999$								number of actuators $\rho = 0.0$							
	2	4	6	8	16	24	32	40	2	4	6	8	16	24	32	40
1						•		•								
2							•	•								
4						•	•	•						•	•	•
5								•						•	•	•
6	•				•								•	•	•	•
7							•	•							•	•
8							•	•							•	•
9						•	•	•							•	•
12								•								
13			•	•												
14						•		•							•	•
15															•	•
16						•										
17						•										
18						•										
20							•									
21						•										
22						•										•
23						•									•	
24														•		
26						•									•	•
28							•								•	•
30				•												
31	•															
32						•										
37						•										
42								•								
44															•	
46																
47						•										
50						•										•
55						•										
56								•								
58																•
61					•											
66						•										
67							•									
68							•									
73						•										
77																•
79						•								•		
81						•									•	•
82						•										
83								•								
91														•		
92								•								
93								•								
95								•								
97			•	•	•		•	•			•	•	•	•	•	•
98			•	•	•		•	•		•	•	•	•	•	•	•
99		•		•	•	•	•	•	•	•	•	•	•	•	•	•
100		•	•	•		•	•	•		•	•	•	•	•	•	•
101		•	•		•	•	•	•		•	•	•	•	•	•	•
102		•	•	•	•	•	•	•		•	•	•	•	•	•	•
103		•	•	•	•	•	•	•		•	•	•	•	•	•	•
104			•	•	•	•	•	•			•	•	•	•	•	•
105				•	•	•	•	•				•	•	•	•	•
106								•								•
107						•	•	•								
109						•	•									
110							•	•								
111							•	•					•		•	
112								•							•	•
113					•			•								
117																•
118																•
120																•
121																•
122															•	•
123																•
124								•						•		•
125														•		•
126							•								•	
127							•	•								
128							•									
129						•	•								•	
130						•										
131							•								•	
132																•
133					•		•									
134								•						•		•
135					•										•	
136															•	•
138																•
139																•
140																
142														•	•	
143						•									•	•
144								•								•
146															•	
148						•		•								
149						•	•	•						•		

THERMAL GRADIENTS FOR DELAMINATION CONTROL

S. Hanagud[*], S.N. Atluri[†], L.N.B. Gummadi[‡] and C.C. McColl[§]
School of Aerospace Engineering
FAA Center of Excellence for Computational Modeling of Aircraft Structures
Georgia Institute of Technology
Atlanta, Ga-30332

ABSTRACT

In this paper, we have studied the feasibility of using thermal gradients to actively control the magnitude of interlaminar stresses. Interlaminar stresses are responsible for the development and growth of a delamination. Closed form analytical solutions have been obtained for interlaminar stresses in a laminated composite plate subjected to a combined mechanical and thermal loading. Thermal gradients required for reducing the magnitude of interlaminar stresses below a prescribed level are determined. Preliminary tests confirm the analytical results.

---

[*]Professor, Member AIAA
[†]Institute Professor and Regents Professor of engineering, Fellow AIAA
[‡]Post Doctoral Fellow
[§]Graduate Student, Student Member AIAA
copyright ©1994 by the American Institute of Aeronautics and Astronautics, Inc., All rights reserved.

INTRODUCTION

In the past, many passive techniques have been proposed to control interlaminar stresses and avoid the formation of delaminations. Pagano and Pipes[1] have suggested the use of a specific stacking order to eliminate the interlaminar tension. This technique can be used in the neighborhood of delamination-prone areas. However, a specific stacking sequence may defeat other design objectives. Pagano and Lackman[2] have shown that by serrating the fiber edges and filling in voids with the matrix material, the on-set of delamination can be prevented. Mignery et. al[3] have suggested sewing or wrapping the laminate in delamination prone areas. Chan et. al.[4] used a thin layer of tough adhesive at the interface of the fiber/matrix to delay the onset of delamination. In reference 5, Chan et. al studied the effect of parameters like the stacking sequence and the poisson's ratio on the interlaminar stresses. The focus of their studies was to prevent or delay the onset of a delamination. They did not consider techniques to arrest the growth of an existing delamination.

The only active control techniques to date, are proposed by Rogers et. al.[6] and Hanagud et. al. [7] and Gummadi[8]. In reference 6, authors have suggested a technique that incorporates strain induced actuators around the damage area. A proper actuation will reduce the stress concentration near the damage, thereby retarding the damage growth. In references 7 and 8, the authors proposed an active control technique that involved detecting the delamination location and then controlling the growth of the detected delamination by using active control techniques.

PROBLEM SETTING

The objective of this paper is to study the feasibility of using smart structure concepts to control delaminations in laminated composite structures. The growth of delaminations in a laminated composite structure can be arrested by reducing the interlaminar stresses that are responsible for the growth. An active control to reduce these interlaminar stresses needs actuators that can be used to counter the interlaminar stresses that are created by applied service loads. Several candidate actuators were considered. These include piezoelectric actuators, electrostrictive actuators, shape memory alloy actuators and imposed thermal gradients. In this paper, we have restricted our studies to the use of imposed thermal gradients.

In order to explore the possible effectiveness of a thermal field in arresting the delaminations, we have selected a simple problem. In this simple problem, a layered composite plate of rectangular shape is considered. It is assumed that the plate is subjected to a uniform axial load along two parallel edges. In order to maintain the simplicity of the feasibility study, no pre-existing delaminations are considered. Instead, the feasibility study is focussed on the reduction of interlaminar stresses at the free edges of the rectangular plate.

In order to understand the structural mechanics of the actuation mechanism, analytical solutions are sought for the interlaminar stresses due to the applied axial forces and control forces transmitted by the actuators. In this study, imposed thermal fields transmit the needed control force.

ANALYTICAL SOLUTIONS FOR INTERLAMINAR STRESSES

Nishioka and Atluri[9] used assumed equilibriated stress shapes and the principle of minimum complementary energy to efficiently estimate the magnitude of interlaminar stresses near the cut-out holes. Kassappogolu and Lagace[10], have essentially used the same method for obtaining analytical solutions for interlaminar stresses at free edges. Their assumed stress field, however, does not satisfy boundary conditions along the loaded edges. We have corrected this situation by assuming a stress field that satisfies boundary conditions along all the free edges.

Webber and Morton[11] described a method for obtaining analytical expressions for free edge stress fields due to thermal effects in a laminated composite plate. They have assumed a uniform temperature increase. They have attempted to obtain an analytical solution to this problem by using a complementary energy approach similar to that described by Kassappogolu and Lagace[10]. They have, however ignored the fact that the thermal expansion can take place in all directions. Interlaminar stresses are calculated by satisfying boundary conditions along only two directions. We also corrected this situation by assuming interlaminar stress fields that satisfy boundary conditions along all the edges.

INTERLAMINAR STRESSES DUE TO MECHANICAL LOADING

In this analysis, we have considered cross ply laminates. A rectangular plate of dimension $2a X 2b X h$ made from orthotropic layers, $(0/90/0/90......90/0)$, as shown in Figure 1, is considered. These layers are assumed to be symmetrically placed about the center plane of the plate. The plate is assumed to be subjected to an axial load of N_x. Using classical laminate plate theory, the stress field in the k th layer $\{\sigma\}^k$ can be determined. But these classical stresses do not satisfy the free surface stress free boundary conditions. To correct this situation, we should consider interlaminar stresses.

An approximate analytical solution has been obtained by assuming a stress field that satisfies equilibrium equations, all boundary conditions and interlaminar traction continuity conditions. The stress field is used to compute the complementary energy in the laminate. By taking the symmetry of the laminate into account, the assumed stress field in the range of $0 \leq x \leq a$ and $0 \leq y \leq b$ can be written as

$$\sigma_{xx}^k = \bar{\sigma}_{xx}^K + A[e^{-\phi_1 x} - e^{-\lambda\phi_1 x}] \quad (1)$$

$$\sigma_{yy}^k = \bar{\sigma}_{yy}^k[1 - e^{-\phi_2 y} - \phi_2 y e^{-\phi_2 y}] \quad (2)$$

$$\sigma_{zz}^k = A[\phi_1 e^{-\phi_1 x}(\phi_1 x - 2) - \quad (3)$$
$$\lambda\phi_1 e^{-\lambda\phi_1 x}(\lambda\phi_1 x - 2)]$$
$$[\frac{z^2}{2} + zB_1^k + B_2^k] +$$
$$\bar{\sigma}_{yy}^k \phi_2^2 e^{-\phi_2 y}[1 - \phi_2 y]$$
$$[\frac{z^2}{2} + zD_1^k + D_2^k]$$

$$\sigma_{xz}^k = A[e^{-\phi_1 x}(1 - \phi_1 x) - e^{-\lambda\phi_1 x} \quad (4)$$
$$(1 - \lambda\phi_1 x)][z + B_1^k]$$

$$\sigma_{yz}^k = \bar{\sigma}_{yy}^k \phi_2^2 y e^{-\phi_2 y}[z + D_1^k] \quad (5)$$

where $\bar{\sigma}_{xx}^k$, and $\bar{\sigma}_{yy}^k$ are the classical laminate solutions [12]. In these equations, exponential functions[10] are used to account for the fact that interlaminar stresses reduce rapidly and the resulting stress field will be the classical laminate plate theory solution away from the free edges. The assumed solution satisfies the boundary condition and the equilibrium equations.

There are $4n$ constants $B_1^k, B_2^k, D_1^k, D_2^k$ for $k = 1, ..., n$. These constants are obtained to satisfy the interlaminar traction conditions similar to reference 10 but in two directions. They are

$$B_1^k = -\frac{1}{\sigma_{xx}^k}\sum_{j=1}^{k-1}\bar{\sigma}_{xx}^j t^j - z^k \quad (6)$$

$$B_2^k = \frac{1}{\sigma_{xx}^k}\sum_{j=1}^{k-1}\bar{\sigma}_{xx}^j t^j z^j - z^{k^2}/2 \quad (7)$$

$$D_1^k = -\frac{1}{\sigma_{yy}^k}\sum_{j=1}^{k-1}\bar{\sigma}_{yy}^j t^j - z^k \quad (8)$$

$$D_2^k = \frac{1}{\sigma_{yy}^k}\sum_{j=1}^{k-1}\bar{\sigma}_{yy}^j t^j z^j - z^{k^2}/2 \quad (9)$$

Constants ϕ_1, ϕ_2, λ and A are determined by minimizing the complementary energy for the laminate so that, the assumed stress field also leads to a compatible displacement field. The expression for the laminate complementary energy is given by the summation

$$\Pi_c = \sum_{k=1}^{n}\frac{1}{2}\int_v\int\int \sigma^T \bar{S}\sigma dV \quad (10)$$
$$-\int_{Au}\int Tu dA$$

Due to geometrical symmetry, only one fourth of the laminate is considered. It is assumed that the laminate is wide enough so that $e^{-a\phi_1}, e^{-a\phi_2}, e^{-b\phi_1}$ and $e^{-b\phi_2}$ are approximated to be zero. The second term in the integration is equal to zero since there is no prescribed displacement in the present problem. Four algebraic nonlinear coupled equations are obtained by taking partial derivatives of the expression for Π_c with respect to the four unknowns ϕ_1, ϕ_2, λ and A. By solving these four equations, $\phi_1, \phi_2, \lambda, A$ can be obtained.

INTERLAMINAR STRESSES DUE TO THERMAL LOADING

A cross-ply rectangular laminate plate of $2aX2bXh$ made from orthotropic layers, (0/90.....90/0), as shown in Figure 1 is considered. Each of the lamina in the plate is assumed to be subjected to a symmetric temperature change of ΔT^k. Using the classical laminate theory, classical stress field in k th layer is determined. To satisfy all the boundary conditions, the stress field is chosen as

$$\sigma_{xx}^k = \bar{\sigma}_{xx}^K[1 - e^{-\phi_1 x} - \phi_1 x e^{-\phi_1 x}] \quad (11)$$
$$\sigma_{yy}^k = \bar{\sigma}_{yy}^k[1 - e^{-\phi_2 y} - \phi_2 y e^{-\phi_2 y}] \quad (12)$$
$$\sigma_{zz}^k = \bar{\sigma}_{xx}^k \phi_1^2 e^{-\phi_1 x}[1 - \phi_1 x] \quad (13)$$
$$[\frac{z^2}{2} + zB_1^k + B_2^k] +$$
$$\bar{\sigma}_{yy}^k \phi_2^2 e^{-\phi_2 y}[1 - \phi_2 y]$$
$$[\frac{z^2}{2} + zD_1^k + D_2^k]$$
$$\sigma_{xz}^k = \bar{\sigma}_{xx}^k \phi_1^2 x e^{-\phi_1 x}[z + B_1^k] \quad (14)$$
$$\sigma_{yz}^k = \bar{\sigma}_{yy}^k \phi_2^2 y e^{-\phi_2 y}[z + D_1^k] \quad (15)$$

where $\bar{\sigma}_{xx}^k$, and $\bar{\sigma}_{yy}^k$ are the classical laminate solutions. Symmetry of the laminate is taken into account. In these equations, exponential functions are used[10,11] to account for the fact that interlaminar stresses reduce rapidly and the resulting stress field will be the classical laminate plate theory solution away from the free edges. The assumed solution satisfy the boundary conditions.

There are $4n$ constants $B_1^k, B_2^k, D_1^k, D_2^k$ for $k = 1, ..., n$. These constants are obtained by satisfying the interlaminar traction conditions and are given by equations 6 to 9. Constants ϕ_1 and ϕ_2 are determined by minimizing the complementary energy for the laminate. The expression for the laminate complementary potential energy under thermal field is given by the summation

$$\Pi_c = \sum_{k=1}^n \frac{1}{2} \int_v \int \int \sigma^T \bar{S} \sigma dV + \quad (16)$$
$$\int_v \int \int \sigma^T e_\theta dV - \int_{Au} \int TudA$$

Again, due to geometrical symmetry, only one fourth of the laminate is considered. The third term in the integration is equal to zero since there is no prescribed displacement in the present problem. Two algebraic nonlinear coupled equations are obtained by taking partial derivatives of the expression for Π_c with respect to the two unknowns ϕ_1 and ϕ_2. By solving these two equations, ϕ_1, ϕ_2 can be obtained and as a result, all the stress filed can be obtained.

The objective of this paper is to use the thermal loading as the actuation mechanism for the reduction of the interlaminar stresses. This actuation mechanism can be used for designing feedback controllers by using control theory.

NUMERICAL RESULTS

To study the accuracy of the assumed stress shapes, a rectangular laminate of CFRP AS-

3201 with [0/90/90/0] orientation is considered. Material properties are $E_{11} = 139$ GPa(20.2MPSI), $E_{22} = 11.7$ GPa(1.7MPSI), $G_{12} = 5.8$ GPa(0.85MPSI), $\mu_{12} = 0.28$. Interlaminar stress with the application of uniform axial stress is shown in Figure 2. Here the interlaminar stresses are shown near the free edge. x-location of the stresses are fixed at four times the thickness where as y location is varied from zero (free edge) to six times the thickness. It can be observed that the interlaminar stress reduces as we go away from the free edge. Dotted line in this figure denote the interlaminar stress estimation using reference 10. Now the x-location is changed to half the thickness and y-location is varied from zero to six times the thickness. Interlaminar normal stresses are shown in Figure 3. This location corresponds to the corner location of the plate. Objective for choosing this location is to study the effect of the unsatisfied boundary condition. Similar variation in the interlaminar stresses is observed between the results of the present approach and that of Reference 10.

To study the effect of unsatisfied boundary condition under thermal loading, we conducted similar studies. Keeping the x-location constant at four times the thickness, interlaminar stresses are plotted varying the y-location from zero to six times the thickness in Figure 4. Results are compared with the results of reference 11. Close agreement is seen between both the results. Now, x-location is changed to half the thickness. This location is close to the boundary in which boundary conditions have not been satisfied in reference 11. Interlaminar stresses are shown in Figure 5 as y location is varied. Significant difference is observed as we move away from the free edge. Results using the approach of reference 11 indicate stresses tend to become zero as we move away from the free edge where as results from the present approach are leading to a constant value. This constant value is due to the interlaminar stresses near the second free boundary. To study the effect of this free boundary, we kept the location of y equal to 4 times the thickness and varied the location of x from zero (free edge) to four times the thickness. Results are shown in Figure 6. Results from the present approach show a large stress at the free edge and reduction in the stresses away from the edge. Results from reference 11 indicate zero interlaminar stresses erroneously. From this, we can conclude that the effect of unsatisfied boundary conditions under thermal loading is significant. Only region this effect is not significant is at the center of the plate.

To study the feasibility of applied thermal gradients, we considered the same rectangular laminate. By applying an axial load of 4375000 N/M (25,000 pounds/inch), interlaminar normal stresses developed at the interface of 0-90 is shown in Figure 7. In order to suppress these interlaminar stresses, we considered a temperature gradient of 25/0/0/25 (^oC) to suppress the interlaminar stresses. Interlaminar normal stress at the interface of 0-90 is shown in Figure 8. It can be seen from the Figures 7 and 8, that the applied thermal field resulted in exactly the same magnitude of interlaminar stresses in the opposite direction by making the effective interlaminar stresses equal to zero. Thus, by using the applied thermal field, we can reduce the high interlaminar stresses by controlling the formation and growth of delamination.

EXPERIMENTAL VERIFICATION

To validate the developed theory, we initiated an experimental investigation with thermal actuators. Experimental techniques for measuring the interlaminar stresses by the use of piezoceramic transducer is first developed.

Test Specimen A 0/90/0 glass epoxy specimen is fabricated using tow pregs supplied by Custom Composites Inc. of Atlanta, Georgia. The dimensions of the specimen are shown in Figure 9. Initially, three separate specimens are fabricated and bonded together, using epoxy resin with embedded piezoceramic transducers for measurement of interlaminar stresses and a thin film of nichrome wire for thermal loading. The thickness of the piezoceramic transducer is 0.0006 inch and the thickness of the nichrome wire is 0.0001 inch. Another piezoceramic transducer is bonded to the upper surface of the specimen for calibration. Dimensions of the overall specimen meet the ASTM standard of Tensile testing [13].

Instrumentation The piezoceramic transducers bonded to the specimen act as sensors. Terminals of the piezoceramic sensor are attached, through a charge amplifier, to a digital oscilloscope. The embedded nichrome wire is connected to a thermal controller for applying the thermal loading. The instrumentation diagram is shown in Figure 10. The mechanical loading is applied through an INSTRON 8501. Since the piezoceramic sensors can not detect a steady load, it was decided to apply a cyclic load of very low frequency that is of the order of 1/2 Hz for mechanical loading and thermal loading. This frequency is significantly below the first harmonic frequency of the beam but above the cutoff frequency of the oscilloscope, at which PZT sensor can detect without errors.

From analysis, it can be shown that when mechanical load is applied, interlaminar stresses can be identified from voltage measurements

$$\sigma_{zz} = \frac{V^2 - V^1}{K_{23}h} \quad (17)$$

†Details of the experimental results will be presented in a separate paper

Here V_1 is the voltage measured by the sensor that is bonded to the surface and V_2 is the voltage measured by the embedded sensor. K_{23} is a function of the piezoceramic material constant and h is the thickness of the layer of the laminate. Similarly, for the specimen that is subjected to thermal loading, interlaminar stresses can be obtained as

$$\sigma_{zz} = \frac{V^2 - V^1 - f(T)}{K_{23}h} \quad (18)$$

Here $f(T)$ is the correction term to account for the different coefficient of thermal expansion of the laminate and thermal effect on transducers.

TESTS Experiments were conducted and interlaminar stresses were determined from the recorded data using appropriate caliberation. Tests were separately conducted under mechanical loading and under both the mechanical and thermal load. The effect of mechanical loading on the interlaminar stress is shown in Figure 11. It can be seen that there is an increase in the interlaminar stress with the increase in the mechanical loading. Magnitudes were compared with analytical results for accuracy.

Results of effect of thermal loading is shown in Figure 12. A mechanical cyclic loading of 0-100 lbs is applied and a thermal loading of 0-15 degree farenheit is applied. The interlaminar stress that was observed to be positive when the mechanical loading is applied is observed to be negative when this thermal loading is applied in addition to the mechanical load indicates that interlaminar tension is changed to interlaminar compression by the application of a relatively small thermal field. The results presented here should be as considered only preliminary. More experiments should be conducted to establish the trend and the amplitude of actual control effort that is needed.

CONCLUSIONS

Interlaminar stresses are determined under axial mechanical loading and under combined axial mechanical and thermal loading. Feasibility of applied thermal gradients as the actuation mechanism for the reduction of the magnitude of interlaminar stresses has been analytically demonstrated. Preliminary experimental investigation validates analytical studies.

ACKNOWLEDGEMENTS

This work was supported by a grant to the Center of Excellence for Computational Modeling of Aircraft Structures from the Federal Aviation Administration.

REFERENCES

[1] Pagano, N.J. and Pipes, R.B., " Some Observations on the Interlaminar strength of composite Laminates", Int.J.Mech. Sci., Vol 15,(1973), pp 679-688.

[2] Pagano, N.J. and Lackman, L.M.," Prevention of Delamination of composite laminates", AIAA Jl Vol.13, 1975,pp 399-401.

[3] Mignery, L.A., Tan, T.M., and Sun, C.T., " The Use of Stitching to Suppress Delamination in Laminated Composites", ASTM Symposium on Delaminations and Debonding of Materials, 1983.

[4] Chan, W.S., Rogers, C. and Aker,S., " Improvement of Edge Delamination Strength of Composite Laminates Using Adhesive Layers", Composite Materials: Testing and Design, ASTM STP 893, 1986.

[5] Chan, W.S., Rogers, C., Cronkhite, J.D and Martin, J, " Delamination Control of Composite Rotor Hubs", Journal of helicopter Society, pp 60-69, 1986.

[6] Rogers, C.A., Liang, C., Li, S,, " Active Damage Control of Hybrid Material Systems Using Induced Strain Actuators", Proc. 32 SDM Conference, pp 1190-1199, 1991.

[7] Hanagud, S., Nagesh Babu, G.L , Roglin, R. and Savanur, S., "Active Control of Delaminations in composite structures" Proc. 33rd AIAA/ASME/ASCE/AHS SDM Conference, pp 1819-1830, 1992.

[8] Gummadi, Lakshmana Nagesh Babu, "Active Control of Delaminations in Smart Composite Structures", Ph.D Thesis presented to school of Aerospace Engineering, Georgia Institute of technology, Atlanta, December 1992.

[9] Nishioka, T. and Atluri, S.N., " Stress Analysis of Holes in Angle- Ply Laminates: An efficient Assumed Stress 'Special - Hole Element' Approach and a simple estimation method", Computers and Structures, Vol. 15, No. 2, 1982, pp 135-147.

[10] Kassapoglou, C. and Lagace, P.A., " An Efficient Method for the Calculation of Interlaminar Stresses in Composite Materials", Journal of Applied Mechanics, Vol 53, 1986. pp 744- 750.

[11] Webber, J.P.H and Morton, S.K., " An Analytical Solution for the Thermal Stresses at the Free Edges of Laminated Plates", Composite Science and Technology, 1993.

[12] Tsai, S.W., " Composites Design", Think Composites, Dayton, 1987.

[13] ASTM Standard D3039 -76 " Standard Test Method for Tensile Properties of Fiber Resin Composites", American Society for Testing and materials, 1993.

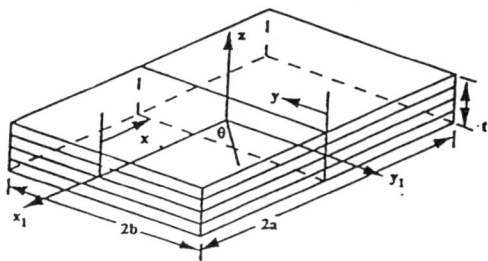

Figure 1: Laminated plate geometry

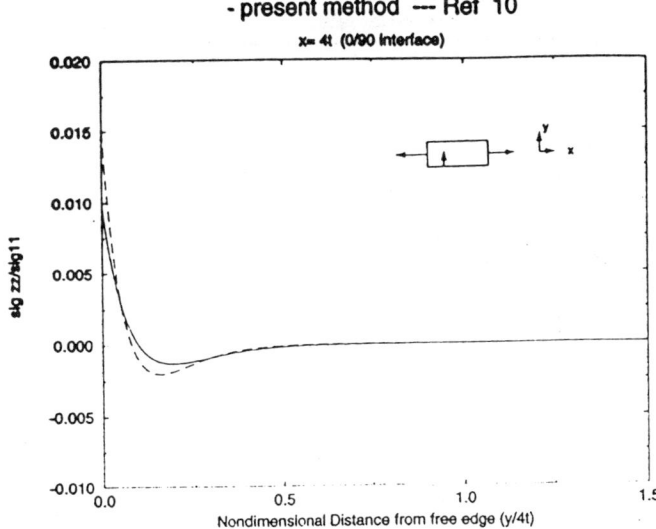

Figure 2: Variation of σ_{zz} due to mechanical load (x=4t)

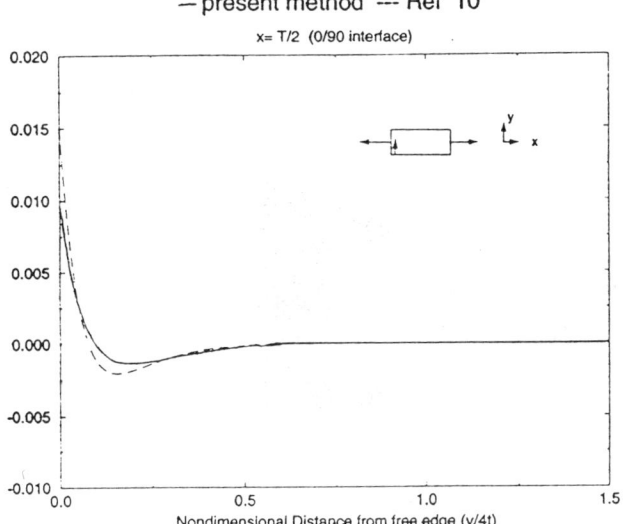

Figure 3: Variation of σ_{zz} due to mechanical load (x=t/2)

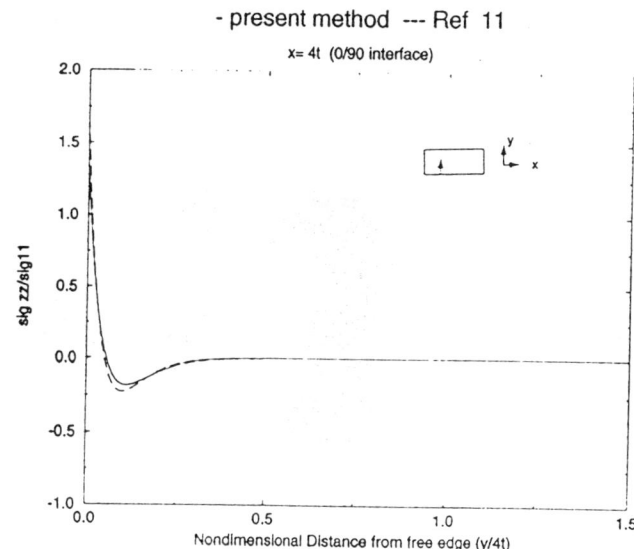

Figure 4: Variation of σ_{zz} due to thermal load (x=4t)

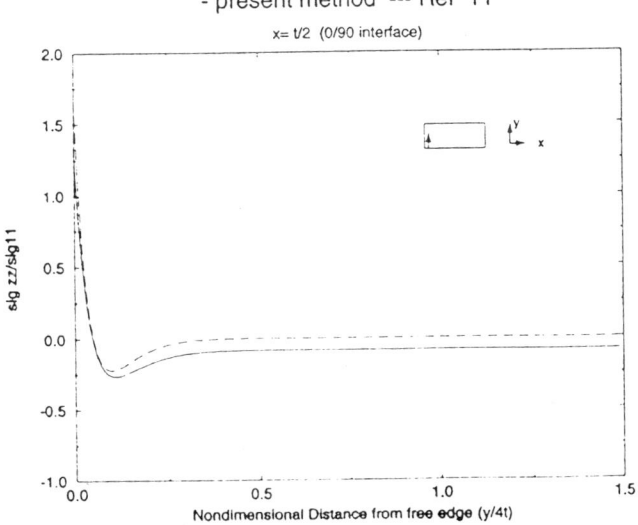

Figure 5: Variation of σ_{zz} due to thermal load (x=t/2)

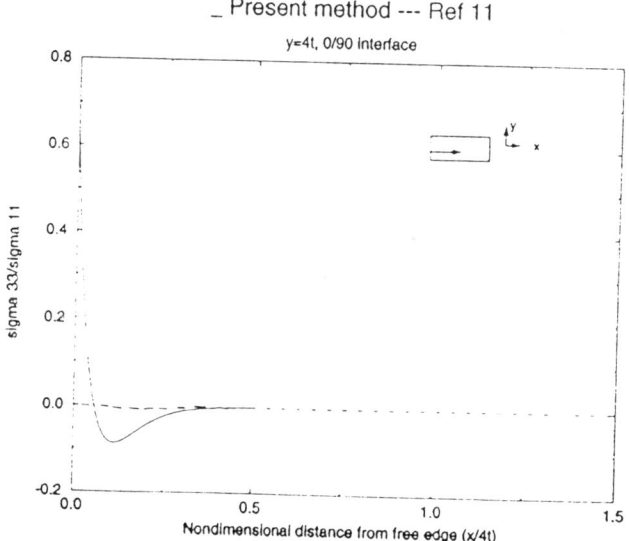

Figure 6: Variation of σ_{zz} due to thermal load (y=4t)

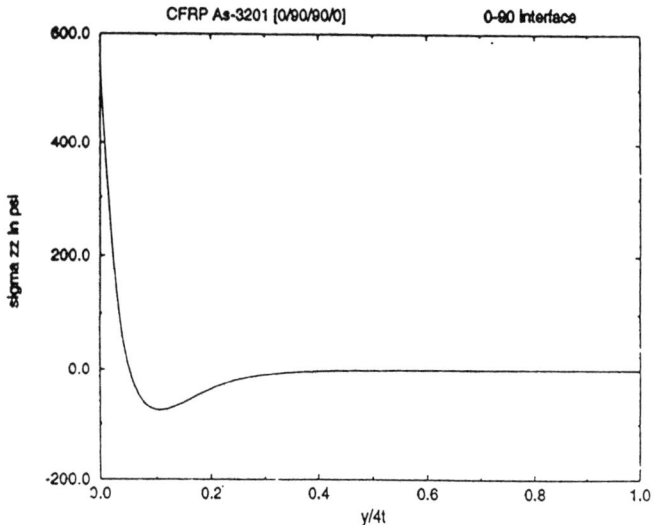

Figure 7: σ_{zz} Distribution due to Mechanical Loading

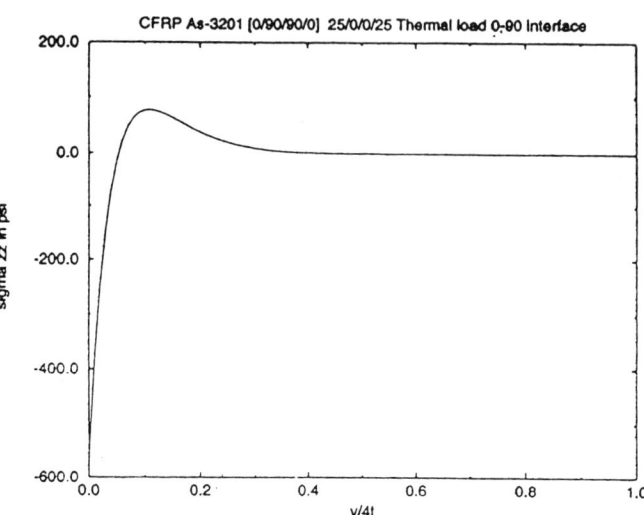

Figure 8: σ_{zz} Distribution due to Thermal Loading

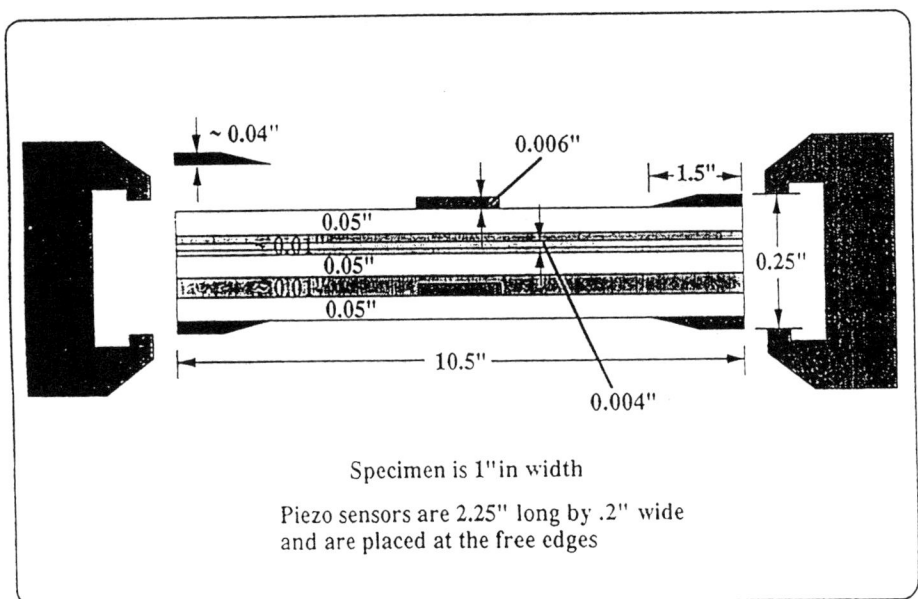

Specimen is 1" in width

Piezo sensors are 2.25" long by .2" wide and are placed at the free edges

Figure 9: Specimen dimensions

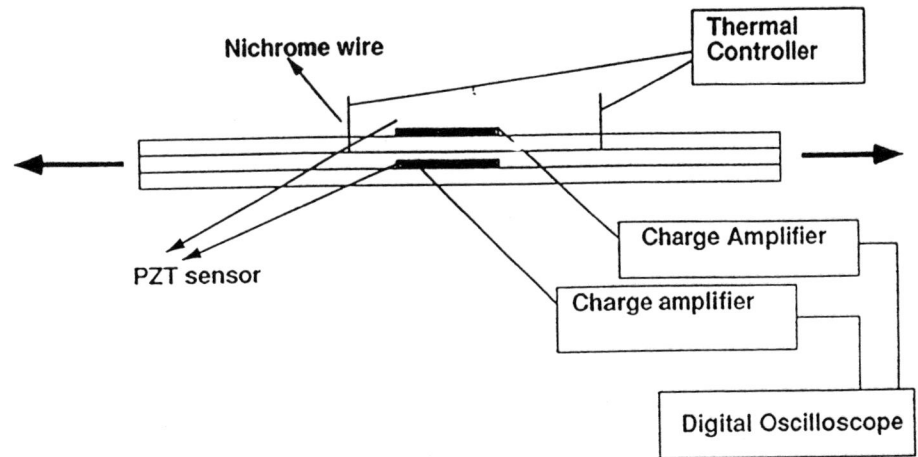

Figure 10: Experimental setup

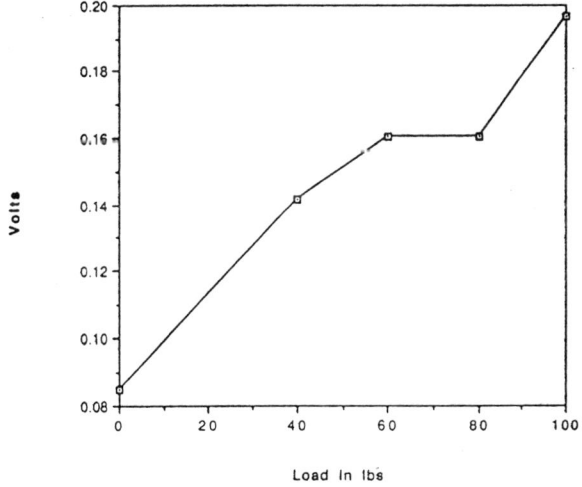

Figure 11: Interlaminar stresses due to mechanical load (σ_{zz})

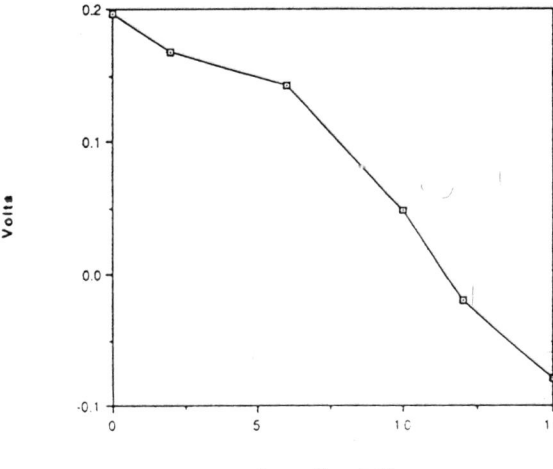

Figure 12: Interlaminar stresses due to mechanical and thermal

OPTIMIZING INDUCED STRAIN ACTUATORS FOR MAXIMUM PANEL DEFLECTION

Tamara J. Leeks[*]

Terrence A. Weisshaar[**]

Purdue University
West Lafayette, Indiana

Abstract

This paper examines the electro-mechanical interaction between a thin self-straining piezoelectric actuator and a simply supported host plate when the actuator is placed on one side of the plate and its objective is to create a large bending deflection. The purpose of the actuator is to create bending deflections to control local aerodynamic pressures and resultant forces such as lift and drag. These studies show that there is a trade-off between the additional stiffening, provided by actuator thickness and the area that it covers on the plate, and the amount of force and moment provided by the actuator. A Rayleigh-Ritz analysis shows that the optimum size, thickness and coverage, of the actuator with respect to the host panel is determined by panel aspect ratio, and relative elastic moduli. The strain energy content of the actuator/plate combination shows that the best combinations of actuator thickness and panel coverage can be identified by plotting strain energy against actuator thickness or area. With an aluminum host plate and a PZT actuator, the best rectangular actuator size is about 0.6 the thickness of the plate and covers about 65% of the host plate.

Introduction

Aerodynamic loads, and the local pressures that create these loads, depend on the surface shape on which they act. Surface panels may be flat or curved and are designed to provide aerodynamic shape and to guarantee structural integrity. Re-shaping smooth aero/structural surfaces to change the pressure distribution is done by bonding or otherwise attaching thin actuators to the inside surfaces of structural panels to create an asymmetric configuration that will bend on command.

Thin plate-like or lattice reinforced panels with embedded self-straining actuators such as shape memory materials or piezoelectric materials have been proposed for aerodynamic control concepts that include actively controlled panels to reduce transonic drag[1,2] and active panel elements to increase supersonic panel flutter speed[3,4,5,6,7]. For transonic drag reduction, the deformation of a panel on the upper surface of a supercritical airfoil can change the flow field and shock wave intensity to reduce drag on command.

---

[*] NSF Graduate Research Fellow, School of Aeronautics and Astronautics, Member AIAA

[**] Professor, School of Aeronautics and Astronautics, Fellow AIAA

Panel flutter suppression with piezoelectric actuators and shape memory alloy actuators is unique in that no articulated device exists to do the same task. In supersonic flow, dynamic oscillations can be reduced by placing thin actuators on the panel surfaces to change the frequencies of the panels on command.

One serious problem with active panel concepts is the difficulty finding a design combination to give large enough panel out-of-plane deformations to create the required changes in aerodynamic forces. Without deflections of the order of a panel thickness (or even more), controlling the size and position of the aerodynamic forces is marginal.

A desirable actuator, such as one using today's piezoelectric materials, can not create significant bending deformation of panels unless the host panel/actuator combination is tailored to extract every bit of electro-mechanical efficiency out of the configuration. An emphasis on efficiency naturally leads to considerations of formal optimization that includes a design objective and design variables. However, before formal optimization can proceed, we must select our design variables and determine the sensitivity of the design objective to these design variables.

This paper is a pre-optimization study that examines the interaction between actuator self-straining ability, bending stiffness, thickness and planform coverage and the host panel bending stiffness and aspect ratio. The purpose of the actuator is to produce large bending deflection. The intent of the study is to identify effective panel/actuator combinations and understand why some combinations are more effective than others.

Background and configuration description

Parameters that affect panel out-of-plane deflection and enter into the optimization process include: the thickness of the actuator compared to the thickness of the panel; the surface area covered by the actuator; the position of the actuator on the panel surface; the boundary conditions at the edges of the panel; the aspect ratio of the panel planform; and, the material properties of the actuator compared to the host panel.

Analytical work described in this paper uses a baseline configuration shown in Figure 1. The size of this panel is consistent with the requirements for an active panel that might be placed between ribs of an active wing. The host plate material is aluminum, with dimensions of 12 in. by 18 in., and a thickness of 0.05

in. A thin lead zirconate titanate (PZT) piezoelectric actuator is attached to the bottom inner surface of the host plate. Expansion or contraction of this actuator will cause panel extension and bending. The area and thickness of the actuator are parameters for the study.

Kim & Jones[8] have studied a similar problem to find the optimal thickness of piezoactuators to maximize the bending moment induced by rectangular actuators surface-bonded to the upper and lower surfaces of a thin flat plate. They found that the best actuator thickness is approximately half the thickness of a steel host substructure and a quarter of the thickness for an aluminum host substructure.

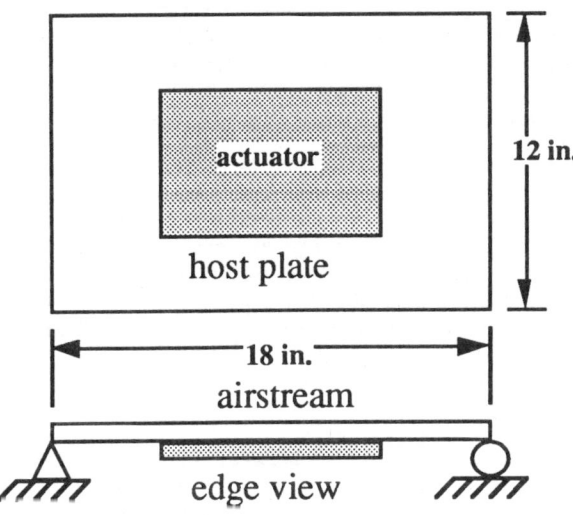

Figure 1 - Panel/actuator geometry

Rogers, Liang, & Jia[9] studied mid-plane symmetric shape memory alloy reinforced plates, using a Rayleigh-Ritz method for their numerical results. They obtained an approximate solution to the plate bending problem, free vibration, buckling, and acoustic transmission loss.

Crawley & Lazarus[10] developed a consistent plate model with embedded actuator stiffness and strain included as laminated plate layers. Their computations were based on a Rayleigh-Ritz model. Wang & Rogers[11] also applied laminate plate theory to a laminated plate containing induced strain actuator "patches" bonded symmetrically to the surface or embedded within the laminate. The thickness and size of these actuator patches are relatively small compared to those of each lamina.

This study differs from previous studies because the actuator is placed on only one side of the host plate. However, we will use laminated plate theory and a Rayleigh-Ritz solution technique, although we will check its accuracy using a finite element analysis.

Unsymmetric host plate/actuator combinations are different than symmetrical combinations in at least two major ways. First of all, nonsymmetry of the material stiffness through the thickness of the plate/actuator combination assures that there will be coupling between in-plane induced strain and panel bending. The bending stiffness of the plate/actuator combination will always be less than a similar combination of material where the total actuator thickness is the same, but distributed symmetrically about the mid-plane[12].

There is a second major difference between our configuration and symmetrical panel actuator combinations. For unsymmetric cross-ply laminates, "large deflection effects" can occur for configuration loading that normally would be regarded as producing deflections in the small plate deflection range. Bending-extension coupling can produce a stiffening or softening effect depending on the direction of the deflection.[13] An unsymmetric cross-ply laminate in cylindrical bending has different apparent bending stiffnesses in the positive and negative deflections. As a result, positive and negative loads of the same magnitude produce different magnitudes of deflection and linear lamination theory may give large differences when the nonlinear effects are ignored.[14]

Nonlinear large deflection effects will most likely increase the bending deflection found from linear theory. Our intent is to provide information as how to maximize panel bending deflections. The additional computational effort required to consider these nonlinearities was not regarded as essential to these results at this time, so nonlinear effects were excluded from this study.

Two computational methods for calculating bending deflection were considered. A finite element program (NASTRAN) was first used to compute the deformations of the plate/actuator combination. In addition, a Rayleigh-Ritz procedure was developed, beginning with a strain energy expression based on laminated plate theory. This expression and the development of the equations of static equilibrium necessary to compute bending deflection are provided in the Appendix. The analysis is based on classical laminated plate theory. The energy expression is useful as a guide to understanding why some plate/actuator combinations are better than others.

Finite Element Model

The panel configuration modeled using the NASTRAN finite element program consisted of quadrilateral plate elements with membrane and bending stiffness. Each of the small elements has the same aspect ratio as the plate, and are arranged 24 along each side, for a total of 576 elements. The actuator elements are directly attached to the host plate with no intervening bonding layer.

For each different actuator/host plate configuration considered, a new model must be created and the finite element program run. While the model computation time is small, the time required to create models and to interpret data was considered to be excessive, given the scope of our study.

The piezoelectric strain is a combination of the piezoelectric constant d_{31} multiplied by the electric field strength E_3. NASTRAN has no piezoelectric finite element capability, so an equivalent thermal element was used. Thermal and piezoelectric strains are both induced strains so that a thermal coefficient of expansion and temperature increase can be assigned to mimic the actuation strain of the piezoelectric material.

Rayleigh-Ritz Approach

An approximate solution for the plate deflections can be obtained using a Rayleigh-Ritz method based on laminate electro-mechanical strain energy and an assumed displacement field for inplane and out-of-plane (bending) deflections. The panel strain energy expression in the Appendix is composed of four basic types of terms and may be written conceptually as:

$U = $ inplane stiffness + bending stiffness
$+$ inplane / bending coupling + induced strain

The stiffnesses involving shear-extension coupling vanish if the laminate is a symmetric laminate with isotropic or specially orthotropic layers, that is

$$A_{16} = A_{26} = B_{16} = B_{26} = D_{16} = D_{26} = 0 \quad (1)$$

The assumed displacements must satisfy the simply-supported plate edge boundary conditions. Two approximate solutions were used. The first is a single polynomial term (a true Rayleigh solution), while the other is a more general series solution (a Ritz solution).

When the actuator is restricted to be at the plate center, a single set of three polynomial assumed displacements, with their origin at the center of the plate, is

$$u_o = A \frac{x}{a} \quad v_o = B \frac{y}{b} \quad (2)$$

$$w = C \left(1 - \frac{x^2}{a^2}\right)\left(1 - \frac{y^2}{b^2}\right) \quad (3)$$

where u, v and w are the inplane displacements in the x,y and bending directions, respectively.

A more general assumed displacement set is[15]

$$u_o = \sum_{m=1}^{M} \sum_{n=1}^{N} A_{mn} \cos\frac{m\pi x}{a} \sin\frac{n\pi y}{b} \quad (4)$$

$$v_o = \sum_{m=1}^{M} \sum_{n=1}^{N} B_{mn} \sin\frac{m\pi x}{a} \cos\frac{n\pi y}{b} \quad (5)$$

$$w = \sum_{m=1}^{M} \sum_{n=1}^{N} C_{mn} \sin\frac{m\pi x}{a} \sin\frac{n\pi y}{b} \quad (6)$$

The reader should note that the deflection coefficients A_{mn} and B_{mn} in Eqns. 4 and 5 are not the same as the laminate stiffness coefficients in Eqn. 1.

When either of the assumed displacement expressions are substituted into the strain energy expression, the Principle of Virtual Work can be used so that the strain energy is minimized with respect to the displacement coefficients. A set of linear equations of static equilibrium results.

When the polynomial deflection is used, this energy minimization results in only three equilibrium conditions for the constants A, B and C, given as follows

$$\frac{\partial U}{\partial A} = 0 \quad \frac{\partial U}{\partial B} = 0 \quad \frac{\partial U}{\partial C} = 0 \quad (7)$$

This set of three simultaneous equation sets must be solved for the deflection coefficient sets A, B, and C. These coefficients are then substituted into the displacement equations to find the plate deflection at the center of the plate.

If only bending of the plate is considered, and the inplane energy and inplane-bending coupling energy is ignored, the strain energy expression contains only bending stiffness matrix terms and the induced piezoelectric moments. To solve for the deflections, only the third of the polynomial equations is substituted into the energy expression. While this reduces the workload, it leads to errors. The coupling terms are very important.

There are differences between retaining the inplane stiffness and inplane/bending stiffness coupling terms and ignoring them. Figure 2 compares the results of the uncoupled bending analysis and the bending-extension coupling analysis using the simple polynomial expression to the finite element solution. All of the cases were run for an actuator to plate thickness ratio of 0.6.

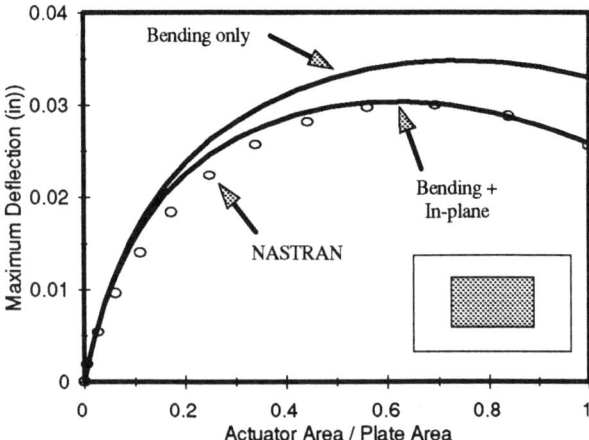

Figure 2 - Plate center bending deflection computed by Ritz method with polynomial approximation.

It can be seen that the uncoupled bending solution over predicts the deflection for the entire range of actuator area ratios, showing that it is important to include the inplane/bending coupling terms. The bending-extension coupling solution is remarkably close to the NASTRAN solution for these simple displacement functions. However, the actuator is restricted to being centered on the plate.

Because we want to assess the effects of changing the actuator location on the plate, and to get a more accurate solution, the trigonometric series displacement functions were used for all other studies. Substituting them into the strain energy expression in the Appendix and minimizing with respect to the undetermined displacement coefficients results in the general conditions

$$\frac{\partial U}{\partial A_{mn}} = 0 \qquad \frac{\partial U}{\partial B_{mn}} = 0 \qquad \frac{\partial U}{\partial C_{mn}} = 0 \qquad (8)$$

The result is a set of $3 M \times N$ simultaneous equations which must be solved for the series coefficients A_{mn}, B_{mn}, and C_{mn}.

The form of the equations for the displacement coefficients is

$$[K_{ij,mn}]\{A_{mn}\} = \{Q_{ij}\} \qquad (9)$$

where $[K_{ij,mn}]$ is a $(3 M \times N) X (3 M \times N)$ matrix of stiffness terms, found from integrating the assumed displacement series strains into the energy expression, $\{A_{mn}\}$ is a $3 M \times N$ vector consisting of the vectors of A_{mn}, B_{mn}, and C_{mn} coefficients. $\{Q_{ij}\}$ is a $3 M \times N$ load vector whose terms represent the induced piezoelectric forces and moments. These latter terms can be thought of as the modal forces and moments for each of the assumed displacement terms.

The stiffness matrix $[K_{ij,mn}]$ is inverted to solve for the displacement coefficients., This matrix becomes large as more terms are added to the series solution. For our results, $M = N = 11$ was used. This results in a stiffness coefficient matrix that is of order 363x363. Figure 3 shows the agreement between this series solution and the finite element results.

Effects of actuator thickness and area

Actuator thickness and area for the baseline plate in Figure 1 were varied to find an actuator/aluminum host plate combination that produced the most panel center bending deflection.

For these studies, the actuator was located at the center of the plate planform and attached to its underside. The maximum bending deflection occurred at the plate center. A range of actuator thickness to plate thickness ratios between 0.1 to 1.0 was examined. For each thickness ratio, a range of actuator planform coverage from 5% to 100% of the plate area was examined. From these results, the combination of actuator thickness ratio and area ratio that produced the largest deflection could be determined.

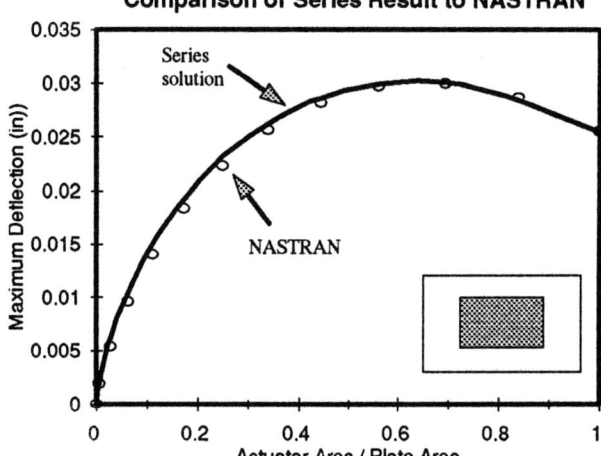

Figure 3 - Plate center bending deflection using trigonometric Ritz approximation with thickness ratio of 0.6 - comparison with NASTRAN (M=N=11)

Figure 4 shows the results of a Rayleigh-Ritz analysis to compute the panel center deflection produced by an actuator with an actuator/host plate thickness ratio between 0.1 and 0.6 for actuator/host plate area ratio between 0.1 and 1. This figure indicates that an actuator/plate thickness ratio of 0.6 produces the largest panel deflection when the actuator/plate area ratio is 65%.

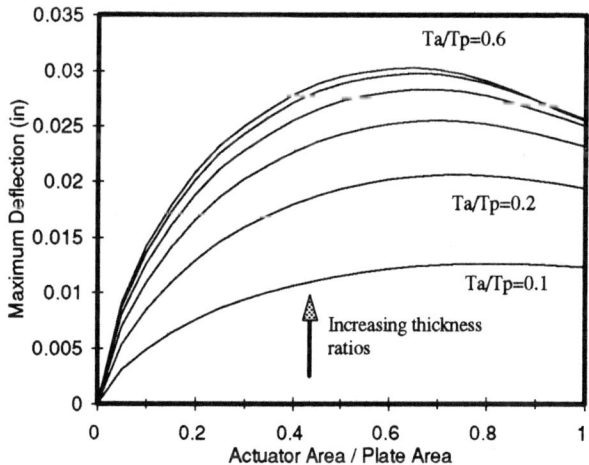

Figure 4 - Maximum panel center bending deflection vs. actuator coverage for thickness ratios between 0.1 and 0.6, in increments of 0.1; aspect ratio 1.5

The curves in Figure 4 show that a slightly smaller area ratio than 0.6 will produce nearly as much panel center deflection. Using a smaller actuator would produce a weight savings with little degradation in performance. An actuator/plate thickness ratio of 0.5 also produces its greatest deflection at 65% area coverage. Note that the deflection produced with an actuator 0.5 as thick as the plate is not much different than that produced by an actuator whose thickness ratio is 0.6.

When the actuator has a thickness ratio of 0.1, a maximum in the curve of deflection vs. actuator coverage occurs at about 80% coverage. However, the deflection does not change much between 50-100% coverage. As the actuator thickness increases, the maximum value in the panel center deflection curve becomes more pronounced and shifts toward a smaller actuator area coverage

Figure 5 shows the results of analyses for actuator/plate thickness ratios between 0.6 and 1 for the entire area ratios between 0.1 and 1. The thickness ratio of 0.6 still produces the most deflection at area ratios greater than 50%. However, thickness ratios of 0.7 and 0.8 produce larger deflections at low actuator area coverage.

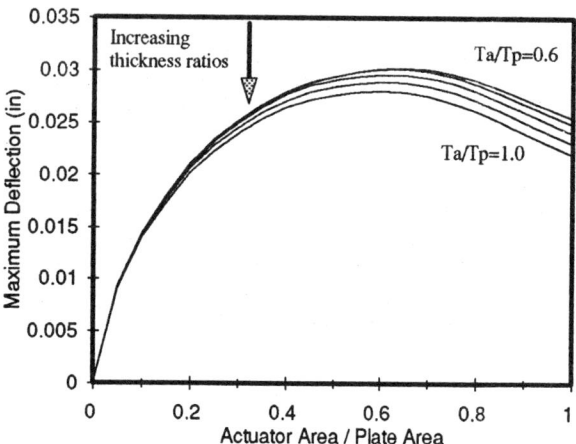

Figure 5 - Plate center bending deflection vs. actuator area coverage for thickness ratios between 0.6 and 1.0 (increments of 0.1); aspect ratio 1.5.

Effects of panel aspect ratio on optimal actuator size

Figure 6 shows the maximum panel center bending deflection obtained at an actuator/host plate thickness ratio of 0.6 for several different panel aspect ratios. Although the aspect ratio can be changed, the total plate area and actuator area remain fixed to be the same as a panel with dimensions 18 in.x 18 in. The square panel with an aspect ratio of 1 has the largest center deflection. The panel with an aspect ratio of 1.5 has a deflection smaller than that of the square panel, but the maximum bending deflection still occurs when the actuator coverage is about 65% of the plate.

The panel with a planform aspect ratio of 3 has a maximum center deflection less than half of the square panel, even though the panel areas are the same. In this case the simple supports are so close to the panel center when the aspect ratio is 3 that the stiffness of the panel is increased with respect to center deflection.

For all of the previous studies, the actuator area and aspect ratio were restrained to be the same as the plate to which it was mounted, although the actuator aspect ratio does not need to be the same. The effect of changing the actuator aspect ratio on the panel center bending deflection was examined using our model.

For the square panel, the best actuator has an actuator thickness ratio of 0.6 and an area coverage of 65%. Any actuator aspect ratio less than or greater than 1 creates a smaller center deflection than the initial configuration, although the differences are small.

When the panel aspect ratio is 1.5, the best actuator also has a thickness ratio of 0.6 and covers an area of 65% of the panel. When the actuator aspect ratio is increased so the actuator spans the long dimension of the panel, the greatest decrease in deflection is about 5.7%.

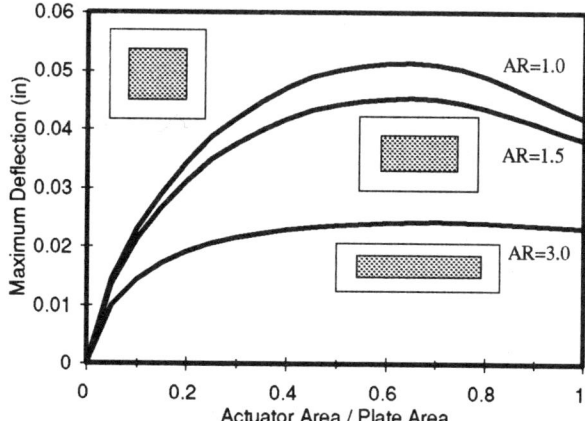

Figure 6 - The effect of actuator coverage on center deflection of three different panels; constant actuator and panel area with actuator/plate thickness ratio 0.6; aspect ratios 1, 1.5 and 3.

For a panel with an aspect ratio of 3, the best actuator has a thickness ratio of 0.6 and an area coverage of 70%. For this case, an actuator with an aspect ratio smaller than the panel produces the greatest deflection, although the increase is only 0.5%.

Non-centrally located actuators

For all previous results, the actuator was centered on the panel planform and the maximum deflection occurred in the center of the panel. For a non-centered actuator, the maximum deflection will not occur in the center of the panel, so a deflection distribution must be examined. This centerline is located in the x-direction, parallel to the long edge of the plate.

The baseline panel with an aspect ratio of 1.5 was used for a study of the effects of non centrally located actuators. The actuator has the same aspect ratio as the panel itself. Two different actuator sizes were used; they covered 6.25% and 25% of the plate area. The actuator was subjected to its maximum electric field throughout and located at one of four different locations along the centerline. In each case, the deflection profile was calculated and compared.

The smallest (6.25% area coverage) actuator was placed in the center of the panel, then shifted toward the left edge of the panel. Figure 7 shows the deflection profile at the centerline of the panel in each case (see inset for positions). The actuator edge can be moved close to the edge of the panel without seeing a large decrease in the maximum deflection. However, once an actuator edge is at or near the panel edge, the peak deflection decreases noticeably. Since the shape and amplitude of the bending deflection on the panel can be changed by relocating the actuator this provides a way to tailor the aerodynamic shape.

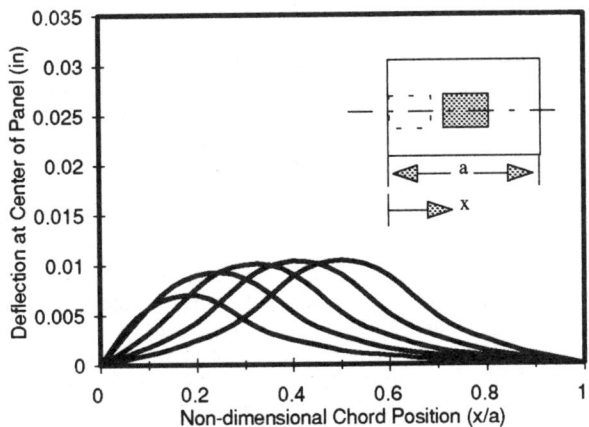

Figure 7 - Bending deflection distribution for a panel with an actuator covering 6.25% of panel.

Figure 8 shows the centerline deflection profile for the actuator covering 25% of the panel area and how this deflection changes as the actuator is repositioned along the panel centerline. The maximum deflection here is about twice that of the smaller 6.25% coverage actuator. The peak deflection when the actuator edge is at the edge of the panel is about 16% less than when the actuator is centered.

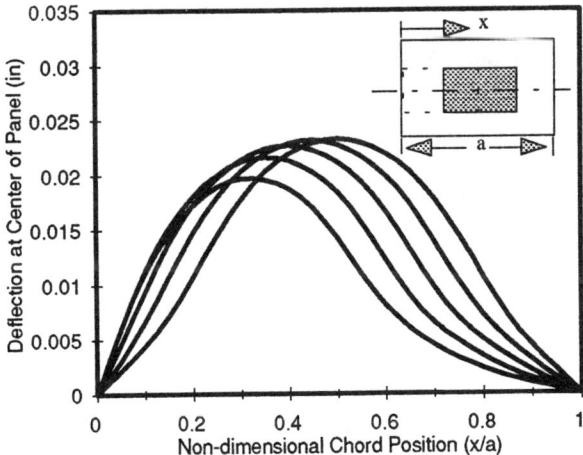

Figure 8 - Bending deflection distributions for a panel with an actuator covering 25% of panel area.

Strain energy density for optimal actuator size

We have observed that actuator thickness and panel area coverage affect the size of the panel bending deflection. When we try to create bending deflection with this type of actuator, two effects are in conflict with each other. First of all, as actuator thickness (or area) increases the actuator is able to create large strains to induce inplane forces and bending moments. On the other hand, as the actuator becomes thicker and covers more area, the panel stiffness also increases so that the plate is more difficult to bend.

When the actuator is small compared to the plate, its attempt to expand is easily thwarted by a relatively massive plate. When a voltage is applied to the actuator, most of the strain energy comes from the actuator being held relatively fixed and expanding only inplane where the inplane stiffness is larger than the actuator bending stiffness. Very little strain energy goes into bending.

On the other hand, as the actuator grows in size, the rate of change of induced bending moment with respect to actuator thickness is relatively large compared to the rate of change of panel bending stiffness. As a result, it is better to increase the size of the actuator and more energy goes into bending. Bending stiffness is much less than inplane stiffness so the rate of change of strain energy with respect to changes in actuator thickness becomes less.

As the actuator thickness increases, the bending stiffness increases more rapidly than the induced moment. At some point, the rate of change of induced bending moment will equal the rate of change in bending stiffness. This is the best actuator for bending deflection. Any increase in actuator size will increase bending stiffness more than induced bending moment and will be counter-productive. The actuator becomes non-optimal.

The dependence of panel strain energy on actuator thickness can be plotted for a panel with two parallel edges free while the other two edges are simply supported. This is the case called cylindrical bending. When the actuator covers the entire plate area, the resulting deflection when the actuator is operated at its full power is a parabola with constant curvature. The expression for strain energy of this special plate is as follows:

$$U = \frac{1}{2}\iint\left[A_{11}\left(\frac{\partial u_o}{\partial x}\right)^2 - 2B_{11}\frac{\partial u_o}{\partial x}\frac{\partial^2 w}{\partial x^2} + D_{11}\left(\frac{\partial^2 w}{\partial x^2}\right)^2 \right.$$
$$\left. -2N_{x\Lambda}\frac{\partial u_o}{\partial x} + 2M_{x\Lambda}\frac{\partial^2 w}{\partial x^2} + \sum_{k=1}^{N} E^{(k)}\left(\Lambda_1^{(k)}\right)^2 t_k\right]dxdy \quad (10)$$

Consider the case of a constant thickness aluminum plate with a PZT actuator covering one 100% of the plate. The strain energy (U divided by the area of the plate) is plotted in Figure 9 against actuator thickness ratio. Also plotted is the curvature of the plate vs. thickness ratio.

The maximum bending deflection occurs when the plate curvature is a maximum and when the actuator thickness ratio is 0.5. The strain energy density has an inflection point at this thickness ratio.

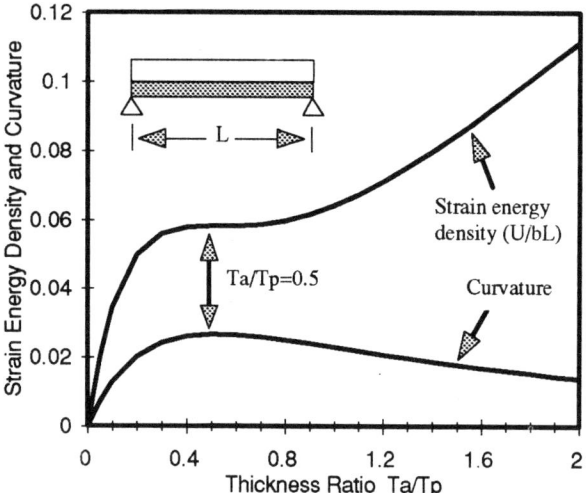

Figure 9 - Beam strain energy density and curvature vs. actuator/plate thickness ratio. Full coverage actuator; cylindrical bending.

The presence of an inflection point is important because it suggests an optimality criterion to select the actuator. The "too small" actuator stores a large part of the mechanical energy because the panel barely deforms. When the actuator reaches a "critical size" the panel bends a great deal, but stores energy in a more "flexible" mode. As actuator size increases further, the dominant energy storage mode again becomes extensional and the slope of the strain energy density curve turns upward and increases rapidly.

The favorable bending energy storage mode is identified by an inflection point or "flat spot" in the strain energy density vs. actuator thickness curve.

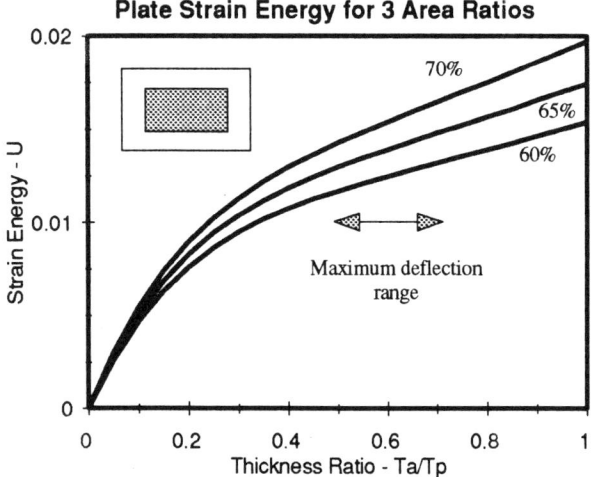

Figure 10 - Plate strain energy vs. actuator thickness ratio for 60%, 65% and 70% area coverage. Panel aspect ratio is 1.5.

The strain energy for the plate can also be plotted versus thickness ratio. Figure 10 shows the strain energy plotted vs. thickness ratio for actuators covering 60%, 65%, and 70% of the plate area for an aluminum plate with an aspect ratio of 1.5. In the cylindrical bending case, it was possible to identify the best thickness ratio by looking for an inflection point in the strain energy plot. However, this inflection point is not so evident for the simply supported panel.

The curves in Figure 10 become nearly linear for larger thickness ratios. Figure 10 indicates that, as long as the slope of the strain energy curve is changing with increases in thickness ratios, adding thickness to the actuator will be beneficial. But once the curve becomes linear, adding actuator thickness will not produce more bending deflection in the panel.

Conclusions

A panel with a self-straining actuator mounted on one side was studied to find the features of the actuator that produces the largest deflection of simply supported rectangular panels. A Rayleigh-Ritz model was developed to compute inplane and bending deflections of a plate with an actuator covering only part of the area.

It was shown that it is important to include the coupling terms and the in-plane actuator forces to accurately model the problem. A simple polynomial assumed displacement field provided good agreement with the finite element results for centrally located actuators with a large area coverage.

To match the finite element results for smaller actuator areas and to provide more accurate results overall, a Rayleigh-Ritz trigonometric series expansion was developed. This model allows the actuator to be placed at any location on the panel. It was found that a series with $M = N = 11$ terms for each of the u, v, and w displacements yielded excellent results, very close to those of the NASTRAN program.

For aluminum panels with aspect ratios between 1.0 and 1.5, the best actuator has a thickness ratio of 0.6 and covers 65% of the panel area. For a panel with an aspect ratio of 3.0, the best actuator has a thickness ratio of 0.6 and covers 70% of the area. For the panels with aspect ratios of 1.0 and 1.5, the actuator with the same aspect ratio as the panel produced the largest deflection. For the panel of aspect ratio 3.0, an actuator with an aspect ratio slightly smaller than that of the panel produced the most deflection.

The effect of actuator location on the deflection profile of a simply supported panel was examined. An actuator with 25% area coverage or less produced the most change in the bending deflection shape.

For a plate with cylindrical bending, when the strain energy is plotted against thickness ratio, the curve shows an inflection point at the best actuator thickness for that configuration. Although the best actuator thickness was obviously at the inflection point in the strain energy curve for the cylindrical bending case, the best actuator thickness for the general plate was less best actuator thickness for the general plate was less obvious.

The plot of strain energy vs. actuator thickness ratio becomes nearly linear at the point corresponding to the best actuator thickness ratio. This indicated that increasing the thickness of the actuator was beneficial until the slope of the strain energy curve reached its smallest value. Adding more thickness to the actuator beyond this point increases stiffness more than it increased the applied moments, and does not increase the deflection that is obtained in the panel.

For this study, actuator thickness, area, and aspect ratio were varied to find which combination produced the largest deflection in a given panel. More precise results for the actuator characteristics could be calculated if an optimization scheme were used in conjunction with the model already developed. Since only a finite number of combinations were tried in this study, the best actuator was found within the limits of the study. Also, the optimal actuator could be found by taking into consideration the weight that is added for a larger actuator. A larger, thicker actuator does not always produce significantly more deflection than a smaller, lighter actuator.

This study was also limited to rectangular actuators placed at any location on the panel. Actuators of shapes other than rectangular should also be considered. Rectangular actuators produce high stresses at the edges and corners[17]. The corner stresses might be avoided if the actuator shape were changed. The deflection produced by elliptical actuators or other actuator shapes should be examined.

The plate deflections obtained here used the PZT actuators to their fullest extent, applying voltages to the actuators that are right at their capabilities. Much needs to be done to improve the efficiency of these materials. On the other hand, we do believe that there is promise for future applications to high speed flow control.

Acknowledgment

This research was supported, in part, by the Air Force Office of Scientific Research under AFOSR Grant 91-0386. Dr. Spencer Wu was contract monitor.

References

1. Muller, M. and Weisshaar, T.A., "Transonic Drag Reduction Through the Use of Induced Strain Actuators to Form an Adaptive Airfoil," AIAA Paper 94-1775, AIAA/ASME Adaptive Structures Forum, Hilton Head, South Carolina, April 1994.

2. Sobieczky, H. and Dulikravich, G.S., "Transonic Airfoil Thickness Variation Requirements for Maintaining Shock-free Flow," in Smart Materials and Structures 1993: Smart Structures and Intelligent Systems, Nesbitt W. Hagood and Gareth J. Knowles, Editors, Proc. SPIE 1917, Vol. 1, pp. 119-124, 1993.

3. Scott, R.C., "Control of Flutter Using Adaptive Materials," M.S. Thesis, Purdue University, West Lafayette, Indiana, May 1990.

4.. Abou-Amer, S.A., "Control of Panel Flutter at High Supersonic Speed," Ph.D. Dissertation, Illinois Institute of Technology, Chicago, Ill., Dec. 1991.

5. Paige, D., "Active Control of Composite Panel Flutter Using Piezoelectric Materials," M.S. Thesis, Purdue University, West Lafayette, Indiana, May 1992.

6. Hajela, P. and Glowasky, R., "Application of Piezoelectric Elements in Supersonic Panel Flutter Suppression," AIAA Paper No. 91-3191, AIAA/AHS/ASEE Aircraft Design Systems and Operations Meeting, Baltimore, Md., Sept., 1991.

7. Scott, R. C. and Weisshaar, T. A., "Controlling Panel Flutter Using Adaptive Materials," Proceedings of the 32nd Structures, Structural Dynamics and Materials Conference, Baltimore, Md., April 1991.

8. Kim, S. J., and Jones, J. D., "Optimal Design of Piezoactuators for Active Noise and Vibration Control," AIAA Journal, Vol. 29, No. 12, December, 1991, pp. 2047-2053.

9. Rogers, C. A., Liang, C., and Jia, J., "Behavior of Shape Memory Alloy Reinforced Composite Plates Part I: Model Formulations and Control Concepts," AIAA Paper No. 89-1389-CP.

10. Crawley, E. F., and Lazarus, K. B., "Induced Strain Actuation of Isotropic and Anisotropic Plates," AIAA Paper No. 89-1326-CP, 1989.

11. Wang, B. T., and Rogers, C. A., "Laminate Plate Theory for Spatially Distributed Induced Strain Actuators," Journal of Composite Materials, Vol. 25, April 1991, pp. 433-452.

12. Vasiliev, V. V., *Mechanics of Composite Structures*, Taylor and Francis, Washington, D.C., 1993

13. Sun, C. T., and Chin, H., On Large Deflection Effects in Unsymmetric Cross-Ply Composite Laminates, *Journal of Composite Materials*, Vol. 22, November, 1988, pp. 1045-1059.

14. Sun, C. T., and Chin, H., Analysis of Asymmetric Composite Laminates, *AIAA Journal*, Vol. 26, NO. 6, June 1988, pp. 714-718.

15. Hetnarski, R.B., Editor, *Thermal Stresses I*, Elsevier Science Publishers, B.V., 1986.

16. Whitney, J. M., *Structural Analysis of Laminated Anisotropic Plates*, Technomic Publishing Co., Inc., 1987.

17. Leeks, T.J., "Optimal Design of Partial-Plate Piezoelectric Actuators for Maximum Panel Deflection," M.S. Thesis, Purdue University, Dec. 1993.

Appendix - Strain Energy Expression

Nomenclature

a	plate length	$N_{x\Lambda}, N_{y\Lambda}, N_{xy\Lambda}$	piezoelectric inplane forces
A_{ij}	extensional stiffness matrix	$M_{x\Lambda}, M_{y\Lambda}, M_{xy\Lambda}$	piezoelectric moments
A, A_{mn}	assumed series solution constants	$\overline{Q}_{ij}^{(k)}$	reduced lamina stiffness for kth layer
b	plate width	T_a	actuator thickness
B_{ij}	coupling stiffness matrix	T_p	host plate thickness
B, B_{mn}	assumed series solution constants	u_o	inplane deflection in x direction
C, C_{mn}	assumed series solution constants	U	panel strain energy
d_{31}	piezoelectric constant	v_o	inplane deflection in y direction
D_{ij}	bending stiffness matrix	w	bending deflection
E_3	electric field strength	$\Lambda_i^{(k)}$	piezoelectric strain of kth lamina

The strain energy expression, U, is used to compute the mechanical strain for a plate/actuator combination in plane stress when the piezoelectric actuator creates induced strain. The mechanical strain is the difference between the total strain and the expansion strain, and is the only strain that creates stress[16]. There is no stress induced if the material is allowed to expand freely. The induced piezoelectric strains are analogous to thermal strains.

To compute strain energy, a reference surface at the mid-plane of the host plate was chosen. This allows the reference surface to remain fixed when the actuator thickness changes thus reducing computational complexity. The last three terms of the equation are independent of the displacements u, v, and w. The integral involving these terms will vanish under the first variation of the strain energy and the terms will not enter the equations when the Rayleigh-Ritz method is applied. For this analysis a rectangular actuator is attached to a plate at an arbitrary position.

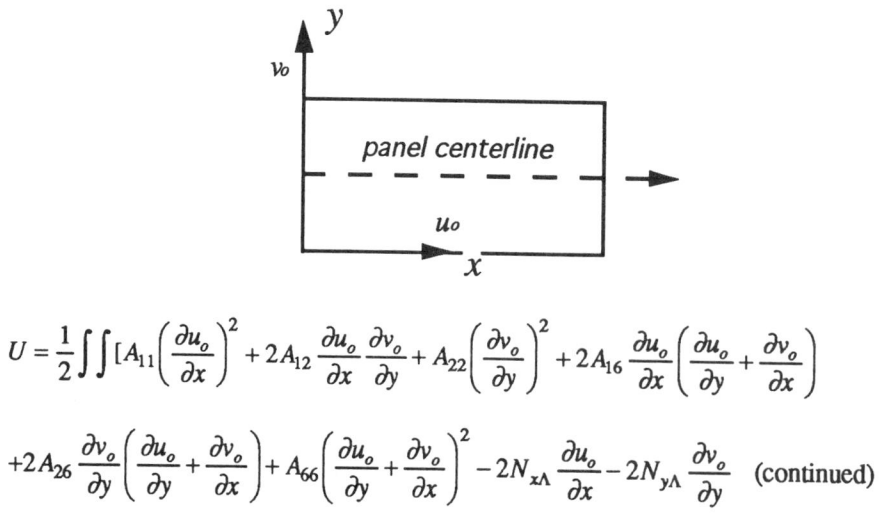

$$U = \frac{1}{2} \int\int [A_{11}\left(\frac{\partial u_o}{\partial x}\right)^2 + 2A_{12}\frac{\partial u_o}{\partial x}\frac{\partial v_o}{\partial y} + A_{22}\left(\frac{\partial v_o}{\partial y}\right)^2 + 2A_{16}\frac{\partial u_o}{\partial x}\left(\frac{\partial u_o}{\partial y} + \frac{\partial v_o}{\partial x}\right)$$

$$+ 2A_{26}\frac{\partial v_o}{\partial y}\left(\frac{\partial u_o}{\partial y} + \frac{\partial v_o}{\partial x}\right) + A_{66}\left(\frac{\partial u_o}{\partial y} + \frac{\partial v_o}{\partial x}\right)^2 - 2N_{x\Lambda}\frac{\partial u_o}{\partial x} - 2N_{y\Lambda}\frac{\partial v_o}{\partial y} \quad \text{(continued)}$$

$$-2N_{xy\Lambda}\left(\frac{\partial u_o}{\partial y}+\frac{\partial v_o}{\partial x}\right)-2B_{11}\frac{\partial u_o}{\partial x}\frac{\partial^2 w}{\partial x^2}-2B_{12}\left(\frac{\partial v_o}{\partial y}\frac{\partial^2 w}{\partial x^2}+\frac{\partial u_o}{\partial x}\frac{\partial^2 w}{\partial y^2}\right)$$

$$-2B_{22}\frac{\partial v_o}{\partial y}\frac{\partial^2 w}{\partial y^2}-2B_{16}\left(\left(\frac{\partial u_o}{\partial y}+\frac{\partial v_o}{\partial x}\right)\frac{\partial^2 w}{\partial x^2}+2\frac{\partial u_o}{\partial x}\frac{\partial^2 w}{\partial x\partial y}\right)$$

$$-2B_{26}\left(\left(\frac{\partial u_o}{\partial y}+\frac{\partial v_o}{\partial x}\right)\frac{\partial^2 w}{\partial y^2}+2\frac{\partial v_o}{\partial y}\frac{\partial^2 w}{\partial x\partial y}\right)-4B_{66}\frac{\partial^2 w}{\partial x\partial y}\left(\frac{\partial u_o}{\partial y}+\frac{\partial v_o}{\partial x}\right)$$

$$+D_{11}\left(\frac{\partial^2 w}{\partial x^2}\right)^2+2D_{12}\frac{\partial^2 w}{\partial x^2}\frac{\partial^2 w}{\partial y^2}+D_{22}\left(\frac{\partial^2 w}{\partial y^2}\right)^2+4D_{16}\frac{\partial^2 w}{\partial x^2}\frac{\partial^2 w}{\partial x\partial y}$$

$$+4D_{26}\frac{\partial^2 w}{\partial y^2}\frac{\partial^2 w}{\partial x\partial y}+4D_{66}\left(\frac{\partial^2 w}{\partial x\partial y}\right)^2+2M_{x\Lambda}\frac{\partial^2 w}{\partial x^2}+2M_{y\Lambda}\frac{\partial^2 w}{\partial y^2}$$

$$+4M_{xy\Lambda}\frac{\partial^2 w}{\partial x\partial y}+\sum_{k=1}^{N}\overline{Q}_{11}^{(k)}\left(\Lambda_1^{(k)}\right)^2(z_k-z_{k-1})$$

$$+2\sum_{k=1}^{N}\overline{Q}_{12}^{(k)}\Lambda_1^{(k)}\Lambda_2^{(k)}(z_k-z_{k-1})+\sum_{k=1}^{N}\overline{Q}_{22}^{(k)}\left(\Lambda_2^{(k)}\right)^2(z_k-z_{k-1})]dxdy \quad \text{(concluded)}$$

MAXIMIZING DEFLECTIONS PRODUCED BY INDUCED STRAIN BEAM-LIKE ACTUATORS FOR TRANSONIC DRAG REDUCTION

Mark B. Muller

Terrence A. Weisshaar

Purdue University
West Lafayette, Indiana 47907-1282

Abstract

This paper examines the use of adaptive materials for transonic drag reduction. Induced strain actuators are used to modify the shape of a transonic airfoil in an attempt to reduce transonic wave drag over a range of transonic flight conditions. The airfoil skin is treated as a beam-like structure spnning between two stringers. A general and analytic structural model based on classical laminated plate theory is developed. The effects of actuator configuration, geometry, and material choices for both the actuatore / host combination and end conditions are investigated. End support conditions are shown to have a significant effect on the amount of deflection that can be produced, but little effect on how the actuator should be sized. Also, large differences are found in desired material properties between symmetric and asymmetric beam/actuator laminates. These results are applied to the problem of reducing transonic airfoil drag via airfoil adaptation. An adaptive airfoil is formed using a supercritical airfoil with an adaptive beam as part of its upper surface. This section's shape is altered by the actuator. Aerodynamic analyses of the modified airfoil shapes are performed to study the effects on transonic drag. These studies show that the induced strain actuator are able to significantly reduce transonic drag for some flight conditions.

Nomenclature

- A = extensional stiffness
- B = extension/bending coupling stiffness
- c = beam width
- D = bending stiffness
- E = Young's modulus
- EI = bending rigidity
- L = beam length
- M = bending moment per unit width
- n = number of layers
- N = in plane force per unit width
- t = material thickness
- w = vertical deflection
- x = horizontal coordinate
- z = vertical coordinate
- α = thickness ratio, t_a/t_b
- β = stiffness ratio, E_a/E_b
- ε = beam extensional strain
- γ = length ratio, L_2/L
- κ = curvature
- Λ = actuator induced strain
- Φ = deflection function

Subscripts and superscripts
- a = actuator
- b = beam
- k = layer
- w = wall
- Λ = actuator induced

Copyright © 1993 American Institute of Aeronautics and Astronautics, Inc. All rights reserved.

* Research Assistant, School of Aeronautics and Astronautics

** Professor, Fellow AIAA, School of Aeronautics and Astronautics

Introduction

Reduction of drag has always been a primary concern of an aircraft designer. This study looks at methods to reduce drag using adaptive materials on aircraft traveling at transonic speeds. Such aircraft include virtually all jet airliners, as well as military fighters and attack aircraft. Specifically, adaptive materials are examined as a method to change an airfoil's shape in flight so as to reduce transonic wave drag over a range of transonic flight conditions. At such speeds, wave drag is a significant source of drag and the flow is very sensitive to small perturbations. It is hoped that this will allow small changes in the airfoil's shape to produce significant reductions in drag.

Supercritical transonic airfoils were developed by Whitcomb and Clark in the mid 1960's[1]. Similar airfoil designs resulted from the work of Pearcey[2]. These airfoils were designed to operate with a large local region of supersonic flow over their upper surfaces. The benefit of this type of design is that the drag rise associated with compressibility occurred at higher Mach numbers than previous airfoils, allowing for increases in aircraft speed or a decrease in aircraft drag at a given transonic speed. A good review of the design and history of NASA supercritical airfoils is given by Harris[3]

Figure 1 shows a typical supercritical airfoil section and a "conventional" airfoil section for comparison. The supercritical airfoil is characterized by a blunt leading edge, a rather flat upper surface, and a large amount of camber towards the rear of the airfoil. The flat upper section allows for the supersonic region to be terminated with a mild shock wave. The rear camber allows for the design lift coefficient to be generated without increasing the upper surface suction, which would increase the strength of the shock wave, thus increasing drag.

Airfoil Adaption

The earliest example of an adaptive wing was the wing warping system employed by the Wright Brothers[4]. The adaption of the wing was used for roll control. More common examples of airfoil adaption are ailerons, flaps, and slats. Ailerons are primarily used for roll control, while flaps and slats are used to adapt airfoils for low speed flight.

Flap scheduling has been proposed for both subsonic and transonic aircraft. The idea is to deflect the flaps either slightly up or down during cruise flight to change the camber of the wing to reduce drag. Somers discusses this idea for a general aviation aircraft using a simple flap[5]. Szodruch and Hilbig[6], as well as Renken[7], have discussed using more complicated camber varying mechanisms for larger aircraft.

Another way for an airfoil to be adaptive is to use actuators placed inside of a wing to change the shape of the airfoil. This concept was investigated in a series of wing tunnel tests by General Dynamics under Navy sponsorship[8,9]. In these models, the wing consisted of a wing box with variable shaped leading and trailing edges. Five actuators were able to change the leading edge incidence, the leading edge radius, lower surface humping, lower surface aft camber, and trailing edge deflection. This effectively allows the airfoil to be reshaped for whatever conditions it is flying at.

In this study, actuators are also placed inside of the airfoil to modify its shape. Instead of mechanical actuators, induced strain actuators made of adaptive materials are used. These materials are bonded to the inside of the wing skin to deflect the wing skin actively in flight by bending it. As actuators, adaptive materials usually have the problem of not being very strong or powerful. Beacuse of this, this study first looks at maximizing the deflections that adaptive materials can produce in beam-like structures. The resulting actuator configurations are then be applied to the aerodynamic problem of adaptive airfoils.

Induced Strain Actuators

An induced strain actuator is a material that induces a strain in a structure to cause deflection in the host structure upon command. It does this by either changing shape or size in response to some stimulus, such as an electric field. One of the most common systems of this type is a bi-metal thermostat consisting of a laminated strip of two metals with different coefficients of thermal expansion. When the temperature changes, the beam bends due to differential expansion of the two sides of the beam.

Several different types of materials may be used for induced strain actuators. These include:

- Piezoelectrics
- Shape Memory Alloys
- Magnetostrictors

These materials are actuated by electric fields, temperature changes, and magnetic fields respectively.

Laminated Beam Model

The moment and deflection of actuators on beams has been modeled by several different methods. Crawley and de Luis[10] introduced the pin force model, which is simple, but does not account for the bending rigidity of the actuator. This makes it inaccurate for thick actuators. Crawley and Anderson[11], as well as Chaudhry and Rogers[12], have used Bernoulli-Euler beam theory models which accurately account for the actuators bending rigidity. Lin and Rogers[13] have used classical theory of elasticity to account for actuator boundary conditions. The approach taken here is based on classical laminated plate theory. A similar approach was taken by Wang and Rogers[14]. For a plane stress beam, this gives the same results as the Bernoulli-Euler beam theory models. It also allows a simple extension from beams to plates, as well as very wide beams, which may be considered as plates or as plane strain beams.

The effect of induced strain actuators on a structure may be modeled in the same way that thermal stresses are modeled. This is done using an induced actuator strain in place of an induced thermal strain. This equivalent thermal loading may be analyzed using classical laminated plate theory (CLPT), as given by Jones[15], with a reduction from plates to beams.

CLPT expresses the relation between the loads and deflection of a plate as

$$\begin{Bmatrix} N \\ M \end{Bmatrix} = \begin{bmatrix} A & B \\ B & D \end{bmatrix} \begin{Bmatrix} \varepsilon \\ \kappa \end{Bmatrix} \quad (1)$$

For a plate, A, B and D are 3x3 matrices. For a plane stress beam, these reduce to scalar quantities given by the following:

$$A = \sum_{k=1}^{n} E_k (z_k - z_{k-1})$$
$$B = \tfrac{1}{2} \sum_{k=1}^{n} E_k (z_k^2 - z_{k-1}^2) \quad (2)$$
$$D = \tfrac{1}{3} \sum_{k=1}^{n} E_k (z_k^3 - z_{k-1}^3)$$

For induced strain loads, there is an applied in plane force and moment, giving

$$\begin{Bmatrix} N + N_\Lambda \\ M + M_\Lambda \end{Bmatrix} = \begin{bmatrix} A & B \\ B & D \end{bmatrix} \begin{Bmatrix} \varepsilon \\ \kappa \end{Bmatrix} \quad (3)$$

Where the induced strain loads are, for a beam, scalar quantities given as:

$$N_\Lambda = \sum_{k=1}^{n} E_k \Lambda_k (z_k - z_{k-1}) \quad (4)$$

$$M_\Lambda = \tfrac{1}{2} \sum_{k=1}^{n} E_k \Lambda_k (z_k^2 - z_{k-1}^2) \quad (5)$$

This allows for the extension and bending of the beam due to the actuators to be found as

$$\begin{Bmatrix} \varepsilon \\ \kappa \end{Bmatrix} = \frac{1}{AD - B^2} \begin{bmatrix} D & -B \\ -B & A \end{bmatrix} \begin{Bmatrix} N_\Lambda \\ M_\Lambda \end{Bmatrix} \quad (6)$$

Note how easily non-symmetric laminated beams of actuators and substrates may be analyzed by this general method. Also note that the matrix inversion is greatly simplified due to A, B and D being scalar quantities. Using this formulation, the following four different configurations are investigated:

- Symmetric laminate, pinned ends
- Asymmetric laminate, pinned ends
- Symmetric laminate, clamped ends
- Asymmetric laminate, clamped ends

Figure 2 shows each of these configurations. The final configuration with clamped ends and an actuator on one side is of most interest, as this is the type that will be used for the airfoil adaption studies.

Symmetrically Laminated Pinned Beams

The symmetrically laminated beam with an actuator patch on both sides of the beam is shown in part a of Figure 2. In this system, one of the actuators extends and the other contracts, to produce a bending moment. If the actuators induce equal but opposite strains, there is no net extension of the beam. Also, the neutral axis for such a beam is at the mid plane.

Applying the laminated beam model gives the following

$$A = E_b t_b (1 + 2\alpha\beta)$$
$$B = 0$$
$$D = \tfrac{1}{12} E_b t_b^3 (1 + 6\alpha\beta + 12\alpha^2\beta + 8\alpha^3\beta) \quad (7)$$
$$N_\Lambda = 0$$
$$M_\Lambda = E_b t_b^3 \Lambda \alpha \beta (1 + \alpha)$$

The beam curvature is

$$\kappa = \frac{12\alpha\beta(1+\alpha)\Lambda}{t_b (1 + 6\alpha\beta + 12\alpha^2\beta + 8\alpha^3\beta)} \quad (8)$$

Using Bernoulli-Euler beam theory, identical results have been derived by both Crawley[11] and Rogers[12].

For comparison of beams with different thickness and stiffness ratios, t_b can be replaced by an expression for total beam bending rigidity. Thus for any combination of stiffness and thickness ratio, the beam will have the same total bending rigidity. The total beam bending rigidity is expressed as

$$\frac{(EI)_{total}}{c} = \frac{AD-B^2}{A} = D$$

$$\frac{(EI)_{total}}{c} = \frac{E_b t_b^3 (1+6\alpha\beta+12\alpha^2\beta+8\alpha^3\beta)}{12} \quad (9)$$

Solving this for t_b and substituting into Equation 8 gives

$$\kappa = \frac{12\alpha\beta(1+\alpha)(cE_b)^{1/3}\Lambda}{[12(EI)_{total}]^{1/3}(1+6\alpha\beta+12\alpha^2\beta+8\alpha^3\beta)^{2/3}} \quad (10)$$

This still contains the beam material modulus. If this is left in, the actuator can be optimized for a given beam material. Instead it can be replaced with the actuator material modulus, in which case the beam is optimized for the actuator material. The latter approach has been chosen, as the choice of actuator materials is more limited than that of beam materials. This is especially true if the beam is a composite material whose stiffness can be tailored to work with the actuator. This gives

$$\kappa = \frac{12\alpha\beta^{2/3}(1+\alpha)(cE_a)^{1/3}\Lambda}{[12(EI)_{total}]^{1/3}(1+6\alpha\beta+12\alpha^2\beta+8\alpha^3\beta)^{2/3}} \quad (11)$$

Solving for the midpoint deflection of the pinned beam gives

$$w\left(\frac{L}{2}\right) = \frac{1}{8}\kappa L^2 \quad (12)$$

$$w\left(\frac{L}{2}\right) = \frac{(cE_a)^{1/3}\Lambda L^2}{[12(EI)_{total}]^{1/3}} \Phi(\alpha,\beta) \quad (13)$$

Where Φ is the deflection function. Φ is a non dimensional function independent of actuator material stiffness, total beam bending rigidity, actuator induced strain, and beam length. Φ can be written as

$$\Phi(\alpha,\beta) = \frac{3\alpha\beta^{2/3}(1+\alpha)}{2(1+6\alpha\beta+12\alpha^2\beta+8\alpha^3\beta)^{2/3}} \quad (14)$$

By maximizing this deflection function, the stiffness and thickness ratios are optimized for the deflection of a pinned beam. Similar formulations will be found for the other beam configurations, allowing for comparison of these configurations.

Asymmetrically Laminated Pinned Beams

The asymmetrically laminated case consists of a beam with an actuator on only one side, as shown in part b of Figure 2. In this case the actuator will extends or contract to produce a net bending moment in the beam. Any actuation of the asymmetric laminate produces bending. This type of system is more desirable for a wing skin, as it does not require an actuator to be on the outside of the wing, where it would be exposed to the environment.

Since this beam actuator combination is asymmetrically laminated, there is bending-extension coupling, which is accounted for by the laminated beam model. The model gives

$$A = t_b E_b (1+\alpha\beta)$$
$$B = \tfrac{1}{2} t_b^2 E_b \alpha (1-\beta)$$
$$D = \tfrac{1}{12} t_b^3 E_b (1+3\alpha\beta+3\alpha^2+\alpha^3\beta) \quad (15)$$
$$N_\Lambda = -t_b E_b \alpha\beta\Lambda$$
$$M_\Lambda = \tfrac{1}{2} t_b^2 E_b \alpha\beta\Lambda$$

This gives a curvature expression of

$$\kappa = \frac{6\alpha\beta(\alpha+1)\Lambda}{t_b(1+4\alpha\beta+6\alpha^2\beta+4\alpha^3\beta+\alpha^4\beta^2)} \quad (16)$$

This result is identical to that obtained from Bernoulli-Euler beam theory by Chaudhry and Rogers [12]. To maximize the beam deflection, the curvature must be maximized by optimizing the stiffness and thickness ratio. As before, the beam bending rigidity can be replaced and the midpoint deflection can be expressed as

$$w\left(\frac{L}{2}\right) = \frac{(cE_a)^{1/3}\Lambda L^2}{[12(EI)_{total}]^{1/3}} \Phi(\alpha,\beta) \quad (17)$$

where the deflection function is given by

$$\Phi(\alpha,\beta) = \frac{\tfrac{3}{4}\alpha\beta^{2/3}(\alpha+1)}{(1+\alpha\beta)^{1/3}(1+4\alpha\beta+6\alpha^2\beta+4\alpha^3\beta+\alpha^4\beta^2)^{2/3}} \quad (18)$$

Symmetrically Laminated Clamped Beams

The symmetrically laminated clamped beam case is pictured in part c of Figure 2. The actuator is centered on the span. This configuration introduces another variable - the length ratio between the actuators and the beam. The actuators do not span the entire beam. If they did the beam would not bend, beacause the moment would be fed directly into the supports.

Noting the symmetry in the system, it can be simplified from three beams to two. For the left end section, the resulting equation for the deflection is

$$w(x) = \frac{6M_w x^2}{E_b b^3} \quad (19)$$

For the center section, laminated beam theory gives

$$\frac{\partial^2 w}{\partial x^2} = \frac{M_w - M_\Lambda}{D} \quad (20)$$

which is integrated twice with respect to x to give

$$w(x) = \frac{(M_w - M_\Lambda)(xL_1)^2}{2D} + C_1(x - L_1) + C_2 \quad (21)$$

Applying boundary conditions solves for the two constants of integration and the wall moment, yielding a midpoint deflection of

$$w\left(\frac{L}{2}\right) = \frac{3(1-\gamma)\gamma L^2 M_\Lambda}{2[\gamma E_b b^3 + 12(1-\gamma)D]} \quad (22)$$

Substituting in the symmetric expressions for M_Λ and D gives

$$w\left(\frac{L}{2}\right) = \frac{3L^2 \Lambda \alpha \beta \gamma (1+\alpha)(1-\gamma)}{t_b[2\gamma + 2(1-\gamma)(1+6\alpha\beta + 12\alpha^2\beta + 8\alpha^3\beta)]} \quad (23)$$

For the pinned beams, the beam thickness was removed by fixing the total beam bending rigidity. Here, a similar approach is used, by setting the total average beam bending rigidity to be constant. This gives

$$\frac{(EI)_{av}}{c} = \frac{E_b t_b^3 [1 - \gamma + \gamma(1 + 6\alpha\beta + 12\alpha^2\beta + 8\alpha^3\beta)]}{12} \quad (24)$$

from which an expression for t_b may be found and substituted into Equation 23, and E_b may be substituted in for as well, giving

$$w\left(\frac{L}{2}\right) = \frac{\Lambda L^2 (cE_a)^{\frac{1}{3}}}{[12(EI)_{av}]^{\frac{1}{3}}} \Phi(\alpha, \beta, \gamma) \quad (25)$$

Where the deflection function may be expressed as

$$\Phi(\alpha, \beta, \gamma) = \Phi_1(\alpha, \beta, \gamma) \Phi_2(\alpha, \beta, \gamma)$$

$$\Phi_1 = \frac{3\alpha\beta^{\frac{2}{3}}\gamma(1+\alpha)(1-\gamma)}{2\gamma + 2(1-\gamma)} \quad (26)$$

$$\Phi_2 = \frac{[1 - \gamma + \gamma(1 + 6\alpha\beta + 12\alpha^2\beta + 8\alpha^3\beta)]^{\frac{1}{3}}}{(1 + 6\alpha\beta + 12\alpha^2\beta + 8\alpha^3\beta)}$$

In contrast with the earlier cases, this deflection function is dependent on three variables. It is, however, still non-dimensional, and the form of the resulting midpoint deflection is still the same.

Asymmetrically Laminated Clamped Beams

The asymmetrically laminated clamped beam case is pictured in part d of Figure 2. It is assumed that the actuator is centered on the beam, that is $L_1 = L_3$. Once again this symmetry is used to reduce the size of the problem.

This case is similar to the symmetrically laminated clamped case except that there is a extension/bending coupling in the actuated cross-section. The deflection for the end section is given by

$$w(x) = \frac{6M_w x}{E_b t_b} \quad (27)$$

For the middle section, laminate theory gives

$$\frac{\partial^2 w}{\partial x^2} = -\kappa = \frac{BN_\Lambda - A(M_\Lambda - M_w)}{AD - B^2} \quad (28)$$

which can be integrated to find the deflection as a function of x, giving two constants of integration.

Applying the appropriate boundary conditions gives a linear system that can be analytically solved to give all of the constants of integration and the wall moment. Doing so gives the midpoint deflection in terms of the beam thickness. Once again the total average beam bending rigidity may be set, giving

$$\frac{(EI)_{av}}{c} = \frac{E_b t_b^3 [1 + \alpha\beta + \alpha\beta\gamma(3 + 6\alpha + 4\alpha^2 + \alpha^3\beta)]}{12(1+\alpha\beta)} \quad (29)$$

From this expression, t_b can be extracted and placed into the integrated deflection expression to give the midpoint deflection as

$$w\left(\frac{L}{2}\right) = \frac{L^2 \Lambda (cE_a)^{\frac{1}{3}}}{[12(EI)_{av}]^{\frac{1}{3}}} \Phi(\alpha, \beta, \gamma) \quad (30)$$

where the deflection function may be written as

$$\Phi(\alpha, \beta, \gamma) = \frac{\Phi_1(\alpha, \beta, \gamma) \Phi_2(\alpha, \beta, \gamma)}{\Phi_3(\alpha, \beta, \gamma) + \Phi_4(\alpha, \beta, \gamma)}$$

$$\Phi_1 = \frac{3\alpha\beta^{\frac{2}{3}}\gamma(1+\alpha)(1-\gamma)}{4(1+\alpha\beta)^{\frac{1}{3}}}$$

$$\Phi_2 = [1 + \alpha\beta + \gamma(3\alpha\beta + 6\alpha^2\beta + 4\alpha^3\beta + \alpha^4\beta^2)]^{\frac{1}{3}} \quad (31)$$

$$\Phi_3 = 1 + 4\alpha\beta + 6\alpha^2\beta + 4\alpha^3\beta + \alpha^4\beta^2$$

$$\Phi_4 = -\gamma(3\alpha\beta + 6\alpha^2\beta + 4\alpha^3\beta + \alpha^4\beta^2)$$

Beam results

For all of these configurations, maximizing the midpoint deflection is achieved by maximizing the deflection function for each case. The deflection functions are non dimensional functions of geometric and structural ratios of the actuator and host structure. For the pin ended cases, a contour plot of Φ can be made over a reasonable design space of stiffness and thickness ratios.

Figure 3 shows what should be obvious; the maximum deflection for a symmetric laminate is achieved by the laminate becoming a bimorph, with a beam thickness of zero. This gives an infinite thickness ratio. In this limiting case, there is no host structure, so the stiffness ratio becomes meaningless. Figure 4 shows the contours of Φ for the asymmetric pin ended case. This plot is less intuitive, since the deflection is maximized by using a low stiffness ratio and high thickness ratio, which corresponds to a thin, high modulus host with a thick actuator. Thus it is better to have the bulk of the beam's bending rigidity come from the actuator, with the host structure serving to offset the neutral axis, allowing the actuator to bend the beam.

For the clamped cases, the deflection function is also dependent on the length ratio. In order to make contour plots similar to those made for the pin ended cases, the length ratios that maximizes Φ were found over the same design space of α and β. Using these ratios, contour plots of Φ were be made. Figure 5 shows the symmetric case, while Figure 6 show the asymmetric case. These plots show the same trends as the pin ended cases. The differences are in the absolute values. Figures 7 and 8 show the optimal values of γ used to make the deflection function plots. These show that as stiffness ratio or thickness ratio increase, the optimal length ratio increases.

Adaptive Airfoils

The idea of an adaptive transonic airfoil is not new. It has been discussed from an aerodynamic point of view by Redeker, Wichmann and Oelker [16], as well as by Sobieczky, Fung, and Seebass [17]. Sobieczky, Fung and Seebass designed a series of transonic airfoils that differed only in shape over a limited portion of the airfoil surface, where each airfoil was shock free at some flight condition. Redeker, Wichmann and Oelker designed three closely related airfoils. The goal of both studies was to have an airfoil that changes shape from one low wave drag shape to another as the flight conditions change.

Concept and Configuration

Analytic models to predict bending have been derived for beams with induced strain actuators attached. By considering a portion of the wing skin to be a beam supported on its ends by stringers, these models can be used to determine deflections. Figure 9 shows the type of configuration considered. An actuator is placed on the inside of the wing skin in between two stringers. This results in an asymmetrically laminated beam with clamped ends, corresponding to the final structural model derived.

In these studies the front end of this actuated panel was placed at 1.5% chord behind the leading edge. The aft end of the actuated panel was varied from chord locations of 10% to 80%. Varying this location produces two effects. One, the farther back the aft end is, the longer the panel, which results in greater overall deflection. Second, the panel length affects the length and location of the deflection the actuator produces in the wing skin.

The base airfoil used was a NASA SC(2) 0612. This airfoil is a phase two NASA supercritical airfoil with a design section C_l of 0.60, and a thickness-to-chord ratio of 12%. Harris gives the design philosophy and coordinate data for this series of airfoils[3]. This airfoil was chosen as it is typical of the type of airfoils used on large transports and jet airliners.

Wing skins are typically made of aluminum, so aluminum host structure was used. Lead zirconium titanate, (PZT), was chosen as the actuator material. It is a peizo-ceramic, which is brittle and nearly as stiff as aluminum. Its combination of high modulus and ability to generate large induced strains make it able to generate more bending deflection than other piezoelectrics[18]. The airfoil is adapted by changing the electric field across the PZT, which induces a strain. Equation 30 shows that the bending of the skin is linearly dependent on the induced strain, which for a piezo-ceramic is linearly dependent on the applied electric field. Thus the bending of the skin is linearly dependent on the applied electric field.

The trade studies with regard to the deflection function can be used to choose the length ratio and thickness ratio of the actuator / host combination to maximize the amount of deflection that may be produced. Table 1 shows the resulting geometry, as well as other configuration data. The actuator and host thicknesses stem from making the combination equivalent in bending rigidity to a 2 mm aluminum wing skin.

Table 1
Adaptive Airfoil characteristics

Base Airfoil	SC (2) 0612
Chord	2 meters
Reynold's Number	20,000,000
Host Material	Aluminum
Actuator Material	PZT

Max Induced Strain	± 260 µstrain
Stiffness Ratio, β	0.875
Thickness Ratio, α	2.238
Length Ratio, γ	0.869
Host Thickness	0.66 mm
Actuator Thickness	1.48 mm
Equivalent Al thickness	2.00 mm

<u>Aerodynamic analysis</u>

All aerodynamic analysis has been performed using the code of Bauer, Garabedian, Korn and Jameson[19,20]. This uses a two dimensional full potential method with a turbulent boundary layer correction. Harris reports this code to give good predictions of drag, as do others[3].

Each adaptive airfoil configuration differed only in the location of the aft end of the actuated panel. To optimize the airfoil at a particular Mach number and lift coefficient, a number of induced strain values (Λ) were used over a range ± 260 µstrain. Of these values of Λ tried, the one producing the least drag was used as the result for the adaptive airfoil. Since this a preliminary study, a formal optimization procedure was not used.

These adaptive airfoils were analyzed at coefficients of lift varying from 0.4 to 0.8 with Mach numbers in the region of Mach drag divergence. This was done so as to investigate the adaptive airfoils' abilities to improve the baseline airfoil in a region near its design C_l of 0.6.

<u>Aerodynamic results</u>

In general, the adaptive airfoils investigated produced small reductions in section drag, with these reductions growing larger as the aft end of the actuated panel moved farther back and lengthened the panel. Figures 10, 11, 12, 13 and 14 show the reduction in drag made by the adaptive airfoils at Mach numbers in the vicinity of drag divergence as a function of the actuated panel aft end location, where zero corresponds to the base line non-adaptive airfoil.

Figure 10 shows a decrease in drag for a Mach number of 0.81 of 12%, but little improvement at any other Mach number. Figure 15 shows only very slight changes in the overall upper surface pressure distributions at this C_l and Mach number for two of the adaptive airfoils as compared to the base airfoil.

A close-up of the shock region is shown in Figure 16, which shows a slight weakening of the shock wave by the adaptive airfoils, which must be the cause of the drag reduction observed in Figure 10. Figure 11 shows only very slight reductions in drag over a range of mach numbers. Figure 12 shows effects similar to the C_l = 0.40 case, except that a much longer actuated panel is needed to reduce drag significantly.

Figure 13 is interesting due to an apparent drag plateau between the Mach numbers of 0.78 and 0.79. Figure 14 shows modest reductions in drag for a number of different Mach numbers, while none of the reductions is as great as that shown in Figure 10 or 12.

Further illustration of the effects described here is shown in Figures 17, 18, 19 and 20. These compare the variation in drag with Mach number for the base line airfoil and the eight adaptive airfoils for a C_l of 0.4. It is apparent that the base line airfoil seems to have an anomaly at Mach 0.81. The reductions in drag at this mach number made by the adaptive airfoils seem to be a result of smoothing out this anomaly. Different values of C_l gave similar results, with large drag reductions stemming from a smoothing out of the base line airfoil curve.

Conclusions

The use of adaptive materials for transonic drag reduction has been investigated. Structural models have been developed and used in conjunction with an established transonic airfoil analysis. This study has shown that there is potential for induced strain actuators bonded to an airfoil skin to reduce transonic wave drag.

The structural model developed is an application of classic laminated plate theory for beam structures. Specifically, three layer symmetric and two layer asymmetric laminations have been studied; however, the general method can be applied to almost any lamination configuration. Also, it may be extended to plates since it is based on plate theory.

This model has been used to study the design trades involved in thickness and stiffness ratios. For symmetric laminates, the result is that a bimorph is desired, with no host structure at all. For asymmetric laminates, the deflection is maximized through the use of very thick, low-modulus actuators.

Neither of these ideal systems may be realistic, due to constraints of weight or strength. Neither of these systems were affected greatly by changing end support conditions. The clamped cases had similar trends to the pinned cases, but with smaller total deflection. For the clamped cases it is also possible to find the length ratio between the actuator and the host beam that maximizes deflections. The symmetric and asymmetric laminates had similar trends and values for optimum length ratios.

These models have been used to calculate deflections of an adaptive transonic airfoil skin. This two dimensional analysis has been done by treating the airfoil skin as a beam supported on its ends by stringers. This has resulted in a series of adaptive airfoils differing only by the length of the actuated skin panel. Using an established airfoil analysis procedure, these adaptive airfoils have shown that they can reduce wave drag by weakening the upper surface shock strength.

It should be noted that these results are based on adaptive airfoils using a single large actuator. Future work should look at using segmented actuators to have more control of the airfoil shape. Also, formal optimization procedures should be used to determine the lowest drag obtainable for a Mach number and lift coefficient.

Acknowledgements

The authors wish to acknowledge the support of this research by NASA Grant NAG-1-157. Mr. Robert C. Scott was grant monitor.

References

1. Whitcomb, R., and Clark, L., "An Airfoil Shape for Efficient Flight at Supercritical Mach Numbers," NASA TM X-1109, 1965.

2. Lock, R., and Fulker, J., "Design of Supercritical Airfoils," *Royal Aircraft Establishment*, Technical Memorandum Aero 1570, May, 1974.

3. Harris, C., "NASA Supercritical Airfoils," NASA Technical Paper 2969, 1990.

4. Cubick, F. E. C., and Jex, H., R., "Aerodynamics, Stability and Control of the 1903 Wright Flyer," National Air and Space Museum, Washington, DC, 1987, pp. 19-43.

5. Somers, Dan M., "Design and Experimental Results for a Flapped Natural-Laminar-Flow Airfoil for General Aviation Applications," NASA Technical Paper 1865, 1981.

6. Szodruch, J., and Hilbig, R., "Variable Wing Camber for Transport Aircraft," Progress in Aerospace Sciences, vol. 25, pp 297-328, 1988.

7. Renken, J. H., "Mission - Adaptive Wing Camber Control Systems for Transport Aircraft," AIAA-85-5006, AIAA 3rd Applied Aerodynamics Conference, October 14-16, 1985.

8. Levinsky, E. S.; Schappelle, R. H.; and Pountney, S., "Airfoil Optimization through the Adaptive Control of Camber and Thickness," CASD-NSC-75-004, September, 1975

9. Levinsky, E. S.; McClain, A. A.; and Schappelle, R. H., "Self Optimizing Flexible Technology Wing Program," Office of Naval Research, ONR-CR212-224-3, February, 1977.

10. Crawley, E., and de Luis, J., "Use of Piezoelectric Actuators as Elements of Intelligent Structures," *AIAA Journal*, Vol. 25, No. 10, 1987, pp. 1371-1385.

11. Crawley, E., and Anderson, E., "Detailed Models of Piezoceramic Actuation of Beams," *Journal of Intelligent Material Systems and Structures*, Vol. 1, January 1990, pp. 4-25.

12. Chaudhry, Z. and Rogers, C., "A Mechanics Approach to Induced Strain Actuation of Structures," *Proceeding, Conference on Adaptive Materials & Smart Structures*, 1993, pp. ??-??

13. Lin, M. and Rogers, C., "Formulation of a Beam Structure with Induced Strain Actuators Based on an Approximated Linear Shear Stress Field," *AIAA*, Paper no. 92-2524; *Proceedings of the 33rd SDM Conference, Dallas, TX*.

14. Wang, B. and Rogers, C., "Laminate Plate Theory for Spatially Distributed Induced Strain Actuators," *Journal of Composite Materials*, Vol. 25, April 1991, pp. 433-452.

15. Jones, R., *Mechanics of Composite Materials*, Scripta Book Company, Washington, 1975.

16. Redeker, G., Wichmann, G., and Oelker, H.-C., "Aerodynamic Investigations Toward an Adaptive Airfoil for a Transonic Transport Aircraft," *Journal of Aircraft*, Vol. 23, No. 5, 1986, pp. 398-405.

17. Sobieczky, H., Fung, K., and Seebass, A., "A New Method for Designing Shock-free Transonic Configurations," *AIAA*, Paper no. 78-1114.

18. Muller, M., "Transonic Drag Reduction Through the Use of Induced Strain Actuators to form an Adaptive Airfoil," Master's Thesis, Purdue University, May 1994.

19 Bauer, F., Garabedian, P., Korn, D., and Jameson, A., Supercritical Wing Sections II. Volume 108 of Lecture Notes in Economics and Mathematical Systems, M. Breckmann and H. P. Künzi, eds., Springer-Verlag, 1975.

20 Bauer, F., Garabedian, P., Korn, D., Supercritical Wing Sections III. Volume 150 of Lecture Notes in Economics and Mathematical Systems, M. Breckmann and H. P. Künzi, eds., Springer-Verlag, 1977.

NASA SC(2) 0612   NACA 23012

Figure 1
Comparison of NASA Supercritical airfoil section (left) with a more conventional section (right)

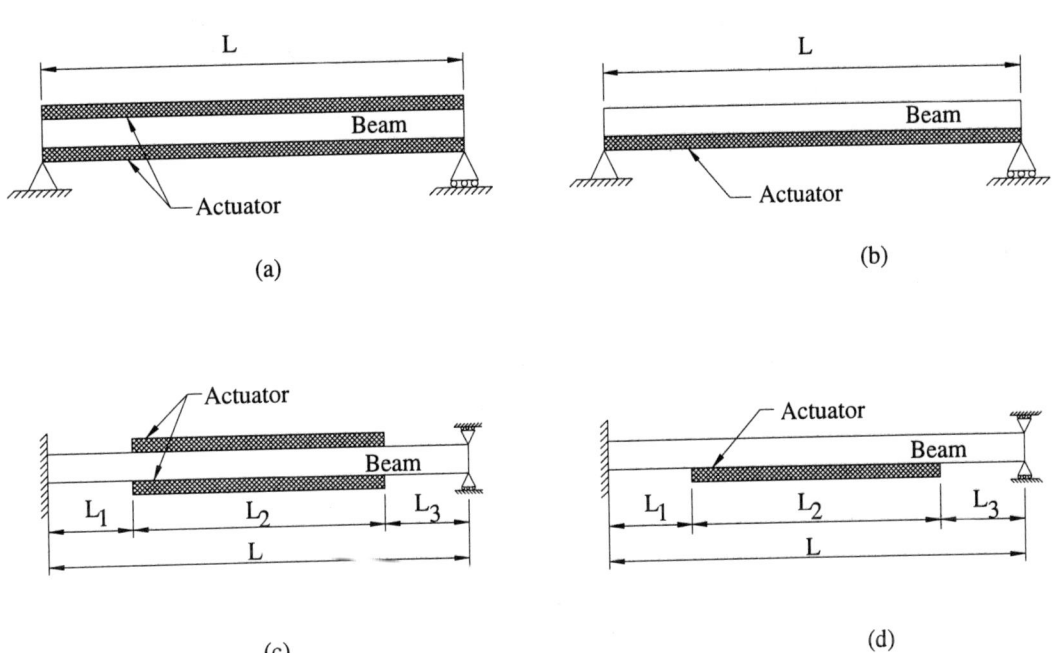

Figure 2
Different beam configurations examined

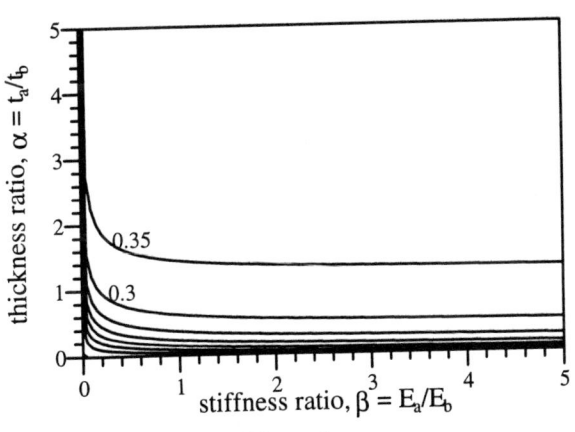

Figure 3
Contours of deflection function for symmetrically laminated pin ended beam

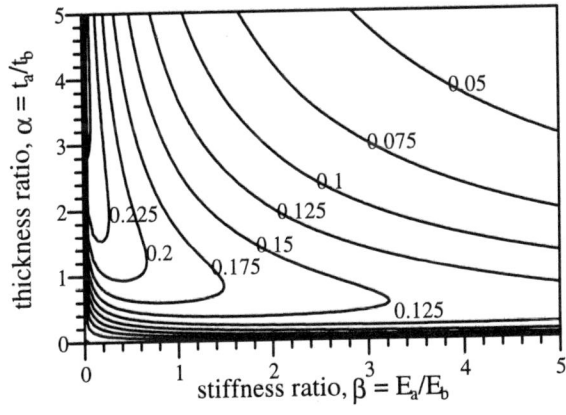

Figure 4
Contours of deflection function for asymmetrically laminated pin ended beam

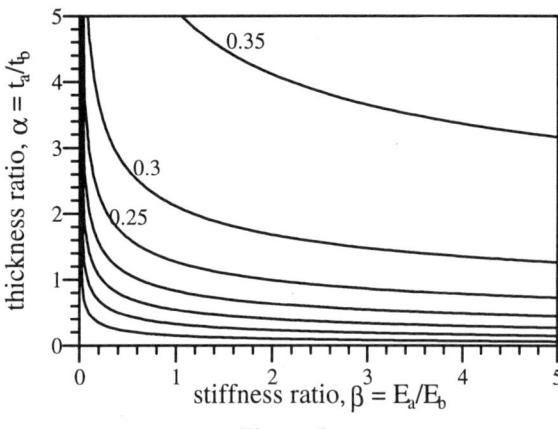

Figure 5
Contours of deflection function for symmetrically laminated clamped beam

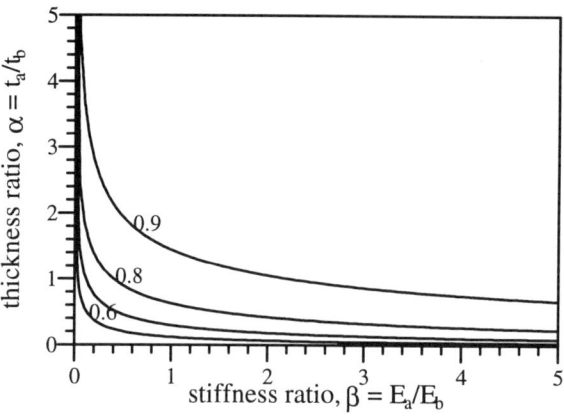

Figure 7
Contours of optimal length ratio for symmetrically laminated clamped beam

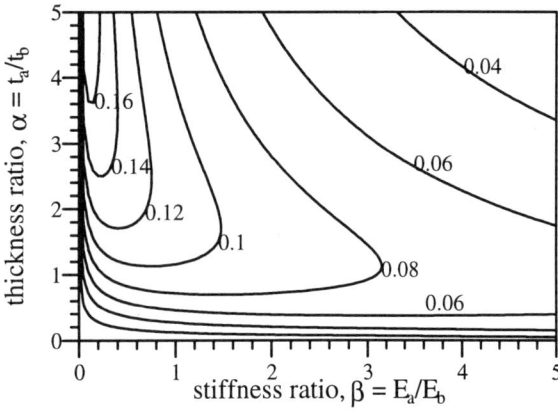

Figure 6
Contours of deflection function for asymmetrically laminated clamped beam

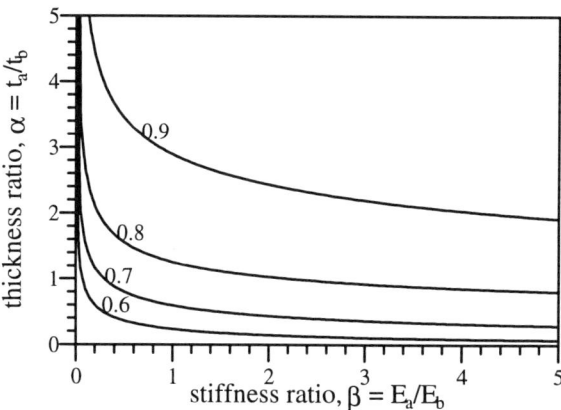

Figure 8
Optimal length ratio contours for asymmetrically laminated clamped beam

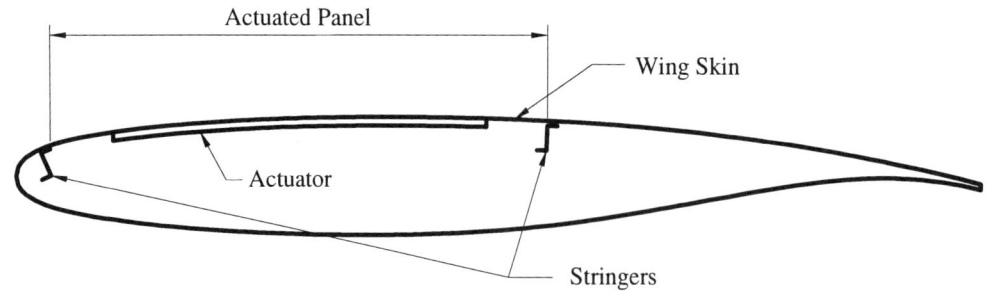

Figure 9
Concept for using induced strain actuator to form an adaptive airfoil

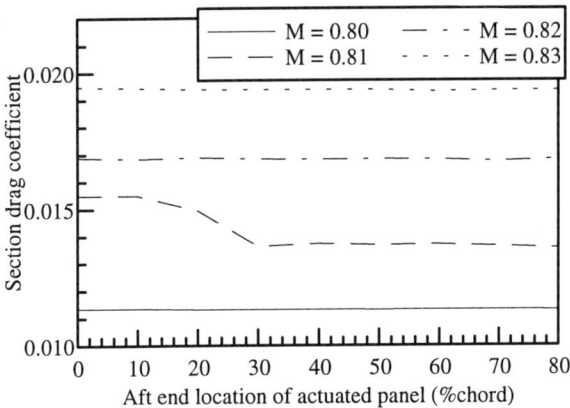

Figure 10
Variation of section drag with adaptive panel length for a section C_l of 0.40

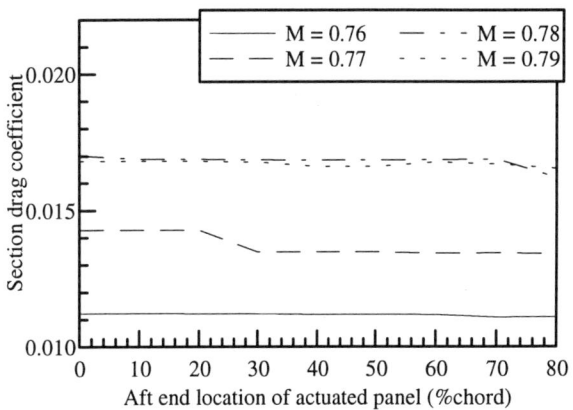

Figure 13
Variation of section drag with adaptive panel length for a section C_l of 0.70

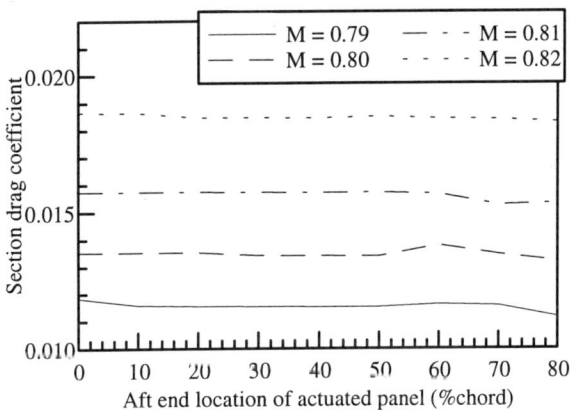

Figure 11
Variation of section drag with adaptive panel length for a section C_l of 0.50

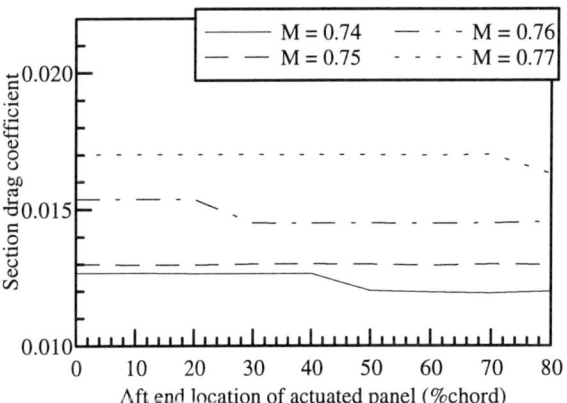

Figure 14
Variation of section drag with adaptive panel length for a section C_l of 0.80

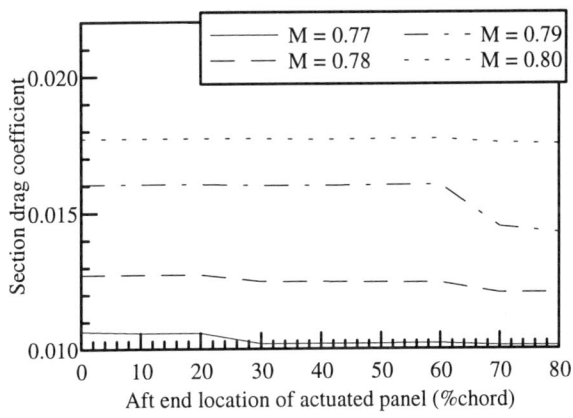

Figure 12
Variation of section drag with adaptive panel length for a section C_l of 0.60

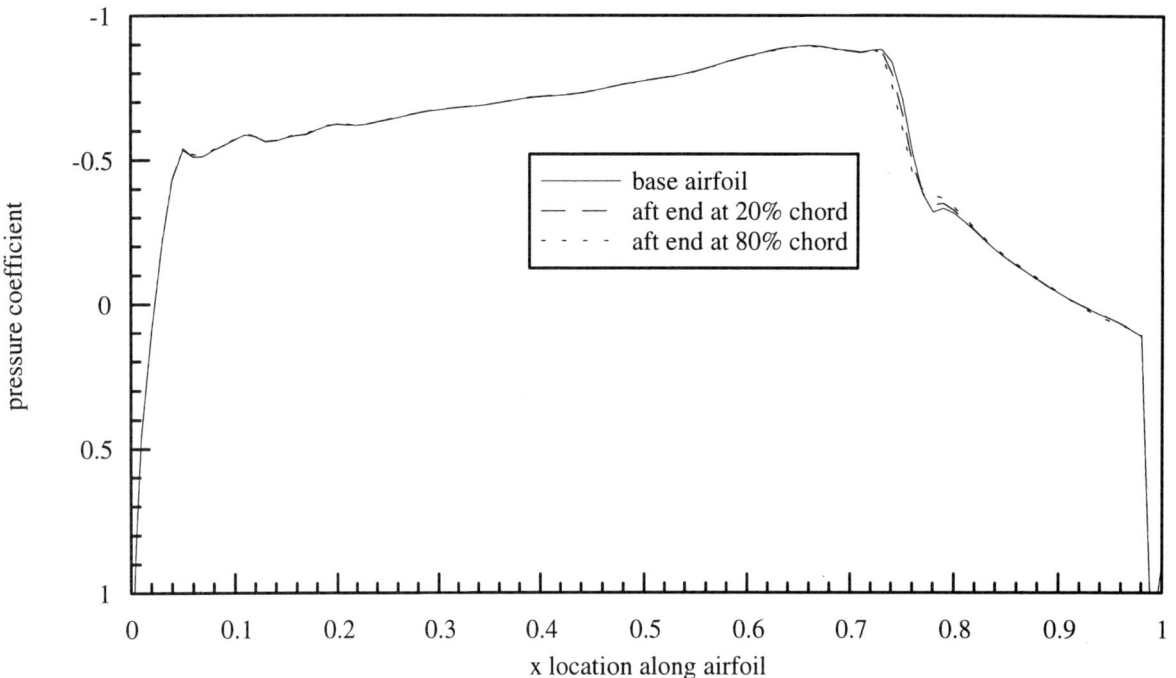

Figure 15
Comparison of upper surface pressure distributions between
base line airfoil and adaptive airfoils

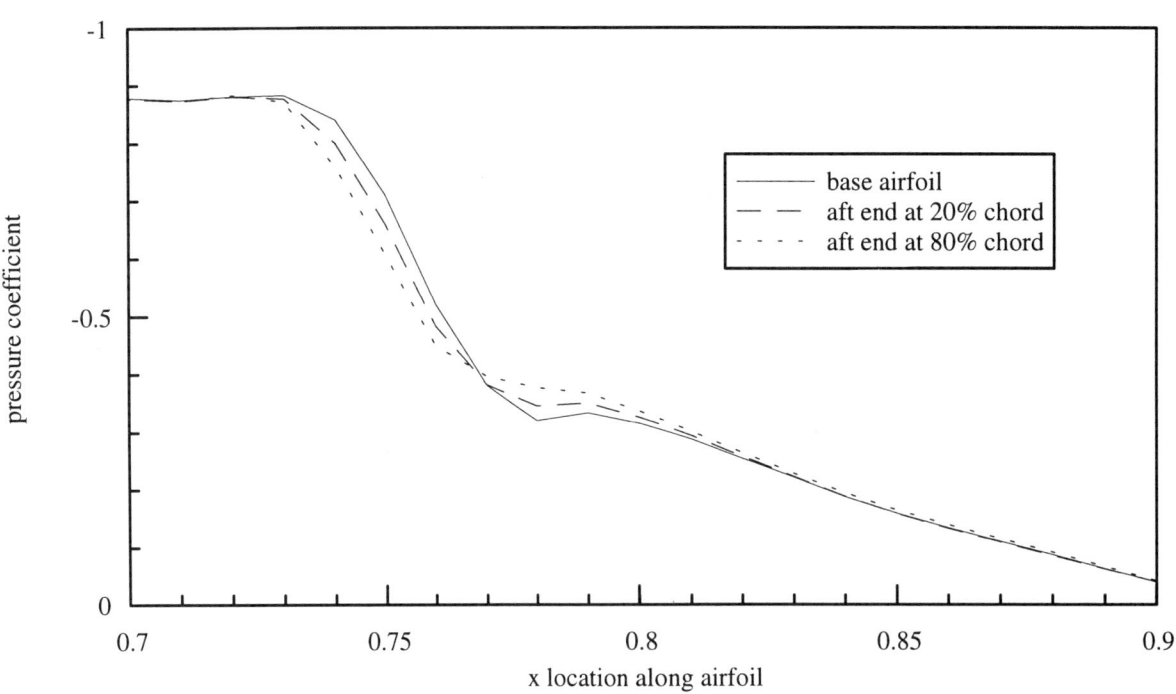

Figure 16
Comparison of upper surface pressure distributions between base line
airfoil and adaptive airfoils in the region of the shock wave

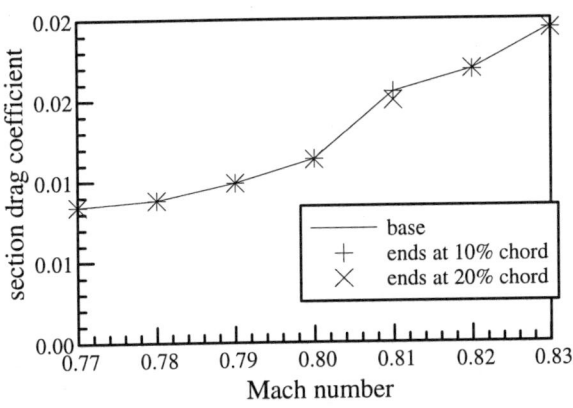

Figure 17
Variation of section drag coefficient with Mach number for base airfoil and adaptive airfoils

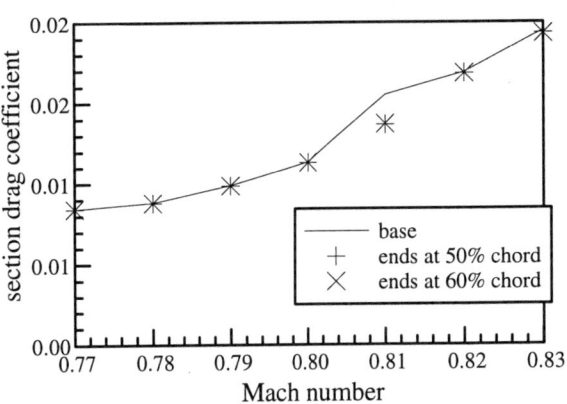

Figure 19
Variation of section drag coefficient with Mach number for base airfoil and adaptive airfoils

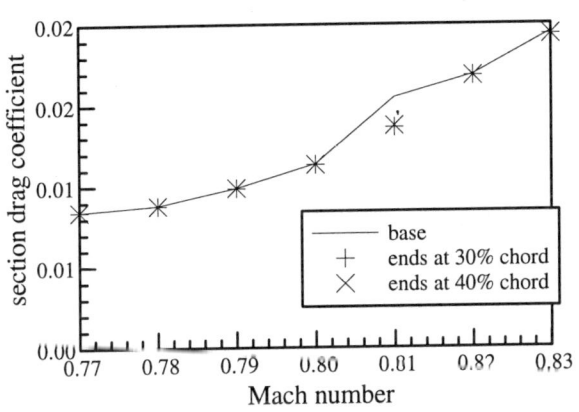

Figure 18
Variation of section drag coefficient with Mach number for base airfoil and adaptive airfoils

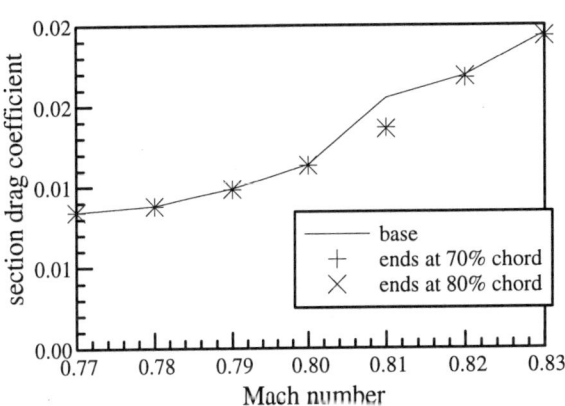

Figure 20
Variation of section drag coefficient with Mach number for base airfoil and adaptive airfoils

IDENTIFICATION OF NONLINEAR JOINTS IN A TRUSS STRUCTURE

Robin J. Bruno
Jet Propulsion Laboratory
California Institute of Technology
Pasadena, California

Abstract

This paper describes a technique to locate and characterize loose joints in a truss structure which are modeled as gaps in the members. The procedure involves prestressing the structure to first eliminate the gaps and then unloading the structure and monitoring a set of displacements. By comparing the calculated displacements to the set of known displacements, the gap location is identified. The loading and unloading is accomplished using a set of actuators whose locations are arbitrary. The sizes of the gaps are determined by monitoring the gap member length changes using the appropriate linear force-displacement relationship for the load level. Three numerical examples are presented to illustrate the procedure assuming ideal displacement data. A last example is presented which uses a displacement set which is corrupted to demonstrate the effect of measurement error.

Introduction

Adaptive structures are attractive alternatives to entirely passive systems in structures which support orbiting space telescopes or optical interferometers. The stringent shape requirements of a structure supporting optical equipment can be maintained with minimal cost and energy output when active members are incorporated into the structure. An accurate structural model is required to perform vibration suppression or shape control however the type of structure under consideration may have a tendency toward joint looseness. The type of structure

---

The research described in this paper was carried out by the Jet Propulsion Laboratory, California Institute of Technology, under a contract with the National Aeronautics and Space Administration.

Copyright © 1994 by the American Institute of Aeronautics and Astronautics, Inc. All rights reserved.

addressed in this paper is of a size which requires a deployable or erectable design to fit the launch vehicle. When final assembly is completed in orbit, there is a possibility that any joints that were not preassembled will not possess the precision fit required. This results in a structure with joint induced nonlinearities and the true response to actuator loading requires the loose joints be located and characterized. The identification of loose joints using active members is addressed in this paper.

Background

Numerous studies have been performed to identify the structural properties of a system. Often, the modal response of the structure is monitored by applying a dynamic excitation to the structure, then the measured frequencies and mode shapes are used to update the mass and stiffness matrices[1,2,3,4]. Chen and Garba[5] used modal information to detect the location and extent of damage in a structure. However the method is only successful for small changes in stiffness. Kim and Bartkowicz[6] also used modal data to detect damage and were able to accurately locate changes of up to 30 percent in the stiffness matrix. Their approach required extensive measurements. Lee, Hossian, and Venkayya[7] investigated correlating the system matrices using a distributed parameter scheme rather than the typical discrete or finite element approach. Their approach depends on a realistic initial estimate of the structures properties.

An interesting method using a neural network approach was developed by Tsou and Shen[8] where the network "learns" the characteristics of the system and is able to generate the appropriate stiffness properties when presented with new response data. This method can detect substantial changes in the beam properties but depends on the analysis of numerous sets of data to train the system. Sanayei and Onipede[9] were able to identify changes in stiffness using a static approach with limited measurements. Using an iterative process of updating the stiffness matrix based on a sensitivity approach, they were able to successfully identify changes in properties when the measurement locations were selected properly.

The approach taken in this study also uses a static approach and requires few measurements and a minimal number of actuators. The current research differs from some of the studies mentioned in that a complete loss of stiffness is detected to determine the location of the loose joints.

Problem Description

The relationship between actuator forces and nodal displacements is clearly defined for a linear structure as well as for a nonlinear structure[10] provided an accurate analysis model is available. The identification of any loose joints is required to develop an accurate force-displacement relationship in the structure. In this paper, the joint looseness is defined as a gap associated with a particular member. The gap member does not develop any internal forces until the structure has been loaded in such a way that the relative displacement, δ, between the endpoints of the gap member changes by an amount exceeding the gap size, ε. This force-displacement relationship of the gap member is shown in Figure 1. While $|\delta| < \varepsilon/2$, the gap remains open, the member is free of internal forces and does not contribute to the system stiffness matrix. When $|\delta| > \varepsilon/2$, the gap is considered closed and the element stiffness is added to the system stiffness.

An accurate model requires knowledge of the gap member locations and the gap sizes, ε. The presence of gaps, which is equivalent to the removal of members, alters the initial stiffness matrix of the unloaded structure significantly. The difference in response between a full structure and a structure with a number of members removed is too great to use the linear structure, that is the structure with no gaps, as an approximation. The number of possible patterns of 5 gaps in 50 possible locations is over 2.1 million. Therefore any attempt to locate the gaps by comparing the structure's response to the response assuming different gap locations would be futile. Further complicating the analysis are the additional unknowns of the number of gaps and the gap sizes. In this paper, the number of gaps, their locations, and their sizes will be determined.

Approach

The number of possible patterns for a known number of gaps in the structure is immense and when adding the additional unknown of the gap size, any attempt to establish an initial structural model becomes unfeasible. Thus the approach taken here is to first completely prestress the structure with a set of actuators to enforce closure of all existing gaps. Once the structure is prestressed, it is then unloaded and the problem is reduced to a series of steps for locating one gap at a time. In the prestressing and subsequent unloading of the structure, a set of actuators at preassigned locations is used and the nodal displacements at a number of degrees of freedom are monitored. To prestress the structure, the structure is loaded until the response of the structure to an additional perturbation is consistent with the known response of the structure with no gaps. When this occurs the gaps have been closed and the actuator loads can be released to begin the detection process. Prior to prestressing the response to an incremental load is calculated which identifies the system behavior when all gaps are open.

At the start of the unloading process, the displacement response under the two conditions of all gaps open and all gaps closed is known. Starting from the prestressed structure, the actuator displacements are decreased incrementally and the nodal displacement response is determined. A vector Δu^n is generated where $\Delta u^n = u^n - u^{n-1}$ and u^n is the vector of displacements at unloading step n. At each step Δu^n is compared to Δu^{n-1}. When $\Delta u^n \neq \Delta u^{n-1}$, a break in linearity has occurred indicating the opening of a gap. The actuator displacements at the break in linearity are stored in a vector δ_m where m indicates the number of gaps detected. The unloading procedure continues until the next break occurs or until the structure is completely unloaded. At the conclusion of the unloading sequence, the number of breaks in linearity in the calculated response indicates the number of gaps present in the structure.

Once the number of gaps has been established, the procedure of determining their locations is initiated. To locate the first gap, displacements of $(\delta_1 + \delta_2)/2$ are applied to the actuators to bring the structure to the point where one gap is open. Then one actuator is perturbed and the difference in the displacement response between the perturbed and unperturbed state at the current load level is calculated and the stored in a vector x. The entries in x are normalized to the maximum value. At this point a matrix, **S**, is created consisting of column entries representing the normalized difference in displacements at the monitored degrees of freedom when one actuator is perturbed and one member has been removed from the structure. The size of **S** is the number of displacement measurements by the number of possible gap locations where the jth column corresponds to the response when the jth possible gap is open and the ith row corresponds to the ith displacement measurement. The gap location is established by calculating the error measurement e_j over all possible gap locations:

$$e_j = \sum_{i=1}^{nd} (x_i - S_{ij})^2 \quad (1)$$
$$j=1,2,3,\ldots,npgap$$

where j represents the gap location, nd is the number of displacement that are monitored, and $npgap$ is the number of possible gap locations. Under ideal conditions with no measurement error, when $e_j = 0$, the gap is in the jth possible location.

If $e_j = 0$ at more than one location, there are multiple columns in S which match x and the gap location has not yet been uniquely identified. This is possible because the number degrees of freedom in the structure almost always exceeds the number of monitored degrees of freedom. If there is more than one matching column, a second actuator in the structure is arbitrarily selected and perturbed. The procedure of perturbing the structure and developing the matrix S is repeated for each new actuator location however the number of columns in S is now reduced to the number of columns in the original matrix which were identical to x. The formulation of S and the error calculation is repeated using different actuators until a unique match is determined.

After the first gap has been identified, the numerical model is updated to reflect the opening of the first gap by removing that member from the stiffness matrix when the actuator displacements fall below δ_1. The actuator loading is then decreased to $(\delta_2 + \delta_3)/2$ and the process is repeated to identify the next gap. In this manner, when each gap location is uniquely identified, the number of static analyses is reduced to $ngap * npgap$ where $ngap$ is the actual number of gaps and $npgap$ is the number of possible gap locations.

At the completion of each stage, the gap location is known and the actuator displacements at which the gap opened is known. To determine the gap size, the linear force-displacement relationships for a truss structure are used. The key equations are taken from the displacement method of analysis and starts with the compatibility equation:

$$\Delta = \beta u \quad (2)$$

where Δ is the member length change, u is the nodal displacements and β is the matrix of direction cosines which relate the two. The next relationship used is the force-displacement relationship for a truss structure:

$$P = K(\Delta - \Delta_0) \quad (3)$$

where P is the vector of member forces, K is a diagonal matrix of member stiffnesses and Δ_0 is the vector of initial member length changes. The final equation used is the force equilibrium relationship:

$$F = \beta^T P \quad (4)$$

where F is the vector of external forces. Combining these three equations and recognizing that the external forces, F, are zero, gives:

$$\Delta = \beta K_G^{-1} \beta^T K \Delta_0 \quad (5)$$

where $K_G = \beta^T K \beta$ is the global stiffness matrix. This provides the relationship between the initial actuator length changes and all member length changes. The independent variable in this formulation is Δ_0, the vector of initial member length changes which is not a measurable quantity. The vector, Δ_0, contains the actuator displacements in a free-free condition. The actual actuator length changes that are generated when the actuators are incorporated into the structure are found in Δ. To relate δ, the actuator length changes to Δ_0, a matrix T_{aa} is created by selecting the rows and columns of $\beta K_G^{-1}\beta^T K$ which correspond to the actuator locations. Then the measured actuator length changes, δ, are expressed by:

$$\delta = T_{aa}\Delta_0 \qquad (6)$$

Using equations (5) and (6) gives the gap member length changes:

$$\Delta_g = T_{ga}T_{aa}^{-1}\delta \qquad (7)$$

where Δ_g is the vector of gap member length changes and T_{ga} is a submatrix of T created by selecting the rows corresponding to the gap locations and the columns corresponding to the actuator locations. Using the actuator displacements, the gap member length changes are monitored and summed to determined the cumulative value of Δ_g at each break in linearity to determine the gap size.

Numerical Examples

A modified version of a support structure for a space-based segmented reflector was used as the basic structure in the numerical examples. This structure consists of 72 members and 63 degrees of freedom with member sizes which range from .77 meters to .92 meters in length. The gap sizes were all set at 100 microns although uniformity in size is not necessary for the procedure. The number of actuators and gaps varies in each example but the set of monitored displacements remains the same. The structural displacements used to generate Δu_n and x are a set of twelve out of plane displacements at the surface. In each case the gap locations and the actuator locations were selected at random. The results are presented in terms of the actuator length changes and the maximum actuator forces generated.

Case 1 - 2 actuators, 4 gaps

In the first case, four gaps were arbitrarily placed in members 10, 42, 53, 62 and five actuators were located in members 15, 23, 43, 60, and 71. The solid circles near the joints represent the gaps and the darkened members indicate the actuator locations in Figure 2. The prestressing was accomplished by exercising only actuators 15 and 23. The prestress actuator displacements and the actuator displacements at which each gap opens are shown in Table 1. The gaps opened in the order 53, 10, 62, and 42, and the all gap sizes were calculated at 100 microns. In this case, each time a gap opened and the response to the actuator perturbation, x, was compared to the columns of S, a unique match was found and no additional actuator perturbations were performed. Therefore in this example, five actuators were selected to perform the gap identification but only two were used with a maximum actuator force of 102 lbs compression. The length changes that occurred at actuators 43, 60 and 71 were in response to the displacements enforced in actuators 15 and 23. Although only two actuators were driven, the remaining three were retained in the analysis to monitor the resultant forces and displacements and insure excessive forces are not generated during the process.

Table 1. Case 1 - Actuator Length Changes for Prestressing and Gap Opening

	actuator length changes, microns				
actuator no.	prestress	gap 53 open	gap 10 open	gap 62 open	gap 42 open
15	473.8	386.1	304.4	245.2	201.4
23	179.3	146.1	115.2	92.8	76.2
43	17.8	14.5	11.4	9.2	7.6
60	-3.5	-2.9	-2.2	-1.8	-1.5
71	-4.9	-4.0	-3.1	-2.5	-2.1

Case 2 - 4 actuators, 5 gaps

In case 2, four actuators were placed in members 18, 23, 45, and 66, and five gaps were located in members 2, 3, 4, 43, and 44 (Figure 3). In this case, the maximum actuator force was 114 lbs in compression and the actuator length changes at prestress and gap openings are shown in Table 2. This case also resulted in a unique match of x to the appropriate column of S at each gap opening. In this case, the gaps sizes were also correctly identified at 100 microns.

Case 3 - 4 actuators, 5 gaps

In case 3, with four actuators and five gaps, the problem of detecting more than one possible gap location was encountered. The gaps are located in members 29, 48, 53, 57, and 62, and the actuators are placed in members 3, 20, 46, and 54 (Figure 4). In this case, the first two gaps, located at members 48 and 29, were identified correctly in terms of location and size. In attempting to locate the third gap, three columns of S provided an error measure of 0 when compared to the displacement vector, x, indicating possible gaps at members 26, 35, or 53. After perturbing each remaining actuator, the same three possibilities remained.

The next step was to take each possible gap location at a time, assume it to be the correct one and continue with the procedure. In doing so, when assuming a gap at location 26, no matches were detected on attempting to locate the fourth gap, thus eliminating the possibility of a gap in member 26. Using 35 and 53 one at a time to detect the fourth gap both resulted in duplicate results of the next gap occurring at member 57. At this point the fourth gap, in member 57, has been determined but the location of the third gap remains undefined.

Once more, two cases were continued, one assuming four gaps at members 48, 29, 35, and 57 and another assuming four gaps at 48, 29, 53, and 57. Both of these cases produced a result of a fifth gap in member 62. At this point, all actuators had been used for perturbations, and two different paths had been followed in the hopes that one would lead to an error. Since gap number three still had not been determined, a new unloading pattern was used.

This time, instead of uniformly decreasing the actuator displacements until the structure is completely unloaded, the actuators were unloaded sequentially. In doing so, different nodal displacements are obtained and all of the gaps were then uniquely identified in location and size. In this case, the maximum actuator force in prestressing was 279 lbs at actuator 20.

Case 4 - include measurement error

The example described here begins with the same gap and actuator placement as that in Case 1. In the present case, however, the displacement vector x, was perturbed with a random error vector to determine when the process of gap identification deteriorates. In the first error analysis, the error vector contains a random distribution of numbers in

Table 2. Case 2 - Actuator Length Changes for Prestressing and Gap Opening

actuator no.	actuator length changes, microns					
	prestress	gap 2 open	gap 3 open	gap 4 open	gap 43 open	gap 44 open
18	198.4	156.4	114.2	112.2	100.8	96.3
23	75.2	83.5	67.5	66.7	62.4	60.5
45	381.4	316.1	235.0	231.2	209.3	199.8
66	200.2	156.0	113.4	111.4	99.9	94.9

the range of 0 to .001 and is added to x, which has been normalized to the maximum displacement. The original vector x, and the displacement vector with measurement error, x_e are:

x	x_e
-0.2230	-0.2228
-0.0857	-0.0856
0.1274	0.1281
-1.0000	-0.9993
-0.7592	-0.7583
-0.0477	-0.0473
-0.3559	-0.3554
-0.2676	-0.2668
-0.0484	-0.0483
0.5771	0.5771
0.3644	0.3650
-0.2303	-0.2297

With this level of error added, the procedure successfully identified all gaps correctly, by selecting j at which the minimum value of e_j occurs. The error level was increased and the gap detection process repeated until the program could no longer accurately identify the gap locations. At an error with a maximum value of .04, the joint identification process remained successful. The original vector and vector with errors in this case were:

x	x_e
-0.2230	-0.2210
-0.0857	-0.0552
0.1274	0.1582
-1.0000	-0.9669
-0.7592	-0.7542
-0.0477	-0.0471
-0.3559	-0.3284
-0.2676	-0.2329
-0.0484	-0.0232
0.5771	0.6065
0.3644	0.3934
-0.2303	-0.1903

The procedure failed at an error distribution with a maximum value of .05. The vectors x and x_e in this case were:

x	x_e
-0.2230	-0.2226
-0.0857	-0.0665
0.1274	0.1307
-1.0000	-0.9791
-0.7592	-0.7249
-0.0477	-0.0182
-0.3559	-0.3094
-0.2676	-0.2253
-0.0484	-0.0220
0.5771	0.5817
0.3644	0.3971
-0.2303	-0.2095

If the error addition is small enough such that the gap location is still identifiable, as when using a maximum error of .04, the gap size will still be calculated correctly. The gap size is dependent on the member length changes at the break in linearity in the unloading process and the gap location. It is unaffected by the error in x.

Conclusions

The locations and sizes of the loose joints in the structure were successfully located using an actuator-induced static loading and unloading procedure. Four numerical examples were presented, three of which assumed no deterioration in the assumed displacement measurements. In one of these three cases, the initial unloading procedure did not provide complete information on the location of the loose joints. However, by altering the unloading sequence, the last gap was correctly located. The last case presents the effect of introducing an error into the structure's displacement vector. The level of error was increased until an error in gap identification occurred.

The procedure described in this paper has the advantage of using a small number of actuators whose positions in the structure are not critical. In each case, the locations of the actuators and gaps were randomly assigned and the gap location and sizes were correctly identified.

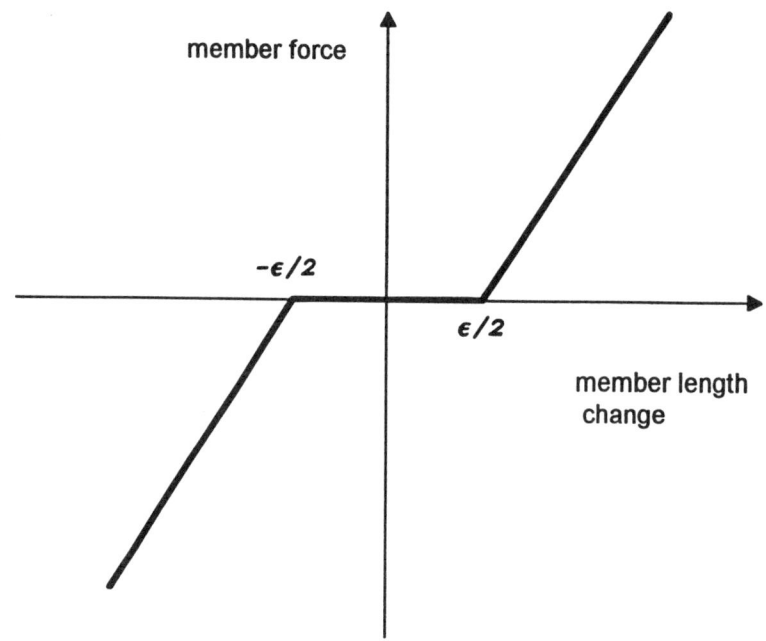

Figure 1. Gap Member Force-Displacement Relationship

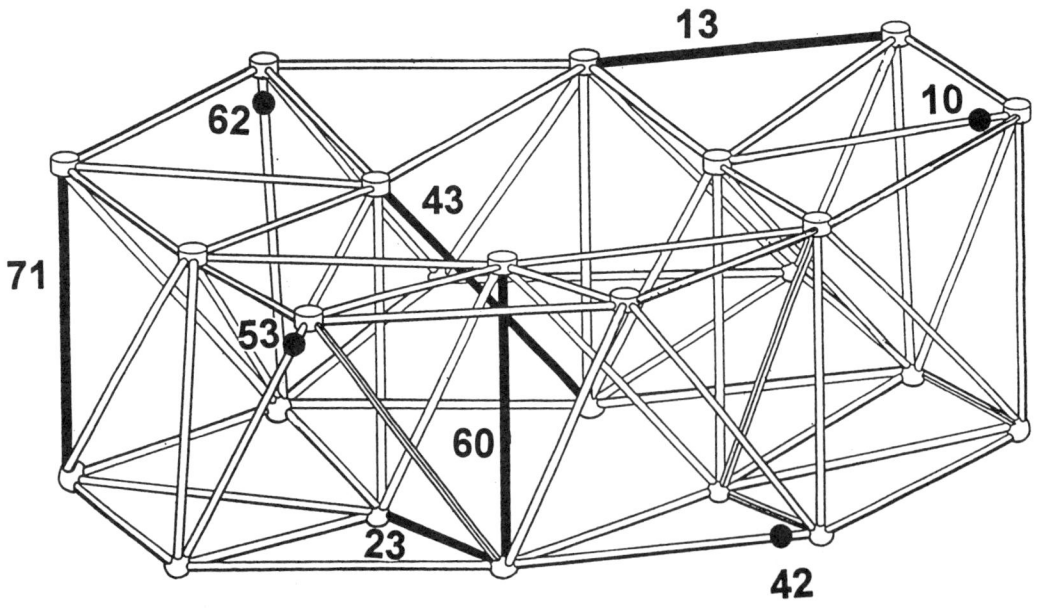

Figure 2. Actuator and Gap Locations for Case 1

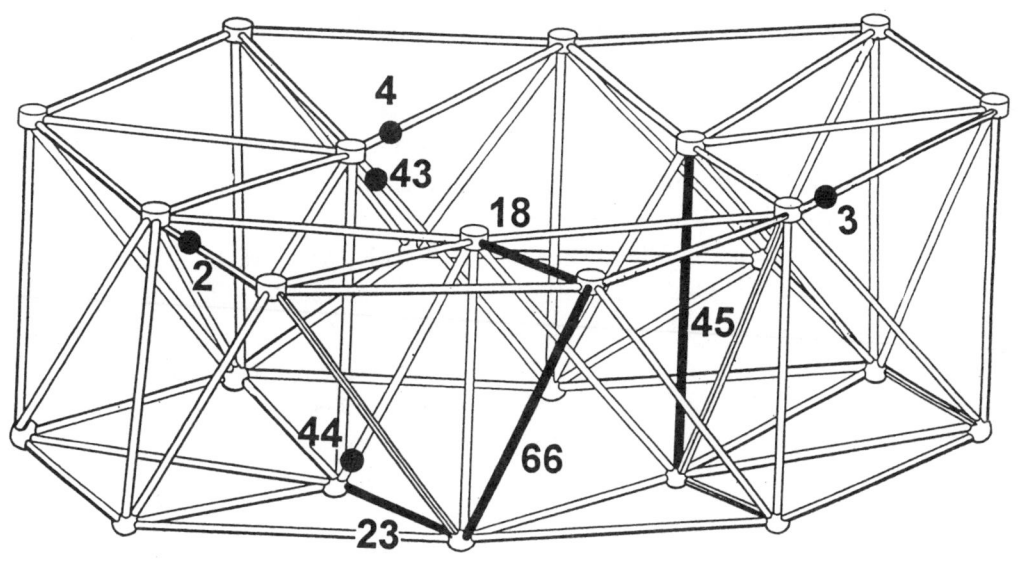

Figure 3. Actuator and Gap Locations for Case 2

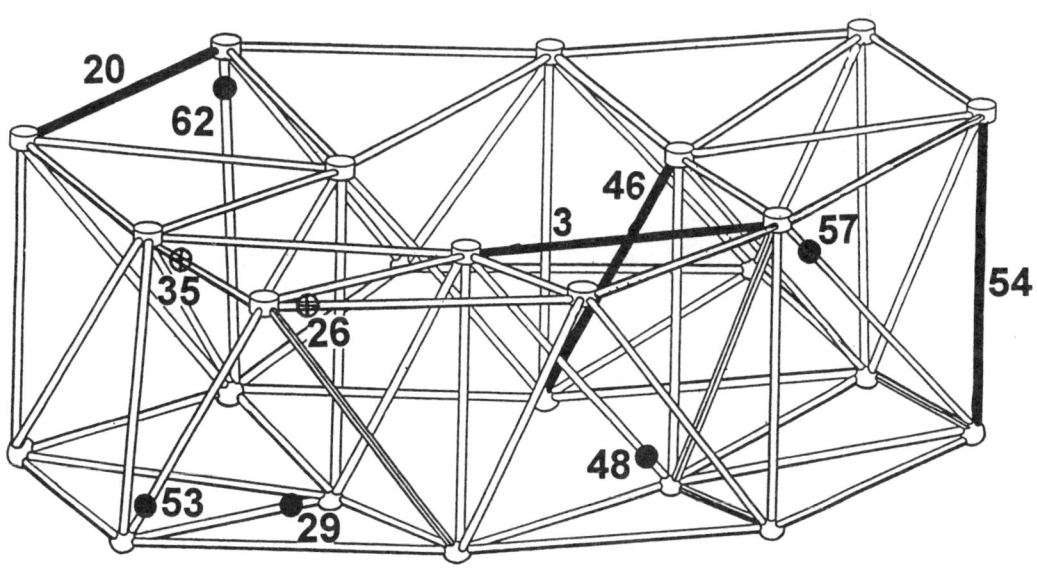

Figure 3. Actuator and Gap Locations for Case 3

References:

1. Hedgepeth, J.M., Critical Requirements for the Design of Large Space Structures. NASA CR-3483,1981.

2. Pearson, J.E., and Hansen, S., "Experimental Studies of a Deformable-Mirror Adaptive Optical System," Journal of Optical Society America, No. 67, pp. 360-369, 1977.

3. Kaouk and Zimmerman, "Structural Damage Assessment Using a Generalized Minimum Rank Perturbation Theory," Proceedings of the 34th AIAA SDM Conference, pp. 1529-1538, La Jolla, California, April, 1993.

4. Kim, H.M. and Doiron, H.H., "On-Orbit Modal Identification of Large Space Structures," Sound and Vibration, Vol. 26, No. 6, pp.24-30, June 1992.

5. Kabe, A.M., "Stiffness Matrix Adjustment Using Mode Data," AIAA Journal, Vol. 23, No. 9, pp. 1431-1436, September, 1985.

6. McGowan, P.E., Smith, S.W., and Javeed, M., "Experiments for Locating Damaged Members in a Truss Structure," Proceedings of the USAF/NASA Conference on System Identification and Health Monitoring of Precision Space Structures, Pasadena, California, 1990, pp. 571-616.

7. Chen, J.-C., Garba, J.A., "On Orbit Damage Assessment for Large Space Structures," Proceedings of the 28th AIAA SDM Conference, 1987.

8. Kim, H.M. and Bartkowicz, T.J., "Damage Detection and Health Monitoring of Large Space Structures," Proceedings of the 34th AIAA SDM Conference pp. 3527-3533, La Jolla, California, April, 1993.

9. Lee, K.Y., Hossian, S.A., Venkayya, V.B., "System Identification of Flexible Structures," Proceedings of the 29th AIAA SDM Conference pp. 1202-1209, Williamsburg, Virginia, 1988.

10. Lindner, D.K., Twitty, G., and Goff, R., "Damage Detection, Location and Estimation for Large Truss Structures," Proceedings of the 34th AIAA SDM Conference pp. 1539-1548, La Jolla, California, April, 1993.

11. Tsou,P., Shen, M.-H., "Structural Damage Detection and Identification using Neural Network,", Proceedings of the 34th AIAA SDM Conference pp. 3551-3560, La Jolla, California, April, 1993.

12. Sanayei, M., Onipede, O., "Damage Assessment of Structures Using Static Test Data," AIAA Journal, Vol. 29, No. 7, July 1991, pp 1174-1179.

AIAA-94-1777-CP

DEVELOPMENT OF INTELLIGENT STRUCTURES USING MULTIOBJECTIVE OPTIMIZATION AND SIMULATED ANNEALING

Charles E. Seeley[†] and Aditi Chattopadhyay[††]
Department of Mechanical and Aerospace Engineering
Arizona State University
Tempe, Arizona 85287

Abstract

A multiobjective optimization procedure is used to address the combined problems of structures/controls synthesis and actuator locations for the design of intelligent structures. Both continuous and discrete design variables are treated equally in the formulation. Multiple and conflicting design objectives such as vibration reduction, dissipated energy, power and a vibration performance index are combined by utilizing an efficient multiobjective optimization formulation. Piezoelectric materials are used as actuators in the control system. A simulated annealing algorithm is used for optimization combined with an approximation technique to reduce computational effort. A numerical example is presented. The efficiency of the procedure is demonstrated by a comparison to a nonlinear programming technique.

Nomenclature

A	state matrix
a	continuous design variable vector
B	control matrix
b	discrete design variable vector
C	capacitance
C	gain matrix
d	prescribed values
e(t)	error
F	objective function
G	scalar feedback gain
g	constraint
I	identity matrix
IAE	integral absolute error
J	dissipated energy
K	stiffness matrix
M	mass matrix
P	disturbance force vector
P	probability factor
ρ	draw-down factor
U	electrical energy
u	control vector
V	voltage
x	displacement vector
y	transverse displacement
z	(0,1) variables
φ	rotation
γ	fraction of initial energy
ω_1	fundamental natural frequency
ζ	actual damping ratio

Subscripts
L   lower bound
U   upper bound

Superscripts
•   time derivative
*   approximate value
-   desired value

Introduction

Aerospace structures must often compromise stiffness to meet stringent weight requirements. The reduced stiffness results in a vulnerability to resonance from relatively minor disturbance forces which would normally not exist. Therefore, it is important that these structures are able to suppress vibrations from an unknown disturbance force, using an active control system. Recently, there has been a significant interest in the design of intelligent structures for vibration control[1-3]. Several different areas are important to include in the design of an intelligent structure. First, the structural design must be considered[4]. This includes selection of a structural material and fabrication of structural members. Second, the design of the control system and actuators must be considered[3,5]. Finally, the location of the sensors and actuators must be included in the design process as well[6,7]. Previously, these problems have been considered separately or sequentially. Since the problem is multidisciplinary in nature, it is desirable to integrate the design objectives from the necessary disciplines inside a closed loop optimization formulation to achieve a more optimal intelligent structure.

In this paper, the integration of structures and controls is addressed using a formal multiobjective formulation technique. The structures/controls problems are associated with continuous design variables such as thicknesses and gains. These variables exhibit derivatives and lend themselves well to gradient based optimization techniques. The actuator location problem involves discrete (0,1) variables which do not possess conventional derivatives. Therefore, their combined presence in the optimization problem poses a formidable difficulty due to the incompatibility of

---

[†] Graduate Research Assistant, Student Member AIAA
[††] Assistant Professor, Senior Member AIAA, Member SPIE, ASME, AHS

Copyright © 1994 by Chattopadhyay. Published by the American Institute of Aeronautics and Astronautics, Inc. with permission.

continuous and discrete variables. However, continuous variables can be represented by a series of discrete (0,1) variables and discrete optimization techniques such as simulated annealing can be used. Simulated annealing algorithms have been shown to be effective in a variety of different engineering applications such as composites[8], actuator locations[6] and circuit board design[9]. In this paper, a procedure is developed to address a multiobjective optimization problem which includes both discrete and continuous design variables using simulated annealing. Since several function evaluations are necessary in such techniques, the procedure can be expensive if an exact analysis is used for each iteration. Therefore, an approximation technique based on a linear Taylor series expansion is used to reduce computational effort. The procedure developed is demonstrated on a structures/controls optimization problem of a cantilever box beam.

Problem Formulation

An aluminum, cantilever beam with square cross section is considered as an example. A piezoceramic material (PZT G-1195)[10] with thickness 0.019mm is used to model the actuators. These are surface bonded to the box beam and act to produce a bending moment about the neutral axis of the beam elements from an applied voltage. Details of this model can be found in refs. 11 and 12. The bending moment provided by the piezoelectric actuators is assumed to be equivalent to an applied moment M at the nodes of each beam element as shown in Fig. 1. This bending moment is the active control force used to control the structure. The dynamic equations of motion, without damping or disturbance force, assume the following standard form:

$$M\ddot{x} + Kx = 0 \qquad (1)$$

where x is the vector of transverse displacements and rotations ($y_1, \phi_1, y_2, \phi_2 \ldots$), M is the augmented mass matrix and K is the augmented stiffness matrix which include mass and stiffness properties of the piezoelectrics only where they exist on the structure. A disturbance force F is applied to the uncontrolled structure over a time interval $0 \leq t \leq t_a$ which leads to the following equation of motion:

$$M\ddot{x} + Kx = P \qquad (2)$$

The actuators become active during a time period $t_a \leq t \leq t_b$ in order to damp out the vibrations caused by the disturbance during the previous time period. The state space equations are written as:

$$\frac{\partial}{\partial t}\begin{bmatrix} x \\ \dot{x} \end{bmatrix} = A \begin{bmatrix} x \\ \dot{x} \end{bmatrix} + Bu \qquad (3)$$

where:

$$A = \begin{bmatrix} 0 & I \\ -MK^{-1} & 0 \end{bmatrix} \qquad (4)$$

$$B = \begin{bmatrix} 0 \\ -M^{-1}C \end{bmatrix} \qquad (5)$$

It is assumed that the actuators and sensors are collocated in order to avoid problems with stability. In the above equations, I is the identity matrix and C is the gain matrix of the piezoelectric actuators. Rate feedback is used in conjunction with the control system so that the effect of the actuators will be similar to structural damping. The feedback control law is written as:

$$u = C\dot{x} \qquad (6)$$

Therefore, the control gain matrix is analogous to a viscous structural damping matrix and the actuators assume the role of active damper elements. Any passive structural damping is assumed to be negligible. Finally, the undamped system is checked during the time interval $t_b \leq t \leq t_c$ to ensure that the total energy of the system and the displacements have been reduced to a specified value[3].

Modal summation techniques are used to reduce the equations of motion to a modal system. This transformation has considerable computational advantages and also eliminates problems related to controlling higher modes which do not have significant contributions on the overall motion. The first four bending modes are retained in this study. The active damping matrix (C) is not proportional to the mass and stiffness matrices, so techniques which take advantage of that situation are not applicable here. As a result, the equations do not decouple.

Optimization Formulation

Objective functions

The optimum design of intelligent structures is associated with several design criteria. In this paper, the objective functions and constraints have been previously formulated[12] using continuous design variables and a nonlinear programming (NLP) technique.

One of the objectives is to minimize the energy dissipated by the piezoelectric actuators. This has the effect of minimizing the work the actuators must do to control the structure. The energy dissipated by each actuator can be expressed as the integral over the time period of active control ($t_a \leq t \leq t_b$) for each actuator. The total energy to be minimized, J, is the sum of these integrals and is given as follows:

$$J = \sum_{i=1}^{IACT} \int_{t_a}^{t_b} \dot{x}_i^T c_i \dot{x}_i \, dt \qquad (7)$$

where IACT is the current number of elements which contain actuators, c_i is the elemental gain matrix and $\dot{x}_i$ are the elemental velocities of the i-th element. The total energy J can also be viewed upon as a quadratic performance index from a controls standpoint using the gain matrix as the arbitrary weight matrix which has physical significance in this case.

Power requirements for a piezoelectric control systems is an important issue in critical aerospace applications[13]. It is desirable to minimize the amount of electrical energy, U, required by the actuators to control the structure. Therefore, it is used as an objective function. For the piezoelectric material used as an actuator, it is assumed that the applied electric field is much greater than the charge generated when the material is deformed. Piezoelectric actuators can be modeled as a parallel RC circuit where R_i is the resistivity and C_i is the capacitance of the i-th piezoelectric actuator. The electrical energy U is an integral of power over time and is summed over all of the actuators to compute the total electrical energy which is expressed as follows:

$$U = \sum_{i=1}^{IACT} \int_{t_a}^{t_b} \left\{ \frac{V_i^2}{R_i} + C_i V_i \frac{dV_i}{dt} \right\} dt \qquad (8)$$

In the above equation, V_i is the voltage of the i-th actuator.

Vibration reduction is an important design criteria. Minimization of a performance index defined by the integral absolute-error (IAE) criterion[14] is suitable for reducing overall vibrational amplitudes and is expressed as follows:

$$\int_{t_a}^{t_b} |e(t)| \, dt \qquad (9)$$

where $e(t)$ is the error of the system. Minimization of the IAE criterion results in a system with reasonable damping and a satisfactory transient-response characteristic. It is also easily evaluated numerically. Therefore, the IAE criteria is used as an objective function and is minimized during optimization to reduce oscillatory motion. Since the desired input of the system approaches zero, the error can be represented simply as the state vector which contains both the displacements and velocities of the dynamic system:

$$e(t) = \begin{bmatrix} x \\ \dot{x} \end{bmatrix} \qquad (10)$$

The integral is evaluated numerically over each component of the state vector and a 2-norm of the resultant vector is used to facilitate its usefulness in the optimization procedure. Finally, the fundamental frequency ω_1 is maximized to avoid possible resonance with the forcing frequency and thereby help reduce vibration.

Constraints

The design criteria that are formulated as constraints are discussed in this section. It is desired that the total energy E of the structure, which is the sum of the potential and the kinetic energies, is reduced to a specified fraction of the original energy after the time interval $T = t_b - t_a$. Therefore, a constraint is imposed as follows:

$$E_{fin} \leq \gamma E_{int} \qquad (11)$$

where E is the energy and the subscripts 'int' and 'fin' refer to the initial and the final states, respectively. The quantity γ is the desired fraction of the initial energy remaining in the system after the specified time T.

A piezoelectric material is associated with a maximum electric field E_{max}. Exceeding this value can result in the loss of the material's special properties. Correspondingly, there exists a maximum voltage V_{max} which can be applied to a piezoelectric actuator and must not be exceeded. The voltage is checked at each time step of the numerical equation solver. Noting that $V_i = G_i \dot{\phi}_i$ for each actuator, when $V_i \geq V_{max}$, saturation of the actuators occurs and the gain for the next time step for that actuator is set to:

$$G_i = \frac{V_{max}}{\dot{\phi}_i} \qquad (12)$$

This ensures that $V_i \leq V_{max}$ without introducing any new design variables. As the rotational velocity increases, the voltage to each actuator remains constant and does not exceed the maximum allowable voltage. As the number of actuators is reduced during the optimization process, the voltage to the remaining actuators increases to redistribute the control forces necessary for satisfying all of the constraints. As a result, this constraint becomes more active.

The desired modal damping ratio corresponding to the j-th mode $(\bar{\zeta})_j$ can be expressed as:

$$\bar{\zeta}_j = \frac{\ln(1/\kappa)}{\omega_j T} \qquad (13)$$

where ω_j is the j-th undamped natural frequency of the structure and κ is the fraction of the initial vibration amplitude which remains after the time interval T. Upon calculation of ζ_j, the actual damping ratio corresponding to the j-th mode, the following constraint is imposed to ensure that the desired damping ratios are achieved.

$$\zeta_j \geq \bar{\zeta}_j \quad (14)$$

The gain matrix C is not constant since it must be altered at each time step to ensure that the voltage does not exceed V_{max}. Therefore, ζ_j is not constant over time either. A minimum gain matrix is obtained at maximum velocity when saturation occurs. The maximum gain matrix is found when V_{max} is not exceeded by any one of the actuators. To evaluate the damping ratio constraint, constant damping ratios are determined by using a gain matrix which is averaged over the time interval when the actuators are active. Although these averaged damping ratios cannot be used to construct an analytical solution of the decaying motion, they approximately describe the character of the solution to ensure that proper damping does occur in the desired modes. It is assumed that the disturbance force has a lower frequency compared to the natural frequency of the structure. By increasing the natural frequency of the structure, the interval between the natural frequency of the structure and the frequency of the disturbance force increases which reduces the response of the structure to the disturbance force. Since the mass can increase in an effort to stiffen the structure and thereby minimize the energy dissipated by the actuators, a constraint is also imposed on the total mass of the beam and the actuators.

Optimization Implementation
Optimization problem

The design variables include web thicknesses of each square box beam element, gains of the actuators and actuator locations. Both continuous (e.g. web thicknesses of each element and actuator gains) and discrete (actuator locations) design variables are used. The discrete (0,1) variables, which are easily adapted to represent the absence/presence of a sensor and actuator pair at a particular location on the structure do not posses conventional derivatives. Therefore, these variables are not compatible with conventional gradient based optimization procedures and require discrete optimization techniques. The combined continuous/discrete optimization problem can be written as follows:

Minimize $\quad F_K(a_I, b_J) \quad K = 1, 2, \ldots \text{NOBJ}$

Subject to $\quad g_M(a_I, b_J) \quad M = 1, 2, \ldots \text{NCON}$

$$I = 1, 2, \ldots \text{NC}$$
$$J = 1, 2, \ldots \text{ND}$$

$$a_L \leq a \leq a_U$$

where NOBJ is the number of objective functions and NCON is the number of constraints. Additionally, **a** is the vector of continuous variables and **b** is the vector of (0,1) variables. The number of continuous design variables is NC and the number of discrete design variables is ND. Upper and lower bounds on the continuous variables are indicated by subscripts L and U respectively. The continuous design variables can be represented by using a linear combination of discrete variables. Specific numerical values which the continuous variables can assume are selected to represent these discrete variables. The transformation from continuous variables to (0,1) variables is as follows[15]:

$$a_I = z_{I1}d_{I1} + z_{I2}d_{I2} + \ldots + z_{IQ}d_{IQ}$$

with

$$z_{I1} + z_{I2} + \ldots + z_{IQ} = 1$$

and

$$z_{IQ} = 0 \text{ or } 1$$

Thus, the I-th continuous design variables can take on the value of any of Q prescribed values d_{IQ}. The new design variable vector consists of the new discrete variables z_{IQ} used to represent the continuous variables in addition to the original discrete variables b_J. Therefore, the objective functions and constraints can be totally represented by discrete (0,1) design variables and the new discrete optimization problem can partially be stated as follows:

Minimize
$$F_K(z_{IQ}, b_J) \quad K = 1, 2, \ldots \text{NOBJ}$$

Subject to
$$g_M(z_{IQ}, b_J) \quad M = 1, 2, \ldots \text{NCON}$$

An optimal solution cannot be guaranteed for a discrete optimization problem without evaluating every possible combination of (0,1) variables which is computationally impractical. Near optimal solutions can be obtained, however, with significant improvements in all objective functions with a reasonable amount of computational effort. Simulated annealing has the ability to climb out of local minima, unlike most NLP techniques. It must be noted that, as with all discrete programming procedures, it is possible that all constraints are not satisfied exactly in the final solution.

Approximate analysis

Exact function evaluations during each search of the design space can be computationally expensive. Therefore, an approximate technique, based on a linear Taylor series expansion, is used and the function is approximated as follows.

$$F^*_K(a,b) = F^*_K(a_0,b_0) + \sum_{I=1}^{NC} \frac{dF_K}{da_I}[a_I - a_{I_0}] + \sum_{J=1}^{ND} \frac{dF_K}{db_J}[b_J - b_{J_0}] \quad (15)$$

where $F^*_K(a,b)$ is the linearized formulation of each objective function. The initial point for the approximation is (a_0,b_0). Gradients of the continuous design variables can be computed using analytical means whenever possible, or by using a finite difference technique. "Gradients" of the discrete variables ($\frac{dF_K}{db_J}$) which represent the sensitivity of the design to the absence/presence of an actuator are computed by calculating the quantity $\Delta F = F(1) - F(0)$ or $\Delta F = F(0) - F(1)$ for each actuator location, depending on whether or not an actuator is present at each location in the initial design. Similar expressions are used for the constraints. During optimization, the approximate problem is solved and the solution is used as a seed point for the next linear approximation. The process is repeated until a satisfactory solution is found. Move limits are imposed on the continuous design variables to protect the validity of the approximation.

Multiobjective formulation

The Kreisselmeier-Steinhauser (K-S) function approach is used to efficiently combine multiple and conflicting design objectives and constraints into a single unconstrained function[16]. Recent studies have successfully demonstrated the usefulness of the K-S function technique in practical design problems[17]. In the K-S formulation, each original objective function is transformed into a reduced objective function as follows:

$$\bar{F}_K = \frac{F^*_K}{F^*_{K_0}} - 1 - g_{max} \leq 0$$

$$K = 1, 2, .. NOBJ \quad (16)$$

where $\bar{F}_K$ are the reduced objective functions, F^*_K are the original linearized objective functions, $F^*_{K_0}$ are their initial values and g_{max} represents the maximum violated constraint. Because these reduced objective functions are analogous to the previous constraints, a new constraint vector g_N is introduced where $N = NCON + NOBJ$. The new K-S objective function to be minimized is then defined as:

$$F = g_{N_{max}} + \frac{1}{\rho}\ln \sum_{N=1}^{NCON+NOBJ} \exp\{\rho(g_N - g_{N_{max}})\} \quad (17)$$

where the multiplier ρ is analogous to a draw down factor controlling the distance from the surface of the K-S function to the surface of the maximum function value. A larger value of ρ moves the K-S function envelope closer to the maximum violated constraint while a smaller value of ρ retains contributions from all of the objective functions and constraints.

Simulated annealing

Simulated annealing algorithms have been applied to a wide variety of engineering problems. Briefly, to minimize an objective function **F**, the algorithm can be stated as follows:

START
Current design is F
Perturb current design F_{new}
If $F_{new} \leq F$ then
F = F_{new}
Else if $P_{acc} \geq P$ then
F = F_{new}
End if
Go to START

The acceptance probability (P_{acc}) of retaining a worse design is computed as follows:

$$P_{acc} = \exp(\frac{1}{T}) \quad (18)$$

where T is the "temperature" which is reduced for successive iterations, thus reducing the probability of accepting a worse design and P is a random number such that $0 \leq P \leq 1$. Occasionally accepting a worse design under the given probability allows the algorithm to climb out of possible local minima. The above loop is repeated a specific number of times for each approximate analysis. The minimized design is then used as a seed point for the new approximation. This process is repeated until a satisfactory solution is obtained as shown in Fig. 2.

Optimization Procedure

Preselected values for the continuous variables, d_Q, can be determined in one of two ways. First, these values can be selected over the entire expected range of the variables corresponding to either standard values or evenly spaced intervals. The second method, which is used in the current study, involves discretization of the interval formed by imposing move limits which are used to protect the validity of the approximation about the seed point a_0. These methods apply to all continuous design variables, so the subscript I has been dropped for clarity. This interval can be divided up into any number of evenly spaced intervals as shown in Fig. 3. The subscripts -ML and +ML indicate the lower and upper bounds formed by the move limit. The expansion point for the Taylor series approximation is a_0 and the constants d_Q represent Q possible values for **a**. The continuous variables can be more accurately represented by increasing the number of points taken inside the interval, or by reducing the move limit interval. These values are determined at each iteration and must be altered as the seed point for each new approximation is relocated. This method increases the efficiency of the simulated annealing algorithm since only those values for the design variables which satisfy the move limits can be selected. This also eliminates the necessity of checking the design variables selected by the simulated annealing algorithm to ensure that the imposed move limits are not exceeded.

Exact function evaluations are the most computationally expensive aspect of the algorithm and the use of the approximate analysis significantly reduces the number of such exact function evaluations necessary. The procedure developed which combines the simulated annealing algorithm with the approximate analysis is computational inexpensive and takes only a small percentage of the total computational effort compared to the exact objective function evaluations required to assemble the approximate analysis. Since the computational effort associated with the simulated annealing algorithm is minimal, there is no reason to be conservative with the number of loops of the algorithm itself. The utility of the algorithm is demonstrated in the results which show improved designs in all cases. The simulated annealing results are compared to those previously obtained when using a nonlinear programming approach[11,12].

Results and Discussion

The procedure developed is demonstrated through a numerical example of a cantilever box beam (Fig. 4). The beam is subjected to a sinusoidal disturbance force of 220N applied to the tip of the box beam over a period of 0.067 sec at a frequency of 25 Hz. Results are presented with the four objective functions previously described. These include the dissipated energy J, the electrical power U and an index of performance IAE which are all minimized. The natural frequency ω_1 is maximized by minimizing its negative value. Constraints are placed on the energy, applied voltage to the actuators, displacements and mass. Damping factor constraints are set such that the initial vibrational amplitudes of the first four modes are reduced to 5% of their original values after one second of active controls. The energy in the system, which is the sum of the potential and kinetic energies, must also be reduced to 5% of its initial value. For the mass constraint, a value of $\bar{m} = 3.5$ kg is used which is the maximum allowable mass of the beam and actuators combined. Design variables include box beam web thicknesses, t_i, actuator locations, α_i and actuator gains. The gain of all of the actuators are set to be equal to reduce the dimension of the design variable vector. Upper and lower bounds are placed on the i-th box beam thickness (0.5mm $\leq t_i \leq$ 2.5mm). The height (h) of the beam is not altered during optimization and is set equal to 10 cm. The gain is constrained to be positive for stability. The absence or presence of the i-th actuator is indicated by $\alpha_i = 0$ or $\alpha_i = 1$, respectively. An initial design is chosen where the web thickness for each box beam element, t_i, is set equal to 1.5 mm as shown in Table 1. The initial design violates both the mass constraint and the damping constraint associated with the first mode.

For the simulated annealing algorithm, the interval formed by the move limits is divided into 50 evenly spaced points (Q=50). Each value of d_Q comprises one of these points for all of the continuous design variables which can be selected. Once the continuous variables are transformed into discrete variables, a total of 510 discrete variables are used during optimization. A random actuator configuration is chosen with actuators located on elements 1, 3, 5, 6, 7, 8 and 10. For this problem, 2000 loops of the simulated annealing algorithm are used for each optimization iteration. It must be noted that these loops are computationally inexpensive due to the implementation of the approximate analysis. Since there are 3.54×10^{23} possible combinations of the discrete variables, 2000 loops represent an extremely small portion of the actual design space. The recommended range of the parameter ρ used in the K-S function is $5 \leq \rho \leq 200$ when used in conjunction with a nonlinear programming technique[18]. When used in the simulated annealing algorithm, a small value of $\rho = 10$ is found to be most appropriate. Larger values of ρ result in numerical instabilities due to jumps in the objective functions and/or constraints due to the use of discrete design variables. Move limits of 10% are placed on each design variable to protect the validity of the approximation. After 10 optimization iterations, ρ is increased to 20 and the move limits are reduced to 5%. The reduced move limit allows smooth convergence of the optimization and a better approximation of the continuous design variables.

The optimum simulated annealing results are obtained after 14 iterations as shown in Fig. 5 where each original objective function is normalized to its initial value. This takes about 47 minutes of CPU time on an IBM RS6000 workstation. The final design consists of actuators at elements 1, 2, 3, 5 and 9 as shown in Table 1. Improvements are made in all of the objective functions. The simulated annealing algorithm results in the dissipated energy, J, being reduced by 98% from the initial to the final design. The index of performance, IAE and the electrical power consumption U are reduced by 93% and 89%, respectively. The natural frequency, ω_1, which is maximized in the optimization formulation, increases by 36%. The damping ratio constraint on the first mode and the voltage constraint represent the most active constraints. The use of different initial actuator configurations also resulted in similar improvements to the objective functions.

The results obtained using the simulated annealing algorithm are compared with those obtained using continuous design variables based on a NLP technique where non optimal actuators were eliminated on the basis of dissipated energy[12]. To briefly describe the algorithm, actuators were initially located at each possible location on the discretized structure which were elements of the finite elements discretization of the box beam. A single optimization iteration consisted of minimizing the objective functions using the current actuator configuration until either convergence or a maximum number of cycles (20) was reached. The actuator which dissipated the least amount of energy, compared to the remaining set of the actuators, was eliminated as a possible actuator location. The new configuration was then re-optimized in the next iteration and a second actuator was eliminated. The procedure was repeated until no more actuators could be eliminated from the set of possible actuator locations without violating one or more constraints. The nonlinear programming technique CONMIN[19] was used in conjunction with the K-S function approach and a Taylor series based approximation was also used to reduce computational effort. The same initial design is used for both the NLP technique and the simulated annealing results except that actuators were initially located on each element for the NLP technique and a random actuator configuration is used for the simulated annealing. Tables 1 - 2 and Figs. 6 - 9 present results of the two optimization approaches.

The optimal configurations obtained using the two techniques differed significantly. From Table 1 it is shown that the NLP technique results in a single actuator at element 1 in contrast to the optimal configuration obtained using the simulated annealing algorithm which comprises five actuators at elements 1, 2, 3, 5 and 9. In general, actuators at the tip of the beam are eliminated first for the NLP technique since they are less effective than those at the root of the box beam. Although the mass constraint is violated in the initial design in both cases, it is not active in the final design in either case. It must be noted that the mass of the optimized beam is slightly higher in the design obtained using simulated annealing due to the increase in the number of actuators compared to the NLP final design. The thickness distribution obtained using the simulated annealing technique is consistent with expected trends. The beam is thickest at the root, and tapered towards the tip except for the last three elements where thickness increases. This is a result of the optimizer's effort to increase the stiffness of the beam at the location where the disturbance force is applied. The thickness distribution obtained using the NLP technique is more nonlinear. This is possibly due to an effort to increase stiffness since a reduced number of actuators are present in the final design. The objective functions obtained using the two approaches are compared in Figs. 6-9 It must be noted that the optimization results depend on the nature and magnitude of the disturbance force. Improvements of the same order are obtained in all of the original objective functions using both techniques although the final actuator configurations are different. This indicates the presence of possible local minima. The major difference is the amount of computational effort required in each case. The simulated annealing algorithm requires about one hour of CPU time whereas the NLP technique requires six hours of CPU time on the same platform to achieve optimum results as shown in Fig. 10. A total of 330 exact function evaluations are required for the simulated annealing while 2200 evaluations are necessary for the NLP technique using CONMIN. The dramatic reduction in required CPU time using the simulated annealing algorithm demonstrates the efficiency of the technique developed

Concluding Remarks

A multiobjective optimization procedure based on a simulated annealing technique is developed to include both discrete and continuous design variables. The Kreisselmeier-Steinhauser multiobjective formulation technique is used to formulate the multiple design objectives optimization problem. An approximate analysis technique is used, based on a linear Taylor series expansion to reduce the computational effort associated with exact function evaluations. A numerical example of a cantilever beam subjected to a sinusoidal disturbance force is presented. Objective functions include energy dissipation, power consumption, an index of performance for vibration control and natural frequency. Constraints are placed on total energy, voltage, displacements and mass. Design variables include actuator gains, box beam web thicknesses, and actuator locations. Significant improvements are obtained in all objective functions and optimal actuator locations are determined inside a closed loop procedure. The results using this technique are compared to a previously used technique based on a nonlinear programming procedure. The following observations can be made from this study.

(1) Both continuous and discrete (0,1) design variables were successfully included in the optimization procedure.

(2) The controls/structures optimization procedure developed using the simulated annealing algorithm resulted in significant improvements in all objective functions while satisfying the imposed constraints.

(3) The procedure resulted in substantial reductions in CPU time compared to the NLP technique.

Acknowledgments

The authors gratefully acknowledge the NASA Space Grant Fellowship program for support of this research.

References

1. Hanagud, S., Obal, M. W., Calise, A. J. "Optimal Vibration Control by the Use of Piezoceramic Sensors and Actuators," *Journal of Guidance, Control, and Dynamics*, Vol. 15, No. 5, 1992, 1199-1205.

2. Miller, D. F. and Shim, J. "Gradient-Based Combined Structural and Control Optimization," *Journal of Guidance*, Vol. 10, No. 3, 1987, 291-298.

3. Horner, G. and Walz, J. "A Design Methodology for Determining Actuator Gains in Spacecraft Vibration Control," *Proc. 26th AIAA/ASME/ASCE/AHS/ASC Structures, Structural Dynamics and Materials Conference*, Apr. 15-17, 1985, Orlando, Florida, 143-151.

4. Sepulveda, A. E., Jin, I. M. and Schmit, L. A. Jr. "Optimal Placement of Active Elements in Control Augmented Structural Synthesis," *Proc. 33rd AIAA/ASME/ASCE/AHS/ASC Structures, Structural Dynamics and Materials Conference*, April 13-15, 1992, Dallas, Tx., 2768-2781.

5. Crawley, E. F. and de Luis, J. "Use of Piezoelectric Actuators as Elements of Intelligent Structures," *AIAA Journal*, Vol. 25, No. 10, 1987, 1373-1385.

6. Onoda, J. and Hanawa, Y. "Optimal Locations of Actuators for Statistical Static Shape Control of Large Space Structures: A Comparison of Approaches," *Proc. 33rd AIAA/ASME/ASCE/AHS/ASC Structures, Structural Dynamics and Materials Conference*, Apr. 13-15, 1992, Dallas, Tx., 2788-2795.

7. Zimmerman, D. C. "A Darwinian Approach to the Actuator Number and Placement Problem with Nonnegligible Actuator Mass," *Proc. 13th Biennial Conference on Mechanical Vibration and Noise, ASME Design Technical Conferences*, DE Vol. 34, 1991, Miami, Fl., 83-88.

8. Lombardi, M., Haftka, R. T. Cinquini, C., "Optimization of Composite Plates for Buckling by Simulated Annealing," *Proc. AIAA/ASME/ASCE/AHS/ASC Structures, Structural Dynamics, and Materials Conference*, April 13-15, 1992, Dallas Tx., 2552-2562.

9. Kirkpatrick, S., Gelatt, C. D. Jr., Vecchi, M. P. "Optimization by Simulated Annealing," *Science* Vol. 220, No. 4598, 1983, 671-680.

10. Piezo Systems Inc. *Product Catalog*, 1993.

11. Seeley, C. E. and Chattopadhyay, A. "The Development of an Optimization Procedure for the Design of Intelligent Structures," *Smart Materials and Structures*, Vol. 2, 1993, 135-146.

12. Seeley, C. E. and Chattopadhyay, A. "A Multiobjective Design Optimization Procedure for Control of Structures Using Piezoelectric Materials," *Journal of Intelligent Material Systems and Structures*, 1994, to be published.

13. Rogers, C. A. "Smart Structures: Where Does the Energy Go?," *Invited Technical Talk, SPIE North American Conference on Smart Structures and Materials*, Feb. 1-4, 1993, Albuquerque, N.M.

14. Ogata, K. *Modern Control Engineering*, 1990, Prentice Hall, N. J.

15. Olsen, G.R. and Vanderplaats, G. N. "Method for Nonlinear Optimization with Discrete Design Variables," *AIAA Journal*, Vol. 27, No.1, 1989, 1584-1589.

16. Kreisselmeir, G. and Steinhauser, R. "Systematic Control Design by Optimizing a Vector Performance Index," *International Federation of Active Controls Symposium on Computer-Aided Design of Control Systems*, Aug. 29-31, 1979, Zurich, Switzerland.

17. Chattopadhyay, A. and McCarthy, T. "Multiobjective Design Optimization of Helicopter Rotor Blades with Multidisciplinary Couplings," *Structural Systems and Industrial Applications*, editors, S. Hernandez and C. A. Brebbia, 1991, 451-461.

18. Wrenn, G. "An Indirect Method for Numerical Optimization Using the Kreisselmier-Steinhayser Function," 1989, NASA contract report 4220.

19. Vanderplaats, G. N. "*CONMIN - A FORTRAN Program for Constrained Function Minimization User's Manual*," 1973, NASA TMX-62,282.

Table 1. Optimization results - design variables.

	Simulated annealing				NLP			
	Initial		Final		Initial		Final	
	t_i (mm)	α_i	t_i (mm)	α_i	t_i (mm)	α_i	t_i (mm)	α_i
Element 1	1.5	1	2.5	1	1.5	1	1.2	1
Element 2	1.5	0	1.7	1	1.5	1	1.7	0
Element 3	1.5	1	1.2	1	1.5	1	2.0	0
Element 4	1.5	0	1.0	0	1.5	1	2.4	0
Element 5	1.5	1	0.7	1	1.5	1	1.7	0
Element 6	1.5	1	0.6	0	1.5	1	1.0	0
Element 7	1.5	1	0.6	0	1.5	1	0.8	0
Element 8	1.5	1	0.9	0	1.5	1	0.7	0
Element 9	1.5	0	0.8	1	1.5	1	0.7	0
Element 10	1.5	1	0.8	0	1.5	1	0.7	0

Table 2. Optimization results - objective functions.

	Simulated annealing		NLP	
	Initial	Final	Initial	Final
Dissipated Energy (Joules)	15.58	0.25	17.65	0.31
Natural frequency (Hz)	29.73	40.47	30.62	39.81
IAE (x10)	86.56	6.27	65.04	5.78
Power (Joules)	6.93	0.74	8.09	0.76
Mass (kg)	4.18	2.96	4.54	3.01

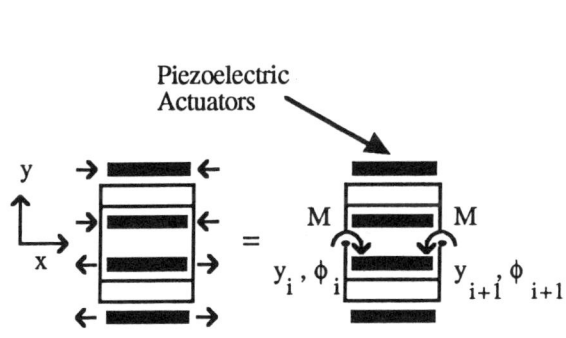

Figure 1. Active piezoelectric beam element.

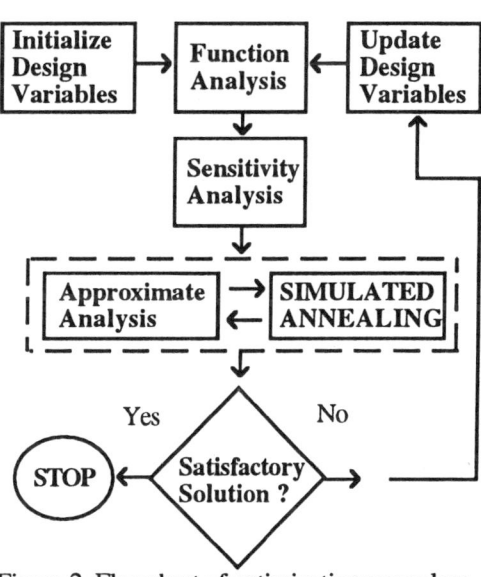

Figure 2. Flowchart of optimization procedure.

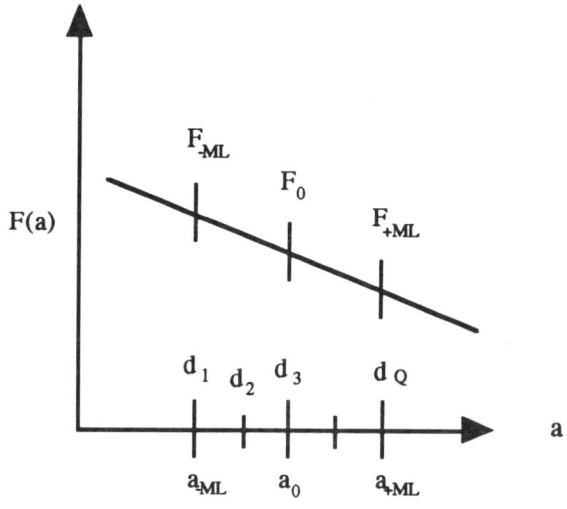

Figure 3. Move limit interval.

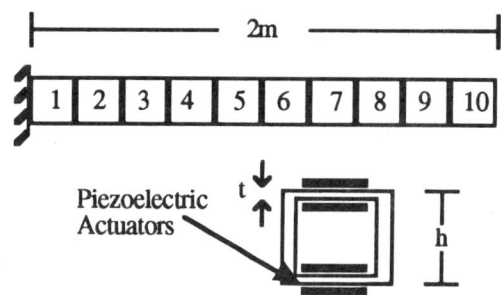

Figure 4. Cantilever box beam with piezoelectric actuators.

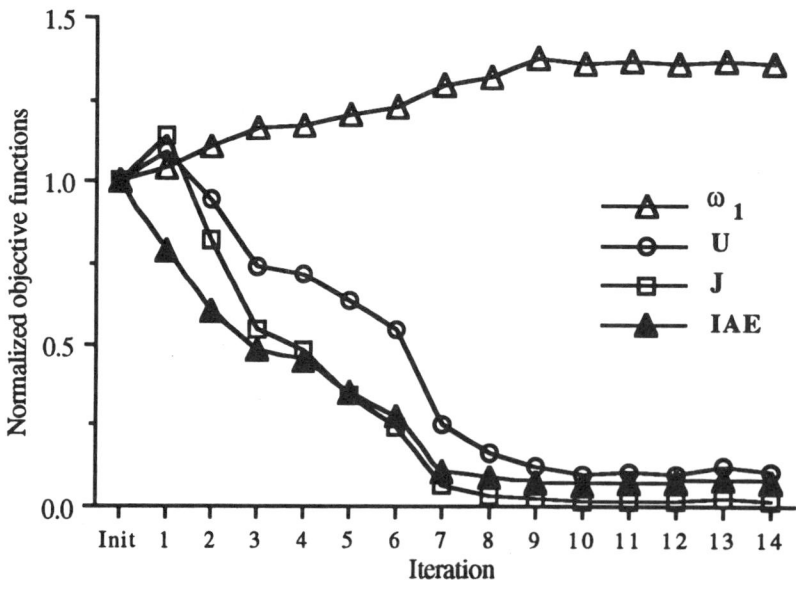

Figure 5. Objective function iteration history.

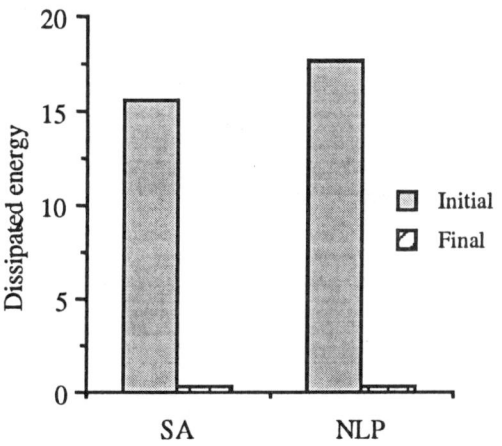

Figure 6. Comparison of dissipated energy (J).

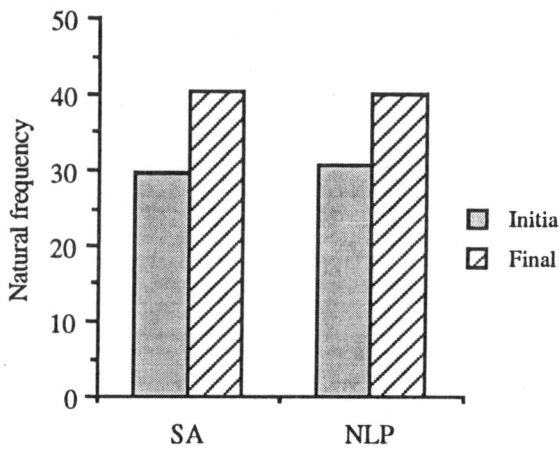

Figure 7. Comparison of natural frequency (ω_1).

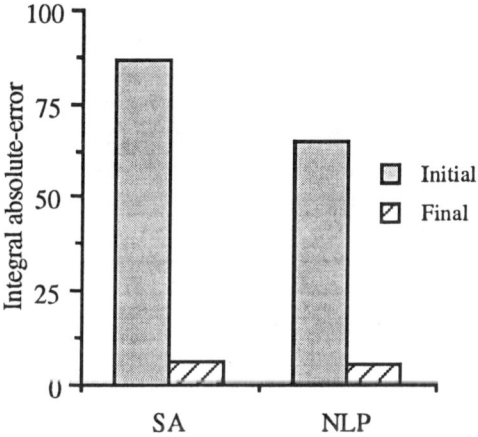

Figure 8. Comparison of IAE index.

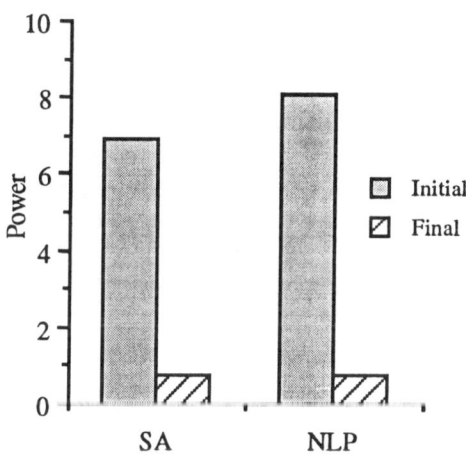

Figure 9. Comparison of power (U).

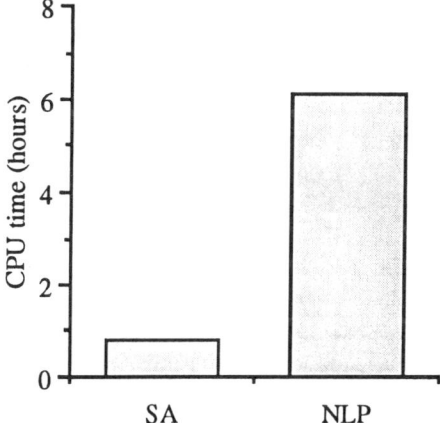

Figure 10. CPU time for SA and NLP.

AIAA-94-1778-CP

A MECHANICAL APPROACH TO INTERFACIAL STRESS ALLEVIATION IN AN INTEGRATED INDUCED STRAIN ACTUATOR/SUBSTRUCTURE SYSTEM

M. W. Lin
C. A. Rogers

Center for Intelligent Material Systems and Structures
Virginia Polytechnic Institute and State University
Blacksburg, Virginia 24061-0261

Abstract

Intensive interfacial shear and peeling stresses localized near the induced strain actuator, i.e., piezoelectric bonded patch, end zones create concerns about structural integrity, particularly interfacial fatigue failure, of integrated induced strain actuator/substructure systems. In this paper, it is demonstrated that the free edge effect of the actuators is responsible for this mechanism. The basis of the mechanics underlining interfacial stress alleviation is discussed based upon actuation force transfer and equilibrium considerations. A new actuator configuration is proposed to eliminate the free edge effect by creating inactive edges of the actuator. Both analytical and finite element models verify that the proposed actuator configuration can significantly reduce the interfacial shear and peeling stresses without sacrificing the effectiveness in transferring the actuation strain from the actuator to the substructure.

Introduction

Recent developments in the implementation of intelligent material systems and structures generally involve highly integrated, surface-bonded or embedded, induced strain actuators as energy input devices or actuating elements in various engineering applications (Bailey and Hubbard, 1985; Crawley and de Luis, 1987; Chaudhry and Rogers, 1991). The means of integrating the actuators and the host substructures is primarily by bonding. The bonding interfaces, therefore, serve as the media for transferring the actuation mechanism. Thus, desirable interfaces must have a high efficiency in transferring the induced actuation strains of the actuators to the substructures, and have a sufficiently high fatigue endurance limit to provide a strong bond to ensure structural integrity.

Two vastly different approaches can be used to achieve these objectives. The first is within the context of material science and engineering, which focuses on methods to increase the stiffness and strength of the interfacial bonds. This approach generally involves the study of the chemical mechanisms of the adhesion between heterogeneous materials. The second approach involves a mechanical approach which seeks the design criteria for optimum configuration of the actuators to increase the efficiency of the actuation transfer and reduce the interfacial stresses. This approach requires an understanding of the mechanics of the mechanical interaction between the actuators and the substructures. The present study will be confined to the mechanical approach to designing an actuator configuration that can alleviate the high stress intensity at the interfaces.

The mechanism of the mechanical interaction between the actuators and the substructures has been modeled analytically by Crawley and de Luis (1987) and Im and Atluri (1989) using one-dimensional analysis based on the shear lag assumption. It was shown that the actuation forces/moments are transferred solely by the interfacial shear stresses localized near the end zones of the actuators. It was concluded that in order to achieve effective induced strain transfer from the actuator to the substructure, a high interfacial shear stress state is desirable. Thus, the issue of interfacial failure due to high intensity of this stress component has never been addressed.

A refined model based on a two-dimensional elasticity formulation was recently presented by Lin and Rogers (1993a, 1993b). This model more accurately describes the stress field in both the actuator and the substructure, particularly near the end zones of the actuators. The interfacial shear and peeling stress distributions were correctly predicted. It was shown that the presence of the free edges of the actuator raises the intensity of the interfacial shear and peeling stresses near the actuator end zones.

Recently, Walkers et al. (1993) conducted a finite element analysis on various actuator edge configurations in an attempt to reduce interfacial shear and peeling stress intensity. It was demonstrated that using various designs of end caps and partial electrode configurations at the ends of the actuators can noticeably reduce the interfacial shear and peeling stresses. Nevertheless, the mechanics underlying this alleviation has not been discussed.

Copyright© 1994 by
Center for Intelligent Material Systems and Structures
Published by the AIAA, Inc. with permission

The objective of the present study is to develop the theoretical basis for the mechanics of the interfacial stress alleviation mechanism. The rationale on the design of actuator configuration to alleviate interfacial shear and peeling stresses is first presented. A new actuator configuration with "inactive" edges is proposed. The issues related to the actuator efficiency brought about by the addition of inactive edges are subsequently discussed. Finally, the effectiveness of this new actuator configuration is characterized by analytical model and finite element analysis.

Theory

Consider a beam structure of length $2L$ with induced strain actuators of length $2l$ symmetrically bonded on the outer surfaces of the beam as shown in Fig. 1. The thickness of the actuator and beam are denoted by a and $2b$, respectively. The actuation mechanism transferred from the actuator to the beam substructure can be easily illustrated using one-dimensional analytical scheme based on the shear lag assumption.

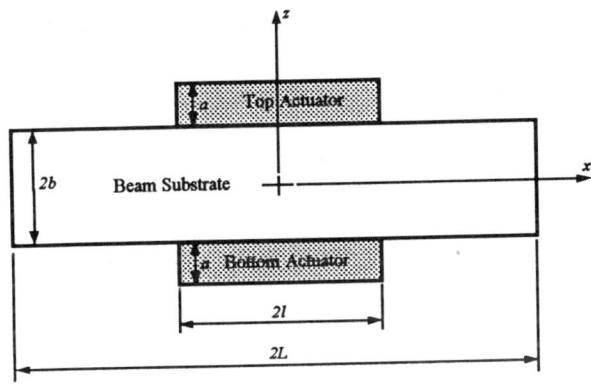

Figure 1. Schematic configuration of the induced strain actuator/beam substructure system.

Consider a free body diagram cut from the top actuator as the actuators are activated. Based on the framework of the one-dimensional shear lag theory, the forces acting on this free body are depicted in Fig. 2(a), where F_a indicates the resulting normal force in the actuator, and S is the shear force on the interface. Note that the normal force F_a is equal in magnitude and opposite in sign to the effective force transferred to the beam substructure. Thus, the quantity F_a can equivalently represent the resulting actuation force in the beam substructure. In this configuration, the free edge effect is apparent, where the ends of the actuator have a force-free boundary condition and the actuation force F_a is solely transferred by the interfacial shear force S. In order to achieve a higher effective force level, a larger interfacial shear force is needed.

Now, if the free edge of the actuator is by some means reconfigured to reduce the free edge effect, the interfacial shear force can be correspondingly reduced, see Fig. 2(b). In other words, part of the force transfer can be accomplished through the ends of the actuator, in addition to the interface. In this configuration, the actuation force is transferred not only by the interfacial shear force, S^*, but also the normal force on the ends of the actuator, F_e. Therefore, a lower interfacial shear force level is needed, i.e., $S^* < S$, to achieve the same level of effective force on the beam substructure F_a as that depicted in Fig. 2(a).

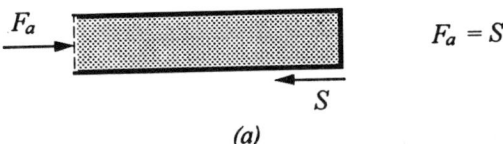

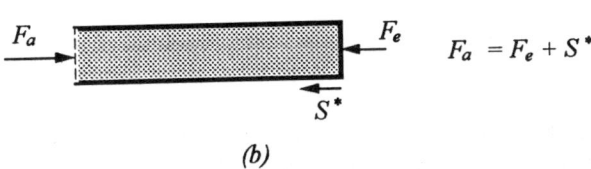

Figure 2. Free-body diagram of the top actuator: (a) with free edge effect, (b) with reduced free edge effect.

The mechanism of alleviating the interfacial stresses by reducing the degree of the free edge effect can be more vigorously analyzed based on equilibrium considerations using a two-dimensional analytical scheme. Consider the integrated induced strain actuator/beam substructure system of Fig. 1. The interfacial stress distributions along the x-axis which satisfy the stress-free boundary condition at the ends of the actuator are illustrated in Fig. 3.

The axial normal stress in the top actuator in general can be expressed as:

$$\sigma_{x(at)}(x,z) = f(z)g(x), \tag{1}$$

where subscript (at) denotes the quantity of the top actuator, and $f(z)$ and $g(x)$ are arbitrary functions depending solely on z and x, respectively. The first equation of equilibrium requires:

$$\sigma_{x,x} + \tau_{xz,z} = 0. \tag{2}$$

Substituting Eq. (1) into (2) for σ_x and integrating with respect to z, the shear stress field is obtained:

$$\tau_{xz(at)}(x,z) = -g'(x)\int f(z)dz + c_1(x), \quad (3)$$

where $c_1(x)$ is an arbitrary function depending on x to be determined by the boundary conditions.

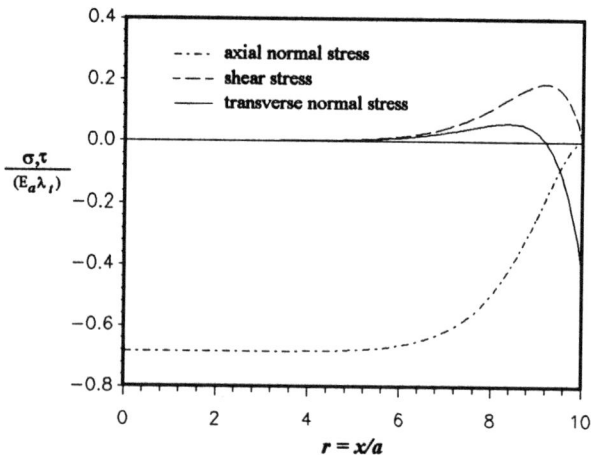

Figure 3. Stress distribution along the x axis at the interface.

The shear stress free boundary condition on the outer lateral surface of the top actuator yields:

$$\tau_{xz(at)}(x, a+b) = 0. \quad (4)$$

Imposing this boundary condition, the shear stress field is defined by:

$$\tau_{xz(at)}(x,z) = g'(x)\{[\int f(z)dz]_{z=a+b} - \int f(z)dz\} = g'(x)F(z). \quad (5)$$

The expression of the shear stress at the interface is then obtained:

$$\tau_{xz(at)}(x,b) = g'(x)F(b). \quad (6)$$

In the above, $g'(x)$, which represents the slope of the axial normal stress in the x direction, can be regarded as a factor that contributes to the variation of the interfacial shear stress in the x axis and its magnitude. It is apparent that the variation of the axial normal stress in the x axis, or more precisely the slope, controls the magnitude and distribution of the interfacial shear stress. In the case of an actuator with free edges, the slope of the axial normal stress increases dramatically near the end zones of the actuator due to the free edge effect. The interfacial shear stress, therefore, yields a noticeable intensity localized at the free end zones. Now, if the increase of the axial normal stress near the actuator edges can be minimized by some means, the intensity of the interfacial shear stress can then be accordingly alleviated. In the extreme, the interfacial shear stress will vanish in the case in which the axial normal stress is uniform in x, i.e., $g'(x) = 0$.

The transverse normal stress field can be obtained likewise by considering the second equilibrium equation:

$$\tau_{xz,x} + \sigma_{z,z} = 0. \quad (7)$$

Substituting Eq. (5) into (7) for τ_{xz} and integrating with respect to z, the transverse normal stress field is obtained:

$$\sigma_{z(at)}(x,z) = -g''(x)\int F(z)dz + c_2(x). \quad (8)$$

Imposing the transverse normal stress free boundary condition on the top outer lateral surface,

$$\sigma_{z(at)}(x, a+b) = 0, \quad (9)$$

the final expression of the transverse normal stress field becomes:

$$\sigma_{z(at)}(x,z) = g''(x)\{[\int F(z)dz]_{z=a+b} - \int F(z)dz\} = g''(x)H(z). \quad (10)$$

The transverse normal stress at the interface, or the peeling stress, is given by:

$$\sigma_{z(at)}(x,b) = g''(x)H(b). \quad (11)$$

Similarly, the slope of the shear stress with respect to x, $g''(x)$, controls the magnitude and variation of the peeling stress in the x direction. As shown in Fig. 3, the interfacial shear stress has the highest slope at the actuator edges, yielding a maximum interfacial peeling stress right at the ends of the actuator. By the same argument, if the free edge effect can be somehow diminished, resulting in a less dramatic change in the slope of the interfacial shear stress, the intensity of the interfacial peeling stress can then be alleviated.

Based on the above analysis, the interfacial shear and peeling stresses can be reduced as the axial normal stress becomes more uniform with respect to x. The non-uniform field of the axial normal stress is shown localized near the end zones of the actuator due to the free edge effect. Thus, as the degree of the free edge effect is reduced, the desired interfacial shear and peeling stress alleviation can be achieved.

One of the most direct approaches to reduce the free edge effect is to incorporate inactive edges at the ends of the actuators, as shown in Fig. 4. The inactive edges have the same thickness as the actuators and a length of e. These inactive edges provide a path for part of the actuation mechanism to be transferred from the ends of the actuators to the beam substructure from a force-transferring point of view. Alternatively, they also can be regarded as blocking elements which will exert axial normal stresses on the actuator edges to reduce the free edge effect.

Actuators with inactive edges can be easily manufactured. The inactive edges can be composed of different material from the actuator with comparable or higher stiffness, but insensitive to the activation stimulus. The inactive edges can also be created by blocking the activation stimulus from the areas near the actuators' edges. For example, in the case of piezoelectric actuators, the inactive edges can be implemented simply by removing the electrode on the surfaces near the end zones of the actuators.

With the addition of inactive edges, the efficiency of the actuators needs to be characterized. In particular, the issues concerning the effectiveness of the actuation force/

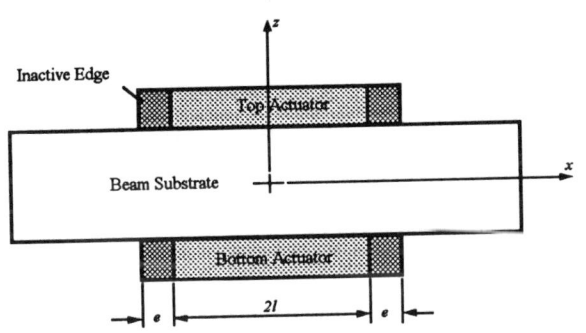

Figure 4. Schematic configuration of the proposed actuator with inactive edges.

moment transfer and the efficiency of the interfacial shear and peeling stress alleviation need to be addressed. Performance characterization of the proposed actuator configuration using both analytical and finite element models is discussed in the following sections.

Analytical modeling

In order to characterize the effectiveness of the proposed actuator configuration, the refined analytical model by Lin and Rogers (1993a) is used. The model was developed based on the plane stress formulation of the theory of elasticity for a beam structure with symmetrically surface-bonded actuator patches, as shown in Fig. 1. The whole-field stress distribution in each constituent was obtained in an approximate manner in closed form by the principle of complementary energy. The model is capable of describing the edge effect and determining the interfacial shear and peeling stress distribution.

The stress field in the top actuator was derived and has the form:

$$\sigma_{x(at)}(x,z) = -E_a \lambda_t + (\frac{a+b-z}{a})^2 \frac{E_a}{E_s} \sigma_{its}(x)$$
$$- \frac{(b-z)(2a+b-z)}{a^2} \sigma_{ot}(x) \quad (12.1)$$

$$\tau_{xz(at)}(x,z) = \frac{1}{3a^2} \{ \frac{E_a(a+b-z)^3}{E_s} \sigma'_{its}(x) + (a+b-z)[2a^2$$
$$-2ab-b^2+2(a+b)z-z^2]\sigma'_{ot}(x) \} \quad (12.2)$$

$$\sigma_{z(at)}(x,z) = \frac{1}{12a^2} \{ \frac{E_a(a+b-z)^4}{E_s} \sigma''_{its}(x) + (a+b$$
$$-z)^2[5a^2-2ab-b^2+2(a+b)z-z^2]\sigma''_{ot}(x) \}, \quad (12.3)$$

where

$$\sigma_{its}(x) = \frac{1}{2(aE_a+bE_s)[aE_a(a+4b)+4b^2E_s]} \{-3a^2 E_a E_s (aE_a$$
$$+2bE_s)\lambda_b + 3aE_a E_s[aE_a(3a+8b)+2bE_s(a+4b)]\lambda_t$$
$$-3a^2 E_a E_s(aE_a+2bE_s)\lambda_b - 4bE_s[aE_a(a+4b)$$
$$+4b^2 E_s]\sigma_c(x) - aE_s[aE_a(7a+16b)$$
$$+bE_s(5a+16b)]\sigma_{ot}(x)+a^2 E_s(3aE_a+5bE_s)\sigma_{ob}(x) \}$$

In the above, E_a and E_s are the Young's modulus of the actuator and the beam substructure, respectively, and λ_t and λ_b are the free induced strains in the top and bottom actuators, respectively. The quantities $\sigma_{ot}(x)$, $\sigma_{ob}(x)$, and $\sigma_c(x)$ are the axial normal stress at the outer fiber of the top and bottom actuator and at the center of the beam, respectively; $\sigma_{its}(x)$ is the axial normal stress of the beam at the top interface; and the prime indicates the derivative with respect to x.

The only unknown quantities in the stress field of Eq. (12) are $\sigma_{ot}(x)$ and its derivatives for the case of pure bending actuation, i.e., $\lambda_b = -\lambda_t$, and has the solution given by:

$$\sigma_{ot}(x) = \frac{D_1^b}{A_3^b} + \sum_{i=1}^{4} p_i^b (\exp)^{a_i^b x}, \quad (13)$$

where $A_i^b (i=1,3)$ and D_1^b are constants related to the material properties and geometry of the actuator and beam, and have been described by Lin and Rogers (1993a). In the above, $\alpha_i^b (i=1,4)$ are the roots of the characteristic equation:

$$A_1^b (\alpha^b)^4 + A_2^b (\alpha^b)^2 + A_3^b = 0, \quad (14)$$

and $P_i^b (i=1,4)$ are arbitrary constants to be determined by the end traction condition at the edges of the actuator.

Since the inactive edges of the actuator were not included in the model, their presence is simulated by prescribing different end traction conditions at the actuators' edges, i.e., $x = \pm l$. Three boundary traction conditions (B.C.) are used:

(1) $\sigma_{x(at)}(\pm l, a+b) = 0; \quad \tau_{xz(at)}(\pm l, z) = 0 \quad (15.1)$

(2) $\sigma_{x(at)}(\pm l, a+b) = 0.5\sigma_{x(at)}(0, a+b);$
$\tau_{xz(at)}(\pm l, z) = 0 \quad (15.2)$

(3) $\sigma_{x(at)}(\pm l, a+b) = \sigma_{x(at)}(0, a+b);$
$\tau_{xz(at)}(\pm l, z) = 0. \quad (15.3)$

B.C. (1) represents the actuator without inactive edges, where the free edge effect is present. B.C. (3) simulates the actuator with inactive edges which have a stiffness capable of transferring the maximum fraction of the actuator's induced strain to the beam substructure over the entire length of the actuator. In other words, the free edge effect is totally eliminated in this case. Finally, B.C. (2) models the actuator with inactive edges which have a stiffness resulting in an average effect of the extreme cases of B.C. (1) and (3).

The effective moments transferred from the actuators to the beam substrate under pure bending actuation for the three cases of boundary traction are shown in Fig. 5. The actuator/beam structure geometric configuration of $b/a = 3$, $l/a = 10$, and $L/a = 20$ is used. The material properties are selected to be $v_s / v_a = 0.73$ and $E_s / E_a = 3.29$, which are comparable to a steel beam with piezoceramic actuators. The quantity of the effective moment is normalized with respect to the moment that the total blocking force of the actuators produce about the neutral axis, i.e., $M^* = 2aE_a\lambda_t (a/2 + b)$, and the x axis is nondimensionalized with respect to the thickness of the actuator.

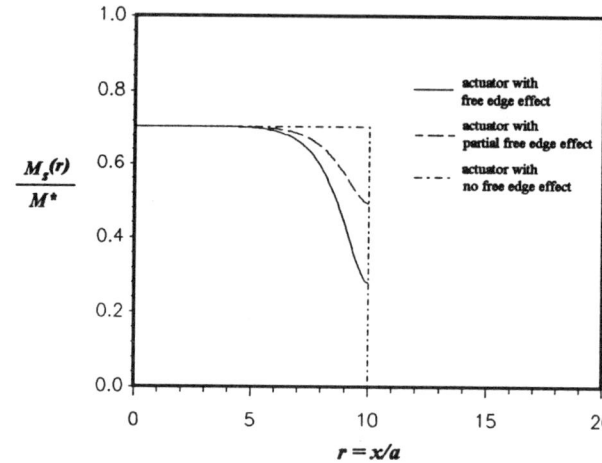

Figure 5. Effective moment distribution along the x axis under different boundary traction conditions as obtained from the analytical model.

It is evident that the magnitude of the effective moment is not affected with the addition of inactive edges in a location away from the ends of the actuator. However, a higher level of the effective moment is shown near the end zones. In the extreme case, i.e., B.C. (3), the free edge effect is totally reduced, resulting in a uniform effective moment along the entire length of the actuator, which represents the maximum achievable actuation transfer mechanism.

Figure 6 shows the corresponding interfacial shear and peeling stress distribution along the x axis. The magnitude of the stresses is normalized with respect to the blocking stress of the actuators, i.e., $E_a\lambda_t$. It is shown that with the inactive edges, the interfacial shear and peeling stresses can be significantly reduced. In the case where a uniform axial normal stress field along the x axis is obtained, the interfacial shear and peeling stresses vanish through the entire interface.

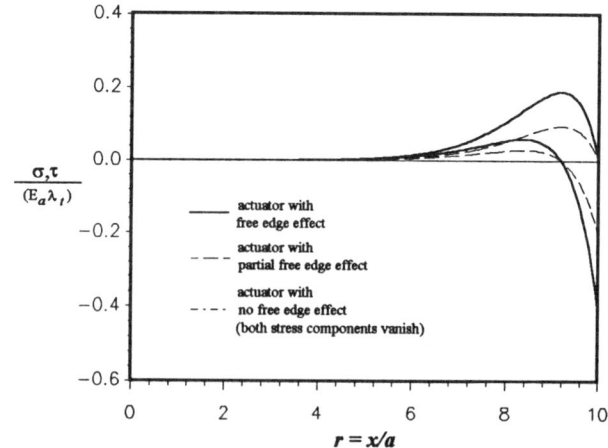

Figure 6. Interfacial shear and peeling stress distribution under different boundary traction conditions as obtained from the analytical model.

In summary, the analytical modeling demonstrates that inactive edges can significantly reduce the interfacial shear and peeling stresses and slightly increase the actuation transfer mechanism near the ends of the actuator. The free edge effect can be reduced by the proposed actuator configuration in which relatively stiffer inactive edges are desirable.

Finite element modeling

The effectiveness of actuators with inactive edges is also characterized using finite element models. Specifically, the geometric effect of the inactive edges is investigated. Figure 7 shows the finite element meshes used to model the actuators with and without inactive edges of length $2a$, where the geometric configuration of $b/a = 3$, $l/a = 10$, and $L/a = 15$ is used. The material properties chosen are $v_s/v_a = 0.73$ and $E_a/E_s = 3.29$. The same material properties as those of the actuator are assigned for the inactive edges. The influence of three different lengths of inactive edges, i.e., $e = a$, $e = 2a$, and $e = 3a$, are investigated. Note that the same element size bias is used for all the models to eliminate mesh and element size effects in the finite element model.

The plane stress linear isoparametric elements were used for all the constituents. Since the characteristic induced strains of the actuator resemble the thermal expansion effects of a structural material, a "fictitious" thermal expansion coefficient was assigned to the actuators, while the beam and inactive edges are insensitive to thermal effects. The desired induced strain level was then obtained by applying a uniform temperature field on the model. The analyses were performed using the ABAQUS finite element package for solution and IDEAS for pre- and post-processing.

The effective moments transferred from the actuators to the beam substructure under pure bending actuation are shown in Fig. 8. No effect is evident for the effective moment level in the area away from the actuator end zones, while the magnitude of the effective moment increases near the actuator ends for the actuators with inactive edges. Further noticeable increase of the effective moment is not shown by increasing the length of e from $2a$ to $3a$, indicating an optimum length of the inactive edge.

Figure 9 depicts the corresponding interfacial shear and peeling stress distribution. It should be noted that the shear stress distribution obtained from the current finite element models does not satisfy the stress-free boundary condition. Nevertheless, for the current study the relative stress level induced by different actuator configurations is of the primary interest. The results should therefore indicate the effectiveness of the interfacial shear stress alleviation. It is shown that both the interfacial shear and peeling stresses are significantly reduced for actuators with inactive edges. Increasing the length of the inactive edges longer than $2a$ does not further reduce the shear stress level. Although further decrease is shown in the peeling stress, the stress level is low enough to be neglected.

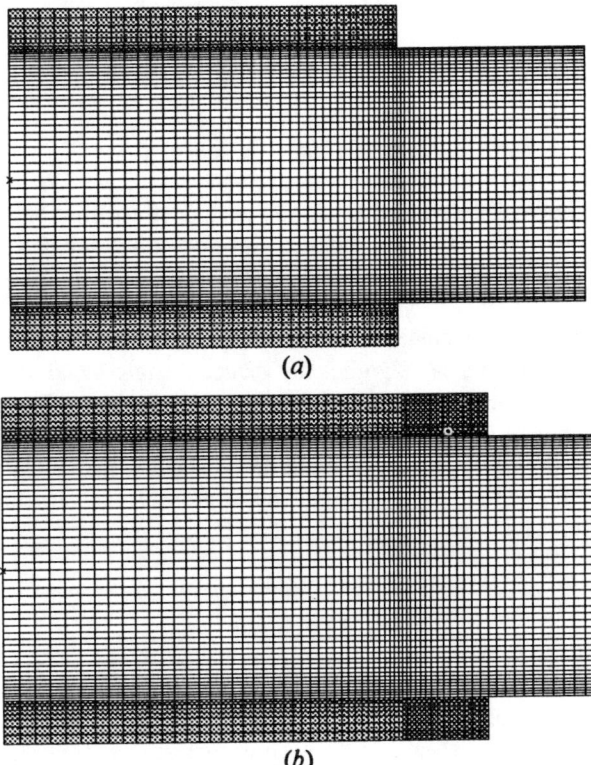

Figure 7. Finite element models for the actuator/beam substructure: (a) without inactive edges, (b) with inactive edges of length $2a$.

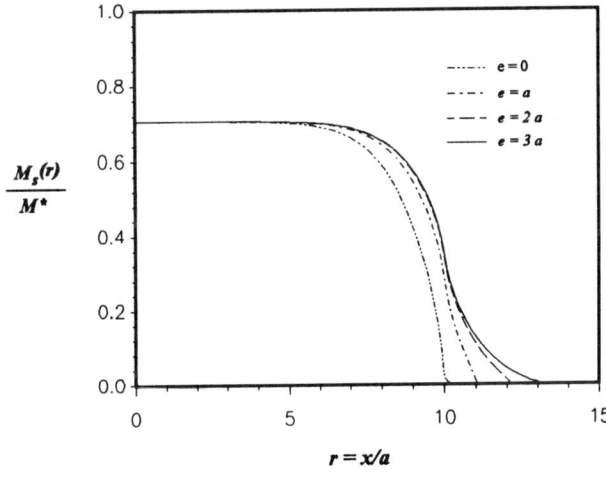

Figure 8. Effective moment distribution along the x axis for cases of different inactive edge lengths as obtained from finite element analysis.

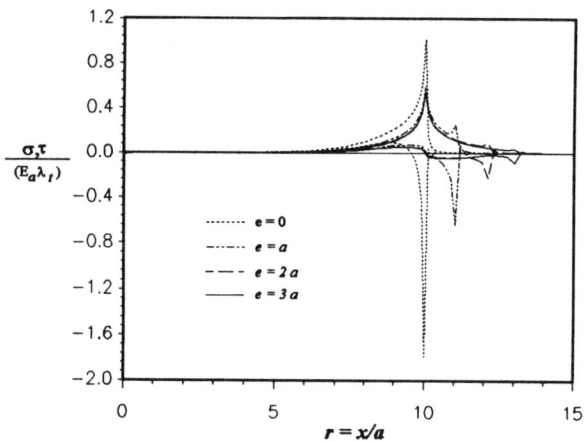

Figure 9. Interfacial shear and peeling stress distribution for cases of different inactive edge lengths as obtained from finite element analysis.

It can be concluded from this analysis that inactive edges on actuators can significantly reduce the interfacial shear and peeling stresses and slightly increase the effective moment in the actuator end zones. An optimum length of the inactive edge is found to be about a two actuator thickness for the present actuator/beam substructure configuration.

Conclusion

The basis of the mechanism underlying interfacial stress alleviation is discussed. It is demonstrated that the interfacial shear and peeling stress concentration localized at the ends of induced strain actuators can be alleviated by reducing the degree of the free edge effect. A new actuator configuration is proposed to eliminate the free edge effect by including inactive edges on the actuators. Both analytical and finite element modeling of the suggested actuator configuration show that the interfacial shear and peeling stresses can be significantly reduced without sacrificing the effectiveness of the transfer of the actuation mechanism. An optimum length of the inactive edges is found from finite element analysis to be about a two-actuator thickness, and inactive edges with a relatively higher stiffness than the actuator are desirable, as shown in the analytical model.

Acknowledgments

The authors gratefully acknowledge the support of the Army Research Office University Research Initiative Center Program, Grant No. DAAL 03-92-6-0181; Dr. Gary Anderson, Program Manger.

References

Bailey, T. and Hubbard Jr., J. E., 1985, "Distributed Piezoelectric-Polymer Active Vibration Control of a Cantilever Beam, "Journal of Guidance and Control, Vol. 8, No. 5, pp. 605-611.

Chaudhry, Z., and Rogers, C. A., 1991, "Bending and Shape Control of Beams Using SMA Actuators," Journal of Intelligent Material Systems and Structures, Vol. 2, No. 4, pp. 581-602.

Crawley, E. F., and de Luis, J., 1987, "Use of Piezoelectric Actuators as Elements of Intelligent Structures," AIAA Journal, Vol. 25, No. 10, pp. 1373-1385.

Im, S,, and Atluri, S. N., 1989, "Effects of a Piezo-Actuator on a Finitely Deformed Beam Subjected to General Loading," AIAA Journal, Vol. 27, No. 12, p.p. 1801-1807.

Lin, M. W., and Rogers, C. A., 1993a, "Modeling of the Actuation Mechanism in a Beam Structure with Induced Strain Actuators," Proceedings of the AIAA/ASME/AHS/ASC 34th SDM Conference, La Jolla, CA, April 19-22, 1993.

Lin, M. W., and Rogers, C. A., 1993b, "Actuation Response of a Beam Structure with Induced Strain Actuators," Proceedings of the Adaptive Structures and Material Systems Symposium, ASME Winter Annual Meeting, New Orleans, LA, November 28-December 3, 1993; in press.

Walker, J., Liang, C., and Rogers, C. A., 1993, "Finite Element Analysis of Adhesively-Bonded Piezoceramic Patches Implementing Modeling Techniques and Design Consideration to Reduce Critical Stresses," Proceedings of the AIAA/ASME/AHS/ASC 34th SDM Conference, La Jolla, CA, April 19-22, 1993.

MODELING CONSIDERATIONS FOR IN-PHASE ACTUATION OF ACTUATORS BONDED TO SHELL STRUCTURES

Frédéric Lalande[†], Zaffir Chaudhry[*] and Craig A. Rogers[**]
Center for Intelligent Material Systems and Structures
Virginia Polytechnic Institute and State University
Blacksburg, VA 24061-0261

Abstract

A model to represent the in-phase actuation of induced strain actuators bonded to the surface of a circular shell is developed. Due to the inherent shell curvature, the equivalent discrete tangential forces generally used to represent the in-phase actuation of the actuators (such as in pin-force models) are not co-linear and result in the application of rigid body forces on the shell. This non-equilibrium state violates the principle of self-equilibrium of fully integrated structures, such as piezoelectrically actuated shells. The solution to this non-equilibrium problem is to apply a uniform transverse pressure over the actuator region to maintain equilibrium. Using this adequate equivalent loading scheme for in-phase actuation, a deformation model for a circular ring is derived based on shell governing equations.

To verify the deformation model, finite element analysis is performed. A perfect match between the in-phase actuation deformation model and the finite element results, when the actuator mass and stiffness are neglected, validates the derived analytical model. A deformed shell comparison between the point force model, often used to represent the actuator in-phase actuation, and the derived analytical model showed major displacement disparities. Thus, by simply applying a uniform transverse pressure along with the discrete tangential forces in order to maintain the self-equilibrium of the shell, the shell deformation can be modeled accurately.

Introduction

Piezoelectrics actuators have been used for active shape, vibration and acoustic control of structures because of their adaptibility and light weight. Their ability to be easily integrated into structures makes them very attractive in structural control since all moving parts encountered with conventional actuators are eliminated. Structural control is implented by simply embedding PZT actuators in the structure or bonding them on the structure. In structural control, the desired deformation in the structure is obtained by the application of localized line forces and moments generated by the expanding or contracting bonded or embedded PZT actuators. In the case of vibration and acoustic control, the piezoelectric actuators will change the impedance of the structure to reduce the unwanted dynamic effects at given frequencies.

Previous research performed on PZT-actuated beam and plate structures has led to models describing their response (Crawley and de Luis, 1987; Crawley and Anderson, 1990; Wang and Rogers, 1991; Dimitriadis et al., 1991; Zhou et al.., 1994; Liang et al., 1993). Simple but efficient models were proposed to describe the response of a plate structure to the piezoelectric actuators (Crawley and Lazarus, 1989). By simply replacing the PZT actuator with line forces and moments on its edges, very accurate results are produced even though this type of model is approximate since the mass and stiffness of the actuator is not considered. However, much less research has been performed on structures with curvature. Some experimental work (Fuller, 1990) and adaptations of flat structure models to curved structures have been made (Sonti and Jones, 1991; Lester and Lefebvre, 1991). Models based on shell equations have also been proposed (Rossi et al., 1993; Sonti, 1993; Zhou et al., 1993; Larson and Vinson, 1993).

In a recent paper, Chaudhry, Lalande and Rogers (1994) considered the modeling of piezoelectric actuator patches on circular cylinders. When the piezoelectric actuators are actuated in-phase, it is found that the point force model used to represent the actuator creates a rigid body motion since the equivalent line forces are not collinear due to the curvature of the shell. Since the PZT actuators are integrated within the structure, self-equilibrium must be satisfied. This equilibrium discrepancy between the actual structure and the equivalent loading scheme will produce serious errors when the shell deformation, based on the line force representation of the actuator, is calculated. Until now, no models take account of this non-equilibrium application of the equivalent line forces. The solution proposed to solve this problem is to apply a uniform transverse pressure over the actuator location to eliminate the rigid body mode. Good agreement between the equivalent loading model and the actual deformation of the piezoelectrically-actuated structure was found.

In this paper, an analytical deformation model of a piezoelectrically-actuated circular ring, which takes into account the non-collinear equivalent line forces, is proposed.

[†] Graduate Research Assistant.
[*] Research Scientist, Member AIAA, ASME.
[**] Professor, Member AIAA, ASME.

Shell equivalent loading model

The first step in this paper is to repeat the conclusions established by Chaudhry, Lalande and Rogers (1994). In that paper, an equivalent loading scheme for shell structures was presented. It was shown that in the case of in-phase actuation, a rigid body mode was present due to the fact that the equivalent line forces L_θ are not collinear (Fig. 1).

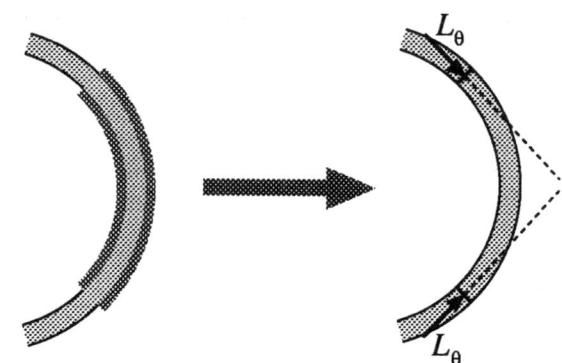

Figure 1. Non-equilibrium of discrete tangential forces in shell structures.

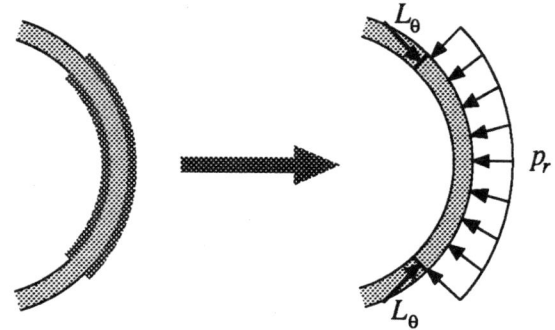

Figure 2. Adequate equivalent loading to maintain equilibrium.

To eliminate this non-equilibrium state of the structure, a transverse uniform pressure is added (Fig. 2). The magnitude of the transverse pressure from simple statics is then:

$$\bar{p}_r = -\frac{\bar{L}_\theta}{a}, \qquad (1)$$

where

$$\bar{L}_\theta = \frac{E_s t_s}{1-\nu} \frac{2}{2+\psi} \Lambda \qquad (2)$$

and

$$\psi = \frac{E_s t_s}{E_a t_a}, \qquad (3)$$

when the Poisson's ratio of the shell and the piezoelectric actuator are assumed to be the same. E, t, L and a are the Young's modulus, the thickness, the free induced strain and the radius of the ring, respectively, while the subscripts s and a stand for shell and actuator, respectively. If a circular ring is considered, the Poisson's effect disappears since there are no constraints in the axial direction. Thus, for the case of a ring, the Poisson's ratio in Eq. (2) is set to zero. Based on this equivalent loading scheme, a deformation model for a circular ring with two discrete tangential forces and a uniform radial pressure will be derived.

Derivation of governing equations

In the current literature (Soedel, 1981; Flugge, 1973), the governing shell equations are limited to shells subjected to external pressure loading only. Since the proposed equivalent loading model involves discrete tangential forces, the available shell equations are not adequate. To simplify the governing equations, the added mass and stiffness of the actuators are neglected. Thus, the governing equations of a uniform stiffness circular ring subjected to radial pressure and discrete tangential loading will be derived using the approach proposed by Soedel (1981). In the case of a thin circular ring, only the in-plane stress resultants N_θ, M_θ and $Q_{\theta r}$ are present. The stress resultants in the axial direction, N_x, M_x, the shear stress resultant $N_{x\theta}$, the twisting moment resultant M_{xq} and the transverse shear force Q_{xr} are all zero since the displacements are constant in the axial direction and the axial displacements are zero.

Starting with the classical theory of elasticity, the polar strain-displacement relations are given by Timoshenko and Goodier (1951):

$$\varepsilon_r = \frac{\partial w}{\partial r} \qquad (4a)$$

$$\varepsilon_\theta = \frac{w}{r} + \frac{\partial v}{r\partial \theta} \qquad (4b)$$

$$\gamma_{\theta r} = \frac{\partial w}{r\partial \theta} + \frac{\partial v}{\partial r} - \frac{v}{r}. \qquad (4c)$$

A neutral surface system of coordinates is used instead of the global polar coordinates to simplify the equations. Thus, the change of variable is:

$$r = a + z = a\left(1 + z/a\right), \qquad (5)$$

where a is the radius of the ring. Combining the polar strain-displacement relations Eq. (4) and the change of coordinate system Eq. (5), the strain-displacement relations in the neutral surface coordinate system become:

$$\varepsilon_r = \frac{\partial w}{\partial z} \tag{6a}$$

$$\varepsilon_\theta = \frac{1}{a(1+z/a)} w + \frac{\partial v}{\partial \theta} \tag{6b}$$

$$\gamma_{\theta r} = \frac{1}{a(1+z/a)} \left[\frac{\partial w}{\partial \theta} - v + a(1+z/a) \frac{\partial v}{\partial z} \right]. \tag{6c}$$

Since the ring is thin, a linear variation in the tangential direction and a constant radial displacement through the thickness are assumed (Kirchhoff's assumption; Soedel, 1981):

$$v = v^o + z\beta \tag{7a}$$

$$w = w^o, \tag{7b}$$

where β is the rotational displacement and v^o and w^o are the neutral surface tangential and radial displacements, respectively. If the shear deflection is neglected ($\gamma_{\theta r} = 0$), the expression of β is obtained from Eqs. (6c) and (7):

$$\beta = \frac{1}{a}\left(v^o - \frac{\partial w^o}{\partial \theta}\right). \tag{8}$$

Even though we assumed the shear strain $\gamma_{\theta r}$ to be zero, this does not imply that the transverse shear force $Q_{\theta r}$ is neglected. When the ring is thin, the z/a term can be neglected since it is much less than one. Thus, the strain-displacement relations become:

$$\varepsilon_r = 0 \tag{9a}$$

$$\varepsilon_\theta = \frac{1}{a}\left(\frac{\partial v^o}{\partial \theta} + w^o\right) + \frac{z}{a^2}\left(\frac{\partial v^o}{\partial \theta} - \frac{\partial^2 w^o}{\partial \theta^2}\right) \tag{9b}$$

$$\gamma_{\theta r} = 0. \tag{9c}$$

The membrane force, bending moment and transverse shear force resultants are obtained by integrating through the thickness of the ring:

$$N_\theta = \int_{-t_s/2}^{t_s/2} \sigma_\theta dz = \int_{-t_s/2}^{t_s/2} E\varepsilon_\theta dz = \frac{K}{a}\left(\frac{\partial v^o}{\partial \theta} + w^o\right) \tag{10a}$$

$$M_\theta = \int_{-t_s/2}^{t_s/2} \sigma_\theta z\, dz = \int_{-t_s/2}^{t_s/2} E\varepsilon_\theta z\, dz = \frac{D}{a^2}\left(\frac{\partial v^o}{\partial \theta} - \frac{\partial^2 w^o}{\partial \theta^2}\right) \tag{10b}$$

$$Q_{\theta r} = \int_{-t_s/2}^{t_s/2} \sigma_{\theta r}\, dz, \tag{10c}$$

where the membrane stiffness is:

$$K = E t_s, \tag{11a}$$

and the bending stiffness is:

$$D = \frac{E t_s^3}{12}. \tag{11b}$$

It must be noted that the Poisson's ratio is not present in the stiffnesses expressions Eq. (11) since the ring is free to deform in the axial direction.

The equilibrium equations derivation is based on the energy method. Using Hamilton's principle, the Love ring equations for the equivalent loading scheme will be developed. Hamilton's principle is given by:

$$\int_{t_o}^{t_1} \left[\delta(U - E_b - E_L) - \delta K\right] dt = 0, \tag{12}$$

where $\delta(U-E_b-E_L)$ is the total variational potential energy and δK is the variational kinetic energy. Since the ring is subjected to static loading, the kinetic energy term is equal to zero.

The first term involved in the total potential energy is the strain energy. The infinitesimal ring element is subjected to tangential stress σ_θ and to $\sigma_{\theta r}$ only. Thus, the strain energy stored in this infinitesimal element is:

$$dU = \frac{1}{2}\left(\sigma_\theta \varepsilon_\theta + \sigma_{\theta r}\gamma_{\theta r}\right) a\, d\theta\, dx\, dz. \tag{13}$$

The transverse shear term is retained in order to obtain an expression for b, even though it was assumed that $\gamma_{\theta r} = 0$ Eq. (9c). The previous equation is integrated over the volume of the ring to obtain the total strain energy:

$$U = \iiint_{x\,\theta\,z} \frac{1}{2}\left(\sigma_\theta \varepsilon_\theta + \sigma_{\theta r}\gamma_{\theta r}\right) a\, d\theta\, dx\, dz. \tag{14}$$

The free body diagram of the ring is shown in Fig. 3. The possible externally applied boundary forces and moment resultants introduce energy into the ring. This second type of potential energy is given by:

$$E_b = \int_x \left(N_\theta^* v^o + Q_{\theta r}^* w^o + M_\theta^* \beta\right) dx. \tag{15}$$

The term of potential energy introduced in the ring due to external loading is due to a radial pressure and a discrete tangential load (Fig. 4). The external loading energy is given by:

$$E_L = \iint_{x\,\theta} p_r w^o \, a d\theta \, dx + \int_x \overline{L}_\theta v^o \, dx. \quad (16)$$

Rewriting the previous equation under a unique double integral by introducing a Dirac function:

$$E_L = \iint_{x\,\theta} \left(\frac{\overline{L}_\theta}{a} \delta(\theta - \theta_p) v^o + p_r w^o \right) a d\theta \, dx, \quad (17)$$

where θ_p is the location of the applied line load. The loads are assumed to be applied on the neutral surface of the ring.

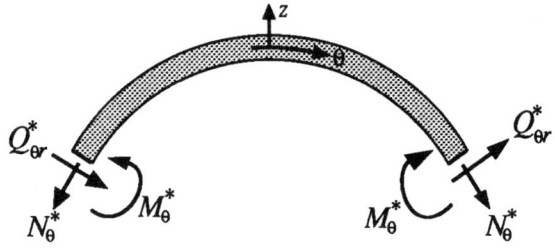

Figure 3. Ring free-body diagram.

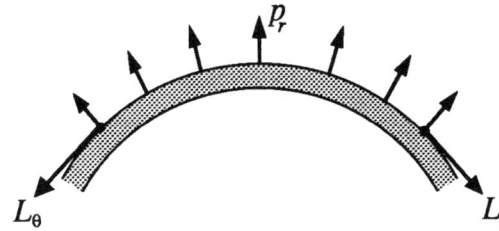

Figure 4. External loading applied to the ring.

Introducing the variational operator in the potential energy expressions, one can obtain the variational strain energy:

$$\delta U = \iiint_{x\,\theta\,z} \left[\sigma_\theta \delta\varepsilon_\theta + \sigma_{\theta r} \delta\varepsilon_{\theta r} \right] a d\theta \, dx \, dz. \quad (18)$$

Substituting the strain-displacement relationships Eq. (9) and using the zero normal shear strain Love simplification, Eq. (18) is rewritten as:

$$\delta U = \iiint_{r\,x\,\theta} \left[\sigma_\theta \left(\frac{\partial \delta v^o}{\partial \theta} + z \frac{\partial \delta \beta}{\partial \theta} + \delta w^o \right) + \sigma_{\theta r} \left(\delta\beta \, a - \frac{\partial \delta v^o}{\partial \theta} - \frac{\partial \delta w^o}{\partial \theta} \right) \right] d\theta \, dx \, dr. \quad (19)$$

Integrating by parts, the above expression for the variational strain energy becomes:

$$\delta U = \left\{ \iint_{x\,\theta} \left[\left(-\frac{\partial N_\theta}{\partial \theta} - Q_{\theta r} \right) \delta v^o + \left(N_\theta - \frac{\partial Q_{\theta r}}{\partial \theta} \right) \delta w^o + \left(-\frac{\partial M_\theta}{\partial \theta} + Q_{\theta r} a \right) \delta\beta \right] d\theta \, dx + \int_x \left[N_\theta \delta v^o + M_\theta \delta\beta + Q_{\theta r} \delta w^o \right] dx \right\} \quad (20a)$$

The variational form of the energy introduced by the external boundary forces is:

$$\delta E_b = \int_x \left(N_\theta^* \delta v^o + Q_{\theta r}^* \delta w^o + M_\theta^* \delta\beta \right) dx, \quad (20b)$$

and the variational external loading energy is:

$$\delta E_L = \iint_{x\,\theta} \left(\frac{\overline{L}_\theta}{a} \delta(\theta - \theta_p) \delta v^o + p_r \delta w^o \right) a d\theta \, dx. \quad (20c)$$

Substituting the expressions of the different types of energy involved in the total variational potential energy Eq. (20) in Hamilton's principle Eq. (12), we obtain

$$\int_{t_o}^{t_1} \left\{ \left\{ \iint_{x\,\theta} \left[\left(-\frac{\partial N_\theta}{\partial \theta} - Q_{\theta r} \right) \delta v^o + \left(N_\theta - \frac{\partial Q_{\theta r}}{\partial \theta} \right) \delta w^o + \left(-\frac{\partial M_\theta}{\partial \theta} + Q_{\theta r} a \right) \delta\beta \right] d\theta \, dx + \int_x \left[N_\theta \delta v^o + M_\theta \delta\beta + Q_{\theta r} \delta w^o \right] dx \right\} - \left\{ \int_x \left[N_\theta^* \delta v^o + Q_{\theta r}^* \delta w^o + M_\theta^* \delta\beta \right] dx \right\} - \left\{ \iint_{x\,\theta} \left[\frac{\overline{L}_\theta}{a} \delta(\theta - \theta_p) \delta v^o + p_r \delta w^o \right] a d\theta \, dx \right\} \right\} dt = 0 \quad (21)$$

Rewriting the previous equation:

$$\int_{t_o}^{t_1}\iint_{x\,\theta}\left[\left(\frac{\partial N_\theta}{\partial \theta}+Q_{\theta r}+\overline{L}_\theta\delta(\theta-\theta_p)\right)\delta v^o+\left(\frac{\partial Q_{\theta r}}{\partial \theta}-N_\theta+a\,p_r\right)\delta w^o+\left(\frac{\partial M_\theta}{\partial \theta}-Q_{\theta r}a\right)\delta\beta\right]d\theta\,dx\,dt+$$
$$\int_{t_o}^{t_1}\int_x\left[(N_\theta{}^*-N_\theta)\delta v^o+(Q_{\theta r}{}^*-Q_{\theta r})\delta w^o+(M_\theta{}^*-M_\theta)\delta\beta\right]dx\,dt=0 \qquad (22)$$

To be satisfied, each integral term of Hamilton's principle must be individually zero. Since the variational displacements cannot be zero due to their arbitrariness, the Love circular ring equations are:

$$\frac{\partial N_\theta}{\partial \theta}+Q_{\theta r}+\overline{L}_\theta\delta(\theta-\theta_p)=0, \qquad (23a)$$

$$\frac{\partial Q_{\theta r}}{\partial \theta}-N_\theta+a\,p_r=0, \qquad (23b)$$

and

$$\frac{\partial M_\theta}{\partial \theta}-Q_{\theta r}a=0. \qquad (23c)$$

Substituting Eq. (23c) in Eqs. (23a) and (23b), the equilibrium equations of the circular ring are found to be:

$$a\frac{dN_\theta(\theta)}{d\theta}+\frac{dM_\theta(\theta)}{d\theta}+a\overline{L}_\theta\delta(\theta-\theta_p)=0, \qquad (24a)$$

$$\frac{d^2M_\theta(\theta)}{d\theta^2}-aN_\theta(\theta)+a^2p_r(\theta)=0. \qquad (24b)$$

The derived equilibrium equations Eq. (24) are very similar to those obtained by Soedel (1981) and Flugge (1973). The difference appears in the tangential loading term $a L_\theta \delta(\theta-\theta_p)$ which replaces the $a^2 p_\theta(\theta)$ term when tangential pressure loading is considered.

By setting the line integrals to zero in Eq. (22), the necessary boundary conditions for the ring are:

$$N_\theta = N_\theta{}^* \text{ or } v^o = v^o{}^*, \qquad (25a)$$

$$M_\theta = M_\theta{}^* \text{ or } \beta = \beta^*, \qquad (25b)$$

and

$$Q = Q^* \text{ or } w^o = w^o{}^*. \qquad (25c)$$

Derivation of the In-Phase Actuation Deformation Model

With the governing equations now derived, the next step is to apply them to the particular problem shown in Fig. 5. To simplify the analytical model derivation, the actuator stiffness will be neglected.

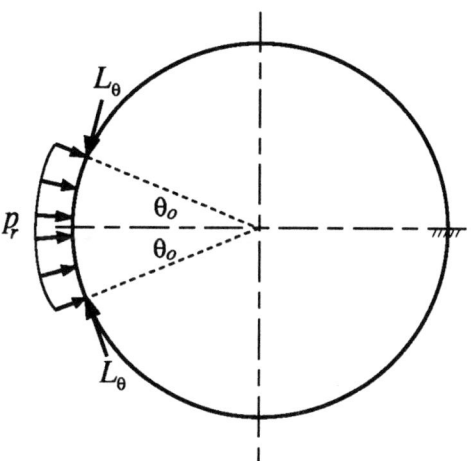

Figure 5. Adequate equivalent actuator loading on the ring.

As established previously, the ring is subjected to discrete tangential forces at the end of the modeled actuator and to a uniform radial pressure of magnitude ($\frac{\overline{L}_\theta}{a}$) Eq. (1), to ensure equilibrium of the ring. The loading of the ring is expressed using Dirac and Heaviside functions:

$$L_\theta(\theta)=\overline{L}_\theta\left[\delta^- - \delta^+\right] \qquad (26)$$

$$p_r(\theta)=\overline{p}_r\left[H^- - H^+\right]=-\frac{\overline{L}_\theta}{a}\left[H^- - H^+\right], \qquad (27)$$

where

$$\theta^- = \theta - (\pi - \theta_o) \qquad (28a)$$

$$\theta^+ = \theta - (\pi + \theta_o) \qquad (28b)$$

$$\delta^- = \delta[\theta^-] \qquad (29a)$$

$$\delta^+ = \delta[\theta^+] \qquad (29b)$$

$$H^- = H[\theta^-] \qquad (30a)$$

$$H^+ = H[\theta^+]. \qquad (30b)$$

The integration constants will be determined from the continuity conditions at $\theta = 0, 2\pi$:

$$w^o(0) = w^o(2\pi) \tag{31a}$$

$$v^o(0) = v^o(2\pi) \tag{31b}$$

$$\beta(0) = \beta(2\pi). \tag{31c}$$

From the rotational displacement expression Eq. (8), it is possible to rewrite the continuity conditions of Eq. (31c), by making use of Eq. (31a), as:

$$w'^o(0) = w'^o(2\pi). \tag{31d}$$

Combining the equilibrium equations Eq. (24), the differential equation for the moment in the ring is obtained as follows:

$$\frac{d^3 M_\theta(\theta)}{d\theta^3} + \frac{dM_\theta(\theta)}{d\theta} = -a^2 \left[\frac{L_\theta(\theta)}{a} + p'_r(\theta) \right] = 0. \tag{32}$$

Substituting the loading expressions Eqs. (26) and (27) in the previous equation (Eq. (32)), it can be seen that the right hand side of the equation will be zero. Solving the differential equation (Eq. (32)), an expression of the moment distribution in the ring is obtained:

$$M_\theta(\theta) = C_1 + C_2 Sin\theta + C_3 Cos\theta. \tag{33}$$

Combining the two stress-displacement equations Eq. (10), the following differential equation is obtained:

$$w^o(\theta) + \frac{d^2 w^o(\theta)}{d\theta^2} = \frac{a^2}{D}\left(\frac{N_\theta(\theta)D}{Ka} - M_\theta(\theta) \right). \tag{34}$$

Rewriting the second equilibrium equation (Eq, (24b)):

$$N_\theta(\theta) = \frac{1}{a} \frac{d^2 M_\theta(\theta)}{d\theta^2} + a\, p_r(\theta). \tag{35}$$

Substituting the expression of the moment Eq. (33) and the tangential force Eq. (35) into Eq. (34), the following differential equation in $w^o(\theta)$ is obtained:

$$w^o(\theta) + \frac{d^2 w(\theta)}{d\theta^2} = -\frac{a^2}{D}\left[\begin{array}{l} C_1 + C_2\left(1 + \frac{D}{Ka^2}\right)Sin\theta + \\ C_3\left(1 + \frac{D}{Ka^2}\right)Cos\theta - \frac{D}{K} p_r(\theta) \end{array} \right]. \tag{36}$$

For thin rings, the $\frac{D}{Ka^2}$ term is neglected since its value is much less than one. The radial displacement equation is obtained using Laplace transform:

$$w^o(\theta) = \frac{a^2}{D}\left\{ \begin{array}{l} C_1(1-Cos\theta) + \frac{C_2}{2}(Sin\theta - \theta Cos\theta) \\ +\frac{C_3}{2}(\theta Sin\theta) + w^o(0)Cos\theta + w^{o'}(0)Sin\theta \\ +\frac{D\bar{p}_r}{K}\left[(1-Cos\theta^-)H^- - (1-Cos\theta^+)H^+\right] \end{array} \right\} \tag{37}$$

Introducing Eq.(37) in Eq. (10a), the tangential displacement differential equation is

$$\frac{dv^o(\theta)}{d\theta} = \frac{a^2}{D}\left\{ \begin{array}{l} -C_1(1-Cos\theta) - \frac{C_2}{2}(Sin\theta - \theta Cos\theta) \\ -\frac{C_3}{2}(\theta Sin\theta) - w^o(0)Cos\theta - w^{o'}(0)Sin\theta \\ +\frac{D\bar{p}_r}{K}\left[Cos\theta^- H^- - Cos\theta^+ H^+\right] \end{array} \right\} \tag{38}$$

Solving this equation using Laplace transformation and applying continuity conditions Eq. (31), the equations of the radial and tangential displacements are found to be:

$$v^o(\theta) = \frac{a^2 \bar{p}_r}{K}\left\{ \begin{array}{l} -\frac{Sin\theta_o}{\pi}(Sin\theta - \theta Cos\theta) \\ +\left[Sin\theta^- H^- - Sin\theta^+ H^+\right] \end{array} \right\}, \tag{39}$$

and

$$w^o(\theta) = \frac{a^2 \bar{p}_r}{K}\left\{ \begin{array}{l} \frac{Sin\theta_o}{\pi}\theta Sin\theta + \\ \left[(1-Cos\theta^-)H^- - (1-Cos\theta^+)H^+\right] \end{array} \right\}. \tag{40}$$

Finite Element Verification

The developed in-phase actuation deformation model is verified using finite element analysis. A ring of 6" radius, 0.032" thickness and 1" deep, and piezoelectric actuators 1/6 of the ring thickness and covering an arc 30° long ($2\theta_o$), are used. Making use of symmetry, the finite element model consists of beam elements only in the upper half, as shown in Fig. 6. Two loading cases are considered: i) temperature contraction equivalent to 1000 μstrain of the beam elements modeling the actuator region; and ii) equivalent discrete forces and uniform pressure loading from Eqs. (1) and (2). The finite element analysis results are shown in Fig. 7, as well as the in-phase actuation deformation model results. A single curve can be observed since the curves match perfectly. Also shown in Fig. 7 are the displacements of the same ring if only discrete tangential forces are applied (point force model). The point force model does not satisfy the ring's self-equilibrium. Major displacement discrepancies between the in-phase actuation deformation and the point force model occur both in shape and magnitude. The point force model overpredicts the displacements by a factor up to 1000. Fig. 8 shows the deformed shape of the self-equilibrium loading and the non-equilibrium loading with the displacements magnified by a 500 and 2 factor, respectively. It can be seen again that when a uniform pressure is not applied to maintain equilibrium, the deformed shape is erroneous. Also, a reaction force in the x-direction at the clamped boundary condition is present if the uniform pressure is not applied. This reaction force should not be present since the actual ring with bonded actuators is in self-equilibrium. The adequate equivalent loading did not show any reaction force in the x-direction at the clamped boundary condition. The verification of the results also have been made with 10° and 60°-long piezoelectric patches, and the coincidence is still perfect between the in-phase actuation deformation model and the finite element analysis.

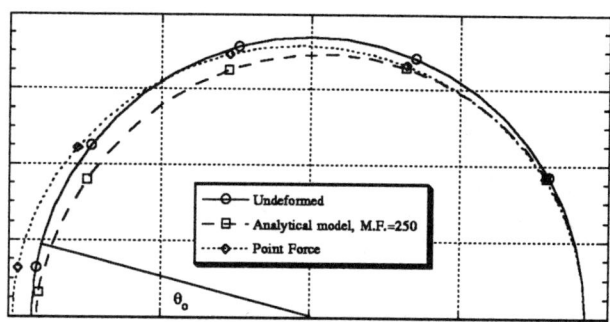

Figure 8. Deformed shape of the ring with and without self-equilibrium loading.

However, it must be mentionned that the deformation of the ring is very sensitive to the applied load in the finite element model. An error of 0.1% in the magnitude of the applied equivalent line force will completely change the response of the ring. This sensitivity of the nodal displacements is due to the low stiffness of the ring (0.032" thick only). The application of a tangential line force of 0.1% magnitude of the applied equivalent line force on the ring will produce nodal displacements of the same order as the self-equilibrium loading nodal displacements.

Also, the pressure loading must be transformed to nodal forces only (lumped loading). The lumped loading is often better for flat elements representing a curved surface (De Salvo and Swanson, 1979).

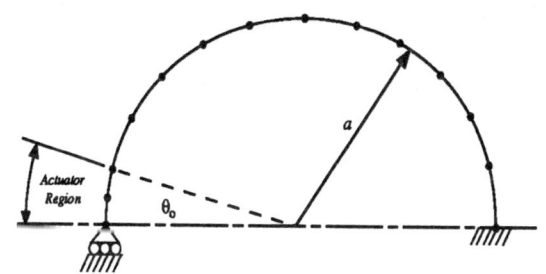

Figure 6. Beam finite element model.

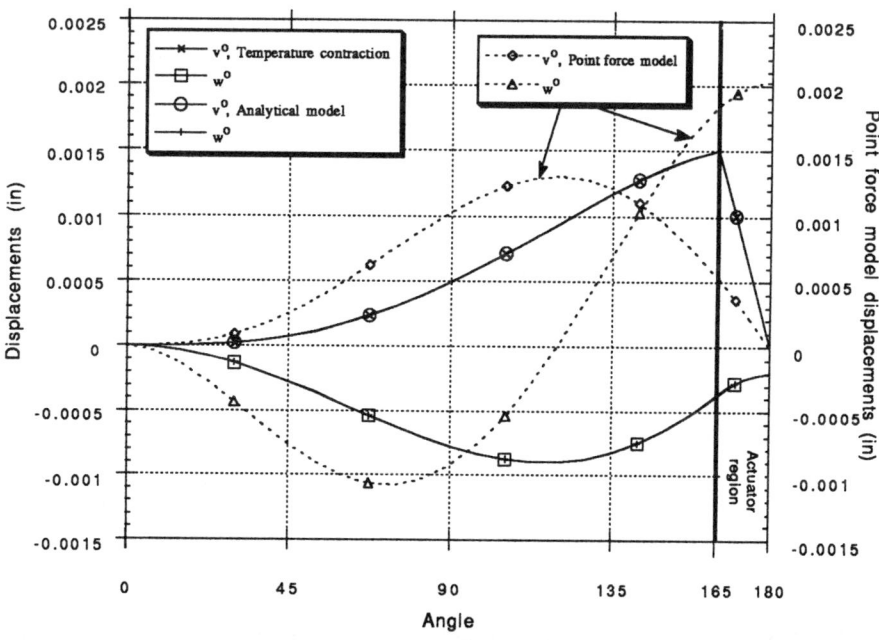

Figure 7. Match of displacements between the analytical model and the beam finite element model.

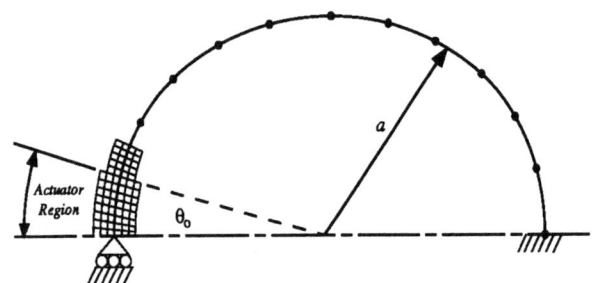

Figure 9. Plane stress finite element model.

Up to this point, the added stiffness of the actuators has been neglected both in the analytical model and the finite element analysis. A second finite element model, using plane stress elements in the actuator region to include the actuator's stiffness, is made to compare the actual behavior of the system to the derived analytical model (Fig. 9). To keep the FE model small, the actuator size is reduced to 10° ($2\theta_o$). The radial and tangential displacements are shown in Fig. 10. Disparities between the analytical model and the plane stress finite element model are present since no assumptions on the actuator stiffness or on the equivalent loading are made on the latter one. Even though displacement differences are present, the plane stress finite element model validates the derived analytical model since it gives results of the same order of magnitude with similar deformed shapes as opposed to the point force model previously discussed. The deformed shape of the analytical model and the plane stress finite element model is shown in Fig. 11.

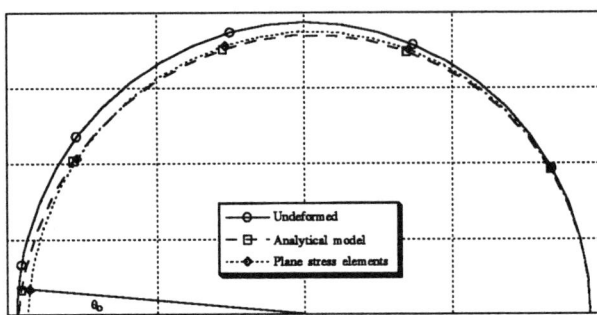

Figure 11. Deformed shape of the ring with and without self-equilibrium loading using plane stress elements.

The discussion of in-phase actuation of induced strain actuators symmetrically bonded on shells can be extended to unsymmetric actuation. Unsymmetric actuation is obtained when the actuators on each side of the shell are submitted to voltages of different magnitudes, or when a single actuator is bonded on one side of the shell. Unsymmetric actuation is a combination of extension and bending of the shell and can be solved using simple superposition. Thus, for unsymmetric actuation, the equivalent loading will consist of discrete tangential forces and moments at the ends of the actuator(s) and a distributed transverse pressure over the actuator(s) footprint to maintain the self-equilibrium of the shell.

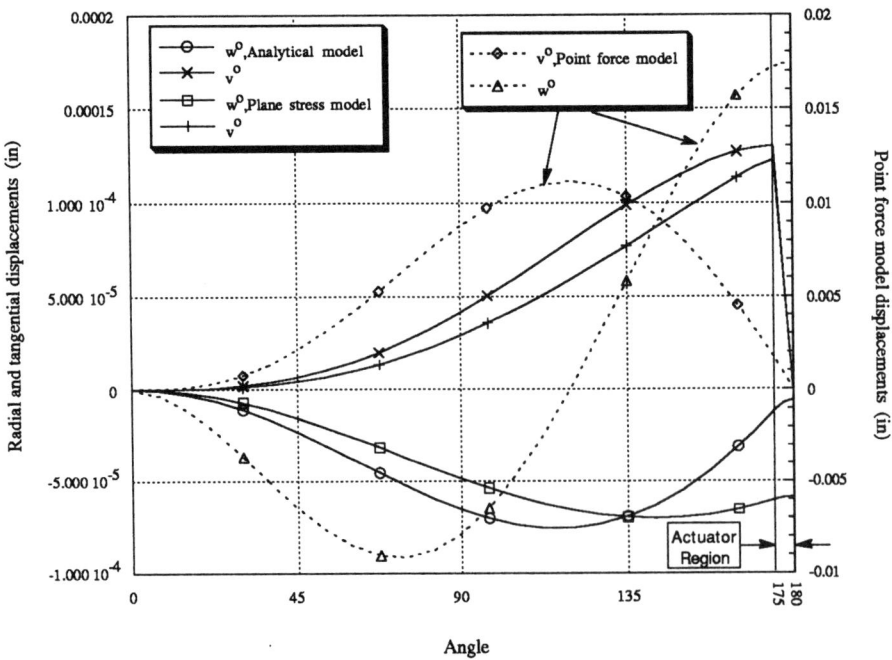

Figure 10. Match of displacements between the analytical model and the plane stress finite element model

Conclusion

In this paper, a deformation model for a circular ring subjected to in-phase actuation was derived. The actuator equivalent loading includes a uniform transverse pressure to maintain the self-equilibrium of the shell structure. The results of the in-phase actuation deformation model are in exact agreement with the finite element results the when actuator stiffness is neglected. If the actuator stiffness is considered, the analytical model gives a good approximation of the shell's deformed shape. If the self-equilibrium is not maintained (point-force model), the predicted deformed shape is completely different from the actual shell response to in-phase actuation. Thus, to obtain accurate results, a uniform transverse pressure should be applied to maintain the shell self-equilibrium.

Acknowledgments

The authors would like to acknowledge the funding support of the Office of Naval Research, Grant ONR N00014-92-J-1170, Dr. Kam Ng, Technical Monitor.

References

1) Chaudhry, Z., Lalande, F. and Rogers, C.A., 1994, "Modeling of Induced Strain Actuator Patches", Proceedings, *SPIE 1994 North American Conference on Smart Structures and Materials*, Orlando, FL, 13-18 February 1994; in press.

2) Crawley, E.F. and Anderson, E.H., 1990, "Detailed Models of Piezoceramic Actuation of Beams", *J. of Intel. Mater. Syst. and Struc.*, Vol. 1, pp. 4-25.

3) Crawley, E.F. and de Luis, J., 1987, "Use of Piezoelectric Actuators as Elements of Intelligent Structures", *AIAA Journal*, 25 (10), pp. 1373-1385.

4) Crawley, E.F. and Lazarus, K.B., 1989, "Induced Strain Actuation of Isotropic and Anisotropic Plates", *AIAA Journal*, Vol. 29, No. 6, pp. 944-951.

5) De Salvo, G.J. and Swanson, J.A., 1979, "ANSYS User's Manual", Swanson analysis systems, Houston, PA.

6) Dimitriadis, E.K., Fuller, C.R. and Rogers, C.A., 1991, "Piezoelectric Actuators for Distributed Vibration Excitation of Thin Plates", *J. of Vibration and Acoustics*, Vol. 113, pp. 100-107.

7) Fuller, C.R., Snyder, S., Hanson, C. and Silcox, R., 1990, "Active Control of Interior Noise in Model Aircraft Fuselages using Piezoceramic Actuators", Proceedings, *AIAA 13th Aeroacoustic Conference*, Vol. 90-3922, Tallahasse, FL, October 22-24 1990.

8) Flugge, W., 1973, "Stresses in shells", Springer-Verlag.

9) Larson, P.H., and Vinson, J.R., 1993, "The Use of Piezoelectric Materials in Curved Beams and rRngs", *Proceedings, ASME Winter Annual Meeting*, New Orleans, LA, 28 Nov-3 Dec 1993; pp. 277-285.

10) Lester, H.C. and Lefebvre, S., 1991, "Piezoelectric Actuator Models for Active Sound and Vibration Control of Cylinders, *Proceedings, Recent Advances in Active Noise and Vibration Control*, Blacksburg, VA, 15-17 April 1991; pp. 3 -26.

11) Liang, C., Sun, F.P., and Rogers, C.A., 1993c, "An Impedance Method for Dynamic Analysis of Active Material Systems", *Proceedings, 34th SDM Conference*, La Jolla, CA, 19-21 April 1993; pp.. 3587-3599.

12) Rossi, A., Liang, C. and Rogers, C.A., 1993, "Impedance Modeling of Piezoelectric Actuator-Driven Systems: an Application to Cylindrical Ring Structures", *Proceedings, 34th SDM Conference*, La Jolla, CA, 19-21 April 1993; pp. 3618-3624

13) Soedel, W., 1981, *Vibrations of Plates and Shells*, Marcel and Decker Inc, New-York.

14) Sonti, V.R. and Jones, J.D., 1991,"Active Vibration Control of Thin Cylindrical Shells using Piezo-electric Actuators", *Proceedings, Recent Advances in Active Noise and Vibration Control*, Blacksburg, VA, 15-17 April 1991; pp. 27-38.

15) Sonti, V.R., 1993, "Curved Piezo-Actuator Models for Active Vibration Control of Cylindrical Shells", *Proceedings, 125th meeting of the Acoustical Society of America*, Ottawa, Canada, 17-21 May .

16) Timoshenko, S. and Goodier, J.N., 1951, *Theory of Elasticity*, McGraw-Hill.

17) Wang, B.T. and Rogers, C.A., 1991, "Modeling of Finite-Length Spatially Distributed Induced Strain Actuators for Laminate Beams and Plates", *Proceedings, 32nd SDM Conference*, Baltimore, MD, 8-10 April 1991; pp. 1511-1520.

18) Zhou, S., Liang, C., and Rogers, C.A., 1993, "Impedance modeling of two dimensional piezoelectric actuators bonded on a cylinder", *Proceedings, 114th ASME Winter Annual Meeting*, New Orleans, LA, 28 Nov -3 Dec 1993, pp. 247-255.

19) Zhou, S., Liang, C. and Rogers, C.A., 1994, "A Dynamic Model of a Piezoelectric Actuator Driven Thin Plate", *Smart Structures and Materials '94*, SPIE, Orlando, FL, 13-18 February 1994; in press.

MODELING PIEZOCERAMIC ACTUATION OF BEAMS IN TORSION

Christopher Park[*] and Inderjit Chopra[**]
Center for Rotorcraft Education and Research
Department of Aerospace Engineering
University of Maryland, College Park, MD 20742

Abstract

This paper develops a one-dimensional model to predict the coupled extension, bending and torsion response of a beam subject to piezoceramic strain actuation. The effects of cross-sectional warping are shown to be negligible for thin rectangular isotropic beams. The impact of adhesive shear lag, on the other hand, is measurable, especially in the torsional response. Experimental test results show that the models are accurate up to a 45° actuator orientation with respect to the beam axis even though detailed strain data indicate that the local strain state is highly two-dimensional.

Nomenclature

A	cross-sectional area
B_λ	warping parameter
C_λ	warping parameter
E	Young's modulus
$\mathcal{E}$	Electric field [V/mil]
G	shear modulus
I	flexural area moment of inertia
J	torsional rigidity constant
L	projected actuator half-length
L_λ	warping parameter
M	bending moment
P	axial force
S	first bending moment of inertia
T	torque
u	axial displacement
v	transverse displacement
w	bending displacement
β	actuator orientation relative to beam axis
γ	shear strain
Γ	shear lag parameter
ε	axial strain
ϕ	twist displacement
κ	bending curvature
λ	warping function
Λ	piezoceramic actuation strain
ν	Poisson's ratio
θ	twist rate
σ	axial stress
τ	shear stress
ψ	stiffness ratio

Superscripts

a	actuator reference
b	beam reference
s	adhesive substrate reference
*	effective value

Subscripts

a	actuator reference
b	beam reference
o	mid-plane reference
s	adhesive substrate reference
κ	bending curvature reference
λ	warping reference

Overbars

~	actuator axes reference

Introduction

Helicopters suffer from excessive vibration, high fatigue loads, poor handling qualities and intolerable noise. The objective is to improve the dynamic performance of the helicopter and reduce vibration to an acceptable level. Research on Higher Harmonic Control (HHC)[1,2,3] and Individual Blade Control (IBC) for helicopter vibration suppression has shown that these concepts can be used to reduce vibration transmitted to the pilot's seat. However HHC and IBC have high weight penalties and are limited in their application to reduce stresses, improve performance and reduce noise. It is envisioned that incorporating smart structure technology in rotor blades can give desirable shape control characteristics to improve the helicopter in all of these areas at a reasonable weight penalty. The need for modeling intelligent structures, in particular beams undergoing combined extension, bending and torsional deflections due to induced strain actuation, is important in the development of smart rotor blades.

Crawley and de Luis[4] presented an analytical uniform strain model of a beam with strain induced actuation by use of surface bonded piezoceramics.

---

[*] Graduate research assistant
[**] Professor and Director, Fellow AIAA

Presented at the 35th Structures, Structural Dynamics and Materials Conference and Adaptive Structures Forum, Hilton Head, NC, April 18-21, 1994.

Copyright 1994 by the American Institute of Aeronautics and Astronautics, Inc. All rights reserved

The model predicted beam extension and bending and included shear lag effects of the adhesive substrate between the piezoceramic and the beam. The dynamic model was experimentally verified for the first two bending modes of a cantilever beam. Crawley and de Luis[5] later presented a uniform strain beam model of both embedded and surface mounted piezoceramic actuators. They validated the dynamic response at resonance using aluminum, glass-epoxy and graphite-epoxy beams. Crawley and Anderson[6] presented a Bernoulli-Euler model and compared it with the previously mentioned uniform strain model, a finite element model and experiment. All of these models assume a pair of actuators aligned with and symmetrically located with respect to the beam axis. The uniform strain model was quite good except for predicting the response of beams with low beam thickness to actuator thickness ratios. The Bernoulli-Euler model was accurate in predicting both bending and extension responses. Park, Walz and Chopra[7] formulated two additional shear lag models. The first investigated coupled extension and bending of a rectangular beam with a single piezoceramic bonded to the beam surface and aligned with the beam axis. The second model permitted an arbitrary orientation of the piezoceramic with respect to the beam axis to predict coupled extension, bending and torsion. Both models utilized a Newtonian shear lag formulation in which the strain was assumed to be constant through the thickness of the actuator and linear through the beam. To maintain a one-dimensional (1-D) characterization of the problem, the actuator was considered a line element and only permitted to induce strain in its lengthwise axis. Both models were correlated with static deflection data and the bending-extension model was further compared to the Bernoulli-Euler model presented by Crawley and Anderson[4]. The extension-bending model was found to adequately predict the experimental results for various geometric parameters. The coupled extension, bending and torsion model, however, was only capable of capturing experimental trends, not magnitudes, especially at high actuator orientation angles relative to the beam axis. Thus, structural modeling of beam extension and bending with embedded and surface mounted actuators has been extensively examined for actuators aligned with the beam axis, however adequate torsion models are not available.

This paper further explores formulation of 1-D torsion models by investigating warping functions, shear lag and experimental strain and deflection data. First, an analytical model is developed with warping considerations but without shear lag effects to illuminate the influence of warping terms on the response. Second, shear lag is introduced to attenuate bending, extension and torsion responses. Configurations with actuators only on one surface (asymmetric) and actuators on both surfaces (symmetric) are developed using both Bernoulli-Euler and uniform strain displacement assumptions for the actuator. Finally, issues associated with constraining the global system response to a 1-D framework are explored and further illustrated by experimental strain and deflection surveys, which also serve to validate the analytical models.

Analytical Models

The Principle of Virtual Work is invoked to construct all of the models presented within this paper. Cross-sectional warping and shear lag effects are incorporated into the models and evaluated relative to those which do not capture such behavior. For illustrative purposes, detailed results are derived for a thin isotropic beam with a surface bonded piezoceramic actuator as shown in Figure 1. All expressions are derived in the absence of external forces and temperature gradients.

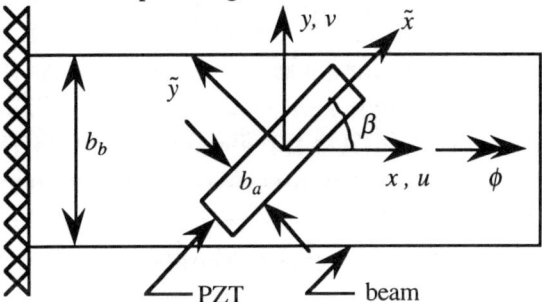

Figure 1 Beam with surface bonded PZT

Bernoulli-Euler Torsion Model

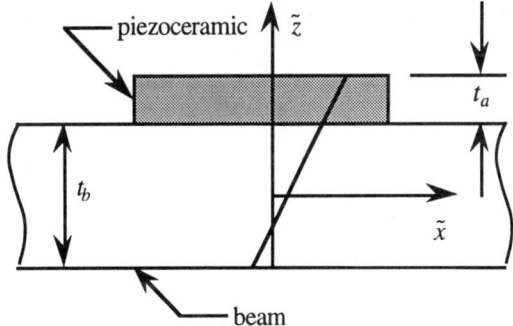

Figure 2 Bernoulli-Euler axial deflections

This model assumes that the structure behaves as a Bernoulli-Euler beam in bending and extension as shown in Figure 2. Torsion is introduced by utilizing a cross-sectional warping function, λ, as suggested by Gjelsvik[8], which is dependent upon the cross-sectional geometry of the structure. Adhesive shear lag effects are not included in this model. The assumed displacement field for the structure is given as

$$u = u_o - zw,_x - \lambda\phi,_x \quad (1a)$$
$$v = -z\phi \quad (1b)$$
$$w = w_o + y\phi \quad (1c)$$

where the warping function $\lambda = \lambda(y,z)$. The non-zero beam strains are

$$\varepsilon_{xx} = u_{o,x} - zw,_{xx} - \lambda\phi,_{xx} \quad (2a)$$
$$\gamma_{zx} = (y - \lambda,_z)\phi,_x \quad (2b)$$
$$\gamma_{xy} = -(z + \lambda,_y)\phi,_x \quad (2c)$$

Assuming no external forces are applied, the Principle of Virtual work reduces to

$$\int_V \{\sigma_{xx}\delta\varepsilon_{xx} + \tau_{zx}\delta\gamma_{zx} + \tau_{xy}\delta\gamma_{xy}\}dV = 0 \quad (3)$$

Further assuming that the actuators are of high aspect ratio, the piezoceramic is approximated as only inducing a strain in its longitudinal axis. The induced strain is thereby transformed to beam axes as

$$\Lambda_{xx} = \Lambda\cos^2\beta \quad \Lambda_{xy} = \Lambda\sin\beta\cos\beta \quad (4)$$

Utilizing the constitutive relationships

$$\sigma_{xx} = E(\varepsilon_{xx} - \Lambda\cos^2\beta) \quad (5a)$$
$$\tau_{zx} = G\gamma_{zx} \quad (5b)$$
$$\tau_{xy} = G(\gamma_{xy} - \Lambda\sin\beta\cos\beta) \quad (5c)$$

and separating the volume integral by sub-components provides the expression

$$\int_{beam}\{E_b\varepsilon_{xx}^b\delta\varepsilon_{xx}^b + G_b\gamma_{zx}^b\delta\gamma_{zx}^b + G_b\gamma_{xy}^b\delta\gamma_{xy}^b\}dV$$
$$+ \int_{actuator}\{E_a(\varepsilon_{xx}^a - \Lambda\cos^2\beta)\delta\varepsilon_{xx}^a + G_a\gamma_{zx}^a\delta\gamma_{zx}^a$$
$$+ G_a(\gamma_{xy}^a - \Lambda\sin\beta\cos\beta)\delta\gamma_{xy}^a\}dV = 0 \quad (6)$$

Substitution of the strain-displacement equations (2) into (6) and integrating over the cross-section gives

$$\int_L\{E_b(A_b u_{o,x} - S_y^b w,_{xx} - B_\lambda^b\phi,_{xx})\delta u_{o,x}$$
$$+ E_b(-S_y^b u_{o,x} + I_{yy}^b w,_{xx} + C_\lambda^b\phi,_{xx})\delta w,_{xx}$$
$$+ E_b(-B_\lambda^b u_{o,x} + C_\lambda^b w,_{xx} + I_{\lambda\lambda}^b\phi,_{xx})\delta\phi,_{xx}$$
$$+ G_b J_b\phi,_x \delta\phi,_x\}dx$$
$$+ \int_L\{E_a(A_a u_{o,x} - S_y^a w,_{xx} - B_\lambda^a\phi,_{xx})\delta u_{o,x}$$
$$+ E_a(-S_y^a u_{o,x} + I_{yy}^a w,_{xx} + C_\lambda^a\phi,_{xx})\delta w,_{xx}$$
$$+ E_a(-B_\lambda^a u_{o,x} + C_\lambda^a w,_{xx} + I_{\lambda\lambda}^a\phi,_{xx})\delta\phi,_{xx}$$
$$+ G_a J_a\phi,_x \delta\phi,_x\}dx$$
$$- \int_L\{E_a A_a\Lambda\cos^2\beta\delta u_{o,x} - E_a S_y^a\Lambda\cos^2\beta\delta w,_{xx}$$
$$- E_a B_\lambda^a\Lambda\cos^2\beta\delta\phi,_{xx}$$
$$- G_a\Lambda(S_y^a + L_\lambda^a)\sin\beta\cos\beta\delta\phi,_x\}dx = 0 \quad (7)$$

where

$$A = \iint dydz \quad (8a)$$
$$S_y = \iint z\, dydz \quad (8b)$$
$$I_{yy} = \iint z^2 dydz \quad (8c)$$
$$J = \iint\left[(y - \lambda,_z)^2 + (z + \lambda,_y)^2\right]dydz \quad (8d)$$
$$B_\lambda = \iint \lambda\, dydz \quad (8e)$$
$$C_\lambda = \iint z\lambda\, dydz \quad (8f)$$
$$I_{\lambda\lambda} = \iint \lambda^2 dydz \quad (8g)$$
$$L_\lambda = \iint \lambda,_y dydz \quad (8h)$$

For a rectangular beam with a piezoceramic bonded to one surface at an orientation angle β, the warping function, shown in Figure 3 is approximated as

$$\lambda = kyz \quad (9)$$

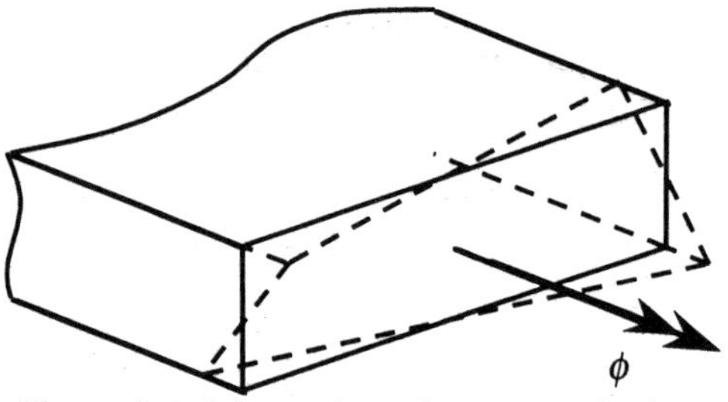

Figure 3 Axial warping of a rectangular bar

The displacement warping formulation (9) exactly recovers the stress function presented by Timoshenko[9] for a thin rectangular bar. For an isotropic beam, $k=1$ and the virtual work equation (7) becomes

$$\int_L\{E_b A_b u_{o,x}\delta u_{o,x} + E_b I_{yy}^b w,_{xx}\delta w,_{xx}$$
$$+ E_b I_{\lambda\lambda}^b\phi,_{xx}\delta\phi,_{xx} + G_b J_b\phi,_x\delta\phi,_x\}dx$$
$$+ \int_L\{E_a[A_a^*(u_{o,x} - \bar{z}_a w,_{xx} - x\bar{z}_a\tan\beta\phi,_{xx})\delta u_{o,x}$$
$$+ I_a^*\left(-\frac{\bar{z}_a A_a^*}{I_a^*}u_{o,x} + w,_{xx} + x\tan\beta\phi,_{xx}\right)\delta w,_{xx}$$
$$+ (-x\bar{z}_a A_a^*\tan\beta u_{o,x} + xI_a^*\tan\beta w,_{xx}$$
$$+ I_{\lambda\lambda}^a\phi,_{xx})\delta\phi,_{xx}] + G_a J_a^*\phi,_x\delta\phi,_x\}dx$$
$$- \int_L\{E_a A_a^*\Lambda\cos^2\beta\delta u,_x - \bar{z}_a E_a A_a^*\Lambda\cos^2\beta\delta w,_{xx}$$
$$- x\bar{z}_a E_a A_a^*\Lambda\sin\beta\cos\beta\delta\phi,_{xx}$$
$$- 2\bar{z}_a G_a A_a^*\Lambda\sin\beta\cos\beta\delta\phi,_x\}dx = 0 \quad (10)$$

where

$$A_a^* = \frac{b_a t_a}{\cos\beta} \tag{11a}$$

$$I_a^* = \frac{1}{\cos\beta}\left(\frac{b_a t_a^3}{12} + \bar{z}_a^2 b_a t_a\right) \tag{11b}$$

$$I_{\lambda\lambda}^a = \left(x^2 \tan^2\beta + \frac{b_a^2}{12\cos^2\beta}\right) I_{yy}^a \tag{11c}$$

$$J_a^* = 4 I_a^* \tag{11d}$$

$$A_b = b_b t_b \tag{11e}$$

$$I_{yy}^b = \frac{1}{12} b_a t_a^3 \tag{11f}$$

$$I_{\lambda\lambda}^b = \frac{1}{144} A_b^3 \tag{11g}$$

$$J_b = \frac{1}{3} b_b t_b^3 \tag{11h}$$

Integrating by parts provides the force boundary conditions and governing partial differential equations, which must be solved numerically. If the warping terms are ignored except to correct the shear strain term and the substitution of variables

$$\varepsilon_o = u_{,x} \qquad \kappa = w_{,xx} \qquad \theta = \phi_{,x} \tag{12}$$

is introduced, the system may be solved directly from

$$\begin{bmatrix} EA & ES & 0 \\ ES & EI & 0 \\ 0 & 0 & GJ \end{bmatrix} \begin{Bmatrix} \varepsilon_o \\ \kappa \\ \theta \end{Bmatrix} = \begin{Bmatrix} P \\ M \\ T \end{Bmatrix} \tag{13}$$

where

$$EA = \iint_A E\,dy\,dz = E_b A_b + E_a A_a^* \tag{14a}$$

$$ES = -\iint_A Ez\,dy\,dz = -\bar{z}_a E_a A_a^* \tag{14b}$$

$$EI = -\iint_A Ez^2\,dy\,dz = E_b I_b + E_a I_a^* \tag{14c}$$

$$GJ = \iint_A G\left[(y - \lambda_{,z})^2 + (z + \lambda_{,y})^2\right] dy\,dz$$
$$= G_b J_b + G_a J_a^* \tag{14d}$$

$$P = \iint_A \Lambda E\,dy\,dz = E_a A_a^* \Lambda \cos^2\beta \tag{14e}$$

$$M = -\iint_A \Lambda Ez\,dy\,dz = -\bar{z}_a E_a A_a^* \Lambda \cos^2\beta \tag{14f}$$

$$T = \iint_A \Lambda G\left[(y - \lambda_{,z}) - (z + \lambda_{,y})\right] dy\,dz$$
$$= -2\bar{z}_a G_a A_a^* \Lambda \sin\beta\cos\beta \tag{14g}$$

The governing equations with warping terms are solved using the finite element method for 1, 10 and 50 elements. The solution exhibits good convergence for a low number of elements as seen in Figure 4. The torsional response has essentially converged for a single element, yet the bending response requires 10 elements. The basic beam element utilized is shown in Figure 5 and the assumed displacements are given as a function of the non-dimensional elemental coordinate, s, as

$$u(s) = (1-s)u_1 + s u_3 \tag{15a}$$

$$w(s) = (2s^3 - 3s^2 + 1)w_1 + l(s^3 - 2s^2 + s)w_1'$$
$$+ (-2s^3 + 3s^2)w_3 + l(s^3 - s^2)w_3' \tag{15b}$$

$$\phi(s) = (2s^2 - 3s + 1)\phi_1 + (-4s^2 + 4s)\phi_2 + (2s^2 - s)\phi_3 \tag{15c}$$

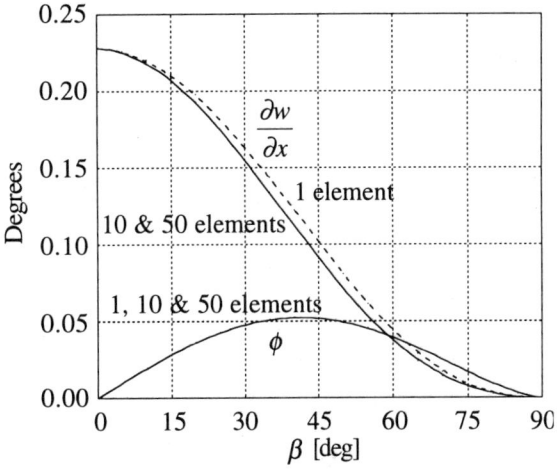

Figure 4 Solution convergence

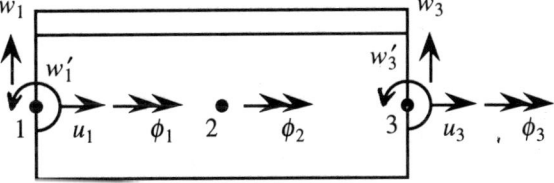

Figure 5 Finite element

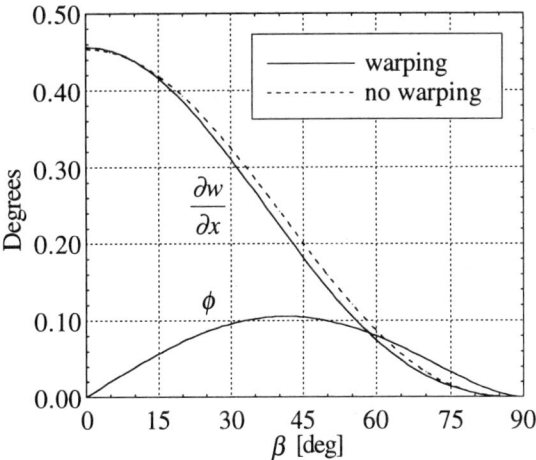

Figure 6 Effects of warping terms

The effects of the warping terms which result in extension-twist and bending-twist elastic couplings are shown in Figure 6 for a $2L^* \times 2.00 \times 0.0305$ inch aluminum beam with two asymmetrically bonded 1.85 x 0.50 x 0.0093 inch G-1195 piezoceramics subjected to a -175 με actuation strain. L^* is the characteristic actuator half-length given by the projection of half of the actuator length onto the beam axis. For this case of a thin rectangular isotropic bar, the warping effects are negligible.

Shear Lag Model

In formulating a shear lag model, the configuration of an isotropic beam with a single surface bonded piezoceramic, such as previously shown in Figure 1, is considered. Since structural couplings due to warping terms are negligible for this type of structure, cross-sectional warping is only considered in as much as to correct the actuator shear strain. The effects of a finite thickness bond layer are considered by allowing the adhesive substrate to react shear loads. Extensional displacements are shown in Figure 7. The displacement field is assumed to be

$$u_b = u_b^o - zw_{b,x}^o - yz\phi_{b,x} \quad (16a)$$
$$v_b = -z\phi_b \quad (16b)$$
$$w_b = w_b^o + y\phi_b \quad (16c)$$
$$u_a = u_a^o - (z-\bar{z}_a)w_{b,x}^o - yz\phi_{a,x} \quad (17a)$$
$$v_a = -z\phi_a \quad (17b)$$
$$w_a = w_b^o + y\phi_a \quad (17c)$$

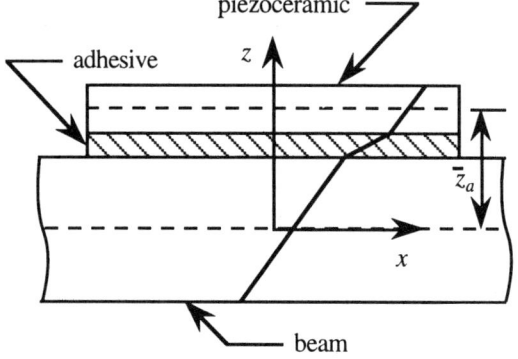

Figure 7 Axial displacement variation

where $\bar{z}_a$ is the location of the actuator mid-plane with reference to the beam mid-plane. Neglecting warping terms in the beam and actuator extensional strains, the non-zero strains become

$$\varepsilon_{xx}^b = u_{b,x}^o - zw_{b,xx}^o \quad (18a)$$
$$\gamma_{xy}^b = -2z\phi_{b,x} \quad (18b)$$
$$\varepsilon_{xx}^a = u_{a,x}^o - (z-\bar{z}_a)w_{b,xx}^o \quad (18c)$$
$$\gamma_{xy}^a = -2z\phi_{a,x} \quad (18d)$$

The adhesive shear strains are related to the differential extension and twist of the beam and actuator. The displacements through the adhesive thickness are linear interpolations between the actuator and beam displacements at the adhesive substrate interfaces. The shear due to bending and extension deformation is

$$\gamma_{zx}^s = \frac{1}{t_s}\left[u_a^o - u_b^o + \frac{t_a+t_b}{2}w_{b,x}^o\right] \quad (19)$$

The torsional shear is assumed to be due to a similar difference in transverse displacement except that both the actuator and beam only undergo rigid body translations as given by the displacement field. Hence, the torsional adhesive shear strain is given by

$$\gamma_{yz}^s = \frac{1}{t_s}\left[\left(t_s+\frac{t_b}{2}\right)\phi_a - \frac{t_b}{2}\phi_b\right] \quad (20)$$

Invoking the Principle of Virtual Work gives

$$\int_{beam}\{E_b\varepsilon_{xx}^b\delta\varepsilon_{xx}^b + G_b\gamma_{xy}^b\delta\gamma_{xy}^b\}dV$$
$$+ \int_{piezo}\{E_a\varepsilon_{xx}^a\delta\varepsilon_{xx}^a + G_a\gamma_{xy}^a\delta\gamma_{xy}^a\}dV$$
$$+ \int_{adhesive}\{G_s\gamma_{yz}^s\delta\gamma_{yz}^s + G_s\gamma_{zx}^s\delta\gamma_{zx}^s\}dV$$
$$= \int_{piezo}\{E_a\Lambda_{xx}^a\delta\varepsilon_{xx}^a + G_a\Lambda_{xy}^a\delta\gamma_{xy}^a\}dV \quad (21)$$

where the induced strains are

$$\Lambda_{xx}^a = \Lambda\cos^2\beta \qquad \Lambda_{xy}^a = \Lambda\sin\beta\cos\beta \quad (22)$$

Introducing the strain-displacements (16) and (17) into (21) and integrating over the volume provides the governing differential equations (23) through (27). Note that the y-axis limits of integration are $y = x\tan\beta \pm b_a/2\cos\beta$ and begin to break down at high orientation angles.

$$E_aA_a^*u_{a,xx}^o - \frac{G_sA_s^*}{t_s^2}\left[u_a^o - u_b^o + \frac{t_a+t_b}{2}w_{b,x}^o\right] = 0 \quad (23)$$

$$E_bA_bu_{b,xx}^o + \frac{G_sA_s^*}{t_s^2}\left[u_a^o - u_b^o + \frac{t_a+t_b}{2}w_{b,x}^o\right] = 0 \quad (24)$$

$$(E_bI_b + E_aI_a^*)w_{b,xxxx}^o$$
$$- \frac{G_sA_s^*}{t_s^2}\frac{t_a+t_b}{2}\left[u_{a,x}^o - u_{b,x}^o + \frac{t_a+t_b}{2}w_{,xx}\right] = 0 \quad (25)$$

$$G_aJ_a^*\phi_{a,xx} - \frac{G_aA_s^*}{t_s^2}\left(t_s+\frac{t_b}{2}\right)\left[\left(t_s+\frac{t_b}{2}\right)\phi_a - \frac{t_b}{2}\phi_b\right] = 0 \quad (26)$$

$$G_bJ_b\phi_{a,xx} + \frac{G_aA_s^*}{t_s^2}\frac{t_b}{2}\left[\left(t_s+\frac{t_b}{2}\right)\phi_a - \frac{t_b}{2}\phi_b\right] = 0 \quad (27)$$

and force boundary conditions at $x = \pm\frac{l}{2}\cos\beta$

$$E_aA_a^*u_{a,x}^o = E_aA_a^*\Lambda\cos^2\beta \quad (28)$$
$$E_bA_bu_{b,x}^o = 0 \quad (29)$$
$$(E_bI_b + E_aI_a^*)w_{,xx} = 0 \quad (30)$$

$$(E_b I_b + E_a I_a^*) w_{,xxx} - \frac{G_s A_s^*}{t_s^2} \frac{t_a+t_b}{2}\left[u_a^o - u_b^o + \frac{t_a+t_b}{2} w_{,x}\right] = 0 \quad (31)$$

$$G_a J_a^* \phi_{a,x} = -2\bar{z}_a G_a A_a^* \Lambda \sin\beta \cos\beta \quad (32)$$

$$G_b J_b \phi_{b,x} = 0 \quad (33)$$

where

$$A_a^* = \frac{b_a t_a}{\cos\beta} \qquad J_a^* = \frac{1}{\cos\beta}\left[\frac{1}{3} b_a t_a^3 + 4\bar{z}_a^2 b_a t_a\right]$$

$$A_b = b_b t_b \qquad I_b = \frac{1}{12} b_b t_b^3 \qquad J_b = \frac{1}{3} b_b t_b^3 \quad (34)$$

$$A_s^* = \frac{b_s t_s}{\cos\beta} \qquad \bar{z}_a = t_s + \frac{t_a+t_b}{2}$$

Note that the torsion equations are not coupled with the bending and extension equations.

Torsional Solution

Rewriting equations (26) and (27) the governing equations become

$$\phi_{a,xx} - \frac{G_s A_s}{t_s^2 G_a J_a}\left(t_s + \frac{t_b}{2}\right)\left[\left(t_s + \frac{t_b}{2}\right)\phi_a - \frac{t_b}{2}\phi_b\right] = 0 \quad (35)$$

$$\phi_{b,xx} + \frac{G_s A_s}{t_s^2 G_b J_b \cos\beta} \frac{t_b}{2}\left[\left(t_s + \frac{t_b}{2}\right)\phi_a - \frac{t_b}{2}\phi_b\right] = 0 \quad (36)$$

Increasing the order, the characteristic equation is

$$D^2\left[D^2 - \frac{G_s A_s}{t_s^2 \cos\beta}\left(\frac{\cos\beta}{G_a J_a}\left(t_s + \frac{t_b}{2}\right)^2 + \frac{1}{G_b J_b}\frac{t_b^2}{4}\right)\right] = 0 \quad (37)$$

The solutions are, therefore, of the form

$$\phi_a = a_0 + a_1 \bar{x} + a_2 \cosh\Gamma_\phi \bar{x} + a_3 \sinh\Gamma_\phi \bar{x} \quad (38)$$

$$\phi_b = b_0 + b_1 \bar{x} + b_2 \cosh\Gamma_\phi \bar{x} + b_3 \sinh\Gamma_\phi \bar{x} \quad (39)$$

where the torsional shear lag parameter, Γ_ϕ, and the characteristic length, L^*, are defined as

$$\Gamma_\phi^2 = (L^*)^2 \frac{G_s A_s^*}{t_s^2}\left[\frac{1}{G_a J_a^*}\left(t_s + \frac{t_b}{2}\right)^2 + \frac{1}{G_b J_b}\frac{t_b^2}{4}\right] \quad (40)$$

$$L^* = \frac{l}{2}\cos\beta \quad (41)$$

Since the original equations are second order, only 4 of the 8 coefficients are independent. Substitution of the solutions into either governing equation provides the dependency relations. Utilizing the force boundary conditions (32) and (33) and the cantilevered geometric boundary condition

$$\phi_b(\bar{x}=-1) = 0 \quad (42)$$

the final solutions are found to be

$$\phi_a = -\frac{TL^*}{GJ^*}\left[\frac{t_b}{2t_s+t_b}(\bar{x}+1) + \frac{2t_s+t_b}{t_b}\frac{G_b J_b}{G_a J_a^*}\left(\frac{\sinh\Gamma_\phi \bar{x}}{\Gamma_\phi \cosh\Gamma_\phi} + \frac{\tanh\Gamma_\phi}{\Gamma_\phi}\right)\right] \quad (43)$$

$$\phi_b = -\frac{TL^*}{GJ^*}\left[\bar{x}+1 - \frac{\sinh\Gamma_\phi \bar{x}}{\Gamma_\phi \cosh\Gamma_\phi} - \frac{\tanh\Gamma_\phi}{\Gamma_\phi}\right] \quad (44)$$

$$GJ^* = \frac{t_b}{2t_s+t_b}G_a J_a^* + \frac{2t_s+t_b}{t_b}G_b J_b \quad (45)$$

$$T = -2\bar{z}_a G_a A_a^* \Lambda \sin\beta \cos\beta \quad (46)$$

Unlike warping, the effects of the shear layer on a rectangular beam may be significant as shown in Figure 8 for a 2.00 x 0.0305 inch aluminum beam with a single 1.85 x 0.50 x 0.0093 inch G-1195 piezoceramics bonded to one surface. Note that the maximum twist occurs when the actuator orientation is near 45°.

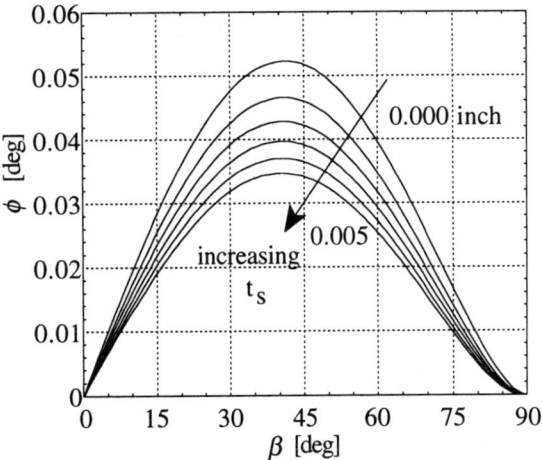

Figure 8 Shear lag effects on twist response

Extension-Bending Solution

Differentiating equations (23) - (25) with respect to x and making the change of variables

$$\varepsilon_a^o = u_{a,x}^o \qquad \varepsilon_b^o = u_{b,x}^o \qquad \kappa = w_{,xx} \quad (47)$$

the system is represented in terms of the actuator mid-plane strain, ε_a^o, beam mid-plane strain, ε_b^o, and bending curvature, κ.

$$\varepsilon_{a,xx}^o - \frac{G_s A_s^*}{t_s^2 E_a A_a^*}\left[\varepsilon_a^o - \varepsilon_b^o + \frac{t_a+t_b}{2}\kappa\right] = 0 \quad (48)$$

$$\varepsilon_{b,xx}^o + \frac{G_s A_s^*}{t_s^2 E_b A_b}\left[\varepsilon_a^o - \varepsilon_b^o + \frac{t_a+t_b}{2}\kappa\right] = 0 \quad (49)$$

$$\kappa_{,xx} - \frac{G_s A_s^*}{t_s^2(E_b I_b + E_a I_a^*)}\frac{t_a+t_b}{2}\left[\varepsilon_a^o - \varepsilon_b^o + \frac{t_a+t_b}{2}\kappa\right] = 0 \quad (50)$$

Solving the transformed system subject to the boundary conditions

$$\varepsilon_a^o\left(\pm\frac{l}{2}\cos\beta\right) = \Lambda \cos^2\beta \quad (51)$$

$$\varepsilon_b^o\left(\pm\frac{l}{2}\cos\beta\right) = 0 \quad (52)$$

$$\kappa\left(\pm\frac{l}{2}\cos\beta\right)=0 \quad (53)$$

provides the solutions

$$\varepsilon_a^o(\bar{x}) = \Lambda\cos^2\beta\left[\psi_a\frac{\cosh\Gamma_u\bar{x}}{\cosh\Gamma_u}+\psi_b+\psi_\kappa\right] \quad (54)$$

$$\varepsilon_b^o(\bar{x}) = \Lambda\psi_b\cos^2\beta\left[1-\frac{\cosh\Gamma_u\bar{x}}{\cosh\Gamma_u}\right] \quad (55)$$

$$\kappa(\bar{x}) = \Lambda\psi_\kappa\frac{2}{t_a+t_b}\cos^2\beta\left[\frac{\cosh\Gamma_u\bar{x}}{\cosh\Gamma_u}-1\right] \quad (56)$$

where the stiffness ratios and axial shear lag parameter are defined as

$$\psi_a = \frac{1}{E_a A_a^*}\left[\frac{1}{E_a A_a^*}+\frac{1}{E_b A_b}+\frac{1}{4}\frac{(t_a+t_b)^2}{E_b I_b+E_a I_a^*}\right]^{-1} \quad (57a)$$

$$\psi_b = \frac{1}{E_b A_b}\left[\frac{1}{E_a A_a^*}+\frac{1}{E_b A_b}+\frac{1}{4}\frac{(t_a+t_b)^2}{E_b I_b+E_a I_a^*}\right]^{-1} \quad (57b)$$

$$\psi_\kappa = \frac{1}{4}\frac{(t_a+t_b)^2}{E_b I_b+E_a I_a^*}\left[\frac{1}{E_a A_a^*}+\frac{1}{E_b A_b}+\frac{1}{4}\frac{(t_a+t_b)^2}{E_b I_b+E_a I_a^*}\right]^{-1} \quad (57c)$$

$$\Gamma_u^2 = (L^*)^2\frac{G_s A_s^*}{t_s^2}\left[\frac{1}{E_a A_a^*}+\frac{1}{E_b A_b}+\frac{1}{4}\frac{(t_a+t_b)^2}{E_b I_b+E_a I_a^*}\right] \quad (57d)$$

$$L^* = \frac{l}{2}\cos\beta \quad (57e)$$

Integrating the beam curvature with respect to $\bar{x}$ and assuming the beam is cantilevered at $\bar{x}=-1$

$$w_{,x}(\bar{x}) = \Lambda\psi_\kappa L^*\frac{2}{t_a+t_b}\cos^2\beta\left[\frac{\sinh\Gamma_u\bar{x}}{\Gamma_u\cosh\Gamma_u}\right.$$
$$\left.+\frac{\tanh\Gamma_u}{\Gamma_u}-1-\bar{x}\right] \quad (58)$$

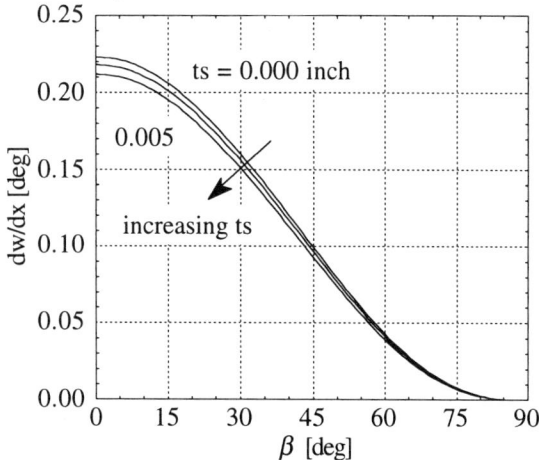

Figure 9 Shear lag effects on extension-bending

Unlike the torsional response, the effects of shear lag are much less on the bending slope as shown in Figure 9 for a 2.00 x 0.0305 inch aluminum beam with a single 1.85 x 0.50 x 0.0093 inch G-1195 piezoceramics bonded to one surface. Adhesive thicknesses are 0.000, 0.001 and 0.005 inches and the actuation strain is 175 µε. The impact of shear lag is less severe on the bending slope than on the twist angle.

If the strain through the actuator is assumed to uniform instead of linear, equations (23), (24) and (25) become

$$E_a A_a^* u_{a,xx}^o - \frac{G_s A_s^*}{t_s^2}\left[u_a^o - u_b^o + \frac{t_b}{2}w_{b,x}^o\right] = 0 \quad (59)$$

$$E_b A_b u_{b,xx}^o + \frac{G_s A_s^*}{t_s^2}\left[u_a^o - u_b^o + \frac{t_b}{2}w_{b,x}^o\right] = 0 \quad (60)$$

$$E_b I_b w_{b,xxxx}^o - \frac{G_s A_s^*}{t_s^2}\frac{t_b}{2}\left[u_{a,x}^o - u_{b,x}^o + \frac{t_b}{2}w_{b,xx}^o\right] = 0 \quad (61)$$

and the corresponding solutions are

$$\varepsilon_a^o(\bar{x}) = \Lambda\cos^2\beta\left[\psi_a\frac{\cosh\Gamma_u\bar{x}}{\cosh\Gamma_u}+\psi_b+\psi_\kappa\right] \quad (62)$$

$$\varepsilon_b^o(\bar{x}) = \Lambda\psi_b\cos^2\beta\left[1-\frac{\cosh\Gamma_u\bar{x}}{\cosh\Gamma_u}\right] \quad (63)$$

$$\kappa(\bar{x}) = -\Lambda\psi_\kappa\cos^2\beta\frac{2}{t_b}\left[1-\frac{\cosh\Gamma_u\bar{x}}{\cosh\Gamma_u}\right] \quad (64)$$

where the stiffness ratios and shear lag parameter are defined by

$$\psi_a = \frac{1}{E_a A_a^*}\left[\frac{1}{E_a A_a^*}+\frac{1}{E_b A_b}+\frac{t_b^2}{4}\frac{1}{E_b I_b}\right]^{-1} \quad (65a)$$

$$\psi_b = \frac{1}{E_b A_b}\left[\frac{1}{E_a A_a^*}+\frac{1}{E_b A_b}+\frac{t_b^2}{4}\frac{1}{E_b I_b}\right]^{-1} \quad (65b)$$

$$\psi_\kappa = \frac{t_b^2}{4}\frac{1}{E_b I_b}\left[\frac{1}{E_a A_a^*}+\frac{1}{E_b A_b}+\frac{t_b^2}{4}\frac{1}{E_b I_b}\right]^{-1} \quad (65c)$$

$$\Gamma_u^2 = (L^*)^2\frac{G_s A_s^*}{t_s^2}\left[\frac{1}{E_a A_a^*}+\frac{1}{E_b A_b}+\frac{t_b^2}{4}\frac{1}{E_b I_b}\right] \quad (66)$$

The torsional response is unaffected by this change in assumed displacement. The effects of a uniform versus linear actuator strain are shown in Figure 10 for two orientation angles for a 2.00 x 0.0305 inch aluminum beam with a surface bonded 1.85 x 0.50 x t_a inch G-1195. For $\beta=0°$, the difference between the two curvatures is greater than 5% for thickness ratios below 18 and greater than 10% below 7. The Bernoulli-Euler strain distribution predicts no bending as the beam thickness vanishes. This result is consistent with the behavior of real beams and the results of Crawley and Anderson[6]. Note that the curvature for the beam is greater for $\beta=0°$ than for $\beta=45°$. The reduction in curvature is attributed to attenuated actuation strain in the beam axis direction.

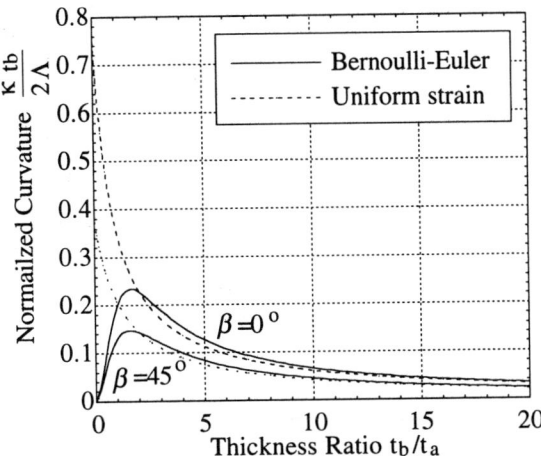

Figure 10 Effects of actuator bending

Symmetric Configuration

If an actuator is added to the bottom surface of the previous configuration and oriented at $-\beta$, then the structural response will be a coupled extension-twist motion provided that both actuators extend or both contract. Using the previous solution methodology, the torsional response is similar to that of equations (43) and (44).

$$\phi_a = -\frac{TL^*}{GJ^*}\left[\frac{t_b}{2t_s+t_b}(\bar{x}+1) \right.$$
$$\left. +\frac{2t_s+t_b}{t_b}\frac{G_bJ_b}{2G_aJ_a^*}\left(\frac{sinh\Gamma_\phi\bar{x}}{\Gamma_\phi cosh\Gamma_\phi}+\frac{tanh\Gamma_\phi}{\Gamma_\phi}\right)\right]$$
(67)

$$\phi_b = -\frac{TL^*}{GJ^*}\left[\bar{x}+1-\frac{sinh\Gamma_\phi\bar{x}}{\Gamma_\phi cosh\Gamma_\phi}-\frac{tanh\Gamma_\phi}{\Gamma_\phi}\right]$$
(68)

The axial response is derived as

$$u_a^o(\bar{x}) = \frac{PL^*}{EA^*}(\bar{x}+1)$$
$$+\frac{PL^*}{EA^*}\frac{E_bA_b}{2E_aA_a^*}\left(\frac{sinh\Gamma_u\bar{x}}{\Gamma_u cosh\Gamma_u}-\frac{tanh\Gamma_u}{\Gamma_u}\right)$$
(69)

$$u_b^o(\bar{x}) = \frac{PL^*}{EA^*}\left[\bar{x}+1-\frac{sinh\Gamma_u\bar{x}}{\Gamma_u cosh\Gamma_u}-\frac{tanh\Gamma_u}{\Gamma_u}\right]$$
(70)

$$\varepsilon_a^o(\bar{x}) = \frac{P}{EA^*}\left[1+\frac{E_bA_b cos\beta}{2E_aA_a}\frac{cosh\Gamma_u\bar{x}}{cosh\Gamma_u}\right]$$
(71)

$$\varepsilon_b^o(\bar{x}) = \frac{P}{EA^*}\left[1-\frac{cosh\Gamma_u\bar{x}}{cosh\Gamma_u}\right]$$
(72)

where

$$GJ^* = \frac{2t_b}{2t_s+t_b}G_aJ_a^* + \frac{2t_s+t_b}{t_b}G_bJ_b$$
(73)

$$EA^* = 2E_aA_a^* + E_bA_b$$
(74)

$$T = -4\bar{z}_aG_aA_a^*\Lambda sin\beta cos\beta$$
(75)

$$P = 2E_aA_a^*\Lambda cos^2\beta$$
(76)

$$\Gamma_u^2 = (L^*)^2\frac{G_sA_s^*}{t_s^2}\left[\frac{1}{2E_aA_a^*}+\frac{1}{E_bA_b}\right]$$
(77)

$$\Gamma_\phi^2 = (L^*)^2\frac{G_sA_s^*}{t_s^2}\left[\frac{1}{2G_aJ_a^*}\left(t_s+\frac{t_b}{2}\right)^2+\frac{1}{G_bJ_b}\frac{t_b^2}{4}\right]$$
(78)

The twist rate for a symmetric specimen is not quite twice that of an asymmetric one as shown in Figure 11 for a 2.00 x 0.0305 inch aluminum beam with three and six 1.85 x 0.25 x 0.0093 inch G-1195 piezoceramics for the asymmetric and symmetric cases, respectively. The induced torque is exactly doubled, however the torsional stiffness exhibits an increase due to the additional piezoceramic stiffness. In the limit of a thick beam with thin piezoceramics, the symmetric twist response approaches twice the asymmetric twist.

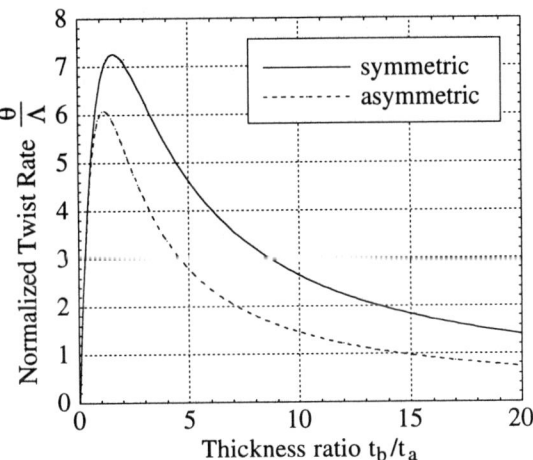

Figure 11 Twist rate comparison

1-D Platform Constraints

The global response of a high aspect ratio structure, such as a rotor blade, is adequately characterized by beam behavior. In addition to mathematical simplicity and physical understanding, modeling induced torsion of beams is preferably constrained to a 1-D response. Since the actuation mechanism is inherently a 2-D phenomenon, means by which to incorporate these 2-D piezoceramic effects into a global 1-D response model are not established, especially for arbitrary actuator geometries. In the limiting case of high aspect ratio actuators with simple geometry, the actuation mechanics appreciably simplify. The limitations of such reductions are difficult to quantify analytically, thereby requiring experimental methods to identify limits of applicability.

There exist several approaches by which to capture piezoceramic actuation in a beam model, each of which has different strengths and weaknesses. The models presented in this paper assume that the piezoceramic only induces strain in its longitudinal axis, which is oriented an angle β with respect to the beam axis. Effectively, this is an assumption on material properties and conflicts with the basic physics of torsion actuation of a beam. Intuitively, torsion should arise due to geometric constraints since the considered piezoceramics exhibit isotropic in-plane characteristics. However, in the limit of high aspect ratio actuators, the utilized modification of material properties is a reasonable compromise for detailed geometric modeling.

The induced force and torque in the presented models are metrics by which to evaluate their effectiveness to actuator orientation. By assuming line element behavior of the piezoceramic, as done in this paper, the resulting force and torque expressions for the asymmetric configuration are

$$P = E_a A_a^* \Lambda \cos^2\beta \tag{79}$$

$$T = -2\bar{z}_a G_a A_a^* \Lambda \sin\beta \cos\beta \tag{80}$$

If, instead, the piezoceramic is permitted full 2-D isotropic material properties, the stress-strain relations are given by

$$\begin{Bmatrix} \sigma_{\bar{x}\bar{x}}^a \\ \sigma_{\bar{y}\bar{y}}^a \\ \tau_{\bar{x}\bar{y}}^a \end{Bmatrix} = \frac{E_a}{1-\nu_a^2} \begin{bmatrix} 1 & \nu_a & 0 \\ \nu_a & 1 & 0 \\ 0 & 0 & \frac{1-\nu_a}{2} \end{bmatrix} \begin{Bmatrix} \varepsilon_{\bar{x}\bar{x}}^a - \Lambda \\ \varepsilon_{\bar{y}\bar{y}}^a - \Lambda \\ \gamma_{\bar{x}\bar{y}}^a \end{Bmatrix} \tag{81}$$

and the internal virtual work becomes

$$\delta U_a = \int_V \frac{E_a}{1-\nu_a^2} \left\{ \left[\varepsilon_{\bar{x}\bar{x}}^a + \nu_a \varepsilon_{\bar{y}\bar{y}}^a - (1+\nu_a)\Lambda \right] \delta\varepsilon_{\bar{x}\bar{x}}^a \right.$$
$$+ \left[\nu_a \varepsilon_{\bar{x}\bar{x}}^a + \varepsilon_{\bar{y}\bar{y}}^a - (1+\nu_a)\Lambda \right] \delta\varepsilon_{\bar{y}\bar{y}}^a$$
$$\left. + \frac{1-\nu_a}{2} \gamma_{\bar{x}\bar{y}}^a \delta\gamma_{\bar{x}\bar{y}}^a \right\} dV \tag{82}$$

Since the strain transverse to the beam axis is identically zero as a result of beam behavior, the strain transformation becomes

$$\begin{Bmatrix} \varepsilon_{\bar{x}\bar{x}}^a \\ \varepsilon_{\bar{y}\bar{y}}^a \\ \gamma_{\bar{x}\bar{y}}^a \end{Bmatrix} = \begin{bmatrix} \cos^2\beta & 2\sin\beta\cos\beta \\ \sin^2\beta & -2\sin\beta\cos\beta \\ -\sin\beta\cos\beta & \cos^2\beta - \sin^2\beta \end{bmatrix} \begin{Bmatrix} \varepsilon_{xx}^a \\ \gamma_{xy}^a \end{Bmatrix} \tag{83}$$

Substituting (83) into the energy expression (82) and regrouping terms produces extension twist coupling which is proportional to $(\cos^2\beta - \sin^2\beta)$. Note that this coupling vanishes exactly when $\beta=45°$. The resulting force and torque are obtained as

$$P = \Lambda \frac{E_a A_a^*}{1-\nu_a} \tag{84}$$

$$T = 0 \tag{85}$$

Even though structural couplings between torsional and bending and extension responses due to the presence of a discrete actuator should be a maximum when $\beta=45°$, no torsion arises. Examining axial force equations (79) and (84), the underlying piezoceramic assumption effects become clear in the limiting configurations of $\beta=0°$ and $\beta=90°$. When $\beta=0°$, the two forces are identical with the exception of the Poisson effect in (84) which is due to plate theory. Hence, both assumptions will result in nearly the same bending response. When $\beta=90°$, the assumption which modifies material properties produces exactly no response whereas the one which utilizes isotropic properties generates the force which both assumptions would produce if the actuator dimensions were reversed and β was set to zero.

If transverse actuator stiffness is neglected, such that the actuator has orthotropic characteristics, the axial force and torque are

$$P = E_a A_a^* \Lambda \cos^2\beta \tag{86}$$

$$T = 4\bar{z}_a E_a A_a^* \Lambda \cos\beta \sin\beta \tag{87}$$

Note that the axial force (86) is identical to (79), yet the induced torque is much higher and intuitively incorrect. This indicates that transverse actuator stiffness is relevant to obtaining sensible predictions.

The importance of including transverse actuator mechanics is clear in terms of the implications on the bending response, yet less so in terms of torsion. Transverse stiffness is essential in obtaining a correct induced torque and 2-D piezoceramic properties are needed to correct the bending response at high orientation angles, $\beta>45°$. Means by which such mechanics can be adequately incorporated into a 1-D framework is not developed. Hence, the bending predictions of the models within this paper are expected to break down above $\beta=45°$. The impact of the underlying assumptions on the torsion response are not obvious and require experimental studies for further insight.

Experimental Results

Strain and deflection data were collected for a variety of specimen parameters to more fully understand the micro mechanics of the torsion problem as well as to provide correlation for the theoretical models. All of the specimens consisted of thin aluminum beams with surfaced bonded piezoceramics. All deflection data were obtained by reflecting a laser beam off the tip of the specimen onto a surface approximately 20 feet away. The vertical and horizontal displacements of the laser beam on this surface were measured and related to the angular rotations. Both G-1195 and PZT-5H piezoceramics were used in the experiments. Note that two separate lots of G-1195 piezoceramics exhibited different characteristics. The free strain of each type was individually characterized and approximated by a polynomial curve fit.

Tip twist angle and bending slope results were obtained for thin aluminum beam configurations with piezoceramics bonded to only one surface (asymmetric case) at an angle β with respect to the beam axis as shown in Figure 12. To obtain measurable results, multiple piezoceramics were required. Local interference effects were minimized by maintaining a 4 inch spacing between piezoceramics and a principle of superposition was assumed. The asymmetric specimens with a piezoceramic aspect ratio of 8 consisted of three 2.00 x 0.25 x 0.0075 inch G-1195(a) piezoceramics bonded to the same surface of a 2.00 x 0.0312 inch aluminum beams. The bond layer thickness was measured to be 0.0015 inches. The asymmetric specimens with 3.7 aspect ratio piezoceramics consisted of two 1.85 x 0.50 x 0.0093 inch G-1195(b) piezoceramics bonded to the same surface of 1.97 x 0.0305 inch aluminum beams. The measure adhesive thickness was 0.004 inches.

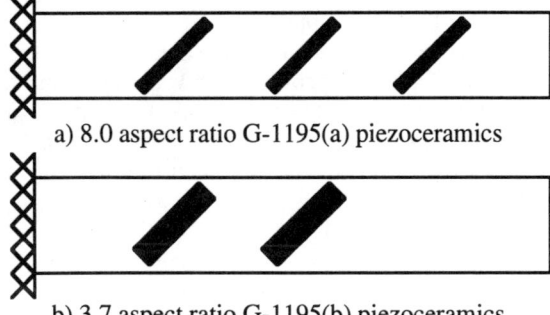

a) 8.0 aspect ratio G-1195(a) piezoceramics

b) 3.7 aspect ratio G-1195(b) piezoceramics

Figure 12 Test specimen configurations

The theoretical shear lag model predictions show good correlation with experimental results for the asymmetric case, shown in Figure 13, in torsional deflections. Bending slope predictions, however, rapidly deteriorate as the orientation angle increases, but are acceptable up to $\beta=45°$ for both sets of specimens. The model considered the actuators as 1-D elements. Hence, the bending slope and twist angle are both expected to approach zero at $\beta=90°$. Since the true bending slope is finite at that limit and the twist angle is zero, the twist predictions are expected to be much better over the range of orientation angles.

To more fully understand the micro mechanics of the beam response, strain gage measurements were taken from several asymmetric test specimens with G-1195 and PZT-5H piezoceramics. Both the piezoceramic and beam surface opposite the PZT were instrumented to obtain detailed strain distributions.

As seen in Figure 14 the lower beam surface axial strain increases towards the edge of the actuator in the lengthwise direction. However, Figure 15 shows that the strain exhibits relatively little decrease in the direction transverse to the actuator. The transverse strain, on the other hand, is approximately constant across the actuator width as shown in Figure 16. All strain distributions were taken from 2.00 x 0.0312 inch aluminum beams with a single 2.00 x 0.50 x 0.0093 inch PZT-5H piezoceramic with $\beta=45°$.

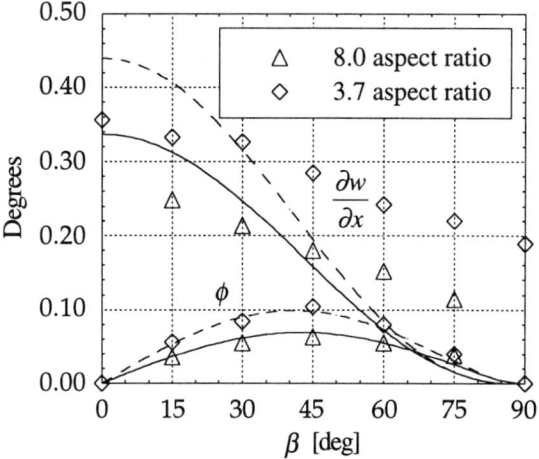

Figure 13 Asymmetric torsion results comparison

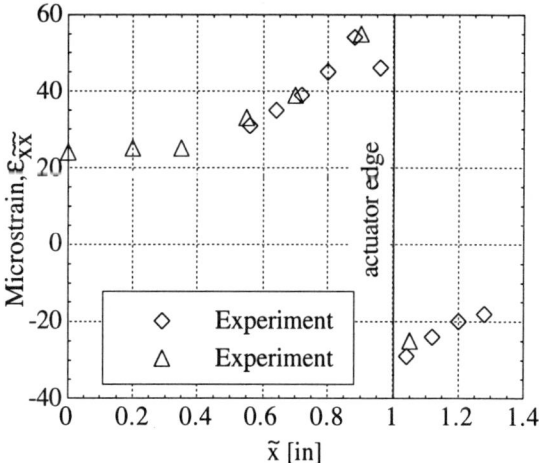

Figure 14 Experimental variation of $\varepsilon_{\tilde{x}\tilde{x}}$

Several shear type strain gages were located at the center of the piezoceramics, shown in Figure 17, on the previous asymmetric specimens with 3.7 PZT aspect ratios. Since the variation in piezoceramic thickness within a given set of specimens is ±0.1 mil and the deflection results exhibit a definable smooth trend, the 4 strain readings from a single specimen are deemed compatible for determining the total strain state for each specimen as if a rosette type gage were used. All measured strains were at 0°, 45°, 90° and 135° relative to the beam axis.

Experimental strain results are shown in Figure

18 for an actuated strain of -175 με. Fourth order polynomial curve fits are overlaid on the test data to illustrate the trends more clearly. As seen, the shear strain is a maximum when $\beta=45°$ and zero at the two limits. The beam extension and transverse strains are mirrored about $\beta=45°$. At the two extremes, the actuated and measured lateral strains are approximately equal, thereby indicating a small energy contribution. However, at 45°, the transverse and longitudinal strains are approximately equal thereby indicating a highly 2-D strain state and a breakdown of the 1-D piezoceramic modeling assumption.

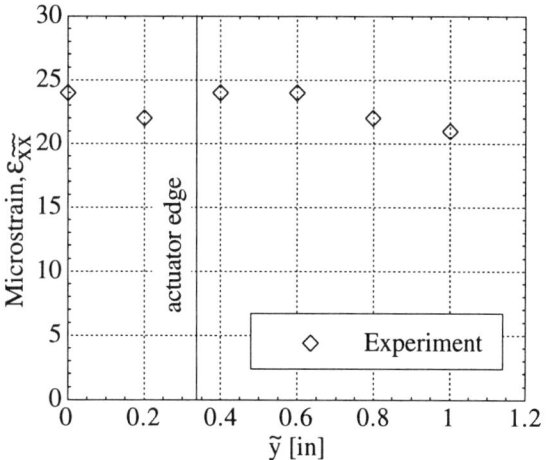

Figure 15 Experimental variation of $\varepsilon_{\tilde{x}\tilde{x}}$

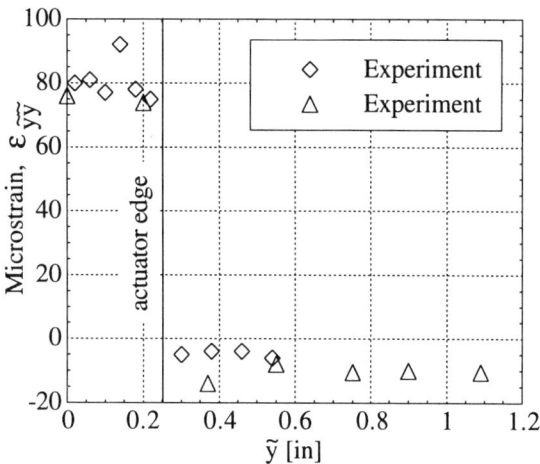

Figure 16 Experimental variation of $\varepsilon_{\tilde{y}\tilde{y}}$

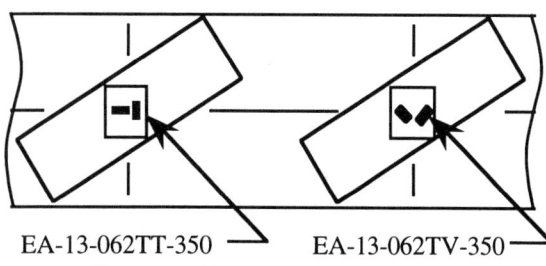

Figure 17 Shear type gage locations

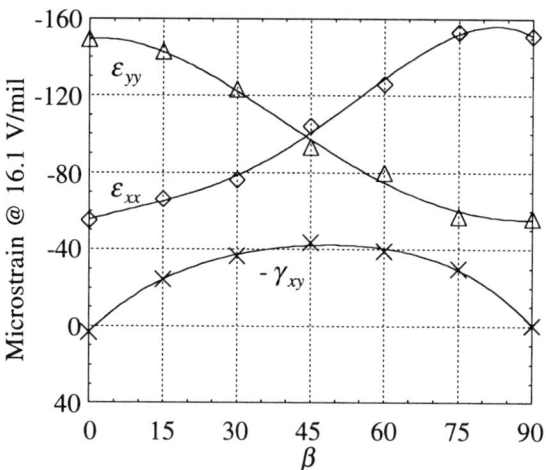

Figure 18 Experimental PZT surface strain

Conclusions

This paper developed several variations of a model to predict the torsion, extension and bending response of a beam subject to piezoceramic actuation and presented strain and deflection data to illustrate the mechanics of the problem and provide model substantiation. In addition, issues associated with reducing 2-D piezoceramic characteristics to a 1-D model were explored. A finite element solution for the model with warping terms is shown to converge for a low number of elements. In the absence of warping terms and shear lag, the model reduces to a system of three equations which may be solved directly.

Torsion is introduced into the models by the application of a cross-sectional warping function which is only dependent upon cross-sectional geometry. For isotropic rectangular beams, the effects of the warping terms in the axial strain expressions were shown to be negligible, however, the warping is important in terms of quantifying the shear strain. Adhesive shear lag effects are also developed and show a measurable impact on the results.

Adequately capturing the 2-D strain induction characteristics of the piezoceramics is critical in obtaining good predictions over a wide range of orientation angles. Modification of the piezoceramic material properties by neglecting transverse induced strain is inconsistent with the basic physics of the system but is a good approximation if high aspect ratios are maintained. Inclusion of the transverse induced strain results in a model which does not twist when $\beta=45°$. Experimentally, the maximum twist occurs at this orientation angle. Completely neglecting transverse actuator stiffness results in adequate axial induced forces but greatly overestimates the induced torque. Therefore, allowing full in-plane isotropic piezoceramic characteristics may more accurately represent the true physics, but such assumptions are incompatible with the utilized beam theory.

Experimental deflection results exhibit fair correlation with model predictions for piezoceramic aspect ratios as low as 3.7. The theoretical bending predictions significantly deviate from experimental data above $\beta=45°$, especially for the specimen with the lower aspect ratio piezoceramics. This divergence is inevitable since theoretical predictions approach zero for $\beta=90°$ as opposed to the finite response obtained experimentally. More sophisticated means of reducing 2-D system behavior to a 1-D global platform, such as soving the full 2-D problem, would improve these results. Twist angle predictions, on the other hand, are accurate for the full range of permissible orientation angles. This is attributed to both theory and experiment being in exact agreement in the 2 limiting actuator orientations, unlike the bending response.

Detailed strain distributions indicate that transverse actuator and beam strain energy contributions are not negligible for all actuator orientations. However, experimental results indicate that 2-D modeling details are only necessary to correct the predictions for high actuator orientations. When $\beta<45°$ the models presented herein are considered sufficient to predict the bending and twist response of rectangular isotropic beams subject to induced strain actuation.

Appendix

This section provides detailed experimental specimen parameters and test results used within this paper. Table 1 provides the geometric parameters of the beam configurations and deflection results and Table 2 contains strain results. Figures 19, 20 and 21 are the free strain characteristics of the piezoceramics used. Note that two separate batches of G-1195 piezoceramics from the same manufacturer exhibit significantly different induced strain characteristics. Corresponding polynomial curve fits are provided in Table 3.

Table 1 Test specimen parameters and deflection data

ID #	β deg	Beam w x l in x in	#	type	Piezoceramics l x w in x in	t mil	Adhesive t mil	Applied Voltage V	Results dw/dx deg	ϕ deg
1	15	2.00 x 0.0312	3 / 0	G-1195	2.00 x 0.25	7.5		+120	0.249	0.0355
2	30	2.00 x 0.0312	3 / 0	G-1195	2.00 x 0.25	7.5	1.7	+120	0.213	0.0558
3	45	2.00 x 0.0312	3 / 0	G-1195	2.00 x 0.25	7.5	1.4	+120	0.180	0.0634
4	60	2.00 x 0.0312	3 / 0	G-1195	2.00 x 0.25	7.5	2.0	+120	0.152	0.0558
5	75	2.00 x 0.0312	3 / 0	G-1195	2.00 x 0.25	7.5		+120	0.114	0.0380
6	0	1.97x 0.0305	2 / 0	G-1195	1.85 x 0.50	9.2 / 9.3	0.7 / 0.4	+150	0.356	0.000
7	15	1.97x 0.0305	2 / 0	G-1195	1.85 x 0.50	9.3 / 9.3	0.7 / 0.7	+150	0.333	0.0561
8	30	1.97x 0.0305	2 / 0	G-1195	1.85 x 0.50	9.3 / 9.2	0.6 / 0.5	+150	0.326	0.0846
9	45	1.97x 0.0305	2 / 0	G-1195	1.85 x 0.50	9.4 / 9.4	0.5 / 0.3	+150	0.285	0.104
10	60	1.97x 0.0305	2 / 0	G-1195	1.85 x 0.50	9.2 / 9.3	0.3 / 0.3	+150	0.242	0.0800
11	75	1.97x 0.0305	2 / 0	G-1195	1.85 x 0.50	9.3 / 9.3	0.5 / 1.0	+150	0.220	0.0406
12	90	1.97x 0.0305	2 / 0	G-1195	1.85 x 0.50	9.2 / 9.2	0.3 / 0.5	+150	0.189	0.000
13	45	2.00 x 0.0312	1 / 0	PZT-5H	2.00 x 0.50	12.5				
14	45	2.00 x 0.0312	1 / 0	PZT-5H	2.00 x 0.50	12.5				
15	45	2.00 x 0.0312	1 / 0	PZT-5H	2.00 x 0.50	12.5	1.5			

Table 2 Microstrain at PZT surface at +150 V

ID #	Gage Orientation			
	0	45	90	135
6	-55	-105	-149	-105
7	-66	-80	-143	-130
8	-76	-63	-123	-148
9	-104	-55	-93	-142
10	-126	-64	-80	-142
11	-153	-75	-57	-128
12	-151	-103	-56	-102

Table 3 Actuation Strain Polynomial coefficients
$\Lambda[\mu\varepsilon] = a_0 + a_1 \mathcal{E} + a_2 \mathcal{E}^2$

PZT	a_0	a_1	a_2
G-1195(a)	2.2439	-10.1561	-0.095227
G-1195(b)	0.34926	-7.72578	-0.18626
PZT-5H	-0.11887	-11.254	-0.34111

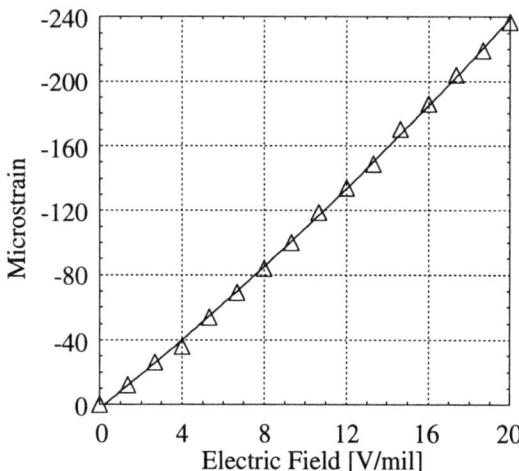

Figure 19 G-1195(a) free strain (specimens 1-5)

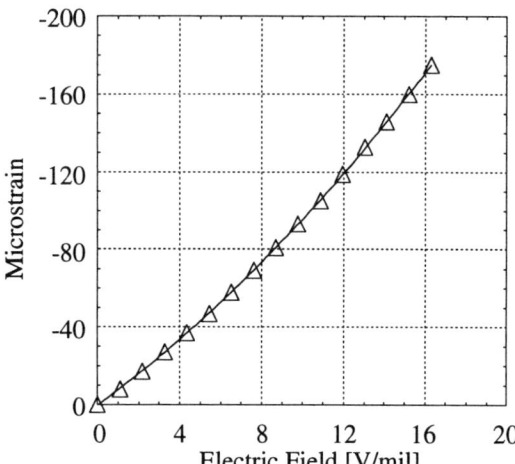

Figure 20 G-1195(b) free strain (specimens 6-12)

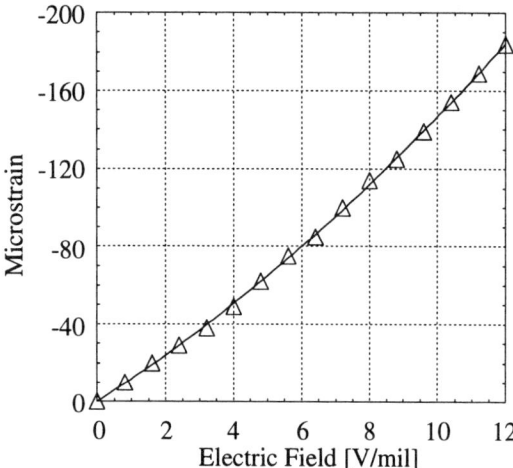

Figure 21 PZT-5H free strain (specimens 13-15)

Acknowledgments

This work is supported by the Army Research Office under the grant DAAL 03-92-G-0121 with Dr. Gary Anderson as Technical Monitor. Experimental data were collected by the authors with assistance from Burtis Spencer.

References

1. Shaw, J., "Higher Harmonic Blade Pitch Control: A System for Helicopter Vibration Control", Ph.D. Dissertation, Department of Aeronautics and Astronautics, Massachusetts Institute of Technology, 1980.

2. Nguyen, K. and Chopra, I., "Application of Higher Harmonic Control to Rotors Operating at High Speed and Thrust," Presented at the 45th Annual Forum of the American Helicopter Society, Boston, Massachusetts, May 1989.

3. Chopra, I. and McCloud III, J.C., "A Numerical Simulation Study of Open Loop, Closed Loop and Adaptive Multicyclic Control Systems," Journal of the American Helicopter Society, Vol. 28, No. 1, January 1983, pp. 311-325.

4. Crawley, E. and de Luis, J., "Use of Piezo-Ceramics As Distributed Actuators in large Space Structures," Proceedings of the 26th AIAA Structures, Structural Dynamics and Materials Conference, Orlando, Florida, April 1985.

5. Crawley, E. and de Luis, J., "Use of Piezoelectric actuators as Elements of Intelligent Structures," AIAA Journal, Vol. 25 No. 10, October 1987, pp. 1373-1385.

6. Crawley, E. and Anderson, E., "Detailed Models of Piezoceramic Actuations of Beams," Proceedings of the 30th AIAA Structures, Structural Dynamics and Materials Conference, Mobile, Alabama, April 1989.

7. Park, C., Walz, C. and Chopra, I., "Bending and Torsion Models of Beams with Induced Strain Actuators", SPIE conference, Albuquerque, NM, 1993.

8. Gjelsvik, A., Theory of Thin Walled Bars, Wiley-Interscience, New York, 1981.

9. Timoshenko, S., Theory of Elasticity, McGraw-Hill, New York, 1934.

TWO-DIMENSIONAL FINITE ELEMENT ANALYSIS OF LAMINATED COMPOSITE PLATES CONTAINING DISTRIBUTED PIEZOELECTRIC ACTUATORS AND SENSORS

Duane T. Detwiler
Altair Engineering, Inc.
Troy, Michigan

M.-H. Herman Shen
Department of Aeronautical and Astronautical Engineering
The Ohio State University, Columbus, Ohio

Vipperla B. Venkayya
Wright Laboratory
Wright Patterson AFB, Ohio

Abstract

A finite element formulation is developed to model the response of laminated composite plates containing distributed piezoceramic actuators and sensors. The equations of motion are derived using the variational principle with respect to the total structural and electrical potential energy. These equations of motion are converted to a two-dimensional finite element equation using a First Order Shear Deformation Laminated Plate Theory. The finite element equation is based on the QUAD4 isoparametric quadrilateral element which is currently used in such FEA codes as COSMIC/NASTRAN and ASTROS. This new piezoelectric finite element formulation was incorporated into an FEA computer code and validated by comparison with published experimental and numerical results.

Introduction

There has been a great deal of interest recently in the concept of active sensing and reactive "smart" structures by the research community. These smart structures contain a system of microsensors which monitor the internal strain of the structure caused by external vibrations, temperatures, or pressures. This information is input into control algorithms programmed into microprocessors which in turn activate a system of microactuators. These microactuators can change the shape of the structure and thereby the mass, stiffness, or energy-dissipation characteristics of the structure which effectively tailors the static or dynamic response of the structure. This type of smart system has applications in large space structures and satellites, aircraft and helicopter structures, automobiles, and automated manufacturing technology.[1-2]

The coupled mechanical and electrical properties of piezoelectric ceramics make them well suited for use as sensors and actuators in smart structures.[2-4] The direct and converse piezoelectric effects govern the interaction between the mechanical and electrical behavior of this type of material. The direct piezoelectric effect states that a strain applied to the material is converted to a charge. On the other hand the converse piezoelectric effect states that an electric potential applied to the material is converted to strain.

The high strength-to-weight ratio of fiber-reinforced laminated composite materials make them attractive for many aerospace applications. The use of these composites grows as advances in

Copyright © 1993 American Institute of Aeronautics and Astronautics, Inc. All rights reserved.

manufacturing technology progress. The construction of a laminated composite by stacking many thin layers together makes the integration of piezoceramic actuators and sensors possible. This integrated smart system would effectively increase the performance of the composite structure. There have been several analytical analysis conducted to demonstrate the feasibility of this integrated concept for simple structures.[4-6]

The design and analysis of large complicated structures with integrated piezoelectric materials requires the development and implementation of finite element methods. Several of these finite element methods have been developed for laminated composite structures with integrated piezoelectric layers.[7-9] Most of these methods use solid three-dimensional hexahedron or brick elements. These elements display excessive shear stiffness as the element thickness decreases. This problem of shear locking was solved by adding three incompatible internal degrees of freedom to the element.[7-8] This makes the analysis of models using these elements more complex and costly to solve. The use of two dimensional plate elements greatly reduces the problem size and the computation time of certain FEA applications.[10] The two-dimensional quadrilateral plate element developed by Hwang and Park is more efficient than solid elements, but it appears to have restricted modeling capabilities.[9]

The purpose of our research is to formulate a laminated composite plate with piezoelectrics as additional layers. This formulation, based on two-dimensional plate theory, is more realistic in the context of current applications. A three-dimensional model may be appealing in research but can be very cumbersome and expensive in modeling practical structures such as wings and control surfaces. This new piezoelectric plate element will be capable of modeling the static and dynamic response of laminated composite plates containing one or more piezoelectric activating or sensing layers subjected to mechanical and electrical loads. Each piezoelectric layer may be at an arbitrary distance from the mid-plane and may have a different applied voltage. (See figures 1. and 2.)

An additional goal of this research was to make it easy to integrate the new finite element formulation into existing finite element codes. Therefore, the new element was based on the QUAD4 isoparametric quadrilateral element from ASTROS and COSMIC/NASTRAN.[12] This element has already been validated for the analysis of laminated composite structures.

The formulation of this new piezoelectric plate finite element will be validated by comparison with three-dimensional formulations and experimental results from literature.[7-8,11]

Derivation of Equations of Motion

The equations of motion for a material with coupled mechanical and electrical properties are based on the stress equations of motion

$$\sigma_{ij,i} + f_j = \rho \ddot{u}_j \quad (1)$$

and the charge equation of electrostatics

$$D_{i,i} = 0 \quad (2)$$

where σ_{ij}, f_i, ρ, u_j, D_i are stresses, body forces, density, displacements, and electric flux density, respectively. These two equations can be combined by multiplying them by the appropriate quantities and integrating over the respective volumes

$$\int_V (\sigma_{ij,i} - \rho \ddot{u}_j) \delta u_j dV + \int_{V_p} D_{j,j} \delta \phi dV_p = 0 \quad (3)$$

where ϕ is the electric potential or voltage. The volume V represents the entire volume of the material while V_p is only the volume of the piezoelectric layers. The body forces are neglected in this equation.

The divergence theorem can be applied to equation (3) to rewrite it in a different form This equation can be further rewritten by using the strain-displacement relationship

$$\int_V \rho \ddot{u}_j \delta u_j dV + \int_V \sigma_{ij} \delta u_{j,i} dV + \int_{V_p} D_i \delta\phi_{,i} dV_p$$
$$= \int_A n_i \sigma_{ij} \delta u_j dA + \int_{A_p} n_i D_i \delta\phi dA_p \quad (4)$$

$$\epsilon_{ij} = \frac{1}{2}(u_{i,j} + u_{j,i}) \quad (5)$$

and the electric field - electric potential relationship

$$E_i = -\phi_{,i} \quad (6)$$

where ϵ_{ij} are the strains and E_i is the electric field strength vector. Substituting equations (5) and (6) into (4) produces

$$\int_V \rho \ddot{u}_i \delta u_i dV + \int_V \sigma_{ij} \delta\epsilon_{ij} dV - \int_{V_p} D_i \delta E_i dV_p$$
$$= \int_A T_i \delta u_i dA + \int_{A_p} Q \delta\phi dA \quad (7)$$

where T_i are the tractions applied on the surface A and Q is the electrical charge applied on the surface of the piezoelectric A_p.

Finite Element Formulation

The finite element formulation is developed based on the QUAD4 plate element currently used in COSMIC/NASTRAN and ASTROS.[12] The QUAD4 is an isoparametric quadrilateral element which uses a bilinear variation of geometry and deformation within the element.

The QUAD4 element has five degrees of freedom (DOF) per node. Two additional DOF per element are included to represent the electrical voltages of two different piezoelectric actuator/sensor layers. (See figure 3.) For every additional piezoelectric layer an additional voltage DOF is needed per element. It is assumed that the strains vary linearly through the thickness of the element. The QUAD4 element may be used to model the coupling of membrane and bending behaviors.

Displacement Functions

Two dimensional interpolation (shape) functions are used to define the geometry field at any point in the element cross-section. These shape functions relate the curvilinear coordinates in the nodal cartesian coordinate system to the element coordinate system.

These shape functions and their derivatives are

$$N_i = \frac{1}{4}(1+\xi\xi_i)(1+\eta\eta_i) \quad (8)$$

$$\frac{\partial N_i}{\partial \xi} = \frac{1}{4}\xi_i(1+\eta\eta_i)$$
$$\frac{\partial N_i}{\partial \eta} = \frac{1}{4}\eta_i(1+\xi\xi_i) \quad (9)$$

Shape functions can be used to express the element deformations in terms of nodal displacements

$$\{u(\xi,\eta)\} = \sum_1^4 N_i(\xi,\eta)\{u\}_i \quad (10)$$

The electric potential is constant throughout the plane of the element

$$\phi_k(\xi,\eta) = \phi_k \quad (11)$$

and therefore the gradient of voltage with respect to x and y is zero.

Strain-Displacement Relationship

The strain-displacement relations for this element is based on the Mindlin first-order shear deformation plate theory[13]. This relation can be written with respect to the global cartesian coordinate system.

This relationship can also be rewritten in the element coordinate system by substituting equation (9) into equation (12)

where $[J]^{-1}$ is the inverse of the Jacobian Matrix

$$\{\epsilon\} = \begin{Bmatrix} \epsilon_x^M \\ \epsilon_y^M \\ \epsilon_{xy}^M \\ \epsilon_x^B \\ \epsilon_y^B \\ \epsilon_{xy}^B \\ \gamma_{yz} \\ \gamma_{zx} \end{Bmatrix} = \begin{bmatrix} \frac{\partial}{\partial x} & 0 & 0 & 0 & 0 \\ 0 & \frac{\partial}{\partial y} & 0 & 0 & 0 \\ \frac{\partial}{\partial y} & \frac{\partial}{\partial x} & 0 & 0 & 0 \\ 0 & 0 & 0 & z\frac{\partial}{\partial x} & 0 \\ 0 & 0 & 0 & 0 & z\frac{\partial}{\partial y} \\ 0 & 0 & 0 & z\frac{\partial}{\partial y} & z\frac{\partial}{\partial x} \\ 0 & 0 & \frac{\partial}{\partial y} & 0 & 1 \\ 0 & 0 & \frac{\partial}{\partial x} & 1 & 0 \end{bmatrix} \begin{Bmatrix} u \\ v \\ w \\ \theta x \\ \theta y \end{Bmatrix} \quad (12)$$

$$\{\epsilon\} = \sum_{i=1}^{4} [J]^{-1}[B]_i \{u\}_i \quad (13)$$

and $[B_u]_i$ is the shape function derivative matrix.

$$[B_u]_i = \begin{bmatrix} N_{i,1} & 0 & 0 & 0 & 0 \\ 0 & N_{i,2} & 0 & 0 & 0 \\ N_{i,2} & N_{i,1} & 0 & 0 & 0 \\ 0 & 0 & 0 & \zeta\frac{t}{2}N_{i,1} & 0 \\ 0 & 0 & 0 & 0 & \zeta\frac{t}{2}N_{i,2} \\ 0 & 0 & 0 & \zeta\frac{t}{2}N_{i,2} & \zeta\frac{t}{2}N_{i,1} \\ 0 & 0 & N_{i,2} & 0 & N_i \\ 0 & 0 & N_{i,1} & N_i & 0 \end{bmatrix} \quad (14)$$

Electric Field Strength - Electric Potential

The electric field strength - electric potential relations for this element are based on equation (6), equation (11), and the assumption that the electric potential varies linearly through the thickness of the piezoelectric layers.

$$\{-E\} = \begin{Bmatrix} -E_{x_1} \\ -E_{y_1} \\ -E_{z_1} \\ -E_{x_2} \\ -E_{y_2} \\ -E_{z_2} \end{Bmatrix} = \begin{bmatrix} \frac{\partial}{\partial x} & 0 \\ \frac{\partial}{\partial y} & 0 \\ \frac{\partial}{\partial z} & 0 \\ 0 & \frac{\partial}{\partial x} \\ 0 & \frac{\partial}{\partial y} \\ 0 & \frac{\partial}{\partial z} \end{bmatrix} \begin{Bmatrix} \phi_1 \\ \phi_2 \end{Bmatrix} = [B_\phi]\{\phi\} = \begin{bmatrix} 0 & 0 \\ 0 & 0 \\ \frac{1}{t_{p_1}} & 0 \\ 0 & 0 \\ 0 & 0 \\ 0 & \frac{1}{t_{p_2}} \end{bmatrix} \begin{Bmatrix} \phi_1 \\ \phi_2 \end{Bmatrix} \quad (15)$$

Stress-Strain Equations

The stress-strain equations are derived using the classical lamination theory which assumes:

(1) Each of the lamina is in a state of plane stress,
(2) All of the lamina are perfectly bonded,
(3) The bonds are infinitesimally thin and non-shear deformable.

This means the laminate behaves as a "single" layer with "special" properties. The stresses and strains are separated into membrane, bending, shear, and electrical components

$$\{\sigma\} = [G]\{\epsilon\}$$

or

$$\begin{Bmatrix} \sigma_M \\ \sigma_B \\ \tau \\ D_1 \\ D_2 \end{Bmatrix} = \begin{bmatrix} [G_1] & [G_4] & 0 & [e]_1 & [e]_2 \\ [G_4] & [G_2] & 0 & [e]_1 & [e]_2 \\ 0 & 0 & [G_3] & 0 & 0 \\ [e]_1 & [e]_1 & 0 & [\xi^s]_1 & 0 \\ [e]_2 & [e]_2 & 0 & 0 & [\xi^s]_2 \end{bmatrix} \begin{Bmatrix} \epsilon_M \\ \epsilon_B \\ \gamma \\ -E_1 \\ -E_2 \end{Bmatrix} \quad (16)$$

where the components for this "single" layer are assembled from the transformed plane stress constitutive relations of each lamina.

$$[G_{ij}]_1 = \frac{1}{t}\sum_{k=1}^{N}[\overline{G_{ij}}]^k (z_k - z_{k-1})$$
$$[G_{ij}]_2 = \frac{1}{3I}\sum_{k=1}^{N}[\overline{G_{ij}}]^k (z_k^3 - z_{k-1}^3) \quad (17)$$
$$[G_{ij}]_4 = \frac{1}{2t^2}\sum_{k=1}^{N}[\overline{G_{ij}}]^k (z_k^2 - z_{k-1}^2)$$

The matrix $[G_{ij}]^k$ is the transformed moduli matrix for each lamina including the piezoelectric layers. The values z_k and z_{k-1} are the distances from the mid-plane of the laminate to the top and bottom of the kth lamina. The electrical-mechanical coupling part of this property matrix is formed from the plane stress piezoelectric coefficient matrix for each piezoelectric layer. The matrix $[\xi^s]_p$ is the dielectric permitivity at constant strain for the pth piezoelectric layer. These plane stress piezoelectric relations must be obtained from the three dimensional relations.[13]

The transverse shear flexibility matrix $[G_3]$ is defined in terms of the transverse shear strain energy through the thickness. The transverse shear stiffness is numerically conditioned to enhance the accuracy of the element for a wide range of modeling practices.[12]

$$[e] = [d][G] = \begin{vmatrix} 0 & 0 & e_{13} \\ 0 & 0 & e_{23} \\ 0 & 0 & 0 \end{vmatrix} \quad (18)$$

$$[\xi^s] = -[\xi^\sigma] + [d][G][d]^T = \begin{vmatrix} \xi^s_{11} & 0 & 0 \\ 0 & \xi^s_{22} & 0 \\ 0 & 0 & \xi^s_{33} \end{vmatrix}$$

Finite Element Equation

The equilibrium equations (7) can now be written for each element in terms of the nodal displacements and the element voltages[14,15]

$$\begin{vmatrix} M & 0 \\ 0 & 0 \end{vmatrix} \begin{vmatrix} \ddot{u} \\ \ddot{\phi} \end{vmatrix} + \begin{vmatrix} K_{uu} & K_{u\phi} \\ K_{\phi u} & K_{\phi\phi} \end{vmatrix} \begin{vmatrix} u \\ \phi \end{vmatrix} = \begin{vmatrix} F \\ Q \end{vmatrix} \quad (19)$$

where [K] is the extended element stiffness matrix and [M] is the element mass matrix. The mechanical stiffness matrix, $[K_{uu}]$ is integrated numerically by (2 x 2 x 2) Gauss-quadrature integration method.[14]

$$[K_{uu}] = \Sigma\Sigma\Sigma [B_u]^T [G][B_u] W_\xi W_\eta W_\zeta \det[J] \quad (20)$$

where (ξ, η, ζ) are the Gaussian integration point coordinates and W_ξ, W_η, W_ζ are the associated weight factors. The electrical-mechanical coupling stiffness matrix, $[K_{u\phi}]$, is integrated in the z direction with respect to the thickness of each piezoelectric layer.

$$[K_{u\phi}] = \int\Sigma\Sigma [B_u]^T [e][B_\phi] W_\xi W_\eta A \, dz_p \quad (21)$$

The piezoelectric permittivity stiffness matrix, $[K_{\phi\phi}]$, is equal to a constant multiplied by the volume of each piezoelectric layer.

$$[K_{\phi\phi}] = [B_\phi]^T [\xi^s][B_\phi] V_p \quad (22)$$

Each of these element stiffness matrices can be assembled into three global matrices and equation (19) can be rewritten to form the actuator and sensor equations for the global static solution.

$$\{u\} = [K_{uu}]^{-1} (\{F\} - [K_{u\phi}]\{\phi_A\}) \quad (23)$$

where ϕ_A is the vector of voltages applied to the piezoelectric actuator layers, and ϕ_S are the

$$\{\phi_s\} = -[K_{\phi\phi}]^{-1} [K_{\phi u}]\{u\} \quad (24)$$

sensor voltages due to the deformation of the structure.

The global dynamic equation can be formed by substituting (24) into the first equation of (19) and moving the actuator voltage to the other side of the equation:

$$[M]\{\ddot{u}\} + ([K_{uu}] - [K_{u\phi}][K_{\phi\phi}][K_{\phi u}])\{u\} = \{F\} - [K_{u\phi}]\{\phi_A\}$$
(25)

This equation can be rewritten into the forced vibration equation by assuming the displacements, forces, and actuator voltages are harmonic variables with different frequencies. This equation can then be reduced to the eigenvalue problem if the right hand side is zero.

Numerical Examples/Model Validation

The proposed piezoelectric QUAD4 finite element formulation was incoporated into a finite element analysis computer program for validation.[13] This FEA program was used to model experiments and numerical simulations performed by Tzou, Crawley and Lazarus, and Ha, Keilers, and Chang.[7-8,11] The numerical results generated from these models were compared with published experimental, theoretical, and numerical results to verify the accuracy of the piezoelectric QUAD4 results.

The first validation case was based on an experiment conducted by Tzou.[7] The experiment used a cantilevered piezoelectric bimorph beam. The experimental apparatus is shown in figure 4. The beam was constructed of two layers of PVDF bonded together and polarized in opposite directions. The finite element model was divided into five equal elements each with two piezoelectric layers. This produced a model with 53 total degrees of freedom. The dimensions of this model are given in figure 5.

The top and bottom surface of the beam were

subjected to an electric potential of 1 volt across the thickness of the beam and the corresponding displacements were determined. The results for the present QUAD4 finite element formulation were then compared to published experimental and theoretical results and the results for a model using beam finite elements. These results show that the QUAD4 element model matches the theoretical results exactly within two decimal places. The Beam element model is slightly more stiff and displays smaller deflections. The experimental value given for tip deflection is less than the others. This is do mostly to shear losses through the bonding layer of the piezoelectrics. Since this bonding is considered perfect in the theoretical and QUAD4 formulations.

The second case was based on a numerical simulation performed by Tzou.[7] The bimorph beam was deflected a prescribed distance by applied tip forces and a sensor voltage across the thickness of each element was determined. These voltages for the QUAD4 element formulation were compared to results for a solid element model which allowed voltage gradients in the x direction. These results show the QUAD4 element voltages to be an average value of the voltage gradients for each solid element. The top and bottom surfaces of each piezoelectric layer is covered by electrode as shown in figure 2. Therefore the voltage is constant in the x-y plane of the element. In reality, if two or more piezoelectric elements are adjacent their voltages are averaged together due to conductivity.

The third validation case was based on experiments conducted by Crawley and Lazarus.[11] The experiment used a cantilevered laminated composite graphite/epoxy plate with distributed G-1195 piezoceramic, (PZT), actuators bonded to the top and bottom surfaces. (See figure 8.) The dimensions for the test setup are shown in the figure.

A constant voltage with an opposite sign was applied to the actuators on each side of the plate. The deflections of the center line and both edges were measured by proximity sensors. The following figures, (9-11 and 12-14), show comparisons of the longitudinal bending, transverse bending, and lateral twisting for a $[0/\pm45]s$ and a $[30_2/0]s$ layup between the QUAD4 finite element model and the results from the experiment and a solid brick finite element model. These deformations were calculated from the lateral deflections along the edges and centerline of the plate.

The first figure in each set indicates the out-of-plane longitudinal bending of the composite plate as a function of position. The solid line with circular symbols represents the QUAD4 numerical results. The dashed line with triangular symbols represents the experimental results. The dashed-dotted line with rectangular symbols represents the solid element numerical results. The figures show all three lines to be close together. The solid element model appears to be more stiff than the QUAD4 model and experimental setup. This is due to the shear locking associated with solid elements. The experimental setup appears more flexible than the solid model but more stiff than the QUAD4 model. This is due to shear losses in the bonding layers for the experimental case.

The second and third figures in each set represent the transverse bending and twisting of the composite plate as a function of position. The experimental results for these cases are more scattered but the QUAD4 element appears reasonably accurate.

The fourth validation case was based on a numerical simulation performed by Ha, Keilers, and Chang.[8] The same cantilevered composite plate model as previously described was used for this case. (See figure 15.) This time the center row of piezoceramics were considered sensors and the other two side rows of piezoceramics were used as actuators on each side of the plate.

A constant voltage of 100 V was applied to the first row of piezoceramics with a positive sign on the top surface and a negative sign on the bottom surface. The same amount of voltage but with opposite signs was applied to the third row. A constant load of 0.2 N was applied at the center tip of the plate. The sensor voltages were numerically determined for this combination of mechanical and electrical loads.

Figure 16 shows a comparison of calculated sensor voltages as a function of position for a $[30_2/0]_s$ layup. The QUAD4 finite element results agree well with the solid finite element simulation. The slight difference is expected due to the difference in the stiffness of models as previously noted.

Conclusions

A new finite element formulation was developed to analyze the mechanical-electrical behavior of laminated composite structures containing distributed piezoelectric actuators and sensors. This formulation was implemented and studied. Several conclusions can be made from the results of this study:

(1) The numerical results generated by the QUAD4 piezoelectric finite element simulations agreed well with experimental data and other solid finite element simulations. This verifies that the electrical-mechanical coupling matrix and the dielectric matrix are formulated correctly.

(2) The finite element model based on QUAD4 elements with one additional degree of freedom per piezoelectric layer is much simpler to formulate and more computationally efficient than models based on solid elements. The total degrees of freedom of the finite element model are reduced by using 2-D QUAD4 elements instead of solid elements. The QUAD4 plate element does not suffer from shear locking. Therefore, incompatible modes are not necessary in the formulation.

(3) Since the structural stiffness and mass matrices are unmodified from the original QUAD4, this new piezoelectric formulation is easily implemented into ASTROS and COSMIC/NASTRAN. The addition of a QUAD4 piezoelectric element into existing FEM codes will provide a versatile tool for the analysis and optimization of large complicated composite structures containing distributed piezoelectric sensing and actuating members.

References

[1] Wada, B.K. "Adaptive Structures," *Proceedings of the AIAA/ASME/ASCE/ASC 30th SDM Conference*, AIAA, New York, pp 1-11.

[2] Gandhi, M.V. and Thompson, B.S., *Smart Materials and Structures*, Chapman & Hall, London, 1992.

[3] Tiersten, H.F., *Linear Piezoelectric Plate Vibrations*, Plenum Press, New York, 1969

[4] Crawley, E.F. and de Luis, J., "Use of Piezoelectric Actuators as Elements of Intelligent Structures," *AIAA Journal*, Vol.25, No.10, 1987 pp.1373-1385

[5] Wang, B.T., and Rogers, C.A., "Modeling of Finite-Length Spatially Distributed Induced Strain Actuators for Laminate Beams and Plates," *Journal of Int. Mat. Sys. and Struc.*, Vol.2, No.1, 1991, pp38-58

[6] Lee, C.-K., O'Sullivan, T.C., and Chiang, W.-W., "Piezoelectric Strain Sensor and Actuator Design for Active Vibration Control," *Proceedings of the AIAA/ASME/ASCE/AHS 32nd SDM Conf.*, AIAA, Washington, DC, 1991

[7] Tzou, H.S., *Piezoelectric Shells Distributed Sensing and Control of Continua*, KAP, Norwell, MA, 1993

[8] Ha, S.K., Keilers, C., and Chang, F.K., "Finite Element Analysis of Composite Structures Containing Distributed Piezoceramic Sensors and Actuators," *AIAA Journal*, Vol.30, No.3, 1992, pp.772-780

[9] Hwang, W.-S. and Park, H.C., "Finite Element Modeling of Piezoelectric Sensors and Actuators," *AIAA Journal*, Vol.31, No.5, 1993, pp.930-937

[10] Hughes, T.J.R., Taylor, R.L., and Kanoknukulchai, W., "A Simple and Efficient Finite Element for Plate Bending," *Int. J. Num. Meth. Engng.*, Vol.11, 1977, pp.1529-1543

[11] Crawley, E.F. and Lazarus, K.B., "Induced Strain Actuation of Isotropic and Anisotropic Plates," *AIAA Journal*, Vol.29, No.6, 1991, pp.944-951

[12] Venkayya, V.B., Tischler, V.A., *QUAD4 Seminar*, WRDC-TR-89-3046, April, 1989

[13] Detwiler, D.T., "Finite Element Modeling of Laminated Composite Plates Containing Distributed Piezoelectric Actuators and Sensors.," M.S. Thesis, The Ohio State Univ., Columbus, OH, December, 1993

[14] Reddy, J.N., *Energy Variational Methods in Applied Mechanics*, John Wiley & Sons, Inc., 1984

[15] Bathe, K.-J., and Wilson, E.L., *Numerical Methods in Finite Element Analysis*, Prentice-Hall, Inc. 1976

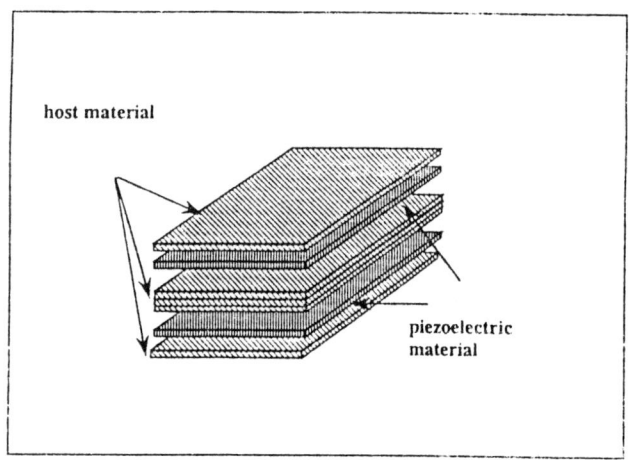

Figure 1 - Schematic of a laminated composite plate with piezoceramic sensor/actuator layers

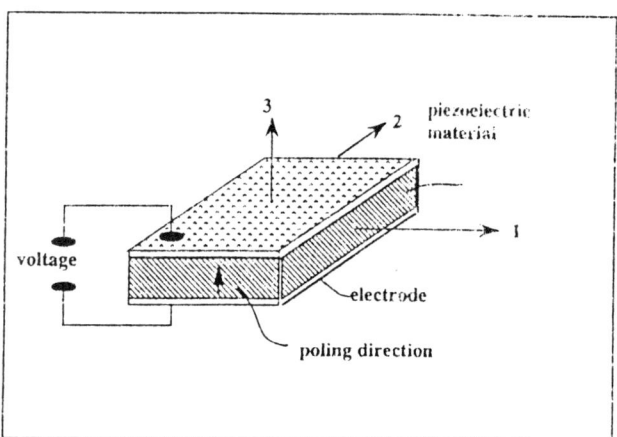

Figure 2 - Schematic of a piezoelectric layer

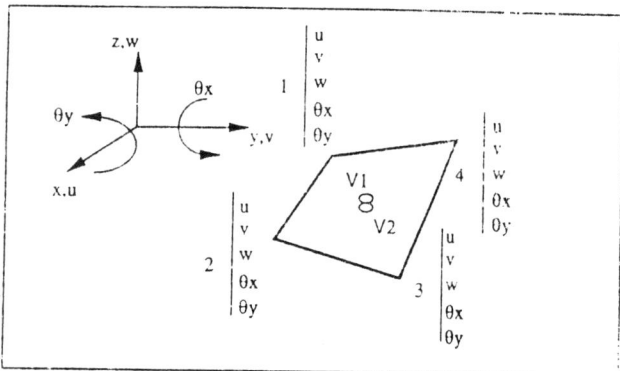

Figure 3 - QUAD4 element with two additional electrical degrees of freedom

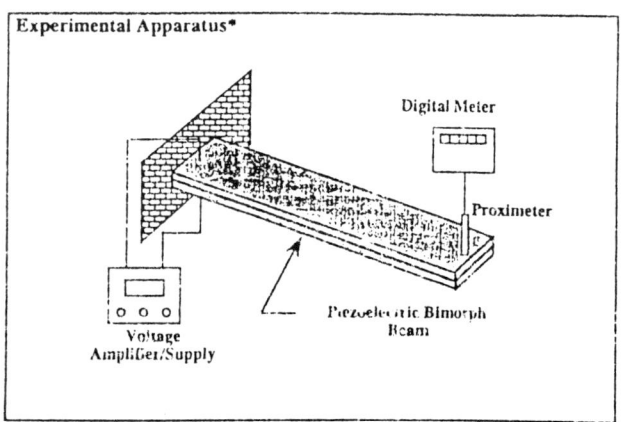

Figure 4 - Experimental setup for piezoelectric bimorph beam

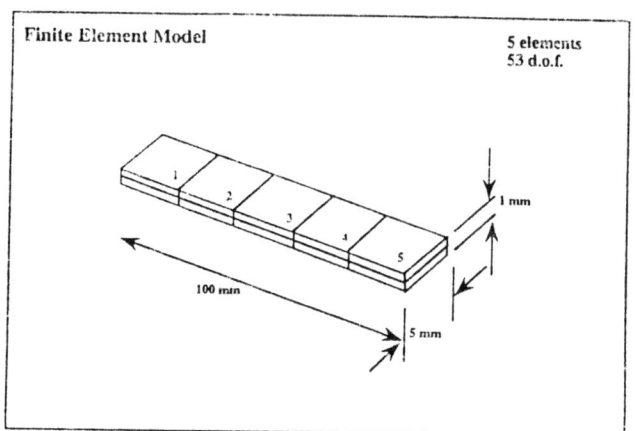

Figure 5 - Finite element model of piezoelectric bimorph beam

Actuation Mechanism

$\times 10^{-7}$ m

Position	1	2	3	4	5
Theory	0.14	0.55	1.24	2.21	3.45
Beam FE*	0.12	0.51	1.16	2.10	3.30
Present	0.14	0.55	1.24	2.21	3.45
EXP*	-	-	-	-	3.15

Figure 6 - Deformations of piezoelectrically actuated bimorph beam

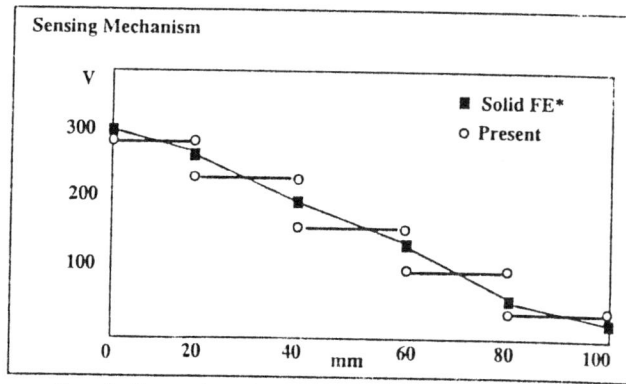

Figure 7 - Sensor voltages for a given displacement of the bimorph beam

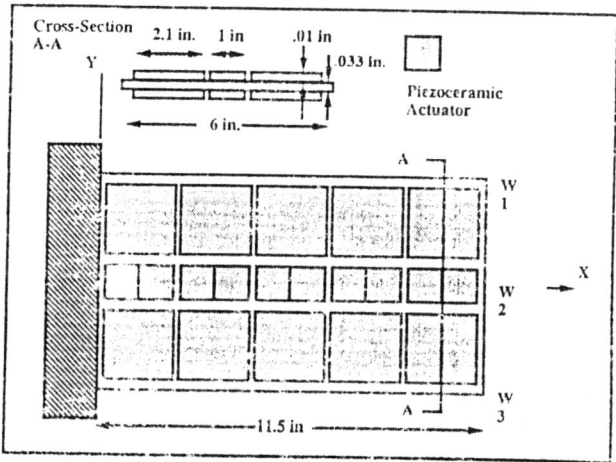

Figure 8 - Cantilevered gr/epoxy composite plate with distributed piezoelectric actuators

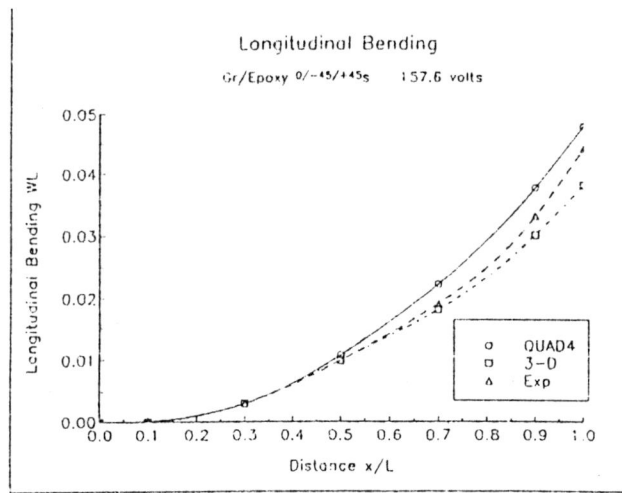

Figure 9 - Longitudinal bending of a cantilevered [0/±45]s composite plate with distributed piezoelectric actuators

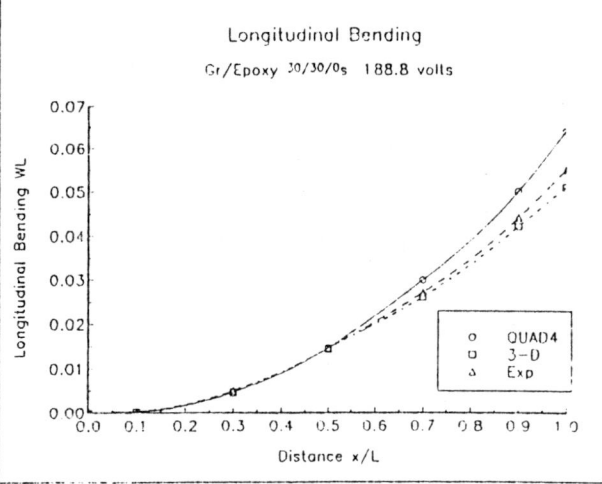

Figure 12 - Longitudinal bending of a cantilevered [30₂/0]s composite plate with distributed piezoelectric actuators

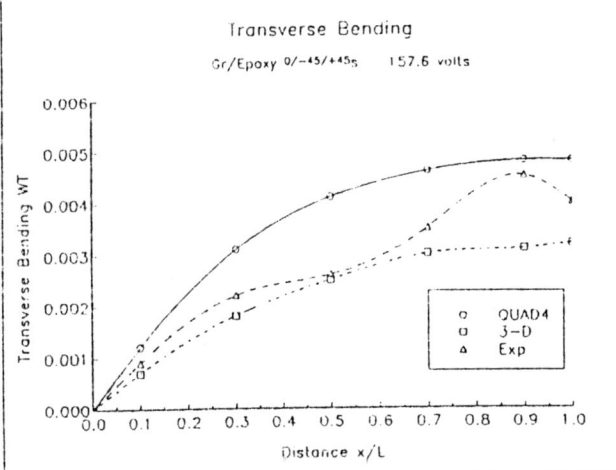

Figure 10 - Transverse bending of a cantilevered [0/±45]s composite plate with distributed piezoelectric actuators

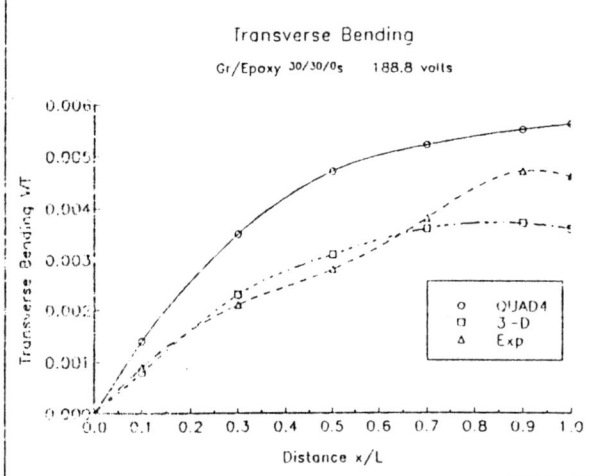

Figure 13 - Transverse bending of a cantilevered [30₂/0]s composite plate with distributed piezoelectric actuators

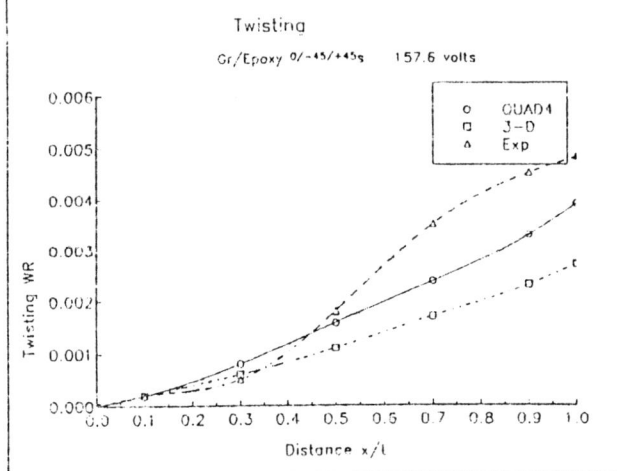

Figure 11 - Twisting of cantilevered [0/±45]s composite plate with distributed piezoelectric actuators

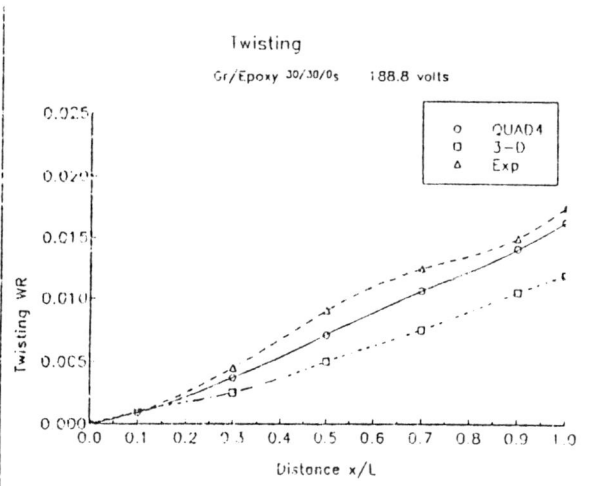

Figure 14 - Lateral twisting of a cantilevered [30₂/0]s composite plate with distributed piezoelectric actuators

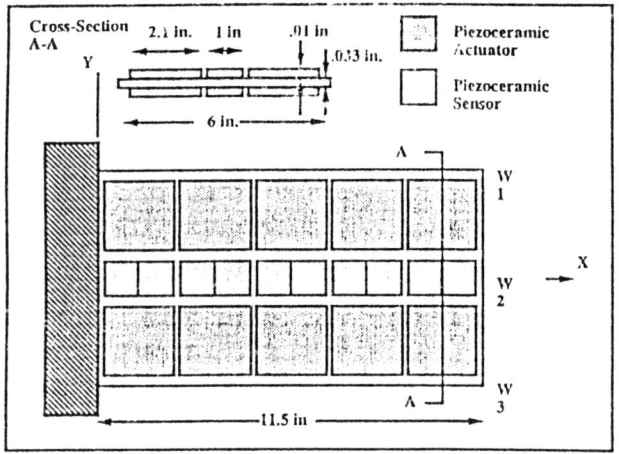

Figure 15 - Cantilevered gr/epoxy composite plate with distributed piezoelectric sensors and actuators

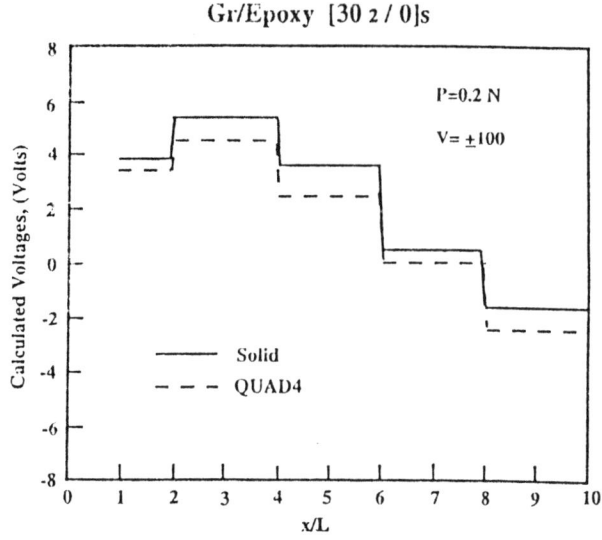

Figure 16 - Calculated sensor voltages due to mechanical and electrical loads

ADAPTIVE ELECTROSTATIC STRUCTURES: A FUNDAMENTAL STUDY OF THE ELECTROSTATIC OSCILLATOR

Larry Silverberg and Ryan McDaniel
Mars Mission Research Center
North Carolina State University
Raleigh, North Carolina 27695-7921

Introduction

Electrostatic structures are a new class of adaptive structures in which the structure is imbedded with electrical charge over its domain, which in turn is controlled by an external electrical field. The external electrical field is obtained by a distribution of electrodes, each charged to a prescribed potential. The result is a structure that can change its shape.

Fundamental experiments in this area accurately predicted equilibrium positions within errors on the order of 0.16% (Ref. 1, 2) and predicted electrostatic frequencies of oscillation to within errors of 10-20%. While the prediction of the equilibrium positions is sufficiently accurate to validate the prediction techniques employed, the prediction of the electrostatic frequencies was insufficiently accurate and raised questions regarding the prediction method employed and several of the associated assumptions. This paper will describe an experiment and prediction method specifically designed to resolve the inaccuracy of the predicted natural frequency obtained in Refs. 1 and 2. In the process, this paper will highlight fundamental issues in predictive modeling of adaptive electrostatic structures.

The field of electrostatics dates back to the classical works of Coloumb and since then has been honored by many historic works. Likewise, the field of mechanics dates back just as far, and within its domain contains a great many classical works. In contrast, the coupling of the two fields is of recent vintage. Coupled problems of electrostatics and mechanics are found in contemporary engineering applications. Examples of these include electrostatic speakers (Refs. 3 and 4), scientific instruments (Refs. 5 and 6) and space-based antennas (Refs. 7 and 8). One way or another, these electrostatic dynamical systems can be regarded as complex electrostatic oscillators. This paper develops an electrostatic oscillator experiment, predicts its behavior and compares the predictions with the measurements.

The electrostatic oscillator experiment presented here will isolate the electrical forces and balance to zero the mechanical forces so that the frequency of the oscillator is due to the electrostatic effects alone. Spherical geometries were selected for the charged surfaces so that the predictions can be made using the method of images (Ref. 9). Furthermore, the parameters were selected to yield a frequency in a range that is easily observed (and measured).

Set Up

The electrostatic oscillator consists of a long slender rod and cross-bar pivoted on two pins as shown in Fig. 1. A conductive sphere of radius R = 1.850 cm is fixed to each end of the rod. Another conductive sphere of identical radius lies below each sphere. The gap between each pair of spheres is G_0 = 2.350 cm when the rod is horizontal. The four spheres are charged to a constant voltage V_0.

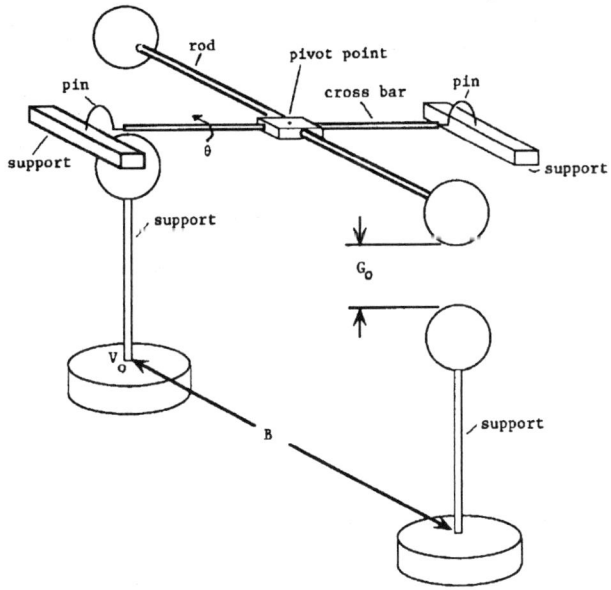

Figure 1. Set up

The rod length B = 37.1 cm is large compared to the sphere radius. Therefore, the electrical forces produced on one side of the rod by the spheres on the other side of the rod are ignored, that is we ignore the electrical forces across the rod, and any effects that these forces would produce. The charges on the spheres are all positive and the electrical force across either gap decreases as the size of the gap increases. Therefore, as the rod rotates either clockwise or counter clockwise, a counteracting moment is exerted on the rod by the electrical forces across the two gaps. This establishes the oscillatory behavior of the rod.

---

[1] Associate Professor, Mechanical and Aerospace Engineering, Member AIAA.
[2] Undergraduate Research Assistant.
"Copyright © 1994 by Larry Silverberg. Published by the American Institute of Aeronautics and Astronautics, Inc. with permission."

Predictions

Electrical charge distributed over the two spheres produces electrical forces across the two gaps. The electrical force across either gap depends on the sphere radius R, the gap length G and the applied voltage V_0. The electrical force is predicted here by the method of images. Other methods would be employed in problems in which the geometry is more complex, (see for example, Refs. 10 and 11). The method of images replaces the two-sphere charge distribution with an infinite series of pairs of fictitious point charges that reproduce the voltage V_0 over the surfaces of the two spheres. Referring to Fig. 2, the first pair of point charges is located at $a_0 = 0$ (letting r = 0) and the value of each charge is $Q_0 = RV_0/k$, where $k = 9 \times 10^9$ Nm²C⁻². The charge Q_0 is identical to the charge on an isolated sphere of radius R and applied voltage V_0. The rth pair of point charges is located at

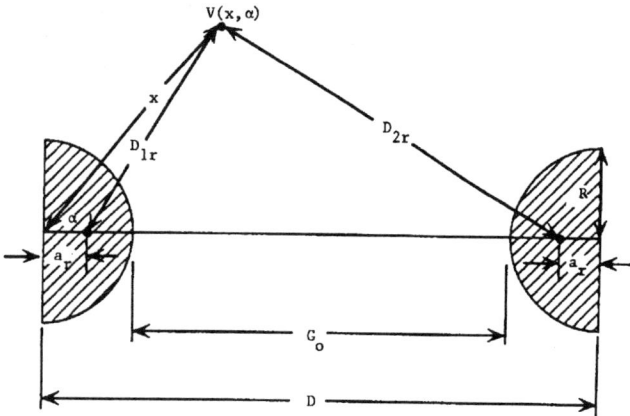

Figure 2. Method of Images

$$a_r = \frac{R^2}{D - a_{r-1}}, \quad (r = 1, 2, ...,) \qquad (1)$$

and the value of each charge is

$$Q_r = -\frac{a_r}{R} Q_{r-1}, \quad (r = 1, 2, ...) \qquad (2)$$

The force across the gap becomes

$$F(D) = k \sum_{r=0}^{\infty} \sum_{s=0}^{\infty} \frac{Q_r Q_s}{(D - (a_r + a_s))^2} \qquad (3)$$

The voltage in free space produced by the point charges becomes

$$V(x, \alpha) = k \sum_{r=0}^{\infty} Q_r \left(\frac{1}{D_{1r}} + \frac{1}{D_{2r}} \right) \qquad (4)$$

in which

$$\begin{aligned} D_{1r}^2 &= x^2 + a_r^2 - 2a_r x \cos \alpha \\ D_{2r}^2 &= x^2 + (D - a_r)^2 - 2(D - a_r) x \cos \alpha \end{aligned} \qquad (5)$$

Note when we let x = R in Eq. (4) that we should obtain the applied voltage on the surface of the sphere $V(R, \alpha) = V_0$. The charge per unit area on the surface of the spheres can also be predicted from Eq. (4). We get (Ref. 9)

$$\sigma(\alpha) = -\frac{1}{4\pi k} \frac{\partial V(x, \alpha)}{\partial x}\bigg|_{x=R} = \sum_{r=0}^{\infty} \Bigg[\frac{R - a_r \cos \alpha}{(R^2 + a_r^2 - 2Ra_r \cos \alpha)^{3/2}}$$
$$+ \frac{R - (D - a_r) \cos \alpha}{(R^2 + (D - a_r)^2 - 2R(D - a_r) \cos \alpha)^{3/2}} \Bigg] Q_r \qquad (6)$$

The oscillatory behavior of the rod can now be established. Summing moments about the pivot point, we obtain the counteracting moment and the predicted natural frequency of oscillation.

$$\omega^2 = -\frac{1}{I} \frac{\partial M_0(\theta)}{\partial \theta}\bigg|_{\theta=0} = -\frac{B^2}{2I} \frac{\partial F(D)}{\partial D}\bigg|_{D=D_0}$$

$$= -\frac{B^2 k}{2I} \sum_{r=0}^{\infty} \sum_{s=0}^{\infty} \Bigg[\frac{-2 Q_r Q_s}{(D - (a_r + a_s))^3} \left(1 - \frac{\partial a_r}{\partial D} - \frac{\partial a_s}{\partial D}\right)$$
$$+ \frac{\frac{\partial Q_r}{\partial D} Q_s + Q_r \frac{\partial Q_s}{\partial D}}{(D - (a_r + a_s))^2} \Bigg]_{D=D_0} \qquad (7)$$

where

$$f(\partial a_r, \partial D) = -\left(\frac{R}{D - a_{r-1}}\right)^2 \left(1 - \frac{\partial a_{r-1}}{\partial D}\right) \qquad (8)$$

$$\frac{\partial Q_r}{\partial D} = -\frac{1}{R} \left(\frac{\partial a_r}{\partial D} Q_{r-1} + a_r \frac{\partial Q_{r-1}}{\partial D} \right)$$

in which I denotes the mass moment of inertia about the pivot point, $\frac{\partial a_0}{\partial D} = \frac{\partial Q_0}{\partial D} = 0$, and ω denotes the predicted natural frequency of oscillation.

Preliminary Results

The natural frequency prediction given in Eq. (7) requires a truncation of the number of pairs of point charges. Denoting this number by n, the accuracy of the approximation can be determined indirectly by checking the accuracy of the voltage at the sphere boundary. Figure 3 shows $V(R, \alpha)$ obtained from Eq. (4) letting $V_0 = 1,000$ V. The natural frequency predictions versus n are shown in Fig. 4. The associated contour plot of lines of constant voltage for n = 10 is shown in Fig. 5.

The above described results are incomplete. Further results will include measured natural frequencies of oscillation, and a description of the sources of errors responsible for the discrepancies between the measured and the predicted natural frequencies.

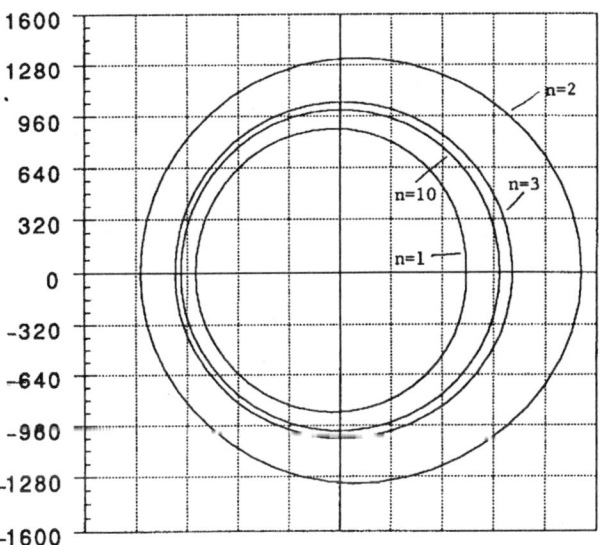

Figure 3. Predicted voltage at the sphere boundary

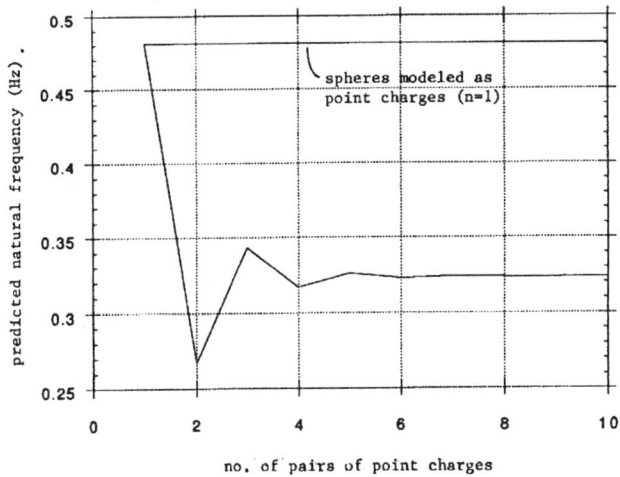

Figure 4. Predicted natural frequency

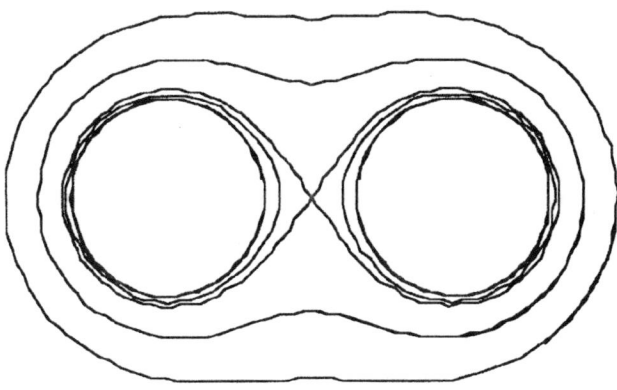

Figure 5. Contour plot of voltage

In addition, a comparison between predicted and measured equilibrium angles will be made using the same test article by making a minor modification to the test article. The rod and two spheres will be deliberately imbalanced causing the rod and two spheres to rotate to a vertical orientation. The imbalance will be measured indirectly from the measured pendulum frequency caused by the imbalance. The conductive coating on the top sphere will be eliminated and a fixed conductive sphere will be placed beside the bottom sphere (attached to the rod). All three conductive spheres will be charged to a voltage V_0, causing the rod to rotate from the vertical axis to a new equilibrium angle. It is this equilibrium angle that will be measured and predicted. These results will be provided along with the presentation of this paper at the Adaptive Structures Forum to be held in Hilton Head, SC on April 21-22, 1994.

References

1. L. Silverberg and W. O. Doggett, "Planar Electrodynamics of Interconnected Charged Particles," *Bulletin of the American Physical Society*, Series II, Vol. 36, No. 10, Nov. 1992.

2. K. Park, "An Investigation of Planar Electrodynamic Structures," Ph.D. Dissertation, North Carolina State University, Raleigh, NC 1992.

3. J. H. Streng, "Charge Movements on the Stretched Membrane in a Circular Electrostatic Push-Pull Loudspeaker," *Journal of the Audio Engineering Society*, Vol. 38, No. 5, May 1990, pp. 331-338.

4. J. H. Streng, "Sound Radiation from Circular Stretched Membranes in Free Space," *Journal of the Audio Engineering Society*, Vol. 37, No. 3, March 1989, pp. 107-118.

5. W. K. Rhim, M. Collender, M. T. Hyson, W. T. Simms, and D. D. Elleman, "Development of an Electrostatic Positioner for Space Material Processing," *Review of*

Scientific Instruments, Vol. 56, No. 2, Feb. 1985, pp. 307-317.

6. N. K. Rhim, S. K. Chung, M. T. Hyson, E. H. Trinh, and D. D. Elleman, "Large Charged drop Levitation Against Gravity," *IEEE Transactions on Industry Applications*, Vol. IA-23, No. 6, Nov/Dec 1987, pp. 975-979.

7. L. Silverberg, "Electrostatically Shaped Membranes," United States Patent, filed Oct. 28, 1992.

8. J. H. Lang, D. H. Staelin, "Electrostatically Figured Reflecting Membrane Antennas for Satellites," *IEEE Transactions on Automatic Control*, Vol. AC-27, June 1982, pp. 666-670.

9. J. D. Jackson, *Classical Electrodynamics*, John Wiley and Sons, Inc., New York, NY, 1975.

10. S. M. Rao, A. W. Glisson, D. R. Wilton, and B. S. Vidula, "A Simple Numerical Solution Procedure for Statics Problems Involving Arbitrary-Shaped Surfaces," *IEEE Transactions on Antennas and Propagation*, Vol. AP-27, No. 5, Sept. 1979, pp. 604-607.

11. P. Dekker, L. Silverberg, and G. Lee, "One-Dimensional Large Deformation theory for Charge Carrying Bodies," presented at the ISCA Conference on Computer Applications in Industry and Engineering, Dec. 15-17, 1993, Honolulu, HI.

AN OBJECT ORIENTED APPROACH TO THE MOTION CONTROL OF A FREE-FLOATING VARIABLE GEOMETRY TRUSS

Shengyang Huang[*], M.C. Natori[**]
Institute of Space and Astronautical Science, 3-1-1, Yoshinodai, Sagamihara 229, Japan

Shoichi Nakai[#], Hiroshi Katsukura[#]
Shimizu Corp., 2-2-2 Uchisaiwai-cho, Chiyoda-ku, Tokyo 110, Japan

Kohichi Miura[✢]
Nihon University, Narashinodai, Funabashi 274, Japan

Abstract

This paper is devoted to the attempt in introducing an object-oriented approach into the motion control of a free-floating variable geometry truss (VGT). Through a static object model, the knowledge base for modeling the construction of VGTs, as well as its motion features and abilities is established. Based upon the knowledge base, a dynamic object model is built up to represent the motion control schema. These two models, along with the functional requirements for control tasks lead to the object-oriented programming for the control system VGTMOT. A motion control experiment for a VGT is also carried out by executing VGTMOT.

I. Introduction

An adaptive structure brings about a new structural concept to the space structure. It can purposefully change its geometric configuration and physical characteristics, so that provides the possibility to optimally adapt the various working environments and tasks[1][2]. To now, the research efforts almost focus on the trying in advancing the concepts of adaptive structure, designing new type of adaptive structures, as well as exploring the kinematics and dynamics of adaptive structures. Also, several works can be found in exploiting the potential applications of adaptive structures. However, few effort to now could be seen in the simulation of control schema for adaptive structures as well as the corresponding implementation in computer. An adaptive structure that is basically intelligent, of course requires the development of computer software to assist in the realization of its various functions. Here, our issue is that, along with the more and more complicated applications of adaptive structures occurring to the space construction so as to lead to the increasing of actuator number, the research about how to effectively achieve the implementing of control schema is becoming to an urgent necessity. A typical example for the complicated applications of adaptive structures is a free-floating variable geometry truss (VGT)[3] that has been investigated by authors in [4][5][6].

Regarding this investigation, rather than a programming method is merely taken into account, the emphasis is more largely placed on modeling and representing an adaptive structure in computer, which leads to a more natural and distinct representation for a complicated control schema and such that results in an easy implementing. Especially, the object oriented model for an adaptive structure might help to explore and so as to make use of the concurrency within a control process. Commonly, a modeling approach is broadly utilized in the design theory. However, the authors believe that is equally important in the structural control if, instead of one or two particular tasks, more highlighted is the adaptivity of structure to the working environment and the structure is maneuvered to adapt a group of task. Obviously, this goal is hard to be achieved only through one or two algorithms. Our argument hereby goes to construct a knowledge base for the problem domain of motion control, which, performing just like the brain of human, contains the inherent motion abilities of structure, as well as a group of control schema that were established upon these abilities.

A knowledge-based control approach has been thought of being fatally weak in the precise control of robotics in that, it is difficult to integrate the computing process into a reasoning-type knowledge base system. Object-oriented approaches (OOAs) bring us a new paradigm to establish integrated knowledge base systems, and an object-oriented programming (OOP) provides the strong computer support to it. The works presented in this paper is devoted to the attempts to construct a knowledge based upon the object oriented approach for the motion control of a representative adaptive structure - VGT structure that is of fully free floating. In this paper, after a brief introduction to the concepts of OOAs, an object-oriented model for a free-floating VGT structure will be outlined, which is utilized to organize the knowledge about the construction and the motion abilities of VGT structures. Then a dynamic model for the objects via the object model is established to deal with the representation of algorithm for a docking task of VGTs. With these two models, we could be facilitated from two aspects: one is it provides the knowledge synthesis to an adaptive structure, which make it available to construct a knowledge-based control schema, the other is it naturally supports concurrent processing to real-time control. Based on these achievements, an object-oriented control system VGTMOT, that is coded by the object-oriented language C++, is developed. Finally, the experiment process carried out by VGTMOT, are shown by pictures.

II. Object-Oriented Approach

An object can be informally defined as a tangible entity

---

[*] Ph. D., Visiting Research Scholar on leave from Dalian University of Technology, P.R.C.
[**] Professor of Spacecraft Engineering, Member AIAA
[#] Ph. D., Chief Reseach Engineer, Ohsaki Research Institute of Shimuzu Corporation
[✢] Research Associate of Department of Junior College

Copyright © 1994 by the American Institute of Aeronautics and Astronautics, Inc. All right reserved.

that exhibits some well-defined behavior. While we start with such definition to represent a physical problem in computer, the term object implies a basic concept that objects serve to unify the ideas of algorithm and data abstraction. Stated differently, there exist invariant properties that characterize an object and its behavior, where, properties are defined to be attributes and behaviors are realized through methods or operations. A model, whose fundamental building blocks are objects that characterize a physical problem domain, is called object model.

The introduction of the concept object-orientation is to cope with the complexity of a physical problem[8]. Concerning the recognition to an objective phenomenon, there in fact exists two orthogonal views. One refers to an emphasis on the events occurring in an objective phenomenon, the other then places the main views on the entities which are involved in the events. The former is obviously oriented to the abstractions for process, namely ordering of events or, say another way, algorithm, while the latter exhibits the abstractions oriented to the objects that either cause action or are the subjects upon which those operations act. A natural problem is, which is the right way to decompose a problem - by algorithms or by objects. Our argument is that the events are superficial, that is just a reflection of internal matters, whereas the entities, or objects, are essential, which are the subjects of events and work through the overall process. Hence, the key to recognize a physical phenomenon is to identify the objects that are characterized by their autonomy, and further to master the inherent relationships among various objects, which leads us to a better choice - to apply the object-oriented decomposition for the problem domain.

An object-oriented paradigm is usually supported by the conceptual framework object models which are broadly thought to mainly encompass the following elements:

Abstraction: An abstraction denotes the essential characteristics of an object that distinguish it from all other kinds of objects and thus provide crisply defined conceptual boundaries, relative to the perspective of the viewer.

Encapsulation: Abstraction and encapsulation are complementary concepts in which, abstraction highlights the outside view of an object, and encapsulation is the process of hiding all of the details of an object that do not contribute to its essential characteristics.

Hierarchy: Hierarchy is a ranking or ordering of abstractions. The two most important hierarchies in a complex system are its class structure and its object structure.

Concurrency: An object model implicitly contains the definitions of both the units of distribution and movement and the entities that communicate, such that easily provides process abstraction and synchronization.

Regarding the engineering applications of OOAs, it is of interest to ask: what exactly do OOAs provide us, to which our issues go to the following statements:

(1) an advanced programming tool;
(2) the strong support to the abstractions of problem domain, that are based not only the static properties but also the dynamic behaviors of objectives;
(3) assisting in the construction of large-scale knowledge bases that integrates both of rule-based knowledge and computing processes;
(4) naturally supporting the parallel computing based upon the distributed treatment of objects.

Say alternately, it offers a brand-new paradigm to the problem solving, the key issues concerning which involve two mutual inverse processes:

(i) to decompose the problem domain into discrete object classes, and then seek for and so as to build up a complete abstractive representation for the problem domain rather than one or two particular tasks or disciplines;
(ii) to establish the synthesis of different objects to serve to the solving of the particular problem upon some control schema.

III. Statements of Problem Background

The problem model concerned with this investigation is an octahedral variable geometry truss (VGT) that was proposed by Miura et al[3] and is set up in the laboratory. A VGT is characterized by its chain feature where the fundamental module of the structure, which consists of a pair of lateral triangular battens and six diagonal members, can be picked up. Two adjacent modules, which share the lateral batten, comprise a repeating unit of the truss. If the actuators are just only equipped in the shared lateral battens (Fig. 1), the geometry and motion of the two adjacent modules can be governed in a unified manner. Thus, through the repetition of units in the longitudinal direction, a VGT can be formed.

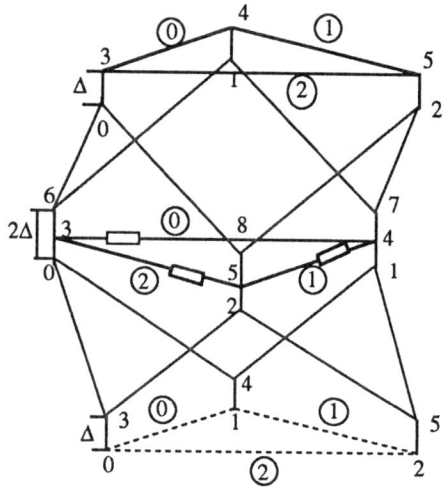

Fig 1. VGT Unit

Concerning this research, we start with our discussion based upon the achievements presented in [4][5][6] that is about the motion control of VGTs in fully free-floating state. As to this case, it is assumed that the structure is under null-gravity space and, for the simplicity, is set within a conservative dynamic system. Besides, the structure moves so slowly that the dynamic effects of structure can be neglected. Consequently, with the conservation of momentum of structure, it is possible to establish a complete mathematics descriptions to govern the motion of structure[5][6].

To aim at the automatic construction of large space structures by using adaptive structures, a free-floating VGT might be expected to finish a series of tasks including docking or dedocking to other structures with its one or two endplanes under complicated environments, as well as obstacle or impact avoidance through autonomous motion of the structure.

IV. Knowledge Organization about a VGT Model

The knowledge about a VGT refers to those that highlights some general abilities oriented to a variety of motion tasks, which are represented by a static object model. A static object model consists of two mutual orthogonal views. One is the abstraction hierarchy of an object class through inheritance to indicate that the low-level object classes are derived and expanded via a fundamental object class. Another is for the constructing description of an enterprise with aggregates of object instances. The unchanged properties within a problem domain refer to the object identity, object relationships, object attributes and the methods. In the later discussion, we use the notations proposed in [9] (also see Appendix) to present the object model of VGTs.

4.1 VGT Bay

A VGT unit (Fig. 1) contains all basic motion features of VGTs. If VGT units, as essential components, are used to simply assembled into together, the constructing redundancy would be caused because of the shared common lateral battens. However, in case that, the three hinges on the bottom plane of a unit are anchored to some platform rather than the bottom triangular batten, obviously the bottom lateral batten would lose its effect, which inspires us to an alternate unit abstraction that is based upon the concept of bay.

A bay differs a little in its construction from a VGT unit, which consists of two lateral triangular trusses (middle and top battens), a virtual bottom plane with the same configuration as its top batten to constrain the three hinges, six diagonal members to connect the top batten with the middle batten, and six diagonal members to link three bottom hinges to the middle battens.

With these bays, a VGT chain structure with one base-fixed endplane can be obtained. As to free-floating VGTs, seeing that the left lateral batten at one end of VGTs also moves in the course of moving of the structure, a virtual bay could be composed upon this left batten. A virtual bay has a triangular batten as its top plane and some virtual diagonal members to link the top plane to the fixed platform. Thus, a generalized bay which is either a bay or the virtual bay, can be defined to be such a variable geometry truss that has three control variables to change its shape, and can be connected to other bays or some fixed base with its top plane or three bottom hinges. A bay is controlled with vector $\mathbf{d}$ (Fig. 2) which stretches between the centroids of the bottom and top triangles[4][5], and the virtual bay employs three Euler angles to control the orientation of its top plane.

4.2 Object Abstraction for VGT Bays

Upon the above discussions, a complex object class 'truss_bay' can be identified for the generalized VGT bay. Also an object class 'virtual_bay' could be abstracted as a child class of 'truss_bay' through inheritance. The basic attributes of 'truss_bay' involve:

(1) organization properties

These properties embody a bay identifier, bay total mass, gravity center of bay within the bay coordinate system, hinge offset Δ (Fig. 1) and control vector $\mathbf{d}$ as well as vector $\mathbf{t}$ which stretches between the gravity centers of the middle and top triangular battens.

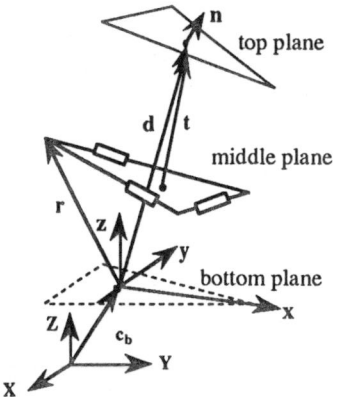

Fig. 2 Bay Coordinate System

(2) geometry and motion attributes

The most significant datum in denoting the geometric and moving behaviors are the rotation transformation matrix $\mathbf{C}_{tb}$ and angular velocity $\boldsymbol{\omega}$ of the top plane that are with respect to the bottom plane. Usually, $\boldsymbol{\omega}$ is expressed as[5]:

$$\boldsymbol{\omega} = \mathbf{P}\dot{\mathbf{d}}$$
$$\mathbf{P} = \frac{2}{\mathbf{d}^T \mathbf{d}} \tilde{\mathbf{d}} \qquad (1)$$

Therefore, project matrix $\mathbf{P}$ rather than $\boldsymbol{\omega}$, as well as $\mathbf{C}_{tb}$ are taken as basic attributes.

(3) angular momentum

The angular momentum of a bay can be obtained by[5]:

$$\mathbf{H} = M\mathbf{g} \times \dot{\mathbf{c}}_b + \mathbf{V}\dot{\mathbf{d}} + \mathbf{Z}\boldsymbol{\omega}_b \qquad (2)$$

where the first term stands for the contribution of gravity center $\mathbf{g}$ of a bay because of the relative translation of the bay coordinate system, the last two terms denote the effects of bay's self-changing as well as the rotation of bay coordinate system with respect to the global reference frame. Hereby the project matrices $\mathbf{V}$ and $\mathbf{Z}$ are picked up as basic attributes.

(4) coupling factors

One coupling stems from the relative movement between the bay coordinate system and the global reference frame, which is denoted by a translation vector $\mathbf{c}_b$ (Fig. 2), and the transformation matrix of bay coordinate system $\mathbf{D}_A$ and $\boldsymbol{\omega}_b$ with respect to the global reference frame. Another refers to the affections to the positions and velocities of other bays of the structure due to the self-changing of vector $\mathbf{d}$.

If there are N bays used to form the structure, then the position vector $\mathbf{r}_i$ of the top plane of a bay is given by[5]

$$\mathbf{r}_i = \sum_{j=1}^{N}(K_{ij}\,{}^A\mathbf{d}_j + \rho_a\,{}^A\mathbf{t}_j) + \mathbf{r}_G \qquad (3)$$

where the left superscripts for vectors **d** and **t** indicate these vectors are defined within the global reference frame, ρ is the mass ratio of bay and $\mathbf{r}_G$ is the gravity center of the structure (Fig. 3). What the attention should be paid on is the coefficient K_{ij} which shows the influence of **d** to the other bays so that is taken to be an attribute.

Another important attribute is the equivalent rigid arm[5] $\pi_{i,j}$, with which the affection to the mass center velocity of the top plane of other bays, that is caused by the rotation of the top plane of bay within the bay coordinate system, can be easily obtained:

$$\dot{\mathbf{r}}_i = {}^A\boldsymbol{\omega}_j \times {}^A\boldsymbol{\pi}_{i,j} \qquad (4)$$

The principal methods concerning the external interface of object class 'truss_bay' are concluded as:

class truss_bay {
  {method group to get the basic attributes};
  {method group to receive the messages such as increment of **d**, K_j and π_{ij}};
  update the geometric and moving state of the bay to be current state upon the input increment of **d**;
  prepare the datum that contribute to the kinematics of structure;
  adjust the geometry and motion of the bay according to the varying of orientation of the top plane; }

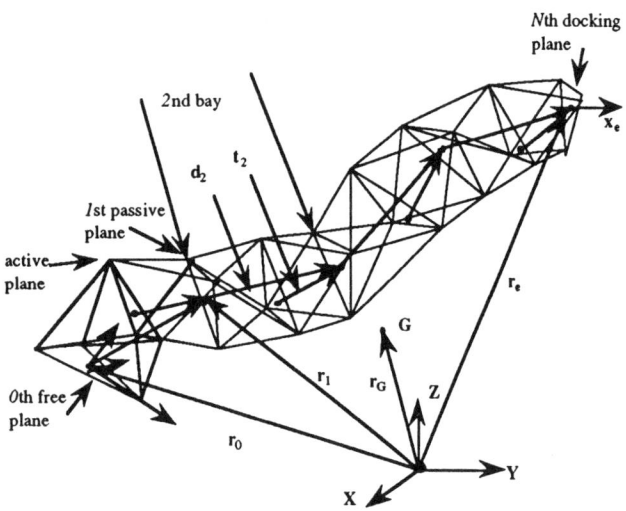

Fig. 3 Chain Construction of a Free-Floating VGT

4.3 Hierarchical Decomposition of an Object Bay

As a complex object, 'truss_bay' can be further decomposed into a series of simple object classes through aggregation.

(1) Object Class 'truss_plane'

Object classes 'truss_plane' could be identified to correspond a lateral batten. The most significant attributes of it are the mass, type, normal vector of the lateral batten, the position vectors of three points (0, 1, 2) (Fig. 1), as well as the datum concerning the velocities of these points for which, it is known[4][5] that the nodal velocity can be expressed as:

$$\dot{\mathbf{r}}_i = \mathbf{R}_i\dot{\mathbf{d}} \qquad (5)$$

so that project matrix **R** is held in. As to the methods of 'truss_plane', we conclude them as follows:

class truss_plane {
  {method set used to get the basic attributes};
  {methods to update the three point coordinate vectors, **R** as well as normal vector of the plane};
  form matrices **V** and **Z** of the batten;
  determine the gravity center of the batten;
  check the length steps of bar according to the maximum stroke length, if it is a length-adjustable bar; }

2) Object Class 'truss_bar'

Seeing that a lateral batten consists of three members, we can further decompose 'truss_plane' into the aggregate of more basic object class 'truss_bar'. The fundamental model of it is a non-uniform length-constant bar[5], whose attributes are the length, type and mass, as well as the eccentric length. The principal methods are:

class truss_bar {
  calculate the inertia of bar;
  locate the gravity center of bar;
  get the non-uniform mass or length ratio of bar;
  get the basic attributes;
  assign or update the length of bar; }

The abstraction for object class 'truss_bar' can further go to the length-adjustable bar, which implies that, under the general abstraction for a member, several varieties of 'truss_bar' can be obtained through inheritance. Subsequently, a child class 'variable_length_bar' of 'truss_bar' is obtained, which would share the attributes of 'truss_bar' and the same operational interface.

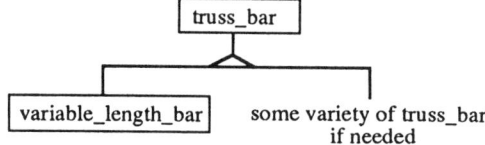

Fig. 4 Abstraction for a VGT Member

4.4 Bay Model

Based upon the abstractions achieved above, a bay object model can be obtained as Fig. 5.

4.5 Object Abstraction for a VGT Structure

A VGT structure is also abstracted to be the object class 'adaptive_truss', which is the highest abstraction on the object model and encompassed by the aggregates of VGT bays (Fig.6).

Object class 'adptive_truss' stands for the synthesis to the overall bays, which is commonly realized through a family of governing equations. The governing equations to achieve the ratio control for free-floating VGTs are expressed to be[6]:

$$\dot{\mathbf{x}}_c = \mathbf{J}_c\,\dot{\mathbf{d}} \qquad (6)$$

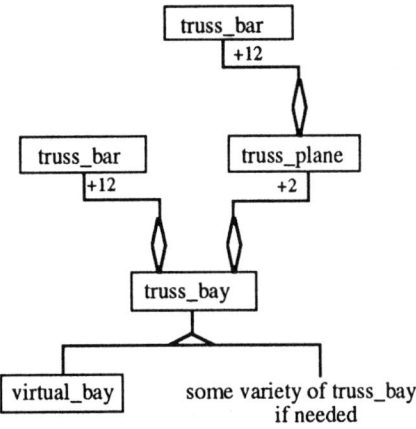

Fig. 5 Scenario of VGT Bay Model

$$\mathbf{H}_0 \dot{\Phi}_0 + \sum_{i=1}^{N} \mathbf{H}_i \dot{\mathbf{d}}_i = const \quad (7)$$

Equation (6) serves for the task control of VGTs by specifying trajectory $\mathbf{x}_e$ for the working endplane, where $\mathbf{J}_e$ is generalized Jacobian matrix[6]. Equation (7) shows the conservation of the angular momentum of structure, where Φ_0 is the orientation of the free endplane. Thus, the basic attributes are easily determined to be matrices $\mathbf{H} = \{\mathbf{H}_1, ..., \mathbf{H}_N\}$, $\mathbf{H}_0$ and $\mathbf{J}_e$, as well as the initial position of structure. The operational interface of 'adaptive_truss' are listed as below:

class adaptive_truss {
   access a bay via the bay list;
   define structural model for control task;
   locate the initial position of the structure;
   simulate the moving of a time step of VGT;
   reset the structure to be the original state; }

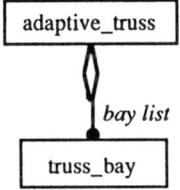

Fig. 6 Object Hierarchy of VGT

V. Object Model of Actuation System

5.1 Two-Level Hierarchy of Control Architecture

An adaptive truss is usually equipped with a lot of actuators, which gives rise to the crisp requirement for computers' ability. Therefore, an urgent theme for the space application of adaptive structures is raised, that is how to construct a control architecture for actuators so as to relieve the excessive demands for the computer conditions. Regarding this problem, a good idea that comes from the thinking of object-oriented problem decomposition, is to make each actuator be an autonomous moving unit, upon whose intelligent abilities some relatively simplified control schema could be established. By using a function-simple CPU board along with a pulse motor and an encoder, an intelligent actuator unit could be obtained, over which a personal computer is then applied to deal with the control of these control units on higher level. A two-level hierarchy of actuation system involved in this investigation has been built up, whose building block diagram is shown as Fig. 7, where PMC-2(98) is a control board for pulse motors, which can simultaneously handle two pulse motors' autonomous motion including treating the feedback from the encoder, check the motor's state, control the motion of the motor according to the commands from the central CPU etc., through motor drive unit HS-330-05.

5.2 Object Abstraction for An Actuator

In response to the composition of an intelligent actuator, conceptually a pulse motor is distinguished from actuators. Firstly a pulse motor, purely as a driving device, is abstracted to be an object class 'pulse_motor', whose attributes includes motor flag, rotary limit-check standard, current status of motor, start pulse ratio and high speed ratio as well as accelerating pulse number etc.. The principal operations of it are concluded to be:

class pulse_motor {
   initialize the LSI of pulse motor;
   wait for motor's stop;
   make a pulse motor rotate to return to its initial state;
   check the state of pulse motor;
   make motor rotate in constant speed;
   immediately make pulse motor stop; }

Secondly an length-type actuator could be constructed by connecting a control board in, for which an object class 'actuator' is identified. The attributes of 'actuator' embodies the port number of personal computer that corresponds to the control board of the motor, the stroke limit, specified length step, rotary direction of the motor and the current state of motor. The methods of it are listed as follows.

class actuator {
   set the initial state of actuator;
   stop actuator;
   make actuator move a length step in constant speed;
   make actuator return to its original position;
   check the moving state of actuator;
   send new moving demands to actuator; }

Object class 'actuator', as a matter of fact is the upgrade of pulse motors from both of function and concept. It needs some basic attributes of 'pulse_motor' to support it, so that naturally keeps the inheriting relationship from 'pulse_motor'.

5.3 Object Abstraction for Actuation System

The control demands to the central CPU from actuators are intermittent, because it usually has some time duration for one actuator's action, within which an actuator autonomously moves without any command from the higher level controller. From the point of computer communication, the actuators just connect with the central controller at some moment for exchanging information, after then they both work in a parallel way. Hence, it needs to abstract the whole actuators that corresponds to the second level control (Fig. 7) into an integrated object class 'motor_drive_system' to represent such independence of actuation system. Object class 'motor_drive_system' is responsible for the management of all actuators, and the communication with the central con-

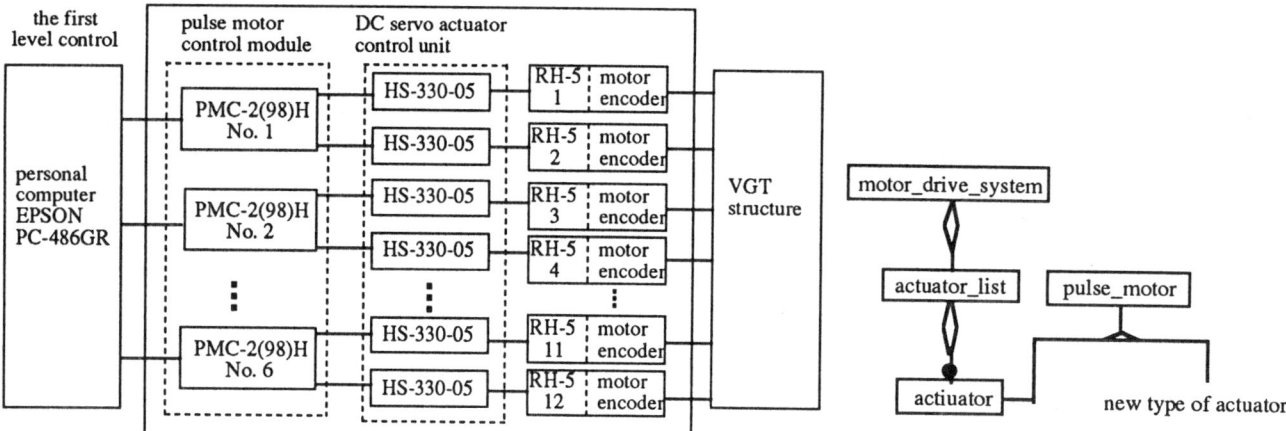

Fig. 7 Two-Level Actuation Control Architecture

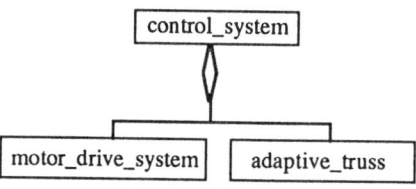

Fig. 8 Object Model of Actuation System

trol. To this end, the fundamental attributes of 'motor_drive_system' are designed to be the speed and length-step cashes, as well as the actuator list, whose integrated operating interface is provided as below:

   class motor_drive_system {
     pack the VGT together;
     create the action data for all actuators;
     modify the running data and make the actuation system work a time step;
     check the moving state of the system;
     reset the system to be the initial state;
     interrupt the running of the system;
     wait for the stop of system;
     set up the running states of all actuators; }

VI. Integration of VGT with Actuation System

While an actuation system is combined with a VGT 'adaptive_truss', an active structural system with adaptivity is then obtained. However, the object 'adaptive_truss' only provides the control simulating for one time step, also 'motor_drive_system' is merely in charge of one actuators' action. Moreover these two activities relate in a 'and-relationship'. Consequently, other object class is needed to coordinate them to complete a task. As an autonomous structural system, an adaptive truss has to recognize the working environment by itself, and design the control tactics upon its knowledge. Due to these reasons, we may define an object class 'control_system' to deal with these jobs, so as to offer a supervisory control on both of 'adaptive_truss' and 'motor_drive_system'. Object class 'control_system', along with 'adaptive_truss' constitute the first level control.

The significant attribute of 'control_system' is the aggregation of 'adaptive_truss' and 'motor_drive_system'. In addition, due to that currently just only a docking task is taken into account, its basic attributes merely contains the task identifier, moving state flag, position of docking port, space constraints etc.. The methods of it are concluded to be:

   class control_system {
     (private methods for internal use)
       set up trajectory for the docking endplane;
       evaluate designed trajectory;
     (public methods for external use)

       build up a control task;
       execute the task;
       interrupt the executing of task;
       reset the control system; }

```
              control_system
             /              \
   motor_drive_system    adaptive_truss
```

Fig. 9 Object Model of an Active VGT

VII. Implementation of Docking Control of VGT

After the object model for VGTs is built up, our issues then turn to the synthesis and control of objects towards to the achievements of particular tasks, which should also be contained in the knowledge base system. The object control oriented to control tasks relies on the algorithms from the problem domain. Its implementation is different from those upon the usual procedural abstractions. It more emphasizes how to optimally put the objects into work from both of the functions and the time sequencing, such that gives rise to an issue to the dynamic analysis of objects which is carried out on the dynamic modeling of objects.

The dynamic analysis of objects examines changes to the objects and their relationships over time, which is denoted on a dynamic model and leads to an easy object-oriented programming. The dynamic model is established upon the concepts of event and state. The attribute values and links held by an object are called its state. Over time, the objects stimulate each other, resulting in a series of changes of their states. An individual stimulus from one object to another is an event. The pattern of events, states, and state transitions for a given class an be abstracted and represented as a state diagram. A state diagram is a network of states and events, whose nodes are states that may be either the attributes' states or child objects, and directed arcs are transition labeled by event names. The state diagrams for the various classes combine into a single dynamics model via shared events.

Below, three diagrams are given out to represent the dynamic model of the docking control for a free-floating VGT. These diagrams, each of which is nested type standing for the generalization of states, are in response to object classes 'control_system', 'adaptive_truss' and 'motor_drive_system' individually. The notation "(...)" following an event name lists the attributes of the event, "[...]" denotes the condition the event occurs, and "/" implies an action aroused by the event. The expression "do: ..." means an activity that characterizes a state. As to those events that can cause an action to be performed without causing a state change, the event name is written inside the state box, followed by a "/" and the name of the action.

It is easy to note that objects 'adaptive_truss' and 'motor_drive_system' are concurrently-active objects during the overall control process. They just exchange messages indirectly at the state '**do**: communicate with ...' through central control 'control_system'. In addition, all bays also show concurrency at the state "**do**: set up kinematics" (Fig. 11).

VIII. Docking Experiment of VGT

Based upon the object model and dynamic model depicted previously, a motion control implementation VGTMOT for a free-floating VGT is developed, which is coded by Turbo C++ and runs on a personal computer EPSON-486. With VGTMOT, a docking experiment[6] for a VGT model is carried out to investigate the applications of adaptive structures in the space construction. The VGT model is suspended to simulate the null-gravity condition, and to be docked to a pre-specified docking port. Fig. 13 shows the initially packed shape of the model, and Fig. 14 gives out the final shape of the model after finishing dock-

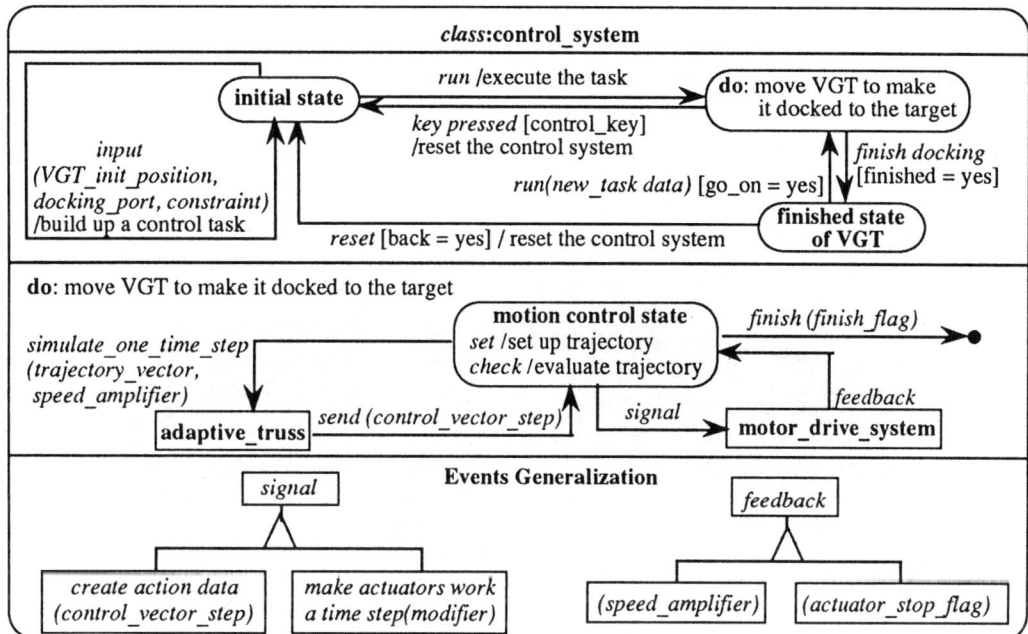

Fig. 10 Diagram of Object Class 'control_system'

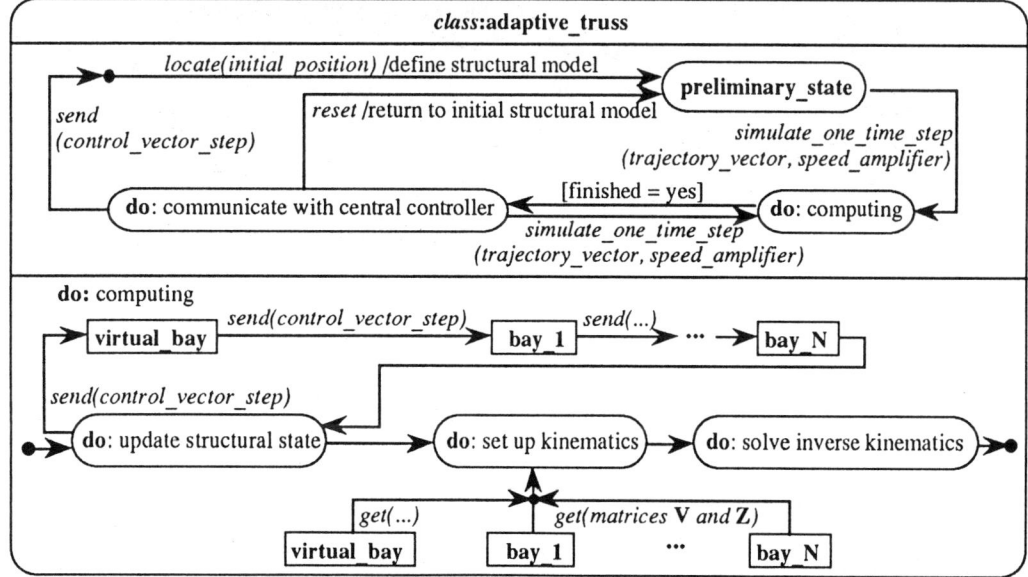

Fig. 11 Diagram of Object Class 'adaptive_truss'

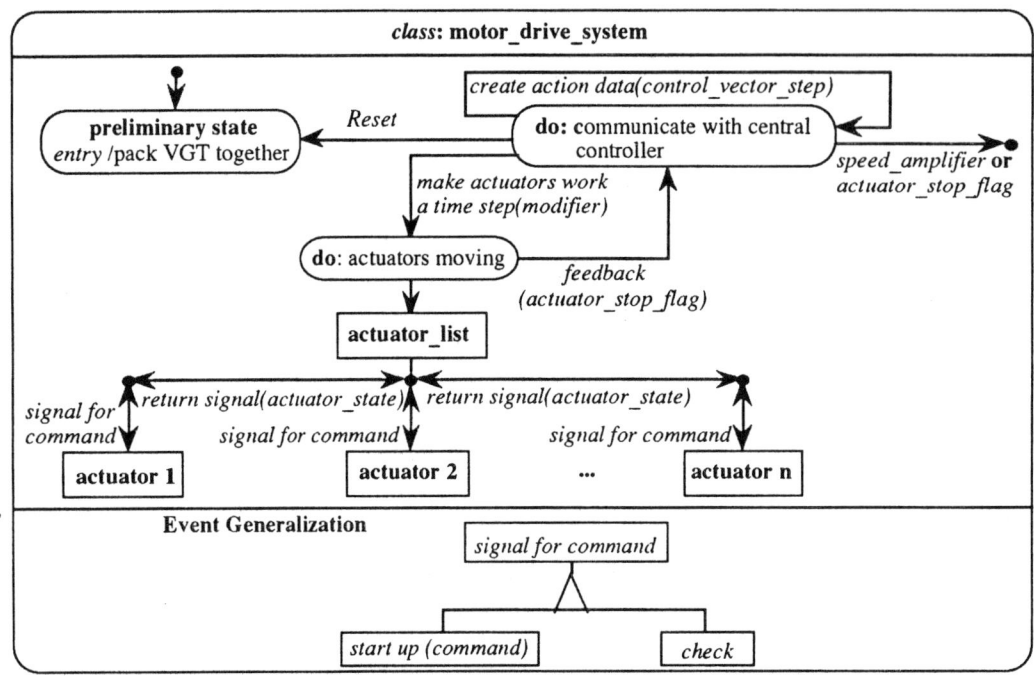

Fig. 12 Diagram of Object Class 'motor_drive_system'

Fig. 13. Initial Shape of a Free-Floating VGT Model

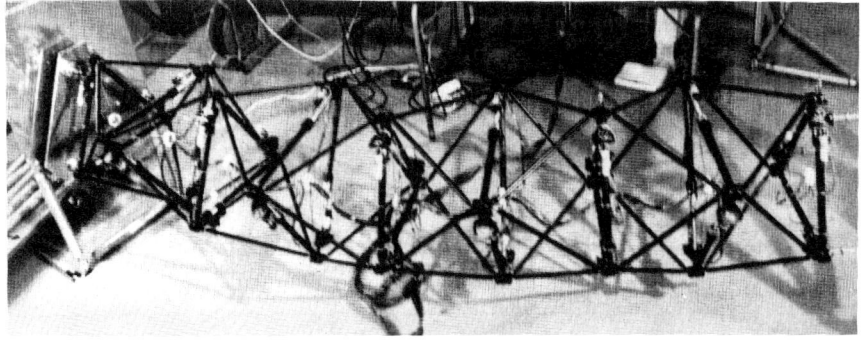

Fig. 14. Final Shape of a Free-Floating VGT Model

ing.

Conclusions

In order to establish a knowledge base for the motion control of a free-floating VGT, the application of an object-oriented approach in the structural control is investigated. Through the problem decomposition oriented to objects, an object model that covers the structural kinematics and actuation system equipped, is built up to represent the motion knowledge held by a free-floating VGT. After then, a synthesis of objects provides us the problem solving for a particular control task. The synthesis process is carried out through a dynamic analysis on the dynamic model, which distinctly exposes the synchronization among various objects over the control process so that results in a concurrent implementing to the control task. These two models lead to a prototype system VGTMOT. To aim at the future space construction by using adaptive structures, a docking experiment is carried out with VGTMOT, the result of the which shows the availability of the developed knowledge base system.

Acknowledgments

The authors would like to acknowledge Profs. K. Higuchi (ISAS) and M. Miki (Univ. Osaka Prefecture) for useful discussion of this study.

References

1. Wade, B.K., J Fanson & E. Crawley, 1990, "Adaptive Structure", *J. Intell. Mater. System and Structure*, Vol. 1, pp. 157-174.
2. Natori, M.C., 1989, "Outline of Space Structure Engineering", *Proc. JSCE*, No. 398/I-10, pp. 1-16 (in Japanese).
3. Miura, K. & H. Furuya, 1988,"Adaptive Structure Concept for Future Space Applications", *AIAA Journal*, Vol. 26, No. 8, Aug., pp. 995-1002.
4. M.C. Natori, S.Y. Huang, K.I. Miura, K.Y. Miura & M.M. Sakamaki, 1992, "Motion Control of A Variable Geometry Truss", *Third Internationl Conference on Adaptive Structures* (San Diego, California, U.S.A.).
5. S.Y. Huang, M.C. Natori, & K.I. Miura, "Kinematics of a Free-Floating Variable Geometry Structure", submitted to *Journal of Guidance, Control, and Dynamics*
6. S.Y. Huang, M.C. Natori & K.I. Miura, 1993, "Inverse Kinematics and Experiment of Docking Control for a Free-Floating Variable Geometry Truss", *Fourth International Conference on Adaptive Structures* (IHK, Cologne, Germany)
7. Coyne, R.D., Rosenman, M.A., Radford, A.D., Balachandran, M., and Gero, J.S., "Knowledge-Based Design System", *Addition-Wesley*, New York, 1990.
8. Grady Booch, "Object Oriented Design with Applications", *The Benjamin/Cummings Publishing Company, Inc.*, 1991.
9. James Rumbaugh, Michael Blaha, William Premerlani, Frederick Eddy & William Lorensen, "Object-Oriented Modeling and Design", Pr*entice-Hall International Editions*. 1991.
10. Nakai, S., Fukuwa, N., Hirose, K., Katukura, H. and Ebihara, M., 1991, "A Knowledge-Based Approach to the Structural Analysis of Large Space Structures", *Proc. of the AIAA/ASME/ASCE/AHS 32nd Structures, Structural Dynamics and Materials Conference*, AIAA, Washington, DC.
11. Darcy M. Bullock & Irving J. Oppenheim, 1992, "Object-Oriented Programming in Robotics Research for Excavation", *Journal of Computing in Civil Engineering*, Vol. 6, No. 3.
12. James H. Garrett Jr. & M. Maher Hakim, 1992, "Object-Oriented Model of Engineering Design Standards", *Journal. of Computing in Civil Engineering*, Vol. 6, No. 3.
13. Mitsunori Miki & Yoshisada Murotsu, 1993, "Object-Oriented Approach to Modeling and Analysis of Truss Structures", *Proc. of the AIAA/ASME/ASCE/AHS 34nd Structures, Structural Dynamics and Materials Conference*, AIAA, La Jolla, California.
14. Richard S. Wiener, Lewis J. Pinson, "An Introduction to Object-Oriented Programming and C++", *Addison-Wesley Publishing Company*, 1988.

Appendix

Object Model Notation[9]

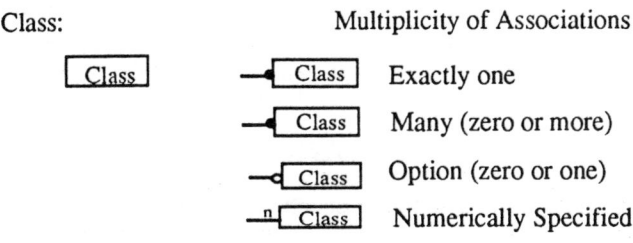

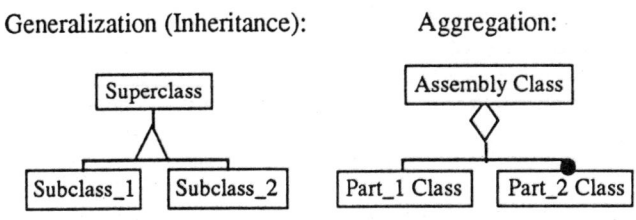

A BIOLOGICALLY INSPIRED CONTROLLER FOR SOUND AND VIBRATION APPLICATIONS

James P. Carneal*

Chris R. Fuller**

Vibrations and Acoustics Laboratories
Virginia Polytechnic Institute and State University
Blacksburg, Virginia 24061-0238

Abstract

A biologically inspired control approach for reducing vibrations in and radiated sound power from distributed elastic systems has been experimentally verified for narrowband excitation. The control paradigm approximates natural biological systems for initiating movement, in that a low number of signals are sent from an advanced, centralized controller (analogous to the brain) and are then distributed by local rules and actions to multiple actuators (analogous to muscle fiber). The controller was applied to attenuate vibrations in a beam and radiated sound power from a plate. Experimental investigations of two different local learning rules were carried out including controller convergence considerations and a comparison with the standard MIMO Filtered-X LMS feedforward control approach. In general the results have demonstrated that the biological control approach has the potential to control multi-modal response in distributed elastic systems using an array of many actuators with a reduced order main controller. Performance was comparable with the Filtered-X LMS approach. Thus significant reductions in control system computational complexity have been realized by this approach.

I. Introduction

Recent work has demonstrated the potential of active control of distributed elastic systems using multiple, independent actuators and sensors. In work concerned with the control of sound radiation from vibrating panels, the importance of number of channels of control and optimization of the transducer position and shape has been demonstrated[1]. However these investigations were carried out for a fixed frequency and it is apparent that for good control over a bandwidth of frequencies, the control actuators and sensors need to be adaptive in shape. At first sight this problem could be solved using an overall transducer broken up into many individual small elements each connected by an individual control channel. In this situation the control transducer would effectively re-optimize its configuration for different conditions by adaptively weighting each transducer segment. Meirovitch and Norris[2] have demonstrated the advantage of such an approach by considering fully distributed control in reducing control spillover. The disadvantage of this approach is that, for systems with a high modal density, the number of actuators and sensors required becomes extremely large. A high number of control channels has a number of problems mainly associated with memory requirements and computational time in the hardware systems used to implement the control. In addition, collinearity of transducer transfer functions causes stability problems in systems with a high number of transducers.

A new approach of controlling distributed elastic systems is presented. The approach is inspired by the action of biological natural systems where a low number of main signals are transmitted from the brain to a large area of muscle tissue to activate many independent segments of muscle. The signals then stimulate local action which is governed by, for example, chemical interaction of locally connected nerves, etc., resulting in multiple subsequent signals for individual muscle cell elongation or contraction. Put simply, a signal is sent from a central complex processor (the brain) and then is broken into multiple signals by local simplified control rules (muscle cells, etc.)[3].

This paper details an experimental implementation of such a process, which has been previously studied in a limited analytical investigation[4]. A distributed elastic system is harmonically excited and controlled by multiple control inputs. In the biologically inspired (BIO) control approach, one control input is chosen as the "master" and is under direction of the main, centralized advanced controller. The other "slave" inputs derive their control laws by localized, simple learning rules related to the behavior of their neighbor actuators and are independent of the main controller direct signal, as seen in Figure 1.

In the following sections, the control algorithm is outlined for the "master" control input and the local learning approaches, the "phase variation" and "optimal distribution" methods. The experimental investigation is discussed including performance metrics, experimental

---

*Ph.D. Candidate, Dept. of Mechanical Engineering
**Professor, Dept. of Mechanical Engineering, Associate Fellow, AIAA.

Copyright © American Institute of Aeronautics and Astronautics, Inc., 1994. All rights reserved.

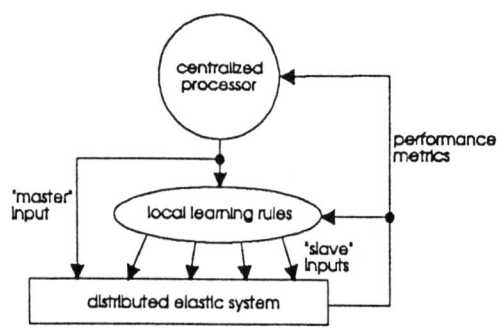

Figure 1. Biological control approach

setup and procedure. Results are presented for the vibration and radiated sound power control experiments, followed by concluding remarks.

II. Controller

The control algorithm is outlined with specifics included for the physical implementation of the BIO controller. For more complete theoretical development and analysis, the reader is referred to Carneal and Fuller[5].

A. Master Control Input

The control algorithm for the master control channel is a single output time averaged gradient (TAG) LMS control algorithm, shown in block diagram form in Figure 2. The TAG algorithm provides a direct cost function measurement which was needed to implement the phase variation local control approach, discussed below. Also, a time averaged cost function has the additional advantage of averaging out the random noise in the error signal which provides a better estimate of the error.

As seen in Figure 2, the error (e) is the sum of the primary disturbance (d) and the secondary control sources where **G** is the transfer function between the reference and the error sensor, and **H** is the transfer function between the control input and the error sensor. By defining the cost function (J) as the square of the error (e), a quadratic function of the FIR filter weights (w) is formed with a unique global minimum. By adaptively changing the FIR filter weight (w), the error (e) can be minimized.

From the standard gradient descent technique, the update equation for the adaptive filter at iteration k is written as:

$$\mathbf{w}_k = \mathbf{w}_{k-1} + \mu(-\nabla_k) \quad (1)$$

where w is the FIR filter coefficient, μ is the convergence coefficient and ∇_k is the gradient of the mean square error performance surface. The gradient is defined as the change in the time averaged cost function with respect to the vector of FIR filter coefficients, written as:

$$\nabla_k = \frac{\partial J}{\partial \mathbf{w}} = \left(\frac{J_k - J_{k-1}}{\mathbf{w}_k - \mathbf{w}_{k-1}} \right). \quad (2)$$

The time averaged cost function is defined as:

$$J_k = \frac{1}{N} \sum_{i=1}^{N} J_i \quad (3)$$

where J_i is the cost function at time step i, and N is the total number of time sample points.

The objective of the TAG algorithm is to minimize the cost function (J). The BIO controller then uses information from the master control channel to implement additional actuators downstream of the master control channel through the use of local control approaches. Note that we have not discussed a procedure for choosing the particular actuator to which the master signal is to be applied. This could be achieved, for example, by a simple search scheme.

B. Local Control Approaches

Once the master control channel is updated, the control inputs to the slave channels are derived using local learning rules. Two local learning rules are first analyzed, the phase variation and the optimal distribution methods. A third local control approach, the moment distribution, is developed from the optimal distribution for the specific application to vibration control.

Mathematically, the local control approach can be expressed as complex gains (γ) applied to the master control channel to determine the control inputs applied to the slave control paths. Note that the method of determining these gains is not limited to the above methods, but could also include experimental measurements, neural network, and genetic algorithm methods.

1. Phase Variation

The following approach for a local learning rule is based on the observation that, for a distributed elastic system with little damping, regions are close to either in-phase or out-of-phase with a fixed point on the structure. Thus the local learning approach is to take the master control input, after reaching an optimal value, and apply it to a slave control path. Figure 3 shows the schematic for the phase variation control system. Using the traditional adaptive feedforward control system schematic (seen in Figure 2) as a baseline, note that H_1 is now defined as the master

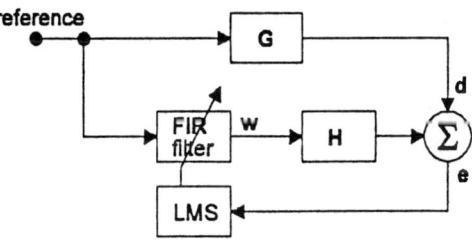

Figure 2. Conventional adaptive feedforward control system schematic

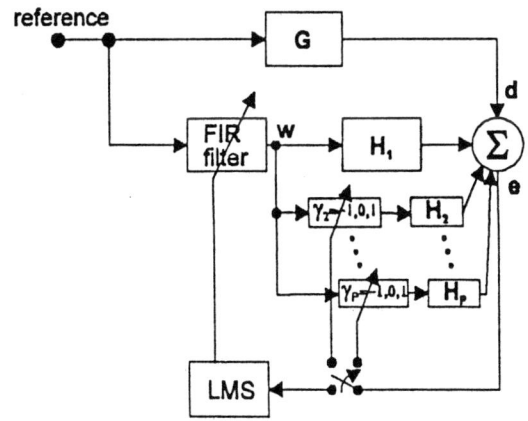

Figure 3. Phase variation control system schematic

control path. Also, note the addition of the slave control paths, H_2 to H_P, along with gains γ_2 to γ_P, where P is the total number of control paths.

In this case, the phase variation algorithm is defined as follows. The master control voltage is applied to the neighbor slave actuator (i.e., the actuator immediately alongside). The slave control voltage is varied in-phase ($\gamma=+1$), out-of-phase ($\gamma=-1$) or turned off ($\gamma=0$) while the cost function, J, is observed for each change. The gain (γ) that results in the lowest cost function is kept and the process is then applied to the next slave control path until all slave control paths are progressively tested. Note that the sequence of activating the slave paths can be varied. By this method, a distributed actuator with a generalized function that drives a response similar to the uncontrolled vibration distribution with low control spillover is constructed. Note that the previous process is sub-optimal as are many biological systems. However, it is believed that good control performance will still be obtained.

2. Optimal Distribution

The optimal distribution local learning rule was solved *a priori* and was developed from adaptive feedforward control theory. Essentially, the measured or calculated control path transfer functions between the actuators and sensors are used to calculate the theoretical optimal gains. These gains are then scaled relative to the master actuator which calibrates the slave gains with respect to the master actuator. The weights of the adaptive filter of the master channel can then be adapted to minimize the cost function, and simultaneously the slave control signals will be derived by proportion.

The implementation of the optimal distribution into the control system is shown in Figure 4. The control signal fed to the master control input (w) is operated on by the optimal complex gains ($\gamma_2..\gamma_P$) and fed to the slave transfer functions ($H_2..H_P$).

Upon application of the optimal distribution to control of total beam out-of-plane vibrational energy density, an interesting phenomenon became apparent. As can be seen in Figure 5, the optimal distribution approximates the static moment distribution in a simply supported beam when a point force is applied. This observation lead to the development of the moment distribution method, described below. Another important observation was that the optimal distribution is *independent* of frequency.

3. Moment Distribution

The optimal distribution was seen to approximate the static moment distribution in a simply supported beam when a point force was applied. Therefore, a local rule was developed based on the internal beam moment, which is zero at the boundary and linear to the maximum moment, which is at the position of the point force. The moment distribution was derived from the position of the control input relative to the point force, defined as:

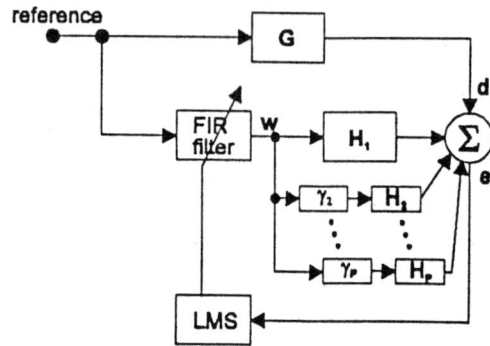

Figure 4. Optimal distribution control system schematic

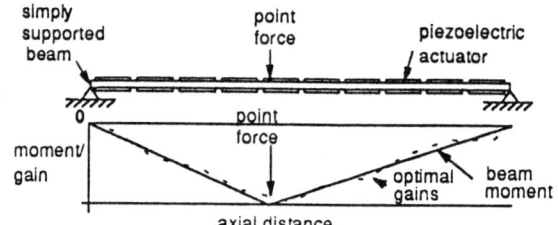

Figure 5. Beam optimal gain distribution

$$\gamma_i = \frac{x_{p,i}}{x_d} \qquad \forall x_p < x_d$$
$$\gamma_i = \frac{x_{p,i} - l}{x_d - l} \qquad \forall x_p > x_d \qquad (4)$$

where x_p is the position of the control input, x_d is the position of the disturbance, and l is the length of the beam. The gains were then scaled so that the gain of the master actuator was 1. As with the optimal distribution, this distribution was *independent* of frequency. However, unlike the optimal distribution, the moment distribution was easily calculable since its form is simply a gain of constant slope. Note that this local rule was applied to the vibration control case only.

III. Experimental Investigation

A. Experimental Setup

The implementation of the biologically inspired (BIO) controller is detailed, followed by a description of the experimental apparatus for the vibration and radiated sound power control experiments.

1. Controller

The BIO controller was programmed on a TMS320C30 digital signal processor (DSP) mounted in a 80386 33 MHz Personal Computer (PC). To increase the maximum sampling rate, the number of operations performed by the DSP was minimized. Therefore, the I/O functions, the cost function and FIR filter calculations were performed on the DSP, while the controller logic was performed on the PC and communicated to the DSP. The sample rate for the controller was set at ten times the excitation frequency with each error signal consisting of 250 time sample points, leading to an update for the FIR filter approximately every 0.05 seconds at a sampling frequency of 5000 Hz. Figures 6b and 7b shows the experimental schematic of the BIO controller for the vibration and radiated sound power experiments, respectively. Note that the structure of the algorithm is similar in the DSP and PC due to the similar definition of cost function, which is discussed later. The only differences are the number of input and output channels.

Again referring to Figure 6b and Figure 7b, the BIO control algorithm is discussed. For the "master channel only" case, the adaptive gains (γ) were set to zero, rendering the control algorithm as a single output system. For the phase variation method, the master channel was allowed to converge, then the FIR filter was locked at its optimum value. The master channel FIR filter coefficients were applied to the slave control channel, one by one, alternating left then right of the master control channel. The filter coefficients were tried in-phase ($\gamma=+1$), out-of-phase ($\gamma=-1$) or turned off ($\gamma=0$) while the cost function (J) was observed for each change. The gain (γ) that resulted in the lowest cost function was kept, the cost function was measured, and the process was applied to the next slave control channel until all slave control paths were progressively tested.

For the optimal and moment distributions, the adaptive gains (γ) were determined *a priori* and then scaled relative to the master actuator. The gains were entered to the control program prior its execution. Once the FIR filter update equation was invoked, the master channel FIR filter was updated as described in Eqs. (1) through (3). However, before the digital to analog conversion process was performed, the master channel FIR filter coefficients were multiplied by the adaptive gains (γ), thereby generating the slave control outputs. When the D/A process was triggered, the master and slave control signals were output simultaneously.

It was stated earlier that the gains (γ) could be determined analytically or experimentally. For vibration control, the optimal distribution gains were determined analytically using adaptive feedforward control theory[5]. However, for the radiated sound power experiment, the optimal distribution gains (γ) were determined experimentally from a full order time averaged gradient (TAG) LMS algorithm, where each output channel was fully independent of any other output channel (as compared to the master/slave relationship of the BIO controller). Once the full order LMS controller had converged, the FIR filter weights were recorded and converted into complex gains. These gains were then scaled so that the master gain was 1. The scaled gains were then input to the BIO control algorithm prior to the invocation of the FIR filter update equation.

2. Vibration Control

To experimentally verify the BIO controller applied to vibration control, an experiment was performed on a steel beam of length 0.380 m, width 0.038 m and thickness 0.0048 m. A photograph and a schematic of the experimental setup is shown in Figures 6a and 6b, respectively. The beam was mounted on a test rig with shims (which allowed rotation of the boundary, but little transverse displacement) in order to approximate simply

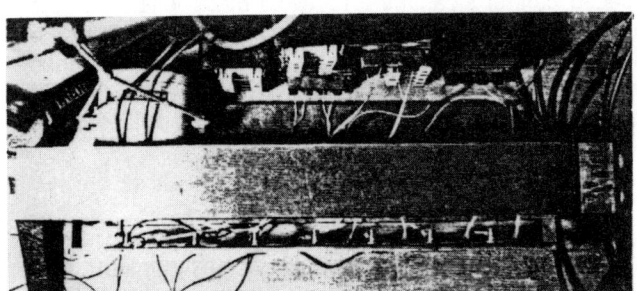

Figure 6a. Photograph of vibration control experiment

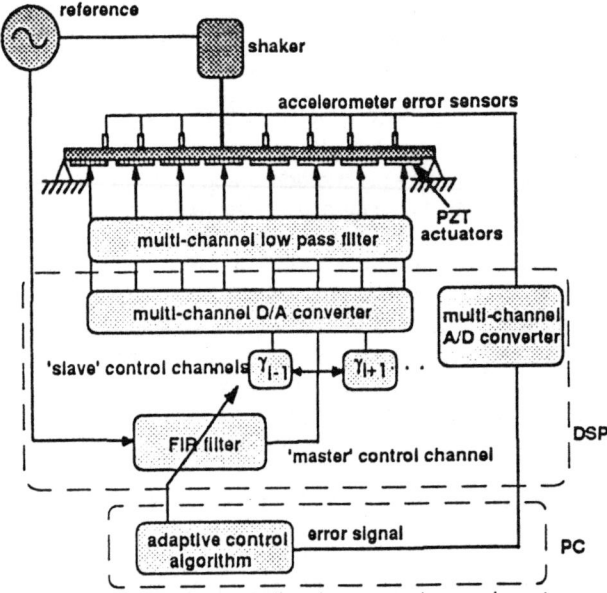

Figure 6b. Schematic of vibration control experiment

supported boundary conditions. Harmonic, narrow band excitation was provided by an approximation of a point force through a shaker mounted at 0.051 m. Experiments were performed at excitation frequencies of 400 Hz and 604 Hz, both of which are off-resonance between the second and third beam modes.

Error signals were provided by the output of seven matched accelerometers located along one side of the beam at evenly spaced locations. To approximate a cost function of beam vibrational energy density, a sum of the mean square voltage read from the accelerometers was provided to the LMS algorithm (and the phase variation method of the BIO controller). The cost function at time step t is defined as:

$$J_t = \sum_{i=1}^{M} |v_i|^2 \qquad (5)$$

where v_i is the accelerometer voltage, subscript i indicates the accelerometer number index, and M is the total number of accelerometers.

Control inputs were provided by eight ceramic piezoelectric (PZT) patches each having dimensions 0.032 m long x 0.038 m wide x 2.59E-04 m thick. The PZT patches were directly bonded to the beam at evenly spaced central locations. To avoid mass loading of the beam, the PZT actuators were bonded to one side of the beam only, instead of the usual PZT pair. It has been experimentally determined that the one-sided PZT actuators were as effective as the two-sided actuators for exciting transverse motion in a beam[6].

3. Radiated Sound Power control

To experimentally verify the BIO controller applied to radiated sound power control, an experiment was performed on plate made of aluminum of length 0.380 m, width 0.300 m and thickness 0.0031 m. A photograph and a schematic of the experimental setup is shown in Figures 7a and 7b, respectively. The plate was mounted in a common wall between a reverberant chamber and an anechoic chamber, which comprises the Transmission Loss Test Facility at Virginia Tech. The mounting apparatus clamped the plate at the boundaries (which allowed negligible rotation of the boundary and negligible transverse displacement) in order to approximate clamped boundary conditions. Harmonic, narrow band excitation was provided by an approximation of a oblique plane wave from a speaker in the source (reverberant) chamber placed at 45 degrees from the plate normal. The speaker was placed at 0.250 m from the plate so that the direct radiated field dominated the input. Experiments were performed at excitation frequencies of 319 Hz, which is off-resonance between plate mode (1,1) and (2,1), and 397 Hz, which is the plate mode (2,1) resonance.

Figure 7a. Photograph of radiated sound power experiment

Error signals were provided by the output of five microphones mounted in a hemispherical configuration in the receiving (anechoic) chamber, which was experimentally determined to have a cutoff frequency of approximately 300 Hz. The microphones were placed according to equal area calculations of a hemisphere. To approximate a cost function of radiated sound power, a sum of the mean square voltage read from the microphones was provided to the LMS algorithm (and the phase variation method of the BIO controller). The cost function at time step t is defined as:

$$J_t = \sum_{i=1}^{M} |v_i|^2 \qquad (6)$$

where v_i is the microphone voltage, subscript i indicates the microphone number index, and M is the total number of microphones.

Control inputs were provided by six ceramic piezoelectric (PZT) patches each having dimensions 0.064 m long x 0.038 m wide x 2.59E-04 m thick. The PZT patches were directly bonded to the plate at evenly spaced central locations forming a 3 by 2 grid on the plate. Difficulties associated with providing a good seal between the source and receiving chambers demanded that the PZT actuators were bonded to one side of the plate only, instead of the usual PZT pair.

B. Performance Metrics

The BIO controller performance was analyzed in terms of reduction in cost function, elastic system vibration, and convergence parameter for a stable system. For a

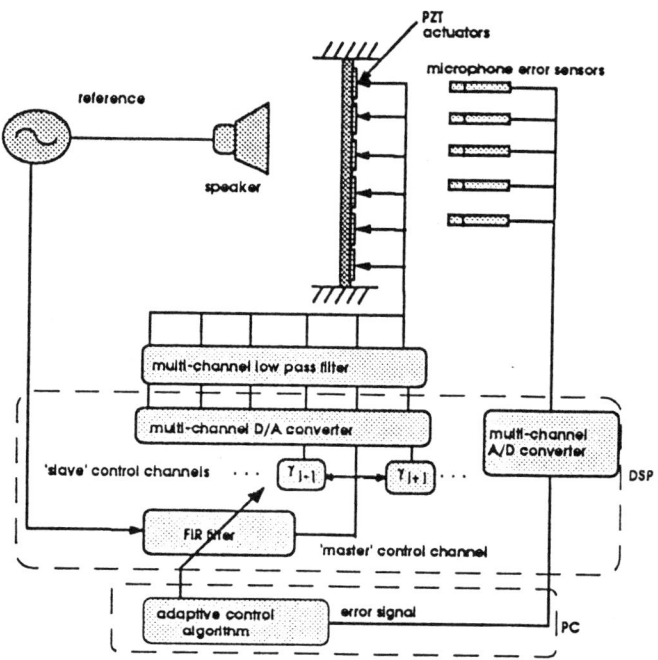

Figure 7b. Schematic of radiated sound power experiment

benchmark comparison, controller performance was compared to the performance of a conventional MIMO Filtered-X LMS controller.

1. Controller Performance

Performance of the control algorithm was defined as the reduction in the cost function as:

$$reduction(dB) = 10 \log_{10} \left(\frac{J_{uncontrolled}}{J_{controlled}} \right) \qquad (7)$$

where $J_{uncontrolled}$ is the uncontrolled cost function and $J_{controlled}$ is the controlled cost function.

For a more detailed analysis of the response of the elastic system, a laser vibrometer was used to determine the out-of-plane vibrational characteristics for the uncontrolled and controlled cases. For the vibration control test, measurements were taken at 21 evenly spaced central axial points on the beam described above. For the radiated sound power test, measurements were taken at

25 evenly spaced locations which formed a 5 by 5 point grid on the plate described above.

2. Controller Convergence

It was predicted in Carneal and Fuller[5] that the maximum convergence parameter (μ_{max}) for stability would be less for a BIO controller compared to one for a single output system. Experiments were performed to verify this trend.

To determine the maximum convergence parameter, the control program was run with the convergence parameter set at different values. The program was allowed to continue until it was determined to be unstable, marginally stable or stable. The maximum convergence parameter was defined as the largest value that resulted in a stable system.

To determine the stability of the system, the variation in cost function was determined, which was defined as difference in the maximum and minimum cost function over several iterations of the update equation (Eq. (1)). The stability of the system was divided into three categories: unstable, marginally stable and stable. A stable algorithm was defined as the variation in cost function being less than 5% of the uncontrolled cost function. A marginally stable algorithm was defined as the variation in cost function being greater than 5% of the uncontrolled cost function, but the maximum cost function did not increase with time. An unstable system was defined as the maximum cost function increasing with time.

3. MIMO Filtered-X LMS Comparison

To provide a benchmark comparison with existing control algorithms used, the BIO controller performance was compared to MIMO Filtered-X LMS controller performance when applied to the same elastic system on the same test rig. Since the cost function was not directly accessible for the Filtered-X LMS, the uncontrolled and controlled cost function (J) was determined by measuring the voltages (v) of the error sensors and summing the square of these voltages:

$$J_t = \sum_{i=1}^{M} |v_i|^2 \qquad (8)$$

where M is the number of error sensors. Note that this process is similar to the determination of cost function for the BIO controller as seen in Eqs. (5) and (6).

C. Experimental Procedure

Several different test cases were performed including off-resonance and on-resonance frequency excitation for the master control channel only case and all of the local learning rules. However, the experimental procedure for a set of tests was the same and is detailed by performance metric.

1. Controller Performance

The experimental procedure for the master and phase variation test case was as follows: First, the disturbance was turned on. The control program was run, the convergence parameter (μ) was set at approximately .01, the master control channel was set to the desired value, and the sampling frequency was set at ten times the excitation frequency. The uncontrolled elastic system out-of-plane vibrational velocity was then measured using the laser vibrometer, and the uncontrolled cost function was measured. After the invocation of the control algorithm, the master control channel was allowed to converge, then the cost function and the elastic system vibration were measured for the "master" case. The phase variation algorithm was invoked, and after completion, the cost function and the elastic system vibration was measured for the "phase variation" case. The controller was then turned off.

For the other local learning rules (the optimal and moment distributions), the disturbance was set at approximately the same level. Once the control programs were run, the convergence parameter (μ) was set at approximately .01, the master control channel was set to the desired value, and the sampling frequency was set at ten times the excitation frequency. Uncontrolled elastic system vibration and cost function were then measured. After the controller was started and allowed to converge, the elastic system vibration and cost function were measured for the controlled case.

2. Controller Convergence

The experimental procedure for the controller convergence test was as follows: The program was executed, the step size, the master actuator, and the sampling frequency were set to desired values. The uncontrolled cost function was recorded, then the control algorithm was invoked. While recording the cost functions at each iteration, the algorithm was allowed to continue until it was determined to be unstable, marginally stable or stable.

3. MIMO Filtered-X LMS Controller

A MIMO Filtered-X LMS controller[7] was applied to control both vibration and radiated sound power using the experimental rig for the beam and plate, respectively. The program was run and the convergence parameter and the sampling frequency were set. A system identification was then performed to determine the transfer functions between the control inputs and the sensors for use in the Filtered-X algorithm. The disturbance was turned on and set at an appropriate level, then the uncontrolled error sensor voltages were recorded and the control algorithm was invoked. After the controller had converged, the controlled error sensor voltages were recorded.

IV. Results

An extensive experimental investigation was performed, however for brevity, only select results are presented. It should be noted that the results presented are indicative of the overall trends observed, and in general, the stated conclusions apply to all the results attained.

A. Vibration Control

The experimental results are presented for the BIO controller applied to vibration control. Two different excitation frequencies were tested, both of which are off-resonance between the second and third modes of the beam and represents a stringent test of the method.

1. Selection of Master Actuator

Analytical models were run to asses the influence of the selection of the master actuator on phase variation performance. Analytical cases[5] were run at a frequency of 400 Hz with each PZT actuator chosen as the master actuator for use in the phase variation algorithm. Table 1 shows the performance of the master actuator, the increase in performance due to the phase variation method, and the control string. The control string of the phase variation method denotes the final implementation of the slave control channels, where 0 indicates off, ⊕ indicates the master control channel (which is by definition in-phase), + indicates channels implemented in-phase, and - indicates channels implemented out-of-phase.

As can be seen in Table 1, the final control performance varied greatly with the choice of master actuator from a low of 0.0 dB for actuators #4 and #8 to a high of 9.8 dB for actuator #2. It is evident that the good performance of actuator #2 was a result of the PZT's co-location with the disturbance force. However, this unique situation of co-located disturbance and actuator is not representative of real control situations, nor is it a true test of distributed control performance.

The increase in control performance achieved with the addition of the phase variation method varied from a low of 0 dB to a high of 6.8 dB, seen in Table 1. Note that for several choices of master actuator, the phase variation method performance increase was less than 3 dB. This was attributed to the limited degrees of freedom associated with fixing the magnitude and varying the phase, as dictated by the phase variation method.

Using the results of Table 1, actuator #7 was chosen to be the master actuator since that actuator best exhibited the capabilities of the phase variation method and had marginal control performance by itself.

As stated previously, the gains determined by the optimal and moment distributions are scaled so that the master actuator gain is 1. The choice of master actuator will not change the relative value of the gain between any one actuator and another and therefore will not influence the optimal and moment distributions.

2. Controller Performance

In this section, the relative control performance for control with the "master" actuator only (referred to as master), then with the local learning rules of the phase variation, the optimal distribution, and the moment distribution methods are compared. For these experiments, the excitation frequency was 400 Hz, the master channel was actuator #7, and the sampling frequency was 4000 Hz. Theoretical[5] and experimental controller performance, seen in Table 2, show the same trends and similar magnitudes, therefore only the experimental results will be discussed.

As seen in Table 2, the master case, where there was only one control input, shows poor control performance of 1.5 dB. Beam vibrational response of the master case, shown in Figure 8, exhibits control spillover in the beam axial location of 160 to 240 mm. The modal decomposition of beam vibrational response, seen in Figure 9, shows a significant reduction in mode 2 with significant spillover into mode 3. It is evident that the master actuator (PZT #7) can effectively couple into mode 2 due to its location along the beam near an anti-node of the second mode. However, this location was also at an anti-node of the third mode, and therefore some of the control energy was used to increase the response of this mode.

While maintaining the control system simplicity of one control channel, the BIO controller using the phase variation method shows admirable control performance of an additional 4.1 dB of attenuation as seen in Table 2. In addition to the master actuator (PZT #7) implemented in-phase, there were three slave control channels implemented out-of-phase and three slave control channels implemented in-phase, as seen in the control string. Beam vibrational response of the phase variation case, seen in Figure 8, shows little control spillover with significant reductions in beam response. The modal decomposition of beam vibrational response, seen in Figure 9, shows reductions in modes 2 and 3, with a slight increase in mode 1 response. It is apparent that the addition of the slave control actuators were able to couple into mode 3 thereby increasing control performance compared to the master case. From these results, it is evident that the phase variation method constructed a distributed actuator (by varying the phase of the master actuator and holding magnitude constant) that coupled effectively into the excited modes of the beam.

The optimal distribution shows good control performance of 24.8 dB, as seen in Table 2 with little or no increase in control system complexity. Beam vibrational response of the optimal distribution method, seen in Figure 8, shows a significant reduction in beam vibrational response with no control spillover. The modal decomposition of beam vibrational response, displayed in Figure 9, shows total reductions in all excited modes. It is obvious that the distributed actuator formed by the optimal distribution method was able to couple effectively into all modes of the beam with no control spillover.

The moment distribution, which was an approximation of the optimal distribution exhibited slightly poorer control performance than the optimal distribution, as expected. As seen in Table 2, control performance was still quite good at 23.6 dB. The modal decomposition of beam vibrational response, seen in Figure 9, shows the same general reduction in modes as the optimal distribution with total reductions in all excited modes. Although the distributed actuator formed by the moment distribution was able to effectively couple into all the modes of the beam, the slight differences due to the approximation (of the optimal distribution) resulted in slightly less reduction of the excited modes.

It was stated earlier that the distributed actuator formed by the optimal and moment distributions was independent of frequency. Tests performed over a wide range of test frequencies and configurations indicated that the optimal distribution provided approximately 24 dB of attenuation for off-resonance harmonic excitation. The moment distribution provided slightly less attenuation of approximately 23 dB. It is evident from the exemplary performance over a wide range of test parameters that the optimal and moment distributions are frequency independent.

3. Controller Convergence and Stability

In Carneal and Fuller[5], it was determined that maximum convergence parameter (μ_{max}) for stability would be less for a BIO controller compared to one for a single output system, due to the additional control inputs. Experiments were run to determine the maximum convergence parameter that could be used for a stable system for the master case and the optimal distribution. For these tests, the system was excited at 604 Hz, the master channel was actuator #7, and the sampling period was 6000 Hz.

Table 3 shows the maximum convergence parameter obtained for a single output system (the master case) and the optimal distribution. The master case has a maximum convergence parameter (μ_{max}) of .07 which is significantly higher than the optimal distribution μ_{max} of .004, which verifies the trends predicted[5].

4. Comparison with 6I6O LMS Control Performance

To gauge the relative magnitude of the BIO controller performance, an experiment was run using an existing 6I6O Filtered-X LMS control system, which used 6 error sensors and 6 control actuators. The 6I6O system had greatly increased complexity over the BIO controller in that the control path transfer function (**H** as defined in Figure 2) was now a matrix and the FIR coefficients were represented by a complex vector, **w**. To determine the updated FIR coefficient vector, matrix calculations were done as per Elliot, et al.[7], resulting in increased control system complexity. For these experiments, the excitation frequency was set at 604 Hz and the sampling frequency was set at 6000 Hz.

Since the existing 6I6O Filtered-X LMS control system had only 6 input channels, accelerometer #4 was not used for an error sensor. In order that the control system was not over determined, only 6 actuators were used. To provide actuator symmetry from the center of the beam, PZT's # 3 and 6 were not utilized as actuators. The uncontrolled and controlled cost function, defined in Eq. (8), was determined by the laser vibrometer measurements at the six accelerometer locations (excluding #4).

As seen in Table 4, the reduction in cost function for 6I6O controller was 19.5 dB which is less than the 25.1 dB obtained by the biological controller using the optimal

Table 1. Influence of master actuator # on phase variation performance (beam; 400 Hz)

Master actuator #	master performance (dB)	phase variation increase (dB)	control string*
1	6.3	0	⊕0000000
2	9.8	0.2	0⊕0+0000
3	8.0	2.7	+0⊕00000
4	0.0	2.5	---⊕++++
5	1.9	1.2	0000⊕+00
6	5.0	2.7	-00+0⊕0+
7	1.6	6.0	-00+++0⊕-
8	0.0	6.8	---++++⊕

Note: ⊕ indicates the master control channel

Table 2. BIO controller performance (beam; 400 Hz)

method	theoretical (dB)	experimental (dB)
master (#7)	1.6	1.5
phase variation control string	7.6 -00+++⊕-*	6.6 -00+++⊕-*
optimal	36.2	24.8
moment	27.5	23.6

Note: ⊕ indicates the master control channel

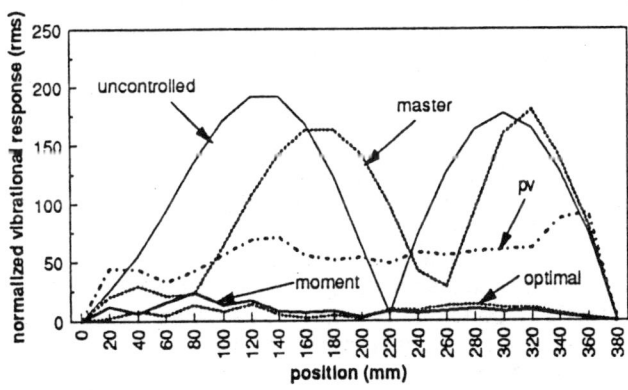

Figure 8. Beam response for uncontrolled and master, phase variation, optimal and moment tests

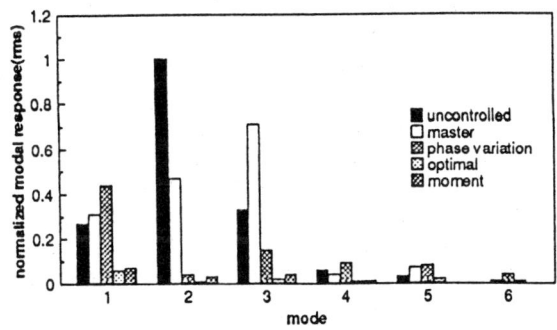

Figure 9. Modal response of beam for uncontrolled, master, phase variation, optimal and moment tests

Table 3. Max. convergence parameter (beam; 604 Hz)

Control method	Max. convergence parameter (μ_{max})
master	0.07
optimal	0.004

Table 4. Controller performance (beam; 604 Hz)

Control method	Performance (dB)
BIO (optimal dist.)	25.1
6I6O Filtered-X LMS	19.5

distribution. The difference in the performance stems from the different number of actuators, where more attenuation is possible with more actuators. (It should be noted that a MIMO control system with the same number of actuators and sensors should theoretically be able to drive the cost function to zero. However, this does not take into account practical considerations such as actuator non-linearities and signal to noise ratios). However, this indicates that comparable performance was attained using the BIO controller with less system complexity than the MIMO Filtered-X LMS algorithm. The results provide a good benchmark comparison for the BIO controller.

B. Radiated Sound Power Control

In this section, the experimental results are presented for the BIO controller applied to radiated sound power control. Experiments were performed at two different excitation frequencies: one off-resonance (319 Hz) and one on-resonance (397 Hz).

1. Selection of Master Actuator

As stated earlier, the selection of the master actuator could be done using a simple search scheme. When the BIO controller was analytically applied to beam vibration control, the influence of the master actuator on phase variation performance was determined analytically. However, for this experimental investigation, the influence of the master actuator on phase variation performance was determined experimentally.

As seen in Table 5, the off-resonance master test performance varies from a low of 0.0 dB for actuator #4 to a high of 3.8 dB for actuator #5. The increase in performance with the phase variation method ranges from a low of 0.7 dB for actuator #4 to a high of 11.8 dB for actuator #1. Note that for 4 actuators the increase in performance was greater than 6 dB, which indicates that the phase variation method is more effective for radiated sound power control (vs. vibration control) due to the increased modal density of the plate. Using the results of Table 5, actuator #1 was chosen to be the master actuator for the off-resonance test.

The on-resonance master test performance, shown in Table 6, varied from a low of 0.0 dB for actuators #2 and #5 to a high of 17.7 dB for actuator #1. The increase in performance with the phase variation method ranged from a low of 0.0 dB for actuators #2 and #5 to a high of 3.6 dB for actuator #6. The above results conclude that the phase variation method was able to increase control performance even when the system was excited on-resonance, which is due to the relatively high modal density of a plate. Note that the poor performance of actuators #2 and #5 was due to their inability to couple into the (2,1) mode due to their placement on a node line of that mode. Using these results, actuator #6 was chosen to be the master actuator for the on-resonance test.

2. Off-resonance Performance

The relative performance for control with the "master" actuator only (referred to as master), the phase variation, and the optimal distribution method are compared for off-resonance excitation. For these experiments, the excitation frequency was 319 Hz, the master actuator was PZT #1, and the sampling frequency was 3000 Hz.

As seen in Table 7, the master case displays poor control performance of 0.6 dB. The modal decomposition, seen in Figure 10, shows no decrease in the (1,1) mode and significant increases in the other modes. It is evident that actuator #1 cannot couple into the plate vibration at this frequency, leading to significant spillover into several modes.

However, the phase variation method is able to construct

Table 5. Influence of master actuator # on phase variation performance (plate; 319 Hz)

Master actuator #	master performance (dB)	phase variation increase (dB)	control string*
1	0.6	11.8	⊕-++-0
2	2.8	7.2	-⊕--+0
3	0.3	4.2	+-⊕+-+
4	0.0	0.7	+-+⊕-+
5	3.8	7.8	-+0-⊕-
6	0.3	7.5	+-++-⊕

Note: ⊕ indicates the master control channel

Table 6. Influence of master actuator # on phase variation performance (plate; 397 Hz)

Master actuator #	master performance (dB)	phase variation increase (dB)	control string*
1	17.7	0.1	⊕-0000
2	0.0	0.0	0⊕0000
3	16.4	2.0	0+⊕0+0
4	17.5	0.0	000⊕00
5	0.0	0.0	0000⊕0
6	17.6	3.6	0+00+⊕

Note: ⊕ indicates the master control channel

a distributed actuator that can couple into the plate vibration. Table 7 shows that control performance increased to 12.4 dB with the phase variation method. As seen in Figure 10, the addition of the slave actuators reduced the response of the (1,1) mode which is the most efficient acoustic radiator below the critical frequency. The increase in performance is a direct result of the reduction in the response of this mode. Increases in response are seen in other modes, however these are significantly less efficient acoustic radiators than the (1,1) mode therefore performance remains good.

The distributed actuator formed by the optimal distribution couples into the plate vibration even better than the phase variation method, although *a priori* knowledge is required. Performance is good at 15.6 dB, as seen in Table 7. The modal decomposition in Figure 10 shows a significant reduction in the (1,1) mode with less spillover into the other modes, which results in increased control performance compared to the phase variation method.

3. On-resonance Performance

In this section, the relative performance for control with the "master" actuator only (referred to as master), the phase variation, and the optimal distribution method are compared for on-resonance excitation. For these experiments, the excitation frequency was 397 Hz, the master actuator was PZT #6, and the sampling frequency was 4000 Hz.

It is not surprising to see the good performance of the master case when the plate is excited on-resonance, which shows a reduction in cost function of 17.6 dB in Table 7. As seen in Figure 11, the master actuator was

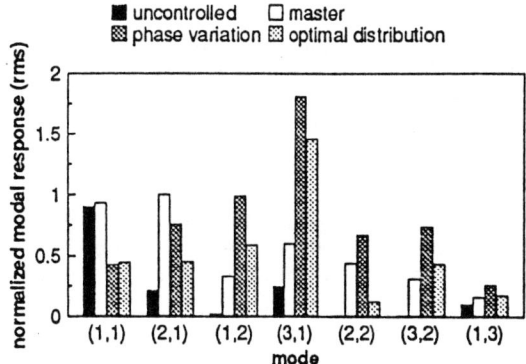

Figure 10. Modal response of plate excited off-resonance

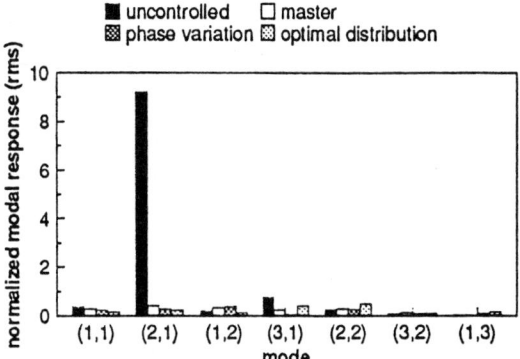

Figure 11. Modal response of plate excited on-resonance

able to reduce the response of the (2,1) mode to the level of the other modes.

With the implementation of the phase variation method, control performance increased 3.6 dB, as shown in Table 7. With the response of the modes nearly equal, the addition of the slave actuators slightly reduced the response of the (1,1) and the (2,1) modes as seen in Figure 11, which resulted in increased performance.

The distributed actuator formed by the optimal distribution has slightly better performance than the phase variation method by 0.4 dB as seen in Table 7. Figure 11 shows the reduction of modes (1,1) and (2,1) was slightly better than the phase variation method.

The above results indicate that even in on-resonance situations, where the resonant mode can theoretically be controlled with one actuator[2], increases in control performance can be attained with the addition of slave actuators. This can be attributed to the high modal density of the plate. However, the addition of the actuators results in diminishing returns in performance.

4. Controller Convergence

For the radiated sound power controller convergence tests, the system was excited on-resonance at 397 Hz, the master channel was actuator #6, and the sampling period was 4000 Hz.

Table 7. BIO controller performance (plate)

Excitation frequency (Hz)	319	397
resonance? mode	no	yes (2,1)
master act. #	1	6
master (dB)	0.6	17.6
phase variation (dB) control string	12.4 ⊕-++-0*	21.2 0+00+⊕*
optimal (dB)	15.6	21.6

*Note: ⊕ indicates the master control channel

Table 8. Maximum convergence parameter (plate; 397 Hz)

Control method	Max. convergence parameter (μ_{max})
master	0.18
optimal	0.02

Table 9. Controller performance (plate; 319 Hz)

Control method	Performance (dB)
BIO (optimal dist.)	15.6
5I5O Filtered-X LMS	10.7

Table 8 shows the maximum convergence parameter obtained for a single output system (the master case) and the optimal distribution. The master case has a maximum convergence parameter (μ_{max}) of 0.18 which is significantly higher than the optimal distribution μ_{max} of 0.02, which verifies the trends predicted[5].

5. Comparison with 5I5O LMS Control Performance

To gauge the relative magnitude of the BIO controller performance, an experiment was run using an existing 5I5O Filtered-X LMS control system, which used 5 error sensors and 5 control actuators. For these experiments, the excitation frequency was 319 Hz and the sampling frequency was 3000 Hz.

Since the existing 5I5O Filtered-X LMS control system had only 5 output channels, PZT 4 was not used due to its poor response at the excitation frequency. The cost function was calculated by the sum of the square of the measured outputs of the microphones. As seen in Table 9, the reduction in cost function for 5I5O controller was 10.7 dB is compared to 15.6 dB obtained by the biological controller using the optimal distribution. Again, The results provide a good benchmark comparison for the BIO controller.

V. Conclusions

A biologically inspired control approach for reducing vibrations in and radiated sound power from distributed elastic systems has been experimentally verified for narrowband excitation. The control approach, inspired by biological systems, approximated the control structure used with biological muscle, where a few main control signals were sent from an advanced, centralized controller (the brain) and distributed by local rules and action to multiple actuators (muscle tissue). Experimental investigations of performance of two different variations of local learning rules were carried out including controller convergence considerations and a comparison with the conventional MIMO Filtered-X LMS control approach.

For vibration control, comparisons with a theoretical analysis showed good agreement of the performance trends. Performance of the various methods decreased vibration levels in a simply supported beam by up to 25 dB for off-resonance harmonic excitation. Also, the distributed actuator formed by the optimal and moment distributions was verified to be independent of frequency over a wide range of test frequencies and configurations.

Performance of the various methods decreased radiated sound power from a clamped plate by up to 16 dB for harmonic off-resonance excitation and up to 22 dB for on-resonance excitation. The addition of the slave control actuators was seen to increase control performance even for on-resonance excitation, due to the high modal density of the plate. However, the addition of actuators resulted in diminishing returns in performance.

For both the vibration and radiated sound power control experiments, the predicted trend of decreasing maximum convergence parameter (for stability) with the additional control inputs of the BIO controller was verified. For the test cases studied, the BIO controller performance was comparable to conventional MIMO Filtered-X LMS performance.

In general the results demonstrated the biological control approach has the potential to control multi-modal response in distributed elastic systems using an array of actuators with a reduced order controller. Thus significant reductions in control system computational complexity have been realized by this approach. Future work will concentrate on additional local learning rules and theoretical application to 2D distributed elastic systems.

Acknowledgments

The authors gratefully acknowledge the support of this work by The Structural Acoustics Branch at NASA Langley Research Center, Dr. Harold Lester, Technical Monitor.

References

[1] R. L. Clark, "Advanced Sensing Techniques for Active Structural Acoustic Control," Ph.D. Dissertation, VPI & SU, 1992.

[2] L. Meirovitch and M. A. Norris, "Vibration Control," Proceedings of Inter-Noise 84, Noise Control Foundation, Poughkeepsie, NY, pp. 477-482, 1984.

[3] B. R. Landau. "Essential Human Anatomy and Physiology," Scott, Foresman and Company, Glenview, Illinois, pp. 191-192, 1976.

[4] C.R. Fuller and J.P. Carneal, "A Biologically Inspired Control Approach for Distributed Elastic Systems," Letter to the Editor, Journal of the Acoustical Society of America, **93** (6), pp. 3511-3513, 1993.

[5] J. P. Carneal and C. R. Fuller, "A Biologically Inspired Controller," to be published in the Journal of the Acoustical Society of America.

[6] G.P. Gibbs and C.R. Fuller, "Excitation of Thin Beams Using Asymmetric Piezoelectric Actuators," Journal of the Acoustical Society of America, **92** (6), pp. 3221-3227, 1992.

[7] S.J. Elliot, I.M. Strothers, and P.A. Nelson, "A Multiple Error LMS Algorithm and Its Application to the Active Control of Sound and Vibration," IEEE Transaction on Acoustic Speech and Signal Processing, Vol. ASSP-35, No. 10, pp. 1423-1434, 1987.

NEURAL NETWORK BASED TIME-OPTIMAL CONTROL OF A MAGNETICALLY LEVITATED PRECISION POSITIONING SYSTEM

Jim Redmond[*] and Susan Tucker[†]

Sandia National Laboratories
PO Box 5800
Albuquerque, NM 87185-0439

Abstract

This paper describes an application of artificial neural networks to the problem of time-optimal control of a magnetically levitated platen. The system of interest is a candidate technology for advanced photolithography machines used in the manufacturing of integrated circuits. The nonlinearities associated with magnetic levitation actuators preclude the direct application of classical time-optimal control methodologies for determining optimal rest-to-rest maneuver strategies. Instead, a computer simulation of the platen system is manipulated to provide a training set for an artificial neural network. The trained network provides optimal switching times for conducting one dimensional rest-to-rest maneuvers of the platen that incorporate the full nonlinear effects of the magnetic levitation actuators. Sample problems illustrate the effectiveness of the neural network based control as compared to traditional proportional derivative control.

1. Introduction

Recent advances in adaptive structures technology have focused on the development of smart materials and their use in active shape control.[1-3] Although these materials are well suited for damping a structure's flexible body modes, they provide no direct control authority over a structure's rigid body motion. Consequently, applications that require rigid body maneuvers must employ alternative sources of actuation. For applications requiring control of a primary structure relative to a fixed structural frame, electromagnetic actuation is gaining in popularity because it provides precision control authority with minimal frictional losses.[4] Recently, this method of actuation was employed to improve the process of integrated circuit (IC) manufacturing.[5] Based on advanced photolithography techniques, IC manufacturing requires precision placement of a platen supporting a silicon wafer. With magnetic levitation actuators, accuracy of better than 10 nm can be achieved for typical rest-to-rest maneuvers of the platen. As with other actuation devices, the use of time-optimal control strategies would improve the speed of the manufacturing process. However, inherent nonlinearities of magnetic levitation devices preclude generating traditional time-optimal switching surfaces.

Previously, an approximate time-optimal feedback strategy was developed for the platen by using linearized estimates of the available control forces to generate simple modal switching curves.[6] Although this technique proved to be useful for relatively large maneuvers, substantial performance gains can be realized by incorporating the actuator nonlinearities and maximizing the control input levels. In this paper, an artificial neural network is trained to provide near optimal switching times for conducting one dimensional rest-to-rest maneuvers. Although they have been successfully applied to a variety of controls problems, artificial neural networks are especially useful for identification and control of nonlinear systems.[7] Their application to time-optimal control problems are becoming more prevalent[8], however no known application exists for systems that include magnetic levitation actuators.

The motion of the platen and wafer containing the ICs is governed by a two stage hierarchial position control scheme. The first (coarse) stage involves controlling large scale two dimensional translations via motor driven screws. The second (fine) stage involves controlling six degree of freedom rigid body motion via magnetic-levitation actuators. This paper focuses on the optimization of the fine stage control system for conducting one dimensional translational maneuvers. The development is based on a

---

[*] Senior Member of Technical Staff, Structural Dynamics Department, Member AIAA
[†] Senior Member of Technical Staff, Manufacturing Applications Department

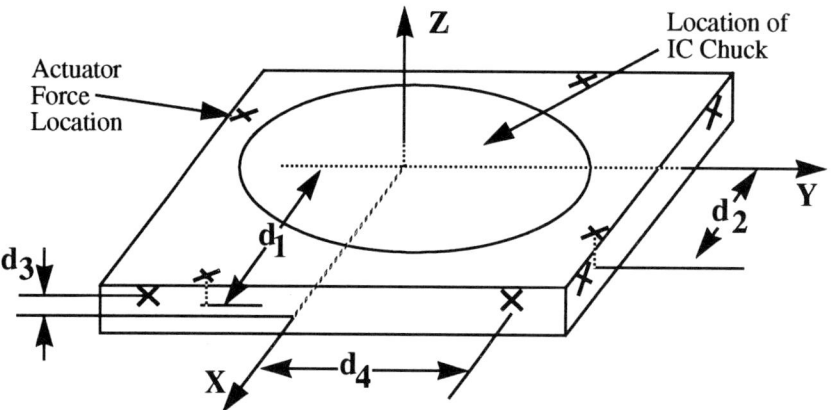

Figure 1. - Fine Stage Platen with Magnetic Levitation Actuators

computer simulation that has been shown to accurately model the behavior of the actual hardware.[9]

A brief description of the system is given in the following section. Time-optimal control is addressed in section 3 through the use of an artificial neural network that accounts for actuator nonlinearities. The neural-network controller is demonstrated on the computer simulation in section 4 and the performance is compared to the performance obtained using a proportional-derivative (PD) controller. Some concluding remarks are given in section 5.

2. System Description

A complete description of the magnetically levitated fine stage is given in Reference 5. However, the details required for controller design are restated here for clarity. A schematic diagram of the fine stage platen is shown in Figure 1 with the inertial coordinate system XYZ nominally located at the platen center of mass when the platen is centered in the fixed coarse stage. The chuck containing the ICs has been omitted for clarity. The fine stage floats within the coarse stage frame which contains the actuation hardware. Three permanent magnets are located below the platen to offset the effects of gravity. Translational motions along the X and Y axes are limited to +/-200 μm before the platen edges contact the coarse stage frame. The platen is composed primarily of aluminum and is assumed to be rigid. For small angles the angular velocities of the body in inertial coordinates are equal to the rate of change of the orientation angles, and the dynamics are described by

$$m\ddot{x}(t) = F_x(t) \quad \text{(EQ 1a)}$$

$$m\ddot{y}(t) = F_y(t) \quad \text{(EQ 1b)}$$

$$m\ddot{z}(t) = F_z(t) \quad \text{(EQ 1c)}$$

$$I_x \ddot{\theta}_x(t) + (I_z - I_y) \dot{\theta}_y(t) \dot{\theta}_z(t) = T_x(t) \quad \text{(EQ 1d)}$$

$$I_y \ddot{\theta}_y(t) + (I_x - I_z) \dot{\theta}_x(t) \dot{\theta}_z(t) = T_y(t) \quad \text{(EQ 1e)}$$

$$I_z \ddot{\theta}_z(t) = T_z(t) \quad \text{(EQ 1f)}$$

in which m is the platen mass, I_x, I_y, I_z are the principal mass moments of inertia, $x(t)$, $y(t)$, $z(t)$, $\theta_x(t)$, $\theta_y(t)$, $\theta_z(t)$ are the modal (rigid body) displacements and rotations, and $F_x(t)$, $F_y(t)$, $F_z(t)$, $T_x(t)$, $T_y(t)$, $T_z(t)$ are the modal control forces and torques. The modal displacements are extracted via a geometric transformation of the signals from six capacitive displacement sensors located in the platen frame. The sensor measurements are processed in a digital signal processor with a sampling rate of 2500 Hz. As shown in equation 1f, the decoupling of the Z axis rotation is a consequence of the symmetry displayed in Figure 1. Although they were neglected in previous control designs[9], the cross-axis coupling of X and Y axes rotations shown in equations 1d and 1e are included for additional accuracy. Note that no damping terms are included in equation 1 since the magnetic levitation minimizes frictional losses.

The purpose of the fine stage control system is to conduct X and Y translational motions in the range +/-100 μm while regulating the other modes. This stage provides positioning accuracies better than 10 nm and is activated

Table 1: Fine Stage Parameters

Platen Mass (m)	0.78 slugs
X-Axis Mass Moment of Inertia (I_x)	8.4 slugs-in^2
Y-Axis Mass Moment of Inertia (I_y)	8.4 slugs-in^2
Z-Axis Mass Moment of Inertia (I_z)	15.6 slugs-in^2
Moment Arm d_1	5.145 in
Moment Arm d_2	2.712 in
Moment Arm d_3	0.658 in
Moment Arm d_4	4.75 in

after the coarse stage directs gross X and Y translations via motor driven ball screws. Sixteen magnetic levitation actuators whose outputs are approximately proportional to the square of the supply currents and inversely proportional to the square of the actuator gaps supply the forces for controlling fine stage motion. Since each magnetic levitation device is capable of producing only unidirectional pulling forces, the sixteen devices are grouped into eight bidirectional actuator pairs. The actuator forces are related to the modal forces via the geometric transformation

$$F(t) = DF_a(t) \quad \text{(EQ 2)}$$

The transformation matrix, the modal force vector, and the actuator force vector are defined as

$$D = \begin{bmatrix} 1 & 1 & 0 & 0 & 0 & 0 & 0 & 0 \\ 0 & 0 & 1 & 1 & 0 & 0 & 0 & 0 \\ 0 & 0 & 0 & 0 & 1 & 1 & 1 & 1 \\ 0 & 0 & d_3 & d_3 & -c & c & c & -c \\ -d_3 & -d_3 & 0 & 0 & c & -c & c & -c \\ d_4 & -d_4 & d_4 & -d_4 & 0 & 0 & 0 & 0 \end{bmatrix} \quad \text{(EQ 3a)}$$

$$F(t) = \begin{bmatrix} F_x(t) & F_y(t) & F_z(t) & T_x(t) & T_y(t) & T_z(t) \end{bmatrix}^T \quad \text{(EQ 3b)}$$

$$F_a(t) = \begin{bmatrix} F_{a1}(t) & F_{a2}(t) & \ldots & F_{a8}(t) \end{bmatrix}^T \quad \text{(EQ 3c)}$$

in which $c = 0.5(d_1 + d_2)$, d_1, d_2, d_3, d_4 represent actuator moment arms, and $F_{ai}(t)$, i=1-8 represent the forces supplied by the actuator pairs. The system parameters are summarized in Table 1.

Each actuator pair is nominally capable of generating +/-15 lbs (66.72 N) of control force when the platen is centered in the coarse stage frame. However, this peak actuator force can vary from 10 to 25 lbs (44.48 to 111.2 N) depending on the platen location. The desired forces from the actuator pairs are based on the desired modal forces from the control algorithm and are computed as the minimum norm solution to equation 2. The actuator gaps and desired actuator forces specify the required supply current for each actuator pair. Since an explicit expression for the required supply current is difficult to obtain due to the actuator nonlinearities, a two dimensional table is used to designate the proper currents for large forces, while a curve fit is used for small forces. The resulting actuator forces are transformed in the simulation to yield the actual modal forces used to drive the table motion. To be consistent with the hardware, a delay of 200 microseconds is included in the simulation to mimic the input to output computation time.

3. Neural Network Controller Development

As previously mentioned, the fine stage positioning system is equipped with a critically damped PD controller developed in reference 9. For nominal translations, the controller bandwidth (undamped natural frequency) is limited to approximately 50 hz to avoid an underdamped response caused by actuator saturation. The relatively slow response of this controller prompted this investigation into the development of a time-optimal control strategy.

3.1 Time-Optimal Control Considerations

Time-optimal control of low-order completely controllable systems is of the classical bang-bang form and can be readily obtained by minimizing the associated

Hamiltonian and subsequently generating state based switching curves.[10] For high-order systems, numerical techniques exist for finding open-loop solutions for a given maneuver. However, several intricacies of the fine stage preclude the direct application of classical techniques. For example, the actuators are incapable of instantaneous switches between full positive and full negative forces as a consequence of the time delays associated with the finite sampling rate, processing operations, and actuator dynamics. Furthermore, the peak actuator forces are not constant but instead are nonlinear functions of the platen position relative to the frame. And although each individual mode is completely controllable with respect to the modal control forces, the coupling of the modes through the actuator forces according to equation 3 renders the overall system singular in the control.[11] Therefore, the time-optimal control is not completely specified by the simple necessary condition of optimality. Although higher order conditions can be considered to derive an optimal control strategy for singular systems, the variation of the peak control force as a function of platen position makes this approach impractical.

The solution developed in this paper uses an artificial neural network to produce optimal switching times that account for nonlinear actuator dynamics. Recall that the main purpose of the fine stage positioning system is to conduct X and Y translations in the range +/-100 μm while regulating the other rigid body modes. As such, the problem is simplified somewhat in that the control for the X and Y translations can be treated independently since these modes do not share common actuators as shown in equation 3. Therefore, we will consider only rest-to-rest translations along the X axis beginning at point x_1 and terminating at point x_2.

Initial investigations revealed that the 2500 Hz sampling rate provides inadequate resolution to determine phase space switching points. Instead, an open-loop control strategy is adopted to alleviate the dependence of the control on the sampling rate. Of course, the inherent uncertainties of open-loop control dictate that the closed-loop control ability must be maintained to address any maneuver errors that can arise as a result of inexact system modelling, external disturbances, or inaccuracies associated with the network. Thus, a hybrid open/closed-loop strategy is proposed for conducting the rest-to-rest maneuvers. Open-loop switching points are used to govern the nominal maneuver based on a 100 kHz clock. After completion of the open-loop phase of the control, the original PD feedback control is used to remove any residual errors in position and velocity.

3.2 Training Set Development

This section presents a method for determining the switch times for maneuvering the fine stage in a time-optimal fashion. Since it is not practical to consider all possible maneuvers in order to generate a table of switch times as a function of initial and final conditions, the goal is to produce a set of optimal maneuvers that can be used to train an artificial neural network. Then, near optimal switch points can be obtained for an arbitrary maneuver by querying the neural network.

Because of the complex actuator dynamics, the open-loop time-optimal control requires two switch times T_{s1} and T_{s2}. The second switch time allows for the finite time required for actuator shutdown. Assuming a positive translation, the time-optimal controller commands a full positive X modal control force for $0<t<T_{s1}$, a full negative force for $T_{s1}<t<T_{s2}$, and a zero modal control force for $T_{s2}<t<T_f$. In order to insure that the full modal control force is utilized, the commanded force should be greater than maximum force available with a minimal actuator gap for the range of motions considered. Since two actuators drive the motion along the X axis as shown in equation 3, the peak force demanded is set nominally to 45 lbs (200.16 N). Note that the final time T_f is not an integral part of the open-loop control strategy and is in general unspecified. If the switching points are valid, the velocity and the control force vanish as the system arrives at x_2. The important parameters for a time-optimal rest-to-rest maneuver are designated by the pattern $\{x_1, x_2, T_{s1}, T_{s2}\}$.

The training set should sufficiently cover the space of interest from -100 to +100 μm and should also include some points beyond the range of interest. But since the endpoints of a training maneuver are arbitrarily chosen, the approach adopted in this study is to specify the initial position and the first switch time. Then, a bisection routine is used to determine the second switch time that produces a near zero velocity as the control force diminishes at the end of the control sequence. When this condition is satisfied, the final position is noted and the training pattern is complete. This procedure is summarized for translation in the positive X direction as follows:

1) Designate starting position x_1 and primary switch time T_{s1}.

2) Initialize two switch times T_{s2a} and T_{s2b} that designate an interval containing the solution T_{s2}. The final velocities associated with the initial

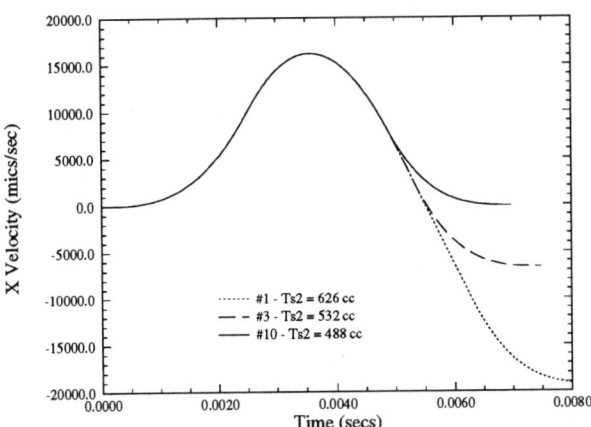

Figure 2. - Convergence to Optimal Velocity Profile for Sample Training Pattern.

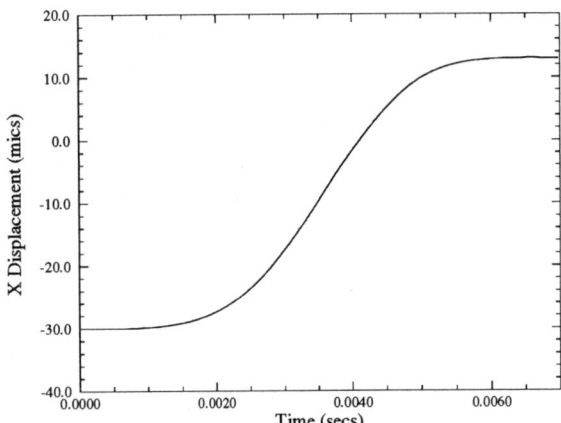

Figure 4. - Optimal Displacement Profile for Sample Training Pattern.

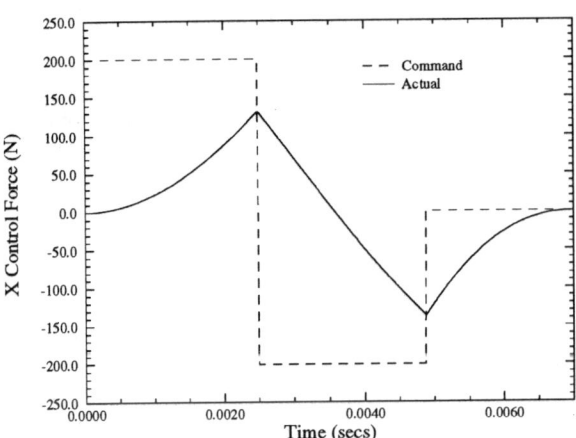

Figure 3. - Optimal Control Force for Sample Training Pattern.

choices should satisfy $V_f(T_{s2a})>0$ and $V_f(T_{s2b})<0$.

3) Let $T_{s2}=(T_{s2a}+T_{s2b})/2$.

4) If $|T_{s2}-T_{s2a}|<\Delta t$, STOP.

5) Command full positive modal force for $0<t<T_{s1}$.

6) Command full negative modal force for $T_{s1}<t<T_{s2}$.

7) Command zero modal force for $T_{s2}<t$.

8) When the modal force is sufficiently small, note the final velocity V_f and the final position x_2.

9) If $V_f>0$, $T_{s2a}=T_{s2}$, GO TO 3.

10) If $V_f<0$, $T_{s2b}=T_{s2}$, GO TO 3.

The convergence criterion Δt is equal to one clock count of the open-loop controller. According to step 4, the second switch time has converged when subsequent adjustments are less than one full clock count.

Assuming a controller clock speed of 100 kHz (1 clock count = 0.00001 sec), this bisection routine is demonstrated in Figure 2 for a positive translation. With an initial position of -30 μm, the primary switch point is set at 250 clock counts (cc) and the algorithm iterates on the second switch time in an effort to minimize the final velocity achieved when the control force drops below 0.00001 N. After ten iterations, the algorithm converges on a secondary switch time of 488 cc. With this switching sequence, the table reaches a peak velocity of approximately 17,000 μm/sec and comes nearly to rest with a residual velocity of approximately 50μm/sec. The commanded and actual modal control forces are shown in Figure 3 and the resulting displacement history is shown in Figure 4. Note that the maneuver is completed too quickly to permit actuator saturation in either the positive or negative direction. This example provides a single training pattern for the artificial neural network representing a maneuver from -30 μm to +13.01 μm. A complete training set is obtained by repeating the procedure for a variety of initial positions and primary switch times. Several example training patterns are given in Table 2. Although it is not part of the training pattern, the final velocity achieved for each maneuver is also provided in Table 2. Final velocities in the range of +/-200 μm/sec are possible due to the increased switching resolution obtained using this open-loop approach.

3.3 Neural Network Description

Although it is theoretically possible to train a network of sufficient size to an arbitrary accuracy in a functional approximation problem, experience indicates that it is difficult and time-prohibitive to train a network approaching 1% RMS error using standard backpropagation techniques. Consequently, the overall network is designed with two internal stages, combining a least mean squares (LMS) estimator with standard feedforward networks trained using a backpropagation technique. In order to produce the two switch times T_{s1} and T_{s2}, the neural network for the open-loop fast clock controller accepts seven inputs, including the starting and stopping positions X_1 and X_2, $sgn(X_2-X_1)$, $|X_2-X_1|^{1/2}sgn(X_2-X_1)$, $log(|X_2-X_1|)sgn(X_2-X_1)$, $(X_2-X_1)^{-1}$, and $(X_2-X_1)^3$. Training data were produced by the method described in section 3.2. First, the LMS algorithm performs linear approximations to the output based on the seven neural network inputs. Then, two internal neural networks approximate the residuals of the least squares fits for switching times T_{s1} and T_{s2}. The two neural networks are configured identically as fully connected feedforward networks with seven inputs, one output, and a single hidden layer of ten nodes using the 'tanh' nonlinearity. The outputs of the LMS estimator and the neural network estimator are combined to produce the final estimates of the switching times.

Training data was divided into two sets for neural network development. The training set consisted of 219 points and was used to determine the network parameters. The test set contained 55 points and was not used for network training. Performance of the network on the two data sets is summarized below. As shown in Table 3, a maximum error of eleven clock counts was obtained over the range of moves within the limits of +/-140μm. Performance of the network on the test data set is an important measure of how well the network generalizes to data not contained in the training set. Since the performance on the test and training data sets is comparable, the network has acquired good generalization capabilities.

4. Control Demonstration

The effectiveness of the neural network based hybrid control scheme becomes apparent when compared to the original PD control for several rest-to-rest maneuvers. As

Table 2. - Sample Neural Network Training Patterns Using 100kHz Clock.

$X_1(\mu m)$	$T_{s1}(cc)$	$T_{s2}(cc)$	$X_2(\mu m)$	$V_f(\mu m/s)$
-50.00	250	463	-14.96	191.58
-40.00	250	476	-14.06	-111.49
-30.00	250	488	13.01	-48.94
-20.00	250	501	27.39	-104.71
-10.00	250	523	42.45	-32.31
0.00	250	552	59.13	56.31
10.00	250	589	77.54	-40.47
20.00	250	631	97.27	-43.96

Table 3. - Neural Network Error Summary (All Data given in Clock Counts)

	Training Set			Test Set		
	RMS	MIN	MAX	RMS	MIN	MAX
T_{s1}	3.2	-10.7	+8.8	2.9	-5.5	+7.9
T_{s2}	3.2	-8.7	+9.5	3.5	-7.5	+8.7

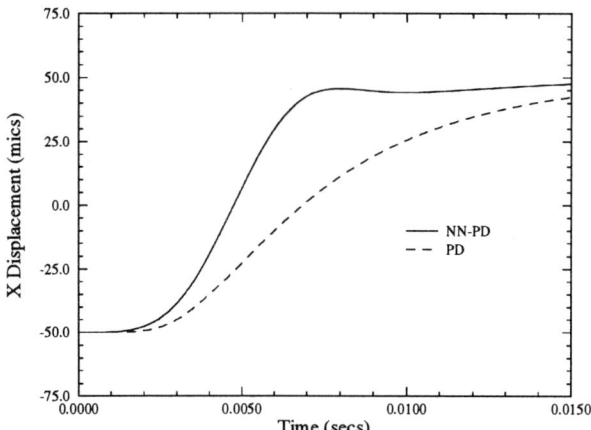

Figure 5. - Comparison of Displacement Profiles for 100 μm Translation.

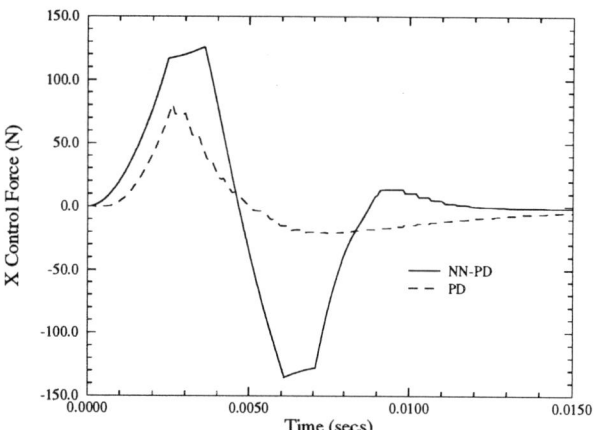

Figure 6. - Comparison of Control Forces for 100 μm Translation.

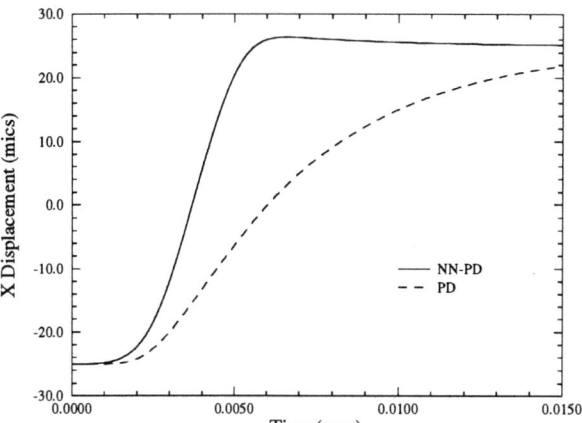

Figure 7. - Comparison of Displacement Profiles for 50 μm Translation.

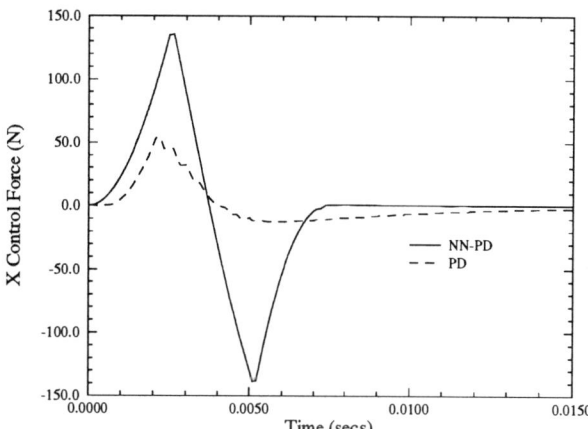

Figure 8. - Comparison of Control Forces for 50 μm Translation.

previously described, the neural network is trained on the basis of a 100 kHz clock and provides primary and secondary switch times in response to inputs of the initial and target locations. The open-loop control force governing the X translation is identical to that described in the previous section while the remaining modal control forces set equal to zero. The closed-loop portion of the hybrid control sequence is activated when the magnitude of the velocity drops below 300 μm/sec at the end of the open-loop phase. The closed-loop controller is identical to the original critically damped PD controller with a 2500 Hz sampling rate, a 200 μsec D/A delay, and a 50 Hz bandwidth.

The displacement profiles resulting from the two control strategies are shown in Figure 5 for a 100μm translation from -50 to +50 μm. As indicated by the control force histories in Figure 6, the neural network produces primary and secondary switch times of 360 and 706 cc,

respectively. The open-loop control force saturates in both the positive and negative directions and the dependence of the maximum actuator force on table position is clearly evident. This portion of the control provides a rapid translation to the vicinity of the target destination. Although the closed-loop control is activated at approximately 0.008 secs, the open-loop control force has not yet diminished. Consequently, the platen moves away from the target point before the actuators catch up to the closed loop control command. At 0.01 secs, the system is within 5% of the target destination as compared to 25% using the PD control alone. The inefficiency associated with the control lag may be alleviated by considering a faster clock and a more accurate network. Such improvements would enhance the ability of the open-loop control to satisfy the end conditions.

The displacement and force profiles for a translation from -25 to +25 μm are shown in Figures 7 and 8.

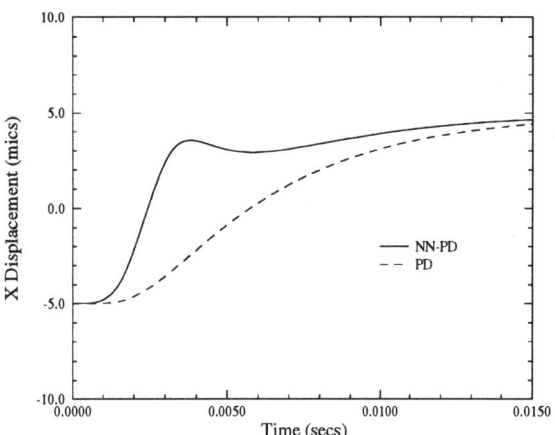

Figure 9. - Comparison of Displacement Profiles for 10 μm Translation.

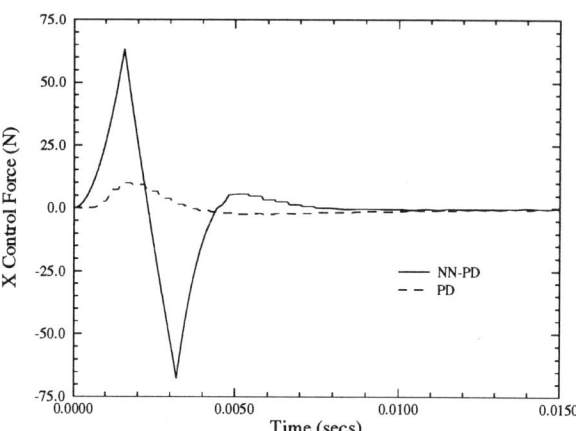

Figure 10. - Comparison of Control Forces for 10 μm Translation.

respectively. Based on switch times of 262 and 521 cc, the open-loop control force barely saturates in the positive and negative directions and produces a displacement profile that slightly overshoots the target destination. The closed-loop control is activated at approximately 0.065 secs. At this point, the neural network maneuver is within 3% of the target point as compared to the 44% percent achieved using feedback control alone.

As shown in Figures 9 and 10, an undershoot response and a residual negative control force result from the open-loop portion of a 10 μm translation. Only fifty percent of the peak control force is utilized due to the rapid switching at 158 and 320 cc. Although the transition to closed-loop control occurs at approximately 0.0035 secs, the actuators require additional time to catch up to the closed loop command. At 0.006 secs, the neural network system is only within 20% of the target destination, still better than the 50% achieved using feedback control alone. As compared to the previous examples, a larger residual error is present at the end of the open-loop control since the neural network error becomes more significant as the switch times decrease. Training the network to a higher degree of accuracy can substantially improve the accuracy obtained with the open-loop control. Preliminary results indicate that reducing the RMS error to within 1 clock count will reduce final position errors to less than 1 μm.

5. Conclusions

A preliminary investigation into the use of an artificial neural network for time-optimal control of the nonlinear photolithography positioning system has been completed. Difficulties associated with existing controller hardware were discussed and a hybrid open/closed-loop control strategy was proposed for one dimensional rest-to-rest maneuvers. A method for determining optimal maneuvers was developed and used to generate a training set for an artificial neural network. Application of the hybrid control approach to the computer simulation revealed a vast improvement in the settling time as compared to the original PD strategy. However, the network error was significant for small maneuvers which required relatively rapid switching.

6. Acknowledgments

This work performed at Sandia National Laboratories is supported by the U.S. Department of Energy under contract DE-AC04-94AL85000. The authors are indebted to Tony Smith and Stew Kohler of Department 2338 for initiating and supporting this project and for supplying the computer model of the mag-lev stage.

7. References

1. Baz, A., Poh, S., Ro, J., Mutua, M., and Gilheany, J., "Active Control of Nitinol-Reinforced Composite Beam," *Intelligent Structural Systems*, 1992, pp. 169-212.

2. Dosch, J.J., Leo, D.J., and Inman, D.J., "Comparison of Control Schemes for a Smart Antenna," *Proceedings of the 31st Conference on Decision and Control*, Tucson Arizona, December 1992, pp. 1815-1820.

3. Segalman, D.J., Witkowski, W.R., Adolf, D.B., and Shahinpoor, M., "Theory and Application of Electrically Controlled Polymeric Gels," *Journal of*

Smart Materials and Structures, Vol. 1, 1992, pp. 95-100.

4. Trumper, D.L., Sanders, J.C., Nguyen, T.H., and Quenn, M.A., "Experimental Results in Nonlinear Compensation of a One-Degree-of-Freedom Magnetic Suspension," *International Symposium on Magnetic Suspension Technology*, NASA Langley Research Center, Hampton, VA, August 1991.

5. Arling, R.W., and Kohler, S.M., "Six Degree of Freedom Fine Motion Positioning Stage Based on Magnetic Levitation," presented at the NASA Conference on Magnetic Levitation, August 1993.

6. Redmond, J., "Nearly Time-Optimal Feedback Control of a Magnetically Levitated Photolithography Positioning System," *Proceedings of the 1994 North American Conference on Smart Structures and Materials*, February 1994.

7. Narendra, K.S. and Parthasarathy, K., "Identification and Control of Dynamical Systems Using Neural Networks," *IEEE Transactions on Neural Networks*, Vol. 1, No. 1, March 1991, pp. 4-27.

8. Swiniarski, R., "Neural Network Application to Adaptive Time-Optimal Control of Nonlinear Systems," *Proceedings of the IEEE International Workshop on Intelligent Motion Control*, Istanbul, Turkey, August 1990, pp.233-238.

9. Kohler, S.M., and Kahle, P.M., "Development of a Control System for the GCA Prototype Stage," Internal Report, Sandia National Labs, 1993.

10. Sage, A.P., White, C.C. III, *Optimum Systems Control*, Prentice Hall, Inc., Englewood Cliffs, NJ, Second Edition, 1977.

11. Athans, M., and Falb, P.L., *Optimal Control*, McGraw Hill Book Company, New York, NY, 1966.

DISCRETE-TIME IMPLEMENTATION OF POSITIVE POSITION FEEDBACK:
ANALYSIS AND DESIGN APPROACHES WITH AN EXPERIMENT *

Gary T. Fagan† and Harry H. Robertshaw‡
Center for Intelligent Materials Systems and Structures
Department of Mechanical Engineering
Virginia Polytechnic Institute and State University
Blacksburg, VA 24061

Abstract

This work discusses the discrete-time implementation of Positive Position Feedback (PPF) for Active Vibration Control. PPF is a collocated direct-output feedback control method that increases the effective damping in a structure. It was introduced as an alternative to direct velocity feedback which can sometimes become unstable due to controller/structure interaction. To date, all work involving PPF revolved around continuous-time systems, therefore, the issue of sample-data systems using PPF are investigated. The issues addressed are: stability of the sampled system, the effects of the sampling rate on the system, and degradation from predicted analog performance. A design procedure for the tuning filters in the Z-plane is also presented. Experimental implementation of PPF on a simply-supported beam resulted in vibration suppression of three modes with a SISO controller. The beam was subjected to both a single-frequency harmonic disturbance and a broadband disturbance. Two- and three-mode controllers were designed with disturbance suppression up to 12dB achieved.

Introduction

The design of a controller for active vibration suppression must not only accomplish the performance objective, but should also be robust in the face of parameter uncertainties and unmodeled dynamics. Collocated control methods have been shown to be more

---

*This research was performed under Army Research Office's University Research Initiative Center Program, Grant No.03-92-6-0181; Dr. Gary Anderson, Program Manager.

†Graduate Research Assistant, Department of Mechanical Engineering, VPI&SU.

‡Associate Professor, Department of Mechanical Engineering, VPI&SU.

Copyright © 1994 by the American Institute of Aeronautics and Astronautics, Inc. All rights reserved.

robust than other traditional control methods, such as LQR or LQG (Balas, 1978; Schafer & Holzach, 1985; Goh & Caughey, 1985) which require full-state estimation or measurement. Collocated control refers to sensors and actuators that are acting at the same physical location on a structure and are 'compatible' such that $C = B^T$ in the state variable description of the system dynamics.

Balas (1978b) was one of the first in the structural controls field to recognize that, in the absence of actuator dynamics, analog velocity feedback with collocated sensors and actuators resulted in a system that was unconditionally stable. This control method is commonly referred to as Direct Velocity Feedback (DVFB). In practice, however, it was observed that this "unconditional global stability" was not always valid. It was assumed that the dynamics of the actuator were sufficiently fast so that they did not "interfere" with the dynamics of the structure being controlled. Unfortunately, all flexible structures have an infinite bandwidth and the dynamics of the actuators will interact with some of the higher modes.

Direct velocity feedback, therefore, is no longer as appealing as it once appeared. The unconditional global stability result vanished with the inclusion of actuator dynamics in the model, and spillover lowered the damping of the higher modes. Various compensation techniques have been suggested that can be used with DVFB to increase the stability margin of the system (e.g. Caughey & Goh, 1985; Balas, 1990; Inman, 1990), but they all increase the complexity of the closed-loop system. As a result, Positive Position Feedback emerged as a more effective alternative to perform vibration suppression.

The idea of Positive Position Feedback (PPF) was first introduced in 1983 by Caughey and Goh in a NASA paper as an alternative to velocity feedback and was shown to have several advantages over DVFB. Conditional global stability can be proven for PPF which include actuator dynamics, and spillover is stabilized. PPF can also be shown to be robust in the face of parameter uncertainties and unmodeled dynamics.

The theory behind analog PPF will be briefly discussed first, followed by discrete implementation concerns and discrete design methods. Lastly, results from initial experiments using discrete PPF on a simply-supported beam for vibration suppression are presented.

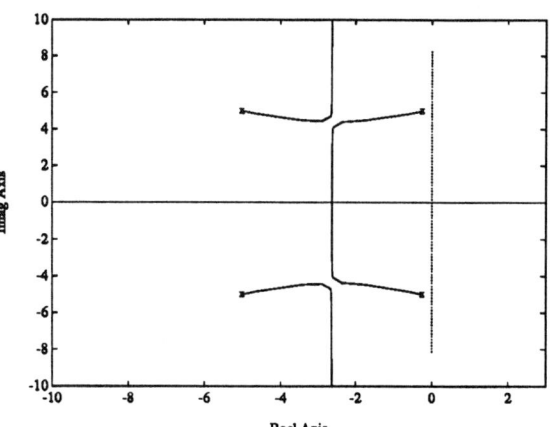

Figure 1: Root Locus of Scalar System Using PPF

Analog PPF

The fundamental theory behind Positive Position Feedback developed by Caughey, Fanson and Goh is best understood when first considering the scalar case. The coupled equations of the structure and actuator are:

$$\ddot{q}(t) + 2\zeta_s\omega_s\dot{q}(t) + \omega_s^2 q(t) = g\omega_s^2 z(t) \\ \ddot{z}(t) + 2\zeta_a\omega_a\dot{z}(t) + \omega_a^2 z(t) = \omega_a^2 q(t) \quad (1)$$

where $q(t)$ and $z(t)$ are the states of the structure and actuator respectively, ζ_s and ζ_a are their associated damping coefficients, ω_s and ω_a are the natural frequencies, and g is a positive scalar gain. The terminology 'Positive Position' refers to the fact that the position state of the system is positively sent to the actuator and the position state of the actuator is positively sent to the system.

The root locus analysis of the scalar system described by Eqn.1 can be seen in Figure 1. The exact form of the root locus will depend on the relative damped natural frequencies of the actuator and structure. Figure 1 shows the root locus when the damped natural frequencies are equal. When the feedback gain equals 1, the locus crosses the imaginary axis, regardless of the particular characteristic values, making the stability criteria independent of the system dynamics. As a result, PPF is robust to the often uncertain parameters associated with structures. In the multivariable system, when there is more than one mode being controlled, it is not possible to have the actuator tuned to all the modes. Also, due to hardware constraints, it may not be possible to find actuators that are "tuned" to the mode to be controlled. It was for these reasons that the concept of "tuning filters" was introduced.

'Tuning Filters' are compensators with second-order dynamics of the same form as that of the actuator. Each filter's frequency is tuned to a specific mode to be controlled in the same manner that the actuator was in the scalar case. By using N_f filters to control N_f modes, the overall structure-actuator-filter equations are:

Structure: $\ddot{q} + D_s\dot{q} + \Omega_s q = \Phi^T u$

Filters: $I_{N_a}(\ddot{Z}_i + 2\zeta_{f_i}\omega_{f_i}\dot{Z}_i + \omega_{f_i}^2 Z_i) = \omega_{f_i} G_i^{\frac{1}{2}}\Phi q,$
$i = 1, ..., N_f$

Actuators: $\ddot{u} + \beta_a\dot{u} + \omega_a^2 u = \sum_{i=1}^{N_f} \omega_{f_i} G_i^{\frac{T}{2}}[\ddot{Z}_i + \beta_a\dot{Z}_i + \omega_a^2 Z_i]$
(2)

where $G_i = G_i^{\frac{T}{2}} G_i^{\frac{1}{2}}$ is the gain for the ith filter, and Φ is the modal participation matrix. The identity matrix at the front of the filter equation is necessary to create symmetry for the stability proof.

With the assumption that the dynamics of the actuator are stable, these dynamics are essentially removed from the overall system by the right hand side of the actuator equation and do not play a part in the stability analysis.

The proof will not be shown here, but it applies Lyapunov's Direct Method for stability. Interested readers are referred to Caughey and Goh (1985) for a detailed analysis of this stability condition.

For the general case of PPF in which several sensor and actuator pairs are used in a global control scheme, each pair has access to information from the other pairs and results in a very large number of coupled equations. The closed-form calculation of the gains for a prescribed amount of damping requires the solution of a complex set of equations which is impractical. A much simpler solution can be found if Local Control PPF (LCPPF) is used instead. Local control means that measurements from a specific sensor on the structure are fed through tuning filters that create control signals for the collocated actuator only. Using this method and assuming that there is one actuator for each filter, the system equations are:

$$\text{Structure: } \ddot{q} + D_s\dot{q} + \Omega_s q = \Phi^T G Z$$
$$\text{Filters: } \ddot{Z} + D_f\dot{Z} + \Omega_f Z = \Omega_f \Phi q \quad (3)$$

where $G \equiv diag(g_1, ..., g_{N_f})$.

The calculation of the control gains for this LCPPF case is accomplished by assuming that for small gains the coupling effects from the uncontrolled modes is negligible. For the simple case when $N_a = N_f$ and each filter is assigned a different mode to control, the system can be decoupled into N_f sets of scalar equations as in the DVFB method. Taking the LaPlace transform of Eqn. 3, the characteristic equation for the ith mode is approximately:

$$(s^2 + 2\zeta_{s_i}\omega_{s_i}s + \omega_{s_i}^2)(s^2 + 2\zeta_{f_i}\omega_{f_i} + \omega_{f_i}^2) - \omega_{s_i}^2\omega_{f_i}^2 g_i = 0$$

This equation can be compared to the closed-loop equation with a prescribed damping:

$$(s^2 + \beta_p s + \omega_p^2)(s^2 + \beta_q + \omega_q^2) = 0$$

By equating coefficients of like powers in s, the specific gain, g_i, can be determined for the prescribed damping in β_p.

While this procedure appears to be a simple method to calculate the gains for a prescribed amount of damping, it becomes exceedingly complicated when $N_a < N_f$, which is usually the case. One actuator can be used with several filters to control more than one mode. Due to the structure of the equations, in this situation, it is no longer possible to separate the equations into individual scalar equations.

Another method to determine the gains for the analog case uses the root locus. For more than one mode, the root locus with one filter is shown in Figure 2. The frequency of the filter is chosen to be slightly higher than the mode being controlled so that the filter pole will be drawn towards the structural zero. This causes the structural pole to be drawn significantly into the left half S-plane before breaking down towards the real axis. The filter gain is chosen as the value at which the filter pole and structural pole are at their closest point. These pole locations are shown

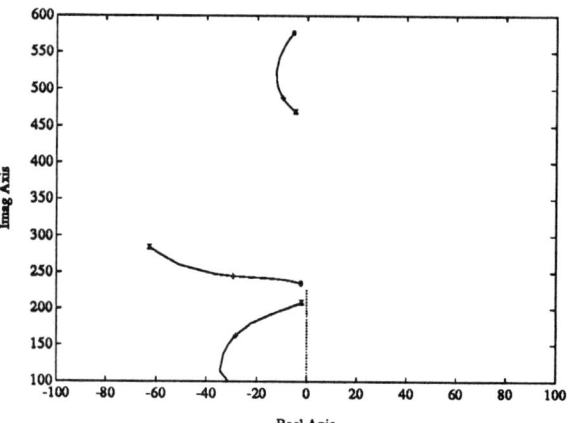

Figure 2: S-plane root locus using PPF with multiple modes and 1 filter

in the figure as crosses.

The design process for when there is more than one filter gain to be determined requires a top-down approach. Since it is not possible to simultaneously vary all the gains of the filters to create the locus, only one filter is tuned at a time. Due to the fact that the filters are second-order and have a two-pole roll off, they affect lower frequency dynamics more than they do the higher frequencies. It is for this reason that the highest frequency filter is designed first. The second highest frequency filter is then designed around the closed loop system of the structure and first filter. It the frequencies of the modes are far enough apart, this second filter should affect the higher frequency dynamics by moving the poles back up the locus only slightly. This effect can be anticipated so that when designing the first filter, the gain is chosen such that the poles are slightly past the coalescent point. The next highest frequency filter is then designed around the closed loop system of the structure and first two filters. If the procedure was performed with the lower frequency filters first, the dynamics of the higher frequency filter would greatly affect the lower frequency filter's designed specifications. The rest of the gains for the filters are determined in this fashion. With this design approach, the lowest frequency structural pole will always be the pole that goes unstable if the gains of the filters are too large.

The root-locus design method for choosing the necessary gains for the tuning filters is much simpler than the analytical method. Therefore, a similar approach will be taken in the discrete-time design procedure.

Discrete-time PPF

There are several matters that need to be considered when implementing PPF with a discrete-time controller: the stability of the sampled system, the sampling rate, and degradation from predicted analog performance. It is the purpose of this section to discuss these issues, as well as addressing a discrete-time design procedure for the tuning filters.

Since discrete-time PPF will be discussed as a root-locus design procedure, sampling and holding inherent in digital control alters the form of the root-locus and needs to be addressed. The input to the system is assumed to be the output of an ideal zero-order hold, and the output of the system is sampled with sampling time Ts. The system is analyzed using Z-transform analysis and the stability region changes from the left half of the S-plane in the continuous time to the inside of the unit circle of the Z-plane. While the poles and zeros still alternate as in the continuous time, instead of being along the inside of the imaginary axis, they are along the inside of the unit circle. In the continuous-time PPF root-locus, there are three zeros at infinity which map to the roots of $(z^3 + 11z^2 + 11z + 1)$ in the Z-plane as the sampling time approaches zero (Åström & Wittenmark, 1990). The location of these zeros as the sampling time is increased will be discussed later. For $T \to 0$, the roots of the polynomial result in zero locations at -9.89, -1.0, and -0.101. This has the direct result of causing two of the higher frequency pole loci in the truncated model to go outside the unit circle. Recall that in the continuous time, it was only the lowest frequency structural poles that would go unstable as the gains are increased. In the digital domain, one of the lowest frequency and two highest frequency structural poles have the potential to go unstable. A typical root locus plot using PPF without filters is shown in Figure 3.

Not only does the sampling itself alter the form of the root locus, the sampling rate has a significant impact on the form. It is the sampling rate that determines the exact location of the open-loop poles and zeros along the unit circle. For a zero-order hold of a system, the poles are mapped into the Z-plane by e^{sT_s}, where s is the complex S-plane pole. It can be seen that as the sampling time T_s approaches zero, all the poles are mapped closer to Z=1. Figure 4 shows the pole locations of a system derived from a simply-supported beam when the sample time is 3000Hz. Figure 5 shows the same system when the

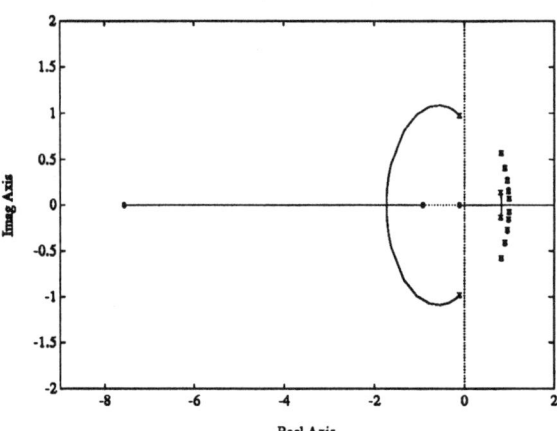

Figure 3: Typical root locus of Discrete PPF with actuator dynamics

sample rate is increased to 17000Hz. As can be seen, fast sampling causes the open-loop poles to move extremely close together.

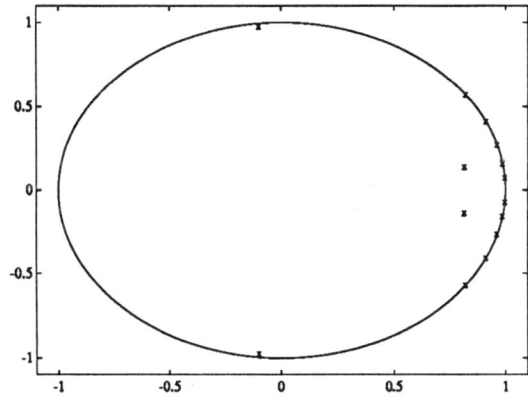

Figure 4: Discrete-time pole locations with a sampling rate of 3000Hz

Not so obvious is the effects of slower sampling rates on the pole and zeros locations. In the controls community, the general rule of thumb is to sample at least 5 to 10 times as fast as the highest frequency of interest. It is well known in signal processing that to avoid aliasing, the system needs to be sampled at least twice as fast as the fastest frequency of interest, also known as the Nyquist frequency. For a stable pole in the Z-plane, as the sampling rate decreases towards twice the pole's natural frequency, the pole ap-

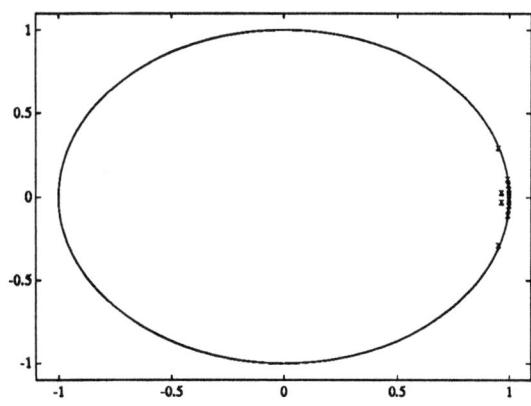

Figure 5: Discrete-time pole locations with a sampling rate of 17000Hz

proaches the negative real axis in a counter-clockwise manner. This effect is represented in Figure 6 for a pole with $\omega_n = 800$Hz.

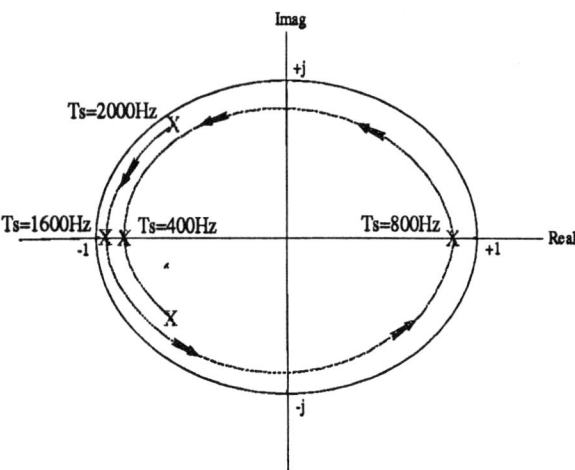

Figure 6: 800Hz-Pole location as Sampling Rate is Decreased

When the sampling rate is exactly twice the pole's natural frequency, the pole lies on the real axis somewhere between 0 and -1 depending on its damping. Since the pole has a complex conjugate, another pole is mirroring the actions of this one. Therefore, the discussion will only focus on one of the poles with the assumption that the other pole is behaving as it's mirror image. As the sampling rate is decreased further, the pole crosses the real axis and proceeds to circle the origin. When the sampling rate is equal to the pole's frequency, it lies on the real axis between 0 and

1. Decreasing the sampling rate even more causes the pole to encircle the origin with an inward spiral. This spiral effect is caused by the transformation $z = e^{sT_s}$. Since s is a complex number, the transformation can be written as $z = e^{T_s(-a \pm bj)} = e^{-aT_s}e^{\pm bjT_s}$. While the absolute value of $e^{\pm bjT_s}$ remains at 1, the reason why the imaginary axis in the S-plane maps to the unit circle in the Z-plane, the value of e^{-aT_s} is decreasing with increasing values of T_s causing the inward spiral effect. As $T \to \infty$, all the poles are mapped to the origin. Since the poles and zeros are alternating, the structural zeros are also follow this same pattern.

The zeros associated with infinity in the S-plane also move as the sampling rate is decreased. The zeros at -9.89 and -1.0 move towards each other along the real axis until they meet at roughly Z=-1.5. At this point, they break apart vertically as shown in Figure 7. When these zeros enter the unit circle , they appear to become paired with the highest frequency pole in the truncated model, and together they spiral inward towards the origin. The zero at -0.101 begins to move to the left along the real axis as the sampling rate is decreased until about Z=-.5, where it then starts back towards the origin as Ts approaches infinity.

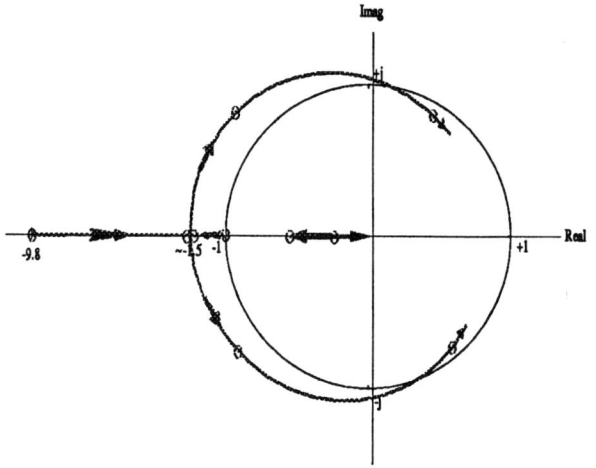

Figure 7: Excess Zero Locations as Sampling Rate is Decreased

Although this analysis followed a fixed-frequency pole as the sampling rate was decreased, it can also be thought of as determining the pole locations of higher frequency poles not included in the model for a fixed sampling rate. In the S-plane, these higher frequency poles were simply further out along the imaginary axis. In the Z-plane, the poles follow a spiraling path towards the origin. With this understanding of the effects of sampling and the sampling rate, the design of the tuning filters will be discussed.

There are two methods which can be used to implement PPF digitally (Fagan, 1993). The first is to design the filters in continuous time, and then use a transformation to convert them to the digital domain. The second method is to design the filters in the Z-plane. Both methods will be discussed here.

The design of the filters in the S-plane has already been discussed, so the focus here will be on the transformation and choice of sampling rates. Each filter can be designed in the continuous time and then individually transformed using a Tustin transformation, $z = \frac{1+\frac{sT_s}{2}}{1-\frac{sT_s}{2}}$. It was found that this transformation correctly mapped the zeros of the filters so as to 'cancel' the actuator poles from the zero-order hold on the system. Once each filter is transformed, they are combined in parallel to determine a single state-space model to be used in the computer control. With this choice of transformation, the effects of the sampling rate were investigated.

It was found that the closed-loop damping was less with a faster sampling rate than with a slower sampling rate. This is due to the interaction of the closely spaced poles and zeros with a fast sampling rate. As discussed earlier, fast sampling causes all the poles to be bunched together near Z=1. In the Z-plane, damping is increased as the poles move towards the center of the unit circle and downward towards the real axis. With a fast sampling rate, the poles are unable to move significantly in the Z-plane to increase their damping. Slower sampling rates spread out the poles and zeros and allow for more movement, thus greater increases in damping.

The question then becomes: How slow should or can the system be sampled? Slower sampling rates are known to cause aliasing and possibly problems with stability, but this does not appear to be a problem with PPF. An analytical model was developed that contained 20 modes, with the highest mode having a frequency of 3320Hz. Two tuning filters were designed in the S-plane to control the first two modes. The system was subjected to a zero-order hold with sampling period of 2000Hz, and the filters were 'Tustin' transformed with the same sampling period. The closed-loop discrete system was formed and the eigenvalues where found to all be stable.

A six-pole model was also analyzed with the highest frequency at 800Hz. The system was sampled at 825Hz and the filters transformed at this same frequency. Once again, all the closed-loop poles were stable with the poles being controlled at a higher damping value than their open-loop values. While this seems promising, the behavior of the system between sampling instants is unknown and may be undesirable. It is therefore recommended to sample at least twice as fast as the fastest mode of interest.

The second method, designing the filters in the digital domain using the root locus, is slightly more complicated than for the continuous time design. Since the sampling rate greatly effects the location of the poles and zeros in the Z-plane, the sampling rate must be known a priori in order to design the filters. Given that the sampling rate is known, there are two different cases of interest: one where there is under-sampling of some of the poles in the model, and one where there is no under-sampling. Under-sampling is when the pole is sampled less than two times per period such that the wrapping effect from the spiraling discussed earlier is occurring.

The latter case will be discussed first. Since the poles are complex conjugates, attention will only be given to the positive imaginary axis with the knowledge that there is symmetry about the real axis. With an appropriate sampling rate so that the poles and zeros are 'evenly' spaced around the unit circle, the filter design will first be focused on the positive real axis side of the unit circle. Representative pole and zero locations are shown in Figure 8.

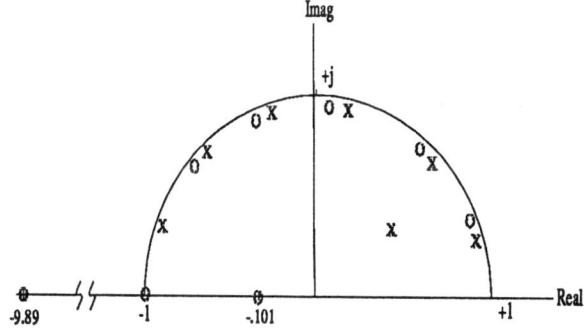

Figure 8: Z-Plane Pole and Zero Locations for Digital Design

With this configuration, it is no longer a simple matter of having the damped frequency of the filters to be greater than that of the mode being controlled as for the continuous locus. Since the structural poles and zeros follow the curvature of the unit circle, the filter poles must be placed relative to the pole and zero being controlled such that the desired locus is formed. The desired locus has the filter pole drawn towards the structural zero, and the structural pole drawn inward so as to increase its damping value, as shown in Figure 9.

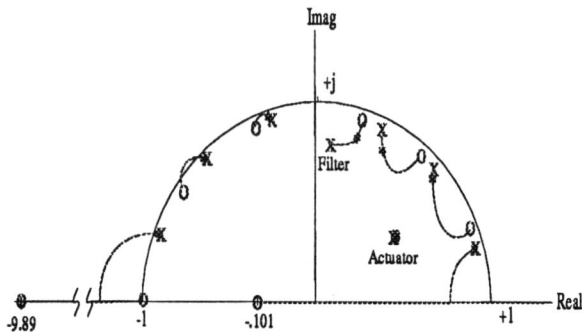

Figure 9: Design of the First Tuning Filter in the Z-Plane

The filter pole location is to the left of the structural pole and zero, and with either a greater or lesser imaginary axis value depending where along the unit circle the controlled pole is located. The filter gain is chosen as the value where the structural pole is slightly past the filter pole. These locations are shown as stars (*) in Figure 9. The gain is chosen so that the structural pole has moved slightly past the filter pole because of spillover. While tuning the next filter, spillover will cause this pole to move slightly back towards its original position.

Even though the locus of two of the higher modes go outside the unit circle, the gains necessary for the design of the filters are not large enough to cause those poles to move so far as to become unstable. Our experience is that it generally requires a gain twice as large as what is necessary in the filter design for this to occur. Unfortunately, unlike analog PPF where spillover resulted in higher than natural damping values for the unmodeled and uncontrolled modes, spillover in digital PPF can result in less than natural damping values for the uncontrolled or unmodeled modes. This is due to the form of the root-locus in the Z-plane. The higher frequency modes loop outward from the origin as they proceed to their respective zeros. This has the result of lowering their damping values. Also, depending on the pole and zero locations, the locus can go outside the unit circle, as seen in the second highest frequency mode in Figure 9, and become unstable. Therefore, care must be taken that the gains for the filter design are not so large as to cause these poles to go outside the unit circle.

When the poles being controlled are in the left half of the unit circle, the design procedure is the same. The filter pole is placed below and slightly to the left of the pole being controlled so that the filter pole is drawn towards the structural zero and the structural pole is drawn inward.

As in the S-plane design, the numerator of the filter contains the pole locations of the actuator so that a zero is placed 'on' the actuator pole and essentially cancelled. Spillover is also greater downward in frequency than upward as in the S-plane, therefore the filters are designed in the same top-down approach as in the continuous time design.

If the poles are spaced far enough apart, the design is not too difficult, but as the sampling rate is increased and the poles move closer together, the design becomes more difficult and the attainable closed-loop damping becomes smaller. This is because of the interaction of the closely spaced poles and zeros discussed earlier.

For the situation where there is under-sampling occurring, there are two different cases that need to be discussed. The first case is when the under-sampled pole lies in quadrant II. Quad II is defined as that region bounded by the negative real axis and positive imaginary axis, and for this analysis, inside the unit circle. The second case is when the under-sampled pole lies in Quad I. Quad I is similarly defined as the region bounded by the positive real axis and positive imaginary axis. This is shown in Figure 10.

For the case when the under-sampled pole was in Quad II, three filters were designed ignoring the fact that there was an under-sampled pole. The filters were tuned to the first three poles of the model. It was determined that the under-sampled pole behaved as any other pole, and did not go unstable any faster than if it was adequately sampled. Like the over-sampled case, this pole required a gain twice as large as the gain necessary in the filter design for it to go unstable.

The last case to be discussed is when the under-

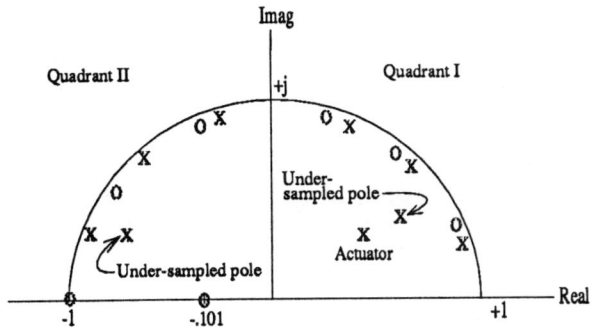

Figure 10: Under-sampled Pole locations, shown in Quad I and Quad II

sampled pole lies in Quad I. At this slow a sampling rate, the zeros associated with infinity have entered the unit circle as discussed earlier and basically became paired with the under-sampled pole. This zero and pole are so close together that the pole is immediately drawn towards the zero. This essentially removes the pole from interfering with the other poles, and the design procedure is the same as if the pole was not there.

The previous discussion was motivated by the fact that a distributed parameter system contains an infinite number of poles. When performing control of such systems, it is necessary to limit the size of the model for the design procedure, which requires truncating. Although the higher frequency poles are not modeled, they still exist. The question was: Where were these poles in the Z-plane? When it was revealed that they were within the unit circle and followed a spiral path towards the origin, the next question was: How do they interact with the modeled poles? Based on analytical models simulated on Matlab, the previous discussion offered a possible answer. The recommended choice of sampling rate is that which evenly spaces the poles to be controlled along the unit circle in Quad I or II without under-sampling the modelled modes. The filters were easily designed in these quadrants, and being evenly spaced allows for greater movement of the structural poles.

When designing the filters in the digital domain, it was necessary to know the sampling frequency to determine the pole locations. This makes the design dependent on the sampling rate; different sampling rates result in different pole locations and thus different filter locations. If the sampling rate changes, the design procedure locating the filter pole location needs to be repeated. If the design is performed in the S-plane and transformed, a different sampling rate simply means the transformation requires a different Ts, without the need to redesign the filters. If the sampling frequency is not going to change, either method can be implemented. If the sampling frequency might change, designing the filters in the S-plane and transforming is the recommended approach. This is the approach used in the next section; the experimental part of this paper. The filters are designed in continuous time and then converted to the Z-plane using the Tustin Transform.

Experimental Results

This section presents the results from a discrete implementation of PPF on a simply-supported beam. These experiments used a single sensor/actuator pair first with two tuning-filters and second with three tuning-filters to perform vibration suppression of a harmonic disturbance.

Two-Mode control

The filters were designed in the S-plane and transformed to the Z-plane. For reasons discussed in the previous section, the highest frequency filter was designed first. For the desired form of the root locus, a filter frequency of 1.3 times the frequency of the pole being controlled and a damping ratio of 0.18 was chosen. This is shown in Figure 11a. A gain of 0.34 resulted in the pole locations shown by a '+' in the figure. This filter and system are closed in a feedback loop to form a new system on which the next filter is designed. The closed-loop poles of the original system and filter are now the open-loop poles of this 'new' system. The frequency of the next filter was chosen to be 1.45 times the frequency of the pole being controlled with a damping ratio of 0.115. This resulted in the root locus shown in Figure 11b. As discussed in the previous section, the filter pole and structural pole previously designed move back up their locus due to the gain of the new filter. A gain of 0.2362 resulted in the pole locations indicated by a '+' on the locus.

To test this controller, the beam was excited by a 32Hz, 300mV harmonic disturbance. Figure 12a shows the response of the beam with the controller turned on after two seconds. There is a drop in the response of the beam by about 12dB. The vibration suppression due to the filter tuned to the second pole is shown in Figure 12b. There is approximately a 5dB reduction in the response of the beam due to the con-

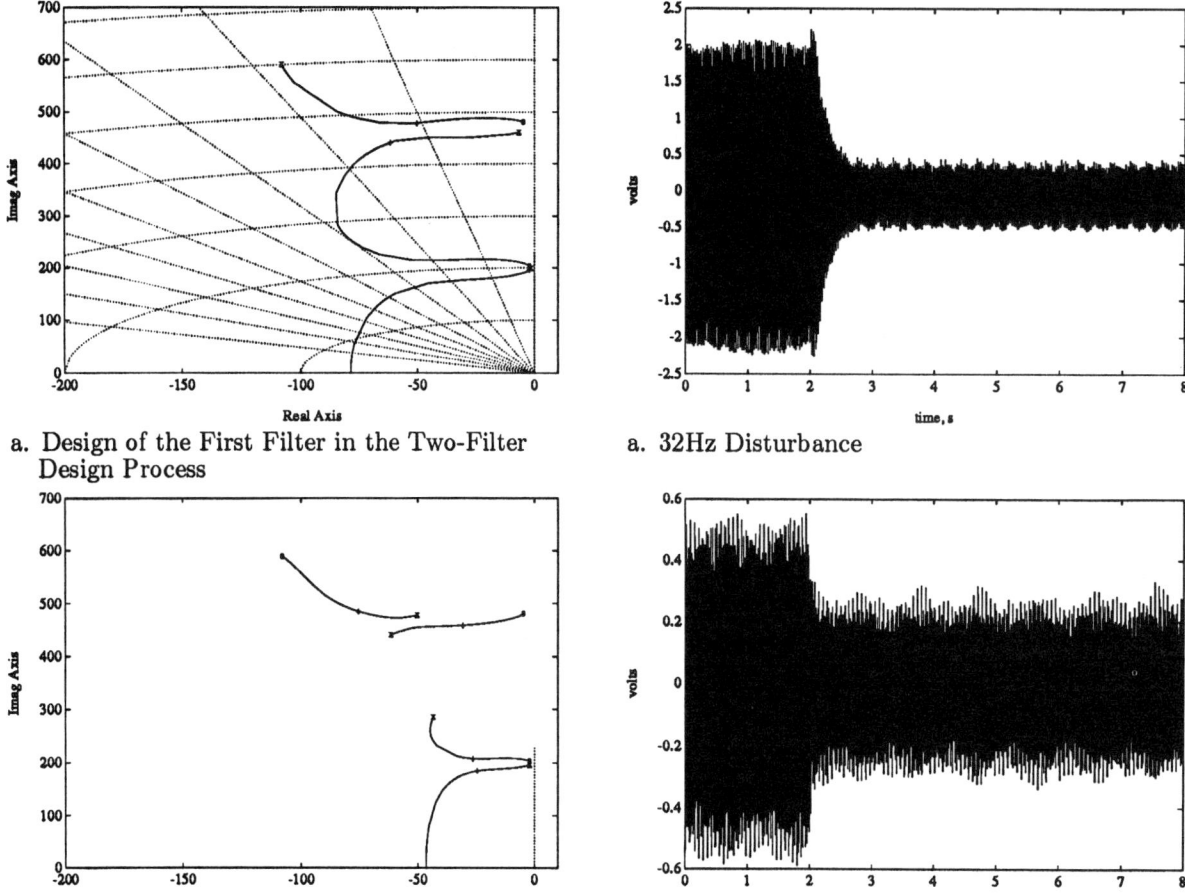

a. Design of the First Filter in the Two-Filter Design Process

b. Design of the Second Filter with the First Filter Already Designed

Figure 11: Two-Filter Design

a. 32Hz Disturbance

b. 74Hz Disturbance

Figure 12: Beam Response to Disturbance with the Two-Filter Controller

troller with a 74Hz, 500mV harmonic disturbance.

To observe the effect of spillover, the beam was excited at the frequency of the next mode. When the controller was turned on, there was an increase in the response. To better understand what is happening, Table 1 lists the open-loop damping values of the system, the predicted damping of the closed-loop system in the S-plane, and the predicted closed-loop damping values of the transformed Z-plane system. The Z-plane values are calculated by using the sampling rate to determine the equivalent S-plane frequencies and damping ratios for comparison.

As can be seen in the table, the predicted closed-loop damping in the Z-plane of the uncontrolled poles is slightly less than their open-loop damping values while the predicted closed-loop damping of the uncontrolled poles in the S-plane is equal to or greater than their open-loop values. This effect was discussed in the discrete PPF section. Therefore, the response of the beam at the uncontrolled modes can increase due to the discrete controller. It should be noted that although the damping of the uncontrolled poles decreased slightly, they remained stable while there was a significant increase in the damping of the two controlled poles.

Three-mode control

The last tests performed simultaneous control of three modes with three tuning filters. The design procedure for the tuning filters is the same as the two-mode controller. The highest frequency filter is designed first. A closed-loop system, CL1, is formed by the open-loop system and this filter. The next highest frequency filter is then designed around CL1. After the second filter is designed, another closed-loop system, CL2, is formed by the first closed-loop system, CL1, and this second filter. The last filter

	Open-Loop freq.(Hz)	damping	S-plane freq.(Hz)	damping	Z-plane freq.(Hz)	damping
pole1	31.7	0.0078	29.6	0.1326	32.0	0.1119
pole2	73.4	0.0141	73.2	0.0669	73.8	0.0619
pole3	127.9	0.0117	129.7	0.0135	129.7	0.0100
pole4	200.7	0.0085	201.1	0.0086	201.1	0.0078
pole5	287.2	0.0030	287.2	0.0030	287.2	0.0029
filter1	46.0	0.15	33.4	0.1268	30.9	0.1830
filter2	95.4	0.18	78.2	0.1545	77.0	0.1750

Table 1: Open-Loop Damping versus Predicted S-plane and Z-plane Closed-Loop Damping Values.

Filter	Freq(Hz)	Damping Ratio	Gain
3	166.0	0.18	0.3779
2	94.0	0.20	0.1465
1	46.0	0.20	0.1436

Table 2: Specifications for the Three-Filter Controller

is then designed around CL2. After the last filter is designed, another closed-loop system, CL3, is formed by CL2 and this last filter. The last closed-loop system, CL3, is the closed-loop system of the original open-loop system and the three filters. The filter frequencies, damping rations, and gains used in the design are listed in Table 2.

The controlled response of the beam are shown in Figures 13 and 14. For Figure 13a, a 34Hz harmonic disturbance was used with the an approximate 8dB reduction in the response due to the controller. A 74Hz harmonic disturbance was used in Figure 13b, with a reduction of 3.5dB due to the controller. When the beam was excited by a 128Hz disturbance, the response decreased by 6dB due to the controller, shown in Figure 14.

Figure 15 shows the open-loop and closed-loop frequency response functions of the system with a three-mode controller. As can be seen in the figure, the responses of the targeted modes have decreased significantly. While the response of the uncontrolled mode at 200Hz decreased slightly, the response at 290Hz has increased by about 3dB. This effect is explained by the discussion in the digital PPF section. Overall, the controller is effective in the vibration suppression of the targeted modes.

Conclusions

This work investigated the use of discretely implemented Positive Position Feedback (PPF) as an effective method to perform Active Vibration Control. A simply-supported beam was used as the testbed which used strain gages as the sensing element and piezoelectric ceramics as the actuator. Using a SISO controller, vibration suppression of several modes was accomplished.

To date, all work involving PPF revolved around continuous-time systems, therefore, the issue of sample-data systems using PPF was investigated. It was shown that unlike the continuous-time PPF control where only the lowest frequency mode had the potential to become unstable, in the discrete-time, some of the higher frequency modes also have the potential to become unstable with large gains.

In addressing the issue of discrete-control of a large number (infinite) of modes with a finite (not infinitesimal) sample period, it was noted that modes existed inside the closed-loop control with a sample period greater than the 'Nyquist' period. It was shown that the S to Z mapping resulted in these undersampled poles (and zeros) following a spiral path towards the origin.

Two design methods for PPF filters for use in discrete control were presented. One method involved designing the filters in the Z-plane. The second method involved designing the filters in the S-plane and converting them to the discrete-time domain. It was shown that if the filters are designed in the continuous-time, the 'Tustin' transformation successfully mapped the filters to the Z-plane for discrete control.

Using the design procedure discussed in the discrete PPF section, tuning filters were designed for use in two different SISO control experiments. In the first

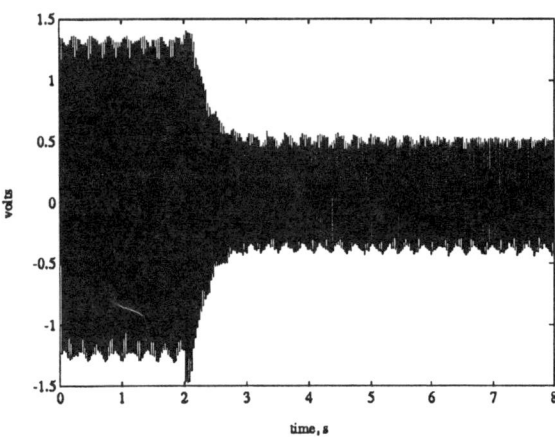

a. 34Hz Disturbance

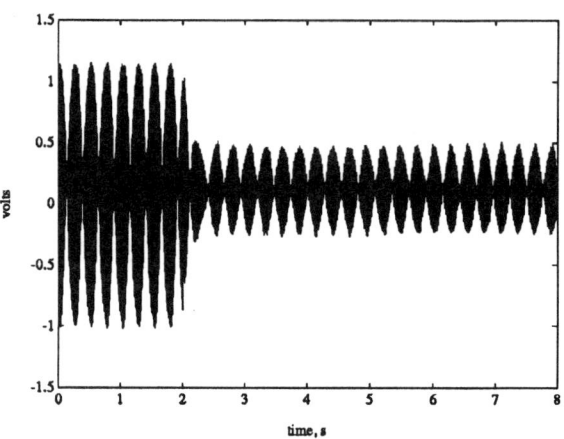

Figure 14: Beam Response to 128Hz Disturbance with Three-Filter Controller

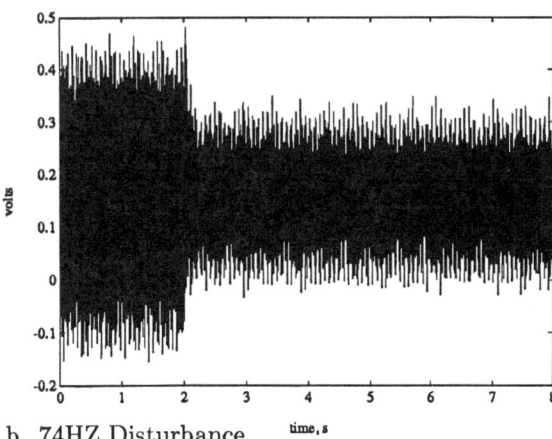

b. 74HZ Disturbance

Figure 13: Beam Responses to 34Hz and 74Hz Disturbance with Three-Filter Controller

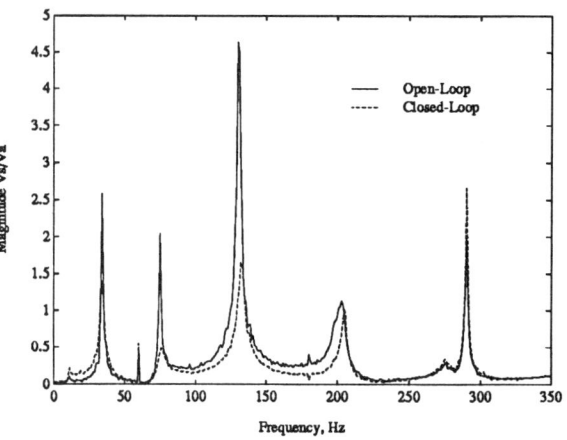

Figure 15: Comparison of Open-loop and Closed-Loop Frequency Response Functions with the Three-Filter Controller

experiment, two tuning filters were used to control two modes with 12dB and 5dB reductions in the beams responses for the 32Hz and 74Hz modes. Similarly, three tuning filters were used to successfully control the 32Hz, 74Hz and 128Hz modes for reductions of 8dB, 4dB, and 6dB respectively. By using more filters to control more modes, there was a reduction in their effectiveness. This is partly due to spillover, discussed in the digital PPF section, that alters the filter's design specifications. It is believed that the fewer tuning filters per sensor/actuator pair results in better vibration suppression of the modes being controlled.

References

Astrom, Karl J., and Bjorn Wittenmark. (1990). Computer-Controlled Systems: Theory and Design, Prentice Hall Publishing Company, Englewood Cliffs, New Jersey.

Balas, Gary J., and John C. Doyle. (1990). "Collocated versus Non-collocated Multivariable Control for Flexible Structure". *Proceedings of the American Controls Conference*, Vol.2, pp 1923-1928.

Balas, Mark J. (1978a). "Active Control of Flexible Systems". *Journal of Optimization Theory and Applications*, Vol.25, no.3, pp. 415–436.

—.(1978b). "Direct Velocity Feedback Control of Large Space Structures". *Journal of Guidance and Control*, Vol.2, no.3, pp. 252–253.

Caughey, T. K., and J. L. Fanson. (1987). "An Experimental Investigation of Vibration Suppression in Large Space Structures Using Positive

Position Feedback". *Dynamics Laboratory Report DYNL-87-1*, California Institute of Technology, Pasadena, CA.

Caughey, T. K., and C. J. Goh. (1983). "Vibration Suppression in Large Space Structures". *NASA Technical Report Number N83-36069*, pp. 119-142.

—. (1982). "Analysis and Control of Quasi-Distributed Parameter Systems". *Dynamics Laboratory Report DYNL-82-3*, California Institute of Technology, Pasadena, CA.

Fagan, Gary T. (1993). "An Experimental Investigation into Active Damage Control Systems Using Positive Position Feedback for AVC", Master's Thesis, VPI&SU.

Fanson, J. L., and T. K. Caughey. (1990). "Positive Position Feedback Control for Large Space Structures". *Proceedings of the 28th AIAA/ASME/ASCE/ AHS/ASC Structures, Structural Dynamics, and Materials Conference*, Monterey, CA, Vol.28, no.4, pp. 717-724.

Goh, C. J., and T. K. Caughey. (1985). "On the stability problem caused by finite actuator dynamics in the collocated control of large space structures". *International Journal of Control*, Vol. 41, no.3, pp. 787–802.

Schafer, Bernd E., and Hans Holzach. (1985). "Experimental Research on Flexible Beam Modal Control". *Journal of Guidance and Control*, Vol.8, no.5, pp. 597–604.

AIAA-94-1788-CP

SPILLOVER AND ENERGY CONSIDERATIONS IN THE OUTPUT FEEDBACK CONTROL OF ADAPTIVE STRUCTURES

C. M. Baycan[*], S. Utku[†]
Duke University
Durham, NC 27708-0288

B. K. Wada[‡]
Jet Propulsion Laboratory
Pasadena, CA 91109

Abstract

In this paper, energy consumption in vibration control of statically indeterminate adaptive structures is investigated. Although, adaptive structures theory is applicable to any type of structure, examples in this paper are based on adaptive truss structures for illustrative purposes. The system is assumed to be autonomous. Vibration control is achieved through the actuators working against element forces. In statically determinate structures, the element forces are generated only by the inertial and dissipative loads, whereas in statically indeterminate structures element stresses are generated not only by inertial and dissipative loads but also by the control elongations. In other words, in statically indeterminate structures, some of the energy supplied by the control is used towards aggravating the controls. In order to overcome this adverse behavior, slave actuators can be used. It is shown that when a sufficient number of slave actuators are employed, element forces due to control elongations are canceled, enabling a statically indeterminate structure to behave as a statically determinate one. It is also demonstrated that the amount of energy required for vibration control is less when slave actuators are used.

1 Introduction

Large space structures are prone to vibrations. Some of the causes for vibrations include meteors, cyclic thermal loading, and docking process in space. Since the space structures in orbit will be carrying equipment that has high precision requirements it is important that the vibrations are eliminated as quickly as possible. One method of suppressing the vibrations is to use passive controls [1, 2, 3]. Since passive control schemes have design limitations, recently more research is directed towards active control. Adaptive truss structures, which have found extensive space applications due to ease of their deployment and assembly, are suitable for active control. An extensive overview on adaptive structures has been given by Wada [4], and Miura and Furuya [5]. In adaptive truss structures, geometry and vibration control is achieved by introducing length changes through the actuators which are embedded in some of the members of the structure [6]. Since all of the control work done is internal, the momentum of the structure does not change. In this work, the vibration control in autonomous systems is considered, i.e., during the control period no external loads are acting on the system.

Although optimal control can be achieved in closed loop linear systems with a quadratic cost functional by the solution of the matrix Riccati equation, the eigenvalue assignment technique with output feedback control will be used here for active vibration control. State feedback control scheme requires the whole state for the control. Due to the limited number of sensors, the practicality of this scheme is restricted to structures with reasonably small number of degrees of freedom or to systems with state estimators, whereas output feedback control scheme can make use of the velocity and/or displacement readings obtained from a small number of sensors, thus eliminating the need for state estimators [7]. Important handicaps of output feedback control scheme are the presence of spillovers, and the possible alteration of the dynamic characteristics of the uncontrolled system. Through, the spillover matrices the controlled and uncontrolled modes are coupled. Also instability, i.e., eigenvalues with positive real parts, may arise. Stability is guaranteed if sensors and actuators are colocated. Also, the first l modes of a structure can be controlled by l actuators [8].

Another important concern is the energy requirement to suppress the vibrations. Main source of energy for space structures is solar energy obtained by photovoltaic cells. Although solar energy is abundant, if a malfunction occurs, the space structure may have to depend on its limited reserve energy for controlling the vibrations. Earth-bound structures, on the other hand have a limited supply of energy when they require it the most i.e. during earthquakes. Total energy required to completely eliminate some of the vibration modes depends on the initial conditions, i.e. the energy at time

---

[*]Graduate Student, Member AIAA
[†]Prof. of Civil Engineering, Prof. of Computer Science
[‡]Deputy Manager, Applied Technologies Section, Fellow AIAA

Copyright ©1994 by American Institute of Aeronautics and Astronautics, Inc. All rights reserved.

$t = 0$, spillover effects, and element forces. Element forces can be broken into two categories: the element forces that may be present before the control actuation, i.e., static loads already on the structure, and the element forces that are due to the motion and the control elongations. The second one is time dependent and the part due to control elongations occur only in statically indeterminate structures. Therefore, for a statically indeterminate structure, control elongations may intensify the vibrations instead of suppressing them by pumping energy to the system. In order to eliminate such adverse behavior, usage of "slave" actuators in statically indeterminate structures have been proposed [9]. The main task of the slave actuators is to relieve stresses, so that a statically indeterminate structure behaves as a determinate one and all of the control energy is directed towards dissipating the vibrations [10].

In this paper, the energy consumption to suppress first l selected modes in a desired manner is investigated for statically determinate, statically indeterminate, and statically indeterminate (but with slave actuators) structures that are in a given initial non-equilibrium state. The location selection of actuators is dealt with first. The criteria for actuator placement is to minimize the spillovers and the energy consumption in the control of the first l number of vibration modes. Although the methodology devised in this paper is applicable to any type of adaptive structure, adaptive truss structures are used in the numerical examples.

2 Dynamics and Control of Adaptive Structures

2.1 Linear Dynamic Excitation-Response Relations

If a structure of M elements, N nodes, a forces per element, e degrees of freedom per node, b deflection constraints, and f force constraints is considered, the number of independent nodal equilibrium equations is given by

$$n = Ne - b \quad (1)$$

On the other hand, the number of element forces (which is the same as the number of element deformations) is

$$m = Ma - f \quad (2)$$

If the structure is statically determinate $m = n$, else it is indeterminate to degree r, where $r = m - n$.

The external nodal loads on the structure can be expressed as an n-tuple $\mathbf{p}$, whereas the initial deformations of the elements, which are also called the control elongations, can be represented by the m-tuple $\mathbf{v}_o$. The response quantities of the structure, namely, nodal displacements, independent element deformations, and the element forces, can be denoted by the n-tuple $\boldsymbol{\xi}$, and the m-tuples $\mathbf{v}$, and $\mathbf{s}$, respectively.

In matrix form the relations between excitations and the response quantities can be shown as below,

$$\begin{bmatrix} \mathrm{diag}(\mathbf{K}^j) & -\mathbf{I} & 0 \\ -\mathbf{I} & 0 & \mathbf{B}^T \\ 0 & \mathbf{B} & 0 \end{bmatrix} \begin{Bmatrix} \mathbf{v} \\ \mathbf{s} \\ \boldsymbol{\xi} \end{Bmatrix} = \begin{Bmatrix} \mathbf{o} \\ -\mathbf{v}_o \\ \mathbf{p} - \mathbf{M}\ddot{\boldsymbol{\xi}} - \hat{\mathbf{C}}\dot{\boldsymbol{\xi}} \end{Bmatrix} \quad (3)$$

where $\mathbf{B}$ is an $(n \times m)$ matrix generated by the direction cosines of bar element axes, $\mathbf{M}$ is the n'th order diagonal mass matrix, and $\hat{\mathbf{C}}$ is the n'th order square damping matrix, $\mathrm{diag}(\mathbf{K}^j)$ is the m'th order block diagonal matrix of element stiffnesses, $\mathbf{I}$ is the identity matrix, and finally $\mathbf{0}$ and $\mathbf{o}$ are zero matrices and vectors of proper dimension. The first row partition of the equation given above presents the constitutive relations, which link the forces in the elements to their deformations. Second row partition is the geometric compatibility equations, which associate nodal deflections to element deformations. Finally, last row partition gives the nodal equilibrium equations.

2.2 Computation of Response due to Controls and Excitations

The response quantities given in Eqn. (3) can be expressed in terms of the nodal loads and controls by inverting the coefficient matrix. Since the coefficient matrix is very sparse, various methods have been devised to compute the response quantities of adaptive structures. If force method of analysis is used, the element forces can be written as,

$$\mathbf{s} = \mathbf{C}\mathbf{x}_r + \overline{\mathbf{C}}\left(\mathbf{p} - \mathbf{M}\ddot{\boldsymbol{\xi}} - \hat{\mathbf{C}}\dot{\boldsymbol{\xi}}\right) \quad (4)$$

where r-tuple $\mathbf{x}_r$ consists of the element forces in the redundant elements. Furthermore, the columns of matrix $\mathbf{C}$ form a basis in the null space of $\mathbf{B}$ and $\overline{\mathbf{C}}$ defines the generalized inverse of $\mathbf{B}$ [11],

Substituting element forces from Eqn. (4) into the first row partition of Eqn. (3), and element deformations from the resulting equation into the second row partition of Eqn. (3), and then multiplying both sides of this equation by $\mathbf{C}^T$, one obtains the relation forces in the redundant elements and excitations and controls,

$$\begin{aligned} \mathbf{F}\mathbf{x}_r &= \mathbf{C}^T \mathbf{v}_o \\ &\quad - \mathbf{C}^T \mathrm{diag}(\mathbf{F}^j)\overline{\mathbf{C}}\left(\mathbf{p} - \mathbf{M}\ddot{\boldsymbol{\xi}} - \hat{\mathbf{C}}\dot{\boldsymbol{\xi}}\right) \end{aligned} \quad (5)$$

where $\mathrm{diag}(\mathbf{F}^j)$ is the block diagonal matrix of element flexibilities and $\mathbf{F}$ is the flexibility matrix of the structure in the direction of the redundant element forces. They are defined as,

$$\mathrm{diag}(\mathbf{F}^j) = \mathrm{diag}(\mathbf{K}^j)^{-1} \quad (6)$$

$$\mathbf{F} = \mathbf{C}^T \mathrm{diag}(\mathbf{F}^j)\mathbf{C} \quad (7)$$

The r-tuple $\mathbf{x}_r$ can be expressed using Eqn. (5). Substituting $\mathbf{x}_r$ into Eqn. (4) enables an expression for element forces, and using $\mathbf{s}$ in first row partition of Eqn. (3) gives a representation for element deformations. Finally, multiplying both sides of the geometric compatibility equations by $\overline{\mathbf{C}}^T$ and using $\mathbf{B}\overline{\mathbf{C}} = \mathbf{I}$, an expression for nodal deflections can be obtained. Consequently, the response quantities can be written in terms of the excitations and controls as follows

$$\left\{ \begin{array}{c} \mathbf{v} \\ \mathbf{s} \\ \boldsymbol{\xi} \end{array} \right\} = \left[\begin{array}{c} \text{diag}(\mathbf{F}^j)\mathbf{K}_c \\ \mathbf{K}_c \\ \overline{\mathbf{C}}^T \left(\text{diag}(\mathbf{F}^j)\mathbf{K}_c - \mathbf{I} \right) \end{array} \right] \mathbf{v}_o$$

$$+ \left[\begin{array}{c} \left(\text{diag}(\mathbf{F}^j) - \mathbf{F}_c \right) \overline{\mathbf{C}} \\ \left(\mathbf{I} - \mathbf{K}_c \text{diag}(\mathbf{F}^j) \right) \overline{\mathbf{C}} \\ \overline{\mathbf{C}}^T \left(\text{diag}(\mathbf{F}^j) - \mathbf{F}_c \right) \overline{\mathbf{C}} \end{array} \right] \mathbf{p}^* \quad (8)$$

where

$$\mathbf{K}_c = \mathbf{C}(\mathbf{C}^T \text{diag}(\mathbf{F}^j)\mathbf{C})^{-1}\mathbf{C}^T \quad (9)$$

$$\mathbf{F}_c = \text{diag}(\mathbf{F}^j)\mathbf{K}_c\text{diag}(\mathbf{F}^j) \quad (10)$$

$$\mathbf{p}^* = \mathbf{p} - \mathbf{M}\ddot{\boldsymbol{\xi}} - \hat{\mathbf{C}}\dot{\boldsymbol{\xi}} \quad (11)$$

If $\mathbf{M}\ddot{\boldsymbol{\xi}} = \hat{\mathbf{C}}\dot{\boldsymbol{\xi}} = \mathbf{o}$ and $\mathbf{p}$ is known (or if $\mathbf{p}^*$ is given), and the controls $\mathbf{v}_o$ are given, then expressions in Eqn. (8) can be used to calculate the complete response of the structure.

2.3 Output Feedback Control and Spillover Matrices

Dynamic equilibrium equation for adaptive structures with controls may be obtained from Eqn. (3) as

$$\mathbf{M}\ddot{\boldsymbol{\xi}} + \hat{\mathbf{C}}\dot{\boldsymbol{\xi}} + \mathbf{K}\boldsymbol{\xi} = -\mathbf{B}\text{diag}\left(\mathbf{K}^j\right)\mathbf{v}_o \quad (12)$$

where, $\mathbf{K}$ is the stiffness of the structure in the direction of nodal degrees of freedom and is,

$$\mathbf{K} = \mathbf{B}\text{diag}(\mathbf{K}^j)\mathbf{B}^T \quad (13)$$

Second order system given above can be converted into a first order one using $\mathbf{x}^T = \begin{bmatrix} \dot{\boldsymbol{\xi}}^T & \boldsymbol{\xi}^T \end{bmatrix}$,

$$\dot{\mathbf{x}} = \mathbf{A}^{(0)}\mathbf{x} + \mathbf{B}_c\mathbf{u} \quad (14)$$

where,

$$\mathbf{A}^{(0)} = \begin{bmatrix} -\mathbf{M}^{-1}\hat{\mathbf{C}} & -\mathbf{M}^{-1}\mathbf{K} \\ \mathbf{I} & 0 \end{bmatrix} \quad (15)$$

$$\mathbf{B}_c = \begin{bmatrix} -\mathbf{M}^{-1}\mathbf{B}\text{diag}\left(\mathbf{K}^j\right) \\ 0 \end{bmatrix} \quad (16)$$

$$\mathbf{u} = \mathbf{v}_o \quad (17)$$

Using Output Feedback Control scheme, control elongations can be computed from the sensor measurements as

$$\mathbf{v}'_o = \mathbf{F}_V\dot{\mathbf{v}}_t + \mathbf{F}_D\mathbf{v}_t \quad (18)$$

where $\mathbf{F}_V$ and $\mathbf{F}_D$ are velocity and displacement feedback gain matrices and $\mathbf{v}_t$ is the q-tuple (p: no. of actuators, l: no. of controlled modes, q: no. of sensors) which specifies the total elongations in the elements with the sensors, and $\mathbf{v}'_o$ lists the non-zero rows of $\mathbf{v}_o$. We assume that the sensors are to measure the length changes in the elements, i.e., $\mathbf{v}_t = \mathbf{v}' - \mathbf{v}'_o$. Then using geometric compatibility equations control elongations can be expressed in terms of the nodal displacements and their rates of change

$$\mathbf{v}_t = \mathbf{B}_s^T\boldsymbol{\xi} \quad (19)$$

$$\dot{\mathbf{v}}_t = \mathbf{B}_s^T\dot{\boldsymbol{\xi}} \quad (20)$$

where $\mathbf{B}_s$ is the portion of the $\mathbf{B}$ matrix associated with the sensor locations. Therefore,

$$\mathbf{v}'_o = \mathbf{F}_V\mathbf{B}_s^T\dot{\boldsymbol{\xi}} + \mathbf{F}_D\mathbf{B}_s^T\boldsymbol{\xi} \quad (21)$$

With this definition, we can rewrite Eqn. (17) as

$$\dot{\mathbf{x}} = \mathbf{A}\mathbf{x} \quad (22)$$

where,

$$\mathbf{A} = \begin{bmatrix} -\mathbf{M}^{-1}\left(\hat{\mathbf{C}} + \mathbf{G}_V\right) & -\mathbf{M}^{-1}\left(\mathbf{K} + \mathbf{G}_D\right) \\ \mathbf{I} & 0 \end{bmatrix} \quad (23)$$

where the matrices $\mathbf{G}_V$ and $\mathbf{G}_D$ are defined as

$$\mathbf{G}_V = \mathbf{B}_s\text{diag}(\mathbf{K}^{j'})\mathbf{F}_V\mathbf{B}_s^T \quad (24)$$

$$\mathbf{G}_D = \mathbf{B}_s\text{diag}(\mathbf{K}^{j'})\mathbf{F}_D\mathbf{B}_s^T \quad (25)$$

The solution to Eqn. (22) given in terms of the initial conditions $\mathbf{x}_o^T = \begin{bmatrix} \dot{\boldsymbol{\xi}}_o^T & \boldsymbol{\xi}_o^T \end{bmatrix}$ is,

$$\mathbf{x}(t) = e^{\mathbf{A}t}\mathbf{x}_o \quad (26)$$

In order to obtain $\mathbf{A}$, one needs the gain matrices. If eigenvalue assignment technique is used, assuming $p = q = l$, the gain matrices can be expressed as [8],

$$\mathbf{F}_D = \text{diag}\left(\mathbf{K}^{j'}\right)^{-1}\boldsymbol{\Psi}^{-1}\text{diag}_c\left(\omega_i^{d\,2} - \omega_i^2\right)\boldsymbol{\Psi}^{-T} \quad (27)$$

$$\mathbf{F}_V = \text{diag}\left(\mathbf{K}^{j'}\right)^{-1}\boldsymbol{\Psi}^{-1}\text{diag}_c\left(2\zeta_i^d\omega_i^d - 2\zeta_i\omega_i\right)\boldsymbol{\Psi}^{-T} \quad (28)$$

where $\boldsymbol{\Psi} = \mathbf{N}_c^T\mathbf{B}_s$, ω_i^d, ζ_i^d, $i = 1, \ldots, q$ are the selected frequencies and the damping ratios for the controlled modes, and ω_i, ζ_i are the inherent frequencies and damping ratios of the structure. The $n \times q$ matrix $\mathbf{N}_c$ lists the controlled modes and it is obtained from the $\mathbf{M}$ normalized modes of the structure $\mathbf{N}$ ($\mathbf{N} = [\mathbf{N}_c|\mathbf{N}_u]$, where $\mathbf{N}_u$ lists the remaining uncontrolled modes). Using the transformation $\boldsymbol{\xi} = \mathbf{N}_c\boldsymbol{\eta}_c + \mathbf{N}_u\boldsymbol{\eta}_u$ and $\dot{\boldsymbol{\xi}} = \mathbf{N}_c\dot{\boldsymbol{\eta}}_c + \mathbf{N}_u\dot{\boldsymbol{\eta}}_u$, in $\boldsymbol{x}^T = \begin{bmatrix} \dot{\boldsymbol{\xi}}^T & \boldsymbol{\xi}^T \end{bmatrix}$, and the latter in Eqn. (22), one can obtain

$$\ddot{\boldsymbol{\eta}}_c + \text{diag}_c\left(2\zeta_i^d\omega_i^d\right)\dot{\boldsymbol{\eta}}_c + \text{diag}_c\left(\omega_i^{d\,2}\right)\boldsymbol{\eta}_c =$$

$$\mathbf{S}^{cud}\boldsymbol{\eta}_u + \mathbf{S}^{cuv}\dot{\boldsymbol{\eta}}_u \quad (29)$$

$$\ddot{\eta}_u + [\mathbf{S}^{uuv} + \text{diag}_u(2\zeta_i\omega_i)]\dot{\eta}_u +$$
$$[\mathbf{S}^{uud} + \text{diag}_u(\omega_i{}^2)]\eta_u = \mathbf{S}^{ucd}\eta_c + \mathbf{S}^{ucv}\dot{\eta}_c \quad (30)$$

where

$$\mathbf{S}^{ijk} = \mathbf{N}_i^T \mathbf{G}_k \mathbf{N}_j, \quad i = c, u; \; j = c, u; \; k = D, V \quad (31)$$

Matrices $\mathbf{S}^{ijk}$, $i = c, u; j = c, u; k = D, V$ are the spillover matrices. If the spillover matrices are all zero, using Eqns. (29) and (30) it can be seen that the controlled modes will damp out with selected frequency ω_i^d and selected damping ratio ζ_i^d, and the uncontrolled ones with frequency ω_i and damping ratio ζ_i. But in output feedback control the spillover matrices are always present. It can be observed from Eqns. (29) and (30) energy is transferred between the controlled and the uncontrolled modes until all motion stops. Since the spillover matrices depend on the selected vibration frequencies and the damping ratios for the controlled modes, extra caution should be taken during the selection process so that the motion of the uncontrolled modes does not become unstable [8].

3 System Energy Considerations

3.1 Control Energy in Active Vibration Control of Statically Determinate Structures

In Fig. 1, a simple statically determinate structure is shown where all mass is lumped at the node, and the member contains an actuator. To explain the basic ideas, it is assumed that the natural damping, $\hat{\mathbf{C}}$, is zero, and the system is autonomous, i.e., the controls $\mathbf{v}_o$ are activated after the transient excitations cease, at time $t = 0$. The state at $t = 0$ is a non-equilibrium state, and it is identified by $\boldsymbol{\xi}(0) = \mathbf{o}$, $\dot{\boldsymbol{\xi}}(0) = \dot{\boldsymbol{\xi}}_o$. The dynamic equilibrium equation of the controlled system may be obtained from Eqn. (12) as,

$$\mathbf{M}\ddot{\boldsymbol{\xi}} + \mathbf{K}\boldsymbol{\xi} = -\mathbf{B}^T \text{diag}(\mathbf{K}^j)\mathbf{v}_o \quad \text{for } t > 0 \quad (32)$$

with the initial conditions given as,

$$\begin{aligned} \boldsymbol{\xi}(0) &= \mathbf{0} \\ \dot{\boldsymbol{\xi}}(0) &= \dot{\boldsymbol{\xi}}_o \\ \mathbf{v}_o(0) &= \mathbf{o} \end{aligned} \quad (33)$$

As discussed in the previous section, the feedback gain matrices $\mathbf{F}_D$ and $\mathbf{F}_V$ of the control law of Eqn. (21) may be obtained either by the "algorithm of optimal linear control with quadratic cost function" [12] or by eigenvalue assignment [8]. Then, the response of the controlled system may be expressed as in Eqn. (26), with $\mathbf{A}$ as in Eqn. (23) with $\hat{\mathbf{C}} = \mathbf{0}$. Once, $\mathbf{x}(t)$ is known, so are $\boldsymbol{\xi}(t)$ and $\dot{\boldsymbol{\xi}}(t)$. Using these, one can compute $\mathbf{v}_o(t)$, $\mathbf{v}(t)$, and $\mathbf{s}(t)$ from Eqn. (21), geometric compatibility relations, and constitutive relations given in Eqn. (3), respectively.

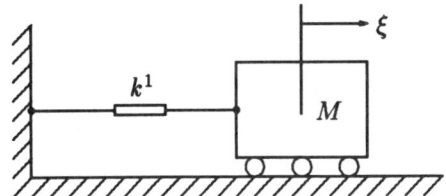

Figure 1: Statically Determinate Single Degree of Freedom Structure

The system's initial energy can be expressed as,

$$E(0) = \frac{1}{2}\dot{\boldsymbol{\xi}}_o^T \mathbf{M} \dot{\boldsymbol{\xi}}_o \quad (34)$$

Let T denote the time when the motion stops. Then, at time $t = T$, the system's energy becomes

$$E(T) = 0 \quad (35)$$

since, $\boldsymbol{\xi}(T) = \mathbf{o}$ and $\dot{\boldsymbol{\xi}}(T) = \mathbf{o}$. The initial energy $E(0)$ is dissipated by the work done in the actuators, E_d. When the motion stops,

$$E_d = E(0) \quad (36)$$

The dissipated energy may be computed alternatively as,

$$E_d = \sum_{i=1}^{n_c} \mathbf{s}^{(i)^T} \mathbf{W}^{(i)} \Delta \mathbf{v}_o^{(i)} \quad (37)$$

where $\mathbf{s}^{(i)} = \mathbf{s}(t_i)$: element forces at time $t_i = \sum_{k=1}^{i} \Delta t$, Δt is the control cycle time ($T = n_c \Delta t$); $\Delta \mathbf{v}_o^{(i)} = \mathbf{v}_o(t_i) - \mathbf{v}_o(t_{i-1})$: incremental actuator induced elongations at time t_i; and $\mathbf{W}^{(i)} = \text{diag}(0/-1)$. Note that $\mathbf{s}^{(i)^T} \mathbf{W}^{(i)} \Delta \mathbf{v}_o^{(i)}$ is the energy required by the actuators in order to insert $\Delta \mathbf{v}_o^{(i)}$ into the structure. Note also that $w_{k,k}^{(i)}$ (k'th diagonal element of $\mathbf{W}^{(i)}$) is zero at time t_i if the k'th component of the required elongation $\Delta \mathbf{v}_o^{(i)}$ and the k'th component of the element force $\mathbf{s}^{(i)}$ are of the same sign, otherwise it is -1. Of course, only elements with an actuator can have a nonzero entry in $\Delta \mathbf{v}_o^{(i)}$. At each control step only when an actuator elongation opposes the corresponding element force, work will be done to resist the motion. In other words, the insertion of $\Delta \mathbf{v}_o^{(i)}$ by the actuator will see no resistance if the components of $\Delta \mathbf{v}_o^{(i)}$ and $\mathbf{s}^{(i)}$ are of the same sign. Therefore, E_d is also the control energy, E_c, i.e., it must be supplied to the actuators in order to drive them during control. Hence the ratio of control energy to the initial energy, e_c, may be expressed as,

$$e_c = \frac{E_c}{E_d} = \frac{E_c}{E(0)} = 1 \quad (38)$$

in statically determinate structures.

3.2 Control Energy in Active Vibration Control of Statically Indeterminate Structures

In Fig. 2, the statically determinate structure of Fig. 1 is made statically indeterminate by adding a second member. Eqn. (32) is applicable to the statically indeterminate structure if one increases $\mathbf{K}$ by $\mathbf{B}\text{diag}\left(\mathbf{K}_r^j\right)\mathbf{B}^T$, the contribution of redundant elements.

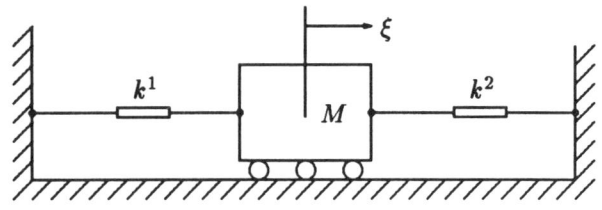

Figure 2: Statically Indeterminate Single Degree of Freedom Structure

Using the same initial conditions displayed in Eqn. (33), it can be seen that the initial energy is as given in Eqn. (34). In order to stop the motion, $E(0)$ must be dissipated completely by the time $t = T$. Unlike the statically determinate case, however the control energy is greater than $E(0)$, i.e., the the energy to be dissipated. In statically indeterminate structures, if the actuators are required to insert incremental control strains $\Delta \mathbf{v}_o^{(i)}$ at time t_i, an actuator will have to work not only against any opposing forces present in its member but also against the resistance of the structure for such insertions. Although, structural resistance against strain insertion does not exist in statically determinate structures, in statically indeterminate structures actuators need additional energy $\Delta E_r^{(i)}$ in order to overcome the structure's resistance. A lower bound of this energy may be expressed as,

$$\Delta E_r^{(i)} = \frac{1}{2}\Delta \mathbf{v}_o^{(i)T} \mathbf{K}_c \Delta \mathbf{v}_o^{(i)} \quad (39)$$

where the last two factors on the right hand side of Eqn. (39) correspond to incremental stresses in the elements due to insertion of $\Delta \mathbf{v}_o^{(i)}$ (see Eqn. (8)), and $\mathbf{K}_c$ is as defined in Eqn. (9). As before, the insertion of strains $\Delta \mathbf{v}_o^{(i)}$ may also be resisted by the element forces $\mathbf{s}^{(i)}$ already present at time t_i, requiring for energy $\Delta E_d^{(i)}$:

$$\Delta E_d^{(i)} = \mathbf{s}^{(i)T} \mathbf{W}^{(i)} \Delta \mathbf{v}_o^{(i)} \quad (40)$$

where the diagonal matrix $\mathbf{W}^{(i)}$ is as defined for Eqn. (37). Using Eqns. (39) and (40), the control energy, E_c, to implement $\mathbf{v}_o(t)$ for $0 \leq t \leq T$ in statically indeterminate structures may be expressed as,

$$E_c = \sum_{i=1}^{n_c} \left(\Delta E_d^{(i)} + \Delta E_r^{(i)} \right) \quad (41)$$

or substituting the definitions from Eqns. (39) and (40),

$$E_c = \sum_{i=1}^{n_c} \mathbf{s}^{(i)T} \mathbf{W}^{(i)} \Delta \mathbf{v}_o^{(i)} + \sum_{i=1}^{n_c} \frac{1}{2}\Delta \mathbf{v}_o^{(i)T} \mathbf{K}_c \Delta \mathbf{v}_o^{(i)} \quad (42)$$

The first term in the above equation is the work done to stop the motion, i.e., the dissipation energy E_d. Hence,

$$\sum_{i=1}^{n_c} \mathbf{s}^{(i)T} \mathbf{W}^{(i)} \Delta \mathbf{v}_o^{(i)} = E_d = E(0) \quad (43)$$

and using the dissipation energy in Eqn. (42),

$$E_c = E(0) + \sum_{i=1}^{n_c} \frac{1}{2}\Delta \mathbf{v}_o^{(i)T} \mathbf{K}_c \Delta \mathbf{v}_o^{(i)} \quad (44)$$

Finally, the ratio of the control energy E_c to the initial energy becomes,

$$e_c = \frac{E_c}{E(0)} = 1 + \frac{\sum_{i=1}^{n_c} \frac{1}{2}\Delta \mathbf{v}_o^{(i)T} \mathbf{K}_c \Delta \mathbf{v}_o^{(i)}}{E(0)} \quad (45)$$

Since the numerator of the second term is a positive definite quantity, $e_c > 1$, i.e., one needs more energy than the initial energy $E(0)$ to stop the vibrations in statically indeterminate structures.

It should be noted that, Eqn. (45) reduces to Eqn. (38), if either the structure is statically determinate (i.e. $\mathbf{C}$ does not exist [11], hence $\mathbf{K}_c = \mathbf{0}$), or $\Delta \mathbf{v}_o^{(i)}$ is in the row space of $\mathbf{B}$ ($\mathbf{K}_c \Delta \mathbf{v}_o^{(i)} = \mathbf{C}(\mathbf{C}^T \text{diag}(\mathbf{F}^j)\mathbf{C})^{-1}\mathbf{C}^T \Delta \mathbf{v}_o^{(i)} = \mathbf{o}$ since $\mathbf{C}^T \Delta \mathbf{v}_o^{(i)} = \mathbf{o}$ because $\mathbf{C}^T \mathbf{B}^T = \mathbf{0}$ [13]). The latter can be achieved by using "slave" actuators [10].

3.3 Spillover of a Different Kind

The extra energy, $(e_c - 1)E(0)$, pumped into statically indeterminate structures is a source of excitation, created by active control itself. This additional energy may be called spillover energy working against the control of vibrations. This is detrimental for the structure, therefore it should be eliminated via employing slave actuators.

4 Numerical Results

In this section numerical results are presented to show the effect of induced strain incompatibilities in active vibration control of space structures. The system is autonomous. The analysis was performed on the planar truss shown in Fig. 3. The initial statically determinate structure has 8 nodes, 13 elements, and 3 deflection constraints. The structure is made of aluminum for which Young's Modulus(E) is 6.8948×10^{10} Pa, and density (ρ) is 2712.64 kg/m^3. All of the elements have the same diameter of 10.0 cm.

It is assumed that the structure does not have any natural damping, i.e., $\hat{\mathbf{C}} = \mathbf{0}$. As stated earlier, output feedback control with eigenvalue assignment is used for the control scheme. The initial disturbance is caused by an impulse load applied in negative y-direction to the tip of the boom at time $t = 0$. It is desired that the first two modes of the structure have 10% damping, i.e., $\zeta_1^d = \zeta_2^d = 0.10$. In order to control the first two vibration modes, two actuators were placed in elements 1 and 2 using the actuator placement criteria given by Lu et al.[7].

The procedure followed is given below,

1. An impulse load of 36.02 N is applied at the tip of the crane. Then the vibrations in the structure are controlled. The energy required to eliminate the vibrations is calculated.

2. Two redundant elements are added to the structure (element 14 is located between nodes 1 and 5, and element 15 is between nodes 5 and 8) for prestressing purposes. The new statically indeterminate structure is shown in Fig. 4. This time, the applied impulse load is 50.0 N. Again, using the actuators located in elements 1 and 2 control of the statically indeterminate structure is achieved.

3. At this step, slave actuators are placed on the redundant elements added to the structure in the previous step (elements 14 and 15). The slave actuators are used to prevent stress buildup due to control actuators. Applying the control law defined earlier, vibrations are eliminated.

In Fig. 5, energy required for vibration control, i.e., E_c is plotted against time for the initial statically determinate structure and the statically indeterminate structure. The difference between the control energies of the statically determinate and the statically indeterminate with the slave actuator cases is due the difference in their initial energies. As it has been stated, the energy requirement is less for statically indeterminate structures when slave actuators are used.

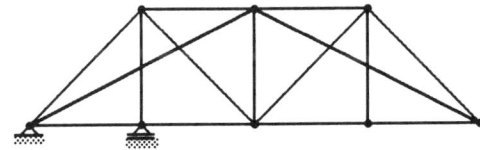

Figure 4: Statically Indeterminate Planar Truss analyzed in Numerical Results, (the heavier lines indicate the redundant elements)

5 Summary and Conclusions

In this paper, control energy for vibration control of statically determinate and indeterminate adaptive structures is investigated. First a formulation for control and dissipation energies for statically determinate and indeterminate structures is given. Then, a ratio of control energy to suppress the vibrations and the initial energy of the structure is presented. It is shown that the ratio, e_c, is greater than 1 for statically indeterminate structures due to the internal stresses created by the control elongations. In order to prevent this consumption, slave actuators are used. Furthermore, it is pointed out that the excess energy supplied to the statically indeterminate structures can be considered as a different kind of spillover energy working against the controls. Finally, a simple planar truss structure is used to demonstrate the theory given in the paper and results obtained from this structures are demonstrated.

The following conclusions are drawn from this work:

- Given the same initial conditions, energy requirement is less for statically determinate structures than for statically indeterminate structures generated by adding redundant elements to the statically determinate ones.

- The extra energy supplied by the actuators in statically indeterminate structures increase the structures' energy state, thus demanding more energy from the controllers.

- Using slave actuators in the redundant elements of statically indeterminate structures, the energy consumption can be reduced.

- Again by employing slave actuators, the transformation of some of the control energy to excitations for the structure is prevented.

References

[1] H. Ashley, "On passive damping mechanisms in large space structures," *J. of Spacecraft and Rockets*, vol. 21, no. 5, pp. 448–455, 1984.

[2] G. S. Chen and B. K. Wada, "Passive damping for space truss structures," in *Paper 88-2469, Proceedings of AIAA/ASME/ASCE/AHS 29th Structures, Structural Dynamics and Materials Conference*, pp. 1742–1749, 1988.

[3] A. R. Ramachandran, Q. C. Xu, L. E. Cross, and R. E. Newnham, "Passive piezoelectric vibration damping," in *Proceedings of First Joint U.S./Japan Conference on Adaptive Structures*, pp. 525–538, 1990.

[4] B. K. Wada, "Adaptive structures: An overview," *AIAA J. of Spacecraft*, vol. 27, no. 3, pp. 330–337, 1990.

[5] K. Miura and H. Furuya, "Adaptive structure concept for future space applications," *AIAA Journal*, vol. 26, no. 8, pp. 995–1002, 1988.

[6] E. H. Anderson, D. M. Moore, J. L. Fanson, and M. A. Ealey, "Development of an active member using piezoelectric and electrostrictive actuation for control of precision structures," in *Paper 90-1085, Proceedings of AIAA/ASME/ASCE/AHS 31st Structures, Structural Dynamics and Materials Conference*, pp. 2221–2233, 1990.

[7] L. Y. Lu, S. Utku, and B. K. Wada, "On the placement of active members in adaptive truss structures for vibration control," *Journal of Smart Materials and Structures*, vol. 1, no. 1, pp. 8–23, 1992.

[8] L. Y. Lu, S. Utku, and B. K. Wada, "Vibration suppression for large scale adaptive truss structures using direct output feedback control," tech. rep., JPL, 1991. see also the Proceedings of The Third International Conference on Adaptive Structures, LaJolla, CA, Nov. 9-11, 1992.

[9] S. Utku, B. Utku, and B. K. Wada, "Vibration inhibition in buildings by the adaptive structures technology and using building's gravitational latent energy," Tech. Rep. 10221, JPL, 1992.

[10] C. M. Baycan, S. Utku, and B. K. Wada, "Vibration control in statically indeterminate adaptive truss structures," in *Paper 93-1691, Proceedings of AIAA/ASME/ASCE/AHS 34th Structures, Structural Dynamics and Materials Conference*, pp. 3289–3296, 1993.

[11] S. Utku, C. H. Norris, and J. B. Wilbur, *Elementary Structural Analysis*. New York, N. Y.: McGraw-Hill, fourth ed., 1991.

[12] M. Athans and P. L. Falb, *Optimal Control*. McGraw Hill, 1966.

[13] A. V. Ramesh, S. Utku, and B. K. Wada, "Real-time control of geometry and stiffness in adaptive structures," *Computer Methods in Applied Mechanics and Engineering*, pp. 761–779, 1991. Also presented in the Second World Congress of Computational Mechanics, Stuttgart, W. Germany.

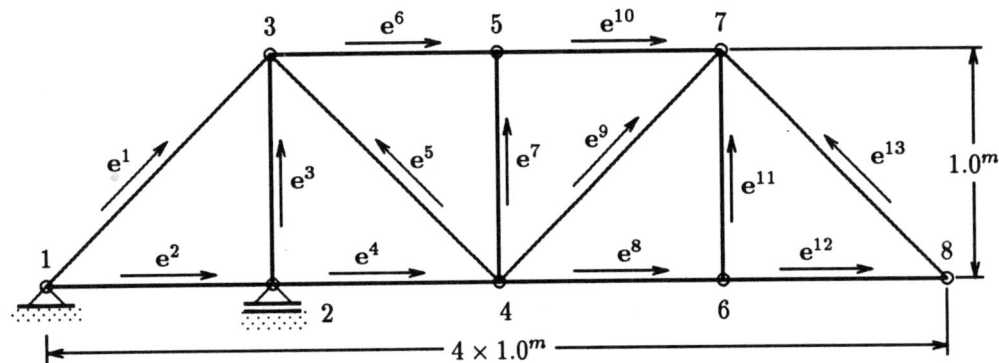

Figure 3: Statically Determinate Planar Truss analyzed in Numerical Results

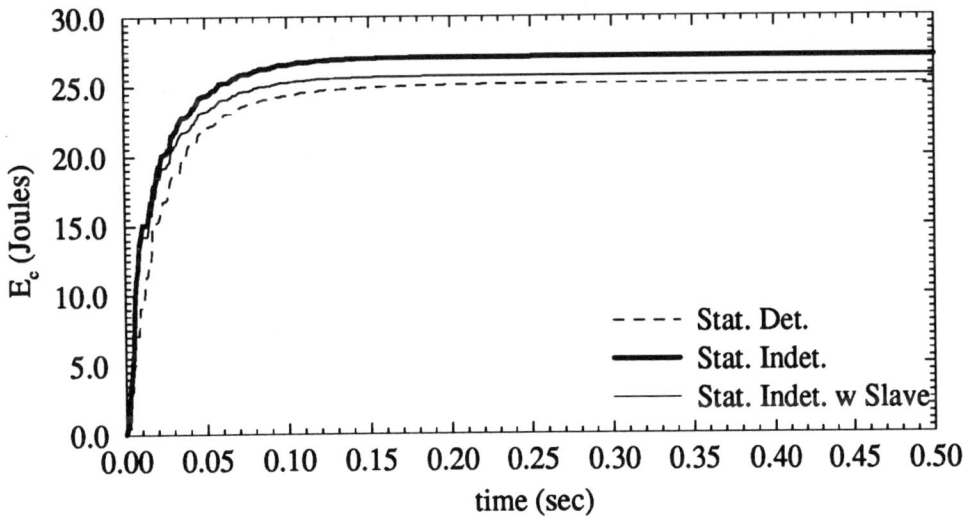

Figure 5: Cotrol Energy (E_c) vs. Time Plots for Vibration Control of the Statically Determinate Truss and Indeterminate Truss with and without Slave Actuators.

DIRECT RATE FEEDBACK FOR PIEZOSTRUCTURES USING SENSORIACTUATORS WITH FEEDTHROUGH DYNAMICS*

Daniel G. Cole[†] and Harry H. Robertshaw[‡]
Center for Intelligent Materials Systems and Structures
Department of Mechanical Engineering
Virginia Polytechnic Institute and State University
Blacksburg, Virginia 24061-0238

Abstract

The control of piezostructures using direct rate feedback (DRFB) is discussed. The collocated condition is achieved using piezoelectric sensoriactuators. The effects of partial compensation on the locus of system zeros is discussed with the result that under-compensation results in an equal number of poles and zeros alternating along the imaginary axis. Global stability of DRFB with partially compensated sensoriactuators is proven, and is shown to be asymptotically stable for structures without rigid-body dynamics. DRFB closed-loop response is discussed, and the trade-offs due to the interaction of mechanical and electrical dynamics are illuminated.

Introduction

The control of complex structures is often not well suited to many fixed gain control techniques (e.g. pole placement, LQR, LQG). This is due to the difficulty of obtaining an accurate plant model and robustness characteristics inherent in the control method. Stability margins and robustness can often times be improved by ensuring appropriate plant transfer functions are minimum phase. Often, the control requirements can be satisfied using reduced-order controllers. Among these, output feedback is easy to implement and under certain conditions can be shown to be globally stable. Very often, control requirements dictate vibration suppression; direct velocity feedback (DVFB) has been shown to be globally stable for positive semi-definite feedback matrices (Balas 1979). Balas investigated structures acted upon by force with outputs proportional to structural velocities.

Structure:

$$\ddot{r}(t) + \mathcal{D}\dot{r}(t) + \Omega r(t) = Bf(t)$$
$$y(t) = C\dot{r}(t)$$

Control Law:

$$f(t) = -Gy(t) = -GC\dot{r}(t)$$

Fundamental to this result is the requirement that $C = B^T$, which describes the collocated condition between the actuators and sensors. In general, this necessary condition is not only contingent on spatial similarity between sensor and actuator but also relies upon the compatibility of the collocated transducer pair. This relationship is best demonstrated by example: a force transducer (shaker) would be compatible with an accelerometer but not with a strain gage. This requirement is discussed rigorously by Burke et al. (1993).

Recent investigations have looked at piezoelectric sensors and actuators, in particular self-sensing actuators or *sensoriactuators* (this term is taken from the similar physiological term sensorimotor), which obtain sensing and excitation from the same piezoelectric element. Sensoriactuators apply structural forces proportional to the applied voltage, and respond with charges (currents) proportional to the structural vibration response. The advantage of

---

*This research was performed under Army Research Office's University Research Initiative Center Program, Grant No.03-92-6-0181; Dr. Gary Anderson, Program Manager.

[†]Graduate Research Assistant, Department of Mechanical Engineering, VPI&SU.

[‡]Associate Professor, Department of Mechanical Engineering, VPI&SU.

Copyright © 1994 by the American Institute of Aeronautics and Astronautics, Inc. All rights reserved.

using sensoriactuators is that the spatial collocation and compatibility relationships are automatically satisfied. However, the electromechanical coupling also includes the electrodynamics of the piezoelectric's capacitance, which acts as a feedthrough term directly coupling the input voltage and output charge (current). Balas' result does not include such a feedthrough term and it is not apparent that a rate feedback law on a system which includes feedthrough dynamics will be globally stable. The feedthrough term can be compensated for using appropriate analog circuitry; however, compensation can only be achieved with prior knowledge of the patch capacitance and, even then, the compensated sensoriactuator problem is difficult (Anderson et al., 1992, Dosch et al., 1992) and perfect compensation is a practical impossibility. One would like to know if the global stability can be achieved even for the uncompensated and partially compensated sensoriactuator configurations and what the effect of inaccurate (or zero-) compensation is on the closed-loop response of the system.

The remainder of this paper discusses the implementation of direct rate feedback (DRFB) laws for piezostructures with the result that such a control law is globally stable for positive semi-definite feedback gain matrix, provided the sensoriactuator is not "over-compensated."

First, a brief discussion sensoriactuator transducer design is provided to illuminate some of the difficulties associated with sensoriactuators and to show that imperfect compensation leads to non-zero feedthrough term. Included here are the effects of varied compensation on the open-loop zeros of the piezostructure.

Second, a proof of global stability of the DRFB problem is shown for positive feedthrough terms (zero and under-compensation). Then, a discussion of the closed-loop behavior using DRFB is provided along with its implications on the dissipation of both electrical and mechanical energy within the structure.

Piezostructures and Sensoriactuators

Piezostructure modelling has been discussed in the literature for a variety of analytical approaches and structural geometries. (Bailey and Hubbard, 1985, Crawley and de Luis, 1987, Crawley and Anderson, 1989, Dimitriadis et al., 1991). For the development that follows the reader is referred to Hagood et al. (1990).

Piezoelectric structures can be described by the following equations:

Actuator eqn: $M\ddot{x} + C\dot{x} + Kx = \Theta v$ (1)
Sensor eqn: $q = \Theta^T x + C_p v$ (2)

where M, C, K are the mass, damping and stiffness matrices of the piezostructure, Θ is the electromechanical coupling matrix for the sensoriactuator and C_p is the piezoelectrics' capacitance matrix. Θ, C_p, are described by the following volume integrals:

$$\Theta = \int_{V_p} \Psi_r^T L_u^T e^T L_\varphi \Psi_\varphi \quad (3)$$

$$C_p = \int_{V_p} \Psi_\varphi^T L_\varphi^T \varepsilon^s L_\varphi \Psi_\varphi \quad (4)$$

where Ψ_r are the displacement distributions of the structure and L_u is the differential operator relating displacements and strain. Ψ_φ are the voltage distributions and L_φ is the gradient operator. e is the piezoelectric material constant relating voltage and stress, and ε^s is the piezoelectric's permitivity at zero strain and is a symmetric, diagonal matrix.

$$\varepsilon^s = \begin{bmatrix} \varepsilon_1^s & 0 & 0 \\ 0 & \varepsilon_1^s & 0 \\ 0 & 0 & \varepsilon_3^s \end{bmatrix}$$

Thus, the piezoelectric capacitance matrix is symmetric and positive definite, $C_p \geq 0$.

The eigenvalue problem can be solved for the homogeneous actuator equation to produce a mass-normalized modal matrix, Φ, of the structure. Letting $x = \Phi r$ we can decouple the equations of motion and rewrite the above relationship as

$$\ddot{r} + \mathcal{D}\dot{r} + \Omega r = \Phi^T \Theta v = \hat{\Theta} v \quad (5)$$

$$q = \hat{\Theta}^T r + C_p v \quad (6)$$

or for rate output

$$\dot{q} = \hat{\Theta}^T \dot{r} + C_p \dot{v} \quad (7)$$

Thus, we notice that the above input-output relationship ($\dot{q}/v$) could be implemented as the DVFB law described by Balas (1979) with the exception of the feedthrough term, $C_p \dot{v}$. The feedthrough term can be compensated for using analog circuitry as shown in Figure 1. The resulting relationship for

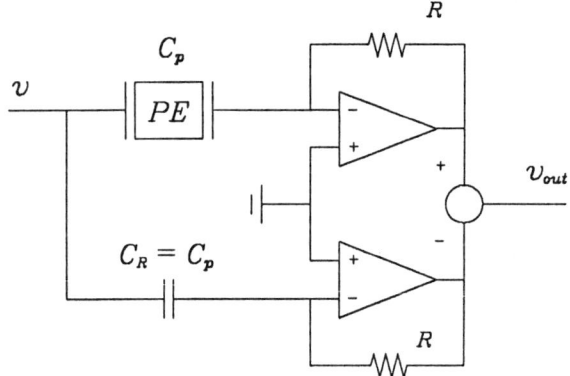

Figure 1: Sensoriactuator compensation circuit

the piezostructure becomes

$$\ddot{r} + \mathcal{D}\dot{r} + \Omega r = \Phi^T \Theta v = \hat{\Theta} v$$

$$\dot{q} = \hat{\Theta}^T \dot{r} + (C_p - C_R)\dot{v} \qquad (8)$$

When the reference capacitor, C_R, is chosen to be equivalent to the capacitance of the piezoelectric then the feedthrough term is zero and DVFB results. If, however, the so called *residual capacitance*

$$C_r = C_p - C_R$$

is not exactly zero (as it is in practice) there would be a change in closed-loop performance; in fact, it is not clear whether DVFB would remain globally stable. Thus, it is necessary to develop stability criteria for the case when the feedthrough term is non-zero. First, we will discuss the effects of partial compensation on the placement of open-loop zeros.

Locus of Open-Loop Zeros for varied C_r

The input-output relationship for a partially compensated sensoriactuator relating charge to voltage is

$$\frac{q(s)}{v(s)} = C_r + \sum_{i=1}^{n} \frac{\hat{\Theta}_i^2}{s^2 + 2\zeta_i \omega_i s + \omega_i^2}$$

This transfer function contains $2n$ poles and $2n$ zeros. Similarly after differentiating, the current-voltage relationship is

$$\frac{\dot{q}(s)}{v(s)} = sC_r + \sum_{i=1}^{n} \frac{s\hat{\Theta}_i^2}{s^2 + 2\zeta_i \omega_i s + \omega_i^2}$$

The zeros of this improper transfer function are those for the q/v transfer function plus a zero at the origin. The values of the zeros, which are important in controller design, depend not only on the poles of the system but also on the values of $\hat{\Theta}_i^2$ and C_r.

The poles of the above transfer function can be seen by inspection to be

$$p_{i,i+1} = -\zeta_i \omega_i \pm j\omega_i \sqrt{1-\zeta_i^2} \qquad i = 1, 2, \ldots, n$$

Martin (1978) showed that undamped collocated structures (no feedthrough dynamics) have alternating poles and zeros on the imaginary axis. Such systems, with no feedthrough term, have two fewer zeros than poles. Including feedthrough dynamics results in an equal number of poles and zeros. The alternating pole/zero feature still occurs, although the zeros are in a different location, and the two extra zeros occur outside the largest poles, as shown in the following theorem. This theorem is similar to Martin's, but also includes effects of feedthrough dynamics on the location of system zeros.

Theorem 1 *The rational transfer function*

$$T(x) = \frac{N(x)}{D(x)} = \frac{a_1}{x-b_1} + \frac{a_2}{x-b_2} + \cdots + \frac{a_n}{x-b_n} + a_{n+1}$$

with $a_i \neq 0$ and $b_1 > b_2 > \cdots > b_n$ will have n alternating poles and zeros on the real axis if and only if

$$\operatorname{sgn}(a_i) = \operatorname{sgn}(a_j) \qquad \forall\, i,j$$

Corollary 1 *If $\operatorname{sgn}(a_i) = \operatorname{sgn}(a_{i+1})$ then $T(x)$ has a real zero between the poles b_i and b_{i+1} if $i \neq n$, or a zero between b_n and $-\infty$ if $i = n$.*

Proof: The numerator of $T(x)$ is

$$N(x) = \sum_{i=1}^{n+1} a_i \prod_{j=1, j\neq i}^{n} (x-b_i)$$

evaluate $N(x)$ at b_1 and b_2 yields

$$N(b_1) = a_1 \prod_{j=2}^{n} (b_1 - b_j)$$

$$N(b_2) = a_2(b_2 - b_1) \prod_{j=3}^{n} (b_2 - b_j)$$

from the ordering of the b_j's

$$\text{sgn}(N(b_1)) = \text{sgn}(a_1)$$
$$\text{sgn}(N(b_2)) = -\text{sgn}(a_2)$$

Since $N(x)$ is continuous and real valued $N(x)$ has a root between b_1 and b_2. Similarly, if all of the a_i's have the same sign then there are zeros between each pole. This accounts for $n-1$ of the roots of $N(x)$. Evaluating the numerator at b_n yields

$$\text{sgn}(N(b_n)) = (-1)^{n-1}\text{sgn}(a_n)$$

and as $x \to -\infty$

$$\text{sgn}(N(x \to -\infty)) = (-1)^n \text{sgn}(a_{n+1})$$

thus, by the same argument used previously, there is a root of $N(x)$ to the left of b_n. □

For the piezostructure mentioned above, this result holds for lightly damped structures ($\zeta \simeq 0$) if we let $x = s^2$, $a_i = \Theta_i^2$, $i = 1, 2, \ldots, n$ $a_{n+1} = C_r$ and $b_i = -\omega_i^2$. Therefore, the system with poles at $s^2 = -\omega_i^2$ has zeros $-z_i^2$ placed such that

$$\omega_1^2 < z_1^2 < \omega_2^2 < z_2^2 < \cdots < \omega_n^2 < z_n^2$$

The location of the open-loop zeros is dependent upon the magnitude of the residual capacitance, C_r (Spangler and Hall, 1992). Thus, *the selection of C_r is another design tool available to the engineer.* The placement of system zeros is easily shown using a root locus analysis.

For the following analysis, we will assume that the structure is lightly damped ($\zeta_i \simeq 0$). We define the open-loop transfer function of the perfectly compensated sensoriactuator system to be

$$G(s) = \sum_{i=1}^{n} \frac{\hat{\Theta}_i^2}{s^2 + \omega_i^2}$$

Then from the previous theorem it has alternating poles and zeros along the imaginary axis. The transfer function for the partially compensated sensoriactuator system is

$$\hat{G}(s) = C_r + G(s)$$

and the zeros of the system are the roots of $\hat{G}(s) = C_r + G(s) = 0$. Since $G(s)$ is not zero at any of the roots of $\hat{G}(s)$, we can multiply by $G^{-1}(s)$ and obtain

$$1 + C_p G^{-1}(s) = 0$$

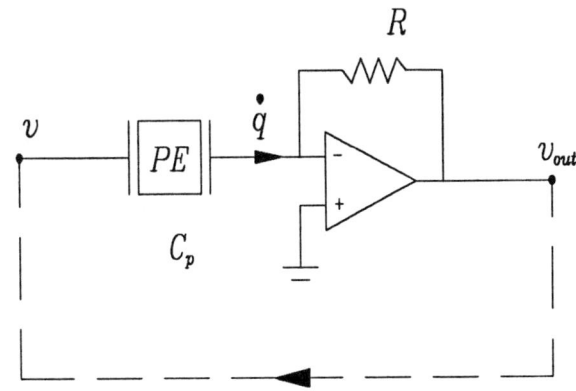

Figure 4: Transresistance amplifier for DRFB

Now, we can find the zeros of the partially compensated piezostructure using a root locus analysis. The loci of system zeros, as C_r is varied, are shown in Figure 2. This effect is also demonstrated in Figure 3 where the capacitance appears to "fill up" and move the zeros of the compensated transfer function.

Direct Rate Feedback

When using direct rate feedback, the control voltage is

$$v = -G\dot{q} = -G(\hat{\Theta}^T \dot{r} + C_p \dot{v}) \quad (9)$$

where $G \geq 0$. The implementation of the DRFB controller uses a transresistance amplifier (I-V converter), as shown in Figure 4, with the relationship

$$\frac{\dot{q}}{v} = -R$$

Here, we are considering uncompensated sensoriactuators but the result holds for partially compensated sensoriactuators, provided $C_r > 0$. The advantage of DRFB is global stability, as shown in the following theorem.

Theorem 2 *The closed-loop piezostructure with direct rate feedback is dissipative, and if $\mathcal{D}$ and Ω are positive definite, it is asymptotically stable.*

Proof: Defining the function

$$E(t) = \frac{1}{2}(\dot{r}^T \dot{r} + r^T \Omega r + v^T C_p v) > 0 \quad (10)$$

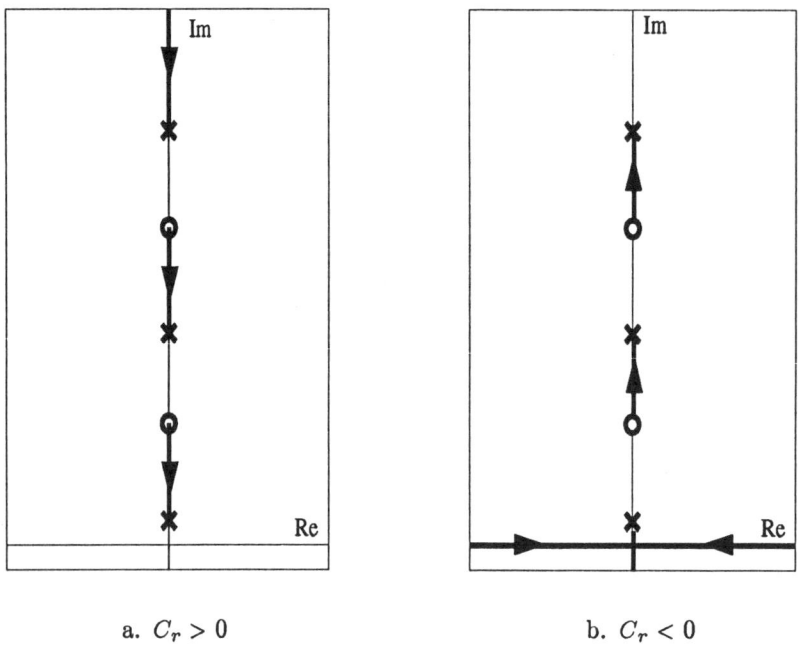

a. $C_r > 0$ b. $C_r < 0$

Figure 2: Zero locus for sensoriactuators

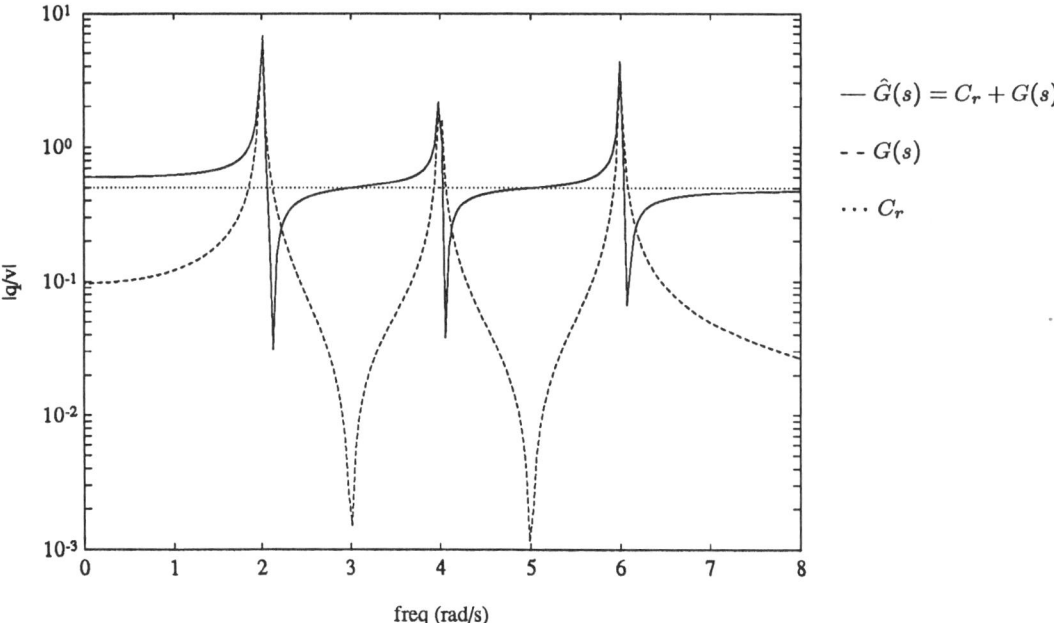

Figure 3: Frequency response for fully and partially compensated sensoriactuators

and taking a derivative with respect to time

$$\dot{E}(t) = \dot{r}^T\ddot{r} + \dot{r}^T\Omega r + \dot{v}^T C_p v \qquad (11)$$

substituting in Equation 5 in the first term

$$\dot{E}(t) = -\dot{r}\mathcal{D}\dot{r} + \dot{r}^T\Theta v + \dot{v}^T C_p v \qquad (12)$$

Now including Equation 9 in the last two terms

$$\begin{aligned}\dot{E}(t) &= -\dot{r}^T\mathcal{D}\dot{r} - \dot{r}^T\hat{\Theta}G(\hat{\Theta}^T\dot{r} + C_p\dot{v}) \\ &\quad -\dot{v}^T C_p G(\hat{\Theta}^T\dot{r} + C_p\dot{v}) \qquad (13) \\ &= -\dot{r}^T\mathcal{D}\dot{r} \\ &\quad -(\hat{\Theta}^T\dot{r} + C_p\dot{v})^T G(\hat{\Theta}^T\dot{r} + C_p\dot{v}) \quad (14) \\ &= -\dot{r}^T\mathcal{D}\dot{r} - \dot{q}G\dot{q} \qquad (15)\end{aligned}$$

Since $\mathcal{D} \geq 0$ and $G \geq 0$ then $\dot{E}(t) \leq 0$ and is dissipative. If, however, $\mathcal{D} > 0$ and $\Omega > 0$ then $E(t)$ is a Lyapunov function and since $\dot{E}(t)$ is negative and $E(t)$ is always decreasing and the system is asymptotically stable. □

This result is only valid if the capacitance is positive definite. For uncompensated piezoelectrics this is always the case; however, if there are sensoriactuator compensation circuits the residual capacitance, C_r, may be negative (over-compensation). In such situations, DRFB is not guaranteed stable. Care should be taken when designing sensoriactuator circuits to avoid over-compensation.

This result does not imply anything about the rate of decay of system energy and states. To discuss the control of specific states (and associated energy) requires an investigation into the location of closed-loop system poles. In fact, there is a complicated exchange of energy between the structure, the piezoelectrics' capacitance and the controller. In general, there is a loss of performance due to the feedthrough term (Spangler and Hall, 1992).

The dissipation of the total energy, E, provided by the controller is a result of sensor current acting in a direction opposite to the applied voltage, a constraint applied by the control law. The result is the controller doing negative work, $W = -\dot{q}v$. Important to the implication of the control law is the selection of the feedback gain matrix G. The gain matrix affects the relationship (phase) between the structures states and the patch voltages which in turn affects the transfer of energy between the structure, piezoelectrics and the controller.

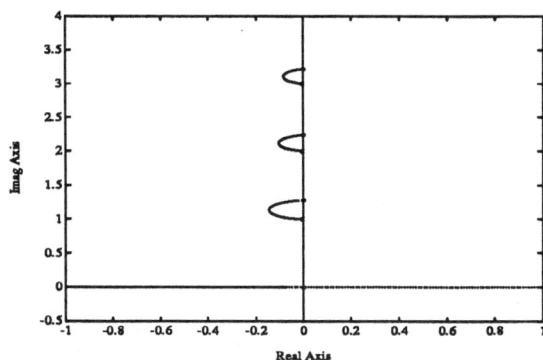

Figure 5: Root Locus for piezoelectric DRFB

Closed-loop Response

The root locus for a piezostructure is shown in Figure 5. We can see that choice of closed loop gain is a compromise between the pole on the real axis which moves to the right as the gain increases and placing the complex poles as far to the left so as to increase damping. If the gain is chosen too large, the pole on the real axis will dominate the response. If, however, the gain is not large enough, the damping in the complex poles will not change sufficiently.

Conclusions

Piezoelectric sensoriactuators can be used to control structures using direct rate feedback (DRFB); additionally, if the gain and feedthrough matrices are positive semi-definite, the closed-loop system can be guaranteed stable. This result is different from direct velocity feedback (DVFB), suggested by Balas (1979), since the plant includes feedthrough dynamics. The collocated condition ($C = B^T$) is satisfied using sensoriactuators, and the effect of compensation on open-loop zeros is discussed. It is shown that, for partially compensated sensoriactuators, there are an equal number of poles and zeros and that they alternate along the imaginary axis; and, the location of system zeros is dependent upon the magnitude of the feedthrough term. It is also suggested that partial compensation could be used as a design tool.

References

1. Anderson, Eric H., Hagood, Nesbitt W., and Goodliffe, Jay M., 1992, "Self-Sensing Piezoelectric Actuation: Analysis and Application to Controlled Structures," AIAA-92-2465-CP.

2. Bailey, T. and Hubbard, J. E., 1985, "Distributed Piezoelectric Polymer Active Vibration Control of a Cantilevered Beam," AIAA *J. of Guidance and Control*, pp. 605.

3. Balas, Mark J., 1979, "Direct Velocity Feedback Control of Large Space Structures," *J. Guidance and Control*, Vol. 2, No. 3, pp. 252–253.

4. Burke, Shawn E., Hubbard, James E. and Meyer, John E., 1993, "Distributed Transducers and Colocation," *Mechanical Systems and Signal Processing*, 7(4), pp. 349–361.

5. Crawley, E. F. and Anderson, E. H., 1989, "Detailed Models of Piezoceramic Actuation of Beams," AIAA-89-1388-CP.

6. Crawley, E. F. and de Luis, J., 1987, "Use of Piezoelectric Actuators as Elements of Intelligent Structures," *AIAA J.*, **25**, 10, pp. 1373–1385.

7. Dimitriadis, E. K., Fuller, C. R. and Rogers, C. A., 1991, "Piezoelectric Actuators for Distrubuted Vibration Excitation of Thin Plates," *J. of Vibration and Acoustics*, Vol. 113, pp. 100-107.

8. Dosch, Jeffrey J., Inman, Daniel J. and Garcia, Ephrahim, 1992, "A Self-Sensing Piezoelectric Actuator for Collocated Control," *J. of Intell. Mater. Syst. and Struct.*, Vol. 3 – January, pp. 166–185.

9. Hagood, Nesbitt W., Chung, Walter H. and von Flotow, Andreas, 1990, "Modelling of Piezoelectric Actuator Dynamics for Active Structural Control," AIAA-90-1087-CP.

10. Martin, G. D., 1978, "On the Control of Flexible Mechanical Systems," Ph.D. Dissertation, Stanford University.

11. Spangler, Ronald L, Jr. and Steven R. Hall, 1992 "Robust Broadband Control of Flexible Structures Using Integral Piezoelectric Elements," Presented at the Third International Conference on Adaptive Structures, San Diego, CA, Nov. 9–11.

A SELF-TUNING PIEZOELECTRIC VIBRATION ABSORBER

Joseph J. Hollkamp*
Wright Laboratory
Wright-Patterson AFB, Ohio

and

Thomas F. Starchville, Jr.**
Dept. of Aerospace Engineering
Penn State University

ABSTRACT

A self-tuning piezoelectric vibration absorber is presented. A piezoelectric absorber, similar to a mechanical vibration absorber, has to be tuned to a particular structural vibration mode in order to be effective. The absorber presented here will tune itself to a particular mode and track that mode if it varies in frequency. Design of the absorber consists of a pair of lead zirconate titanate (PZT) tiles attached to the structure and shunted by an inductor-resistor circuit. This produces an electrical resonance that can be tuned to the desired structural mode by a simple control system. The absorber is experimentally demonstrated on a cantilevered beam. The experiments include an examination of the response of the absorber to an abrupt change in system parameters.

INTRODUCTION

In the past several years, researchers have begun using piezoelectric devices for structural vibration suppression. These devices transform electrical energy to mechanical energy and vice versa. Vibration absorbers using piezoelectric devices have been presented by Hagood and von Flotow (1991), Hagood and Crawley (1989), Edberg, Bicos, and Fechter (1991) and Hollkamp (1994). The absorber is created by connecting a simple inductor-resistor network across a piezoelectric actuator. The network combined with the inherent capacitance of the piezoelectric produces a damped electrical resonance which when properly *tuned* to a vibration mode will absorb energy from the structure.

Also recently, Smith, Maly, and Johnson (1991) have demonstrated an adaptive absorber using a viscoelastic material (VEM) as the spring for a single-degree-of-freedom mechanical absorber. The stiffness of the VEM can be varied with temperature, thereby varying the tuning of the absorber. A controller tunes the absorber to suppress a vibration mode and keeps the absorber tuned to that mode even if its frequency shifts.

In this paper, a self-tuning *piezoelectric* absorber will be presented. The theory behind the device will be discussed and a simple control scheme presented. Finally, the absorber will be demonstrated experimentally on a cantilevered beam with attached lead zirconate titanate (PZT) tiles.

BACKGROUND THEORY

Piezoelectric materials are useful for vibration control because they have the unique ability to strain when an electrical voltage is applied and produce an electrical voltage when strained. In short, they have the ability to transform electrical energy to mechanical energy and vice versa. The piezoelectric material described in this paper are sheets of PZT. The PZT sheet is poled along one axis, usually across the thickness with electrodes covering the top and

---

* Research Engineer, Member AIAA
**Graduate Student, Member AIAA

This paper is declared a work of the U.S. Government and is not subject to copyright protection in the United States.

bottom of the sheet. An applied voltage across the thickness produces strain in the other two dimensions. Detailed models for piezoelectric materials can be found in Crawley and DeLuis (1987).

Piezoelectrics can be used as passive energy dissipation devices by using an electrical impedance as a shunt. If the shunt consists of an inductor and resistor in series, the shunt combined with the inherent capacitance of the piezoelectric creates a damped electrical resonance. The resonance can be tuned so that the piezoelectric device acts as a damped vibration absorber. A damped vibration absorber replaces a single structural mode with two highly damped modes. Hagood and von Flotow (1991) give a detailed analytical derivation of a piezoelectric damped vibration absorber applied to an undamped mechanical structure.

The transfer function of the absorber can be shown to be

$$\frac{x}{x^{ST}} = \frac{\gamma^2 + \delta^2 r\gamma + \delta^2}{(\gamma^2 + 2\zeta\gamma + 1)(\gamma^2 + \delta^2 r\gamma + \delta^2) + K_{ij}^2(\gamma^2 + \delta^2 r\gamma)} \quad (1)$$

where

$$\begin{aligned} \delta &= \omega_e / \omega^E \\ \gamma &= s / \omega^E \\ r &= R\, C_{pi}^S\, \omega^E \\ \omega_e &= 1/\sqrt{LC_{pi}^S} \end{aligned} \quad (2)$$

and where x is displacement; s is the Laplace domain variable; ζ is the modal damping factor in the original structural; C_{pi}^S is the inherent capacitance of the piezoelectric (measured at constant strain); ω^E is the natural frequency of the structural mode when the piezoelectric device is shorted; ω_e is the electrical resonant frequency; and R and L are the resistance and inductance of the shunt. Also, K_{ij} is the generalized electromechanical coupling coefficient defined by

$$K_{ij}^2 = \frac{K_{jj}^E}{K + K_{jj}^E} \frac{k_{ij}^2}{1 - k_{ij}^2} \quad (3)$$

where k_{ij} is the electromechanical coupling coefficient of the piezoelectric material and K_{jj}^E represents the mechanical stiffness of the shorted piezoelectric. The static displacement to a given force, F, is defined by x^{ST} as

$$x^{ST} = F/(K + K_{jj}^E) \quad (4)$$

where K is the mechanical stiffness of the structure. Note that the preceding parameters are modal quantities.

Optimal inductance and resistance values for an undamped system can be found based upon the analogy with the damped mechanical vibration absorber (Hagood and von Flotow, 1991). The analysis uses transfer function intersection points (Den Hartog, 1956; Timoshenko, Young, and Weaver, 1974). The optimal tuning values are

$$\begin{aligned} \delta_{opt} &= \sqrt{1 + K_{ij}^2} \\ r_{opt} &= \sqrt{2}\, \frac{K_{ij}}{1 + K_{ij}^2} \end{aligned} \quad (5)$$

For lightly damped systems, these optimal values are approximations. For demonstration purposes, Figure 1 shows the effect of optimal absorbers on a lightly damped structure for various generalized electromechanical coupling coefficients. Note that there are other optimal solutions for the piezoelectric absorber (Hagood and von Flotow, 1991) and for damped vibration absorbers in general (Den Hartog, 1956; Timoshenko, Young, and Weaver, 1974; Juang, 1984).

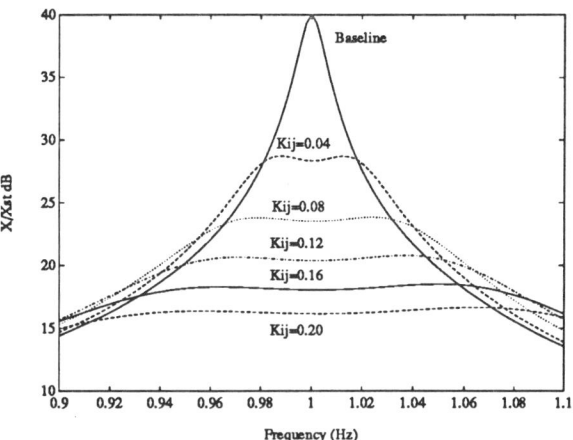

Figure 1. The performance of the optimally tuned piezoelectric absorber as Kij varies in comparison to a baseline structure with a 0.5% viscous damping ratio.

THE SELF-TUNING PIEZOELECTRIC ABSORBER

The design of the vibration absorber is straightforward if the frequency of a vibration mode is known; however if the frequency is unknown or varying, an adaptive control scheme must be used. A *self-tuning* piezoelectric absorber can be made by adaptively tuning the electrical resonance (i.e. by adjusting the shunt inductance and resistance). A performance criterion must be chosen for the control scheme. A natural choice would be the RMS response of the structure, but this is a good choice only if the magnitude of the disturbance is consistent. Smith, Maly, and Johnson (1991) in the design of a self-tuning *mechanical* absorber, used the ratio of RMS response of the absorber to the RMS response of the structure as the performance criterion. This RMS ratio performance criterion eliminates the need for a consistent disturbance. The self-tuning piezoelectric absorber presented here will use a similar performance criterion. Here, the voltage across the shunt will be used as the measure of the absorber response.

The transfer function for a mechanical absorber is easy to derive. Hagood and von Flotow (1991) provide a nondimensional form for an undamped structure

$$\frac{x}{x^{ST}} = \frac{\gamma^2 + \delta^2 r\gamma + \delta^2}{(\gamma^2+1)(\gamma^2+\delta^2 r\gamma+\delta^2)+\beta(\delta^2\gamma^2+\delta^2 r\gamma^3)} \quad (6)$$

where the parameters now correspond to their mechanical analogies and β is the ratio of the absorber mass to the system mass. The transfer function for the absorber mass is easily derived as

$$\frac{x_{abs}}{x^{ST}} = \frac{\delta^2 r\gamma + \delta^2}{(\gamma^2+1)(\gamma^2+\delta^2 r\gamma+\delta^2)+\beta(\delta^2\gamma^2+\delta^2 r\gamma^3)} \quad (7)$$

where x_{abs} is the displacement of the absorber mass.

The piezoelectric absorber control scheme uses the voltage across the shunt. If there is no external current supplied and only a uniaxial load in the transverse direction is applied, the voltage across the shunt can be shown to be

$$V_i = C_1 \frac{\gamma^2 + \delta^2 r\gamma}{\gamma^2 + \delta^2 r\gamma + \delta^2} S_{jj} \quad (8)$$

where C_1 is a proportionality constant based upon the size and piezoelectric properties of the material and S_{jj} is the strain in the transverse direction. Since the strain is proportional to displacement, the transfer function for the voltage across the shunt becomes

$$\frac{V_i}{x^{ST}} = \frac{C(\gamma^2+\delta^2 r\gamma)}{(\gamma^2+2\zeta\gamma+1)(\gamma^2+\delta^2 r\gamma+\delta^2)+K_{ij}^2(\gamma^2+\delta^2 r\gamma)} \quad (9)$$

where C is a proportionality constant. The form of this transfer function is slightly different than that of the mechanical absorber transfer function (Equation 7). Thus the voltage across the shunt is not *directly* analogous to measuring the mechanical absorber displacement, but it is a useful measure of the absorber response. The piezoelectric absorber transfer function, Equation 9, is used to demonstrate the analytical prediction of the RMS ratio as the shunt is tuned. In Figure 2, the RMS ratio is a maximum ($\delta=0.9975$) near the minimum RMS structural response ($\delta=1.025$). Thus a control law based on maximizing the RMS ratio will not allow us to provide the minimum structural response, but it will be very close. A simple control law based on this performance

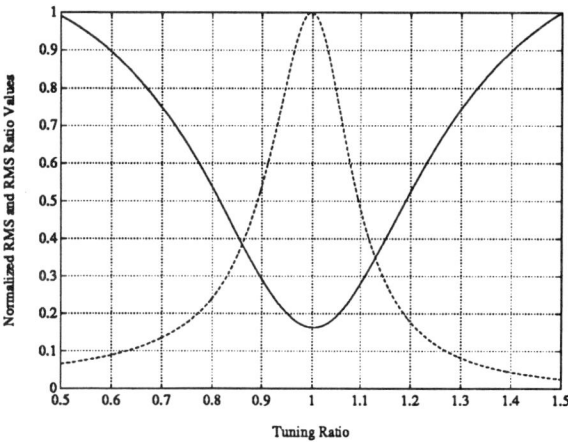

Figure 2. The analytical performance of the piezoelectric vibration absorber as the tuning ratio varies. Shown is the RMS response of the system (solid) and the ratio of the absorber RMS response to the system RMS response (dashed).

function can be stated as: if the estimation of the slope of the RMS ratio is positive, increase the electrical frequency, otherwise decrease it by a set amount.

THE EXPERIMENTS

A set of experiments was conducted to verify the control theory. A cantilevered aluminum beam shown in Figure 3 was constructed for this purpose. Six pairs of PZT tiles, a tile on each side of the beam, are attached at the locations shown in the figure. The type of PZT is G-1195. One side of each tile is grounded to the beam and the other side is connected in parallel to the other tile in each pair. Synthetic inductors (Chen 1986), schematic in Figure 4, are used in the shunt; these were chosen for ease of use. The effective inductance is linearly proportional to the value of an adjustable resistor which in this study was a motorized potentiometer. A PZT tile (one side of pair #5) was used as a disturbance; the input to the tile was a band limited (0-100 Hz) gaussian random signal. The disturbance signal is not used in the tuning scheme, but is used to form transfer functions for demonstration. An additional mass can be added to the beam to provide an abrupt change in the system.

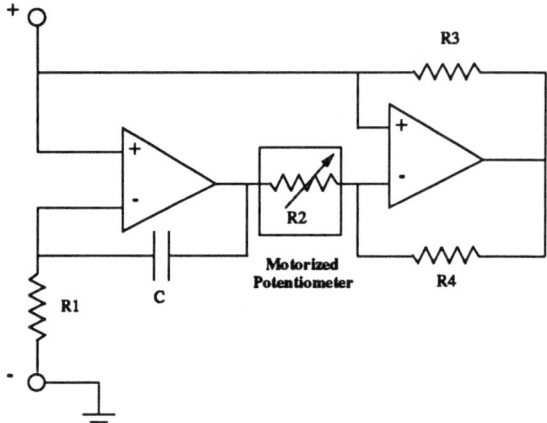

Figure 4. A synthetic inductor design using two operational amplifiers. The inductance is controlled by selecting R2.

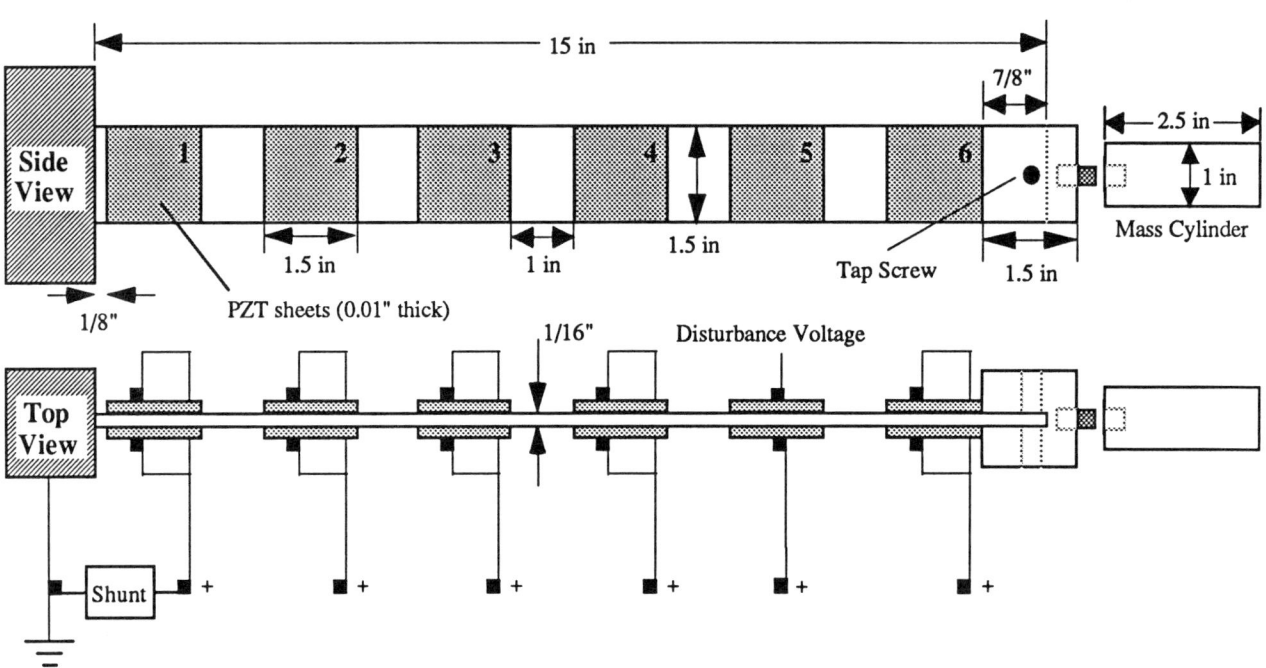

Figure 3. A schematic of the experimental cantilevered beam. Six pairs of PZT tiles are bonded to the structure. The absorber or shunt response is measured from PZT pair #1 and the system response is measured from PZT pair #4. One PZT sheet of pair #5 is used for excitation. An additional mass can be added to the tip of the beam.

An absorber was created using the PZT tiles at location 1. The second bending mode (40.8 Hz, 0.3% damping) was chosen as the mode to suppress. The synthetic inductor provided both the inductance and resistance of the shunt. (Ideally, the synthetic inductor can create high inductance values without an associated resistance. Practically, there is an effective resistance). A personal computer with data acquisition cards was used to measure the voltage across the synthetic inductor (the absorber response) and the voltage across the PZT tiles at location 4 (the system response). The signals from each of the two PZT pairs were sent through band pass filters to eliminate other structural modes from the measurement signals. The filters passed data in the 10 to 70 Hz frequency range. Software was written for the personal computer to acquire the filtered data, estimate the RMS response, and command the motorized potentiometer to change according to the control law. A Hewlett Packard Paragon system served as the source driver to produce the random input and measure the frequency response functions during the experiments. Figure 5 illustrates a schematic of the experimental setup.

A preliminary experiment was conducted to validate the performance function. Figure 6 shows the results of this preliminary experiment. Here, the tuning of the absorber was manually varied and RMS ratio estimated. Since there was a large amount of uncertainty in the estimation of the RMS ratio, a polynomial fit to the data around the peak was used to more accurately predict the maximum tuning ratio. The RMS ratio is a maximum when the tuning ratio is very near unity ($\delta = 0.99$, in this case). Recall that the RMS response of system was said to be poor choice for the absorber performance criterion. Figure 6 confirms this statement. This plot shows that the RMS response of the system is very inconsistent. Also in the preliminary experiments the amount of data used to estimate the RMS ratio was varied. While ten seconds of data provided good estimates, this seemed too long to wait for each tuning update. Two seconds of data provided noisy estimates; five seconds of data was considerably better. Of course, the amount of data necessary to form good estimates depends on the modal frequency and the quality of data. In all the experiments reported in this paper, five seconds of data was used to form the estimates.

After the preliminary experiments confirmed that the choice of the performance criterion was acceptable, experiments were conducted to allow the absorber to self-tune. A change in voltage to the motorized potentiometer resulted in a shift in the electrical resonance of the shunt; however, a linear change in command voltage corresponds to a linear change in inductance not a linear change in frequency. In the preliminary experiments, a

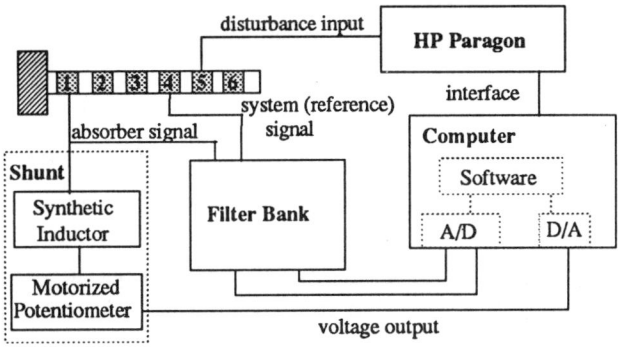

Figure 5. Schematic of the experimental setup. Signals from the absorber piezoelectric sheets (pair #1) and the system piezoelectric sheets (pair #4) are passed through a filter bank with a frequency window of 10-70 Hz. The computer estimates the RMS ratio and controls the motorized potentiometer.

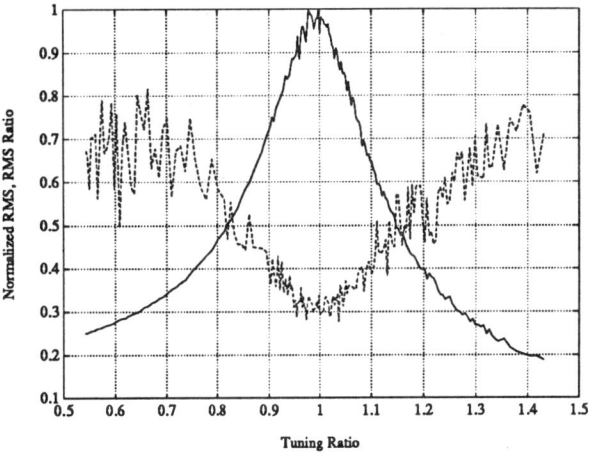

Figure 6. The measured performance of the piezoelectric vibration absorber. Shown is the normalized RMS ratio (solid) and RMS of the system response (dashed).

linear change in frequency was found to be more effective than a linear change in inductance. To achieve a constant frequency increment, the change in command voltage in the control scheme is nonlinear.

A series of fine (0.8 Hz) and coarse frequency (4.0 Hz) increment experiments were conducted to determine the ability of the absorber to self-tune. The response for each case is shown in Figure 7. Note in the figure that the rise time of the absorber for the coarse tuning is fast (approximately 10 sec) due to the larger step-size in frequency. The response afterwards, however, dithers around the "optimal" tuning ratio by what would seem a significant amount. The rise time for the fine tuning case is very slow, but the dither is smaller.

To better understand the implications of the dither, Figures 8 and 9 should be considered. Four different frequency response functions (FRFs) are presented to illustrate when the absorber is: 1) optimally tuned ($\delta \approx 1.0$), 2) tuned to the minimum of the dithering, 3) tuned to the maximum of the dithering, and 4) turned off. Each of the three cases involving the absorber resulted in a decrease of the resonant peak. A comparison of the power from these FRFs normalized to the baseline (no shunt) is shown in Table 1. The power was calculated by the following equation:

$$Power = \int_{2\pi 10}^{2\pi 70} |H(j\omega)|^2 \, d\omega \quad (10)$$

where $H(j\omega)$ is a experimental FRF. Notice in Table 1 for the coarse and fine frequency changes, the highest reduction in power (83%) occurs when the absorber is optimally tuned; this was expected. Comparing this value to the cases when the absorber was allowed to self-tune using a coarse change, the range of power reduction is between 61% and 83%. This shows that even though the absorber is not able to stay exactly tuned throughout the experiment (due to RMS estimation uncertainty), it still significantly reduces the vibration energy. The range of power reduction for the fine frequency change is noticeably tighter (81% to 83%). This shows that even though the

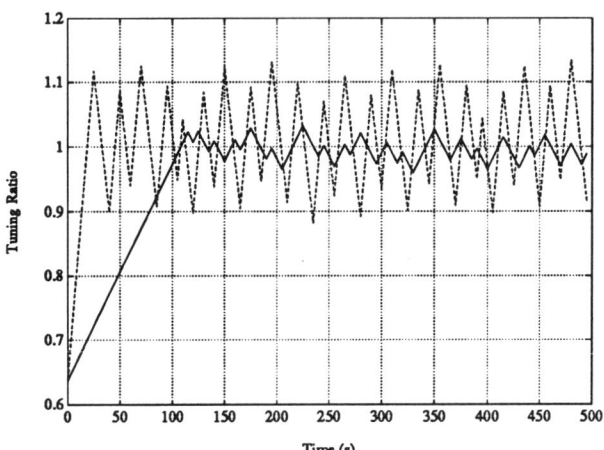

Figure 7. Experimental results for the piezoelectric vibration absorber when a coarse frequency adjustment of 4 Hz (dashed) and a fine frequency adjustment of 0.8 Hz (solid) are used.

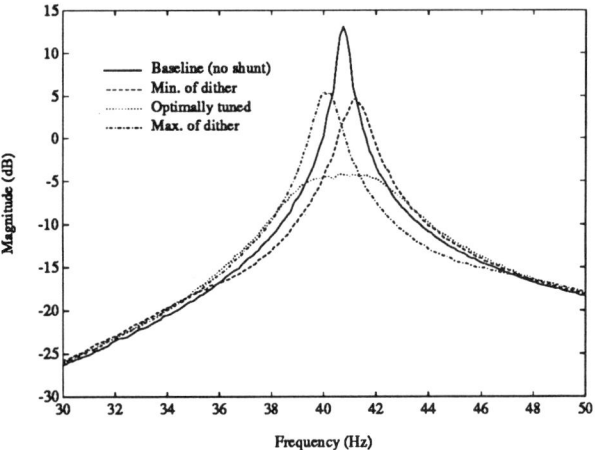

Figure 8. Measured frequency response functions for four conditions for a coarse frequency tuning test.

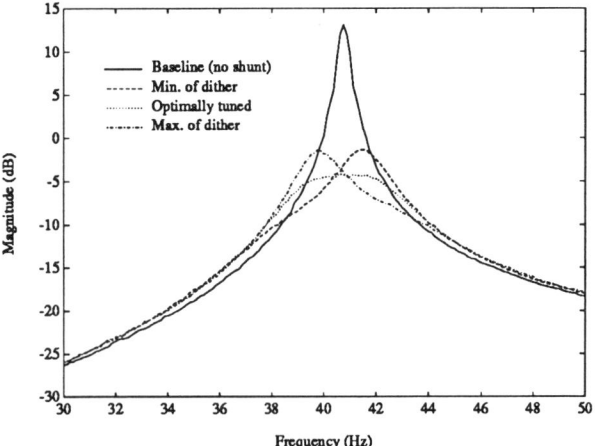

Figure 9. Measured frequency response functions for four conditions for a fine frequency tuning test.

Table 1. Comparison of normalized power for four conditions for coarse tuning (4 Hz) and fine tuning (0.8 Hz).

Tuning Condition	Coarse Frequency Tuning (4 Hz)		Fine Frequency Tuning (0.8 Hz)	
	Tuning Ratio (δ)	Normalized Power(%)	Tuning Ratio (δ)	Normalized Power (%)
Optimally tuned	1.00	17	1.00	17
Min. of dithering	0.88	35	0.96	19
Max. of dithering	1.13	39	1.03	19
Baseline (no shunt)	-----	100	-----	100

rise time for this case is longer, the overall performance of the absorber as it begins to tune is very good compared to a coarse frequency change. Figure 10 shows the experimentally measured time response of the baseline system due to an impulse in comparison to the fine tuning worst case (the maximum of the dither).

Now that it has been demonstrated that the piezoelectric vibration absorber can self-tune to a particular structural mode, the system was subjected to an abrupt change to cause the resonant frequency to shift. Causing an abrupt change in system parameters at a point during the experiment was accomplished by attaching a cylindrical mass to the end of the beam as shown in Figure 3. The cylinder was made of aluminum, had a mass of 114.75 g, and reduced the frequency of the second mode from 40.8 Hz to 30.55 Hz, while the damping remained constant at 0.3% of critical. It was decided that this was a large enough change in frequency to test the self-tuning of the absorber. Figure 11 shows the frequency response of the baseline structure with and without the added mass.

The self-tuning results for coarse and fine frequency adjustments are presented in Figure 12. The mass cylinder was added to the structure at 500 seconds into the experiment. Notice that the response for the first portion of the test is similar to that in Figure 7, the absorber does a better job self-tuning with the smaller frequency step-size. When the mass is applied and the frequency shifts, the absorber, in both cases, tracks the shift with excellent results.

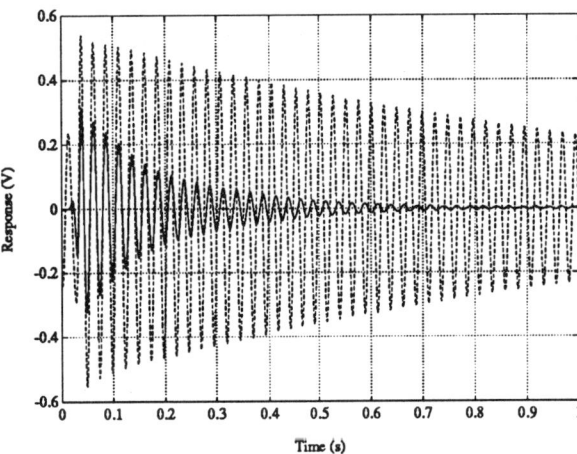

Figure 10. Free system response without the absorber (dashed) and with the absorber tuned to the fine tuning dither maximum(solid). An impulse was used for excitation.

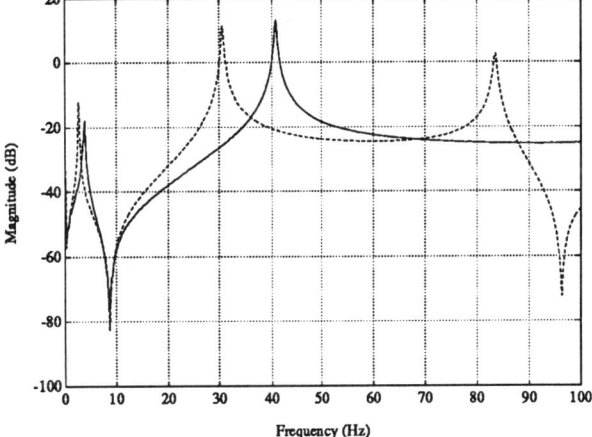

Figure 11. Frequency response comparison for the baseline system with (dashed) and without (solid) the additional mass.

Ideally, one would want a compromise between the coarse and fine tuning. At the start of an experiment, a fast rise time is desirable (coarse tuning), after the absorber begins to tune, a small amount of variation (fine tuning) is desired. But we would like to track the abrupt change with coarse tuning and later switch back to fine tuning. One way to achieve this is by monitoring the RMS ratio. Notice the abrupt change in the RMS ratio, shown in Figure 13. The control law can be modified to use coarse tuning if the change in RMS ratio is greater than 20%, otherwise use fine tuning. The results for the modified control law are shown in Figure 14. Now we get the fast rise time associated with the coarse tuning and the smaller dither associated with fine tuning. Another way to adjust the frequency increments is to make it inversely proportional to the RMS ratio. Figure 15 shows the results for this case.

CONCLUSIONS

The design of a self-tuning piezoelectric vibration absorber has been presented. It was demonstrated that the absorber could tune itself to a structural mode using both a coarse (4 Hz) and a fine (0.8 Hz) frequency step-size in the control law. The control law itself was designed for simplicity using the change in the RMS ratio slope to increase or decrease the electrical resonance of the absorber. Although there was some dither in the absorber tuning, this could be attributed to a

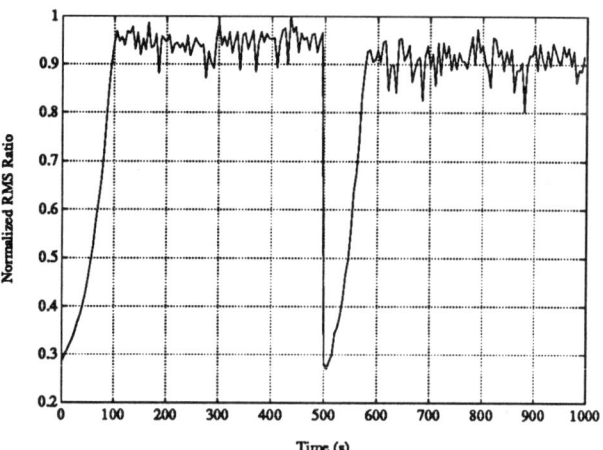

Figure 13. Time history of the RMS ratio for an abrupt change in system properties.

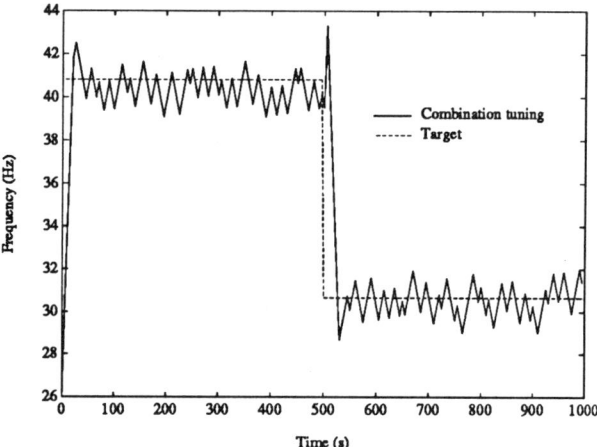

Figure 14. Results for an abrupt system change using a combination of coarse and fine frequency tuning.

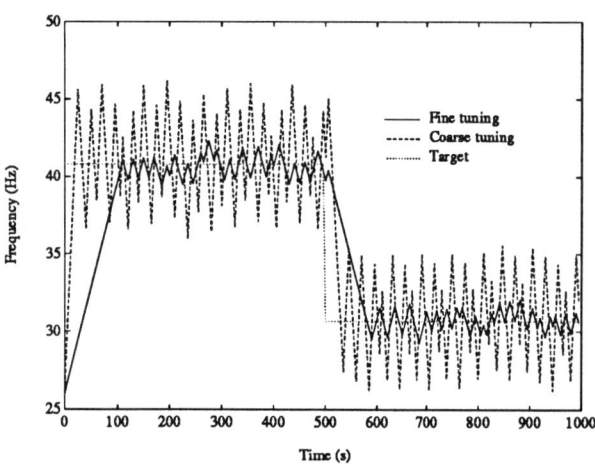

Figure 12. Results with an abrupt change in system properties for coarse and fine frequency tuning. A tip mass was added to the structure after 500 seconds.

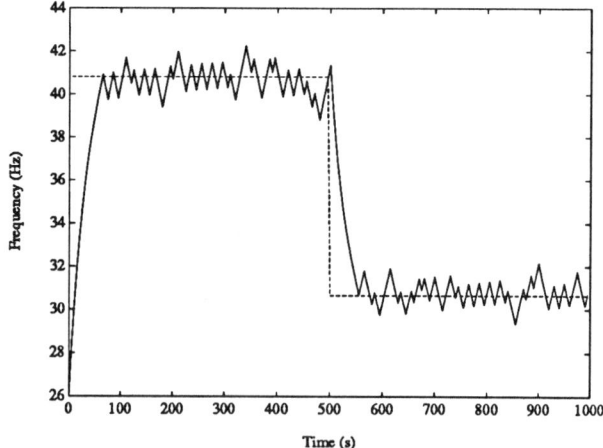

Figure 15. Results for an abrupt system change. The frequency adjustment is inversely proportional to RMS ratio.

fair amount of uncertainty in estimating the RMS of the absorber and system. One way to improve this measure could be to use a longer data stream when performing the RMS calculation. This, however, would greatly increase the amount of time for the absorber to properly tune and as seen in the FRF power comparisons, the resulting reduction in vibration energy is still very favorable.

The robustness or "tunability" of the absorber was examined by abruptly changing the system parameters to determine if once tuned and a change occurred, could it retune to the new frequency. A mass was added to the structure to reduce the modal frequency and the absorber demonstrated that it could retune itself after this abrupt change.

Other results obtained in experiments, but not presented here, include reducing the size of the tuning increment as time progresses. This control scheme smoothed out the tuning dither. But in order to track system changes, the scheme has to monitor the RMS ratio. If the RMS ratio significantly changes, the tuning increment reverts back to a large size. Also, the ability of the absorber to tune to an off resonance harmonic excitation was tested. The absorber tuned to the excitation frequency but since the absorber is a *modal* device, it was unable to extract any off resonance energy from the system.

ACKNOWLEDGMENTS

This work was sponsored by the Air Force Office of Scientific Research under the 2302AW Task (monitored by Dr Spencer Wu) and the Graduate Student Research Program.

REFERENCES

Chen, W.-K., 1986, *Passive and Active Filters*, John Wiley and Sons, Inc., New York.

Crawley, E.F., and DeLuis, J., 1987, "Use of Piezoelectric Actuators as Elements of Intelligent Structures," *AIAA Journal*, Vol. 25, No. 10, pp. 1373-1385.

Den Hartog, J.P., 1956. *Mechanical Vibrations*, McGraw-Hill Book Co., New York,.

Dosch, J.J., Inman, D.J., and Garcia, E., 1992, "A Self-Sensing Piezoelectric Actuator for Collocated Control," *Journal of Intelligent Material Systems and Structures*, Vol. 3, No. 1, pp. 166-185.

Edberg, D.L., Bicos, A.S., and Fechter, J.S., 1991, "On Piezoelectric Energy Conversion for Electronic Passive Damping Enhancement," *Proceedings of Damping '91*, San Diego, Paper GBA-1.

Hagood, N.W., and Crawley, E.F., 1989, "Experimental Investigation into Passive Damping Enhancement for Space Structures," *Proc. of the 30th AIAA/ASME/ASC/AHS Structures Structural Dynamics and Materials Conference*, Mobile, Alabama, AIAA Paper 89-3436, pp. 97-109.

Hagood, N.W., and von Flotow, A., 1991. "Damping of Structural Vibrations with Piezoelectric Materials and Passive Electrical Networks," *Journal of Sound and Vibration*, Vol. 146, No. 2, pp. 243-268.

Hollkamp J.J., 1994, "Multimodal Passive Vibration Suppression with Piezoelectric Materials and Resonant Shunts," *The Journal of Intelligent Material Systems and Structures*, to be published.

Juang, J.-N., 1984, "Optimal Design of a Passive Vibration Absorber for a Truss Beam," *Journal of Guidance, Control and Dynamics*, Vol. 7, No. 6, pp. 733-739.

Juang, J.-N., and Pappa, R.S., 1985, "An Eigensystem Realization Algorithm for Modal Parameter Identification and Model Reduction, *Journal of Guidance, Control and Dynamics*, Vol. 8, No. 5, pp. 620-627.

Smith K.E., Maly J.R., and Johnson C.D., 1991, "Smart Tuned-Mass Dampers," *Proceedings of the ADPA/AIAA/ASME/SPIE Conference on Active Materials and Adaptive Structures,"* Alexandria, VA, pp. 19-22.

Timoshenko, S., Young, D.H., and Weaver, W., 1974, *Vibration Problems in Engineering*, John Wiley and Sons, Inc., New York.

AUTHOR INDEX

Abdallah, M. MAT-35
Aboutorabi, H. STR-15
Adamson, J. STR-22
Agnes, G. SDN-54
Agrawal, B. SDN-49
Ahmad, M. STR-50
Ahmadi, G. SDN-49
Ahn, S. STR-31
Akgün, M. STR-23
Akl, J. SDN-38
Albinger, J. MAT-24
Alichanbari, H. SDN-37
Allaei, D. DYN-8, WIP-25
Allen, J. MDO-36
Alvarez, M. STR-13
Alvin, K. DYN-4, SDN-40
Ambur, D. MAT-16, 35, STR-07, 10
Aminpour, M. STR-05
Amirouche, F. DYN-3
Amos, A. SDN-38
Anderson, E. SDN-54, DYN-2, 4, MDO-09
Angrilli, F. DES-26
Annis Jr., C. STR-15
Arbocz, J. STR-10, 50
Ari-Gur, J. SDN-21
Armanios, E. SDN-04
Arora, J. MDO-09
Atluri, S. ASF-7, MAT-24, STR-50
Ausman, J. MDO-27
Austin, F. SDN-33
Avva, V. STR-11
Babu, G. ASF-7, MAT-34
Backman, B. MAT-35
Badir, A. SDN-04
Bai, J. SDN-39
Baker, D. WIP-25, STR-11
Balachandran, B. DYN-2
Balakrishna, S. SDN-48
Balmés, E. DYN-5
Banerjee, J. SDN-04, STR-14
Bang, H. SDN-49
Barlow, R. STR-31
Bartkowicz, T. DYN-8
Barton, O. STR-41
Baruch, M. STR-14, 50
Baruh, H. ASF-2
Basci, M. STR-53
Baycan, C. ASF-9
Bechtel, G. STR-31
Benaroya, H. SDN-21, 30
Bendiksen, O. SDN-19
Bennett, G. MAT-44, MDO-27
Berdichevsky, V. SDN-30
Bergman, L. ASF-5
Berke, L. STR-32
Bert, C. STR-02
Bhumbla, R. STR-05
Bianchini, E. SDN-54
Bies, M. DYN-7
Birman, V. ASF-4
Blackburn, C. DYN-5
Blair, M. WIP-25
Blelloch, P. SDN-47
Bloebaum, C. MAT-44, MDO-09, 18, 27
Bohorquez, L. DES-46
Boivin, N. DYN-2
Bookout, P. SDN-40
Bostic, S. MAT-44
Bothwell, C. ASF-5
Bouadi, H. WIP-25
Brainin, L. MDO-27
Braisted, W. DES-08
Brent, D. MAT-12
Brillhart, R. ASF-1
Brinson, C. MAT-42
Brockman, R. DES-08
Bruno, R. ASF-7
Brunty, J. DYN-7
Buchholz, R. DYN-1
Buehrle, R. SDN-48
Bullock, S. STR-07
Bunakov, V. DES-46
Bushnell, D. STR-41
Bushnell, W. STR-41
Butler, B. SDN-04
Byun, C. SDN-28
Cairns, D. MAT-35
Callahan, J. ASF-2
Campbell, M. SDN-39
Canfield, R. MDO-36
Cao, T. DYN-4
Capitanio, R. SDN-40
Carapella, E. WIP-25
Carne, T. SDN-33
Castanier, M. SDN-52
Celi, R. MDO-36
Cerro, J. STR-07
Cesnik, C. STR-23
Chamis, C. MAT-24
Chamis, C. MDO-18, STR-13, 22, 43
Chanda, H. STR-32
Chandra, R. SDN-01

Chang, C. MDO-45
Chang, F-K. DES-46
Chang, K. STR-03
Changhe, K. DES-08
Chao, R. DYN-1
Chattopadhyay, A. MDO-36, STR-10
Chaudhry, Z. ASF-6, 8
Chawla, V. ASF-3
Chen, H. STR-05
Chen, P. MAT-24
Chen, R. DYN-9
Chen, W-L. STR-32
Chen, W. MAT-44, MDO-36
Chen, X. STR-23
Cheng, S. SDN-47
Chi, H-W. MDO-09
Chick, S. STR-31
Chin, C. DYN-2, SDN-19
Chini, G. MAT-12
Cho, M. STR-14, 23
Choi, K. STR-03
Chopra, I. MDO-18, SDN-01
Christensen, E. DYN-7, SDN-52
Chrostowski, J. SDN-02
Clark, R. DYN-7
Clendenin, B. SDN-49
Cole, D. ASF-9
Collier, C. MAT-42
Cornuault, C. SDN-47
Cox, D. SDN-33
Cox, K. DYN-7
Craddock, J. MAT-34, STR-50
Crassidis, A. SDN-39
Crawley, E. SDN-33, 37, 39
Cuda, S. STR-11
Cudney, H. ASF-5, DYN-8
Cudney, H. WIP-25
Cui, K. DYN-3
Cunningham, D. SDN-54
Cunningham, T. SDN-48
Cusano, C. DES-17
Cutchins, M. DYN-3
Dambski, J. DES-26
Das, A. DYN-4
David, K. DES-26
Davidson, B. MAT-16
Davila, C. STR-05, 14
Davis, L. SDN-54
Dawson, R. DYN-7
Dayal, V. MAT-24
Demuts, E. STR-03
Detwiler, D. ASF-8
Di Sciuva, M. STR-06
Di, R. STR-07
Dickson, J. STR-07
Diehl, C. ASF-5
Dinkler, D. DYN-2
Ditman, J. ASF-5
Dong, S. STR-50
Doorbar, P. SDN-04
Dopker, B. MAT-16
Dowell, E. SDN-28
Eason, T. STR-23
Eastep, F. SDN-19
Eberhardt, J. MAT-12
Eberle, R. DYN-7
Ehlers, S. ASF-6
El-Wakeil, O. MDO-09
Elishakoff, I. STR-06
Ellison, B. DYN-7
Ellison, J. SDN-49
Ema, M. STR-10
Enomoto, N. SDN-48
Epps, J. SDN-01
Epstein, R. SDN-28
Ernst, E. STR-07
Fadel, G. MDO-09
Fagan, G. ASF-9
Fanti, F. DES-26
Farouk, A. STR-11
Farris, T. MAT-44
Feldes, R. STR-11
Fenn, R. ASF-5
Fiore, J. STR-03
Fish, J. WIP-25
Flanagan, G. DES-08
Flitter, L. SDN-01
Flowers, G. ASF-3
Flynn, B. DES-29
Folkman, S. DYN-4
Foster, L. SDN-21
Foster, R. MAT-12
Fox, E. STR-22
Frady, G. DYN-7
Frazier, W. DES-17
Freed, A. MAT-42, STR-23
French, M. SDN-19
Frengley, M. DES-46
Freymann, R. SDN-38
Friedmann, P. DYN-9, SDN-01
Frostig, Y. STR-14
Fu, L. STR-06

Fuchs, H. DES-46
Fuller, C. ASF-9
Furuya, H. ASF-7
Gagliardi, J. DES-08
Gandhi, F. SDN-01
Ganguli, R. MDO-18
Garcia, E. ASF-1, 2
Gates, T. MAT-42
Gennady, K. SDN-04
George, D. SDN-33
Ghandi, K. ASF-5
Ghaessgen, E. DES-08
Giurgiutiu, V. ASF-6, MAT-34
Glaessgen, E. DES-08
Gordis, J. SDN-30
Gordon, R. WIP-25
Gorrell, S. DES-08
Gozu, K. STR-10
Graesser, D. DES-29
Gramoll, K. MAT-35, STR-20
Grandhi, R. MDO-18
Grandt Jr., A. MAT-24
Grasso, S. SDN-48
Gray, G. DYN-2
Grebenstein, E. STR-03
Greer Jr., J. SDN-38
Greschik, G. DES-26
Griffin Jr., O. DES-08
Griffin, S. ASF-3
Grimsley, F. STR-20
Grodsinsky, C. SDN-49
Gu, H. STR-10
Gu, Y. ASF-2
Guillot, D. DYN-9, SDN-38
Guo, R. MDO-36
Gürdal, Z. MDO-45, STR-41
Guruswamy, G. DYN-9, SDN-28, STR-43
Gute, D. STR-32
Haas, D. SDN-01
Haensel, C. MAT-24
Haftka, R. ASF-7, DYN-8, MDO-09, 45, STR-43
Hagood, N. ASF-5, SDN-54
Hajela, P. MDO-45, SDN-04
Hale, M. SDN-48
Halford, G. STR-31
Hall, K. SDN-19
Hamilton, B. DES-46
Hamilton, D. DYN-7
Hamling, S. STR-15
Hanagud, S. ASF-3, 6, 7, 9
Hänle, U. DYN-2
Haque, I. DYN-3
Harrell, L. SDN-21
Harvey, M. MDO-27
Haryadi, S. STR-43
Hasselman, T. SDN-02
Hawwa, M. DYN-3, STR-43
Hayman, G. DES-17
Hedgepeth, J. DES-17, STR-53
Heightland, C. WIP-25
Heppler, G. DES-08, SDN-21
Hernandez, S. MDO-36
Heylinger, P. ASF-4
Higuchi, K. DES-26, SDN-48
Hilton, H. STR-50
Hinada, M. DYN-1
Hodges, D. SDN-02, STR-23
Hol, J. STR-10
Holkamp, J. ASF-2, 9
Honlinger, H. ASF-4
Hong, S. STR-06
Hong, Z. ASF-3
Hooper, S. MAT-35, SDN-30
Hopkins, D. ASF-4, MDO-18, STR-32
Hopkins, M. SDN-28
Horner, G. ASF-2
Hoyniak, D. SDN-52
Hsiao, M-H. SDN-33
Hsias, M. ASF-3
Hu, T-H. SDN-21
Huang, J-K. ASF-3, SDN-33
Huang, Y. STR-32
Hufford, L. ASF-1
Hung, C-L. DES-46
Huseyin, K. SDN-21
Huston, R. MAT-16
Hwang, Y-T. MDO-45
Hyams, D. MDO-09
Hyer, M. DES-46, WIP-25
Icardi, U. STR-06
Ilcewicz, L. DES-29, MAT-16
Inman, D. DYN-6
Ishikawa, T. STR-10
Jackson, A. STR-06
Jacoby, M. SDN-02
Jacques, D. MDO-36
Jaihal, P. ASF-3
James, G. SDN-33
Jaunky, N. STR-10
Jeans, L. STR-20
Jegley, D. MAT-35
Jensen, D. DES-08

Jiangning, Q. SDN-02
Jin, J-D. DYN-9
Johanson, R. MDO-45
Johnson, C. DYN-7
Johnson, E. MDO-45, STR-53
Juang, J-N. SDN-21
Juhasz, B. DYN-1
Juneja, V. DYN-8
Kaljevic, I. STR-32
Kaminsky, J. DYN-7
Kammer, D. ASF-1, DYN-6
Kandil, H. SDN-19
Kandil, O. SDN-19
Kao, P-J. MDO-27, SDN-30
Kapania, R. STR-43
Kar, d. MDO-27
Karpel, M. MDO-27
Karpouzian, G. SDN-28
Kasagi, A. STR-06
Katayama, K. SDN-48
Katsube, N. MAT-42
Kawiecki, G. ASF-2
Keel, L. ASF-1
Keilers, d. STR-03
Kennedy, D. STR-07
Kerdjoudj, M. DYN-3
Keremes, J. STR-31
Khalessi, M. STR-13, STR-31
Khalessi, R. STR-13
Khamseh, A. MAT-24
Khot, N. DYN-6
Kikuchi, N. MDO-45
Kilgore, W. SDN-48
Kim, C. DYN-9
Kim, H. DYN-8
Kim, I. DES-29
Kim, K. DYN-9
Kim, T. ASF-7, STR-50
Kim, W. ASF-3
Kim, Y. DYN-9
Kinoi, S. SDN-48
Kirsch, U. MDO-36
Kishi, K. DYN-7
Kistler, L. DES-46
Knight, N. STR-10
Knowles, G. SDN-30
Ko, J. STR-14, 23
Koch, R. SDN-52
Kodiyalam, S. MDO-27, SDN-30
Koenig, M. MAT-24
Kogiso, N. MDO-45
Kohsetsu, Y. SDN-48
Kokan, D. STR-20
Kolonay, R. SDN-37
Kondor, S. ASF-6
Kosmatka, J. STR-05, 50
Kramer, R. STR-03
Kreider, W. DYN-4
Kristinsdottir, B. DES-29
Kriz, R. MAT-34
Kröplin, B. DYN-2
Krueger, R. MAT-24
Kumar, S. WIP-25
Kurdila, A. DYN-9, STR-05, 14, 23
Kuwao, F. DYN-7
Kwak, M. DYN-4
Kweon, J. STR-06
Lagace, P. STR-15
Lalande, F. ASF-8
Lamson, S. MDO-27
Lang, E. ASF-2
Larkin, P. STR-03
Larson, C. DYN-5
Layton, J. SDN-37, 39
Lee, B. SDN-37
Lee, C. STR-53
Lee, I. STR-06
Lee, J. STR-41
Lee, S. ASF-4, STR-05
Lee, U. DYN-5
Lee-Glauser, G. SDN-49
Leeks, T. ASF-7
Leiva, J. MDO-36
Leo, D. DYN-6
Lesieutre, G. SDN-54
Leung, M. MDO-45
Levraea, J. SDN-54
Lew, J-S. ASF-1
Li, C. DYN-8
Li, D. SDN-21
Li, E. SDN-04
Li, M. SDN-02
Li, W-L. MDO-27
Li, Y. STR-06
Liang, C. ASF-2
Liang, X. DES-08
Liaw, D. STR-22
Libai, A. DYN-2
Librescu, L. SDN-28, STR-06
Lim, K. DYN-6
Lim, T. ASF-4, DYN-8, SDN-54
Lin, C. DYN-8, SDN-37, STR-05

Lin, H-Z. STR-13, 31
Lin, K. MAT-35, STR-15
Lin, M. ASF-8
Lin, W. STR-06
Lin, Z. STR-13
Lindenmoyer, A. DYN-1
Lindner, D. ASF-4
Link, T. ASF-1
Liu, C. DYN-5
Liu, W. STR-05
Livine, E. MDO-27
Locke, J. MAT-12, SDN-04
Lomenzo, R. WIP-25
Lou, M. MAT-12
Louis, t. MDO-27
Lu, C. MDO-36
Lucas, T. MDO-36
Luo, J. MAT-34
Lykins, C. STR-22
Mabson, G. DES-29, STR-20
Mack, N. MDO-45
MacMurdy, D. STR-43
Majed, A. DYN-1
Majima, O. STR-10
Mak, W. DES-17
Malla, R. DYN-4
Malloy, J. MDO-18
Malone Jr., J. STR-22
Manning, R. ASF-4
Mantena, P. WIP-25
Mapar, J. SDN-21
Marcucelli, K. WIP-25
Marek, E. DYN-5
Marquart, E. SDN-54
Martin, R. DYN-4, MAT-42
Massey, S. SDN-19
Masters, B. SDN-33
Matsuike, J. SDN-48
Matsuzaki, Y. DYN-9, SDN-47
Mayne, R. SDN-39
McCarty, C. DES-08
McCling, R. STR-31
McColl, C. ASF-7
McCulloch, A. SDN-49
McGowan, D. MAT-44
McInerney, K. SDN-39
McKnight, R. STR-31
McManus, H. STR-15
McManus, M. DYN-7
Mei, C. ASF-3, SDN-04, 19, 38, 47
Meirovitch, L. SDN-28
Mendel, M. STR-31
Messac, A. SDN-47
Mester, S. SDN-30
Metschan, S. DES-29
Meyer, T. SDN-39
Meyers, C. WIP-25
Michael, C. STR-20
Mignolet, M. SDN-33
Mignosa, L. STR-53
Miki, M. MDO-45
Milano, J. SDN-01
Miller, R. DES-17
Millott, T. DYN-9, SDN-01
Millwater, H. STR-31
Minesugi, K. ASF-06, DYN-7
Minguet, P. MAT-35, STR-10
Minnetyan, L. MAT-16, 24
Mistree, F. MDO-36
Miura, K. ASF-8
Moffat, J. DYN-9
Mohammed, I. MAT-24
Mook, D. SDN-19, 39
Mordfin, T. DYN-3
Morel, M. SDN-52
Morita, Y. DYN-1
Morton, S. DES-17
Moukawsher, E. MAT-24
Mulubagal, G. MDO-18
Murphy, D. MAT-16, SDN-52
Murthy, P. MAT-16, 24
Mushung, L. DYN-7
Nagar, A. MAT-44
Nakai, S. ASF-8
Nakamura, M. DYN-7
Nam, C. ASF-3
Namburu, R. SDN-38
Narayanan, G. MAT-42, STR-23
Nast, T. SDN-02
Natori, M. ASF-8, DES-26
Nayfeh, A. DYN-2, 3, 4, SDN-19, STR-43
Nayfeh, S. DYN-2
Nemeth, M. STR-06
Neussl, M. MAT-24
Nevill Jr., G. MDO-45
Newton, D. ASF-2
Nguyen, D. MDO-09, SDN-02
Nielsen, D. STR-53
Niitsu, M. DYN-7, SDN-48
Nikolaidis, E. STR-22
Nishimura, H. SDN-39
Nishio, M. DYN-1
Nishmura, H. WIP-25
Nissley, D. MAT-42

Nitzsche, F. ASF-6
Nonami, K. WIP-25
Noor, A. STR-10
Norden, C. DES-08
Norris, M. ASF-2
Nozue, T. DYN-1
O'Connor, D. STR-31
Ochoa, O. STR-23
Oh, S. ASF-3
Okazaki, K. DES-26
Onoda, J. ASF-6, DYN-7
Ordonneau, J. SDN-47
Ottarsson, G. SDN-52
Owen, B. STR-41
Oz, H. DYN-6
Ozbek, A. SDN-02, 30
Pai, S. DYN-4, STR-32
Palazotto, A. SDN-38, STR-07
Pandiyan, R. SDN-49
Papalambros, P. MDO-45
Park, C. ASF-8
Park, H. STR-05
Park, J. MAT-24, MDO-09
Park, K. DES-17, 26, SDN-40
Park, K.C. DYN-4
Park, Y. STR-05
Parmerter, R. STR-14, 23
Parthasarathy, V. MDO-27
Pates, C. SDN-38
Pelessone, D. STR-43
Pellegrino, S. MAT-35
Pendleton, E. SDN-28
Percheron, T. SDN-47
Peters, J. STR-10
Peterson, L. ASF-1, DYN-4, SDN-39, 40, STR-07
Petra, J-M. MAT-16
Pham, Q. DYN-7
Pierre, C. DYN-2, SDN-52
Pilant, M. STR-14
Polaha, J. MAT-35
Pomfret, C. STR-22
Poole, E. MAT-35
Povich, C. ASF-4
Powell, B. MDO-27
Prabhakar, A. DYN-7
Prakash, B. DES-17, 26, MAT-34
Pramono, E. SDN-37
Prasad, C. MAT-16
Prasad, G. ASF-3
Price, S. SDN-37
Protasov, V. DES-46
Pyo, C. MAT-24
Qin, J. MDO-09
Quinn, G. DYN-7
Quinn, R. DYN-5
Ragon, S. DES-17
Rahematpura, M. SDN-30
Rais-Rohani, M. MDO-27, WIP-25
Raju, I. STR-14
Ramaprasad, S. MAT-35
Rao, K. MAT-34
Rao, S. STR-53
Raouf, R. STR-07
Rastogi, N. STR-53
Ratliff, G. SDN-54
Ravindranath, D. DES-17, 26
Rayburn, J. DYN-7
Reccanello, F. DES-26
Red-Horse, J. SDN-33
Reddy, T. SDN-37, 52
Redmond, J. ASF-9
Rehfield, L. DES-17
Reid, R. STR-14
Reifsnider, K. MAT-34
Reiss, R. STR-41
Renaud, J. MDO-27
Rezaeepazhand, J. STR-41
Rhodes, M. MDO-18
Riddick, J. STR-11, 43
Robertshaw, H. ASF-9
Robinson, M. STR-11
Rockoff, L. DES-26
Rodrigues, E. SDN-33
Rogers, C. ASF-2, 6, 8
Roglin, R. ASF-6
Ronnau, A. MAT-24
Rosetti, D. ASF-2
Rosner, A. MDO-18
Ross, T. SDN-02
Rouse, M. MAT-35
Rowley, R. MAT-44
Ruckman, C. DYN-6
Run, Z. ASF-3
Ryu, B. SDN-48
Sadler, R. STR-11
Saigal, S. STR-32
Salajegheh, E. MDO-09
Salama, M. ASF-7
Salas, R. DYN-7
Sankar, B. STR-15
Saravanos, D. ASF-4
Sato, N. DYN-7
Sawicki, A. STR-20

Schaeffel Jr., J. SDN-47
Scharton, T. SDN-40
Schlaegel, W. DYN-4
Schneiderman, S. MAT-34
Sciulli, D. DYN-4
Seeley, C. ASF-7
Sekine, K. DES-26
Sensburg, O. ASF-4
Sepulveda, A. MDO-18
Serhan, H. MAT-44, MDO-27
Sha, D. SDN-02, STR-23
Sharp, D. STR-10
Shastry, K. DES-17, 26
Shaw, S. DYN-2
Shekthman, I. SDN-30
Shen, M. ASF-8
Shenyang, H. ASF-8
Shiao, M. MDO-18, STR-13
Shieh, R. STR-32
Shih, C. MDO-45
Shih, F. STR-03
Shih, I-C. MDO-36
Shinji, A. SDN-47
Shirahatti, U. SDN-38
Shiraki, K. DYN-7
Shivakumar, K. STR-11, 43
Shyy, J-K. MDO-36
Silverberg, L. ASF-8
Simha, T. DES-26
Simitses, G. STR-41
Simmermacher, T. SDN-39
Simonetta, S. SDN-21
Simpson, P. STR-53
Singh, R. MAT-12, 24, SDN-02
Singhal, S. MDO-18
Sinha, S. SDN-49
Skopp, G. STR-31
Slater, J. MAT-16
Sleight, D. STR-14
Smith, B. MAT-16
Smith, C. SDN-01
Smith, H. SDN-02
Smith, J. STR-14, 53
Smith, S. DYN-8
Smith, W. ASF-2
Sobel, L. STR-10
Song, O. SDN-28
Sonik, V. STR-15
Sorini, P. MAT-44
Spanos, P. DYN-1, 7
Spence, A. MDO-36
Sridharan, S. STR-06
Srinivasan, R. SDN-28, 37
Srivastava, R. SDN-52
Stabb, M. DYN-2
Staley, L. STR-03
Starchville Jr., T. ASF-9
Starnes Jr., J. DES-17, 46, MAT-16, STR-06, 10
Stavrinidis, C. SDN-40
Steadman, D. ASF-3
Steinberg, E. STR-13
Stoyack, J. MAT-44
Straube, T. SDN-40
Strganac, T. DYN-9
Striz, A. STR-32
Stroud, W. STR-22
Su, J. DYN-6
Su, T-J. SDN-21
Suemasu, H. STR-10
Sues, R. MDO-18
Sugavanam, S. SDN-49
Sugiyama, Y. SDN-48
Suleman, A. ASF-3
Sullivan, R. MAT-12
Sumali, H. WIP-25
Sun, C. MAT-34
Sun, F. ASF-2
Suryanarayan, S. DYN-5
Sutjahjo, E. STR-43
Swanson, G. DES-29
Szewczyk, Z. MDO-45
Szolwinski, M. MAT-44
Tadikonda, S. DYN-3
Takemoto, Y. DYN-1
Tamma, K. SDN-02, STR-23, 32
Tan, P. MAT-24
Tandara, M. SDN-47
Tang, S. DYN-9
Tasker, F. DYN-5, SDN-01
Taylor, R. STR-53
Teboub, Y. SDN-04
Tefend, M. DYN-6
Telford, K. STR-11
Tengler, N. DYN-1, SDN-47
Teranishi, K. STR-41
Thatker, A. SDN-04
Thomas, D. DES-17
Thomas, H. MDO-36
Thompson, L. STR-11
Thomson, D. DES-08
Thornton, E. MAT-12, 44
Tianlu, S. DES-08
Tich, E. MDO-27

Tinker, M. SDN-40
Tomlinson, G. ASF-4
Torng, T. MDO-18
Torng, T. STR-31
Tracy Sr., J. STR-11
Triller, M. DYN-6
Tsai, P. STR-13
Tsang, P. STR-15
Tsuchiya, M. DYN-1
Tsujihata, A. DYN-1
Tsukashima, T. DYN-1
Tu, E. DYN-9
Tucker, S. ASF-9
Turner, D. SDN-38
Turner, T. SDN-04
Tuttle, M. DES-29
Tzou, H. ASF-2
Ujino, I. SDN-48
Utku, S. ASF-3, 9
Vadali, R. STR-05
Vadde, S. MDO-36
van Schoor, M. SDN-33
Vanderplaats, G. MDO-09
Vasilev, V. DES-46
Veley, D. STR-53
Venkatesan, C. DYN-9, SDN-01
Venkayya, V. ASF-3, SDN-37
Verderaime, V. STR-13
Vizzini, A. DES-08
Volk, J. MDO-27
Volovoi, V. SDN-30
Waas, A. MAT-24, STR-06
Wada, B. ASF-3, 9
Walker, T. MAT-16
Wallace, C. WIP-25
Waltz, C. ASF-6
Wang, B. MDO-36
Wang, G. SDN-30
Wang, J. STR-14
Wang, L. MDO-18
Watanabe, N. STR-41
Waters Jr., W. MAT-16
Watkins, W. STR-07
Watson, K. DES-08
Watson, L. MDO-45
Webber, J. DES-17
Webster, R. STR-03
Weeratunga, S. SDN-37
Weidong, Y. DES-08
Weisshaar, T. ASF-7, SDN-33
Wells, R. STR-23
Wereley, N. ASF-5, DES-46, STR-32
Whitcomb, J. MAT-34
Whitney, S. STR-14
Williams, F. STR-07
Williams, T. SDN-39
Worden, K. ASF-4
Wu, K. SDN-47
Wu, Y-T. STR-31
Wu, Y. MAT-42
Xiaodong, L. DES-08
Xie, Y. DYN-5
Xue, D. ASF-3, MAT-44, SDN-19, 47
Yamanaka, K. SDN-21
Yan, Y. DES-46
Yang, H. SDN-37
Yang, R. MDO-18
Yao, S. STR-11
Yehuda, S. SDN-04
Yen, S. STR-50
Yi, S. STR-50
Yin, W-L. STR-50
Yiu, Y. STR-32
You, Z. SDN-30
Young Jr., C. SDN-48
Youssef, H. SDN-33
Yu, Z. SDN-01
Yuan, A. SDN-01
Yuan, K. DYN-9
Yurkovich, R. MDO-27
Zabinsky, Z. DES-29
Zahrah, T. DES-46
Zee, R. SDN-21
Zhang, d. STR-50
Zhong, Z. SDN-04
Zhou, M. MDO-09
Zhou, R. SDN-19
Zhou, S. ASF-5
Zhou, X. STR-23
Zimmerman, D. DYN-5, 8, SDN-39, 49